Chemistry Complete Solution Manual

Santa Barbara City College

9e

Steven S. Zumdahl | Susan A. Zumdahl

CENGAGE Learning

Australia • Brazil • Japan • Korea • Mexico • Singapore • Spain • United Kingdom • United States

CENGAGE Learning

Chemistry Complete Solution Manual
Santa Barbara City College
9e

Senior Project Development Manager:
Linda deStefano

Market Development Manager:
Heather Kramer

Senior Production/Manufacturing Manager:
Donna M. Brown

Production Editorial Manager:
Kim Fry

Sr. Rights Acquisition Account Manager:
Todd Osborne

Chemistry, 9th Edition
Steven S. Zumdahl | Susan A. Zumdahl

For product information and technology assistance, contact us at
Cengage Learning Customer & Sales Support, 1-800-354-9706
For permission to use material from this text or product,
submit all requests online at **cengage.com/permissions**
Further permissions questions can be emailed to
permissionrequest@cengage.com

ISBN-13: 978-1-305-01012-3

ISBN-10: 1-305-01012-4

Cengage Learning
5191 Natorp Boulevard
Mason, Ohio 45040
USA

Cengage Learning is a leading provider of customized learning solutions with office locations around the globe, including Singapore, the United Kingdom, Australia, Mexico, Brazil, and Japan. Locate your local office at:
international.cengage.com/region.

Cengage Learning products are represented in Canada by Nelson Education, Ltd.
For your lifelong learning solutions, visit **www.cengage.com/custom.**
Visit our corporate website at **www.cengage.com.**

Printed in the United States of America

TABLE OF CONTENTS

CHAPTER 1

CHEMICAL FOUNDATIONS

Questions

17. A law summarizes what happens, e.g., law of conservation of mass in a chemical reaction or the ideal gas law, PV = nRT. A theory (model) is an attempt to explain why something happens. Dalton's atomic theory explains why mass is conserved in a chemical reaction. The kinetic molecular theory explains why pressure and volume are inversely related at constant temperature and moles of gas present, as well as explaining the other mathematical relationships summarized in PV = nRT.

18. A dynamic process is one that is active as opposed to static. In terms of the scientific method, scientists are always performing experiments to prove or disprove a hypothesis or a law or a theory. Scientists do not stop asking questions just because a given theory seems to account satisfactorily for some aspect of natural behavior. The key to the scientific method is to continually ask questions and perform experiments. Science is an active process, not a static one.

19. The fundamental steps are

    (1) making observations;
    (2) formulating hypotheses;
    (3) performing experiments to test the hypotheses.

    The key to the scientific method is performing experiments to test hypotheses. If after the test of time the hypotheses seem to account satisfactorily for some aspect of natural behavior, then the set of tested hypotheses turns into a theory (model). However, scientists continue to perform experiments to refine or replace existing theories.

20. A random error has equal probability of being too high or too low. This type of error occurs when estimating the value of the last digit of a measurement. A systematic error is one that always occurs in the same direction, either too high or too low. For example, this type of error would occur if the balance you were using weighed all objects 0.20 g too high, that is, if the balance wasn't calibrated correctly. A random error is an indeterminate error, whereas a systematic error is a determinate error.

21. A qualitative observation expresses what makes something what it is; it does not involve a number; e.g., the air we breathe is a mixture of gases, ice is less dense than water, rotten milk stinks.

    The SI units are mass in kilograms, length in meters, and volume in the derived units of m^3. The assumed uncertainty in a number is ±1 in the last significant figure of the number. The precision of an instrument is related to the number of significant figures associated with an experimental reading on that instrument. Different instruments for measuring mass, length, or volume have varying degrees of precision. Some instruments only give a few significant figures for a measurement, whereas others will give more significant figures.

22. Precision: reproducibility; accuracy: the agreement of a measurement with the true value.

a. Imprecise and inaccurate data: 12.32 cm, 9.63 cm, 11.98 cm, 13.34 cm

b. Precise but inaccurate data: 8.76 cm, 8.79 cm, 8.72 cm, 8.75 cm

c. Precise and accurate data: 10.60 cm, 10.65 cm, 10.63 cm, 10.64 cm

Data can be imprecise if the measuring device is imprecise as well as if the user of the measuring device has poor skills. Data can be inaccurate due to a systematic error in the measuring device or with the user. For example, a balance may read all masses as weighing 0.2500 g too high or the user of a graduated cylinder may read all measurements 0.05 mL too low.

A set of measurements that are imprecise implies that all the numbers are not close to each other. If the numbers aren't reproducible, then all the numbers can't be very close to the true value. Some say that if the average of imprecise data gives the true value, then the data are accurate; a better description is that the data takers are extremely lucky.

23. Significant figures are the digits we associate with a number. They contain all of the certain digits and the first uncertain digit (the first estimated digit). What follows is one thousand indicated to varying numbers of significant figures: 1000 or 1×10^3 (1 S.F.); 1.0×10^3 (2 S.F.); 1.00×10^3 (3 S.F.); 1000. or 1.000×10^3 (4 S.F.).

To perform the calculation, the addition/subtraction significant figure rule is applied to 1.5 − 1.0. The result of this is the one-significant-figure answer of 0.5. Next, the multiplication/division rule is applied to 0.5/0.50. A one-significant-figure number divided by a two-significant-figure number yields an answer with one significant figure (answer = 1).

24. From Figure 1.9 of the text, a change in temperature of 180°F is equal to a change in temperature of 100°C and 100 K. A degree unit on the Fahrenheit scale is not a large as a degree unit on the Celsius or Kelvin scales. Therefore, a 20° change in the Celsius or Kelvin temperature would correspond to a larger temperature change than a 20° change in the Fahrenheit scale. The 20° temperature change on the Celsius and Kelvin scales are equal to each other.

25. Straight line equation: $y = mx + b$, where m is the slope of the line and b is the y-intercept. For the T_F vs. T_C plot:

$$T_F = (9/5)T_C + 32$$
$$y = \quad m \quad x + b$$

The slope of the plot is 1.8 (= 9/5) and the y-intercept is 32°F.

For the T_C vs. T_K plot:

$$T_C = \quad T_K \; - 273$$
$$y = m\,x \; + b$$

The slope of the plot is 1, and the y-intercept is −273°C.

26. a. coffee; saltwater; the air we breathe ($N_2 + O_2$ + others); brass (Cu + Zn)

b. book; human being; tree; desk

c. sodium chloride (NaCl); water (H_2O); glucose ($C_6H_{12}O_6$); carbon dioxide (CO_2)

d. nitrogen (N_2); oxygen (O_2); copper (Cu); zinc (Zn)

e. boiling water; freezing water; melting a popsicle; dry ice subliming

f. Electrolysis of molten sodium chloride to produce sodium and chlorine gas; the explosive reaction between oxygen and hydrogen to produce water; photosynthesis, which converts H_2O and CO_2 into $C_6H_{12}O_6$ and O_2; the combustion of gasoline in our car to produce CO_2 and H_2O

Exercises

Significant Figures and Unit Conversions

27. a. exact b. inexact

c. exact d. inexact (π has an infinite number of decimal places.)

28. a. one significant figure (S.F.). The implied uncertainty is ±1000 pages. More significant figures should be added if a more precise number is known.

b. two S.F. c. four S.F.

d. two S.F. e. infinite number of S.F. (exact number) f. one S.F.

29. a. $\underline{6.07} \times 10^{-15}$; 3 S.F. b. $0.00\underline{3840}$; 4 S.F. c. $\underline{17.00}$; 4 S.F.

d. $\underline{8} \times 10^8$; 1 S.F. e. $\underline{463.8052}$; 7 S.F. f. $\underline{3}00$; 1 S.F.

g. $\underline{301}$; 3 S.F. h. $\underline{300.}$; 3 S.F.

30. a. $\underline{1}00$; 1 S.F. b. $\underline{1.0} \times 10^2$; 2 S.F. c. $\underline{1.00} \times 10^3$; 3 S.F.

d. $\underline{100.}$; 3 S.F. e. $0.00\underline{48}$; 2 S.F. f. $0.00\underline{480}$; 3 S.F.

g. $\underline{4.80} \times 10^{-3}$; 3 S.F. h. $\underline{4.800} \times 10^{-3}$; 4 S.F.

31. When rounding, the last significant figure stays the same if the number after this significant figure is less than 5 and increases by one if the number is greater than or equal to 5.

a. 3.42×10^{-4} b. 1.034×10^4 c. 1.7992×10^1 d. 3.37×10^5

32. a. 4×10^5 b. 3.9×10^5 c. 3.86×10^5 d. 3.8550×10^5

33. Volume measurements are estimated to one place past the markings on the glassware. The first graduated cylinder is labeled to 0.2 mL volume increments, so we estimate volumes to

the hundredths place. Realistically, the uncertainty in this graduated cylinder is ±0.05 mL. The second cylinder, with 0.02 mL volume increments, will have an uncertainty of ±0.005 mL. The approximate volume in the first graduated cylinder is 2.85 mL, and the volume in the other graduated cylinder is approximately 0.280 mL. The total volume would be:

$$\begin{array}{r} 2.85\text{ mL} \\ \underline{+0.280\text{ mL}} \\ 3.13\text{ mL} \end{array}$$

We should report the total volume to the hundredths place because the volume from the first graduated cylinder is only read to the hundredths (read to two decimal places). The first graduated cylinder is the least precise volume measurement because the uncertainty of this instrument is in the hundredths place, while the uncertainty of the second graduated cylinder is to the thousandths place. It is always the lease precise measurement that limits the precision of a calculation.

34. a. Volumes are always estimated to one position past the marked volume increments. The estimated volume of the first beaker is 32.7 mL, the estimated volume of the middle beaker is 33 mL, and the estimated volume in the last beaker is 32.73 mL.

b. Yes, all volumes could be identical to each other because the more precise volume readings can be rounded to the other volume readings. But because the volumes are in three different measuring devices, each with its own unique uncertainty, we cannot say with certainty that all three beakers contain the same amount of water.

c. $$\begin{array}{l} 32.7\text{ mL} \\ 33\text{ mL} \\ \underline{32.73\text{ mL}} \\ 98.43\text{ mL} = 98\text{ mL} \end{array}$$

The volume in the middle beaker can only be estimated to the ones place, which dictates that the sum of the volume should be reported to the ones place. As is always the case, the least precise measurement determines the precision of a calculation.

35. For addition and/or subtraction, the result has the same number of decimal places as the number in the calculation with the fewest decimal places. When the result is rounded to the correct number of significant figures, the last significant figure stays the same if the number after this significant figure is less than 5 and increases by one if the number is greater than or equal to 5. The underline shows the last significant figure in the intermediate answers.

a. $212.2 + 26.7 + 402.09 = 640.\underline{9}9 = 641.0$

b. $1.0028 + 0.221 + 0.10337 = 1.32\underline{7}17 = 1.327$

c. $52.331 + 26.01 - 0.9981 = 77.3\underline{4}29 = 77.34$

d. $2.01 \times 10^2 + 3.014 \times 10^3 = 2.01 \times 10^2 + 30.14 \times 10^2 = 32.1\underline{5} \times 10^2 = 3215$

When the exponents are different, it is easiest to apply the addition/subtraction rule when all numbers are based on the same power of 10.

e. $7.255 - 6.8350 = 0.42 = 0.420$ (first uncertain digit is in the third decimal place).

36. For multiplication and/or division, the result has the same number of significant figures as the number in the calculation with the fewest significant figures.

a. $\dfrac{0.102 \times 0.0821 \times 273}{1.01} = \underline{2.26}35 = 2.26$

b. $0.14 \times 6.022 \times 10^{23} = \underline{8.4}31 \times 10^{22} = 8.4 \times 10^{22}$; since 0.14 only has two significant figures, the result should only have two significant figures.

c. $4.0 \times 10^4 \times 5.021 \times 10^{-3} \times 7.34993 \times 10^2 = \underline{1.4}76 \times 10^5 = 1.5 \times 10^5$

d. $\dfrac{2.00 \times 10^6}{3.00 \times 10^{-7}} = \underline{6.66}67 \times 10^{12} = 6.67 \times 10^{12}$

37. a. Here, apply the multiplication/division rule first; then apply the addition/subtraction rule to arrive at the one-decimal-place answer. We will generally round off at intermediate steps in order to show the correct number of significant figures. However, you should round off at the end of all the mathematical operations in order to avoid round-off error. The best way to do calculations is to keep track of the correct number of significant figures during intermediate steps, but round off at the end. For this problem, we underlined the last significant figure in the intermediate steps.

$$\frac{2.526}{3.1} + \frac{0.470}{0.623} + \frac{80.705}{0.4326} = 0.8\underline{1}48 + 0.75\underline{4}4 + 186.\underline{5}58 = 188.1$$

b. Here, the mathematical operation requires that we apply the addition/subtraction rule first, then apply the multiplication/division rule.

$$\frac{6.404 \times 2.91}{18.7 - 17.1} = \frac{6.404 \times 2.91}{1.\underline{6}} = 12$$

c. $6.071 \times 10^{-5} - 8.2 \times 10^{-6} - 0.521 \times 10^{-4} = 60.71 \times 10^{-6} - 8.2 \times 10^{-6} - 52.1 \times 10^{-6}$
$= 0.\underline{4}1 \times 10^{-6} = 4 \times 10^{-7}$

d. $\dfrac{3.8 \times 10^{-12} + 4.0 \times 10^{-13}}{4 \times 10^{12} + 6.3 \times 10^{13}} = \dfrac{38 \times 10^{-13} + 4.0 \times 10^{-13}}{4 \times 10^{12} + 63 \times 10^{12}} = \dfrac{4\underline{2} \times 10^{-13}}{6\underline{7} \times 10^{12}} = 6.3 \times 10^{-26}$

e. $\dfrac{9.5 + 4.1 + 2.8 + 3.175}{4} = \dfrac{19.\underline{5}75}{4} = 4.89 = 4.9$

Uncertainty appears in the first decimal place. The average of several numbers can only be as precise as the least precise number. Averages can be exceptions to the significant figure rules.

f. $\dfrac{8.925 - 8.905}{8.925} \times 100 = \dfrac{0.02\underline{0}}{8.925} \times 100 = 0.22$

38. a. $6.022 \times 10^{23} \times 1.05 \times 10^2 = 6.32 \times 10^{25}$

b. $\dfrac{6.6262 \times 10^{-34} \times 2.998 \times 10^{8}}{2.54 \times 10^{-9}} = 7.82 \times 10^{-17}$

c. $1.285 \times 10^{-2} + 1.24 \times 10^{-3} + 1.879 \times 10^{-1}$

$= 0.1285 \times 10^{-1} + 0.0124 \times 10^{-1} + 1.879 \times 10^{-1} = 2.020 \times 10^{-1}$

When the exponents are different, it is easiest to apply the addition/subtraction rule when all numbers are based on the same power of 10.

d. $\dfrac{(1.00866 - 1.00728)}{6.02205 \times 10^{23}} = \dfrac{0.00138}{6.02205 \times 10^{23}} = 2.29 \times 10^{-27}$

e. $\dfrac{9.875 \times 10^{2} - 9.795 \times 10^{2}}{9.875 \times 10^{2}} \times 100 = \dfrac{0.080 \times 10^{2}}{9.875 \times 10^{2}} \times 100 = 8.1 \times 10^{-1}$

f. $\dfrac{9.42 \times 10^{2} + 8.234 \times 10^{2} + 1.625 \times 10^{3}}{3} = \dfrac{0.942 \times 10^{3} + 0.824 \times 10^{3} + 1.625 \times 10^{3}}{3}$

$= 1.130 \times 10^{3}$

39. a. $8.43 \text{ cm} \times \dfrac{1 \text{ m}}{100 \text{ cm}} \times \dfrac{1000 \text{mm}}{\text{m}} = 84.3 \text{ mm}$ b. $2.41 \times 10^{2} \text{ cm} \times \dfrac{1 \text{ m}}{100 \text{ cm}} = 2.41 \text{ m}$

c. $294.5 \text{ nm} \times \dfrac{1 \text{ m}}{1 \times 10^{9} \text{ nm}} \times \dfrac{100 \text{ cm}}{\text{m}} = 2.945 \times 10^{-5} \text{ cm}$

d. $1.445 \times 10^{4} \text{ m} \times \dfrac{1 \text{ km}}{1000 \text{m}} = 14.45 \text{ km}$ e. $235.3 \text{ m} \times \dfrac{1000 \text{mm}}{\text{m}} = 2.353 \times 10^{5} \text{ mm}$

f. $903.3 \text{ nm} \times \dfrac{1 \text{ m}}{1 \times 10^{9} \text{ nm}} \times \dfrac{1 \times 10^{6} \text{ µm}}{\text{m}} = 0.9033 \text{µm}$

40. a. $1 \text{ Tg} \times \dfrac{1 \times 10^{12} \text{ g}}{\text{Tg}} \times \dfrac{1 \text{kg}}{1000 \text{g}} = 1 \times 10^{9} \text{kg}$

b. $6.50 \times 10^{2} \text{ Tm} \times \dfrac{1 \times 10^{12} \text{ m}}{\text{Tm}} \times \dfrac{1 \times 10^{9} \text{ nm}}{\text{m}} = 6.50 \times 10^{23} \text{nm}$

c. $25 \text{ fg} \times \dfrac{1 \text{g}}{1 \times 10^{15} \text{ fg}} \times \dfrac{1 \text{kg}}{1000 \text{g}} = 25 \times 10^{-18} \text{kg} = 2.5 \times 10^{-17} \text{kg}$

d. $8.0 \text{ dm}^3 \times \dfrac{1 \text{ L}}{\text{dm}^3} = 8.0 \text{ L}$ $(1 \text{ L} = 1 \text{ dm}^3 = 1000 \text{ cm}^3 = 1000 \text{ mL})$

e. $1 \text{ mL} \times \dfrac{1 \text{ L}}{1000 \text{mL}} \times \dfrac{1 \times 10^{6} \text{ µL}}{\text{L}} = 1 \times 10^{3} \text{ µL}$

f. $1 \text{ µg} \times \dfrac{1 \text{g}}{1 \times 10^{6} \text{ µg}} \times \dfrac{1 \times 10^{12} \text{ pg}}{\text{g}} = 1 \times 10^{6} \text{ pg}$

41. a. Appropriate conversion factors are found in Appendix 6. In general, the number of significant figures we use in the conversion factors will be one more than the number of significant figures from the numbers given in the problem. This is usually sufficient to avoid round-off error.

$$3.91\text{ kg} \times \frac{1\text{ lb}}{0.4536\text{kg}} = 8.62\text{ lb};\ 0.62\text{ lb} \times \frac{16\text{ oz}}{\text{lb}} = 9.9\text{ oz}$$

Baby's weight = 8 lb and 9.9 oz or, to the nearest ounce, 8 lb and 10. oz.

$$51.4\text{ cm} \times \frac{1\text{ in}}{2.54\text{ cm}} = 20.2\text{ in} \approx 20\ 1/4\text{ in} = \text{baby's height}$$

b. $$25{,}000\text{ mi} \times \frac{1.61\text{ km}}{\text{mi}} = 4.0 \times 10^4\text{ km};\ 4.0 \times 10^4\text{ km} \times \frac{1000\text{m}}{\text{km}} = 4.0 \times 10^7\text{ m}$$

c. $$V = l \times w \times h = 1.0\text{ m} \times \left(5.6\text{ cm} \times \frac{1\text{ m}}{100\text{cm}}\right) \times \left(2.1\text{ dm} \times \frac{1\text{ m}}{10\text{ dm}}\right) = 1.2 \times 10^{-2}\text{ m}^3$$

$$1.2 \times 10^{-2}\text{ m}^3 \times \left(\frac{10\text{ dm}}{\text{m}}\right)^3 \times \frac{1\text{ L}}{\text{dm}^3} = 12\text{ L}$$

$$12\text{ L} \times \frac{1000\text{cm}^3}{\text{L}} \times \left(\frac{1\text{ in}}{2.54\text{ cm}}\right)^3 = 730\text{ in}^3;\ 730\text{ in}^3 \times \left(\frac{1\text{ ft}}{12\text{ in}}\right)^3 = 0.42\text{ ft}^3$$

42. a. $$908\text{ oz} \times \frac{1\text{ lb}}{16\text{ oz}} \times \frac{0.4536\text{kg}}{\text{lb}} = 25.7\text{ kg}$$

b. $$12.8\text{ L} \times \frac{1\text{ qt}}{0.9463\text{L}} \times \frac{1\text{ gal}}{4\text{ qt}} = 3.38\text{ gal}$$

c. $$125\text{ mL} \times \frac{1\text{ L}}{1000\text{mL}} \times \frac{1\text{ qt}}{0.9463\text{L}} = 0.132\text{ qt}$$

d. $$2.89\text{ gal} \times \frac{4\text{ qt}}{1\text{ gal}} \times \frac{1\text{ L}}{1.057\text{ qt}} \times \frac{1000\text{mL}}{1\text{ L}} = 1.09 \times 10^4\text{ mL}$$

e. $$4.48\text{ lb} \times \frac{453.6\text{ g}}{1\text{ lb}} = 2.03 \times 10^3\text{ g}$$

f. $$550\text{ mL} \times \frac{1\text{ L}}{1000\text{mL}} \times \frac{1.06\text{ qt}}{\text{L}} = 0.58\text{ qt}$$

43. a. $$1.25\text{ mi} \times \frac{8\text{ furlongs}}{\text{mi}} = 10.0\text{ furlongs};\ 10.0\text{ furlongs} \times \frac{40\text{ rods}}{\text{furlong}} = 4.00 \times 10^2\text{ rods}$$

$$4.00 \times 10^2\text{ rods} \times \frac{5.5\text{ yd}}{\text{rod}} \times \frac{36\text{ in}}{\text{yd}} \times \frac{2.54\text{ cm}}{\text{in}} \times \frac{1\text{ m}}{100\text{cm}} = 2.01 \times 10^3\text{ m}$$

$$2.01 \times 10^3 \text{ m} \times \frac{1\text{ km}}{1000\text{m}} = 2.01 \text{ km}$$

b. Let's assume we know this distance to ±1 yard. First, convert 26 miles to yards.

$$26 \text{ mi} \times \frac{5280\text{ft}}{\text{mi}} \times \frac{1\text{ yd}}{3\text{ ft}} = 45{,}760.\text{ yd}$$

$$26 \text{ mi} + 385 \text{ yd} = 45{,}760.\text{ yd} + 385 \text{ yd} = 46{,}145 \text{ yards}$$

$$46{,}145 \text{ yard} \times \frac{1\text{ rod}}{5.5\text{ yd}} = 8390.0 \text{ rods};\ 8390.0 \text{ rods} \times \frac{1\text{ furlong}}{40\text{ rods}} = 209.75 \text{ furlongs}$$

$$46{,}145 \text{ yard} \times \frac{36\text{ in}}{\text{yd}} \times \frac{2.54\text{ cm}}{\text{in}} \times \frac{1\text{ m}}{100\text{cm}} = 42{,}195 \text{ m};\ 42{,}195 \text{ m} \times \frac{1\text{ km}}{1000\text{m}}$$

$$= 42.195 \text{ km}$$

44. a. $1 \text{ ha} \times \frac{10{,}000\text{m}^2}{\text{ha}} \times \left(\frac{1\text{km}}{1000\text{m}}\right)^2 = 1 \times 10^{-2} \text{ km}^2$

b. $5.5 \text{ acre} \times \frac{160\text{rod}^2}{\text{acre}} \times \left(\frac{5.5\text{ yd}}{\text{rod}} \times \frac{36\text{in}}{\text{yd}} \times \frac{2.54\text{cm}}{\text{in}} \times \frac{1\text{ m}}{100\text{cm}}\right)^2 = 2.2 \times 10^4 \text{ m}^2$

$$2.2 \times 10^4 \text{ m}^2 \times \frac{1\text{ ha}}{1 \times 10^4\text{ m}^2} = 2.2 \text{ ha};\ 2.2 \times 10^4 \text{ m}^2 \times \left(\frac{1\text{ km}}{1000\text{m}}\right)^2 = 0.022 \text{ km}^2$$

c. Area of lot = 120 ft × 75 ft = $9.0 \times 10^3 \text{ ft}^2$

$$9.0 \times 10^3 \text{ ft}^2 \times \left(\frac{1\text{ yd}}{3\text{ ft}} \times \frac{1\text{ rod}}{5.5\text{ yd}}\right)^2 \times \frac{1\text{ acre}}{160\text{rod}^2} = 0.21 \text{ acre};\ \frac{\$6{,}500}{0.21\text{acre}} = \frac{\$31{,}000}{\text{acre}}$$

We can use our result from (b) to get the conversion factor between acres and hectares (5.5 acre = 2.2 ha.). Thus 1 ha = 2.5 acre.

$$0.21 \text{ acre} \times \frac{1\text{ ha}}{2.5\text{ acre}} = 0.084 \text{ ha};\ \text{the price is: } \frac{\$6{,}500}{0.084\text{ha}} = \frac{\$77{,}000}{\text{ha}}$$

45. a. $1 \text{ troy lb} \times \frac{12\text{ troyoz}}{\text{troylb}} \times \frac{20\text{pw}}{\text{troyoz}} \times \frac{24\text{ grains}}{\text{pw}} \times \frac{0.0648\text{g}}{\text{grain}} \times \frac{1\text{kg}}{1000\text{g}} = 0.373 \text{ kg}$

$$1 \text{ troy lb} = 0.373 \text{ kg} \times \frac{2.205\text{lb}}{\text{kg}} = 0.822 \text{ lb}$$

b. $1 \text{ troy oz} \times \frac{20\text{ pw}}{\text{troyoz}} \times \frac{24\text{ grains}}{\text{pw}} \times \frac{0.0648\text{g}}{\text{grain}} = 31.1 \text{ g}$

$$1 \text{ troy oz} = 31.1 \text{ g} \times \frac{1\text{ carat}}{0.200\text{g}} = 156 \text{ carats}$$

c. 1 troy lb = 0.373 kg; $0.373\text{ kg} \times \frac{1000\text{g}}{\text{kg}} \times \frac{1\text{ cm}^3}{19.3\text{ g}} = 19.3\text{ cm}^3$

46. a. $1\text{ grain ap} \times \frac{1\text{ scruple}}{20\text{ grain ap}} \times \frac{1\text{ dram ap}}{3\text{ scruples}} \times \frac{3.888\text{g}}{\text{dram ap}} = 0.06480\text{ g}$

From the previous question, we are given that 1 grain troy = 0.0648 g = 1 grain ap. So the two are the same.

b. $1\text{ oz ap} \times \frac{8\text{ dram ap}}{\text{oz ap}} \times \frac{3.888\text{g}}{\text{dram ap}} \times \frac{1\text{ oz troy*}}{31.1\text{ g}} = 1.00\text{ oz troy}$; *see Exercise 45b.

c. $5.00 \times 10^2\text{ mg} \times \frac{1\text{ g}}{1000\text{mg}} \times \frac{1\text{ dram ap}}{3.888\text{g}} \times \frac{3\text{ scruples}}{\text{dram ap}} = 0.386\text{ scruple}$

$0.386\text{ scruple} \times \frac{20\text{ grains ap}}{\text{scruple}} = 7.72\text{ grains ap}$

d. $1\text{ scruple} \times \frac{1\text{ dram ap}}{3\text{ scruples}} \times \frac{3.888\text{g}}{\text{dram ap}} = 1.296\text{ g}$

47. $15.6\text{ g} \times \frac{1\text{ capsule}}{0.65\text{ g}} = 24\text{ capsules}$

48. $1.5\text{ teaspoons} \times \frac{80.\text{ mg acet}}{0.50\text{ teaspoon}} = 240\text{ mg acetaminophen}$

$\frac{240\text{mg acet}}{24\text{ lb}} \times \frac{1\text{ lb}}{0.454\text{kg}} = 22\text{ mg acetaminophen/kg}$

$\frac{240\text{mg acet}}{35\text{ lb}} \times \frac{1\text{ lb}}{0.454\text{kg}} = 15\text{ mg acetaminophen/kg}$

The range is from 15 to 22 mg acetaminophen per kg of body weight.

49. $\text{warp } 1.71 = \left(5.00 \times \frac{3.00 \times 10^8\text{ m}}{\text{s}}\right) \times \frac{1.094\text{yd}}{\text{m}} \times \frac{60\text{ s}}{\text{min}} \times \frac{60\text{ min}}{\text{h}} \times \frac{1\text{ knot}}{2030\text{yd/h}}$

$= 2.91 \times 10^9\text{ knots}$

$\left(5.00 \times \frac{3.00 \times 10^8\text{ m}}{\text{s}}\right) \times \frac{1\text{ km}}{1000\text{m}} \times \frac{1\text{ mi}}{1.609\text{km}} \times \frac{60\text{ s}}{\text{min}} \times \frac{60\text{ min}}{\text{h}} = 3.36 \times 10^9\text{ mi/h}$

50. $\frac{100.\text{ m}}{9.58\text{ s}} = 10.4\text{ m/s}$; $\frac{100.\text{ m}}{9.58\text{ s}} \times \frac{1\text{ km}}{1000\text{m}} \times \frac{60\text{ s}}{\text{min}} \times \frac{60\text{ min}}{\text{h}} = 37.6\text{ km/h}$

$$\frac{100.\text{ m}}{9.58\text{ s}} \times \frac{1.0936\text{ yd}}{\text{m}} \times \frac{3\text{ ft}}{\text{yd}} = 34.2\text{ ft/s}; \quad \frac{34.2\text{ ft}}{\text{s}} \times \frac{1\text{ mi}}{5280\text{ ft}} \times \frac{60\text{ s}}{\text{min}} \times \frac{60\text{ min}}{\text{h}} = 23.3\text{ mi/h}$$

$$1.00 \times 10^2\text{ yd} \times \frac{1\text{ m}}{1.0936\text{ yd}} \times \frac{9.58\text{ s}}{100.\text{ m}} = 8.76\text{ s}$$

51. $$\frac{65\text{ km}}{\text{h}} \times \frac{0.6214\text{ mi}}{\text{km}} = 40.4 = 40.\text{ mi/h}$$

To the correct number of significant figures (2), 65 km/h does not violate a 40 mi/h speed limit.

52. $$112\text{ km} \times \frac{0.6214\text{ mi}}{\text{km}} \times \frac{1\text{ h}}{65\text{ mi}} = 1.1\text{ h} = 1\text{ h and } 6\text{ min}$$

$$112\text{ km} \times \frac{0.6214\text{ mi}}{\text{km}} \times \frac{1\text{ gal}}{28\text{ mi}} \times \frac{3.785\text{ L}}{\text{gal}} = 9.4\text{ L of gasoline}$$

53. $$\frac{2.45\text{ euros}}{\text{kg}} \times \frac{1\text{ kg}}{2.2046\text{ lb}} \times \frac{\$1.32}{\text{euro}} = \$1.47/\text{lb}$$

One pound of peaches costs $1.47.

54. For the gasoline car:

$$500.\text{ mi} \times \frac{1\text{ gal}}{28.0\text{ mi}} \times \frac{\$3.50}{\text{gal}} = \$62.5$$

For the E85 car:

$$500.\text{ mi} \times \frac{1\text{ gal}}{22.5\text{ mi}} \times \frac{\$2.85}{\text{gal}} = \$63.3$$

The E85 vehicle would cost slightly more to drive 500. miles as compared to the gasoline vehicle ($63.3 versus $62.5).

55. $$\text{Volume of lake} = 100\text{ mi}^2 \times \left(\frac{5280\text{ ft}}{\text{mi}}\right)^2 \times 20\text{ ft} = 6 \times 10^{10}\text{ ft}^3$$

$$6 \times 10^{10}\text{ ft}^3 \times \left(\frac{12\text{ in}}{\text{ft}} \times \frac{2.54\text{ cm}}{\text{in}}\right)^3 \times \frac{1\text{ mL}}{\text{cm}^3} \times \frac{0.4\ \mu\text{g}}{\text{mL}} = 7 \times 10^{14}\ \mu\text{g mercury}$$

$$7 \times 10^{14}\ \mu\text{g} \times \frac{1\text{ g}}{1 \times 10^6\ \mu\text{g}} \times \frac{1\text{ kg}}{1 \times 10^3\text{ g}} = 7 \times 10^5\text{ kg of mercury}$$

56. Volume of room = 18 ft × 12 ft × 8 ft = 1700 ft^3 (carrying one extra significant figure)

$$1700\ \text{ft}^3 \times \left(\frac{12\ \text{in}}{\text{ft}}\right)^3 \times \left(\frac{2.54\ \text{cm}}{\text{in}}\right)^3 \times \left(\frac{1\ \text{m}}{100\ \text{cm}}\right)^3 = 48\ \text{m}^3$$

$$48\ \text{m}^3 \times \frac{400{,}000\ \mu\text{g CO}}{\text{m}^3} \times \frac{1\ \text{g CO}}{1 \times 10^6\ \mu\text{g CO}} = 19\ \text{g} = 20\ \text{g CO (to 1 sig. fig.)}$$

Temperature

57. a. $T_C = \frac{5}{9}(T_F - 32) = \frac{5}{9}(-459°F - 32) = -273°C$; $T_K = T_C + 273 = -273°C + 273 = 0\ K$

b. $T_C = \frac{5}{9}(-40.°F - 32) = -40.°C$; $T_K = -40.°C + 273 = 233\ K$

c. $T_C = \frac{5}{9}(68°F - 32) = 20.°C$; $T_K = 20.°C + 273 = 293\ K$

d. $T_C = \frac{5}{9}(7 \times 10^7°F - 32) = 4 \times 10^7°C$; $T_K = 4 \times 10^7°C + 273 = 4 \times 10^7\ K$

58. 96.1°F ±0.2°F; first, convert 96.1°F to °C. $T_C = \frac{5}{9}(T_F - 32) = \frac{5}{9}(96.1 - 32) = 35.6°C$

A change in temperature of 9°F is equal to a change in temperature of 5°C. So the uncertainty is:

$$\pm 0.2°F \times \frac{5°\ C}{9°\ F} = \pm 0.1°C.$$ Thus 96.1 ±0.2°F = 35.6 ±0.1°C.

59. a. $T_F = \frac{9}{5} \times T_C + 32 = \frac{9}{5} \times 39.2°C + 32 = 102.6°F$ (*Note*: 32 is exact.)

$T_K = T_C + 273.2 = 39.2 + 273.2 = 312.4\ K$

b. $T_F = \frac{9}{5} \times (-25) + 32 = -13°F$; $T_K = -25 + 273 = 248\ K$

c. $T_F = \frac{9}{5} \times (-273) + 32 = -459°F$; $T_K = -273 + 273 = 0\ K$

d. $T_F = \frac{9}{5} \times 801 + 32 = 1470°F$; $T_K = 801 + 273 = 1074\ K$

60. a. $T_C = T_K - 273 = 233 - 273 = -40.°C$

$T_F = \frac{9}{5} \times T_C + 32 = \frac{9}{5} \times (-40.) + 32 = -40.°F$

b. $T_C = 4 - 273 = -269°C$; $T_F = \frac{9}{5} \times (-269) + 32 = -452°F$

c. $T_C = 298 - 273 = 25°C$; $T_F = \frac{9}{5} \times 25 + 32 = 77°F$

d. $T_C = 3680 - 273 = 3410°C$; $T_F = \frac{9}{5} \times 3410 + 32 = 6170°F$

61. $T_F = \frac{9}{5} \times T_C + 32$; from the problem, we want the temperature where $T_F = 2T_C$.

Substituting:

$$2T_C = \frac{9}{5} \times T_C + 32,\ (0.2)T_C = 32,\ T_C = \frac{32}{0.2} = 160°C$$

$T_F = 2T_C$ when the temperature in Fahrenheit is 2(160) = 320°F. Because all numbers when solving the equation are exact numbers, the calculated temperatures are also exact numbers.

62. $T_C = \frac{5}{9}(T_F - 32) = \frac{5}{9}(72 - 32) = 22°C$

$T_C = T_K - 273 = 313 - 273 = 40.°C$

The difference in temperature between Jupiter at 313 K and Earth at 72°F is 40.°C – 22 °C = 18°C.

63. a. A change in temperature of 140°C is equal to 50°X. Therefore, $\frac{140°C}{50°X}$ is the unit conversion between a degree on the X scale to a degree on the Celsius scale. To account for the different zero points, –10° must be subtracted from the temperature on the X scale to get to the Celsius scale. The conversion between °X to °C is:

$$T_C = T_X \times \frac{140°C}{50°X} - 10°C,\ T_C = T_X \times \frac{14°C}{5°X} - 10°C$$

The conversion between °C to °X would be:

$$T_X = (T_C + 10°C)\ \frac{5°X}{14°C}$$

b. Assuming 10°C and $\frac{5°X}{14°C}$ are exact numbers:

$$T_X = (22.0°C + 10°C)\ \frac{5°X}{14°C} = 11.4°X$$

c. Assuming exact numbers in the temperature conversion formulas:

$$T_C = 58.0°X \times \frac{14°C}{5°X} - 10°C = 152°C$$

$$T_K = 152°C + 273 = 425\ K$$

$$T_F = \frac{9°F}{5°C} \times 152°C + 32°F = 306°F$$

64. a.

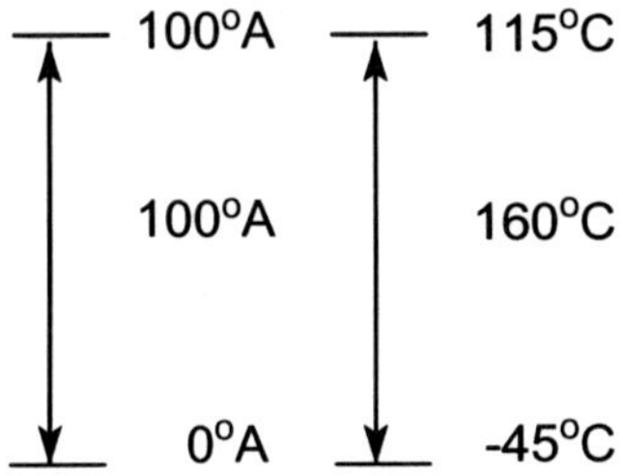

A change in temperature of 160°C equals a change in temperature of 100°A.

So $\frac{160°C}{100°A}$ is our unit conversion for a degree change in temperature.

At the freezing point: 0°A = –45°C

Combining these two pieces of information:

$$T_A = (T_C + 45°C) \times \frac{100°A}{160°C} = (T_C + 45°C) \times \frac{5°A}{8°C} \text{ or } T_C = T_A \times \frac{8°C}{5°A} - 45°C$$

b. $T_C = (T_F - 32) \times \frac{5}{9};\ T_C = T_A \times \frac{8}{5} - 45 = (T_F - 32) \times \frac{5}{9}$

$$T_F - 32 = \frac{9}{5} \times \left(T_A \times \frac{8}{5} - 45\right) = T_A \times \frac{72}{25} - 81,\ T_F = T_A \times \frac{72°F}{25°A} - 49°F$$

c. $T_C = T_A \times \frac{8}{5} - 45$ and $T_C = T_A$; so $T_C = T_C \times \frac{8}{5} - 45,\ \frac{3T_C}{5} = 45,\ T_C = 75°C = 75°A$

d. $T_C = 86°A \times \frac{8°C}{5°A} - 45°C = 93°C;\ T_F = 86°A \times \frac{72°F}{25°A} - 49°F = 199°F = 2.0 \times 10^2 °F$

e. $T_A = (45°C + 45°C) \times \frac{5°A}{8°C} = 56°A$

Density

65. Mass = $350\ lb \times \frac{453.6\ g}{lb} = 1.6 \times 10^5\ g;\ V = 1.2 \times 10^4\ in^3 \times \left(\frac{2.54\ cm}{in}\right)^3 = 2.0 \times 10^5\ cm^3$

$$\text{Density} = \frac{\text{mass}}{\text{volume}} = \frac{1 \times 10^5\ g}{2.0 \times 10^5\ cm^3} = 0.80\ g/cm^3$$

Because the material has a density less than water, it will float in water.

66. $V = \frac{4}{3}\pi r^3 = \frac{4}{3} \times 3.14 \times (0.50\ cm)^3 = 0.52\ cm^3;\ d = \frac{2.0\ g}{0.52\ cm^3} = 3.8\ g/cm^3$

The ball will sink.

67. $V = \frac{4}{3}\pi r^3 = \frac{4}{3} \times 3.14 \times \left(7.0 \times 10^5 \text{ km} \times \frac{1000\text{m}}{\text{km}} \times \frac{100\text{cm}}{\text{m}}\right)^3 = 1.4 \times 10^{33} \text{ cm}^3$

$$\text{Density} = \frac{\text{mass}}{\text{volume}} = \frac{2 \times 10^{36} \text{ kg} \times \frac{1000\text{g}}{\text{kg}}}{1.4 \times 10^{33} \text{ cm}^3} = 1.4 \times 10^6 \text{ g/cm}^3 = 1 \times 10^6 \text{ g/cm}^3$$

68. $V = l \times w \times h = 2.9 \text{ cm} \times 3.5 \text{ cm} \times 10.0 \text{ cm} = 1.0 \times 10^2 \text{ cm}^3$

$$d = \text{density} = \frac{615.0 \text{ g}}{1.0 \times 10^2 \text{ cm}^3} = \frac{6.2 \text{ g}}{\text{cm}^3}$$

69. a. $5.0 \text{ carat} \times \frac{0.200\text{g}}{\text{carat}} \times \frac{1 \text{ cm}^3}{3.51\text{g}} = 0.28 \text{ cm}^3$

b. $2.8 \text{ mL} \times \frac{1 \text{ cm}^3}{\text{mL}} \times \frac{3.51\text{g}}{\text{cm}^3} \times \frac{1 \text{ carat}}{0.200\text{g}} = 49 \text{ carats}$

70. For ethanol: $100. \text{ mL} \times \frac{0.789\text{g}}{\text{mL}} = 78.9 \text{ g}$

For benzene: $1.00 \text{ L} \times \frac{1000\text{mL}}{\text{L}} \times \frac{0.880\text{g}}{\text{mL}} = 880. \text{ g}$

Total mass = 78.9 g + 880. g = 959 g

71. $V = 21.6 \text{ mL} - 12.7 \text{ mL} = 8.9 \text{ mL}$; density $= \frac{33.42 \text{ g}}{8.9 \text{ mL}} = 3.8 \text{ g/mL} = 3.8 \text{ g/cm}^3$

72. $5.25 \text{ g} \times \frac{1 \text{ cm}^3}{10.5 \text{ g}} = 0.500 \text{ cm}^3 = 0.500 \text{ mL}$

The volume in the cylinder will rise to 11.7 mL (11.2 mL + 0.500 mL = 11.7 mL).

73. a. Both have the same mass of 1.0 kg.

b. 1.0 mL of mercury; mercury is more dense than water. *Note*: 1 mL = 1 cm^3.

$1.0 \text{ mL} \times \frac{13.6 \text{ g}}{\text{mL}} = 14 \text{ g of mercury}$; $1.0 \text{ mL} \times \frac{0.998\text{g}}{\text{mL}} = 1.0 \text{ g of water}$

c. Same; both represent 19.3 g of substance.

$19.3 \text{ mL} \times \frac{0.9982\text{g}}{\text{mL}} = 19.3 \text{ g of water}$; $1.00 \text{ mL} \times \frac{19.32 \text{ g}}{\text{mL}} = 19.3 \text{ g of gold}$

d. 1.0 L of benzene (880 g versus 670 g)

$75 \text{ mL} \times \frac{8.96 \text{ g}}{\text{mL}} = 670 \text{ g of copper}$; $1.0 \text{ L} \times \frac{1000\text{mL}}{\text{L}} \times \frac{0.880\text{g}}{\text{mL}} = 880 \text{ g of benzene}$

74. a. $1.50 \text{ qt} \times \frac{1 \text{ L}}{1.0567 \text{qt}} \times \frac{1000 \text{mL}}{\text{L}} \times \frac{0.789 \text{g}}{\text{mL}} = 1120 \text{ g ethanol}$

b. $3.5 \text{ in}^3 \times \left(\frac{2.54 \text{ cm}}{\text{in}}\right)^3 \times \frac{13.6 \text{ g}}{\text{cm}^3} = 780 \text{ g mercury}$

75. a. 1.0 kg feather; feathers are less dense than lead.

b. 100 g water; water is less dense than gold.

c. Same; both volumes are 1.0 L.

76. a. $H_2(g)$: $V = 25.0 \text{ g} \times \frac{1 \text{ cm}^3}{0.000084 \text{g}} = 3.0 \times 10^5 \text{ cm}^3$ [$H_2(g)$ = hydrogen gas.]

b. $H_2O(l)$: $V = 25.0 \text{ g} \times \frac{1 \text{ cm}^3}{0.9982 \text{g}} = 25.0 \text{ cm}^3$ [$H_2O(l)$ = water.]

c. $Fe(s)$: $V = 25.0 \text{ g} \times \frac{1 \text{ cm}^3}{7.87 \text{ g}} = 3.18 \text{ cm}^3$ [$Fe(s)$ = iron.]

Notice the huge volume of the gaseous H_2 sample as compared to the liquid and solid samples. The same mass of gas occupies a volume that is over 10,000 times larger than the liquid sample. Gases are indeed mostly empty space.

77. $V = 1.00 \times 10^3 \text{ g} \times \frac{1 \text{ cm}^3}{22.57 \text{ g}} = 44.3 \text{ cm}^3$

$44.3 \text{ cm}^3 = l \times w \times h = 4.00 \text{ cm} \times 4.00 \text{ cm} \times h, \ h = 2.77 \text{ cm}$

78. $V = 22 \text{ g} \times \frac{1 \text{ cm}^3}{8.96 \text{ g}} = 2.5 \text{ cm}^3$; $V = \pi r^2 \times l$, where l = length of the wire

$2.5 \text{ cm}^3 = \pi \times \left(\frac{0.25 \text{ mm}}{2}\right)^2 \times \left(\frac{1 \text{ cm}}{10 \text{ mm}}\right)^2 \times l, \ l = 5.1 \times 10^3 \text{ cm} = 170 \text{ ft}$

Classification and Separation of Matter

79. A gas has molecules that are very far apart from each other, whereas a solid or liquid has molecules that are very close together. An element has the same type of atom, whereas a compound contains two or more different elements. Picture i represents an element that exists as two atoms bonded together (like H_2 or O_2 or N_2). Picture iv represents a compound (like CO, NO, or HF). Pictures iii and iv contain representations of elements that exist as individual atoms (like Ar, Ne, or He).

a. Picture iv represents a gaseous compound. Note that pictures ii and iii also contain a gaseous compound, but they also both have a gaseous element present.

b. Picture vi represents a mixture of two gaseous elements.

c. Picture v represents a solid element.

d. Pictures ii and iii both represent a mixture of a gaseous element and a gaseous compound.

80. Solid: rigid; has a fixed volume and shape; slightly compressible

Liquid: definite volume but no specific shape; assumes shape of the container; slightly Compressible

Gas: no fixed volume or shape; easily compressible

Pure substance: has constant composition; can be composed of either compounds or elements

Element: substances that cannot be decomposed into simpler substances by chemical or physical means.

Compound: a substance that can be broken down into simpler substances (elements) by chemical processes.

Homogeneous mixture: a mixture of pure substances that has visibly indistinguishable parts.

Heterogeneous mixture: a mixture of pure substances that has visibly distinguishable parts.

Solution: a homogeneous mixture; can be a solid, liquid or gas

Chemical change: a given substance becomes a new substance or substances with different properties and different composition.

Physical change: changes the form (g, l, or s) of a substance but does no change the chemical composition of the substance.

81. Homogeneous: Having visibly indistinguishable parts (the same throughout).

Heterogeneous: Having visibly distinguishable parts (not uniform throughout).

a. heterogeneous (due to hinges, handles, locks, etc.)

b. homogeneous (hopefully; if you live in a heavily polluted area, air may be heterogeneous.)

c. homogeneous

d. homogeneous (hopefully, if not polluted)

e. heterogeneous

f. heterogeneous

82. a. heterogeneous

b. homogeneous

c. heterogeneous

d. homogeneous (assuming no imperfections in the glass)

e. heterogeneous (has visibly distinguishable parts)

83. a. pure b. mixture c. mixture d. pure e. mixture (copper and zinc)

f. pure g. mixture h. mixture i. mixture

Iron and uranium are elements. Water (H_2O) is a compound because it is made up of two or more different elements. Table salt is usually a homogeneous mixture composed mostly of sodium chloride (NaCl), but will usually contain other substances that help absorb water vapor (an anticaking agent).

84. Initially, a mixture is present. The magnesium and sulfur have only been placed together in the same container at this point, but no reaction has occurred. When heated, a reaction occurs. Assuming the magnesium and sulfur had been measured out in exactly the correct ratio for complete reaction, the remains after heating would be a pure compound composed of magnesium and sulfur. However, if there were an excess of either magnesium or sulfur, the remains after reaction would be a mixture of the compound produced and the excess reactant.

85. Chalk is a compound because it loses mass when heated and appears to change into another substance with different physical properties (the hard chalk turns into a crumbly substance).

86. Because vaporized water is still the *same substance* as solid water (H_2O), no chemical reaction has occurred. Sublimation is a physical change.

87. A physical change is a change in the state of a substance (solid, liquid, and gas are the three states of matter); a physical change does not change the chemical composition of the substance. A chemical change is a change in which a given substance is converted into another substance having a different formula (composition).

a. Vaporization refers to a liquid converting to a gas, so this is a physical change. The formula (composition) of the moth ball does not change.

b. This is a chemical change since hydrofluoric acid (HF) is reacting with glass (SiO_2) to form new compounds that wash away.

c. This is a physical change because all that is happening during the boiling process is the conversion of liquid alcohol to gaseous alcohol. The alcohol formula (C_2H_5OH) does not change.

d. This is a chemical change since the acid is reacting with cotton to form new compounds.

88. a. Distillation separates components of a mixture, so the orange liquid is a mixture (has an average color of the yellow liquid and the red solid). Distillation utilizes boiling point differences to separate out the components of a mixture. Distillation is a physical change because the components of the mixture do not become different compounds or elements.

b. Decomposition is a type of chemical reaction. The crystalline solid is a compound, and decomposition is a chemical change where new substances are formed.

c. Tea is a mixture of tea compounds dissolved in water. The process of mixing sugar into tea is a physical change. Sugar doesn't react with the tea compounds, it just makes the solution sweeter.

Additional Exercises

89. Because each pill is 4.0% Lipitor by mass, for every 100.0 g of pills, there are 4.0 g of Lipitor present. Note that 100 pills is assumed to be an exact number.

$$100 \text{ pills} \times \frac{2.5 \text{ g}}{\text{pill}} \times \frac{4.0 \text{ g Lipitor}}{100.0 \text{ g pills}} \times \frac{1 \text{ kg}}{1000 \text{g}} = 0.010 \text{ kg Lipitor}$$

90. $126 \text{ gal} \times \frac{4 \text{ qt}}{\text{gal}} \times \frac{1 \text{ L}}{1.057 \text{qt}} = 477 \text{ L}$

91. $\text{Total volume} = \left(200.\text{ m} \times \frac{100 \text{cm}}{\text{m}}\right) \times \left(300.\text{ m} \times \frac{100 \text{cm}}{\text{m}}\right) \times 4.0 \text{ cm} = 2.4 \times 10^9 \text{ cm}^3$

Volume of topsoil covered by 1 bag =

$$\left[10.\text{ ft}^2 \times \left(\frac{12 \text{ in}}{\text{ft}}\right)^2 \times \left(\frac{2.54 \text{ cm}}{\text{in}}\right)^2\right] \times \left(1.0 \text{ in} \times \frac{2.54 \text{ cm}}{\text{in}}\right) = 2.4 \times 10^4 \text{ cm}^3$$

$$2.4 \times 10^9 \text{ cm}^3 \times \frac{1 \text{ bag}}{2.4 \times 10^4 \text{ cm}^3} = 1.0 \times 10^5 \text{ bags topsoil}$$

92. a. No; if the volumes were the same, then the gold idol would have a much greater mass because gold is much more dense than sand.

b. $\text{Mass} = 1.0 \text{ L} \times \frac{1000 \text{cm}^3}{\text{L}} \times \frac{19.32 \text{ g}}{\text{cm}^3} \times \frac{1 \text{ kg}}{1000 \text{g}} = 19.32 \text{ kg} \ (= 42.59 \text{ lb})$

It wouldn't be easy to play catch with the idol because it would have a mass of over 40 pounds.

93. $1 \text{ light year} = 1 \text{ yr} \times \frac{365 \text{day}}{\text{yr}} \times \frac{24 \text{ h}}{\text{day}} \times \frac{60 \text{ min}}{\text{h}} \times \frac{60 \text{ s}}{\text{min}} \times \frac{186{,}000 \text{mi}}{\text{s}}$

$= 5.87 \times 10^{12} \text{ miles}$

$$9.6 \text{ parsecs} \times \frac{3.26 \text{ light yr}}{\text{parsec}} \times \frac{5.87 \times 10^{12} \text{ mi}}{\text{light yr}} \times \frac{1.609 \text{km}}{\text{mi}} \times \frac{1000 \text{m}}{\text{km}} = 3.0 \times 10^{17} \text{ m}$$

94. $1 \text{ s} \times \frac{1 \text{ min}}{60 \text{s}} \times \frac{1 \text{ h}}{60 \text{ min}} \times \frac{65 \text{ mi}}{\text{h}} \times \frac{5280 \text{ft}}{\text{mi}} = 95.3 \text{ ft} = 100 \text{ ft}$

If you take your eyes off the road for one second traveling at 65 mph, your car travels approximately 100 feet.

95. a. $0.25 \text{ lb} \times \frac{453.6 \text{ g}}{\text{lb}} \times \frac{1.0 \text{ g trytophan}}{100.0 \text{ g turkey}} = 1.1 \text{ g tryptophan}$

b. $0.25 \text{ qt} \times \frac{0.9463 \text{L}}{\text{qt}} \times \frac{1.04 \text{kg}}{\text{L}} \times \frac{1000 \text{kg}}{\text{kg}} \times \frac{2.0 \text{ g tryptophan}}{100.0 \text{ g milk}} = 4.9 \text{ g tryptophan}$

96. A chemical change involves the change of one or more substances into other substances through a reorganization of the atoms. A physical change involves the change in the form of a substance, but not its chemical composition.

a. physical change (Just smaller pieces of the same substance.)

b. chemical change (Chemical reactions occur.)

c. chemical change (Bonds are broken.)

d. chemical change (Bonds are broken.)

e. physical change (Water is changed from a liquid to a gas.)

f. physical change (Chemical composition does not change.)

97. $18.5 \text{ cm} \times \frac{10.0^\circ \text{F}}{5.25 \text{ cm}} = 35.2^\circ\text{F}$ increase; $T_{\text{final}} = 98.6 + 35.2 = 133.8^\circ\text{F}$

$T_c = 5/9\ (133.8 - 32) = 56.56^\circ\text{C}$

98. $\text{Mass}_{\text{benzene}} = 58.80 \text{ g} - 25.00 \text{ g} = 33.80 \text{ g}$; $V_{\text{benzene}} = 33.80 \text{ g} \times \frac{1 \text{ cm}^3}{0.880 \text{g}} = 38.4 \text{ cm}^3$

$V_{\text{solid}} = 50.0 \text{ cm}^3 - 38.4 \text{ cm}^3 = 11.6 \text{ cm}^3$; density $= \frac{25.00 \text{ g}}{11.6 \text{ cm}^3} = 2.16 \text{ g/cm}^3$

99. a. Volume × density = mass; the orange block is more dense. Because mass (orange) > mass (blue) and because volume (orange) < volume (blue), the density of the orange block must be greater to account for the larger mass of the orange block.

b. Which block is more dense cannot be determined. Because mass (orange) > mass (blue) and because volume (orange) > volume (blue), the density of the orange block may or may not be larger than the blue block. If the blue block is more dense, its density cannot be so large that its mass is larger than the orange block's mass.

c. The blue block is more dense. Because mass (blue) = mass (orange) and because volume (blue) < volume (orange), the density of the blue block must be larger in order to equate the masses.

d. The blue block is more dense. Because mass (blue) > mass (orange) and because the volumes are equal, the density of the blue block must be larger in order to give the blue block the larger mass.

100. Circumference = c = 2πr; $V = \frac{4\pi r^3}{3} = \frac{4\pi}{3}\left(\frac{c}{2\pi}\right)^3 = \frac{c^3}{6\pi^2}$

$$\text{Largest density} = \frac{5.25\text{ oz}}{\frac{(9.00\text{ in})^3}{6\pi^2}} = \frac{5.25\text{ oz}}{12.3\text{ in}^3} = \frac{0.427\text{ oz}}{\text{in}^3}$$

$$\text{Smallest density} = \frac{5.00\text{ oz}}{\frac{(9.25\text{ in})^3}{6\pi^2}} = \frac{5.00\text{ oz}}{13.4\text{ in}^3} = \frac{0.73\text{ oz}}{\text{in}^3}$$

Maximum range is: $\frac{(0.373 - 0.427)\text{ oz}}{\text{in}^3}$ or 0.40 ±0.03 oz/in^3 (Uncertainty is in 2nd decimal place.)

101. $V = V_{final} - V_{initial}$; $d = \frac{28.90\text{ g}}{9.8\text{ cm}^3 - 6.4\text{ cm}^3} = \frac{28.90\text{ g}}{3.4\text{ cm}^3} = 8.5\text{ g/cm}^3$

$d_{max} = \frac{mass_{max}}{V_{min}}$; we get V_{min} from 9.7 cm^3 – 6.5 cm^3 = 3.2 cm^3.

$d_{max} = \frac{28.93\text{ g}}{3.2\text{ cm}^3} = \frac{9.0\text{ g}}{\text{cm}^3}$; $d_{min} = \frac{mass_{min}}{V_{max}} = \frac{28.87\text{ g}}{9.9\text{ cm}^3 - 6.3\text{ cm}^3} = \frac{8.0\text{ g}}{\text{cm}^3}$

The density is 8.5 ±0.5 g/cm^3.

102. We need to calculate the maximum and minimum values of the density, given the uncertainty in each measurement. The maximum value is:

$$d_{max} = \frac{19.625\text{ g} + 0.002\text{ g}}{25.00\text{ cm}^3 - 0.03\text{ cm}^3} = \frac{19.627\text{ g}}{24.97\text{ cm}^3} = 0.7860\text{ g/cm}^3$$

The minimum value of the density is:

$$d_{min} = \frac{19.625\text{ g} - 0.002\text{ g}}{25.00\text{ cm}^3 + 0.03\text{ cm}^3} = \frac{19.623\text{ g}}{25.03\text{ cm}^3} = 0.7840\text{ g/cm}^3$$

The density of the liquid is between 0.7840 and 0.7860 g/cm^3. These measurements are sufficiently precise to distinguish between ethanol (d = 0.789 g/cm^3) and isopropyl alcohol (d = 0.785 g/cm^3).

ChemWork Problems

The answers to the problems 103-108 (or a variation to these problems) are found in OWL. These problems are also assignable in OWL.

Challenge Problems

109. In a subtraction, the result gets smaller, but the uncertainties add. If the two numbers are very close together, the uncertainty may be larger than the result. For example, let's assume we want to take the difference of the following two measured quantities, 999,999 ±2 and 999,996 ±2. The difference is 3 ±4. Because of the uncertainty, subtracting two similar numbers is poor practice.

110. In general, glassware is estimated to one place past the markings.

a. 128.7 mL glassware   b. 18 mL glassware   c. 23.45 mL glassware

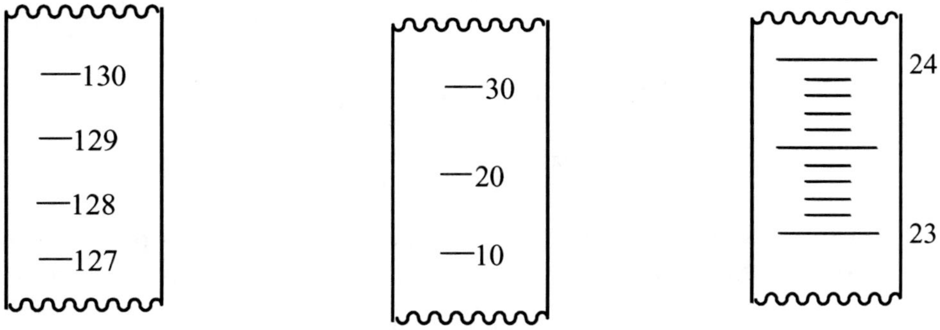

read to tenth's place   read to one's place   read to two decimal places

Total volume = 128.7 + 18 + 23.45 = 170.15 = 170. (Due to 18, the sum would be known only to the ones place.)

111. a. $\frac{2.70 - 2.64}{2.70} \times 100 = 2\%$ b. $\frac{|16.12 - 16.48|}{16.12} \times 100 = 2.2\%$

c. $\frac{1.000 - 0.9981}{1.000} \times 100 = \frac{0.002}{1.000} \times 100 = 0.2\%$

112. a. At some point in 1982, the composition of the metal used in minting pennies was changed because the mass changed during this year (assuming the volume of the pennies were constant).

b. It should be expressed as 3.08 ±0.05 g. The uncertainty in the second decimal place will swamp any effect of the next decimal places.

113. Heavy pennies (old): mean mass = 3.08 ±0.05 g

Light pennies (new): mean mass = $\frac{(2.467 + 2.545 + 2.518)}{3}$ = 2.51 ±0.04 g

Because we are assuming that volume is additive, let's calculate the volume of 100.0 g of each type of penny, then calculate the density of the alloy. For 100.0 g of the old pennies, 95 g will be Cu (copper) and 5 g will be Zn (zinc).

$V = 95 \text{ g Cu} \times \frac{1\text{ cm}^3}{8.96\text{g}} + 5 \text{ g Zn} \times \frac{1\text{ cm}^3}{7.14\text{g}} = 11.3 \text{ cm}^3$ (carrying one extra sig. fig.)

Density of old pennies $= \frac{100.\text{ g}}{11.3\text{ cm}^3} = 8.8 \text{ g/cm}^3$

For 100.0 g of new pennies, 97.6 g will be Zn and 2.4 g will be Cu.

$V = 2.4 \text{ g Cu} \times \frac{1\text{ cm}^3}{8.96\text{g}} + 97.6 \text{ g Zn} \times \frac{1\text{ cm}^3}{7.14\text{g}} = 13.94 \text{ cm}^3$ (carrying one extra sig. fig.)

Density of new pennies $= \frac{100.\text{ g}}{13.94\text{ cm}^3} = 7.17 \text{ g/cm}^3$

$d = \dfrac{\text{mass}}{\text{volume}}$; because the volume of both types of pennies are assumed equal, then:

$$\frac{d_{new}}{d_{old}} = \frac{\text{mass}_{new}}{\text{mass}_{old}} = \frac{7.17\ g/cm^3}{8.8\ g/cm^3} = 0.81$$

The calculated average mass ratio is: $\dfrac{\text{mass}_{new}}{\text{mass}_{old}} = \dfrac{2.51\ g}{3.08\ g} = 0.815$

To the first two decimal places, the ratios are the same. If the assumptions are correct, then we can reasonably conclude that the difference in mass is accounted for by the difference in alloy used.

114. a. At 8 a.m., approximately 57 cars pass through the intersection per hour.

b. At 12 a.m. (midnight), only 1 or 2 cars pass through the intersection per hour.

c. Traffic at the intersection is limited to less than 10 cars per hour from 8 p.m. to 5 a.m. Starting at 6 a.m., there is a steady increase in traffic through the intersection, peaking at 8 a.m. when approximately 57 cars pass per hour. Past 8 a.m. traffic moderates to about 40 cars through the intersection per hour until noon, and then decreases to 21 cars per hour by 3 p.m. Past 3 p.m. traffic steadily increases to a peak of 52 cars per hour at 5 p.m., and then steadily decreases to the overnight level of less than 10 cars through the intersection per hour.

d. The traffic pattern through the intersection is directly related to the work schedules of the general population as well as to the store hours of the businesses in downtown.

e. Run the same experiment on a Sunday, when most of the general population doesn't work and when a significant number of downtown stores are closed in the morning.

115. Let x = mass of copper and y = mass of silver.

$105.0\ g = x + y$ and $10.12\ mL = \dfrac{x}{8.96} + \dfrac{y}{10.5}$; solving:

$$\left(10.12 = \frac{x}{8.96} + \frac{105.0 - x}{10.5}\right) \times 8.96 \times 10.5,\ 952.1 = (10.5)x + 940.8 - (8.96)x$$

(carrying 1 extra sig. fig.)

$$11.3 = (1.54)x,\ x = 7.3\ g;\ \text{mass \% Cu} = \frac{7.3\ g}{105.0\ g} \times 100 = 7.0\%\ \text{Cu}$$

116. a.

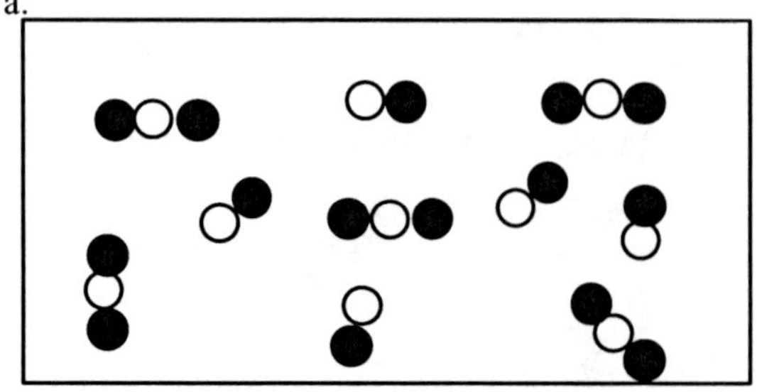

2 compounds

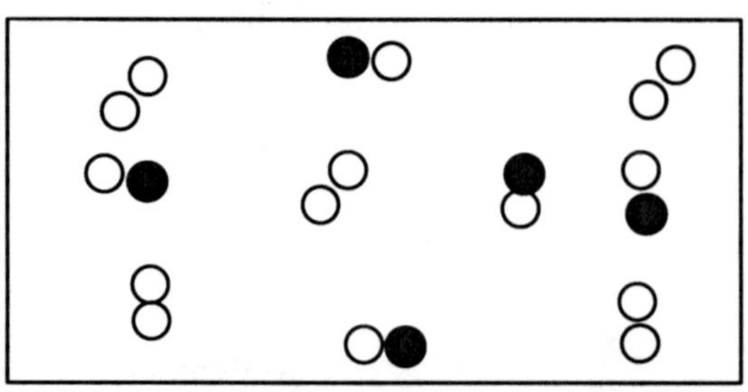

compound and element (diatomic)

b.

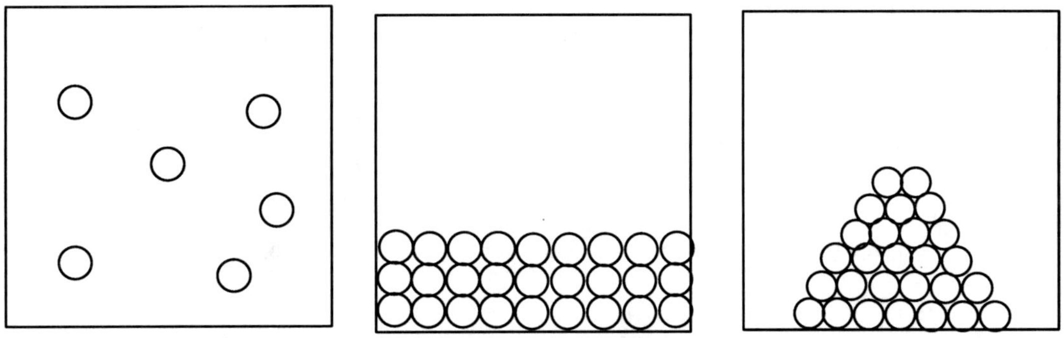

gas element (monoatomic)

atoms/molecules far apart; random order; takes volume of container

liquid element

atoms/molecules close together; somewhat ordered arrangement; takes volume of container

solid element

atoms/molecules close together; ordered arrangement; has its own volume

117. a. One possibility is that rope B is not attached to anything and rope A and rope C are connected via a pair of pulleys and/or gears.

b. Try to pull rope B out of the box. Measure the distance moved by C for a given movement of A. Hold either A or C firmly while pulling on the other rope.

118. The bubbles of gas is air in the sand that is escaping; methanol and sand are not reacting. We will assume that the mass of trapped air is insignificant.

Mass of dry sand = 37.3488 g – 22.8317 g = 14.5171 g

Mass of methanol = 45.2613 g – 37.3488 g = 7.9125 g

Volume of sand particles (air absent) = volume of sand and methanol – volume of methanol

Volume of sand particles (air absent) = 17.6 mL – 10.00 mL = 7.6 mL

$$\text{Density of dry sand (air present)} = \frac{14.5171\text{g}}{10.0\text{ mL}} = 1.45\text{ g/mL}$$

$$\text{Density of methanol} = \frac{7.9125\text{g}}{10.00\text{ mL}} = 0.7913\text{ g/mL}$$

$$\text{Density of sand particles (air absent)} = \frac{14.5171\text{g}}{7.6\text{ mL}} = 1.9\text{ g/mL}$$

Integrative Problems

119. 2.97×10^8 persons $\times$ 0.0100 = 2.97×10^6 persons contributing

$$\frac{\$4.75 \times 10^8}{2.97 \times 10^6\text{ persons}} = \$160./\text{person};\ \frac{\$160.}{\text{person}} \times \frac{20\text{ nickels}}{\$1} = 3.20 \times 10^3\text{ nickels/person}$$

$$\frac{\$160.}{\text{person}} \times \frac{1 £}{\$1.869} = 85.6 \text{ £/person}$$

120. $\frac{22610\text{kg}}{\text{m}^3} \times \frac{1000\text{g}}{\text{kg}} \times \frac{1\text{ m}^3}{1 \times 10^6 \text{ cm}^3} = 22.61 \text{ g/cm}^3$

Volume of block = 10.0 cm × 8.0 cm × 9.0 cm = 720 cm^3; $\frac{22.61\text{g}}{\text{cm}^3} \times 720 \text{ cm}^3 = 1.6 \times 10^4$ g

121. At 200.0°F: $T_C = \frac{5}{9}(200.0°\text{F} - 32°\text{F}) = 93.33°\text{C}$; $T_K = 93.33 + 273.15 = 366.48$ K

At −100.0°F: $T_C = \frac{5}{9}(-100.0°\text{F} - 32°\text{F}) = -73.33°\text{C}$; $T_K = -73.33°\text{C} + 273.15 = 199.82$ K

$\Delta T(°\text{C}) = [93.33°\text{C} - (-73.33°\text{C})] = 166.66°\text{C}$; $\Delta T(\text{K}) = (366.48 \text{ K} - 199.82 \text{ K}) = 166.66$ K

The "300 Club" name only works for the Fahrenheit scale; it does not hold true for the Celsius and Kelvin scales.

CHAPTER 2

ATOMS, MOLECULES, AND IONS

Questions

16. Some elements exist as molecular substances. That is, hydrogen normally exists as H_2 molecules, not single hydrogen atoms. The same is true for N_2, O_2, F_2, Cl_2, etc.

17. A compound will always contain the same numbers (and types) of atoms. A given amount of hydrogen will react only with a specific amount of oxygen. Any excess oxygen will remain unreacted.

18. The halogens have a high affinity for electrons, and one important way they react is to form anions of the type X^-. The alkali metals tend to give up electrons easily and in most of their compounds exist as M^+ cations. *Note*: These two very reactive groups are only one electron away (in the periodic table) from the least reactive family of elements, the noble gases.

19. Law of conservation of mass: Mass is neither created nor destroyed. The total mass before a chemical reaction always equals the total mass after a chemical reaction.

    Law of definite proportion: A given compound always contains exactly the same proportion of elements by mass. For example, water is always 1 g H for every 8 g oxygen.

    Law of multiple proportions: When two elements form a series of compounds, the ratios of the mass of the second element that combine with 1 g of the first element always can be reduced to small whole numbers: For CO_2 and CO discussed in Section 2.2, the mass ratios of oxygen that react with 1 g carbon in each compound are in a 2 : 1 ratio.

20. a. The smaller parts are electrons and the nucleus. The nucleus is broken down into protons and neutrons, which can be broken down into quarks. For our purpose, electrons, neutrons, and protons are the key smaller parts of an atom.

    b. All atoms of hydrogen have 1 proton in the nucleus. Different isotopes of hydrogen have 0, 1, or 2 neutrons in the nucleus. Because we are talking about atoms, this implies a neutral charge, which dictates 1 electron present for all hydrogen atoms. If charged ions were included, then different ions/atoms of H could have different numbers of electrons.

    c. Hydrogen atoms always have 1 proton in the nucleus, and helium atoms always have 2 protons in the nucleus. The number of neutrons can be the same for a hydrogen atom and a helium atom. Tritium (^3H) and ^4He both have 2 neutrons. Assuming neutral atoms, then the number of electrons will be 1 for hydrogen and 2 for helium.

    d. Water (H_2O) is always 1 g hydrogen for every 8 g of O present, whereas H_2O_2 is always 1 g hydrogen for every 16 g of O present. These are distinctly different compounds, each with its own unique relative number and types of atoms present.

e. A chemical equation involves a reorganization of the atoms. Bonds are broken between atoms in the reactants, and new bonds are formed in the products. The number and types of atoms between reactants and products do not change. Because atoms are conserved in a chemical reaction, mass is also conserved.

21. J. J. Thomson's study of cathode-ray tubes led him to postulate the existence of negatively charged particles that we now call electrons. Thomson also postulated that atoms must contain positive charge in order for the atom to be electrically neutral. Ernest Rutherford and his alpha bombardment of metal foil experiments led him to postulate the nuclear atom–an atom with a tiny dense center of positive charge (the nucleus) with electrons moving about the nucleus at relatively large distances away; the distance is so large that an atom is mostly empty space.

22. The atom is composed of a tiny dense nucleus containing most of the mass of the atom. The nucleus itself is composed of neutrons and protons. Neutrons have a mass slightly larger than that of a proton and have no charge. Protons, on the other hand, have a 1+ relative charge as compared to the 1– charged electrons; the electrons move about the nucleus at relatively large distances. The volume of space that the electrons move about is so large, as compared to the nucleus, that we say an atom is mostly empty space.

23. The number and arrangement of electrons in an atom determine how the atom will react with other atoms, i.e., the electrons determine the chemical properties of an atom. The number of neutrons present determines the isotope identity and the mass number.

24. Density = mass/volume; if the volumes are assumed equal, then the much more massive proton would have a much larger density than the relatively light electron.

25. For lighter, stable isotopes, the number of protons in the nucleus is about equal to the number of neutrons. When the number of protons and neutrons is equal to each other, the mass number (protons + neutrons) will be twice the atomic number (protons). Therefore, for lighter isotopes, the ratio of the mass number to the atomic number is close to 2. For example, consider ^{28}Si, which has 14 protons and (28 – 14 =) 14 neutrons. Here, the mass number to atomic number ratio is 28/14 = 2.0. For heavier isotopes, there are more neutrons than protons in the nucleus. Therefore, the ratio of the mass number to the atomic number increases steadily upward from 2 as the isotopes get heavier and heavier. For example, ^{238}U has 92 protons and (238 – 92 =) 146 neutrons. The ratio of the mass number to the atomic number for ^{238}U is 238/92 = 2.6.

26. Some properties of metals are

(1) conduct heat and electricity;
(2) malleable (can be hammered into sheets);
(3) ductile (can be pulled into wires);
(4) lustrous appearance;
(5) form cations when they form ionic compounds.

Nonmetals generally do not have these properties, and when they form ionic compounds, nonmetals always form anions.

27. Carbon is a nonmetal. Silicon and germanium are called metalloids because they exhibit both metallic and nonmetallic properties. Tin and lead are metals. Thus metallic character increases as one goes down a family in the periodic table. The metallic character decreases from left to right across the periodic table.

28. a. A molecule has no overall charge (an equal number of electrons and protons are present). Ions, on the other hand, have extra electrons added or removed to form anions (negatively charged ions) or cations (positively charged ions).

b. The sharing of electrons between atoms is a covalent bond. An ionic bond is the force of attraction between two oppositely charged ions.

c. A molecule is a collection of atoms held together by covalent bonds. A compound is composed of two or more different elements having constant composition. Covalent and/or ionic bonds can hold the atoms together in a compound. Another difference is that molecules do not necessarily have to be compounds. H_2 is two hydrogen atoms held together by a covalent bond. H_2 is a molecule, but it is not a compound; H_2 is a diatomic element.

d. An anion is a negatively charged ion; e.g., Cl^-, O^{2-}, and SO_4^{2-} are all anions. A cation is a positively charged ion, e.g., Na^+, Fe^{3+}, and NH_4^+ are all cations.

29. a. This represents ionic bonding. Ionic bonding is the electrostatic attraction between anions and cations.

b. This represents covalent bonding where electrons are shared between two atoms. This could be the space-filling model for H_2O or SF_2 or NO_2, etc.

30. Natural niacin and commercially produced niacin have the exact same formula of $C_6H_5NO_2$. Therefore, both sources produce niacin having an identical nutritional value. There may be other compounds present in natural niacin that would increase the nutritional value, but the nutritional value due to just niacin is identical to the commercially produced niacin.

31. Statements a and b are true. Counting over in the periodic table, element 118 will be the next noble gas (a nonmetal). For statement c, hydrogen has mostly nonmetallic properties. For statement d, a family of elements is also known as a group of elements. For statement e, two items are incorrect. When a metal reacts with a nonmetal, an ionic compound is produced, and the formula of the compound would be AX_2 (alkaline earth metals form 2+ ions and halogens form 1− ions in ionic compounds). The correct statement would be: When an alkaline earth metal, A, reacts with a halogen, X, the formula of the ionic compound formed should be AX_2.

32. a. Dinitrogen monoxide is correct. N and O are both nonmetals, resulting in a covalent compound. We need to use the covalent rules of nomenclature. The other two names are for ionic compounds.

b. Copper(I) oxide is correct. With a metal in a compound, we have an ionic compound. Because copper, like most transition metals, forms at least a couple of different stable charged ions in compounds, we must indicate the charge on copper in the name. Copper oxide could be CuO or Cu_2O, hence why we must give the charge of most transition metal compounds. Dicopper monoxide is the name if this were a covalent compound, which it is not.

c. Lithium oxide is correct. Lithium forms 1+ charged ions in stable ionic compounds. Because lithium is assumed to form 1+ ions in compounds, we do not need to indicate the charge of the metal ion in the compound. Dilithium monoxide would be the name if Li_2O were a covalent compound (a compound composed of only nonmetals).

Exercises

Development of the Atomic Theory

33. a. The composition of a substance depends on the numbers of atoms of each element making up the compound (depends on the formula of the compound) and not on the composition of the mixture from which it was formed.

    b. Avogadro's hypothesis (law) implies that volume ratios are proportional to molecule ratios at constant temperature and pressure. $H_2(g) + Cl_2(g) \rightarrow 2\ HCl(g)$. From the balanced equation, the volume of HCl produced will be twice the volume of H_2 (or Cl_2) reacted.

34. Avogadro's hypothesis (law) implies that volume ratios are equal to molecule ratios at constant temperature and pressure. Here, 1 volume of N_2 reacts with 3 volumes of H_2 to produce 2 volumes of the gaseous product or in terms of molecule ratios:

$$1\ N_2 + 3\ H_2 \rightarrow 2\ \text{product}$$

In order for the equation to be balanced, the product must be NH_3.

35. From the law of definite proportions, a given compound always contains exactly the same proportion of elements by mass. The first sample of chloroform has a total mass of 12.0 g C + 106.4 g Cl + 1.01 g H = 119.41 g (carrying extra significant figures). The mass percent of carbon in this sample of chloroform is:

$$\frac{12.0\ \text{g C}}{119.41\ \text{g total}} \times 100 = 10.05\%\ \text{C by mass}$$

From the law of definite proportions, the second sample of chloroform must also contain 10.05% C by mass. Let x = mass of chloroform in the second sample:

$$\frac{30.0\ \text{g C}}{x} \times 100 = 10.05, \quad x = 299\ \text{g chloroform}$$

36. A compound will always have a constant composition by mass. From the initial data given, the mass ratio of H : S : O in sulfuric acid (H_2SO_4) is:

$$\frac{2.02}{2.02} : \frac{32.07}{2.02} : \frac{64.00}{2.02} = 1 : 15.9 : 31.7$$

If we have 7.27 g H, then we will have 7.27 × 15.9 = 116 g S and 7.27 × 31.7 = 230. g O in the second sample of H_2SO_4.

37. Hydrazine: 1.44×10^{-1} g H/g N; ammonia: 2.16×10^{-1} g H/g N; hydrogen azide: 2.40×10^{-2} g H/g N. Let's try all of the ratios:

$$\frac{0.144}{0.0240} = 6.00; \quad \frac{0.216}{0.0240} = 9.00; \quad \frac{0.0240}{0.0240} = 1.00; \quad \frac{0.216}{0.144} = 1.50 = \frac{3}{2}$$

All the masses of hydrogen in these three compounds can be expressed as simple whole-number ratios. The g H/g N in hydrazine, ammonia, and hydrogen azide are in the ratios 6 : 9 : 1.

38. The law of multiple proportions does not involve looking at the ratio of the mass of one element with the total mass of the compounds. To illustrate the law of multiple proportions, we compare the mass of carbon that combines with 1.0 g of oxygen in each compound:

compound 1: 27.2 g C and 72.8 g O (100.0 – 27.2 = mass O)

compound 2: 42.9 g C and 57.1 g O (100.0 – 42.9 = mass O)

The mass of carbon that combines with 1.0 g of oxygen is:

compound 1: $\frac{27.2 \text{ g C}}{72.8 \text{ g O}} = 0.374 \text{ g C/g O}$

compound 2: $\frac{42.9 \text{ g C}}{57.1 \text{ g O}} = 0.751 \text{ g C/g O}$

$\frac{0.751}{0.374} = \frac{2}{1}$; this supports the law of multiple proportions because this carbon ratio is a whole number.

39. For CO and CO_2, it is easiest to concentrate on the mass of oxygen that combines with 1 g of carbon. From the formulas (two oxygen atoms per carbon atom in CO_2 versus one oxygen atom per carbon atom in CO), CO_2 will have twice the mass of oxygen that combines per gram of carbon as compared to CO. For CO_2 and C_3O_2, it is easiest to concentrate on the mass of carbon that combines with 1 g of oxygen. From the formulas (three carbon atoms per two oxygen atoms in C_3O_2 versus one carbon atom per two oxygen atoms in CO_2), C_3O_2 will have three times the mass of carbon that combines per gram of oxygen as compared to CO_2. As expected, the mass ratios are whole numbers as predicted by the law of multiple proportions.

40. Compound I: $\frac{14.0 \text{ g R}}{3.00 \text{ g Q}} = \frac{4.67 \text{ g R}}{1.00 \text{ g Q}}$; compound II: $\frac{7.00 \text{ g R}}{4.50 \text{ g Q}} = \frac{1.56 \text{ g R}}{1.00 \text{ g Q}}$

The ratio of the masses of R that combine with 1.00 g Q is: $\frac{4.67}{1.56} = 2.99 \approx 3$

As expected from the law of multiple proportions, this ratio is a small whole number.

Because compound I contains three times the mass of R per gram of Q as compared with compound II (RQ), the formula of compound I should be R_3Q.

41. Mass is conserved in a chemical reaction because atoms are conserved. Chemical reactions involve the reorganization of atoms, so formulas change in a chemical reaction, but the number and types of atoms do not change. Because the atoms do not change in a chemical reaction, mass must not change. In this equation we have two oxygen atoms and four hydrogen atoms both before and after the reaction occurs.

42. Mass is conserved in a chemical reaction.

	ethanol +	oxygen $\rightarrow$	water +	carbon dioxide
Mass:	46.0 g	96.0 g	54.0 g	?

Mass of reactants = 46.0 + 96.0 = 142.0 g = mass of products

142.0 g = 54.0 g + mass of CO_2, mass of CO_2 = 142.0 – 54.0 = 88.0 g

43. To get the atomic mass of H to be 1.00, we divide the mass of hydrogen that reacts with 1.00 g of oxygen by 0.126; that is, $\frac{0.126}{0.126} = 1.00$. To get Na, Mg, and O on the same scale, we do the same division.

$$\text{Na: } \frac{2.875}{0.126} = 22.8; \ \text{Mg: } \frac{1.500}{0.126} = 11.9; \ \text{O: } \frac{1.00}{0.126} = 7.94$$

	H	O	Na	Mg
Relative value	1.00	7.94	22.8	11.9
Accepted value	1.008	16.00	22.99	24.31

For your information, the atomic masses of O and Mg are incorrect. The atomic masses of H and Na are close to the values given in the periodic table. Something must be wrong about the assumed formulas of the compounds. It turns out the correct formulas are H_2O, Na_2O, and MgO. The smaller discrepancies result from the error in the assumed atomic mass of H.

44. If the formula is InO, then one atomic mass of In would combine with one atomic mass of O, or:

$$\frac{A}{16.00} = \frac{4.784\,\text{g In}}{1.000\,\text{g O}}, \ \text{A = atomic mass of In = 76.54}$$

If the formula is In_2O_3, then two times the atomic mass of In will combine with three times the atomic mass of O, or:

$$\frac{2A}{(3)16.00} = \frac{4.784\,\text{g In}}{1.000\,\text{g O}}, \ \text{A = atomic mass of In = 114.8}$$

The latter number is the atomic mass of In used in the modern periodic table.

The Nature of the Atom

45. From section 2.5, the nucleus has "a diameter of about 10^{-13} cm" and the electrons "move about the nucleus at an average distance of about 10^{-8} cm from it." We will use these

statements to help determine the densities. Density of hydrogen nucleus (contains one proton only):

$$V_{nucleus} = \frac{4}{3}\pi r^3 = \frac{4}{3}(3.14)(5 \times 10^{-14}\ cm)^3 = 5 \times 10^{-40}\ cm^3$$

$$d = density = \frac{1.67 \times 10^{-24}\ g}{5 \times 10^{-40}\ cm^3} = 3 \times 10^{15}\ g/cm^3$$

Density of H atom (contains one proton and one electron):

$$V_{atom} = \frac{4}{3}(3.14)(1 \times 10^{-8}\ cm)^3 = 4 \times 10^{-24}\ cm^3$$

$$d = \frac{1.67 \times 10^{-24}\ g + 9 \times 10^{-28} g}{4 \times 10^{-24}\ cm^3} = 0.4\ g/cm^3$$

46. Because electrons move about the nucleus at an average distance of about 1×10^{-8} cm, the diameter of an atom will be about 2×10^{-8} cm. Let's set up a ratio:

$$\frac{\text{diameter of nucleus}}{\text{diameter of atom}} = \frac{1\ mm}{\text{diameter of model}} = \frac{1 \times 10^{-13}\ cm}{2 \times 10^{-8}\ cm}; \text{ solving:}$$

diameter of model = 2×10^5 mm = 200 m

47. $5.93 \times 10^{-18}\ C \times \frac{1\ \text{electron charge}}{1.602 \times 10^{-19}\ C}$ = 37 negative (electron) charges on the oil drop

48. First, divide all charges by the smallest quantity, 6.40×10^{-13}.

$$\frac{2.56 \times 10^{-12}}{6.40 \times 10^{-13}} = 4.00; \quad \frac{7.68}{0.640} = 12.0; \quad \frac{3.84}{0.640} = 6.00$$

Because all charges are whole-number multiples of 6.40×10^{-13} zirkombs, the charge on one electron could be 6.40×10^{-13} zirkombs. However, 6.40×10^{-13} zirkombs could be the charge of two electrons (or three electrons, etc.). All one can conclude is that the charge of an electron is 6.40×10^{-13} zirkombs or an integer fraction of 6.40×10^{-13} zirkombs.

49. sodium–Na; radium–Ra; iron–Fe; gold–Au; manganese–Mn; lead–Pb

50. fluorine–F; chlorine–Cl; bromine–Br; sulfur–S; oxygen–O; phosphorus–P

51. Sn–tin; Pt–platinum; Hg–mercury; Mg–magnesium; K–potassium; Ag–silver

52. As–arsenic; I–iodine; Xe–xenon; He–helium; C–carbon; Si–silicon

53. a. Metals: Mg, Ti, Au, Bi, Ge, Eu, and Am. Nonmetals: Si, B, At, Rn, and Br.

b. Si, Ge, B, and At. The elements at the boundary between the metals and the nonmetals are B, Si, Ge, As, Sb, Te, Po, and At. Aluminum has mostly properties of metals, so it is generally not classified as a metalloid.

54. a. The noble gases are He, Ne, Ar, Kr, Xe, and Rn (helium, neon, argon, krypton, xenon, and radon). Radon has only radioactive isotopes. In the periodic table, the whole number enclosed in parentheses is the mass number of the longest-lived isotope of the element.

b. Promethium (Pm) has only radioactive isotopes.

55. a. transition metals b. alkaline earth metals c. alkali metals

d. noble gases e. halogens

56. Use the periodic table to identify the elements.

a. Cl; halogen b. Be; alkaline earth metal

c. Eu; lanthanide metal d. Hf; transition metal

e. He; noble gas f. U; actinide metal

g. Cs; alkali metal

57. a. Element 8 is oxygen. A = mass number = 9 + 8 = 17; $^{17}_{8}O$

b. Chlorine is element 17. $^{37}_{17}Cl$ c. Cobalt is element 27. $^{60}_{27}Co$

d. Z = 26; A = 26 + 31 = 57; $^{57}_{26}Fe$ e. Iodine is element 53. $^{131}_{53}I$

f. Lithium is element 3. $^{7}_{3}Li$

58. a. Cobalt is element 27. A = mass number = 27 + 31 = 58; $^{58}_{27}Co$

b. $^{10}_{5}B$ c. $^{23}_{12}Mg$ d. $^{132}_{53}I$ e. $^{47}_{20}Ca$ f. $^{65}_{29}Cu$

59. Z is the atomic number and is equal to the number of protons in the nucleus. A is the mass number and is equal to the number of protons plus neutrons in the nucleus. X is the symbol of the element. See the front cover of the text which has a listing of the symbols for the various elements and corresponding atomic number or see the periodic table on the cover to determine the identity of the various atoms. Because all of the atoms have equal numbers of protons and electrons, each atom is neutral in charge.

a. $^{23}_{11}Na$ b. $^{19}_{9}F$ c. $^{16}_{8}O$

60. The atomic number for carbon is 6. ^{14}C has 6 protons, 14 – 6 = 8 neutrons, and 6 electrons in the neutral atom. ^{12}C has 6 protons, 12 – 6 = 6 neutrons, and 6 electrons in the neutral atom. The only difference between an atom of ^{14}C and an atom of ^{12}C is that ^{14}C has two additional neutrons.

61. a. $^{79}_{35}Br$: 35 protons, 79 – 35 = 44 neutrons. Because the charge of the atom is neutral, the number of protons = the number of electrons = 35.

b. $^{81}_{35}Br$: 35 protons, 46 neutrons, 35 electrons

c. $^{239}_{94}Pu$: 94 protons, 145 neutrons, 94 electrons

d. $^{133}_{55}Cs$: 55 protons, 78 neutrons, 55 electrons

e. $^{3}_{1}H$: 1 proton, 2 neutrons, 1 electron

f. $^{56}_{26}Fe$: 26 protons, 30 neutrons, 26 electrons

62. a. $^{235}_{92}U$: 92 p, 143 n, 92 e b. $^{27}_{13}Al$: 13 p, 14 n, 13 e c. $^{57}_{26}Fe$: 26 p, 31 n, 26 e

d. $^{208}_{82}Pb$: 82 p, 126 n, 82 e e. $^{86}_{37}Rb$: 37 p, 49 n, 37 e f. $^{41}_{20}Ca$: 20 p, 21 n, 20 e

63. a. Ba is element 56. Ba^{2+} has 56 protons, so Ba^{2+} must have 54 electrons in order to have a net charge of 2+.

b. Zn is element 30. Zn^{2+} has 30 protons and 28 electrons.

c. N is element 7. N^{3-} has 7 protons and 10 electrons.

d. Rb is element 37, Rb^{+} has 37 protons and 36 electrons.

e. Co is element 27. Co^{3+} has 27 protons and 24 electrons.

f. Te is element 52. Te^{2-} has 52 protons and 54 electrons.

g. Br is element 35. Br^{-} has 35 protons and 36 electrons.

64. a. $^{24}_{12}Mg$: 12 protons, 12 neutrons, 12 electrons

b. $^{24}_{12}Mg^{2+}$: 12 p, 12 n, 10 e c. $^{59}_{27}Co^{2+}$: 27 p, 32 n, 25 e

d. $^{59}_{27}Co^{3+}$: 27 p, 32 n, 24 e e. $^{59}_{27}Co$: 27 p, 32 n, 27 e

f. $^{79}_{34}Se$: 34 p, 45 n, 34 e g. $^{79}_{34}Se^{2-}$: 34 p, 45 n, 36 e

h. $^{63}_{28}Ni$: 28 p, 35 n, 28 e i. $^{59}_{28}Ni^{2+}$: 28 p, 31 n, 26 e

65. Atomic number = 63 (Eu); net charge = +63 – 60 = 3+; mass number = 63 + 88 = 151; symbol: $^{151}_{63}Eu^{3+}$

Atomic number = 50 (Sn); mass number = 50 + 68 = 118; net charge = +50 – 48 = 2+; symbol: $^{118}_{50}Sn^{2+}$

66. Atomic number = 16 (S); net charge = +16 − 18 = 2−; mass number = 16 + 18 = 34; symbol: $^{34}_{16}S^{2-}$

Atomic number = 16 (S); net charge = +16 − 18 = 2−; mass number = 16 + 16 = 32; symbol: $^{32}_{16}S^{2-}$

67.

Symbol	Number of protons in nucleus	Number of neutrons in nucleus	Number of electrons	Net charge
$^{238}_{92}U$	92	146	92	0
$^{40}_{20}Ca^{2+}$	20	20	18	2+
$^{51}_{23}V^{3+}$	23	28	20	3+
$^{89}_{39}Y$	39	50	39	0
$^{79}_{35}Br^{-}$	35	44	36	1−
$^{31}_{15}P^{3-}$	15	16	18	3−

68.

Symbol	Number of protons in nucleus	Number of neutrons in nucleus	Number of electrons	Net charge
$^{53}_{26}Fe^{2+}$	26	27	24	2+
$^{59}_{26}Fe^{3+}$	26	33	23	3+
$^{210}_{85}At^{-}$	85	125	86	1−
$^{27}_{13}Al^{3+}$	13	14	10	3+
$^{128}_{52}Te^{2-}$	52	76	54	2−

69. In ionic compounds, metals lose electrons to form cations, and nonmetals gain electrons to form anions. Group 1A, 2A, and 3A metals form stable 1+, 2+, and 3+ charged cations, respectively. Group 5A, 6A, and 7A nonmetals form 3–, 2–, and 1– charged anions, respectively.

a. Lose 2 e^- to form Ra^{2+}. b. Lose 3 e^- to form In^{3+}. c. Gain 3 e^- to form P^{3-}.

d. Gain 2 e^- to form Te^{2-}. e. Gain 1 e^- to form Br^-. f. Lose 1 e^- to form Rb^+.

70. See Exercise 69 for a discussion of charges various elements form when in ionic compounds.

a. Element 13 is Al. Al forms 3+ charged ions in ionic compounds. Al^{3+}

b. Se^{2-} c. Ba^{2+} d. N^{3-} e. Fr^+ f. Br^-

Nomenclature

71. a. sodium bromide b. rubidium oxide

c. calcium sulfide d. aluminum iodide

e. SrF_2 f. Al_2Se_3

g. K_3N h. Mg_3P_2

72. a. mercury(I) oxide b. iron(III) bromide

c. cobalt(II) sulfide d. titanium(IV) chloride

e. Sn_3N_2 f. CoI_3

g. HgO h. CrS_3

73. a. cesium fluoride b. lithium nitride

c. silver sulfide (Silver only forms stable 1+ ions in compounds, so no Roman numerals are needed.)

d. manganese(IV) oxide e. titanium(IV) oxide f. strontium phosphide

74. a. $ZnCl_2$ (Zn only forms stable +2 ions in compounds, so no Roman numerals are needed.)

b. SnF_4 c. Ca_3N_2 d. Al_2S_3

e. Hg_2Se f. AgI (Ag only forms stable +1 ions in compounds.)

75. a. barium sulfite b. sodium nitrite

c. potassium permanganate d. potassium dichromate

76. a. $Cr(OH)_3$ b. $Mg(CN)_2$

c. $Pb(CO_3)_2$ d. $NH_4C_2H_3O_2$

77. a. dinitrogen tetroxide b. iodine trichloride

c. sulfur dioxide d. diphosphorus pentasulfide

78. a. B_2O_3 b. AsF_5
c. N_2O d. SCl_6

79. a. copper(I) iodide b. copper(II) iodide c. cobalt(II) iodide
d. sodium carbonate e. sodium hydrogen carbonate or sodium bicarbonate
f. tetrasulfur tetranitride g. selenium tetrachloride h. sodium hypochlorite
i. barium chromate j. ammonium nitrate

80. a. acetic acid b. ammonium nitrite c. cobalt(III) sulfide
d. iodine monochloride e. lead(II) phosphate f. potassium chlorate
g. sulfuric acid h. strontium nitride i. aluminum sulfite
j. tin(IV) oxide k. sodium chromate l. hypochlorous acid

Note: For the compounds named as acids, we assume these are dissolved in water.

81. In the case of sulfur, SO_4^{2-} is sulfate, and SO_3^{2-} is sulfite. By analogy:

SeO_4^{2-}: selenate; SeO_3^{2-}: selenite; TeO_4^{2-}: tellurate; TeO_3^{2-}: tellurite

82. From the anion names of hypochlorite (ClO^-), chlorite (ClO_2^-), chlorate (ClO_3^-), and perchlorate (ClO_4^-), the oxyanion names for similar iodine ions would be hypoiodite (IO^-), iodite (IO_2^-), iodate (IO_3^-), and periodate (IO_4^-). The corresponding acids would be hypoiodous acid (HIO), iodous acid (HIO_2), iodic acid (HIO_3), and periodic acid (HIO_4).

83. a. SF_2 b. SF_6 c. NaH_2PO_4
d. Li_3N e. $Cr_2(CO_3)_3$ f. SnF_2
g. $NH_4C_2H_3O_2$ h. NH_4HSO_4 i. $Co(NO_3)_3$
j. Hg_2Cl_2; mercury(I) exists as Hg_2^{2+} ions. k. $KClO_3$ l. NaH

84. a. CrO_3 b. S_2Cl_2 c. NiF_2
d. K_2HPO_4 e. AlN
f. NH_3 (Nitrogen trihydride is the systematic name.) g. MnS_2
h. $Na_2Cr_2O_7$ i. $(NH_4)_2SO_3$ j. CI_4

85. a. Na_2O b. Na_2O_2 c. KCN
d. $Cu(NO_3)_2$ e. $SeBr_4$ f. HIO_2
g. PbS_2 h. CuCl
i. GaAs (We would predict the stable ions to be Ga^{3+} and As^{3-}.)
j. CdSe (Cadmium only forms 2+ charged ions in compounds.)
k. ZnS (Zinc only forms 2+ charged ions in compounds.)
l. HNO_2 m. P_2O_5

86. a. $(NH_4)_2HPO_4$ b. Hg_2S c. SiO_2

d. Na_2SO_3 e. $Al(HSO_4)_3$ f. NCl_3

g. HBr h. $HBrO_2$ i. $HBrO_4$

j. KHS k. CaI_2 l. $CsClO_4$

87. a. nitric acid, HNO_3 b. perchloric acid, $HClO_4$ c. acetic acid, $HC_2H_3O_2$

d. sulfuric acid, H_2SO_4 e. phosphoric acid, H_3PO_4

88. a. Iron forms 2+ and 3+ charged ions; we need to include a Roman numeral for iron. Iron(III) chloride is correct.

b. This is a covalent compound, so use the covalent rules. Nitrogen dioxide is correct.

c. This is an ionic compound, so use the ionic rules. Calcium oxide is correct. Calcium only forms stable 2+ ions when in ionic compounds, so no Roman numeral is needed.

d. This is an ionic compound, so use the ionic rules. Aluminum sulfide is correct.

e. This is an ionic compound, so use the ionic rules. Mg is magnesium. Magnesium acetate is correct.

f. Phosphide is P^{3-}, while phosphate is PO_4^{3-}. Because phosphate has a 3− charge, the charge on iron is 3+. Iron(III) phosphate is correct.

g. This is a covalent compound, so use the covalent rules. Diphosphorus pentasulfide is correct.

h. Because each sodium is 1+ charged, we have the O_2^{2-} (peroxide) ion present. Sodium peroxide is correct. Note that sodium oxide would be Na_2O.

i. HNO_3 is nitric acid, not nitrate acid. Nitrate acid does not exist.

j. H_2S is hydrosulfuric acid or dihydrogen sulfide or just hydrogen sulfide (common name). H_2SO_4 is sulfuric acid.

Additional Exercises

89. Yes, 1.0 g H would react with 37.0 g ^{37}Cl, and 1.0 g H would react with 35.0 g ^{35}Cl.

No, the mass ratio of H/Cl would always be 1 g H/37 g Cl for ^{37}Cl and 1 g H/35 g Cl for ^{35}Cl. As long as we had pure ^{37}Cl or pure ^{35}Cl, the ratios will always hold. If we have a mixture (such as the natural abundance of chlorine), the ratio will also be constant as long as the composition of the mixture of the two isotopes does not change.

90. Carbon (C); hydrogen (H); oxygen (O); nitrogen (N); phosphorus (P); sulfur (S)

For lighter elements, stable isotopes usually have equal numbers of protons and neutrons in the nucleus; these stable isotopes are usually the most abundant isotope for each element. Therefore, a predicted stable isotope for each element is ^{12}C, ^{2}H, ^{16}O, ^{14}N, ^{30}P, and ^{32}S. These are stable isotopes except for ^{30}P, which is radioactive. The most stable (and most abundant) isotope of phosphorus is ^{31}P. There are exceptions. Also, the most abundant isotope for hydrogen is ^{1}H; this has just a proton in the nucleus. ^{2}H (deuterium) is stable (not radioactive), but ^{1}H is also stable as well as most abundant.

91. $^{53}_{26}Fe^{2+}$ has 26 protons, 53 − 26 = 27 neutrons, and two fewer electrons than protons (24 electrons) in order to have a net charge of 2+.

92. a. False. Neutrons have no charge; therefore, all particles in a nucleus are not charged.

b. False. The atom is best described as having a tiny dense nucleus containing most of the mass of the atom with the electrons moving about the nucleus at relatively large distances away; so much so that an atom is mostly empty space.

c. False. The mass of the nucleus makes up most of the mass of the entire atom.

d. True.

e. False. The number of protons in a neutral atom must equal the number of electrons.

93. From the Na_2X formula, X has a 2− charge. Because 36 electrons are present, X has 34 protons and 79 − 34 = 45 neutrons, and is selenium.

a. True. Nonmetals bond together using covalent bonds and are called covalent compounds.

b. False. The isotope has 34 protons.

c. False. The isotope has 45 neutrons.

d. False. The identity is selenium, Se.

94. a. Fe^{2+}: 26 protons (Fe is element 26.); protons − electrons = net charge, 26 − 2 = 24 electrons; FeO is the formula since the oxide ion has a 2− charge, and the name is iron(II) oxide.

b. Fe^{3+}: 26 protons; 23 electrons; Fe_2O_3; iron(III) oxide

c. Ba^{2+}: 56 protons; 54 electrons; BaO; barium oxide

d. Cs^{+}: 55 protons; 54 electrons; Cs_2O; cesium oxide

e. S^{2-}: 16 protons; 18 electrons; Al_2S_3; aluminum sulfide

f. P^{3-}: 15 protons; 18 electrons; AlP; aluminum phosphide

g. Br^{-}: 35 protons; 36 electrons; $AlBr_3$; aluminum bromide

h. N^{3-}: 7 protons; 10 electrons; AlN; aluminum nitride

95. a. $Pb(C_2H_3O_2)_2$: lead(II) acetate

b. $CuSO_4$: copper(II) sulfate

c. CaO: calcium oxide

d. $MgSO_4$: magnesium sulfate

e. $Mg(OH)_2$: magnesium hydroxide

f. $CaSO_4$: calcium sulfate

g. N_2O: dinitrogen monoxide or nitrous oxide (common name)

96. a. This is element 52, tellurium. Te forms stable 2! charged ions in ionic compounds (like other oxygen family members).

b. Rubidium. Rb, element 37, forms stable 1+ charged ions.

c. Argon. Ar is element 18.

d. Astatine. At is element 85.

97. From the XBr_2 formula, the charge on element X is 2+. Therefore, the element has 88 protons, which identifies it as radium, Ra. 230 – 88 = 142 neutrons.

98. Because this is a relatively small number of neutrons, the number of protons will be very close to the number of neutrons present. The heavier elements have significantly more neutrons than protons in their nuclei. Because this element forms anions, it is a nonmetal and will be a halogen because halogens form stable 1– charged ions in ionic compounds. From the halogens listed, chlorine, with an average atomic mass of 35.45, fits the data. The two isotopes are ^{35}Cl and ^{37}Cl, and the number of electrons in the 1– ion is 18. Note that because the atomic mass of chlorine listed in the periodic table is closer to 35 than 37, we can assume that ^{35}Cl is the more abundant isotope. This is discussed in Chapter 3.

99. a. Ca^{2+} and N^{3-}: Ca_3N_2, calcium nitride

b. K^+ and O^{2-}: K_2O, potassium oxide

c. Rb^+ and F^-: RbF, rubidium fluoride

d. Mg^{2+} and S^{2-}: MgS, magnesium sulfide

e. Ba^{2+} and I^-: BaI_2, barium iodide

f. Al^{3+} and Se^{2-}: Al_2Se_3, aluminum selenide

g. Cs^+ and P^{3-}: Cs_3P, cesium phosphide

h. In^{3+} and Br^-: $InBr_3$, indium(III) bromide. In also forms In^+ ions, but one would predict In^{3+} ions from its position in the periodic table.

100. These compounds are similar to phosphate (PO_4^{3-}) compounds. Na_3AsO_4 contains Na^+ ions and AsO_4^{3-} ions. The name would be sodium arsenate. H_3AsO_4 is analogous to phosphoric acid, H_3PO_4. H_3AsO_4 would be arsenic acid. $Mg_3(SbO_4)_2$ contains Mg^{2+} ions and SbO_4^{3-} ions, and the name would be magnesium antimonate.

101. a. Element 15 is phosphorus, P. This atom has 15 protons and 31 – 15 = 16 neutrons.

b. Element 53 is iodine, I. 53 protons; 74 neutrons

c. Element 19 is potassium, K. 19 protons; 20 neutrons

d. Element 70 is ytterbium, Yb. 70 protons; 103 neutrons

102. Mass is conserved in a chemical reaction.

	chromium(III) oxide	+	aluminum	→	chromium	+	aluminum oxide
Mass:	34.0 g		12.1 g		23.3 g		?

Mass aluminum oxide produced = (34.0 + 12.1) − 23.3 = 22.8 g

ChemWork Problems

The answers to the problems 103-108 (or a variation to these problems) are found in OWL. These problems are also assignable in OWL.

Challenge Problems

109. Copper (Cu), silver (Ag), and gold (Au) make up the coinage metals.

110. Because the gases are at the same temperature and pressure, the volumes are directly proportional to the number of molecules present. Let's assume hydrogen and oxygen to be monatomic gases and that water has the simplest possible formula (HO). We have the equation:

$$H + O \rightarrow HO$$

But the volume ratios are also equal to the molecule ratios, which correspond to the coefficients in the equation:

$$2\ H + O \rightarrow 2\ HO$$

Because atoms cannot be created nor destroyed in a chemical reaction, this is not possible. To correct this, we can make oxygen a diatomic molecule:

$$2\ H + O_2 \rightarrow 2\ HO$$

This does not require hydrogen to be diatomic. Of course, if we know water has the formula H_2O, we get:

$$2\ H + O_2 \rightarrow 2\ H_2O$$

The only way to balance this is to make hydrogen diatomic:

$$2\ H_2 + O_2 \rightarrow 2\ H_2O$$

111. Avogadro proposed that equal volumes of gases (at constant temperature and pressure) contain equal numbers of molecules. In terms of balanced equations, Avogadro's hypothesis (law) implies that volume ratios will be identical to molecule ratios. Assuming one molecule of octane reacting, then 1 molecule of C_xH_y produces 8 molecules of CO_2 and 9 molecules of H_2O. $C_xH_y + n\ O_2 \rightarrow 8\ CO_2 + 9\ H_2O$. Because all the carbon in octane ends up as carbon in

CO_2, octane must contain 8 atoms of C. Similarly, all hydrogen in octane ends up as hydrogen in H_2O, so one molecule of octane must contain 9 × 2 = 18 atoms of H. Octane formula = C_8H_{18}, and the ratio of C : H = 8 : 18 or 4 : 9.

112. From Section 2.5 of the text, the average diameter of the nucleus is about 10^{-13} cm, and the electrons move about the nucleus at an average distance of about 10^{-8} cm. From this, the diameter of an atom is about 2×10^{-8} cm.

$$\frac{2 \times 10^{-8}\ \text{cm}}{1 \times 10^{-13}\ \text{cm}} = 2 \times 10^5;\quad \frac{1\ \text{mi}}{1\ \text{grape}} = \frac{5280\ \text{ft}}{1\ \text{grape}} = \frac{63{,}360\ \text{in}}{1\ \text{grape}}$$

Because the grape needs to be 2×10^5 times smaller than a mile, the diameter of the grape would need to be $63{,}360/(2 \times 10^5) \approx 0.3$ in. This is a reasonable size for a small grape.

113. The alchemists were incorrect. The solid residue must have come from the flask.

114. The equation for the reaction would be 2 Na(s) + Cl_2(g) → 2 NaCl(s). The sodium reactant exists as singular sodium atoms packed together very tightly and in a very organized fashion. This type of packing of atoms represents the solid phase. The chlorine reactant exists as Cl_2 molecules. In the picture of chlorine, there is a lot of empty space present. This only occurs in the gaseous phase. When sodium and chlorine react, the ionic compound NaCl forms. NaCl exists as separate Na^+ and Cl^- ions. Because the ions are packed very closely together and are packed in a very organized fashion, NaCl is depicted in the solid phase.

115. a. Both compounds have C_2H_6O as the formula. Because they have the same formula, their mass percent composition will be identical. However, these are different compounds with different properties because the atoms are bonded together differently. These compounds are called isomers of each other.

b. When wood burns, most of the solid material in wood is converted to gases, which escape. The gases produced are most likely CO_2 and H_2O.

c. The atom is not an indivisible particle but is instead composed of other smaller particles, called electrons, neutrons, and protons.

d. The two hydride samples contain different isotopes of either hydrogen and/or lithium. Although the compounds are composed of different isotopes, their properties are similar because different isotopes of the same element have similar properties (except, of course, their mass).

116. Let X_a be the formula for the atom/molecule X, Y_b be the formula for the atom/molecule Y, X_cY_d be the formula of compound I between X and Y, and X_eY_f be the formula of compound II between X and Y. Using the volume data, the following would be the balanced equations for the production of the two compounds.

$X_a + 2\ Y_b \rightarrow 2\ X_cY_d;\ 2\ X_a + Y_b \rightarrow 2\ X_eY_f$

From the balanced equations, $a = 2c = e$ and $b = d = 2f$.

Substituting into the balanced equations:

$$X_{2c} + 2\ Y_{2f} \rightarrow 2\ X_cY_{2f};\ \ 2\ X_{2c} + Y_{2f} \rightarrow 2\ X_{2c}Y_f$$

For simplest formulas, assume that $c = f = 1$. Thus:

$$X_2 + 2\ Y_2 \rightarrow 2\ XY_2 \text{ and } 2\ X_2 + Y_2 \rightarrow 2\ X_2Y$$

Compound I = XY_2: If X has relative mass of 1.00, $\dfrac{1.00}{1.00 + 2y} = 0.3043,\ y = 1.14.$

Compound II = X_2Y: If X has relative mass of 1.00, $\dfrac{2.00}{2.00 + y} = 0.6364,\ y = 1.14.$

The relative mass of Y is 1.14 times that of X. Thus, if X has an atomic mass of 100, then Y will have an atomic mass of 114.

117. Most of the mass of the atom is due to the protons and the neutrons in the nucleus, and protons and neutrons have about the same mass (1.67×10^{-24} g). The ratio of the mass of the molecule to the mass of a nuclear particle will give a good approximation of the number of nuclear particles (protons and neutrons) present.

$$\frac{7.31 \times 10^{-23}\text{ g}}{1.67 \times 10^{-24}\text{ g}} = 43.8 \approx 44 \text{ nuclear particles}$$

Thus there are 44 protons and neutrons present. If the number of protons equals the number of neutrons, we have 22 protons in the molecule. One possibility would be the molecule CO_2 [6 + 2(8) = 22 protons].

118. For each experiment, divide the larger number by the smaller. In doing so, we get:

experiment 1	X = 1.0	Y = 10.5
experiment 2	Y = 1.4	Z = 1.0
experiment 3	X = 1.0	Y = 3.5

Our assumption about formulas dictates the rest of the solution. For example, if we assume that the formula of the compound in experiment 1 is XY and that of experiment 2 is YZ, we get relative masses of:

X = 2.0; Y = 21; Z = 15 (= 21/1.4)

and a formula of X_3Y for experiment 3 [three times as much X must be present in experiment 3 as compared to experiment 1 (10.5/3.5 = 3)].

However, if we assume the formula for experiment 2 is YZ and that of experiment 3 is XZ, then we get:

X = 2.0; Y = 7.0; Z = 5.0 (= 7.0/1.4)

and a formula of XY_3 for experiment 1. Any answer that is consistent with your initial assumptions is correct.

The answer to part d depends on which (if any) of experiments 1 and 3 have a formula of XY in the compound. If the compound in experiment 1 has a formula of XY, then:

$$21 \text{ g XY} \times \frac{4.2 \text{ g Y}}{(4.2 + 0.4) \text{ g XY}} = 19.2 \text{ g Y (and 1.8 g X)}$$

If the compound in experiment 3 has the XY formula, then:

$$21 \text{ g XY H } \frac{7.0 \text{ g Y}}{(7.0 + 2.0) \text{ g XY}} = 16.3 \text{ g Y (and 4.7 g X)}$$

Note that it could be that neither experiment 1 nor experiment 3 has XY as the formula. Therefore, there is no way of knowing an absolute answer here.

Integrated Problems

119. The systematic name of Ta_2O_5 is tantalum(V) oxide. Tantalum is a transition metal and requires a Roman numeral. Sulfur is in the same group as oxygen, and its most common ion is S^{2-}. There-fore, the formula of the sulfur analogue would be Ta_2S_5.

Total number of protons in Ta_2O_5:

Ta, Z = 73, so 73 protons × 2 = 146 protons; O, Z = 8, so 8 protons × 5 = 40 protons

Total protons = 186 protons

Total number of protons in Ta_2S_5:

Ta, Z = 73, so 73 protons × 2 = 146 protons; S, Z = 16, so 16 protons × 5 = 80 protons

Total protons = 226 protons

Proton difference between Ta_2S_5 and Ta_2O_5: 226 protons – 186 protons = 40 protons

120. The cation has 51 protons and 48 electrons. The number of protons corresponds to the atomic number. Thus this is element 51, antimony. There are 3 fewer electrons than protons. Therefore, the charge on the cation is 3+. The anion has one-third the number of protons of the cation, which corresponds to 17 protons; this is element 17, chlorine. The number of electrons in this anion of chlorine is 17 + 1 = 18 electrons. The anion must have a charge of 1−.

The formula of the compound formed between Sb^{3+} and Cl^- is $SbCl_3$. The name of the compound is antimony(III) chloride. The Roman numeral is used to indicate the charge on Sb because the predicted charge is not obvious from the periodic table.

121. Number of electrons in the unknown ion:

$$2.55 \times 10^{-26} \text{ g} \times \frac{1 \text{ kg}}{1000 \text{g}} \times \frac{1 \text{ electron}}{9.11 \times 10^{-31} \text{ kg}} = 28 \text{ electrons}$$

Number of protons in the unknown ion:

$$5.34 \times 10^{-23}\ \text{g} \times \frac{1\,\text{kg}}{1000\text{g}} \times \frac{1\,\text{proton}}{1.67\times10^{-27}\,\text{kg}} = 32\ \text{protons}$$

Therefore, this ion has 32 protons and 28 electrons. This is element number 32, germanium (Ge). The net charge is 4+ because four electrons have been lost from a neutral germanium atom.

The number of electrons in the unknown atom:

$$3.92 \times 10^{-26}\ \text{g} \times \frac{1\,\text{kg}}{1000\text{g}} \times \frac{1\,\text{electron}}{9.11\times0^{-31}\,\text{kg}} = 43\ \text{electrons}$$

In a neutral atom, the number of protons and electrons is the same. Therefore, this is element 43, technetium (Tc).

The number of neutrons in the technetium atom:

$$9.35 \times 10^{-23}\ \text{g} \times \frac{1\,\text{kg}}{1000\text{g}} \times \frac{1\,\text{proton}}{1.67\times10^{-27}\,\text{kg}} = 56\ \text{neutrons}$$

The mass number is the sum of the protons and neutrons. In this atom, the mass number is 43 protons + 56 neutrons = 99. Thus this atom and its mass number is ^{99}Tc.

Marathon Problem

122. a. For each set of data, divide the larger number by the smaller number to determine relative masses.

$$\frac{0.602}{0.295} = 2.04;\quad A = 2.04 \text{ when } B = 1.00$$

$$\frac{0.401}{0.172} = 2.33;\quad C = 2.33 \text{ when } B = 1.00$$

$$\frac{0.374}{0.320} = 1.17;\quad C = 1.17 \text{ when } A = 1.00$$

To have whole numbers, multiply the results by 3.

Data set 1: A = 6.1 and B = 3.0

Data set 2: C = 7.0 and B = 3.0

Data set 3: C = 3.5 and A = 3.0 or C = 7.0 and A = 6.0

Assuming 6.0 for the relative mass of A, the relative masses would be A = 6.0, B = 3.0, and C = 7.0 (if simplest formulas are assumed).

b. Gas volumes are proportional to the number of molecules present. There are many possible correct answers for the balanced equations. One such solution that fits the gas volume data is:

$$6\,A_2 + B_4 \rightarrow 4\,A_3B$$

$$B_4 + 4\,C_3 \rightarrow 4\,BC_3$$

$$3\,A_2 + 2\,C_3 \rightarrow 6\,AC$$

In any correct set of reactions, the calculated mass data must match the mass data given initially in the problem. Here, the new table of relative masses would be:

$$\frac{6\,(\text{mass } A_2)}{\text{mass } B_4} = \frac{0.602}{0.295};\ \text{mass } A_2 = 0.340(\text{mass } B_4)$$

$$\frac{4\,(\text{mass } C_3)}{\text{mass } B_4} = \frac{0.401}{0.172};\ \text{mass } C_3 = 0.583(\text{mass } B_4)$$

$$\frac{2\,(\text{mass } C_3)}{3\,(\text{mass } A_2)} = \frac{0.374}{0.320};\ \text{mass } A_2 = 0.570(\text{mass } C_3)$$

Assume some relative mass number for any of the masses. We will assume that mass B = 3.0, so mass B_4 = 4(3.0) = 12.

Mass C_3 = 0.583(12) = 7.0, mass C = 7.0/3

Mass A_2 = 0.570(7.0) = 4.0, mass A = 4.0/2 = 2.0

When we assume a relative mass for B = 3.0, then A = 2.0 and C = 7.0/3. The relative masses having all whole numbers would be A = 6.0, B = 9.0, and C = 7.0.

Note that any set of balanced reactions that confirms the initial mass data is correct. This is just one possibility.

CHAPTER 3

STOICHIOMETRY

Questions

23.

Isotope	Mass	Abundance
^{12}C	$12.000\bar{0}$ u	98.89%
^{13}C	13.034 u	1.11%

Average mass = $0.9889(12.000\bar{0}) + 0.0111(13.034) = 12.01$ u

Note: u is an abbreviation for amu (atomic mass units).

From the relative abundances, there would be 9889 atoms of ^{12}C and 111 atoms of ^{13}C in the 10,000 atom sample. The average mass of carbon is independent of the sample size; it will always be 12.01 u.

$$\text{Total mass} = 10{,}000 \text{ atoms} \times \frac{12.01\text{u}}{\text{atom}} = 1.201 \times 10^5 \text{ u}$$

For 1 mole of carbon (6.0221×10^{23} atoms C), the average mass would still be 12.01 u.

The number of ^{12}C atoms would be $0.9889(6.0221 \times 10^{23}) = 5.955 \times 10^{23}$ atoms ^{12}C, and the number of ^{13}C atoms would be $0.0111(6.0221 \times 10^{23}) = 6.68 \times 10^{21}$ atoms ^{13}C.

$$\text{Total mass} = 6.0221 \times 10^{23} \text{ atoms} \times \frac{12.01\text{u}}{\text{atom}} = 7.233 \times 10^{24} \text{ u}$$

$$\text{Total mass in g} = 6.0221 \times 10^{23} \text{ atoms} \times \frac{12.01\text{u}}{\text{atom}} \times \frac{1\text{ g}}{6.0221 \times 10^{23} \text{ u}} = 12.01 \text{ g/mol}$$

By using the carbon-12 standard to define the relative masses of all of the isotopes, as well as to define the number of things in a mole, then each element's average atomic mass in units of grams is the mass of a mole of that element as it is found in nature.

24. Consider a sample of glucose, $C_6H_{12}O_6$. The molar mass of glucose is 180.16 g/mol. The chemical formula allows one to convert from molecules of glucose to atoms of carbon, hydrogen, or oxygen present and vice versa. The chemical formula also gives the mole relationship in the formula. One mole of glucose contains 6 mol C, 12 mol H, and 6 mol O. Thus mole conversions between molecules and atoms are possible using the chemical formula. The molar mass allows one to convert between mass and moles of compound, and Avogadro's number (6.022×10^{23}) allows one to convert between moles of compound and number of molecules.

25. Avogadro's number of dollars = 6.022×10^{23} dollars/mol dollars

$$\frac{1 \text{ mol dollars} \times \dfrac{6.022 \times 10^{23} \text{ dollars}}{\text{mol dollars}}}{7 \times 10^9 \text{ people}} = 8.6 \times 10^{13} = 9 \times 10^{13} \text{ dollars/person}$$

1 trillion = 1,000,000,000,000 = 1×10^{12}; each person would have 90 trillion dollars.

26. Molar mass of CO_2 = 12.01 + 2(16.00) = 44.01 g/mol

One mol of CO_2 contains 6.022×10^{23} molecules of CO_2, 6.022×10^{23} atoms of C, and 1.204×10^{24} atoms of O. We could also break down 1 mol of CO_2 into the number of protons and the number of electrons present (1.325×10^{25} protons and 1.325×10^{25} electrons). In order to determine the number of neutrons present, we would need to know the isotope abundances for carbon and oxygen.

The mass of 1 mol of CO_2 would be 44.01 g. From the molar mass, one mol of CO_2 would contain 12.01 g C and 32.00 g O. We could also break down 1 mol of CO_2 into the mass of protons and mass of electrons present (22.16 g protons and 1.207×10^{-2} g electrons). This assumes no mass loss when the individual particles come together to form the atom. This is not a great assumption as will be discussed in Chapter 19 on Nuclear Chemistry.

27. Only in b are the empirical formulas the same for both compounds illustrated. In b, general formulas of X_2Y_4 and XY_2 are illustrated, and both have XY_2 for an empirical formula.

For a, general formulas of X_2Y and X_2Y_2 are illustrated. The empirical formulas for these two compounds are the same as the molecular formulas. For c, general formulas of XY and XY_2 are illustrated; these general formulas are also the empirical formulas. For d, general formulas of XY_4 and X_2Y_6 are illustrated. XY_4 is also the molecular formula, but X_2Y_6 has the empirical formula of XY_3.

28. The molar mass is the mass of 1 mole of the compound. The empirical mass is the mass of 1 mole of the empirical formula. The molar mass is a whole-number multiple of the empirical mass. The masses are the same when the molecular formula = empirical formula, and the masses are different when the two formulas are different. When different, the empirical mass must be multiplied by the same whole number used to convert the empirical formula to the molecular formula. For example, $C_6H_{12}O_6$ is the molecular formula for glucose, and CH_2O is the empirical formula. The whole-number multiplier is 6. This same factor of 6 is the multiplier used to equate the empirical mass (30 g/mol) of glucose to the molar mass (180 g/mol).

29. The mass percent of a compound is a constant no matter what amount of substance is present. Compounds always have constant composition.

30. A balanced equation starts with the correct formulas of the reactants and products. The coefficients necessary to balance the equation give molecule relationships as well as mole relationships between reactants and products. The state (phase) of the reactants and products is also given. Finally, special reaction conditions are sometimes listed above or below the arrow. These can include special catalysts used and/or special temperatures required for a reaction to occur.

31. Only one product is formed in this representation. This product has two Y atoms bonded to an X. The other substance present in the product mixture is just the excess of one of the reactants (Y). The best equation has smallest whole numbers. Here, answer c would be this smallest whole number equation ($X + 2\ Y \rightarrow XY_2$). Answers a and b have incorrect products listed, and for answer d, an equation only includes the reactants that go to produce the product; excess reactants are not shown in an equation.

32. A balanced equation must have the same number and types of atoms on both sides of the equation, but it also needs to have correct formulas. The illustration has the equation as:

$$H + O \rightarrow H_2O$$

Under normal conditions, hydrogen gas and oxygen gas exist as diatomic molecules. So the first change to make is to change H + O to $H_2 + O_2$. To balance this equation, we need one more oxygen atom on the product side. Trial and error eventually gives the correct balanced equation of:

$$2\ H_2 + O_2 \rightarrow 2\ H_2O$$

This equation uses the smallest whole numbers and has the same number of oxygen atoms and hydrogen atoms on both sides of the equation (4 H + 2 O atoms). So in your drawing, there should be two H_2 molecules, 1 O_2 molecule, and 2 H_2O molecules.

33. The theoretical yield is the stoichiometric amount of product that should form if the limiting reactant is completely consumed and the reaction has 100% yield.

34. A reactant is present in excess if there is more of that reactant present than is needed to combine with the limiting reactant for the process. By definition, the limiting reactant cannot be present in excess. An excess of any reactant does not affect the theoretical yield for a process; the theoretical yield is determined by the limiting reactant.

35. The specific information needed is mostly the coefficients in the balanced equation and the molar masses of the reactants and products. For percent yield, we would need the actual yield of the reaction and the amounts of reactants used.

a. Mass of CB produced = 1.00×10^4 molecules A_2B_2

$$\times \frac{1\ \text{mol}\ A_2B_2}{6.022 \times 10^{23}\ \text{molecules}\ A_2B_2} \times \frac{2\ \text{mol CB}}{1\ \text{mol}\ A_2B_2} \times \frac{\text{molar mass of CB}}{\text{mol CB}}$$

b. Atoms of A produced = 1.00×10^4 molecules $A_2B_2 \times \dfrac{2\ \text{atoms A}}{1\ \text{molecule}\ A_2B_2}$

c. Moles of C reacted = 1.00×10^4 molecules $A_2B_2 \times \dfrac{1\ \text{mol}\ A_2B_2}{6.022 \times 10^{23}\ \text{molecules}\ A_2B_2}$

$$\times \frac{2\ \text{mol C}}{1\ \text{mol}\ A_2B_2}$$

d. Percent yield = $\dfrac{\text{actual mass}}{\text{theoretical mass}} \times 100$; the theoretical mass of CB produced was calculated in part a. If the actual mass of CB produced is given, then the percent yield can be determined for the reaction using the percent yield equation.

36. One method is to assume each quantity of reactant is limiting, then calculate the amount of product that could be produced from each reactant. This gives two possible answers (assuming two reactants). The correct answer (the amount of product that could be produced) is always the smaller number. Even though there is enough of the other reactant to form more product, once the smaller quantity is reached, the limiting reactant runs out, and the reaction cannot continue.

A second method would be to pick one of the reactants and then calculate how much of the other reactant would be required to react with all of it. How the answer compares to the actual amount of that reactant present allows one to deduce the identity of the limiting reactant. Once the identity is known, one would take the limiting reactant and convert it to mass of product formed.

Exercises

Atomic Masses and the Mass Spectrometer

37. Let A = average atomic mass

$$A = 0.0140(203.973) + 0.2410(205.9745) + 0.2210(206.9759) + 0.5240(207.9766)$$

$A = 2.86 + 49.64 + 45.74 + 109.0 = 207.2$ u; from the periodic table, the element is Pb.

Note: u is an abbreviation for amu (atomic mass units).

38. Average atomic mass = A = 0.0800(45.952632) + 0.0730(46.951764) + 0.7380(47.947947) + 0.0550(48.947841) + 0.0540(49.944792) = 47.88 amu

This is element Ti (titanium).

39. Let A = mass of ^{185}Re:

$$186.207 = 0.6260(186.956) + 0.3740(A),\ 186.207 - 117.0 = 0.3740(A)$$

$A = \dfrac{69.2}{0.3740} = 185$ u (A = 184.95 u without rounding to proper significant figures.)

40. Abundance ^{28}Si = 100.00 − (4.70 + 3.09) = 92.21%; from the periodic table, the average atomic mass of Si is 28.09 u.

$$28.09 = 0.9221(27.98) + 0.0470(\text{atomic mass } ^{29}\text{Si}) + 0.0309(29.97)$$

Atomic mass ^{29}Si = 29.01 u

The mass of ^{29}Si is actually a little less than 29 u. There are other isotopes of silicon that are considered when determining the 28.09 u average atomic mass of Si listed in the atomic table.

41. Let x = % of ^{151}Eu and y = % of ^{153}Eu, then $x + y = 100$ and $y = 100 - x$.

$$151.96 = \frac{x(150.9196) + (100 - x)(152.9209)}{100}$$

$15196 = (150.9196)x + 15292.09 - (152.9209)x,\ -96 = -(2.0013)x$

$x = 48\%$; 48% ^{151}Eu and 100 – 48 = 52% ^{153}Eu

42. If silver is 51.82% ^{107}Ag, then the remainder is ^{109}Ag (48.18%). Determining the atomic mass (A) of ^{109}Ag:

$$107.868 = \frac{51.82(106.905) + 48.18(A)}{100}$$

10786.8 = 5540. + (48.18)A, A = 108.9 u = atomic mass of ^{109}Ag

43. There are three peaks in the mass spectrum, each 2 mass units apart. This is consistent with two isotopes differing in mass by two mass units. The peak at 157.84 corresponds to a Br_2 molecule composed of two atoms of the lighter isotope. This isotope has mass equal to 157.84/2 or 78.92. This corresponds to ^{79}Br. The second isotope is ^{81}Br with mass equal to 161.84/2 = 80.92. The peaks in the mass spectrum correspond to $^{79}Br_2$, $^{79}Br^{81}Br$, and $^{81}Br_2$ in order of increasing mass. The intensities of the highest and lowest masses tell us the two isotopes are present in about equal abundance. The actual abundance is 50.68% ^{79}Br and 49.32% ^{81}Br.

44. Because we are not given the relative masses of the isotopes, we need to estimate the masses of the isotopes. A good estimate is to assume that only the protons and neutrons contribute to the overall mass of the atom and that the atomic mass of a proton and neutron are each 1.00 u. So the masses are about: ^{54}Fe, 54.00 u; ^{56}Fe, 56.00 u; ^{57}Fe, 57.00 u; ^{58}Fe, 58.00 u. Using these masses, the calculated average atomic mass would be:

$$0.0585(54.00) + 0.9175(56.00) + 0.0212(57.00) + 0.0028(58.00) = 55.91 \text{ u}$$

The average atomic mass listed in the periodic table is 55.85 u.

Moles and Molar Masses

45. When more than one conversion factor is necessary to determine the answer, we will usually put all the conversion factors into one calculation instead of determining intermediate answers. This method reduces round-off error and is a time saver.

$$500.\text{ atoms Fe} \times \frac{1\text{ mol Fe}}{6.022 \times 10^{23}\text{ atoms Fe}} \times \frac{55.85\text{ g Fe}}{\text{mol Fe}} = 4.64 \times 10^{-20}\text{ g Fe}$$

46. $500.0\text{ g Fe} \times \frac{1\text{ mol Fe}}{55.85\text{ g Fe}} = 8.953\text{ mol Fe}$

$$8.953\text{ mol Fe} \times \frac{6.022 \times 10^{23}\text{ atoms Fe}}{\text{mol Fe}} = 5.391 \times 10^{24}\text{ atoms Fe}$$

47. $1.00 \text{ carat} \times \frac{0.200 \text{ g C}}{\text{carat}} \times \frac{1 \text{ mol C}}{12.01 \text{ g C}} \times \frac{6.022 \times 10^{23} \text{ atoms C}}{\text{mol C}} = 1.00 \times 10^{22} \text{ atoms C}$

48. $5.0 \times 10^{21} \text{ atoms C} \times \frac{1 \text{ mol C}}{6.022 \times 10^{23} \text{ atoms C}} = 8.3 \times 10^{-3} \text{ mol C}$

$8.3 \times 10^{-3} \text{ mol C} \times \frac{12.01 \text{ g C}}{\text{mol C}} = 0.10 \text{ g C}$

49. Al_2O_3: 2(26.98) + 3(16.00) = 101.96 g/mol

Na_3AlF_6: 3(22.99) + 1(26.98) + 6(19.00) = 209.95 g/mol

50. HFC−134a, CH_2FCF_3: 2(12.01) + 2(1.008) + 4(19.00) = 102.04 g/mol

HCFC−124, $CHClFCF_3$: 2(12.01) + 1(1.008) + 1(35.45) + 4(19.00) = 136.48 g/mol

51. a. The formula is NH_3. 14.01 g/mol + 3(1.008 g/mol) = 17.03 g/mol

b. The formula is N_2H_4. 2(14.01) + 4(1.008) = 32.05 g/mol

c. $(NH_4)_2Cr_2O_7$: 2(14.01) + 8(1.008) + 2(52.00) + 7(16.00) = 252.08 g/mol

52. a. The formula is P_4O_6. 4(30.97 g/mol) + 6(16.00 g/mol) = 219.88 g/mol

b. $Ca_3(PO_4)_2$: 3(40.08) + 2(30.97) + 8(16.00) = 310.18 g/mol

c. Na_2HPO_4: 2(22.99) + 1(1.008) + 1(30.97) + 4(16.00) = 141.96 g/mol

53. a. $1.00 \text{ g } NH_3 \times \frac{1 \text{ mol } NH_3}{17.03 \text{ g } NH_3} = 0.0587 \text{ mol } NH_3$

b. $1.00 \text{ g } N_2H_4 \times \frac{1 \text{ mol } N_2H_4}{32.05 \text{ g } N_2H_4} = 0.0312 \text{ mol } N_2H_4$

c. $1.00 \text{ g } (NH_4)_2Cr_2O_7 \times \frac{1 \text{ mol } (NH_4)_2Cr_2O_7}{252.08 \text{ g } (NH_4)_2Cr_2O_7} = 3.97 \times 10^{-3} \text{ mol } (NH_4)_2Cr_2O_7$

54. a. $1.00 \text{ g } P_4O_6 \times \frac{1 \text{ mol } P_4O_6}{219.88 \text{ g}} = 4.55 \times 10^{-3} \text{ mol } P_4O_6$

b. $1.00 \text{ g } Ca_3(PO_4)_2 \times \frac{1 \text{ mol } Ca_3(PO_4)_2}{310.18 \text{ g}} = 3.22 \times 10^{-3} \text{ mol } Ca_3(PO_4)_2$

c. $1.00 \text{ g } Na_2HPO_4 \times \frac{1 \text{ mol } Na_2HPO_4}{141.96 \text{ g}} = 7.04 \times 10^{-3} \text{ mol } Na_2HPO_4$

55. a. $5.00 \text{ mol NH}_3 \times \frac{17.03\text{g NH}_3}{\text{mol NH}_3} = 85.2 \text{ g NH}_3$

b. $5.00 \text{ mol N}_2\text{H}_4 \times \frac{32.05 \text{ g N}_2\text{H}_4}{\text{mol N}_2\text{H}_4} = 160. \text{ g N}_2\text{H}_4$

c. $5.00 \text{ mol (NH}_4)_2\text{Cr}_2\text{O}_7 \times \frac{252.08 \text{ g (NH}_4)_2\text{Cr}_2\text{O}_7}{1 \text{ mol (NH}_4)_2\text{Cr}_2\text{O}_7} = 1260 \text{ g (NH}_4)_2\text{Cr}_2\text{O}_7$

56. a. $5.00 \text{ mol P}_4\text{O}_6 \times \frac{219.88 \text{ g}}{1 \text{ mol P}_4\text{O}_6} = 1.10 \times 10^3 \text{ g P}_4\text{O}_6$

b. $5.00 \text{ mol Ca}_3(\text{PO}_4)_2 \times \frac{310.18 \text{ g}}{\text{mol Ca}_3(\text{PO}_4)_2} = 1.55 \times 10^3 \text{ g Ca}_3(\text{PO}_4)_2$

c. $5.00 \text{ mol Na}_2\text{HPO}_4 \times \frac{141.96 \text{ g}}{\text{mol Na}_2\text{HPO}_4} = 7.10 \times 10^2 \text{ g Na}_2\text{HPO}_4$

57. Chemical formulas give atom ratios as well as mole ratios.

a. $5.00 \text{ mol NH}_3 \times \frac{1 \text{ mol N}}{\text{mol NH}_3} \times \frac{14.01 \text{ g N}}{\text{mol N}} = 70.1 \text{ g N}$

b. $5.00 \text{ mol N}_2\text{H}_4 \times \frac{2 \text{ mol N}}{\text{mol N}_2\text{H}_4} \times \frac{14.01 \text{ g N}}{\text{mol N}} = 140. \text{ g N}$

c. $5.00 \text{ mol (NH}_4)_2\text{Cr}_2\text{O}_7 \times \frac{2 \text{ mol N}}{\text{mol (NH}_4)_2\text{Cr}_2\text{O}_7} \times \frac{14.01 \text{ g N}}{\text{mol N}} = 140. \text{ g N}$

58. a. $5.00 \text{ mol P}_4\text{O}_6 \times \frac{4 \text{ mol P}}{\text{mol P}_4\text{O}_6} \times \frac{30.97 \text{ g P}}{\text{mol P}} = 619 \text{ g P}$

b. $5.00 \text{ mol Ca}_3(\text{PO}_4)_2 \times \frac{2 \text{ mol P}}{\text{mol Ca}_3(\text{PO}_4)_2} \times \frac{30.97 \text{ g P}}{\text{mol P}} = 310. \text{ g P}$

c. $5.00 \text{ mol Na}_2\text{HPO}_4 \times \frac{1 \text{ mol P}}{\text{mol Na}_2\text{HPO}_4} \times \frac{30.97 \text{ g P}}{\text{mol P}} = 155 \text{ g P}$

59. a. $1.00 \text{ g NH}_3 \times \frac{1 \text{ mol NH}_3}{17.03 \text{ g NH}_3} \times \frac{6.022 \times 10^{23} \text{ molecules NH}_3}{\text{mol NH}_3}$

$= 3.54 \times 10^{22} \text{ molecules NH}_3$

b. $1.00 \text{ g N}_2\text{H}_4 \times \frac{1 \text{ mol N}_2\text{H}_4}{32.05 \text{ g N}_2\text{H}_4} \times \frac{6.022 \times 10^{23} \text{ molecules N}_2\text{H}_4}{\text{mol N}_2\text{H}_4}$

$= 1.88 \times 10^{22} \text{ molecules N}_2\text{H}_4$

c. $1.00 \text{ g } (NH_4)_2Cr_2O_7 \times \frac{1 \text{ mol } (NH_4)_2Cr_2O_7}{252.08 \text{ g } (NH_4)_2Cr_2O_7}$

$\times \frac{6.022 \times 10^{23} \text{ formula units } (NH_4)_2Cr_2O_7}{\text{mol } (NH_4)_2Cr_2O_7} = 2.39 \times 10^{21} \text{ formula units } (NH_4)_2Cr_2O_7$

60. a. $1.00 \text{ g } P_4O_6 \times \frac{1 \text{ mol } P_4O_6}{219.88 \text{ g}} \times \frac{6.022 \times 10^{23} \text{ molecules}}{\text{mol } P_4O_6} = 2.74 \times 10^{21} \text{ molecules } P_4O_6$

b. $1.00 \text{ g } Ca_3(PO_4)_2 \times \frac{1 \text{ mol } Ca_3(PO_4)_2}{310.18 \text{ g}} \times \frac{6.022 \times 10^{23} \text{ formula units}}{\text{mol } Ca_3(PO_4)_2}$

$= 1.94 \times 10^{21} \text{ formula units } Ca_3(PO_4)_2$

c. $1.00 \text{ g } Na_2HPO_4 \times \frac{1 \text{ mol } Na_2HPO_4}{141.96 \text{ g}} \times \frac{6.022 \times 10^{23} \text{ formula units}}{\text{mol } Na_2HPO_4}$

$= 4.24 \times 10^{21} \text{ formula units } Na_2HPO_4$

61. Using answers from Exercise 59:

a. $3.54 \times 10^{22} \text{ molecules } NH_3 \times \frac{1 \text{ atom N}}{\text{molecule } NH_3} = 3.54 \times 10^{22} \text{ atoms N}$

b. $1.88 \times 10^{22} \text{ molecules } N_2H_4 \times \frac{2 \text{ atoms N}}{\text{molecule } N_2H_4} = 3.76 \times 10^{22} \text{ atoms N}$

c. $2.39 \times 10^{21} \text{ formula units } (NH_4)_2Cr_2O_7 \times \frac{2 \text{ atoms N}}{\text{formula unit } (NH_4)_2Cr_2O_7}$

$= 4.78 \times 10^{21} \text{ atoms N}$

62. Using answers from Exercise 60:

a. $2.74 \times 10^{21} \text{ molecules } P_4O_6 \times \frac{4 \text{ atoms P}}{\text{molecule } P_4O_6} = 1.10 \times 10^{22} \text{ atoms P}$

b. $1.94 \times 10^{21} \text{ formula units } Ca_3(PO_4)_2 \times \frac{2 \text{ atoms P}}{\text{formula unit } Ca_3(PO_4)_2} = 3.88 \times 10^{21} \text{ atoms P}$

c. $4.24 \times 10^{21} \text{ formula units } Na_2HPO_4 \times \frac{1 \text{ atom P}}{\text{formula unit } Na_2HPO_4} = 4.24 \times 10^{21} \text{ atoms P}$

63. Molar mass of CCl_2F_2 = 12.01 + 2(35.45) + 2(19.00) = 120.91 g/mol

$5.56 \text{ mg } CCl_2F_2 \times \frac{1 \text{ g}}{1000 \text{ mg}} \times \frac{1 \text{ mol}}{120.91 \text{ g}} \times \frac{6.022 \times 10^{23} \text{ molecules}}{\text{mol}}$

$= 2.77 \times 10^{19} \text{ molecules } CCl_2F_2$

$$5.56 \times 10^{-3}\text{ g } CCl_2F_2 \times \frac{1\text{ mol } CCl_2F_2}{120.91\text{ g}} \times \frac{2\text{ mol Cl}}{1\text{ mol } CCl_2F} \times \frac{35.45\text{ g Cl}}{\text{mol Cl}}$$

$$= 3.26 \times 10^{-3}\text{ g} = 3.26\text{ mg Cl}$$

64. The $\bullet 2H_2O$ is part of the formula of bauxite (they are called waters of hydration). Combining elements together, the chemical formula for bauxite would be $Al_2O_5H_4$.

a. Molar mass = 2(26.98) + 5(16.00) + 4(1.008) = 137.99 g/mol

b. $$0.58\text{ mol } Al_2O_3\bullet 2H_2O \times \frac{2\text{ mol Al}}{\text{mol } Al_2O_3\bullet 2H_2O} \times \frac{26.98\text{g Al}}{\text{mol Al}} = 31\text{ g Al}$$

c. $$0.58\text{ mol } Al_2O_3\bullet 2H_2O \times \frac{2\text{ mol Al}}{\text{mol } Al_2O_3\bullet 2H_2O} \times \frac{6.022 \times 10^{23}\text{ atoms}}{\text{mol Al}}$$

$$= 7.0 \times 10^{23}\text{ atoms Al}$$

d. $$2.1 \times 10^{24}\text{ formula units } Al_2O_3\bullet 2H_2O \times \frac{1\text{ mol } Al_2O_3\bullet 2H_2O}{6.022 \times 10^{23}\text{ formula units}} \times \frac{137.99\text{ g}}{\text{mol}}$$

$$= 480\text{ g } Al_2O_3\bullet 2H_2O$$

65. a. $$150.0\text{ g } Fe_2O_3 \times \frac{1\text{ mol}}{159.70\text{ g}} = 0.9393\text{ mol } Fe_2O_3$$

b. $$10.0\text{ mg } NO_2 \times \frac{1\text{ g}}{1000\text{mg}} \times \frac{1\text{ mol}}{46.01\text{ g}} = 2.17 \times 10^{-4}\text{ mol } NO_2$$

c. $$1.5 \times 10^{16}\text{ molecules } BF_3 \times \frac{1\text{ mol}}{6.02 \times 10^{23}\text{ molecules}} = 2.5 \times 10^{-8}\text{ mol } BF_3$$

66. a. $$20.0\text{ mg } C_8H_{10}N_4O_2 \times \frac{1\text{ g}}{1000\text{mg}} \times \frac{1\text{ mol}}{194.20\text{ g}} = 1.03 \times 10^{-4}\text{ mol } C_8H_{10}N_4O_2$$

b. $$2.72 \times 10^{21}\text{ molecules } C_2H_5OH \times \frac{1\text{ mol}}{6.022 \times 10^{23}\text{ molecules}}$$

$$= 4.52 \times 10^{-3}\text{ mol } C_2H_5OH$$

c. $$1.50\text{ g } CO_2 \times \frac{1\text{ mol}}{44.01\text{ g}} = 3.41 \times 10^{-2}\text{ mol } CO_2$$

67. a. A chemical formula gives atom ratios as well as mole ratios. We will use both ideas to show how these conversion factors can be used.

Molar mass of $C_2H_5O_2N$ = 2(12.01) + 5(1.008) + 2(16.00) + 14.01 = 75.07 g/mol

$$5.00 \text{ g } C_2H_5O_2N \times \frac{1 \text{ mol } C_2H_5O_2N}{75.07 \text{ g } C_2H_5O_2N} \times \frac{6.022 \times 10^{23} \text{ molecules } C_2H_5O_2N}{\text{mol } C_2H_5O_2N}$$

$$\times \frac{1 \text{ atom N}}{\text{molecule } C_2H_5O_2N} = 4.01 \times 10^{22} \text{ atoms N}$$

b. Molar mass of Mg_3N_2 = 3(24.31) + 2(14.01) = 100.95 g/mol

$$5.00 \text{ g } Mg_3N_2 \times \frac{1 \text{ mol } Mg_3N_2}{100.95 \text{ g } Mg_3N_2} \times \frac{6.022 \times 10^{23} \text{ formula units } Mg_3N_2}{\text{mol } Mg_3N_2}$$

$$\times \frac{2 \text{ atoms N}}{\text{mol } Mg_3N_2} = 5.97 \times 10^{22} \text{ atoms N}$$

c. Molar mass of $Ca(NO_3)_2$ = 40.08 + 2(14.01) + 6(16.00) = 164.10 g/mol

$$5.00 \text{ g } Ca(NO_3)_2 \times \frac{1 \text{ mol } Ca(NO_3)_2}{164.10 \text{ g } Ca(NO_3)_2} \times \frac{2 \text{ mol N}}{\text{mol } Ca(NO_3)_2} \times \frac{6.022 \times 10^{23} \text{ atoms N}}{\text{mol N}}$$

$$= 3.67 \times 10^{22} \text{ atoms N}$$

d. Molar mass of N_2O_4 = 2(14.01) + 4(16.00) = 92.02 g/mol

$$5.00 \text{ g } N_2O_4 \times \frac{1 \text{ mol } N_2O_4}{92.02 \text{ g } N_2O_4} \times \frac{2 \text{ mol N}}{\text{mol } N_2O_4} \times \frac{6.022 \times 10^{23} \text{ atoms N}}{\text{mol N}}$$

$$= 6.54 \times 10^{22} \text{ atoms N}$$

68. $4.24 \text{ g } C_6H_6 \times \frac{1 \text{ mol}}{78.11 \text{ g}} = 5.43 \times 10^{-2} \text{ mol } C_6H_6$

$$5.43 \times 10^{-2} \text{ mol } C_6H_6 \times \frac{6.022 \times 10^{23} \text{ molecules}}{\text{mol}} = 3.27 \times 10^{22} \text{ molecules } C_6H_6$$

Each molecule of C_6H_6 contains 6 atoms C + 6 atoms H = 12 atoms total.

$$3.27 \times 10^{22} \text{ molecules } C_6H_6 \times \frac{12 \text{ atoms total}}{\text{molecule}} = 3.92 \times 10^{23} \text{ atoms total}$$

$$0.224 \text{ mol } H_2O \times \frac{18.02 \text{ g}}{\text{mol}} = 4.04 \text{ g } H_2O$$

$$0.224 \text{ mol } H_2O \times \frac{6.022 \times 10^{23} \text{ molecules}}{\text{mol}} = 1.35 \times 10^{23} \text{ molecules } H_2O$$

$$1.35 \times 10^{23} \text{ molecules } H_2O \times \frac{3 \text{ atoms total}}{\text{molecule}} = 4.05 \times 10^{23} \text{ atoms total}$$

$$2.71 \times 10^{22} \text{ molecules } CO_2 \times \frac{1\text{ mol}}{6.022 \times 10^{23}\text{ molecules}} = 4.50 \times 10^{-2} \text{ mol } CO_2$$

$$4.50 \times 10^{-2} \text{ mol } CO_2 \times \frac{44.01\text{ g}}{\text{mol}} = 1.98 \text{ g } CO_2$$

$$2.71 \times 10^{22} \text{ molecules } CO_2 \times \frac{3\text{ atoms total}}{\text{molecule } CO_2} = 8.13 \times 10^{22} \text{ atoms total}$$

$$3.35 \times 10^{22} \text{ atoms total} \times \frac{1\text{ molecule}}{6\text{ atoms total}} = 5.58 \times 10^{21} \text{ molecules } CH_3OH$$

$$5.58 \times 10^{21} \text{ molecules } CH_3OH \times \frac{1\text{ mol}}{6.022 \times 10^{23}\text{ molecules}} = 9.27 \times 10^{-3} \text{ mol } CH_3OH$$

$$9.27 \times 10^{-3} \text{ mol } CH_3OH \times \frac{32.04\text{ g}}{\text{mol}} = 0.297 \text{ g } CH_3OH$$

69. Molar mass of $C_6H_8O_6$ = 6(12.01) + 8(1.008) + 6(16.00) = 176.12 g/mol

$$500.0 \text{ mg} \times \frac{1\text{ g}}{1000\text{ mg}} \times \frac{1\text{ mol}}{176.12\text{ g}} = 2.839 \times 10^{-3} \text{ mol } C_6H_8O_6$$

$$2.839 \times 10^{-3} \text{ mol} \times \frac{6.022 \times 10^{23}\text{ molecules}}{\text{mol}} = 1.710 \times 10^{21} \text{ molecules } C_6H_8O_6$$

70. a. 9(12.01) + 8(1.008) + 4(16.00) = 180.15 g/mol

b. $500.\text{ mg} \times \frac{1\text{ g}}{1000\text{ mg}} \times \frac{1\text{ mol}}{180.15\text{ g}} = 2.78 \times 10^{-3} \text{ mol } C_9H_8O_4$

$$2.78 \times 10^{-3} \text{ mol} \times \frac{6.022 \times 10^{23}\text{ molecules}}{\text{mol}} = 1.67 \times 10^{21} \text{ molecules } C_9H_8O_4$$

71. a. 2(12.01) + 3(1.008) + 3(35.45) + 2(16.00) = 165.39 g/mol

b. $500.0 \text{ g} \times \frac{1\text{ mol}}{165.39\text{ g}} = 3.023 \text{ mol } C_2H_3Cl_3O_2$

c. $2.0 \times 10^{-2} \text{ mol} \times \frac{165.39\text{ g}}{\text{mol}} = 3.3 \text{ g } C_2H_3Cl_3O_2$

d. $5.0 \text{ g } C_2H_3Cl_3O_2 \times \frac{1\text{ mol}}{165.39\text{ g}} \times \frac{6.022 \times 10^{23}\text{ molecules}}{\text{mol}} \times \frac{3\text{ atoms Cl}}{\text{molecule}}$

$= 5.5 \times 10^{22}$ atoms of chlorine

e. $1.0 \text{ g Cl} \times \frac{1 \text{ mol Cl}}{35.45 \text{ g}} \times \frac{1 \text{ mol } C_2H_3Cl_3O_2}{3 \text{ mol Cl}} \times \frac{165.39 \text{ g } C_2H_3Cl_3O_2}{\text{mol } C_2H_3Cl_3O_2} = 1.6 \text{ g chloral hydrate}$

f. $500 \text{ molecules} \times \frac{1 \text{ mol}}{6.022 \times 10^{23} \text{ molecules}} \times \frac{165.39 \text{ g}}{\text{mol}} = 1.373 \times 10^{-19} \text{ g } C_2H_3Cl_3O_2$

72. As we shall see in later chapters, the formula written as $(CH_3)_2N_2O$ tries to tell us something about how the atoms are attached to each other. For our purposes in this problem, we can write the formula as $C_2H_6N_2O$.

a. $2(12.01) + 6(1.008) + 2(14.01) + 1(16.00) = 74.09 \text{ g/mol}$

b. $250 \text{ mg} \times \frac{1 \text{ g}}{1000 \text{ mg}} \times \frac{1 \text{ mol}}{74.09 \text{ g}} = 3.4 \times 10^{-3} \text{ mol}$ c. $0.050 \text{ mol} \times \frac{74.09 \text{ g}}{\text{mol}} = 3.7 \text{ g}$

d. $1.0 \text{ mol } C_2H_6N_2O \times \frac{6.022 \times 10^{23} \text{ molecules } C_2H_6N_2O}{\text{mol } C_2H_6N_2O} \times \frac{6 \text{ atoms of H}}{\text{molecule } C_2H_6N_2O}$

$= 3.6 \times 10^{24}$ atoms of hydrogen

e. $1.0 \times 10^6 \text{ molecules} \times \frac{1 \text{ mol}}{6.022 \times 10^{23} \text{ molecules}} \times \frac{74.09 \text{ g}}{\text{mol}} = 1.2 \times 10^{-16} \text{ g}$

f. $1 \text{ molecule} \times \frac{1 \text{ mol}}{6.022 \times 10^{23} \text{ molecules}} \times \frac{74.09 \text{ g}}{\text{mol}} = 1.230 \times 10^{-22} \text{ g } C_2H_6N_2O$

Percent Composition

73. a. $C_3H_4O_2$: Molar mass $= 3(12.01) + 4(1.008) + 2(16.00) = 36.03 + 4.032 + 32.00$
$= 72.06 \text{ g/mol}$

$\text{Mass \% C} = \frac{36.03 \text{ g C}}{72.06 \text{ g compound}} \times 100 = 50.00\% \text{ C}$

$\text{Mass \% H} = \frac{4.032 \text{ g H}}{72.06 \text{ g compound}} \times 100 = 5.595\% \text{ H}$

Mass % O = 100.00 – (50.00 + 5.595) = 44.41% O or:

$\% \text{ O} = \frac{32.00 \text{ g}}{72.06 \text{ g}} \times 100 = 44.41\% \text{ O}$

b. $C_4H_6O_2$: Molar mass $= 4(12.01) + 6(1.008) + 2(16.00) = 48.04 + 6.048 + 32.00$
$= 86.09 \text{ g/mol}$

$\text{Mass \% C} = \frac{48.04 \text{ g}}{86.09 \text{ g}} \times 100 = 55.80\% \text{ C}$; mass $\% \text{ H} = \frac{6.048 \text{ g}}{86.09 \text{ g}} \times 100 = 7.025\% \text{ H}$

Mass % O = 100.00 – (55.80 + 7.025) = 37.18% O

c. C_3H_3N: Molar mass = 3(12.01) + 3(1.008) + 1(14.01) = 36.03 + 3.024 + 14.01 = 53.06 g/mol

$$\text{Mass \% C} = \frac{36.03\text{ g}}{53.06\text{ g}} \times 100 = 67.90\%\text{ C};\quad \text{mass \% H} = \frac{3.024\text{ g}}{53.06\text{ g}} \times 100 = 5.699\%\text{ H}$$

$$\text{Mass \% N} = \frac{14.01\text{ g}}{53.06\text{ g}} \times 100 = 26.40\%\text{ N or \% N} = 100.00 - (67.90 + 5.699) = 26.40\%\text{ N}$$

74. In 1 mole of $YBa_2Cu_3O_7$, there are 1 mole of Y, 2 moles of Ba, 3 moles of Cu, and 7 moles of O.

$$\text{Molar mass} = 1\text{ mol Y}\left(\frac{88.91\text{ g Y}}{\text{mol Y}}\right) + 2\text{ mol Ba}\left(\frac{137.3\text{ g Ba}}{\text{mol Ba}}\right) + 3\text{ mol Cu}\left(\frac{63.55\text{ g Cu}}{\text{mol Cu}}\right) + 7\text{ mol O}\left(\frac{16.00\text{ g O}}{\text{mol O}}\right)$$

Molar mass = 88.91 + 274.6 + 190.65 + 112.00 = 666.2 g/mol

$$\text{Mass \% Y} = \frac{88.91\text{ g}}{666.2\text{ g}} \times 100 = 13.35\%\text{ Y};\quad \text{mass \% Ba} = \frac{274.6\text{ g}}{666.2\text{ g}} \times 100 = 41.22\%\text{ Ba}$$

$$\text{Mass \% Cu} = \frac{190.65\text{ g}}{666.2\text{ g}} \times 100 = 28.62\%\text{ Cu};\quad \text{mass \% O} = \frac{112.0\text{ g}}{666.2\text{ g}} \times 100 = 16.81\%\text{ O}$$

75. NO: $\text{Mass \% N} = \dfrac{14.01\text{ g N}}{30.01\text{ g NO}} \times 100 = 46.68\%\text{ N}$

NO_2: $\text{Mass \% N} = \dfrac{14.01\text{ g N}}{46.01\text{ g NO}_2} \times 100 = 30.45\%\text{ N}$

N_2O: $\text{Mass \% N} = \dfrac{2(14.01)\text{ g N}}{44.02\text{ g N}_2\text{O}} \times 100 = 63.65\%\text{ N}$

From the calculated mass percents, only NO is 46.7% N by mass, so NO could be this species. Any other compound having NO as an empirical formula could also be the compound.

76. a. $C_8H_{10}N_4O_2$: Molar mass = 8(12.01) + 10(1.008) + 4(14.01) + 2(16.00) = 194.20 g/mol

$$\text{Mass \% C} = \frac{8(12.01)\text{ g C}}{194.20\text{ g C}_8\text{H}_{10}\text{N}_4\text{O}_2} \times 100 = \frac{96.08\text{ g}}{194.20\text{ g}} \times 100 = 49.47\%\text{ C}$$

b. $C_{12}H_{22}O_{11}$: Molar mass = 12(12.01) + 22(1.008) + 11(16.00) = 342.30 g/mol

$$\text{Mass \% C} = \frac{12(12.01)\text{ g C}}{342.30\text{ g C}_{12}\text{H}_{22}\text{O}_{11}} \times 100 = 42.10\%\text{ C}$$

c. C_2H_5OH: Molar mass = 2(12.01) + 6(1.008) + 1(16.00) = 46.07 g/mol

$$\text{Mass \% C} = \frac{2(12.01)\text{ g C}}{46.07\text{ g } C_2H_5OH} \times 100 = 52.14\%\text{ C}$$

The order from lowest to highest mass percentage of carbon is:

sucrose ($C_{12}H_{22}O_{11}$) < caffeine ($C_8H_{10}N_4O_2$) < ethanol (C_2H_5OH)

77. There are 0.390 g Cu for every 100.000 g of fungal laccase. Assuming 100.00 g fungal laccase:

$$\text{Mol fungal laccase} = 0.390\text{ g Cu} \times \frac{1\text{ mol Cu}}{63.55\text{ g Cu}} \times \frac{1\text{ mol fungal laccase}}{4\text{ mol Cu}} = 1.53 \times 10^{-3}\text{ mol}$$

$$\frac{x\text{ g fungal laccase}}{\text{mol fungal laccase}} = \frac{100.000\text{ g}}{1.53 \times 10^{-3}\text{ mol}},\ x = \text{molar mass} = 6.54 \times 10^4\text{ g/mol}$$

78. There are 0.347 g Fe for every 100.000 g hemoglobin (Hb). Assuming 100.000 g hemoglobin:

$$\text{Mol Hb} = 0.347\text{ g Fe} \times \frac{1\text{ mol Fe}}{55.85\text{ g Fe}} \times \frac{1\text{ mol Hb}}{4\text{ mol Fe}} = 1.55 \times 10^{-3}\text{ mol Hb}$$

$$\frac{x\text{ g Hb}}{\text{mol Hb}} = \frac{100.000\text{ g Hb}}{1.55 \times 10^{-3}\text{ mol Hb}},\ x = \text{molar mass} = 6.45 \times 10^4\text{ g/mol}$$

Empirical and Molecular Formulas

79. a. Molar mass of CH_2O = $1\text{ mol C}\left(\frac{12.01\text{ g C}}{\text{mol C}}\right) + 2\text{ mol H}\left(\frac{1.008\text{ g H}}{\text{mol H}}\right)$

$$+ 1\text{ mol O}\left(\frac{16.00\text{ g O}}{\text{mol O}}\right) = 30.03\text{ g/mol}$$

$$\%\text{ C} = \frac{12.01\text{ g C}}{30.03\text{ g } CH_2O} \times 100 = 39.99\%\text{ C};\ \%\text{ H} = \frac{2.016\text{ g H}}{30.03\text{ g } CH_2O} \times 100 = 6.713\%\text{ H}$$

$$\%\text{ O} = \frac{16.00\text{ g O}}{30.03\text{ g } CH_2O} \times 100 = 53.28\%\text{ O or \% O} = 100.00 - (39.99 + 6.713) = 53.30\%$$

b. Molar mass of $C_6H_{12}O_6$ = 6(12.01) + 12(1.008) + 6(16.00) = 180.16 g/mol

$$\%\text{ C} = \frac{76.06\text{ g C}}{180.16\text{ g } C_6H_{12}O_6} \times 100 = 40.00\%;\quad \%\text{ H} = \frac{12.(1.008)\text{ g}}{180.16\text{ g}} \times 100 = 6.714\%$$

% O = 100.00 – (40.00 + 6.714) = 53.29%

c. Molar mass of $HC_2H_3O_2$ = 2(12.01) + 4(1.008) + 2(16.00) = 60.05 g/mol

$$\% \text{ C} = \frac{24.02 \text{ g}}{60.05 \text{ g}} \times 100 = 40.00\%; \quad \% \text{ H} = \frac{4.032 \text{ g}}{60.05 \text{ g}} \times 100 = 6.714\%$$

% O = 100.00 – (40.00 + 6.714) = 53.29%

80. All three compounds have the same empirical formula, CH_2O, and different molecular formulas. The composition of all three in mass percent is also the same (within rounding differences). Therefore, elemental analysis will give us only the empirical formula.

81. a. The molecular formula is N_2O_4. The smallest whole number ratio of the atoms (the empirical formula) is NO_2.

b. Molecular formula: C_3H_6; empirical formula: CH_2

c. Molecular formula: P_4O_{10}; empirical formula: P_2O_5

d. Molecular formula: $C_6H_{12}O_6$; empirical formula: CH_2O

82. a. SNH: Empirical formula mass = 32.07 + 14.01 + 1.008 = 47.09 g/mol

$\frac{188.35 \text{ g}}{47.09 \text{ g}}$ = 4.000; so the molecular formula is $(SNH)_4$ or $S_4N_4H_4$.

b. $NPCl_2$: Empirical formula mass = 14.01 + 30.97 + 2(35.45) = 115.88 g/mol

$\frac{347.64 \text{ g}}{115.88 \text{ g}}$ = 3.0000; molecular formula is $(NPCl_2)_3$ or $N_3P_3Cl_6$.

c. CoC_4O_4: 58.93 + 4(12.01) + 4(16.00) = 170.97 g/mol

$\frac{341.94 \text{ g}}{170.97 \text{ g}}$ = 2.0000; molecular formula: $Co_2C_8O_8$

d. SN: 32.07 + 14.01 = 46.08 g/mol; $\frac{184.32 \text{ g}}{46.08 \text{ g}}$ = 4.000; molecular formula: S_4N_4

83. Out of 100.00 g of compound, there are:

$$48.64 \text{ g C} \times \frac{1 \text{ mol C}}{12.01 \text{ g C}} = 4.050 \text{ mol C}; \; 8.16 \text{ g H} \times \frac{1 \text{ mol H}}{1.008 \text{ g H}} = 8.10 \text{ mol H}$$

$$\% \text{ O} = 100.00 - 48.64 - 8.16 = 43.20\%; \; 43.20 \text{ g O} \times \frac{1 \text{ mol O}}{16.00 \text{ g O}} = 2.700 \text{ mol O}$$

Dividing each mole value by the smallest number:

$$\frac{4.050}{2.700} = 1.500; \quad \frac{8.10}{2.700} = 3.00; \quad \frac{2.700}{2.700} = 1.000$$

Because a whole number ratio is required, the C : H : O ratio is 1.5 : 3 : 1 or 3 : 6 : 2. So the empirical formula is $C_3H_6O_2$.

84. Assuming 100.00 g of nylon-6:

$$63.68 \text{ g C} \times \frac{1 \text{ mol C}}{12.01 \text{ g C}} = 5.302 \text{ mol C}; \quad 12.38 \text{ g N} \times \frac{1 \text{ mol N}}{14.01 \text{ g N}} = 0.8837 \text{ mol N}$$

$$9.80 \text{ g H} \times \frac{1 \text{ mol H}}{1.008 \text{ g H}} = 9.72 \text{ mol H}; \quad 14.14 \text{ g O} \times \frac{1 \text{ mol O}}{16.00 \text{ g O}} = 0.8838 \text{ mol O}$$

Dividing each mole value by the smallest number:

$$\frac{5.302}{0.8837} = 6.000; \quad \frac{9.72}{0.8837} = 11.0; \quad \frac{0.8838}{0.8837} = 1.000$$

The empirical formula for nylon-6 is $C_6H_{11}NO$

85. Compound I: Mass O = 0.6498 g Hg_xO_y – 0.6018 g Hg = 0.0480 g O

$$0.6018 \text{ g Hg} \times \frac{1 \text{ mol Hg}}{200.6 \text{ g Hg}} = 3.000 \times 10^{-3} \text{ mol Hg}$$

$$0.0480 \text{ g O} \times \frac{1 \text{ mol O}}{16.00 \text{ g O}} = 3.00 \times 10^{-3} \text{ mol O}$$

The mole ratio between Hg and O is 1 : 1, so the empirical formula of compound I is HgO.

Compound II: Mass Hg = 0.4172 g Hg_xO_y – 0.016 g O = 0.401 g Hg

$$0.401 \text{ g Hg} \times \frac{1 \text{ mol Hg}}{200.6 \text{ g Hg}} = 2.00 \times 10^{-3} \text{ mol Hg}; \quad 0.016 \text{ g O} \times \frac{1 \text{ mol O}}{16.00 \text{ g O}} = 1.0 \times 10^{-3} \text{ mol O}$$

The mole ratio between Hg and O is 2 : 1, so the empirical formula is Hg_2O.

86. $$1.121 \text{ g N} \times \frac{1 \text{ mol N}}{14.01 \text{ g N}} = 8.001 \times 10^{-2} \text{ mol N}; \quad 0.161 \text{ g H} \times \frac{1 \text{ mol H}}{1.008 \text{ g H}} = 1.60 \times 10^{-1} \text{ mol H}$$

$$0.480 \text{ g C} \times \frac{1 \text{ mol C}}{12.01 \text{ g C}} = 4.00 \times 10^{-2} \text{ mol C}; \quad 0.640 \text{ g O} \times \frac{1 \text{ mol O}}{16.00 \text{ g O}} = 4.00 \times 10^{-2} \text{ mol O}$$

Dividing all mole values by the smallest number:

$$\frac{8.001 \times 10^{-2}}{4.00 \times 10^{-2}} = 2.00; \quad \frac{1.60 \times 10^{-1}}{4.00 \times 10^{-2}} = 4.00; \quad \frac{4.00 \times 10^{-2}}{4.00 \times 10^{-2}} = 1.00$$

The empirical formula is N_2H_4CO.

87. Out of 100.0 g, there are:

$$69.6 \text{ g S} \times \frac{1 \text{ mol S}}{32.07 \text{ g S}} = 2.17 \text{ mol S}; \quad 30.4 \text{ g N} \times \frac{1 \text{ mol N}}{14.01 \text{ g N}} = 2.17 \text{ mol N}$$

The empirical formula is SN because the mole values are in a 1 : 1 mole ratio.

The empirical formula mass of SN is ~ 46 g/mol. Because 184/46 = 4.0, the molecular formula is S_4N_4.

88. Assuming 100.0 g of compound:

$$26.7 \text{ g P} \times \frac{1 \text{ mol P}}{30.97 \text{ g P}} = 0.862 \text{ mol P}; \quad 12.1 \text{ g N} \times \frac{1 \text{ mol N}}{14.01 \text{ g N}} = 0.864 \text{ mol N}$$

$$61.2 \text{ g Cl} \times \frac{1 \text{ mol Cl}}{35.45 \text{ g Cl}} = 1.73 \text{ mol Cl}$$

$\frac{1.73}{0.862} = 2.01$; the empirical formula is $PNCl_2$.

The empirical formula mass is ≈ 31.0 + 14.0 + 2(35.5) = 116 g/mol.

$\frac{\text{Molar mass}}{\text{Empirical formula mass}} = \frac{580}{116} = 5.0$; the molecular formula is $(PNCl_2)_5 = P_5N_5Cl_{10}$.

89. Assuming 100.00 g of compound:

$$47.08 \text{ g C} \times \frac{1 \text{ mol C}}{12.01 \text{ g C}} = 3.920 \text{ mol C}; \quad 6.59 \text{ g H} \times \frac{1 \text{ mol H}}{1.008 \text{ g H}} = 6.54 \text{ mol H}$$

$$46.33 \text{ g Cl} \times \frac{1 \text{ mol Cl}}{35.45 \text{ g Cl}} = 1.307 \text{ mol Cl}$$

Dividing all mole values by 1.307 gives:

$$\frac{3.920}{1.307} = 2.999; \quad \frac{6.54}{1.307} = 5.00; \quad \frac{1.307}{1.307} = 1.000$$

The empirical formula is C_3H_5Cl.

The empirical formula mass is 3(12.01) + 5(1.008) + 1(35.45) = 76.52 g/mol.

$$\frac{\text{Molar mass}}{\text{Empirical formula mass}} = \frac{153}{76.52} = 2.00$$; the molecular formula is $(C_3H_5Cl)_2 = C_6H_{10}Cl_2$.

90. Assuming 100.00 g of compound (mass oxygen = 100.00 g − 41.39 g C − 3.47 g H = 55.14 g O):

$$41.39 \text{ g C} \times \frac{1 \text{ mol C}}{12.01 \text{ g C}} = 3.446 \text{ mol C}; \quad 3.47 \text{ g H} \times \frac{1 \text{ mol H}}{1.008 \text{ g H}} = 3.44 \text{ mol H}$$

$$55.14 \text{ g O} \times \frac{1 \text{ mol O}}{16.00 \text{ g O}} = 3.446 \text{ mol O}$$

All are the same mole values, so the empirical formula is CHO. The empirical formula mass is 12.01 + 1.008 + 16.00 = 29.02 g/mol.

$$\text{Molar mass} = \frac{15.0 \text{ g}}{0.129 \text{ mol}} = 116 \text{ g/mol}$$

$$\frac{\text{Molar mass}}{\text{Empirical mass}} = \frac{116}{29.02} = 4.00; \quad \text{molecular formula} = (CHO)_4 = C_4H_4O_4$$

91. When combustion data are given, it is assumed that all the carbon in the compound ends up as carbon in CO_2 and all the hydrogen in the compound ends up as hydrogen in H_2O. In the sample of fructose combusted, the masses of C and H are:

$$\text{mass C} = 2.20 \text{ g } CO_2 \times \frac{1 \text{ mol } CO_2}{44.01 \text{ g } CO_2} \times \frac{1 \text{ mol C}}{\text{mol } CO_2} \times \frac{12.01 \text{ g C}}{\text{mol C}} = 0.600 \text{ g C}$$

$$\text{mass H} = 0.900 \text{ g } H_2O \times \frac{1 \text{ mol } H_2O}{18.02 \text{ g } H_2O} \times \frac{2 \text{ mol H}}{\text{mol } H_2O} \times \frac{1.008 \text{ g H}}{\text{mol H}} = 0.101 \text{ g H}$$

Mass O = 1.50 g fructose − 0.600 g C − 0.101 g H = 0.799 g O

So, in 1.50 g of the fructose, we have:

$$0.600 \text{ g C} \times \frac{1 \text{ mol C}}{12.01 \text{ g C}} = 0.0500 \text{ mol C}; \quad 0.101 \text{ g H} \times \frac{1 \text{ mol H}}{1.008 \text{ g H}} = 0.100 \text{ mol H}$$

$$0.799 \text{ g O} \times \frac{1 \text{ mol O}}{16.00 \text{ g O}} = 0.0499 \text{ mol O}$$

Dividing by the smallest number: $\frac{0.100}{0.0499} = 2.00$; the empirical formula is CH_2O.

92. This compound contains nitrogen, and one way to determine the amount of nitrogen in the compound is to calculate composition by mass percent. We assume that all the carbon in 33.5 mg CO_2 came from the 35.0 mg of compound and all the hydrogen in 41.1 mg H_2O came from the 35.0 mg of compound.

$$3.35 \times 10^{-2}\ \text{g CO}_2 \times \frac{1\ \text{mol CO}_2}{44.01\ \text{g CO}_2} \times \frac{1\ \text{mol C}}{\text{mol CO}_2} \times \frac{12.01\ \text{g C}}{\text{mol C}} = 9.14 \times 10^{-3}\ \text{g C}$$

$$\text{Mass \% C} = \frac{9.14 \times 10^{-3}\ \text{g C}}{3.50 \times 10^{-2}\ \text{g compound}} \times 100 = 26.1\%\ \text{C}$$

$$4.11 \times 10^{-2}\ \text{g H}_2\text{O} \times \frac{1\ \text{mol H}_2\text{O}}{18.02\ \text{g H}_2\text{O}} \times \frac{2\ \text{mol H}}{\text{mol H}_2\text{O}} \times \frac{1.008\ \text{g H}}{\text{mol H}} = 4.60 \times 10^{-3}\ \text{g H}$$

$$\text{Mass \% H} = \frac{4.60 \times 10^{-3}\ \text{g H}}{3.50 \times 10^{-2}\ \text{g compound}} \times 100 = 13.1\%\ \text{H}$$

The mass percent of nitrogen is obtained by difference:

Mass % N = 100.0 – (26.1 + 13.1) = 60.8% N

Now perform the empirical formula determination by first assuming 100.0 g of compound. Out of 100.0 g of compound, there are:

$$26.1\ \text{g C} \times \frac{1\ \text{mol C}}{12.01\ \text{g C}} = 2.17\ \text{mol C};\quad 13.1\ \text{g H} \times \frac{1\ \text{mol H}}{1.008\ \text{g H}} = 13.0\ \text{mol H}$$

$$60.8\ \text{g N} \times \frac{1\ \text{mol N}}{14.01\ \text{g N}} = 4.34\ \text{mol N}$$

Dividing all mole values by 2.17 gives: $\frac{2.17}{2.17} = 1.00$; $\frac{13.0}{2.17} = 5.99$; $\frac{4.34}{2.17} = 2.00$

The empirical formula is CH_6N_2.

93. The combustion data allow determination of the amount of hydrogen in cumene. One way to determine the amount of carbon in cumene is to determine the mass percent of hydrogen in the compound from the data in the problem; then determine the mass percent of carbon by difference (100.0 – mass % H = mass % C).

$$42.8\ \text{mg H}_2\text{O} \times \frac{1\ \text{g}}{1000\ \text{mg}} \times \frac{2.016\ \text{g H}}{18.02\ \text{g H}_2\text{O}} \times \frac{1000\ \text{mg}}{\text{g}} = 4.79\ \text{mg H}$$

$$\text{Mass \% H} = \frac{4.79\ \text{mg H}}{47.6\ \text{mg cumene}} \times 100 = 10.1\%\ \text{H};\ \text{mass \% C} = 100.0 - 10.1 = 89.9\%\ \text{C}$$

Now solve the empirical formula problem. Out of 100.0 g cumene, we have:

$$89.9\ \text{g C} \times \frac{1\ \text{mol C}}{12.01\ \text{g C}} = 7.49\ \text{mol C};\quad 10.1\ \text{g H} \times \frac{1\ \text{mol H}}{1.008\ \text{g H}} = 10.0\ \text{mol H}$$

$\frac{10.0}{7.49} = 1.34 \approx \frac{4}{3}$; the mole H to mole C ratio is 4 : 3. The empirical formula is C_3H_4.

Empirical formula mass ≈ 3(12) + 4(1) = 40 g/mol.

The molecular formula must be $(C_3H_4)_3$ or C_9H_{12} because the molar mass of this formula will be between 115 and 125 g/mol (molar mass ≈ 3 × 40 g/mol = 120 g/mol).

94. There are several ways to do this problem. We will determine composition by mass percent:

$$16.01 \text{ mg } CO_2 \times \frac{1\text{ g}}{1000\text{mg}} \times \frac{12.01\text{ g C}}{44.01\text{ g } CO_2} \times \frac{1000\text{mg}}{\text{g}} = 4.369 \text{ mg C}$$

$$\% \text{ C} = \frac{4.369 \text{ mg C}}{10.68 \text{ mg compound}} \times 100 = 40.91\% \text{ C}$$

$$4.37 \text{ mg } H_2O \times \frac{1\text{ g}}{1000\text{mg}} \times \frac{2.016\text{ g H}}{18.02\text{ g } H_2O} \times \frac{1000\text{mg}}{\text{g}} = 0.489 \text{ mg H}$$

$$\% \text{ H} = \frac{0.489 \text{ mg}}{10.68 \text{ mg}} \times 100 = 4.58\% \text{ H}; \ \% \text{ O} = 100.00 - (40.91 + 4.58) = 54.51\% \text{ O}$$

So, in 100.00 g of the compound, we have:

$$40.91 \text{ g C} \times \frac{1 \text{ mol C}}{12.01 \text{ g C}} = 3.406 \text{ mol C}; \ 4.58 \text{ g H} \times \frac{1 \text{ mol H}}{1.008 \text{ g H}} = 4.54 \text{ mol H}$$

$$54.51 \text{ g O} \times \frac{1 \text{ mol O}}{16.00 \text{ g O}} = 3.407 \text{ mol O}$$

Dividing by the smallest number: $\frac{4.54}{3.406} = 1.33 \approx \frac{4}{3}$; the empirical formula is $C_3H_4O_3$.

The empirical formula mass of $C_3H_4O_3$ is ≈ 3(12) + 4(1) + 3(16) = 88 g/mol.

Because $\frac{176.1}{88} = 2.0$, the molecular formula is $C_6H_8O_6$.

Balancing Chemical Equations

95. When balancing reactions, start with elements that appear in only one of the reactants and one of the products, and then go on to balance the remaining elements.

a. $C_6H_{12}O_6(s) + O_2(g) \rightarrow CO_2(g) + H_2O(g)$

Balance C atoms: $C_6H_{12}O_6 + O_2 \rightarrow 6\ CO_2 + H_2O$

Balance H atoms: $C_6H_{12}O_6 + O_2 \rightarrow 6\ CO_2 + 6\ H_2O$

Lastly, balance O atoms: $C_6H_{12}O_6(s) + 6\ O_2(g) \rightarrow 6\ CO_2(g) + 6\ H_2O(g)$

b. $Fe_2S_3(s) + HCl(g) \rightarrow FeCl_3(s) + H_2S(g)$

Balance Fe atoms: $Fe_2S_3 + HCl \rightarrow 2\ FeCl_3 + H_2S$

Balance S atoms: $Fe_2S_3 + HCl \rightarrow 2\ FeCl_3 + 3\ H_2S$

There are 6 H and 6 Cl on right, so balance with 6 HCl on left:

$$Fe_2S_3(s) + 6\ HCl(g) \rightarrow 2\ FeCl_3(s) + 3\ H_2S(g).$$

c. $CS_2(l) + NH_3(g) \rightarrow H_2S(g) + NH_4SCN(s)$

C and S balanced; balance N:

$$CS_2 + 2\ NH_3 \rightarrow H_2S + NH_4SCN$$

H is also balanced. $CS_2(l) + 2\ NH_3(g) \rightarrow H_2S(g) + NH_4SCN(s)$

96. An important part to this problem is writing out correct formulas. If the formulas are incorrect, then the balanced reaction is incorrect.

a. $C_2H_5OH(l) + 3\ O_2(g) \rightarrow 2\ CO_2(g) + 3\ H_2O(g)$

b. $3\ Pb(NO_3)_2(aq) + 2\ Na_3PO_4(aq) \rightarrow Pb_3(PO_4)_2(s) + 6\ NaNO_3(aq)$

c. $Zn(s) + 2\ HCl(aq) \rightarrow ZnCl_2(aq) + H_2(g)$

d. $Sr(OH)_2(aq) + 2\ HBr(aq) \rightarrow 2H_2O(l) + SrBr_2(aq)$

97. $2\ H_2O_2(aq) \xrightarrow[\text{catalyst}]{MnO_2} 2\ H_2O(l) + O_2(g)$

98. $Fe_3O_4(s) + 4\ H_2(g) \rightarrow 3\ Fe(s) + 4\ H_2O(g)$

$Fe_3O_4(s) + 4\ CO(g) \rightarrow 3\ Fe(s) + 4\ CO_2(g)$

99. a. $3\ Ca(OH)_2(aq) + 2\ H_3PO_4(aq) \rightarrow 6\ H_2O(l) + Ca_3(PO_4)_2(s)$

b. $Al(OH)_3(s) + 3\ HCl(aq) \rightarrow AlCl_3(aq) + 3\ H_2O(l)$

c. $2\ AgNO_3(aq) + H_2SO_4(aq) \rightarrow Ag_2SO_4(s) + 2\ HNO_3(aq)$

100. a. $2\ KO_2(s) + 2\ H_2O(l) \rightarrow 2\ KOH(aq) + O_2(g) + H_2O_2(aq)$ or

$4\ KO_2(s) + 6\ H_2O(l) \rightarrow 4\ KOH(aq) + O_2(g) + 4\ H_2O_2(aq)$

b. $Fe_2O_3(s) + 6\ HNO_3(aq) \rightarrow 2\ Fe(NO_3)_3(aq) + 3\ H_2O(l)$

c. $4\ NH_3(g) + 5\ O_2(g) \rightarrow 4\ NO(g) + 6\ H_2O(g)$

d. $PCl_5(l) + 4\ H_2O(l) \rightarrow H_3PO_4(aq) + 5\ HCl(g)$

e. $2\ CaO(s) + 5\ C(s) \rightarrow 2\ CaC_2(s) + CO_2(g)$

f. $2\ MoS_2(s) + 7\ O_2(g) \rightarrow 2\ MoO_3(s) + 4\ SO_2(g)$

g. $FeCO_3(s) + H_2CO_3(aq) \rightarrow Fe(HCO_3)_2(aq)$

101. a. The formulas of the reactants and products are $C_6H_6(l) + O_2(g) \rightarrow CO_2(g) + H_2O(g)$. To balance this combustion reaction, notice that all of the carbon in C_6H_6 has to end up as carbon in CO_2 and all of the hydrogen in C_6H_6 has to end up as hydrogen in H_2O. To balance C and H, we need 6 CO_2 molecules and 3 H_2O molecules for every 1 molecule of C_6H_6. We do oxygen last. Because we have 15 oxygen atoms in 6 CO_2 molecules and 3 H_2O molecules, we need 15/2 O_2 molecules in order to have 15 oxygen atoms on the reactant side.

$C_6H_6(l) + \frac{15}{2} O_2(g) \rightarrow 6\ CO_2(g) + 3\ H_2O(g)$; multiply by two to give whole numbers.

$2\ C_6H_6(l) + 15\ O_2(g) \rightarrow 12\ CO_2(g) + 6\ H_2O(g)$

b. The formulas of the reactants and products are $C_4H_{10}(g) + O_2(g) \rightarrow CO_2(g) + H_2O(g)$.

$C_4H_{10}(g) + \frac{13}{2} O_2(g) \rightarrow 4\ CO_2(g) + 5\ H_2O(g)$; multiply by two to give whole numbers.

$2\ C_4H_{10}(g) + 13\ O_2(g) \rightarrow 8\ CO_2(g) + 10\ H_2O(g)$

c. $C_{12}H_{22}O_{11}(s) + 12\ O_2(g) \rightarrow 12\ CO_2(g) + 11\ H_2O(g)$

d. $2\ Fe(s) + \frac{3}{2} O_2(g) \rightarrow Fe_2O_3(s)$; for whole numbers: $4\ Fe(s) + 3\ O_2(g) \rightarrow 2\ Fe_2O_3(s)$

e. $2\ FeO(s) + \frac{1}{2} O_2(g) \rightarrow Fe_2O_3(s)$; for whole numbers, multiply by two.

$4\ FeO(s) + O_2(g) \rightarrow 2\ Fe_2O_3(s)$

102. a. $16\ Cr(s) + 3\ S_8(s) \rightarrow 8\ Cr_2S_3(s)$

b. $2\ NaHCO_3(s) \rightarrow Na_2CO_3(s) + CO_2(g) + H_2O(g)$

c. $2\ KClO_3(s) \rightarrow 2\ KCl(s) + 3\ O_2(g)$

d. $2\ Eu(s) + 6\ HF(g) \rightarrow 2\ EuF_3(s) + 3\ H_2(g)$

103. a. $SiO_2(s) + C(s) \rightarrow Si(s) + CO(g)$; Si is balanced.

Balance oxygen atoms: $SiO_2 + C \rightarrow Si + 2\ CO$

Balance carbon atoms: $SiO_2(s) + 2\ C(s) \rightarrow Si(s) + 2\ CO(g)$

b. $SiCl_4(l) + Mg(s) \rightarrow Si(s) + MgCl_2(s)$; Si is balanced.

Balance Cl atoms: $SiCl_4 + Mg \rightarrow Si + 2\ MgCl_2$

Balance Mg atoms: $SiCl_4(l) + 2\ Mg(s) \rightarrow Si(s) + 2\ MgCl_2(s)$

c. $Na_2SiF_6(s) + Na(s) \rightarrow Si(s) + NaF(s)$; Si is balanced.

Balance F atoms: $Na_2SiF_6 + Na \rightarrow Si + 6\ NaF$

Balance Na atoms: $Na_2SiF_6(s) + 4\ Na(s) \rightarrow Si(s) + 6\ NaF(s)$

104. $CaSiO_3(s) + 6\ HF(aq) \rightarrow CaF_2(aq) + SiF_4(g) + 3\ H_2O(l)$

Reaction Stoichiometry

105. The stepwise method to solve stoichiometry problems is outlined in the text. Instead of calculating intermediate answers for each step, we will combine conversion factors into one calculation. This practice reduces round-off error and saves time.

$Fe_2O_3(s) + 2\ Al(s) \rightarrow 2\ Fe(l) + Al_2O_3(s)$

$$15.0\text{ g Fe} \times \frac{1\text{ mol Fe}}{55.85\text{ g Fe}} = 0.269\text{ mol Fe; } 0.269\text{ mol Fe} \times \frac{2\text{ mol Al}}{2\text{ mol Fe}} \times \frac{26.98\text{ g Al}}{\text{mol Al}} = 7.26\text{ g Al}$$

$$0.269\text{ mol Fe} \times \frac{1\text{ mol Fe}_2\text{O}_3}{2\text{ mol Fe}} \times \frac{159.70\text{ g Fe}_2\text{O}_3}{\text{mol Fe}_2\text{O}_3} = 21.5\text{ g Fe}_2\text{O}_3$$

$$0.269\text{ mol Fe} \times \frac{1\text{ mol Al}_2\text{O}_3}{2\text{ mol Fe}} \times \frac{101.96\text{ g Al}_2\text{O}_3}{\text{mol Al}_2\text{O}_3} = 13.7\text{ g Al}_2\text{O}_3$$

106. $10\ KClO_3(s) + 3\ P_4(s) \rightarrow 3\ P_4O_{10}(s) + 10\ KCl(s)$

$$52.9\text{ g KClO}_3 \times \frac{1\text{ mol KClO}_3}{122.55\text{ g KClO}_3} \times \frac{3\text{ mol P}_4\text{O}_{10}}{10\text{ mol KClO}_3} \times \frac{283.88\text{ g P}_4\text{O}_{10}}{\text{mol P}_4\text{O}_{10}} = 36.8\text{ g P}_4\text{O}_{10}$$

107. $$1.000\text{ kg Al} \times \frac{1000\text{g Al}}{\text{kg Al}} \times \frac{1\text{ mol Al}}{26.98\text{ g Al}} \times \frac{3\text{ mol NH}_4\text{ClO}_4}{3\text{ mol Al}} \times \frac{117.49\text{ g NH}_4\text{ClO}_4}{\text{mol NH}_4\text{ClO}_4}$$

$$= 4355\text{ g} = 4.355\text{ kg NH}_4\text{ClO}_4$$

108. a. $Ba(OH)_2{\bullet}8H_2O(s) + 2\ NH_4SCN(s) \rightarrow Ba(SCN)_2(s) + 10\ H_2O(l) + 2\ NH_3(g)$

b. $$6.5\text{ g Ba(OH)}_2{\bullet}8\text{H}_2\text{O} \times \frac{1\text{ mol Ba(OH)}_2 \bullet 8\text{H}_2\text{O}}{315.4\text{g}} = 0.0206\text{ mol} = 0.021\text{ mol}$$

$$0.021 \text{ mol Ba(OH)}_2\text{•8H}_2\text{O} \times \frac{2 \text{ mol NH}_4\text{SCN}}{1 \text{ mol Ba(OH)}_2 \bullet 8\text{H}_2\text{O}} \times \frac{76.13 \text{ g NH}_4\text{SCN}}{\text{mol NH}_4\text{SCN}}$$

$$= 3.2 \text{ g NH}_4\text{SCN}$$

109. a. $$1.0 \times 10^2 \text{ mg NaHCO}_3 \times \frac{1 \text{ g}}{1000 \text{mg}} \times \frac{1 \text{ mol NaHCO}_3}{84.01 \text{ g NaHCO}_3} \times \frac{1 \text{ mol C}_6\text{H}_8\text{O}_7}{3 \text{ mol NaHCO}_3}$$

$$\times \frac{192.12 \text{ g C}_6\text{H}_8\text{O}_7}{\text{mol C}_6\text{H}_8\text{O}_7} = 0.076 \text{ g or } 76 \text{ mg C}_6\text{H}_8\text{O}_7$$

b. $$0.10 \text{ g NaHCO}_3 \times \frac{1 \text{ mol NaHCO}_3}{84.01 \text{ g NaHCO}_3} \times \frac{3 \text{ mol CO}_2}{3 \text{ mol NaHCO}_3} \times \frac{44.01 \text{ g CO}_2}{\text{mol CO}_2}$$

$$= 0.052 \text{ g or } 52 \text{ mg CO}_2$$

110. a. $$1.00 \times 10^2 \text{ g C}_7\text{H}_6\text{O}_3 \times \frac{1 \text{ mol C}_7\text{H}_6\text{O}_3}{138.12 \text{ g C}_7\text{H}_6\text{O}_3} \times \frac{1 \text{ mol C}_4\text{H}_6\text{O}_3}{1 \text{ mol C}_7\text{H}_6\text{O}_3} \times \frac{102.09 \text{ g C}_4\text{H}_6\text{O}_3}{1 \text{ mol C}_4\text{H}_6\text{O}_3}$$

$$= 73.9 \text{ g C}_4\text{H}_6\text{O}_3$$

b. $$1.00 \times 10^2 \text{ g C}_7\text{H}_6\text{O}_3 \times \frac{1 \text{ mol C}_7\text{H}_6\text{O}_3}{138.12 \text{ g C}_7\text{H}_6\text{O}_3} \times \frac{1 \text{ mol C}_9\text{H}_8\text{O}_4}{1 \text{ mol C}_7\text{H}_6\text{O}_3} \times \frac{180.15 \text{ g C}_9\text{H}_8\text{O}_4}{\text{mol C}_9\text{H}_8\text{O}_4}$$

$$= 1.30 \times 10^2 \text{ g aspirin}$$

111. $$1.0 \times 10^4 \text{ kg waste} \times \frac{3.0 \text{ kg NH}_4^+}{100 \text{ kg waste}} \times \frac{1000 \text{g}}{\text{kg}} \times \frac{1 \text{ mol NH}_4^+}{18.04 \text{ g NH}_4^+} \times \frac{1 \text{ mol C}_5\text{H}_7\text{O}_2\text{N}}{55 \text{ mol NH}_4^+}$$

$$\times \frac{113.12 \text{ g C}_5\text{H}_7\text{O}_2\text{N}}{\text{mol C}_5\text{H}_7\text{O}_2\text{N}} = 3.4 \times 10^4 \text{ g tissue if all NH}_4^+ \text{ converted}$$

Because only 95% of the NH_4^+ ions react:

mass of tissue = $(0.95)(3.4 \times 10^4 \text{ g}) = 3.2 \times 10^4$ g or 32 kg bacterial tissue

112. $$1.0 \times 10^3 \text{ g phosphorite} \times \frac{75 \text{ g Ca}_3(\text{PO}_4)_2}{100 \text{ g phosphorite}} \times \frac{1 \text{ mol Ca}_3(\text{PO}_4)_2}{310.18 \text{ g Ca}_3(\text{PO}_4)_2}$$

$$\times \frac{1 \text{ mol P}_4}{2 \text{ mol Ca}_3(\text{PO}_4)_2} \times \frac{123.88 \text{ g P}_4}{\text{mol P}_4} = 150 \text{ g P}_4$$

113. $$1.0 \text{ ton CuO} \times \frac{907 \text{ kg}}{\text{ton}} \times \frac{1000 \text{g}}{\text{kg}} \times \frac{1 \text{ mol CuO}}{79.55 \text{ g CuO}} \times \frac{1 \text{ mol C}}{2 \text{ mol CuO}} \times \frac{12.01 \text{ g C}}{\text{mol C}} \times \frac{100. \text{ g coke}}{95 \text{ g C}}$$

$$= 7.2 \times 10^4 \text{ g or } 72 \text{ kg coke}$$

114. $2 \text{ LiOH(s)} + \text{CO}_2\text{(g)} \rightarrow \text{Li}_2\text{CO}_3\text{(aq)} + \text{H}_2\text{O(l)}$

The total volume of air exhaled each minute for the 7 astronauts is 7 × 20. = 140 L/min.

$$25{,}000 \text{ g LiOH} \times \frac{1 \text{ mol LiOH}}{23.95 \text{ g LiOH}} \times \frac{1 \text{ mol CO}_2}{2 \text{ mol LiOH}} \times \frac{44.01 \text{ g CO}_2}{\text{mol CO}_2} \times \frac{100 \text{ g air}}{4.0 \text{ g CO}_2}$$

$$\times \frac{1 \text{ mL air}}{0.0010 \text{ g air}} \times \frac{1 \text{ L}}{1000 \text{ mL}} \times \frac{1 \text{ min}}{140 \text{ L air}} \times \frac{1 \text{ h}}{60 \text{ min}} = 68 \text{ h} = 2.8 \text{ days}$$

Limiting Reactants and Percent Yield

115. The product formed in the reaction is NO_2; the other species present in the product representation is excess O_2. Therefore, NO is the limiting reactant. In the pictures, 6 NO molecules react with 3 O_2 molecules to form 6 NO_2 molecules.

$$6 \text{ NO(g)} + 3 \text{ O}_2\text{(g)} \rightarrow 6 \text{ NO}_2\text{(g)}$$

For smallest whole numbers, the balanced reaction is:

$$2 \text{ NO(g)} + \text{O}_2\text{(g)} \rightarrow 2 \text{ NO}_2\text{(g)}$$

116. In the following table we have listed three rows of information. The "Initial" row is the number of molecules present initially, the "Change" row is the number of molecules that react to reach completion, and the "Final" row is the number of molecules present at completion. To determine the limiting reactant, let's calculate how much of one reactant is necessary to react with the other.

$$10 \text{ molecules O}_2 \times \frac{4 \text{ molecules NH}_3}{5 \text{ molecules O}_2} = 8 \text{ molecules NH}_3 \text{ to react with all of the O}_2$$

Because we have 10 molecules of NH_3 and only 8 molecules of NH_3 are necessary to react with all of the O_2, O_2 is limiting. Now use the 10 molecules of O_2 and the molecule relationships given in the balanced equation to determine the number of molecules of each product formed, then complete the table.

	$4 \text{ NH}_3\text{(g)}$	+	$5 \text{ O}_2\text{(g)}$	→	4 NO(g)	+	$6 \text{ H}_2\text{O(g)}$
Initial	10 molecules		10 molecules		0		0
Change	−8 molecules		−10 molecules		+8 molecules		+12 molecules
Final	2 molecules		0		8 molecules		12 molecules

The total number of molecules present after completion = 2 molecules NH_3 + 0 molecules O_2 + 8 molecules NO + 12 molecules H_2O = 22 molecules.

117. a. The strategy we will generally use to solve limiting reactant problems is to assume each reactant is limiting, and then calculate the quantity of product each reactant could produce if it were limiting. The reactant that produces the smallest quantity of product is the limiting reactant (runs out first) and therefore determines the mass of product that can be produced.

Assuming N_2 is limiting:

$$1.00 \times 10^3 \text{ g N}_2 \times \frac{1 \text{ mol N}_2}{28.02 \text{ g N}_2} \times \frac{2 \text{ mol NH}_3}{\text{mol N}_2} \times \frac{17.03 \text{ g NH}_3}{\text{mol NH}_3} = 1.22 \times 10^3 \text{ g NH}_3$$

Assuming H_2 is limiting:

$$5.00 \times 10^2 \text{ g } H_2 \times \frac{1 \text{ mol } H_2}{2.016 \text{ g } H_2} \times \frac{2 \text{ mol } NH_3}{3 \text{ mol } H_2} \times \frac{17.03 \text{ g } NH_3}{\text{mol } NH_3} = 2.82 \times 10^3 \text{ g } NH_3$$

Because N_2 produces the smaller mass of product (1220 g vs. 2820 g NH_3), N_2 is limiting and 1220 g NH_3 can be produced. As soon as 1220 g of NH_3 is produced, all of the N_2 has run out. Even though we have enough H_2 to produce more product, there is no more N_2 present as soon as 1220 g of NH_3 have been produced.

b. $$1.00 \times 10^3 \text{ g } N_2 \times \frac{1 \text{ mol } N_2}{28.02 \text{ g } N_2} \times \frac{3 \text{ mol } H_2}{\text{mol } N_2} \times \frac{2.016 \text{ g } H_2}{\text{mol } H_2} = 216 \text{ g } H_2 \text{ reacted}$$

Excess H_2 = 500. g H_2 initially – 216 g H_2 reacted = 284 g H_2 in excess (unreacted)

118. $Ca_3(PO_4)_2 + 3\ H_2SO_4 \rightarrow 3\ CaSO_4 + 2\ H_3PO_4$

Assuming $Ca_3(PO_4)_2$ is limiting:

$$1.0 \times 10^3 \text{ g } Ca_3(PO_4)_2 \times \frac{1 \text{ mol } Ca_3(PO_4)_2}{310.18 \text{ g } Ca_3(PO_4)_2} \times \frac{3 \text{ mol } CaSO_4}{\text{mol } Ca_3(PO_4)_2} \times \frac{136.15 \text{ g } CaSO_4}{\text{mol } CaSO_4}$$

$$= 1300 \text{ g } CaSO_4$$

Assuming concentrated H_2SO_4 reagent is limiting:

$$1.0 \times 10^3 \text{ g conc. } H_2SO_4 \times \frac{98 \text{ g } H_2SO_4}{100 \text{ g conc. } H_2SO_4} \times \frac{1 \text{ mol } H_2SO_4}{98.09 \text{ g } H_2SO_4}$$

$$\times \frac{3 \text{ mol } CaSO_4}{3 \text{ mol } H_2SO_4} \times \frac{136.15 \text{ g } CaSO_4}{\text{mol } CaSO_4} = 1400 \text{ g } CaSO_4$$

Because $Ca_3(PO_4)_2$ produces the smaller quantity of product, $Ca_3(PO_4)_2$ is limiting and 1300 g $CaSO_4$ can be produced.

$$1.0 \times 10^3 \text{ g } Ca_3(PO_4)_2 \times \frac{1 \text{ mol } Ca_3(PO_4)_2}{310.18 \text{ g } Ca_3(PO_4)_2} \times \frac{2 \text{ mol } H_3PO_4}{\text{mol } Ca_3(PO_4)_2} \times \frac{97.99 \text{ g } H_3PO_4}{\text{mol } H_3PO_4}$$

$$= 630 \text{ g } H_3PO_4 \text{ produced}$$

119. Assuming BaO_2 is limiting:

$$1.50 \text{ g } BaO_2 \times \frac{1 \text{ mol } BaO_2}{169.3 \text{ g } BaO_2} \times \frac{1 \text{ mol } H_2O_2}{\text{mol } BaO_2} \times \frac{34.02 \text{ g } H_2O_2}{\text{mol } H_2O_2} = 0.301 \text{ g } H_2O_2$$

Assuming HCl is limiting:

$$25.0 \text{ mL} \times \frac{0.0272 \text{ g HCl}}{\text{mL}} \times \frac{1 \text{ mol HCl}}{36.46 \text{ g HCl}} \times \frac{1 \text{ mol } H_2O_2}{2 \text{ mol HCl}} \times \frac{34.02 \text{ g } H_2O_2}{\text{mol } H_2O_2} = 0.317 \text{ g } H_2O_2$$

BaO_2 produces the smaller amount of H_2O_2, so it is limiting and a mass of 0.301 g of H_2O_2 can be produced.

Initial mol HCl present: $25.0 \text{ mL} \times \dfrac{0.0272 \text{ g HCl}}{\text{mL}} \times \dfrac{1 \text{ mol HCl}}{36.46 \text{ g HCl}} = 1.87 \times 10^{-2} \text{ mol HCl}$

The amount of HCl reacted:

$$1.50 \text{ g } BaO_2 \times \frac{1 \text{ mol } BaO_2}{169.3 \text{ g } BaO_2} \times \frac{2 \text{ mol HCl}}{\text{mol } BaO_2} = 1.77 \times 10^{-2} \text{ mol HCl}$$

Excess mol HCl = $1.87 \times 10^{-2} \text{ mol} - 1.77 \times 10^{-2} \text{ mol} = 1.0 \times 10^{-3} \text{ mol HCl}$

Mass of excess HCl = $1.0 \times 10^{-3} \text{ mol HCl} \times \dfrac{36.46 \text{ g HCl}}{\text{mol HCl}} = 3.6 \times 10^{-2} \text{ g HCl unreacted}$

120. Assuming Ag_2O is limiting:

$$25.0 \text{ g } Ag_2O \times \frac{1 \text{ mol } Ag_2O}{231.8 \text{ g } Ag_2O} \times \frac{2 \text{ mol } AgC_{10}H_9N_4SO_2}{\text{mol } Ag_2O} \times \frac{357.18 \text{ g } AgC_{10}H_9N_4SO_2}{\text{mol } AgC_{10}H_9N_4SO_2}$$

$$= 77.0 \text{ g } AgC_{10}H_9N_4SO_2$$

Assuming $C_{10}H_{10}N_4SO_2$ is limiting:

$$50.0 \text{ g } C_{10}H_{10}N_4SO_2 \times \frac{1 \text{ mol } C_{10}H_{10}N_4SO_2}{250.29 \text{ g } C_{10}H_{10}N_4SO_2} \times \frac{2 \text{ mol } AgC_{10}H_9N_4SO_2}{2 \text{ mol } C_{10}H_{10}N_4SO_2}$$

$$\times \frac{357.18 \text{ g } AgC_{10}H_9N_4SO_2}{\text{mol } AgC_{10}H_9N_4SO_2} = 71.4 \text{ g } AgC_{10}H_9N_4SO_2$$

Because $C_{10}H_{10}N_4SO_2$ produces the smaller amount of product, it is limiting and 71.4 g of silver sulfadiazine can be produced.

121. To solve limiting-reagent problems, we will generally assume each reactant is limiting and then calculate how much product could be produced from each reactant. The reactant that produces the smallest amount of product will run out first and is the limiting reagent.

$$5.00 \times 10^6 \text{ g } NH_3 \times \frac{1 \text{ mol } NH_3}{17.03 \text{ g } NH_3} \times \frac{2 \text{ mol HCN}}{2 \text{ mol } NH_3} = 2.94 \times 10^5 \text{ mol HCN}$$

$$5.00 \times 10^6 \text{ g } O_2 \times \frac{1 \text{ mol } O_2}{32.00 \text{ g } O_2} \times \frac{2 \text{ mol HCN}}{3 \text{ mol } O_2} = 1.04 \times 10^5 \text{ mol HCN}$$

$$5.00 \times 10^6 \text{ g } CH_4 \times \frac{1 \text{ mol } CH_4}{16.04 \text{ g } CH_4} \times \frac{2 \text{ mol HCN}}{2 \text{ mol } CH_4} = 3.12 \times 10^5 \text{ mol HCN}$$

O_2 is limiting because it produces the smallest amount of HCN. Although more product could be produced from NH_3 and CH_4, only enough O_2 is present to produce 1.04×10^5 mol HCN. The mass of HCN produced is:

$$1.04 \times 10^5 \text{ mol HCN} \times \frac{27.03 \text{ g HCN}}{\text{mol HCN}} = 2.81 \times 10^6 \text{ g HCN}$$

$$5.00 \times 10^6 \text{ g } O_2 \times \frac{1 \text{ mol } O_2}{32.00 \text{ g } O_2} \times \frac{6 \text{ mol } H_2O}{3 \text{ mol } O_2} \times \frac{18.02 \text{ g } H_2O}{1 \text{ mol } H_2O} = 5.63 \times 10^6 \text{ g } H_2O$$

122. If C_3H_6 is limiting:

$$15.0 \text{ g } C_3H_6 \times \frac{1 \text{ mol } C_3H_6}{42.08 \text{ g } C_3H_6} \times \frac{2 \text{ mol } C_3H_3N}{2 \text{ mol } C_3H_6} \times \frac{53.06 \text{ g } C_3H_3N}{\text{mol } C_3H_3N} = 18.9 \text{ g } C_3H_3N$$

If NH_3 is limiting:

$$5.00 \text{ g } NH_3 \times \frac{1 \text{ mol } NH_3}{17.03 \text{ g } NH_3} \times \frac{2 \text{ mol } C_3H_3N}{2 \text{ mol } NH_3} \times \frac{53.06 \text{ g } C_3H_3N}{\text{mol } C_3H_3N} = 15.6 \text{ g } C_3H_3N$$

If O_2 is limiting:

$$10.0 \text{ g } O_2 \times \frac{1 \text{ mol } O_2}{32.00 \text{ g } O_2} \times \frac{2 \text{ mol } C_3H_3N}{3 \text{ mol } O_2} \times \frac{53.06 \text{ g } C_3H_3N}{\text{mol } C_3H_3N} = 11.1 \text{ g } C_3H_3N$$

O_2 produces the smallest amount of product; thus O_2 is limiting, and 11.1 g C_3H_3N can be produced.

123. $C_2H_6(g) + Cl_2(g) \rightarrow C_2H_5Cl(g) + HCl(g)$

If C_2H_6 is limiting:

$$300. \text{ g } C_2H_6 \times \frac{1 \text{ mol } C_2H_6}{30.07 \text{ g } C_2H_6} \times \frac{1 \text{ mol } C_2H_5Cl}{\text{mol } C_2H_6} \times \frac{64.51 \text{ g } C_2H_5Cl}{\text{mol } C_2H_5Cl} = 644 \text{ g } C_2H_5Cl$$

If Cl_2 is limiting:

$$650. \text{ g } Cl_2 \times \frac{1 \text{ mol } Cl_2}{70.90 \text{ g } Cl_2} \times \frac{1 \text{ mol } C_2H_5Cl}{\text{mol } Cl_2} \times \frac{64.51 \text{ g } C_2H_5Cl}{\text{mol } C_2H_5Cl} = 591 \text{ g } C_2H_5Cl$$

Cl_2 is limiting because it produces the smaller quantity of product. Hence, the theoretical yield for this reaction is 591 g C_2H_5Cl. The percent yield is:

$$\text{percent yield} = \frac{\text{actual}}{\text{theoretical}} \times 100 = \frac{490. \text{ g}}{591 \text{ g}} \times 100 = 82.9\%$$

124. a. $$1142 \text{ g } C_6H_5Cl \times \frac{1 \text{ mol } C_6H_5Cl}{112.55 \text{ g } C_6H_5Cl} \times \frac{1 \text{ mol } C_{14}H_9Cl_5}{2 \text{ mol } C_6H_5Cl} \times \frac{354.46 \text{ g } C_{14}H_9Cl_5}{\text{mol } C_{14}H_9Cl_5}$$

$$= 1798 \text{ } C_{14}H_9Cl_5$$

$$485 \text{ g } C_2HOCl_3 \times \frac{1 \text{ mol } C_2HOCl_3}{147.38 \text{ g } C_2HOCl_3} \times \frac{1 \text{ mol } C_{14}H_9Cl_5}{\text{mol } C_2HOCl_3} \times \frac{354.46 \text{ g } C_{14}H_9Cl_5}{\text{mol } C_{14}H_9Cl_5}$$

$$= 1170 \text{ g } C_{14}H_9Cl_5$$

From the masses of product calculated, C_2HOCl_3 is limiting and 1170 g $C_{14}H_9Cl_5$ can be produced.

b. C_2HOCl_3 is limiting, and C_6H_5Cl is in excess.

c. $$485 \text{ g } C_2HOCl_3 \times \frac{1 \text{ mol } C_2HOCl_3}{147.38 \text{ g } C_2HOCl_3} \times \frac{2 \text{ mol } C_6H_5Cl}{\text{mol } C_2HOCl_3} \times \frac{112.55 \text{ g } C_6H_5Cl}{\text{mol } C_6H_5Cl}$$

$$= 741 \text{ g } C_6H_5Cl \text{ reacted}$$

1142 g − 741 g = 401 g C_6H_5Cl in excess

d. $$\text{Percent yield} = \frac{200.0 \text{ g DDT}}{1170 \text{g DDT}} \times 100 = 17.1\%$$

125. $$2.50 \text{ metric tons } Cu_3FeS_3 \times \frac{1000 \text{kg}}{\text{metricton}} \times \frac{1000 \text{g}}{\text{kg}} \times \frac{1 \text{ mol } Cu_3FeS_3}{342.71 \text{ g}} \times \frac{3 \text{ mol Cu}}{1 \text{ mol } Cu_3FeS_3}$$

$$\times \frac{63.55 \text{ g}}{\text{mol Cu}} = 1.39 \times 10^6 \text{ g Cu (theoretical)}$$

$$1.39 \times 10^6 \text{ g Cu (theoretical)} \times \frac{86.3 \text{ g Cu (actual)}}{100. \text{ g Cu (theoretical)}} = 1.20 \times 10^6 \text{ g Cu} = 1.20 \times 10^3 \text{ kg Cu}$$

$$= 1.20 \text{ metric tons Cu (actual)}$$

126. $P_4(s) + 6\ F_2(g) \rightarrow 4\ PF_3(g)$; the theoretical yield of PF_3 is:

$$120. \text{ g } PF_3 \text{ (actual)} \times \frac{100.0 \text{ g } PF_3 \text{ (theoretical)}}{78.1 \text{ g } PF_3 \text{ (actual)}} = 154 \text{ g } PF_3 \text{ (theoretical)}$$

$$154 \text{ g } PF_3 \times \frac{1 \text{ mol } PF_3}{87.97 \text{ g } PF_3} \times \frac{6 \text{ mol } F_2}{4 \text{ mol } PF_3} \times \frac{38.00 \text{ g } F_2}{\text{mol } F_2} = 99.8 \text{ g } F_2$$

99.8 g F_2 is needed to actually produce 120. g of PF_3 if the percent yield is 78.1%.

Additional Exercises

127. $^{12}C_2{}^1H_6$: 2(12.000000) + 6(1.007825) = 30.046950 u

$^{12}C^1H_2{}^{16}O$: 1(12.000000) + 2(1.007825) + 1(15.994915) = 30.010565 u

$^{14}N^{16}O$: 1(14.003074) + 1(15.994915) = 29.997989 u

The peak results from $^{12}C^1H_2{}^{16}O$.

128. We would see the peaks corresponding to:

$^{10}B^{35}Cl_3$ [mass ≈ 10 + 3(35) = 115 u], $^{10}B^{35}Cl_2{}^{37}Cl$ (117), $^{10}B^{35}Cl^{37}Cl_2$ (119),
$^{10}B^{37}Cl_3$ (121), $^{11}B^{35}Cl_3$ (116), $^{11}B^{35}Cl_2{}^{37}Cl$ (118), $^{11}B^{35}Cl^{37}Cl_2$ (120), $^{11}B^{37}Cl_3$ (122)

We would see a total of eight peaks at approximate masses of 115, 116, 117, 118, 119, 120, 121, and 122.

129. Molar mass XeF_n = $$\frac{0.368\text{ g } XeF_n}{9.03 \times 10^{20}\text{ molecules } XeF_n \times \dfrac{1\text{ mol } XeF_n}{6.022 \times 10^{23}\text{ molecules}}} = 245\text{ g/mol}$$

245 g = 131.3 g + n(19.00 g), n = 5.98; formula = XeF_6

130. a. $$14\text{ mol C} \times \frac{12.01\text{ g}}{\text{mol C}} + 18\text{ mol H} \times \frac{1.008\text{ g}}{\text{mol H}} + 2\text{ mol N} \times \frac{14.01\text{ g}}{\text{mol N}}$$
$$+ 5\text{ mol O} \times \frac{16.00\text{ g}}{\text{mol O}} = 294.30\text{ g}$$

b. $$10.0\text{ g } C_{14}H_{18}N_2O_5 \times \frac{1\text{ mol } C_{14}H_{18}N_2O_5}{294.30\text{ g } C_{14}H_{18}N_2O_5} = 3.40 \times 10^{-2}\text{ mol } C_{14}H_{18}N_2O_5$$

c. $$1.56\text{ mol} \times \frac{294.3\text{ g}}{\text{mol}} = 459\text{ g } C_{14}H_{18}N_2O_5$$

d. $$5.0\text{ mg} \times \frac{1\text{ g}}{1000\text{ mg}} \times \frac{1\text{ mol}}{294.30\text{ g}} \times \frac{6.022 \times 10^{23}\text{ molecules}}{\text{mol}}$$
$$= 1.0 \times 10^{19}\text{ molecules } C_{14}H_{18}N_2O_5$$

e. The chemical formula tells us that 1 molecule of $C_{14}H_{18}N_2O_5$ contains 2 atoms of N. If we have 1 mole of $C_{14}H_{18}N_2O_5$ molecules, then 2 moles of N atoms are present.

$$1.2\text{ g } C_{14}H_{18}N_2O_5 \times \frac{1\text{ mol } C_{14}H_{18}N_2O_5}{294.30\text{ g } C_{14}H_{18}N_2O_5} \times \frac{2\text{ mol N}}{\text{mol } C_{14}H_{18}N_2O_5}$$
$$\times \frac{6.022 \times 10^{23}\text{ atoms N}}{\text{mol N}} = 4.9 \times 10^{21}\text{ atoms N}$$

f. $$1.0 \times 10^{9}\text{ molecules} \times \frac{1\text{ mol}}{6.022 \times 10^{23}\text{ atoms}} \times \frac{294.30\text{ g}}{\text{mol}} = 4.9 \times 10^{-13}\text{ g}$$

g. $$1\text{ molecule} \times \frac{1\text{ mol}}{6.022 \times 10^{23}\text{ atoms}} \times \frac{294.30\text{ g}}{\text{mol}} = 4.887 \times 10^{-22}\text{ g } C_{14}H_{18}N_2O_5$$

131. Molar mass = 20(12.01) + 29(1.008) + 19.00 + 3(16.00) = 336.43 g/mol

$$\text{Mass \% C} = \frac{20(12.01)\text{ g C}}{336.43\text{ g compound}} \times 100 = 71.40\%\text{ C}$$

$$\text{Mass \% H} = \frac{29(1.008)\text{ g H}}{336.43\text{ g compound}} \times 100 = 8.689\%\text{ H}$$

$$\text{Mass \% F} = \frac{19.00\text{ g F}}{336.43\text{ g compound}} \times 100 = 5.648\%\text{ F}$$

Mass % O = 100.00 – (71.40 + 8.689 + 5.648) = 14.26% O or:

$$\text{\% O} = \frac{3(16.00)\text{ g O}}{336.43\text{ g compound}} \times 100 = 14.27\%\text{ O}$$

132. In 1 hour, the 1000. kg of wet cereal produced contains 580 kg H_2O and 420 kg of cereal. We want the final product to contain 20.% H_2O. Let x = mass of H_2O in final product.

$$\frac{x}{420 + x} = 0.20,\ \ x = 84 + (0.20)x,\ \ x = 105 \approx 110\text{ kg } H_2O$$

The amount of water to be removed is 580 – 110 = 470 kg/h.

133. Out of 100.00 g of adrenaline, there are:

$$56.79\text{ g C} \times \frac{1\text{ mol C}}{12.01\text{ g C}} = 4.729\text{ mol C};\ \ 6.56\text{ g H} \times \frac{1\text{ mol H}}{1.008\text{ g H}} = 6.51\text{ mol H}$$

$$28.37\text{ g O} \times \frac{1\text{ mol O}}{16.00\text{ g O}} = 1.773\text{ mol O};\ \ 8.28\text{ g N} \times \frac{1\text{ mol N}}{14.01\text{ g N}} = 0.591\text{ mol N}$$

Dividing each mole value by the smallest number:

$$\frac{4.729}{0.591} = 8.00;\ \ \frac{6.51}{0.591} = 11.0;\ \ \frac{1.773}{0.591} = 3.00;\ \ \frac{0.591}{0.591} = 1.00$$

This gives adrenaline an empirical formula of $C_8H_{11}O_3N$.

134. Assuming 100.00 g of compound (mass hydrogen = 100.00 g – 49.31 g C – 43.79 g O = 6.90 g H):

$$49.31\text{ g C} \times \frac{1\text{ mol C}}{12.01\text{ g C}} = 4.106\text{ mol C};\ \ 6.90\text{ g H} \times \frac{1\text{ mol H}}{1.008\text{ g H}} = 6.85\text{ mol H}$$

$$43.79\text{ g O} \times \frac{1\text{ mol O}}{16.00\text{ g O}} = 2.737\text{ mol O}$$

Dividing all mole values by 2.737 gives:

$$\frac{4.106}{2.737} = 1.500;\ \ \frac{6.85}{2.737} = 2.50;\ \ \frac{2.737}{2.737} = 1.000$$

Because a whole number ratio is required, the empirical formula is $C_3H_5O_2$.

Empirical formula mass: 3(12.01) + 5(1.008) +2(16.00) = 73.07 g/mol

$$\frac{\text{Molar mass}}{\text{Empirical formula mass}} = \frac{146.1}{73.07} = 1.999;\quad \text{molecular formula} = (C_3H_5O_2)_2 = C_6H_{10}O_4$$

135. There are many valid methods to solve this problem. We will assume 100.00 g of compound, and then determine from the information in the problem how many moles of compound equals 100.00 g of compound. From this information, we can determine the mass of one mole of compound (the molar mass) by setting up a ratio. Assuming 100.00 g cyanocobalamin:

$$\text{mol cyanocobalamin} = 4.34\text{ g Co} \times \frac{1\text{ mol Co}}{58.93\text{ g Co}} \times \frac{1\text{ mol cyanocobalamin}}{\text{mol Co}}$$

$$= 7.36 \times 10^{-2}\text{ mol cyanocobalamin}$$

$$\frac{x\text{ g cyanocobalamin}}{1\text{ mol cyanocobalamin}} = \frac{100.00\text{ g}}{7.36\times10^{-2}\text{ mol}},\ x = \text{molar mass} = 1.36 \times 10^3\text{ g/mol}$$

136. $$2\text{ tablets} \times \frac{0.262\text{ g } C_7H_5BiO_4}{\text{tablet}} \times \frac{1\text{ mol } C_7H_5BiO_4}{362.11\text{ g } C_7H_5BiO_4} \times \frac{1\text{ mol Bi}}{1\text{ mol } C_7H_5BiO_4} \times \frac{209.0\text{ g Bi}}{\text{mol Bi}}$$

$$= 0.302\text{ g Bi consumed}$$

137. Empirical formula mass = 12.01 + 1.008 = 13.02 g/mol; because 104.14/13.02 = 7.998 ≈ 8, the molecular formula for styrene is $(CH)_8 = C_8H_8$.

$$2.00\text{ g } C_8H_8 \times \frac{1\text{ mol } C_8H_8}{104.14\text{ g } C_8H_8} \times \frac{8\text{ mol H}}{\text{mol } C_8H_8} \times \frac{6.022\times10^{23}\text{ atoms H}}{\text{mol H}} = 9.25 \times 10^{22}\text{ atoms H}$$

138. $$41.98\text{ mg } CO_2 \times \frac{12.01\text{ mg C}}{44.01\text{ mg } CO_2} = 11.46\text{ mg C};\ \%\text{ C} = \frac{11.46\text{ mg}}{19.81\text{ mg}} \times 100 = 57.85\%\text{ C}$$

$$6.45\text{ mg } H_2O \times \frac{2.016\text{ mg H}}{18.02\text{ mg } H_2O} = 0.722\text{ mg H};\ \%\text{ H} = \frac{0.772\text{ mg}}{19.81\text{ mg}} \times 100 = 3.64\%\text{ H}$$

% O = 100.00 – (57.85 + 3.64) = 38.51% O

Out of 100.00 g terephthalic acid, there are:

$$57.85\text{ g C} \times \frac{1\text{ mol C}}{12.01\text{ g C}} = 4.817\text{ mol C};\ 3.64\text{ g H} \times \frac{1\text{ mol H}}{1.008\text{ g H}} = 3.61\text{ mol H}$$

$$38.51\text{ g O} \times \frac{1\text{ mol O}}{16.00\text{ g O}} = 2.407\text{ mol O}$$

$$\frac{4.817}{2.407} = 2.001;\quad \frac{3.61}{2.407} = 1.50;\quad \frac{2.407}{2.407} = 1.000$$

The C : H : O mole ratio is 2 : 1.5 : 1 or 4 : 3 : 2. The empirical formula is $C_4H_3O_2$.

Mass of $C_4H_3O_2 \approx 4(12) + 3(1) + 2(16) = 83$g/mol.

$$\text{Molar mass} = \frac{41.5\text{ g}}{0.250\text{ mol}} = 166\text{ g/mol};\ \frac{166}{83} = 2.0;\ \text{the molecular formula is } C_8H_6O_4.$$

139. $$17.3\text{ g H} \times \frac{1\text{ mol H}}{1.008\text{ g H}} = 17.2\text{ mol H};\ \ 82.7\text{ g C} \times \frac{1\text{ mol C}}{12.01\text{ g C}} = 6.89\text{ mol C}$$

$$\frac{17.2}{6.89} = 2.50;\ \text{the empirical formula is } C_2H_5.$$

The empirical formula mass is ~29 g/mol, so two times the empirical formula would put the compound in the correct range of the molar mass. Molecular formula = $(C_2H_5)_2 = C_4H_{10}$.

$$2.59 \times 10^{23}\text{ atoms H} \times \frac{1\text{ molecule } C_4H_{10}}{10\text{ atoms H}} \times \frac{1\text{ mol } C_4H_{10}}{6.022 \times 10^{23}\text{ molecules}} = 4.30 \times 10^{-2}\text{ mol } C_4H_{10}$$

$$4.30 \times 10^{-2}\text{ mol } C_4H_{10} \times \frac{58.12\text{ g}}{\text{mol } C_4H_{10}} = 2.50\text{ g } C_4H_{10}$$

140. Assuming 100.00 g E_3H_8:

$$\text{mol E} = 8.73\text{ g H} \times \frac{1\text{ mol H}}{1.008\text{ g H}} \times \frac{3\text{ mol E}}{8\text{ mol H}} = 3.25\text{ mol E}$$

$$\frac{x\text{ g E}}{1\text{ mol E}} = \frac{91.27\text{ g E}}{3.25\text{ mol E}},\ x = \text{molar mass of E} = 28.1\text{ g/mol};\ \text{atomic mass of E} = 28.1\text{ u}$$

Note: From the periodic table, element E is silicon, Si.

141. Mass of H_2O = 0.755 g $CuSO_4 \cdot xH_2O$ – 0.483 g $CuSO_4$ = 0.272 g H_2O

$$0.483\text{ g } CuSO_4 \times \frac{1\text{ mol } CuSO_4}{159.62\text{ g } CuSO_4} = 0.00303\text{ mol } CuSO_4$$

$$0.272\text{ g } H_2O \times \frac{1\text{ mol } H_2O}{18.02\text{ g } H_2O} = 0.0151\text{ mol } H_2O$$

$$\frac{0.0151\text{ mol } H_2O}{0.00303\text{ g } CuSO_4} = \frac{4.98\text{ mol } H_2O}{1\text{ mol } CuSO_4};\ \text{compound formula} = CuSO_4 \cdot 5H_2O,\ x = 5$$

142. a. Only acrylonitrile contains nitrogen. If we have 100.00 g of polymer:

$$8.80\text{ g N} \times \frac{1\text{ mol } C_3H_3N}{14.01\text{ g N}} \times \frac{53.06\text{ g } C_3H_3N}{1\text{ mol } C_3H_3N} = 33.3\text{ g } C_3H_3N$$

$$\% \ C_3H_3N = \frac{33.3 \text{ g } C_3H_3N}{100.00 \text{ g polymer}} = 33.3\% \ C_3H_3N$$

Only butadiene in the polymer reacts with Br_2:

$$0.605 \text{ g } Br_2 \times \frac{1 \text{ mol } Br_2}{159.8 \text{ g } Br_2} \times \frac{1 \text{ mol } C_4H_6}{\text{mol } Br_2} \times \frac{54.09 \text{ g } C_4H_6}{\text{mol } C_4H_6} = 0.205 \text{ g } C_4H_6$$

$$\% \ C_4H_6 = \frac{0.205 \text{ g}}{1.20 \text{ g}} \times 100 = 17.1\% \ C_4H_6$$

b. If we have 100.0 g of polymer:

$$33.3 \text{ g } C_3H_3N \times \frac{1 \text{ mol } C_3H_3N}{53.06 \text{ g}} = 0.628 \text{ mol } C_3H_3N$$

$$17.1 \text{ g } C_4H_6 \times \frac{1 \text{ mol } C_4H_6}{54.09 \text{ g } C_4H_6} = 0.316 \text{ mol } C_4H_6$$

$$49.6 \text{ g } C_8H_8 \times \frac{1 \text{ mol } C_8H_8}{104.14 \text{ g } C_8H_8} = 0.476 \text{ mol } C_8H_8$$

Dividing by 0.316: $\frac{0.628}{0.316} = 1.99$; $\frac{0.316}{0.316} = 1.00$; $\frac{0.476}{0.316} = 1.51$

This is close to a mole ratio of 4 : 2 : 3. Thus there are 4 acrylonitrile to 2 butadiene to 3 styrene molecules in the polymer, or $(A_4B_2S_3)_n$.

143. $1.20 \text{ g } CO_2 \times \frac{1 \text{ mol } CO_2}{44.01 \text{ g}} \times \frac{1 \text{ mol } C}{\text{mol } CO_2} \times \frac{1 \text{ mol } C_{24}H_{30}N_3O}{24 \text{ mol } C} \times \frac{376.51 \text{ g}}{\text{mol } C_{24}H_{30}N_3O}$

$= 0.428 \text{ g } C_{24}H_{30}N_3O$

$$\frac{0.428 \text{ g } C_{24}H_{30}N_3O}{1.00 \text{ g sample}} \times 100 = 42.8\% \ C_{24}H_{30}N_3O \text{ (LSD)}$$

144. a. $CH_4(g) + 4\ S(s) \rightarrow CS_2(l) + 2\ H_2S(g)$ or $2\ CH_4(g) + S_8(s) \rightarrow 2\ CS_2(l) + 4\ H_2S(g)$

b. $120. \text{ g } CH_4 \times \frac{1 \text{ mol } CH_4}{16.04 \text{ g } CH_4} \times \frac{1 \text{ mol } CS_2}{\text{mol } CH_4} \times \frac{76.15 \text{ g } CS_2}{\text{mol } CS_2} = 570. \text{ g } CS_2$

$$120. \text{ g S} \times \frac{1 \text{ mol S}}{32.07 \text{ g S}} \times \frac{1 \text{ mol } CS_2}{4 \text{ mol S}} \times \frac{76.15 \text{ g } CS_2}{\text{mol } CS_2} = 71.2 \text{ g } CS_2$$

Because S produces the smaller quantity of CS_2, sulfur is the limiting reactant and 71.2 g CS_2 can be produced. The same amount of CS_2 would be produced using the balanced equation with S_8.

145. $$126 \text{ g } B_5H_9 \times \frac{1 \text{ mol } B_5H_9}{63.12 \text{ g } B_5H_9} \times \frac{9 \text{ mol } H_2O}{2 \text{ mol } B_5H_9} \times \frac{18.02 \text{ g } H_2O}{\text{mol } H_2O} = 162 \text{ g } H_2O$$

$$192 \text{ g } O_2 \times \frac{1 \text{ mol } O_2}{32.00 \text{ g } O_2} \times \frac{9 \text{ mol } H_2O}{12 \text{ mol } O_2} \times \frac{18.02 \text{ g } H_2O}{\text{mol } H_2O} = 81.1 \text{ g } H_2O$$

Because O_2 produces the smallest quantity of product, O_2 is limiting and 81.1g H_2O can be produced.

146. $2\ NaNO_3(s) \rightarrow 2\ NaNO_2(s) + O_2(g)$; the amount of $NaNO_3$ in the impure sample is:

$$0.2864 \text{ g } NaNO_2 \times \frac{1 \text{ mol } NaNO_2}{69.00 \text{ g } NaNO_2} \times \frac{2 \text{ mol } NaNO_3}{2 \text{ mol } NaNO_2} \times \frac{85.00 \text{ g } NaNO_3}{\text{mol } NaNO_3}$$

$$= 0.3528 \text{ g } NaNO_3$$

$$\text{Mass percent } NaNO_3 = \frac{0.3528 \text{ g } NaNO_3}{0.4230 \text{ g sample}} \times 100 = 83.40\%$$

147. $$453 \text{ g Fe} \times \frac{1 \text{ mol Fe}}{55.85 \text{ g Fe}} \times \frac{1 \text{ mol } Fe_2O_3}{2 \text{ mol Fe}} \times \frac{159.70 \text{ g } Fe_2O_3}{\text{mol } Fe_2O_3} = 648 \text{ g } Fe_2O_3$$

$$\text{Mass percent } Fe_2O_3 = \frac{648 \text{ g } Fe_2O_3}{752 \text{ g ore}} \times 100 = 86.2\%$$

148. a. $$\text{Mass of Zn in alloy} = 0.0985 \text{ g } ZnCl_2 \times \frac{65.38 \text{ g Zn}}{136.28 \text{ g } ZnCl_2} = 0.0473 \text{ g Zn}$$

$$\%\ Zn = \frac{0.0473 \text{ g Zn}}{0.5065 \text{ g brass}} \times 100 = 9.34\%\ Zn; \quad \%\ Cu = 100.00 - 9.34 = 90.66\%\ Cu$$

b. The Cu remains unreacted. After filtering, washing, and drying, the mass of the unreacted copper could be measured.

149. Assuming 1 mole of vitamin A (286.4 g vitamin A):

$$\text{mol C} = 286.4 \text{ g vitamin A} \times \frac{0.8386 \text{ g C}}{\text{g vitamin A}} \times \frac{1 \text{ mol C}}{12.01 \text{ g C}} = 20.00 \text{ mol C}$$

$$\text{mol H} = 286.4 \text{ g vitamin A} \times \frac{0.1056 \text{ g H}}{\text{g vitamin A}} \times \frac{1 \text{ mol H}}{1.008 \text{ g H}} = 30.00 \text{ mol H}$$

Because 1 mole of vitamin A contains 20 mol C and 30 mol H, the molecular formula of vitamin A is $C_{20}H_{30}E$. To determine E, let's calculate the molar mass of E:

$$286.4 \text{ g} = 20(12.01) + 30(1.008) + \text{molar mass E}, \quad \text{molar mass E} = 16.0 \text{ g/mol}$$

From the periodic table, E = oxygen, and the molecular formula of vitamin A is $C_{20}H_{30}O$.

150. a. At 40.0 g of Na added, Cl_2 and Na both run out at the same time (both are limiting reactants). Past 40.0 g of Na added, Cl_2 is limiting, and because the amount of Cl_2 present in each experiment was the same quantity, no more NaCl can be produced. Before 40.0 g of Na added, Na was limiting. As more Na was added (up to 40.0 g Na), more NaCl was produced.

b. $20.0 \text{ g Na} \times \frac{1 \text{ mol Na}}{22.99 \text{ g Na}} \times \frac{2 \text{ mol NaCl}}{2 \text{ mol Na}} \times \frac{58.44 \text{ g NaCl}}{\text{mol NaCl}} = 50.8 \text{ g NaCl}$

c. At 40.0 g Na added, both Cl_2 and Na are present in stoichiometric amounts.

$$40.0 \text{ g Na} \times \frac{1 \text{ mol Na}}{22.99 \text{ g Na}} \times \frac{1 \text{ mol } Cl_2}{2 \text{ mol Na}} \times \frac{70.90 \text{ g } Cl_2}{\text{mol } Cl_2} = 61.7 \text{ g } Cl_2$$

61.7 g Cl_2 was present at 40.0 g Na added, and from the problem, the same 61.7 g Cl_2 was present in each experiment.

d. At 50.0 g Na added, Cl_2 is limiting:

$$61.7 \text{ g } Cl_2 \times \frac{1 \text{ mol } Cl_2}{70.90 \text{ g } Cl_2} \times \frac{2 \text{ mol NaCl}}{1 \text{ mol } Cl_2} \times \frac{58.44 \text{ g NaCl}}{\text{mol NaCl}} = 101.7 \text{ g} = 102 \text{ g NaCl}$$

e. $20.0 \text{ g Na} \times \frac{1 \text{ mol Na}}{22.99 \text{ g Na}} \times \frac{1 \text{ mol } Cl_2}{2 \text{ mol Na}} \times \frac{70.90 \text{ g } Cl_2}{\text{mol } Cl_2} = 30.8 \text{ g } Cl_2 \text{ reacted}$

Excess Cl_2 = 61.7 g Cl_2 initially – 30.8 g Cl_2 reacted = 30.9 g Cl_2 in excess

Note: We know that 40.0 g Na is the point where Na and the 61.7 g of Cl_2 run out at the same time. So if 20.0 g of Na are reacted, one-half of the Cl_2 that was present at 40.0 g Na reacted will be in excess. The previous calculation confirms this.

For 50.0 g Na reacted, Cl_2 is limiting and 40.0 g Na will react as determined previously.

Excess Na = 50.0 g Na initially – 40.0 g Na reacted = 10.0 g Na in excess.

151. X_2Z: 40.0% X and 60.0% Z by mass; $\frac{\text{mol X}}{\text{mol Z}} = 2 = \frac{40.0/A_x}{60.0/A_z} = \frac{(40.0)A_z}{(60.0)A_x}$ or $A_z = 3A_x$, where A = molar mass.

For XZ_2, molar mass = $A_x + 2A_z = A_x + 2(3A_x) = 7A_x$.

Mass percent X = $\frac{A_x}{7A_x} \times 100 = 14.3\%$ X; % Z = 100.0 – 14.3 = 85.7% Z

ChemWork Problems

The answers to the problems 152-159 (or a variation to these problems) are found in OWL. These problems are also assignable in OWL.

Challenge Problems

160. GaAs can be either ^{69}GaAs or ^{71}GaAs. The mass spectrum for GaAs will have 2 peaks at 144 (= 69 + 75) and 146 (= 71 + 75) with intensities in the ratio of 60 : 40 or 3 : 2.

Ga_2As_2 can be $^{69}Ga_2As_2$, $^{69}Ga^{71}GaAs_2$, or $^{71}Ga_2As_2$. The mass spectrum will have 3 peaks at 288, 290, and 292 with intensities in the ratio of 36 : 48 : 16 or 9 : 12 : 4. We get this ratio from the following probability table:

	^{69}Ga (0.60)	^{71}Ga (0.40)
^{69}Ga (0.60)	0.36	0.24
^{71}Ga (0.40)	0.24	0.16

161. The volume of a gas is proportional to the number of molecules of gas. Thus the formulas are:

I: NH_3; II: N_2H_4; III: HN_3

The mass ratios are:

$$\text{I: } \frac{82.25\text{ g N}}{17.75\text{ g H}} = \frac{4.634\text{g N}}{\text{g H}}; \quad \text{II: } \frac{6.949\text{g N}}{\text{g H}}; \quad \text{III: } \frac{41.7\text{ g N}}{\text{g H}}$$

If we set the atomic mass of H equal to 1.008, then the atomic mass, A, for nitrogen is:

I: 14.01; II: 14.01; III. 14.0

For example, for compound I: $\frac{A}{3(1.008)} = \frac{4.634}{1}$, A = 14.01

162. $\frac{^{85}\text{Rb atoms}}{^{87}\text{Rb atoms}} = 2.591$; if we had exactly 100 atoms, x = number of ^{85}Rb atoms, and $100 - x$ = number of ^{87}Rb atoms.

$$\frac{x}{100-x} = 2.591,\ x = 259.1 - (2.591)x,\ x = \frac{259.1}{3.591} = 72.15;\ 72.15\%\ ^{85}\text{Rb}$$

$$0.7215(84.9117) + 0.2785(\text{A}) = 85.4678,\quad \text{A} = \frac{85.4678 - 61.26}{0.2785} = 86.92\ \text{u}$$

163. First, we will determine composition in mass percent. We assume that all the carbon in the 0.213 g CO_2 came from the 0.157 g of the compound and that all the hydrogen in the 0.0310 g H_2O came from the 0.157 g of the compound.

$$0.213\ \text{g CO}_2 \times \frac{12.01\ \text{g C}}{44.01\ \text{g CO}_2} = 0.0581\ \text{g C};\ \%\ \text{C} = \frac{0.0581\ \text{g C}}{0.157\ \text{g compound}} \times 100 = 37.0\%\ \text{C}$$

$$0.0310\ \text{g H}_2\text{O} \times \frac{2.016\ \text{g H}}{18.02\ \text{g H}_2\text{O}} = 3.47 \times 10^{-3}\ \text{g H};\ \%\ \text{H} = \frac{3.47 \times 10^{-3}\ \text{g}}{0.157\ \text{g}} \times 100 = 2.21\%\ \text{H}$$

We get the mass percent of N from the second experiment:

$$0.0230\ \text{g NH}_3\ \text{H}\ \frac{14.01\ \text{g N}}{17.03\ \text{g NH}_3} = 1.89 \times 10^{-2}\ \text{g N}$$

$$\%\ \text{N} = \frac{1.89 \times 10^{-2}\ \text{g}}{0.103\ \text{g}} \times 100 = 18.3\%\ \text{N}$$

The mass percent of oxygen is obtained by difference:

$$\%\ \text{O} = 100.00 - (37.0 + 2.21 + 18.3) = 42.5\%\ \text{O}$$

So, out of 100.00 g of compound, there are:

$$37.0\ \text{g C} \times \frac{1\ \text{mol C}}{12.01\ \text{g C}} = 3.08\ \text{mol C};\quad 2.21\ \text{g H} \times \frac{1\ \text{mol H}}{1.008\ \text{g H}} = 2.19\ \text{mol H}$$

$$18.3\ \text{g N} \times \frac{1\ \text{mol N}}{14.01\ \text{g N}} = 1.31\ \text{mol N};\quad 42.5\ \text{g O} \times \frac{1\ \text{mol O}}{16.00\ \text{g O}} = 2.66\ \text{mol O}$$

Lastly, and often the hardest part, we need to find simple whole-number ratios. Divide all mole values by the smallest number:

$$\frac{3.08}{1.31} = 2.35;\quad \frac{2.19}{1.31} = 1.67;\quad \frac{1.31}{1.31} = 1.00;\quad \frac{2.66}{1.31} = 2.03$$

Multiplying all these ratios by 3 gives an empirical formula of $C_7H_5N_3O_6$.

164. $1.0 \times 10^6 \text{ kg HNO}_3 \times \frac{1000\text{g HNO}_3}{\text{kg HNO}_3} \times \frac{1\text{mol HNO}_3}{63.02\text{ g HNO}_3} = 1.6 \times 10^7 \text{ mol HNO}_3$

We need to get the relationship between moles of HNO_3 and moles of NH_3. We have to use all three equations.

$$\frac{2\text{ mol HNO}_3}{3\text{ mol NO}_2} \times \frac{2\text{ mol NO}_2}{2\text{ mol NO}} \times \frac{4\text{ mol NO}}{4\text{ mol NH}_3} = \frac{16\text{ mol HNO}_3}{24\text{ mol NH}_3}$$

Thus we can produce 16 mol HNO_3 for every 24 mol NH_3, we begin with:

$$1.6 \times 10^7 \text{ mol HNO}_3 \times \frac{24\text{ mol NH}_3}{16\text{ mol HNO}_3} \times \frac{17.03\text{ g NH}_3}{\text{mol NH}_3} = 4.1 \times 10^8 \text{ g or } 4.1 \times 10^5 \text{ kg NH}_3$$

This is an oversimplified answer. In practice, the NO produced in the third step is recycled back continuously into the process in the second step. If this is taken into consideration, then the conversion factor between mol NH_3 and mol HNO_3 turns out to be 1 : 1; that is, 1 mole of NH_3 produces 1 mole of HNO_3. Taking into consideration that NO is recycled back gives an answer of 2.7×10^5 kg NH_3 reacted.

165. $\text{Fe(s)} + \frac{1}{2}\text{O}_2\text{(g)} \rightarrow \text{FeO(s)};\ 2\text{ Fe(s)} + \frac{3}{2}\text{O}_2\text{(g)} \rightarrow \text{Fe}_2\text{O}_3\text{(s)}$

$$20.00\text{ g Fe} \times \frac{1\text{ mol Fe}}{55.85\text{ g}} = 0.3581\text{ mol}$$

$$(11.20 - 3.24)\text{ g O}_2 \times \frac{1\text{ mol O}_2}{32.00\text{ g}} = 0.2488\text{ mol O}_2\text{ consumed (1 extra sig. fig.)}$$

Let's assume x moles of Fe reacts to form x moles of FeO. Then $0.3581 - x$, the remaining moles of Fe, reacts to form Fe_2O_3. Balancing the two equations in terms of x:

$$x\text{ Fe} + \frac{1}{2}x\text{ O}_2 \rightarrow x\text{ FeO}$$

$$(0.3581 - x)\text{ mol Fe} + \frac{3}{2}\left(\frac{0.3581 - x}{2}\right)\text{mol O}_2 \rightarrow \left(\frac{0.3581 - x}{2}\right)\text{mol Fe}_2\text{O}_3$$

Setting up an equation for total moles of O_2 consumed:

$$\frac{1}{2}x + \frac{3}{4}(0.3581 - x) = 0.2488\text{ mol O}_2, \quad x = 0.0791 = 0.079\text{ mol FeO}$$

$$0.079\text{ mol FeO} \times \frac{71.85\text{ g FeO}}{\text{mol}} = 5.7\text{ g FeO produced}$$

$$\text{Mol Fe}_2\text{O}_3\text{ produced} = \frac{0.3581 - 0.079}{2} = 0.140\text{ mol Fe}_2\text{O}_3$$

$$0.140\text{ mol Fe}_2\text{O}_3 \times \frac{159.70\text{ g Fe}_2\text{O}_3}{\text{mol}} = 22.4\text{ g Fe}_2\text{O}_3\text{ produced}$$

166. $2\ C_2H_6(g) + 7\ O_2(g) \rightarrow 4\ CO_2(g) + 6\ H_2O(l)$; $C_3H_8(g) + 5\ O_2(g) \rightarrow 3\ CO_2(g) + 4\ H_2O(l)$
30.07 g/mol 44.09 g/mol

Let x = mass C_2H_6, so $9.780 - x$ = mass C_3H_8. Use the balanced equation to set up a mathematical expression for the moles of O_2 required.

$$\frac{x}{30.07} \times \frac{7}{2} + \frac{9.780 - x}{44.09} \times \frac{5}{1} = 1.120 \text{ mol } O_2$$

Solving: $x = 3.7$ g C_2H_6; $\frac{3.7 \text{ g}}{9.780 \text{ g}} \times 100 = 38\%$ C_2H_6 by mass

167. The two relevant equations are:

$$Zn(s) + 2\ HCl(aq) \rightarrow ZnCl_2(aq) + H_2(g) \text{ and } Mg(s) + 2\ HCl(aq) \rightarrow MgCl_2(aq) + H_2(g)$$

Let x = mass Mg, so $10.00 - x$ = mass Zn. From the balanced equations, moles H_2 produced = moles Zn reacted + moles Mg reacted.

$$\text{Mol } H_2 = 0.5171 \text{ g } H_2 \times \frac{1 \text{ mol } H_2}{2.016 \text{ g } H_2} = 0.2565 \text{ mol } H_2$$

$$0.2565 = \frac{x}{24.31} + \frac{10.00 - x}{65.38}; \text{ solving: } x = 4.008 \text{ g Mg}$$

$$\frac{4.008 \text{ g}}{10.00 \text{ g}} \times 100 = 40.08\% \text{ Mg}$$

168. $a\ N_2H_4 + b\ NH_3 + (10.00 - 4.062)\ O_2 \rightarrow c\ NO_2 + d\ H_2O$

Setting up four equations to solve for the four unknowns:

$2a + b = c$ (N mol balance)

$2c + d = 2(10.00 - 4.062)$ (O mol balance)

$4a + 3b = 2d$ (H mol balance)

$a(32.05) + b(17.03) = 61.00$ (mass balance)

Solving the simultaneous equations gives $a = 1.12 = 1.1$ mol N_2H_4.

$$\frac{1.1 \text{ mol } N_2H_4 \times 32.05 \text{ g/mol } N_2H_4}{61.00 \text{ g}} \times 100 = 58\% \ N_2H_4$$

169. We know that water is a product, so one of the elements in the compound is hydrogen.

$X_aH_b + O_2 \rightarrow H_2O + ?$

To balance the H atoms, the mole ratio between X_aH_b and $H_2O = \frac{2}{b}$.

Mol compound = $\frac{1.39\text{ g}}{62.09\text{ g/mol}}$ = 0.0224 mol; mol $H_2O = \frac{1.21\text{ g}}{18.02\text{ g/mol}}$ = 0.0671 mol

$\frac{2}{b} = \frac{0.0224}{0.0671}$, $b = 6$; X_aH_6 has a molar mass of 62.09 g/mol.

62.09 = a(molar mass of X) + 6(1.008), a(molar mass of X) = 56.04

Some possible identities for X could be Fe (a = 1), Si (a = 2), N (a = 4), and Li (a = 8). N fits the data best, so N_4H_6 is the most likely formula.

170. The balanced equation is 2 Sc(s) + 2x HCl(aq) $\longrightarrow$ 2 $ScCl_x$(aq)+ x H_2(g)

The mole ratio of Sc : $H_2 = \frac{2}{x}$.

Mol Sc = 2.25 g Sc × $\frac{1\text{ mol Sc}}{44.96\text{ g Sc}}$ = 0.0500 mol Sc

Mol H_2 = 0.1502 g H_2 × $\frac{1\text{ mol }H_2}{2.016\text{ g }H_2}$ = 0.07450 mol H_2

$\frac{2}{x} = \frac{0.0500}{0.07450}$, $x = 3$; the formula is $ScCl_3$.

171. Total mass of copper used:

$$10{,}000\text{ boards} \times \frac{(8.0\text{ cm} \times 16.0\text{ cm} \times 0.060\text{ cm})}{\text{board}} \times \frac{8.96\text{ g}}{\text{cm}^3} = 6.9 \times 10^5\text{ g Cu}$$

Amount of Cu to be recovered = 0.80 × (6.9 × 10^5 g) = 5.5 × 10^5 g Cu.

$$5.5 \times 10^5\text{ g Cu} \times \frac{1\text{ mol Cu}}{63.55\text{ g Cu}} \times \frac{1\text{ mol Cu(NH}_3)_4\text{Cl}_2}{\text{mol Cu}} \times \frac{202.59\text{ g Cu(NH}_3)_4\text{Cl}_2}{\text{mol Cu(NH}_3)_4\text{Cl}_2}$$

$$= 1.8 \times 10^6\text{ g Cu(NH}_3)_4\text{Cl}_2$$

$$5.5 \times 10^5\text{ g Cu} \times \frac{1\text{ mol Cu}}{63.55\text{ g Cu}} \times \frac{4\text{ mol NH}_3}{\text{mol Cu}} \times \frac{17.03\text{ g NH}_3}{\text{mol NH}_3} = 5.9 \times 10^5\text{ g NH}_3$$

172. a. From the reaction stoichiometry we would expect to produce 4 mol of acetaminophen for every 4 mol of $C_6H_5O_3N$ reacted. The actual yield is 3 mol of acetaminophen compared to a theoretical yield of 4 mol of acetaminophen. Solving for percent yield by mass (where M = molar mass acetaminophen):

$$\text{percent yield} = \frac{3\text{ mol} \times \text{M}}{4\text{ mol} \times \text{M}} \times 100 = 75\%$$

b. The product of the percent yields of the individual steps must equal the overall yield, 75%.

$(0.87)(0.98)(x) = 0.75$, $x = 0.88$; step III has a percent yield of 88%.

173. 10.00 g XCl_2 + excess $Cl_2 \rightarrow$ 12.55 g XCl_4; 2.55 g Cl reacted with XCl_2 to form XCl_4. XCl_4 contains 2.55 g Cl and 10.00 g XCl_2. From the mole ratios, 10.00 g XCl_2 must also contain 2.55 g Cl; mass X in XCl_2 = 10.00 – 2.55 = 7.45 g X.

$$2.55 \text{ g Cl} \times \frac{1 \text{ mol Cl}}{35.45 \text{ g Cl}} \times \frac{1 \text{ mol XCl}_2}{2 \text{ mol Cl}} \times \frac{1 \text{ mol X}}{\text{mol XCl}_2} = 3.60 \times 10^{-2} \text{ mol X}$$

So 3.60×10^{-2} mol X has a mass equal to 7.45 g X. The molar mass of X is:

$$\frac{7.45 \text{ g X}}{3.60 \times 10^{-2} \text{ mol X}} = 207 \text{ g/mol X; atomic mass} = 207 \text{ u, so X is Pb.}$$

174. 4.000 g $M_2S_3 \rightarrow$ 3.723 g MO_2

There must be twice as many moles of MO_2 as moles of M_2S_3 in order to balance M in the reaction. Setting up an equation for 2(mol M_2S_3) = mol MO_2 where A = molar mass M:

$$2\left[\frac{4.000 \text{ g}}{2A + 3(32.07)}\right] = \frac{3.723 \text{ g}}{A + 2(16.00)}, \quad \frac{8.000}{2A + 96.21} = \frac{3.723}{A + 32.00}$$

$(8.000)A + 256.0 = (7.446)A + 358.2$, $(0.554)A = 102.2$, A = 184 g/mol; atomic mass = 184 u

Note: From the periodic table, M is tungsten, W.

175. Consider the case of aluminum plus oxygen. Aluminum forms Al^{3+} ions; oxygen forms O^{2-} anions. The simplest compound between the two elements is Al_2O_3. Similarly, we would expect the formula of any Group 6A element with Al to be Al_2X_3. Assuming this, out of 100.00 g of compound, there are 18.56 g Al and 81.44 g of the unknown element, X. Let's use this information to determine the molar mass of X, which will allow us to identify X from the periodic table.

$$18.56 \text{ g Al} \times \frac{1 \text{ mol Al}}{26.98 \text{ g Al}} \times \frac{3 \text{ mol X}}{2 \text{ mol Al}} = 1.032 \text{ mol X}$$

81.44 g of X must contain 1.032 mol of X.

$$\text{Molar mass of X} = \frac{81.44 \text{ g X}}{1.032 \text{ mol X}} = 78.91 \text{ g/mol X.}$$

From the periodic table, the unknown element is selenium, and the formula is Al_2Se_3.

176. Let x = mass KCl and y = mass KNO_3. Assuming 100.0 g of mixture, $x + y = 100.0$ g.

Molar mass KCl = 74.55 g/mol; molar mass KNO_3 = 101.11 g/mol

$$\text{Mol KCl} = \frac{x}{74.55};\ \text{mol } KNO_3 = \frac{y}{101.11}$$

Knowing that the mixture is 43.2% K, then in the 100.0 g mixture, an expression for the mass of K is:

$$39.10\left(\frac{x}{74.55} + \frac{y}{101.11}\right) = 43.2$$

We have two equations and two unknowns:

$$\begin{aligned} (0.5245)x + (0.3867)y &= 43.2 \\ x + \quad\quad y &= 100.0 \end{aligned}$$

Solving, $x = 32.9$ g KCl; $\dfrac{32.9\text{ g}}{100.0\text{ g}} \times 100 = 32.9\%$ KCl

177. The balanced equations are:

$$4\ NH_3(g) + 5\ O_2(g) \rightarrow 4\ NO(g) + 6\ H_2O(g) \text{ and } 4\ NH_3(g) + 7\ O_2(g) \rightarrow 4\ NO_2(g) + 6\ H_2O(g)$$

Let $4x$ = number of moles of NO formed, and let $4y$ = number of moles of NO_2 formed. Then:

$$4x\ NH_3 + 5x\ O_2 \rightarrow 4x\ NO + 6x\ H_2O \text{ and } 4y\ NH_3 + 7y\ O_2 \rightarrow 4y\ NO_2 + 6y\ H_2O$$

All the NH_3 reacted, so $4x + 4y = 2.00$. $10.00 - 6.75 = 3.25$ mol O_2 reacted, so $5x + 7y = 3.25$.

Solving by the method of simultaneous equations:

$$\begin{aligned} 20x + 28y &= 13.0 \\ \underline{-20x - 20y} &\underline{= -10.0} \\ 8y &= 3.0,\quad y = 0.38;\ 4x + 4 \times 0.38 = 2.00,\ x = 0.12 \end{aligned}$$

Mol NO = $4x = 4 \times 0.12 = 0.48$ mol NO formed

178. $C_xH_yO_z$ + oxygen $\rightarrow x\ CO_2 + y/2\ H_2O$

$$\text{Mass \% C in aspirin} = \frac{2.20\text{ g } CO_2 \times \dfrac{1\text{ mol } CO_2}{44.01\text{ g } CO_2} \times \dfrac{1\text{ mol C}}{\text{mol } CO_2} \times \dfrac{12.01\text{ g C}}{\text{mol C}}}{1.00\text{ g aspirin}} = 60.0\%\text{ C}$$

$$\text{Mass \% H in aspirin} = \frac{0.400\,\text{g H}_2\text{O} \times \dfrac{1\,\text{mol H}_2\text{O}}{18.02\,\text{g H}_2\text{O}} \times \dfrac{2\,\text{mol H}}{\text{mol H}_2\text{O}} \times \dfrac{1.008\,\text{g H}}{\text{mol H}}}{1.00\,\text{g aspirin}} = 4.48\%\ \text{H}$$

Mass % O = 100.00 – (60.0 + 4.48) = 35.5% O

Assuming 100.00 g aspirin:

$$60.0\ \text{g C} \times \frac{1\,\text{mol C}}{12.01\,\text{g C}} = 5.00\ \text{mol C};\quad 4.48\ \text{g H} \times \frac{1\,\text{mol H}}{1.008\,\text{g H}} = 4.44\ \text{mol H}$$

$$35.5\ \text{g O} \times \frac{1\,\text{mol O}}{16.00\,\text{g O}} = 2.22\ \text{mol O}$$

Dividing by the smallest number: $\frac{5.00}{2.22} = 2.25;\ \frac{4.44}{2.22} = 2.00$

Empirical formula: $(C_{2.25}H_{2.00}O)_4 = C_9H_8O_4$. Empirical mass ≈ 9(12) + 8(1) + 4(16) = 180 g/mol; this is in the 170–190 g/mol range, so the molecular formula is also $C_9H_8O_4$.

Balance the aspirin synthesis reaction to determine the formula for salicylic acid.

$C_aH_bO_c + C_4H_6O_3 \rightarrow C_9H_8O_4 + C_2H_4O_2$, $C_aH_bO_c$ = salicylic acid = $C_7H_6O_3$

Integrative Problems

179. a. $1.05 \times 10^{-20}\ \text{g Fe} \times \frac{1\,\text{mol Fe}}{55.85\,\text{g Fe}} \times \frac{6.022\times 10^{23}\ \text{atoms Fe}}{\text{mol Fe}} = 113\ \text{atoms Fe}$

b. The total number of platinum atoms is 14 × 20 = 280 atoms (exact number). The mass of these atoms is:

$$280\ \text{atoms Pt} \times \frac{1\,\text{mol Pt}}{6.022 \times 10^{23}\ \text{atoms Pt}} \times \frac{195.1\,\text{g Pt}}{\text{mol Pt}} = 9.071 \times 10^{-20}\ \text{g Pt}$$

c. $9.071 \times 10^{-20}\ \text{g Ru} \times \frac{1\,\text{mol Ru}}{101.1\,\text{g Ru}} \times \frac{6.022 \times 10^{23}\ \text{atoms Ru}}{\text{mol Ru}} = 540.3 = 540\ \text{atoms Ru}$

180. Assuming 100.00 g of tetrodotoxin:

$$41.38\ \text{g C} \times \frac{1\,\text{mol C}}{12.01\,\text{g C}} = 3.445\ \text{mol C};\quad 13.16\ \text{g N} \times \frac{1\,\text{mol N}}{14.01\,\text{g N}} = 0.9393\ \text{mol N}$$

$$5.37\ \text{g H} \times \frac{1\,\text{mol H}}{1.008\,\text{g H}} = 5.33\ \text{mol H};\quad 40.09\ \text{g O} \times \frac{1\,\text{mol O}}{16.00\,\text{g O}} = 2.506\ \text{mol O}$$

Divide by the smallest number:

$$\frac{3.445}{0.9393} = 3.668; \quad \frac{5.33}{0.9393} = 5.67; \quad \frac{2.506}{0.9393} = 2.668$$

To get whole numbers for each element, multiply through by 3.

Empirical formula: $(C_{3.668}H_{5.67}NO_{2.668})_3 = C_{11}H_{17}N_3O_8$; the mass of the empirical formula is 319.3 g/mol.

$$\text{Molar mass tetrodotoxin} = \frac{1.59 \times 10^{-21}\ \text{g}}{3\ \text{molecules} \times \dfrac{1\ \text{mol}}{6.022 \times 10^{23}\ \text{molecules}}} = 319\ \text{g/mol}$$

Because the empirical mass and molar mass are the same, the molecular formula is the same as the empirical formula, $C_{11}H_{17}N_3O_8$.

$$165\ \text{lb} \times \frac{1\ \text{kg}}{2.2046\ \text{lb}} \times \frac{10.\ \mu\text{g}}{\text{kg}} \times \frac{1 \times 10^{-6}\ \text{g}}{\mu\text{g}} \times \frac{1\ \text{mol}}{319.3\ \text{g}} \times \frac{6.022 \times 10^{23}\ \text{molecules}}{1\ \text{mol}}$$

$= 1.4 \times 10^{18}$ molecules tetrodotoxin is the LD_{50} dosage

181. $$\text{Molar mass } X_2 = \frac{0.105\ \text{g}}{8.92 \times 10^{20}\ \text{molecules} \times \dfrac{1\ \text{mol}}{6.022 \times 10^{23}\ \text{molecules}}} = 70.9\ \text{g/mol}$$

The mass of X = 1/2(70.9 g/mol) = 35.5 g/mol. This is the element chlorine.

Assuming 100.00 g of MX_3 (= MCl_3) compound:

$$54.47\ \text{g Cl} \times \frac{1\ \text{mol}}{35.45\ \text{g}} = 1.537\ \text{mol Cl}$$

$$1.537\ \text{mol Cl} \times \frac{1\ \text{mol M}}{3\ \text{mol Cl}} = 0.5123\ \text{mol M}$$

$$\text{Molar mass of M} = \frac{45.53\ \text{g M}}{0.5123\ \text{mol M}} = 88.87\ \text{g/mol M}$$

M is the element yttrium (Y), and the name of YCl_3 is yttrium(III) chloride.

The balanced equation is $2\ Y + 3\ Cl_2 \rightarrow 2\ YCl_3$.

Assuming Cl_2 is limiting:

$$1.00\ \text{g Cl}_2 \times \frac{1\ \text{mol Cl}_2}{70.90\ \text{g Cl}_2} \times \frac{2\ \text{mol YCl}_3}{3\ \text{mol Cl}_2} \times \frac{195.26\ \text{g YCl}_3}{1\ \text{mol YCl}_3} = 1.84\ \text{g YCl}_3$$

Assuming Y is limiting:

$$1.00 \text{ g Y} \times \frac{1 \text{ mol Y}}{88.91 \text{ g Y}} \times \frac{2 \text{ mol } YCl_3}{2 \text{ mol Y}} \times \frac{195.26 \text{ g } YCl_3}{1 \text{ mol } YCl_3} = 2.20 \text{ g } YCl_3$$

Because Cl_2, when it all reacts, produces the smaller amount of product, Cl_2 is the limiting reagent, and the theoretical yield is 1.84 g YCl_3.

182. $2 \text{ As} + 4 \text{ AsI}_3 \rightarrow 3 \text{ As}_2\text{I}_4$

Volume of As cube = $(3.00 \text{ cm})^3 = 27.0 \text{ cm}^3$

$$27.0 \text{ cm}^3 \times \frac{5.72 \text{ g As}}{\text{cm}^3} \times \frac{1 \text{ mol As}}{74.92 \text{ g As}} \times \frac{3 \text{ mol As}_2\text{I}_4}{2 \text{ mol As}} \times \frac{657.44 \text{ g As}_2\text{I}_4}{\text{mol As}_2\text{I}_4} = 2030 \text{ g As}_2\text{I}_4$$

$$1.01 \times 10^{24} \text{ molecules AsI}_3 \times \frac{1 \text{ mol AsI}_3}{6.022 \times 10^{23} \text{ molecules AsI}_3} \times \frac{3 \text{ mol As}_2\text{I}_4}{4 \text{ mol AsI}_3}$$

$$\times \frac{657.44 \text{ g As}_2\text{I}_4}{\text{mol As}_2\text{I}_4} = 827 \text{ g As}_2\text{I}_4$$

Because the reactant AsI_3 produces the smaller quantity of product, then AsI_3 is the limiting reactant and 827 g As_2I_4 is the theoretical yield.

$$0.756 = \frac{\text{actual yield}}{827 \text{ g}}, \text{ actual yield} = 0.756 \times 827 \text{ g} = 625 \text{ g As}_2\text{I}_4$$

Marathon Problems

183. Let M = unknown element; mass O in oxide = 3.708 g – 2.077 g = 1.631 g O

In 3.708 g of compound:

$$1.631 \text{ g O} \times \frac{1 \text{ mol O}}{16.00 \text{ g O}} = 0.1019 \text{ g mol O}$$

If MO is the formula of the oxide, then M has a molar mass of $\frac{2.077 \text{ g M}}{0.1019 \text{ mol M}} = 20.38$ g/mol.

This is too low for the molar mass. We must have fewer moles of M than moles O present in the formula. Some possibilities are MO_2, M_2O_3, MO_3, etc. It is a guessing game as to which to try. Let's assume an MO_2 formula. Then the molar mass of M is:

$$\frac{2.077 \text{ g M}}{0.1019 \text{ mol O} \times \frac{1 \text{ mol M}}{2 \text{ mol O}}} = 40.77 \text{ g/mol}$$

This is close to calcium, but calcium forms an oxide having the CaO formula, not CaO_2.

If MO_3 is assumed to be the formula, then the molar mass of M calculates to be 61.10 g/mol, which is too large. Therefore, the mol O to mol M ratio must be between 2 and 3. Some reasonable possibilities are 2.25, 2.33, 2.5, 2.67, and 2.75 (these are reasonable because they will lead to whole number formulas). Trying a mol O to mol M ratio of 2.5 : 1 gives a molar mass of:

$$\frac{2.077 \text{ g M}}{0.1019 \text{ mol O} \times \dfrac{1 \text{ mol M}}{2.5 \text{ mol O}}} = 50.96 \text{ g/mol}$$

This is the molar mass of vanadium, and V_2O_5 is a reasonable formula for an oxide of vanadium. The other choices for the O : M mole ratios between 2 and 3 do not give as reasonable results. Therefore, M is most likely vanadium, and the formula is V_2O_5.

184. a. i. If the molar mass of A is greater than the molar mass of B, then we cannot determine the limiting reactant because, while we have a fewer number of moles of A, we also need fewer moles of A (from the balanced reaction).

ii. If the molar mass of B is greater than the molar mass of A, then B is the limiting reactant because we have a fewer number of moles of B and we need more B (from the balanced reaction).

b. $A + 5\ B \rightarrow 3\ CO_2 + 4\ H_2O$

To conserve mass: 44.01 + 5(B) = 3(44.01) + 4(18.02); solving: B = 32.0 g/mol

Because B is diatomic, the best choice for B is O_2.

c. We can solve this without mass percent data simply by balancing the equation:

$$A + 5\ O_2 \rightarrow 3\ CO_2 + 4\ H_2O$$

A must be C_3H_8 (which has a similar molar mass to CO_2). This is also the empirical formula.

Note: $\dfrac{3(12.01)}{3(12.01) + 8(1.008)} \times 100 = 81.71\%$ C. So this checks.

CHAPTER 4

TYPES OF CHEMICAL REACTIONS AND SOLUTION STOICHIOMETRY

Questions

13. a. Polarity is a term applied to covalent compounds. Polar covalent compounds have an unequal sharing of electrons in bonds that results in unequal charge distribution in the overall molecule. Polar molecules have a partial negative end and a partial positive end. These are not full charges as in ionic compounds but are charges much smaller in magnitude. Water is a polar molecule and dissolves other polar solutes readily. The oxygen end of water (the partial negative end of the polar water molecule) aligns with the partial positive end of the polar solute, whereas the hydrogens of water (the partial positive end of the polar water molecule) align with the partial negative end of the solute. These opposite charge attractions stabilize polar solutes in water. This process is called hydration. Nonpolar solutes do not have permanent partial negative and partial positive ends; nonpolar solutes are not stabilized in water and do not dissolve.

    b. KF is a soluble ionic compound, so it is a strong electrolyte. KF(aq) actually exists as separate hydrated K^+ ions and hydrated F^- ions in solution: $C_6H_{12}O_6$ is a polar covalent molecule that is a nonelectrolyte. $C_6H_{12}O_6$ is hydrated as described in part a.

    c. RbCl is a soluble ionic compound, so it exists as separate hydrated Rb^+ ions and hydrated Cl^- ions in solution. AgCl is an insoluble ionic compound, so the ions stay together in solution and fall to the bottom of the container as a precipitate.

    d. HNO_3 is a strong acid and exists as separate hydrated H^+ ions and hydrated NO_3^- ions in solution. CO is a polar covalent molecule and is hydrated as explained in part a.

14. 2.0 L × 3.0 mol/L = 6.0 mol HCl; the 2.0 L of solution contains 6.0 mol of the solute. HCl is a strong acid; it exists in aqueous solution as separate hydrated H^+ ions and hydrated Cl^- ions. So the solution will contain 6.0 mol of H^+(aq) and 6.0 mol of Cl^- (aq). For the acetic acid solution, $HC_2H_3O_2$ is a weak acid instead of a strong acid. Only some of the 6.0 moles of $HC_2H_3O_2$ molecules will dissociate into H^+(aq) and $C_2H_3O_2^-$(aq). The 2.0 L of 3.0 *M* $HC_2H_3O_2$ solution will contain mostly hydrated $HC_2H_3O_2$ molecules but will also contain some hydrated H^+ ions and hydrated $C_2H_3O_2^-$ ions.

15. Only statement b is true. A concentrated solution can also contain a nonelectrolyte dissolved in water, e.g., concentrated sugar water. Acids are either strong or weak electrolytes. Some ionic compounds are not soluble in water, so they are not labeled as a specific type of electrolyte.

16. One mole of NaOH dissolved in 1.00 L of solution will produce 1.00 *M* NaOH. First, weigh out 40.00 g of NaOH (1.000 mol). Next, add some water to a 1-L volumetric flask (an instrument that is precise to 1.000 L). Dissolve the NaOH in the flask, add some more water, mix, add more water, mix, etc. until water has been added to 1.000-L mark of the volumetric flask. The result is 1.000 L of a 1.000 *M* NaOH solution. Because we know the volume to four significant figures as well as the mass, the molarity will be known to four significant figures. This is good practice, if you need a three-significant-figure molarity, your measurements should be taken to four significant figures.

When you need to dilute a more concentrated solution with water to prepare a solution, again make all measurements to four significant figures to ensure three significant figures in the molarity. Here, we need to cut the molarity in half from 2.00 *M* to 1.00 *M*. We would start with 1 mole of NaOH from the concentrated solution. This would be 500.0 mL of 2.00 *M* NaOH. Add this to a 1-L volumetric flask with addition of more water and mixing until the 1.000-L mark is reached. The resulting solution would be 1.00 *M*.

17. Use the solubility rules in Table 4.1. Some soluble bromides by Rule 2 would be NaBr, KBr, and NH_4Br (there are others). The insoluble bromides by Rule 3 would be AgBr, $PbBr_2$, and Hg_2Br_2. Similar reasoning is used for the other parts to this problem.

Sulfates: Na_2SO_4, K_2SO_4, and $(NH_4)_2SO_4$ (and others) would be soluble, and $BaSO_4$, $CaSO_4$, and $PbSO_4$ (or Hg_2SO_4) would be insoluble.

Hydroxides: NaOH, KOH, $Ca(OH)_2$ (and others) would be soluble, and $Al(OH)_3$, $Fe(OH)_3$, and $Cu(OH)_2$ (and others) would be insoluble.

Phosphates: Na_3PO_4, K_3PO_4, $(NH_4)_3PO_4$ (and others) would be soluble, and Ag_3PO_4, $Ca_3(PO_4)_2$, and $FePO_4$ (and others) would be insoluble.

Lead: $PbCl_2$, $PbBr_2$, PbI_2, $Pb(OH)_2$, $PbSO_4$, and PbS (and others) would be insoluble. $Pb(NO_3)_2$ would be a soluble Pb^{2+} salt.

18. $Pb(NO_3)_2(aq) + 2\ KI(aq) \rightarrow PbI_2(s) + 2\ KNO_3(aq)$ (formula equation)

$Pb^{2+}(aq) + 2\ NO_3^-(aq) + 2\ K^+(aq) + 2\ I^-(aq) \rightarrow PbI_2(s) + 2\ K^+(aq) + 2\ NO_3^-(aq)$
(complete ionic equation)

The 1.0 mol of Pb^{2+} ions would react with the 2.0 mol of I^- ions to form 1.0 mol of the PbI_2 precipitate. Even though the Pb^{2+} and I^- ions are removed, the spectator ions K^+ and NO_3^- are still present. The solution above the precipitate will conduct electricity because there are plenty of charge carriers present in solution.

19. The Brønsted-Lowry definitions are best for our purposes. An acid is a proton donor, and a base is a proton acceptor. A proton is an H^+ ion. Neutral hydrogen has 1 electron and 1 proton, so an H^+ ion is just a proton. An acid-base reaction is the transfer of an H^+ ion (a proton) from an acid to a base.

20. The acid is a diprotic acid (H_2A), meaning that it has two H^+ ions in the formula to donate to a base. The reaction is $H_2A(aq) + 2\ NaOH(aq) \rightarrow 2\ H_2O(l) + Na_2A(aq)$, where A^{2-} is what is left over from the acid formula when the two protons (H^+ ions) are reacted.

For the HCl reaction, the base has the ability to accept two protons. The most common examples are $Ca(OH)_2$, $Sr(OH)_2$, and $Ba(OH)_2$. A possible reaction would be $2\ HCl(aq) + Ca(OH)_2(aq) \rightarrow 2\ H_2O(l) + CaCl_2(aq)$.

21. a. The species reduced is the element that gains electrons. The reducing agent causes reduction to occur by itself being oxidized. The reducing agent generally refers to the entire formula of the compound/ion that contains the element oxidized.

 b. The species oxidized is the element that loses electrons. The oxidizing agent causes oxidation to occur by itself being reduced. The oxidizing agent generally refers to the entire formula of the compound/ion that contains the element reduced.

 c. For simple binary ionic compounds, the actual charges on the ions are the same as the oxidation states. For covalent compounds, nonzero oxidation states are imaginary charges the elements would have if they were held together by ionic bonds (assuming the bond is between two different nonmetals). Nonzero oxidation states for elements in covalent compounds are not actual charges. Oxidation states for covalent compounds are a bookkeeping method to keep track of electrons in a reaction.

22. Reference the Problem Solving Strategy box in Section 4.10 of the text for the steps involved in balancing redox reactions by oxidation states. The key to the oxidation states method is to balance the electrons gained by the species reduced with the number of electrons lost from the species oxidized. This is done by assigning oxidation states and, from the change in oxidation states, determining the coefficients necessary to balance electrons gained with electrons lost. After the loss and gain of electrons is balanced, the remainder of the equation is balanced by inspection.

Exercises

Aqueous Solutions: Strong and Weak Electrolytes

23. a. $NaBr(s) \rightarrow Na^+(aq) + Br^-(aq)$

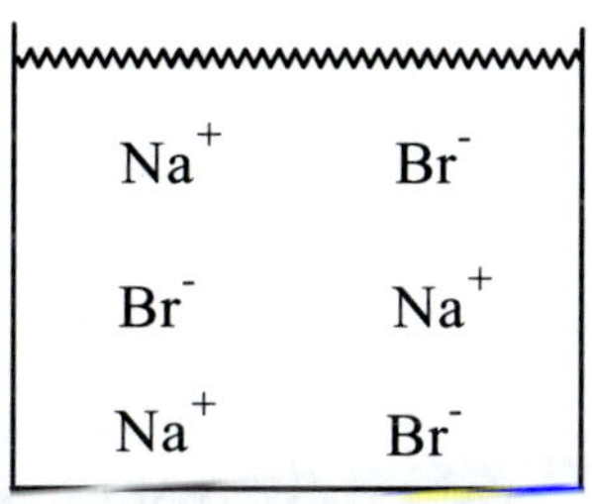

Your drawing should show equal number of Na^+ and Br^- ions.

 b. $MgCl_2(s) \rightarrow Mg^{2+}(aq) + 2\ Cl^-(aq)$

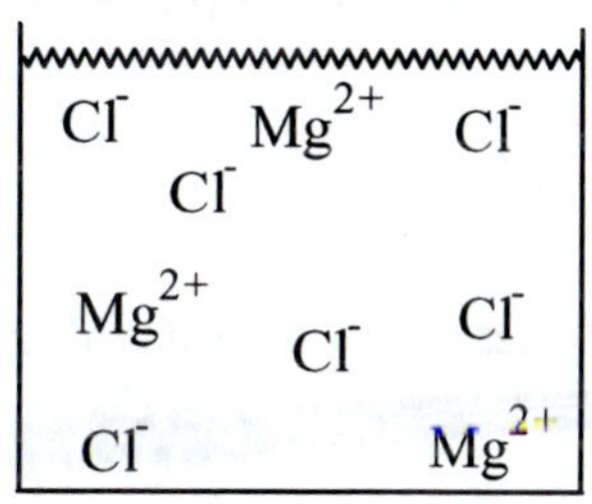

Your drawing should show twice the number of Cl^- ions as Mg^{2+} ions.

c. $Al(NO_3)_3(s) \rightarrow Al^{3+}(aq) + 3\ NO_3^-(aq)$

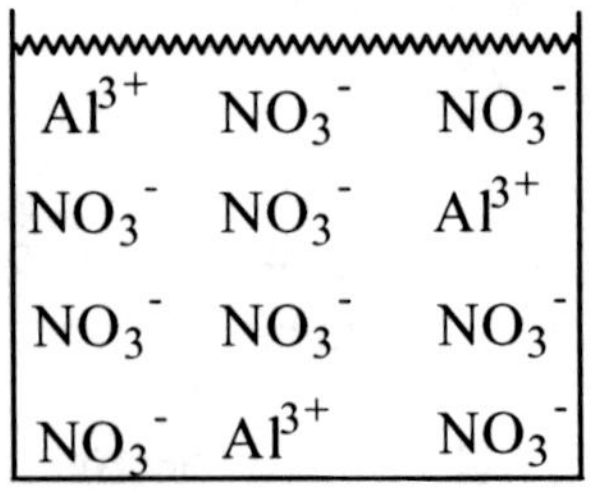

d. $(NH_4)_2SO_4(s) \rightarrow 2\ NH_4^+(aq) + SO_4^{2-}(aq)$

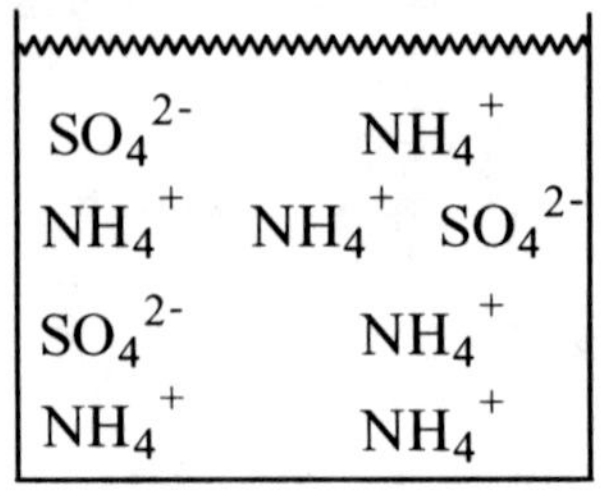

For e-i, your drawings should show equal numbers of the cations and anions present because each salt is a 1 : 1 salt. The ions present are listed in the following dissolution reactions.

e. $NaOH(s) \rightarrow Na^+(aq) + OH^-(aq)$

f. $FeSO_4(s) \rightarrow Fe^{2+}(aq) + SO_4^{2-}(aq)$

g. $KMnO_4(s) \rightarrow K^+(aq) + MnO_4^-\ (aq)$

h. $HClO_4(aq) \rightarrow H^+(aq) + ClO_4^-(aq)$

i. $NH_4C_2H_3O_2(s) \rightarrow NH_4^+(aq) + C_2H_3O_2^-(aq)$

24. a. $Ba(NO_3)_2(aq) \rightarrow Ba^{2+}(aq) + 2\ NO_3^-(aq)$; picture iv represents the Ba^{2+} and NO_3^- ions present in $Ba(NO_3)_2(aq)$.

b. $NaCl(aq) \rightarrow Na^+(aq) + Cl^-(aq)$; picture ii represents NaCl(aq).

c. $K_2CO_3(aq) \rightarrow 2\ K^+(aq) + CO_3^{2-}(aq)$; picture iii represents $K_2CO_3(aq)$.

d. $MgSO_4(aq) \rightarrow Mg^{2+}(aq) + SO_4^{2-}(aq)$; picture i represents $MgSO_4(aq)$.

$HNO_3(aq) \rightarrow H^+(aq) + NO_3^-(aq)$. Picture ii best represents the strong acid HNO_3. Strong acids are strong electrolytes. $HC_2H_3O_2$ only partially dissociates in water; acetic acid is a weak electrolyte. None of the pictures represent weak electrolyte solutions; they all are representations of strong electrolytes.

25. $CaCl_2(s) \rightarrow Ca^{2+}(aq) + 2\ Cl^-(aq)$

26. $MgSO_4(s) \rightarrow Mg^{2+}(aq) + SO_4^{2-}(aq)$; $NH_4NO_3(s) \rightarrow NH_4^+(aq) + NO_3^-(aq)$

Solution Concentration: Molarity

27. a. $5.623\text{ g NaHCO}_3 \times \dfrac{1\text{ mol NaHCO}_3}{84.01\text{ g NaHCO}_3} = 6.693 \times 10^{-2}\text{ mol NaHCO}_3$

$$M = \frac{6.693 \times 10^{-2}\text{ mol}}{250.0\text{ mL}} \times \frac{1000\text{mL}}{\text{L}} = 0.2677\ M\text{ NaHCO}_3$$

b. $0.1846 \text{ g } K_2Cr_2O_7 \times \frac{1 \text{ mol } K_2Cr_2O_7}{294.20 \text{ g } K_2Cr_2O_7} = 6.275 \times 10^{-4} \text{ mol } K_2Cr_2O_7$

$$M = \frac{6.275 \times 10^{-4} \text{ mol}}{500.0 \times 10^{-3} \text{ L}} = 1.255 \times 10^{-3}\ M\ K_2Cr_2O_7$$

c. $0.1025 \text{ g Cu} \times \frac{1 \text{ mol Cu}}{63.55 \text{ g Cu}} = 1.613 \times 10^{-3} \text{ mol Cu} = 1.613 \times 10^{-3} \text{ mol } Cu^{2+}$

$$M = \frac{1.613 \times 10^{-3} \text{ mol } Cu^{2+}}{200.0 \text{ mL}} \times \frac{1000 \text{ mL}}{\text{L}} = 8.065 \times 10^{-3}\ M\ Cu^{2+}$$

28. $75.0 \text{ mL} \times \frac{0.79 \text{ g}}{\text{mL}} \times \frac{1 \text{ mol}}{46.07 \text{ g}} = 1.3 \text{ mol } C_2H_5OH$; molarity $= \frac{1.3 \text{ mol}}{0.250 \text{ L}} = 5.2\ M\ C_2H_5OH$

29. a. $M_{Ca(NO_3)_2} = \frac{0.100 \text{ mol } Ca(NO_3)_2}{0.100 \text{ L}} = 1.00\ M$

$Ca(NO_3)_2(s) \rightarrow Ca^{2+}(aq) + 2\ NO_3^-(aq)$; $M_{Ca^{2+}} = 1.00\ M$; $M_{NO_3^-} = 2(1.00) = 2.00\ M$

b. $M_{Na_2SO_4} = \frac{2.5 \text{ mol } Na_2SO_4}{1.25 \text{ L}} = 2.0\ M$

$Na_2SO_4(s) \rightarrow 2\ Na^+(aq) + SO_4^{2-}(aq)$; $M_{Na^+} = 2(2.0) = 4.0\ M$; $M_{SO_4^{2-}} = 2.0\ M$

c. $5.00 \text{ g } NH_4Cl \times \frac{1 \text{ mol } NH_4Cl}{53.49 \text{ g } NH_4Cl} = 0.0935 \text{ mol } NH_4Cl$

$$M_{NH_4Cl} = \frac{0.0935 \text{ mol } NH_4Cl}{0.5000 \text{ L}} = 0.187\ M$$

$NH_4Cl(s) \rightarrow NH_4^+(aq) + Cl^-(aq)$; $M_{NH_4^+} = M_{Cl^-} = 0.187\ M$

d. $1.00 \text{ g } K_3PO_4 \times \frac{1 \text{ mol } K_3PO_4}{212.27 \text{ g}} = 4.71 \times 10^{-3} \text{ mol } K_3PO_4$

$$M_{K_3PO_4} = \frac{4.71 \times 10^{-3} \text{ mol}}{0.2500 \text{ L}} = 0.0188\ M$$

$K_3PO_4(s) \rightarrow 3\ K^+(aq) + PO_4^{3-}(aq)$; $M_{K^+} = 3(0.0188) = 0.0564\ M$; $M_{PO_4^{3-}} = 0.0188\ M$

30. a. $M_{Na_3PO_4} = \frac{0.0200 \text{ mol}}{0.0100 \text{ L}} = 2.00\ M$

$Na_3PO_4(s) \rightarrow 3\ Na^+(aq) + PO_4^{3-}(aq)$; $M_{Na^+} = 3(2.00) = 6.00\ M$; $M_{PO_4^{3-}} = 2.00\ M$

b. $M_{Ba(NO_3)_2} = \frac{0.300 \text{ mol}}{0.6000 \text{ L}} = 0.500\ M$

$Ba(NO_3)_2(s) \rightarrow Ba^{2+}(aq) + 2\ NO_3^-(aq)$; $M_{Ba^{2+}} = 0.500\ M$; $M_{NO_3^-} = 2(0.500) = 1.00\ M$

c. $$M_{KCl} = \frac{1.00 \text{ g KCl} \times \frac{1 \text{ mol KCl}}{74.55 \text{ g KCl}}}{0.5000 \text{ L}} = 0.0268\ M$$

$KCl(s) \rightarrow K^+(aq) + Cl^-(aq)$; $M_{K^+} = M_{Cl^-} = 0.0268\ M$

d. $$M_{(NH_4)_2SO_4} = \frac{132 \text{ g } (NH_4)_2SO_4 \times \frac{1 \text{ mol } (NH_4)_2SO_4}{132.15 \text{ g}}}{1.50 \text{ L}} = 0.666\ M$$

$(NH_4)_2SO_4(s) \rightarrow 2\ NH_4^+(aq) + SO_4^{2-}(aq)$

$M_{NH_4^+} = 2(0.666) = 1.33\ M$; $M_{SO_4^{2-}} = 0.666\ M$

31. Mol solute = volume (L) × molarity$\left(\frac{\text{mol}}{\text{L}}\right)$; $AlCl_3(s) \rightarrow Al^{3+}(aq) + 3\ Cl^-(aq)$

$$\text{Mol } Cl^- = 0.1000 \text{ L} \times \frac{0.30 \text{ mol } AlCl_3}{\text{L}} \times \frac{3 \text{ mol } Cl^-}{\text{mol } AlCl_3} = 9.0 \times 10^{-2} \text{ mol } Cl^-$$

$MgCl_2(s) \rightarrow Mg^{2+}(aq) + 2\ Cl^-(aq)$

$$\text{Mol } Cl^- = 0.0500 \text{ L} \times \frac{0.60 \text{ mol } MgCl_2}{\text{L}} \times \frac{2 \text{ mol } Cl^-}{\text{mol } MgCl_2} = 6.0 \times 10^{-2} \text{ mol } Cl^-$$

$NaCl(s) \rightarrow Na^+(aq) + Cl^-(aq)$

$$\text{Mol } Cl^- = 0.2000 \text{ L} \times \frac{0.40 \text{ mol NaCl}}{\text{L}} \times \frac{1 \text{ mol } Cl^-}{\text{mol NaCl}} = 8.0 \times 10^{-2} \text{ mol } Cl^-$$

100.0 mL of 0.30 M $AlCl_3$ contains the most moles of Cl^- ions.

32. $NaOH(s) \rightarrow Na^+(aq) + OH^-(aq)$, 2 total mol of ions (1 mol Na^+ and 1 mol Cl^-) per mol NaOH.

$$0.1000 \text{ L} \times \frac{0.100 \text{ mol NaOH}}{\text{L}} \times \frac{2 \text{ mol ions}}{\text{mol NaOH}} = 2.0 \times 10^{-2} \text{ mol ions}$$

$BaCl_2(s) \rightarrow Ba^{2+}(aq) + 2\ Cl^-(aq)$, 3 total mol of ions per mol $BaCl_2$.

$$0.0500 \text{ L} \times \frac{0.200 \text{ mol}}{\text{L}} \times \frac{3 \text{ mol ions}}{\text{mol BaCl}_2} = 3.0 \times 10^{-2} \text{ mol ions}$$

$Na_3PO_4(s) \rightarrow 3\ Na^+(aq) + PO_4^{3-}(aq)$, 4 total mol of ions per mol Na_3PO_4.

$$0.0750 \text{ L} \times \frac{0.150 \text{ mol Na}_3\text{PO}_4}{\text{L}} \times \frac{4 \text{ mol ions}}{\text{mol Na}_3\text{PO}_4} = 4.50 \times \text{mol } 10^{-2} \text{ ions}$$

75.0 mL of 0.150 *M* Na_3PO_4 contains the largest number of ions.

33. Molar mass of NaOH = 22.99 + 16.00 + 1.008 = 40.00 g/mol

$$\text{Mass NaOH} = 0.2500 \text{ L} \times \frac{0.400 \text{ mol NaOH}}{\text{L}} \times \frac{40.00 \text{ g NaOH}}{\text{mol NaOH}} = 4.00 \text{ g NaOH}$$

34. $$10. \text{ g AgNO}_3 \times \frac{1 \text{ mol AgNO}_3}{169.9 \text{ g}} \times \frac{1 \text{ L}}{0.25 \text{ mol AgNO}_3} = 0.24 \text{ L} = 240 \text{ mL}$$

35. a. $$2.00 \text{ L} \times \frac{0.250 \text{ mol NaOH}}{\text{L}} \times \frac{40.00 \text{ g NaOH}}{\text{mol NaOH}} = 20.0 \text{ g NaOH}$$

Place 20.0 g NaOH in a 2-L volumetric flask; add water to dissolve the NaOH, and fill to the mark with water, mixing several times along the way.

b. $$2.00 \text{ L} \times \frac{0.250 \text{ mol NaOH}}{\text{L}} \times \frac{1 \text{ L stock}}{1.00 \text{ mol NaOH}} = 0.500 \text{ L}$$

Add 500. mL of 1.00 *M* NaOH stock solution to a 2-L volumetric flask; fill to the mark with water, mixing several times along the way.

c. $$2.00 \text{ L} \times \frac{0.100 \text{ mol K}_2\text{CrO}_4}{\text{L}} \times \frac{194.20 \text{ g K}_2\text{CrO}_4}{\text{mol K}_2\text{CrO}_4} = 38.8 \text{ g K}_2\text{CrO}_4$$

Similar to the solution made in part a, instead using 38.8 g K_2CrO_4.

d. $$2.00 \text{ L} \times \frac{0.100 \text{ mol K}_2\text{CrO}_4}{\text{L}} \times \frac{1 \text{ L stock}}{1.75 \text{ mol K}_2\text{CrO}_4} = 0.114 \text{ L}$$

Similar to the solution made in part b, instead using 114 mL of the 1.75 *M* K_2CrO_4 stock solution.

36. a. $1.00 \text{ L solution} \times \dfrac{0.50 \text{ mol } H_2SO_4}{\text{L}} = 0.50 \text{ mol } H_2SO_4$

$$0.50 \text{ mol } H_2SO_4 \times \frac{1 \text{ L}}{18 \text{ mol } H_2SO_4} = 2.8 \times 10^{-2} \text{ L conc. } H_2SO_4 \text{ or } 28 \text{ mL}$$

Dilute 28 mL of concentrated H_2SO_4 to a total volume of 1.00 L with water. The resulting 1.00 L of solution will be a 0.50 *M* H_2SO_4 solution.

b. We will need 0.50 mol HCl.

$$0.50 \text{ mol HCl} \times \frac{1 \text{ L}}{12 \text{ mol HCl}} = 4.2 \times 10^{-2} \text{ L} = 42 \text{ mL}$$

Dilute 42 mL of concentrated HCl to a final volume of 1.00 L.

c. We need 0.50 mol $NiCl_2$.

$$0.50 \text{ mol } NiCl_2 \times \frac{1 \text{ mol } NiCl_2 \bullet 6H_2O}{\text{mol } NiCl_2} \times \frac{237.69 \text{ g } NiCl_2 \bullet 6H_2O}{\text{mol } NiCl_2 \bullet 6H_2O}$$

$$= 118.8 \text{ g } NiCl_2 \bullet 6H_2O \approx 120 \text{ g}$$

Dissolve 120 g $NiCl_2 \bullet 6H_2O$ in water, and add water until the total volume of the solution is 1.00 L.

d. $1.00 \text{ L} \times \dfrac{0.50 \text{ mol } HNO_3}{\text{L}} = 0.50 \text{ mol } HNO_3$

$$0.50 \text{ mol } HNO_3 \times \frac{1 \text{ L}}{16 \text{ mol } HNO_3} = 0.031 \text{ L} = 31 \text{ mL}$$

Dissolve 31 mL of concentrated reagent in water. Dilute to a total volume of 1.00 L.

e. We need 0.50 mol Na_2CO_3.

$$0.50 \text{ mol } Na_2CO_3 \times \frac{105.99 \text{ g } Na_2CO_3}{\text{mol}} = 53 \text{ g } Na_2CO_3$$

Dissolve 53 g Na_2CO_3 in water, dilute to 1.00 L.

37. $10.8 \text{ g } (NH_4)_2SO_4 \times \dfrac{1 \text{ mol}}{132.15 \text{ g}} = 8.17 \times 10^{-2} \text{ mol } (NH_4)_2SO_4$

$$\text{Molarity} = \frac{8.17 \times 10^{-2} \text{ mol}}{100.0 \text{ mL}} \times \frac{1000 \text{ mL}}{\text{L}} = 0.817 \, M \, (NH_4)_2SO_4$$

Moles of $(NH_4)_2SO_4$ in final solution:

$$10.00 \times 10^{-3}\ \text{L} \times \frac{0.817\ \text{mol}}{\text{L}} = 8.17 \times 10^{-3}\ \text{mol}$$

$$\text{Molarity of final solution} = \frac{8.17 \times 10^{-3}\ \text{mol}}{(10.00 + 50.00)\ \text{mL}} \times \frac{1000\ \text{mL}}{\text{L}} = 0.136\ M\ (NH_4)_2SO_4$$

$(NH_4)_2SO_4(s) \rightarrow 2\ NH_4^+(aq) + SO_4^{2-}(aq)$; $M_{NH_4^+} = 2(0.136) = 0.272\ M$; $M_{SO_4^{2-}} = 0.136\ M$

38. $\text{Molarity} = \dfrac{\text{total mol } HNO_3}{\text{total volume}}$; total volume = 0.05000 L + 0.10000 L = 0.15000 L

$$\text{Total mol } HNO_3 = 0.05000\ \text{L} \times \frac{0.100\ \text{mol } HNO_3}{\text{L}} + 0.10000\ \text{L} \times \frac{0.200\ \text{mol } HNO_3}{\text{L}}$$

Total mol HNO_3 = 5.00×10^{-3} mol + 2.00×10^{-2} mol = 2.50×10^{-2} mol HNO_3

$$\text{Molarity} = \frac{2.50 \times 10^{-2}\ \text{mol } HNO_3}{0.15000\ \text{L}} = 0.167\ M\ HNO_3$$

As expected, the molarity of HNO_3 is between 0.100 M and 0.200 M.

39. $\text{Mol } Na_2CO_3 = 0.0700\ \text{L} \times \dfrac{3.0\ \text{mol } Na_2CO_3}{\text{L}} = 0.21\ \text{mol } Na_2CO_3$

$Na_2CO_3(s) \rightarrow 2\ Na^+(aq) + CO_3^{2-}(aq)$; mol Na^+ = 2(0.21 mol) = 0.42 mol

$$\text{Mol } NaHCO_3 = 0.0300\ \text{L} \times \frac{1.0\ \text{mol } NaHCO_3}{\text{L}} = 0.030\ \text{mol } NaHCO_3$$

$NaHCO_3(s) \rightarrow Na^+(aq) + HCO_3^-(aq)$; mol Na^+ = 0.030 mol

$$M_{Na^+} = \frac{\text{total mol } Na^+}{\text{total volume}} = \frac{0.42\ \text{mol} + 0.030\ \text{mol}}{0.0700\ \text{L} + 0.0300\ \text{L}} = \frac{0.45\ \text{mol}}{0.1000\ \text{L}} = 4.5\ M\ Na^+$$

40. $\text{Mol } CoCl_2 = 0.0500\ \text{L} \times \dfrac{0.250\ \text{mol } CoCl_2}{\text{L}} = 0.0125\ \text{mol}$

$$\text{Mol } NiCl_2 = 0.0250\ \text{L} \times \frac{0.350\ \text{mol } NiCl_2}{\text{L}} = 0.00875\ \text{mol}$$

Both $CoCl_2$ and $NiCl_2$ are soluble chloride salts by the solubility rules. A 0.0125-mol aqueous sample of $CoCl_2$ is actually 0.0125 mol Co^{2+} and 2(0.0125 mol) = 0.0250 mol Cl^-. A 0.00875-mol aqueous sample of $NiCl_2$ is actually 0.00875 mol Ni^{2+} and 2(0.00875) = 0.0175 mol Cl^-. The total volume of solution that these ions are in is 0.0500 L + 0.0250 L = 0.0750 L.

$$M_{Co^{2+}} = \frac{0.0125 \text{mol} Co^{2+}}{0.0750 L} = 0.167\ M;\ M_{Ni^{2+}} = \frac{0.00875 \text{mol} Ni^{2+}}{0.0750 L} = 0.117\ M$$

$$M_{Cl^-} = \frac{0.0250 \text{mol} Cl^- + 0.0175 \text{mol} Cl^-}{0.0750 L} = 0.567\ M$$

41. Stock solution = $\frac{10.0 \text{ mg}}{500.0 \text{ mL}} = \frac{10.0 \times 10^{-3} \text{ g}}{500.0 \text{ mL}} = \frac{2.00 \times 10^{-5} \text{ g steroid}}{\text{mL}}$

$$100.0 \times 10^{-6} \text{ L stock} \times \frac{1000 \text{mL}}{\text{L}} \times \frac{2.00 \times 10^{-5} \text{ g steroid}}{\text{mL}} = 2.00 \times 10^{-6} \text{ g steroid}$$

This is diluted to a final volume of 100.0 mL.

$$\frac{2.00 \times 10^{-6} \text{ g steroid}}{100.0 \text{ mL}} \times \frac{1000 \text{mL}}{\text{L}} \times \frac{1 \text{ mol steroid}}{336.43 \text{ g steroid}} = 5.94 \times 10^{-8}\ M \text{ steroid}$$

42. Stock solution:

$$1.584 \text{ g } Mn^{2+} \times \frac{1 \text{ mol } Mn^{2+}}{54.94 \text{ g } Mn^{2+}} = 2.883 \times 10^{-2} \text{ mol } Mn^{2+}$$

$$\text{Molarity} = \frac{2.833 \times 10^{-2} \text{ mol } Mn^{2+}}{1.000 L} = 2.883 \times 10^{-2}\ M$$

Solution A:

$$50.00 \text{ mL} \times \frac{1 \text{ L}}{1000 \text{mL}} \times \frac{2.833 \times 10^{-2} \text{ mol}}{\text{L}} = 1.442 \times 10^{-3} \text{ mol } Mn^{2+}$$

$$\text{Molarity} = \frac{1.442 \times 10^{-3} \text{ mol}}{1000.0 \text{ mL}} \times \frac{1000 \text{mL}}{1 \text{ L}} = 1.442 \times 10^{-3}\ M$$

Solution B:

$$10.0 \text{ mL} \times \frac{1 \text{ L}}{1000 \text{mL}} \times \frac{1.442 \times 10^{-3} \text{ mol}}{\text{L}} = 1.442 \times 10^{-5} \text{ mol } Mn^{2+}$$

$$\text{Molarity} = \frac{1.442 \times 10^{-5} \text{ mol}}{0.2500 L} = 5.768 \times 10^{-5}\ M$$

Solution C:

$$10.00 \times 10^{-3} \text{ L} \times \frac{5.768 \times 10^{-5} \text{ mol}}{\text{L}} = 5.768 \times 10^{-7} \text{ mol } Mn^{2+}$$

$$\text{Molarity} = \frac{5.768 \times 10^{-7} \text{ mol}}{0.5000 L} = 1.154 \times 10^{-6}\ M$$

Precipitation Reactions

43. The solubility rules referenced in the following answers are outlined in Table 4.1 of the text.

a. Soluble: Most nitrate salts are soluble (Rule 1).

b. Soluble: Most chloride salts are soluble except for Ag^+, Pb^{2+}, and Hg_2^{2+} (Rule 3).

c. Soluble: Most sulfate salts are soluble except for $BaSO_4$, $PbSO_4$, Hg_2SO_4, and $CaSO_4$ (Rule 4.)

d. Insoluble: Most hydroxide salts are only slightly soluble (Rule 5).
Note: We will interpret the phrase "slightly soluble" as meaning insoluble and the phrase "marginally soluble" as meaning soluble. So the marginally soluble hydroxides $Ba(OH)_2$, $Sr(OH)_2$, and $Ca(OH)_2$ will be assumed soluble unless noted otherwise.

e. Insoluble: Most sulfide salts are only slightly soluble (Rule 6). Again, "slightly soluble" is interpreted as "insoluble" in problems like these.

f. Insoluble: Rule 5 (see answer d).

g. Insoluble: Most phosphate salts are only slightly soluble (Rule 6).

44. The solubility rules referenced in the following answers are from Table 4.1 of the text. The phrase "slightly soluble" is interpreted to mean insoluble, and the phrase "marginally soluble" is interpreted to mean soluble.

a. Soluble (Rule 3)

b. Soluble (Rule 1)

c. Inoluble (Rule 4)

d. Soluble (Rules 2 and 3)

e. Insoluble (Rule 6)

f. Insoluble (Rule 5)

g. Insoluble (Rule 6)

h. Soluble (Rule 2)

45. In these reactions, soluble ionic compounds are mixed together. To predict the precipitate, switch the anions and cations in the two reactant compounds to predict possible products; then use the solubility rules in Table 4.1 to predict if any of these possible products are insoluble (are the precipitate). Note that the phrase "slightly soluble" in Table 4.1 is interpreted to mean insoluble, and the phrase "marginally soluble" is interpreted to mean soluble.

a. Possible products = $FeCl_2$ and K_2SO_4; both salts are soluble, so no precipitate forms.

b. Possible products = $Al(OH)_3$ and $Ba(NO_3)_2$; precipitate = $Al(OH)_3(s)$

c. Possible products = $CaSO_4$ and NaCl; precipitate = $CaSO_4(s)$

d. Possible products = KNO_3 and NiS; precipitate = NiS(s)

46. Use Table 4.1 to predict the solubility of the possible products.

a. Possible products = Hg_2SO_4 and $Cu(NO_3)_2$; precipitate = Hg_2SO_4

b. Possible products = $NiCl_2$ and $Ca(NO_3)_2$; both salts are soluble so no precipitate forms.

c. Possible products = KI and $MgCO_3$; precipitate = $MgCO_3$

d. Possible products = NaBr and $Al_2(CrO_4)_3$; precipitate = $Al_2(CrO_4)_3$

47. For the following answers, the balanced formula equation is first, followed by the complete ionic equation, then the net ionic equation.

a. No reaction occurs since all possible products are soluble salts.

b. $2\ Al(NO_3)_3(aq) + 3\ Ba(OH)_2(aq) \rightarrow 2\ Al(OH)_3(s) + 3\ Ba(NO_3)_2(aq)$

$2\ Al^{3+}(aq) + 6\ NO_3^-(aq) + 3\ Ba^{2+}(aq) + 6\ OH^-(aq) \rightarrow$
$2\ Al(OH)_3(s) + 3\ Ba^{2+}(aq) + 6\ NO_3^-(aq)$

$Al^{3+}(aq) + 3\ OH^-(aq) \rightarrow Al(OH)_3(s)$

c. $CaCl_2(aq) + Na_2SO_4(aq) \rightarrow CaSO_4(s) + 2\ NaCl(aq)$

$Ca^{2+}(aq) + 2\ Cl^-(aq) + 2\ Na^+(aq) + SO_4^{2-}(aq) \rightarrow CaSO_4(s) + 2\ Na^+(aq) + 2\ Cl^-(aq)$

$Ca^{2+}(aq) + SO_4^{2-}(aq) \rightarrow CaSO_4(s)$

d. $K_2S(aq) + Ni(NO_3)_2(aq) \rightarrow 2\ KNO_3(aq) + NiS(s)$

$2\ K^+(aq) + S^{2-}(aq) + Ni^{2+}(aq) + 2\ NO_3^-(aq) \rightarrow 2\ K^+(aq) + 2\ NO_3^-(aq) + NiS(s)$

$Ni^{2+}(aq) + S^{2-}(aq) \rightarrow NiS(s)$

48. a. $Hg_2(NO_3)_2(aq) + CuSO_4(aq) \rightarrow Hg_2SO_4(s) + Cu(NO_3)_2(aq)$

$Hg_2^{2+}(aq) + 2\ NO_3^-(aq) + Cu^{2+}(aq) + SO_4^{2-}(aq) \rightarrow Hg_2SO_4(s) + Cu^{2+}(aq) + 2\ NO_3^-(aq)$

$Hg_2^{2+}(aq) + SO_4^{2-}(aq) \rightarrow Hg_2SO_4(s)$

b. No reaction occurs since both possible products are soluble.

c. $K_2CO_3(aq) + MgI_2(aq) \rightarrow 2\ KI(aq) + MgCO_3(s)$

$2\ K^+(aq) + CO_3^{2-}(aq) + Mg^{2+}(aq) + 2I^-(aq) \rightarrow 2\ K^+(aq) + 2\ I^-(aq) + MgCO_3(s)$

$Mg^{2+}(aq) + CO_3^{2-}(aq) \rightarrow MgCO_3(s)$

d. $3\ Na_2CrO_4(aq) + 2\ Al(Br)_3(aq) \rightarrow 6\ NaBr(aq) + Al_2(CrO_4)_3(s)$

$6\ Na^+(aq) + 3\ CrO_4^{2-}(aq) + 2\ Al^{3+}(aq) + 6\ Br^-(aq) \rightarrow 6\ Na^+(aq) + 6\ Br^-(aq) +$
$Al_2(CrO_4)_3(s)$

$2\ Al^{3+}(aq) + 3\ CrO_4^{2-}(aq) \rightarrow Al_2(CrO_4)_3(s)$

49. a. When $CuSO_4(aq)$ is added to $Na_2S(aq)$, the precipitate that forms is CuS(s). Therefore, Na^+ (the gray spheres) and SO_4^{2-} (the bluish green spheres) are the spectator ions.

$CuSO_4(aq) + Na_2S(aq) \rightarrow CuS(s) + Na_2SO_4(aq)$; $Cu^{2+}(aq) + S^{2-}(aq) \rightarrow CuS(s)$

b. When $CoCl_2(aq)$ is added to $NaOH(aq)$, the precipitate that forms is $Co(OH)_2(s)$. Therefore, Na^+ (the gray spheres) and Cl^- (the green spheres) are the spectator ions.

$CoCl_2(aq) + 2\ NaOH(aq) \rightarrow Co(OH)_2(s) + 2\ NaCl(aq)$

$Co^{2+}(aq) + 2\ OH^-(aq) \rightarrow Co(OH)_2(s)$

c. When $AgNO_3(aq)$ is added to $KI(aq)$, the precipitate that forms is $AgI(s)$. Therefore, K^+ (the red spheres) and NO_3^- (the blue spheres) are the spectator ions.

$AgNO_3(aq) + KI(aq) \rightarrow AgI(s) + KNO_3(aq)$; $Ag^+(aq) + I^-(aq) \rightarrow AgI(s)$

50. There are many acceptable choices for spectator ions. We will generally choose Na^+ and NO_3^- as the spectator ions because sodium salts and nitrate salts are usually soluble in water.

a. $Fe(NO_3)_3(aq) + 3\ NaOH(aq) \rightarrow Fe(OH)_3(s) + 3\ NaNO_3(aq)$

b. $Hg_2(NO_3)_2(aq) + 2\ NaCl(aq) \rightarrow Hg_2Cl_2(s) + 2\ NaNO_3(aq)$

c. $Pb(NO_3)_2(aq) + Na_2SO_4(aq) \rightarrow PbSO_4(s) + 2\ NaNO_3(aq)$

d. $BaCl_2(aq) + Na_2CrO_4(aq) \rightarrow BaCrO_4(s) + 2\ NaCl(aq)$

51. a. $(NH_4)_2SO_4(aq) + Ba(NO_3)_2(aq) \rightarrow 2\ NH_4NO_3(aq) + BaSO_4(s)$

$Ba^{2+}(aq) + SO_4^{2-}(aq) \rightarrow BaSO_4(s)$

b. $Pb(NO_3)_2(aq) + 2\ NaCl(aq) \rightarrow PbCl_2(s) + 2\ NaNO_3(aq)$

$Pb^{2+}(aq) + 2\ Cl^-(aq) \rightarrow PbCl_2(s)$

c. Potassium phosphate and sodium nitrate are both soluble in water. No reaction occurs.

d. No reaction occurs because all possible products are soluble.

e. $CuCl_2(aq) + 2\ NaOH(aq) \rightarrow Cu(OH)_2(s) + 2\ NaCl(aq)$

$Cu^{2+}(aq) + 2\ OH^-(aq) \rightarrow Cu(OH)_2(s)$

52. a. $CrCl_3(aq) + 3\ NaOH(aq) \rightarrow Cr(OH)_3(s) + 3\ NaCl(aq)$

$Cr^{3+}(aq) + 3\ OH^-(aq) \rightarrow Cr(OH)_3(s)$

b. $2\ AgNO_3(aq) + (NH_4)_2CO_3(aq) \rightarrow Ag_2CO_3(s) + 2\ NH_4NO_3(aq)$

$2\ Ag^+(aq) + CO_3^{2-}(aq) \rightarrow Ag_2CO_3(s)$

c. $CuSO_4(aq) + Hg_2(NO_3)_2(aq) \rightarrow Cu(NO_3)_2(aq) + Hg_2SO_4(s)$

$Hg_2^{2+}(aq) + SO_4^{2-}(aq) \rightarrow Hg_2SO_4(s)$

d. No reaction occurs because all possible products (SrI_2 and KNO_3) are soluble.

53. Because a precipitate formed with Na_2SO_4, the possible cations are Ba^{2+}, Pb^{2+}, Hg_2^{2+}, and Ca^{2+} (from the solubility rules). Because no precipitate formed with KCl, Pb^{2+} and Hg_2^{2+} cannot be present. Because both Ba^{2+} and Ca^{2+} form soluble chlorides and soluble hydroxides, both these cations could be present. Therefore, the cations could be Ba^{2+} and Ca^{2+} (by the solubility rules in Table 4.1). For students who do a more rigorous study of solubility, Sr^{2+} could also be a possible cation (it forms an insoluble sulfate salt, whereas the chloride and hydroxide salts of strontium are soluble).

54. Because no precipitates formed upon addition of NaCl or Na_2SO_4, we can conclude that Hg_2^{2+} and Ba^{2+} are not present in the sample because Hg_2Cl_2 and $BaSO_4$ are insoluble salts. However, Mn^{2+} may be present since Mn^{2+} does not form a precipitate with either NaCl or Na_2SO_4. A precipitate formed with NaOH; the solution must contain Mn^{2+} because it forms a precipitate with OH^- [$Mn(OH)_2(s)$].

55. $2\ AgNO_3(aq) + Na_2CrO_4(aq) \rightarrow Ag_2CrO_4(s) + 2\ NaNO_3(aq)$

$$0.0750\text{ L} \times \frac{0.100\text{ mol AgNO}_3}{\text{L}} \times \frac{1\text{ mol Na}_2\text{CrO}_4}{2\text{ mol AgNO}_3} \times \frac{161.98\text{ g Na}_2\text{CrO}_4}{\text{mol Na}_2\text{CrO}_4} = 0.607\text{ g Na}_2\text{CrO}_4$$

56. $2\ Na_3PO_4(aq) + 3\ Pb(NO_3)_2(aq) \rightarrow Pb_3(PO_4)_2(s) + 6\ NaNO_3(aq)$

$$0.1500\text{ L} \times \frac{0.250\text{ mol Pb(NO}_3)_2}{\text{L}} \times \frac{2\text{ mol Na}_3\text{PO}_4}{3\text{ mol Pb(NO}_3)_2} \times \frac{1\text{ L Na}_3\text{PO}_4}{0.100\text{ mol Na}_3\text{PO}_4} = 0.250\text{ L}$$

$$= 250.\text{ mL Na}_3\text{PO}_4$$

57. $Al(NO_3)_3(aq) + 3\ KOH(aq) \rightarrow Al(OH)_3(s) + 3\ KNO_3(aq)$

Assuming $Al(NO_3)_3$ is limiting:

$$0.0500\text{ L} \times \frac{0.200\text{ mol Al(NO}_3)_3}{\text{L}} \times \frac{1\text{ mol Al(OH)}_3}{\text{mol Al(NO}_3)_3} \times \frac{78.00\text{ g Al(OH)}_3}{\text{mol Al(OH)}_3} = 0.780\text{ g Al(OH)}_3$$

Assuming KOH is limiting:

$$0.2000\text{ L} \times \frac{0.100\text{ mol KOH}}{\text{L}} \times \frac{1\text{ mol Al(OH)}_3}{3\text{ mol KOH}} \times \frac{78.00\text{ g Al(OH)}_3}{\text{mol Al(OH)}_3} = 0.520\text{ g Al(OH)}_3$$

Because KOH produces the smaller mass of the $Al(OH)_3$ precipitate, KOH is the limiting reagent and 0.520 g $Al(OH)_3$ can form.

58. The balanced equation is $3\ BaCl_2(aq) + Fe_2(SO_4)_3(aq) \rightarrow 3\ BaSO_4(s) + 2\ FeCl_3(aq)$.

$$100.0\text{ mL BaCl}_2 \times \frac{1\text{ L}}{1000\text{ mL}} \times \frac{0.100\text{ mol BaCl}_2}{\text{L}} \times \frac{3\text{ mol BaSO}_4}{3\text{ mol BaCl}_2} \times \frac{233.4\text{ g BaSO}_4}{\text{mol BaSO}_4}$$

$$= 2.33\text{ g BaSO}_4$$

$$100.0 \text{ mL } Fe_2(SO_4)_3 \times \frac{1\text{ L}}{1000\text{mL}} \times \frac{0.100\text{mol } Fe_2(SO_4)_3}{\text{L}} \times \frac{3\text{ mol } BaSO_4}{\text{mol } Fe_2(SO_4)_3} \times \frac{233.4\text{ g } BaSO_4}{\text{mol } BaSO_4} = 7.00\text{ g } BaSO_4$$

The $BaCl_2$ reagent produces the smaller quantity of the $BaSO_4$ precipitate, so $BaCl_2$ is limiting and 2.33 g $BaSO_4$ can form.

59. The reaction is $AgNO_3(aq) + NaBr(aq) \rightarrow AgBr(s) + NaNO_3(aq)$.

Assuming $AgNO_3$ is limiting:

$$100.0\text{ mL } AgNO_3 \times \frac{1\text{ L}}{1000\text{mL}} \times \frac{0.150\text{mol } AgNO_3}{\text{L } AgNO_3} \times \frac{1\text{ mol } AgBr}{\text{mol } AgNO_3} \times \frac{187.8\text{ g } AgBr}{\text{mol } AgBr}$$

$$= 2.82\text{ g } AgBr$$

Assuming NaBr is limiting:

$$20.0\text{ mL NaBr} \times \frac{1\text{ L}}{1000\text{mL}} \times \frac{1.00\text{ mol NaBr}}{\text{L NaBr}} \times \frac{1\text{ mol } AgBr}{\text{mol NaBr}} \times \frac{187.8\text{ g } AgBr}{\text{mol } AgBr}$$

$$= 3.76\text{ g } AgBr$$

The $AgNO_3$ reagent produces the smaller quantity of AgBr, so $AgNO_3$ is limiting and 2.82 g AgBr can form.

60. $2\ AgNO_3(aq) + CaCl_2(aq) \rightarrow 2\ AgCl(s) + Ca(NO_3)_2(aq)$

$$0.1000\text{ L} \times \frac{0.20\text{ mol } AgNO_3}{\text{L}} \times \frac{2\text{ mol } AgCl}{2\text{ mol } AgNO_3} \times \frac{143.4\text{ g } AgCl}{\text{mol } AgCl} = 2.9\text{ g } AgCl$$

$$0.1000\text{ L} \times \frac{0.15\text{ mol } CaCl_2}{\text{L}} \times \frac{2\text{ mol } AgCl}{\text{mol } CaCl_2} \times \frac{143.4\text{ g } AgCl}{\text{mol } AgCl} = 4.3\text{ g } AgCl$$

$AgNO_3$ is limiting (it produces the smaller mass of AgCl) and 2.9 g AgCl can form.

The net ionic equation is $Ag^+(aq) + Cl^-(aq) \rightarrow AgCl(s)$. The ions remaining in solution are the unreacted Cl^- ions and the spectator ions NO_3^- and Ca^{2+} (all Ag^+ is used up in forming AgCl). The moles of each ion present initially (before reaction) can be easily determined from the moles of each reactant. We have 0.1000 L(0.20 mol $AgNO_3$/L) = 0.020 mol $AgNO_3$, which dissolves to form 0.020 mol Ag^+ and 0.020 mol NO_3^-. We also have 0.1000 L(0.15 mol $CaCl_2$/L) = 0.015 mol $CaCl_2$, which dissolves to form 0.015 mol Ca^{2+} and 2(0.015) = 0.030 mol Cl^-. To form the 2.9 g of AgCl precipitate, 0.020 mol Ag^+ will react with 0.020 mol of Cl^- to form 0.020 mol AgCl (which has a mass of 2.9 g).

Mol unreacted Cl^- = 0.030 mol Cl^- initially – 0.020 mol Cl^- reacted

Mol unreacted Cl^- = 0.010 mol Cl^-

$$M_{Cl^-} = \frac{0.010 \text{ mol Cl}^-}{\text{total volume}} = \frac{0.010 \text{ mol Cl}^-}{0.1000 \text{ L} + 0.1000 \text{ L}} = 0.050\ M\ \text{Cl}^-$$

The molarities of the spectator ions are:

$$\frac{0.020 \text{ mol NO}_3^-}{0.2000 \text{ L}} = 0.10\ M\ \text{NO}_3^-;\quad \frac{0.015 \text{ mol Ca}^{2+}}{0.2000 \text{ L}} = 0.075\ M\ \text{Ca}^{2+}$$

61. a. The balanced reaction is 2 KOH(aq) + $Mg(NO_3)_2$(aq) $\rightarrow$ $Mg(OH)_2$(s) + 2 KNO_3(aq).

b. The precipitate is magnesium hydroxide.

c. Assuming KOH is limiting:

$$0.1000 \text{ L KOH} \times \frac{0.200 \text{ mol KOH}}{\text{L KOH}} \times \frac{1 \text{ mol Mg(OH)}_2}{2 \text{ mol KOH}} \times \frac{58.33 \text{ g Mg(OH)}_2}{\text{mol Mg(OH)}_2} = 0.583 \text{ g Mg(OH)}_2$$

Assuming $Mg(NO_3)_2$ is limiting:

$$0.1000 \text{ L Mg(NO}_3)_2 \times \frac{0.200 \text{ mol Mg(NO}_3)_2}{\text{L Mg(NO}_3)_2} \times \frac{1 \text{ mol Mg(OH)}_2}{\text{mol Mg(NO}_3)_2} \times \frac{58.33 \text{ g Mg(OH)}_2}{\text{mol Mg(OH)}_2} = 1.17 \text{ g Mg(OH)}_2$$

The KOH reagent is limiting because it produces the smaller quantity of the $Mg(OH)_2$ precipitate. So 0.583 g $Mg(OH)_2$ can form.

d. The net ionic equation for this reaction is Mg^{2+}(aq) + 2 OH^-(aq) $\rightarrow$ $Mg(OH)_2$(s).

Because KOH is the limiting reagent, all of the OH^- is used up in the reaction. So M_{OH^-} = 0 M. Note that K^+ is a spectator ion, so it is still present in solution after precipitation was complete. Also present will be the excess Mg^{2+} and NO_3^- (the other spectator ion).

$$\text{Total Mg}^{2+} = 0.1000 \text{ L Mg(NO}_3)_2 \times \frac{0.200 \text{ mol Mg(NO}_3)_2}{\text{L Mg(NO}_3)_2} \times \frac{1 \text{ mol Mg}^{2+}}{\text{mol Mg(NO}_3)_2} = 0.0200 \text{ mol Mg}^{2+}$$

$$\text{Mol Mg}^{2+} \text{ reacted} = 0.1000 \text{ L KOH} \times \frac{0.200 \text{ mol KOH}}{\text{L KOH}} \times \frac{1 \text{ mol Mg(NO}_3)_2}{2 \text{ mol KOH}} \times \frac{1 \text{ mol Mg}^{2+}}{\text{mol Mg(NO}_3)_2} = 0.0100 \text{ mol Mg}^{2+}$$

$$M_{Mg^{2+}} = \frac{\text{mol excess Mg}^{2+}}{\text{total volume}} = \frac{(0.0200 - 0.0100) \text{ mol Mg}^{2+}}{0.1000 \text{ L} + 0.1000 \text{ L}} = 5.00 \times 10^{-2}\ M\ \text{Mg}^{2+}$$

The spectator ions are K^+ and NO_3^-. The moles of each are:

$$\text{mol } K^+ = 0.1000 \text{ L KOH} \times \frac{0.200 \text{ mol KOH}}{\text{L KOH}} \times \frac{1 \text{ mol } K^+}{\text{mol KOH}} = 0.0200 \text{ mol } K^+$$

$$\text{mol } NO_3^- = 0.1000 \text{ L } Mg(NO_3)_2 \times \frac{0.200 \text{ mol } Mg(NO_3)_2}{\text{L } Mg(NO_3)_2} \times \frac{2 \text{ mol } NO_3^-}{\text{mol } Mg(NO_3)_2}$$

$$= 0.0400 \text{ mol } NO_3^-$$

The concentrations are:

$$\frac{0.0200 \text{ mol } K^+}{0.2000 \text{ L}} = 0.100\ M\ K^+;\quad \frac{0.0400 \text{ mol } NO_3^-}{0.2000 \text{ L}} = 0.200\ M\ NO_3^-$$

62. a. $Cu(NO_3)_2(aq) + 2\ KOH(aq) \rightarrow Cu(OH)_2(s) + 2\ KNO_3(aq)$

Solution A contains 2.00 L × 2.00 mol/L = 4.00 mol $Cu(NO_3)_2$, and solution B contains 2.00 L × 3.00 mol/L = 6.00 mol KOH. In the picture in the problem, we have 4 formula units of $Cu(NO_3)_2$ (4 Cu^{2+} ions and 8 NO_3^- ions) and 6 formula units of KOH (6 K^+ ions and 6 OH^- ions). With 4 Cu^{2+} ions and 6 OH^- ions present, OH^- is limiting (when all 6 molecules of OH^- react, we only need 3 of the 4 Cu^{2+} ions to react with all of the OH^- present). After reaction, one Cu^{2+} ion remains as 3 $Cu(OH)_2(s)$ formula units form as precipitate. The following drawing summarizes the ions that remain in solution and the relative amount of precipitate that forms. Note that K^+ and NO_3^- ions are spectator ions. In the drawing, V_1 is the volume of solution A or B, and V_2 is the volume of the combined solutions, with $V_2 = 2V_1$. The drawing exaggerates the amount of precipitate that would actually form.

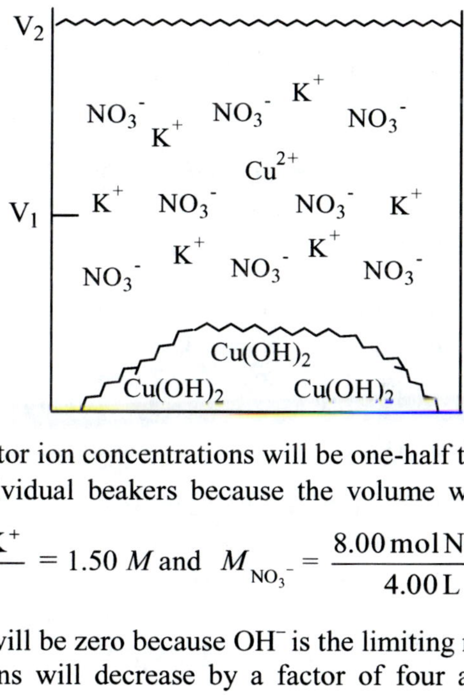

b. The spectator ion concentrations will be one-half the original spectator ion concentrations in the individual beakers because the volume was doubled. Or using moles, $M_{K^+} =$

$\dfrac{6.00 \text{ mol } K^+}{4.00 \text{ L}} = 1.50\ M$ and $M_{NO_3^-} = \dfrac{8.00 \text{ mol } NO_3^-}{4.00 \text{ L}} = 2.00\ M$. The concentration of

OH^- ions will be zero because OH^- is the limiting reagent. From the drawing, the number of Cu^{2+} ions will decrease by a factor of four as the precipitate forms. Because the volume of solution doubled, the concentration of Cu^{2+} ions will decrease by a factor of eight after the two beakers are mixed:

$$M_{Cu^+} = 2.00\ M\left(\frac{1}{8}\right) = 0.250\ M$$

Alternately, one could certainly use moles to solve for $M_{Cu^{2+}}$:

$$\text{Mol } Cu^{2+} \text{ reacted} = 2.00\text{ L} \times \frac{3.00\text{ mol } OH^-}{L} \times \frac{1\text{ mol } Cu^{2+}}{2\text{ mol } OH^-} = 3.00\text{ mol } Cu^{2+} \text{ reacted}$$

$$\text{Mol } Cu^{2+} \text{ present initially} = 2.00\text{ L} \times \frac{2.00\text{ mol } Cu^{2+}}{L} = 4.00\text{ mol } Cu^{2+} \text{ present initially}$$

Excess Cu^{2+} present after reaction = 4.00 mol – 3.00 mol = 1.00 mol Cu^{2+} excess

$$M_{Cu^{2+}} = \frac{1.00\text{ mol } Cu^{2+}}{2.00\text{ L} + 2.00\text{ L}} = 0.250\ M$$

$$\text{Mass of precipitate} = 6.00\text{ mol KOH} \times \frac{1\text{ mol } Cu(OH)_2}{2\text{ mol KOH}} \times \frac{97.57\text{ g } Cu(OH)_2}{\text{mol } Cu(OH)_2} = 293\text{ g } Cu(OH)_2$$

63. $M_2SO_4(aq) + CaCl_2(aq) \rightarrow CaSO_4(s) + 2\ MCl(aq)$

$$1.36\text{ g } CaSO_4 \times \frac{1\text{ mol } CaSO_4}{136.15\text{ g } CaSO_4} \times \frac{1\text{ mol } M_2SO_4}{\text{mol } CaSO_4} = 9.99 \times 10^{-3}\text{ mol } M_2SO_4$$

From the problem, 1.42 g M_2SO_4 was reacted, so:

$$\text{molar mass} = \frac{1.42\text{ g } M_2SO_4}{9.99 \times 10^{-3}\text{ mol } M_2SO_4} = 142\text{ g/mol}$$

142 u = 2(atomic mass M) + 32.07 + 4(16.00), atomic mass M = 23 u

From periodic table, M = Na (sodium).

64. a. Na^+, NO_3^-, Cl^-, and Ag^+ ions are present before any reaction occurs. The excess Ag^+ added will remove all of the Cl^- ions present. Therefore, Na^+, NO_3^-, and the excess Ag^+ ions will all be present after precipitation of AgCl is complete.

b. $Ag^+(aq) + Cl^-(aq) \rightarrow AgCl(s)$

c. $$\text{Mass NaCl} = 0.641\text{ g AgCl} \times \frac{1\text{ mol AgCl}}{143.4\text{ g}} \times \frac{1\text{ mol } Cl^-}{\text{mol AgCl}} \times \frac{1\text{ mol NaCl}}{\text{mol } Cl^-} \times \frac{58.44\text{ g}}{\text{mol NaCl}} = 0.261\text{ g NaCl}$$

$$\text{Mass \% NaCl} = \frac{0.261\text{ g NaCl}}{1.50\text{ g mixture}} \times 100 = 17.4\%\text{ NaCl}$$

Acid-Base Reactions

65. All the bases in this problem are ionic compounds containing OH^-. The acids are either strong or weak electrolytes. The best way to determine if an acid is a strong or weak electrolyte is to memorize all the strong electrolytes (strong acids). Any other acid you encounter that is not a strong acid will be a weak electrolyte (a weak acid), and the formula should be left unaltered in the complete ionic and net ionic equations. The strong acids to recognize are HCl, HBr, HI, HNO_3, $HClO_4$, and H_2SO_4. For the following answers, the order of the equations are formula, complete ionic, and net ionic.

a. $2\ HClO_4(aq) + Mg(OH)_2(s) \rightarrow 2\ H_2O(l) + Mg(ClO_4)_2(aq)$

$2\ H^+(aq) + 2\ ClO_4^-(aq) + Mg(OH)_2(s) \rightarrow 2\ H_2O(l) + Mg^{2+}(aq) + 2\ ClO_4^-(aq)$

$2\ H^+(aq) + Mg(OH)_2(s) \rightarrow 2\ H_2O(l) + Mg^{2+}(aq)$

b. $HCN(aq) + NaOH(aq) \rightarrow H_2O(l) + NaCN(aq)$

$HCN(aq) + Na^+(aq) + OH^-(aq) \rightarrow H_2O(l) + Na^+(aq) + CN^-(aq)$

$HCN(aq) + OH^-(aq) \rightarrow H_2O(l) + CN^-(aq)$

c. $HCl(aq) + NaOH(aq) \rightarrow H_2O(l) + NaCl(aq)$

$H^+(aq) + Cl^-(aq) + Na^+(aq) + OH^-(aq) \rightarrow H_2O(l) + Na^+(aq) + Cl^-(aq)$

$H^+(aq) + OH^-(aq) \rightarrow H_2O(l)$

66. a. $3\ HNO_3(aq) + Al(OH)_3(s) \rightarrow 3\ H_2O(l) + Al(NO_3)_3(aq)$

$3\ H^+(aq) + 3\ NO_3^-(aq) + Al(OH)_3(s) \rightarrow 3\ H_2O(l) + Al^{3+}(aq) + 3\ NO_3^-(aq)$

$3\ H^+(aq) + Al(OH)_3(s) \rightarrow 3\ H_2O(l) + Al^{3+}(aq)$

b. $HC_2H_3O_2(aq) + KOH(aq) \rightarrow H_2O(l) + KC_2H_3O_2(aq)$

$HC_2H_3O_2(aq) + K^+(aq) + OH^-(aq) \rightarrow H_2O(l) + K^+(aq) + C_2H_3O_2^-(aq)$

$HC_2H_3O_2(aq) + OH^-(aq) \rightarrow H_2O(l) + C_2H_3O_2^-(aq)$

c. $Ca(OH)_2(aq) + 2\ HCl(aq) \rightarrow 2\ H_2O(l) + CaCl_2(aq)$

$Ca^{2+}(aq) + 2\ OH^-(aq) + 2\ H^+(aq) + 2\ Cl^-(aq) \rightarrow 2\ H_2O(l) + Ca^{2+}(aq) + 2\ Cl^-(aq)$

$2\ H^+(aq) + 2\ OH^-(aq) \rightarrow 2\ H_2O(l)$ or $H^+(aq) + OH^-(aq) \rightarrow H_2O(l)$

67. All the acids in this problem are strong electrolytes (strong acids). The acids to recognize as strong electrolytes are HCl, HBr, HI, HNO_3, $HClO_4$, and H_2SO_4.

a. $KOH(aq) + HNO_3(aq) \rightarrow H_2O(l) + KNO_3(aq)$

b. $Ba(OH)_2(aq) + 2\ HCl(aq) \rightarrow 2\ H_2O(l) + BaCl_2(aq)$

c. $3\ HClO_4(aq) + Fe(OH)_3(s) \rightarrow 3\ H_2O(l) + Fe(ClO_4)_3(aq)$

d. $AgOH(s) + HBr(aq) \rightarrow AgBr(s) + H_2O(l)$

e. $Sr(OH)_2(aq) + 2\ HI(aq) \rightarrow 2\ H_2O(l) + SrI_2(aq)$

68. a. Perchloric acid plus potassium hydroxide is a possibility.

$HClO_4(aq) + KOH(aq) \rightarrow H_2O(l) + KClO_4(aq)$

b. Nitric acid plus cesium hydroxide is a possibility.

$HNO_3(aq) + CsOH(aq) \rightarrow H_2O(l) + CsNO_3(aq)$

c. Hydroiodic acid plus calcium hydroxide is a possibility.

$2\ HI(aq) + Ca(OH)_2(aq) \rightarrow 2\ H_2O(l) + CaI_2(aq)$

69. If we begin with 50.00 mL of 0.200 *M* NaOH, then:

$$50.00 \times 10^{-3}\,\text{L} \times \frac{0.200\,\text{mol}}{\text{L}} = 1.00 \times 10^{-2}\ \text{mol NaOH is to be neutralized}$$

a. $NaOH(aq) + HCl(aq) \rightarrow NaCl(aq) + H_2O(l)$

$$1.00 \times 10^{-2}\ \text{mol NaOH} \times \frac{1\,\text{mol HCl}}{\text{mol NaOH}} \times \frac{1\,\text{L}}{0.100\,\text{mol}} = 0.100\ \text{L or 100. mL}$$

b. $HNO_3(aq) + NaOH(aq) \rightarrow H_2O(l) + NaNO_3(aq)$

$$1.00 \times 10^{-2}\ \text{mol NaOH} \times \frac{1\,\text{mol HNO}_3}{\text{mol NaOH}} \times \frac{1\,\text{L}}{0.150\,\text{mol HNO}_3} = 6.67 \times 10^{-2}\ \text{L or 66.7 mL}$$

c. $HC_2H_3O_2(aq) + NaOH(aq) \rightarrow H_2O(l) + NaC_2H_3O_2(aq)$

$$1.00 \times 10^{-2}\ \text{mol NaOH} \times \frac{1\,\text{mol HC}_2\text{H}_3\text{O}_2}{\text{mol NaOH}} \times \frac{1\,\text{L}}{0.200\,\text{mol HC}_2\text{H}_3\text{O}_2} = 5.00 \times 10^{-2}\ \text{L}$$

$$= 50.0\ \text{mL}$$

70. We begin with 25.00 mL of 0.200 *M* HCl or 25.00×10^{-3} L × 0.200 mol/L $= 5.00 \times 10^{-3}$ mol HCl.

a. $HCl(aq) + NaOH(aq) \rightarrow H_2O(l) + NaCl(aq)$

$$5.00 \times 10^{-3}\ \text{mol HCl} \times \frac{1\,\text{mol HCl}}{\text{mol NaOH}} \times \frac{1\,\text{L}}{0.100\,\text{mol NaOH}} = 5.00 \times 10^{-2}\ \text{L or 50.0 mL}$$

b. $2\ HCl(aq) + Sr(OH)_2(aq) \rightarrow 2\ H_2O(l) + SrCl_2(aq)$

$$5.00 \times 10^{-3} \text{ mol HCl} \times \frac{1 \text{ mol Sr(OH)}_2}{2 \text{ mol HCl}} \times \frac{1 \text{ L}}{0.0500 \text{ mol Sr(OH)}_2} = 5.00 \times 10^{-2} \text{ L}$$

$$= 50.0 \text{ mL}$$

c. $HCl(aq) + KOH(aq) \rightarrow H_2O(l) + KCl(aq)$

$$5.00 \times 10^{-3} \text{ mol HCl} \times \frac{1 \text{ mol KOH}}{\text{mol HCl}} \times \frac{1 \text{ L}}{0.250 \text{ mol KOH}} = 2.00 \times 10^{-2} \text{ L} = 20.0 \text{ mL}$$

71. $Ba(OH)_2(aq) + 2\ HCl(aq) \rightarrow BaCl_2(aq) + 2\ H_2O(l)$; $H^+(aq) + OH^-(aq) \rightarrow H_2O(l)$

$$75.0 \times 10^{-3} \text{ L} \times \frac{0.250 \text{ mol HCl}}{\text{L}} = 1.88 \times 10^{-2} \text{ mol HCl} = 1.88 \times 10^{-2} \text{ mol H}^+$$

$$+ 1.88 \times 10^{-2} \text{ mol Cl}^-$$

$$225.0 \times 10^{-3} \text{ L} \times \frac{0.0550 \text{ mol Ba(OH)}_2}{\text{L}} = 1.24 \times 10^{-2} \text{ mol Ba(OH)}_2$$

$$= 1.24 \times 10^{-2} \text{ mol Ba}^{2+} + 2.48 \times 10^{-2} \text{ mol OH}^-$$

The net ionic equation requires a 1 : 1 mole ratio between OH^- and H^+. The actual mole OH^- to mole H^+ ratio is greater than 1 : 1, so OH^- is in excess. Because 1.88×10^{-2} mol OH^- will be neutralized by the H^+, we have $(2.48 - 1.88) \times 10^{-2} = 0.60 \times 10^{-2}$ mol OH^- in excess.

$$M_{OH^-} = \frac{\text{mol OH}^- \text{ excess}}{\text{total volume}} = \frac{6.0 \times 10^{-3} \text{ mol OH}^-}{0.0750 \text{ L} + 0.2250 \text{ L}} = 2.0 \times 10^{-2}\ M \text{ OH}^-$$

72. HCl and HNO_3 are strong acids; $Ca(OH)_2$ and RbOH are strong bases. The net ionic equation that occurs is $H^+(aq) + OH^-(aq) \rightarrow H_2O(l)$.

$$\text{Mol H}^+ = 0.0500 \text{ L} \times \frac{0.100 \text{ mol HCl}}{\text{L}} \times \frac{1 \text{ mol H}^+}{\text{mol HCl}}$$

$$+ 0.1000 \text{ L} \times \frac{0.200 \text{ mol HNO}_3}{\text{L}} \times \frac{1 \text{ mol H}^+}{\text{mol HNO}_3} = 0.00500 + 0.0200 = 0.0250 \text{ mol H}^+$$

$$\text{Mol OH}^- = 0.5000 \text{ L} \times \frac{0.0100 \text{ mol Ca(OH)}_2}{\text{L}} \times \frac{2 \text{ mol OH}^-}{\text{mol Ca(OH)}_2}$$

$$+ 0.2000 \text{ L} \times \frac{0.100 \text{ mol RbOH}}{\text{L}} \times \frac{1 \text{ mol OH}^-}{\text{mol RbOH}} = 0.0100 + 0.0200 = 0.0300 \text{ mol OH}^-$$

We have an excess of OH^-, so the solution is basic (not neutral). The moles of excess OH^- = 0.0300 mol OH^- initially – 0.0250 mol OH^- reacted (with H^+) = 0.0050 mol OH^- excess.

$$M_{OH^-} = \frac{0.0050 \text{ mol OH}^-}{(0.0500 + 0.1000 + 0.5000 + 0.2000) \text{ L}} = \frac{0.0050 \text{ mol}}{0.8500 \text{ L}} = 5.9 \times 10^{-3}\ M$$

73. $HCl(aq) + NaOH(aq) \rightarrow H_2O(l) + NaCl(aq)$

$$24.16 \times 10^{-3}\text{ L NaOH} \times \frac{0.106\text{ mol NaOH}}{\text{L NaOH}} \times \frac{1\text{ mol HCl}}{\text{mol NaOH}} = 2.56 \times 10^{-3}\text{ mol HCl}$$

$$\text{Molarity of HCl} = \frac{2.56 \times 10^{-3}\text{ mol}}{25.00 \times 10^{-3}\text{ L}} = 0.102\ M\text{ HCl}$$

74. $HC_2H_3O_2(aq) + NaOH(aq) \rightarrow H_2O(l) + NaC_2H_3O_2(aq)$

a. $$16.58 \times 10^{-3}\text{ L soln} \times \frac{0.5062\text{ mol NaOH}}{\text{L soln}} \times \frac{1\text{ mol }HC_2H_3O_2}{\text{mol NaOH}}$$

$$= 8.393 \times 10^{-3}\text{ mol }HC_2H_3O_2$$

$$\text{Concentration of }HC_2H_3O_2(aq) = \frac{8.393 \times 10^{-3}\text{ mol}}{0.01000\text{L}} = 0.8393\ M$$

b. If we have 1.000 L of solution: Total mass = $1000.\text{ mL} \times \frac{1.006\text{ g}}{\text{mL}} = 1006\text{ g solution}$

$$\text{Mass of }HC_2H_3O_2 = 0.8393\text{ mol} \times \frac{60.05\text{ g}}{\text{mol}} = 50.40\text{ g }HC_2H_3O_2$$

$$\text{Mass \% acetic acid} = \frac{50.40\text{ g}}{1006\text{g}} \times 100 = 5.010\%$$

75. $2\ HNO_3(aq) + Ca(OH)_2(aq) \rightarrow 2\ H_2O(l) + Ca(NO_3)_2(aq)$

$$35.00 \times 10^{-3}\text{ L }HNO_3 \times \frac{0.0500\text{ mol }HNO_3}{\text{L }HNO_3} \times \frac{1\text{ mol }Ca(OH)_2}{2\text{ mol }HNO_3} \times \frac{1\text{ L }Ca(OH)_2}{0.0200\text{ mol }Ca(OH)_2}$$

$$= 0.0438\text{ L} = 43.8\text{ mL }Ca(OH)_2$$

76. Strong bases contain the hydroxide ion (OH^-). The reaction that occurs is $H^+ + OH^- \rightarrow H_2O$.

$$0.0120\text{ L} \times \frac{0.150\text{ mol }H^+}{\text{L}} \times \frac{1\text{ mol }OH^-}{\text{mol }H^+} = 1.80 \times 10^{-3}\text{ mol }OH^-$$

The 30.0 mL of the unknown strong base contains 1.80×10^{-3} mol OH^-.

$$\frac{1.80 \times 10^{-3}\text{ mol }OH^-}{0.0300\text{L}} = 0.0600\text{ M }OH^-$$

The unknown base concentration is one-half the concentration of OH^- ions produced from the base, so the base must contain 2 OH^- in each formula unit. The three soluble strong bases that have two OH^- ions in the formula are $Ca(OH)_2$, $Sr(OH)_2$, and $Ba(OH)_2$. These are all possible identities for the strong base.

77. KHP is a monoprotic acid: $NaOH(aq) + KHP(aq) \rightarrow H_2O(l) + NaKP(aq)$

$$\text{Mass KHP} = 0.02046 \text{ L NaOH} \times \frac{0.1000 \text{ mol NaOH}}{\text{L NaOH}} \times \frac{1 \text{ mol KHP}}{\text{mol NaOH}} \times \frac{204.22 \text{ g KHP}}{\text{mol KHP}}$$

$$= 0.4178 \text{ g KHP}$$

78. Because KHP is a monoprotic acid, the reaction is (KHP is an abbreviation for potassium hydrogen phthalate):

$$NaOH(aq) + KHP(aq) \rightarrow NaKP(aq) + H_2O(l)$$

$$0.1082 \text{ g KHP} \times \frac{1 \text{ mol KHP}}{204.22 \text{ g KHP}} \times \frac{1 \text{ mol NaOH}}{\text{mol KHP}} = 5.298 \times 10^{-4} \text{ mol NaOH}$$

There are 5.298×10^{-4} mol of sodium hydroxide in 34.67 mL of solution. Therefore, the concentration of sodium hydroxide is:

$$\frac{5.298 \times 10^{-4} \text{ mol}}{34.67 \times 10^{-3} \text{ L}} = 1.528 \times 10^{-2}\ M \text{ NaOH}$$

Oxidation-Reduction Reactions

79. Apply the rules in Table 4.2.

a. $KMnO_4$ is composed of K^+ and MnO_4^- ions. Assign oxygen an oxidation state of −2, which gives manganese a +7 oxidation state because the sum of oxidation states for all atoms in MnO_4^- must equal the 1− charge on MnO_4^-. K, +1; O, −2; Mn, +7.

b. Assign O a −2 oxidation state, which gives nickel a +4 oxidation state. Ni, +4; O, −2.

c. $Na_4Fe(OH)_6$ is composed of Na^+ cations and $Fe(OH)_6^{4-}$ anions. $Fe(OH)_6^{4-}$ is composed of an iron cation and 6 OH^- anions. For an overall anion charge of 4−, iron must have a +2 oxidation state. As is usually the case in compounds, assign O a −2 oxidation state and H a +1 oxidation state. Na, +1; Fe, +2; O, −2; H, +1.

d. $(NH_4)_2HPO_4$ is made of NH_4^+ cations and HPO_4^{2-} anions. Assign +1 as the oxidation state of H and −2 as the oxidation state of O. In NH_4^+, $x + 4(+1) = +1$, $x = -3$ = oxidation state of N. In HPO_4^{2-}, $+1 + y + 4(-2) = -2$, $y = +5$ = oxidation state of P.

e. O, −2; P, +3

f. O, −2; $3x + 4(-2) = 0$, $x = +8/3$ = oxidation state of Fe; this is the average oxidation state of the three iron ions in Fe_3O_4. In the actual formula unit, there are two Fe^{3+} ions and one Fe^{2+} ion.

g. O, −2; F, −1; Xe, +6

h. F, −1; S, +4

i. O, −2; C, +2

j. H, +1; O, −2; C, 0

80. a. UO_2^{2+}: O, −2; for U, $x + 2(-2) = +2$, $x = \underline{+6}$

b. As_2O_3: O, −2; for As, $2(x) + 3(-2) = 0$, $x = \underline{+3}$

c. $NaBiO_3$: Na, +1; O, −2; for Bi, $+1 + x + 3(-2) = 0$, $x = \underline{+5}$

d. As_4: As, 0

e. $HAsO_2$: Assign H = +1 and O = −2; for As, $+1 + x + 2(-2) = 0$, $x = \underline{+3}$

f. $Mg_2P_2O_7$: Composed of Mg^{2+} ions and $P_2O_7^{4-}$ ions. Mg, +2; O, −2; P, +5

g. $Na_2S_2O_3$: Composed of Na^+ ions and $S_2O_3^{2-}$ ions. Na, +1; O, −2; S, +2

h. Hg_2Cl_2: Hg, +1; Cl, −1

i. $Ca(NO_3)_2$: Composed of Ca^{2+} ions and NO_3^- ions. Ca, +2; O, −2; N, +5

81. a. −3 b. −3 c. $2(x) + 4(+1) = 0$, $x = -2$

d. +2 e. +1 f. +4

g. +3 h. +5 i. 0

82. a. $SrCr_2O_7$: Composed of Sr^{2+} and $Cr_2O_7^{2-}$ ions. Sr, +2; O, −2; Cr, $2x + 7(-2) = -2$, $x = +6$

b. Cu, +2; Cl, −1 c. O, 0; d. H, +1; O, −1

e. Mg^{2+} and CO_3^{2-} ions present. Mg, +2; O, −2; C, +4; f. Ag, 0

g. Pb^{2+} and SO_3^{2-} ions present. Pb, +2; O, −2; S, +4; h. O, −2; Pb, +4

i. Na^+ and $C_2O_4^{2-}$ ions present. Na, +1; O, −2; C, $2x + 4(-2) = -2$, $x = +3$

j. O, −2; C, +4

k. Ammonium ion has a 1+ charge (NH_4^+), and sulfate ion has a 2− charge (SO_4^{2-}). Therefore, the oxidation state of cerium must be +4 (Ce^{4+}). H, +1; N, −3; O, −2; S, +6

l. O, −2; Cr, +3

83. To determine if the reaction is an oxidation-reduction reaction, assign oxidation states. If the oxidation states change for some elements, then the reaction is a redox reaction. If the oxidation states do not change, then the reaction is not a redox reaction. In redox reactions, the species oxidized (called the reducing agent) shows an increase in oxidation states, and the species reduced (called the oxidizing agent) shows a decrease in oxidation states.

	Redox?	Oxidizing Agent	Reducing Agent	Substance Oxidized	Substance Reduced
a.	Yes	Ag^+	Cu	Cu	Ag^+
b.	No	–	–	–	–
c.	No	–	–	–	–
d.	Yes	$SiCl_4$	Mg	Mg	$SiCl_4$ (Si)
e.	No	–	–	–	–

In b, c, and e, no oxidation numbers change.

84. The species oxidized shows an increase in oxidation states and is called the reducing agent. The species reduced shows a decrease in oxidation states and is called the oxidizing agent. The pertinent oxidation states are listed by the substance oxidized and the substance reduced.

	Redox?	Oxidizing Agent	Reducing Agent	Substance Oxidized	Substance Reduced
a.	Yes	H_2O	CH_4	CH_4 (C, $-4 \rightarrow +2$)	H_2O (H, $+1 \rightarrow 0$)
b	Yes	$AgNO_3$	Cu	Cu ($0 \rightarrow +2$)	$AgNO_3$ (Ag, $+1 \rightarrow 0$)
c.	Yes	HCl	Zn	Zn ($0 \rightarrow +2$)	HCl (H, $+1 \rightarrow 0$)

d. No; there is no change in any of the oxidation numbers.

85. Each sodium atom goes from the 0 oxidation state in Na to the +1 oxidation state in NaF. Each Na atom loses one electron. Each fluorine atom goes from the 0 oxidation state in F_2 to the −1 state in NaF. In order to match electrons gained by fluorine with electrons lost by sodium, 1 F atom is needed for every Na atom in the balanced equation. Because F_2 contains two fluorine atoms, two sodium atoms will be needed to balance the electrons. The following balanced equation makes sense from an atom standpoint but also makes sense from an electron standpoint.

$$2\,Na(s) + F_2(g) \rightarrow 2\,NaF(s)$$

86. Each oxygen atom goes from the 0 oxidation state in O_2 to the −2 oxidation state in MgO. Each magnesium atom goes from the 0 oxidation state in Mg to the +2 oxidation state in MgO. To match electron gain with electron loss, 1 atom of O is needed for each atom of Mg in the balanced equation. Because two oxygen atoms are in each O_2 molecule, we will need two Mg atoms for every O_2 molecule. The balanced equation below balances atoms but also balances electrons, which must always be the case in any correctly balanced equation.

$$2\,Mg(s) + O_2(g) \rightarrow 2\,MgO(s)$$

87. a. The first step is to assign oxidation states to all atoms (see numbers above the atoms).

$$\overset{-3\ +1}{C_2H_6} + \overset{0}{O_2} \rightarrow \overset{+4\ -2}{CO_2} + \overset{+1\ -2}{H_2O}$$

Each carbon atom changes from −3 to +4, an increase of 7. Each oxygen atom changes from 0 to −2, a decrease of 2. We need 7/2 O atoms for every C atom in order to balance electron gain with electron loss.

$C_2H_6 + 7/2\ O_2 \rightarrow CO_2 + H_2O$

Balancing the remainder of the equation by inspection:

$$C_2H_6(g) + 7/2\ O_2(g) \rightarrow 2\ CO_2(g) + 3\ H_2O(g)$$

or

$$2\ C_2H_6(g) + 7\ O_2(g) \rightarrow 4\ CO_2(g) + 6\ H_2O(g)$$

b. The oxidation state of magnesium changes from 0 to +2, an increase of 2. The oxidation state of hydrogen changes from +1 to 0, a decrease of 1. We need 2 H atoms for every Mg atom in order to balance the electrons transferred. The balanced equation is:

$$Mg(s) + 2\ HCl(aq) \rightarrow Mg^{2+}(aq) + 2\ Cl^-(aq) + H_2(g)$$

c. The oxidation state of nickel increases by 2 (0 to +2), and the oxidation state of cobalt decreases by 1 (+3 to +2). We need 2 Co^{3+} ions for every Ni atom in order to balance electron gain with electron loss. The balanced equation is:

$$Ni(s) + 2\ Co^{3+}(aq) \rightarrow Ni^{2+}(aq) + 2\ Co^{2+}(aq)$$

d. The equation is balanced (mass and charge balanced). Each hydrogen atom gains one electron (+1 → 0), and each zinc atom loses two electrons (0 → +2). We need 2 H atoms for every Zn atom in order to balance the electrons transferred. This is the ratio in the given equation:

$$Zn(s) + H_2SO_4(aq) \rightarrow ZnSO_4(aq) + H_2(g)$$

88. a. The first step is to assign oxidation states to all atoms (see numbers above the atoms).

$$\overset{0}{Cl_2} + \overset{0}{Al} \rightarrow \overset{+3}{Al^{3+}} + \overset{-1}{Cl^-}$$

Each aluminum atom changes in oxidation state from 0 to +3, an increase of 3. Each chlorine atom changes from 0 to −1, a decrease of 1. We need 3 Cl atoms for every Al atom in the balanced equation in order to balance electron gain with electron loss.

$3/2\ Cl_2 + Al \rightarrow Al^{3+} + 3\ Cl^-$

For whole numbers, multiply through by two. The balanced equation is:

$$3\ Cl_2(g) + 2\ Al(s) \rightarrow 2Al^{3+}(aq) + 6\ Cl^-(aq)$$

b. $$\overset{0}{O_2} + \overset{+1\ -2}{H_2O} + \overset{0}{Pb} \rightarrow \overset{+2\ -2\ +1}{Pb(OH)_2}$$

From the oxidation states written above the elements, lead is oxidized, and oxygen in O_2 is reduced. Each lead atom changes from 0 to +2, an increase of 2, and each O atom in

O_2 changes from 0 to −2, a decrease of 2. We need 1 Pb atom for each O atom in O_2 to balance the electrons transferred. Balancing the electrons:

$$O_2 + H_2O + 2\ Pb \rightarrow 2\ Pb(OH)_2$$

The last step is to balance the rest of the equation by inspection. In this reaction, when the H atoms become balanced, the entire equation is balanced. The balanced overall equation is:

$$O_2(g) + 2\ H_2O(l) + 2\ Pb(s) \rightarrow 2\ Pb(OH)_2(s)$$

c. $\overset{+1}{H^+} + \overset{+7\ -2}{MnO_4^-} + \overset{+2}{Fe^{2+}} \rightarrow \overset{+2}{Mn^{2+}} + \overset{+3}{Fe^{3+}} + \overset{+1\ -2}{H_2O}$

From the oxidation states written above each element, manganese is reduced (goes from +7 to +3), and Fe is oxidized (goes from +2 to +3). In order to balance the electrons transferred, we need 5 Fe atoms for every Mn atom. Balancing the electrons gives:

$$H^+ + MnO_4^- + 5\ Fe^{2+} \rightarrow Mn^{2+} + 5\ Fe^{3+} + H_2O$$

Balancing the O atoms, then the H atoms by inspection, leads to the following overall balanced equation.

$$8\ H^+(aq) + MnO_4^-(aq) + 5\ Fe^{2+}(aq) \rightarrow Mn^{2+}(aq) + 5\ Fe^{3+}(aq) + 4\ H_2O(l)$$

Additional Exercises

89. Desired uncertainty is 1% of 0.02, or ±0.0002. So we want the solution to be 0.0200 ± 0.0002 *M,* or the concentration should be between 0.0198 and 0.0202 *M*. We should use a 1-L volumetric flask to make the solution. They are good to ±0.1%. We want to weigh out between 0.0198 mol and 0.0202 mol of KIO_3.

Molar mass of KIO_3 = 39.10 + 126.9 + 3(16.00) = 214.0 g/mol

$$0.0198 \text{ mol} \times \frac{214.0 \text{ g}}{\text{mol}} = 4.237 \text{ g}; \quad 0.0202 \text{ mol} \times \frac{214.0 \text{ g}}{\text{mol}} = 4.323 \text{ g (carrying extra sig. figs.)}$$

We should weigh out between 4.24 and 4.32 g of KIO_3. We should weigh it to the nearest milligram, or nearest 0.1 mg. Dissolve the KIO_3 in water, and dilute (with mixing along the way) to the mark in a 1-L volumetric flask. This will produce a solution whose concentration is within the limits and is known to at least the fourth decimal place.

90. Solution A: $\frac{4 \text{ molecules}}{1.0 \text{ L}}$; solution B: $\frac{6 \text{ molecules}}{4.0 \text{ L}} = \frac{1.5 \text{ molecules}}{1.0 \text{ L}}$

Solution C: $\frac{4 \text{ molecules}}{2.0 \text{ L}} = \frac{2 \text{ molecules}}{1.0 \text{ L}}$; solution D: $\frac{6 \text{ molecules}}{2.0 \text{ L}} = \frac{3 \text{ molecules}}{1.0 \text{ L}}$

Solution A has the most molecules per unit volume so solution A is most concentrated. This is followed by solution D, then solution C. Solution B has the fewest molecules per unit volume, so solution B is least concentrated.

91. $32.0 \text{ g } C_{12}H_{22}O_{11} \times \dfrac{1 \text{ mol } C_{12}H_{22}O_{11}}{342.30 \text{ g}} = 0.0935 \text{ mol } C_{12}H_{22}O_{11}$ added to blood

The blood sugar level would increase by:

$$\frac{0.0935 \text{ mol } C_{12}H_{22}O_{11}}{5.0 \text{ L}} = 0.019 \text{ mol/L}$$

92. Mol $CaCl_2$ present $= 0.230 \text{ L } CaCl_2 \times \dfrac{0.275 \text{ mol } CaCl_2}{\text{L } CaCl_2} = 6.33 \times 10^{-2} \text{ mol } CaCl_2$

The volume of $CaCl_2$ solution after evaporation is:

$$6.33 \times 10^{-2} \text{ mol } CaCl_2 \times \frac{1 \text{ L } CaCl_2}{1.10 \text{ mol } CaCl_2} = 5.75 \times 10^{-2} \text{ L} = 57.5 \text{ mL } CaCl_2$$

Volume H_2O evaporated = 230. mL – 57.5 mL = 173 mL H_2O evaporated

93. There are other possible correct choices for most of the following answers. We have listed only three possible reactants in each case.

a. $AgNO_3$, $Pb(NO_3)_2$, and $Hg_2(NO_3)_2$ would form precipitates with the Cl^- ion.

$Ag^+(aq) + Cl^-(aq) \rightarrow AgCl(s)$; $Pb^{2+}(aq) + 2\ Cl^-(aq) \rightarrow PbCl_2(s)$

$Hg_2^{2+}(aq) + 2\ Cl^-(aq) \rightarrow Hg_2Cl_2(s)$

b. Na_2SO_4, Na_2CO_3, and Na_3PO_4 would form precipitates with the Ca^{2+} ion.

$Ca^{2+}(aq) + SO_4^{2-}(aq) \rightarrow CaSO_4(s)$; $Ca^{2+}(aq) + CO_3^{2-}(aq) \rightarrow CaCO_3(s)$

$3\ Ca^{2+}(aq) + 2\ PO_4^{3-}(aq) \rightarrow Ca_3(PO_4)_2(s)$

c. NaOH, Na_2S, and Na_2CO_3 would form precipitates with the Fe^{3+} ion.

$Fe^{3+}(aq) + 3\ OH^-(aq) \rightarrow Fe(OH)_3(s)$; $2\ Fe^{3+}(aq) + 3\ S^{2-}(aq) \rightarrow Fe_2S_3(s)$

$2\ Fe^{3+}(aq) + 3\ CO_3^{2-}(aq) \rightarrow Fe_2(CO_3)_3(s)$

d. $BaCl_2$, $Pb(NO_3)_2$, and $Ca(NO_3)_2$ would form precipitates with the SO_4^{2-} ion.

$Ba^{2+}(aq) + SO_4^{2-}(aq) \rightarrow BaSO_4(s)$; $Pb^{2+}(aq) + SO_4^{2-}(aq) \rightarrow PbSO_4(s)$

$Ca^{2+}(aq) + SO_4^{2-}(aq) \rightarrow CaSO_4(s)$

e. Na_2SO_4, NaCl, and NaI would form precipitates with the Hg_2^{2+} ion.

$Hg_2^{2+}(aq) + SO_4^{2-}(aq) \rightarrow Hg_2SO_4(s)$; $Hg_2^{2+}(aq) + 2\ Cl^-(aq) \rightarrow Hg_2Cl_2(s)$

$Hg_2^{2+}(aq) + 2\ I^-(aq) \rightarrow Hg_2I_2(s)$

f. NaBr, Na_2CrO_4, and Na_3PO_4 would form precipitates with the Ag^+ ion.

$Ag^+(aq) + Br^-(aq) \rightarrow AgBr(s)$; $2\ Ag^+(aq) + CrO_4^{2-}(aq) \rightarrow Ag_2CrO_4(s)$

$3\ Ag^+(aq) + PO_4^{3-}(aq) \rightarrow Ag_3PO_4(s)$

94. a. $MgCl_2(aq) + 2\ AgNO_3(aq) \rightarrow 2\ AgCl(s) + Mg(NO_3)_2(aq)$

$$0.641\text{ g AgCl} \times \frac{1\text{ mol AgCl}}{143.4\text{ g AgCl}} \times \frac{1\text{ mol MgCl}_2}{2\text{ mol AgCl}} \times \frac{95.21\text{ g}}{\text{mol MgCl}_2} = 0.213\text{ g MgCl}_2$$

$$\frac{0.213\text{ g MgCl}_2}{1.50\text{ g mixture}} \times 100 = 14.2\%\ \text{MgCl}_2$$

b. $$0.213\text{ g MgCl}_2 \times \frac{1\text{ mol MgCl}_2}{95.21\text{ g}} \times \frac{2\text{ mol AgNO}_3}{\text{mol MgCl}_2} \times \frac{1\text{ L}}{0.500\text{ mol AgNO}_3} \times \frac{1000\text{ mL}}{1\text{ L}}$$

$$= 8.95\text{ mL AgNO}_3$$

95. $XCl_2(aq) + 2\ AgNO_3(aq) \rightarrow 2\ AgCl(s) + X(NO_3)_2(aq)$

$$1.38\text{ g AgCl} \times \frac{1\text{ mol AgCl}}{143.4\text{ g}} \times \frac{1\text{ mol XCl}_2}{2\text{ mol AgCl}} = 4.81 \times 10^{-3}\text{ mol XCl}_2$$

$$\frac{1.00\text{ g XCl}_2}{4.91 \times 10^{-3}\text{ mol XCl}_2} = 208\text{ g/mol};\ x + 2(35.45) = 208,\ x = 137\text{ g/mol}$$

From the periodic table, the metal X is barium (Ba).

96. From the periodic table, use aluminum in the formulas to convert from mass of $Al(OH)_3$ to mass of $Al_2(SO_4)_3$ in the mixture.

$$0.107\text{ g Al(OH)}_3 \times \frac{1\text{ mol Al(OH)}_3}{78.00\text{ g}} \times \frac{1\text{ mol Al}^{3+}}{\text{mol Al(OH)}_3} \times \frac{1\text{ mol Al}_2(\text{SO}_4)_3}{2\text{ mol Al}^{3+}}$$

$$\times \frac{342.17\text{ g Al}_2(\text{SO}_4)_3}{\text{mol Al}_2(\text{SO}_4)_3} = 0.235\text{ g Al}_2(\text{SO}_4)_3$$

$$\text{Mass \% Al}_2(\text{SO}_4)_3 = \frac{0.235\text{ g}}{1.45\text{ g}} \times 100 = 16.2\%$$

97. All the Tl in TlI came from Tl in Tl_2SO_4. The conversion from TlI to Tl_2SO_4 uses the molar masses and formulas of each compound.

$$0.1824\text{ g TlI} \times \frac{204.4\text{ g Tl}}{331.3\text{ g TlI}} \times \frac{504.9\text{ g Tl}_2\text{SO}_4}{408.8\text{ g Tl}} = 0.1390\text{ g Tl}_2\text{SO}_4$$

$$\text{Mass \% Tl}_2\text{SO}_4 = \frac{0.1390\text{ g Tl}_2\text{SO}_4}{9.486\text{ g pesticide}} \times 100 = 1.465\%\ \text{Tl}_2\text{SO}_4$$

98. a. $Fe^{3+}(aq) + 3\ OH^-(aq) \rightarrow Fe(OH)_3(s)$

$Fe(OH)_3$: 55.85 + 3(16.00) + 3(1.008) = 106.87 g/mol

$$0.107\text{ g Fe(OH)}_3 \times \frac{55.85\text{ g Fe}}{106.87\text{ g Fe(OH)}_3} = 0.0559\text{ g Fe}$$

b. $Fe(NO_3)_3$: 55.85 + 3(14.01) + 9(16.00) = 241.86 g/mol

$$0.0559\text{ g Fe} \times \frac{241.86\text{ g Fe(NO}_3)_3}{55.85\text{ g Fe}} = 0.242\text{ g Fe(NO}_3)_3$$

c. $$\text{Mass \% Fe(NO}_3)_3 = \frac{0.242\text{ g}}{0.456\text{ g}} \times 100 = 53.1\%$$

99. With the ions present, the only possible precipitate is $Cr(OH)_3$.

$Cr(NO_3)_3(aq) + 3\ NaOH(aq) \rightarrow Cr(OH)_3(s) + 3\ NaNO_3(aq)$

$$\begin{array}{l}\text{Mol NaOH used}\\ \text{to form precipitate}\end{array} = 2.06\text{ g Cr(OH)}_3 \times \frac{1\text{ mol Cr(OH)}_3}{103.02\text{ g}} \times \frac{3\text{ mol NaOH}}{\text{mol Cr(OH)}_3} = 6.00 \times 10^{-2}\text{ mol}$$

$NaOH(aq) + HCl(aq) \rightarrow NaCl(aq) + H_2O(l)$

$$\begin{array}{l}\text{Mol NaOH used}\\ \text{to react with HCl}\end{array} = 0.1000\text{ L} \times \frac{0.400\text{ mol HCl}}{\text{L}} \times \frac{1\text{ mol NaOH}}{\text{mol HCl}} = 4.00 \times 10^{-2}\text{ mol}$$

$$M_{NaOH} = \frac{\text{total mol NaOH}}{\text{volume}} = \frac{6.00 \times 10^{-2}\text{ mol} + 4.00 \times 10^{-2}\text{ mol}}{0.0500\text{ L}} = 2.00\ M\text{ NaOH}$$

100. a. $MgO(s) + 2\ HCl(aq) \rightarrow MgCl_2(aq) + H_2O(l)$

$Mg(OH)_2(s) + 2\ HCl(aq) \rightarrow MgCl_2(aq) + 2\ H_2O(l)$

$Al(OH)_3(s) + 3\ HCl(aq) \rightarrow AlCl_3(aq) + 3\ H_2O(l)$

b. Let's calculate the number of moles of HCl neutralized per gram of substance. We can get these directly from the balanced equations and the molar masses of the substances.

$$\frac{2\text{ mol HCl}}{\text{mol MgO}} \times \frac{1\text{ mol MgO}}{40.31\text{ g MgO}} = \frac{4.962 \times 10^{-2}\text{ mol HCl}}{\text{g MgO}}$$

$$\frac{2\text{ mol HCl}}{\text{mol Mg(OH)}_2} \times \frac{1\text{ mol Mg(OH)}_2}{58.33\text{ g Mg(OH)}_2} = \frac{3.429 \times 10^{-2}\text{ mol HCl}}{\text{g Mg(OH)}_2}$$

$$\frac{3 \text{ mol HCl}}{\text{mol Al(OH)}_3} \times \frac{1 \text{ mol Al(OH)}_3}{78.00 \text{ g Al(OH)}_3} = \frac{3.846 \times 10^{-2} \text{ mol HCl}}{\text{g Al(OH)}_3}$$

Therefore, 1 gram of magnesium oxide would neutralize the most 0.10 *M* HCl.

101. Using HA as an abbreviation for the monoprotic acid acetylsalicylic acid:

$$\text{HA(aq)} + \text{NaOH(aq)} \rightarrow \text{H}_2\text{O(l)} + \text{NaA(aq)}$$

$$\text{Mol HA} = 0.03517 \text{ L NaOH} \times \frac{0.5065 \text{ mol NaOH}}{\text{L NaOH}} \times \frac{1 \text{ mol HA}}{\text{mol NaOH}} = 1.781 \times 10^{-2} \text{ mol HA}$$

Fom the problem, 3.210 g HA was reacted, so:

$$\text{molar mass} = \frac{3.210 \text{ g HA}}{1.781 \times 10^{-2} \text{ mol HA}} = 180.2 \text{ g/mol}$$

102. $\text{Mg(s)} + 2 \text{ HCl(aq)} \rightarrow \text{MgCl}_2\text{(aq)} + \text{H}_2\text{(g)}$

$$3.00 \text{ g Mg} \times \frac{1 \text{ mol Mg}}{24.31 \text{ g Mg}} \times \frac{2 \text{ mol HCl}}{\text{mol Mg}} \times \frac{1 \text{ L}}{5.0 \text{ mol HCl}} = 0.049 \text{ L} = 49 \text{ mL HCl}$$

103. Let HA = unknown monoprotic acid; $\text{HA(aq)} + \text{NaOH(aq)} \rightarrow \text{NaA(aq)} + \text{H}_2\text{O(l)}$

$$\text{Mol HA present} = 0.0250 \text{ L} \times \frac{0.500 \text{ mol NaOH}}{\text{L}} \times \frac{1 \text{ mol HA}}{1 \text{ mol NaOH}} = 0.0125 \text{ mol HA}$$

$$\frac{x \text{ g HA}}{\text{mol HA}} = \frac{2.20 \text{ g HA}}{0.0125 \text{ mol HA}}, \quad x = \text{molar mass of HA} = 176 \text{ g/mol}$$

Empirical formula weight ≈ 3(12) + 4(1) + 3(16) = 88 g/mol.

Because 176/88 = 2.0, the molecular formula is $(\text{C}_3\text{H}_4\text{O}_3)_2 = \text{C}_6\text{H}_8\text{O}_6$.

104. We get the empirical formula from the elemental analysis. Out of 100.00 g carminic acid, there are:

$$53.66 \text{ g C} \times \frac{1 \text{ mol C}}{12.01 \text{ g C}} = 4.468 \text{ mol C}; \quad 4.09 \text{ g H} \times \frac{1 \text{ mol H}}{1.008 \text{ g H}} = 4.06 \text{ mol H}$$

$$42.25 \text{ g O} \times \frac{1 \text{ mol O}}{16.00 \text{ g O}} = 2.641 \text{ mol O}$$

Dividing the moles by the smallest number gives:

$$\frac{4.468}{2.641} = 1.692; \quad \frac{4.06}{2.641} = 1.54$$

These numbers don't give obvious mole ratios. Let's determine the mol C to mol H ratio:

$$\frac{4.468}{4.06} = 1.10 = \frac{11}{10}$$

So let's try $\frac{4.06}{10} = 0.406$ as a common factor: $\frac{4.468}{0.406} = 11.0$; $\frac{4.06}{0.406} = 10.0$; $\frac{2.641}{0.406} = 6.50$

Therefore, $C_{22}H_{20}O_{13}$ is the empirical formula.

We can get molar mass from the titration data. The balanced reaction is $HA(aq) + OH^-(aq) \rightarrow H_2O(l) + A^-(aq)$, where HA is an abbreviation for carminic acid, an acid with one acidic proton (H^+).

$$18.02 \times 10^{-3}\text{ L soln} \times \frac{0.0406\text{ mol NaOH}}{\text{L soln}} \times \frac{1\text{ mol carminic acid}}{\text{mol NaOH}}$$
$$= 7.32 \times 10^{-4}\text{ mol carminic acid}$$

$$\text{Molar mass} = \frac{0.3602\text{ g}}{7.32 \times 10^{-4}\text{ mol}} = \frac{492\text{ g}}{\text{mol}}$$

The empirical formula mass of $C_{22}H_{20}O_{13} \approx 22(12) + 20(1) + 13(16) = 492$ g.

Therefore, the molecular formula of carminic acid is also $C_{22}H_{20}O_{13}$.

105. $$0.104\text{ g AgCl} \times \frac{1\text{ mol AgCl}}{143.4\text{ g AgCl}} \times \frac{1\text{ mol Cl}^-}{\text{mol AgCl}} \times \frac{35.45\text{ g Cl}^-}{\text{mol Cl}^-} = 2.57 \times 10^{-2}\text{ g Cl}^-$$

All of the Cl^- in the AgCl precipitate came from the chlorisondamine chloride compound in the medication. So we need to calculate the quantity of $C_{14}H_{20}Cl_6N_2$ which contains 2.57×10^{-2} g Cl^-.

Molar mass of $C_{14}H_{20}Cl_6N_2 = 14(12.01) + 20(1.008) + 6(35.45) + 2(14.01) = 429.02$ g/mol

There are 6(35.45) = 212.70 g chlorine for every mole (429.02 g) of $C_{14}H_{20}Cl_6N_2$.

$$2.57 \times 10^{-2}\text{ g Cl}^- \times \frac{429.02\text{ g } C_{14}H_{20}Cl_6N_2}{212.70\text{ g Cl}^-} = 5.18 \times 10^{-2}\text{ g } C_{14}H_{20}Cl_6N_2$$

$$\text{Mass \% chlorisondamine chloride} = \frac{5.18 \times 10^{-2}\text{ g}}{1.28\text{ g}} \times 100 = 4.05\%$$

106. All the sulfur in $BaSO_4$ came from the saccharin. The conversion from $BaSO_4$ to saccharin utilizes the molar masses of each compound.

$$0.5032\text{ g } BaSO_4 \times \frac{32.07\text{ g S}}{233.4\text{ g } BaSO_4} \times \frac{183.19\text{ g } C_7H_5NO_3S}{32.07\text{ g S}} = 0.3949\text{ g } C_7H_5NO_3S$$

$$\frac{\text{Average mass}}{\text{Tablet}} = \frac{0.3949\text{ g}}{10\text{ tablets}} = \frac{3.949 \times 10^{-2}\text{ g}}{\text{tablet}} = \frac{39.49\text{ mg}}{\text{tablet}}$$

$$\text{Average mass \%} = \frac{0.3949\text{ g } C_7H_5NO_3S}{0.5894\text{ g}} \times 100 = 67.00\% \text{ saccharin by mass}$$

107. Use the silver nitrate data to calculate the mol Cl^- present, then use the formula of douglasite ($2KCl\bullet FeCl_2\bullet 2H_2O$) to convert from Cl^- to douglasite (1 mole of douglasite contains 4 moles of Cl^-). The net ionic equation is $Ag^+ + Cl^- \rightarrow AgCl(s)$.

$$0.03720\text{ L} \times \frac{0.1000\text{ mol }Ag^+}{\text{L}} \times \frac{1\text{ mol }Cl^-}{\text{mol }Ag^+} \times \frac{1\text{ mol douglasite}}{4\text{ mol }Cl^-} \times \frac{311.88\text{ g douglasite}}{\text{mol}}$$
$$= 0.2900\text{ g douglasite}$$

$$\text{Mass \% douglasite} = \frac{0.2900\text{ g}}{0.4550\text{ g}} \times 100 = 63.74\%$$

108. a. $Al(s) + 3\ HCl(aq) \rightarrow AlCl_3(aq) + 3/2\ H_2(g)$ or $2\ Al(s) + 6\ HCl(aq) \rightarrow 2\ AlCl_3(aq) + 3\ H_2(g)$

Hydrogen is reduced (goes from the +1 oxidation state to the 0 oxidation state), and aluminum Al is oxidized (0 → +3).

b. Balancing S is most complicated since sulfur is in both products. Balance C and H first; then worry about S.

$CH_4(g) + 4\ S(s) \rightarrow CS_2(l) + 2\ H_2S(g)$

Sulfur is reduced (0 → −2), and carbon is oxidized (−4 → +4).

c. Balance C and H first; then balance O.

$C_3H_8(g) + 5\ O_2(g) \rightarrow 3\ CO_2(g) + 4\ H_2O(l)$

Oxygen is reduced (0 → −2), and carbon is oxidized (−8/3 → +4).

d. Although this reaction is mass balanced, it is not charge balanced. We need 2 moles of silver on each side to balance the charge.

$Cu(s) + 2\ Ag^+(aq) \rightarrow 2\ Ag(s) + Cu^{2+}(aq)$

Silver is reduced (+1 → 0), and copper is oxidized (0 → +2).

109. $Cr_2O_7^{2-}$: $2(x) + 7(-2) = -2,\ x = +6$

C_2H_5OH (C_2H_6O): $2(y) + 6(+1) + (-2) = 0,\ y = -2$

CO_2: $z + 2(-2) = 0,\ z = +4$

Each chromium atom goes from the oxidation state of +6 in $Cr_2O_7^{2-}$ to +3 in Cr^{3+}. Each chromium atom gains three electrons; chromium is the species reduced. Each carbon atom goes from the oxidation state of −2 in C_2H_5OH to +4 in CO_2. Each carbon atom loses six electrons; carbon is the species oxidized. From the balanced equation, we have four chromium atoms and two carbon atoms. With each chromium atom gaining three electrons, a total of 4(3) = 12 electrons are transferred in the balanced reaction. This is confirmed from the 2 carbon atoms in the balanced equation, where each carbon atom loses six electrons [2(6) = 12 electrons transferred].

ChemWork Problems

The answers to the problems 110-119 (or a variation to these problems) are found in OWL. These problems are also assignable in OWL.

Challenge Problems

120. Let x = mass of NaCl, and let y = mass K_2SO_4. So $x + y = 10.00$.

Two reactions occur: $Pb^{2+}(aq) + 2\ Cl^-(aq) \rightarrow PbCl_2(s)$ and

$Pb^{2+}(aq) + SO_4^{2-}(aq) \rightarrow PbSO_4(s)$

Molar mass of NaCl = 58.44 g/mol; molar mass of K_2SO_4 = 174.27 g/mol; molar mass of $PbCl_2$ = 278.1 g/mol; molar mass of $PbSO_4$ = 303.3 g/mol

$$\frac{x}{58.44} = \text{moles NaCl}; \quad \frac{y}{174.27} = \text{moles } K_2SO_4$$

mass of $PbCl_2$ + mass $PbSO_4$ = total mass of solid

$$\frac{x}{58.44}(1/2)(278.1) + \frac{y}{174.27}(303.3) = 21.75$$

We have two equations:

$(2.379)x + (1.740)y = 21.75$ and $x + y = 10.00$. Solving:

$$x = 6.81 \text{ g NaCl}; \quad \frac{6.81 \text{ g NaCl}}{10.00 \text{ g mixture}} \times 100 = 68.1\% \text{ NaCl}$$

121. a. $5.0 \text{ ppb Hg in water} = \dfrac{5.0 \text{ ng Hg}}{\text{g soln}} = \dfrac{5.0 \times 10^{-9} \text{ g Hg}}{\text{mL soln}}$

$$\frac{5.0 \times 10^{-9} \text{ g Hg}}{\text{mL}} \times \frac{1 \text{ mol Hg}}{200.6 \text{ g Hg}} \times \frac{1000 \text{ mL}}{\text{L}} = 2.5 \times 10^{-8}\ M \text{ Hg}$$

b. $\dfrac{1.0 \times 10^{-9} \text{ g CHCl}_3}{\text{mL}} \times \dfrac{1 \text{ mol CHCl}_3}{119.37 \text{ g CHCl}_3} \times \dfrac{1000 \text{ mL}}{\text{L}} = 8.4 \times 10^{-9}\ M \text{ CHCl}_3$

c. $10.0 \text{ ppm As} = \dfrac{10.0\ \mu\text{g As}}{\text{g soln}} = \dfrac{10.0 \times 10^{-6} \text{ g As}}{\text{mL soln}}$

$$\frac{10.0 \times 10^{-6}\text{ g As}}{\text{mL}} \times \frac{1\text{ mol As}}{74.92\text{ g As}} \times \frac{1000\text{mL}}{\text{L}} = 1.33 \times 10^{-4}\ M\text{ As}$$

d. $$\frac{0.10 \times 10^{-6}\text{ g DDT}}{\text{mL}} \times \frac{1\text{ mol DDT}}{354.46\text{ g DDT}} \times \frac{1000\text{mL}}{\text{L}} = 2.8 \times 10^{-7}\ M\text{ DDT}$$

122. We want 100.0 mL of each standard. To make the 100. ppm standard:

$$\frac{100.\ \mu\text{g Cu}}{\text{mL}} \times 100.0\text{ mL solution} = 1.00 \times 10^{4}\ \mu\text{g Cu needed}$$

$$1.00 \times 10^{4}\ \mu\text{g Cu} \times \frac{1\text{ mL stock}}{1000.0\ \mu\text{g Cu}} = 10.0\text{ mL of stock solution}$$

Therefore, to make 100.0 mL of 100. ppm solution, transfer 10.0 mL of the 1000.0 ppm stock solution to a 100-mL volumetric flask, and dilute to the mark.

Similarly:

75.0 ppm standard, dilute 7.50 mL of the 1000.0 ppm stock to 100.0 mL.

50.0 ppm standard, dilute 5.00 mL of the 1000.0 ppm stock to 100.0 mL.

25.0 ppm standard, dilute 2.50 mL of the 1000.0 ppm stock to 100.0 mL.

10.0 ppm standard, dilute 1.00 mL of the 1000.0 ppm stock to 100.0 mL.

123. a. $$0.308\text{ g AgCl} \times \frac{35.45\text{ g Cl}}{143.4\text{ g AgCl}} = 0.0761\text{ g Cl};\ \%\text{ Cl} = \frac{0.0761\text{g}}{0.256\text{g}} \times 100 = 29.7\%\text{ Cl}$$

Cobalt(III) oxide, Co_2O_3: 2(58.93) + 3(16.00) = 165.86 g/mol

$$0.145\text{ g } Co_2O_3 \times \frac{117.86\text{ g Co}}{165.86\text{ g } Co_2O_3} = 0.103\text{ g Co};\ \%\text{ Co} = \frac{0.103\text{g}}{0.416\text{g}} \times 100 = 24.8\%\text{ Co}$$

The remainder, 100.0 – (29.7 + 24.8) = 45.5%, is water.

Assuming 100.0 g of compound:

$$45.5\text{ g } H_2O \times \frac{2.016\text{ g H}}{18.02\text{ g } H_2O} = 5.09\text{ g H};\ \%\text{ H} = \frac{5.09\text{ g H}}{100.0\text{ g compound}} \times 100 = 5.09\%\text{ H}$$

$$45.5\text{ g } H_2O \times \frac{16.00\text{ g O}}{18.02\text{ g } H_2O} = 40.4\text{ g O};\ \%\text{ O} = \frac{40.4\text{ g O}}{100.0\text{ g compound}} \times 100 = 40.4\%\text{ O}$$

The mass percent composition is 24.8% Co, 29.7% Cl, 5.09% H, and 40.4% O.

b. Out of 100.0 g of compound, there are:

$$24.8 \text{ g Co H } \frac{1 \text{ mol}}{58.93 \text{ g Co}} = 0.421 \text{ mol Co}; \ 29.7 \text{ g Cl} \times \frac{1 \text{ mol}}{35.45 \text{ g Cl}} = 0.838 \text{ mol Cl}$$

$$5.09 \text{ g H} \times \frac{1 \text{ mol}}{1.008 \text{ g H}} = 5.05 \text{ mol H}; \ 40.4 \text{ g O} \times \frac{1 \text{ mol}}{16.00 \text{ g O}} = 2.53 \text{ mol O}$$

Dividing all results by 0.421, we get $CoCl_2 \cdot 6H_2O$ for the empirical formula, which is also the actual formula given the information in the problem. The $\cdot 6H_2O$ represent six waters of hydration in the chemical formula.

c. $CoCl_2 \cdot 6H_2O(aq) + 2\ AgNO_3(aq) \rightarrow 2\ AgCl(s) + Co(NO_3)_2(aq) + 6\ H_2O(l)$

$CoCl_2 \cdot 6H_2O(aq) + 2\ NaOH(aq) \rightarrow Co(OH)_2(s) + 2\ NaCl(aq) + 6\ H_2O(l)$

$Co(OH)_2 \rightarrow Co_2O_3$ This is an oxidation-reduction reaction. Thus we also need to include an oxidizing agent. The obvious choice is O_2.

$4\ Co(OH)_2(s) + O_2(g) \rightarrow 2\ Co_2O_3(s) + 4\ H_2O(l)$

124. a. $C_{12}H_{10-n}Cl_n + n\ Ag^+ \rightarrow n\ AgCl$; molar mass of AgCl = 143.4 g/mol

Molar mass of PCB = $12(12.01) + (10 - n)(1.008) + n(35.45) = 154.20 + (34.44)n$

Because n mol AgCl is produced for every 1 mol PCB reacted, $n(143.4)$ g of AgCl will be produced for every $[154.20 + (34.44)n]$ g of PCB reacted.

$$\frac{\text{Mass of AgCl}}{\text{Mass of PCB}} = \frac{(143.4)n}{154.20 + (34.44)n} \text{ or } \text{mass}_{AgCl}[154.20 + (34.44)n] = \text{mass}_{PCB}(143.4)n$$

b. $0.4971[154.20 + (34.44)n] = 0.1947(143.4)n, \ 76.65 + (17.12)n = (27.92)n$

$76.65 = (10.80)n, \ n = 7.097$

125. $Zn(s) + 2\ AgNO_2(aq) \rightarrow 2\ Ag(s) + Zn(NO_2)_2(aq)$

Let x = mass of Ag and y = mass of Zn after the reaction has stopped. Then $x + y = 29.0$ g. Because the moles of Ag produced will equal two times the moles of Zn reacted:

$$(19.0 - y) \text{ g Zn} \times \frac{1 \text{ mol Zn}}{65.38 \text{ g Zn}} \times \frac{2 \text{ mol Ag}}{1 \text{ mol Zn}} = x \text{ g Ag} \times \frac{1 \text{ mol Ag}}{107.9 \text{ g Ag}}$$

Simplifying:

$$3.059 \times 10^{-2}(19.0 - y) = (9.268 \times 10^{-3})x$$

Substituting $x = 29.0 - y$ into the equation gives:

$$3.059 \times 10^{-2}(19.0 - y) = 9.268 \times 10^{-3}(29.0 - y)$$

Solving:

$$0.581 - (3.059 \times 10^{-2})y = 0.269 - (9.268 \times 10^{-3})y,\ (2.132 \times 10^{-2})y = 0.312,\ y = 14.6 \text{ g Zn}$$

14.6 g Zn is present, and 29.0 – 14.6 = 14.4 g Ag is also present after the reaction is stopped.

126. $Ag^+(aq) + Cl^-(aq) \rightarrow AgCl(s)$; let x = mol NaCl and y = mol KCl.

$(22.90 \times 10^{-3}\ L) \times 0.1000\ mol/L = 2.290 \times 10^{-3}\ mol\ Ag^+ = 2.290 \times 10^{-3}\ mol\ Cl^-$ total

$x + y = 2.290 \times 10^{-3}\ mol\ Cl^-$, $x = 2.290 \times 10^{-3} - y$

Because the molar mass of NaCl is 58.44 g/mol and the molar mass of KCl is 74.55 g/mol:

$$(58.44)x + (74.55)y = 0.1586\ g$$

$58.44(2.290 \times 10^{-3} - y) + (74.55)y = 0.1586$, $(16.11)y = 0.0248$, $y = 1.54 \times 10^{-3}$ mol KCl

$$\text{Mass \% KCl} = \frac{1.54 \times 10^{-3}\ \text{mol} \times 74.55\ \text{g/mol}}{0.1586\text{g}} \times 100 = 72.4\%\ \text{KCl}$$

% NaCl = 100.0 – 72.4 = 27.6% NaCl

127. $0.298\ g\ BaSO_4 \times \frac{96.07\ g\ SO_4^{2-}}{233.4\ g\ BaSO_4} = 0.123\ g\ SO_4^{2-}$; % sulfate $= \frac{0.123\ g\ SO_4^{2-}}{0.205\ g} = 60.0\%$

Assume we have 100.0 g of the mixture of Na_2SO_4 and K_2SO_4. There are:

$$60.0\ g\ SO_4^{2-} \times \frac{1\ mol}{96.07\ g} = 0.625\ mol\ SO_4^{2-}$$

There must be 2 × 0.625 = 1.25 mol of 1+ cations to balance the 2– charge of SO_4^{2-}.

Let x = number of moles of K^+ and y = number of moles of Na^+; then $x + y = 1.25$.

The total mass of Na^+ and K^+ must be 40.0 g in the assumed 100.0 g of mixture. Setting up an equation:

$$x\ mol\ K^+ \times \frac{39.10\ g}{mol} + y\ mol\ Na^+ \times \frac{22.99\ g}{mol} = 40.0\ g$$

So we have two equations with two unknowns: $x + y = 1.25$ and $(39.10)x + (22.99)y = 40.0$

$x = 1.25 - y$, so $39.10(1.25 - y) + (22.99)y = 40.0$

$48.9 - (39.10)y + (22.99)y = 40.0$, $-(16.11)y = -8.9$

$y = 0.55$ mol Na^+ and $x = 1.25 - 0.55 = 0.70$ mol K^+

Therefore:

$$0.70 \text{ mol K}^+ \times \frac{1 \text{ mol } K_2SO_4}{2 \text{ mol K}^+} = 0.35 \text{ mol } K_2SO_4;\ 0.35 \text{ mol } K_2SO_4 \times \frac{174.27 \text{ g}}{\text{mol}} = 61 \text{ g } K_2SO_4$$

We assumed 100.0 g; therefore, the mixture is 61% K_2SO_4 and 39% Na_2SO_4.

128. a. Let x = mass of Mg, so $10.00 - x$ = mass of Zn. $Ag^+(aq) + Cl^-(aq) \rightarrow AgCl(s)$.

From the given balanced equations, there is a 2 : 1 mole ratio between mol Mg and mol Cl^-. The same is true for Zn. Because mol Ag^+ = mol Cl^- present, one can set up an equation relating mol Cl^- present to mol Ag^+ added.

$$x \text{ g Mg} \times \frac{1 \text{ mol Mg}}{24.31 \text{ g Mg}} \times \frac{2 \text{ mol Cl}^-}{\text{mol Mg}} + (10.00 - x) \text{ g Zn} \times \frac{1 \text{ mol Zn}}{65.38 \text{ g Zn}} \times \frac{2 \text{ mol Cl}^-}{\text{mol Zn}}$$

$$= 0.156 \text{ L} \times \frac{3.00 \text{ mol Ag}^+}{\text{L}} \times \frac{1 \text{ mol Cl}^-}{\text{mol Ag}^+} = 0.468 \text{ mol Cl}^-$$

$$\frac{2x}{24.31} + \frac{2(10.00 - x)}{65.38} = 0.468,\ 24.31 \times 65.38 \left(\frac{2x}{24.31} + \frac{20.00 - 2x}{65.38} = 0.468\right)$$

$(130.8)x + 486.2 - (48.62)x = 743.8$ (carrying 1 extra sig. fig.)

$$(82.2)x = 257.6,\quad x = 3.13 \text{ g Mg};\qquad \% \text{ Mg} = \frac{3.13 \text{ g Mg}}{10.00 \text{ g mixture}} \times 100 = 31.3\% \text{ Mg}$$

b. $0.156 \text{ L} \times \frac{3.00 \text{ mol Ag}^+}{\text{L}} \times \frac{1 \text{ mol Cl}^-}{\text{mol Ag}^+} = 0.468 \text{ mol Cl}^- = 0.468 \text{ mol HCl added}$

$$M_{HCl} = \frac{0.468 \text{ mol}}{0.0780 \text{ L}} = 6.00\ M \text{ HCl}$$

129. $Pb^{2+}(aq) + 2\ Cl^-(aq) \rightarrow PbCl_2(s)$

$$3.407 \text{ g } PbCl_2 \times \frac{1 \text{ mol } PbCl_2}{278.1 \text{ g } PbCl_2} \times \frac{1 \text{ mol } Pb^{2+}}{\text{mol } PbCl_2} = 0.01225 \text{ mol } Pb^{2+}$$

$$\frac{0.01225 \text{ mol}}{2.00 \times 10^{-3} \text{ L}} = 6.13\ M\ Pb^{2+} = 6.13\ M\ Pb(NO_3)_2$$

This is also the $Pb(NO_3)_2$ concentration in the 80.0 mL of evaporated solution.

$$\text{Original concentration} = \frac{\text{moles } Pb(NO_3)_2}{\text{original volume}} = \frac{0.0800 \text{ L} \times 6.13 \text{ mol/L}}{0.1000 \text{ L}} = 4.90\ M\ Pb(NO_3)_2$$

130. $\text{Mol } CuSO_4 = 87.7 \text{ mL} \times \frac{1 \text{ L}}{1000 \text{ mL}} \times \frac{0.500 \text{ mol}}{\text{L}} = 0.0439 \text{ mol}$

$$\text{Mol Fe} = 2.00 \text{ g} \times \frac{1 \text{ mol Fe}}{55.85 \text{ g}} = 0.0358 \text{ mol}$$

The two possible reactions are:

I. $CuSO_4(aq) + Fe(s) \rightarrow Cu(s) + FeSO_4(aq)$

II. $3\ CuSO_4(aq) + 2\ Fe(s) \rightarrow 3\ Cu(s) + Fe_2(SO_4)_3(aq)$

If reaction I occurs, Fe is limiting, and we can produce:

$$0.0358 \text{ mol Fe} \times \frac{1 \text{ mol Cu}}{\text{mol Fe}} \times \frac{63.55 \text{ g Cu}}{\text{mol Cu}} = 2.28 \text{ g Cu}$$

If reaction II occurs, $CuSO_4$ is limiting, and we can produce:

$$0.0439 \text{ mol } CuSO_4 \times \frac{3 \text{ mol Cu}}{3 \text{ mol } CuSO_4} \times \frac{63.55 \text{ g Cu}}{\text{mol Cu}} = 2.79 \text{ g Cu}$$

Assuming 100% yield, reaction I occurs because it fits the data best.

131. 0.2750 L × 0.300 mol/L = 0.0825 mol H^+; let y = volume (L) delivered by Y and z = volume (L) delivered by Z.

$$H^+(aq) + OH^-(aq) \rightarrow H_2O(l);\quad \underbrace{y(0.150 \text{ mol/L}) + z(0.250 \text{ mol/L})}_{\text{mol } OH^-} = 0.0825 \text{ mol } H^+$$

$$0.2750 \text{ L} + y + z = 0.655 \text{ L},\quad y + z = 0.380,\quad z = 0.380 - y$$

$$y(0.150) + (0.380 - y)(0.250) = 0.0825, \text{ solving: } y = 0.125 \text{ L},\ z = 0.255 \text{ L}$$

$$\text{Flow rate for Y} = \frac{125 \text{ mL}}{60.65 \text{ min}} = 2.06 \text{ mL/min};\ \text{flow rate for Z} = \frac{255 \text{ mL}}{60.65 \text{ min}} = 4.20 \text{ mL/min}$$

132. a. $H_3PO_4(aq) + 3\ NaOH(aq) \rightarrow 3\ H_2O(l) + Na_3PO_4(aq)$

b. $3\ H_2SO_4(aq) + 2\ Al(OH)_3(s) \rightarrow 6\ H_2O(l) + Al_2(SO_4)_3(aq)$

c. $H_2Se(aq) + Ba(OH)_2(aq) \rightarrow 2\ H_2O(l) + BaSe(s)$

d. $H_2C_2O_4(aq) + 2\ NaOH(aq) \rightarrow 2\ H_2O(l) + Na_2C_2O_4(aq)$

133. $2\ H_3PO_4(aq) + 3\ Ba(OH)_2(aq) \rightarrow 6\ H_2O(l) + Ba_3(PO_4)_2(s)$

$$0.01420 \text{ L} \times \frac{0.141 \text{ mol } H_3PO_4}{\text{L}} \times \frac{3 \text{ mol } Ba(OH)_2}{2 \text{ mol } H_3PO_4} \times \frac{1 \text{ L } Ba(OH)_2}{0.0521 \text{ mol } Ba(OH)_2} = 0.0576 \text{ L}$$

$$= 57.6 \text{ mL } Ba(OH)_2$$

134. $35.08 \text{ mL NaOH} \times \frac{1\text{ L}}{1000\text{mL}} \times \frac{2.12 \text{ mol NaOH}}{\text{L NaOH}} \times \frac{1 \text{ mol } H_2SO_4}{2 \text{ mol NaOH}} = 3.72 \times 10^{-2} \text{ mol } H_2SO_4$

$$\text{Molarity} = \frac{3.72 \times 10^{-2} \text{ mol}}{10.00 \text{ mL}} \times \frac{1000\text{mL}}{\text{L}} = 3.72\ M\ H_2SO_4$$

135. The pertinent equations are:

$$2\ NaOH(aq) + H_2SO_4(aq) \rightarrow Na_2SO_4(aq) + 2\ H_2O(l)$$

$$HCl(aq) + NaOH(aq) \rightarrow NaCl(aq) + H_2O(l)$$

$$\text{Amount of NaOH added} = 0.0500 \text{ L} \times \frac{0.213\text{mol}}{\text{L}} = 1.07 \times 10^{-2} \text{ mol NaOH}$$

Amount of NaOH neutralized by HCl:

$$0.01321 \text{ L HCl H } \frac{0.103 \text{mol HCl}}{\text{L HCl}} \times \frac{1 \text{ mol NaOH}}{\text{mol HCl}} = 1.36 \times 10^{-3} \text{ mol NaOH}$$

The difference, 9.3×10^{-3} mol, is the amount of NaOH neutralized by the sulfuric acid.

$$9.3 \times 10^{-3} \text{ mol NaOH} \times \frac{1 \text{ mol } H_2SO_4}{2 \text{ mol NaOH}} = 4.7 \times 10^{-3} \text{ mol } H_2SO_4$$

$$\text{Concentration of } H_2SO_4 = \frac{4.7 \times 10^{-3} \text{ mol}}{0.1000\text{L}} = 4.7 \times 10^{-2}\ M\ H_2SO_4$$

136. Let H_2A = formula for the unknown diprotic acid.

$$H_2A(aq) + 2\ NaOH(aq) \rightarrow 2\ H_2O(l) + Na_2A(aq)$$

$$\text{Mol } H_2A = 0.1375 \text{ L} \times \frac{0.750 \text{mol NaOH}}{\text{L}} \times \frac{1 \text{mol } H_2A}{2 \text{ mol NaOH}} = 0.0516 \text{ mol}$$

$$\text{Molar mass of } H_2A = \frac{6.50\text{ g}}{0.0516\text{mol}} = 126 \text{ g/mol}$$

137. $\text{Mol } C_6H_8O_7 = 0.250 \text{ g } C_6H_8O_7 \times \frac{1 \text{mol } C_6H_8O_7}{192.12 \text{ g } C_6H_8O_7} = 1.30 \times 10^{-3} \text{ mol } C_6H_8O_7$

Let H_xA represent citric acid, where x is the number of acidic hydrogens. The balanced neutralization reaction is:

$$H_xA(aq) + x\ OH^-(aq) \rightarrow x\ H_2O(l) + A^{x-}(aq)$$

$$\text{Mol } OH^- \text{ reacted} = 0.0372 \text{ L} \times \frac{0.105 \text{mol } OH^-}{\text{L}} = 3.91 \times 10^{-3} \text{ mol } OH^-$$

$$x = \frac{\text{mol OH}^-}{\text{mol citric acid}} = \frac{3.91 \times 10^{-3}\ \text{mol}}{1.30 \times 10^{-3}\ \text{mol}} = 3.01$$

Therefore, the general acid formula for citric acid is H_3A, meaning that citric acid has three acidic hydrogens per citric acid molecule (citric acid is a triprotic acid).

138. a. Flow rate = 5.00×10^4 L/s + 3.50×10^3 L/s = 5.35×10^4 L/s

b. $C_{HCl} = \frac{3.50 \times 10^3 (65.0)}{5.35 \times 10^4} = 4.25$ ppm HCl

c. 1 ppm = 1 mg/kg H_2O = 1 mg/L (assuming density = 1.00 g/mL)

$$8.00\ \text{h} \times \frac{60\ \text{min}}{\text{h}} \times \frac{60\ \text{s}}{\text{min}} \times \frac{1.80 \times 10^4\ \text{L}}{\text{s}} \times \frac{4.25\ \text{mg HCl}}{\text{L}} \times \frac{1\ \text{g}}{1000\text{mg}} = 2.20 \times 10^6\ \text{g HCl}$$

$$2.20 \times 10^6\ \text{g HCl} \times \frac{1\ \text{mol HCl}}{36.46\ \text{g HCl}} \times \frac{1\ \text{mol CaO}}{2\ \text{mol HCl}} \times \frac{56.08\ \text{g Ca}}{\text{mol CaO}} = 1.69 \times 10^6\ \text{g CaO}$$

d. The concentration of Ca^{2+} going into the second plant was:

$$\frac{5.00 \times 10^4 (10.2)}{5.35 \times 10^4} = 9.53\ \text{ppm}$$

The second plant used: 1.80×10^4 L/s × (8.00 × 60 × 60) s = 5.18×10^8 L of water.

$$1.69 \times 10^6\ \text{g CaO} \times \frac{40.08\ \text{g Ca}^{2+}}{56.08\ \text{g CaO}} = 1.21 \times 10^6\ \text{g Ca}^{2+}\ \text{was added to this water.}$$

$$C_{Ca^{2+}}\ \text{(plant water)} = 9.53 + \frac{1.21 \times 10^9\ \text{mg}}{5.18 \times 10^8\ \text{L}} = 9.53 + 2.34 = 11.87\ \text{ppm}$$

Because 90.0% of this water is returned, $(1.80 \times 10^4) \times 0.900 = 1.62 \times 10^4$ L/s of water with 11.87 ppm Ca^{2+} is mixed with $(5.35 - 1.80) \times 10^4 = 3.55 \times 10^4$ L/s of water containing 9.53 ppm Ca^{2+}.

$$C_{Ca^{2+}}\ \text{(final)} = \frac{(1.62 \times 10^4\ \text{L/s})(11.87\ \text{ppm}) + (3.55 \times 10^4\ \text{L/s})(9.53\ \text{ppm})}{1.62 \times 10^4\ \text{L/s} + 3.55 \times 10^4\ \text{L/s}} = 10.3\ \text{ppm}$$

139. Mol KHP used = $0.4016\ \text{g} \times \frac{1\ \text{mol}}{204.22\ \text{g}} = 1.967 \times 10^{-3}$ mol KHP

Because 1 mole of NaOH reacts completely with 1 mole of KHP, the NaOH solution contains 1.967×10^{-3} mol NaOH.

$$\text{Molarity of NaOH} = \frac{1.967 \times 10^{-3}\ \text{mol}}{25.06 \times 10^{-3}\ \text{L}} = \frac{7.849 \times 10^{-2}\ \text{mol}}{\text{L}}$$

$$\text{Maximum molarity} = \frac{1.967 \times 10^{-3}\ \text{mol}}{25.01 \times 10^{-3}\ \text{L}} = \frac{7.865 \times 10^{-2}\ \text{mol}}{\text{L}}$$

$$\text{Minimum molarity} = \frac{1.967 \times 10^{-3}\ \text{mol}}{25.11 \times 10^{-3}\ \text{L}} = \frac{7.834 \times 10^{-2}\ \text{mol}}{\text{L}}$$

We can express this as 0.07849 ±0.00016 *M*. An alternative way is to express the molarity as 0.0785 ±0.0002 *M*. This second way shows the actual number of significant figures in the molarity. The advantage of the first method is that it shows that we made all our individual measurements to four significant figures.

Integrative Problems

140. a. Assume 100.00 g of material.

$$42.23\ \text{g C} \times \frac{1\ \text{mol C}}{12.01\ \text{g C}} = 3.516\ \text{mol C};\ \ 55.66\ \text{g F} \times \frac{1\ \text{mol F}}{19.00\ \text{g F}} = 2.929\ \text{mol F}$$

$$2.11\ \text{g B} \times \frac{1\ \text{mol B}}{10.81\ \text{g B}} = 0.195\ \text{mol B}$$

Dividing by the smallest number: $\frac{3.516}{0.195} = 18.0;\ \frac{2.929}{0.195} = 15.0$

The empirical formula is $C_{18}F_{15}B$.

b. $0.3470\ \text{L} \times \frac{0.01267\ \text{mol}}{\text{L}} = 4.396 \times 10^{-3}\ \text{mol BARF}$

$$\text{Molar mass of BARF} = \frac{2.251\ \text{g}}{4.396 \times 10^{-3}\ \text{mol}} = 512.1\ \text{g/mol}$$

The empirical formula mass of BARF is 511.99 g. Therefore, the molecular formula is the same as the empirical formula, $C_{18}F_{15}B$.

141. $3\ (NH_4)_2CrO_4(aq) + 2\ Cr(NO_2)_3(aq) \rightarrow 6\ NH_4NO_2(aq) + Cr_2(CrO_4)_3(s)$

$$0.203\ \text{L} \times \frac{0.307\ \text{mol}\ (NH_4)_2CrO_4}{\text{L}} \times \frac{1\ \text{mol}\ Cr_2(CrO_4)_3}{3\ \text{mol}\ (NH_4)_2CrO_4} \times \frac{452.00\ \text{g}\ Cr_2(CrO_4)_3}{\text{mol}\ Cr_2(CrO_4)_3}$$

$$= 9.39\ \text{g}\ Cr_2(CrO_4)_3$$

$$0.137\ \text{L} \times \frac{0.269\ \text{mol}\ Cr(NO_2)_3}{\text{L}} \times \frac{1\ \text{mol}\ Cr_2(CrO_4)_3}{2\ \text{mol}\ Cr(NO_2)_3} \times \frac{452.00\ \text{g}\ Cr_2(CrO_4)_3}{\text{mol}\ Cr_2(CrO_4)_3}$$

$$= 8.33\ \text{g}\ Cr_2(CrO_4)_3$$

The $Cr(NO_2)_3$ reagent produces the smaller amount of product, so $Cr(NO_2)_3$ is limiting and the theoretical yield of $Cr_2(CrO_4)_3$ is 8.33 g.

$$0.880 = \frac{\text{actual yield}}{8.33\text{ g}},\ \text{actual yield} = (8.33\text{ g})(0.880) = 7.33\text{ g } Cr_2(CrO_4)_3 \text{ isolated}$$

142. The oxidation states of the elements in the various ions are:

VO^{2+}: O, −2; V, $x + (-2) = +2$, $x = +4$

MnO_4^-: O, −2; Mn, $x + 4(-2) = -1$, $x = +7$

$V(OH)_4^+$: O, −2, H, +1; V, $x + 4(-2) + 4(+1) = +1$, $x = +5$

Mn^{2+}: Mn, +2

Vanadium goes from the +4 oxidation state in VO^{2+} to the +5 oxidation state in $V(OH)_4^+$. Manganese goes from the +7 oxidation state in MnO_4^- to the +2 oxidation state in Mn^{2+}. We need 5 V atoms for every Mn atom in order to balance the electrons transferred. Balancing the electrons transferred, then balancing the rest by inspection gives:

$$MnO_4^-(aq) + 5\ VO^{2+}(aq) + 11\ H_2O(l) \rightarrow 5\ V(OH)_4^+(aq) + Mn^{2+}(aq) + 22\ H^+(aq)$$

$$0.02645\text{ L} \times \frac{0.02250\text{ mol } MnO_4^-}{\text{L}} \times \frac{5\text{ mol } VO^{2+}}{\text{mol } MnO_4^-} \times \frac{1\text{ mol V}}{\text{mol } VO^{2+}} \times \frac{50.94\text{ g V}}{\text{mol V}} = 0.1516\text{ g V}$$

$$0.581 = \frac{0.1516\text{ g V}}{\text{mass of ore sample}},\ \text{mass of ore sample} = 0.1516/0.581 = 0.261\text{ g}$$

143. X^{2-} contains 36 electrons, so X^{2-} has 34 protons, which identifies X as selenium (Se). The name of H_2Se would be hydroselenic acid following the conventions described in Chapter 2.

$$H_2Se(aq) + 2\ OH^-(aq) \rightarrow Se^{2-}(aq) + 2\ H_2O(l)$$

$$0.0356\text{ L} \times \frac{0.175\text{ mol } OH^-}{\text{L}} \times \frac{1\text{ mol } H_2Se}{2\text{ mol } OH^-} \times \frac{80.98\text{ g } H_2Se}{\text{mol } H_2Se} = 0.252\text{ g } H_2Se$$

Marathon Problems

144. Mol $BaSO_4$ = $0.2327\text{ g} \times \frac{1\text{ mol}}{233.4\text{ g}} = 9.970 \times 10^{-4}$ mol $BaSO_4$

The moles of the sulfate salt depend on the formula of the salt. The general equation is:

$$M_x(SO_4)_y(aq) + y\ Ba^{2+}(aq) \rightarrow y\ BaSO_4(s) + x\ M^{z+}$$

Depending on the value of y, the mole ratio between the unknown sulfate salt and $BaSO_4$ varies. For example, if Pat thinks the formula is $TiSO_4$, the equation becomes:

$$TiSO_4(aq) + Ba^{2+}(aq) \rightarrow BaSO_4(s) + Ti^{2+}(aq)$$

Because there is a 1 : 1 mole ratio between mol $BaSO_4$ and mol $TiSO_4$, you need 9.970×10^{-4} mol of $TiSO_4$. Because 0.1472 g of salt was used, the compound would have a molar mass of (assuming the $TiSO_4$ formula):

$$0.1472 \text{ g}/9.970 \times 10^{-4} \text{ mol} = 147.6 \text{ g/mol}$$

From atomic masses in the periodic table, the molar mass of $TiSO_4$ is 143.95 g/mol. From just these data, $TiSO_4$ seems reasonable.

Chris thinks the salt is sodium sulfate, which would have the formula Na_2SO_4. The equation is:

$$Na_2SO_4(aq) + Ba^{2+}(aq) \rightarrow BaSO_4(s) + 2\ Na^+(aq)$$

As with $TiSO_4$, there is a 1:1 mole ratio between mol $BaSO_4$ and mol Na_2SO_4. For sodium sulfate to be a reasonable choice, it must have a molar mass of about 147.6 g/mol. Using atomic masses, the molar mass of Na_2SO_4 is 142.05 g/mol. Thus Na_2SO_4 is also reasonable.

Randy, who chose gallium, deduces that gallium should have a 3+ charge (because it is in column 3A), and the formula of the sulfate would be $Ga_2(SO_4)_3$. The equation would be:

$$Ga_2(SO_4)_3(aq) + 3\ Ba^{2+}(aq) \rightarrow 3\ BaSO_4(s) + 2\ Ga^{3+}(aq)$$

The calculated molar mass of $Ga_2(SO_4)_3$ would be:

$$\frac{0.1472\text{g } Ga_2(SO_4)_3}{9.970 \times 10^{-4} \text{ mol } BaSO_4} \times \frac{3 \text{ mol } BaSO_4}{\text{mol } Ga_2(SO_4)_3} = 442.9 \text{ g/mol}$$

Using atomic masses, the molar mass of $Ga_2(SO_4)_3$ is 427.65 g/mol. Thus $Ga_2(SO_4)_3$ is also reasonable.

Looking in references, sodium sulfate (Na_2SO_4) exists as a white solid with orthorhombic crystals, whereas gallium sulfate [$Ga_2(SO_4)_3$] is a white powder. Titanium sulfate exists as a green powder, but its formula is $Ti_2(SO_4)_3$. Because this has the same formula as gallium sulfate, the calculated molar mass should be around 443 g/mol. However, the molar mass of $Ti_2(SO_4)_3$ is 383.97 g/mol. It is unlikely, then, that the salt is titanium sulfate.

To distinguish between Na_2SO_4 and $Ga_2(SO_4)_3$, one could dissolve the sulfate salt in water and add NaOH. Ga^{3+} would form a precipitate with the hydroxide, whereas Na_2SO_4 would not. References confirm that gallium hydroxide is insoluble in water.

145. a. Compound A = $M(NO_3)_x$; in 100.00 g of compd.: $8.246 \text{ g N} \times \dfrac{48.00 \text{ g O}}{14.01 \text{ g N}} = 28.25 \text{ g O}$

Thus the mass of nitrate in the compound = 8.246 + 28.25 g = 36.50 g (if x = 1).

If x = 1: mass of M = 100.00 − 36.50 g = 63.50 g

$$\text{Mol M} = \text{mol N} = \frac{8.246 \text{g}}{14.01 \text{g/mol}} = 0.5886 \text{ mol}$$

$$\text{Molar mass of metal M} = \frac{63.50\text{ g}}{0.5886\text{ mol}} = 107.9 \text{ g/mol (This is silver, Ag.)}$$

If $x = 2$: mass of M = 100.00 − 2(36.50) = 27.00 g

$$\text{Mol M} = \tfrac{1}{2} \text{ mol N} = \frac{0.5886\text{ mol}}{2} = 0.2943 \text{ mol}$$

$$\text{Molar mass of metal M} = \frac{27.00\text{ g}}{0.2943\text{ mol}} = 91.74 \text{ g/mol}$$

This is close to Zr, but Zr does not form stable 2+ ions in solution; it forms stable 4+ ions. Because we cannot have $x = 3$ or more nitrates (three nitrates would have a mass greater than 100.00 g), compound A must be $AgNO_3$.

Compound B: K_2CrO_x is the formula. This salt is composed of K^+ and CrO_x^{2-} ions. Using oxidation states, $6 + x(-2) = -2$, $x = 4$. Compound B is K_2CrO_4 (potassium chromate).

b. The reaction is:

$$2\ AgNO_3(aq) + K_2CrO_4(aq) \rightarrow Ag_2CrO_4(s) + 2\ KNO_3(aq)$$

The blood red precipitate is $Ag_2CrO_4(s)$.

c. 331.8 g Ag_2CrO_4 formed; this is equal to the molar mass of Ag_2CrO_4, so 1 mole of precipitate formed. From the balanced reaction, we need 2 mol $AgNO_3$ to react with 1 mol K_2CrO_4 to produce 1 mol (331.8 g) of Ag_2CrO_4.

$$2.000 \text{ mol } AgNO_3 \times \frac{169.9\text{ g}}{\text{mol}} = 339.8 \text{ g } AgNO_3$$

$$1.000 \text{ mol } K_2CrO_4 \times \frac{194.2\text{ g}}{\text{mol}} = 194.2 \text{ g } K_2CrO_4$$

The problem says that we have equal masses of reactants. Our two choices are 339.8 g $AgNO_3$ + 339.8 g K_2CrO_4 or 194.2 g $AgNO_3$ + 194.2 g K_2CrO_4. If we assume the 194.2-g quantities are correct, then when 194.2 g K_2CrO_4 (1 mol) reacts, 339.8 g $AgNO_3$ (2.0 mol) must be present to react with all the K_2CrO_4. We only have 194.2 g $AgNO_3$ present; this cannot be correct. Instead of K_2CrO_4 limiting, $AgNO_3$ must be limiting, and we have reacted 339.8 g $AgNO_3$ and 339.8 g K_2CrO_4.

$$\text{Solution A: } \frac{2.000\text{ mol Ag}^+}{0.5000\text{ L}} = 4.000\ M\ Ag^+;\ \frac{2.000\text{ mol NO}_3^-}{0.5000\text{ L}} = 4.000\ M\ NO_3^-$$

$$\text{Solution B: } 339.8 \text{ g } K_2CrO_4 \times \frac{1\text{ mol}}{194.2\text{ g}} = 1.750 \text{ mol } K_2CrO_4$$

$$\frac{2 \times 1.750\text{ mol K}^+}{0.5000\text{ L}} = 7.000\ M\ K^+;\ \frac{1.750\text{ mol CrO}_4^{2-}}{0.5000\text{ L}} = 3.500\ M\ CrO_4^{2-}$$

d. After the reaction, moles of K^+ and moles of NO_3^- remain unchanged because they are spectator ions. Because Ag^+ is limiting, its concentration will be 0 *M* after precipitation is complete. The following summarizes the changes that occur as the precipitate forms.

$$2\,Ag^+(aq) + CrO_4^{2-}(aq) \rightarrow Ag_2CrO_4(s)$$

	$2\,Ag^+(aq)$	$CrO_4^{2-}(aq)$	$Ag_2CrO_4(s)$
Initial	2.000 mol	1.750 mol	0
Change	−2.000 mol	−1.000 mol	+1.000 mol
After rxn	0	0.750 mol	1.000 mol

$$M_{K^+} = \frac{2 \times 1.750\,\text{mol}}{1.0000\,\text{L}} = 3.500\,M\,K^+; \quad M_{NO_3^-} = \frac{2.000\,\text{mol}}{1.0000\,\text{L}} = 2.000\,M\,NO_3^-$$

$$M_{CrO_4^{2-}} = \frac{0.750\,\text{mol}}{1.0000\,\text{L}} = 0.750\,M\,CrO_4^{2-}; \quad M_{Ag^+} = 0\,M \text{ (the limiting reagent)}$$

CHAPTER 5

GASES

Questions

20. Molecules in the condensed phases (liquids and solids) are very close together. Molecules in the gaseous phase are very far apart. A sample of gas is mostly empty space. Therefore, one would expect 1 mole of $H_2O(g)$ to occupy a huge volume as compared to 1 mole of $H_2O(l)$.

21. The column of water would have to be 13.6 times taller than a column of mercury. When the pressure of the column of liquid standing on the surface of the liquid is equal to the pressure of air on the rest of the surface of the liquid, then the height of the column of liquid is a measure of atmospheric pressure. Because water is 13.6 times less dense than mercury, the column of water must be 13.6 times longer than that of mercury to match the force exerted by the columns of liquid standing on the surface.

22. A bag of potato chips is a constant-pressure container. The volume of the bag increases or decreases in order to keep the internal pressure equal to the external (atmospheric) pressure. The volume of the bag increased because the external pressure decreased. This seems reasonable as atmospheric pressure is lower at higher altitudes than at sea level. We ignored n (moles) as a possibility because the question said to concentrate on external conditions. It is possible that a chemical reaction occurred that would increase the number of gas molecules inside the bag. This would result in a larger volume for the bag of potato chips. The last factor to consider is temperature. During ski season, one would expect the temperature of Lake Tahoe to be colder than Los Angeles. A decrease in T would result in a decrease in the volume of the potato chip bag. This is the exact opposite of what actually happened, so apparently the temperature effect is not dominant.

23. The P versus 1/V plot is incorrect. The plot should be linear with <u>positive</u> slope and a y-intercept of zero. PV = k, so P = k(1/V). This is in the form of the straight-line equation $y = mx + b$. The y-axis is pressure, and the x-axis is 1/V.

24. The decrease in temperature causes the balloon to contract (V and T are directly related). Because weather balloons do expand, the effect of the decrease in pressure must be dominant.

25. d = (molar mass)P/RT; density is directly proportional to the molar mass of a gas. Helium, with the smallest molar mass of all the noble gases, will have the smallest density.

26. Rigid container: As temperature is increased, the gas molecules move with a faster average velocity. This results in more frequent and more forceful collisions, resulting in an increase in pressure. Density = mass/volume; the moles of gas are constant, and the volume of the container is constant, so density in this case must be temperature-independent (density is constant).

Flexible container: The flexible container is a constant-pressure container. Therefore, the final internal pressure will be unaffected by an increase in temperature. The density of the gas, however, will be affected because the container volume is affected. As T increases, there is an immediate increase in P inside the container. The container expands its volume to reduce the internal pressure back to the external pressure. We have the same mass of gas in a larger volume. Gas density will decrease in the flexible container as T increases.

27. At STP (T = 273.2 K and P = 1.000 atm), the volume of 1.000 mol of gas is:

$$V = \frac{nRT}{P} = \frac{1.000\text{mol} \times \dfrac{0.08206\text{L atm}}{\text{K mol}} \times 273.2\text{K}}{1.000\text{atm}} = 22.42 \text{ L}$$

At STP, the volume of 1.000 mole of any gas is 22.42 L, assuming the gas behaves ideally. Therefore, the molar volume of He(g) and N_2(g) at STP both equal 22.42 L/mol. If the temperature increases to 25.0°C (298.2 K), the volume of 1.000 mole of a gas will be larger than 22.42 L/mole because molar volume is directly related to the temperature at constant pressure. If 1.000 mole of a gas is collected over water at a total pressure of 1.000 atm, the partial pressure of the collected gas will be less than 1.000 atm because water vapor is present ($P_{total} = P_{gas} + P_{H_2O}$). At some partial pressure below 1.000 atm, the volume of 1.000 mole of a gas will be larger than 22.42 L/mol because molar volume is inversely related to the pressure at constant temperature.

28. For the first diagram, there is a total volume of 3X after the stopcock is open. The six total gas particles will be equally distributed (on average) over the entire volume (3X). So per X volume, there will be two gas particles. Your first drawing should have four gas particles in the 2X volume flask and two gas particles in the X volume flask.

Applying Boyle's law, the pressure in the two flasks after the stopcock is opened is:

$$P_1V_1 = P_2V_2,\ P_2 = \frac{P_1V_1}{V_2} = \frac{P_1 \times 2X}{3X} = \frac{2}{3}P_1$$

The final pressure in both flasks will be two-thirds that of the initial pressure in the left flask.

For the second diagram, there is a total volume of 2X after the stopcock is opened. The gas particles will be equally distributed (on average) so that your drawing should have three gas particles in each flask. The final pressure is:

$$P_2 = \frac{P_1V_1}{V_2} = \frac{P_1 \times X}{2X} = \frac{P_1}{2}$$

The final pressure in both flasks will be one-half that of the initial pressure in the left flask.

29. No; at any nonzero Kelvin temperature, there is a distribution of kinetic energies. Similarly, there is a distribution of velocities at any nonzero Kelvin temperature. The reason there is a distribution of kinetic energies at any specific temperature is because there is a distribution of velocities for any gas sample at any specific temperature.

30. a. Containers ii, iv, vi, and viii have volumes twice those of containers i, iii, v, and vii. Containers iii, iv, vii, and viii have twice the number of molecules present than containers i, ii, v, and vi. The container with the lowest pressure will be the one that has the fewest moles of gas present in the largest volume (containers ii and vi both have the lowest P). The smallest container with the most moles of gas present will have the highest pressure (containers iii and vii both have the highest P). All the other containers (i, iv, v, and viii) will have the same pressure between the two extremes. The order is ii = vi < i = iv = v = viii < iii = vii.

b. All have the same average kinetic energy because the temperature is the same in each container. Only temperature determines the average kinetic energy.

c. The least dense gas will be in container ii because it has the fewest of the lighter Ne atoms present in the largest volume. Container vii has the most dense gas because the largest number of the heavier Ar atoms are present in the smallest volume. To determine the ordering for the other containers, we will calculate the relative density of each. In the table below, m_1 equals the mass of Ne in container i, V_1 equals the volume of container i, and d_1 equals the density of the gas in container i.

Container	i	ii	iii	iv	v	vi	vii	viii
mass, volume	m_1, V_1	m_1, $2V_1$	$2m_1$, V_1	$2m_1$, $2V_1$	$2m_1$, V_1	$2m_1$, $2V_1$	$4m_1$, V_1	$4m_1$, $2V_1$
density $\left(\frac{\text{mass}}{\text{volume}}\right)$	$\frac{m_1}{V_1} = d_1$	$\frac{m_1}{2V_1} = \frac{1}{2}d_1$	$\frac{2\,m_1}{V_1} = 2d_1$	$\frac{2\,m_1}{2\,V_1} = d_1$	$\frac{2\,m_1}{V_1} = 2d_1$	$\frac{2\,m_1}{2\,V_1} = d_1$	$\frac{4\,m_1}{V_1} = 4d_1$	$\frac{4\,m_1}{2\,V_1} = 2d_1$

From the table, the order of gas density is ii < i = iv = vi < iii = v = viii < vii.

d. $\mu_{rms} = (3RT/M)^{1/2}$; the root mean square velocity only depends on the temperature and the molar mass. Because T is constant, the heavier argon molecules will have a slower root mean square velocity than the neon molecules. The order is v = vi = vii = viii < i = ii = iii = iv.

31. $2\ NH_3(g) \rightarrow N_2(g) + 3\ H_2(g)$; as reactants are converted into products, we go from 2 moles of gaseous reactants to 4 moles of gaseous products (1 mol N_2 + 3 mol H_2). Because the moles of gas doubles as reactants are converted into products, the volume of the gases will double (at constant P and T).

$PV = nRT,\ P = \left(\frac{RT}{V}\right)n = (\text{constant})n$; pressure is directly related to n at constant T and V.

As the reaction occurs, the moles of gas will double, so the pressure will double. Because 1 mole of N_2 is produced for every 2 moles of NH_3 reacted, $P_{N_2} = (1/2)P^{o}_{NH_3}$. Owing to the 3 : 2 mole ratio in the balanced equation, $P_{H_2} = (3/2)P^{o}_{NH_3}$.

Note: $P_{total} = P_{H_2} + P_{N_2} = (3/2)P^{o}_{NH_3} + (1/2)P^{o}_{NH_3} = 2P^{o}_{NH_3}$. As we said earlier, the total pressure will double from the initial pressure of NH_3 as reactants are completely converted into products.

32. Statements a, c, and e are true. For statement b, if temperature is constant, then the average kinetic energy will be constant no matter what the identity of the gas (KE_{ave} = 3/2 RT). For statement d, as T increases, the average velocity of the gas particles increases. When gas particles are moving faster, the effect of interparticle interactions is minimized. For statement f, the KMT predicts that P is directly related to T at constant V and n. As T increases, the gas molecules move faster, on average, resulting in more frequent and more forceful collisions. This leads to an increase in P.

33. The values of *a* are: H_2, $\frac{0.244\,\text{atm L}^2}{\text{mol}^2}$; CO_2, 3.59; N_2, 1.39; CH_4, 2.25

Because *a* is a measure of intermolecular attractions, the attractions are greatest for CO_2.

34. The van der Waals constant *b* is a measure of the size of the molecule. Thus C_3H_8 should have the largest value of *b* because it has the largest molar mass (size).

35. PV = nRT; Figure 5.6 is illustrating how well Boyle's law works. Boyle's law studies the pressure-volume relationship for a gas at constant moles of gas (n) and constant temperature (T). At constant n and T, the PV product for an ideal gas equals a constant value of nRT, no matter what the pressure of the gas. Figure 5.6 plots the PV product versus P for three different gases. The ideal value for the PV product is shown with a dotted line at about a value of 22.41 L atm. From the plot, it looks like the plot for Ne is closest to the dotted line, so we can conclude that of the three gases in the plot, Ne behaves most ideally. The O_2 plot is also fairly close to the dotted line, so O_2 also behaves fairly ideally. CO_2, on the other hand, has a plot farthest from the ideal plot; hence CO_2 behaves least ideally.

36. Dalton's law of partial pressures holds if the total pressure of a mixture of gases depends only on the total moles of gas particles present and not on the identity of the gases in the mixtures. If the total pressure of a mixture of gases were to depend on the identities of the gases, then each gas would behave differently at a certain set of conditions, and determining the pressure of a mixture of gases would be very difficult. All ideal gases are assumed volumeless and are assumed to exert no forces among the individual gas particles. Only in this scenario can Dalton's law of partial pressure hold true for an ideal gas. If gas particles did have a volume and/or did exert forces among themselves, then each gas, with its own identity and size, would behave differently. This is not observed for ideal gases.

Exercises

Pressure

37. a. $4.8\text{ atm} \times \frac{760\text{ mm Hg}}{\text{atm}} = 3.6 \times 10^3\text{ mm Hg}$

b. $3.6 \times 10^3\text{ mm Hg} \times \frac{1\text{ torr}}{\text{mm Hg}} = 3.6 \times 10^3\text{ torr}$

c. $4.8\text{ atm} \times \frac{1.013 \times 10^5\text{ Pa}}{\text{atm}} = 4.9 \times 10^5\text{ Pa}$

d. $4.8\text{ atm} \times \frac{14.7\text{ psi}}{\text{atm}} = 71\text{ psi}$

38. a. $2200 \text{ psi} \times \frac{1 \text{ atm}}{14.7 \text{ psi}} = 150 \text{ atm}$

b. $150 \text{ atm} \times \frac{1.013 \times 10^5 \text{ Pa}}{\text{atm}} \times \frac{1 \text{ MPa}}{1 \times 10^6 \text{ Pa}} = 15 \text{ MPa}$

c. $150 \text{ atm} \times \frac{760 \text{ torr}}{\text{atm}} = 1.1 \times 10^5 \text{ torr}$

39. $6.5 \text{ cm} \times \frac{10 \text{ mm}}{\text{cm}} = 65 \text{ mm Hg} = 65 \text{ torr};\ 65 \text{ torr} \times \frac{1 \text{ atm}}{760 \text{ torr}} = 8.6 \times 10^{-2} \text{ atm}$

$8.6 \times 10^{-2} \text{ atm} \times \frac{1.013 \times 10^5 \text{ Pa}}{\text{atm}} = 8.7 \times 10^3 \text{ Pa}$

40. $20.0 \text{ in Hg} \times \frac{2.54 \text{ cm}}{\text{in}} \times \frac{10 \text{ mm}}{\text{cm}} = 508 \text{ mm Hg} = 508 \text{ torr};\ 508 \text{ torr} \times \frac{1 \text{ atm}}{760 \text{ torr}} = 0.668 \text{ atm}$

41. If the levels of mercury in each arm of the manometer are equal, then the pressure in the flask is equal to atmospheric pressure. When they are unequal, the difference in height in millimeters will be equal to the difference in pressure in millimeters of mercury between the flask and the atmosphere. Which level is higher will tell us whether the pressure in the flask is less than or greater than atmospheric.

a. $P_{flask} < P_{atm}$; $P_{flask} = 760. - 118 = 642$ torr

$642 \text{ torr} \times \frac{1 \text{ atm}}{760 \text{ torr}} = 0.845 \text{ atm}$

$0.845 \text{ atm} \times \frac{1.013 \times 10^5 \text{ Pa}}{\text{atm}} = 8.56 \times 10^4 \text{ Pa}$

b. $P_{flask} > P_{atm}$; $P_{flask} = 760.$ torr $+ 215$ torr $= 975$ torr

$975 \text{ torr} \times \frac{1 \text{ atm}}{760 \text{ torr}} = 1.28 \text{ atm}$

$1.28 \text{ atm} \times \frac{1.013 \times 10^5 \text{ Pa}}{\text{atm}} = 1.30 \times 10^5 \text{ Pa}$

c. $P_{flask} = 635 - 118 = 517$ torr; $P_{flask} = 635 + 215 = 850.$ torr

42. a. The pressure is proportional to the mass of the fluid. The mass is proportional to the volume of the column of fluid (or to the height of the column assuming the area of the column of fluid is constant).

$d = \text{density} = \dfrac{\text{mass}}{\text{volume}}$; in this case, the volume of silicon oil will be the same as the volume of mercury in Exercise 41.

$$V = \frac{m}{d};\ V_{Hg} = V_{oil};\quad \frac{m_{Hg}}{d_{Hg}} = \frac{m_{oil}}{d_{oil}},\ m_{oil} = \frac{m_{Hg}d_{oil}}{d_{Hg}}$$

Because P is proportional to the mass of liquid:

$$P_{oil} = P_{Hg}\left(\frac{d_{oil}}{d_{Hg}}\right) = P_{Hg}\left(\frac{1.30}{13.6}\right) = (0.0956)P_{Hg}$$

This conversion applies only to the column of silicon oil.

$P_{flask} = 760.\ \text{torr} - (0.0956 \times 118)\ \text{torr} = 760. - 11.3 = 749\ \text{torr}$

$$749\ \text{torr} \times \frac{1\ \text{atm}}{760\ \text{torr}} = 0.986\ \text{atm};\ 0.986\ \text{atm} \times \frac{1.013 \times 10^5\ \text{Pa}}{\text{atm}} = 9.99 \times 10^4\ \text{Pa}$$

$P_{flask} = 760.\ \text{torr} + (0.0956 \times 215)\ \text{torr} = 760. + 20.6 = 781\ \text{torr}$

$$781\ \text{torr} \times \frac{1\ \text{atm}}{760\ \text{torr}} = 1.03\ \text{atm};\ 1.03\ \text{atm} \times \frac{1.013 \times 10^5\ \text{Pa}}{\text{atm}} = 1.04 \times 10^5\ \text{Pa}$$

b. If we are measuring the same pressure, the height of the silicon oil column would be 13.6 ÷ 1.30 = 10.5 times the height of a mercury column. The advantage of using a less dense fluid than mercury is in measuring small pressures. The height difference measured will be larger for the less dense fluid. Thus the measurement will be more precise.

Gas Laws

43. At constant n and T, PV = nRT = constant, so $P_1V_1 = P_2V_2$; at sea level, P = 1.00 atm = 760. mm Hg.

$$V_2 = \frac{P_1V_1}{P_2} = \frac{760.\ \text{mm} \times 2.0\ \text{L}}{500.\ \text{mm Hg}} = 3.0\ \text{L}$$

The balloon will burst at this pressure because the volume must expand beyond the 2.5 L limit of the balloon.

Note: To solve this problem, we did not have to convert the pressure units into atm; the units of mm Hg canceled each other. In general, only convert units if you have to. Whenever the gas constant R is not used to solve a problem, pressure and volume units must only be consistent and not necessarily in units of atm and L. The exception is temperature, which must always be converted to the Kelvin scale.

44. The pressure exerted on the balloon is constant, and the moles of gas present is constant. From Charles's law, $V_1/T_1 = V_2/T_2$ at constant P and n.

$$V_2 = \frac{V_1T_2}{T_1} = \frac{700.\text{ mL} \times 100.\text{ K}}{(273.2 + 20.0)\text{ K}} = 239\text{ mL}$$

As expected, as temperature decreases, the volume decreases.

45. At constant T and P, Avogadro's law holds ($V \propto n$).

$$\frac{V_1}{n_1} = \frac{V_2}{n_2},\ n_2 = \frac{V_2n_1}{V_1} = \frac{20.\text{ L} \times 0.50\text{ mol}}{11.2\text{ L}} = 0.89\text{ mol}$$

As expected, as V increases, n increases.

46. As NO_2 is converted completely into N_2O_4, the moles of gas present will decrease by one-half (from the 2 : 1 mole ratio in the balanced equation). Using Avogadro's law:

$$\frac{V_1}{n_1} = \frac{V_2}{n_2},\ V_2 = V_1 \times \frac{n_2}{n_1} = 25.0\text{ mL} \times \frac{1}{2} = 12.5\text{ mL}$$

$N_2O_4(g)$ will occupy one-half the original volume of $NO_2(g)$. This is expected because the moles of gas present decrease by one-half when NO_2 is converted into N_2O_4.

47. a. $$PV = nRT,\ V = \frac{nRT}{P} = \frac{2.00\text{ mol} \times \frac{0.08206\text{ L atm}}{\text{K mol}} \times (155 + 273)\text{ K}}{5.00\text{ atm}} = 14.0\text{ L}$$

b. $$PV = nRT,\ n = \frac{PV}{RT} = \frac{0.300\text{ atm} \times 2.00\text{ L}}{\frac{0.08206\text{ L atm}}{\text{K mol}} \times 155\text{ K}} = 4.72 \times 10^{-2}\text{ mol}$$

c. $$PV = nRT,\ T = \frac{PV}{nR} = \frac{4.47\text{ atm} \times 25.0\text{ L}}{2.01\text{ mol} \times \frac{0.08206\text{ L atm}}{\text{K mol}}} = 678\text{ K} = 405°\text{C}$$

d. $$PV = nRT,\ P = \frac{nRT}{V} = \frac{10.5\text{ mol} \times \frac{0.08206\text{ L atm}}{\text{K mol}} \times (273 + 75)\text{ K}}{2.25\text{ L}} = 133\text{ atm}$$

48. a. $P = 7.74 \times 10^3\text{ Pa} \times \frac{1\text{ atm}}{1.013 \times 10^5\text{ Pa}} = 0.0764\text{ atm};\ T = 25 + 273 = 298\text{ K}$

$$PV = nRT,\ n = \frac{PV}{RT} = \frac{0.0764\text{ atm} \times 0.0122\text{ L}}{\frac{0.08206\text{ L atm}}{\text{K mol}} \times 298\text{ K}} = 3.81 \times 10^{-5}\text{ mol}$$

b. $PV = nRT,\ P = \dfrac{nRT}{V} = \dfrac{0.421\,\text{mol} \times \dfrac{0.08206\,\text{L atm}}{\text{K mol}} \times 223\,\text{K}}{0.0430\,\text{L}} = 179\ \text{atm}$

c. $V = \dfrac{nRT}{P} = \dfrac{4.4 \times 10^{-2}\ \text{mol} \times \dfrac{0.08206\,\text{L atm}}{\text{K mol}} \times (331 + 273)\,\text{K}}{455\,\text{torr} \times \dfrac{1\,\text{atm}}{760\,\text{torr}}} = 3.6\ \text{L}$

d. $T = \dfrac{PV}{nR} = \dfrac{\left(745\,\text{mm Hg} \times \dfrac{1\,\text{atm}}{760\,\text{mm Hg}}\right) \times 11.2\ \text{L}}{0.401\,\text{mol} \times \dfrac{0.08206\,\text{L atm}}{\text{K mol}}} = 334\ \text{K} = 61°\text{C}$

49. $n = \dfrac{PV}{RT} = \dfrac{2.70\,\text{atm} \times 200.0\,\text{L}}{\dfrac{0.08206\,\text{L atm}}{\text{K mol}} \times (273 + 24)\,\text{K}} = 22.2\ \text{mol}$

For He: $22.2\ \text{mol} \times \dfrac{4.003\,\text{g He}}{\text{mol}} = 88.9\ \text{g He}$

For H_2: $22.2\ \text{mol} \times \dfrac{2.016\,\text{g}\ H_2}{\text{mol}} = 44.8\ \text{g}\ H_2$

50. a. $n = \dfrac{PV}{RT} = \dfrac{1.00\,\text{atm} \times 6.0\,\text{L}}{\dfrac{0.08206\,\text{L atm}}{\text{K mol}} \times 298\,\text{K}} = 0.25\,\text{mol air}$

b. $n = \dfrac{1.97\,\text{atm} \times 6.0\,\text{L}}{0.08206\,\text{L atm/K mol} \times 298\,\text{K}} = 0.48\,\text{mol air}$

c. $n = \dfrac{0.296\,\text{atm} \times 6.0\,\text{L}}{0.08206\,\text{L atm/K mol} \times 200.\,\text{K}} = 0.11\,\text{mol air}$

Air is indeed "thinner" at high elevations.

51. $PV = nRT,\ n = \dfrac{PV}{RT} = \dfrac{14.5\,\text{atm} \times (75.0 \times 10^{-3}\ \text{L})}{\dfrac{0.08206\,\text{L atm}}{\text{K mol}} \times 295\,\text{K}} = 0.0449\ \text{mol}\ O_2$

52. $P = \dfrac{nRT}{V} = \dfrac{\left(0.60\,\text{g} \times \dfrac{1\,\text{mol}}{32.00\,\text{g}}\right) \times \dfrac{0.08206\,\text{L atm}}{\text{K mol}} \times (273 + 22)\,\text{K}}{5.0\,\text{L}} = 0.091\ \text{atm}$

53. a. $PV = nRT$; $175 \text{ g Ar} \times \dfrac{1 \text{ mol Ar}}{39.95 \text{ g Ar}} = 4.38 \text{ mol Ar}$

$$T = \frac{PV}{nR} = \frac{10.0 \text{ atm} \times 2.50 \text{ L}}{4.38 \text{ mol} \times \dfrac{0.08206 \text{ L atm}}{\text{K mol}}} = 69.6 \text{ K}$$

b. $PV = nRT$, $P = \dfrac{nRT}{V} = \dfrac{4.38 \text{ mol} \times \dfrac{0.08206 \text{ L atm}}{\text{K mol}} \times 255 \text{ K}}{2.50 \text{ L}} = 32.3 \text{ atm}$

54. $0.050 \text{ mL} \times \dfrac{1.149 \text{ g}}{\text{mL}} \times \dfrac{1 \text{ mol } O_2}{32.00 \text{ g}} = 1.8 \times 10^{-3} \text{ mol } O_2$

$$V = \frac{nRT}{P} = \frac{1.8 \times 10^{-3} \text{ mol} \times \dfrac{0.08206 \text{ L atm}}{\text{K mol}} \times 310. \text{ K}}{1.0 \text{ atm}} = 4.6 \times 10^{-2} \text{ L} = 46 \text{ mL}$$

55. For a gas at two conditions: $\dfrac{P_1V_1}{n_1T_1} = \dfrac{P_2V_2}{n_2T_2}$

Because V is constant: $\dfrac{P_1}{n_1T_1} = \dfrac{P_2}{n_2T_2}$, $n_2 = \dfrac{n_1P_2T_1}{P_1T_2}$

$$n_2 = \frac{1.50 \text{ mol} \times 800. \text{ torr} \times 298 \text{ K}}{400. \text{ torr} \times 323 \text{ K}} = 2.77 \text{ mol}$$

Moles of gas added $= n_2 - n_1 = 2.77 - 1.50 = 1.27$ mol

For two-condition problems, units for P and V just need to be the same units for both conditions, not necessarily atm and L. The unit conversions from other P or V units would cancel when applied to both conditions. However, temperature always must be converted to the Kelvin scale. The temperature conversions between other units and Kelvin will not cancel each other.

56. $PV = nRT$, n is constant. $\dfrac{PV}{T} = nR = \text{constant}$, $\dfrac{P_1V_1}{T_1} = \dfrac{P_2V_2}{T_2}$

$V_2 = (1.040)V_1$, $\dfrac{V_1}{V_2} = \dfrac{1.000}{1.040}$

$$P_2 = \frac{P_1V_1T_2}{V_2T_1} = 75 \text{ psi} \times \frac{1.000}{1.040} \times \frac{(273 + 58) \text{ K}}{(273 + 19) \text{ K}} = 82 \text{ psi}$$

57. At two conditions: $\dfrac{P_1V_1}{n_1T_1} = \dfrac{P_2V_2}{n_2T_2}$; all gases are assumed to follow the ideal gas law. The identity of the gas in container B is unimportant as long as we know the moles of gas present.

$$\frac{P_B}{P_A} = \frac{V_A n_B T_B}{V_B n_A T_A} = \frac{1.0\ \text{L} \times 2.0\ \text{mol} \times 560.\ \text{K}}{2.0\ \text{L} \times 1.0\ \text{mol} \times 280.\ \text{K}} = 2.0$$

The pressure of the gas in container B is twice the pressure of the gas in container A.

58. The pressure is doubled so $P_2 = 2P_1$ and the absolute temperature is halved so $T_2 = ½T_1$ (or $T_1 = 2T_2$). The moles of gas did not change, so $n_2 = n_1$. The volume effect of these changes is:

$$\frac{P_1V_1}{n_1T_1} = \frac{P_2V_2}{n_2T_2},\ \frac{V_2}{V_1} = \frac{P_1n_2T_2}{P_2n_1T_1} = \frac{P_1T_2}{P_2T_1} = \frac{P_1 \times T_2}{2P_1 \times 2T_2} = 1/4$$

The volume of the gas decreases by a factor of four when the pressure is doubled and the absolute temperature is halved.

59. a. At constant n and V, $\frac{P_1}{T_1} = \frac{P_2}{T_2}$, $P_2 = \frac{P_1T_2}{T_1} = 11.0\ \text{atm} \times \frac{318\ \text{K}}{273\ \text{K}} = 12.8\ \text{atm}$

b. $\frac{P_1}{T_1} = \frac{P_2}{T_2}$, $T_2 = \frac{T_1P_2}{P_1} = 273\ \text{K} \times \frac{6.50\ \text{atm}}{11.0\ \text{atm}} = 161\ \text{K}$

c. $T_2 = \frac{T_1P_2}{P_1} = 273\ \text{K} \times \frac{25.0\ \text{atm}}{11.0\ \text{atm}} = 620.\ \text{K}$

60. Because the container is flexible, P is assumed constant. The moles of gas present are also constant.

$$\frac{P_1V_1}{n_1T_1} = \frac{P_2V_2}{n_2T_2},\ \frac{V_1}{T_1} = \frac{V_2}{T_2};\ V_{sphere} = 4/3\ \pi r^3$$

$$V_2 = \frac{V_1T_2}{T_1},\ 4/3\ \pi(r_2)^3 = \frac{4/3\ \pi(1.00\ \text{cm})^3 \times 361\ \text{K}}{280.\ \text{K}}$$

$$r_2^3 = \frac{361\ \text{K}}{280.\ \text{K}} = 1.29,\ r_2 = (1.29)^{1/3} = 1.09\ \text{cm} = \text{radius of sphere after heating}$$

61. $\frac{PV}{T} = nR = \text{constant}$, $\frac{P_1V_1}{T_1} = \frac{P_2V_2}{T_2}$

$$P_2 = \frac{P_1V_1T_2}{V_2T_1} = 710.\ \text{torr} \times \frac{5.0 \times 10^2\ \text{mL}}{25\ \text{mL}} \times \frac{(273 + 820.)\ \text{K}}{(273 + 30.)\ \text{K}} = 5.1 \times 10^4\ \text{torr}$$

62. $PV = nRT$, $\frac{nT}{P} = \frac{V}{R} = \text{constant}$, $\frac{n_1T_1}{P_1} = \frac{n_2T_2}{P_2}$; moles × molar mass = mass

$$\frac{n_1(\text{molar mass})T_1}{P_1} = \frac{n_2(\text{molar mass})T_2}{P_2},\ \frac{\text{mass}_1 \times T_1}{P_1} = \frac{\text{mass}_2 \times T_2}{P_2}$$

$$\text{Mass}_2 = \frac{\text{mass}_1 \times T_1P_2}{T_2P_1} = \frac{1.00 \times 10^3 \text{ g} \times 291\text{K} \times 650.\text{ psi}}{299\text{K} \times 2050.\text{ psi}} = 309 \text{ g}$$

63. PV = nRT, n is constant. $\frac{PV}{T} = nR = \text{constant}$, $\frac{P_1V_1}{T_1} = \frac{P_2V_2}{T_2}$, $V_2 = \frac{V_1P_1T_2}{V_2T_1}$

$$V_2 = 1.00 \text{ L} \times \frac{760.\text{torr}}{220.\text{ torr}} \times \frac{(273-31)\text{ K}}{(273+23)\text{ K}} = 2.82 \text{ L}; \ \Delta V = 2.82 - 1.00 = 1.82 \text{ L}$$

64. PV = nRT, P is constant. $\frac{nT}{V} = \frac{P}{R} = \text{constant}$, $\frac{n_1T_1}{V_1} = \frac{n_2T_2}{V_2}$

$$\frac{n_2}{n_1} = \frac{T_1V_2}{T_2V_1} = \frac{294\text{K}}{335\text{K}} \times \frac{4.20 \times 10^3 \text{ m}^3}{4.00 \times 10^3 \text{ m}^3} = 0.921$$

Gas Density, Molar Mass, and Reaction Stoichiometry

65. STP: T = 273 K and P = 1.00 atm; at STP, the molar volume of a gas is 22.42 L.

$$2.00 \text{ L } O_2 \times \frac{1 \text{ mol } O_2}{22.42 \text{ L}} \times \frac{4 \text{ mol Al}}{3 \text{ mol } O_2} \times \frac{26.98 \text{ g Al}}{\text{mol Al}} = 3.21 \text{ g Al}$$

Note: We could also solve this problem using PV = nRT, where n_{O_2} = PV/RT. You don't have to memorize 22.42 L/mol at STP.

66. $CO_2(s) \rightarrow CO_2(g)$; $4.00 \text{ g } CO_2 \times \frac{1 \text{ mol } CO_2}{44.01 \text{g } CO_2} = 9.09 \times 10^{-2} \text{ mol } CO_2$

At STP, the molar volume of a gas is 22.42 L. $9.09 \times 10^{-2} \text{ mol } CO_2 \times \frac{22.42 \text{ L}}{\text{mol } CO_2} = 2.04 \text{ L}$

67. $2 \text{ NaN}_3(s) \rightarrow 2 \text{ Na}(s) + 3 \text{ N}_2(g)$

$$n_{N_2} = \frac{PV}{RT} = \frac{1.00 \text{ atm} \times 70.0 \text{ L}}{\frac{0.08206 \text{L atm}}{\text{K mol}} \times 273\text{K}} = 3.12 \text{ mol } N_2 \text{ needed to fill air bag.}$$

$$\text{Mass NaN}_3 \text{ reacted} = 3.12 \text{ mol } N_2 \times \frac{2 \text{ mol NaN}_3}{3 \text{ mol } N_2} \times \frac{65.02 \text{ g NaN}_3}{\text{mol NaN}_3} = 135 \text{ g NaN}_3$$

68. Because the solution is 50.0% H_2O_2 by mass, the mass of H_2O_2 decomposed is 125/2 = 62.5 g.

$$62.5 \text{ g } H_2O_2 \times \frac{1 \text{ mol } H_2O_2}{34.02 \text{ g } H_2O_2} \times \frac{1 \text{ mol } O_2}{2 \text{ mol } H_2O_2} = 0.919 \text{ mol } O_2$$

$$V = \frac{nRT}{P} = \frac{0.919\,\text{mol} \times \dfrac{0.08206\,\text{L atm}}{\text{K mol}} \times 300.\,\text{K}}{746\,\text{torr} \times \dfrac{1\,\text{atm}}{760\,\text{torr}}} = 23.0\ \text{L } O_2$$

69. $$n_{H_2} = \frac{PV}{RT} = \frac{1.0\,\text{atm} \times \left[4800\,\text{m}^3 \times \left(\dfrac{100\,\text{cm}}{\text{m}}\right)^3 \times \dfrac{1\,\text{L}}{1000\,\text{cm}^3}\right]}{\dfrac{0.08206\,\text{L atm}}{\text{K mol}} \times 273\,\text{K}} = 2.1 \times 10^5\ \text{mol}$$

2.1×10^5 mol H_2 is in the balloon. This is 80.% of the total amount of H_2 that had to be generated:

$0.80(\text{total mol } H_2) = 2.1 \times 10^5$, total mol $H_2 = 2.6 \times 10^5$ mol

$$2.6 \times 10^5\ \text{mol } H_2 \times \frac{1\,\text{mol Fe}}{\text{mol } H_2} \times \frac{55.85\,\text{g Fe}}{\text{mol Fe}} = 1.5 \times 10^7\ \text{g Fe}$$

$$2.6 \times 10^5\ \text{mol } H_2 \times \frac{1\,\text{mol } H_2SO_4}{\text{mol } H_2} \times \frac{98.09\,\text{g } H_2SO_4}{\text{mol } H_2SO_4} \times \frac{100\,\text{g reagent}}{98\,\text{g } H_2SO_4}$$
$$= 2.6 \times 10^7\ \text{g of 98\% sulfuric acid}$$

70. $$5.00\ \text{g S} \times \frac{1\,\text{mol S}}{32.07\,\text{g}} = 0.156\ \text{mol S}$$

0.156 mol S will react with 0.156 mol O_2 to produce 0.156 mol SO_2. More O_2 is required to convert SO_2 into SO_3.

$$0.156\ \text{mol } SO_2 \times \frac{1\,\text{mol } O_2}{2\,\text{mol } SO_2} = 0.0780\ \text{mol } O_2$$

Total mol O_2 reacted = 0.156 + 0.0780 = 0.234 mol O_2

$$V = \frac{nRT}{P} = \frac{0.234\,\text{mol} \times \dfrac{0.08206\,\text{L atm}}{\text{K mol}} \times 623\,\text{K}}{5.25\,\text{atm}} = 2.28\ \text{L } O_2$$

71. $Kr(g) + 2\,Cl_2(g) \rightarrow KrCl_4(s)$; $$n_{Kr} = \frac{PV}{RT} = \frac{0.500\,\text{atm} \times 15.0\,\text{L}}{\dfrac{0.08206\,\text{L atm}}{\text{K mol}} \times 623\,\text{K}} = 0.147\ \text{mol Kr}$$

We could do the same calculation for Cl_2. However, the only variable that changed is the pressure. Because the partial pressure of Cl_2 is triple that of Kr, moles of Cl_2 = 3(0.147) = 0.441 mol Cl_2. The balanced equation requires 2 moles of Cl_2 to react with every mole of Kr. However, we actually have three times as many moles of Cl_2 as we have of Kr. So Cl_2 is in excess and Kr is the limiting reagent.

$$0.147 \text{ mol Kr} \times \frac{1 \text{ mol KrCl}_4}{\text{mol Kr}} \times \frac{225.60 \text{ g KrCl}_4}{\text{mol Kr}} = 33.2 \text{ g KrCl}_4$$

72. PV = nRT, V and T are constant. $\frac{P_1}{n_1} = \frac{P_2}{n_2}, \ \frac{P_2}{P_1} = \frac{n_2}{n_1}$

Let's calculate the partial pressure of C_3H_3N that can be produced from each of the starting materials assuming each reactant is limiting. The reactant that produces the smallest amount of product will run out first and is the limiting reagent.

$$P_{C_3H_3N} = 0.500 \text{ MPa} \times \frac{2 \text{ MPa } C_3H_3N}{2 \text{ MPa } C_3H_6} = 0.500 \text{ MPa if } C_3H_6 \text{ is limiting}$$

$$P_{C_3H_3N} = 0.800 \text{ MPa} \times \frac{2 \text{ MPa } C_3H_3N}{2 \text{ MPa } NH_3} = 0.800 \text{ MPa if } NH_3 \text{ is limiting}$$

$$P_{C_3H_3N} = 1.500 \text{ MPa} \times \frac{2 \text{ MPa } C_3H_3N}{3 \text{ MPa } O_2} = 1.000 \text{ MPa if } O_2 \text{ is limiting}$$

C_3H_6 is limiting. Although more product could be produced from NH_3 and O_2, there is only enough C_3H_6 to produce 0.500 MPa of C_3H_3N. The partial pressure of C_3H_3N in atmospheres after the reaction is:

$$0.500 \times 10^6 \text{ Pa} \times \frac{1 \text{ atm}}{1.013 \times 10^5 \text{ Pa}} = 4.94 \text{ atm}$$

$$n = \frac{PV}{RT} = \frac{4.94 \text{ atm} \times 150. \text{ L}}{\frac{0.08206 \text{ L atm}}{\text{K mol}} \times 298 \text{ K}} = 30.3 \text{ mol } C_3H_3N$$

$$30.3 \text{ mol} \times \frac{53.06 \text{ g}}{\text{mol}} = 1.61 \times 10^3 \text{ g } C_3H_3N \text{ can be produced.}$$

73. $CH_3OH + 3/2\ O_2 \rightarrow CO_2 + 2\ H_2O$ or $2\ CH_3OH(l) + 3\ O_2(g) \rightarrow 2\ CO_2(g) + 4\ H_2O(g)$

$$50.0 \text{ mL} \times \frac{0.850 \text{ g}}{\text{mL}} \times \frac{1 \text{ mol}}{32.04 \text{ g}} = 1.33 \text{ mol } CH_3OH(l) \text{ available}$$

$$n_{O_2} = \frac{PV}{RT} = \frac{2.00 \text{ atm} \times 22.8 \text{ L}}{\frac{0.08206 \text{ L atm}}{\text{K mol}} \times 300. \text{ K}} = 1.85 \text{ mol } O_2 \text{ available}$$

Assuming CH_3OH is limiting:

$$1.33 \text{ mol } CH_3OH \times \frac{4 \text{ mol } H_2O}{2 \text{ mol } CH_3OH} = 2.66 \text{ mol } H_2O$$

Assuming O_2 is limiting:

$$1.85 \text{ mol } O_2 \times \frac{4 \text{ mol} H_2O}{3 \text{ mol} O_2} = 2.47 \text{ mol } H_2O$$

Because the O_2 reactant produces the smaller quantity of H_2O, O_2 is limiting and 2.47 mol of H_2O can be produced.

74. For ammonia (in 1 minute):

$$n_{NH_3} = \frac{P_{NH_3} \times V_{NH_3}}{RT} = \frac{90. \text{ atm} \times 500. \text{ L}}{\frac{0.08206 \text{ L atm}}{\text{K mol}} \times 496 \text{ K}} = 1.1 \times 10^3 \text{ mol } NH_3$$

NH_3 flows into the reactor at a rate of 1.1×10^3 mol/min.

For CO_2 (in 1 minute):

$$n_{CO_2} = \frac{P_{CO_2} \times V_{CO_2}}{RT} = \frac{45 \text{ atm} \times 600. \text{ L}}{\frac{0.08206 \text{ L atm}}{\text{K mol}} \times 496 \text{ K}} = 6.6 \times 10^2 \text{ mol } CO_2$$

CO_2 flows into the reactor at 6.6×10^2 mol/min.

If NH_3 is limiting:

$$\frac{1.1 \times 10^3 \text{ mol} NH_3}{\text{min}} \times \frac{1 \text{ mol urea}}{2 \text{ mol} NH_3} \times \frac{60.06 \text{ g urea}}{\text{mol urea}} = 3.3 \times 10^4 \text{ g urea/min}$$

If CO_2 is limiting:

$$\frac{660 \text{ mol} CO_2}{\text{min}} \times \frac{1 \text{ mol urea}}{\text{mol} CO_2} \times \frac{60.06 \text{ g urea}}{\text{mol urea}} = 4.0 \times 10^4 \text{ g urea/min}$$

Because the NH_3 reactant produces the smaller quantity of product, NH_3 is limiting and 3.3×10^4 g urea/min can be formed.

75. a. $CH_4(g) + NH_3(g) + O_2(g) \rightarrow HCN(g) + H_2O(g)$; balancing H first, then O, gives:

$$CH_4 + NH_3 + \frac{3}{2} O_2 \rightarrow HCN + 3\ H_2O \text{ or } 2\ CH_4(g) + 2\ NH_3(g) + 3\ O_2(g) \rightarrow 2\ HCN(g) + 6\ H_2O(g)$$

b. PV = nRT, T and P constant; $\frac{V_1}{n_1} = \frac{V_2}{n_2}, \frac{V_1}{V_2} = \frac{n_1}{n_2}$

The volumes are all measured at constant T and P, so the volumes of gas present are directly proportional to the moles of gas present (Avogadro's law). Because Avogadro's law applies, the balanced reaction gives mole relationships as well as volume relationships.

If CH_4 is limiting: $20.0\text{ L }CH_4 \times \dfrac{2\text{ L HCN}}{2\text{ L }CH_4} = 20.0\text{ L HCN}$

If NH_3 is limiting: $20.0\text{ L }NH_3 \times \dfrac{2\text{ L HCN}}{2\text{ L }NH_3} = 20.0\text{ L HCN}$

If O_2 is limiting: $20.0\text{ L }O_2 \times \dfrac{2\text{ L HCN}}{3\text{ L }O_2} = 13.3\text{ L HCN}$

O_2 produces the smallest quantity of product, so O_2 is limiting and 13.3 L HCN can be produced.

76. From the balanced equation, ethene reacts with hydrogen in a 1 : 1 mole ratio. Because T and P are constant, a greater volume of H_2 and thus more moles of H_2 are flowing into the reaction container than moles of ethene. So ethene is the limiting reagent.

In 1 minute:

$$n_{C_2H_4} = \frac{PV}{RT} = \frac{25.0\text{ atm} \times 1000.\text{ L}}{\dfrac{0.08206\text{L atm}}{\text{K mol}} \times 573\text{K}} = 532\text{ mol }C_2H_4\text{ reacted}$$

$$\text{Theoretical yield} = \frac{532\text{ mol }C_2H_4}{\text{min}} \times \frac{1\text{ mol }C_2H_6}{\text{mol }C_2H_4} \times \frac{30.07\text{ g }C_2H_6}{\text{mol }C_2H_6} \times \frac{1\text{ kg}}{1000\text{ g}} = 16.0\text{ kg }C_2H_6/\text{min}$$

$$\text{Percent yield} = \frac{15.0\text{kg/min}}{16.0\text{kg/min}} \times 100 = 93.8\%$$

77. Molar mass $= \dfrac{dRT}{P}$, where d = density of gas in units of g/L.

$$\text{Molar mass} = \frac{3.164\text{g/L} \times \dfrac{0.08206\text{L atm}}{\text{K mol}} \times 273.2\text{K}}{1.000\text{atm}} = 70.98\text{ g/mol}$$

The gas is diatomic, so the average atomic mass = 70.93/2 = 35.47 u. From the periodic table, this is chlorine, and the identity of the gas is Cl_2.

78. P × (molar mass) = dRT, $d = \dfrac{\text{mass}}{\text{volume}}$, P × (molar mass) $= \dfrac{\text{mass}}{V} \times RT$

$$\text{Molar mass} = \frac{\text{mass} \times RT}{PV} = \frac{0.800\text{ g} \times \dfrac{0.08206\text{L atm}}{\text{K mol}} \times 373\text{K}}{(750.\text{ torr} \times \dfrac{1\text{ atm}}{760\text{ torr}}) \times 0.256\text{L}} = 96.9\text{ g/mol}$$

Mass of CHCl ≈ 12.0 + 1.0 + 35.5 = 48.5 g/mol; $\dfrac{96.9}{48.5} = 2.00$; molecular formula is $C_2H_2Cl_2$.

79. $$d_{UF_6} = \frac{P \times (\text{molar mass})}{RT} = \frac{\left(745\,\text{torr} \times \frac{1\,\text{atm}}{760\,\text{torr}}\right) \times 352.0\ \text{g/mol}}{\frac{0.08206\,\text{L atm}}{\text{K mol}} \times 333\,\text{K}} = 12.6\ \text{g/L}$$

80. d = P × (molar mass)/RT; we need to determine the average molar mass of air. We get this by using the mole fraction information to determine the weighted value for the molar mass. If we have 1.000 mol of air:

$$\text{average molar mass} = 0.78\ \text{mol}\ N_2 \times \frac{28.02\ \text{g}\ N_2}{\text{mol}\ N_2} + 0.21\ \text{mol}\ O_2 \times \frac{32.00\ \text{g}\ O_2}{\text{mol}\ O_2}$$
$$+ 0.010\ \text{mol Ar} \times \frac{39.95\ \text{g Ar}}{\text{mol Ar}} = 28.98 = 29\ \text{g}$$

$$d_{air} = \frac{1.00\,\text{atm} \times 29\,\text{g/mol}}{\frac{0.08206\,\text{L atm}}{\text{K mol}} \times 273\,\text{K}} = 1.3\ \text{g/L}$$

Partial Pressure

81. The container has 5 He atoms, 3 Ne atoms, and 2 Ar atoms for a total of 10 atoms. The mole fractions of the various gases will be equal to the molecule fractions.

$$\chi_{He} = \frac{5\ \text{He atoms}}{10\ \text{total atoms}} = 0.50;\quad \chi_{Ne} = \frac{3\ \text{Ne atoms}}{10\ \text{total atoms}} = 0.30$$

$$\chi_{Ar} = 1.00 - 0.50 - 0.30 = 0.20$$

$$P_{He} = \chi_{He} \times P_{total} = 0.50(1.00\ \text{atm}) = 0.50\ \text{atm}$$

$$P_{Ne} = \chi_{Ne} \times P_{Total} = 0.30(1.00\text{atm}) = 0.30\ \text{atm}$$

$$P_{Ar} = 1.00\ \text{atm} - 0.50\ \text{atm} - 0.30\ \text{atm} = 0.20\ \text{atm}$$

82. a. There are 6 He atoms and 4 Ne atoms, and each flask has the same volume. The He flask has 1.5 times as many atoms of gas present as the Ne flask, so the pressure in the He flask will be 1.5 times greater (assuming a constant temperature).

b. Because the flask volumes are the same, your drawing should have the various atoms equally distributed between the two flasks. So each flask should have 3 He atoms and 2 Ne atoms.

c. After the stopcock is opened, each flask will have 5 total atoms and the pressures will be equal. If six atoms of He gave an initial pressure of $P_{He,\ initial}$, then 5 total atoms will have a pressure of $5/6 \times P_{He,\ initial}$.

Using similar reasoning, 4 atoms of Ne gave an initial pressure of $P_{Ne, initial}$, so 5 total atoms will have a pressure of $5/4 \times P_{Ne, initial}$. Summarizing:

$$P_{final} = \frac{5}{6}P_{He, initial} = \frac{5}{4}P_{Ne, initial}$$

d. For the partial pressures, treat each gas separately. For helium, when the stopcock is opened, the six atoms of gas are now distributed over a larger volume. To solve for the final partial pressures, use Boyle's law for each gas.

$$\text{For He: } P_2 = \frac{P_1V_1}{V_2} = P_{He, initial} \times \frac{X}{2X} = \frac{P_{He, initial}}{2}$$

The partial pressure of helium is exactly halved. The same result occurs with neon so that when the volume is doubled, the partial pressure is halved. Summarizing:

$$P_{He, final} = \frac{P_{He, initial}}{2}; \; P_{Ne, final} = \frac{P_{Ne, initial}}{2}$$

83. $$P_{CO_2} = \frac{nRT}{V} = \frac{\left(7.8\text{ g} \times \frac{1\text{ mol}}{44.01\text{ g}}\right) \times \frac{0.08206\text{ L atm}}{\text{K mol}} \times 300.\text{ K}}{4.0\text{ L}} = 1.1\text{ atm}$$

With air present, the partial pressure of CO_2 will still be 1.1 atm. The total pressure will be the sum of the partial pressures, $P_{total} = P_{CO_2} + P_{air}$.

$$P_{total} = 1.1\text{ atm} + \left(740\text{ torr} \times \frac{1\text{ atm}}{760\text{ torr}}\right) = 1.1 + 0.97 = 2.1\text{ atm}$$

84. $$n_{H_2} = 1.00\text{ g } H_2 \times \frac{1\text{ mol } H_2}{2.016\text{ g } H_2} = 0.496\text{ mol } H_2; \; n_{He} = 1.00\text{ g He} \times \frac{1\text{ mol He}}{4.003\text{ g He}}$$
$$= 0.250\text{ mol He}$$

$$P_{H_2} = \frac{n_{H_2} \times RT}{V} = \frac{0.496\text{ mol} \times \frac{0.08206\text{ L atm}}{\text{K mol}} \times (273 + 27)\text{ K}}{1.00\text{ L}} = 12.2\text{ atm}$$

$$P_{He} = \frac{n_{He} \times RT}{V} = 6.15\text{ atm}; \; P_{total} = P_{H_2} + P_{He} = 12.2\text{ atm} + 6.15\text{ atm} = 18.4\text{ atm}$$

85. Treat each gas separately and determine how the partial pressure of each gas changes when the container volume increases. Once the partial pressures of H_2 and N_2 are determined, the total pressure will be the sum of these two partial pressures. At constant n and T, the relationship $P_1V_1 = P_2V_2$ holds for each gas.

For H_2: $P_2 = \dfrac{P_1V_1}{V_2} = 475 \text{ torr} \times \dfrac{2.00\text{ L}}{3.00\text{ L}} = 317 \text{ torr}$

For N_2: $P_2 = 0.200 \text{ atm} \times \dfrac{1.00\text{ L}}{3.00\text{ L}} = 0.0667 \text{ atm}$; $0.0667 \text{ atm} \times \dfrac{760\text{ torr}}{\text{atm}} = 50.7 \text{ torr}$

$P_{total} = P_{H_2} + P_{N_2} = 317 + 50.7 = 368 \text{ torr}$

86. For H_2: $P_2 = \dfrac{P_1V_1}{V_2} = 360. \text{ torr} \times \dfrac{2.00\text{ L}}{3.00\text{ L}} = 240. \text{ torr}$

$P_{total} = P_{H_2} + P_{N_2}$, $P_{N_2} = P_{total} - P_{H_2} = 320. \text{ torr} - 240. \text{ torr} = 80. \text{ torr}$

For N_2: $P_1 = \dfrac{P_2V_2}{V_1} = 80. \text{ torr} \times \dfrac{3.00\text{ L}}{1.00\text{ L}} = 240 \text{ torr}$

87. $P_1V_1 = P_2V_2$; the total volume is 1.00 L + 1.00 L + 2.00 L = 4.00 L.

For He: $P_2 = \dfrac{P_1V_1}{V_2} = 200. \text{ torr} \times \dfrac{1.00\text{ L}}{4.00\text{ L}} = 50.0 \text{ torr He}$

For Ne: $P_2 = 0.400 \text{ atm} \times \dfrac{1.00\text{ L}}{4.00\text{ L}} = 0.100 \text{ atm}$; $0.100 \text{ atm} \times \dfrac{760\text{ torr}}{\text{atm}} = 76.0 \text{ torr Ne}$

For Ar: $P_2 = 24.0 \text{ kPa} \times \dfrac{2.00\text{ L}}{4.00\text{ L}} = 12.0 \text{ kPa}$; $12.0 \text{ kPa} \times \dfrac{1\text{ atm}}{101.3\text{ kPa}} \times \dfrac{760\text{ torr}}{\text{atm}}$

$= 90.0 \text{ torr Ar}$

$P_{total} = 50.0 + 76.0 + 90.0 = 216.0 \text{ torr}$

88. We can use the ideal gas law to calculate the partial pressure of each gas or to calculate the total pressure. There will be less math if we calculate the total pressure from the ideal gas law.

$n_{O_2} = 1.5 \times 10^2 \text{ mg } O_2 \times \dfrac{1\text{ g}}{1000\text{ mg}} \times \dfrac{1\text{ mol } O_2}{32.00\text{ g } O_2} = 4.7 \times 10^{-3} \text{ mol } O_2$

$n_{NH_3} = 5.0 \times 10^{21} \text{ molecules } NH_3 \times \dfrac{1\text{ mol } NH_3}{6.022 \times 10^{23}\text{ molecules } NH_3} = 8.3 \times 10^{-3} \text{ mol } NH_3$

$n_{total} = n_{N_2} + n_{O_2} + n_{NH_3} = 5.0 \times 10^{-2} + 4.7 \times 10^{-3} + 8.3 \times 10^{-3} = 6.3 \times 10^{-2} \text{ mol total}$

$$P_{total} = \frac{n_{total} \times RT}{V} = \frac{6.3 \times 10^{-2}\text{ mol} \times \frac{0.08206\text{ L atm}}{\text{K mol}} \times 273\text{ K}}{1.0\text{ L}} = 1.4\text{ atm}$$

$P_{N_2} = \chi_{N_2} \times P_{total}$, $\chi_{N_2} = \dfrac{n_{N_2}}{n_{total}}$; $P_{N_2} = \dfrac{5.0 \times 10^{-2}\text{ mol}}{6.3 \times 10^{-2}\text{ mol}} \times 1.4\text{ atm} = 1.1\text{ atm}$

$$P_{O_2} = \frac{4.7 \times 10^{-3}}{6.3 \times 10^{-2}} \times 1.4\text{ atm} = 0.10\text{ atm};\ P_{NH_3} = \frac{8.3 \times 10^{-3}}{6.3 \times 10^{-2}} \times 1.4\text{ atm} = 0.18\text{ atm}$$

89. a. Mole fraction $CH_4 = \chi_{CH_4} = \frac{P_{CH_4}}{P_{total}} = \frac{0.175\text{ atm}}{0.175\text{ atm} + 0.250\text{ atm}} = 0.412$

$\chi_{O_2} = 1.000 - 0.412 = 0.588$

b. $PV = nRT,\ n_{total} = \frac{P_{total} \times V}{RT} = \frac{0.425\text{ atm} \times 10.5\text{ L}}{\frac{0.08206\text{ L atm}}{\text{K mol}} \times 338\text{ K}} = 0.161\text{ mol}$

c. $\chi_{CH_4} = \frac{n_{CH_4}}{n_{total}},\ n_{CH_4} = \chi_{CH_4} \times n_{total} = 0.412 \times 0.161\text{ mol} = 6.63 \times 10^{-2}\text{ mol } CH_4$

$6.63 \times 10^{-2}\text{ mol } CH_4 \times \frac{16.04\text{ g } CH_4}{\text{mol } CH_4} = 1.06\text{ g } CH_4$

$n_{O_2} = 0.588 \times 0.161\text{ mol} = 9.47 \times 10^{-2}\text{ mol } O_2;\ 9.47 \times \text{mol } O_2 \times \frac{32.00\text{ g } O_2}{\text{mol } O_2}$

$= 3.03\text{ g } O_2$

90. $52.5\text{ g } O_2 \times \frac{1\text{ mol } O_2}{32.00\text{ g } O_2} = 1.64\text{ mol } O_2;\ 65.1\text{ g } CO_2 \times \frac{1\text{ mol } CO_2}{44.01\text{ g } CO_2} = 1.48\text{ mol } CO_2$

$\chi_{O_2} = \frac{n_{O_2}}{n_{total}} = \frac{1.64\text{ mol}}{(1.64 + 1.48)\text{ mol}} = 0.526$

$P_{O_2} = \chi_{O_2} \times P_{total} = 0.526 \times 9.21\text{ atm} = 4.84\text{ atm}$

$P_{CO_2} = 9.21 - 4.84 = 4.37\text{ atm}$

91. $P_{total} = P_{H_2} + P_{H_2O},\ 1.032\text{ atm} = P_{H_2} + 32\text{ torr} \times \frac{1\text{ atm}}{760\text{ torr}},\ P_{H_2} = 1.032 - 0.042 = 0.990\text{ atm}$

$n_{H_2} = \frac{P_{H_2}V}{RT} = \frac{0.990\text{ atm} \times 0.240\text{ L}}{\frac{0.08206\text{ L atm}}{\text{K mol}} \times 303\text{ K}} = 9.56 \times 10^{-3}\text{ mol } H_2$

$9.56 \times 10^{-3}\text{ mol } H_2 \times \frac{1\text{ mol Zn}}{\text{mol } H_2} \times \frac{65.38\text{ g Zn}}{\text{mol Zn}} = 0.625\text{ g Zn}$

92. To calculate the volume of gas, we can use P_{total} and n_{total} ($V = n_{total}RT/P_{total}$), or we can use P_{He} and n_{He} ($V = n_{He}RT/P_{He}$). Because n_{H_2O} is unknown, we will use P_{He} and n_{He}.

$P_{He} + P_{H_2O} = 1.00\text{ atm} = 760.\text{ torr},\ P_{He} + 23.8\text{ torr} = 760.\text{ torr},\ P_{He} = 736\text{ torr}$

$$n_{He} = 0.586\ g \times \frac{1\ mol}{4.003\ g} = 0.146\ mol\ He$$

$$V = \frac{n_{He}RT}{P_{He}} = \frac{0.146\ mol \times \frac{0.08206\ L\ atm}{K\ mol} \times 298\ K}{736\ torr \times \frac{1\ atm}{760\ torr}} = 3.69\ L$$

93. $2\ NaClO_3(s) \rightarrow 2\ NaCl(s) + 3\ O_2(g)$

$P_{total} = P_{O_2} + P_{H_2O},\ P_{O_2} = P_{total} - P_{H_2O} = 734\ torr - 19.8\ torr = 714\ torr$

$$n_{O_2} = \frac{P_{O_2} \times V}{RT} = \frac{\left(714\ torr \times \frac{1\ atm}{760\ torr}\right) \times 0.0572\ L}{\frac{0.08206\ L\ atm}{K\ mol} \times (273 + 22)\ K} = 2.22 \times 10^{-3}\ mol\ O_2$$

$$\text{Mass } NaClO_3 \text{ decomposed} = 2.22 \times 10^{-3}\ mol\ O_2 \times \frac{2\ mol\ NaClO_3}{3\ mol\ O_2} \times \frac{106.44\ g\ NaClO_3}{mol\ NaClO_3} = 0.158\ g\ NaClO_3$$

$$\text{Mass \% } NaClO_3 = \frac{0.158\ g}{0.8765\ g} \times 100 = 18.0\%$$

94. 10.10 atm − 7.62 atm = 2.48 atm is the pressure of the amount of F_2 reacted.

PV = nRT, V and T are constant. $\frac{P}{n}$ = constant, $\frac{P_1}{n_1} = \frac{P_2}{n_2}$ or $\frac{P_1}{P_2} = \frac{n_1}{n_2}$

$$\frac{\text{Moles } F_2 \text{ reacted}}{\text{Moles Xe reacted}} = \frac{2.48\ atm}{1.24\ atm} = 2.00;\ \text{so } Xe + 2\ F_2 \rightarrow XeF_4$$

95. Because P and T are constant, V and n are directly proportional. The balanced equation requires 2 L of H_2 to react with 1 L of CO (2 : 1 volume ratio due to 2 : 1 mole ratio in the balanced equation). If in 1 minute all 16.0 L of H_2 react, only 8.0 L of CO are required to react with it. Because we have 25.0 L of CO present in that 1 minute, CO is in excess and H_2 is the limiting reactant. The volume of CH_3OH produced at STP will be one-half the volume of H_2 reacted due to the 1 : 2 mole ratio in the balanced equation. In 1 minute, 16.0 L/2 = 8.00 L CH_3OH is produced (theoretical yield).

$$n_{CH_3OH} = \frac{PV}{RT} = \frac{1.00\ atm \times 8.00\ L}{\frac{0.08206\ L\ atm}{K\ mol} \times 273\ K} = 0.357\ mol\ CH_3OH \text{ in 1 minute}$$

$$0.357\ mol\ CH_3OH \times \frac{32.04\ g\ CH_3OH}{mol\ CH_3OH} = 11.4\ g\ CH_3OH \text{ (theoretical yield per minute)}$$

$$\text{Percent yield} = \frac{\text{actual yield}}{\text{theoretical yield}} \times 100 = \frac{5.30\text{ g}}{11.4\text{ g}} \times 100 = 46.5\%\text{ yield}$$

96. $$750.\text{ mL juice} \times \frac{12\text{ mL } C_2H_5OH}{100\text{ mL juice}} = 90.\text{ mL } C_2H_5OH\text{ present}$$

$$90.\text{ mL } C_2H_5OH \times \frac{0.79\text{ g } C_2H_5OH}{\text{mL } C_2H_5OH} \times \frac{1\text{ mol } C_2H_5OH}{46.07\text{ g } C_2H_5OH} \times \frac{2\text{ mol } CO_2}{2\text{ mol } C_2H_5OH} = 1.5\text{ mol } CO_2$$

The CO_2 will occupy (825 – 750. =) 75 mL not occupied by the liquid (headspace).

$$P_{CO_2} = \frac{n_{CO_2}RT}{V} = \frac{1.5\text{ mol} \times \frac{0.08206\text{ L atm}}{\text{K mol}} \times 298\text{ K}}{75 \times 10^{-3}\text{ L}} = 490\text{ atm}$$

Actually, enough CO_2 will dissolve in the wine to lower the pressure of CO_2 to a much more reasonable value.

97. $2\ HN_3(g) \rightarrow 3\ N_2(g) + H_2(g)$; at constant V and T, P is directly proportional to n. In the reaction, we go from 2 moles of gaseous reactants to 4 moles of gaseous products. Because moles doubled, the final pressure will double (P_{total} = 6.0 atm). Similarly, from the 2 : 1 mole ratio between HN_3 and H_2, the partial pressure of H_2 will be 3.0/2 = 1.5 atm. The partial pressure of N_2 will be (3/2)3.0 atm = 4.5 atm. This is from the 2 : 3 mole ratio between HN_3 and N_2.

98. $2\ SO_2(g) + O_2(g) \rightarrow 2\ SO_3(g)$; because P and T are constant, volume ratios will equal mole ratios ($V_f/V_i = n_f/n_i$). Let x = mol SO_2 = mol O_2 present initially. From the balanced equation, 2 mol of SO_2 react for every 1 mol of O_2 that reacts. Because we have equal moles of SO_2 and O_2 present initially, and because SO_2 is used up twice as fast as O_2, SO_2 is the limiting reagent. Therefore, no SO_2 will be present after the reaction goes to completion. However, excess $O_2(g)$ will be present as well as the $SO_3(g)$ produced.

$$\text{Mol } O_2 \text{ reacted} = x\text{ mol } SO_2 \times \frac{1\text{ mol } O_2}{2\text{ mol } SO_2} = x/2\text{ mol } O_2$$

Mol O_2 remaining = x mol O_2 initially – $x/2$ mol O_2 reacted = $x/2$ mol O_2

$$\text{Mol } SO_3 \text{ produced} = x\text{ mol } SO_2 \times \frac{2\text{ mol } SO_3}{2\text{ mol } SO_2} = x\text{ mol } SO_3$$

Total moles gas initially = x mol SO_2 + x mol O_2 = $2x$

Total moles gas after reaction = $x/2$ mol O_2 + x mol SO_3 = $(3/2)x$ = $(1.5)x$

$$\frac{n_f}{n_i} = \frac{V_f}{V_i} = \frac{(1.5)x}{2x} = \frac{1.5}{2} = 0.75;\ V_f/V_i = 0.75 : 1 \text{ or } 3 : 4$$

The volume of the reaction container shrinks to 75% of the initial volume.

99. $150 \text{ g } (CH_3)_2N_2H_2 \times \frac{1 \text{ mol}(CH_3)_2N_2H_2}{60.10 \text{ g}} \times \frac{3 \text{ mol } N_2}{\text{mol}(CH_3)_2N_2H_2} = 7.5 \text{ mol } N_2$ produced

$$P_{N_2} = \frac{nRT}{V} = \frac{7.5 \text{ mol} \times \frac{0.08206 \text{ L atm}}{\text{K mol}} \times 400.\text{K}}{250 \text{ L}} = 0.98 \text{ atm}$$

We could do a similar calculation for P_{H_2O} and P_{CO_2} and then calculate P_{total} (= $P_{N_2} + P_{H_2O} + P_{CO_2}$). Or we can recognize that 9 total moles of gaseous products form for every mole of $(CH_3)_2N_2H_2$ reacted (from the balanced equation given in the problem). This is three times the moles of N_2 produced. Therefore, P_{total} will be three times larger than P_{N_2}.

$P_{total} = 3 \times P_{N_2} = 3 \times 0.98 \text{ atm} = 2.9 \text{ atm}$.

100. The partial pressure of CO_2 that reacted is 740. − 390. = 350. torr. Thus the number of moles of CO_2 that react is given by:

$$n = \frac{PV}{RT} = \frac{\frac{350.}{760} \text{ atm} \times 3.00 \text{ L}}{\frac{0.08206 \text{ L atm}}{\text{K mol}} \times 293 \text{K}} = 5.75 \times 10^{-2} \text{ mol } CO_2$$

$$5.75 \times 10^{-2} \text{ mol } CO_2 \times \frac{1 \text{ mol MgO}}{1 \text{ mol } CO_2} \times \frac{40.31 \text{ g MgO}}{\text{mol MgO}} = 2.32 \text{ g MgO}$$

$$\text{Mass \% MgO} = \frac{2.32 \text{ g}}{2.85 \text{ g}} \times 100 = 81.4\% \text{ MgO}$$

Kinetic Molecular Theory and Real Gases

101. $KE_{avg} = (3/2)RT$; the average kinetic energy depends only on temperature. At each temperature, CH_4 and N_2 will have the same average KE. For energy units of joules (J), use R = 8.3145 J/K•mol. To determine average KE per molecule, divide the molar KE_{avg} by Avogadro's number, 6.022×10^{23} molecules/mol.

$$\text{At 273 K: } KE_{avg} = \frac{3}{2} \times \frac{8.3145 \text{ J}}{\text{K mol}} \times 273 \text{ K} = 3.40 \times 10^3 \text{ J/mol} = 5.65 \times 10^{-21} \text{ J/molecule}$$

$$\text{At 546 K: } KE_{avg} = \frac{3}{2} \times \frac{8.3145 \text{ J}}{\text{K mol}} \times 546 \text{ K} = 6.81 \times 10^3 \text{ J/mol} = 1.13 \times 10^{-20} \text{ J/molecule}$$

102. $n_{Ar} = \frac{228 \text{ g}}{39.95 \text{ g/mol}} = 5.71 \text{ mol Ar}; \ \chi_{CH_4} = \frac{n_{CH_4}}{n_{CH_4} + n_{Ar}} = 0.650 = \frac{n_{CH_4}}{n_{CH_4} + 5.71}$

$0.650(n_{CH_4} + 5.71) = n_{CH_4}, \ 3.71 = (0.350)n_{CH_4}, \ n_{CH_4} = 10.6 \text{ mol } CH_4$

$KE_{avg} = \frac{3}{2}RT$ for 1 mole of gas

$KE_{total} = (10.6 + 5.71)\text{ mol} \times 3/2 \times 8.3145\text{ J/K•mol} \times 298\text{ K} = 6.06 \times 10^4\text{ J} = 60.6\text{ kJ}$

103. $\mu_{rms} = \left(\frac{3RT}{M}\right)^{1/2}$, where $R = \frac{8.3145 J}{K\ mol}$ and M = molar mass in kg.

For CH_4, $M = 1.604 \times 10^{-2}$ kg, and for N_2, $M = 2.802 \times 10^{-2}$ kg.

For CH_4 at 273 K: $\mu_{rms} = \left(\frac{3 \times \frac{8.3145 J}{K\ mol} \times 273 K}{1.604 \times 10^{-2}\ kg/mol}\right)^{1/2} = 652$ m/s

Similarly, μ_{rms} for CH_4 at 546 K is 921 m/s.

For N_2 at 273 K: $\mu_{rms} = \left(\frac{3 \times \frac{8.3145 J}{K\ mol} \times 273 K}{2.802 \times 10^{-2}\ kg/mol}\right)^{1/2} = 493$ m/s

Similarly, for N_2 at 546 K, $\mu_{rms} = 697$ m/s.

104. $\mu_{rms} = \left(\frac{3RT}{M}\right)^{1/2}$; $\frac{\mu_{UF_6}}{\mu_{He}} = \frac{\left(\frac{3RT_{UF_6}}{M_{UF_6}}\right)^{1/2}}{\left(\frac{3RT_{He}}{M_{He}}\right)^{1/2}} = \left(\frac{M_{He}T_{UF_6}}{M_{UF_6}T_{He}}\right)^{1/2}$

We want the root mean square velocities to be equal, and this occurs when:

$$M_{He}T_{UF_6} = M_{UF_6}T_{He}$$

The ratio of the temperatures is: $\frac{T_{UF_6}}{T_{He}} = \frac{M_{UF_6}}{M_{He}} = \frac{352.0}{4.003} = 87.93$

The heavier UF_6 molecules would need a temperature 87.93 times that of the He atoms in order for the root mean square velocities to be equal.

105. The number of gas particles is constant, so at constant moles of gas, either a temperature change or a pressure change results in the smaller volume. If the temperature is constant, an increase in the external pressure would cause the volume to decrease. Gases are mostly empty space so gases are easily compressible.

If the pressure is constant, a decrease in temperature would cause the volume to decrease. As the temperature is lowered, the gas particles move with a slower average velocity and don't collide with the container walls as frequently and as forcefully. As a result, the internal pressure decreases. In order to keep the pressure constant, the volume of the container must decrease in order to increase the gas particle collisions per unit area.

106. In this situation, the volume has increased by a factor of two. One way to double the volume of a container at constant pressure and temperature is to double the number of moles of gas particles present. As gas particles are added, more collisions per unit area occur and the internal pressure increases. In order to keep the pressure constant, the container volume must increase.

Another way to double the volume of a container at constant pressure and moles of gas is to double the absolute temperature. As temperature increases, the gas molecules collide more frequently with the walls of the container. In order to keep pressure constant, the container volume must increase.

The last variable which can be changed is pressure. If the external pressure exerted on the container is halved, the volume will double (assuming constant temperature and moles). As the external pressure applied is reduced, the volume of the container must increase in order to equalize the higher internal pressure with the lower external applied pressure.

107.

	a	b	c	d
Avg. KE	increase	decrease	same (KE $\propto$ T)	same
Avg. velocity	increase	decrease	same ($\frac{1}{2}$ mv^2 = KE $\propto$ T)	same
Wall coll. freq	increase	decrease	increase	increase

Average kinetic energy and average velocity depend on T. As T increases, both average kinetic energy and average velocity increase. At constant T, both average kinetic energy and average velocity are constant. The collision frequency is proportional to the average velocity (as velocity increases, it takes less time to move to the next collision) and to the quantity n/V (as molecules per volume increase, collision frequency increases).

108. V, T, and P are all constant, so n must be constant. Because we have equal moles of gas in each container, gas B molecules must be heavier than gas A molecules.

a. Both gas samples have the same number of molecules present (n is constant).

b. Because T is constant, KE_{avg} must be the same for both gases [KE_{avg} = (3/2)RT].

c. The lighter gas A molecules will have the faster average velocity.

d. The heavier gas B molecules do collide more forcefully, but gas A molecules, with the faster average velocity, collide more frequently. The end result is that P is constant between the two containers.

109. a. They will all have the same average kinetic energy because they are all at the same temperature [KE_{avg} = (3/2)RT].

b. Flask C; H_2 has the smallest molar mass. At constant T, the lighter molecules have the faster average velocity. This must be true for the average kinetic energies to be the same.

110. a. All the gases have the same average kinetic energy since they are all at the same temperature [$KE_{avg} = (3/2)RT$].

b. At constant T, the lighter the gas molecule, the faster the average velocity [$\mu_{avg} \propto \mu_{rms} \propto (1/M)^{1/2}$].

Xe (131.3 g/mol) < Cl_2 (70.90 g/mol) < O_2 (32.00 g/mol) < H_2 (2.016 g/mol)
slowest fastest

c. At constant T, the lighter H_2 molecules have a faster average velocity than the heavier O_2 molecules. As temperature increases, the average velocity of the gas molecules increases. Separate samples of H_2 and O_2 can only have the same average velocities if the temperature of the O_2 sample is greater than the temperature of the H_2 sample.

111. Graham's law of effusion: $\frac{\text{Rate}_1}{\text{Rate}_2} = \left(\frac{M_2}{M_1}\right)^{1/2}$

Let Freon-12 = gas 1 and Freon-11 = gas 2:

$$\frac{1.07}{1.00} = \left(\frac{137.4}{M_1}\right)^{1/2}, \quad 1.14 = \frac{137.4}{M_1}, \quad M_1 = 121 \text{ g/mol}$$

The molar mass of CF_2Cl_2 is equal to 121 g/mol, so Freon-12 is CF_2Cl_2.

112. $\frac{\text{Rate}_1}{\text{Rate}_2} = \left(\frac{M_2}{M_1}\right)^{1/2}$; $\text{rate}_1 = \frac{24.0 \text{ mL}}{\text{min}}$; $\text{rate}_2 = \frac{47.8 \text{ mL}}{\text{min}}$; $M_2 = \frac{16.04 \text{ g}}{\text{mol}}$; $M_1 = ?$

$$\frac{24.0}{47.8} = \left(\frac{16.04}{M_1}\right)^{1/2} = 0.502, \quad 16.04 = (0.502)^2 \times M_1, \quad M_1 = \frac{16.04}{0.252} = \frac{63.7 \text{ g}}{\text{mol}}$$

113. $\frac{\text{Rate}_1}{\text{Rate}_2} = \left(\frac{M_2}{M_1}\right)^{1/2}$, $\frac{\text{rate}(^{12}C^{17}O)}{\text{rate}(^{12}C^{18}O)} = \left(\frac{30.0}{29.0}\right)^{1/2} = 1.02$; $\frac{\text{Rate}(^{12}C^{16}O)}{\text{Rate}(^{12}C^{18}O)} = \left(\frac{30.0}{28.0}\right)^{1/2} = 1.04$

The relative rates of effusion of $^{12}C^{16}O$ to $^{12}C^{17}O$ to $^{12}C^{18}O$ are 1.04 : 1.02 : 1.00.

Advantage: CO_2 isn't as toxic as CO.

Major disadvantages of using CO_2 instead of CO:

1. Can get a mixture of oxygen isotopes in CO_2.

2. Some species, for example, $^{12}C^{16}O^{18}O$ and $^{12}C^{17}O_2$, would effuse (gaseously diffuse) at about the same rate because the masses are about equal. Thus some species cannot be separated from each other.

114. $\frac{\text{Rate}_1}{\text{Rate}_2} = \left(\frac{M_2}{M_1}\right)^{1/2}$, where M = molar mass; let gas (1) = He and gas (2) = Cl_2.

Effusion rates in this problem are equal to the volume of gas that effuses per unit time (L/min). Let t = time in the following expression.

$$\frac{\frac{1.0\text{ L}}{4.5\text{ min}}}{\frac{1.0\text{ L}}{t}} = \left(\frac{70.90}{4.003}\right)^{1/2}, \quad \frac{t}{4.5\text{ min}} = 4.209, \ t = 19\text{ min}$$

115. a. PV = nRT

$$P = \frac{nRT}{V} = \frac{0.5000\text{mol} \times \frac{0.08206\text{L atm}}{\text{K mol}} \times (25.0 + 273.2)\text{ K}}{1.0000\text{L}} = 12.24\text{ atm}$$

b. $\left[P + a\left(\frac{n}{V}\right)^2\right](V - nb) = nRT$; for N_2: a = 1.39 atm L^2/mol^2 and b = 0.0391 L/mol

$$\left[P + 1.39\left(\frac{0.5000}{1.0000}\right)^2 \text{atm}\right](1.0000\text{ L} - 0.5000 \times 0.0391\text{ L}) = 12.24\text{ L atm}$$

(P + 0.348 atm)(0.9805 L) = 12.24 L atm

$$P = \frac{12.24\text{ L atm}}{0.9805\text{L}} - 0.348\text{ atm} = 12.48 - 0.348 = 12.13\text{ atm}$$

c. The ideal gas law is high by 0.11 atm, or $\frac{0.11}{12.13} \times 100 = 0.91\%$.

116. a. PV = nRT

$$P = \frac{nRT}{V} = \frac{0.5000\text{mol} \times \frac{0.08206\text{L atm}}{\text{K mol}} \times 298.2\text{ K}}{10.000\text{L}} = 1.224\text{ atm}$$

b. $\left[P + a\left(\frac{n}{V}\right)^2\right](V - nb) = nRT$; for N_2: a = 1.39 atm L^2/mol^2 and b = 0.0391 L/mol

$$\left[P + 1.39\left(\frac{0.5000}{10.000}\right)^2 \text{atm}\right](10.000\text{ L} - 0.5000 \times 0.0391\text{ L}) = 12.24\text{ L atm}$$

$(P + 0.00348 \text{ atm})(10.000 \text{ L} - 0.0196 \text{ L}) = 12.24 \text{ L atm}$

$$P + 0.00348 \text{ atm} = \frac{12.24 \text{ L atm}}{9.980 \text{ L}} = 1.226 \text{ atm}, \quad P = 1.226 - 0.00348 = 1.223 \text{ atm}$$

c. The results agree to ±0.001 atm (0.08%).

d. In Exercise 115, the pressure is relatively high, and there is significant disagreement. In Exercise 116, the pressure is around 1 atm, and both gas laws show better agreement. The ideal gas law is valid at relatively low pressures.

Atmospheric Chemistry

117. $\chi_{He} = 5.24 \times 10^{-6}$ from Table 5.4. $P_{He} = \chi_{He} \times P_{total} = 5.24 \times 10^{-6} \times 1.0 \text{ atm} = 5.2 \times 10^{-6} \text{ atm}$

$$\frac{n}{V} = \frac{P}{RT} = \frac{5.2 \times 10^{-6} \text{ atm}}{\frac{0.08206 \text{ L atm}}{\text{K mol}} \times 298 \text{ K}} = 2.1 \times 10^{-7} \text{ mol He/L}$$

$$\frac{2.1 \times 10^{-7} \text{ mol}}{\text{L}} \times \frac{1 \text{ L}}{1000 \text{ cm}^3} \times \frac{6.022 \times 10^{23} \text{ atoms}}{\text{mol}} = 1.3 \times 10^{14} \text{ atoms He/cm}^3$$

118. At 15 km, $T \approx -60°C$ and $P = 0.1$ atm. Use $\frac{P_1V_1}{T_1} = \frac{P_2V_2}{T_2}$ since n is constant.

$$V_2 = \frac{V_1P_1T_2}{P_2T_1} = \frac{1.0 \text{ L} \times 1.00 \text{ atm} \times 213 \text{ K}}{0.1 \text{ atm} \times 298 \text{ K}} = 7 \text{ L}$$

119. $S(s) + O_2(g) \rightarrow SO_2(g)$, combustion of coal

$2\ SO_2(g) + O_2(g) \rightarrow 2\ SO_3(g)$, reaction with atmospheric O_2

$SO_3(g) + H_2O(l) \rightarrow H_2SO_4(aq)$, reaction with atmospheric H_2O

120. $H_2SO_4(aq) + CaCO_3(s) \rightarrow CaSO_4(aq) + H_2O(l) + CO_2(g)$

121. a. If we have 1.0×10^6 L of air, then there are 3.0×10^2 L of CO.

$P_{CO} = \chi_{CO}P_{total}$; $\chi_{CO} = \frac{V_{CO}}{V_{total}}$ because $V \propto n$; $P_{CO} = \frac{3.0 \times 10^2}{1.0 \times 10^6} \times 628 \text{ torr} = 0.19 \text{ torr}$

b. $n_{CO} = \frac{P_{CO}V}{RT}$; assuming 1.0 m³ air, 1 m³ = 1000 L:

$$n_{CO} = \frac{\frac{0.19}{760}\text{ atm} \times (1.0 \times 10^3 \text{ L})}{\frac{0.08206 \text{ L atm}}{\text{K mol}} \times 273 \text{ K}} = 1.1 \times 10^{-2} \text{ mol CO}$$

$$1.1 \times 10^{-2} \text{ mol} \times \frac{6.02 \times 10^{23} \text{ molecules}}{\text{mol}} = 6.6 \times 10^{21} \text{ CO molecules in } 1.0 \text{ m}^3 \text{ of air}$$

c. $$\frac{6.6 \times 10^{21} \text{ molecules}}{\text{m}^3} \times \left(\frac{1 \text{ m}}{100 \text{ cm}}\right)^3 = \frac{6.6 \times 10^{15} \text{ molecules CO}}{\text{cm}^3}$$

122. For benzene:

$$89.6 \times 10^{-9} \text{ g} \times \frac{1 \text{ mol}}{78.11 \text{ g}} = 1.15 \times 10^{-9} \text{ mol benzene}$$

$$V_{benzene} = \frac{n_{benzene}RT}{P} = \frac{1.15 \times 10^{-9} \text{ mol} \times \frac{0.08206 \text{ L atm}}{\text{K mol}} \times 296 \text{ K}}{748 \text{ torr} \times \frac{1 \text{ atm}}{760 \text{ torr}}} = 2.84 \times 10^{-8} \text{ L}$$

$$\text{Mixing ratio} = \frac{2.84 \times 10^{-8} \text{ L}}{3.00 \text{ L}} \times 10^6 = 9.47 \times 10^{-3} \text{ ppmv}$$

$$\text{Or ppbv} = \frac{\text{vol. of X} \times 10^9}{\text{total vol.}} = \frac{2.84 \times 10^{-8} \text{ L}}{3.00 \text{ L}} \times 10^9 = 9.47 \text{ ppbv}$$

$$\frac{1.15 \times 10^{-9} \text{ mol benzene}}{3.00 \text{ L}} \times \frac{1 \text{ L}}{1000 \text{ cm}^3} \times \frac{6.022 \times 10^{23} \text{ molecules}}{\text{mol}}$$
$$= 2.31 \times 10^{11} \text{ molecules benzene/cm}^3$$

For toluene:

$$153 \times 10^{-9} \text{ g } C_7H_8 \times \frac{1 \text{ mol}}{92.13 \text{ g}} = 1.66 \times 10^{-9} \text{ mol toluene}$$

$$V_{toluene} = \frac{n_{toluene}RT}{P} = \frac{1.66 \times 10^{-9} \text{ mol} \times \frac{0.08206 \text{ L atm}}{\text{K mol}} \times 296 \text{ K}}{748 \text{ torr} \times \frac{1 \text{ atm}}{760 \text{ torr}}} = 4.10 \times 10^{-8} \text{ L}$$

$$\text{Mixing ratio} = \frac{4.10 \times 10^{-8} \text{ L}}{3.00 \text{ L}} \times 10^6 = 1.37 \times 10^{-2} \text{ ppmv (or 13.7 ppbv)}$$

$$\frac{1.66 \times 10^{-9} \text{ mol toluene}}{3.00 \text{ L}} \times \frac{1 \text{ L}}{1000 \text{ cm}^3} \times \frac{6.022 \times 10^{23} \text{ molecules}}{\text{mol}}$$
$$= 3.33 \times 10^{11} \text{ molecules toluene/cm}^3$$

Additional Exercises

123. a. PV = nRT

PV = constant

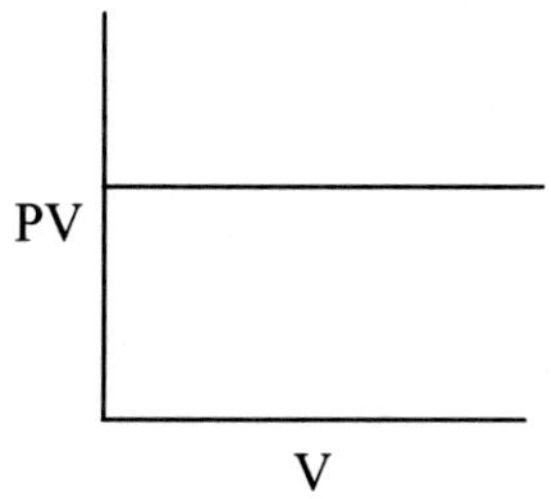

b. PV = nRT

$$P = \left(\frac{nR}{V}\right) \times T = \text{const} \times T$$

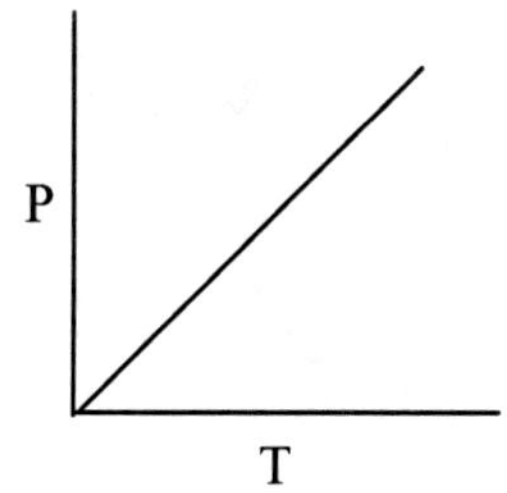

c. PV = nRT

$$T = \left(\frac{P}{nR}\right) \times V = \text{const} \times V$$

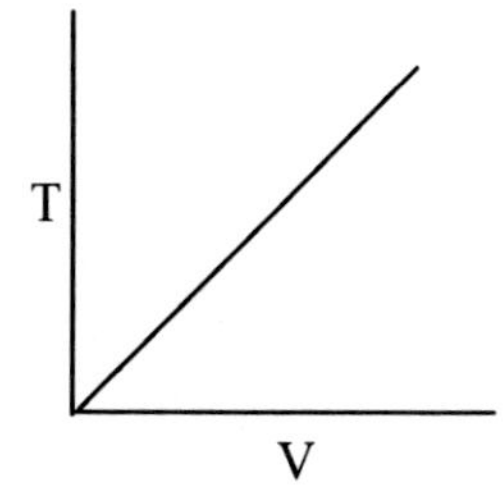

d. PV = nRT

PV = constant

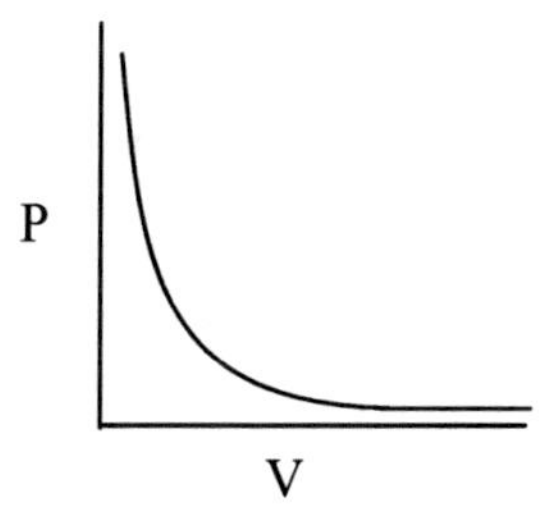

e. $P = \frac{nR}{V} = \frac{\text{constant}}{V}$

$$P = \text{constant} \times \frac{1}{V}$$

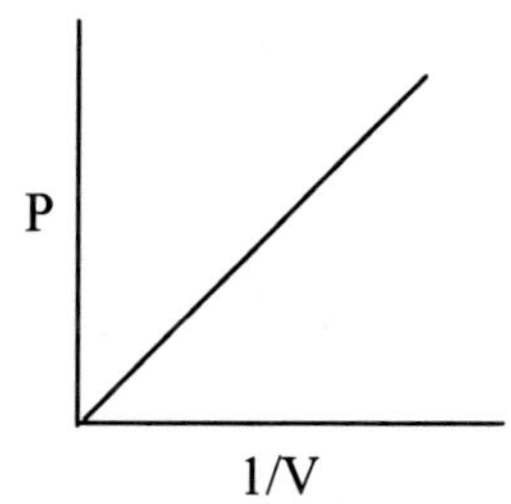

f. PV = nRT

$$\frac{PV}{T} = nR = \text{constant}$$

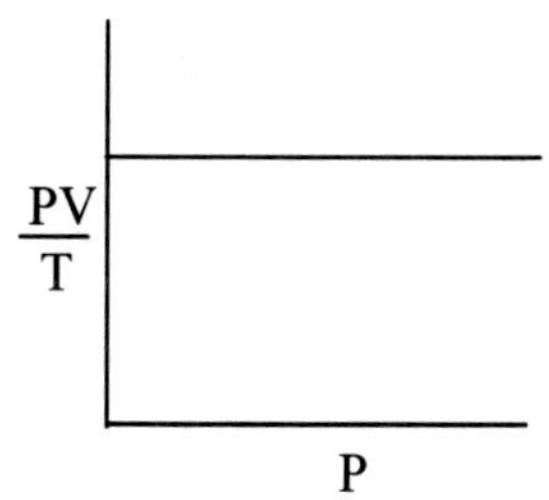

Note: The equation for a straight line is $y = mx + b$, where y is the y-axis and x is the x-axis. Any equation that has this form will produce a straight line with slope equal to m and a y intercept equal to b. Plots b, c, and e have this straight-line form.

124. At constant T and P, Avogadro's law applies; that is, equal volumes contain equal moles of molecules. In terms of balanced equations, we can say that mole ratios and volume ratios between the various reactants and products will be equal to each other. $Br_2 + 3\ F_2 \rightarrow 2\ X$; 2 moles of X must contain 2 moles of Br and 6 moles of F; X must have the formula BrF_3 for a balanced equation.

125. $$14.1 \times 10^2 \text{ in Hg} \bullet \text{in}^3 \times \frac{2.54 \text{ cm}}{\text{in}} \times \frac{10 \text{ mm}}{1 \text{ cm}} \times \frac{1 \text{ atm}}{760 \text{ mm}} \times \left(\frac{2.54 \text{ cm}}{\text{in}}\right)^3 \times \frac{1 \text{ L}}{1000 \text{cm}^3}$$
$$= 0.772 \text{ atm} \bullet \text{L}$$

Boyle's law: PV = k, where k = nRT; from Example 5.3, the k values are around 22 atm•L. Because k = nRT, we can assume that Boyle's data and the Example 5.3 data were taken at different temperatures and/or had different sample sizes (different moles).

126. $Mn(s) + x\,HCl(g) \rightarrow MnCl_x(s) + \frac{x}{2}\,H_2(g)$

$$n_{H_2} = \frac{PV}{RT} = \frac{0.951\,\text{atm} \times 3.22\,\text{L}}{\frac{0.08206\,\text{L atm}}{\text{K mol}} \times 373\,\text{K}} = 0.100 \text{ mol } H_2$$

$$\text{Mol Cl in compound} = \text{mol HCl} = 0.100 \text{ mol } H_2 \times \frac{x \text{ mol Cl}}{\frac{x}{2} \text{ mol } H_2} = 0.200 \text{ mol Cl}$$

$$\frac{\text{Mol Cl}}{\text{Mol Mn}} = \frac{0.200 \text{ mol Cl}}{2.747 \text{ g Mn} \times \frac{1 \text{ mol Mn}}{54.94 \text{ g Mn}}} = \frac{0.200 \text{ mol Cl}}{0.05000 \text{ mol Mn}} = 4.00$$

The formula of compound is $MnCl_4$.

127. Assume some mass of the mixture. If we had 100.0 g of the gas, we would have 50.0 g He and 50.0 g Xe.

$$\chi_{He} = \frac{n_{He}}{n_{He} + n_{Xe}} = \frac{\frac{50.0\text{g}}{4.003\text{g/mol}}}{\frac{50.0\text{g}}{4.003\text{g/mol}} + \frac{50.0\text{g}}{131.3\text{g/mol}}} = \frac{12.5 \text{ mol He}}{12.5 \text{ mol He} + 0.381 \text{ mol Xe}} = 0.970$$

No matter what the initial mass of mixture is assumed, the mole fraction of helium will always be 0.970.

$P_{He} = \chi_{He}P_{total} = 0.970 \times 600.$ torr = 582 torr; $P_{Xe} = 600. - 582 = 18$ torr

128. Assuming 100.0 g of cyclopropane:

$$85.7 \text{ g C} \times \frac{1 \text{ mol C}}{12.01 \text{ g}} = 7.14 \text{ mol C}$$

$$14.3 \text{ g H} \times \frac{1 \text{ mol H}}{1.008 \text{ g}} = 14.2 \text{ mol H}; \quad \frac{14.2}{7.14} = 1.99$$

The empirical formula for cyclopropane is CH_2, which has an empirical mass ≈ 12.0 + 2(1.0) = 14.0 g/mol.

$$P \times (\text{molar mass}) = dRT, \quad \text{molar mass} = \frac{dRT}{P} = \frac{1.88 \text{ g/L} \times \frac{0.08206 \text{ L atm}}{\text{K mol}} \times 273\text{K}}{1.00 \text{ atm}}$$

$$= 42.1 \text{g/mol}$$

Because 42.1/14.0 ≈ 3.0, the molecular formula for cyclopropane is $(CH_2)_{\times 3} = C_3H_6$.

129. $P_{total} = P_{N_2} + P_{H_2O}$, P_{N_2} = 726 torr – 23.8 torr = 702 torr × $\frac{1\,\text{atm}}{760\,\text{torr}}$ = 0.924 atm

$$n_{N_2} = \frac{P_{N_2} \times V}{RT} = \frac{0.924\,\text{atm} \times 31.8 \times 10^{-3}\ \text{L}}{\frac{0.08206\,\text{L atm}}{\text{K mol}} \times 298\,\text{K}} = 1.20 \times 10^{-3}\ \text{mol}\ N_2$$

Mass of N in compound = 1.20×10^{-3} mol N_2 × $\frac{28.02\ \text{g}\ N_2}{\text{mol}}$ = 3.36×10^{-2} g nitrogen

Mass % N = $\frac{3.36 \times 10^{-2}\ \text{g}}{0.253\,\text{g}}$ × 100 = 13.3% N

130. 33.5 mg CO_2 × $\frac{12.01\,\text{mg C}}{44.01\,\text{mg}\ CO_2}$ = 9.14 mg C; % C = $\frac{9.14\,\text{mg}}{35.0\ \text{mg}}$ × 100 = 26.1% C

41.1 mg H_2O × $\frac{2.016\,\text{mg H}}{18.02\,\text{mg}\ H_2O}$ = 4.60 mg H; % H = $\frac{4.60\,\text{mg}}{35.0\ \text{mg}}$ × 100 = 13.1% H

$$n_{N_2} = \frac{P_{N_2} V}{RT} = \frac{\frac{740.}{760}\,\text{atm} \times 35.6 \times 10^{-3}\ \text{L}}{\frac{0.08206\,\text{L atm}}{\text{K mol}} \times 298\,\text{K}} = 1.42 \times 10^{-3}\ \text{mol}\ N_2$$

1.42×10^{-3} mol N_2 × $\frac{28.02\ \text{g}\ N_2}{\text{mol}\ N_2}$ = 3.98×10^{-2} g nitrogen = 39.8 mg nitrogen

Mass % N = $\frac{39.8\ \text{mg}}{65.2\ \text{mg}}$ × 100 = 61.0% N

Or we can get % N by difference: % N = 100.0 – (26.1 + 13.1) = 60.8%

Out of 100.0 g:

26.1 g C × $\frac{1\,\text{mol}}{12.01\,\text{g}}$ = 2.17 mol C; $\frac{2.17}{2.17}$ = 1.00

13.1 g H × $\frac{1\,\text{mol}}{1.008\,\text{g}}$ = 13.0 mol H; $\frac{13.0}{2.17}$ = 5.99

60.8 g N × $\frac{1\,\text{mol}}{14.01\,\text{g}}$ = 4.34 mol N; $\frac{4.34}{2.17}$ = 2.00; empirical formula is CH_6N_2.

$$\frac{\text{Rate}_1}{\text{Rate}_2} = \left(\frac{M}{39.95}\right)^{1/2} = \frac{26.4}{24.6} = 1.07,\ M = (1.07)^2 \times 39.95 = 45.7\ \text{g/mol}$$

Empirical formula mass of CH_6N_2 ≈ 12 + 6 + 28 = 46 g/mol. Thus the molecular formula is also CH_6N_2.

131. We will apply Boyle's law to solve. PV = nRT = constant, $P_1V_1 = P_2V_2$

Let condition (1) correspond to He from the tank that can be used to fill balloons. We must leave 1.0 atm of He in the tank, so P_1 = 200. – 1.00 = 199 atm and V_1 = 15.0 L. Condition (2) will correspond to the filled balloons with P_2 = 1.00 atm and V_2 = N(2.00 L), where N is the number of filled balloons, each at a volume of 2.00 L.

199 atm × 15.0 L = 1.00 atm × N(2.00 L), N = 1492.5; we can't fill 0.5 of a balloon, so N = 1492 balloons or, to 3 significant figures, 1490 balloons.

132. Mol of He removed $= \frac{PV}{RT} = \frac{1.00\,\text{atm} \times 1.75 \times 10^{-3}\text{ L}}{\frac{0.08206\text{L atm}}{\text{K mol}} \times 298\text{K}} = 7.16 \times 10^{-5}$ mol

In the original flask, 7.16×10^{-5} mol of He exerted a partial pressure of 1.960 – 1.710 = 0.250 atm.

$$V = \frac{nRT}{P} = \frac{7.16 \times 10^{-5}\text{mol} \times 0.08206 \times 298\text{K}}{0.250\text{atm}} = 7.00 \times 10^{-3}\text{ L} = 7.00\text{ mL}$$

133. For O_2, n and T are constant, so $P_1V_1 = P_2V_2$.

$$P_1 = \frac{P_2V_2}{V_1} = 785\text{ torr} \times \frac{1.94\text{ L}}{2.00\text{ L}} = 761\text{ torr} = P_{O_2}$$

$P_{total} = P_{O_2} + P_{H_2O}$, P_{H_2O} = 785 – 761 = 24 torr

134. PV = nRT, V and T are constant. $\frac{P_1}{n_1} = \frac{P_2}{n_2}$ or $\frac{P_1}{P_2} = \frac{n_1}{n_2}$

When V and T are constant, then pressure is directly proportional to moles of gas present, and pressure ratios are identical to mole ratios.

At 25°C: $2\ H_2(g) + O_2(g) \rightarrow 2\ H_2O(l)$, $H_2O(l)$ is produced at 25°C.

The balanced equation requires 2 mol H_2 for every mol O_2 reacted. The same ratio (2 : 1) holds true for pressure units. So if all 2.00 atm of H_2 react, only 1.00 atm of O_2 will react with it. Because we have 3.00 atm of O_2 present, oxygen is in excess and hydrogen is the limiting reactant. The only gas present at 25°C after the reaction goes to completion will be the excess O_2.

$$P_{O_2}\text{ (reacted)} = 2.00\text{ atm } H_2 \times \frac{1\text{ atm } O_2}{2\text{ atm } H_2} = 1.00\text{ atm } O_2$$

P_{O_2} (excess) = P_{O_2} (initial) – P_{O_2} (reacted) = 3.00 atm - 1.00 atm = 2.00 atm O_2 = P_{total}

At 125°C: $2\ H_2(g) + O_2(g) \rightarrow 2\ H_2O(g)$, $H_2O(g)$ is produced at 125°C.

The major difference in the problem at 125°C versus 25°C is that gaseous water is now a product (instead of liquid H_2O), which will increase the total pressure because an additional gas is present. *Note*: For this problem, it is assumed that 2.00 atm of H_2 and 3.00 atm of O_2 are reacted at 125°C instead of 25°C.

$$P_{H_2O}\text{(produced)} = 2.00\text{ atm }H_2 \times \frac{2\text{ atm }H_2O}{2\text{ atm }H_2} = 2.00\text{ atm }H_2O$$

$$P_{total} = P_{O_2}\text{(excess)} + P_{H_2O}\text{(produced)} = 2.00\text{ atm }O_2 + 2.00\text{ atm }H_2O = 4.00\text{ atm} = P_{total}$$

135. $$1.00 \times 10^3\text{ kg Mo} \times \frac{1000\text{g}}{\text{kg}} \times \frac{1\text{ mol Mo}}{95.94\text{ g Mo}} = 1.04 \times 10^4\text{ mol Mo}$$

$$1.04 \times 10^4\text{ mol Mo} \times \frac{1\text{ mol }MoO_3}{\text{mol Mo}} \times \frac{7/2\text{ mol }O_2}{\text{mol }MoO_3} = 3.64 \times 10^4\text{ mol }O_2$$

$$V_{O_2} = \frac{n_{O_2}RT}{P} = \frac{3.64 \times 10^4\text{ mol} \times \frac{0.08206\text{L atm}}{\text{K mol}} \times 290.\text{ K}}{1.00\text{ atm}} = 8.66 \times 10^5\text{ L of }O_2$$

$$8.66 \times 10^5\text{ L }O_2 \times \frac{100\text{ L air}}{21\text{ L }O_2} = 4.1 \times 10^6\text{ L air}$$

$$1.04 \times 10^4\text{ mol Mo} \times \frac{3\text{ mol }H_2}{\text{mol Mo}} = 3.12 \times 10^4\text{ mol }H_2$$

$$V_{H_2} = \frac{3.12 \times 10^4\text{ mol} \times \frac{0.08206\text{L atm}}{\text{K mol}} \times 290.\text{ K}}{1.00\text{ atm}} = 7.42 \times 10^5\text{ L of }H_2$$

136. For NH_3: $P_2 = \frac{P_1V_1}{V_2} = 0.500\text{ atm} \times \frac{2.00\text{ L}}{3.00\text{ L}} = 0.333\text{ atm}$

For O_2: $P_2 = \frac{P_1V_1}{V_2} = 1.50\text{ atm} \times \frac{1.00\text{ L}}{3.00\text{ L}} = 0.500\text{ atm}$

After the stopcock is opened, V and T will be constant, so $P \propto n$.

Assuming NH_3 is limiting: $0.333\text{ atm }NH_3 \times \frac{4\text{ atm NO}}{4\text{ atm }NH_3} = 0.333\text{ atm NO}$

Assuming O_2 is limiting: $0.500\text{ atm }O_2 \times \frac{4\text{ atm NO}}{5\text{ atm }O_2} = 0.400\text{ atm NO}$

NH_3 produces the smaller amount of product, so NH_3 is limiting and 0.333 atm of NO can be produced.

137. Out of 100.00 g of compound there are:

$$58.51 \text{ g C} \times \frac{1 \text{ mol C}}{12.01 \text{ g C}} = 4.872 \text{ mol C}; \quad \frac{4.872}{2.435} = 2.001$$

$$7.37 \text{ g H} \times \frac{1 \text{ mol H}}{1.008 \text{ g H}} = 7.31 \text{ mol H}; \quad \frac{7.31}{2.435} = 3.00$$

$$34.12 \text{ g N} \times \frac{1 \text{ mol N}}{14.01 \text{ g N}} = 2.435 \text{ mol N}; \quad \frac{2.435}{2.435} = 1.000$$

The empirical formula is C_2H_3N.

$$\frac{\text{Rate}_1}{\text{Rate}_2} = \left(\frac{M_2}{M_1}\right)^{1/2}; \text{ let gas (1) = He; } 3.20 = \left(\frac{M_2}{4.003}\right)^{1/2}, \ M_2 = 41.0 \text{ g/mol}$$

The empirical formula mass of $C_2H_3N \approx 2(12.0) + 3(1.0) + 1(14.0) = 41.0$ g/mol. So the molecular formula is also C_2H_3N.

138. If Be^{3+}, the formula is $Be(C_5H_7O_2)_3$ and molar mass $\approx 13.5 + 15(12) + 21(1) + 6(16)$ $= 311$ g/mol. If Be^{2+}, the formula is $Be(C_5H_7O_2)_2$ and molar mass $\approx 9.0 + 10(12) + 14(1) + 4(16) = 207$ g/mol.

Data set I (molar mass = dRT/P and d = mass/V):

$$\text{molar mass} = \frac{\text{mass} \times RT}{PV} = \frac{0.2022 \text{ g} \times \frac{0.08206 \text{ L atm}}{\text{K mol}} \times 286 \text{ K}}{(765.2 \text{ torr} \times \frac{1 \text{ atm}}{760 \text{ torr}}) \times (22.6 \times 10^{-3} \text{ L})} = 209 \text{ g/mol}$$

Data set II:

$$\text{molar mass} = \frac{\text{mass} \times RT}{PV} = \frac{0.2224 \text{ g} \times \frac{0.08206 \text{ L atm}}{\text{K mol}} \times 290. \text{ K}}{(764.6 \text{ torr} \times \frac{1 \text{ atm}}{760 \text{ torr}}) \times (26.0 \times 10^{-3} \text{ L})} = 202 \text{ g/mol}$$

These results are close to the expected value of 207 g/mol for $Be(C_5H_7O_2)_2$. Thus we conclude from these data that beryllium is a divalent element with an atomic weight (mass) of 9.0 u.

139. $$0.2766 \text{ g } CO_2 \times \frac{12.01 \text{ g C}}{44.01 \text{ g } CO_2} = 7.548 \times 10^{-2} \text{ g C}; \ \% \text{ C} = \frac{7.548 \times 10^{-2} \text{ g}}{0.1023 \text{ g}} \times 100 = 73.78\% \text{ C}$$

$$0.0991 \text{ g } H_2O \times \frac{2.016 \text{ g H}}{18.02 \text{ g } H_2O} = 1.11 \times 10^{-2} \text{ g H}; \ \% \text{ H} = \frac{1.11 \times 10^{-2} \text{ g}}{0.1023 \text{ g}} \times 100 = 10.9\% \text{ H}$$

$$PV = nRT, \ n_{N_2} = \frac{PV}{RT} = \frac{1.00 \text{ atm} \times 27.6 \times 10^{-3} \text{ L}}{\frac{0.08206 \text{ L atm}}{\text{K mol}} \times 273 \text{ K}} = 1.23 \times 10^{-3} \text{ mol } N_2$$

$$1.23 \times 10^{-3} \text{ mol } N_2 \times \frac{28.02 \text{ g } N_2}{\text{mol } N_2} = 3.45 \times 10^{-2} \text{ g nitrogen}$$

$$\text{Mass \% N} = \frac{3.45 \times 10^{-2} \text{ g}}{0.4831 \text{ g}} \times 100 = 7.14\% \text{ N}$$

$$\text{Mass \% O} = 100.00 - (73.78 + 10.9 + 7.14) = 8.2\% \text{ O}$$

Out of 100.00 g of compound, there are:

$$73.78 \text{ g C} \times \frac{1 \text{ mol}}{12.01 \text{ g}} = 6.143 \text{ mol C}; \ 7.14 \text{ g N} \times \frac{1 \text{ mol}}{14.01 \text{ g}} = 0.510 \text{ mol N}$$

$$10.9 \text{ g H} \times \frac{1 \text{ mol}}{1.008 \text{ g}} = 10.8 \text{ mol H}; \ 8.2 \text{ g O} \times \frac{1 \text{ mol}}{16.00 \text{ g}} = 0.51 \text{ mol O}$$

Dividing all values by 0.51 gives an empirical formula of $C_{12}H_{21}NO$.

$$\text{Molar mass} = \frac{dRT}{P} = \frac{\frac{4.02 \text{ g}}{\text{L}} \times \frac{0.08206 \text{ L atm}}{\text{K mol}} \times 400. \text{ K}}{256 \text{ torr} \times \frac{1 \text{ atm}}{760 \text{ torr}}} = 392 \text{ g/mol}$$

Empirical formula mass of $C_{12}H_{21}NO \approx 195$ g/mol; $\frac{392}{195} \approx 2$

Thus the molecular formula is $C_{24}H_{42}N_2O_2$.

140. At constant T, the lighter the gas molecules, the faster the average velocity. Therefore, the pressure will increase initially because the lighter H_2 molecules will effuse into container A faster than air will escape. However, the pressures will eventually equalize once the gases have had time to mix thoroughly.

ChemWork Problems

The answers to the problems 141-148 (or a variation to these problems) are found in OWL. These problems are also assignable in OWL.

Challenge Problems

149. $BaO(s) + CO_2(g) \rightarrow BaCO_3(s)$; $CaO(s) + CO_2(g) \rightarrow CaCO_3(s)$

$$n_i = \frac{P_i V}{RT} = \text{initial moles of } CO_2 = \frac{\frac{750.}{760} \text{ atm} \times 1.50 \text{ L}}{\frac{0.08206 \text{ L atm}}{\text{K mol}} \times 303.2 \text{ K}} = 0.0595 \text{ mol } CO_2$$

$$n_f = \frac{P_f V}{RT} = \text{final moles of } CO_2 = \frac{\frac{230.}{760}\text{atm} \times 1.50\,\text{L}}{\frac{0.08206\,\text{L atm}}{\text{K mol}} \times 303.2\text{ K}} = 0.0182 \text{ mol } CO_2$$

$0.0595 - 0.0182 = 0.0413$ mol CO_2 reacted

Because each metal reacts 1 : 1 with CO_2, the mixture contains a total of 0.0413 mol of BaO and CaO. The molar masses of BaO and CaO are 153.3 and 56.08 g/mol, respectively.

Let x = mass of BaO and y = mass of CaO, so:

$$x + y = 5.14 \text{ g and } \frac{x}{153.3} + \frac{y}{56.08} = 0.0413 \text{ mol or } x + (2.734)y = 6.33$$

Solving by simultaneous equations:

$$\begin{aligned} x + (2.734)y &= 6.33 \\ -x \quad\quad -y &= -5.14 \\ \hline (1.734)y &= 1.19, \quad y - 1.19/1.734 = 0.686 \end{aligned}$$

$y = 0.686$ g CaO and $5.14 - y = x = 4.45$ g BaO

$$\text{Mass \% BaO} = \frac{4.45\text{ g BaO}}{5.14\text{ g}} \times 100 = 86.6\%\text{ BaO; } \%\text{CaO} = 100.0 - 86.6 = 13.4\%\text{ CaO}$$

150. $Cr(s) + 3\ HCl(aq) \rightarrow CrCl_3(aq) + 3/2\ H_2(g)$; $Zn(s) + 2\ HCl(aq) \rightarrow ZnCl_2(aq) + H_2(g)$

$$\text{Mol } H_2 \text{ produced} = n = \frac{PV}{RT} = \frac{\left(750.\text{ torr} \times \frac{1\text{ atm}}{760\text{ torr}}\right) \times 0.225\text{ L}}{\frac{0.08206\text{ L atm}}{\text{K mol}} \times (273 + 27)\text{ K}} = 9.02 \times 10^{-3} \text{ mol } H_2$$

9.02×10^{-3} mol H_2 = mol H_2 from Cr reaction + mol H_2 from Zn reaction

From the balanced equation: 9.02×10^{-3} mol H_2 = mol Cr × (3/2) + mol Zn × 1

Let x = mass of Cr and y = mass of Zn, then:

$$x + y = 0.362 \text{ g and } 9.02 \times 10^{-3} = \frac{(1.5)x}{52.00} + \frac{y}{65.38}$$

We have two equations and two unknowns. Solving by simultaneous equations:

$$\begin{aligned} 9.02 \times 10^{-3} &= (0.02885)x + (0.01530)y \\ -0.01530 \times 0.362 &= -(0.01530)x - (0.01530)y \\ \hline 3.48 \times 10^{-3} &= (0.01355)x, \quad x = \text{mass of Cr} = \frac{3.48 \times 10^{-3}}{0.01355} = 0.257 \text{ g} \end{aligned}$$

$$y = \text{mass of Zn} = 0.362\text{ g} - 0.257\text{ g} = 0.105\text{ g Zn; mass \% Zn} = \frac{0.105\text{ g}}{0.362\text{ g}} \times 100 = 29.0\%\text{ Zn}$$

151. Assuming 1.000 L of the hydrocarbon (C_xH_y), then the volume of products will be 4.000 L, and the mass of products ($H_2O + CO_2$) will be:

$$1.391 \text{ g/L} \times 4.000 \text{ L} = 5.564 \text{ g products}$$

$$\text{Mol } C_xH_y = n_{C_xH_y} = \frac{PV}{RT} = \frac{0.959 \text{ atm} \times 1.000 \text{ L}}{\frac{0.08206 \text{ L atm}}{\text{K mol}} \times 298 \text{ K}} = 0.0392 \text{ mol}$$

$$\text{Mol products} = n_p = \frac{PV}{RT} = \frac{1.51 \text{ atm} \times 4.000 \text{ L}}{\frac{0.08206 \text{ L atm}}{\text{K mol}} \times 375 \text{ K}} = 0.196 \text{ mol}$$

C_xH_y + oxygen $\rightarrow x\, CO_2 + y/2\, H_2O$

Setting up two equations:

$(0.0392)x + 0.0392(y/2) = 0.196$ (moles of products)

$(0.0392)x(44.01 \text{ g/mol}) + 0.0392(y/2)(18.02 \text{ g/mol}) = 5.564 \text{ g}$ (mass of products)

Solving: $x = 2$ and $y = 6$, so the formula of the hydrocarbon is C_2H_6.

152. a. Let x = moles SO_2 = moles O_2 and z = moles He.

$\frac{P \bullet MM}{RT}$, where MM = molar mass

$$1.924 \text{ g/L} = \frac{1.000 \text{ atm} \times MM}{\frac{0.08206 \text{ L atm}}{\text{K mol}} \times 273.2 \text{ K}}, \quad MM_{mixture} = 43.13 \text{ g/mol}$$

Assuming 1.000 total moles of mixture is present, then: $x + x + z = 1.000$ and:

$$64.07 \text{ g/mol} \times x + 32.00 \text{ g/mol} \times x + 4.003 \text{ g/mol} \times z = 43.13 \text{ g}$$

$2x + z = 1.000$ and $(96.07)x + (4.003)z = 43.13$

Solving: $x = 0.4443$ mol and $z = 0.1114$ mol

Thus: $\chi_{He} = 0.1114 \text{ mol}/1.000 \text{ mol} = 0.1114$

b. $2\, SO_2(g) + O_2(g) \rightarrow 2\, SO_3(g)$

Initially, assume 0.4443 mol SO_2, 0.4443 mol O_2, and 0.1114 mol He. Because SO_2 is limiting, we end up with 0.2222 mol O_2, 0.4443 mol SO_3, and 0.1114 mol He in the gaseous product mixture. This gives $n_{initial}$ = 1.0000 mol and n_{final} = 0.7779 mol.

In a reaction, mass is constant. $d = \frac{mass}{V}$ and $V \propto n$ at constant P and T, so $d \propto \frac{1}{n}$.

$$\frac{n_{initial}}{n_{final}} = \frac{1.0000}{0.7779} = \frac{d_{final}}{d_{initial}},\ d_{final} = \left(\frac{1.0000}{0.7779}\right) \times 1.924 \text{ g/L},\ d_{final} = 2.473 \text{ g/L}$$

153. a. The reaction is $CH_4(g) + 2\ O_2(g) \rightarrow CO_2(g) + 2\ H_2O(g)$.

$$PV = nRT,\ \frac{PV}{n} = RT = \text{constant},\ \frac{P_{CH_4}V_{CH_4}}{n_{CH_4}} = \frac{P_{air}V_{air}}{n_{air}}$$

The balanced equation requires 2 mol O_2 for every mole of CH_4 that reacts. For three times as much oxygen, we would need 6 mol O_2 per mole of CH_4 reacted ($n_{O_2} = 6n_{CH_4}$). Air is 21% mole percent O_2, so $n_{O_2} = (0.21)n_{air}$. Therefore, the moles of air we would need to deliver the excess O_2 are:

$$n_{O_2} = (0.21)n_{air} = 6n_{CH_4},\ n_{air} = 29n_{CH_4},\ \frac{n_{air}}{n_{CH_4}} = 29$$

In 1 minute:

$$V_{air} = V_{CH_4} \times \frac{n_{air}}{n_{CH_4}} \times \frac{P_{CH_4}}{P_{air}} = 200.\text{ L} \times 29 \times \frac{1.50\text{ atm}}{1.00\text{ atm}} = 8.7 \times 10^3 \text{ L air/min}$$

b. If x mol of CH_4 were reacted, then $6x$ mol O_2 were added, producing $(0.950)x$ mol CO_2 and $(0.050)x$ mol of CO. In addition, $2x$ mol H_2O must be produced to balance the hydrogens.

$$CH_4(g) + 2\ O_2(g) \rightarrow CO_2(g) + 2\ H_2O(g);\ CH_4(g) + 3/2\ O_2(g) \rightarrow CO(g) + 2\ H_2O(g)$$

Amount O_2 reacted:

$$(0.950)x \text{ mol } CO_2 \times \frac{2 \text{ mol } O_2}{\text{mol } CO_2} = (1.90)x \text{ mol } O_2$$

$$(0.050)x \text{ mol CO} \times \frac{1.5 \text{ mol } O_2}{\text{mol CO}} = (0.075)x \text{ mol } O_2$$

Amount of O_2 left in reaction mixture $= (6.00)x - (1.90)x - (0.075)x = (4.03)x$ mol O_2

Amount of $N_2 = (6.00)x \text{ mol } O_2 \times \frac{79 \text{ mol } N_2}{21 \text{ mol } O_2} = (22.6)x \approx 23x$ mol N_2

The reaction mixture contains:

$(0.950)x$ mol CO_2 + $(0.050)x$ mol CO + $(4.03)x$ mol O_2 + $(2.00)x$ mol H_2O
+ $23x$ mol N_2 = $(30.)x$ mol of gas total

$$\chi_{CO} = \frac{(0.050)x}{(30.)x} = 0.0017;\ \chi_{CO_2} = \frac{(0.950)x}{(30.)x} = 0.032;\ \chi_{O_2} = \frac{(4.03)x}{(30.)x} = 0.13$$

$$\chi_{H_2O} = \frac{(2.00)x}{(30.)x} = 0.067;\quad \chi_{N_2} = \frac{23x}{(30.)x} = 0.77$$

154. The reactions are:

$$C(s) + 1/2\ O_2(g) \rightarrow CO(g) \text{ and } C(s) + O_2(g) \rightarrow CO_2(g)$$

$$PV = nRT,\ P = n\left(\frac{RT}{V}\right) = n(\text{constant})$$

Because the pressure has increased by 17.0%, the number of moles of gas has also increased by 17.0%.

$$n_{final} = (1.170)n_{initial} = 1.170(5.00) = 5.85 \text{ mol gas} = n_{O_2} + n_{CO} + n_{CO_2}$$

$n_{CO} + n_{CO_2} = 5.00$ (balancing moles of C). Solving by simultaneous equations:

$$\begin{array}{rl} n_{O_2} + n_{CO} + n_{CO_2} & = 5.85 \\ -(n_{CO} + n_{CO_2} & = 5.00) \\ \hline n_{O_2} & = 0.85 \end{array}$$

If all C were converted to CO_2, no O_2 would be left. If all C were converted to CO, we would get 5 mol CO and 2.5 mol excess O_2 in the reaction mixture. In the final mixture, moles of CO equals twice the moles of O_2 present ($n_{CO} = 2n_{O_2}$).

$$n_{CO} = 2n_{O_2} = 1.70 \text{ mol CO};\ 1.70 + n_{CO_2} = 5.00,\ n_{CO_2} = 3.30 \text{ mol } CO_2$$

$$\chi_{CO} = \frac{1.70}{5.85} = 0.291;\ \chi_{CO_2} = \frac{3.30}{5.85} = 0.564;\ \chi_{O_2} = \frac{0.85}{5.85} = 0.145 \approx 0.15$$

155. a. Volume of hot air: $V = \frac{4}{3}\pi r^3 = \frac{4}{3}\pi(2.50\text{ m})^3 = 65.4\text{ m}^3$

(*Note*: Radius = diameter/2 = 5.00/2 = 2.50 m)

$$65.4\text{ m}^3 \times \left(\frac{10\text{ dm}}{\text{m}}\right)^3 \times \frac{1\text{ L}}{\text{dm}^3} = 6.54 \times 10^4\text{ L}$$

$$n = \frac{PV}{RT} = \frac{\left(745\text{ torr} \times \frac{1\text{ atm}}{760\text{ torr}}\right) \times 6.54 \times 10^4\text{ L}}{\frac{0.08206\text{ L atm}}{\text{K mol}} \times (273 + 65)\text{ K}} = 2.31 \times 10^3\text{ mol air}$$

$$\text{Mass of hot air} = 2.31 \times 10^3 \text{ mol} \times \frac{29.0 \text{ g}}{\text{mol}} = 6.70 \times 10^4 \text{ g}$$

$$\text{Air displaced: } n = \frac{PV}{RT} = \frac{\frac{745}{760} \text{ atm} \times 6.54 \times 10^4 \text{ L}}{\frac{0.08206 \text{ L atm}}{\text{K mol}} \times (273 + 21) \text{ K}} = 2.66 \times 10^3 \text{ mol air}$$

$$\text{Mass of air displaced} = 2.66 \times 10^3 \text{ mol} \times \frac{29.0 \text{ g}}{\text{mol}} = 7.71 \times 10^4 \text{ g}$$

$$\text{Lift} = 7.71 \times 10^4 \text{ g} - 6.70 \times 10^4 \text{ g} = 1.01 \times 10^4 \text{ g}$$

b. Mass of air displaced is the same, 7.71×10^4 g. Moles of He in balloon will be the same as moles of air displaced, 2.66×10^3 mol, because P, V, and T are the same.

$$\text{Mass of He} = 2.66 \times 10^3 \text{ mol} \times \frac{4.003 \text{ g}}{\text{mol}} = 1.06 \times 10^4 \text{ g}$$

$$\text{Lift} = 7.71 \times 10^4 \text{ g} - 1.06 \times 10^4 \text{ g} = 6.65 \times 10^4 \text{ g}$$

c. Hot air: $n = \frac{PV}{RT} = \frac{\frac{630.}{760} \text{ atm} \times (6.54 \times 10^4 \text{ L})}{\frac{0.08206 \text{ L atm}}{\text{K mol}} \times 338 \text{ K}} = 1.95 \times 10^3 \text{ mol air}$

$$1.95 \times 10^3 \text{ mol} \times \frac{29.0 \text{ g}}{\text{mol}} = 5.66 \times 10^4 \text{ g of hot air}$$

$$\text{Air displaced: } n = \frac{PV}{RT} = \frac{\frac{630.}{760} \text{ atm} \times (6.54 \times 10^4 \text{ L})}{\frac{0.08206 \text{ L atm}}{\text{K mol}} \times 294 \text{ K}} = 2.25 \times 10^3 \text{ mol air}$$

$$2.25 \times 10^3 \text{ mol} \times \frac{29.0 \text{ g}}{\text{mol}} = 6.53 \times 10^4 \text{ g of air displaced}$$

$$\text{Lift} = 6.53 \times 10^4 \text{ g} - 5.66 \times 10^4 \text{ g} = 8.7 \times 10^3 \text{ g}$$

156. a. When the balloon is heated, the balloon will expand (P and n remain constant). The mass of the balloon is the same, but the volume increases, so the density of the argon in the balloon decreases. When the density is less than that of air, the balloon will rise.

b. Assuming the balloon has no mass, when the density of the argon equals the density of air, the balloon will float in air. Above this temperature, the balloon will rise.

$d_{air} = \frac{P \cdot MM_{air}}{RT}$, where MM_{air} = average molar mass of air

$$MM_{air} = 0.790 \times 28.02 \text{ g/mol} + 0.210 \times 32.00 \text{ g/mol} = 28.9 \text{ g/mol}$$

$$d_{air} = \frac{1.00\,\text{atm} \times 28.9\,\text{g/mol}}{\frac{0.08206\,\text{L atm}}{\text{K mol}} \times 298\,\text{K}} = 1.18\ \text{g/L}$$

$$d_{argon} = \frac{1.00\,\text{atm} \times 39.95\,\text{g/mol}}{\frac{0.08206\,\text{L atm}}{\text{K mol}} \times \text{T}} = 1.18\ \text{g/L},\ \ \text{T} = 413\ \text{K}$$

Heat the Ar above 413 K or 140.°C, and the balloon would float.

157. a. Average molar mass of air = 0.790 × 28.02 g/mol + 0.210 × 32.00 g/mol = 28.9 g/mol

Molar mass of helium = 4.003 g/mol

A given volume of air at a given set of conditions has a larger density than helium at those conditions due to the larger average molar mass of air. We need to heat the air to a temperature greater than 25°C in order to lower the air density (by driving air molecules out of the hot air balloon) until the density is the same as that for helium (at 25°C and 1.00 atm).

b. To provide the same lift as the helium balloon (assume V = 1.00 L), the mass of air in the hot air balloon (V = 1.00 L) must be the same as that in the helium balloon. Let MM = molar mass:

$$\text{P}\bullet\text{MM} = \text{dRT},\ \ \text{mass} = \frac{\text{MM}\bullet\text{PV}}{\text{RT}};\ \ \text{solving: mass He} = 0.164\ \text{g}$$

$$\text{Mass air} = 0.164\ \text{g} = \frac{28.9\,\text{g/mol} \times 1.00\,\text{atm} \times 1.00\,\text{L}}{\frac{0.08206\,\text{L atm}}{\text{K mol}} \times \text{T}}$$

T = 2150 K (a very high temperature)

158. $$\left(\text{P} + \frac{\text{an}^2}{\text{V}^2}\right) \times (\text{V} - \text{nb}) = \text{nRT},\ \ \text{PV} + \frac{\text{an}^2\,\text{V}}{\text{V}^2} - \text{nbP} - \frac{\text{an}^3\,\text{b}}{\text{V}^2} = \text{nRT}$$

$$\text{PV} + \frac{\text{an}^2}{\text{V}} - \text{nbP} - \frac{\text{an}^3\,\text{b}}{\text{V}^2} = \text{nRT}$$

At low P and high T, the molar volume of a gas will be relatively large. Thus the an^2/V and an^3b/V^2 terms become negligible at low P and high T because V is large. Because nb is the actual volume of the gas molecules themselves, nb << V and the –nbP term will be negligible as compared to PV. Thus PV = nRT.

159. d = molar mass(P/RT); at constant P and T, the density of gas is directly proportional to the molar mass of the gas. Thus the molar mass of the gas has a value which is 1.38 times that of the molar mass of O_2.

Molar mass = 1.38(32.00 g/mol) = 44.2 g/mol

Because H_2O is produced when the unknown binary compound is combusted, the unknown must contain hydrogen. Let A_xH_y be the formula for unknown compound.

$$\text{Mol } A_xH_y = 10.0 \text{ g } A_xH_y \times \frac{1 \text{ mol } A_xH_y}{44.2 \text{ g}} = 0.226 \text{ mol } A_xH_y$$

$$\text{Mol H} = 16.3 \text{ g } H_2O \times \frac{1 \text{ mol } H_2O}{18.02 \text{ g}} \times \frac{2 \text{ mol H}}{\text{mol } H_2O} = 1.81 \text{ mol H}$$

$$\frac{1.81 \text{ mol H}}{0.226 \text{ mol } A_xH_y} = 8 \text{ mol H/mol } A_xH_y;\ A_xH_y = A_xH_8$$

The mass of the x moles of A in the A_xH_8 formula is:

$$44.2 \text{ g} - 8(1.008 \text{ g}) = 36.1 \text{ g}$$

From the periodic table and by trial and error, some possibilities for A_xH_8 are ClH_8, F_2H_8, C_3H_8, and Be_4H_8. C_3H_8 and Be_4H_8 fit the data best, and because C_3H_8 (propane) is a known substance, C_3H_8 is the best possible identity from the data in this problem.

160. a. Initially $P_{N_2} = P_{H_2} = 1.00$ atm, and the total pressure is 2.00 atm ($P_{total} = P_{N_2} + P_{H_2}$). The total pressure after reaction will also be 2.00 atm because we have a constant-pressure container. Because V and T are constant before the reaction takes place, there must be equal moles of N_2 and H_2 present initially. Let x = mol N_2 = mol H_2 that are present initially. From the balanced equation, $N_2(g) + 3\ H_2(g) \longrightarrow 2\ NH_3(g)$, H_2 will be limiting because three times as many moles of H_2 are required to react as compared to moles of N_2. After the reaction occurs, none of the H_2 remains (it is the limiting reagent).

$$\text{Mol } NH_3 \text{ produced} = x \text{ mol } H_2 \times \frac{2 \text{ mol } NH_3}{3 \text{ mol } H_2} = 2x/3$$

$$\text{Mol } N_2 \text{ reacted} = x \text{ mol } H_2 \times \frac{1 \text{ mol } N_2}{3 \text{ mol } H_2} = x/3$$

Mol N_2 remaining = x mol N_2 present initially – $x/3$ mol N_2 reacted = $2x/3$ mol N_2

After the reaction goes to completion, equal moles of $N_2(g)$ and $NH_3(g)$ are present ($2x/3$). Because equal moles are present, the partial pressure of each gas must be equal ($P_{N_2} = P_{NH_3}$).

$$P_{total} = 2.00 \text{ atm} = P_{N_2} + P_{NH_3};\ \text{solving: } P_{N_2} = 1.00 \text{ atm} = P_{NH_3}$$

b. V ∝ n because P and T are constant. The moles of gas present initially are:

$$n_{N_2} + n_{H_2} = x + x = 2x \text{ mol}$$

After reaction, the moles of gas present are:

$$n_{N_2} + n_{NH_3} = \frac{2x}{3} + \frac{2x}{3} = 4x/3 \text{ mol}$$

$$\frac{V_{after}}{V_{initial}} = \frac{n_{after}}{n_{initial}} = \frac{4x/3}{2x} = \frac{2}{3}$$

The volume of the container will be two-thirds the original volume, so:

$$V = 2/3(15.0 \text{ L}) = 10.0 \text{ L}$$

Integrative Problems

161. The redox equation must be balanced. Each uranium atom changes oxidation sates from +4 in UO^{2+} to +6 in UO_2^{2+} (a loss of two electrons for each uranium atom). Each nitrogen atom changes oxidation states from +5 in NO_3^- to +2 in NO (a gain of three electrons for each nitrogen atom). To balance the electrons transferred, we need two N atoms for every three U atoms. The balanced equation is:

$$2\,H^+(aq) + 2\,NO_3^-(aq) + 3\,UO^{2+}(aq) \rightarrow 3\,UO_2^{2+}(aq) + 2\,NO(g) + H_2O(l)$$

$$n_{NO} = \frac{PV}{RT} = \frac{1.5 \text{ atm} \times 0.255 \text{ L}}{\frac{0.08206 \text{ L atm}}{\text{K mol}} \times 302 \text{ K}} = 0.015 \text{ mol NO}$$

$$0.015 \text{ mol NO} \times \frac{3 \text{ mol } UO^{2+}}{2 \text{ mol NO}} = 0.023 \text{ mol } UO^{2+}$$

162. a. $156 \text{ mL} \times \frac{1.34 \text{ g}}{\text{mL}} = 209 \text{ g } HSiCl_3$ = actual yield of $HSiCl_3$

$$n_{HCl} = \frac{PV}{RT} = \frac{10.0 \text{ atm} \times 15.0 \text{ L}}{\frac{0.08206 \text{ L atm}}{\text{K mol}} \times 308 \text{ K}} = 5.93 \text{ mol HCl}$$

$$5.93 \text{ mol HCl} \times \frac{1 \text{ mol } HSiCl_3}{3 \text{ mol HCl}} \times \frac{135.45 \text{ g } HSiCl_3}{1 \text{ mol } HSiCl_3} = 268 \text{ g } HSiCl_3$$

$$\text{Percent yield} = \frac{\text{actual yield}}{\text{theoretical yield}} \times 100 = \frac{209 \text{ g}}{268 \text{ g}} \times 100 = 78.0\%$$

b. $209 \text{ g } HiSCl_3 \times \frac{1 \text{ mol } HSiCl_3}{135.45 \text{ g } HSiCl_3} \times \frac{1 \text{ mol } SiH_4}{4 \text{ mol } HSiCl_3} = 0.386 \text{ mol } SiH_4$

This is the theoretical yield. If the percent yield is 93.1%, then the actual yield is:

$$0.386 \text{ mol } SiH_4 \times 0.931 = 0.359 \text{ mol } SiH_4$$

$$V_{SiH_4} = \frac{nRT}{P} = \frac{0.359\,\text{mol} \times \frac{0.08206\,\text{L atm}}{\text{K mol}} \times 308\,\text{K}}{10.0\,\text{atm}} = 0.907\text{ L} = 907\text{ mL } SiH_4$$

163. ThF_4, 232.0 + 4(19.00) = 308.0 g/mL

$$d = \frac{\text{molar mass} \times P}{RT} = \frac{308.0\text{ g/mol} \times 2.5\text{ atm}}{\frac{0.08206\,\text{L atm}}{\text{K mol}} \times (1680 + 273)\text{ K}} = 4.8\text{ g/L}$$

The gas with the smaller molar mass will effuse faster. Molar mass of ThF_4 = 308.0 g/mol; molar mass of UF_3 = 238.0 + 3(19.00) = 295.0 g/mol. Therefore, UF_3 will effuse faster.

$$\frac{\text{Rate of effusion of } UF_3}{\text{Rate of effusion of } ThF_4} = \sqrt{\frac{\text{molar mass of } ThF_4}{\text{molar mass of } UF_3}} = \sqrt{\frac{308.0\text{g/mol}}{295.0\text{g/mol}}} = 1.02$$

UF_3 effuses 1.02 times faster than ThF_4.

164. The partial pressures can be determined by using the mole fractions.

$P_{methane} = P_{total} \times \chi_{methane}$ = 1.44 atm × 0.915 = 1.32 atm; P_{ethane} = 1.44 – 1.32 = 0.12 atm

Determining the number of moles of natural gas combusted:

$$n_{natural\ gas} = \frac{PV}{RT} = \frac{1.44\,\text{atm} \times 15.00\,\text{L}}{\frac{0.08206\,\text{L atm}}{\text{K mol}} \times 293\,\text{K}} = 0.898\text{ mol natural gas}$$

$n_{methane} = n_{natural\ gas} \times \chi_{methane}$ = 0.898 mol × 0.915 = 0.822 mol methane

n_{ethane} = 0.898 – 0.822 = 0.076 mol ethane

$CH_4(g) + 2\ O_2(g) \rightarrow CO_2(g) + 2\ H_2O(l)$; $2\ C_2H_6 + 7\ O_2(g) \rightarrow 4\ CO_2(g) + 6\ H_2O(l)$

$$0.822\text{ mol } CH_4 \times \frac{2\,\text{mol}\,H_2O}{1\,\text{mol}\,CH_4} \times \frac{18.02\,\text{g}\,H_2O}{\text{mol}\,H_2O} = 29.6\text{ g } H_2O$$

$$0.076\text{ mol } C_2H_6 \times \frac{6\,\text{mol}\,H_2O}{2\,\text{mol}\,C_2H_6} \times \frac{18.02\,\text{g}\,H_2O}{\text{mol}\,H_2O} = 4.1\text{ g } H_2O$$

The total mass of H_2O produced = 29.6 g + 4.1 g = 33.7 g H_2O.

Marathon Problem

165. a. The formula of the compound A_xB_y depends on which gas is limiting, A_2 or B_2. We need to determine both possible products. The procedure we will use is to assume one reactant is limiting, and then determine what happens to the initial total moles of gas as it is converted into the product. Because P and T are constant, volume $\propto$ n. Because mass is conserved in a chemical reaction, any change in density must be due to a change in volume of the container as the reaction goes to completion.

$$\text{Density} = d \propto \frac{1}{V} \quad \text{and} \quad V \propto n, \quad \text{so:} \quad \frac{d_{after}}{d_{initial}} = \frac{n_{initial}}{n_{after}}$$

Assume the molecular formula of the product is A_xB_y where x and y are whole numbers. First, let's consider when A_2 is limiting with x moles each of A_2 and B_2 in our equimolar mixture. Note that the coefficient in front of A_xB_y in the equation must be 2 for a balanced reaction.

	$x\,A_2(g)$	$+\ y\,B_2(g)$	$\rightarrow$ $2\,A_xB_y(g)$
Initial	x mol	x mol	0 mol
Change	$-x$ mol	$-y$ mol	+2 mol
Final	0	$(x - y)$ mol	2 mol

$$\frac{d_{after}}{d_{initial}} = 1.50 = \frac{n_{initial}}{n_{after}} = \frac{2x}{x - y + 2}$$

$(1.50)x - (1.50)y + 3.00 = 2x,\ \ 3.00 - (1.50)y = (0.50)x$

Because x and y are whole numbers, y must be 1 because the above equation does not allow y to be 2 or greater. When $y = 1$, $x = 3$ giving a formula of A_3B if A_2 is limiting.

Assuming B_2 is limiting with y moles in the equimolar mixture:

	$x\,A_2(g)$	$+\ y\,B_2(g)$	$\rightarrow$ $2\,A_xB_y(g)$
Initial	y	y	0
Change	$-x$	$-y$	+2
After	$y - x$	0	2

$$\frac{\text{density}_{after}}{\text{density}_{before}} = 1.50 = \frac{n_{initial}}{n_{after}} = \frac{2y}{y - x + 2}$$

Solving gives $x = 1$ and $y = 3$ for a molecular formula of AB_3 when B_2 is limiting.

b. In both possible products, the equations dictated that only one mole of either A or B had to be present in the formula. Any number larger than 1 would not fit the data given in the problem. Thus the two formulas determined are both molecular formulas and not just empirical formulas.

CHAPTER 6

THERMOCHEMISTRY

Questions

11. Path-dependent functions for a trip from Chicago to Denver are those quantities that depend on the route taken. One can fly directly from Chicago to Denver, or one could fly from Chicago to Atlanta to Los Angeles and then to Denver. Some path-dependent quantities are miles traveled, fuel consumption of the airplane, time traveling, airplane snacks eaten, etc. State functions are path-independent; they only depend on the initial and final states. Some state functions for an airplane trip from Chicago to Denver would be longitude change, latitude change, elevation change, and overall time zone change.

12. Products have a lower potential energy than reactants when the bonds in the products are stronger (on average) than in the reactants. This occurs generally in exothermic processes. Products have a higher potential energy than reactants when the reactants have the stronger bonds (on average). This is typified by endothermic reactions.

13. $2\ C_8H_{18}(l) + 25\ O_2(g) \rightarrow 16\ CO_2(g) + 18\ H_2O(g)$; the combustion of gasoline is exothermic (as is typical of combustion reactions). For exothermic reactions, heat is released into the surroundings giving a negative q value. To determine the sign of w, concentrate on the moles of gaseous reactants versus the moles of gaseous products. In this combustion reaction, we go from 25 moles of reactant gas molecules to 16 + 18 = 34 moles of product gas molecules. As reactants are converted to products, an expansion will occur because the moles of gas increase. When a gas expands, the system does work on the surroundings, and w is a negative value.

14. $\Delta H = \Delta E + P\Delta V$ at constant P; from the definition of enthalpy, the difference between ΔH and ΔE, at constant P, is the quantity $P\Delta V$. Thus, when a system at constant P can do pressure-volume work, then $\Delta H \neq \Delta E$. When the system cannot do PV work, then $\Delta H = \Delta E$ at constant pressure. An important way to differentiate ΔH from ΔE is to concentrate on q, the heat flow; the heat flow by a system at constant pressure equals ΔH, and the heat flow by a system at constant volume equals ΔE.

15. a. The ΔH value for a reaction is specific to the coefficients in the balanced equation. Because the coefficient in front of H_2O is a 2, 891 kJ of heat is released when 2 mol of H_2O is produced. For 1 mol of H_2O formed, 891/2 = 446 kJ of heat is released.

    b. 891/2 = 446 kJ of heat released for each mol of O_2 reacted.

16. Use the coefficients in the balanced rection to determine the heat required for the various quantities.

    a. $1 \text{ mol Hg} \times \dfrac{90.7 \text{ kJ}}{\text{mol Hg}} = 90.7 \text{ kJ required}$

b. $1 \text{ mol } O_2 \times \dfrac{90.7 \text{ kJ}}{1/2 \text{ mol } O_2} = 181.4 \text{ kJ required}$

c. When an equation is reversed, $\Delta H_{new} = -\Delta H_{old}$. When an equation is multiplied by some integer *n*, then $\Delta H_{new} = n(\Delta H_{old})$.

$$Hg(l) + 1/2\ O_2(g) \rightarrow HgO(s) \qquad \Delta H = -90.7 \text{ kJ}$$

$$2Hg(l) + O_2(g) \rightarrow 2HgO(s) \qquad \Delta H = 2(-90.7 \text{ kJ}) = -181.4 \text{ kJ};\ 181.4 \text{ kJ released}$$

17. Given:

$$CH_4(g) + 2\ O_2(g) \rightarrow CO_2(g) + 2\ H_2O(l) \qquad \Delta H = -891 \text{ kJ}$$
$$CH_4(g) + 2\ O_2(g) \rightarrow CO_2(g) + 2\ H_2O(g) \qquad \Delta H = -803 \text{ kJ}$$

Using Hess's law:

$$H_2O(l) + 1/2\ CO_2(g) \rightarrow 1/2\ CH_4(g) + O_2(g) \qquad \Delta H_1 = -1/2(-891 \text{ kJ})$$
$$1/2\ CH_4(g) + O_2(g) \rightarrow 1/2\ CO_2(g) + H_2O(g) \qquad \Delta H_2 = 1/2(-803 \text{ kJ})$$
$$\overline{H_2O(l) \rightarrow H_2O(g) \qquad \Delta H = \Delta H_1 + \Delta H_2 = 44 \text{ kJ}}$$

The enthalpy of vaporization of water is 44 kJ/mol.

Note: When an equation is reversed, the sign on ΔH is reversed. When the coefficients in a balanced equation are multiplied by an integer, then the value of ΔH is multiplied by the same integer.

18. A state function is a function whose change depends only on the initial and final states and not on how one got from the initial to the final state. An extensive property depends on the amount of substance. Enthalpy changes for a reaction are path-independent, but they do depend on the quantity of reactants consumed in the reaction. Therefore, enthalpy changes are a state function and an extensive property.

19. The zero point for ΔH_f° values are elements in their standard state. All substances are measured in relationship to this zero point.

20. a. $CH_4(g) + 2\ O_2(g) \rightarrow CO_2(g) + 2\ H_2O(l) \qquad \Delta H = ?$

Utilizing Hess's law:

$$\text{Reactants} \rightarrow \text{Standard State Elements} \quad \Delta H = \Delta H_a + \Delta H_b = 75 + 0 = 75 \text{ kJ}$$
$$\text{Standard State Elements} \rightarrow \text{Products} \quad \Delta H = \Delta H_c + \Delta H_d = -394 - 572 = -966 \text{ kJ}$$
$$\overline{\text{Reactants} \rightarrow \text{Products} \quad \Delta H = 75 - 966 = -891 \text{ kJ}}$$

b. The standard enthalpy of formation for an element in its standard state is given a value of zero. To assign standard enthalpy of formation values for all other substances, there needs to be a reference point from which all enthalpy changes are determined. This reference point is the elements in their standard state which is defined as the zero point.

So when using standard enthalpy values, a reaction is broken up into two steps. The first step is to calculate the enthalpy change necessary to convert the reactants to the elements in their standard state. The second step is to determine the enthalpy change that occurs when the elements in their standard state go to form the products. When these two steps are added together, the reference point (the elements in their standard state) cancels out and we are left with the enthalpy change for the reaction.

c. This overall reaction is just the reverse of all the steps in the part a answer. So $\Delta H° = +966 - 75 = 891$ kJ. Products are first converted to the elements in their standard state which requires 966 kJ of heat. Next, the elements in the standard states go to form the original reactants [$CH_4(g) + 2\ O_2(g)$] which has an enthalpy change of −75 kJ. All of the signs are reversed because the entire process is reversed.

21. No matter how insulated your thermos bottle, some heat will always escape into the surroundings. If the temperature of the thermos bottle (the surroundings) is high, less heat initially will escape from the coffee (the system); this results in your coffee staying hotter for a longer period of time.

22. From the photosynthesis reaction, $CO_2(g)$ is used by plants to convert water into glucose and oxygen. If the plant population is significantly reduced, not as much CO_2 will be consumed in the photosynthesis reaction. As the CO_2 levels of the atmosphere increase, the greenhouse effect due to excess CO_2 in the atmosphere will become worse.

23. Fossil fuels contain carbon; the incomplete combustion of fossil fuels produces CO(g) instead of $CO_2(g)$. This occurs when the amount of oxygen reacting is not sufficient to convert all the carbon to CO_2. Carbon monoxide is a poisonous gas to humans.

24. Advantages: H_2 burns cleanly (less pollution) and gives a lot of energy per gram of fuel. Water as a source of hydrogen is abundant and cheap.

Disadvantages: Expensive and gas storage and safety issues

Exercises

Potential and Kinetic Energy

25. $KE = \frac{1}{2}mv^2$; convert mass and velocity to SI units. $1\ J = \frac{1\ kg\ m^2}{s^2}$

$$\text{Mass} = 5.25\ \text{oz} \times \frac{1\ \text{lb}}{16\ \text{oz}} \times \frac{1\ \text{kg}}{2.205\ \text{lb}} = 0.149\ \text{kg}$$

$$\text{Velocity} = \frac{1.0 \times 10^2\ \text{mi}}{\text{h}} \times \frac{1\ \text{h}}{60\ \text{min}} \times \frac{1\ \text{min}}{60\ \text{s}} \times \frac{1760\ \text{yd}}{\text{mi}} \times \frac{1\ \text{m}}{1.094\ \text{yd}} = \frac{45\ \text{m}}{\text{s}}$$

$$KE = \frac{1}{2}mv^2 = \frac{1}{2} \times 0.149\ \text{kg} \times \left(\frac{45\ \text{m}}{\text{s}}\right)^2 = 150\ \text{J}$$

26. $KE = \frac{1}{2}mv^2 = \frac{1}{2} \times 2.0\text{ kg} \times \left(\frac{1.0\text{ m}}{\text{s}}\right)^2 = 1.0\text{ J}$; $KE = \frac{1}{2}mv^2 = \frac{1}{2} \times 1.0\text{ kg} \times \left(\frac{2.0\text{ m}}{\text{s}}\right)^2$

$= 2.0\text{ J}$

The 1.0-kg object with a velocity of 2.0 m/s has the greater kinetic energy.

27. a. Potential energy is energy due to position. Initially, ball A has a higher potential energy than ball B because the position of ball A is higher than the position of ball B. In the final position, ball B has the higher position so ball B has the higher potential energy.

b. As ball A rolled down the hill, some of the potential energy lost by A has been converted to random motion of the components of the hill (frictional heating). The remainder of the lost potential energy was added to B to initially increase its kinetic energy and then to increase its potential energy.

28. Ball A: $PE = mgz = 2.00\text{ kg} \times \frac{9.81\text{m}}{\text{s}^2} \times 10.0\text{ m} = \frac{196\text{kg m}^2}{\text{s}^2} = 196\text{ J}$

At point I: All this energy is transferred to ball B. All of B's energy is kinetic energy at this point. $E_{total} = KE = 196\text{ J}$. At point II, the sum of the total energy will equal 196 J.

At point II: $PE = mgz = 4.00\text{ kg} \times \frac{9.81\text{m}}{\text{s}^2} \times 3.00\text{ m} = 118\text{ J}$

$KE = E_{total} - PE = 196\text{ J} - 118\text{ J} = 78\text{ J}$

Heat and Work

29. $\Delta E = q + w = 45\text{ kJ} + (-29\text{ kJ}) = 16\text{ kJ}$

30. $\Delta E = q + w = -125 + 104 = -21\text{ kJ}$

31. a. $\Delta E = q + w = -47\text{ kJ} + 88\text{ kJ} = 41\text{ kJ}$

b. $\Delta E = 82 - 47 = 35\text{ kJ}$

c. $\Delta E = 47 + 0 = 47\text{ kJ}$

d. When the surroundings do work on the system, $w > 0$. This is the case for a.

32. Step 1: $\Delta E_1 = q + w = 72\text{ J} + 35\text{ J} = 107\text{ J}$, step 2: $\Delta E_2 = 35\text{ J} - 72\text{ J} = -37\text{ J}$

$\Delta E_{overall} = \Delta E_1 + \Delta E_2 = 107\text{ J} - 37\text{ J} = 70.\text{ J}$

33. $\Delta E = q + w$; work is done by the system on the surroundings in a gas expansion; w is negative.

$300.\text{ J} = q - 75\text{ J}$, $q = 375\text{ J}$ of heat transferred to the system

34. a. $\Delta E = q + w = -23 \text{ J} + 100. \text{ J} = 77 \text{ J}$

b. $w = -P\Delta V = -1.90 \text{ atm}(2.80 \text{ L} - 8.30 \text{ L}) = 10.5 \text{ L atm} \times \frac{101.3 \text{ J}}{\text{L atm}} = 1060 \text{ J}$

$\Delta E = q + w = 350. \text{ J} + 1060 = 1410 \text{ J}$

c. $w = -P\Delta V = -1.00 \text{ atm}(29.1 \text{ L} - 11.2 \text{ L}) = -17.9 \text{ L atm} \times \frac{101.3 \text{ J}}{\text{L atm}} = -1810 \text{ J}$

$\Delta E = q + w = 1037 \text{ J} - 1810 \text{ J} = -770 \text{ J}$

35. $w = -P\Delta V$; we need the final volume of the gas. Because T and n are constant, $P_1V_1 = P_2V_2$.

$$V_2 = \frac{V_1 P_1}{P_2} = \frac{10.0 \text{ L}(15.0 \text{ atm})}{2.00 \text{ atm}} = 75.0 \text{ L}$$

$$w = -P\Delta V = -2.00 \text{ atm}(75.0 \text{ L} - 10.0 \text{ L}) = -130. \text{ L atm} \times \frac{101.3 \text{ J}}{\text{L atm}} \times \frac{1 \text{ kJ}}{1000 \text{ J}}$$

$= -13.2 \text{ kJ} = \text{work}$

36. $w = -210. \text{ J} = -P\Delta V$, $-210 \text{ J} = -P(25 \text{ L} - 10. \text{ L})$, $P = 14 \text{ atm}$

37. In this problem, $q = w = -950. \text{ J}$.

$-950. \text{ J} \times \frac{1 \text{ L atm}}{101.3 \text{ J}} = -9.38 \text{ L atm}$ of work done by the gases

$w = -P\Delta V$, $-9.38 \text{ L atm} = \frac{-650.}{760} \text{ atm} \times (V_f - 0.040 \text{ L})$, $V_f - 0.040 = 11.0 \text{ L}$, $V_f = 11.0 \text{ L}$

38. $\Delta E = q + w$, $-102.5 \text{ J} = 52.5 \text{ J} + w$, $w = -155.0 \text{ J} \times \frac{1 \text{ L atm}}{101.3 \text{ J}} = -1.530 \text{ L atm}$

$w = -P\Delta V$, $-1.530 \text{ L atm} = -0.500 \text{ atm} \times \Delta V$, $\Delta V = 3.06 \text{ L}$

$\Delta V = V_f - V_i$, $3.06 \text{ L} = 58.0 \text{ L} - V_i$, $V_i = 54.9 \text{ L}$ = initial volume

39. $q = \text{molar heat capacity} \times \text{mol} \times \Delta T = \frac{20.8 \text{ J}}{^\circ\text{C mol}} \times 39.1 \text{ mol} \times (38.0 - 0.0)^\circ\text{C} = 30{,}900 \text{ J}$

$= 30.9 \text{ kJ}$

$w = -P\Delta V = -1.00 \text{ atm} \times (998 \text{ L} - 876 \text{ L}) = -122 \text{ L atm} \times \frac{101.3 \text{ J}}{\text{L atm}} = -12{,}400 \text{ J} = -12.4 \text{ kJ}$

$\Delta E = q + w = 30.9 \text{ kJ} + (-12.4 \text{ kJ}) = 18.5 \text{ kJ}$

40. $H_2O(g) \rightarrow H_2O(l)$; $\Delta E = q + w$; $q = -40.66$ kJ; $w = -P\Delta V$

$$\text{Volume of 1 mol } H_2O(l) = 1.000 \text{ mol } H_2O(l) \times \frac{18.02\text{ g}}{\text{mol}} \times \frac{1\text{ cm}^3}{0.996\text{ g}} = 18.1\text{ cm}^3 = 18.1\text{ mL}$$

$$w = -P\Delta V = -1.00\text{ atm} \times (0.0181\text{ L} - 30.6\text{ L}) = 30.6\text{ L atm} \times \frac{101.3\text{ J}}{\text{L atm}} = 3.10 \times 10^3\text{ J}$$

$$= 3.10\text{ kJ}$$

$$\Delta E = q + w = -40.66\text{ kJ} + 3.10\text{ kJ} = -37.56\text{ kJ}$$

Properties of Enthalpy

41. This is an endothermic reaction, so heat must be absorbed in order to convert reactants into products. The high-temperature environment of internal combustion engines provides the heat.

42. One should try to cool the reaction mixture or provide some means of removing heat because the reaction is very exothermic (heat is released). The $H_2SO_4(aq)$ will get very hot and possibly boil unless cooling is provided.

43. a. Heat is absorbed from the water (it gets colder) as KBr dissolves, so this is an endothermic process.

    b. Heat is released as CH_4 is burned, so this is an exothermic process.

    c. Heat is released to the water (it gets hot) as H_2SO_4 is added, so this is an exothermic process.

    d. Heat must be added (absorbed) to boil water, so this is an endothermic process.

44. a. The combustion of gasoline releases heat, so this is an exothermic process.

    b. $H_2O(g) \longrightarrow H_2O(l)$; heat is released when water vapor condenses, so this is an exothermic process.

    c. To convert a solid to a gas, heat must be absorbed, so this is an endothermic process.

    d. Heat must be added (absorbed) in order to break a bond, so this is an endothermic process.

45. $4\text{ Fe}(s) + 3\text{ O}_2(g) \rightarrow 2\text{ Fe}_2\text{O}_3(s)$ $\Delta H = -1652$ kJ; note that 1652 kJ of heat is released when 4 mol Fe reacts with 3 mol O_2 to produce 2 mol Fe_2O_3.

    a. $4.00\text{ mol Fe} \times \dfrac{-1652\text{kJ}}{4\text{ mol Fe}} = -1650$ kJ; 1650 kJ of heat released

    b. $1.00\text{ mol Fe}_2\text{O}_3 \times \dfrac{-1652\text{kJ}}{2\text{ mol Fe}_2\text{O}_3} = -826$ kJ; 826 kJ of heat released

c. $1.00\text{ g Fe} \times \frac{1\text{ mol Fe}}{55.85\text{ g}} \times \frac{-1652\text{ kJ}}{4\text{ mol Fe}} = -7.39\text{ kJ}$; 7.39 kJ of heat released

d. $10.0\text{ g Fe} \times \frac{1\text{ mol Fe}}{55.85\text{ g Fe}} \times \frac{-1652\text{ kJ}}{4\text{ mol Fe}} = -73.9\text{ kJ}$

$$2.00\text{ g O}_2 \times \frac{1\text{ mol O}_2}{32.00\text{ g O}_2} \times \frac{-1652\text{ kJ}}{3\text{ mol O}_2} = -34.4\text{ kJ}$$

Because 2.00 g O_2 releases the smaller quantity of heat, O_2 is the limiting reactant and 34.4 kJ of heat can be released from this mixture.

46. a. $1.00\text{ mol H}_2\text{O} \times \frac{-572\text{ kJ}}{2\text{ mol H}_2\text{O}} = -286\text{ kJ}$; 286 kJ of heat released

b. $4.03\text{ g H}_2 \times \frac{1\text{ mol H}_2}{2.016\text{ g H}_2} \times \frac{-572\text{ kJ}}{2\text{ mol H}_2} = -572\text{ kJ}$; 572 kJ of heat released

c. $186\text{ g O}_2 \times \frac{1\text{ mol O}_2}{32.00\text{ g O}_2} \times \frac{-572\text{ kJ}}{\text{mol O}_2} = -3320\text{ kJ}$; 3320 kJ of heat released

d. $n_{H_2} = \frac{PV}{RT} = \frac{1.0\text{ atm} \times 2.0 \times 10^8\text{ L}}{\frac{0.08206\text{ L atm}}{\text{K mol}} \times 298\text{ K}} = 8.2 \times 10^6\text{ mol H}_2$

$$8.2 \times 10^6\text{ mol H}_2 \times \frac{-572\text{ kJ}}{2\text{ mol H}_2} = -2.3 \times 10^9\text{ kJ};\ 2.3 \times 10^9\text{ kJ of heat released}$$

47. From Example 6.3, $q = 1.3 \times 10^8$ J. Because the heat transfer process is only 60.% efficient, the total energy required is $1.3 \times 10^8\text{ J} \times \frac{100.\text{ J}}{60.\text{ J}} = 2.2 \times 10^8\text{ J}$.

$$\text{Mass C}_3\text{H}_8 = 2.2 \times 10^8\text{ J} \times \frac{1\text{ mol C}_3\text{H}_8}{2221 \times 10^3\text{ J}} \times \frac{44.09\text{ g C}_3\text{H}_8}{\text{mol C}_3\text{H}_8} = 4.4 \times 10^3\text{ g C}_3\text{H}_8$$

48. a. $1.00\text{ g CH}_4 \times \frac{1\text{ mol CH}_4}{16.04\text{ g CH}_4} \times \frac{-891\text{ kJ}}{\text{mol CH}_4} = -55.5\text{ kJ}$

b. $n = \frac{PV}{RT} = \frac{740.\text{ torr} \times \frac{1\text{ atm}}{760\text{ torr}} \times 1.00 \times 10^3\text{ L}}{\frac{0.08206\text{ L atm}}{\text{K mol}} \times 298\text{ K}} = 39.8\text{ mol CH}_4$

$$39.8\text{ mol CH}_4 \times \frac{-891\text{ kJ}}{\text{mol CH}_4} = -3.55 \times 10^4\text{ kJ}$$

49. When a liquid is converted into gas, there is an increase in volume. The 2.5 kJ/mol quantity is the work done by the vaporization process in pushing back the atmosphere.

50. $\Delta H = \Delta E + P\Delta V$; from this equation, $\Delta H > \Delta E$ when $\Delta V > 0$, $\Delta H < \Delta E$ when $\Delta V < 0$, and $\Delta H = \Delta E$ when $\Delta V = 0$. Concentrate on the moles of gaseous products versus the moles of gaseous reactants to predict ΔV for a reaction.

a. There are 2 moles of gaseous reactants converting to 2 moles of gaseous products, so $\Delta V = 0$. For this reaction, $\Delta H = \Delta E$.

b. There are 4 moles of gaseous reactants converting to 2 moles of gaseous products, so $\Delta V < 0$ and $\Delta H < \Delta E$.

c. There are 9 moles of gaseous reactants converting to 10 moles of gaseous products, so $\Delta V > 0$ and $\Delta H > \Delta E$.

Calorimetry and Heat Capacity

51. Specific heat capacity is defined as the amount of heat necessary to raise the temperature of one gram of substance by one degree Celsius. Therefore, $H_2O(l)$ with the largest heat capacity value requires the largest amount of heat for this process. The amount of heat for $H_2O(l)$ is:

$$\text{energy} = s \times m \times \Delta T = \frac{4.18\ \text{J}}{^\circ\text{C g}} \times 25.0\ \text{g} \times (37.0^\circ\text{C} - 15.0^\circ\text{C}) = 2.30 \times 10^3\ \text{J}$$

The largest temperature change when a certain amount of energy is added to a certain mass of substance will occur for the substance with the smallest specific heat capacity. This is Hg(l), and the temperature change for this process is:

$$\Delta T = \frac{\text{energy}}{s \times m} = \frac{10.7\ \text{kJ} \times \frac{1000\text{J}}{\text{kJ}}}{\frac{0.14\ \text{J}}{^\circ\text{C g}} \times 550.\ \text{g}} = 140^\circ\text{C}$$

52. a. $s = \text{specific heat capacity} = \frac{0.24\ \text{J}}{^\circ\text{C g}} = \frac{0.24\ \text{J}}{\text{K g}}$ since $\Delta T(\text{K}) = \Delta T(^\circ\text{C})$

$$\text{Energy} = s \times m \times \Delta T = \frac{0.24\ \text{J}}{^\circ\text{C g}} \times 150.0\ \text{g} \times (298\ \text{K} - 273\ \text{K}) = 9.0 \times 10^2\ \text{J}$$

b. $\text{Molar heat capacity} = \frac{0.24\ \text{J}}{^\circ\text{C g}} \times \frac{107.9\ \text{g Ag}}{\text{mol Ag}} = \frac{26\ \text{J}}{^\circ\text{C mol}}$

c. $1250\ \text{J} = \frac{0.24\ \text{J}}{^\circ\text{C g}} \times m \times (15.2^\circ\text{C} - 12.0^\circ\text{C}),\ m = \frac{1250}{0.24 \times 3.2} = 1.6 \times 10^3\ \text{g Ag}$

53. $s = \text{specific heat capacity} = \frac{q}{m \times \Delta T} = \frac{133\ \text{J}}{5.00\ \text{g} \times (55.1 - 25.2)^\circ\text{C}} = 0.890\ \text{J/}^\circ\text{C}\cdot\text{g}$

From Table 6.1, the substance is solid aluminum.

54. $s = \dfrac{585\text{ J}}{125.6\text{ g} \times (53.5 - 20.0)^\circ\text{C}} = 0.139\text{ J/°C•g}$

Molar heat capacity $= \dfrac{0.139\text{ J}}{^\circ\text{C g}} \times \dfrac{200.6\text{ g}}{\text{molHg}} = \dfrac{27.9\text{ J}}{^\circ\text{C mol}}$

55. | Heat loss by hot water | = | heat gain by cooler water |

The magnitudes of heat loss and heat gain are equal in calorimetry problems. The only difference is the sign (positive or negative). To avoid sign errors, keep all quantities positive and, if necessary, deduce the correct signs at the end of the problem. Water has a specific heat capacity = s = 4.18 J/°C•g = 4.18 J/K•g (ΔT in °C = ΔT in K).

Heat loss by hot water $= s \times m \times \Delta T = \dfrac{4.18\text{ J}}{\text{K g}} \times 50.0\text{ g} \times (330.\text{ K} - T_f)$

Heat gain by cooler water $= \dfrac{4.18\text{ J}}{\text{K g}} \times 30.0\text{ g} \times (T_f - 280.\text{ K})$; heat loss = heat gain, so:

$$\frac{209\text{ J}}{\text{K}} \times (330.\text{ K} - T_f) = \frac{125\text{ J}}{\text{K}} \times (T_f - 280.\text{ K})$$

$$6.90 \times 10^4 - 209T_f = 125T_f - 3.50 \times 10^4,\ 334T_f = 1.040 \times 10^5,\ T_f = 311\text{ K}$$

Note that the final temperature is closer to the temperature of the more massive hot water, which is as it should be.

56. Heat loss by hot water = heat gain by cold water; keeping all quantities positive helps to avoid sign errors:

$$\frac{4.18\text{ J}}{^\circ\text{C g}} \times m_{hot} \times (55.0^\circ\text{C} - 37.0^\circ\text{C}) = \frac{4.18\text{ J}}{^\circ\text{C g}} \times 90.0\text{ g} \times (37.0^\circ\text{C} - 22.0^\circ\text{C})$$

$$m_{hot} = \frac{90.0\text{ g} \times 15.0^\circ\text{C}}{18.0^\circ\text{C}} = 75.0\text{ g hot water needed}$$

57. Heat loss by Al + heat loss by Fe = heat gain by water; keeping all quantities positive to avoid sign error:

$$\frac{0.89\text{ J}}{^\circ\text{C g}} \times 5.00\text{ g Al} \times (100.0^\circ\text{C} - T_f) + \frac{0.45\text{ J}}{^\circ\text{C g}} \times 10.00\text{ g Fe} \times (100.0 - T_f)$$
$$= \frac{4.18\text{ J}}{^\circ\text{C g}} \times 97.3\text{ g } H_2O \times (T_f - 22.0^\circ\text{C})$$

$$4.5(100.0 - T_f) + 4.5(100.0 - T_f) = 407(T_f - 22.0),\ 450 - (4.5)T_f + 450 - (4.5)T_f = 407T_f - 8950$$

$$416T_f = 9850,\ T_f = 23.7^\circ\text{C}$$

58. Heat released to water = $5.0\text{ g }H_2 \times \dfrac{120.\text{J}}{\text{g }H_2} + 10.\text{ g methane} \times \dfrac{50.\text{J}}{\text{g methane}} = 1.10 \times 10^3\text{ J}$

Heat gain by water = $1.10 \times 10^3\text{ J} = \dfrac{4.18\text{ J}}{^\circ\text{C g}} \times 50.0\text{ g} \times \Delta T$

$\Delta T = 5.26°\text{C}$, $5.26°\text{C} = T_f - 25.0°\text{C}$, $T_f = 30.3°\text{C}$

59. Heat gain by water = heat loss by metal = s × m × ΔT, where s = specific heat capacity.

Heat gain = $\dfrac{4.18\text{ J}}{^\circ\text{C g}} \times 150.0\text{ g} \times (18.3°\text{C} - 15.0°\text{C}) = 2100\text{ J}$

A common error in calorimetry problems is sign errors. Keeping all quantities positive helps to eliminate sign errors.

Heat loss = $2100\text{ J} = s \times 150.0\text{ g} \times (75.0°\text{C} - 18.3°\text{C})$, $s = \dfrac{2100\text{J}}{150.0\text{ g} \times 56.7\,^\circ\text{C}} = 0.25\text{ J/°C•g}$

60. Heat gain by water = heat loss by Cu; keeping all quantities positive helps to avoid sign errors:

$$\frac{4.18\text{ J}}{^\circ\text{C g}} \times \text{mass} \times (24.9°\text{C} - 22.3°\text{C}) = \frac{0.20\text{ J}}{^\circ\text{C g}} \times 110.\text{ g Cu} \times (82.4°\text{C} - 24.9°\text{C})$$

11(mass) = 1300, mass = 120 g H_2O

61. $50.0 \times 10^{-3}\text{ L} \times 0.100\text{ mol/L} = 5.00 \times 10^{-3}$ mol of both $AgNO_3$ and HCl are reacted. Thus 5.00×10^{-3} mol of AgCl will be produced because there is a 1 : 1 mole ratio between reactants.

Heat lost by chemicals = heat gained by solution

Heat gain = $\dfrac{4.18\text{ J}}{^\circ\text{C g}} \times 100.0\text{ g} \times (23.40 - 22.60)°\text{C} = 330\text{ J}$

Heat loss = 330 J; this is the heat evolved (exothermic reaction) when 5.00×10^{-3} mol of AgCl is produced. So q = –330 J and ΔH (heat per mol AgCl formed) is negative with a value of:

$$\Delta H = \frac{-330\text{J}}{5.00 \times 10^{-3}\text{ mol}} \times \frac{1\text{ kJ}}{1000\text{J}} = -66\text{ kJ/mol}$$

Note: Sign errors are common with calorimetry problems. However, the correct sign for ΔH can be determined easily from the ΔT data; i.e., if ΔT of the solution increases, then the reaction is exothermic because heat was released, and if ΔT of the solution decreases, then the reaction is endothermic because the reaction absorbed heat from the water. For calorimetry problems, keep all quantities positive until the end of the calculation and then decide the sign for ΔH. This will help eliminate sign errors.

62. $NaOH(aq) + HCl(aq) \rightarrow NaCl(aq) + H_2O(l)$

We have a stoichiometric mixture. All of the NaOH and HCl will react.

$0.10 \text{ L} \times \frac{1.0 \text{ mol}}{\text{L}} = 0.10$ mol of HCl is neutralized by 0.10 mol NaOH.

Heat lost by chemicals = heat gained by solution

Volume of solution = 100.0 + 100.0 = 200.0 mL

$$\text{Heat gain} = \frac{4.18 \text{ J}}{{}^\circ\text{C g}} \times \left(200.0 \text{ mL} \times \frac{1.0 \text{ g}}{\text{mL}}\right) \times (31.3 - 24.6)^\circ\text{C} = 5.6 \times 10^3 \text{ J} = 5.6 \text{ kJ}$$

Heat loss = 5.6 kJ; this is the heat released by the neutralization of 0.10 mol HCl. Because the temperature increased, the sign for ΔH must be negative, i.e., the reaction is exothermic. For calorimetry problems, keep all quantities positive until the end of the calculation and then decide the sign for ΔH. This will help eliminate sign errors.

$$\Delta H = \frac{-5.6 \text{ kJ}}{0.10 \text{ mol}} = -56 \text{ kJ/mol}$$

63. Heat lost by solution = heat gained by KBr; mass of solution = 125 g + 10.5 g = 136 g

Note: Sign errors are common with calorimetry problems. However, the correct sign for ΔH can easily be obtained from the ΔT data. When working calorimetry problems, keep all quantities positive (ignore signs). When finished, deduce the correct sign for ΔH. For this problem, T decreases as KBr dissolves, so ΔH is positive; the dissolution of KBr is endothermic (absorbs heat).

$$\text{Heat lost by solution} = \frac{4.18 \text{ J}}{{}^\circ\text{C g}} \times 136 \text{ g} \times (24.2^\circ\text{C} - 21.1^\circ\text{C}) = 1800 \text{ J} = \text{heat gained by KBr}$$

$$\Delta H \text{ in units of J/g} = \frac{1800 \text{ J}}{10.5 \text{ g KBr}} = 170 \text{ J/g}$$

$$\Delta H \text{ in units of kJ/mol} = \frac{170 \text{ J}}{\text{g KBr}} \times \frac{119.0 \text{ g KBr}}{\text{mol KBr}} \times \frac{1 \text{ kJ}}{1000 \text{ J}} = 20. \text{ kJ/mol}$$

64. $NH_4NO_3(s) \rightarrow NH_4^+(aq) + NO_3^-(aq)$ $\Delta H = ?$; mass of solution = 75.0 g + 1.60 g = 76.6 g

Heat lost by solution = heat gained as NH_4NO_3 dissolves. To help eliminate sign errors, we will keep all quantities positive (q and ΔT) and then deduce the correct sign for ΔH at the end of the problem. Here, because temperature decreases as NH_4NO_3 dissolves, heat is absorbed as NH_4NO_3 dissolves, so this is an endothermic process (ΔH is positive).

$$\text{Heat lost by solution} = \frac{4.18 \text{ J}}{{}^\circ\text{C g}} \times 76.6 \text{ g} \times (25.00 - 23.34)^\circ\text{C} = 532 \text{ J} = \text{heat gained as } NH_4NO_3 \text{ dissolves}$$

$$\Delta H = \frac{532\text{ J}}{1.60\text{ g } NH_4NO_3} \times \frac{80.05\text{ g } NH_4NO_3}{\text{mol } NH_4NO_3} \times \frac{1\text{ kJ}}{1000\text{ J}} = 26.6\text{ kJ/mol } NH_4NO_3 \text{ dissolving}$$

65. Because ΔH is exothermic, the temperature of the solution will increase as $CaCl_2(s)$ dissolves. Keeping all quantities positive:

$$\text{heat loss as } CaCl_2 \text{ dissolves} = 11.0\text{ g } CaCl_2 \times \frac{1\text{ mol } CaCl_2}{110.98\text{ g } CaCl_2} \times \frac{81.5\text{ kJ}}{\text{mol } CaCl_2} = 8.08\text{ kJ}$$

$$\text{heat gained by solution} = 8.08 \times 10^3\text{ J} = \frac{4.18\text{ J}}{°\text{C g}} \times (125 + 11.0)\text{ g} \times (T_f - 25.0°\text{C})$$

$$T_f - 25.0°\text{C} = \frac{8.08 \times 10^3}{4.18 \times 136} = 14.2°\text{C}, \; T_f = 14.2°\text{C} + 25.0°\text{C} = 39.2°\text{C}$$

66. $$0.1000\text{ L} \times \frac{0.500\text{ mol HCl}}{\text{L}} \times \frac{118\text{ kJ heat released}}{2\text{ mol HCl}} = 2.95\text{ kJ of heat released if HCl limiting}$$

$$0.3000\text{ L} \times \frac{0.100\text{ mol } Ba(OH)_2}{\text{L}} \times \frac{118\text{ kJ heat released}}{\text{mol } Ba(OH)_2} = 3.54\text{ kJ heat released if } Ba(OH)_2 \text{ limiting}$$

Because the HCl reagent produces the smaller amount of heat released, HCl is limiting and 2.95 kJ of heat are released by this reaction.

$$\text{Heat gained by solution} = 2.95 \times 10^3\text{ J} = \frac{4.18\text{ J}}{°\text{C g}} \times 400.0\text{ g} \times \Delta T$$

$$\Delta T = 1.76°\text{C} = T_f - T_i = T_f - 25.0°\text{C}, \; T_f = 26.8°\text{C}$$

67. a. Heat gain by calorimeter = heat loss by CH_4 = $6.79\text{ g } CH_4 \times \frac{1\text{ mol } CH_4}{16.04\text{ g}} \times \frac{802\text{ kJ}}{\text{mol}}$ = 340. kJ

$$\text{Heat capacity of calorimeter} = \frac{340.\text{ kJ}}{10.8\text{ °C}} = 31.5\text{ kJ/°C}$$

b. Heat loss by C_2H_2 = heat gain by calorimeter = $16.9°\text{C} \times \frac{31.5\text{ kJ}}{°\text{C}} = 532\text{ kJ}$

A bomb calorimeter is at constant volume, so the heat released/gained = $q_V = \Delta E$:

$$\Delta E_{comb} = \frac{-532\text{ kJ}}{12.6\text{ g } C_2H_2} \times \frac{26.04\text{ g}}{\text{mol } C_2H_2} = -1.10 \times 10^3\text{ kJ/mol}$$

68. First, we need to get the heat capacity of the calorimeter from the combustion of benzoic acid. Heat lost by combustion = heat gained by calorimeter.

$$\text{Heat loss} = 0.1584\text{ g} \times \frac{26.42\text{ kJ}}{\text{g}} = 4.185\text{ kJ}$$

Heat gain = 4.185 kJ = $C_{cal} \times \Delta T$, $C_{cal} = \dfrac{4.185\,\text{kJ}}{2.54^\circ\text{C}} = 1.65$ kJ/°C

Now we can calculate the heat of combustion of vanillin. Heat loss = heat gain.

$$\text{Heat gain by calorimeter} = \frac{1.65\,\text{kJ}}{^\circ\text{C}} \times 3.25^\circ\text{C} = 5.36\text{ kJ}$$

Heat loss = 5.36 kJ, which is the heat evolved by combustion of the vanillin.

$$\Delta E_{comb} = \frac{-5.36\,\text{kJ}}{0.2130\,\text{g}} = -25.2\text{ kJ/g};\quad \Delta E_{comb} = \frac{-25.2\,\text{kJ}}{\text{g}} \times \frac{152.14\,\text{g}}{\text{mol}} = -3830\text{ kJ/mol}$$

Hess's Law

69. Information given:

$$C(s) + O_2(g) \rightarrow CO_2(g) \qquad \Delta H = -393.7\text{ kJ}$$
$$CO(g) + 1/2\, O_2(g) \rightarrow CO_2(g) \qquad \Delta H = -283.3\text{ kJ}$$

Using Hess's law:

$$2\,C(s) + 2\,O_2(g) \rightarrow 2\,CO_2(g) \qquad \Delta H_1 = 2(-393.7\text{ kJ})$$
$$2\,CO_2(g) \rightarrow 2\,CO(g) + O_2(g) \qquad \Delta H_2 = -2(-283.3\text{ kJ})$$
$$\overline{2\,C(s) + O_2(g) \rightarrow 2\,CO(g) \qquad \Delta H = \Delta H_1 + \Delta H_2 = -220.8\text{ kJ}}$$

Note: When an equation is reversed, the sign on ΔH is reversed. When the coefficients in a balanced equation are multiplied by an integer, then the value of ΔH is multiplied by the same integer.

70. Given:

$$C_4H_4(g) + 5\,O_2(g) \rightarrow 4\,CO_2(g) + 2\,H_2O(l) \qquad \Delta H_{comb} = -2341\text{ kJ}$$
$$C_4H_8(g) + 6\,O_2(g) \rightarrow 4\,CO_2(g) + 4\,H_2O(l) \qquad \Delta H_{comb} = -2755\text{ kJ}$$
$$H_2(g) + 1/2\,O_2(g) \rightarrow H_2O(l) \qquad \Delta H_{comb} = -286\text{ kJ}$$

By convention, $H_2O(l)$ is produced when enthalpies of combustion are given, and because per-mole quantities are given, the combustion reaction refers to 1 mole of that quantity reacting with $O_2(g)$.

Using Hess's law to solve:

$$C_4H_4(g) + 5\,O_2(g) \rightarrow 4\,CO_2(g) + 2\,H_2O(l) \qquad \Delta H_1 = -2341\text{ kJ}$$
$$4\,CO_2(g) + 4\,H_2O(l) \rightarrow C_4H_8(g) + 6\,O_2(g) \qquad \Delta H_2 = -\,(-2755\text{ kJ})$$
$$2\,H_2(g) + O_2(g) \rightarrow 2\,H_2O(l) \qquad \Delta H_3 = 2(-286\text{ kJ})$$
$$\overline{C_4H_4(g) + 2\,H_2(g) \rightarrow C_4H_8(g) \qquad \Delta H = \Delta H_1 + \Delta H_2 + \Delta H_3 = -158\text{ kJ}}$$

71. $2\ N_2(g) + 6\ H_2(g) \rightarrow 4\ NH_3(g)$ $\Delta H = -2(92\ kJ)$

$6\ H_2O(g) \rightarrow 6\ H_2(g) + 3\ O_2(g)$ $\Delta H = -3(-484\ kJ)$

$2\ N_2(g) + 6\ H_2O(g) \rightarrow 3\ O_2(g) + 4\ NH_3(g)$ $\Delta H = 1268\ kJ$

No, because the reaction is very endothermic (requires a lot of heat to react), it would not be a practical way of making ammonia because of the high energy costs required.

72. $ClF + 1/2\ O_2 \rightarrow 1/2\ Cl_2O + 1/2\ F_2O$ $\Delta H = 1/2(167.4\ kJ)$

$1/2\ Cl_2O + 3/2\ F_2O \rightarrow ClF_3 + O_2$ $\Delta H = -1/2(341.4\ kJ)$

$F_2 + 1/2\ O_2 \rightarrow F_2O$ $\Delta H = 1/2(-43.4\ kJ)$

$ClF(g) + F_2(g) \rightarrow ClF_3$ $\Delta H = -108.7\ kJ$

73. $NO + O_3 \rightarrow NO_2 + O_2$ $\Delta H = -199\ kJ$

$3/2\ O_2 \rightarrow O_3$ $\Delta H = -1/2(-427\ kJ)$

$O \rightarrow 1/2\ O_2$ $\Delta H = -1/2(495\ kJ)$

$NO(g) + O(g) \rightarrow NO_2(g)$ $\Delta H = -233\ kJ$

74. We want ΔH for $N_2H_4(l) + O_2(g) \rightarrow N_2(g) + 2\ H_2O(l)$. It will be easier to calculate ΔH for the combustion of four moles of N_2H_4 because we will avoid fractions.

$9\ H_2 + 9/2\ O_2 \rightarrow 9\ H_2O$ $\Delta H = 9(-286\ kJ)$

$3\ N_2H_4 + 3\ H_2O \rightarrow 3\ N_2O + 9\ H_2$ $\Delta H = -3(-317\ kJ)$

$2\ NH_3 + 3\ N_2O \rightarrow 4\ N_2 + 3\ H_2O$ $\Delta H = -1010.\ kJ$

$N_2H_4 + H_2O \rightarrow 2\ NH_3 + 1/2\ O_2$ $\Delta H = -\ (-143\ kJ)$

$4\ N_2H_4(l) + 4\ O_2(g) \rightarrow 4\ N_2(g) + 8\ H_2O(l)$ $\Delta H = -2490.\ kJ$

For $N_2H_4(l) + O_2(g) \rightarrow N_2(g) + 2\ H_2O(l)$ $\Delta H = \dfrac{-2490.\ kJ}{4} = -623\ kJ$

Note: By the significant figure rules, we could report this answer to four significant figures. However, because the ΔH values given in the problem are only known to ±1 kJ, our final answer will at best be ±1 kJ.

75. $CaC_2 \rightarrow Ca + 2\ C$ $\Delta H = -\ (-62.8\ kJ)$

$CaO + H_2O \rightarrow Ca(OH)_2$ $\Delta H = -653.1\ kJ$

$2\ CO_2 + H_2O \rightarrow C_2H_2 + 5/2\ O_2$ $\Delta H = -\ (-1300.\ kJ)$

$Ca + 1/2\ O_2 \rightarrow CaO$ $\Delta H = -635.5\ kJ$

$2\ C + 2\ O_2 \rightarrow 2\ CO_2$ $\Delta H = 2(-393.5\ kJ)$

$CaC_2(s) + 2\ H_2O(l) \rightarrow Ca(OH)_2(aq) + C_2H_2(g)$ $\Delta H = -713\ kJ$

76.

Reaction	ΔH
$P_4O_{10} \rightarrow P_4 + 5\ O_2$	$\Delta H = -(-2967.3\ kJ)$
$10\ PCl_3 + 5\ O_2 \rightarrow 10\ Cl_3PO$	$\Delta H = 10(-285.7\ kJ)$
$6\ PCl_5 \rightarrow 6\ PCl_3 + 6\ Cl_2$	$\Delta H = -6(-84.2\ kJ)$
$P_4 + 6\ Cl_2 \rightarrow 4\ PCl_3$	$\Delta H = -1225.6$
$P_4O_{10}(s) + 6\ PCl_5(g) \rightarrow 10\ Cl_3PO(g)$	$\Delta H = -610.1\ kJ$

Standard Enthalpies of Formation

77. The change in enthalpy that accompanies the formation of 1 mole of a compound from its elements, with all substances in their standard states, is the standard enthalpy of formation for a compound. The reactions that refer to ΔH_f° are:

$Na(s) + 1/2\ Cl_2(g) \rightarrow NaCl(s)$; $H_2(g) + 1/2\ O_2(g) \rightarrow H_2O(l)$

$6\ C(\text{graphite, s}) + 6\ H_2(g) + 3\ O_2(g) \rightarrow C_6H_{12}O_6(s)$

$Pb(s) + S(\text{rhombic, s}) + 2\ O_2(g) \rightarrow PbSO_4(s)$

78. a. Aluminum oxide = Al_2O_3; $2\ Al(s) + 3/2\ O_2(g) \rightarrow Al_2O_3(s)$

b. $C_2H_5OH(l) + 3\ O_2(g) \rightarrow 2\ CO_2(g) + 3\ H_2O(l)$

c. $NaOH(aq) + HCl(aq) \rightarrow H_2O(l) + NaCl(aq)$

d. $2\ C(\text{graphite, s}) + 3/2\ H_2(g) + 1/2\ Cl_2(g) \rightarrow C_2H_3Cl(g)$

e. $C_6H_6(l) + 15/2\ O_2(g) \rightarrow 6\ CO_2(g) + 3\ H_2O(l)$

Note: ΔH_{comb} values assume 1 mole of compound combusted.

f. $NH_4Br(s) \rightarrow NH_4^+(aq) + Br^-(aq)$

79. In general, $\Delta H^\circ = \sum n_p \Delta H^\circ_{f,\ products} - \sum n_r \Delta H^\circ_{f,\ reactants}$, and all elements in their standard state have $\Delta H^\circ_f = 0$ by definition.

a. The balanced equation is $2\ NH_3(g) + 3\ O_2(g) + 2\ CH_4(g) \rightarrow 2\ HCN(g) + 6\ H_2O(g)$.

$$\Delta H^\circ = (2\ \text{mol HCN} \times \Delta H^\circ_{f,\ HCN} + 6\ \text{mol}\ H_2O(g) \times \Delta H^\circ_{f,\ H_2O}) - (2\ \text{mol}\ NH_3 \times \Delta H^\circ_{f,\ NH_3} + 2\ \text{mol}\ CH_4 \times \Delta H^\circ_{f,CH_4})$$

$$\Delta H^\circ = [2(135.1) + 6(-242)] - [2(-46) + 2(-75)] = -940.\ kJ$$

b. $Ca_3(PO_4)_2(s) + 3\ H_2SO_4(l) \rightarrow 3\ CaSO_4(s) + 2\ H_3PO_4(l)$

$$\Delta H° = \left[3\ \text{mol}\ CaSO_4(s)\left(\frac{-1433\ \text{kJ}}{\text{mol}}\right) + 2\ \text{mol}\ H_3PO_4(l)\left(\frac{-1267\ \text{kJ}}{\text{mol}}\right)\right] - \left[1\ \text{mol}\ Ca_3(PO_4)_2(s)\left(\frac{-4126\ \text{kJ}}{\text{mol}}\right) + 3\ \text{mol}\ H_2SO_4(l)\left(\frac{-814\ \text{kJ}}{\text{mol}}\right)\right]$$

$\Delta H° = -6833\ \text{kJ} - (-6568\ \text{kJ}) = -265\ \text{kJ}$

c. $NH_3(g) + HCl(g) \rightarrow NH_4Cl(s)$

$\Delta H° = (1\ \text{mol}\ NH_4Cl \times \Delta H°_{f,\ NH_4Cl}) - (1\ \text{mol}\ NH_3 \times \Delta H°_{f,\ NH_3} + 1\ \text{mol}\ HCl \times \Delta H°_{f,\ HCl})$

$$\Delta H° = \left[1\ \text{mol}\left(\frac{-314\ \text{kJ}}{\text{mol}}\right)\right] - \left[1\ \text{mol}\left(\frac{-46\ \text{kJ}}{\text{mol}}\right) + 1\ \text{mol}\left(\frac{-92\ \text{kJ}}{\text{mol}}\right)\right]$$

$\Delta H° = -314\ \text{kJ} + 138\ \text{kJ} = -176\ \text{kJ}$

80. a. The balanced equation is $C_2H_5OH(l) + 3\ O_2(g) \rightarrow 2\ CO_2(g) + 3\ H_2O(g)$.

$$\Delta H° = \left[2\ \text{mol}\left(\frac{-393.5\ \text{kJ}}{\text{mol}}\right) + 3\ \text{mol}\left(\frac{-242\ \text{kJ}}{\text{mol}}\right)\right] - \left[1\ \text{mol}\left(\frac{-278\ \text{kJ}}{\text{mol}}\right)\right]$$

$\Delta H° = -1513\ \text{kJ} - (-278\ \text{kJ}) = -1235\ \text{kJ}$

b. $SiCl_4(l) + 2\ H_2O(l) \rightarrow SiO_2(s) + 4\ HCl(aq)$

Because HCl(aq) is $H^+(aq) + Cl^-(aq)$, $\Delta H°_f = 0 - 167 = -167$ kJ/mol.

$$\Delta H° = \left[4\ \text{mol}\left(\frac{-167\ \text{kJ}}{\text{mol}}\right) + 1\ \text{mol}\left(\frac{-911\ \text{kJ}}{\text{mol}}\right)\right] - \left[1\ \text{mol}\left(\frac{-687\ \text{kJ}}{\text{mol}}\right) + 2\ \text{mol}\left(\frac{-286\ \text{kJ}}{\text{mol}}\right)\right]$$

$\Delta H° = -1579\ \text{kJ} - (-1259\ \text{kJ}) = -320.\ \text{kJ}$

c. $MgO(s) + H_2O(l) \rightarrow Mg(OH)_2(s)$

$$\Delta H° = \left[1\ \text{mol}\left(\frac{-925\ \text{kJ}}{\text{mol}}\right)\right] - \left[1\ \text{mol}\left(\frac{-602\ \text{kJ}}{\text{mol}}\right) + 1\ \text{mol}\left(\frac{-286\ \text{kJ}}{\text{mol}}\right)\right]$$

$\Delta H° = -925\ \text{kJ} - (-888\ \text{kJ}) = -37\ \text{kJ}$

81. a. $4\ NH_3(g) + 5\ O_2(g) \rightarrow 4\ NO(g) + 6\ H_2O(g)$; $\Delta H° = \sum n_p \Delta H°_{f,\ \text{products}} - \sum n_r \Delta H°_{f,\ \text{reactants}}$

$$\Delta H° = \left[4\ \text{mol}\left(\frac{90.\ \text{kJ}}{\text{mol}}\right) + 6\ \text{mol}\left(\frac{-242\ \text{kJ}}{\text{mol}}\right)\right] - \left[4\ \text{mol}\left(\frac{-46\ \text{kJ}}{\text{mol}}\right)\right] = -908\ \text{kJ}$$

$2\ NO(g) + O_2(g) \rightarrow 2\ NO_2(g)$

$$\Delta H° = \left[2\ \text{mol}\left(\frac{34\ \text{kJ}}{\text{mol}}\right)\right] - \left[2\ \text{mol}\left(\frac{90.\ \text{kJ}}{\text{mol}}\right)\right] = -112\ \text{kJ}$$

$3\ NO_2(g) + H_2O(l) \rightarrow 2\ HNO_3(aq) + NO(g)$

$$\Delta H° = \left[2\ \text{mol}\left(\frac{-207\ \text{kJ}}{\text{mol}}\right) + 1\ \text{mol}\left(\frac{90.\ \text{kJ}}{\text{mol}}\right)\right] - \left[3\ \text{mol}\left(\frac{34\ \text{kJ}}{\text{mol}}\right) + 1\ \text{mol}\left(\frac{-286\ \text{kJ}}{\text{mol}}\right)\right]$$

$$-140.\ \text{kJ}$$

Note: All ΔH_f° values are assumed ±1 kJ.

b. $12\ NH_3(g) + 15\ O_2(g) \rightarrow 12\ NO(g) + 18\ H_2O(g)$
$12\ NO(g) + 6\ O_2(g) \rightarrow 12\ NO_2(g)$
$12\ NO_2(g) + 4\ H_2O(l) \rightarrow 8\ HNO_3(aq) + 4\ NO(g)$
$4\ H_2O(g) \rightarrow 4\ H_2O(l)$

$12\ NH_3(g) + 21\ O_2(g) \rightarrow 8\ HNO_3(aq) + 4\ NO(g) + 14\ H_2O(g)$

The overall reaction is exothermic because each step is exothermic.

82. $4\ Na(s) + O_2(g) \rightarrow 2\ Na_2O(s)$ $\quad \Delta H° = 2\ \text{mol}\left(\frac{-416\ \text{kJ}}{\text{mol}}\right) = -832\ \text{kJ}$

$2\ Na(s) + 2\ H_2O(l) \rightarrow 2\ NaOH(aq) + H_2(g)$

$$\Delta H° = \left[2\ \text{mol}\left(\frac{-470.\ \text{kJ}}{\text{mol}}\right)\right] - \left[2\ \text{mol}\left(\frac{-286\ \text{kJ}}{\text{mol}}\right)\right] = -368\ \text{kJ}$$

$2Na(s) + CO_2(g) \rightarrow Na_2O(s) + CO(g)$

$$\Delta H° = \left[1\ \text{mol}\left(\frac{-416\ \text{kJ}}{\text{mol}}\right) + 1\ \text{mol}\left(\frac{-110.5\ \text{kJ}}{\text{mol}}\right)\right] - \left[1\ \text{mol}\left(\frac{-393.5\ \text{kJ}}{\text{mol}}\right)\right] = -133\ \text{kJ}$$

In Reactions 2 and 3, sodium metal reacts with the "extinguishing agent." Both reactions are exothermic, and each reaction produces a flammable gas, H_2 and CO, respectively.

83. $3\ Al(s) + 3\ NH_4ClO_4(s) \rightarrow Al_2O_3(s) + AlCl_3(s) + 3\ NO(g) + 6\ H_2O(g)$

$$\Delta H° = \left[6\ \text{mol}\left(\frac{-242\ \text{kJ}}{\text{mol}}\right) + 3\ \text{mol}\left(\frac{90.\ \text{kJ}}{\text{mol}}\right) + 1\ \text{mol}\left(\frac{-704\ \text{kJ}}{\text{mol}}\right) + 1\ \text{mol}\left(\frac{-1676\ \text{kJ}}{\text{mol}}\right)\right]$$

$$- \left[3\ \text{mol}\left(\frac{-295\ \text{kJ}}{\text{mol}}\right)\right] = -2677\ \text{kJ}$$

84. $5\ N_2O_4(l) + 4\ N_2H_3CH_3(l) \rightarrow 12\ H_2O(g) + 9\ N_2(g) + 4\ CO_2(g)$

$$\Delta H° = \left[12\,\text{mol}\left(\frac{-242\,\text{kJ}}{\text{mol}}\right) + 4\,\text{mol}\left(\frac{-393.5\,\text{kJ}}{\text{mol}}\right)\right]$$

$$- \left[5\,\text{mol}\left(\frac{-20.\,\text{kJ}}{\text{mol}}\right) + 4\,\text{mol}\left(\frac{54\,\text{kJ}}{\text{mol}}\right)\right] = -4594\ \text{kJ}$$

85. $2\ ClF_3(g) + 2\ NH_3(g) \rightarrow N_2(g) + 6\ HF(g) + Cl_2(g) \quad \Delta H° = -1196\ \text{kJ}$

$$\Delta H° = (6\,\Delta H^\circ_{f,\ HF}) - (2\,\Delta H^\circ_{f,\ ClF_3} + 2\Delta H^\circ_{f,\ NH_3})$$

$$-1196\ \text{kJ} = 6\ \text{mol}\left(\frac{-271\,\text{kJ}}{\text{mol}}\right) - 2\,\Delta H^\circ_{f,\ ClF_3} - 2\ \text{mol}\left(\frac{-46\,\text{kJ}}{\text{mol}}\right)$$

$$-1196\ \text{kJ} = -1626\ \text{kJ} - 2\,\Delta H^\circ_{f,\ ClF_3} + 92\ \text{kJ},\quad \Delta H^\circ_{f,\ ClF_3} = \frac{(-1626 + 92 + 1196)\,\text{kJ}}{2\,\text{mol}} = \frac{-169\,\text{kJ}}{\text{mol}}$$

86. $C_2H_4(g) + 3\ O_2(g) \rightarrow 2\ CO_2(g) + 2\ H_2O(l) \quad \Delta H° = -1411.1\ \text{kJ}$

$$\Delta H° = -1411.1\ \text{kJ} = 2(-393.5)\ \text{kJ} + 2(-285.8)\ \text{kJ} - \Delta H^\circ_{f,C_2H_4}$$

$$-1411.1\ \text{kJ} = -1358.6\ \text{kJ} - \Delta H^\circ_{f,C_2H_4},\quad \Delta H^\circ_{f,C_2H_4} = 52.5\ \text{kJ/mol}$$

Energy Consumption and Sources

87. $C(s) + H_2O(g) \rightarrow H_2(g) + CO(g) \quad \Delta H° = -110.5\ \text{kJ} - (-242\ \text{kJ}) = 132\ \text{kJ}$

88. $CO(g) + 2\ H_2(g) \rightarrow CH_3OH(l) \quad \Delta H° = -239\ \text{kJ} - (-110.5\ \text{kJ}) = -129\ \text{kJ}$

89. $C_2H_5OH(l) + 3\ O_2(g) \rightarrow 2\ CO_2(g) + 3\ H_2O(l)$

$$\Delta H° = [2(-393.5\ \text{kJ}) + 3(-286\ \text{kJ})] - (-278\ \text{kJ}) = -1367\ \text{kJ/mol ethanol}$$

$$\frac{-1367\,\text{kJ}}{\text{mol}} \times \frac{1\,\text{mol}}{46.07\,\text{g}} = -29.67\ \text{kJ/g}$$

90. $CH_3OH(l) + 3/2\ O_2(g) \rightarrow CO_2(g) + 2\ H_2O(l)$

$$\Delta H° = [-393.5\ \text{kJ} + 2(-286\ \text{kJ})] - (-239\ \text{kJ}) = -727\ \text{kJ/mol}\ CH_3OH$$

$$\frac{-727\,\text{kJ}}{\text{mol}} \times \frac{1\,\text{mol}}{32.04\,\text{g}} = -22.7\ \text{kJ/g versus } -29.67\ \text{kJ/g for ethanol (from Exercise 89)}$$

Ethanol has a slightly higher fuel value per gram than methanol.

91. $C_3H_8(g) + 5\ O_2(g) \rightarrow 3\ CO_2(g) + 4\ H_2O(l)$

$\Delta H° = [3(-393.5\text{ kJ}) + 4(-286\text{ kJ})] - (-104\text{ kJ}) = -2221\text{ kJ/mol } C_3H_8$

$\frac{-2221\text{kJ}}{\text{mol}} \times \frac{1\text{ mol}}{44.09\text{ g}} = \frac{-50.37\text{kJ}}{\text{g}}$ versus -47.7 kJ/g for octane (Example 6.11)

The fuel values are very close. An advantage of propane is that it burns more cleanly. The boiling point of propane is –42°C. Thus it is more difficult to store propane, and there are extra safety hazards associated with using high-pressure compressed-gas tanks.

92. 1 mole of $C_2H_2(g)$ and 1 mole of $C_4H_{10}(g)$ have equivalent volumes at the same T and P.

$$\frac{\text{Enthalpy of combustion per volume of } C_2H_2}{\text{Enthalpy of combustion per volume of } C_4H_{10}} = \frac{\text{enthalpy of combustion per mol of } C_2H_2}{\text{enthalpy of combustion per mol of } C_4H_{10}}$$

$$\frac{\text{Enthalpy of combustion per volume of } C_2H_2}{\text{Enthalpy of combustion per volume of } C_4H_{10}} = \frac{\frac{-49.9\text{ kJ}}{\text{g } C_2H_2} \times \frac{26.04\text{ g } C_2H_2}{\text{mol } C_2H_2}}{\frac{-49.5\text{ kJ}}{\text{g } C_4H_{10}} \times \frac{58.12\text{ g } C_4H_{10}}{\text{mol } C_4H_{10}}} = 0.452$$

More than twice the volume of acetylene is needed to furnish the same energy as a given volume of butane.

93. The molar volume of a gas at STP is 22.42 L (from Chapter 5).

$$4.19 \times 10^6\text{ kJ} \times \frac{1\text{ mol } CH_4}{891\text{ kJ}} \times \frac{22.42\text{ L } CH_4}{\text{mol } CH_4} = 1.05 \times 10^5\text{ L } CH_4$$

94. Mass of $H_2O = 1.00\text{ gal} \times \frac{3.785\text{ L}}{\text{gal}} \times \frac{1000\text{ mL}}{\text{L}} \times \frac{1.00\text{ g}}{\text{mL}} = 3790\text{ g } H_2O$

Energy required (theoretical) $= s \times m \times \Delta T = \frac{4.18\text{ J}}{°\text{C g}} \times 3790\text{ g} \times 10.0\text{ °C} = 1.58 \times 10^5\text{ J}$

For an actual (80.0% efficient) process, more than this quantity of energy is needed since heat is always lost in any transfer of energy. The energy required is:

$$1.58 \times 10^5\text{ J} \times \frac{100.\text{ J}}{80.0\text{ J}} = 1.98 \times 10^5\text{ J}$$

Mass of $C_2H_2 = 1.98 \times 10^5\text{ J} \times \frac{1\text{ mol } C_2H_2}{1300. \times 10^3\text{ J}} \times \frac{26.04\text{ g } C_2H_2}{\text{mol } C_2H_2} = 3.97\text{ g } C_2H_2$

Additional Exercises

95. $2.0\text{ h} \times \frac{5500\text{kJ}}{\text{h}} \times \frac{1\text{ mol } H_2O}{40.6\text{ kJ}} \times \frac{18.02\text{ g } H_2O}{\text{mol}} = 4900\text{ g} = 4.9\text{ kg } H_2O$

96. From the problem, walking 4.0 miles consumes 400 kcal of energy.

$$1 \text{ lb fat} \times \frac{454\text{g}}{\text{lb}} \times \frac{7.7\text{ kcal}}{\text{g}} \times \frac{4\text{ mi}}{400\text{kcal}} \times \frac{1\text{ h}}{4\text{ mi}} = 8.7\text{ h} = 9\text{ h}$$

97. a. $2\ SO_2(g) + O_2(g) \rightarrow 2\ SO_3(g)$; $w = -P\Delta V$; because the volume of the piston apparatus decreased as reactants were converted to products ($\Delta V < 0$), w is positive ($w > 0$).

b. $COCl_2(g) \rightarrow CO(g) + Cl_2(g)$; because the volume increased ($\Delta V > 0$), w is negative ($w < 0$).

c. $N_2(g) + O_2(g) \rightarrow 2\ NO(g)$; because the volume did not change ($\Delta V = 0$), no PV work is done ($w = 0$).

In order to predict the sign of w for a reaction, compare the coefficients of all the product gases in the balanced equation to the coefficients of all the reactant gases. When a balanced reaction has more moles of product gases than moles of reactant gases (as in b), the reaction will expand in volume (ΔV positive), and the system does work on the surroundings. When a balanced reaction has a decrease in the moles of gas from reactants to products (as in a), the reaction will contract in volume (ΔV negative), and the surroundings will do compression work on the system. When there is no change in the moles of gas from reactants to products (as in c), $\Delta V = 0$ and $w = 0$.

98. a. $N_2(g) + 3\ H_2(g) \rightarrow 2\ NH_3(g)$; from the balanced equation, 1 molecule of N_2 will react with 3 molecules of H_2 to produce 2 molecules of NH_3. So the picture after the reaction should only have 2 molecules of NH_3 present. Another important part of your drawing will be the relative volume of the product container. The volume of a gas is directly proportional to the number of gas molecules present (at constant T and P). In this problem, 4 total molecules of gas were present initially ($1\ N_2 + 3\ H_2$). After reaction, only 2 molecules are present ($2\ NH_3$). Because the number of gas molecules decreases by a factor of 2 (from 4 total to 2 total), the volume of the product gas must decrease by a factor of 2 as compared to the initial volume of the reactant gases. Summarizing, the picture of the product container should have 2 molecules of NH_3 and should be at a volume which is one-half the original reactant container volume.

b. $w = -P\Delta V$; here the volume decreased, so ΔV is negative. When ΔV is negative, w is positive. As the reactants were converted to products, a compression occurred which is associated with work flowing into the system (w is positive).

99. a. $C_{12}H_{22}O_{11}(s) + 12\ O_2(g) \rightarrow 12\ CO_2(g) + 11\ H_2O(l)$

b. A bomb calorimeter is at constant volume, so heat released = $q_V = \Delta E$:

$$\Delta E = \frac{-24.00\text{ kJ}}{1.46\text{ g}} \times \frac{342.30\text{ g}}{\text{mol}} = -5630\text{ kJ/mol } C_{12}H_{22}O_{11}$$

c. $PV = nRT$; at constant P and T, $P\Delta V = RT\Delta n$, where Δn = moles of gaseous products − moles of gaseous reactants.

At constant P and T: $\Delta H = \Delta E + P\Delta V = \Delta E + RT\Delta n$

For this reaction, $\Delta n = 12 - 12 = 0$, so $\Delta H = \Delta E = -5630$ kJ/mol.

100. $w = -P\Delta V$; Δn = moles of gaseous products – moles of gaseous reactants. Only gases can do PV work (we ignore solids and liquids). When a balanced reaction has more moles of product gases than moles of reactant gases (Δn positive), the reaction will expand in volume (ΔV positive), and the system will do work on the surroundings. For example, in reaction c, $\Delta n = 2 - 0 = 2$ moles, and this reaction would do expansion work against the surroundings. When a balanced reaction has a decrease in the moles of gas from reactants to products (Δn negative), the reaction will contract in volume (ΔV negative), and the surroundings will do compression work on the system, e.g., reaction a, where $\Delta n = 0 - 1 = -1$. When there is no change in the moles of gas from reactants to products, $\Delta V = 0$ and $w = 0$, e.g., reaction b, where $\Delta n = 2 - 2 = 0$.

When $\Delta V > 0$ ($\Delta n > 0$), then $w < 0$, and the system does work on the surroundings (c and e).

When $\Delta V < 0$ ($\Delta n < 0$), then $w > 0$, and the surroundings do work on the system (a and d).

When $\Delta V = 0$ ($\Delta n = 0$), then $w = 0$ (b).

101. $\Delta E_{overall} = \Delta E_{step\ 1} + \Delta E_{step\ 2}$; this is a cyclic process, which means that the overall initial state and final state are the same. Because ΔE is a state function, $\Delta E_{overall} = 0$ and $\Delta E_{step\ 1} = -\Delta E_{step\ 2}$.

$\Delta E_{step\ 1} = q + w = 45\text{ J} + (-10.\text{ J}) = 35\text{ J}$

$\Delta E_{step\ 2} = -\Delta E_{step\ 1} = -35\text{ J} = q + w,\ -35\text{ J} = -60\text{ J} + w,\ w = 25\text{ J}$

102. $2\text{ K(s)} + 2\text{ H}_2\text{O(l)} \rightarrow 2\text{ KOH(aq)} + \text{H}_2\text{(g)}\quad \Delta H° = 2(-481\text{ kJ}) - 2(-286\text{ kJ}) = -390.\text{ kJ}$

$$5.00\text{ g K} \times \frac{1\text{ mol K}}{39.10\text{ g K}} \times \frac{-390.\text{ kJ}}{2\text{ mol K}} = -24.9\text{ kJ}$$

24.9 kJ of heat is released on reaction of 5.00 g K.

$$24{,}900\text{ J} = \frac{4.18\text{ J}}{\text{g °C}} \times (1.00 \times 10^3\text{ g}) \times \Delta T,\ \Delta T = \frac{24{,}900}{4.18 \times 1.00 \times 10^3} = 5.96°\text{C}$$

Final temperature = 24.0 + 5.96 = 30.0°C

103. $\text{HNO}_3\text{(aq)} + \text{KOH(aq)} \rightarrow \text{H}_2\text{O(l)} + \text{KNO}_3\text{(aq)}\quad \Delta H = -56\text{ kJ}$

$$0.2000\text{ L} \times \frac{0.400\text{ mol HNO}_3}{\text{L}} \times \frac{56\text{ kJ heat released}}{\text{mol HNO}_3} = 4.5\text{ kJ heat released if HNO}_3\text{ limiting}$$

$$0.1500\text{ L} \times \frac{0.500\text{ mol KOH}}{\text{L}} \times \frac{56\text{ kJ heat released}}{\text{mol KOH}} = 4.2\text{ kJ heat released if KOH limiting}$$

Because the KOH reagent produces the smaller quantity of heat released, KOH is limiting and 4.2 kJ of heat released.

104. $Na_2SO_4(aq) + Ba(NO_3)_2(aq) \rightarrow BaSO_4(s) + 2\ NaNO_3(aq) \qquad \Delta H = ?$

$$1.00\ L \times \frac{2.00\ mol\ Na_2SO_4}{L} \times \frac{1\ mol\ BaSO_4}{mol\ Na_2SO_4} = 2.00\ mol\ BaSO_4 \text{ if } Na_2SO_4 \text{ limiting}$$

$$2.00\ L \times \frac{0.750\ mol\ Ba(NO_3)_2}{L} \times \frac{1\ mol\ BaSO_4}{mol\ Ba(NO_3)_2} = 1.50\ mol\ BaSO_4 \text{ if } Ba(NO_3)_2 \text{ limiting}$$

The $Ba(NO_3)_2$ reagent produces the smaller quantity of product, so $Ba(NO_3)_2$ is limiting and 1.50 mol $BaSO_4$ can form.

Heat gain by solution = heat loss by reaction

$$\text{Mass of solution} = 3.00\ L \times \frac{1000\ mL}{L} \times \frac{2.00\ g}{mL} = 6.00 \times 10^3\ g$$

$$\text{Heat gain by solution} = \frac{6.37\ J}{^\circ C\ g} \times 6.00 \times 10^3\ g \times (42.0 - 30.0)^\circ C = 4.59 \times 10^5\ J$$

Because the solution gained heat, the reaction is exothermic; $q = -4.59 \times 10^5$ J for the reaction.

$$\Delta H = \frac{-4.59 \times 10^5\ J}{1.50\ mol\ BaSO_4} = -3.06 \times 10^5\ J/mol = -306\ kJ/mol$$

105. $|q_{surr}| = |q_{solution} + q_{cal}|$; we normally assume that q_{cal} is zero (no heat gain/loss by the calorimeter). However, if the calorimeter has a nonzero heat capacity, then some of the heat absorbed by the endothermic reaction came from the calorimeter. If we ignore q_{cal}, then q_{surr} is too small, giving a calculated ΔH value that is less positive (smaller) than it should be.

106. The specific heat of water is 4.18 J/°C•g, which is equal to 4.18 kJ/°C•kg.

We have 1.00 kg of H_2O, so: $1.00\ kg \times \dfrac{4.18\ kJ}{^\circ C\ kg} = 4.18\ kJ/^\circ C$

This is the portion of the heat capacity that can be attributed to H_2O.

Total heat capacity = $C_{cal} + C_{H_2O}$, $C_{cal} = 10.84 - 4.18 = 6.66$ kJ/°C

107. Heat released = 1.056 g × 26.42 kJ/g = 27.90 kJ = heat gain by water and calorimeter

$$\text{Heat gain} = 27.90\ kJ = \left(\frac{4.18\ kJ}{^\circ C\ kg} \times 0.987\ kg \times \Delta T\right) + \left(\frac{6.66\ kJ}{^\circ C} \times \Delta T\right)$$

$27.90 = (4.13 + 6.66)\Delta T = (10.79)\Delta T,\ \Delta T = 2.586^\circ C$

$2.586^\circ C = T_f - 23.32^\circ C,\ T_f = 25.91^\circ C$

108. For Exercise 83, a mixture of 3 mol Al and 3 mol NH_4ClO_4 yields 2677 kJ of energy. The mass of the stoichiometric reactant mixture is:

$$\left(3\text{ mol} \times \frac{26.98\text{ g}}{\text{mol}}\right) + \left(3\text{ mol} \times \frac{117.49\text{ g}}{\text{mol}}\right) = 433.41\text{ g}$$

For 1.000 kg of fuel: $1.000 \times 10^3\text{ g} \times \frac{-2677\text{kJ}}{433.41\text{g}} = -6177\text{ kJ}$

In Exercise 84, we get 4594 kJ of energy from 5 mol of N_2O_4 and 4 mol of $N_2H_3CH_3$. The mass is $\left(5\text{ mol} \times \frac{92.02\text{ g}}{\text{mol}}\right) + \left(4\text{ mol} \times \frac{46.08\text{ g}}{\text{mol}}\right) = 644.42\text{ kJ}.$

For 1.000 kg of fuel: $1.000 \times 10^3\text{ g} \times \frac{-4594\text{kJ}}{644.42\text{ g}} = -7129\text{ kJ}$

Thus we get more energy per kilogram from the $N_2O_4/N_2H_3CH_3$ mixture.

109.

$1/2\text{ D} \rightarrow 1/2\text{ A} + \text{B}$	$\Delta H = -1/6(-403\text{ kJ})$
$1/2\text{ E} + \text{F} \rightarrow 1/2\text{ A}$	$\Delta H = 1/2(-105.2\text{ kJ})$
$1/2\text{ C} \rightarrow 1/2\text{ E} + 3/2\text{ D}$	$\Delta H = 1/2(64.8\text{ kJ})$
$\text{F} + 1/2\text{ C} \rightarrow \text{A} + \text{B} + \text{D}$	$\Delta H = 47.0\text{ kJ}$

110. To avoid fractions, let's first calculate ΔH for the reaction:

$$6\text{ FeO(s)} + 6\text{ CO(g)} \rightarrow 6\text{ Fe(s)} + 6\text{ CO}_2\text{(g)}$$

$6\text{ FeO} + 2\text{ CO}_2 \rightarrow 2\text{ Fe}_3\text{O}_4 + 2\text{ CO}$	$\Delta H° = -2(18\text{ kJ})$
$2\text{ Fe}_3\text{O}_4 + \text{CO}_2 \rightarrow 3\text{ Fe}_2\text{O}_3 + \text{CO}$	$\Delta H° = -(-39\text{ kJ})$
$3\text{ Fe}_2\text{O}_3 + 9\text{ CO} \rightarrow 6\text{ Fe} + 9\text{ CO}_2$	$\Delta H° = 3(-23\text{ kJ})$
$6\text{ FeO(s)} + 6\text{ CO(g)} \rightarrow 6\text{ Fe(s)} + 6\text{ CO}_2\text{(g)}$	$\Delta H° = -66\text{ kJ}$

So for $\text{FeO(s)} + \text{CO(g)} \rightarrow \text{Fe(s)} + \text{CO}_2\text{(g)}$, $\Delta H° = \frac{-66\text{ kJ}}{6} = -11\text{ kJ}.$

111. a. $\Delta H° = 3\text{ mol}(227\text{ kJ/mol}) - 1\text{ mol}(49\text{ kJ/mol}) = 632\text{ kJ}$

b. Because $3\text{ C}_2\text{H}_2\text{(g)}$ is higher in energy than $C_6H_6(l)$, acetylene will release more energy per gram when burned in air. Note that 3 moles of C_2H_2 has the same mass as 1 mole of C_6H_6.

112.

$$\begin{array}{ll} I(g) + Cl(g) \rightarrow ICl(g) & \Delta H = -(211.3\text{ kJ}) \\ 1/2\ Cl_2(g) \rightarrow Cl(g) & \Delta H = 1/2(242.3\text{ kJ}) \\ 1/2\ I_2(g) \rightarrow I(g) & \Delta H = 1/2(151.0\text{ kJ}) \\ 1/2\ I_2(s) \rightarrow 1/2\ I_2(g) & \Delta H = 1/2(62.8\text{ kJ}) \\ \hline 1/2\ I_2(s) + 1/2\ Cl_2(g) \rightarrow ICl(g) & \Delta H = 16.8\text{ kJ/mol} = \Delta H^\circ_{f,\ ICl} \end{array}$$

113. Heat gained by water = heat lost by nickel = s × m × ΔT, where s = specific heat capacity.

$$\text{Heat gain} = \frac{4.18\text{ J}}{^\circ\text{C g}} \times 150.0\text{ g} \times (25.0^\circ\text{C} - 23.5^\circ\text{C}) = 940\text{ J}$$

A common error in calorimetry problems is sign errors. Keeping all quantities positive helps to eliminate sign errors.

$$\text{Heat loss} = 940\text{ J} = \frac{0.444\text{ J}}{^\circ\text{C g}} \times \text{mass} \times (99.8 - 25.0)\ ^\circ\text{C},\ \text{mass} = \frac{940}{0.444 \times 74.8} = 28\text{ g}$$

114. $$\text{Heat gain by calorimeter} = \frac{1.56\text{ kJ}}{^\circ\text{C}} \times 3.2^\circ\text{C} = 5.0\text{ kJ} = \text{heat loss by quinine}$$

Heat loss = 5.0 kJ, which is the heat evolved (exothermic reaction) by the combustion of 0.1964 g of quinone. Because we are at constant volume, $q_V = \Delta E$.

$$\Delta E_{comb} = \frac{-5.0\text{ kJ}}{0.1964\text{g}} = -25\text{ kJ/g}; \qquad \Delta E_{comb} = \frac{-25\text{ kJ}}{\text{g}} \times \frac{108.09\text{ g}}{\text{mol}} = -2700\text{ kJ/mol}$$

115. a. $C_2H_4(g) + O_3(g) \rightarrow CH_3CHO(g) + O_2(g)$ $\Delta H^\circ = -166\text{ kJ} - [143\text{ kJ} + 52\text{ kJ}] = -361\text{ kJ}$

b. $O_3(g) + NO(g) \rightarrow NO_2(g) + O_2(g)$ $\Delta H^\circ = 34\text{ kJ} - [90.\text{ kJ} + 143\text{ kJ}] = -199\text{ kJ}$

c. $SO_3(g) + H_2O(l) \rightarrow H_2SO_4(aq)$ $\Delta H^\circ = -909\text{ kJ} - [-396\text{ kJ} + (-286\text{ kJ})] = -227\text{ kJ}$

d. $2\ NO(g) + O_2(g) \rightarrow 2\ NO_2(g)$ $\Delta H^\circ = 2(34)\text{ kJ} - 2(90.)\text{ kJ} = -112\text{ kJ}$

ChemWork Problems

The answers to the problems 116-123 (or a variation to these problems) are found in OWL. These problems are also assignable in OWL.

Challenge Problems

124. Only when there is a volume change can PV work be done. In pathway 1 (steps 1 + 2), only the first step does PV work (step 2 has a constant volume of 30.0 L). In pathway 2 (steps 3 + 4), only step 4 does PV work (step 3 has a constant volume of 10.0 L).

Pathway 1: $w = -P\Delta V = -2.00 \text{ atm}(30.0 \text{ L} - 10.0 \text{ L}) = -40.0 \text{ L atm} \times \dfrac{101.3 \text{ J}}{\text{L atm}}$

$$= -4.05 \times 10^3 \text{ J}$$

Pathway 2: $w = -P\Delta V = -1.00 \text{ atm}(30.0 \text{ L} - 10.0 \text{ L}) = -20.0 \text{ L atm} \times \dfrac{101.3 \text{ J}}{\text{L atm}}$

$$= -2.03 \times 10^3 \text{ J}$$

Note: The sign is negative because the system is doing work on the surroundings (an expansion). We get different values of work for the two pathways; both pathways have the same initial and final states. Because w depends on the pathway, work cannot be a state function.

125. $A(l) \rightarrow A(g)$ $\Delta H_{vap} = 30.7$ kJ; at constant pressure, $\Delta H = q_p = 30.7$ kJ

Because PV = nRT, at constant pressure and temperature: $w = -P\Delta V = -RT\Delta n$, where:

Δn = moles of gaseous products − moles of gaseous reactants = 1 − 0 = 1

$w = -RT\Delta n = -8.3145 \text{ J/K•mol}(80. + 273 \text{ K})(1 \text{ mol}) = -2940 \text{ J} = -2.94 \text{ kJ}$

$\Delta E = q + w = 30.7 \text{ kJ} + (-2.94 \text{ kJ}) = 27.8 \text{ kJ}$

126. Energy needed = $\dfrac{20. \times 10^3 \text{ g } C_{12}H_{22}O_{11}}{\text{h}} \times \dfrac{1 \text{ mol } C_{12}H_{22}O_{11}}{342.3 \text{ g } C_{12}H_{22}O_{11}} \times \dfrac{5640 \text{ kJ}}{\text{mol}} = 3.3 \times 10^5 \text{ kJ/h}$

Energy from sun = $1.0 \text{ kW/m}^2 = 1000 \text{ W/m}^2 = \dfrac{1000 \text{ J}}{\text{s m}^2} = \dfrac{1.0 \text{ kJ}}{\text{s m}^2}$

$10{,}000 \text{ m}^2 \times \dfrac{1.0 \text{ kJ}}{\text{s m}^2} \times \dfrac{60 \text{ s}}{\text{min}} \times \dfrac{60 \text{ min}}{\text{h}} = 3.6 \times 10^7 \text{ kJ/h}$

Percent efficiency = $\dfrac{\text{energy used per hour}}{\text{total energy per hour}} \times 100 = \dfrac{3.3 \times 10^5 \text{ kJ}}{3.6 \times 10^7 \text{ kJ}} \times 100 = 0.92\%$

127. Energy used in 8.0 hours = 40. kWh = $\dfrac{40. \text{ kJ h}}{\text{s}} \times \dfrac{3600 \text{ s}}{\text{h}} = 1.4 \times 10^5 \text{ kJ}$

Energy from the sun in 8.0 hours = $\dfrac{10. \text{ kJ}}{\text{s m}^2} \times \dfrac{60 \text{ s}}{\text{min}} \times \dfrac{60 \text{ min}}{\text{h}} \times 8.0 \text{ h} = 2.9 \times 10^4 \text{ kJ/m}^2$

Only 19% of the sunlight is converted into electricity:

$0.19 \times (2.9 \times 10^4 \text{ kJ/m}^2) \times \text{area} = 1.4 \times 10^5 \text{ kJ}$, area = 25 m^2

128. a. $2 \text{ HNO}_3(aq) + \text{Na}_2\text{CO}_3(s) \rightarrow 2 \text{ NaNO}_3(aq) + \text{H}_2\text{O}(l) + \text{CO}_2(g)$

$\Delta H° = [2(-467 \text{ kJ}) + (-286 \text{ kJ}) + (-393.5 \text{ kJ})] - [2(-207 \text{ kJ}) + (-1131 \text{ kJ})] = -69 \text{ kJ}$

$$2.0 \times 10^4 \text{ gallons} \times \frac{4 \text{ qt}}{\text{gal}} \times \frac{946 \text{ mL}}{\text{qt}} \times \frac{1.42 \text{ g}}{\text{mL}} = 1.1 \times 10^8 \text{ g of concentrated nitric acid solution}$$

$$1.1 \times 10^8 \text{ g solution} \times \frac{70.0 \text{ g } HNO_3}{100.0 \text{ g solution}} = 7.7 \times 10^7 \text{ g } HNO_3$$

$$7.7 \times 10^7 \text{ g } HNO_3 \times \frac{1 \text{ mol } HNO_3}{63.02 \text{ g } HNO_3} \times \frac{1 \text{ mol } Na_2CO_3}{2 \text{ mol } HNO_3} \times \frac{105.99 \text{ g } Na_2CO_3}{\text{mol } Na_2CO_3} = 6.5 \times 10^7 \text{ g } Na_2CO_3$$

$$7.7 \times 10^7 \text{ g } HNO_3 \times \frac{1 \text{ mol } HNO_3}{63.02 \text{ g } HNO_3} \times \frac{-69 \text{ kJ}}{2 \text{ mol } HNO_3} = -4.2 \times 10^7 \text{ kJ}$$

4.2×10^7 kJ of heat was released.

b. They feared the heat generated by the neutralization reaction would vaporize the unreacted nitric acid, causing widespread airborne contamination.

129. $$400 \text{ kcal} \times \frac{4.18 \text{ kJ}}{\text{kcal}} = 1.7 \times 10^3 \text{ kJ} \approx 2 \times 10^3 \text{ kJ}$$

$$PE = mgz = \left(180 \text{ lb} \times \frac{1 \text{ kg}}{2.205 \text{ lb}}\right) \times \frac{9.81 \text{ m}}{s^2} \times \left(8 \text{ in} \times \frac{2.54 \text{ cm}}{\text{in}} \times \frac{1 \text{ m}}{100 \text{ cm}}\right) = 160 \text{ J} \approx 200 \text{ J}$$

200 J of energy is needed to climb one step. The total number of steps to climb are:

$$2 \times 10^6 \text{ J} \times \frac{1 \text{ step}}{200 \text{ J}} = 1 \times 10^4 \text{ steps}$$

130. $H_2(g) + 1/2\ O_2(g) \rightarrow H_2O(l)$ $\Delta H° = \Delta H^\circ_{f,\,H_2O(l)} = -285.8$ kJ; we want the reverse reaction:

$H_2O(l) \rightarrow H_2(g) + 1/2\ O_2(g)$ $\Delta H° = 285.8$ kJ

$w = -P\Delta V$; because $PV = nRT$, at constant T and P, $P\Delta V = RT\Delta n$, where Δn = moles of gaseous products – moles of gaseous reactants. Here, $\Delta n = (1 \text{ mol } H_2 + 0.5 \text{ mol } O_2) - (0) =$ 1.50 mol.

$$\Delta E° = \Delta H° - P\Delta V = \Delta H° - RT\Delta n$$

$$\Delta E° = 285.8 \text{ kJ} - \left(8.3145 \text{ J/K} \bullet \text{mol} \times 298 \text{ K} \times \frac{1 \text{ kJ}}{1000 \text{ J}} \times 1.50 \text{ mol}\right)$$

$$\Delta E° = 285.8 \text{ kJ} - 3.72 \text{ kJ} = 282.1 \text{ kJ}$$

131. There are five parts to this problem. We need to calculate:

(1) q required to heat $H_2O(s)$ from –30. °C to 0°C; use the specific heat capacity of $H_2O(s)$

(2) q required to convert 1 mol $H_2O(s)$ at 0°C into 1 mol $H_2O(l)$ at 0°C; use ΔH_{fusion}

(3) q required to heat $H_2O(l)$ from 0°C to 100.°C; use the specific heat capacity of $H_2O(l)$

(4) q required to convert 1 mol $H_2O(l)$ at 100.°C into 1 mol $H_2O(g)$ at 100.°C; use $\Delta H_{vaporization}$

(5) q required to heat $H_2O(g)$ from 100.°C to 140.°C; use the specific heat capacity of $H_2O(g)$

We will sum up the heat required for all five parts, and this will be the total amount of heat required to convert 1.00 mol of $H_2O(s)$ at –30.°C to $H_2O(g)$ at 140.°C.

$q_1 = 2.03\ J/°C{\bullet}g \times 18.02\ g \times [0 - (-30.)]°C = 1.1 \times 10^3\ J$

$q_2 = 1.00\ mol \times 6.02 \times 10^3\ J/mol = 6.02 \times 10^3\ J$

$q_3 = 4.18\ J/°C{\bullet}g \times 18.02\ g \times (100. - 0)°C = 7.53 \times 10^3\ J$

$q_4 = 1.00\ mol \times 40.7 \times 10^3\ J/mol = 4.07 \times 10^4\ J$

$q_5 = 2.02\ J/°C{\bullet}g \times 18.02\ g \times (140. - 100.)°C = 1.5 \times 10^3\ J$

$q_{total} = q_1 + q_2 + q_3 + q_4 + q_5 = 5.69 \times 10^4\ J = 56.9\ kJ$

132. When a mixture of ice and water exists, the temperature of the mixture remains at 0°C until all of the ice has melted. Because an ice-water mixture exists at the end of the process, the temperature remains at 0°C. All of the energy released by the element goes to convert ice into water. The energy required to do this is related to $\Delta H_{fusion} = 6.02$ kJ/mol (from Exercise 131).

Heat loss by element = heat gain by ice cubes at 0°C

$$\text{Heat gain} = 109.5\ g\ H_2O \times \frac{1\ mol\ H_2O}{18.02\ g} \times \frac{6.02\ kJ}{mol\ H_2O} = 36.6\ kJ$$

$$\text{Specific heat of element} = \frac{q}{mass \times \Delta T} = \frac{36{,}600\ J}{500.0\ g \times (195 - 0)°C} = 0.375\ J/°C{\bullet}g$$

Integrative Problems

133. $N_2(g) + 2\ O_2(g) \rightarrow 2\ NO_2(g) \qquad \Delta H = 67.7\ kJ$

$$n_{N_2} = \frac{PV}{RT} = \frac{3.50\ atm \times 0.250\ L}{\frac{0.08206\ L\ atm}{K\ mol} \times 373\ K} = 2.86 \times 10^{-2}\ mol\ N_2$$

$$n_{O_2} = \frac{PV}{RT} = \frac{3.50\ atm \times 0.450\ L}{\frac{0.08206\ L\ atm}{K\ mol} \times 373\ K} = 5.15 \times 10^{-2}\ mol\ O_2$$

$$2.86 \times 10^{-2} \text{ mol } N_2 \times \frac{2 \text{ mol } NO_2}{1 \text{ mol } N_2} = 5.72 \times 10^{-2} \text{ mol } NO_2 \text{ produced if } N_2 \text{ is limiting.}$$

$$5.15 \times 10^{-2} \text{ mol } O_2 \times \frac{2 \text{ mol } NO_2}{2 \text{ mol } O_2} = 5.15 \times 10^{-2} \text{ mol } NO_2 \text{ produced if } O_2 \text{ is limiting.}$$

O_2 is limiting because it produces the smaller quantity of product. The heat required is:

$$5.15 \times 10^{-2} \text{ mol } NO_2 \times \frac{67.7 \text{ kJ}}{2 \text{ mol } NO_2} = 1.74 \text{ kJ}$$

134. a. $4\ CH_3NO_2(l) + 3\ O_2(g) \rightarrow 4\ CO_2(g) + 2\ N_2(g) + 6\ H_2O(g)$

$$\Delta H^\circ_{rxn} = -1288.5 \text{ kJ} = [4 \text{ mol}(-393.5 \text{ kJ/mol}) + 6 \text{ mol}(-242 \text{ kJ/mol})] - [4 \text{ mol}(\Delta H^\circ_{f,\ CH_3NO_2})]$$

Solving: $\Delta H^\circ_{f,\ CH_3NO_2} = -434$ kJ/mol

b. $P_{total} = 950.\ \text{torr} \times \frac{1 \text{ atm}}{760 \text{ torr}} = 1.25 \text{ atm};\ P_{N_2} = P_{total} \times \chi_{N_2} = 1.25 \text{ atm} \times 0.134 = 0.168 \text{ atm}$

$$n_{N_2} = \frac{0.168 \text{ atm} \times 15.0 \text{ L}}{\frac{0.08206 \text{ L atm}}{\text{K mol}} \times 373 \text{ K}} = 0.0823 \text{ mol } N_2$$

$$0.0823 \text{ mol } N_2 \times \frac{28.02 \text{ g } N_2}{1 \text{ mol } N_2} = 2.31 \text{ g } N_2$$

135. Heat loss by U = heat gain by heavy water; volume of cube = (cube edge)^3

$$\text{Mass of heavy water} = 1.00 \times 10^3 \text{ mL} \times \frac{1.11 \text{ g}}{\text{mL}} = 1110 \text{ g}$$

$$\text{Heat gain by heavy water} = \frac{4.211 \text{ J}}{^\circ\text{C g}} \times 1110 \text{ g} \times (28.5 - 25.5)^\circ\text{C} = 1.4 \times 10^4 \text{ J}$$

$$\text{Heat loss by U} = 1.4 \times 10^4 \text{ J} = \frac{0.117 \text{ J}}{^\circ\text{C g}} \times \text{mass} \times (200.0 - 28.5)^\circ\text{C},\quad \text{mass} = 7.0 \times 10^2 \text{ g U}$$

$$7.0 \times 10^2 \text{ g U} \times \frac{1 \text{ cm}^3}{19.05 \text{ g}} = 37 \text{ cm}^3;\ \text{cube edge} = (37 \text{ cm}^3)^{1/3} = 3.3 \text{ cm}$$

Marathon Problems

136. Pathway I:

Step 1: (5.00 mol, 3.00 atm, 15.0 L) → (5.00 mol, 3.00 atm, 55.0 L)

$$w = -P\Delta V = -(3.00 \text{ atm})(55.0 \text{ L} - 15.0 \text{ L}) = -120. \text{ L atm}$$

$$w = -120. \text{ L atm} \times \frac{101.3 \text{ J}}{\text{L atm}} \times \frac{1 \text{ kJ}}{1000 \text{J}} = -12.2 \text{ kJ}$$

Step 1 is at constant pressure. The heat released/gained at constant pressure = q_p = ΔH. From the problem, $\Delta H = nC_p\Delta T$ for an ideal gas. Using the ideal gas law, let's substitute for ΔT.

$\Delta(PV) = \Delta(nRT) = nR\Delta T$ for a specific gas sample. So: $\Delta T = \dfrac{\Delta(PV)}{nR}$

$\Delta H = q_p = nC_p\Delta T = nC_p \times \dfrac{\Delta(PV)}{nR} = \dfrac{C_p\Delta(PV)}{R}$; *Note*: $\Delta(PV) = (P_2V_2 - P_1V_1)$

For an ideal monatomic gas, $C_p = \dfrac{5}{2}R$; substituting into the above equation:

$$\Delta H = \left(\frac{5}{2}R\right)\frac{\Delta(PV)}{R} = \frac{5}{2}\Delta(PV)$$

$$\Delta H = q_p = \frac{5}{2}\Delta(PV) = \frac{5}{2}(3.00 \text{ atm} \times 55.0 \text{ L} - 3.00 \text{ atm} \times 15.0 \text{ L}) = 300. \text{ L atm}$$

$$\Delta H = q_p = 300. \text{ L atm} \times \frac{101.3 \text{ J}}{\text{L atm}} \times \frac{1 \text{ kJ}}{1000 \text{J}} = 30.4 \text{ kJ}$$

$$\Delta E = q + w = 30.4 \text{ kJ} - 12.2 \text{ kJ} = 18.2 \text{ kJ}$$

Note: We could have used $\Delta E = nC_v\Delta T$ to calculate the same answer (ΔE = 18.2 kJ).

Step 2: (5.00 mol, 3.00 atm, 55.0 L) → (5.00 mol, 6.00 atm, 20.0 L)

In this step, neither pressure nor volume are constant. We will need to determine q in a different way. However, it will always hold for an ideal gas that $\Delta E = nC_v\Delta T$ and $\Delta H = nC_p\Delta T$.

$$\Delta E = nC_v\Delta T = n\left(\frac{3}{2}R\right)\left(\frac{\Delta(PV)}{nR}\right) = \frac{3}{2}\Delta PV$$

$$\Delta E = \frac{3}{2}(120. - 165) \text{ L atm} = -67.5 \text{ L atm} \quad \text{(Carry an extra significant figure.)}$$

$$\Delta E = -67.5 \text{ L atm} \times \frac{101.3 \text{ J}}{\text{L atm}} \times \frac{1 \text{ kJ}}{1000 \text{J}} = -6.8 \text{ kJ}$$

$$\Delta H = nC_p\Delta T = n\left(\frac{5}{2}R\right)\left(\frac{\Delta(PV)}{nR}\right) = \frac{5}{2}\Delta PV$$

$$\Delta H = \frac{5}{2}(120. - 165)\text{ L atm} = -113\text{ L atm}$$ (Carry an extra significant figure.)

$$w = -P\Delta V = -(6.00\text{ atm})(20.0 - 55.0)\text{ L} = 210.\text{ L atm}$$

$$w = 210.\text{ L atm} \times \frac{101.3\text{ J}}{\text{L atm}} \times \frac{1\text{ kJ}}{1000\text{J}} = 21.3\text{ kJ}$$

$$\Delta E = q + w,\ -6.8\text{ kJ} = q + 21.3\text{ kJ},\ q = -28.1\text{ kJ}$$

Summary:

Path I	Step 1	Step 2	Total
q	30.4 kJ	−28.1 kJ	2.3 kJ
w	−12.2 kJ	21.3 kJ	9.1 kJ
ΔE	18.2 kJ	−6.8 kJ	11.4 kJ
ΔH	30.4 kJ	−11 kJ	19 kJ

Pathway II:

Step 3: (5.00 mol, 3.00 atm, 15.0 L) → (5.00 mol, 6.00 atm, 15.0 L)

Step 3 is at constant volume. The heat released/gained at constant volume = $q_v = \Delta E$.

$$\Delta E = nC_v\Delta T = n\left(\frac{3}{2}R\right)\left(\frac{\Delta(PV)}{nR}\right) = \frac{3}{2}\Delta PV$$

$$\Delta E = q_v = \frac{3}{2}\Delta(PV) = \frac{3}{2}(6.00\text{ atm} \times 15.0\text{ L} - 3.00\text{ atm} \times 15.0\text{ L})$$

$$\Delta E = q_v = \frac{3}{2}(90.0 - 45.0)\text{ L atm} = 67.5\text{ L atm}$$

$$\Delta E = q_v = 67.5\text{ L atm} \times \frac{101.3\text{ J}}{\text{L atm}} \times \frac{1\text{ kJ}}{1000\text{J}} = 6.84\text{ kJ}$$

$$w = -P\Delta V = 0 \text{ because } \Delta V = 0$$

$$\Delta H = \Delta E + \Delta(PV) = 67.5\text{ L atm} + 45.0\text{ L atm} = 112.5\text{ L atm} = 11.40\text{ kJ}$$

Step 4: (5.00 mol, 6.00 atm, 15.0 L) → (5.00 mol, 6.00 atm, 20.0 L)

Step 4 is at constant pressure so $q_p = \Delta H$.

$$\Delta H = q_p = nC_p\Delta T = \left(\frac{5}{2}R\right)\left(\frac{\Delta(PV)}{nR}\right) = \frac{5}{2}\Delta PV$$

$$\Delta H = \frac{5}{2}(120. - 90.0)\text{ L atm} = 75\text{ L atm}$$

$$\Delta H = q_p = 75 \text{ L atm} \times \frac{101.3 \text{ J}}{\text{L atm}} \times \frac{1 \text{ kJ}}{1000 \text{J}} = 7.6 \text{ kJ}$$

$$w = -P\Delta V = -(6.00 \text{ atm})(20.0 - 15.0) \text{ L} = -30. \text{ L atm}$$

$$w = -30. \text{ L atm} \times \frac{101.3 \text{ J}}{\text{L atm}} \times \frac{1 \text{ kJ}}{1000 \text{J}} = -3.0 \text{ kJ}$$

$$\Delta E = q + w = 7.6 \text{ kJ} - 3.0 \text{ kJ} = 4.6 \text{ kJ}$$

Summary:

Path II	**Step 3**	**Step 4**	**Total**
q	6.84 kJ	7.6 kJ	14.4 kJ
w	0	−3.0 kJ	−3.0 kJ
ΔE	6.84 kJ	4.6 kJ	11.4 kJ
ΔH	11.40 kJ	7.6 kJ	19.0 kJ

State functions are independent of the particular pathway taken between two states; path functions are dependent on the particular pathway. In this problem, the overall values of ΔH and ΔE for the two pathways are the same (see the two summaries of results); hence ΔH and ΔE are state functions. However, the overall values of q and w for the two pathways are different; hence q and w are path functions.

137. $C_xH_y + \left(\frac{2x + y/2}{2}\right) O_2 \rightarrow x\, CO_2 + y/2\, H_2O$

$$[x(-393.5) + y/2\,(-242)] - \Delta H^\circ_{C_xH_y} = -2044.5, \quad -(393.5)x - 121y - \Delta H_{C_xH_y} = -2044.5$$

$$d_{gas} = \frac{P \bullet MM}{RT}, \text{ where MM = average molar mass of } CO_2/H_2O \text{ mixture}$$

$$0.751 \text{ g/L} = \frac{1.00 \text{ atm} \times MM}{\frac{0.08206 \text{L atm}}{\text{K mol}} \times 473 \text{K}}, \text{ MM of } CO_2/H_2O \text{ mixture} = 29.1 \text{ g/mol}$$

Let a = mol CO_2 and $1.00 - a$ = mol H_2O (assuming 1.00 total moles of mixture)

$(44.01)a + (1.00 - a) \times 18.02 = 29.1$; solving: $a = 0.426$ mol CO_2 ; mol $H_2O = 0.574$ mol

$$\text{Thus: } \frac{0.574}{0.426} = \frac{\frac{y}{2}}{x}, \quad 2.69 = \frac{y}{x}, \quad y = (2.69)x$$

For whole numbers, multiply by three, which gives $y = 8$, $x = 3$. Note that $y = 16$, $x = 6$ is possible, along with other combinations. Because the hydrocarbon has a lower density than Kr, the molar mass of C_xH_y must be less than the molar mass of Kr (83.80 g/mol). Only C_3H_8 works.

$$-2044.5 = -393.5(3) - 121(8) - \Delta H^\circ_{C_3H_8}, \quad \Delta H^\circ_{C_3H_8} = -104 \text{ kJ/mol}$$

CHAPTER 7

ATOMIC STRUCTURE AND PERIODICITY

Questions

19. The equations relating the terms are $\nu\lambda = c$, $E = h\nu$, and $E = hc/\lambda$. From the equations, wavelength and frequency are inversely related, photon energy and frequency are directly related, and photon energy and wavelength are inversely related. The unit of 1 Joule (J) = 1 kg m^2/s^2. This is why you must change mass units to kg when using the deBroglie equation.

20. Frequency is the number of waves (cycles) of electromagnetic radiation per second that pass a given point in space. Speed refers to the distance a wave travels per unit time. All electromagnetic radiation (EMR) travels at the same speed (c, the speed of light = 2.998×10^8 m/s). However, each wavelength of EMR has its own unique frequency,

21. The photoelectric effect refers to the phenomenon in which electrons are emitted from the surface of a metal when light strikes it. The light must have a certain minimum frequency (energy) in order to remove electrons from the surface of a metal. Light having a frequency below the minimum value results in no electrons being emitted, whereas light at or higher than the minimum frequency does cause electrons to be emitted. For light having a frequency higher than the minimum frequency, the excess energy is transferred into kinetic energy for the emitted electron. Albert Einstein explained the photoelectric effect by applying quantum theory.

22. The emission of light by excited atoms has been the key interconnection between the macroscopic world we can observe and measure, and what is happening on a microscopic basis within an atom. Excited atoms emit light (which we can observe and measure) because of changes in the microscopic structure of the atom. By studying the emissions of atoms, we can trace back to what happened inside the atom. Specifically, our current model of the atom relates the energy of light emitted to electrons in the atom moving from higher allowed energy states to lower allowed energy states.

23. Example 7.3 calculates the deBroglie wavelength of a ball and of an electron. The ball has a wavelength on the order of 10^{-34} m. This is incredibly short and, as far as the wave-particle duality is concerned, the wave properties of large objects are insignificant. The electron, with its tiny mass, also has a short wavelength; on the order of 10^{-10} m. However, this wavelength is significant because it is on the same order as the spacing between atoms in a typical crystal. For very tiny objects like electrons, the wave properties are important. The wave properties must be considered, along with the particle properties, when hypothesizing about the electron motion in an atom.

24. a. For hydrogen (Z = 1), the energy levels in units of joules are given by the equation $E_n = -2.178 \times 10^{-18}(1/n^2)$. As n increases, the differences between $1/n^2$ for consecutive energy levels becomes smaller and smaller. Consider the difference between $1/n^2$ values for $n = 1$ and $n = 2$ as compared to $n = 3$ and $n = 4$.

For $n = 1$ and $n = 2$:

$$\frac{1}{1^2} - \frac{1}{2^2} = 1 - 0.25 = 0.75$$

For $n = 3$ and $n = 4$:

$$\frac{1}{3^2} - \frac{1}{4^2} = 0.1111 - 0.0625 = 0.0486$$

Because the energy differences between $1/n^2$ values for consecutive energy levels decrease as n increases, the energy levels get closer together as n increases.

b. For a spectral transition for hydrogen, $\Delta E = E_f - E_i$:

$$\Delta E = -2.178 \times 10^{-18}\ \text{J}\left(\frac{1}{n_f^2} - \frac{1}{n_i^2}\right)$$

where n_i and n_f are the levels of the initial and final states, respectively. A positive value of ΔE always corresponds to an absorption of light, and a negative value of ΔE always corresponds to an emission of light.

In the diagram, the red line is for the $n_i = 3$ to $n_f = 2$ transition.

$$\Delta E = -2.178 \times 10^{-18}\ \text{J}\left(\frac{1}{2^2} - \frac{1}{3^2}\right) = -2.178 \times 10^{-18}\ \text{J}\left(\frac{1}{4} - \frac{1}{9}\right)$$

$$\Delta E = -2.178 \times 10^{-18}\ \text{J} \times (0.2500 - 0.1111) = -3.025 \times 10^{-19}\ \text{J}$$

The photon of light must have precisely this energy (3.025×10^{-19} J).

$$|\Delta E| = E_{photon} = h\nu = \frac{hc}{\lambda}$$

$$\lambda = \frac{hc}{|\Delta E|} = \frac{6.6261 \times 10^{-34}\ \text{J s} \times 2.9979 \times 10^8\ \text{m/s}}{3.025 \times 10^{-19}\ \text{J}} = 6.567 \times 10^{-7}\ \text{m} = 656.7\ \text{nm}$$

From Figure 7.2, λ = 656.7 nm is red light so the diagram is correct for the red line.

In the diagram, the green line is for the $n_i = 4$ to $n_f = 2$ transition.

$$\Delta E = -2.178 \times 10^{-18}\ \text{J}\left(\frac{1}{2^2} - \frac{1}{4^2}\right) = -4.084 \times 10^{-19}\ \text{J}$$

$$\lambda = \frac{hc}{|\Delta E|} = \frac{6.6261 \times 10^{-34}\ \text{J s} \times 2.9979 \times 10^8\ \text{m/s}}{4.084 \times 10^{-19}\ \text{J}} = 4.864 \times 10^{-7}\ \text{m} = 486.4\ \text{nm}$$

From Figure 7.2, λ = 486.4 nm is green-blue light. The diagram is consistent with this line.

In the diagram, the blue line is for the $n_i = 5$ to $n_f = 2$ transition.

$$\Delta E = -2.178 \times 10^{-18}\ \text{J}\left(\frac{1}{2^2} - \frac{1}{5^2}\right) = -4.574 \times 10^{-19}\ \text{J}$$

$$\lambda = \frac{hc}{|\Delta E|} = \frac{6.6261 \times 10^{-34}\ \text{J s} \times 2.9979 \times 10^8\ \text{m/s}}{4.574 \times 10^{-19}\ \text{J}} = 4.343 \times 10^{-7}\ \text{m} = 434.3\ \text{nm}$$

From Figure 7.2, λ = 434.3 nm is blue or blue-violet light. The diagram is consistent with this line also.

25. The Bohr model was an important step in the development of the current quantum mechanical model of the atom. The idea that electrons can only occupy certain, allowed energy levels is illustrated nicely (and relatively easily). We talk about the Bohr model to present the idea of quantized energy levels.

26. The figure on the left tells us that the probability of finding the electron in the 1s orbital at points along a line drawn outward from the nucleus in any direction. This probability is greatest close to the nucleus and drops off rapidly as the distance from the nucleus increases.

    The figure on the right represents the total probability of finding the electron at a particular distance from the nucleus for a 1s hydrogen orbital. For this distribution, the hydrogen 1s orbital is divided into successive thin spherical shells and the total probability of finding the electron in each spherical shell is plotted versus distance from the nucleus. This graph is called the radial probability distribution.

    The radial probability distribution initially shows a steady increase with distance from the nucleus, reaches a maximum, then shows a steady decrease. Even though it is likely to find an electron near the nucleus, the volume of the spherical shell close to the nucleus is tiny, resulting in a low radial probability. The maximum radial probability distribution occurs at a distance of 5.29×10^{-2} nm from the nucleus; the electron is most likely to be found in the volume of the shell centered at this distance from the nucleus. The 5.29×10^{-2} nm distance is the exact radius of innermost ($n = 1$) orbit in the Bohr model.

27. When the p and d orbital functions are evaluated at various points in space, the results sometimes have positive values and sometimes have negative values. The term phase is often associated with the + and – signs. For example, a sine wave has alternating positive and negative phases. This is analogous to the positive and negative values (phases) in the p and d orbitals.

28. The widths of the various blocks in the periodic table are determined by the number of electrons that can occupy the specific orbital(s). In the s block, we have one orbital ($\ell = 0$, $m_\ell = 0$) that can hold two electrons; the s block is two elements wide. For the f block, there are 7 degenerate f orbitals ($\ell = 3$, $m_\ell = -3, -2, -1, 0, 1, 2, 3$), so the f block is 14 elements wide. The g block corresponds to $\ell = 4$. The number of degenerate g orbitals is 9. This comes from the 9 possible m_ℓ values when $\ell = 4$ ($m_\ell = -4, -3, -2, -1, 0, 1, 2, 3, 4$). With 9 orbitals, each orbital holding two electrons, the g block would be 18 elements wide. The h block has $\ell = 5$, $m_\ell = -5, -4, -3, -2, -1, 0, 1, 2, 3, 4, 5$. With 11 degenerate h orbitals, the h block would be 22 elements wide.

29. If one more electron is added to a half-filled subshell, electron-electron repulsions will increase because two electrons must now occupy the same atomic orbital. This may slightly decrease the stability of the atom.

30. Size decreases from left to right and increases going down the periodic table. Thus, going one element right and one element down would result in a similar size for the two elements diagonal to each other. The ionization energies will be similar for the diagonal elements since the periodic trends also oppose each other. Electron affinities are harder to predict, but atoms with similar sizes and ionization energies should also have similar electron affinities.

31. The valence electrons are strongly attracted to the nucleus for elements with large ionization energies. One would expect these species to readily accept another electron and have very exothermic electron affinities. The noble gases are an exception; they have a large ionization energy but have an endothermic electron affinity. Noble gases have a filled valence shell of electrons. The added electron in a noble gas must go into a higher n value atomic orbital, having a significantly higher energy, and this is very unfavorable.

32. Electron-electron repulsions become more important when we try to add electrons to an atom. From the standpoint of electron-electron repulsions, larger atoms would have more favorable (more exothermic) electron affinities. Considering only electron-nucleus attractions, smaller atoms would be expected to have the more favorable (more exothermic) electron affinities. These trends are exactly the opposite of each other. Thus the overall variation in electron affinity is not as great as ionization energy in which attractions to the nucleus dominate.

33. For hydrogen and hydrogen-like (one-electron ions), all atomic orbitals with the same n value have the same energy. For polyatomic atoms/ions, the energy of the atomic orbitals also depends on ℓ. Because there are more nondegenerate energy levels for polyatomic atoms/ions as compared to hydrogen, there are many more possible electronic transitions resulting in more complicated line spectra.

34. Each element has a characteristic spectrum because each element has unique energy levels. Thus the presence of the characteristic spectral lines of an element confirms its presence in any particular sample.

35. Yes, the maximum number of unpaired electrons in any configuration corresponds to a minimum in electron-electron repulsions.

36. The electron is no longer part of that atom. The proton and electron are completely separated.

37. Ionization energy applies to the removal of the electron from an atom in the gas phase. The work function applies to the removal of an electron from the surface of a solid element.

$M(g) \rightarrow M^+(g) + e^-$ ionization energy; $M(s) \rightarrow M^+(s) + e^-$ work function

38. Li^+ ions are the smallest of the alkali metal cations and will be most strongly attracted to the water molecules.

Exercises

Light and Matter

39. $\nu = \dfrac{c}{\lambda} = \dfrac{2.998 \times 10^8 \text{ m/s}}{780. \times 10^{-9} \text{ m}} = 3.84 \times 10^{14} \text{ s}^{-1}$

40. $99.5 \text{ MHz} = 99.5 \times 10^6 \text{ Hz} = 99.5 \times 10^6 \text{ s}^{-1}$; $\lambda = \dfrac{c}{\nu} = \dfrac{2.998 \times 10^8 \text{ m/s}}{99.5 \times 10^6 \text{ s}^{-1}} = 3.01 \text{ m}$

41. $\nu = \dfrac{c}{\lambda} = \dfrac{3.00 \times 10^8 \text{ m/s}}{1.0 \times 10^{-2} \text{ m}} = 3.0 \times 10^{10} \text{ s}^{-1}$

$E = h\nu = 6.63 \times 10^{-34} \text{ J s} \times 3.0 \times 10^{10} \text{ s}^{-1} = 2.0 \times 10^{-23} \text{ J/photon}$

$$\frac{2.0 \times 10^{-23} \text{ J}}{\text{photon}} \times \frac{6.02 \times 10^{23} \text{ photons}}{\text{mol}} = 12 \text{ J/mol}$$

42. $E = h\nu = \dfrac{hc}{\lambda} = \dfrac{6.63 \times 10^{-34} \text{ J s} \times 3.00 \times 10^8 \text{ m/s}}{25 \text{ nm} \times \dfrac{1 \text{ m}}{1 \times 10^9 \text{ nm}}} = 8.0 \times 10^{-18} \text{ J/photon}$

$$\frac{8.0 \times 10^{-18} \text{ J}}{\text{photon}} \times \frac{6.02 \times 10^{23} \text{ photons}}{\text{mol}} = 4.8 \times 10^6 \text{ J/mol}$$

43. 280 nm: $\nu = \dfrac{c}{\lambda} = \dfrac{3.00 \times 10^8 \text{ m/s}}{280 \text{nm} \times \dfrac{1 \text{ m}}{1 \times 10^9 \text{ nm}}} = 1.1 \times 10^{15} \text{ s}^{-1}$

320 nm: $\nu = \dfrac{3.00 \times 10^8 \text{ m/s}}{320 \times 10^{-9} \text{ nm}} = 9.4 \times 10^{14} \text{ s}^{-1}$

The compounds in the sunscreen absorb ultraviolet B (UVB) electromagnetic radiation having a frequency from $9.4 \times 10^{14} \text{ s}^{-1}$ to $1.1 \times 10^{15} \text{ s}^{-1}$.

44. S-type cone receptors: $\lambda = \dfrac{c}{\nu} = \dfrac{2.998 \times 10^8 \text{ m/s}}{6.00 \times 10^{14} \text{ s}^{-1}} = 5.00 \times 10^{-7} \text{ m} = 500. \text{ nm}$

$$\lambda = \frac{2.998 \times 10^8 \text{ m/s}}{7.49 \times 10^{14} \text{ s}^{-1}} = 4.00 \times 10^{-7} \text{ m} = 400. \text{ nm}$$

S-type cone receptors detect 400-500 nm light. From Figure 7.2 in the text, this is violet to green light, respectively.

M-type cone receptors: $\lambda = \dfrac{2.998 \times 10^8 \text{ m/s}}{4.76 \times 10^{14} \text{ s}^{-1}} = 6.30 \times 10^{-7} \text{ m} = 630. \text{ nm}$

$$\lambda = \frac{2.998 \times 10^8 \text{ m/s}}{6.62 \times 10^{14} \text{ s}^{-1}} = 4.53 \times 10^{-7} \text{ m} = 453 \text{ nm}$$

M-type cone receptors detect 450-630 nm light. From Figure 7.2 in the text, this is blue to orange light, respectively.

L-type cone receptors: $\lambda = \dfrac{2.998 \times 10^8 \text{ m/s}}{4.28 \times 10^{14} \text{ s}^{-1}} = 7.00 \times 10^{-7} \text{ m} = 700. \text{ nm}$

$$\lambda = \frac{2.998 \times 10^8 \text{ m/s}}{6.00 \times 10^{14} \text{ s}^{-1}} = 5.00 \times 10^{-7} \text{ m} = 500. \text{ nm}$$

L-type cone receptors detect 500-700 nm light. This represents green to red light, respectively.

45. The wavelength is the distance between consecutive wave peaks. Wave *a* shows 4 wavelengths, and wave *b* shows 8 wavelengths.

Wave *a*: $\lambda = \dfrac{1.6 \times 10^{-3} \text{ m}}{4} = 4.0 \times 10^{-4} \text{ m}$

Wave *b*: $\lambda = \dfrac{1.6 \times 10^{-3} \text{ m}}{8} = 2.0 \times 10^{-4} \text{ m}$

Wave *a* has the longer wavelength. Because frequency and photon energy are both inversely proportional to wavelength, wave *b* will have the higher frequency and larger photon energy since it has the shorter wavelength.

$$\nu = \frac{c}{\lambda} = \frac{2.998 \times 10^8 \text{ m/s}}{2.0 \times 10^{-4} \text{ m}} = 1.5 \times 10^{12} \text{ s}^{-1}$$

$$E = \frac{hc}{\lambda} = \frac{6.626 \times 10^{-34} \text{ J s} \times 2.998 \times 10^8 \text{ m/s}}{2.0 \times 10^{-4} \text{ m}} = 9.9 \times 10^{-22} \text{ J}$$

Because both waves are examples of electromagnetic radiation, both waves travel at the same speed, c, the speed of light. From Figure 7.2 of the text, both of these waves represent infrared electromagnetic radiation.

46. Referencing Figure 7.2 of the text, 2.12×10^{-10} m electromagnetic radiation is X rays.

$$\lambda = \frac{c}{\nu} = \frac{2.9979 \times 10^8 \text{ m/s}}{107.1 \times 10^6 \text{ s}^{-1}} = 2.799 \text{ m}$$

From the wavelength calculated above, 107.1 MHz electromagnetic radiation is FM radiowaves.

$$\lambda = \frac{hc}{E} = \frac{6.626 \times 10^{-34}\ \text{J s} \times 2.998 \times 10^{8}\ \text{m/s}}{3.97 \times 10^{-19}\ \text{J}} = 5.00 \times 10^{-7}\ \text{m}$$

The 3.97×10^{-19} J/photon electromagnetic radiation is visible (green) light.

The photon energy and frequency order will be the exact opposite of the wavelength ordering because E and ν are both inversely related to λ. From the previously calculated wavelengths, the order of photon energy and frequency is:

FM radiowaves < visible (green) light < X rays	
longest λ	shortest λ
lowest ν	highest ν
smallest E	largest E

47. $$E_{photon} = \frac{hc}{\lambda} = \frac{6.626 \times 10^{-34}\ \text{J s} \times 2.998 \times 10^{8}\ \text{m/s}}{150.\ \text{nm} \times \dfrac{1\ \text{m}}{1 \times 10^{9}\ \text{nm}}} = 1.32 \times 10^{-18}\ \text{J}$$

$$1.98 \times 10^{5}\ \text{J} \times \frac{1\ \text{photon}}{1.32 \times 10^{-18}\ \text{J}} \times \frac{1\ \text{atom C}}{\text{photon}} = 1.50 \times 10^{23}\ \text{atoms C}$$

48. $$E_{photon} = h\nu = \frac{hc}{\lambda},\ E_{photon} = \frac{6.626 \times 10^{-34}\ \text{J s} \times 2.998 \times 10^{8}\ \text{m/s}}{1.0 \times 10^{-10}\ \text{m}} = 2.0 \times 10^{-15}\ \text{J}$$

$$\frac{2.0 \times 10^{-15}\ \text{J}}{\text{photon}} \times \frac{6.02 \times 10^{23}\ \text{photons}}{\text{mol}} \times \frac{1\ \text{kJ}}{1000\ \text{J}} = 1.2 \times 10^{6}\ \text{kJ/mol}$$

$$E_{photon} = \frac{6.626 \times 10^{-34}\ \text{J s} \times 2.998 \times 10^{8}\ \text{m/s}}{1.0 \times 10^{4}\ \text{m}} = 2.0 \times 10^{-29}\ \text{J}$$

$$\frac{2.0 \times 10^{-29}\ \text{J}}{\text{photon}} \times \frac{6.02 \times 10^{23}\ \text{photons}}{\text{mol}} \times \frac{1\ \text{kJ}}{1000\ \text{J}} = 1.2 \times 10^{-8}\ \text{kJ/mol}$$

X rays do have an energy greater than the carbon-carbon bond energy. Therefore, X rays could conceivably break carbon-carbon bonds in organic compounds and thereby disrupt the function of an organic molecule. Radiowaves, however, do not have sufficient energy to break carbon-carbon bonds and are therefore relatively harmless.

49. The energy needed to remove a single electron is:

$$\frac{279.7\ \text{kJ}}{\text{mol}} \times \frac{1\ \text{mol}}{6.0221 \times 10^{23}} = 4.645 \times 10^{-22}\ \text{kJ} = 4.645 \times 10^{-19}\ \text{J}$$

$$E = \frac{hc}{\lambda},\ \lambda = \frac{hc}{E} = \frac{6.6261 \times 10^{-34}\ \text{J s} \times 2.9979 \times 10^{8}\ \text{m/s}}{4.645 \times 10^{-19}\ \text{J}} = 4.277 \times 10^{-7}\ \text{m} = 427.7\ \text{nm}$$

50. $\frac{208.4 \text{ kJ}}{\text{mol}} \times \frac{1 \text{ mol}}{6.0221 \times 10^{23}} = 3.461 \times 10^{-22} \text{ kJ} = 3.461 \times 10^{-19}$ J to remove one electron

$$E = \frac{hc}{\lambda}, \; \lambda = \frac{hc}{E} = \frac{6.6261 \times 10^{-34} \text{ J s} \times 2.9979 \times 10^{8} \text{ m/s}}{3.461 \times 10^{-19} \text{ J}} = 5.739 \times 10^{-7} \text{ m} = 573.9 \text{ nm}$$

51. Ionization energy = energy to remove an electron = $7.21 \times 10^{-19} = E_{photon}$

$E_{photon} = h\nu$ and $\lambda\nu = c$. So $\nu = \frac{c}{\lambda}$ and $E = \frac{hc}{\lambda}$.

$$\lambda = \frac{hc}{E_{photon}} = \frac{6.626 \times 10^{-34} \text{ J s} \times 2.998 \times 10^{8} \text{ m/s}}{7.21 \times 10^{-19} \text{ J}} = 2.76 \times 10^{-7} \text{ m} = 276 \text{ nm}$$

52. $\frac{890.1 \text{ kJ}}{\text{mol}} \times \frac{1 \text{ mol}}{6.0221 \times 10^{23} \text{ atoms}} = \frac{1.478 \times 10^{-21} \text{ kJ}}{\text{atom}} = \frac{1.478 \times 10^{-18} \text{ J}}{\text{atom}}$

= ionization energy per atom

$$E = \frac{hc}{\lambda}, \; \lambda = \frac{hc}{E} = \frac{6.626 \times 10^{-34} \text{ J s} \times 2.9979 \times 10^{8} \text{ m/s}}{1.478 \times 10^{-18} \text{ J}} = 1.344 \times 10^{-7} \text{ m} = 134.4 \text{ nm}$$

No, it will take light having a wavelength of 134.4 nm or less to ionize gold. A photon of light having a wavelength of 225 nm is longer wavelength and thus lower energy than 134.4 nm light.

53. a. 10.% of speed of light = $0.10 \times 3.00 \times 10^{8}$ m/s = 3.0×10^{7} m/s

$$\lambda = \frac{h}{mv}, \; \lambda = \frac{6.63 \times 10^{-34} \text{ J s}}{9.11 \times 10^{-31} \text{ kg} \times 3.0 \times 10^{7} \text{ m/s}} = 2.4 \times 10^{-11} \text{ m} = 2.4 \times 10^{-2} \text{ nm}$$

Note: For units to come out, the mass must be in kg because $1 \text{ J} = \frac{1 \text{ kg m}^2}{\text{s}^2}$.

b. $\lambda = \frac{h}{mv} = \frac{6.63 \times 10^{-34} \text{ J s}}{0.055 \text{kg} \times 35 \text{ m/s}} = 3.4 \times 10^{-34} \text{ m} = 3.4 \times 10^{-25}$ nm

This number is so small that it is insignificant. We cannot detect a wavelength this small. The meaning of this number is that we do not have to worry about the wave properties of large objects.

54. a. $\lambda = \frac{h}{mv} = \frac{6.626 \times 10^{-34} \text{ J s}}{1.675 \times 10^{-27} \text{ kg} \times (0.0100 \times 2.998 \times 10^{8} \text{ m/s})} = 1.32 \times 10^{-13} \text{m}$

b. $\lambda = \frac{h}{mv}, \; v = \frac{h}{\lambda m} = \frac{6.626 \times 10^{-34} \text{ J s}}{75 \times 10^{-12} \text{ m} \times 1.675 \times 10^{-27} \text{ kg}} = 5.3 \times 10^{3}$ m/s

55. $\lambda = \dfrac{h}{mv}$, $m = \dfrac{h}{\lambda v} = \dfrac{6.63 \times 10^{-34}\ \text{J s}}{1.5 \times 10^{-15}\ \text{m} \times (0.90 \times 3.00 \times 10^{8}\ \text{m/s})} = 1.6 \times 10^{-27}$ kg

This particle is probably a proton or a neutron.

56. $\lambda = \dfrac{h}{mv}$, $v = \dfrac{h}{\lambda m}$; for $\lambda = 1.0 \times 10^{2}$ nm $= 1.0 \times 10^{-7}$ m:

$$v = \frac{6.63 \times 10^{-34}\ \text{J s}}{9.11 \times 10^{-31}\ \text{kg} \times 1.0 \times 10^{-7}\ \text{m}} = 7.3 \times 10^{3}\ \text{m/s}$$

For $\lambda = 1.0$ nm $= 1.0 \times 10^{-9}$ m: $v = \dfrac{6.63 \times 10^{-34}\ \text{J s}}{9.11 \times 10^{-31}\ \text{kg} \times 1.0 \times 10^{-9}\ \text{m}} = 7.3 \times 10^{5}$ m/s

Hydrogen Atom: The Bohr Model

57. For the H atom (Z = 1): $E_n = -2.178 \times 10^{-18}\ \text{J}/n^2$; for a spectral transition, $\Delta E = E_f - E_i$:

$$\Delta E = -2.178 \times 10^{-18}\ \text{J}\left(\frac{1}{n_f^2} - \frac{1}{n_i^2}\right)$$

where n_i and n_f are the levels of the initial and final states, respectively. A positive value of ΔE always corresponds to an absorption of light, and a negative value of ΔE always corresponds to an emission of light.

a. $\Delta E = -2.178 \times 10^{-18}\ \text{J}\left(\dfrac{1}{2^2} - \dfrac{1}{3^2}\right) = -2.178 \times 10^{-18}\ \text{J}\left(\dfrac{1}{4} - \dfrac{1}{9}\right)$

$\Delta E = -2.178 \times 10^{-18}\ \text{J} \times (0.2500 - 0.1111) = -3.025 \times 10^{-19}\ \text{J}$

The photon of light must have precisely this energy (3.025×10^{-19} J).

$$|\Delta E| = E_{photon} = h\nu = \frac{hc}{\lambda},\ \lambda = \frac{hc}{|\Delta E|} = \frac{6.6261 \times 10^{-34}\ \text{J s} \times 2.9979 \times 10^{8}\ \text{m/s}}{3.025 \times 10^{-19}\ \text{J}}$$

$$= 6.567 \times 10^{-7}\ \text{m} = 656.7\ \text{nm}$$

From Figure 7.2, this is visible electromagnetic radiation (red light).

b. $\Delta E = -2.178 \times 10^{-18}\ \text{J}\left(\dfrac{1}{2^2} - \dfrac{1}{4^2}\right) = -4.084 \times 10^{-19}\ \text{J}$

$$\lambda = \frac{hc}{|\Delta E|} = \frac{6.6261 \times 10^{-34}\ \text{J s} \times 2.9979 \times 10^{8}\ \text{m/s}}{4.084 \times 10^{-19}\ \text{J}} = 4.864 \times 10^{-7}\ \text{m} = 486.4\ \text{nm}$$

This is visible electromagnetic radiation (green-blue light).

c. $\Delta E = -2.178 \times 10^{-18}\text{J}\left(\frac{1}{1^2} - \frac{1}{2^2}\right) = -1.634 \times 10^{-18}\text{J}$

$$\lambda = \frac{6.6261 \times 10^{-34}\ \text{J s} \times 2.9979 \times 10^{8}\ \text{m/s}}{1.634 \times 10^{-18}\ \text{J}} = 1.216 \times 10^{-7}\ \text{m} = 121.6\ \text{nm}$$

This is ultraviolet electromagnetic radiation.

58. a. $\Delta E = -2.178 \times 10^{-18}\text{J}\left(\frac{1}{3^2} - \frac{1}{4^2}\right) = -1.059 \times 10^{-19}\ \text{J}$

$$\lambda = \frac{hc}{|\Delta E|} = \frac{6.6261 \times 10^{-34}\ \text{J s} \times 2.9979 \times 10^{8}\ \text{m/s}}{1.059 \times 10^{-19}\ \text{J}} = 1.876 \times 10^{-6}\ \text{m} = 1876\ \text{nm}$$

From Figure 7.2, this is infrared electromagnetic radiation.

b. $\Delta E = -2.178 \times 10^{-18}\text{J}\left(\frac{1}{4^2} - \frac{1}{5^2}\right) = -4.901 \times 10^{-20}\ \text{J}$

$$\lambda = \frac{hc}{|\Delta E|} = \frac{6.6261 \times 10^{-34}\ \text{J s} \times 2.9979 \times 10^{8}\ \text{m/s}}{4.901 \times 10^{-20}\ \text{J}} = 4.053 \times 10^{-6}\ \text{m}$$

$= 4053$ nm (infrared)

c. $\Delta E = -2.178 \times 10^{-18}\ \text{J}\left(\frac{1}{3^2} - \frac{1}{5^2}\right) = -1.549 \times 10^{-19}\ \text{J}$

$$\lambda = \frac{hc}{|\Delta E|} = \frac{6.6261 \times 10^{-34}\ \text{J s} \times 2.9979 \times 10^{8}\ \text{m/s}}{1.549 \times 10^{-19}\ \text{J}} = 1.282 \times 10^{-6}\ \text{m}$$

$= 1282$ nm (infrared)

59.

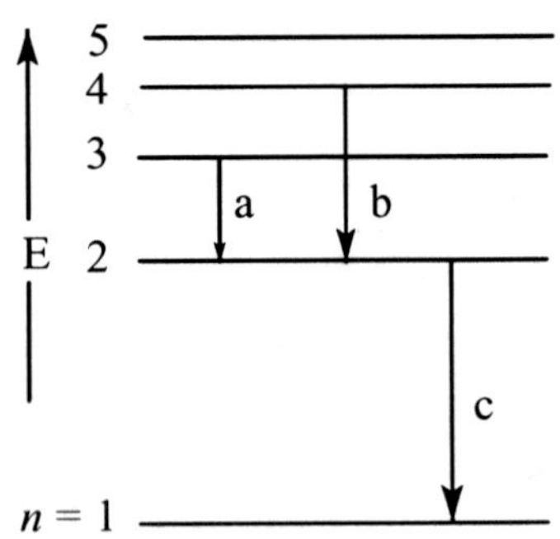

a. $3 \rightarrow 2$

b. $4 \rightarrow 2$

c. $2 \rightarrow 1$

Energy levels are not to scale.

60.

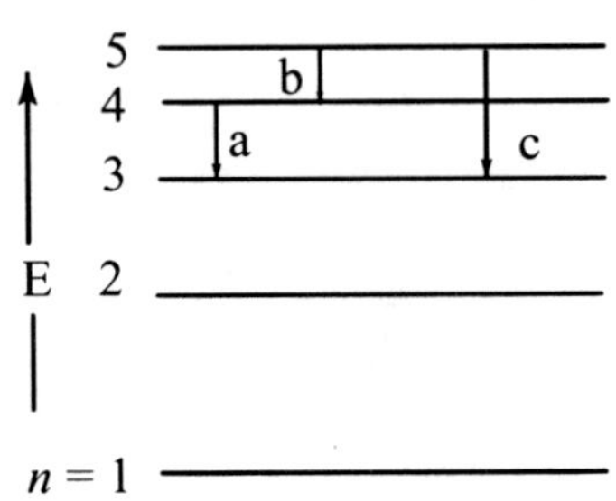

a. $4 \rightarrow 3$

b. $5 \rightarrow 4$

c. $5 \rightarrow 3$

Energy levels are not to scale.

61. The longest wavelength light emitted will correspond to the transition with the smallest energy change (smallest ΔE). This is the transition from n = 6 to n = 5.

$$\Delta E = -2.178 \times 10^{-18}\ \text{J}\left(\frac{1}{5^2} - \frac{1}{6^2}\right) = -2.662 \times 10^{-20}\ \text{J}$$

$$\lambda = \frac{hc}{|\Delta E|} = \frac{6.6261 \times 10^{-34}\ \text{J s} \times 2.9979 \times 10^{8}\ \text{m/s}}{2.662 \times 10^{-20}\ \text{J}} = 7.462 \times 10^{-6}\ \text{m} = 7462\ \text{nm}$$

The shortest wavelength emitted will correspond to the largest ΔE; this is n = 6 → n = 1.

$$\Delta E = -2.178 \times 10^{-18}\ \text{J}\left(\frac{1}{1^2} - \frac{1}{6^2}\right) = -2.118 \times 10^{-18}\ \text{J}$$

$$\lambda = \frac{hc}{|\Delta E|} = \frac{6.6261 \times 10^{-34}\ \text{J s} \times 2.9979 \times 10^{8}\ \text{m/s}}{2.118 \times 10^{-18}\ \text{J}} = 9.379 \times 10^{-8}\ \text{m} = 93.79\ \text{nm}$$

62. There are 4 possible transitions for an electron in the $n = 5$ level (5 → 4, 5 → 3, 5 → 2, and 5 → 1). If an electron initially drops to the $n = 4$ level, three additional transitions can occur (4 → 3, 4 → 2, and 4 → 1). Similarly, there are two more transitions from the $n = 3$ level (3 → 2, 3 → 1) and one more transition for the $n = 2$ level (2 → 1). There are a total of 10 possible transitions for an electron in the $n = 5$ level for a possible total of 10 different wavelength emissions.

63. $$\Delta E = -2.178 \times 10^{-18}\ \text{J}\left(\frac{1}{n_f^2} - \frac{1}{n_i^2}\right) = -2.178 \times 10^{-18}\ \text{J}\left(\frac{1}{5^2} - \frac{1}{1^2}\right) = 2.091 \times 10^{-18}\ \text{J} = E_{photon}$$

$$\lambda = \frac{hc}{E} = \frac{6.6261 \times 10^{-34}\ \text{J s} \times 2.9979 \times 10^{8}\ \text{m/s}}{2.091 \times 10^{-18}\ \text{J}} = 9.500 \times 10^{-8}\ \text{m} = 95.00\ \text{nm}$$

Because wavelength and energy are inversely related, visible light (λ ≈ 400–700 nm) is not energetic enough to excite an electron in hydrogen from $n = 1$ to $n = 5$.

$$\Delta E = -2.178 \times 10^{-18}\ \text{J}\left(\frac{1}{6^2} - \frac{1}{2^2}\right) = 4.840 \times 10^{-19}\ \text{J}$$

$$\lambda = \frac{hc}{E} = \frac{6.6261 \times 10^{-34}\ \text{J s} \times 2.9979 \times 10^{8}\ \text{m/s}}{4.840 \times 10^{-19}\ \text{J}} = 4.104 \times 10^{-7}\ \text{m} = 410.4\ \text{nm}$$

Visible light with λ = 410.4 nm will excite an electron from the $n = 2$ to the $n = 6$ energy level.

64. a. False; it takes less energy to ionize an electron from $n = 3$ than from the ground state.

b. True

c. False; the energy difference between $n = 3$ and $n = 2$ is smaller than the energy difference between $n = 3$ and $n = 1$; thus the wavelength is larger for the $n = 3 \rightarrow n = 2$ electronic transition than for the $n = 3 \rightarrow n = 1$ transition. E and λ are inversely proportional to each other (E = hc/λ).

d. True

e. False; $n = 2$ is the first excited state, and $n = 3$ is the second excited state.

65. Ionization from $n = 1$ corresponds to the transition $n_i = 1 \rightarrow n_f = \infty$, where $E_\infty = 0$.

$$\Delta E = E_\infty - E_1 = -E_1 = 2.178 \times 10^{-18}\left(\frac{1}{1^2}\right) = 2.178 \times 10^{-18}\text{ J} = E_{photon}$$

$$\lambda = \frac{hc}{E} = \frac{6.6261 \times 10^{-34}\text{ J s} \times 2.9979 \times 10^{8}\text{ m/s}}{2.178 \times 10^{-18}\text{ J}} = 9.120 \times 10^{-8}\text{ m} = 91.20\text{ nm}$$

To ionize from $n = 2$, $\Delta E = E_\infty - E_2 = -E_2 = 2.178 \times 10^{-18}\left(\frac{1}{2^2}\right) = 5.445 \times 10^{-19}\text{ J}$

$$\lambda = \frac{6.6261 \times 10^{-34}\text{ J s} \times 2.9979 \times 10^{8}\text{ m/s}}{5.445 \times 10^{-19}\text{ J}} = 3.648 \times 10^{-7}\text{ m} = 364.8\text{ nm}$$

66. $\Delta E = E_\infty - E_n = -E_n = 2.178 \times 10^{-18}\text{ J}\left(\frac{1}{n^2}\right)$

$$E_{photon} = \frac{hc}{\lambda} = \frac{6.626 \times 10^{-34}\text{ J s} \times 2.9979 \times 10^{8}\text{ m/s}}{1460 \times 10^{-9}\text{ m}} = 1.36 \times 10^{-19}\text{ J}$$

$$E_{photon} = \Delta E = 1.36 \times 10^{-19}\text{ J} = 2.178 \times 10^{-18}\left(\frac{1}{n^2}\right),\quad n^2 = 16.0,\ n = 4$$

67. $|\Delta E| = E_{photon} = h\nu = 6.662 \times 10^{-34}\text{ J s} \times 6.90 \times 10^{14}\text{ s}^{-1} = 4.57 \times 10^{-19}\text{ J}$

$\Delta E = -4.57 \times 10^{-19}$ J because we have an emission.

$$-4.57 \times 10^{-19}\text{ J} = E_n - E_5 = -2.178 \times 10^{-18}\text{ J}\left(\frac{1}{n^2} - \frac{1}{5^2}\right)$$

$$\frac{1}{n^2} - \frac{1}{25} = 0.210,\quad \frac{1}{n^2} = 0.250,\quad n^2 = 4,\ n = 2$$

The electronic transition is from $n = 5$ to $n = 2$.

68. $|\Delta E| = E_{photon} = \dfrac{hc}{\lambda} = \dfrac{6.6261 \times 10^{-34}\text{ J s} \times 2.9979 \times 10^{8}\text{ m/s}}{397.2 \times 10^{-9}\text{ m}} = 5.001 \times 10^{-19}\text{ J}$

$\Delta E = -5.001 \times 10^{-19}$ J because we have an emission.

$$-5.001 \times 10^{-19}\,\text{J} = \text{E}_2 - \text{E}_n = -2.178 \times 10^{-18}\,\text{J}\left(\frac{1}{2^2} - \frac{1}{n^2}\right)$$

$$0.2296 = \frac{1}{4} - \frac{1}{n^2},\ \frac{1}{n^2} = 0.0204,\ n = 7$$

Quantum Mechanics, Quantum Numbers, and Orbitals

69. a. $\Delta p = m\Delta v = 9.11 \times 10^{-31}\,\text{kg} \times 0.100\,\text{m/s} = \dfrac{9.11 \times 10^{-32}\,\text{kg m}}{\text{s}}$

$$\Delta p\Delta x \geq \frac{h}{4\pi},\ \Delta x = \frac{h}{4\pi\Delta p} = \frac{6.626 \times 10^{-34}\,\text{J s}}{4 \times 3.142 \times (9.11 \times 10^{-32}\,\text{kg m/s})} = 5.79 \times 10^{-4}\,\text{m}$$

b. $\Delta x = \dfrac{h}{4\pi\Delta p} = \dfrac{6.626 \times 10^{-34}\,\text{J s}}{4 \times 3.142 \times 0.145\,\text{kg} \times 0.100\,\text{m/s}} = 3.64 \times 10^{-33}\,\text{m}$

c. The diameter of an H atom is roughly $\sim 10^{-8}$ cm. The uncertainty in position is much larger than the size of the atom.

d. The uncertainty is insignificant compared to the size of a baseball.

70. Units of $\Delta E \cdot \Delta t = J \times s$, the same as the units of Planck's constant.

Units of $\Delta(mv) \cdot \Delta x = \text{kg} \times \dfrac{\text{m}}{\text{s}} \times \text{m} = \dfrac{\text{kg m}^2}{\text{s}} = \dfrac{\text{kg m}^2}{\text{s}^2} \times \text{s} = \text{J} \times \text{s}$

71. $n = 1, 2, 3, \ldots;\quad \ell = 0, 1, 2, \ldots (n-1);\quad m_\ell = -\ell \ldots -2, -1, 0, 1, 2, \ldots +\ell$

72. a. This general shape represents a p orbital ($\ell = 1$) and because there is a node in each of the lobes, this figure represents a 3p orbital ($n = 3$, $\ell = 1$)

b. This is an s orbital ($\ell = 0$). And because there is one node present, this is a 2s orbital ($n = 2$, $\ell = 0$).

c. This is the shape of a specific d oriented orbital ($\ell = 2$). This orbital is designated as a d_{z^2}. Because no additional nodes are present inside any of the boundary surfaces, this is a $3d_{z^2}$ orbital ($n = 3$, $\ell = 2$).

73. a. allowed

b. For $\ell = 3$, m_ℓ can range from -3 to $+3$; thus $+4$ is not allowed.

c. n cannot equal zero.

d. ℓ cannot be a negative number.

74. a. For $n = 3$, $\ell = 3$ is not possible.

d. m_s cannot equal -1.

e. ℓ cannot be a negative number.

f. For $\ell = 1$, m_ℓ cannot equal 2.

The quantum numbers in parts b and c are allowed.

75. ψ^2 gives the probability of finding the electron at that point.

76. The diagrams of the orbitals in the text give only 90% probabilities of where the electron may reside. We can never be 100% certain of the location of the electrons due to Heisenberg's uncertainty principle.

Polyelectronic Atoms

77. He: $1s^2$; Ne: $1s^22s^22p^6$; Ar: $1s^22s^22p^63s^23p^6$; each peak in the diagram corresponds to a subshell with different values of n. Corresponding subshells are closer to the nucleus for heavier elements because of the increased nuclear charge.

78. In polyelectronic atoms, the orbitals of a given principal quantum level are not degenerate. In polyelectronic atoms, the energy order of the n = 1, 2, and 3 orbitals are (not to scale):

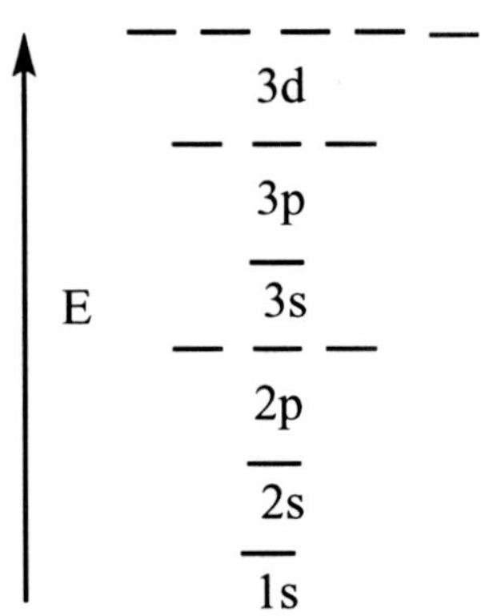

In general, the lower the n value for an orbital, the closer on average the electron can be to the nucleus, and the lower the energy. Within a specific n value orbital (like 2s vs. 2p or 3s vs. 3p vs. 3d), it is generally true that $E_{ns} < E_{np} < E_{nd} < E_{nf}$.

To rationalize this order, we utilize the radial probability distributions. In the 2s and 2p distribution, notice that the 2s orbital has a small hump of electron density very near the nucleus. This indicates that an electron in the 2s orbital can be very close to the nucleus some of the time. The 2s electron penetrates to the nucleus more than a 2p electron, and with this penetration comes a lower overall energy for the 2s orbital as compared to the 2p orbital.

In the n = 3 radial probability distribution, the 3s electron has two humps of electron density very close to the nucleus, and the 3p orbital has one hump very close to the nucleus. The 3s orbital electron is most penetrating, with the 3p orbital electron the next most penetrating, followed by the least penetrating 3d orbital electron. The more penetrating the electron, the lower the overall energy. Hence the 3s orbital is lower energy than the 3p orbitals which is lower energy than the 3d orbitals.

79. 5p: three orbitals $3d_{z^2}$: one orbital 4d: five orbitals

$n = 5$: $\ell = 0$ (1 orbital), $\ell = 1$ (3 orbitals), $\ell = 2$ (5 orbitals), $\ell = 3$ (7 orbitals), $\ell = 4$ (9 orbitals); total for $n = 5$ is 25 orbitals.

$n = 4$: $\ell = 0$ (1), $\ell = 1$ (3), $\ell = 2$ (5), $\ell = 3$ (7); total for $n = 4$ is 16 orbitals.

80. 1p, 0 electrons ($\ell \neq 1$ when $n = 1$); $6d_{x^2-y^2}$, 2 electrons (specifies one atomic orbital); 4f, 14 electrons (7 orbitals have 4f designation); $7p_y$, 2 electrons (specifies one atomic orbital); 2s, 2 electrons (specifies one atomic orbital); $n = 3$, 18 electrons (3s, 3p, and 3d orbitals are possible; there are one 3s orbital, three 3p orbitals, and five 3d orbitals).

81. a. $n = 4$: ℓ can be 0, 1, 2, or 3. Thus we have s (2 e^-), p (6 e^-), d (10 e^-), and f (14 e^-) orbitals present. Total number of electrons to fill these orbitals is 32.

b. $n = 5$, $m_\ell = +1$: For $n = 5$, $\ell = 0, 1, 2, 3, 4$. For $\ell = 1, 2, 3, 4$, all can have $m_\ell = +1$. There are four distinct orbitals having these quantum numbers, which can hold 8 electrons.

c. $n = 5$, $m_s = +1/2$: For $n = 5$, $\ell = 0, 1, 2, 3, 4$. Number of orbitals = 1, 3, 5, 7, 9 for each value of ℓ, respectively. There are 25 orbitals with $n = 5$. They can hold 50 electrons, and 25 of these electrons can have $m_s = +1/2$.

d. $n = 3$, $\ell = 2$: These quantum numbers define a set of 3d orbitals. There are 5 degenerate 3d orbitals that can hold a total of 10 electrons.

e. $n = 2$, $\ell = 1$: These define a set of 2p orbitals. There are 3 degenerate 2p orbitals that can hold a total of 6 electrons.

82. a. It is impossible to have $n = 0$. Thus no electrons can have this set of quantum numbers.

b. The four quantum numbers completely specify a single electron in a 2p orbital.

c. $n = 3$, $m_s = +1/2$: 3s, 3p, and 3d orbitals all have $n = 3$. These nine orbitals can each hold one electron with $m_s = +1/2$; 9 electrons can have these quantum numbers

d. $n = 2$, $\ell = 2$: this combination is not possible ($\ell \neq 2$ for $n = 2$). Zero electrons in an atom can have these quantum numbers.

e. $n = 1$, $\ell = 0$, $m_\ell = 0$: these define a 1s orbital that can hold 2 electrons.

83. a. Na: $1s^2 2s^2 2p^6 3s^1$; Na has 1 unpaired electron.

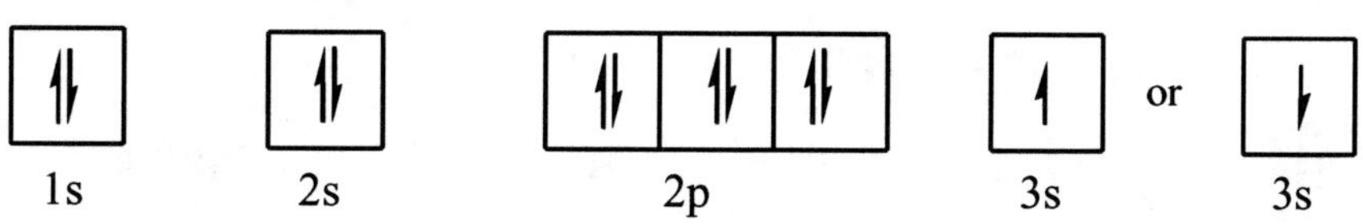

b. Co: $1s^22s^22p^63s^23p^64s^23d^7$; Co has 3 unpaired electrons.

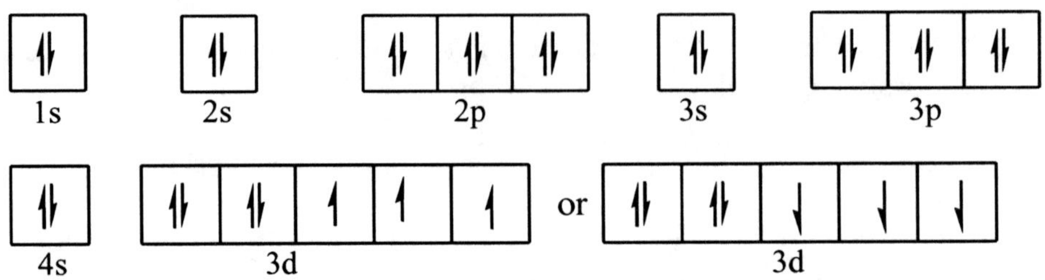

c. Kr: $1s^22s^22p^63s^23p^64s^23d^{10}4p^6$; Kr has 0 unpaired electrons.

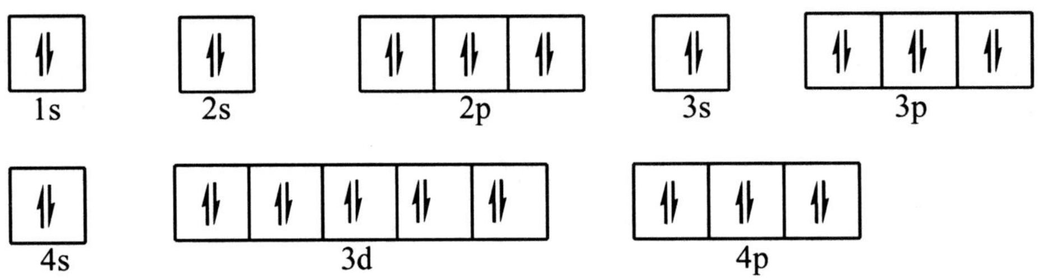

84. The two exceptions are Cr and Cu.

Cr: $1s^22s^22p^63s^23p^64s^13p^5$; Cr has 6 unpaired electrons.

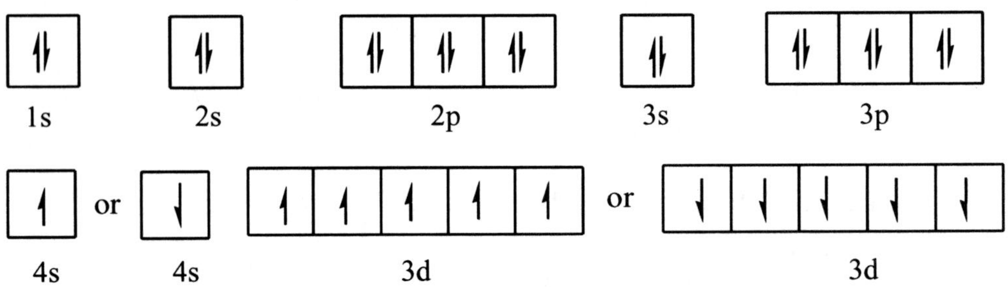

Cu: $1s^22s^22p^63s^23p^64s^13d^{10}$; Cu has 1 unpaired electron.

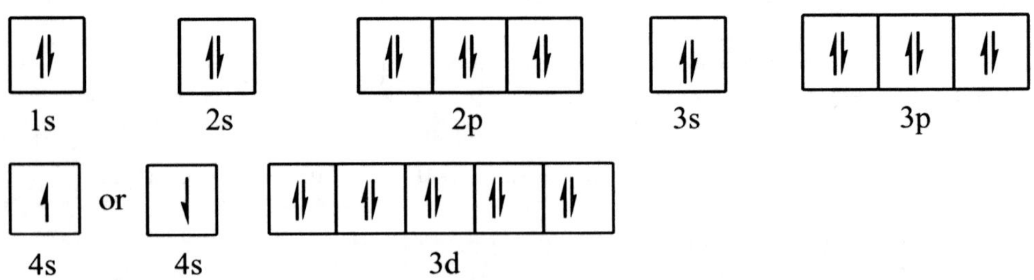

85. Si: $1s^22s^22p^63s^23p^2$ or $[Ne]3s^23p^2$; Ga: $1s^22s^22p^63s^23p^64s^23d^{10}4p^1$ or $[Ar]4s^23d^{10}4p^1$

As: $[Ar]4s^23d^{10}4p^3$; Ge: $[Ar]4s^23d^{10}4p^2$; Al: $[Ne]3s^23p^1$; Cd: $[Kr]5s^24d^{10}$

S: $[Ne]3s^23p^4$; Se: $[Ar]4s^23d^{10}4p^4$

86. Cu: $[Ar]4s^2 3d^9$ (using periodic table), $[Ar]4s^1 3d^{10}$ (actual)

O: $1s^2 2s^2 2p^4$; La: $[Xe]6s^2 5d^1$; Y: $[Kr]5s^2 4d^1$; Ba: $[Xe]6s^2$

Tl: $[Xe]6s^2 4f^{14} 5d^{10} 6p^1$; Bi: $[Xe]6s^2 4f^{14} 5d^{10} 6p^3$

87. The following are complete electron configurations. Noble gas shorthand notation could also be used.

Sc: $1s^2 2s^2 2p^6 3s^2 3p^6 4s^2 3d^1$; Fe: $1s^2 2s^2 2p^6 3s^2 3p^6 4s^2 3d^6$

P: $1s^2 2s^2 2p^6 3s^2 3p^3$; Cs: $1s^2 2s^2 2p^6 3s^2 3p^6 4s^2 3d^{10} 4p^6 5s^2 4d^{10} 5p^6 6s^1$

Eu: $1s^2 2s^2 2p^6 3s^2 3p^6 4s^2 3d^{10} 4p^6 5s^2 4d^{10} 5p^6 6s^2 4f^6 5d^1$*

Pt: $1s^2 2s^2 2p^6 3s^2 3p^6 4s^2 3d^{10} 4p^6 5s^2 4d^{10} 5p^6 6s^2 4f^{14} 5d^8$*

Xe: $1s^2 2s^2 2p^6 3s^2 3p^6 4s^2 3d^{10} 4p^6 5s^2 4d^{10} 5p^6$; Br: $1s^2 2s^2 2p^6 3s^2 3p^6 4s^2 3d^{10} 4p^5$

**Note*: These electron configurations were predicted using only the periodic table. The actual electron configurations are: Eu: $[Xe]6s^2 4f^7$ and Pt: $[Xe]6s^1 4f^{14} 5d^9$

88. Cl: $1s^2 2s^2 2p^6 3s^2 3p^5$ or $[Ne]3s^2 3p^5$ Sb: $[Kr]5s^2 4d^{10} 5p^3$

Sr: $1s^2 2s^2 2p^6 3s^2 3p^6 4s^2 3d^{10} 4p^6 5s^2$ or $[Kr]5s^2$ W: $[Xe]6s^2 4f^{14} 5d^4$

Pb: $[Xe]6s^2 4f^{14} 5d^{10} 6p^2$ Cf: $[Rn]7s^2 5f^{10}$*

**Note*: Predicting electron configurations for lanthanide and actinide elements is difficult since they have 0, 1, or 2 electrons in d orbitals. This is the actual Cf electron configuration.

89. a. Both In and I have one unpaired 5p electron, but only the nonmetal I would be expected to form a covalent compound with the nonmetal F. One would predict an ionic compound to form between the metal In and the nonmetal F.

I: $[Kr]5s^2 4d^{10} 5p^5$ $\underline{\uparrow\downarrow}\ \underline{\uparrow\downarrow}\ \underline{\uparrow}$
5p

b. From the periodic table, this will be element 120. Element 120: $[Rn]7s^2 5f^{14} 6d^{10} 7p^6 8s^2$

c. Rn: $[Xe]6s^2 4f^{14} 5d^{10} 6p^6$; note that the next discovered noble gas will also have 4f electrons (as well as 5f electrons).

d. This is chromium, which is an exception to the predicted filling order. Cr has 6 unpaired electrons, and the next most is 5 unpaired electrons for Mn.

Cr: $[Ar]4s^1 3d^5$ $\underline{\uparrow}$ (4s) $\underline{\uparrow}\ \underline{\uparrow}\ \underline{\uparrow}\ \underline{\uparrow}\ \underline{\uparrow}$ (3d)

90. a. As: $1s^22s^22p^63s^23p^64s^23d^{10}4p^3$

b. Element 116 will be below Po in the periodic table: $[Rn]7s^25f^{14}6d^{10}7p^4$

c. Ta: $[Xe]6s^24f^{14}5d^3$ or Ir: $[Xe]6s^24f^{14}5d^7$

d. At: $[Xe]6s^24f^{14}5d^{10}6p^5$; note that element 117 will also have electrons in the 6p atomic orbitals (as well as electrons in the 7p orbitals).

91. a. The complete ground state electron for this neutral atom is $1s^22s^22p^63s^23p^4$. This atom has 2 + 2 + 6 + 2 + 4 = 16 electrons. Because the atom is neutral, it also has 16 protons, making the atom sulfur, S.

b. Complete excited state electron configuration: $1s^22s^12p^4$; this neutral atom has 2 + 1 + 4 = 7 electrons, which means it has 7 protons, which identifies it as nitrogen, N.

c. Complete ground state electron configuration: $1s^22s^22p^63s^23p^64s^23d^{10}4p^5$; this 1– charged ion has 35 electrons. Because the overall charge is 1–, this ion has 34 protons which identifies it as selenium. The ion is Se^-.

92. a. This atom has 10 electrons. Ne b. S

c. The predicted ground state configuration is $[Kr]5s^24d^9$. From the periodic table, the element is Ag. *Note*: $[Kr]5s^14d^{10}$ is the actual ground state electron configuration for Ag.

d. Bi: $[Xe]6s^24f^{14}5d^{10}6p^3$; the three unpaired electrons are in the 6p orbitals.

93. Hg: $1s^22s^22p^63s^23p^64s^23d^{10}4p^65s^24d^{10}5p^66s^24f^{14}5d^{10}$

a. From the electron configuration for Hg, we have $3s^2$, $3p^6$, and $3d^{10}$ electrons; 18 total electrons with $n = 3$.

b. $3d^{10}$, $4d^{10}$, $5d^{10}$; 30 electrons are in the d atomic orbitals.

c. $2p^6$, $3p^6$, $4p^6$, $5p^6$; each set of np orbitals contain one p_z atomic orbital. Because we have 4 sets of np orbitals and two electrons can occupy the p_z orbital, there are 4(2) = 8 electrons in p_z atomic orbitals.

d. All the electrons are paired in Hg, so one-half of the electrons are spin up ($m_s = +1/2$) and the other half are spin down ($m_s = -1/2$). 40 electrons have spin up.

94. Element 115, Uup, is in Group 5A under Bi (bismuth):

Uup: $1s^22s^22p^63s^23p^64s^23d^{10}4p^65s^24d^{10}5p^66s^24f^{14}5d^{10}6p^67s^25f^{14}6d^{10}7p^3$

a. $5s^2$, $5p^6$, $5d^{10}$, and $5f^{14}$; 32 electrons have $n = 5$ as one of their quantum numbers.

b. $\ell = 3$ are f orbitals. $4f^{14}$ and $5f^{14}$ are the f orbitals used. They are all filled, so 28 electrons have $\ell = 3$.

c. p, d, and f orbitals all have one of the degenerate orbitals with $m_\ell = 1$. There are 6 orbitals with $m_\ell = 1$ for the various p orbitals used; there are 4 orbitals with $m_\ell = 1$ for the various d orbitals used; and there are 2 orbitals with $m_\ell = 1$ for the various f orbitals used. We have a total of $6 + 4 + 2 = 12$ orbitals with $m_\ell = 1$. Eleven of these orbitals are filled with 2 electrons, and the 7p orbitals are only half-filled. The number of electrons with $m_\ell = 1$ is $11 \times (2\ e^-) + 1 \times (1\ e^-) = 23$ electrons.

d. The first 112 electrons are all paired; one-half of these electrons (56 e^-) will have $m_s = -1/2$. The 3 electrons in the 7p orbitals singly occupy each of the three degenerate 7p orbitals; the three electrons are spin parallel, so the 7p electrons either have $m_s = +1/2$ or $m_s = -1/2$. Therefore, either 56 electrons have $m_s = -1/2$ or 59 electrons have $m_s = -1/2$.

95. B: $1s^2 2s^2 2p^1$

	n	ℓ	m_ℓ	m_s
1s	1	0	0	+1/2
1s	1	0	0	−1/2
2s	2	0	0	+1/2
2s	2	0	0	−1/2
2p*	2	1	−1	+1/2

*This is only one of several possibilities for the 2p electron. The 2p electron in B could have $m_\ell = -1$, 0 or +1 and $m_s = +1/2$ or $-1/2$ for a total of six possibilities.

N: $1s^2 2s^2 2p^3$

	n	ℓ	m_ℓ	m_s	
1s	1	0	0	+1/2	
1s	1	0	0	−1/2	
2s	2	0	0	+1/2	
2s	2	0	0	−1/2	
2p	2	1	−1	+1/2	
2p	2	1	0	+1/2	(Or all 2p electrons could have $m_s = -1/2$.)
2p	2	1	+1	+1/2	

96. Ti : $[Ar]4s^2 3d^2$

	n	ℓ	m_ℓ	m_s	
4s	4	0	0	+1/2	
4s	4	0	0	−1/2	
3d	3	2	−2	+1/2	Only one of 10 possible combinations of m_ℓ and m_s for the first d electron. For the ground state, the second d electron should be in a different orbital with spin parallel; 4 possibilities.
3d	3	2	−1	+1/2	

97. Group 1A: 1 valence electron; ns^1; Li: $[He]2s^1$; $2s^1$ is the valence electron configuration for Li.

Group 2A: 2 valence electrons; ns^2; Ra: $[Rn]7s^2$; $7s^2$ is the valence electron configuration for Ra.

Group 3A: 3 valence electrons; ns^2np^1; Ga: $[Ar]4s^23d^{10}4p^1$; $4s^24p^1$ is the valence electron configuration for Ga. Note that valence electrons for the representative elements of Groups 1A-8A are considered those electrons in the highest n value, which for Ga is $n = 4$. We do not include the 3d electrons as valence electrons because they are not in $n = 4$ level.

Group 4A: 4 valence electrons; ns^2np^2; Si: $[Ne]3s^23p^2$; $3s^23p^2$ is the valence electron configuration for Si.

Group 5A: 5 valence electrons; ns^2np^3; Sb: $[Kr]5s^24d^{10}5p^3$; $5s^25p^3$ is the valence electron configuration for Sb.

Group 6A: 6 valence electrons; ns^2np^4; Po: $[Xe]6s^24f^{14}5d^{10}6p^4$; $6s^26p^4$ is the valence electron configuration for Po.

Group 7A: 7 valence electrons; ns^2np^5; 117: $[Rn]7s^25f^{14}6d^{10}7p^5$; $7s^27p^5$ is the valence electron configuration for 117.

Group 8A: 8 valence electrons; ns^2np^6; Ne: $[He]2s^22p^6$; $2s^22p^6$ is the valence electron configuration for Ne.

98. a. 2 valence electrons; $4s^2$ b. 6 valence electrons; $2s^22p^4$

c. 7 valence electrons; $7s^27p^5$ d. 3 valence electrons; $5s^25p^1$

e. 8 valence electrons; $3s^23p^6$ f. 5 valence electrons; $6s^26p^3$

99. O: $1s^22s^22p_x^22p_y^2$ ($\underline{\uparrow\downarrow}$ $\underline{\uparrow\downarrow}$ $\underline{\quad}$); there are no unpaired electrons in this oxygen atom. This configuration would be an excited state, and in going to the more stable ground state ($\underline{\uparrow\downarrow}$ $\underline{\uparrow}$ $\underline{\uparrow}$), energy would be released.

100. The number of unpaired electrons is in parentheses.

a. excited state of boron (1)

B ground state: $1s^22s^22p^1$ (1)

b. ground state of neon (0)

Ne ground state: $1s^22s^22p^6$ (0)

c. exited state of fluorine (3)

F ground state: $1s^22s^22p^5$ (1)

$\underline{\uparrow\downarrow}$ $\underline{\uparrow\downarrow}$ $\underline{\uparrow}$

2p

d. excited state of iron (6)

Fe ground state: $[Ar]4s^23d^6$ (4)

$\underline{\uparrow\downarrow}$ $\underline{\uparrow}$ $\underline{\uparrow}$ $\underline{\uparrow}$ $\underline{\uparrow}$

3d

101. None of the s block elements have 2 unpaired electrons. In the p block, the elements with either ns^2np^2 or ns^2np^4 valence electron configurations have 2 unpaired electrons. For elements 1-36, these are elements C, Si, and Ge (with ns^2np^2) and elements O, S, and Se (with ns^2np^4). For the d block, the elements with configurations nd^2 or nd^8 have two unpaired electrons. For elements 1-36, these are Ti ($3d^2$) and Ni ($3d^8$). A total of 8 elements from the first 36 elements have two unpaired electrons in the ground state.

102. The s block elements with ns^1 for a valence electron configuration have one unpaired electron. These are elements H, Li, Na, and K for the first 36 elements. The p block elements with ns^2np^1 or ns^2np^5 valence electron configurations have one unpaired electron. These are elements B, Al, and Ga (ns^2np^1) and elements F, Cl, and Br (ns^2np^5) for the first 36 elements. In the d block, Sc ($[Ar]4s^23d^1$) and Cu ($[Ar]4s^13d^{10}$) each have one unpaired electron. A total of 12 elements from the first 36 elements have one unpaired electron in the ground state.

103. We get the number of unpaired electrons by examining the incompletely filled subshells. The paramagnetic substances have unpaired electrons, and the ones with no unpaired electrons are not paramagnetic (they are called diamagnetic).

Li: $1s^22s^1$ ↑ ; paramagnetic with 1 unpaired electron.
2s

N: $1s^22s^22p^3$ ↑ ↑ ↑ ; paramagnetic with 3 unpaired electrons.
2p

Ni: $[Ar]4s^23d^8$ ↑↓ ↑↓ ↑↓ ↑ ↑ ; paramagnetic with 2 unpaired electrons.
3d

Te: $[Kr]5s^24d^{10}5p^4$ ↑↓ ↑ ↑ ; paramagnetic with 2 unpaired electrons.
5p

Ba: $[Xe]6s^2$ ↑↓ ; not paramagnetic because no unpaired electrons are present.
6s

Hg: $[Xe]6s^24f^{14}5d^{10}$ ↑↓ ↑↓ ↑↓ ↑↓ ↑↓ ; not paramagnetic because no unpaired electrons.
5d

104. We get the number of unpaired electrons by examining the incompletely filled subshells.

O: $[He]2s^22p^4$	$2p^4$:	↑↓ ↑ ↑	two unpaired e^-
O^+: $[He]2s^22p^3$	$2p^3$:	↑ ↑ ↑	three unpaired e^-
O^-: $[He]2s^22p^5$	$2p^5$:	↑↓ ↑↓ ↑	one unpaired e^-
Os: $[Xe]6s^24f^{14}5d^6$	$5d^6$:	↑↓ ↑ ↑ ↑ ↑	four unpaired e^-
Zr: $[Kr]5s^24d^2$	$4d^2$:	↑ ↑ __ __ __	two unpaired e^-
S: $[Ne]3s^23p^4$	$3p^4$:	↑↓ ↑ ↑	two unpaired e^-

F: $[He]2s^22p^5$ $2p^5$: ↿⇂ ↿⇂ ↿ one unpaired e^-

Ar: $[Ne]3s^23p^6$ $3p^6$ ↿⇂ ↿⇂ ↿⇂ zero unpaired e^-

The Periodic Table and Periodic Properties

105. Size (radius) decreases left to right across the periodic table, and size increases from top to bottom of the periodic table.

 a. S < Se < Te b. Br < Ni < K c. F < Si < Ba

 All follow the general radius trend.

106. a. Be < Na < Rb b. Ne < Se < Sr c. O < P < Fe

 All follow the general radius trend.

107. The ionization energy trend is the opposite of the radius trend; ionization energy (IE), in general, increases left to right across the periodic table and decreases from top to bottom of the periodic table.

 a. Te < Se < S b. K < Ni < Br c. Ba < Si < F

 All follow the general ionization energy trend.

108. a. Rb < Na < Be b. Sr < Se < Ne c. Fe < P < O

 All follow the general ionization energy trend.

109. a. He (From the general radius trend.) b. Cl

 c. Element 116 is the next oxygen family member to be discovered (under Po), element 119 is the next alkali metal to be discovered (under Fr), and element 120 is the next alkaline earth metal to be discovered (under Ra). From the general radius trend, element 116 will be the smallest.

 d. Si

 e. Na^+; this ion has the fewest electrons as compared to the other sodium species present. Na^+ has the smallest number of electron-electron repulsions, which makes it the smallest ion with the largest ionization energy.

110. a. Ba (From the general ionization energy trend.) b. K

 c. O; in general, Group 6A elements have a lower ionization energy than neighboring Group 5A elements. This is an exception to the general ionization energy trend across the periodic table.

d. S^{2-}; this ion has the most electrons compared to the other sulfur species present. S^{2-} has the largest number of electron-electron repulsions, which leads to S^{2-} having the largest size and smallest ionization energy.

e. Cs; this follows the general ionization energy trend.

111. a. Sg: $[Rn]7s^25f^{14}6d^4$ b. W

c. Sg is in the same group as chromium and would be expected to form compounds and ions similar to that of chromium. CrO_3, Cr_2O_3, CrO_4^{2-}, $Cr_2O_7^{2-}$ are some know chromium compounds/ions, so SgO_3, Sg_2O_3, SgO_4^{2-}, and $Sg_2O_7^{2-}$ are some likely possibilities.

112. a. Uus will have 117 electrons. $[Rn]7s^25f^{14}6d^{10}7p^5$

b. It will be in the halogen family and will be most similar to astatine (At).

c. Uus should form 1– charged anions like the other halogens. Like the other halogens, some possibilities are NaUus, $Mg(Uus)_2$, $C(Uus)_4$, and $O(Uus)_2$

d. Assuming Uus is like the other halogens, some possibilities are $UusO^-$, $UusO_2^-$, $UusO_3^-$, and $UusO_4^-$.

113. As: $[Ar]4s^23d^{10}4p^3$; Se: $[Ar]4s^23d^{10}4p^4$; the general ionization energy trend predicts that Se should have a higher ionization energy than As. Se is an exception to the general ionization energy trend. There are extra electron-electron repulsions in Se because two electrons are in the same 4p orbital, resulting in a lower ionization energy for Se than predicted.

114. Expected order from the ionization energy trend: Be < B < C < N < O

B and O are exceptions to the general ionization energy trend. The ionization energy of O is lower because of the extra electron-electron repulsions present when two electrons are paired in the same orbital. This makes it slightly easier to remove an electron from O compared to N. B is an exception because of the smaller penetrating ability of the 2p electron in B compared to the 2s electrons in Be. The smaller penetrating ability makes it slightly easier to remove an electron from B compared to Be. The correct ionization energy ordering, taking into account the two exceptions, is B < Be < C < O < N.

115. a. As we remove succeeding electrons, the electron being removed is closer to the nucleus, and there are fewer electrons left repelling it. The remaining electrons are more strongly attracted to the nucleus, and it takes more energy to remove these electrons.

b. Al: $1s^22s^22p^63s^23p^1$; for I_4, we begin removing an electron with $n = 2$. For I_3, we remove an electron with $n = 3$ (the last valence electron). In going from $n = 3$ to $n = 2$, there is a big jump in ionization energy because the $n = 2$ electrons are closer to the nucleus on average than the $n = 3$ electrons. Since the $n = 2$ electrons are closer, on average, to the nucleus, they are held more tightly and require a much larger amount of energy to remove compared to the $n = 3$ electrons. In general, valence electrons are much easier to remove than inner-core electrons.

116. The general ionization energy trend says that ionization energy increases going left to right across the periodic table. However, one of the exceptions to this trend occurs between Groups 2A and 3A. Between these two groups, Group 3A elements usually have a lower ionization energy than Group 2A elements. Therefore, Al should have the lowest first ionization energy value, followed by Mg, with Si having the largest ionization energy. Looking at the values for the first ionization energy in the graph, the green plot is Al, the blue plot is Mg, and the red plot is Si.

Mg (the blue plot) is the element with the huge jump between I_2 and I_3. Mg has two valence electrons, so the third electron removed is an inner core electron. Inner core electrons are always much more difficult to remove than valence electrons since they are closer to the nucleus, on average, than the valence electrons.

117. a. More favorable electron affinity: C and Br; the electron affinity trend is very erratic. Both N and Ar have positive electron affinity values (unfavorable) due to their electron configurations (see text for detailed explanation).

b. Higher ionization energy: N and Ar (follows the ionization energy trend)

c. Larger size: C and Br (follows the radius trend)

118. a. More favorable electron affinity: K and Cl; Mg has a positive electron affinity value, and F has a more positive electron affinity value than expected from its position relative to Cl.

b. Higher ionization energy: Mg and F   c. Larger radius: K and Cl

119. Al(−44), Si(−120), P(−74), S(−200.4), Cl(−348.7); based on the increasing nuclear charge, we would expect the electron affinity values to become more exothermic as we go from left to right in the period. Phosphorus is out of line. The reaction for the electron affinity of P is:

$$\underset{[Ne]3s^2 3p^3}{P(g)} + e^- \rightarrow \underset{[Ne]3s^2 3p^4}{P^-(g)}$$

The additional electron in P^- will have to go into an orbital that already has one electron. There will be greater repulsions between the paired electrons in P^-, causing the electron affinity of P to be less favorable than predicted based solely on attractions to the nucleus.

120. Electron affinity refers to the energy associated with the process of adding an electron to a gaseous substance. Be, N, and Ne all have endothermic (unfavorable) electron affinity values. In order to add an electron to Be, N, or Ne, energy must be added. Another way of saying this is that Be, N, and Ne become less stable (have a higher energy) when an electron is added to each. To rationalize why those three atoms have endothermic (unfavorable) electron affinity values, let's see what happens to the electron configuration as an electron is added.

$$\underset{[He]2s^2}{Be(g)} + e^- \rightarrow \underset{[He]2s^2 2p^1}{Be^-(g)} \qquad \underset{[He]2s^2 2p^3}{N(g)} + e^- \rightarrow \underset{[He]2s^2 2p^4}{N^-(g)}$$

$$\underset{[He]2s^2 2p^6}{Ne(g)} + e^- \rightarrow \underset{[He]2s^2 2p^6 3s^1}{Ne^-(g)}$$

In each case something energetically unfavorable occurs when an electron is added. For Be, the added electron must go into a higher-energy 2p atomic orbital because the 2s orbital is full. In N, the added electron must pair up with another electron in one of the 2p atomic orbitals; this adds electron-electron repulsions. In Ne, the added electron must be added to a much higher 3s atomic orbital because the n = 2 orbitals are full.

121. The electron affinity trend is very erratic. In general, electron affinity decreases down the periodic table, and the trend across the table is too erratic to be of much use.

a. Se < S; S is most exothermic.

b. I < Br < F < Cl; Cl is most exothermic. (F is an exception).

122. a. N < O < F, F is most exothermic.

b. Al < P < Si; Si is most exothermic.

123. Electron-electron repulsions are much greater in O^- than in S^- because the electron goes into a smaller 2p orbital versus the larger 3p orbital in sulfur. This results in a more favorable (more exothermic) electron affinity for sulfur.

124. O; the electron-electron repulsions will be much more severe for $O^- + e^- \rightarrow O^{2-}$ than for $O + e^- \rightarrow O^-$, resulting in O having the more exothermic (favorable) electron affinity.

125. a. $Se^{3+}(g) \rightarrow Se^{4+}(g) + e^-$

b. $S^-(g) + e^- \rightarrow S^{2-}(g)$

c. $Fe^{3+}(g) + e^- \rightarrow Fe^{2+}(g)$

d. $Mg(g) \rightarrow Mg^+(g) + e^-$

126. a. The electron affinity (EA) of Mg^{2+} is ΔH for $Mg^{2+}(g) + e^- \rightarrow Mg^+(g)$; this is just the reverse of the second ionization energy (I_2) for Mg. $EA(Mg^{2+}) = -I_2(Mg) = -1445$ kJ/mol (Table 7.5). Note that when an equation is reversed, the sign on the equation is also reversed.

b. I_1 of Cl^- is ΔH for $Cl^-(g) \rightarrow Cl(g) + e^-$; $I_1(Cl^-) = -EA(Cl) = 348.7$ kJ/mol (Table 7.7)

c. $Cl^+(g) + e^- \rightarrow Cl(g)$ $\quad \Delta H = -I_1(Cl) = -1255$ kJ/mol $= EA(Cl^+)$ (Table 7.5)

d. $Mg^-(g) \rightarrow Mg(g) + e^-$ $\quad \Delta H = -EA(Mg) = -230$ kJ/mol $= I_1(Mg^-)$

Alkali Metals

127. It should be potassium peroxide (K_2O_2) because K^+ ions are stable in ionic compounds. K^{2+} ions are not stable; the second ionization energy of K is very large compared to the first.

128. a. Li_3N; lithium nitride b. NaBr; sodium bromide c. K_2S; potassium sulfide

129. $$\nu = \frac{c}{\lambda} = \frac{2.9979 \times 10^8 \text{ m/s}}{455.5 \times 10^{-9} \text{ m}} = 6.582 \times 10^{14} \text{ s}^{-1}$$

$$E = h\nu = 6.6261 \times 10^{-34} \text{ J s} \times 6.582 \times 10^{14} \text{ s}^{-1} = 4.361 \times 10^{-19} \text{ J}$$

130. For 589.0 nm: $\nu = \frac{c}{\lambda} = \frac{2.9979 \times 10^8 \text{ m/s}}{589.0 \times 10^{-9} \text{ m}} = 5.090 \times 10^{14} \text{ s}^{-1}$

$$E = h\nu = 6.6261 \times 10^{-34} \text{ J s} \times 5.090 \times 10^{14} \text{ s}^{-1} = 3.373 \times 10^{-19} \text{ J}$$

For 589.6 nm: $\nu = c/\lambda = 5.085 \times 10^{14} \text{ s}^{-1}$; $E = h\nu = 3.369 \times 10^{-19}$ J

The energies in kJ/mol are:

$$3.373 \times 10^{-19} \text{ J} \times \frac{1 \text{ kJ}}{1000 \text{ J}} \times \frac{6.0221 \times 10^{23}}{\text{mol}} = 203.1 \text{ kJ/mol}$$

$$3.369 \times 10^{-19} \text{ J} \times \frac{1 \text{ kJ}}{1000 \text{ J}} \times \frac{6.0221 \times 10^{23}}{\text{mol}} = 202.9 \text{ kJ/mol}$$

131. Yes; the ionization energy general trend is to decrease down a group, and the atomic radius trend is to increase down a group. The data in Table 7.8 confirm both of these general trends.

132. It should be element 119 with the ground state electron configuration $[Rn]7s^25f^{14}6d^{10}7p^68s^1$.

133. a. $6 \text{ Li(s)} + N_2\text{(g)} \rightarrow 2 \text{ Li}_3\text{N(s)}$ b. $2 \text{ Rb(s)} + \text{S(s)} \rightarrow \text{Rb}_2\text{S(s)}$

134. a. $2 \text{ Cs(s)} + 2 \text{ H}_2\text{O(l)} \rightarrow 2 \text{ CsOH(aq)} + \text{H}_2\text{(g)}$ b. $2 \text{ Na(s)} + \text{Cl}_2\text{(g)} \rightarrow 2 \text{ NaCl(s)}$

Additional Exercises

135. No; lithium metal is very reactive. It will react somewhat violently with water, making it completely unsuitable for human consumption. Lithium has a small first ionization energy, so it is more likely that the lithium prescribed will be in the form of a soluble lithium salt (a soluble ionic compound with Li^+ as the cation).

136. a. $\lambda = \frac{c}{\nu} = \frac{3.00 \times 10^8 \text{ m/s}}{6.0 \times 10^{13} \text{ s}^{-1}} = 5.0 \times 10^{-6} \text{ m}$

b. From Figure 7.2, this is infrared electromagnetic radiation.

c. $E = h\nu = 6.63 \times 10^{-34} \text{ J s} \times 6.0 \times 10^{13} \text{ s}^{-1} = 4.0 \times 10^{-20}$ J/photon

$$\frac{4.0 \times 10^{-20} \text{ J}}{\text{photon}} \times \frac{6.022 \times 10^{23} \text{ photons}}{\text{mol}} = 2.4 \times 10^4 \text{ J/mol}$$

d. Frequency and photon energy are directly related ($E = h\nu$). Because $5.4 \times 10^{13} \text{ s}^{-1}$ electromagnetic radiation (EMR) has a lower frequency than $6.0 \times 10^{13} \text{ s}^{-1}$ EMR, the $5.4 \times 10^{13} \text{ s}^{-1}$ EMR will have less energetic photons.

137. $E = \frac{310\text{ kJ}}{\text{mol}} \times \frac{1\text{ mol}}{6.022 \times 10^{23}} = 5.15 \times 10^{-22}\text{ kJ} = 5.15 \times 10^{-19}\text{ J}$

$$E = \frac{hc}{\lambda},\ \lambda = \frac{hc}{E} = \frac{6.626 \times 10^{-34}\text{ J s} \times 2.998 \times 10^{8}\text{ m/s}}{5.15 \times 10^{-19}\text{ J}} = 3.86 \times 10^{-7}\text{ m} = 386\text{ nm}$$

138. Energy to make water boil = s × m × ΔT = $\frac{4.18\text{ J}}{°\text{C g}} \times 50.0\text{ g} \times 75.0°\text{C} = 1.57 \times 10^4\text{ J}$

$$E_{photon} = \frac{hc}{\lambda} = \frac{6.626 \times 10^{-34}\text{ J s} \times 2.998 \times 10^{8}\text{ m/s}}{9.75 \times 10^{-2}\text{ m}} = 2.04 \times 10^{-24}\text{ J}$$

$$1.57 \times 10^4\text{ J} \times \frac{1\text{ s}}{750.\text{ J}} = 20.9\text{ s};\ \ 1.57 \times 10^4\text{ J} \times \frac{1\text{ photon}}{2.04 \times 10^{-24}\text{ J}} = 7.70 \times 10^{27}\text{ photons}$$

139. $60 \times 10^6\text{ km} \times \frac{1000\text{ m}}{\text{km}} \times \frac{1\text{ s}}{3.00 \times 10^8\text{ m}} = 200\text{ s}$ (about 3 minutes)

140. $\lambda = \frac{hc}{E} = \frac{6.626 \times 10^{-34}\text{ J s} \times 2.998 \times 10^{8}\text{ m/s}}{3.59 \times 10^{-19}\text{ J}} = 5.53 \times 10^{-7}\text{ m} \times \frac{100\text{ cm}}{\text{m}}$

$$= 5.53 \times 10^{-5}\text{ cm}$$

From the spectrum, $\lambda = 5.53 \times 10^{-5}$ cm is greenish-yellow light.

141. $\Delta E = -R_H\left(\frac{1}{n_f^2} - \frac{1}{n_i^2}\right) = -2.178 \times 10^{-18}\text{ J}\left(\frac{1}{2^2} - \frac{1}{6^2}\right) = -4.840 \times 10^{-19}\text{ J}$

$$\lambda = \frac{hc}{|\Delta E|} = \frac{6.6261 \times 10^{-34}\text{ J s} \times 2.9979 \times 10^{8}\text{ m/s}}{4.840 \times 10^{-19}\text{ J}} = 4.104 \times 10^{-7}\text{ m} \times \frac{100\text{ cm}}{\text{m}}$$

$$= 4.104 \times 10^{-5}\text{ cm}$$

From the spectrum, $\lambda = 4.104 \times 10^{-5}$ cm is violet light, so the n = 6 to n = 2 visible spectrum line is violet.

142. Exceptions: Cr, Cu, Nb, Mo, Tc, Ru, Rh, Pd, Ag, Pt, and Au; Tc, Ru, Rh, Pd, and Pt do not correspond to the supposed extra stability of half-filled and filled subshells as normally applied.

143. a. True for H only. b. True for all atoms. c. True for all atoms.

144. $n = 5$; $m_\ell = -4, -3, -2, -1, 0, 1, 2, 3, 4$; 18 electrons

145. 1p: $n = 1$, $\ell = 1$ is not possible; 3f: $n = 3$, $\ell = 3$ is not possible; 2d: $n = 2$, $\ell = 2$ is not possible; in all three incorrect cases, $n = \ell$. The maximum value ℓ can have is $n - 1$, not n.

146. O: $1s^22s^22p^4$; C: $1s^22s^22p^2$; H: $1s^1$; N: $1s^22s^22p^3$; Ca: $[Ar]4s^2$; P: $[Ne]3s^23p^3$

Mg: $[Ne]3s^2$; K: $[Ar]4s^1$

147. From the radii trend, the smallest-size element (excluding hydrogen) would be the one in the most upper right corner of the periodic table. This would be O. The largest-size element would be the one in the most lower left of the periodic table. Thus K would be the largest. The ionization energy trend is the exact opposite of the radii trend. So K, with the largest size, would have the smallest ionization energy. From the general ionization energy trend, O should have the largest ionization energy. However, there is an exception to the general ionization energy trend between N and O. Due to this exception, N would have the largest ionization energy of the elements examined.

148. a. The 4+ ion contains 20 electrons. Thus the electrically neutral atom will contain 24 electrons. The atomic number is 24, which identifies it as chromium.

b. The ground state electron configuration of the ion must be $1s^22s^22p^63s^23p^64s^03d^2$; there are 6 electrons in s orbitals.

c. 12 d. 2

e. From the mass, this is the isotope $^{50}_{24}Cr$. There are 26 neutrons in the nucleus.

f. $1s^22s^22p^63s^23p^64s^13d^5$ is the ground state electron configuration for Cr. Cr is an exception to the normal filling order.

149. Valence electrons are easier to remove than inner-core electrons. The large difference in energy between I_2 and I_3 indicates that this element has two valence electrons. This element is most likely an alkaline earth metal since alkaline earth metal elements all have two valence electrons.

150. All oxygen family elements have ns^2np^4 valence electron configurations, so this nonmetal is from the oxygen family.

a. 2 + 4 = 6 valence electrons.

b. O, S, Se, and Te are the nonmetals from the oxygen family (Po is a metal).

c. Because oxygen family nonmetals form 2– charged ions in ionic compounds, K_2X would be the predicted formula, where X is the unknown nonmetal.

d. From the size trend, this element would have a smaller radius than barium.

e. From the ionization energy trend, this element would have a smaller ionization energy than fluorine.

151. a.

$$\begin{array}{ll} Na(g) \rightarrow Na^+(g) + e^- & I_1 = 495 \text{ kJ} \\ Cl(g) + e^- \rightarrow Cl^-(g) & EA = -348.7 \text{ kJ} \\ \hline Na(g) + Cl(g) \rightarrow Na^+(g) + Cl^-g) & \Delta H = 146 \text{ kJ} \end{array}$$

b.

$Mg(g) \rightarrow Mg^{+}(g) + e^{-}$	$I_1 = 735$ kJ
$F(g) + e^{-} \rightarrow F^{-}(g)$	EA = −327.8 kJ
$Mg(g) + F(g) \rightarrow Mg^{+}(g) + F^{-}(g)$	ΔH = 407 kJ

c.

$Mg^{+}(g) \rightarrow Mg^{2+}(g) + e^{-}$	$I_2 = 1445$ kJ
$F(g) + e^{-} \rightarrow F^{-}(g)$	EA = −327.8 kJ
$Mg^{+}(g) + F(g) \rightarrow Mg^{2+}(g) + F^{-}(g)$	ΔH = 1117 kJ

d. Using parts b and c, we get:

$Mg(g) + F(g) \rightarrow Mg^{+}(g) + F^{-}(g)$	ΔH = 407 kJ
$Mg^{+}(g) + F(g) \rightarrow Mg^{2+}(g) + F^{-}(g)$	ΔH = 1117 kJ
$Mg(g) + 2\ F(g) \rightarrow Mg^{2+}(g) + 2\ F^{-}(g)$	ΔH = 1524 kJ

152. Sc: $[Ar]4s^23d^1$, 1 unpaired electron; Ti: $[Ar]4s^23d^2$; 2 unpaired e^-; V: $[Ar]4s^23d^3$, 3 unpaired e^-; Cr: $[Ar]4s^13d^5$, 6 unpaired electrons (Cr is an exception to the normal filling order); Mn: $[Ar]4s^23d^5$, 5 unpaired e^-; Fe: $[Ar]4s^23d^6$, 4 unpaired e^-; Co: $[Ar]4s^23d^7$, 3 unpaired e^-; Ni: $[Ar]4s^23d^8$, 2 unpaired e^-; Cu: $[Ar]\ 4s^13d^{10}$, 1 unpaired e^- (Cu is also an exception to the normal filling order); Zn: $[Ar]4s^23d^{10}$, 0 unpaired e^-.

ChemWork Problems

The answers to the problems 153-162 (or a variation to these problems) are found in OWL. These problems are also assignable in OWL.

Challenge Problems

163. $\lambda = \dfrac{h}{mv}$, where m = mass and v = velocity; $v_{rms} = \sqrt{\dfrac{3RT}{m}}$, $\lambda = \dfrac{h}{m\sqrt{\dfrac{3RT}{m}}} = \dfrac{h}{\sqrt{3RTm}}$

$$\text{For one atom, } R = \frac{8.3145\text{J}}{\text{K mol}} \times \frac{1\text{ mol}}{6.022 \times 10^{23}\text{ atoms}} = 1.381 \times 10^{-23}\text{ J/K•atom}$$

$$2.31 \times 10^{-11}\text{ m} = \frac{6.626 \times 10^{-34}\text{ J s}}{\sqrt{m}\sqrt{3(1.381 \times 10^{-23})(373\text{K})}},\quad m = 5.32 \times 10^{-26}\text{ kg} = 5.32 \times 10^{-23}\text{ g}$$

$$\text{Molar mass} = \frac{5.32 \times 10^{-23}\text{ g}}{\text{atom}} \times \frac{6.022 \times 10^{23}\text{ atoms}}{\text{mol}} = 32.0\text{ g/mol}$$

The atom is sulfur (S).

164. $E_{photon} = \dfrac{hc}{\lambda} = \dfrac{6.6261 \times 10^{-34}\ \text{J s} \times 2.9979 \times 10^{8}\ \text{m/s}}{253.4 \times 10^{-9}\ \text{m}} = 7.839 \times 10^{-19}\ \text{J}$

$\Delta E = 7.839 \times 10^{-19}$ J; the general energy equation for one-electron ions is $E_n = -2.178 \times 10^{-18}\ \text{J}\ (Z^2)/n^2$, where Z = atomic number.

$$\Delta E = -2.178 \times 10^{-18}\ \text{J}\ (Z)^2 \left(\frac{1}{n_f^2} - \frac{1}{n_i^2}\right),\quad Z = 4 \text{ for } Be^{3+}$$

$$\Delta E = -7.839 \times 10^{-19}\ \text{J} = -2.178 \times 10^{-18}\,(4)^2 \left(\frac{1}{n_f^2} - \frac{1}{5^2}\right)$$

$$\frac{7.839 \times 10^{-19}}{2.178 \times 10^{-18} \times 16} + \frac{1}{25} = \frac{1}{n_f^2},\quad \frac{1}{n_f^2} = 0.06249,\quad n_f = 4$$

This emission line corresponds to the $n = 5 \rightarrow n = 4$ electronic transition.

165. a. Because wavelength is inversely proportional to energy, the spectral line to the right of B (at a larger wavelength) represents the lowest possible energy transition; this is $n = 4$ to $n = 3$. The B line represents the next lowest energy transition, which is $n = 5$ to $n = 3$, and the A line corresponds to the $n = 6$ to $n = 3$ electronic transition.

b. Because this spectrum is for a one-electron ion, $E_n = -2.178 \times 10^{-18}\ \text{J}\ (Z^2/n^2)$. To determine ΔE and, in turn, the wavelength of spectral line A, we must determine Z, the atomic number of the one electron species. Use spectral line B data to determine Z.

$$\Delta E_{5 \rightarrow 3} = -2.178 \times 10^{-18}\ \text{J} \left(\frac{Z^2}{3^2} - \frac{Z^2}{5^2}\right) = -2.178 \times 10^{-18} \left(\frac{16Z^2}{9 \times 25}\right)$$

$$E = \frac{hc}{\lambda} = \frac{6.6261 \times 10^{-34}\ \text{J s}(2.9979 \times 10^{8}\ \text{m/s})}{142.5 \times 10^{-9}\ \text{m}} = 1.394 \times 10^{-18}\ \text{J}$$

Because an emission occurs, $\Delta E_{5 \rightarrow 3} = -1.394 \times 10^{-18}$ J.

$$\Delta E = -1.394 \times 10^{-18}\ \text{J} = -2.178 \times 10^{-18}\ \text{J} \left(\frac{16Z^2}{9 \times 25}\right),\quad Z^2 = 9.001,\ Z = 3;\ \text{the ion is } Li^{2+}.$$

Solving for the wavelength of line A:

$$\Delta E_{6 \rightarrow 3} = -2.178 \times 10^{-18}(3)^2 \left(\frac{1}{3^2} - \frac{1}{6^2}\right) = -1.634 \times 10^{-18}\ \text{J}$$

$$\lambda = \frac{hc}{|\Delta E|} = \frac{6.6261 \times 10^{-34}\ \text{J s}(2.9979 \times 10^{8}\ \text{m/s})}{1.634 \times 10^{-18}\ \text{J}} = 1.216 \times 10^{-7}\ \text{m} = 121.6\ \text{nm}$$

166. For hydrogen: $\Delta E = -2.178 \times 10^{-18}\ J \left(\frac{1}{2^2} - \frac{1}{5^2}\right) = -4.574 \times 10^{-19}\ J$

For a similar blue light emission, He^+ will need about the same ΔE value.

For He^+: $E_n = -2.178 \times 10^{-18} J\ (Z^2/n^2)$, where $Z = 2$:

$$\Delta E = -4.574 \times 10^{-19}\ J = -2.178 \times 10^{-18}\ J \left(\frac{2^2}{n_f^2} - \frac{2^2}{4^2}\right)$$

$$0.2100 = \frac{4}{n_f^2} - \frac{4}{16},\ 0.4600 = \frac{4}{n_f^2},\quad n_f = 2.949$$

The transition from $n = 4$ to $n = 3$ for He^+ should emit a similar colored blue light as the $n = 5$ to $n = 2$ hydrogen transition; both these transitions correspond to very nearly the same energy change.

167. For one-electron species, $E_n = -R_H Z^2/n^2$. The ground state ionization energy is the energy change for the $n = 1 \rightarrow n = \infty$ transition. So:

$$\text{ionization energy} = E_\infty - E_1 = -E_1 = R_H Z^2/n^2 = R_H Z^2$$

$$\frac{4.72 \times 10^4\ kJ}{mol} \times \frac{1\,mol}{6.022 \times 10^{23}} \times \frac{1000 J}{kJ} = 2.178 \times 10^{-18}\ J\ (Z^2);\ \text{solving: } Z = 6$$

Element 6 is carbon (X = carbon), and the charge for a one-electron carbon ion is 5+ ($m = 5$). The one-electron ion is C^{5+}.

168. A node occurs when $\psi = 0$. $\psi_{300} = 0$ when $27 - 18\sigma + 2\sigma^2 = 0$.

Solving using the quadratic formula: $\sigma = \frac{18 \pm \sqrt{(18)^2 - 4(2)(27)}}{4} = \frac{18 \pm \sqrt{108}}{4}$

$\sigma = 7.10$ or $\sigma = 1.90$; because $\sigma = r/a_o$, the nodes occur at $r = (7.10)a_o = 3.76 \times 10^{-10}$ m and at $r = (1.90)a_o = 1.01 \times 10^{-10}$ m, where r is the distance from the nucleus.

169. For $r = a_o$ and $\theta = 0°$ (Z = 1 for H):

$$\psi_{2p_z} = \frac{1}{4(2\pi)^{1/2}}\left(\frac{1}{5.29 \times 10^{-11}}\right)^{3/2} (1)\ e^{-1/2} \cos 0 = 1.57 \times 10^{14};\ \psi^2 = 2.46 \times 10^{28}$$

For $r = a_o$ and $\theta = 90°$, $\psi_{2p_z} = 0$ since $\cos 90° = 0$; $\psi^2 = 0$; there is no probability of finding an electron in the $2p_z$ orbital with $\theta = 0°$. As expected, the xy plane, which corresponds to $\theta = 0°$, is a node for the $2p_z$ atomic orbital.

170. a. Each orbital could hold 3 electrons.

b. The first period corresponds to $n = 1$ which can only have 1s orbitals. The 1s orbital could hold 3 electrons; hence the first period would have three elements. The second period corresponds to $n = 2$, which has 2s and 2p orbitals. These four orbitals can each hold three electrons. A total of 12 elements would be in the second period.

c. 15 d. 21

171. a. 1st period: $p = 1,\ q = 1,\ r = 0,\ s = \pm 1/2$ (2 elements)

2nd period: $p = 2,\ q = 1,\ r = 0,\ s = \pm 1/2$ (2 elements)

3rd period: $p = 3,\ q = 1,\ r = 0,\ s = \pm 1/2$ (2 elements)

$p = 3,\ q = 3,\ r = -2,\ s = \pm 1/2$ (2 elements)

$p = 3,\ q = 3,\ r = 0,\ s = \pm 1/2$ (2 elements)

$p = 3,\ q = 3,\ r = +2,\ s = \pm 1/2$ (2 elements)

4th period: $p = 4$; q and r values are the same as with $p = 3$ (8 total elements)

1							2
3							4
5	6	7	8	9	10	11	12
13	14	15	16	17	18	19	20

b. Elements 2, 4, 12, and 20 all have filled shells and will be least reactive.

c. Draw similarities to the modern periodic table.

XY could be X^+Y^-, $X^{2+}Y^{2-}$, or $X^{3+}Y^{3-}$. Possible ions for each are:

X^+ could be elements 1, 3, 5, or 13; Y^- could be 11 or 19.

X^{2+} could be 6 or 14; Y^{2-} could be 10 or 18.

X^{3+} could be 7 or 15; Y^{3-} could be 9 or 17.

Note: X^{4+} and Y^{4-} ions probably won't form.

XY_2 will be $X^{2+}(Y^-)_2$; See above for possible ions.

X_2Y will be $(X^+)_2Y^{2-}$ See above for possible ions.

XY_3 will be $X^{3+}(Y^-)_3$; See above for possible ions.

X_2Y_3 will be $(X^{3+})_2(Y^{2-})_3$; See above for possible ions.

d. $p = 4,\ q = 3,\ r = -2,\ s = \pm 1/2$ (2 electrons)

$p = 4,\ q = 3,\ r = 0,\ s = \pm 1/2$ (2 electrons)

$p = 4,\ q = 3,\ r = +2,\ s = \pm 1/2$ (2 electrons)

A total of 6 electrons can have $p = 4$ and $q = 3$.

e. $p = 3,\ q = 0,\ r = 0$; this is not allowed; q must be odd. Zero electrons can have these quantum numbers.

f. $p = 6,\ q = 1,\ r = 0,\ s = \pm 1/2$ (2 electrons)

$p = 6,\ q = 3,\ r = -2, 0, +2;\ s = \pm 1/2$ (6 electrons)

$p = 6,\ q = 5,\ r = -4, -2, 0, +2, +4;\ s = \pm 1/2$ (10 electrons)

Eighteen electrons can have $p = 6$.

172. The third ionization energy refers to the following process: $E^{2+}(g) \rightarrow E^{3+}(g) + e^-$ $\Delta H = I_3$. The electron configurations for the 2+ charged ions of Na to Ar are:

Na^{2+}: $1s^22s^22p^5$
Mg^{2+}: $1s^22s^22p^6$
Al^{2+}: $[Ne]3s^1$
Si^{2+}: $[Ne]3s^2$
P^{2+}: $[Ne]3s^23p^1$
S^{2+}: $[Ne]3s^23p^2$
Cl^{2+}: $[Ne]3s^23p^3$
Ar^{2+}: $[Ne]3s^23p^4$

I_3 for sodium and magnesium should be extremely large compared with the others because $n = 2$ electrons are much more difficult to remove than $n = 3$ electrons. Between Na^{2+} and Mg^{2+}, one would expect to have the same trend as seen with I_1(F) versus I_1(Ne); these neutral atoms have identical electron configurations to Na^{2+} and Mg^{2+}. Therefore, the $1s^22s^22p^5$ ion (Na^{2+}) should have a lower ionization energy than the $1s^22s^22p^6$ ion (Mg^{2+}).

The remaining 2+ ions (Al^{2+} to Ar^{2+}) should follow the same trend as the neutral atoms having the same electron configurations. The general ionization energy trend predicts an increase from $[Ne]3s^1$ to $[Ne]3s^23p^4$. The exceptions occur between $[Ne]3s^2$ and $[Ne]3s^23p^1$ and between $[Ne]3s^23p^3$ and $[Ne]3s^23p^4$. $[Ne]3s^23p^1$ is out of order because of the small penetrating ability of the 3p electron as compared with the 3s electrons. $[Ne]3s^23p^4$ is out of order because of the extra electron-electron repulsions present when two electrons are paired in the same orbital. Therefore, the correct ordering for Al^{2+} to Ar^{2+} should be $Al^{2+} < P^{2+} < Si^{2+} < S^{2+} < Ar^{2+} < Cl^{2+}$, where P^{2+} and Ar^{2+} are out of line for the same reasons that Al and S are out of line in the general ionization energy trend for neutral atoms. A qualitative plot of the third ionization energies for elements Na through Ar follows.

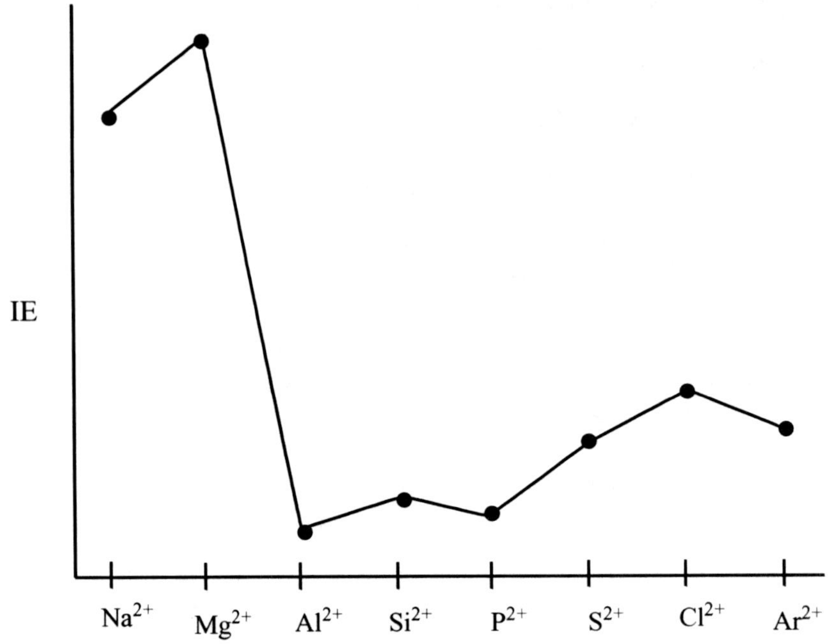

Note: The actual numbers in Table 7.5 support most of this plot. No I_3 is given for Na^{2+}, so you cannot check this. The only deviation from our discussion is I_3 for Ar^{2+} which is greater than I_3 for Cl^{2+} instead of less than.

173. The ratios for Mg, Si, P, Cl, and Ar are about the same. However, the ratios for Na, Al, and S are higher. For Na, the second ionization energy is extremely high because the electron is taken from $n = 2$ (the first electron is taken from $n = 3$). For Al, the first electron requires a bit less energy than expected by the trend due to the fact it is a 3p electron versus a 3s electron. For S, the first electron requires a bit less energy than expected by the trend due to electrons being paired in one of the p orbitals.

174. Size also decreases going across a period. Sc and Ti along with Y and Zr are adjacent elements. There are 14 elements (the lanthanides) between La and Hf, making Hf considerably smaller.

175. a. Assuming the Bohr model applies to the 1s electron, $E_{1s} = -R_H Z^2/n^2 = -R_H Z^2_{eff}$, where $n = 1$. Ionization energy $= E_\infty - E_{1s} = 0 - E_{1s} = R_H Z^2_{eff}$.

$$\frac{2.462 \times 10^6 \text{ kJ}}{\text{mol}} \times \frac{1 \text{ mol}}{6.0221 \times 10^{23}} \times \frac{1000 \text{J}}{\text{kJ}} = 2.178 \times 10^{-18} \text{ J } (Z_{eff})^2, \ Z_{eff} = 43.33$$

b. Silver is element 47, so Z = 47 for silver. Our calculated Z_{eff} value is less than 47. Electrons in other orbitals can penetrate the 1s orbital. Thus a 1s electron can be slightly shielded from the nucleus by these penetrating electrons, giving a Z_{eff} close to but less than Z.

176. None of the noble gases and no subatomic particles had been discovered when Mendeleev published his periodic table. Thus there was no known element out of place in terms of reactivity and there was no reason to predict an entire family of elements. Mendeleev ordered his table by mass; he had no way of knowing there were gaps in atomic numbers (they hadn't been discovered yet).

177. $m = \dfrac{h}{\lambda v} = \dfrac{6.626 \times 10^{-34}\ \text{kg m}^2/\text{s}}{3.31 \times 10^{-15}\ \text{m} \times (0.0100 \times 2.998 \times 10^8\ \text{m/s})} = 6.68 \times 10^{-26}\ \text{kg/atom}$

$$\frac{6.68 \times 10^{-26}\ \text{kg}}{\text{atom}} \times \frac{6.022 \times 10^{23}\ \text{atoms}}{\text{mol}} \times \frac{1000\text{g}}{1\ \text{kg}} = 40.2\ \text{g/mol}$$

The element is calcium, Ca.

Integrated Problems

178. a. $\nu = \dfrac{E}{h} = \dfrac{7.52 \times 10^{-19}\ \text{J}}{6.626 \times 10^{-34}\ \text{J s}} = 1.13 \times 10^{15}\ \text{s}^{-1}$

$$\lambda = \frac{c}{\nu} = \frac{2.998 \times 10^8\ \text{m/s}}{1.13 \times 10^{15}\text{s}^{-1}} = 2.65 \times 10^{-7}\ \text{m} = 265\ \text{nm}$$

b. E_{photon} and λ are inversely related ($E = hc/\lambda$). Any wavelength of electromagnetic radiation less than or equal to 265 nm ($\lambda \leq 265$) will have sufficient energy to eject an electron. So, yes, 259-nm EMR will eject an electron.

c. This is the electron configuration for copper, Cu, an exception to the expected filling order.

179. a. An atom of francium has 87 protons and 87 electrons. Francium is an alkali metal and forms stable 1+ cations in ionic compounds. This cation would have 86 electrons. The electron configurations will be:

Fr: $[\text{Rn}]7s^1$; Fr^+: $[\text{Rn}] = [\text{Xe}]6s^24f^{14}5d^{10}6p^6$

b. $1.0\ \text{oz Fr} \times \dfrac{1\ \text{lb}}{16\ \text{oz}} \times \dfrac{1\ \text{kg}}{2.205\ \text{lb}} \times \dfrac{1000\text{g}}{1\ \text{kg}} \times \dfrac{1\ \text{mol Fr}}{223\ \text{g Fr}} \times \dfrac{6.02 \times 10^{23}\ \text{atoms}}{1\ \text{mol Fr}}$

$= 7.7 \times 10^{22}$ atoms Fr

c. ^{223}Fr is element 87, so it has 223 – 87 = 136 neutrons.

$$136\ \text{neutrons} \times \frac{1.67493 \times 10^{-27}\ \text{kg}}{1\ \text{neutron}} \times \frac{1000\text{g}}{1\ \text{kg}} = 2.27790 \times 10^{-22}\ \text{g neutrons}$$

180. a. $[\text{Kr}]5s^24d^{10}5p^6$ = Xe; $[\text{Kr}]5s^24d^{10}5p^1$ = In; $[\text{Kr}]5s^24d^{10}5p^3$ = Sb

From the general radii trend, the increasing size order is Xe < Sb < In.

b. $[\text{Ne}]3s^23p^5$ = Cl; $[\text{Ar}]4s^23d^{10}4p^3$ = As; $[\text{Ar}]4s^23d^{10}4p^5$ = Br

From the general ionization energy trend, the decreasing ionization energy order is: Cl > Br > As.

CHAPTER 8

BONDING: GENERAL CONCEPTS

Questions

15. a. This diagram represents a polar covalent bond as in HCl. In a polar covalent bond, there is an electron rich region (indicated by the red color) and an electron poor region (indicated by the blue color). In HCl, the more electronegative Cl atom (on the red side of the diagram) has a slightly greater ability to attract the bonding electrons than does H (on the blue side of the diagram), which in turn produces a dipole moment.

b. This diagram represents an ionic bond as in NaCl. Here, the electronegativity differences between the Na and Cl are so great that the valence electron of sodium is transferred to the chlorine atom. This results in the formation of a cation, an anion, and an ionic bond.

c. This diagram represents a pure covalent bond as in H_2. Both atoms attract the bonding electrons equally, so there is no bond dipole formed. This is illustrated in the electrostatic potential diagram as the various red and blue colors are equally distributed about the molecule. The diagram shows no one region that is red nor one region that is blue (there is no specific partial negative end and no specific partial positive end), so the molecule is nonpolar.

16. In F_2 the bonding is pure covalent, with the bonding electrons shared equally between the two fluorine atoms. In HF, there is also a shared pair of bonding electrons, but the shared pair is drawn more closely to the fluorine atom. This is called a polar covalent bond as opposed to the pure covalent bond in F_2.

17. Of the compounds listed, P_2O_5 is the only compound containing only covalent bonds. $(NH_4)_2SO_4$, $Ca_3(PO_4)_2$, K_2O, and KCl are all compounds composed of ions, so they exhibit ionic bonding. The polyatomic ions in $(NH_4)_2SO_4$ are NH_4^+ and SO_4^{2-}. Covalent bonds exist between the N and H atoms in NH_4^+ and between the S and O atoms in SO_4^{2-}. Therefore, $(NH_4)_2SO_4$ contains both ionic and covalent bonds. The same is true for $Ca_3(PO_4)_2$. The bonding is ionic between the Ca^{2+} and PO_4^{3-} ions and covalent between the P and O atoms in PO_4^{3-}. Therefore, $(NH_4)_2SO_4$ and $Ca_3(PO_4)_2$ are the compounds with both ionic and covalent bonds.

18. Ionic solids are held together by strong electrostatic forces that are omnidirectional.

i. For electrical conductivity, charged species must be free to move. In ionic solids, the charged ions are held rigidly in place. Once the forces are disrupted (melting or dissolution), the ions can move about (conduct).

ii. Melting and boiling disrupts the attractions of the ions for each other. Because these electrostatic forces are strong, it will take a lot of energy (high temperature) to accomplish this.

iii. If we try to bend a piece of material, the ions must slide across each other. For an ionic solid the following might happen:

strong attraction strong repulsion

Just as the layers begin to slide, there will be very strong repulsions causing the solid to snap across a fairly clean plane.

iv. Polar molecules are attracted to ions and can break up the lattice.

These properties and their correlation to chemical forces will be discussed in detail in Chapters 10 and 11.

19. Electronegativity increases left to right across the periodic table and decreases from top to bottom. Hydrogen has an electronegativity value between B and C in the second row and identical to P in the third row. Going further down the periodic table, H has an electronegativity value between As and Se (row 4) and identical to Te (row 5). It is important to know where hydrogen fits into the electronegativity trend, especially for rows 2 and 3. If you know where H fits into the trend, then you can predict bond dipole directions for nonmetals bonded to hydrogen.

20. Linear structure (180° bond angle)

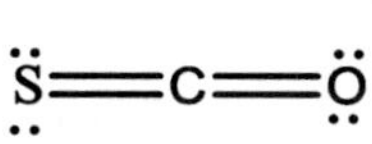

Polar; bond dipoles do not cancel.

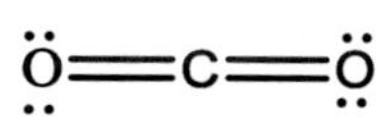

Nonpolar; bond dipoles cancel.

Trigonal planar structure (120° bond angle)

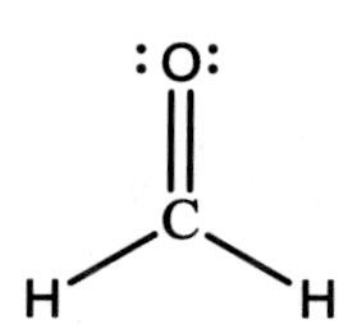

Polar; bond dipoles do not cancel.

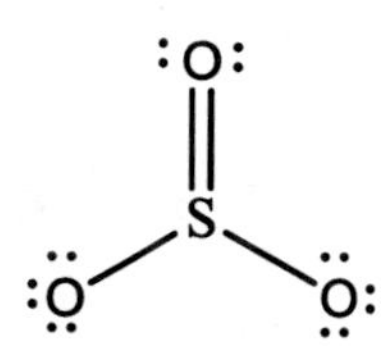

+ 2 other resonance structures

Nonpolar; bond dipoles cancel.

Tetrahedral structure (109.5° bond angles)

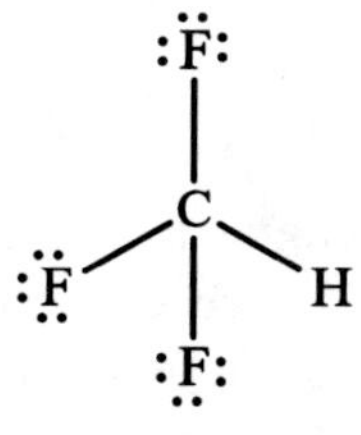

Polar; bond dipoles do not cancel.

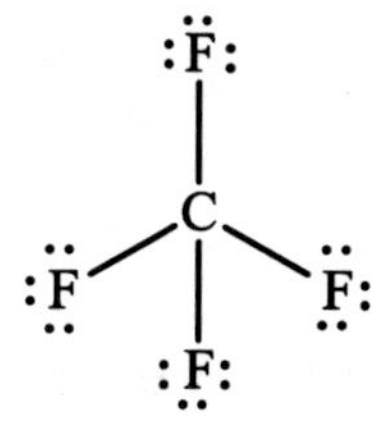

Nonpolar; bond dipoles cancel.

21. For ions, concentrate on the number of protons and the number of electrons present. The species whose nucleus holds the electrons most tightly will be smallest. For example, anions are larger than the neutral atom. The anion has more electrons held by the same number of protons in the nucleus. These electrons will not be held as tightly, resulting in a bigger size for the anion as compared to the neutral atom. For isoelectronic ions, the same number of electrons are held by different numbers of protons in the various ions. The ion with the most protons holds the electrons tightest and is smallest in size.

22. Two other factors that must be considered are the ionization energy needed to produce more positively charged ions and the electron affinity needed to produce more negatively charged ions. The favorable lattice energy more than compensates for the unfavorable ionization energy of the metal and for the unfavorable electron affinity of the nonmetal as long as electrons are added to or removed from the valence shell. Once the valence shell is empty, the ionization energy required to remove the next (inner-core) electron is extremely unfavorable; the same is true for electron affinity when an electron is added to a higher *n* shell. These two quantities are so unfavorable after the valence shell is complete that they overshadow the favorable lattice energy, and the higher charged ionic compounds do not form.

23. Fossil fuels contain a lot of carbon and hydrogen atoms. Combustion of fossil fuels (reaction with O_2) produces CO_2 and H_2O. Both these compounds have very strong bonds. Because stronger product bonds are formed than reactant bonds broken, combustion reactions are very exothermic.

24. Statements a and c are true. For statement a, XeF_2 has 22 valence electrons, and it is impossible to satisfy the octet rule for all atoms with this number of electrons. The best Lewis structure is:

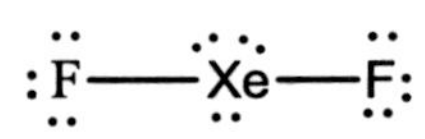

For statement c, NO^+ has 10 valence electrons, whereas NO^- has 12 valence electrons. The Lewis structures are:

Because a triple bond is stronger than a double bond, NO^+ has a stronger bond.

For statement b, SF_4 has five electron pairs around the sulfur in the best Lewis structure; it is an exception to the octet rule. Because OF_4 has the same number of valence electrons as SF_4, OF_4 would also have to be an exception to the octet rule. However, row 2 elements such as O never have more than 8 electrons around them, so OF_4 does not exist. For statement d, two resonance structures can be drawn for ozone:

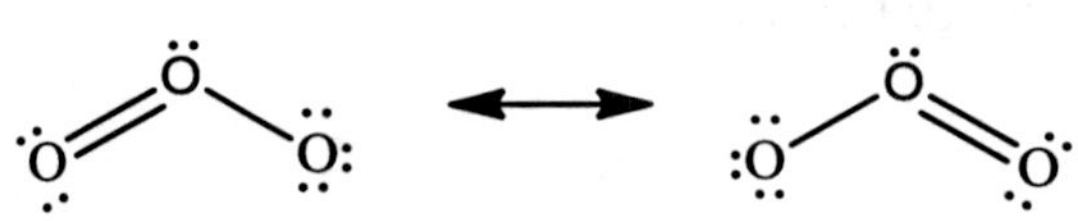

When resonance structures can be drawn, the actual bond lengths and strengths are all equal to each other. Even though each Lewis structure implies the two O–O bonds are different, this is not the case in real life. In real life, both of the O–O bonds are equivalent. When resonance structures can be drawn, you can think of the bonding as an average of all of the resonance structures.

25. CO_2, 4 + 2(6) = 16 valence electrons

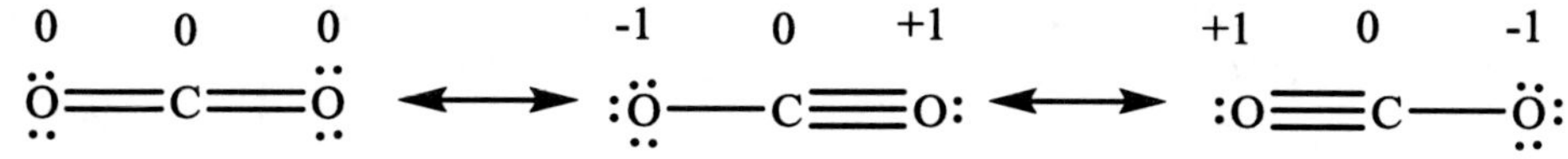

The formal charges are shown above the atoms in the three Lewis structures. The best Lewis structure for CO_2 from a formal charge standpoint is the first structure having each oxygen double bonded to carbon. This structure has a formal charge of zero on all atoms (which is preferred). The other two resonance structures have nonzero formal charges on the oxygens, making them less reasonable. For CO_2, we usually ignore the last two resonance structures and think of the first structure as the true Lewis structure for CO_2.

26. Only statement c is true. The bond dipoles in CF_4 and KrF_4 are arranged in a manner that they all cancel each other out, making them nonpolar molecules (CF_4 has a tetrahedral molecular structure, whereas KrF_4 has a square planar molecular structure). In SeF_4, the bond dipoles in this see-saw molecule do not cancel each other out, so SeF_4 is polar. For statement a, all the molecules have either a trigonal planar geometry or a trigonal bipyramid geometry, both of which have 120° bond angles. However, $XeCl_2$ has three lone pairs and two bonded chlorine atoms around it. $XeCl_2$ has a linear molecular structure with a 180° bond angle. With three lone pairs, we no longer have a 120° bond angle in $XeCl_2$. For statement b, SO_2 has a V-shaped molecular structure with a bond angle of about 120°. CS_2 is linear with a 180° bond angle, and SCl_2 is V-shaped but with an approximate 109.5° bond angle. The three compounds do not have the same bond angle. For statement d, central atoms adopt a geometry to minimize electron repulsions, not maximize them.

Exercises

Chemical Bonds and Electronegativity

27. Using the periodic table, the general trend for electronegativity is:

(1) Increase as we go from left to right across a period

(2) Decrease as we go down a group

Using these trends, the expected orders are:

a. C < N < O b. Se < S < Cl c. Sn < Ge < Si d. Tl < Ge < S

28. a. Rb < K < Na b. Ga < B < O c. Br < Cl < F d. S < O < F

29. The most polar bond will have the greatest difference in electronegativity between the two atoms. From positions in the periodic table, we would predict:

a. Ge–F b. P–Cl c. S–F d. Ti–Cl

30. a. Sn–H b. Tl–Br c. Si–O d. O–F

31. The general trends in electronegativity used in Exercises 27 and 29 are only rules of thumb. In this exercise, we use experimental values of electronegativities and can begin to see several exceptions. The order of EN from Figure 8.3 is:

a. C (2.5) < N (3.0) < O (3.5) same as predicted

b. Se (2.4) < S (2.5) < Cl (3.0) same

c. Si = Ge = Sn (1.8) different

d. Tl (1.8) = Ge (1.8) < S (2.5) different

Most polar bonds using actual EN values:

a. Si–F and Ge–F have equal polarity (Ge–F predicted).

b. P–Cl (same as predicted)

c. S–F (same as predicted)

d. Ti–Cl (same as predicted)

32. The order of EN from Figure 8.3 is:

a. Rb (0.8) = K (0.8) < Na (0.9), different

b. Ga (1.6) < B (2.0) < O (3.5), same

c. Br (2.8) < Cl (3.0) < F (4.0), same

d. S (2.5) < O (3.5) < F (4.0), same

Most polar bonds using actual EN values:

a. C–H most polar (Sn–H predicted)

b. Al–Br most polar (Tl–Br predicted).

c. Si–O (same as predicted).

d. Each bond has the same polarity, but the bond dipoles point in opposite directions. Oxygen is the positive end in the O–F bond dipole, and oxygen is the negative end in the O–Cl bond dipole (O–F predicted).

33. Use the electronegativity trend to predict the partial negative end and the partial positive end of the bond dipole (if there is one). To do this, you need to remember that H has electronegativity between B and C and identical to P. Answers b, d, and e are incorrect. For d (Br_2), the bond between two Br atoms will be a pure covalent bond, where there is equal sharing of the bonding electrons, and no dipole moment exists. For b and e, the bond polarities are reversed. In Cl–I, the more electronegative Cl atom will be the partial negative end of the bond dipole, with I having the partial positive end. In O–P, the more electronegative oxygen will be the partial negative end of the bond dipole, with P having the partial positive end. In the following, we used arrows to indicate the bond dipole. The arrow always points to the partial negative end of a bond dipole (which always is the most electronegative atom in the bond).

34. See Exercise 33 for a discussion on bond dipoles. We will use arrows to indicate the bond dipoles. The arrow always points to the partial negative end of the bond dipole, which will always be to the more electronegative atom. The tail of the arrow indicates the partial positive end of the bond dipole.

a. $\overset{\longrightarrow}{\text{C—O}}$

b. P–H is a pure covalent (nonpolar) bond because P and H have identical electronegativities.

c. $\overset{\longrightarrow}{\text{H—Cl}}$

d. $\overset{\longleftarrow}{\text{Br—Te}}$

e. $\overset{\longrightarrow}{\text{Se—S}}$ The actual electronegativity difference between Se and S is so small that this bond is probably best characterized as a pure covalent bond having no bond dipole.

35. Bonding between a metal and a nonmetal is generally ionic. Bonding between two nonmetals is covalent, and in general, the bonding between two different nonmetals is usually polar covalent. When two different nonmetals have very similar electronegativities, the bonding is pure covalent or just covalent.

a. ionic b. covalent c. polar covalent

d. ionic e. polar covalent f. covalent

36. The possible ionic bonds that can form are between the metal Cs and the nonmetals P, O, and H. These ionic compounds are Cs_3P, Cs_2O, and CsH. The bonding between the various nonmetals will be covalent. P_4, O_2, and H_2 are all pure covalent (or just covalent) with equal sharing of the bonding electrons. P–H will also be a covalent bond because P and H have identical electronegativities. The other possible covalent bonds that can form will all be polar covalent because the nonmetals involved in the bonds all have intermediate differences in electronegativities. The possible polar covalent bonds are P–O and O–H.

Note: The bonding among cesium atoms is called metallic. This type of bonding between metals will be discussed in Chapter 10.

37. Electronegativity values increase from left to right across the periodic table. The order of electronegativities for the atoms from smallest to largest electronegativity will be H = P < C < N < O < F. The most polar bond will be F–H since it will have the largest difference in electronegativities, and the least polar bond will be P–H since it will have the smallest difference in electronegativities ($\Delta EN = 0$). The order of the bonds in decreasing polarity will be F–H > O–H > N–H > C–H > P–H.

38. Ionic character is proportional to the difference in electronegativity values between the two elements forming the bond. Using the trend in electronegativity, the order will be:

Br–Br < N–O < C–F < Ca–O < K–F

least ionic character — most ionic character

Note that Br–Br, N–O, and C–F bonds are all covalent bonds since the elements are all nonmetals. The Ca–O and K–F bonds are ionic, as is generally the case when a metal forms a bond with a nonmetal.

39. A permanent dipole moment exists in a molecule if the molecule has one specific area with a partial negative end (a red end in an electrostatic potential diagram) and a different specific region with a partial positive end (a blue end in an electrostatic potential diagram). If the blue and red colors are equally distributed in the electrostatic potential diagrams, then no permanent dipole exists.

a. Has a permanent dipole. b. Has no permanent dipole.

c. Has no permanent dipole. d. Has a permanent dipole.

e. Has no permanent dipole. f. Has no permanent dipole.

40. a. H_2O; both H_2O and NH_3 have permanent dipole moments in part due to the polar O–H and N–H bonds. But because oxygen is more electronegative than nitrogen, one would expect H_2O to have a slightly greater dipole moment. This diagram has the more intense red color on one end and the more intense blue color at the other end indicating a larger dipole moment.

b. NH_3; this diagram is for a polar molecule, but the colors are not as intense as the diagram in part a. Hence, this diagram is for a molecule which is not as polar as H_2O. Since N is less electronegative than O, NH_3 will not be as polar as H_2O.

c. CH_4; this diagram has no one specific red region and has four blue regions arranged symmetrically about the molecule. This diagram is for a molecule which has no dipole moment. This is only true for CH_4. The C–H bonds are at best, slightly polar because carbon and hydrogen have similar electronegativity values. In addition, the slightly polar C–H bond dipoles are arranged about carbon so that they cancel each other out, making CH_4 a nonpolar molecule. See Example 8.2.

Ions and Ionic Compounds

41. Al^{3+}: $[He]2s^22p^6$; Ba^{2+}: $[Kr]5s^24d^{10}5p^6$; Se^{2-}: $[Ar]4s^23d^{10}4p^6$; I^-: $[Kr]5s^24d^{10}5p^6$

42. Te^{2-}: $[Kr]5s^24d^{10}5p^6$; Cl^-: $[Ne]3s^23p^6$; Sr^{2+}: $[Ar]4s^23d^{10}4p^6$; Li^+: $1s^2$

43. a. Li^+ and N^{3-} are the expected ions. The formula of the compound would be Li_3N (lithium nitride).

b. Ga^{3+} and O^{2-}; Ga_2O_3, gallium(III) oxide or gallium oxide

c. Rb^+ and Cl^-; RbCl, rubidium chloride d. Ba^{2+} and S^{2-}; BaS, barium sulfide

44. a. Al^{3+} and Cl^-; $AlCl_3$, aluminum chloride b. Na^+ and O^{2-}; Na_2O, sodium oxide

c. Sr^{2+} and F^-; SrF_2, strontium fluoride d. Ca^{2+} and S^{2-}; CaS, calcium sulfide

45. a. Mg^{2+}: $1s^22s^22p^6$; K^+: $1s^22s^22p^63s^23p^6$; Al^{3+}: $1s^22s^22p^6$

b. N^{3-}, O^{2-}, and F^-: $1s^22s^22p^6$; Te^{2-}: $[Kr]5s^24d^{10}5p^6$

46. a. Sr^{2+}: $[Ar]4s^23d^{10}4p^6$; Cs^+: $[Kr]5s^24d^{10}5p^6$; In^+: $[Kr]5s^24d^{10}$; Pb^{2+}: $[Xe]6s^24f^{14}5d^{10}$

b. P^{3-} and S^{2-}: $[Ne]3s^23p^6$; Br^-: $[Ar]4s^23d^{10}4p^6$

47. a. Sc^{3+}: [Ar] b. Te^{2-}: [Xe] c. Ce^{4+}: [Xe] and Ti^{4+}: [Ar] d. Ba^{2+}: [Xe]

All these ions have the noble gas electron configuration shown in brackets.

48. a. Cs_2S is composed of Cs^+ and S^{2-}. Cs^+ has the same electron configuration as Xe, and S^{2-} has the same configuration as Ar.

b. SrF_2; Sr^{2+} has the Kr electron configuration, and F^- has the Ne configuration.

c. Ca_3N_2; Ca^{2+} has the Ar electron configuration, and N^{3-} has the Ne configuration.

d. $AlBr_3$; Al^{3+} has the Ne electron configuration, and Br^- has the Kr configuration.

49. a. Na^+ has 10 electrons. F^-, O^{2-}, and N^{3-} are some possible anions also having 10 electrons.

b. Ca^{2+} has 18 electrons. Cl^-, S^{2-}, and P^{3-} also have 18 electrons.

c. Al^{3+} has 10 electrons. F^-, O^{2-}, and N^{3-} also have 10 electrons.

d. Rb^+ has 36 electrons. Br^-, Se^{2-}, and As^{3-} also have 36 electrons.

50. a. Ne has 10 electrons. AlN, MgF_2, and Na_2O are some possible ionic compounds where each ion has 10 electrons.

b. CaS, K_3P, and KCl are some examples where each ion is isoelectronic with Ar; i.e., each ion has 18 electrons.

c. Each ion in Sr_3As_2, $SrBr_2$, and Rb_2Se is isoelectronic with Kr.

d. Each ion in BaTe and CsI is isoelectronic with Xe.

51. Neon has 10 electrons; there are many possible ions with 10 electrons. Some are N^{3-}, O^{2-}, F^-, Na^+, Mg^{2+}, and Al^{3+}. In terms of size, the ion with the most protons will hold the electrons the tightest and will be the smallest. The largest ion will be the ion with the fewest protons. The size trend is:

$Al^{3+} < Mg^{2+} < Na^+ < F^- < O^{2-} < N^{3-}$
smallest largest

52. All these ions have 18 e^-; the smallest ion (Sc^{3+}) has the most protons attracting the 18 e^-, and the largest ion has the fewest protons (S^{2-}). The order in terms of increasing size is $Sc^{3+} < Ca^{2+} < K^+ < Cl^- < S^{2-}$. In terms of the atom size indicated in the question:

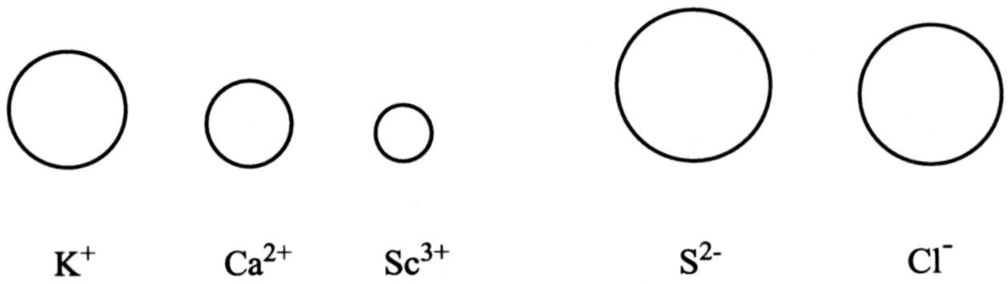

53. a. $Cu > Cu^+ > Cu^{2+}$ b. $Pt^{2+} > Pd^{2+} > Ni^{2+}$ c. $O^{2-} > O^- > O$

d. $La^{3+} > Eu^{3+} > Gd^{3+} > Yb^{3+}$ e. $Te^{2-} > I^- > Cs^+ > Ba^{2+} > La^{3+}$

For answer a, as electrons are removed from an atom, size decreases. Answers b and d follow the radius trend. For answer c, as electrons are added to an atom, size increases. Answer e follows the trend for an isoelectronic series; i.e., the smallest ion has the most protons.

54. a. $V > V^{2+} > V^{3+} > V^{5+}$ b. $Cs^+ > Rb^+ > K^+ > Na^+$ c. $Te^{2-} > I^- > Cs^+ > Ba^{2+}$

d. $P^{3-} > P^{2-} > P^- > P$ e. $Te^{2-} > Se^{2-} > S^{2-} > O^{2-}$

55. Lattice energy is proportional to $-Q_1Q_2/r$, where Q is the charge of the ions and r is the distance between the centers of the ions. The more negative the lattice energy, the more stable the ionic compound. So greater charged ions as well as smaller sized ions lead to more negative lattice energy values and more stable ionic compounds.

a. NaCl; Na^+ is smaller than K^+. b. LiF; F^- is smaller than Cl^-.

c. MgO; O^{2-} has a greater charge than OH^-. d. $Fe(OH)_3$; Fe^{3+} has a greater charge than Fe^{2+}.

e. Na_2O; O^{2-} has a greater charge than Cl^-. f. MgO; both ions are smaller in MgO.

56. a. LiF; Li^+ is smaller than Cs^+. b. NaBr; Br^- is smaller than I^-.

c. BaO; O^{2-} has a greater charge than Cl^-. d. $CaSO_4$; Ca^{2+} has a greater charge than Na^+.

e. K_2O; O^{2-} has a greater charge than F^-. f. Li_2O; both ions are smaller in Li_2O.

57.

$K(s) \rightarrow K(g)$	$\Delta H = 90.$ kJ (sublimation)
$K(g) \rightarrow K^+(g) + e^-$	$\Delta H = 419$ kJ (ionization energy)
$1/2\ Cl_2(g) \rightarrow Cl(g)$	$\Delta H = 239/2$ kJ (bond energy)
$Cl(g) + e^- \rightarrow Cl^-(g)$	$\Delta H = -349$ kJ (electron affinity)
$K^+(g) + Cl^-(g) \rightarrow KCl(s)$	$\Delta H = -690.$ kJ (lattice energy)
$K(s) + 1/2\ Cl_2(g) \rightarrow KCl(s)$	$\Delta H_f^\circ = -411$ kJ/mol

58.

$$\begin{array}{rll}
Mg(s) \rightarrow Mg(g) & \Delta H = 150.\ kJ & \text{(sublimation)} \\
Mg(g) \rightarrow Mg^{+}(g) + e^{-} & \Delta H = 735\ kJ & (IE_1) \\
Mg^{+}(g) \rightarrow Mg^{2+}(g) + e^{-} & \Delta H = 1445\ kJ & (IE_2) \\
F_2(g) \rightarrow 2\ F(g) & \Delta H = 154\ kJ & (BE) \\
2\ F(g) + 2\ e^{-} \rightarrow 2\ F^{-}(g) & \Delta H = 2(-328)\ kJ & (EA) \\
Mg^{2+}(g) + 2\ F^{-}(g) \rightarrow MgF_2(s) & \Delta H = -2913\ kJ & (LE) \\
\hline
Mg(s) + F_2(g) \rightarrow MgF_2(s) & \Delta H_f^\circ = -1085\ kJ/mol &
\end{array}$$

59. From the data given, it takes less energy to produce $Mg^{+}(g) + O^{-}(g)$ than to produce $Mg^{2+}(g) + O^{2-}(g)$. However, the lattice energy for $Mg^{2+}O^{2-}$ will be much more exothermic than that for $Mg^{+}O^{-}$ due to the greater charges in $Mg^{2+}O^{2-}$. The favorable lattice energy term dominates, and $Mg^{2+}O^{2-}$ forms.

60.

$$\begin{array}{rl}
Na(g) \rightarrow Na^{+}(g) + e^{-} & \Delta H = IE = 495\ kJ\ \text{(Table 7.5)} \\
F(g) + e^{-} \rightarrow F^{-}(g) & \Delta H = EA = -327.8\ kJ\ \text{(Table 7.7)} \\
\hline
Na(g) + F(g) \rightarrow Na^{+}(g) + F^{-}(g) & \Delta H = 167\ kJ
\end{array}$$

The described process is endothermic. What we haven't accounted for is the extremely favorable lattice energy. Here, the lattice energy is a large negative (exothermic) value, making the overall formation of NaF a favorable exothermic process.

61. Use Figure 8.11 as a template for this problem.

$$\begin{array}{rl}
Li(s) \rightarrow Li(g) & \Delta H_{sub} = ? \\
Li(g) \rightarrow Li^{+}(g) + e^{-} & \Delta H = 520.\ kJ \\
1/2\ I_2(g) \rightarrow I(g) & \Delta H = 151/2\ kJ \\
I(g) + e^{-} \rightarrow I^{-}(g) & \Delta H = -295\ kJ \\
Li^{+}(g) + I^{-}(g) \rightarrow LiI(s) & \Delta H = -753\ kJ \\
\hline
Li(s) + 1/2\ I_2(g) \rightarrow LiI(s) & \Delta H = -292\ kJ
\end{array}$$

$\Delta H_{sub} + 520. + 151/2 - 295 - 753 = -292,\ \Delta H_{sub} = 161\ kJ$

62. Let us look at the complete cycle for Na_2S.

$$\begin{array}{rl}
2\ Na(s) \rightarrow 2\ Na(g) & 2\Delta H_{sub,\ Na} = 2(109)\ kJ \\
2\ Na(g) \rightarrow 2\ Na^{+}(g) + 2\ e^{-} & 2IE = 2(495)\ kJ \\
S(s) \rightarrow S(g) & \Delta H_{sub,\ S} = 277\ kJ \\
S(g) + e^{-} \rightarrow S^{-}(g) & EA_1 = -200.\ kJ \\
S^{-}(g) + e^{-} \rightarrow S^{2-}(g) & EA_2 = ? \\
2\ Na^{+}(g) + S^{2-}(g) \rightarrow Na_2S(s) & LE = -2203\ kJ \\
\hline
2\ Na(s) + S(s) \rightarrow Na_2S(s) & \Delta H_f^\circ = -365\ kJ
\end{array}$$

$\Delta H_f^\circ = 2\Delta H_{sub,\ Na} + 2IE + \Delta H_{sub,\ S} + EA_1 + EA_2 + LE,\ -365 = -918 + EA_2,\ EA_2 = 553\ kJ$

For each salt: $\Delta H_f^\circ = 2\Delta H_{sub,\,M} + 2IE + 277 - 200. + LE + EA_2$

K_2S: $-381 = 2(90.) + 2(419) + 277 - 200. - 2052 + EA_2$, $EA_2 = 576$ kJ

Rb_2S: $-361 = 2(82) + 2(409) + 277 - 200. - 1949 + EA_2$, $EA_2 = 529$ kJ

Cs_2S: $-360. = 2(78) + 2(382) + 277 - 200. - 1850. + EA_2$, $EA_2 = 493$ kJ

We get values from 493 to 576 kJ.

The mean value is: $\dfrac{553 + 576 + 529 + 493}{4} = 538$ kJ. We can represent the results as EA_2 = 540 ±50 kJ.

63. Ca^{2+} has a greater charge than Na^+, and Se^{2-} is smaller than Te^{2-}. The effect of charge on the lattice energy is greater than the effect of size. We expect the trend from most exothermic lattice energy to least exothermic to be:

$$\underset{(-2862)}{CaSe} > \underset{(-2721)}{CaTe} > \underset{(-2130)}{Na_2Se} > \underset{(-2095)}{Na_2Te}$$

This is what we observe.

64. Lattice energy is proportional to the charge of the cation times the charge of the anion Q_1Q_2.

Compound	Q_1Q_2	Lattice Energy
$FeCl_2$	$(+2)(-1) = -2$	−2631 kJ/mol
$FeCl_3$	$(+3)(-1) = -3$	−5359 kJ/mol
Fe_2O_3	$(+3)(-2) = -6$	−14,744 kJ/mol

Bond Energies

65. a. H—H + Cl—Cl ⟶ 2 H—Cl

Bonds broken:

1 H–H (432 kJ/mol)
1 Cl–Cl (239 kJ/mol)

Bonds formed:

2 H–Cl (427 kJ/mol)

$\Delta H = \Sigma D_{broken} - \Sigma D_{formed}$, $\Delta H = 432 \text{ kJ} + 239 \text{ kJ} - 2(427) \text{ kJ} = -183$ kJ

b. N≡N + 3 H—H ⟶ 2 H—N(—H)—H

Bonds broken:

1 N≡N (941 kJ/mol)
3 H–H (432 kJ/mol)

Bonds formed:

6 N–H (391 kJ/mol)

$\Delta H = 941 \text{ kJ} + 3(432) \text{ kJ} - 6(391) \text{ kJ} = -109$ kJ

66. Sometimes some of the bonds remain the same between reactants and products. To save time, only break and form bonds that are involved in the reaction.

a.

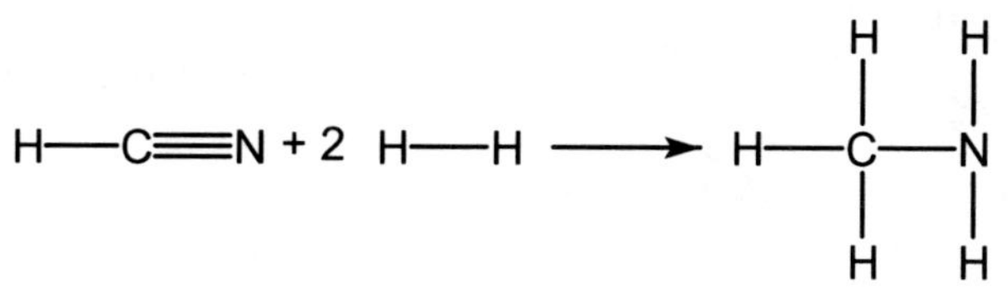

Bonds broken:

1 C≡N (891 kJ/mol)
2 H–H (432 kJ/mol)

Bonds formed:

1 C–N (305 kJ/mol)
2 C–H (413 kJ/mol)
2 N–H (391 kJ/mol)

ΔH = 891 kJ + 2(432 kJ) – [305 kJ + 2(413 kJ) + 2(391 kJ)] = –158 kJ

b.

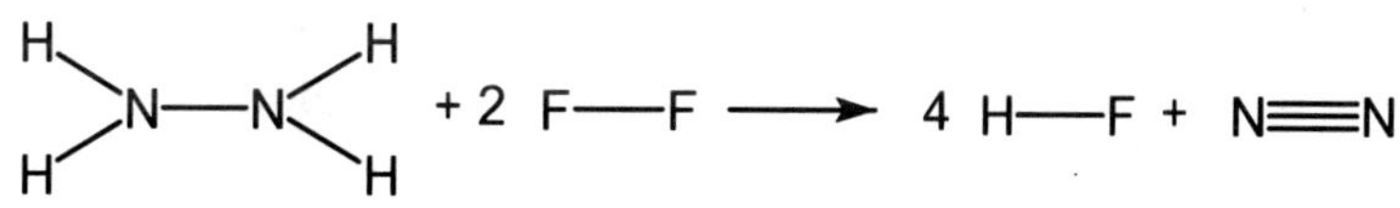

Bonds broken:

1 N–N (160. kJ/mol)
4 N–H (391 kJ/mol)
2 F–F (154 kJ/mol)

Bonds formed:

4 H–F (565 kJ/mol)
1 N ≡ N (941 kJ/mol)

ΔH = 160. kJ + 4(391 kJ) + 2(154 kJ) – [4(565 kJ) + 941 kJ] = –1169 kJ

67.

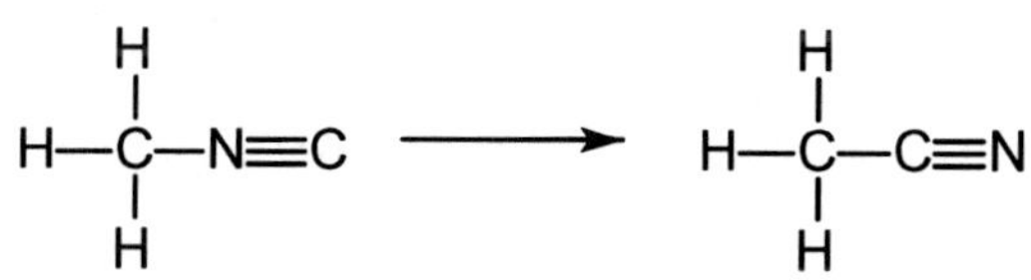

Bonds broken: 1 C–N (305 kJ/mol) Bonds formed: 1 C–C (347 kJ/mol)

$\Delta H = \Sigma D_{broken} - \Sigma D_{formed}$, ΔH = 305 – 347 = –42 kJ

Note: Sometimes some of the bonds remain the same between reactants and products. To save time, only break and form bonds that are involved in the reaction.

68.

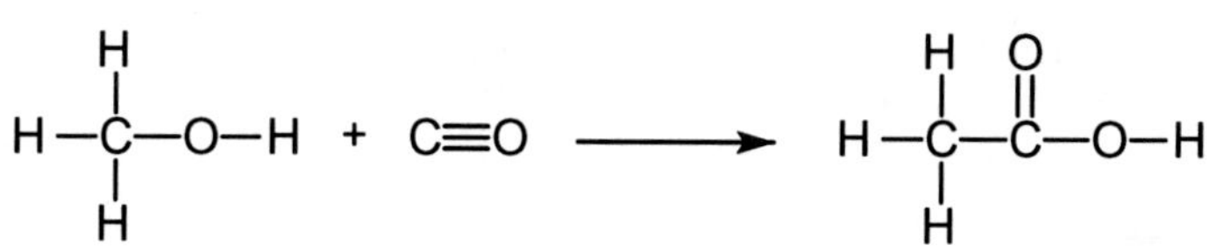

Bonds broken:

1 C≡O (1072 kJ/mol)
1 C–O (358 kJ/mol)

Bonds formed:

1 C–C (347 kJ/mol)
1 C=O (745 kJ/mol)
1 C–O (358 kJ/mol)

ΔH = 1072 + 358 – [347 + 745 + 358] = –20. kJ

69.

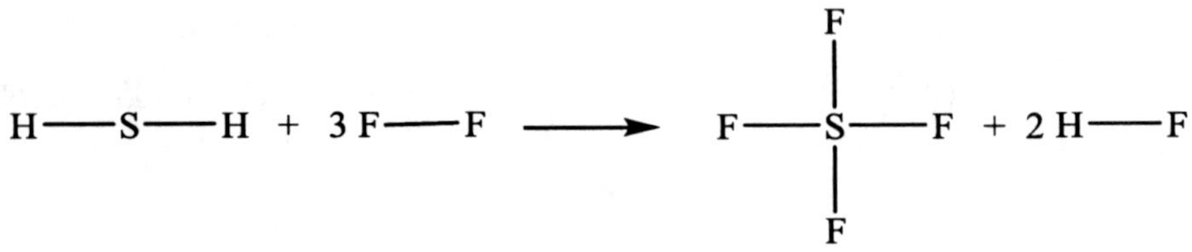

Bonds broken:

2 S–H (347 kJ/mol)
3 F–F (154 kJ/mol)

Bonds formed:

4 S–F (327 kJ/mol)
2 H–F (565 kJ/mol)

ΔH = 2(347) + 3(154) – [4(327) + 2(565)] = –1282 kJ

70.

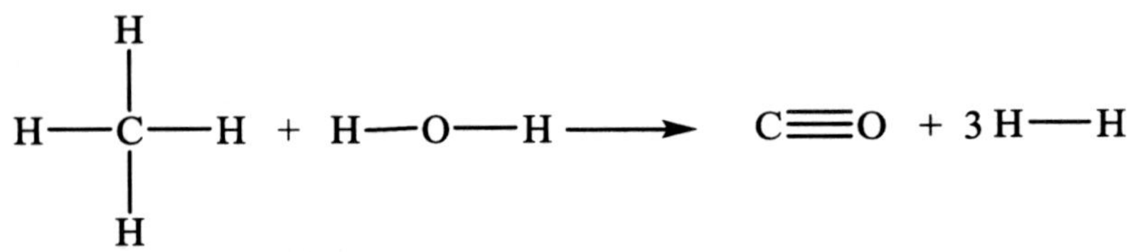

Bonds broken:

4 C–H (413 kJ/mol)
2 O–H (467 kJ/mol)

Bonds formed:

1 C≡O (1072 kJ/mol)
3 H–H (432 kJ/mol)

ΔH = 4(413) + 2(467) – [1072 + 3(432)] = 218 kJ

71. H–C≡C–H + 5/2 O=O → 2 O=C=O + H–O–H

Bonds broken:

2 C–H (413 kJ/mol)
1 C≡C (839 kJ/mol)
5/2 O=O (495 kJ/mol)

Bonds formed:

2 × 2 C=O (799 kJ/mol)
2 O–H (467 kJ/mol)

ΔH = 2(413 kJ) + 839 kJ + 5/2 (495 kJ) – [4(799 kJ) + 2(467 kJ)] = –1228 kJ

72. CH_4 + 2 O=O → O=C=O + 2 H–O–H

Bonds broken:

- 4 C–H (413 kJ/mol)
- 2 O=O (495 kJ/mol)

Bonds formed:

- 2 C=O (799 kJ/mol)
- 2 × 2 O–H (467 kJ/mol)

$\Delta H = 4(413\text{ kJ}) + 2(495\text{ kJ}) - [2(799\text{ kJ}) + 4(467\text{ kJ})] = -824\text{ kJ}$

73.

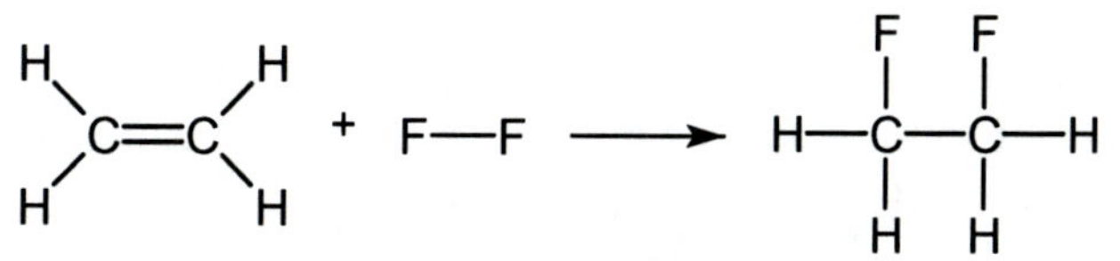

ΔH = -549 kJ

Bonds broken:

- 1 C=C (614 kJ/mol)
- 1 F–F (154 kJ/mol)

Bonds formed:

- 1 C–C (347 kJ/mol)
- 2 C–F (D_{CF} = C–F bond energy)

$\Delta H = -549\text{ kJ} = 614\text{ kJ} + 154\text{ kJ} - [347\text{ kJ} + 2D_{CF}]$, $2D_{CF} = 970.$, $D_{CF} = 485\text{ kJ/mol}$

74. Let x = bond energy for A_2, so $2x$ = bond energy for AB.

$\Delta H = -285\text{ kJ} = x + 432\text{ kJ} - [2(2x)]$, $3x = 717$, $x = 239\text{ kJ/mol}$

The bond energy for A_2 is 239 kJ/mol.

75. a. $\Delta H° = 2\Delta H^{o}_{f,\,HCl} = 2\text{ mol}(-92\text{ kJ/mol}) = -184\text{ kJ}$ (–183 kJ from bond energies)

b. $\Delta H° = 2\Delta H^{o}_{f,\,NH_3} = 2\text{ mol}(-46\text{ kJ/mol}) = -92\text{ kJ}$ (–109 kJ from bond energies)

Comparing the values for each reaction, bond energies seem to give a reasonably good estimate for the enthalpy change of a reaction. The estimate is especially good for gas phase reactions.

76. $CH_3OH(g) + CO(g) \rightarrow CH_3COOH(l)$ $\Delta H° = -484\text{ kJ} - [(-201\text{ kJ}) + (-110.5\text{ kJ})] = -173\text{ kJ}$

Using bond energies, ΔH = –20. kJ. For this reaction, bond energies give a much poorer estimate for ΔH as compared with gas-phase reactions (see Exercise 75). The major reason for the large discrepancy is that not all species are gases in Exercise 68. Bond energies do not account for the energy changes that occur when liquids and solids form instead of gases. These energy changes are due to intermolecular forces and will be discussed in Chapter 10.

77. a. Using SF_4 data: $SF_4(g) \rightarrow S(g) + 4\ F(g)$

$$\Delta H° = 4D_{SF} = 278.8 + 4\,(79.0) - (-775) = 1370.\text{ kJ}$$

$$D_{SF} = \frac{1370.\text{ kJ}}{4\text{ mol SF bonds}} = 342.5\text{ kJ/mol} = \text{S–F bond energy}$$

Using SF_6 data: $SF_6(g) \rightarrow S(g) + 6\ F(g)$

$$\Delta H° = 6D_{SF} = 278.8 + 6\ (79.0) - (-1209) = 1962\ kJ$$

$$D_{SF} = \frac{1962\,kJ}{6\ mol} = 327.0\ kJ/mol = S\text{–}F\ bond\ energy$$

b. The S–F bond energy in Table 8.4 is 327 kJ/mol. The value in the table was based on the S–F bond in SF_6.

c. S(g) and F(g) are not the most stable forms of the elements at 25°C and 1 atm. The most stable forms are $S_8(s)$ and $F_2(g)$; $\Delta H_f^\circ = 0$ for these two species.

78. $NH_3(g) \rightarrow N(g) + 3\ H(g)$ $\Delta H° = 3D_{NH} = 472.7 + 3(216.0) - (-46.1) = 1166.8\ kJ$

$$D_{NH} = \frac{1166.8\ kJ}{3\ mol\ NH\ bonds} = 388.93\ kJ/mol \approx 389\ kJ/mol$$

D_{calc} = 389 kJ/mol as compared with 391 kJ/mol in Table 8.4. There is good agreement.

79.

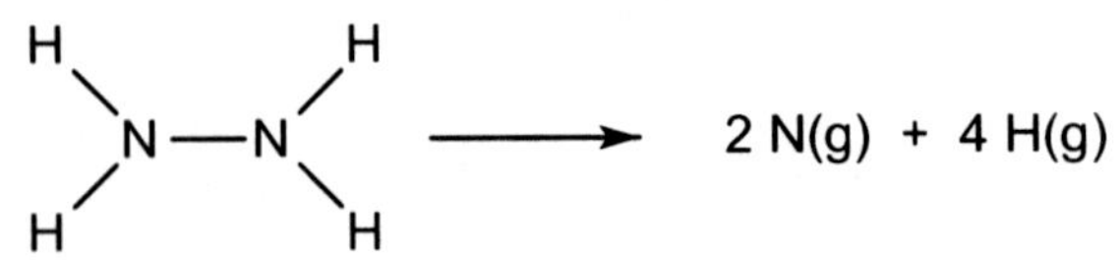

$\Delta H = D_{N\text{–}N} + 4D_{N\text{–}H} = D_{N\text{–}N} + 4(388.9)$

$$\Delta H° = 2\Delta H^\circ_{f,\,N} + 4\Delta H^\circ_{f,\,H} - \Delta H^\circ_{f,\,N_2H_4} = 2(472.7\ kJ) + 4(216.0\ kJ) - 95.4\ kJ$$

$$\Delta H° = 1714.0\ kJ = D_{N\text{–}N} + 4(388.9)$$

$D_{N\text{–}N}$ = 158.4 kJ/mol (versus 160. kJ/mol in Table 8.4)

80. $1/2\ N_2(g) + 1/2\ O_2(g) \rightarrow NO(g)$ $\Delta H = 90.\ kJ$

Bonds broken:

1/2 N≡N (941 kJ/mol)
1/2 O=O (495 kJ/mol)

Bonds formed:

1 NO (D_{NO} = NO bond energy)

$$\Delta H = 90.\ kJ = 1/2(941) + 1/2(495) - (D_{NO}),\ D_{NO} = 628\ kJ/mol$$

From this data, the calculated NO bond energy is 628 kJ/mol.

Lewis Structures and Resonance

81. Drawing Lewis structures is mostly trial and error. However, the first two steps are always the same. These steps are (1) count the valence electrons available in the molecule/ion, and (2) attach all atoms to each other with single bonds (called the skeletal structure). Unless noted otherwise, the atom listed first is assumed to be the atom in the middle, called the central atom, and all other atoms in the formula are attached to this atom. The most notable exceptions to the rule are formulas that begin with H, e.g., H_2O, H_2CO, etc. Hydrogen can never be a central atom since this would require H to have more than two electrons. In these compounds, the atom listed second is assumed to be the central atom.

After counting valence electrons and drawing the skeletal structure, the rest is trial and error. We place the remaining electrons around the various atoms in an attempt to satisfy the octet rule (or duet rule for H). Keep in mind that practice makes perfect. After practicing, you can (and will) become very adept at drawing Lewis structures.

a. F_2 has 2(7) = 14 valence electrons.

Skeletal structure

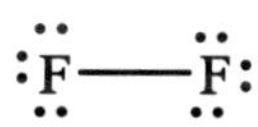

Lewis structure

b. O_2 has 2(6) = 12 valence electrons.

Skeletal structure

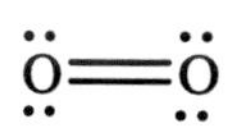

Lewis structure

c. CO has 4 + 6 = 10 valence electrons.

Skeletal structure

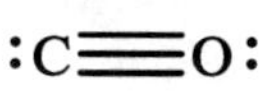

Lewis structure

d. CH_4 has 4 + 4(1) = 8 valence electrons.

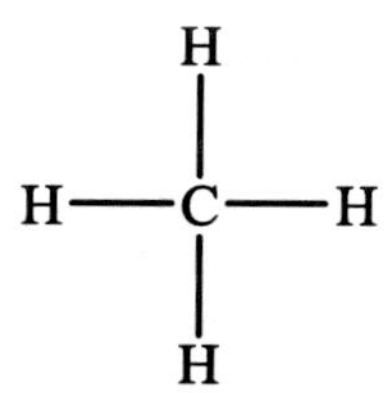

Skeletal structure

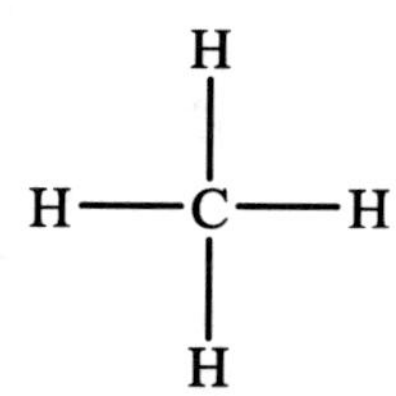

Lewis structure

e. NH_3 has 5 + 3(1) = 8 valence electrons.

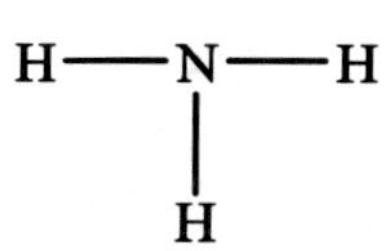

Skeletal structure

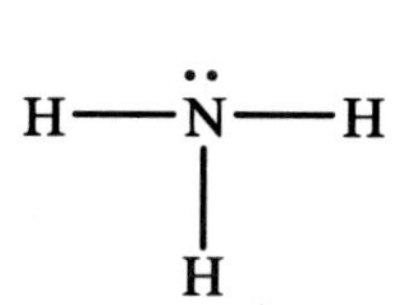

Lewis structure

f. H_2O has 2(1) + 6 = 8 valence electrons.

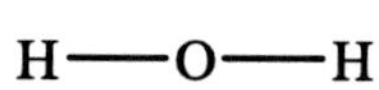

Skeletal structure

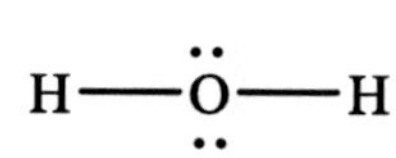

Lewis structure

g. HF has 1 + 7 = 8 valence electrons.

Skeletal structure

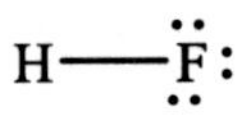

Lewis structure

82. a. H_2CO has 2(1) + 4 + 6 = 12 valence electrons.

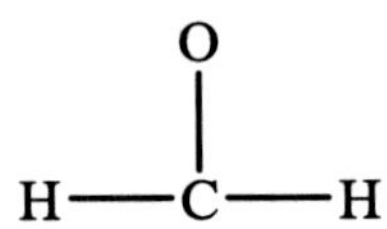

Skeletal structure

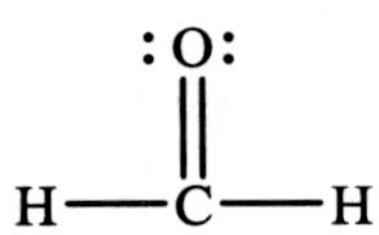

Lewis structure

b. CO_2 has 4 + 2(6) = 16 valence electrons.

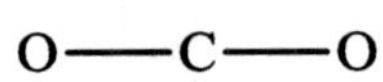

Skeletal structure

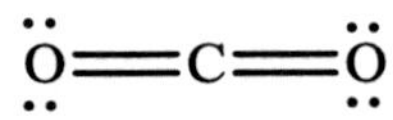

Lewis structure

c. HCN has 1 + 4 + 5 = 10 valence electrons.

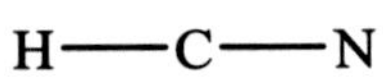

Skeletal structure

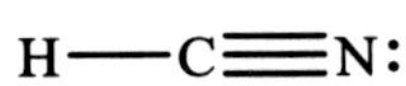

Lewis structure

83. Drawing Lewis structures is mostly trial and error. However, the first two steps are always the same. These steps are (1) count the valence electrons available in the molecule/ion, and (2) attach all atoms to each other with single bonds (called the skeletal structure). Unless noted otherwise, the atom listed first is assumed to be the atom in the middle, called the central atom, and all other atoms in the formula are attached to this atom. The most notable exceptions to the rule are formulas that begin with H, e.g., H_2O, H_2CO, etc. Hydrogen can never be a central atom since this would require H to have more than two electrons. In these compounds, the atom listed second is assumed to be the central atom.

After counting valence electrons and drawing the skeletal structure, the rest is trial and error. We place the remaining electrons around the various atoms in an attempt to satisfy the octet rule (or duet rule for H).

a. CCl_4 has 4 + 4(7) = 32 valence electrons.

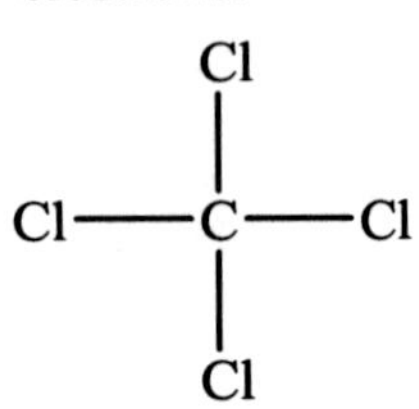

Skeletal structure

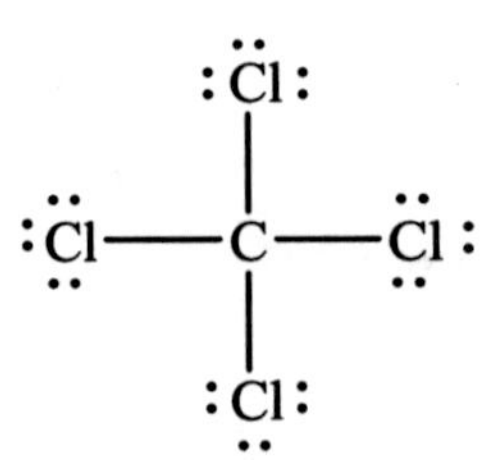

Lewis structure

b. NCl_3 has 5 + 3(7) = 26 valence electrons.

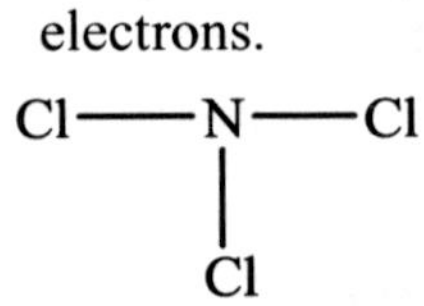

Skeletal structure

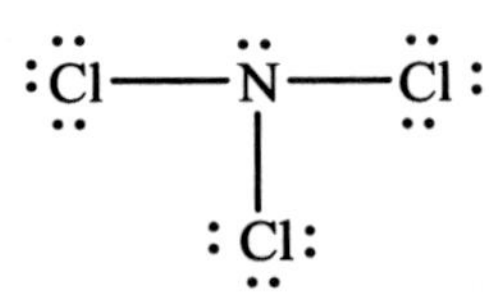

Lewis structure

c. $SeCl_2$ has 6 + 2(7) = 20 valence electrons.

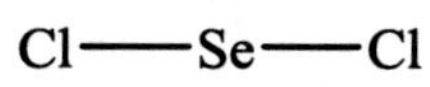

Skeletal structure

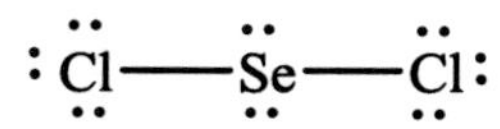

Lewis structure

d. ICl has 7 + 7 = 14 valence electrons.

Skeletal structure

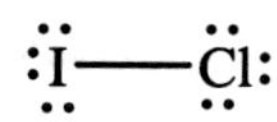

Lewis structure

84. a. $POCl_3$ has 5 + 6 + 3(7) = 32 valence electrons.

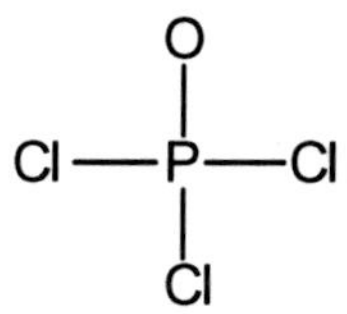

Skeletal structure

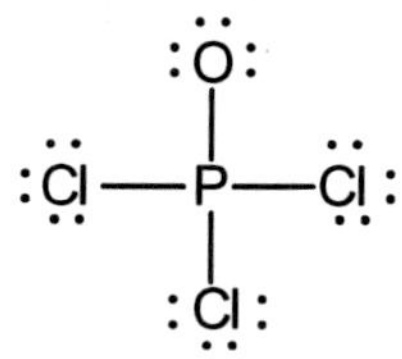

Lewis structure

Note: This structure uses all 32 e^- while satisfying the octet rule for all atoms. This is a valid Lewis structure.

SO_4^{2-} has 6 + 4(6) + 2 = 32 valence electrons.

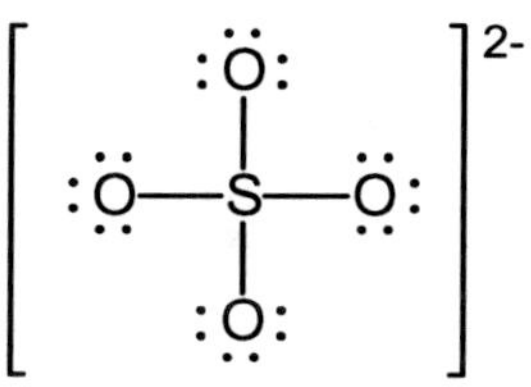

Note: A negatively charged ion will have additional electrons to those that come from the valence shell of the atoms. The magnitude of the negative charge indicates the number of extra electrons to add in.

XeO_4, 8 + 4(6) = 32 e^-

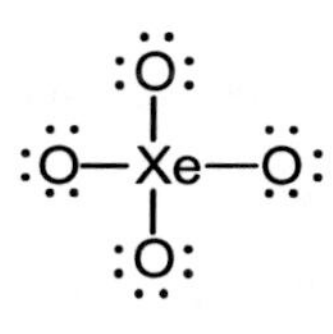

PO_4^{3-}, 5 + 4(6) + 3 = 32 e^-

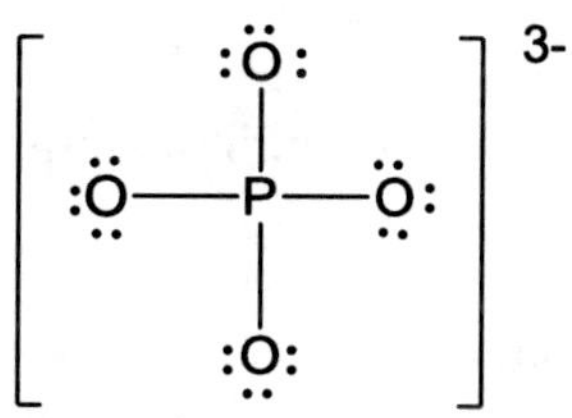

ClO_4^- has 7 + 4(6) + 1 = 32 valence electrons

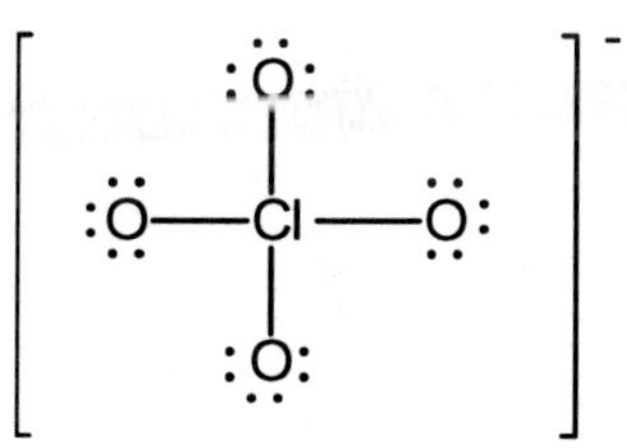

Note: All of these species have the same number of atoms and the same number of valence electrons. They also have the same Lewis structure.

b. NF_3 has 5 + 3(7) = 26 valence electrons. SO_3^{2-}, 6 + 3(6) + 2 = 26 e^-

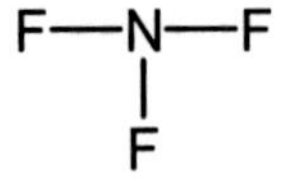

Skeletal structure

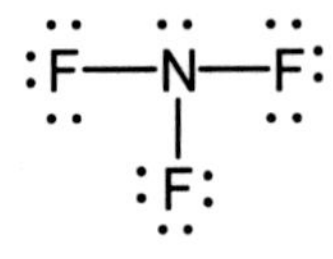

Lewis structure

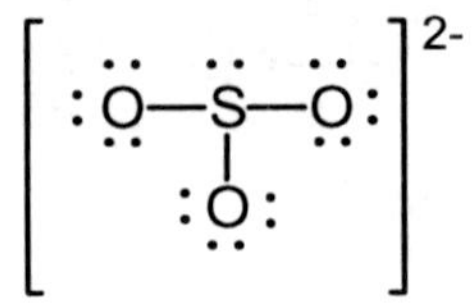

PO_3^{3-}, 5 + 3(6) + 3 = 26 e^-

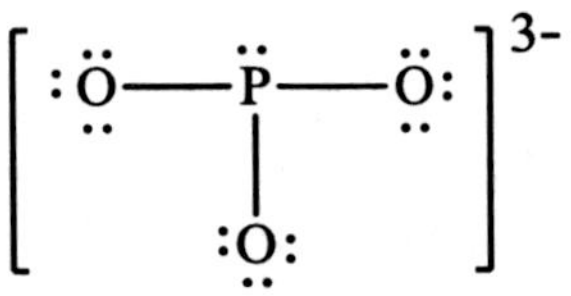

ClO_3^-, 7 + 3(6) + 1 = 26 e^-

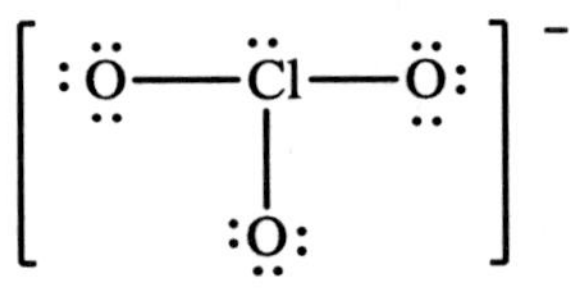

Note: Species with the same number of atoms and valence electrons have similar Lewis structures.

c. ClO_2^- has 7 + 2(6) + 1 = 20 valence

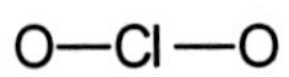

Skeletal structure

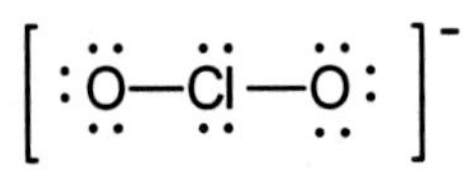

Lewis structure

SCl_2, 6 + 2(7) = 20 e^-

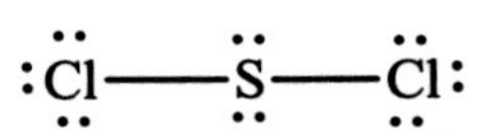

PCl_2^-, 5 + 2(7) + 1 = 20 e^-

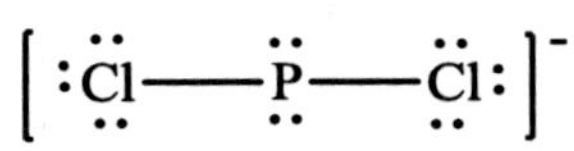

Note: Species with the same number of atoms and valence electrons have similar Lewis structures.

d. Molecules ions that have the same number of valence electrons and the same number of atoms will have similar Lewis structures.

85. BeH_2, 2 + 2(1) = 4 valence electrons BH_3, 3 + 3(1) = 6 valence electrons

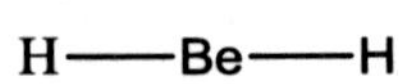

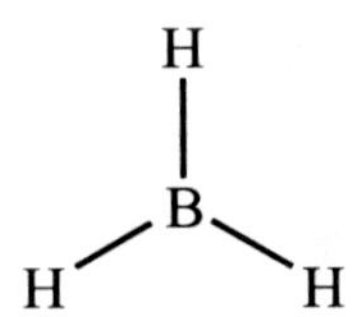

86. a. NO_2, $5 + 2(6) = 17\ e^-$

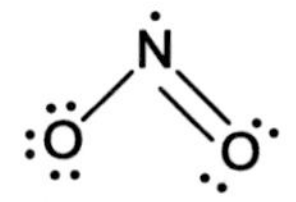

Plus others

N_2O_4, $2(5) + 4(6) = 34\ e^-$

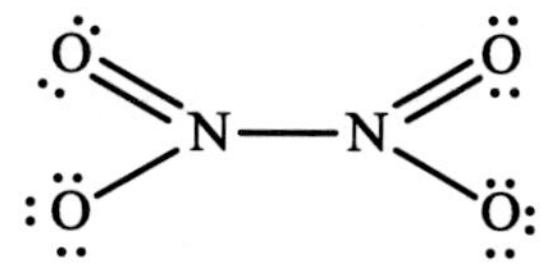

Plus other resonance structures

b. BH_3, $3 + 3(1) = 6\ e^-$

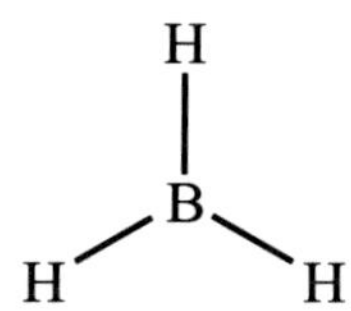

NH_3, $5 + 3(1) = 8\ e^-$

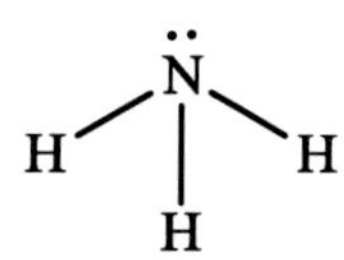

BH_3NH_3, $6 + 8 = 14\ e^-$

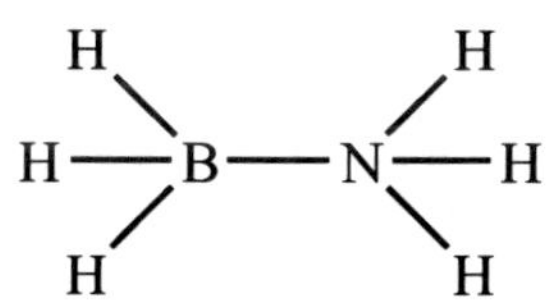

In reaction a, NO_2 has an odd number of electrons, so it is impossible to satisfy the octet rule. By dimerizing to form N_2O_4, the odd electron on two NO_2 molecules can pair up, giving a species whose Lewis structure can satisfy the octet rule. In general, odd-electron species are very reactive. In reaction b, BH_3 is electron-deficient. Boron has only six electrons around it. By forming BH_3NH_3, the boron atom satisfies the octet rule by accepting a lone pair of electrons from NH_3 to form a fourth bond.

87. PF_5, $5 + 5(7) = 40$ valence electrons

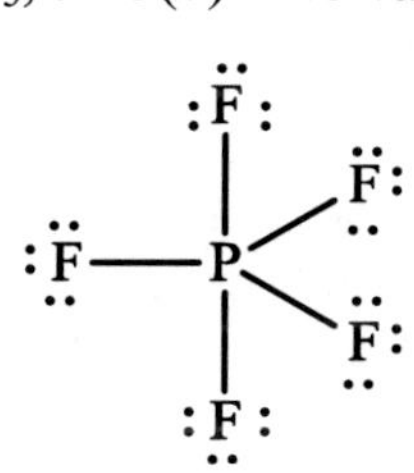

SF_4, $6 + 4(7) = 34\ e^-$

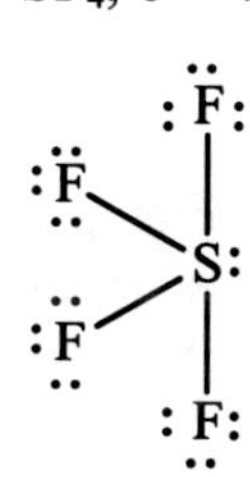

ClF_3, $7 + 3(7) = 28\ e^-$

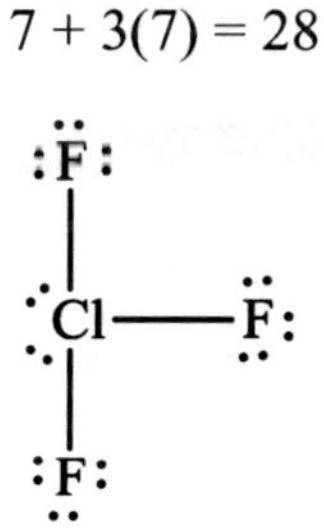

Br_3^-, $3(7) + 1 = 22\ e^-$

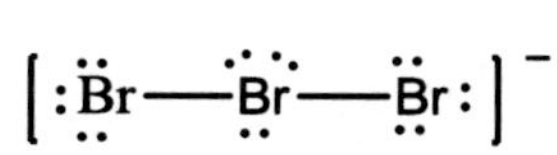

Row 3 and heavier nonmetals can have more than 8 electrons around them when they have to. Row 3 and heavier elements have empty d orbitals that are close in energy to valence s and p orbitals. These empty d orbitals can accept extra electrons.

For example, P in PF_5 has its five valence electrons in the 3s and 3p orbitals. These s and p orbitals have room for three more electrons, and if it has to, P can use the empty 3d orbitals for any electrons above 8.

88. SF_6, $6 + 6(7) = 48\ e^-$

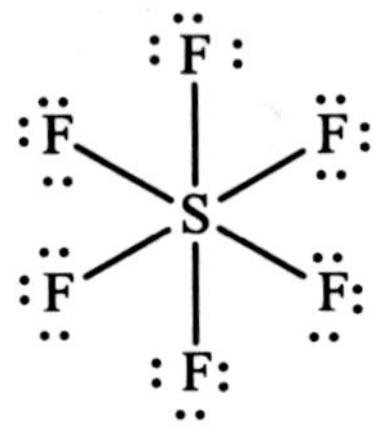

ClF_5, $7 + 5(7) = 42\ e^-$

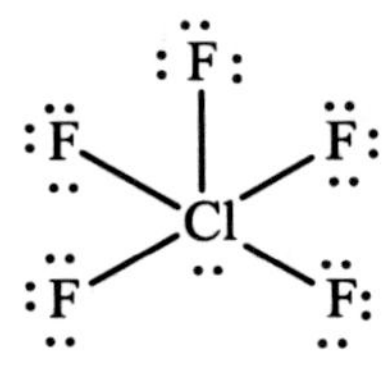

XeF_4, $8 + 4(7) = 36\ e^-$

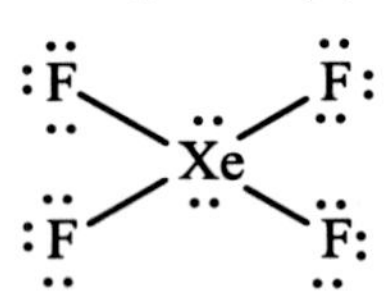

89. a. NO_2^- has $5 + 2(6) + 1 = 18$ valence electrons. The skeletal structure is O–N–O.

To get an octet about the nitrogen and only use 18 e^-, we must form a double bond to one of the oxygen atoms.

Because there is no reason to have the double bond to a particular oxygen atom, we can draw two resonance structures. Each Lewis structure uses the correct number of electrons and satisfies the octet rule, so each is a valid Lewis structure. Resonance structures occur when you have multiple bonds that can be in various positions. We say the actual structure is an average of these two resonance structures.

NO_3^- has $5 + 3(6) + 1 = 24$ valence electrons. We can draw three resonance structures for NO_3^-, with the double bond rotating among the three oxygen atoms.

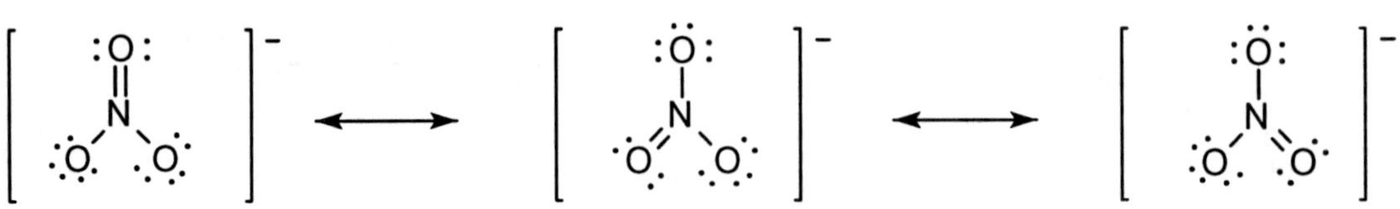

N_2O_4 has 2(5) + 4(6) = 34 valence electrons. We can draw four resonance structures for N_2O_4.

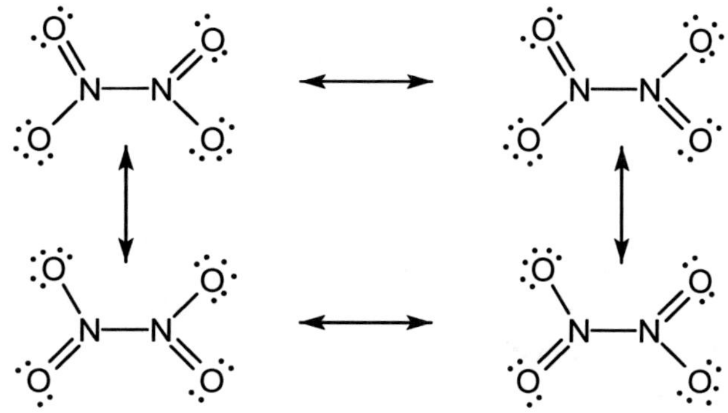

b. OCN^- has 6 + 4 + 5 + 1 = 16 valence electrons. We can draw three resonance structures for OCN^-.

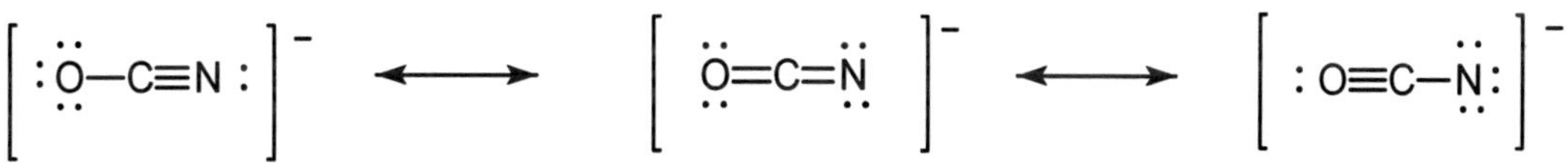

SCN^- has 6 + 4 + 5 + 1 = 16 valence electrons. Three resonance structures can be drawn.

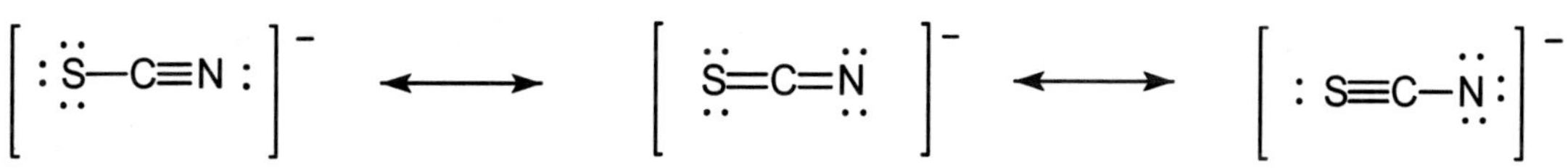

N_3^- has 3(5) + 1 = 16 valence electrons. As with OCN^- and SCN^-, three different resonance structures can be drawn.

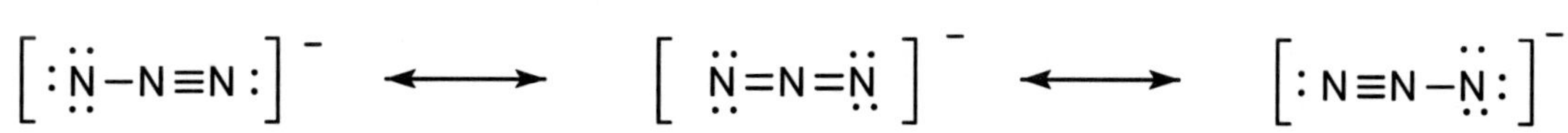

90. Ozone: O_3 has 3(6) = 18 valence electrons.

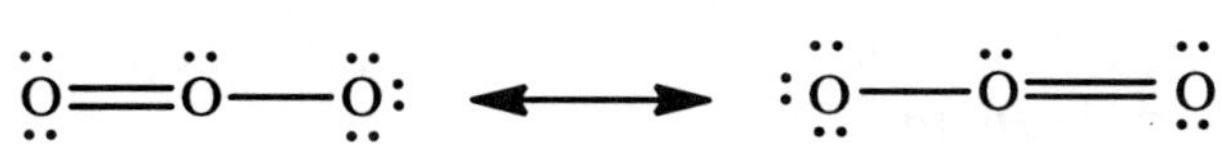

Sulfur dioxide: SO_2 has 6 + 2(6) = 18 valence electrons.

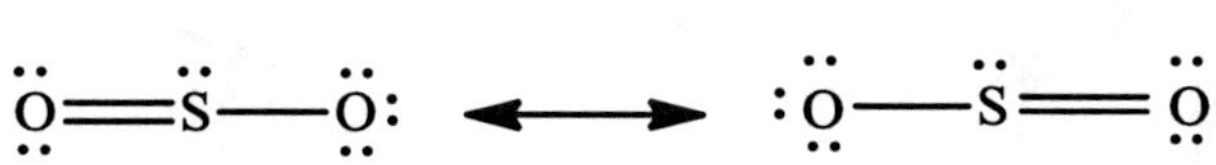

Sulfur trioxide: SO_3 has 6 + 3(6) = 24 valence electrons.

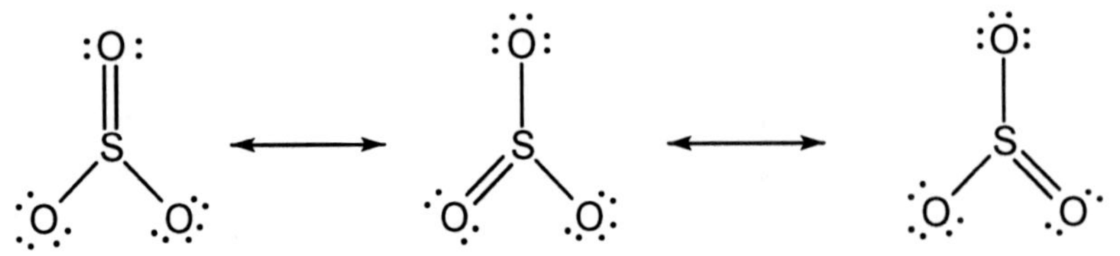

91. Benzene has 6(4) + 6(1) = 30 valence electrons. Two resonance structures can be drawn for benzene. The actual structure of benzene is an average of these two resonance structures; i.e., all carbon-carbon bonds are equivalent with a bond length and bond strength somewhere between a single and a double bond.

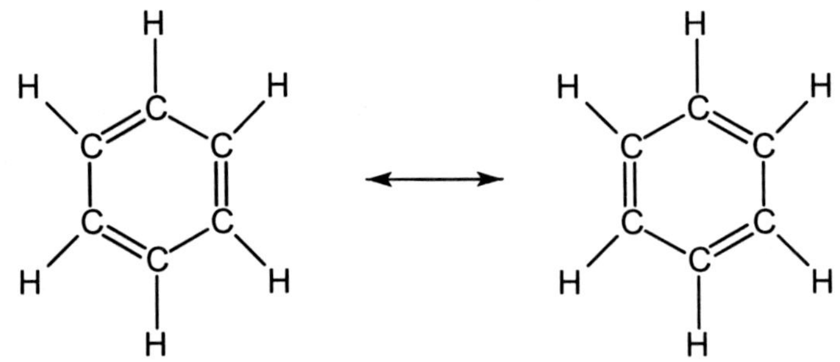

92. Borazine ($B_3N_3H_6$) has 3(3) + 3(5) + 6(1) = 30 valence electrons. The possible resonance structures are similar to those of benzene in Exercise 91.

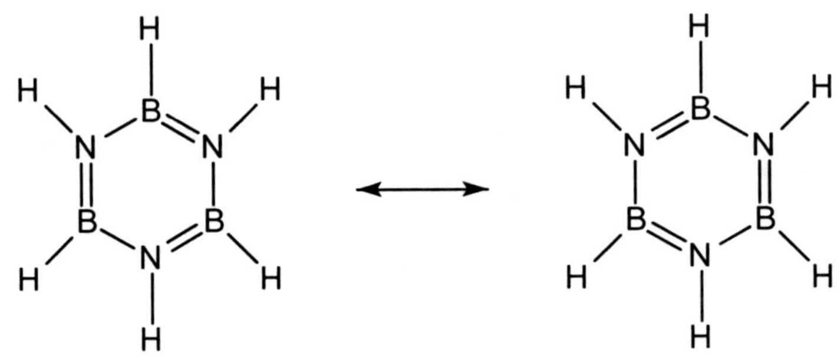

93. We will use a hexagon to represent the six-member carbon ring, and we will omit the four hydrogen atoms and the three lone pairs of electrons on each chlorine. If no resonance existed, we could draw four different molecules:

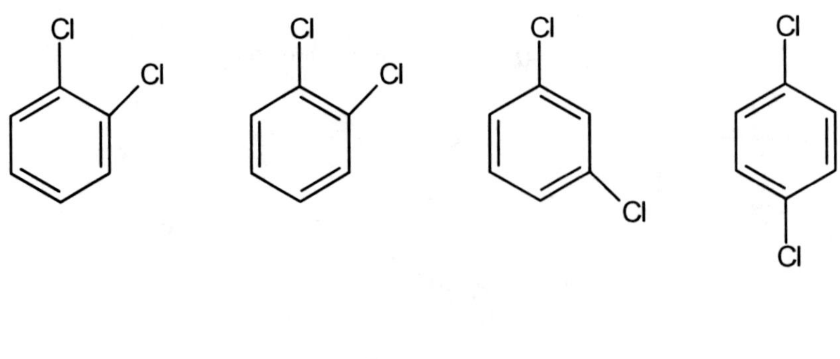

If the double bonds in the benzene ring exhibit resonance, then we can draw only three different dichlorobenzenes. The circle in the hexagon represents the delocalization of the three double bonds in the benzene ring (see Exercise 91).

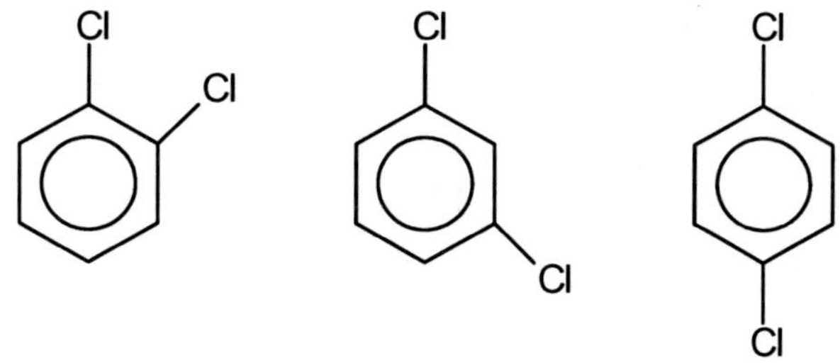

With resonance, all carbon-carbon bonds are equivalent. We can't distinguish between a single and double bond between adjacent carbons that have a chlorine attached. That only three isomers are observed supports the concept of resonance.

94. CO_3^{2-} has 4 + 3(6) + 2 = 24 valence electrons.

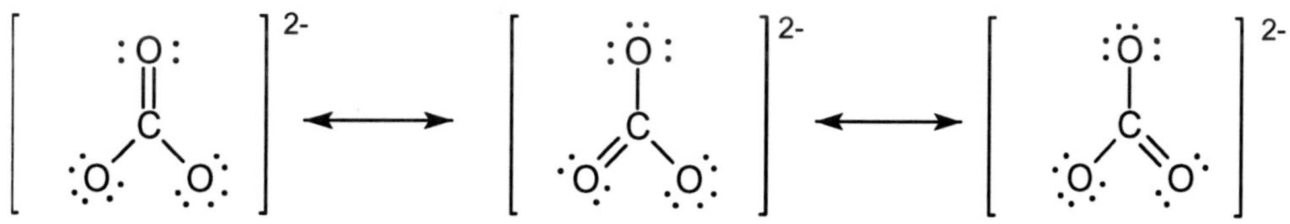

Three resonance structures can be drawn for CO_3^{2-}. The actual structure for CO_3^{2-} is an average of these three resonance structures. That is, the three C–O bond lengths are all equivalent, with a length somewhere between a single and a double bond. The actual bond length of 136 pm is consistent with this resonance view of CO_3^{2-}.

95. CH_3NCO has 4 + 3(1) + 5 + 4 + 6 = 22 valence electrons. Three resonance structures can be drawn for methyl isocyanate.

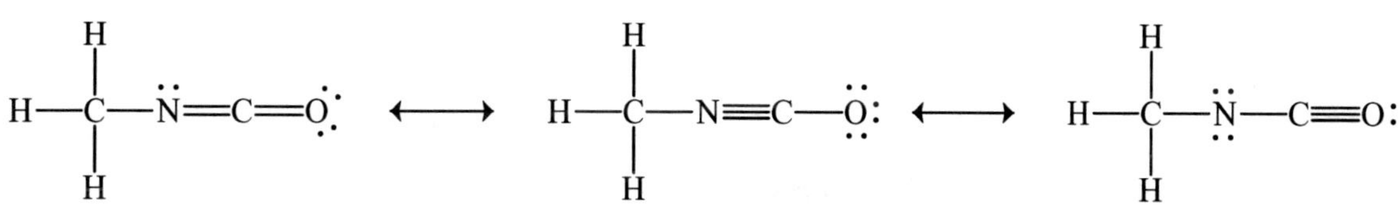

96. PAN ($H_3C_2NO_5$) has 3(1) + 2(4) + 5 + 5(6) = 46 valence electrons.

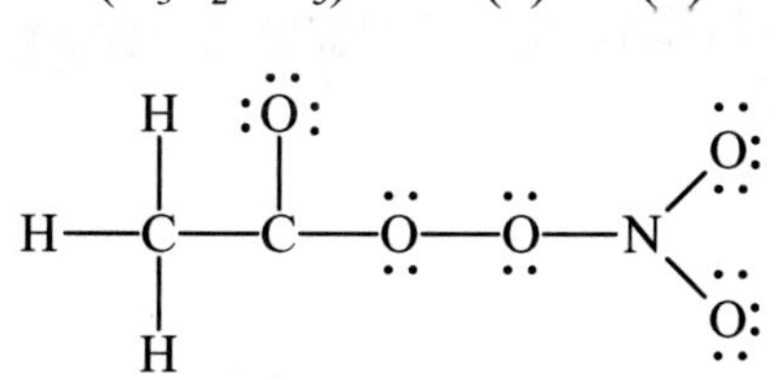

This is the skeletal structure with complete octets about oxygen atoms (46 electrons used).

This structure has used all 46 electrons, but there are only six electrons around one of the carbon atoms and the nitrogen atom. Two unshared pairs must become shared; i.e., we must form double bonds. The three possible resonance structures for PAN are:

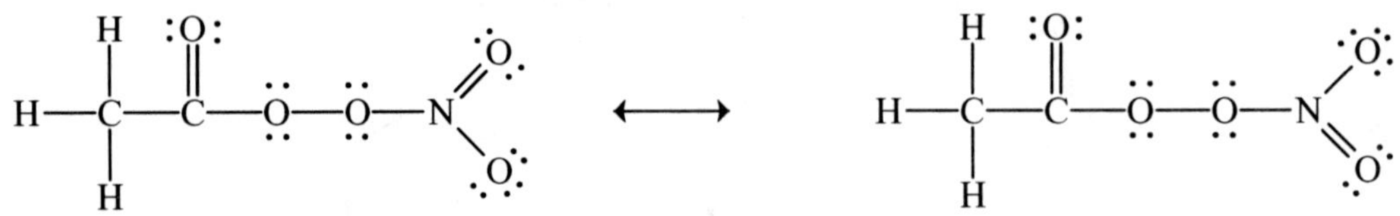

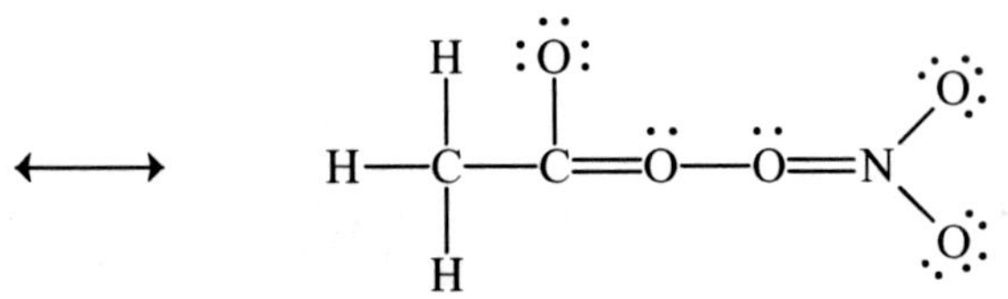

(last form not important)

97. The Lewis structures for the various species are:

CO (10 e^-): 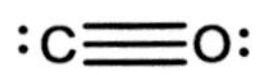 Triple bond between C and O.

CO_2 (16 e^-): 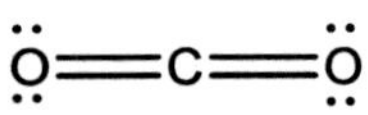 Double bond between C and O.

CO_3^{2-} (24 e^-): 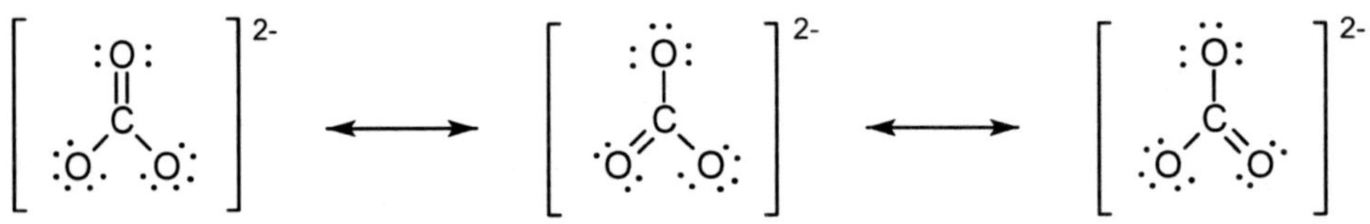

Average of 1 1/3 bond between C and O in CO_3^{2-}.

CH_3OH (14 e^-): 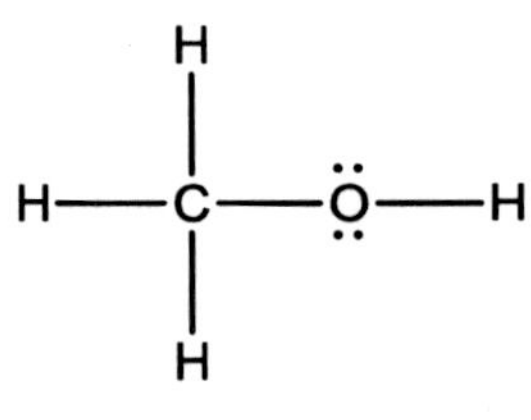 Single bond between C and O.

As the number of bonds increases between two atoms, bond strength increases, and bond length decreases. With this in mind, then:

Longest → shortest C – O bond: $CH_3OH > CO_3^{2-} > CO_2 > CO$

Weakest → strongest C – O bond: $CH_3OH < CO_3^{2-} < CO_2 < CO$

98. H_2NOH (14 e^-) 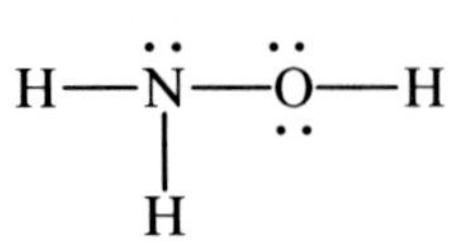 Single bond between N and O

N_2O (16 e^-): 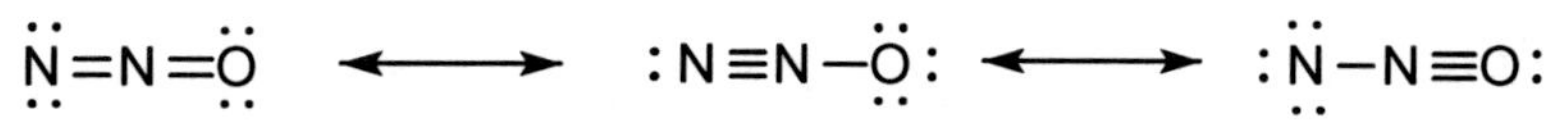

Average of a double bond between N and O

NO^+ (10 e^-): 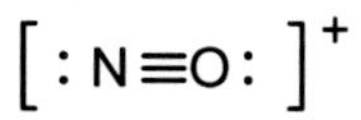 Triple bond between N and O

NO_2^- (18 e^-): 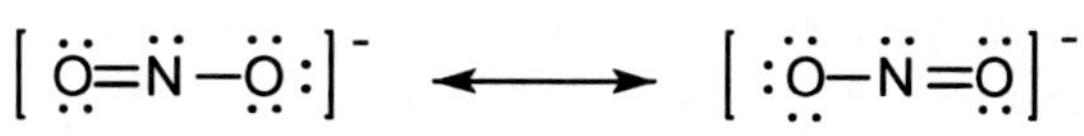

Average of 1 1/2 bond between N and O

NO_3^- (24 e^-): 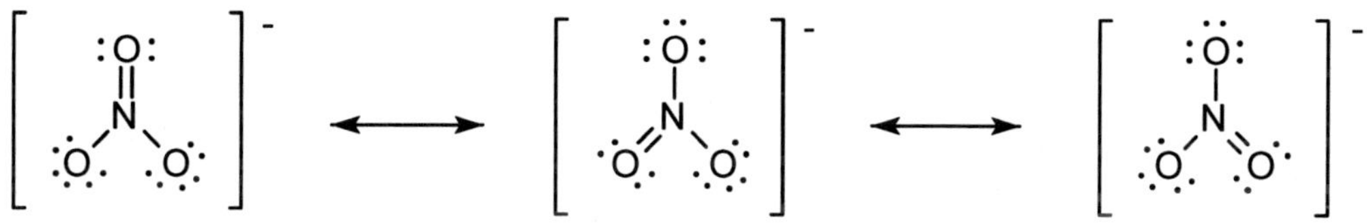

Average of 1 1/3 bond between N and O

From the Lewis structures, the order from shortest → longest N–O bond is:

$NO^+ < N_2O < NO_2^- < NO_3^- < H_2NOH$

Formal Charge

99. BF_3 has 3 + 3(7) = 24 valence electrons. The two Lewis structures to consider are:

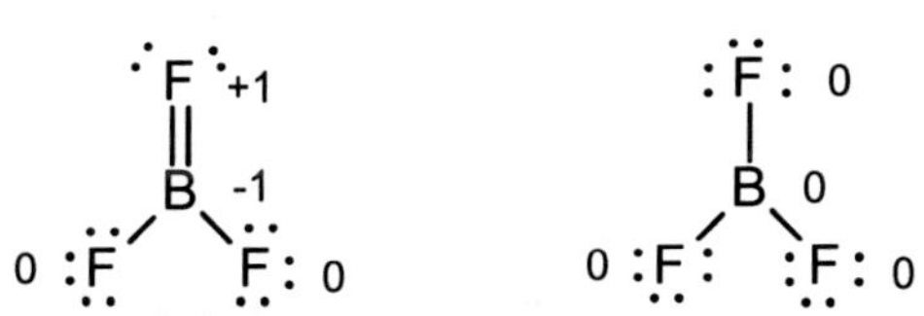

The formal charges for the various atoms are assigned in the Lewis structures. Formal charge = number of valence electrons on free atom – number of lone pair electrons on atoms – 1/2 (number of shared electrons of atom). For B in the first Lewis structure, formal charge (FC) = 3 – 0 – 1/2(8) = –1. For F in the first structure with the double bond, FC = 7 – 4 – 1/2(4) = +1. The others all have a formal charge equal to zero [FC = 7 – 6 – 1/2(2) – 0].

The first Lewis structure obeys the octet rule but has a +1 formal charge on the most electronegative element there is, fluorine, and a negative formal charge on a much less electronegative element, boron. This is just the opposite of what we expect: negative formal charge on F and positive formal charge on B. The other Lewis structure does not obey the octet rule for B but has a zero formal charge on each element in BF_3. Because structures generally want to minimize formal charge, then BF_3 with only single bonds is best from a formal charge point of view.

100. :C≡O: Carbon: FC = 4 – 2 – 1/2(6) = –1; oxygen: FC = 6 – 2 – 1/2(6) = +1

Electronegativity predicts the opposite polarization. The two opposing effects seem to partially cancel to give a much less polar molecule than expected.

101. See Exercise 84 for the Lewis structures of $POCl_3$, SO_4^{2-}, ClO_4^- and PO_4^{3-}. All these compounds/ions have similar Lewis structures to those of SO_2Cl_2 and XeO_4 shown below. Formal charge = [number of valence electrons on free atom] – [number of lone pair electrons on atom + 1/2(number of shared electrons of atom)].

a. $POCl_3$: P, FC = 5 – 1/2(8) = +1

b. SO_4^{2-}: S, FC = 6 – 1/2(8) = +2

c. ClO_4^-: Cl, FC = 7 – 1/2(8) = +3

d. PO_4^{3-}: P, FC = 5 – 1/2(8) = +1

e. SO_2Cl_2, 6 + 2(6) + 2(7) = 32 e^-

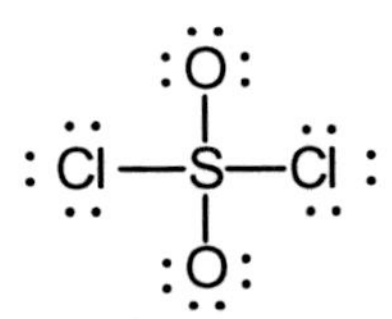

S, FC = 6 – 1/2(8) = +2

f. XeO_4, 8 + 4(6) = 32 e^-

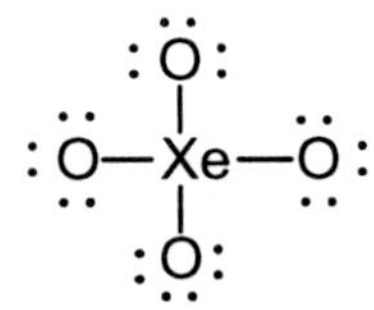

Xe, FC = 8 – 1/2(8) = +4

g. ClO_3^-, 7 + 3(6) + 1 = 26 e^-

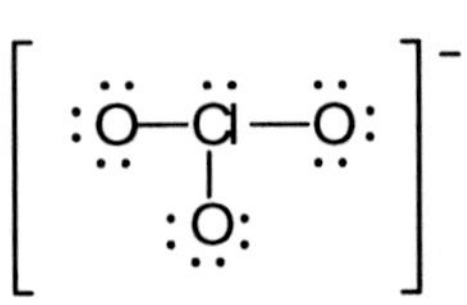

Cl, FC = 7 – 2 – 1/2(6) = +2

h. NO_4^{3-}, 5 + 4(6) + 3 = 32 e^-

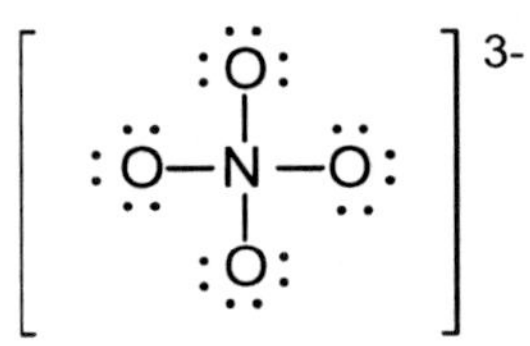

N, FC = 5 – 1/2(8) = +1

102. For SO_4^{2-}, ClO_4^-, PO_4^{3-} and ClO_3^-, only one of the possible resonance structures is drawn.

a. Must have five bonds to P to minimize formal charge of P. The best choice is to form a double bond to O since this will give O a formal charge of zero, and single bonds to Cl for the same reason.

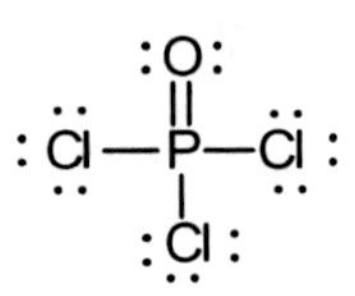

P, FC = 0

b. Must form six bonds to S to minimize formal charge of S.

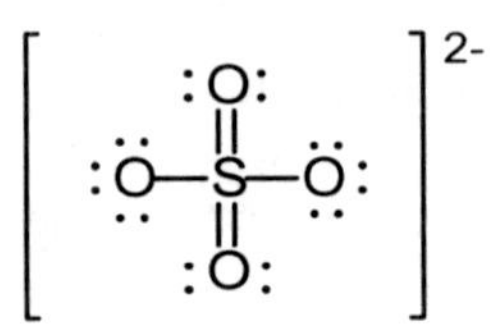

S, FC = 0

c. Must form seven bonds to Cl to minimize formal charge.

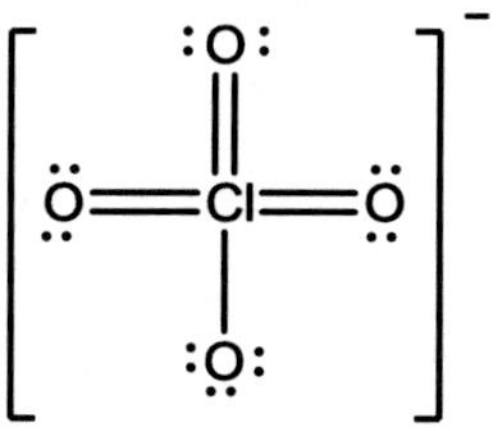

Cl, FC = O

d. Must form five bonds to P to to minimize formal charge.

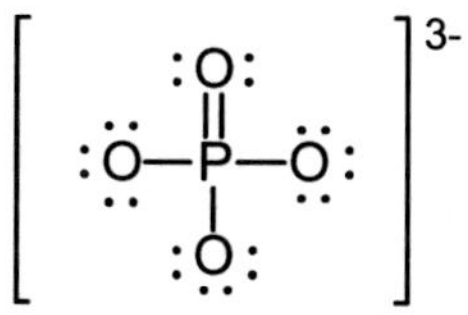

P, FC = 0

e.

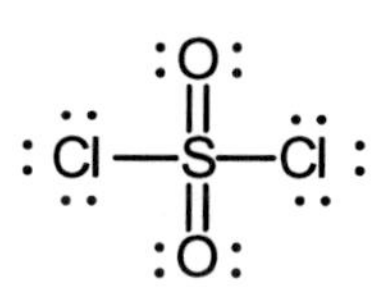

S, FC = 0
Cl, FC = 0
O, FC = 0

f.

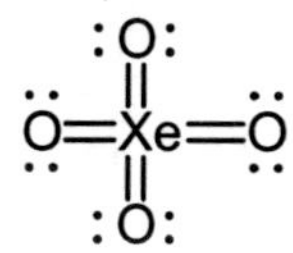

Xe, FC = 0

g.

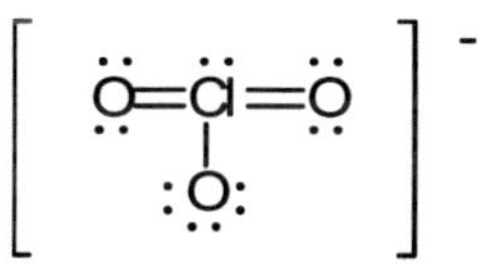

Cl, FC = 0

h. We can't. The following structure has a zero formal charge for N:

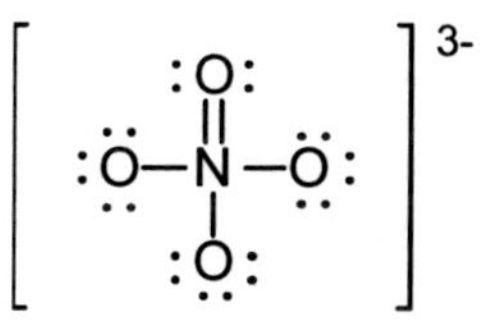

but N does not expand its octet. We wouldn't expect this resonance form to exist.

103. O_2F_2 has 2(6) + 2(7) = 26 valence e^-. The formal charge and oxidation number (state) of each atom is below the Lewis structure of O_2F_2.

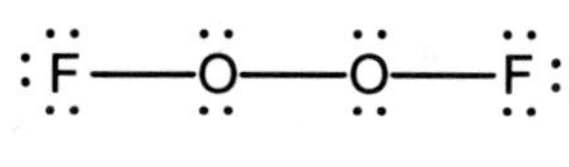

	F	O	O	F
Formal Charge	0	0	0	0
Oxid. Number	-1	+1	+1	-1

Oxidation states are more useful when accounting for the reactivity of O_2F_2. We are forced to assign +1 as the oxidation state for oxygen due to the bonding to fluorine. Oxygen is very electronegative, and +1 is not a stable oxidation state for this element.

104. OCN^- has 6 + 4 + 5 + 1 = 16 valence electrons.

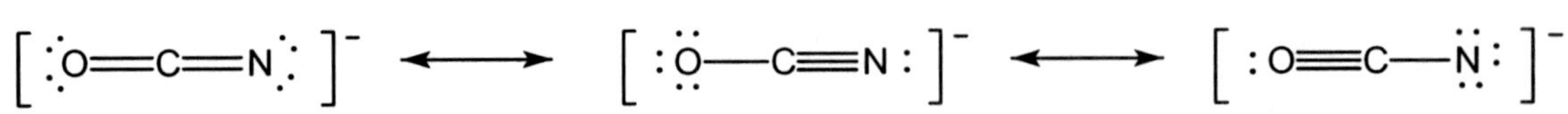

Formal charge: 0 0 -1 | -1 0 0 | +1 0 -2

Only the first two resonance structures should be important. The third places a positive formal charge on the most electronegative atom in the ion and a –2 formal charge on N.

CNO^- will also have 16 valence electrons.

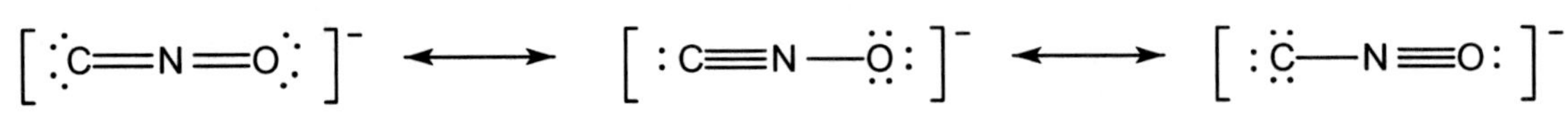

Formal charge: -2 +1 0 | -1 +1 -1 | -3 +1 +1

All the resonance structures for fulminate (CNO^-) involve greater formal charges than in cyanate (OCN^-), making fulminate more reactive (less stable).

105. SCl, 6 + 7 = 13; the formula could be SCl (13 valence electrons), S_2Cl_2 (26 valence electrons), S_3Cl_3 (39 valence electrons), etc. For a formal charge of zero on S, we will need each sulfur in the Lewis structure to have two bonds to it and two lone pairs [FC = 6 – 4 – 1/2(4) = 0]. Cl will need one bond and three lone pairs for a formal charge of zero [FC = 7 – 6 – 1/2(2) = 0]. Since chlorine wants only one bond to it, it will not be a central atom here. With this in mind, only S_2Cl_2 can have a Lewis structure with a formal charge of zero on all atoms. The structure is:

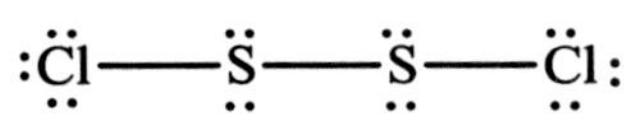

106. The nitrogen-nitrogen bond length of 112 pm is between a double (120 pm) and a triple (110 pm) bond. The nitrogen-oxygen bond length of 119 pm is between a single (147 pm) and a double bond (115 pm). The third resonance structure shown below doesn't appear to be as important as the other two since there is no evidence from bond lengths for a nitrogen-oxygen triple bond or a nitrogen-nitrogen single bond as in the third resonance form. We can adequately describe the structure of N_2O using the resonance forms:

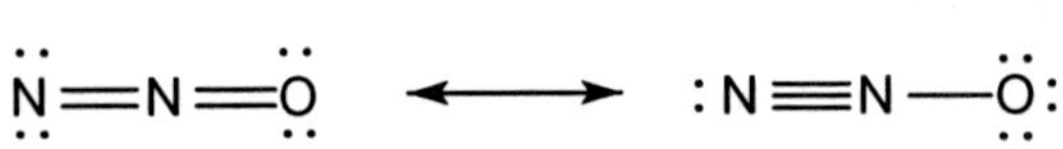

Assigning formal charges for all three resonance forms:

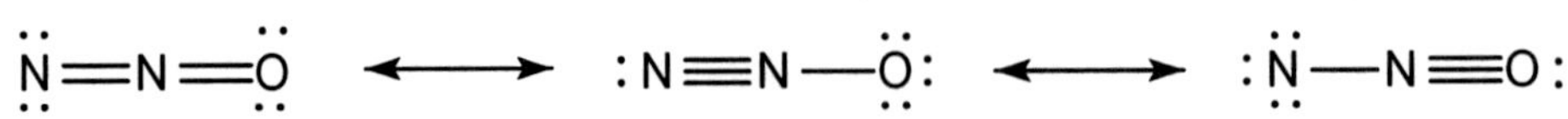

-1 +1 0 | 0 +1 -1 | -2 +1 +1

For:

(:N=), FC = 5 - 4 - 1/2(4) = -1

(=N=), FC = 5 - 1/2(8) = +1 , Same for (≡N—) and (—N≡)

(:N—), FC = 5 - 6 - 1/2(2) = -2 ; (:N≡), FC = 5 - 2 - 1/2(6) = 0

(=O:), FC = 6 - 4 - 1/2(4) = 0 ; (—O:), FC = 6 - 6 - 1/2(2) = -1

(≡O:), FC = 6 - 2 - 1/2(6) = +1

We should eliminate N–N≡O because it has a formal charge of +1 on the most electronegative element (O). This is consistent with the observation that the N–N bond is between a double and triple bond and that the N–O bond is between a single and double bond.

107. For formal charge values of zero:

(1) each carbon in the structure has 4 bonding pairs of electrons and no lone pairs;

(2) each N has 3 bonding pairs of electrons and 1 lone pair of electrons;

(3) each O has 2 bonding pairs of electrons and 2 lone pairs of electrons;

(4) each H is attached by only a single bond (1 bonding pair of electrons).

Following these guidelines, the Lewis structure is:

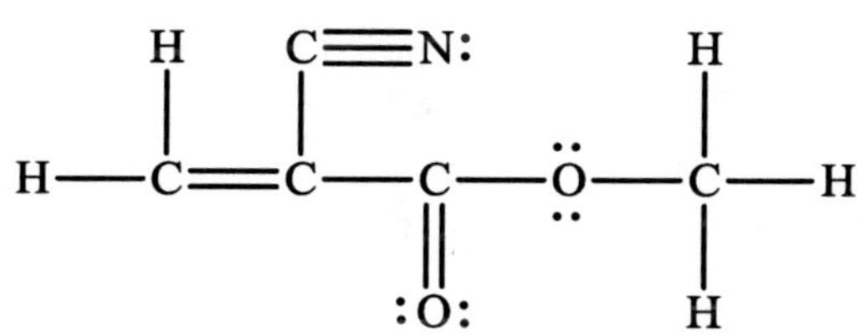

108. For a formal charge of zero, carbon atoms in the structure will all satisfy the octet rule by forming four bonds (with no lone pairs). Oxygen atoms have a formal charge of zero by forming two bonds and having two lone pairs of electrons. Hydrogen atoms have a formal charge of zero by forming a single bond (with no lone pairs). Following these guidelines, two resonance structures can be drawn for benzoic acid (see next page).

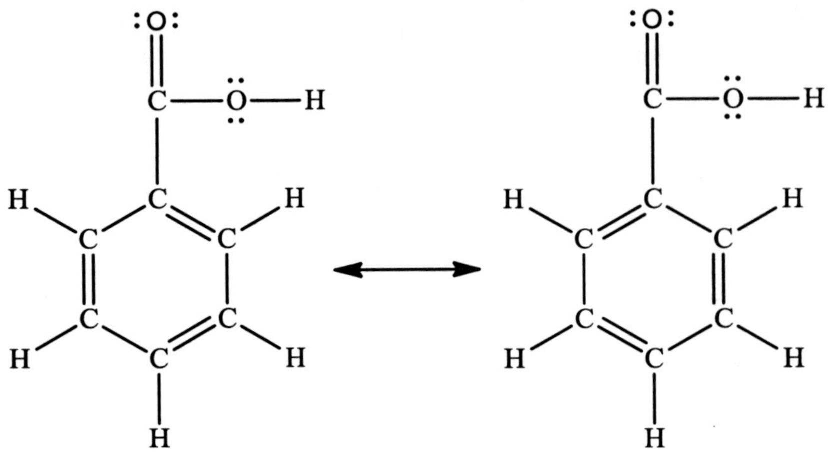

Molecular Structure and Polarity

109. The first step always is to draw a valid Lewis structure when predicting molecular structure. When resonance is possible, only one of the possible resonance structures is necessary to predict the correct structure because all resonance structures give the same structure. The Lewis structures are in Exercises 83 and 89. The structures and bond angles for each follow.

83: a. CCl_4: tetrahedral, 109.5°
b. NCl_3: trigonal pyramid, <109.5°
c. $SeCl_2$: V-shaped or bent, <109.5°
d. ICl: linear, but there is no bond angle present

Note: NCl_3 and $SeCl_2$ both have lone pairs of electrons on the central atom that result in bond angles that are something less than predicted from a tetrahedral arrangement (109.5°). However, we cannot predict the exact number. For the solutions manual, we will insert a less than sign to indicate this phenomenon. For bond angles equal to 120°, the lone pair phenomenon isn't as significant as compared to smaller bond angles. For these molecules, for example, NO_2^-, we will insert an approximate sign in front of the 120° to note that there may be a slight distortion from the VSEPR predicted bond angle.

89: a. NO_2^-: V-shaped, ≈120°; NO_3^-: trigonal planar, 120°
N_2O_4: trigonal planar, 120° about both N atoms
b. OCN^-, SCN^-, and N_3^- are all linear with 180° bond angles.

110. See Exercises 84 and 90 for the Lewis structures.

84: a. All are tetrahedral; 109.5°
b. All are trigonal pyramid; <109.5°
c. All are V-shaped; <109.5°

90: O_3 and SO_2 are V-shaped (or bent) with a bond angle ≈120°. SO_3 is trigonal planar with 120° bond angles.

111. From the Lewis structures (see Exercise 87), Br_3^- would have a linear molecular structure, ClF_3 would have a T-shaped molecular structure, and SF_4 would have a see-saw molecular structure. For example, consider ClF_3 (28 valence electrons):

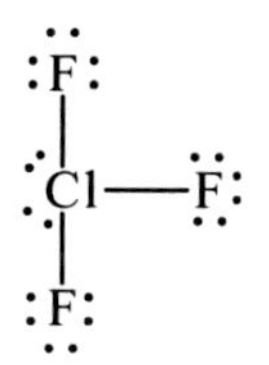

The central Cl atom is surrounded by five electron pairs, which requires a trigonal bipyramid geometry. Since there are three bonded atoms and two lone pairs of electrons about Cl, we describe the molecular structure of ClF_3 as T-shaped with predicted bond angles of about 90°. The actual bond angles will be slightly less than 90° due to the stronger repulsive effect of the lone-pair electrons as compared to the bonding electrons.

112. From the Lewis structures (see Exercise 88), XeF_4 would have a square planar molecular structure, and ClF_5 would have a square pyramid molecular structure.

113. a. SeO_3, $6 + 3(6) = 24\ e^-$

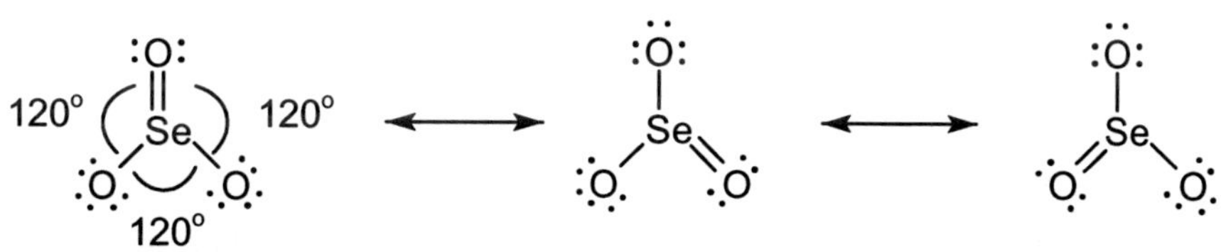

SeO_3 has a trigonal planar molecular structure with all bond angles equal to 120°. Note that any one of the resonance structures could be used to predict molecular structure and bond angles.

b. SeO_2, $6 + 2(6) = 18\ e^-$

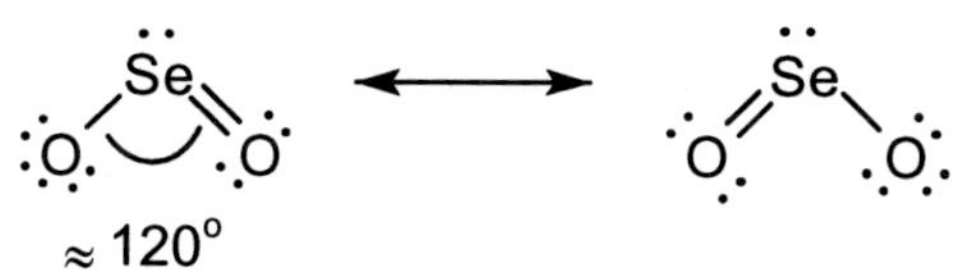

SeO_2 has a V-shaped molecular structure. We would expect the bond angle to be approximately 120° as expected for trigonal planar geometry.

Note: Both SeO_3 and SeO_2 structures have three effective pairs of electrons about the central atom. All of the structures are based on a trigonal planar geometry, but only SeO_3 is described as having a trigonal planar structure. Molecular structure always describes the relative positions of the atoms.

114. a. PCl_3 has $5 + 3(7) = 26$ valence electrons.

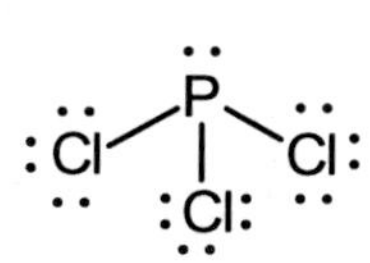

Trigonal pyramid; all angles are <109.5°.

b. SCl_2 has $6 + 2(7) = 20$ valence electrons.

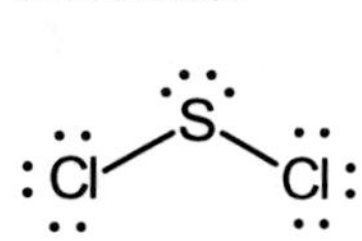

V-shaped; angle is <109.5°.

c. SiF_4 has 4 + 4(7) = 32 valence electrons.

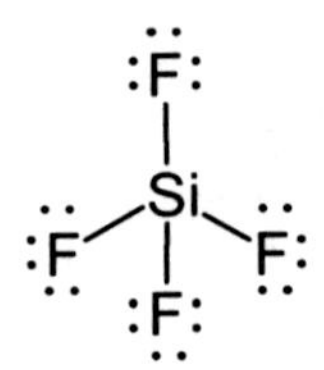

Tetrahedral; all angles are 109.5°.

Note: In PCl_3, SCl_2, and SiF_4, there are four pairs of electrons about the central atom in each case in this exercise. All of the structures are based on a tetrahedral geometry, but only SiF_4 has a tetrahedral structure. We consider only the relative positions of the atoms when describing the molecular structure.

115. a. $XeCl_2$ has 8 + 2(7) = 22 valence electrons.

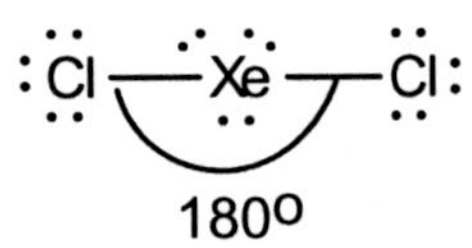

There are five pairs of electrons about the central Xe atom. The structure will be based on a trigonal bipyramid geometry. The most stable arrangement of the atoms in $XeCl_2$ is a linear molecular structure with a 180° bond angle.

b. ICl_3 has 7 + 3(7) = 28 valence electrons.

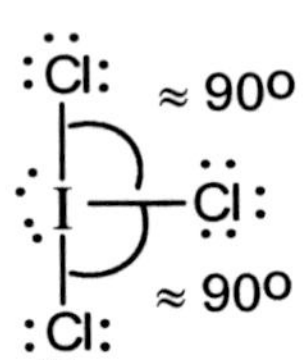

T-shaped; the ClICl angles are ≈90°. Since the lone pairs will take up more space, the ClICl bond angles will probably be slightly less than 90°.

c. TeF_4 has 6 + 4(7) = 34 valence electrons.

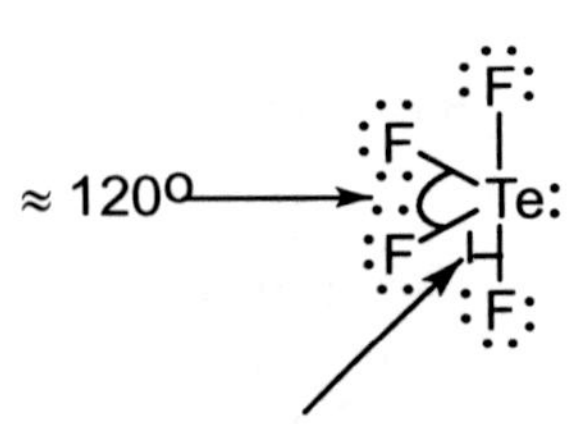

See-saw or teeter-totter or distorted tetrahedron

d. PCl_5 has 5 + 5(7) = 40 valence electrons.

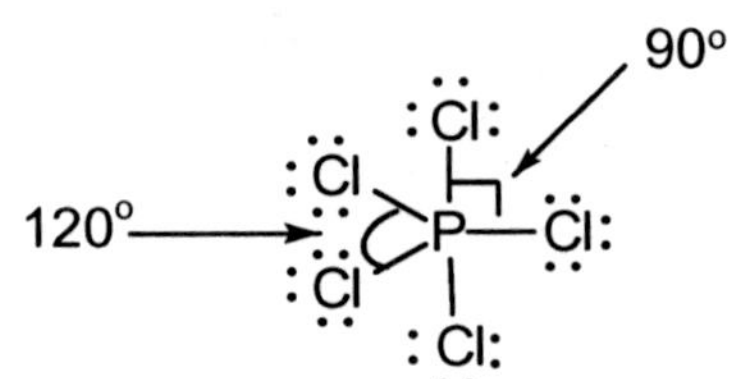

Trigonal bipyramid

All the species in this exercise have five pairs of electrons around the central atom. All the structures are based on a trigonal bipyramid geometry, but only in PCl_5 are all the pairs, bonding pairs. Thus PCl_5 is the only one for which we describe the molecular structure as trigonal bipyramid. Still, we had to begin with the trigonal bipyramid geometry to get to the structures (and bond angles) of the others.

116. a. ICl_5, 7 + 5(7) = 42 e^-

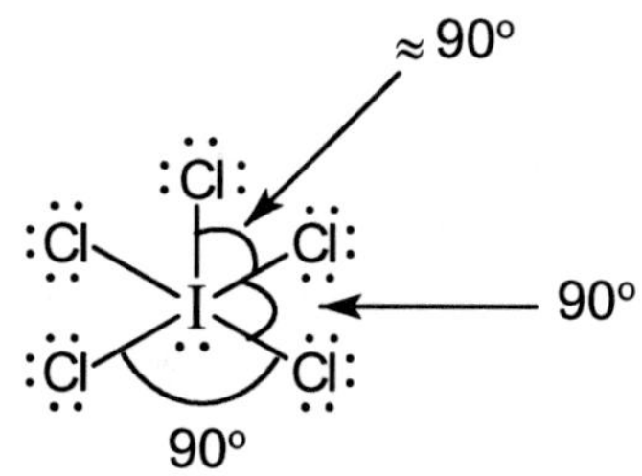

Square pyramid, ≈90° bond angles

b. $XeCl_4$, 8 + 4(7) = 36 e^-

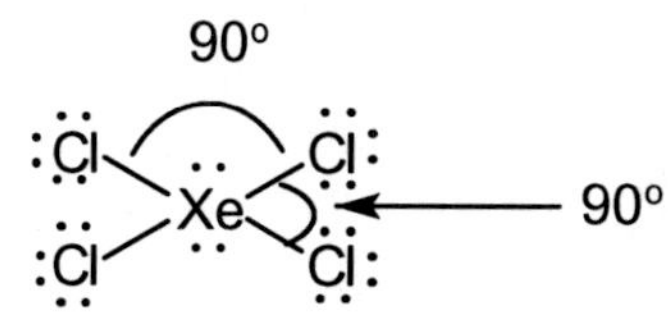

Square planar, 90° bond angles

c. $SeCl_6$ has 6 + 6(7) = 48 valence electrons.

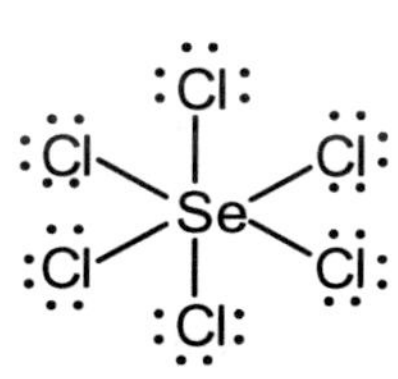

Octahedral, 90° bond angles

Note: All these species have six pairs of electrons around the central atom. All three structures are based on the octahedron, but only $SeCl_6$ has an octahedral molecular structure.

117. SeO_3 and SeO_2 both have polar bonds, but only SeO_2 has a dipole moment. The three bond dipoles from the three polar Se–O bonds in SeO_3 will all cancel when summed together. Hence SeO_3 is nonpolar since the overall molecule has no resulting dipole moment. In SeO_2, the two Se–O bond dipoles do not cancel when summed together; hence SeO_2 has a net dipole moment (is polar). Since O is more electronegative than Se, the negative end of the dipole moment is between the two O atoms, and the positive end is around the Se atom. The arrow in the following illustration represents the overall dipole moment in SeO_2. Note that to predict polarity for SeO_2, either of the two resonance structures can be used.

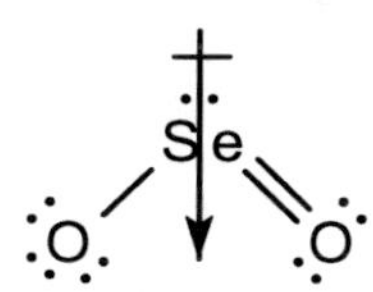

118. All have polar bonds; in SiF_4, the individual bond dipoles cancel when summed together, and in PCl_3 and SCl_2, the individual bond dipoles do not cancel. Therefore, SiF_4 has no net dipole moment (is nonpolar), and PCl_3 and SCl_2 have net dipole moments (are polar). For PCl_3, the negative end of the dipole moment is between the more electronegative chlorine atoms, and the positive end is around P. For SCl_2, the negative end is between the more electronegative Cl atoms, and the positive end of the dipole moment is around S.

119. All have polar bonds, but only TeF_4 and ICl_3 have dipole moments. The bond dipoles from the five P–Cl bonds in PCl_5 cancel each other when summed together, so PCl_5 has no net dipole moment. The bond dipoles in $XeCl_2$ also cancel:

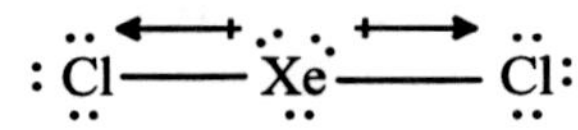

Because the bond dipoles from the two Xe–Cl bonds are equal in magnitude but point in opposite directions, they cancel each other, and $XeCl_2$ has no net dipole moment (is nonpolar). For TeF_4 and ICl_3, the arrangement of these molecules is such that the individual bond dipoles do *not* all cancel, so each has an overall net dipole moment (is polar).

120. All have polar bonds, but only ICl_5 has an overall net dipole moment. The six bond dipoles in $SeCl_6$ all cancel each other, so $SeCl_6$ has no net dipole moment. The same is true for $XeCl_4$:

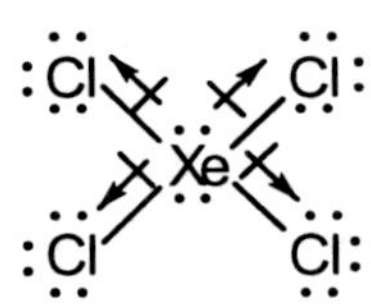

When the four bond dipoles are added together, they all cancel each other, resulting in $XeCl_4$ having no overall dipole moment (is nonpolar). ICl_5 has a structure in which the individual bond dipoles do *not* all cancel, hence ICl_5 has a dipole moment (is polar)

121. Molecules that have an overall dipole moment are called polar molecules, and molecules that do not have an overall dipole moment are called nonpolar molecules.

a. OCl_2, $6 + 2(7) = 20\ e^-$

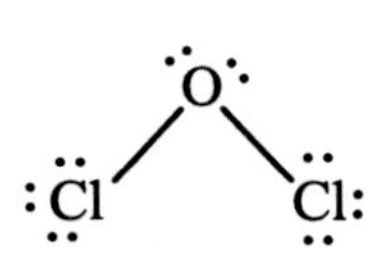

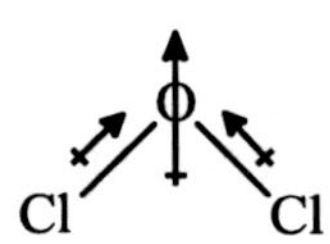

V-shaped, polar; OCl_2 is polar because the two O–Cl bond dipoles don't cancel each other. The resulting dipole moment is shown in the drawing.

KrF_2, $8 + 2(7) = 22\ e^-$

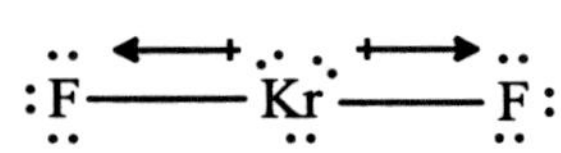

Linear, nonpolar; the molecule is nonpolar because the two Kr–F bond dipoles cancel each other.

BeH_2, $2 + 2(1) = 4\ e^-$

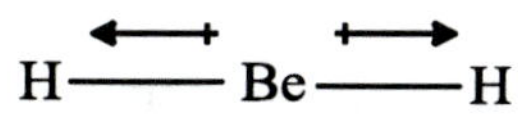

Linear, nonpolar; Be–H bond dipoles are equal and point in opposite directions. They cancel each other. BeH_2 is nonpolar.

SO_2, $6 + 2(6) = 18\ e^-$

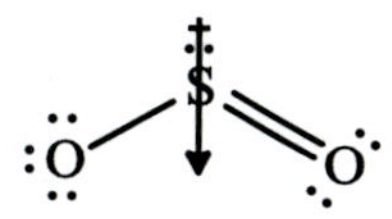

V-shaped, polar; the S–O bond dipoles do not cancel, so SO_2 is polar (has a net dipole moment). Only one resonance structure is shown.

Note: All four species contain three atoms. They have different structures because the number of lone pairs of electrons around the central atom are different in each case.

b. SO_3, $6 + 3(6) = 24\ e^-$

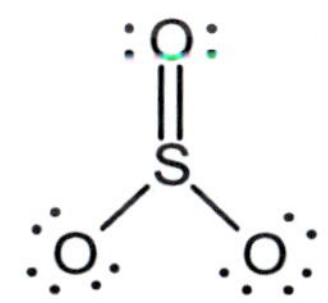

Trigonal planar, nonpolar; bond dipoles cancel. Only one resonance structure is shown.

NF_3, $5 + 3(7) = 26\ e^-$

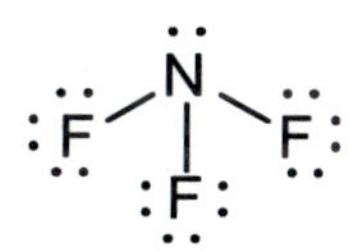

Trigonal pyramid, polar; bond dipoles do not cancel.

IF_3 has $7 + 3(7) = 28$ valence electrons.

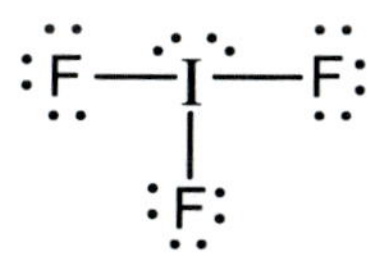

T-shaped, polar; bond dipoles do not cancel.

Note: Each molecule has the same number of atoms but different structures because of differing numbers of lone pairs around each central atom.

c. CF_4, $4 + 4(7) = 32\ e^-$

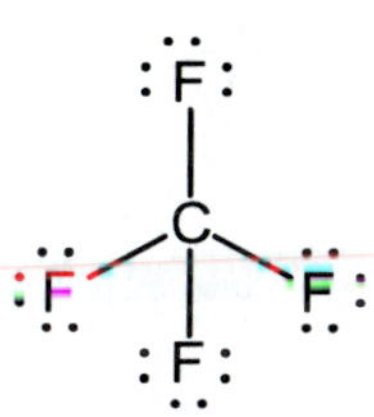

Tetrahedral, nonpolar; bond dipoles cancel.

SeF_4, $6 + 4(7) = 34\ e^-$

See-saw, polar; bond dipoles do not cancel.

KrF_4, 8 + 4(7) = 36 valence electrons

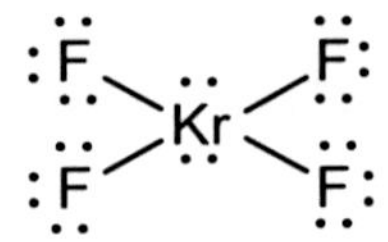

Square planar, nonpolar; bond dipoles cancel.

Note: Again, each molecule has the same number of atoms but different structures because of differing numbers of lone pairs around the central atom.

d. IF_5, 7 + 5(7) = 42 e^-

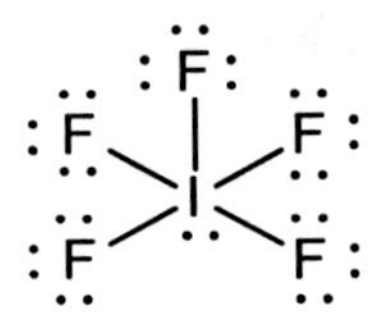

Square pyramid, polar; bond dipoles do not cancel.

AsF_5, 5 + 5(7) = 40 e^-

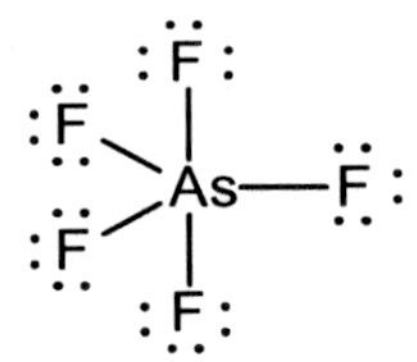

Trigonal bipyramid, nonpolar; bond dipoles cancel.

Note: Yet again, the molecules have the same number of atoms but different structures because of the presence of differing numbers of lone pairs.

122. a.

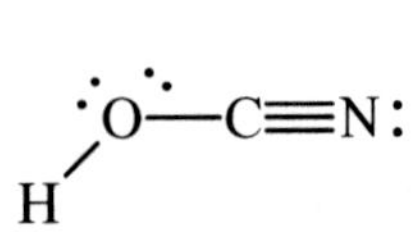

Polar; the bond dipoles do not cancel.

b.

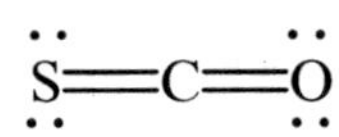

Polar; the C–O bond is a more polar bond than the C–S bond, so the two bond dipoles do not cancel each other.

c.

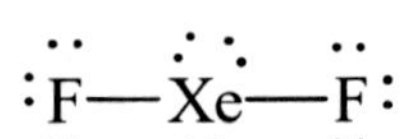

Nonpolar; the two Xe–F bond dipoles cancel each other.

d.

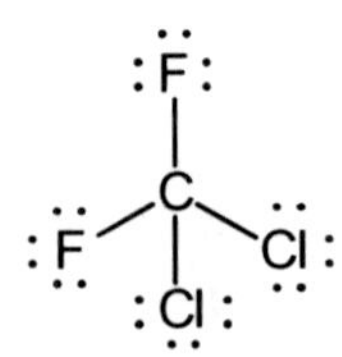

Polar; all the bond dipoles are not equivalent, and they don't cancel each other.

e.

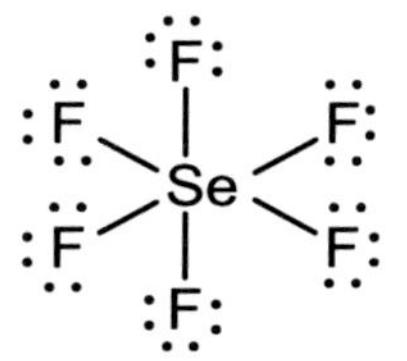

Nonpolar; the six Se–F bond dipoles cancel each other.

f.

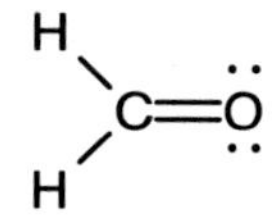

Polar; the bond dipoles are not equivalent, and they don't cancel

123. EO_3^- is the formula of the ion. The Lewis structure has 26 valence electrons. Let x = number of valence electrons of element E.

$26 = x + 3(6) + 1,\ x = 7$ valence electrons

Element E is a halogen because halogens have seven valence electrons. Some possible identities are F, Cl, Br, and I. The EO_3^- ion has a trigonal pyramid molecular structure with bond angles of less than 109.5° (<109.5°).

124. The formula is EF_2O^{2-}, and the Lewis structure has 28 valence electrons.

$28 = x + 2(7) + 6 + 2,\ x = 6$ valence electrons for element E

Element E must belong to the Group 6A elements since E has six valence electrons. E must also be a row 3 or heavier element since this ion has more than eight electrons around the central E atom (row 2 elements never have more than eight electrons around them). Some possible identities for E are S, Se, and Te. The ion has a T-shaped molecular structure with bond angles of $\approx$90°.

125. All these molecules have polar bonds that are symmetrically arranged about the central atoms. In each molecule, the individual bond dipoles cancel each other out to give no net overall dipole moment. All these molecules are nonpolar even though they all contain polar bonds.

126. XeF_2Cl_2, $8 + 2(7) + 2(7) = 36\ e^-$

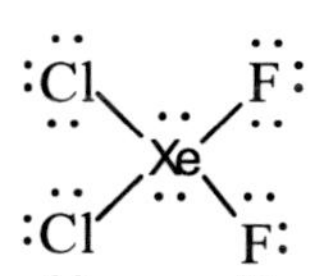

polar

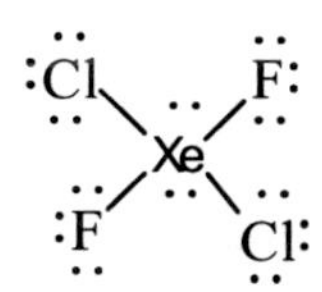

nonpolar

The two possible structures for XeF_2Cl_2 are above. In the first structure, the F atoms are 90° apart from each other, and the Cl atoms are also 90° apart. The individual bond dipoles would not cancel in this molecule, so this molecule is polar. In the second possible structure, the F atoms are 180° apart, as are the Cl atoms. Here, the bond dipoles are symmetrically arranged so they do cancel each other out, and this molecule is nonpolar. Therefore, measurement of the dipole moment would differentiate between the two compounds. These are different compounds and not resonance structures.

Additional Exercises

127. a. Radius: $N^+ < N < N^-$; IE: $N^- < N < N^+$

N^+ has the fewest electrons held by the seven protons in the nucleus, whereas N^- has the most electrons held by the seven protons. The seven protons in the nucleus will hold the electrons most tightly in N^+ and least tightly in N^-. Therefore, N^+ has the smallest radius with the largest ionization energy (IE), and N^- is the largest species with the smallest IE.

b. Radius: $Cl^+ < Cl < Se < Se^-$; IE: $Se^- < Se < Cl < Cl^+$

The general trends tell us that Cl has a smaller radius than Se and a larger IE than Se. Cl^+, with fewer electron-electron repulsions than Cl, will be smaller than Cl and have a larger IE. Se^-, with more electron-electron repulsions than Se, will be larger than Se and have a smaller IE.

c. Radius: $Sr^{2+} < Rb^+ < Br^-$; IE: $Br^- < Rb^+ < Sr^{2+}$

These ions are isoelectronic. The species with the most protons (Sr^{2+}) will hold the electrons most tightly and will have the smallest radius and largest IE. The ion with the fewest protons (Br^-) will hold the electrons least tightly and will have the largest radius and smallest IE.

128. a. $Na^+(g) + Cl^-(g) \rightarrow NaCl(s)$ b. $NH_4^+(g) + Br^-(g) \rightarrow NH_4Br(s)$

c. $Mg^{2+}(g) + S^{2-}(g) \rightarrow MgS(s)$ d. $O_2(g) \rightarrow 2\ O(g)$

129. a.

$HF(g) \rightarrow H(g) + F(g)$	$\Delta H = 565$ kJ
$H(g) \rightarrow H^+(g) + e^-$	$\Delta H = 1312$ kJ
$F(g) + e^- \rightarrow F^-(g)$	$\Delta H = -327.8$ kJ
$HF(g) \rightarrow H^+(g) + F^-(g)$	$\Delta H = 1549$ kJ

b.

$HCl(g) \rightarrow H(g) + Cl(g)$	$\Delta H = 427$ kJ
$H(g) \rightarrow H^+(g) + e^-$	$\Delta H = 1312$ kJ
$Cl(g) + e^- \rightarrow Cl^-(g)$	$\Delta H = -348.7$ kJ
$HCl(g) \rightarrow H^+(g) + Cl^-(g)$	$\Delta H = 1390.$ kJ

c.

$HI(g) \rightarrow H(g) + I(g)$	$\Delta H = 295$ kJ
$H(g) \rightarrow H^+(g) + e^-$	$\Delta H = 1312$ kJ
$I(g) + e^- \rightarrow I^-(g)$	$\Delta H = -295.2$ kJ
$HI(g) \rightarrow H^+(g) + I^-(g)$	$\Delta H = 1312$ kJ

d.

$H_2O(g) \rightarrow OH(g) + H(g)$	$\Delta H = 467$ kJ
$H(g) \rightarrow H^+(g) + e^-$	$\Delta H = 1312$ kJ
$OH(g) + e^- \rightarrow OH^-(g)$	$\Delta H = -180.$ kJ
$H_2O(g) \rightarrow H^+(g) + OH^-(g)$	$\Delta H = 1599$ kJ

130. CO_3^{2-} has 4 + 3(6) + 2 = 24 valence electrons.

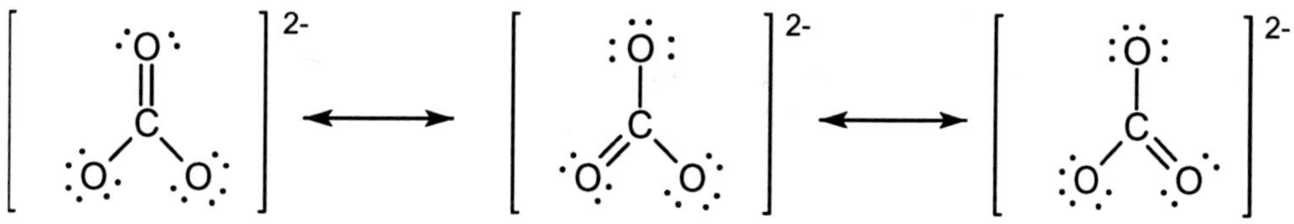

HCO_3^- has 1 + 4 + 3(6) + 1 = 24 valence electrons.

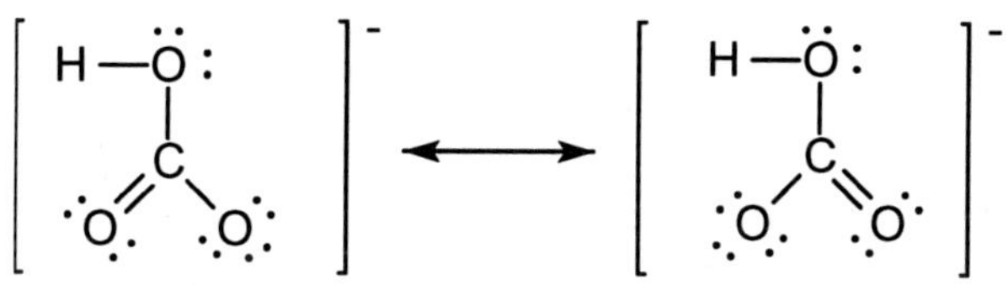

H_2CO_3 has 2(1) + 4 + 3(6) = 24 valence electrons.

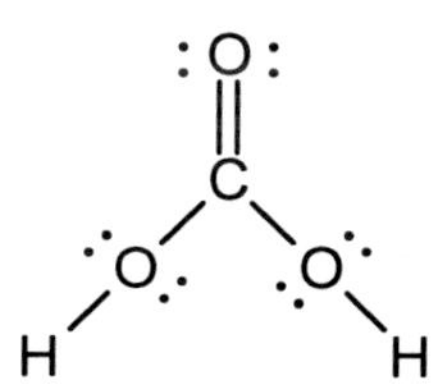

The Lewis structures for the reactants and products are:

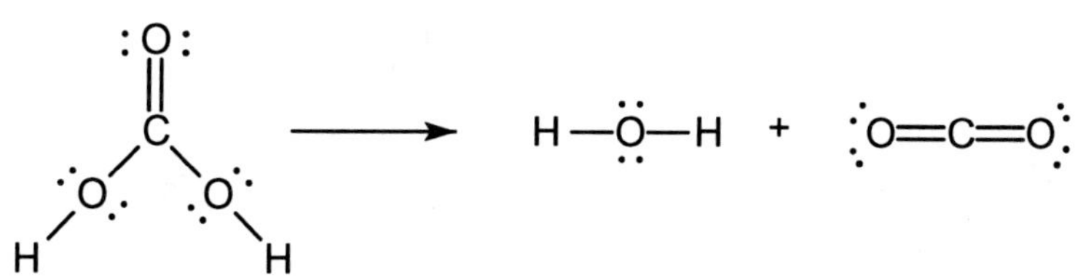

Bonds broken:	Bonds formed:
2 C–O (358 kJ/mol)	1 C=O (799 kJ/mol)
1 O–H (467 kJ/mol)	1 O–H (467 kJ/mol)

$\Delta H = 2(358) + 467 - [799 + 467] = -83$ kJ; the carbon-oxygen double bond is stronger than two carbon-oxygen single bonds; hence CO_2 and H_2O are more stable than H_2CO_3.

131. The stable species are:

a. NaBr: In $NaBr_2$, the sodium ion would have a 2+ charge, assuming that each bromine has a 1– charge. Sodium doesn't form stable Na^{2+} ionic compounds.

b. ClO_4^-: ClO_4 has 31 valence electrons, so it is impossible to satisfy the octet rule for all atoms in ClO_4. The extra electron from the 1– charge in ClO_4^- allows for complete octets for all atoms.

c. XeO_4: We can't draw a Lewis structure that obeys the octet rule for SO_4 (30 electrons), unlike XeO_4 (32 electrons).

d. SeF_4: Both compounds require the central atom to expand its octet. O is too small and doesn't have low-energy d orbitals to expand its octet (which is true for all row 2 elements).

132. a. All have 24 valence electrons and the same number of atoms in the formula. All have the same resonance Lewis structures; the structures are all trigonal planar with 120° bond angles. The Lewis structures for NO_3^- and CO_3^{2-} will be the same as the three SO_3 Lewis structures shown below.

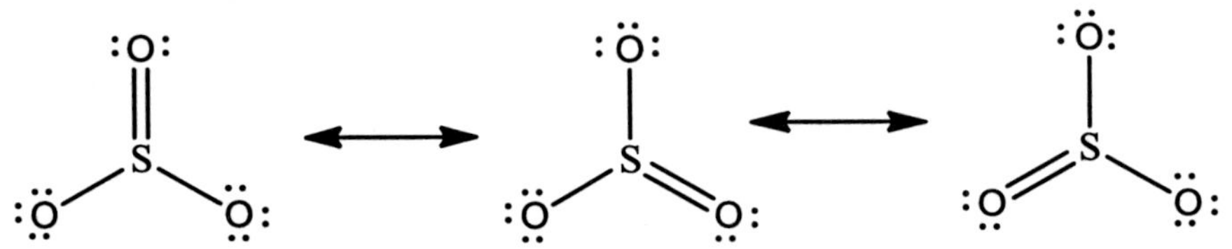

b. All have 18 valence electrons and the same number of atoms. All have the same resonance Lewis structures; the molecular structures are all V-shaped with ≈120° bond angles. O_3 and SO_2 have the same two Lewis structures as is shown for NO_2^-.

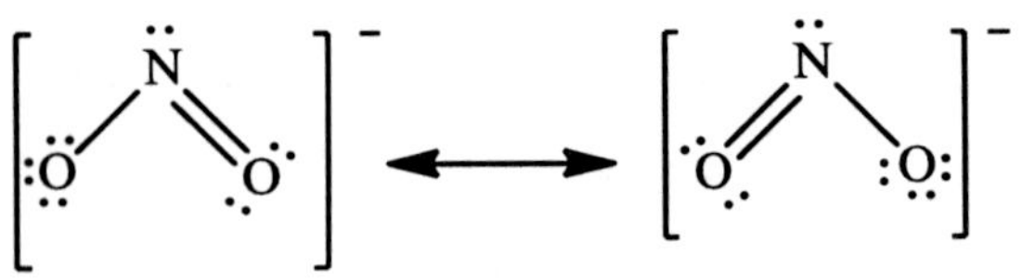

133. a. $XeCl_4$, $8 + 4(7) = 36\ e^-$ $XeCl_2$, $8 + 2(7) = 22\ e^-$

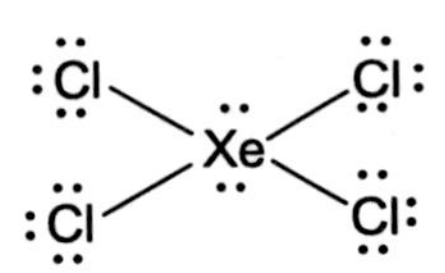

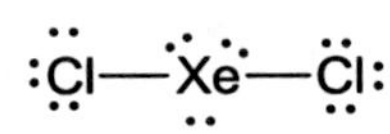

Square planar, 90°, nonpolar Linear, 180°, nonpolar

Both compounds have a central Xe atom with lone pairs and terminal Cl atoms, and both compounds do not satisfy the octet rule. In addition, both are nonpolar because the Xe–Cl bond dipoles and lone pairs around Xe are arranged in such a manner that they cancel each other out. The last item in common is that both have 180° bond angles. Although we haven't emphasized this, the bond angles between the Cl atoms on the diagonal in $XeCl_4$ are 180° apart from each other.

b. All of these are polar covalent compounds. The bond dipoles do not cancel out each other when summed together. The reason the bond dipoles are not symmetrically arranged in these compounds is that they all have at least one lone pair of electrons on the central atom, which disrupts the symmetry. Note that there are molecules that have lone pairs and are nonpolar, e.g., $XeCl_4$ and $XeCl_2$ in the preceding problem. A lone pair on a central atom does not guarantee a polar molecule.

134. The general structure of the trihalide ions is:

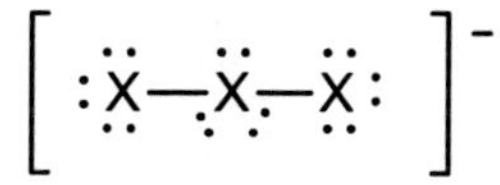

Bromine and iodine are large enough and have low-energy, empty d orbitals to accommodate the expanded octet. Fluorine is small, and its valence shell contains only 2s and 2p orbitals (four orbitals) and cannot expand its octet. The lowest-energy d orbitals in F are 3d; they are too high in energy compared with 2s and 2p to be used in bonding.

135. Yes, each structure has the same number of effective pairs around the central atom, giving the same predicted molecular structure for each compound/ion. (A multiple bond is counted as a single group of electrons.)

136. a.

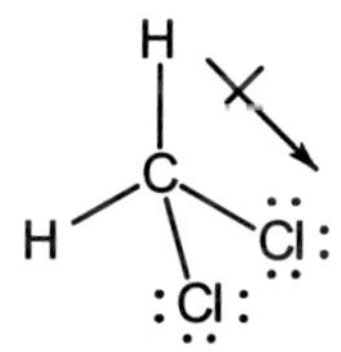

The C–H bonds are assumed nonpolar since the electronegativities of C and H are about equal.

$\delta+$ $\delta-$
C–Cl is the charge distribution for each C–Cl bond. In CH_2Cl_2, the two individual C–Cl bond dipoles add together to give an overall dipole moment for the molecule. The overall dipole will point from C (positive end) to the midpoint of the two Cl atoms (negative end).

In $CHCl_3$, the C–H bond is essentially nonpolar. The three C–Cl bond dipoles in $CHCl_3$ add together to give an overall dipole moment for the molecule. The overall dipole will have the negative end at the midpoint of the three chlorines and the positive end around the carbon.

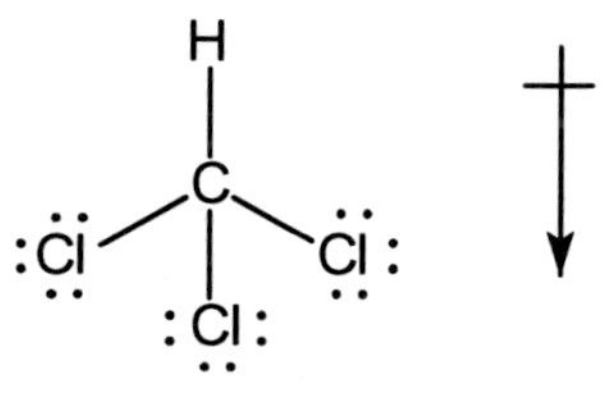

CCl_4 is nonpolar. CCl_4 is a tetrahedral molecule where all four C–Cl bond dipoles cancel when added together. Let's consider just the C and two of the Cl atoms. There will be a net dipole pointing in the direction of the middle of the two Cl atoms.

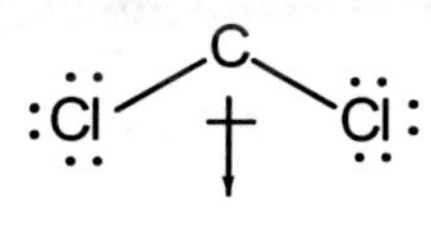

There will be an equal and opposite dipole arising from the other two Cl atoms.

Combining:

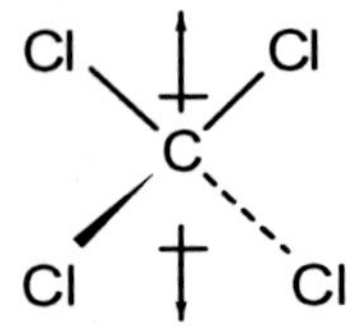

The two dipoles cancel, and CCl_4 is nonpolar.

b. CO_2 is nonpolar. CO_2 is a linear molecule with two equivalence bond dipoles that cancel. N_2O, which is also a linear molecule, is polar because the nonequivalent bond dipoles do not cancel.

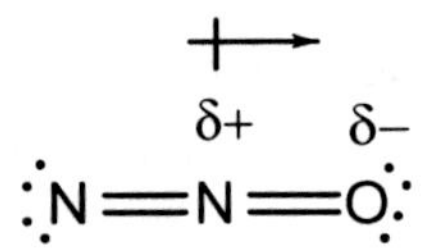

c. NH_3 is polar. The 3 N–H bond dipoles add together to give a net dipole in the direction of the lone pair. We would predict PH_3 to be nonpolar on the basis of electronegativitity, i.e., P–H bonds are nonpolar. However, the presence of the lone pair makes the PH_3 molecule slightly polar. The net dipole is in the direction of the lone pair and has a magnitude about one third that of the NH_3 dipole.

$\delta-$ $\delta+$
N – H

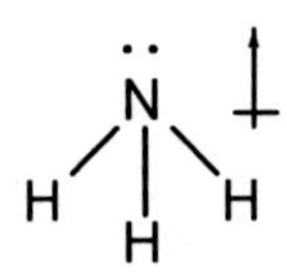

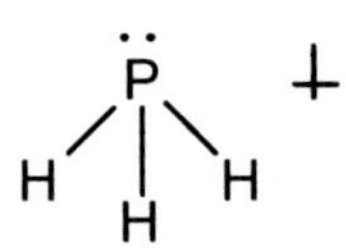

137. TeF_5^- has 6 + 5(7) + 1 = 42 valence electrons.

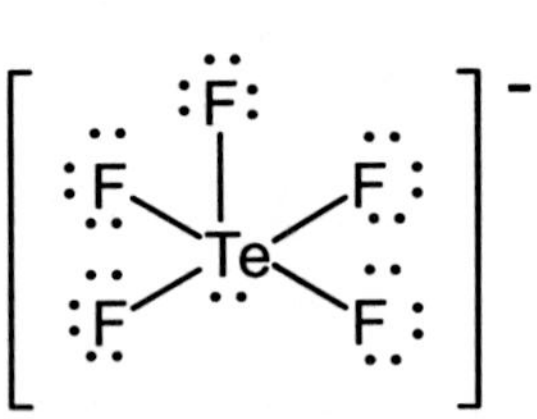

The lone pair of electrons around Te exerts a stronger repulsion than the bonding pairs of electrons. This pushes the four square-planar F atoms away from the lone pair and reduces the bond angles between the axial F atom and the square-planar F atoms.

138. C ≡ O (1072 kJ/mol); N ≡ N (941 kJ/mol); CO is polar, whereas N_2 is nonpolar. This may lead to a great reactivity for the CO bond.

ChemWork Problems

The answers to the problems 139-146 (or a variation to these problems) are found in OWL. These problems are also assignable in OWL.

Challenge Problems

147. a. There are two attractions of the form $\frac{(+1)(-1)}{r}$, where r = 1 × 10^{-10} m = 0.1 nm.

$$V = 2 \times (2.31 \times 10^{-19}\ \text{J nm})\left[\frac{(+1)(-1)}{0.1\ \text{nm}}\right] = -4.62 \times 10^{-18}\ \text{J} = -5 \times 10^{-18}\ \text{J}$$

b. There are four attractions of +1 and −1 charges at a distance of 0.1 nm from each other. The two negative charges and the two positive charges repel each other across the diagonal of the square. This is at a distance of $\sqrt{2}$ × 0.1 nm.

$$V = 4 \times (2.31 \times 10^{-19})\left[\frac{(+1)(-1)}{0.1}\right] + 2.31 \times 10^{-19}\left[\frac{(+1)(+1)}{\sqrt{2}\,(0.1)}\right] + 2.31 \times 10^{-19}\left[\frac{(-1)(-1)}{\sqrt{2}\,(0.1)}\right]$$

$$V = -9.24 \times 10^{-18}\ \text{J} + 1.63 \times 10^{-18}\ \text{J} + 1.63 \times 10^{-18}\ \text{J} = -5.98 \times 10^{-18}\ \text{J} = -6 \times 10^{-18}\ \text{J}$$

Note: There is a greater net attraction in arrangement b than in a.

148.

	(IE – EA)	(IE – EA)/502	EN (text)	2006/502 = 4.0
F	2006 kJ/mol	4.0	4.0	
Cl	1604 kJ/mol	3.2	3.0	
Br	1463 kJ/mol	2.9	2.8	
I	1302 kJ/mol	2.6	2.5	

The values calculated from ionization energies and electron affinities show the same trend as (and agree fairly closely) with the values given in the text.

149. The reaction is:

$$1/2\ I_2(g) + 1/2\ Cl_2(g) \rightarrow ICl(g) \qquad \Delta H_f^\circ = ?$$

Using Hess's law:

$1/2\ I_2(s) \rightarrow 1/2\ I_2(g)$	ΔH = 1/2(62 kJ)	(Appendix 4)
$1/2\ I_2(g) \rightarrow I(g)$	ΔH = 1/2(149 kJ)	(Table 8.4)
$1/2\ Cl_2(g) \rightarrow Cl(g)$	ΔH = 1/2(239 kJ)	(Table 8.4)
$I(g) + Cl(g) \rightarrow ICl(g)$	ΔH = −208 kJ	(Table 8.4)
$1/2\ I_2(s) + 1/2\ Cl_2(g) \rightarrow ICl(g)$	ΔH = 17 kJ so ΔH_f° = 17 kJ/mol	

150.

Reaction	ΔH
$2\ Li^+(g) + 2\ Cl^-(g) \rightarrow 2\ LiCl(s)$	$\Delta H = 2(-829\ kJ)$
$2\ Li(g) \rightarrow 2\ Li^+(g) + 2\ e^-$	$\Delta H = 2(520.\ kJ)$
$2\ Li(s) \rightarrow 2\ Li(g)$	$\Delta H = 2(166\ kJ)$
$2\ HCl(g) \rightarrow 2\ H(g) + 2\ Cl(g)$	$\Delta H = 2(427\ kJ)$
$2\ Cl(g) + 2\ e^- \rightarrow 2\ Cl^-(g)$	$\Delta H = 2(-349\ kJ)$
$2\ H(g) \rightarrow H_2(g)$	$\Delta H = -(432\ kJ)$
$2\ Li(s) + 2\ HCl(g) \rightarrow 2\ LiCl(s) + H_2(g)$	$\Delta H = -562\ kJ$

151. See Figure 8.11 to see the data supporting MgO as an ionic compound. Note that the lattice energy is large enough to overcome all of the other processes (removing two electrons from Mg, etc.). The bond energy for O_2 (247 kJ/mol) and electron affinity (737 kJ/mol) are the same when making CO. However, ionizing carbon to form a C^{2+} ion must be too large. See Figure 7.32 to see that the first ionization energy for carbon is about 350 kJ/mol greater than the first ionization energy for magnesium. If all other numbers were equal, the overall energy change would be down to ~250 kJ/mol (see Figure 8.11). It is not unreasonable to assume that the second ionization energy for carbon is more than 250 kJ/mol greater than the second ionization energy of magnesium. This would result in a positive ΔH value for the formation of CO as an ionic compound. One wouldn't expect CO to be ionic if the energetics were unfavorable.

152. a. (1) Removing an electron from the metal: ionization energy, positive ($\Delta H > 0$)

(2) Adding an electron to the nonmetal: electron affinity, often negative ($\Delta H < 0$)

(3) Allowing the metal cation and nonmetal anion to come together: lattice energy, negative ($\Delta H < 0$)

b. Often the sign of the sum of the first two processes is positive (or unfavorable). This is especially true due to the fact that we must also vaporize the metal and often break a bond on a diatomic gas. For example, the ionization energy for Na is +495 kJ/mol, and the electron affinity for F is −328 kJ/mol. Overall, the energy change is +167 kJ/mol (unfavorable).

c. For an ionic compound to form, the sum must be negative (exothermic).

d. The lattice energy must be favorable enough to overcome the endothermic process of forming the ions; i.e., the lattice energy must be a large negative quantity.

e. While Na_2Cl (or $NaCl_2$) would have a greater lattice energy than NaCl, the energy to make a Cl^{2-} ion (or Na^{2+} ion) must be larger (more unfavorable) than what would be gained by the larger lattice energy. The same argument can be made for MgO compared to MgO_2 or Mg_2O. The energy to make the ions is too unfavorable or the lattice energy is not favorable enough, and the compounds do not form.

153. As the halogen atoms get larger, it becomes more difficult to fit three halogen atoms around the small nitrogen atom, and the NX_3 molecule becomes less stable.

154. a. I.

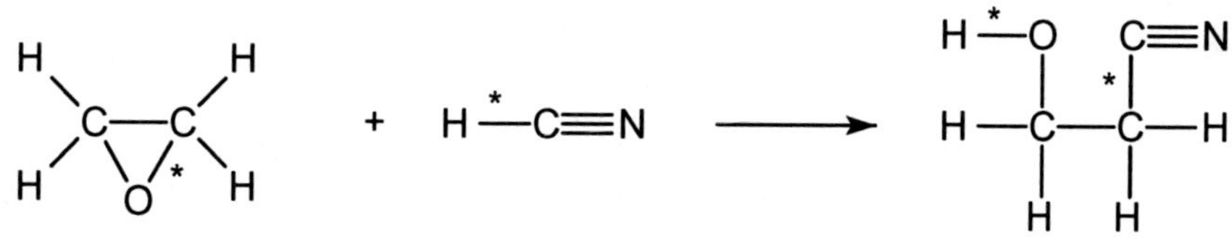

Bonds broken (*):

1 C–O (358 kJ)
1 C–H (413 kJ)

Bonds formed (*):

1 O–H (467 kJ)
1 C–C (347 kJ)

ΔH_I = 358 kJ + 413 kJ – (467 kJ + 347 kJ) = –43 kJ

II.

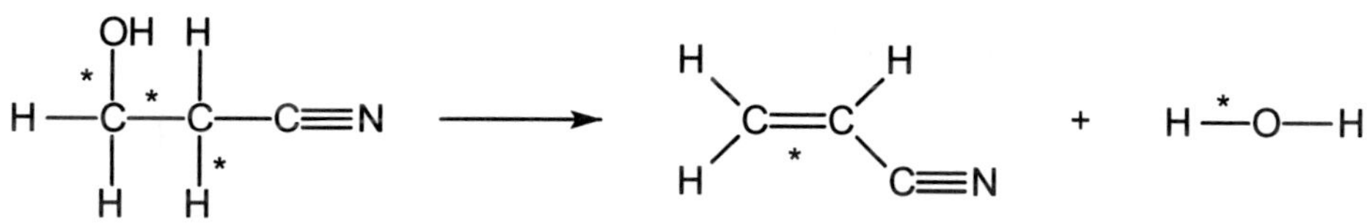

Bonds broken (*):

1 C–O (358 kJ/mol)
1 C–H (413 kJ/mol)
1 C–C (347 kJ/mol)

Bonds formed (*):

1 H–O (467 kJ/mol)
1 C=C (614 kJ/mol)

ΔH_{II} = 358 kJ + 413 kJ + 347 kJ – [467 kJ + 614 kJ] = +37 kJ

$\Delta H_{overall} = \Delta H_I + \Delta H_{II}$ = –43 kJ + 37 kJ = –6 kJ

b.

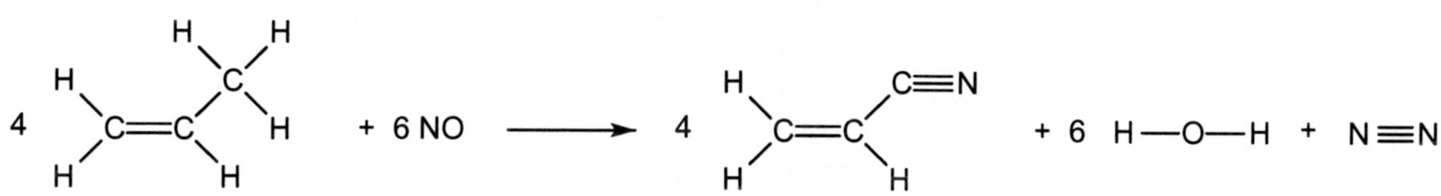

Bonds broken:

4 × 3 C–H (413 kJ/mol)
6 N=O (630. kJ/mol)

Bonds formed:

4 C≡N (891 kJ/mol)
6 × 2 H–O (467 kJ/mol)
1 N≡N (941 kJ/mol)

ΔH = 12(413) + 6(630.) – [4(891) + 12(467) + 941] = –1373 kJ

c.

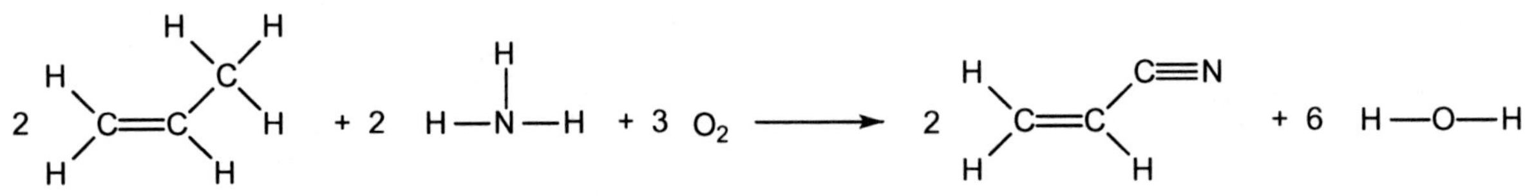

Bonds broken:

2 × 3 C–H (413 kJ/mol)
2 × 3 N–H (391 kJ/mol)
3 O=O (495 kJ/mol)

Bonds formed:

2 C≡N (891 kJ/mol)
6 × 2 O–H (467 kJ/mol)

$\Delta H = 6(413) + 6(391) + 3(495) - [2(891) + 12(467)] = -1077$ kJ

d. Because both reactions are highly exothermic, the high temperature is not needed to provide energy. It must be necessary for some other reason. The reason is to increase the speed of the reaction. This is discussed in Chapter 12 on kinetics.

155. a. i. $C_6H_6N_{12}O_{12} \rightarrow 6\ CO + 6\ N_2 + 3\ H_2O + 3/2\ O_2$

The NO_2 groups are assumed to have one N–O single bond and one N=O double bond, and each carbon atom has one C–H single bond. We must break and form all bonds.

Bonds broken:

3 C–C (347 kJ/mol)
6 C–H (413 kJ/mol)
12 C–N (305 kJ/mol)
6 N–N (160. kJ/mol)
6 N–O (201 kJ/mol)
6 N=O (607 kJ/mol)

$\Sigma D_{broken} = 12{,}987$ kJ

Bonds formed:

6 C≡O (1072 kJ/mol)
6 N≡N (941 kJ/mol)
6 H–O (467 kJ/mol)
3/2 O=O (495 kJ/mol)

$\Sigma D_{formed} = 15{,}623$ kJ

$\Delta H = \Sigma D_{broken} - \Sigma D_{formed} = 12{,}987\text{ kJ} - 15{,}623\text{ kJ} = -2636$ kJ

ii. $C_6H_6N_{12}O_{12} \rightarrow 3\ CO + 3\ CO_2 + 6\ N_2 + 3\ H_2O$

Note: The bonds broken will be the same for all three reactions.

Bonds formed:

3 C≡O (1072 kJ/mol)
6 C=O (799 kJ/mol)
6 N≡N (941 kJ/mol)
6 H–O (467 kJ/mol)

$\Sigma D_{formed} = 16{,}458$ kJ

$\Delta H = 12{,}987\text{ kJ} - 16{,}458\text{ kJ} = -3471$ kJ

iii. $C_6H_6N_{12}O_{12} \rightarrow 6\ CO_2 + 6\ N_2 + 3\ H_2$

Bonds formed:

12 C=O (799 kJ/mol)
6 N≡N (941 kJ/mol)
3 H–H (432 kJ/mol)

$\Sigma D_{formed} = 16{,}530.$ kJ

$\Delta H = 12{,}987$ kJ – 16,530. kJ = –3543 kJ

b. Reaction iii yields the most energy per mole of CL-20, so it will yield the most energy per kilogram.

$$\frac{-3543\text{kJ}}{\text{mol}} \times \frac{1\,\text{mol}}{438.23\,\text{g}} \times \frac{1000\text{g}}{\text{kg}} = -8085 \text{ kJ/kg}$$

156. We can draw resonance forms for the anion after the loss of H^+, we can argue that the extra stability of the anion causes the proton to be more readily lost, i.e., makes the compound a better acid.

a.

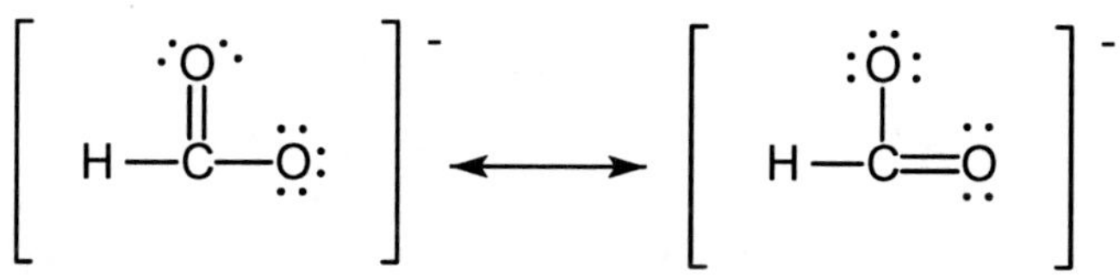

b.

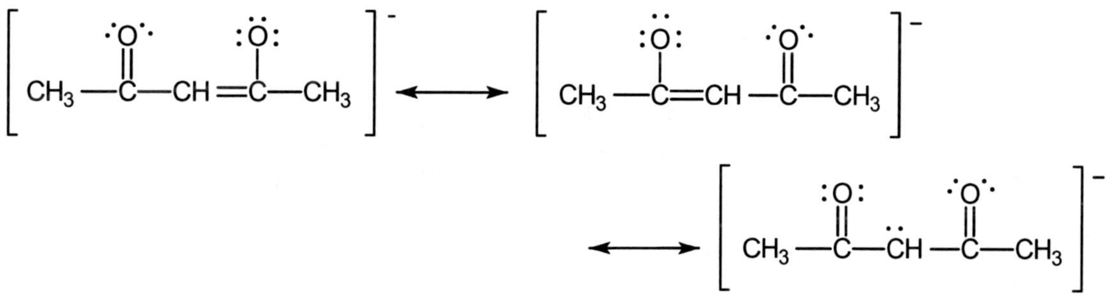

c.

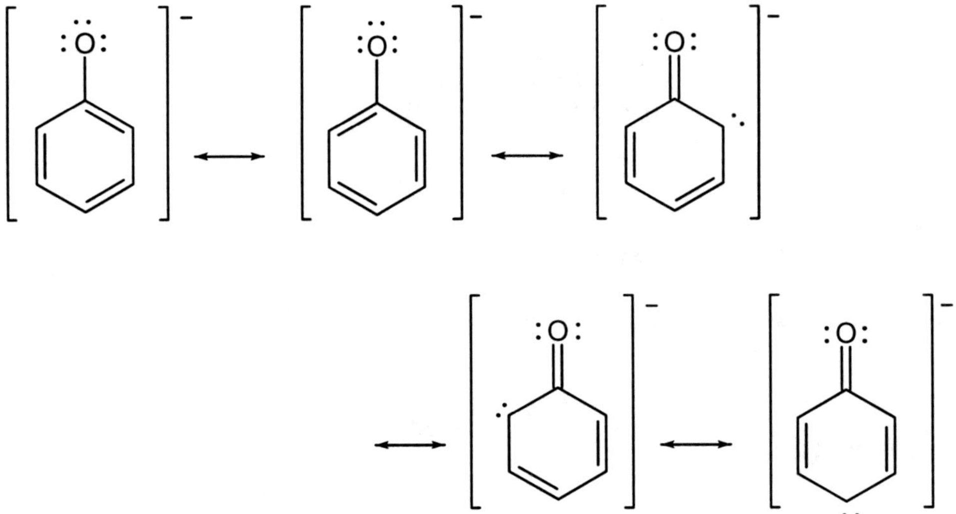

In all three cases, extra resonance forms can be drawn for the anion that are not possible when the H^+ is present, which leads to enhanced stability.

157. For carbon atoms to have a formal charge of zero, each C atom must satisfy the octet rule by forming four bonds (with no lone pairs). For nitrogen atoms to have a formal charge of zero, each N atom must satisfy the octet rule by forming three bonds and have one lone pair of electrons. For oxygen atoms to have a formal charge of zero, each O atom must satisfy the octet rule by forming two bonds and have two lone pairs of electrons. With these bonding requirements in mind, then the Lewis structure of histidine, where all atoms have a formal charge of zero, is:

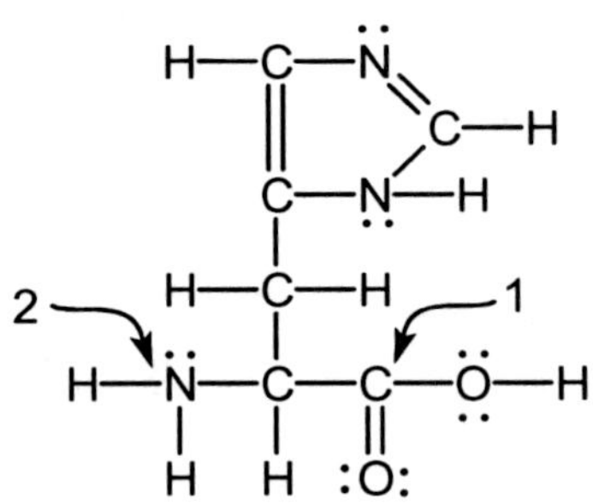

We would expect 120° bond angles about the carbon atom labeled 1 and ≈109.5° bond angles about the nitrogen atom labeled 2. The nitrogen bond angles should be slightly smaller than 109.5° due to the lone pair of electrons on nitrogen.

158. This molecule has 30 valence electrons. The only C–N bond that can possibly have a double-bond character is the N bound to the C with O attached. Double bonds to the other two C–N bonds would require carbon in each case to have 10 valence electrons (which carbon never does). The resonance structures are:

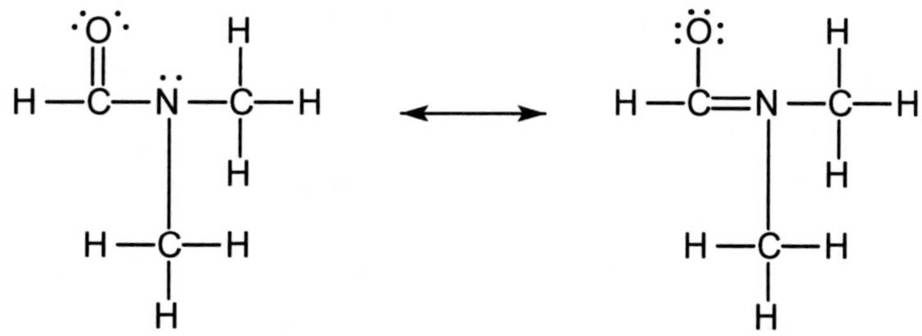

159. a. $BrFI_2$, 7 +7 + 2(7) = 28 e^-; two possible structures exist with Br as the central atom; each has a T-shaped molecular structure.

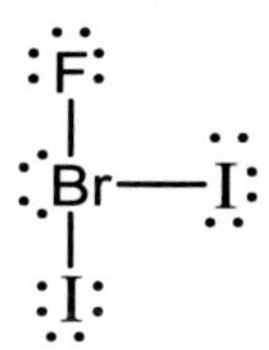

90° bond angles between I atoms

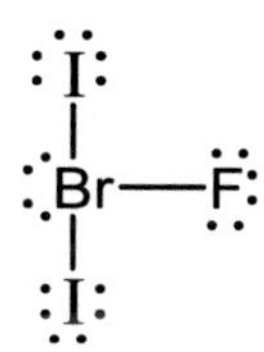

180° bond angles between I atoms

b. XeO_2F_2, 8 + 2(6) + 2(7) = 34 e^-; three possible structures exist with Xe as the central atom; each has a see-saw molecular structure.

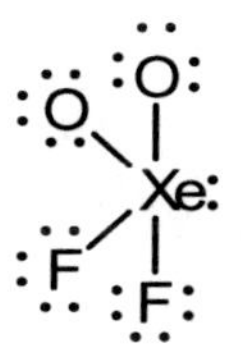

90° bond angle between O atoms

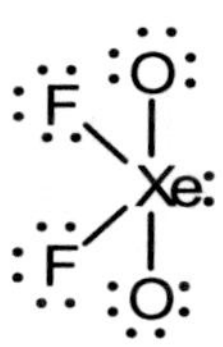

180° bond angle between O atoms

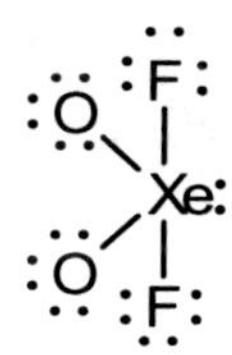

120° bond angle between O atoms

c. $TeF_2Cl_3^-$; 6 + 2(7) + 3(7) + 1 = 42 e^-; three possible structures exist with Te as the central atom; each has a square pyramid molecular structure.

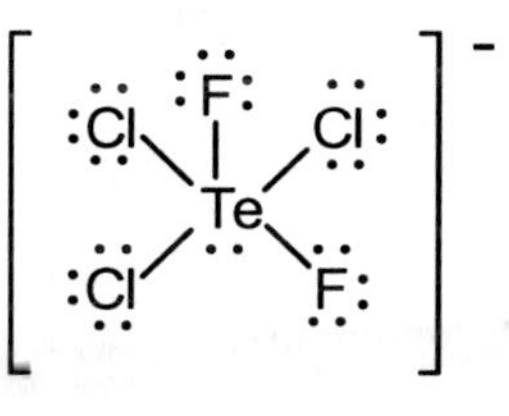

One F is 180° from the lone pair.

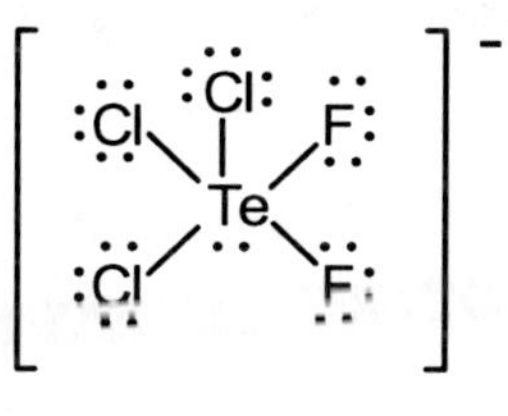

Both F atoms are 90° from the lone pair and 90° from each other.

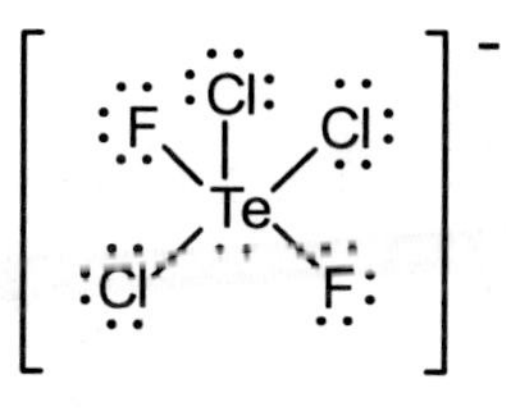

Both F atoms are 90° from the lone pair and 180° from each other.

160. The skeletal structure of caffeine is:

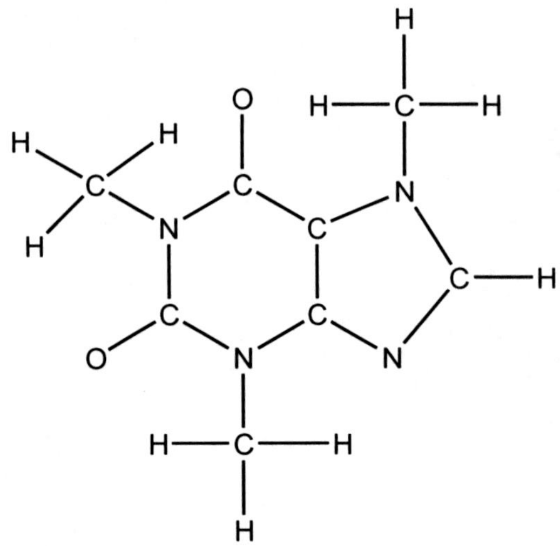

For a formal charge of zero on all atoms, the bonding requirements are:

(1) four bonds and no lone pairs for each carbon atom;

(2) three bonds and one lone pair for each nitrogen atom;

(3) two bonds and two lone pairs for each oxygen atom;

(4) one bond and no lone pairs for each hydrogen atom.

Following these guidelines gives a Lewis structure that has a formal charge of zero for all the atoms in the molecule. The Lewis structure is:

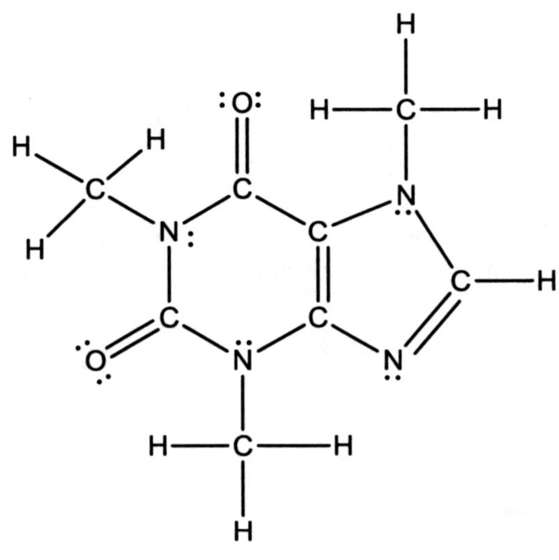

Integrative Problems

161. Assuming 100.00 g of compound: $42.81 \text{ g F} = \frac{1 \text{ mol X}}{19.00 \text{ g F}} = 2.253 \text{ mol F}$

The number of moles of X in XF_5 is: $2.53 \text{ mol F} \times \frac{1 \text{ mol X}}{5 \text{ mol F}} = 0.4506 \text{ mol X}$

This number of moles of X has a mass of 57.19 g (= 100.00 g – 42.81 g). The molar mass of X is:

$$\frac{57.19 \text{ g X}}{0.4506 \text{ mol X}} = 126.9 \text{ g/mol};$$ this is element I and the compound is IF_5.

IF_5, $7 + 5(7) = 42\ e^-$

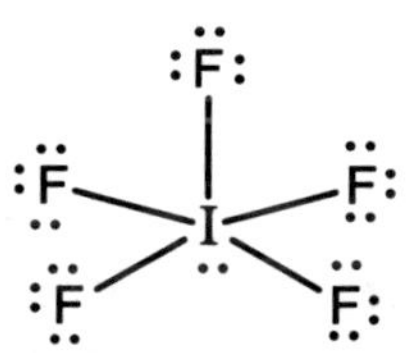

The molecular structure is square pyramid.

162. If X^{2-} has a configuration of $[Ar]4s^23d^{10}4p^6$, then X must have a configuration with two fewer electrons, $[Ar]4s^23d^{10}4p^4$. This is element Se.

$SeCN^-$, $6 + 4 + 5 + 1 = 16\ e^-$

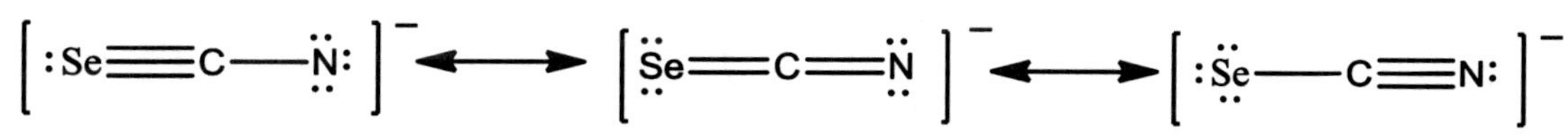

163. The elements are identified by their electron configurations:

$[Ar]4s^13d^5$ = Cr; $[Ne]3s^23p^3$ = P; $[Ar]4s^23d^{10}4p^3$ = As; $[Ne]3s^23p^5$ = Cl

Following the electronegativity trend, the order is Cr < As < P < Cl.

Marathon Problem

164. <u>Compound A</u>: This compound is a strong acid (part g). HNO_3 is a strong acid and is available in concentrated solutions of 16 *M* (part c). The highest possible oxidation state of nitrogen is +5, and in HNO_3, the oxidation state of nitrogen is +5 (part b). Therefore, compound A is most likely HNO_3. The Lewis structures for HNO_3 are:

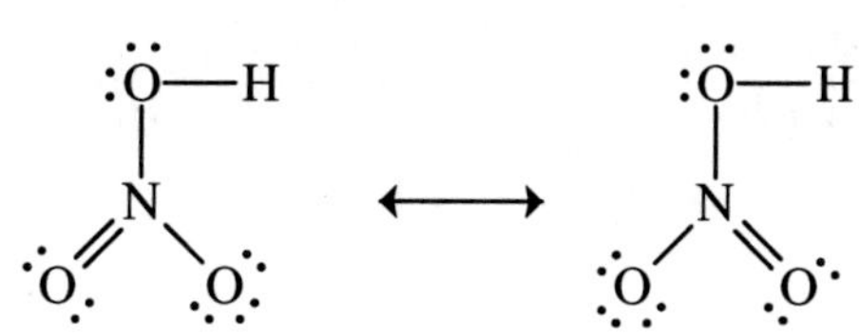

Compound B: This compound is basic (part g) and has one nitrogen (part b). The formal charge of zero (part b) tells us that there are three bonds to the nitrogen and that the nitrogen has one lone pair. Assuming compound B is monobasic, then the data in part g tell us that the molar mass of B is 33.0 g/mol (21.98 mL of 1.000 *M* HCl = 0.02198 mol HCl; thus there are 0.02198 mol of B; 0.726 g/0.02198 mol = 33.0 g/mol). Because this number is rather small, it limits the possibilities. That is, there is one nitrogen, and the remainder of the atoms are O and H. Since the molar mass of B is 33.0 g/mol, then only one O oxygen atom can be present. The N and O atoms have a combined molar mass of 30.0 g/mol; the rest is made up of hydrogens (3 H atoms), giving the formula NH_3O. From the list of K_b values for weak bases in Appendix 5 of the text, compound B is most likely NH_2OH. The Lewis structure is:

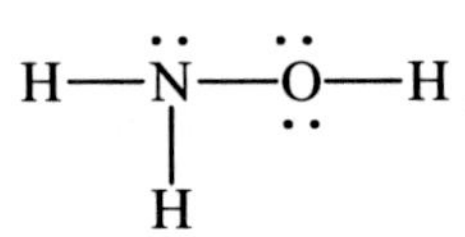

Compound C: From parts a and f and assuming compound A is HNO_3 , then compound C contains the nitrate ion, NO_3^-. Because part b tells us that there are two nitrogens, the other ion needs to have one N atom and some H atoms. In addition, compound C must be a weak acid (part g), which must be due to the other ion since NO_3^- has no acidic properties. Also, the nitrogen atom in the other ion must have an oxidation state of -3 (part b) and a formal charge of +1. The ammonium ion fits the data. Thus compound C is most likely NH_4NO_3. A Lewis structure is:

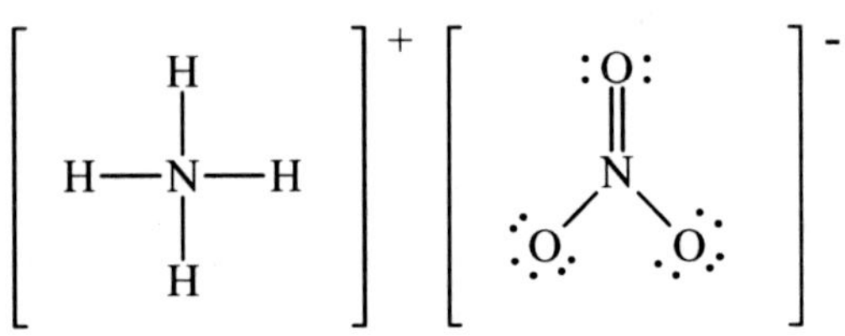

Note: Two more resonance structures can be drawn for NO_3^-.

Compound D: From part f, this compound has one less oxygen atom than compound C; thus NH_4NO_2 is a likely formula. Data from part e confirm this. Assuming 100.0 g of compound, we have:

$$43.7 \text{ g N} \times 1 \text{ mol}/14.01 \text{ g} = 3.12 \text{ mol N}$$
$$50.0 \text{ g O} \times 1 \text{ mol}/16.00 \text{ g} = 3.12 \text{ mol O}$$
$$6.3 \text{ g H} \times 1 \text{ mol}/1.008 \text{ g} = 6.25 \text{ mol H}$$

There is a 1 : 1 : 2 mole ratio among N to O to H. The empirical formula is NOH_2, which has an empirical formula mass of 32.0 g/mol.

$$\text{Molar mass} = \frac{dRT}{P} = \frac{2.86\text{g/L}(0.08206\text{L atm/K}\bullet\text{mol})(273\text{K})}{1.00\text{atm}} = 64.1 \text{ g/mol}$$

For a correct molar mass, the molecular formula of compound D is $N_2O_2H_4$ or NH_4NO_2. A Lewis structure is:

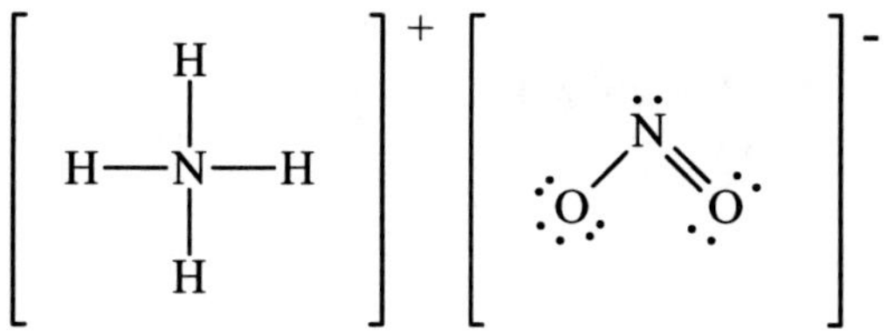

Note: One more resonance structure for NO_2^- can be drawn.

<u>Compound E:</u> A basic solution (part g) that is commercially available at 15 *M* (part c) is ammonium hydroxide (NH_4OH). This is also consistent with the information given in parts b and d. The Lewis structure for NH_4OH is:

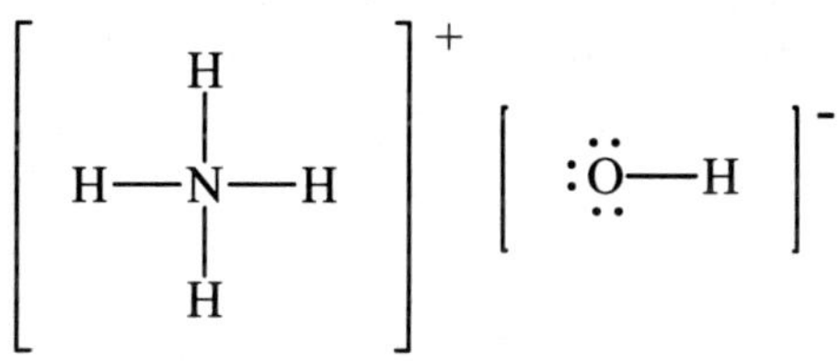

CHAPTER 9

COVALENT BONDING: ORBITALS

Questions

9. In hybrid orbital theory, some or all of the valence atomic orbitals of the central atom in a molecule are mixed together to form hybrid orbitals; these hybrid orbitals point to where the bonded atoms and lone pairs are oriented. The sigma bonds are formed from the hybrid orbitals overlapping head to head with an appropriate orbital from the bonded atom. The π bonds, in hybrid orbital theory, are formed from unhybridized p atomic orbitals. The p orbitals overlap side to side to form the π bond, where the π electrons occupy the space above and below a line joining the atoms (the internuclear axis). Assuming the z-axis is the internuclear axis, then the p_z atomic orbital will always be hybridized whether the hybridization is sp, sp^2, sp^3, dsp^3 or d^2sp^3. For sp hybridization, the p_x and p_y atomic orbitals are unhybridized; they are used to form two π bonds to the bonded atom(s). For sp^2 hybridization, either the p_x or the p_y atomic orbital is hybridized (along with the s and p_z orbitals); the other p orbital is used to form a π bond to a bonded atom. For sp^3 hybridization, the s and all the p orbitals are hybridized; no unhybridized p atomic orbitals are present, so no π bonds form with sp^3 hybridization. For dsp^3 and d^2sp^3 hybridization, we just mix in one or two d orbitals into the hybridization process. Which specific d orbitals are used is not important to our discussion.

10. The MO theory is a mathematical model. The allowed electron energy levels (molecular orbitals) in a molecule are solutions to the mathematical problem. The square of the solutions gives the shapes of the molecular orbitals. A sigma bond is an allowed energy level where the greatest electron probability is between the nuclei forming the bond. Valence s orbitals form sigma bonds, and if the z-axis is the internuclear axis, then valence p_z orbitals also form sigma bonds. For a molecule like HF, a sigma-bonding MO results from the combination of the H 1s orbital and the F $2p_z$ atomic orbital.

    For π bonds, the electron density lies above and below the internuclear axis. The π bonds are formed when p_x orbitals are combined (side-to-side overlap) and when p_y orbitals are combined.

11. We use d orbitals when we have to; i.e., we use d orbitals when the central atom on a molecule has more than eight electrons around it. The d orbitals are necessary to accommodate the electrons over eight. Row 2 elements never have more than eight electrons around them, so they never hybridize d orbitals. We rationalize this by saying there are no d orbitals close in energy to the valence 2s and 2p orbitals (2d orbitals are forbidden energy levels). However, for row 3 and heavier elements, there are 3d, 4d, 5d, etc. orbitals that will be close in energy to the valence s and p orbitals. It is row 3 and heavier nonmetals that hybridize d orbitals when they have to.

For sulfur, the valence electrons are in 3s and 3p orbitals. Therefore, 3d orbitals are closest in energy and are available for hybridization. Arsenic would hybridize 4d orbitals to go with the valence 4s and 4p orbitals, whereas iodine would hybridize 5d orbitals since the valence electrons are in $n = 5$.

12. Rotation occurs in a bond as long as the orbitals that go to form that bond still overlap when the atoms are rotating. Sigma bonds, with the head-to-head overlap, remain unaffected by rotating the atoms in the bonds. Atoms that are bonded together by only a sigma bond (single bond) exhibit this rotation phenomenon. The π bonds, however, cannot be rotated. The p orbitals must be parallel to each other to form the π bond. If we try to rotate the atoms in a π bond, the p orbitals would no longer have the correct alignment necessary to overlap. Because π bonds are present in double and triple bonds (a double bond is composed of 1 σ and 1 π bond, and a triple bond is always 1 σ and 2 π bonds), the atoms in a double or triple bond cannot rotate (unless the bond is broken).

13. Bonding and antibonding molecular orbitals are both solutions to the quantum mechanical treatment of the molecule. Bonding orbitals form when in-phase orbitals combine to give constructive interference. This results in enhanced electron probability located between the two nuclei. The end result is that a bonding MO is lower in energy than the atomic orbitals from which it is composed. Antibonding orbitals form when out-of-phase orbitals combine. The mismatched phases produce destructive interference leading to a node of electron probability between the two nuclei. With electron distribution pushed to the outside, the energy of an antibonding orbital is higher than the energy of the atomic orbitals from which it is composed.

14. From experiment, B_2 is paramagnetic. If the σ_{2p} MO is lower in energy than the two degenerate π_{2p} MOs, the electron configuration for B_2 would have all electrons paired. Experiment tells us we must have unpaired electrons. Therefore, the MO diagram is modified to have the π_{2p} orbitals lower in energy than the σ_{2p} orbitals. This gives two unpaired electrons in the electron configuration for B_2, which explains the paramagnetic properties of B_2. The model allowed for s and p orbitals to mix, which shifted the energy of the σ_{2p} orbital to above that of the π_{2p} orbitals.

15. The localized electron model does not deal effectively with molecules containing unpaired electrons. We can draw all of the possible structures for NO with its odd number of valence electrons but still not have a good feel for whether the bond in NO is weaker or stronger than the bond in NO^-. MO theory can handle odd electron species without any modifications. From the MO electron configurations, the bond order is 2.5 for NO and 2 for NO^-. Therefore, NO should have the stronger bond (and it does). In addition, hybrid orbital theory does not predict that NO^- is paramagnetic. The MO theory correctly makes this prediction.

16. NO_3^-, $5 + 3(6) + 1 = 24\ e^-$

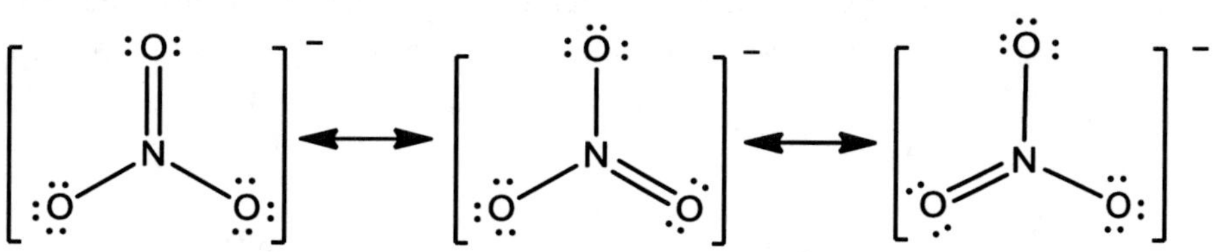

When resonance structures can be drawn, it is usually due to a multiple bond that can be in different positions. This is the case for NO_3^-. Experiment tells us that the three N–O bonds are equivalent in length and strength. To explain this, we say the π electrons are delocalized in the molecule. For NO_3^-, the π bonding system is composed of an unhybridized p atomic orbital from all the atoms in NO_3^-. These p orbitals are oriented perpendicular to the plane of the atoms in NO_3^-. The π bonding system consists of all of the perpendicular p orbitals overlapping forming a diffuse electron cloud above and below the entire surface of the NO_3^- ion. Instead of having the π electrons situated above and below two specific nuclei, we think of the π electrons in NO_3^- as extending over the entire surface of the molecule (hence the term delocalized). See Figure 9.48 for an illustration of the π bonding system in NO_3^-.

Exercises

The Localized Electron Model and Hybrid Orbitals

17. H_2O has 2(1) + 6 = 8 valence electrons.

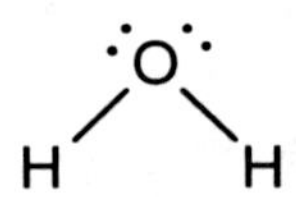

H_2O has a tetrahedral arrangement of the electron pairs about the O atom that requires sp^3 hybridization. Two of the four sp^3 hybrid orbitals are used to form bonds to the two hydrogen atoms, and the other two sp^3 hybrid orbitals hold the two lone pairs on oxygen. The two O–H bonds are formed from overlap of the sp^3 hybrid orbitals from oxygen with the 1s atomic orbitals from the hydrogen atoms. Each O–H covalent bond is called a sigma (σ) bond since the shared electron pair in each bond is centered in an area on a line running between the two atoms.

18. CCl_4 has 4 + 4(7) = 32 valence electrons.

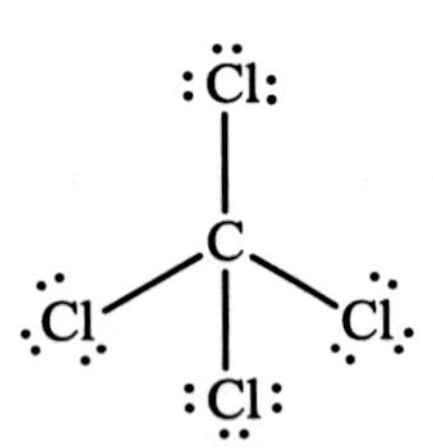

CCl_4 has a tetrahedral arrangement of the electron pairs about the carbon atom that requires sp^3 hybridization. The four sp^3 hybrid orbitals from carbon are used to form the four bonds to chlorine. The chlorine atoms also have a tetrahedral arrangement of electron pairs, and we will assume that they are also sp^3 hybridized. The C–Cl sigma bonds are all formed from overlap of sp^3 hybrid orbitals from carbon with sp^3 hybrid orbitals from each chlorine atom.

19. H_2CO has 2(1) + 4 + 6 = 12 valence electrons.

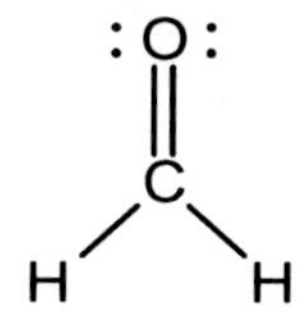

The central carbon atom has a trigonal planar arrangement of the electron pairs that requires sp^2 hybridization. The two C–H sigma bonds are formed from overlap of the sp^2 hybrid orbitals from carbon with the hydrogen 1s atomic orbitals. The double bond between carbon and oxygen consists of one σ and one π bond. The oxygen atom, like the carbon atom, also has a trigonal planar arrangement of the electrons that requires sp^2 hybridization. The σ bond in the double bond is formed from overlap of a carbon sp^2 hybrid orbital with an oxygen sp^2 hybrid orbital. The π bond in the double bond is formed from overlap of the unhybridized p atomic orbitals. Carbon and oxygen each has one unhybridized p atomic orbital that is parallel with the other. When two parallel p atomic orbitals overlap, a π bond results where the shared electron pair occupies the space above and below a line joining the atoms in the bond.

20. C_2H_2 has 2(4) + 2(1) = 10 valence electrons.

Each carbon atom in C_2H_2 is sp hybridized since each carbon atom is surrounded by two effective pairs of electrons; i.e., each carbon atom has a linear arrangement of the electrons. Since each carbon atom is sp hybridized, then each carbon atom has two unhybridized p atomic orbitals. The two C–H sigma bonds are formed from overlap of carbon sp hybrid orbitals with hydrogen 1s atomic orbitals. The triple bond is composed of one σ bond and two π bonds. The sigma bond between to the carbon atoms is formed from overlap of sp hybrid orbitals from each carbon atom. The two π bonds of the triple bond are formed from parallel overlap of the two unhybridized p atomic orbitals from each carbon.

21. Ethane, C_2H_6, has 2(4) + 6(1) = 14 valence electrons.

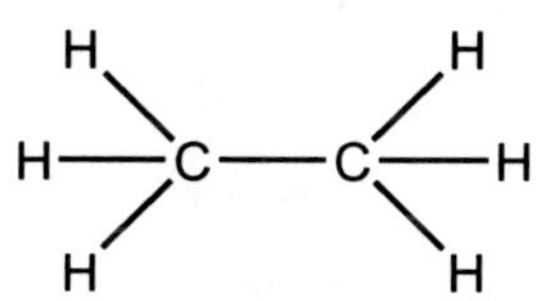

The carbon atoms are sp^3 hybridized. The six C–H sigma bonds are formed from overlap of the sp^3 hybrid orbitals from C with the 1s atomic orbitals from the hydrogen atoms. The carbon-carbon sigma bond is formed from overlap of an sp^3 hybrid orbital from each C atom.

Ethanol, C_2H_6O has 2(4) + 6(1) + 6 = 20 e^-

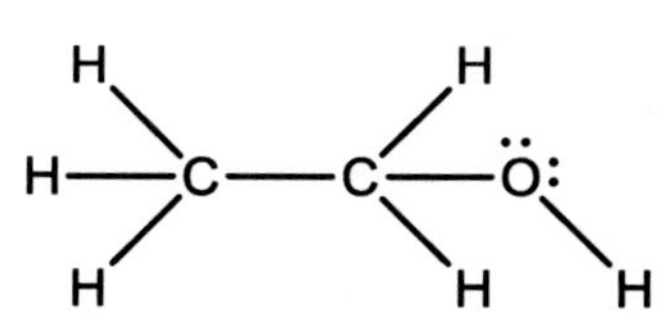

The two C atoms and the O atom are sp^3 hybridized. All bonds are formed from overlap with these sp^3 hybrid orbitals. The C–H and O–H sigma bonds are formed from overlap of sp^3 hybrid orbitals with hydrogen 1s atomic orbitals. The C–C and C–O sigma bonds are formed from overlap of the sp^3 hybrid orbitals from each atom.

22. HCN, 1 + 4 + 5 = 10 valence electrons

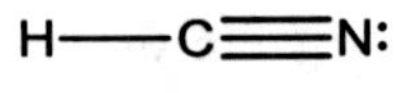

Assuming N is hybridized, both C and N atoms are sp hybridized. The C–H σ bond is formed from overlap of a carbon sp^3 hybrid orbital with a hydrogen 1s atomic orbital. The triple bond is composed of one σ bond and two π bonds. The sigma bond is formed from head-to-head overlap of the sp hybrid orbitals from the C and N atoms. The two π bonds in the triple bond are formed from overlap of the two unhybridized p atomic orbitals from each C and N atom.

$COCl_2$, 4 + 6 + 2(7) = 24 valence electrons

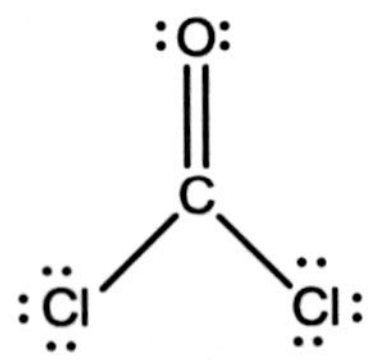

Assuming all atoms are hybridized, the carbon and oxygen atoms are sp^2 hybridized, and the two chlorine atoms are sp^3 hybridized. The two C–Cl σ bonds are formed from overlap of sp^2 hybrids from C with sp^3 hybrid orbitals from Cl. The double bond between the carbon and oxygen atoms consists of one σ and one π bond. The σ bond in the double bond is formed from head-to-head overlap of an sp^2 orbital from carbon with an sp^2 hybrid orbital from oxygen. The π bond is formed from parallel overlap of the unhybridized p atomic orbitals from each atom of C and O.

23. See Exercises 8.83 and 8.89 for the Lewis structures. To predict the hybridization, first determine the arrangement of electron pairs about each central atom using the VSEPR model; then use the information in Figure 9.24 of the text to deduce the hybridization required for that arrangement of electron pairs.

8.83 a. CCl_4: C is sp^3 hybridized. b. NCl_3: N is sp^3 hybridized.

c. $SeCl_2$: Se is sp^3 hybridized. d. ICl: Both I and Cl are sp^3 hybridized.

8.89 a. The central N atom is sp^2 hybridized in NO_2^- and NO_3^-. In N_2O_4, both central N atoms are sp^2 hybridized.

b. In OCN^- and SCN^-, the central carbon atoms in each ion are sp hybridized, and in N_3^-, the central N atom is also sp hybridized.

24. See Exercises 8.84 and 8.90 for the Lewis structures.

8.84 a. All the central atoms are sp^3 hybridized.

b. All the central atoms are sp^3 hybridized.

c. All the central atoms are sp^3 hybridized.

8.90 In O_3 and in SO_2, the central atoms are sp^2 hybridized, and in SO_3, the central sulfur atom is also sp^2 hybridized.

25. All exhibit dsp^3 hybridization. All of these molecules/ions have a trigonal bipyramid arrangement of electron pairs about the central atom; all have central atoms with dsp^3 hybridization. See Exercise 8.87 for the Lewis structures.

26. All these molecules have an octahedral arrangement of electron pairs about the central atom; all have central atoms with d^2sp^3 hybridization. See Exercise 8.88 for the Lewis structures.

27. The molecules in Exercise 8.113 all have a trigonal planar arrangement of electron pairs about the central atom, so all have central atoms with sp^2 hybridization. The molecules in Exercise 8.114 all have a tetrahedral arrangement of electron pairs about the central atom, so all have central atoms with sp^3 hybridization. See Exercises 8.113 and 8.114 for the Lewis structures.

28. The molecules in Exercise 8.115 all have central atoms with dsp^3 hybridization because all are based on the trigonal bipyramid arrangement of electron pairs. The molecules in Exercise 8.116 all have central atoms with d^2sp^3 hybridization because all are based on the octahedral arrangement of electron pairs. See Exercises 8.115 and 8.116 for the Lewis structures.

29. a.

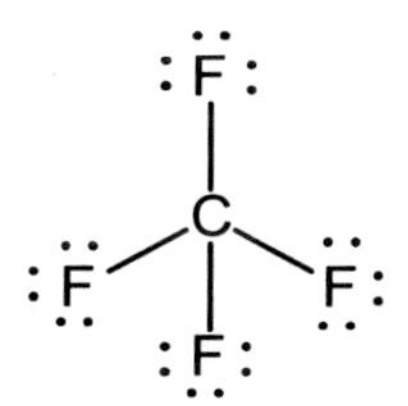

tetrahedral sp^3
109.5° nonpolar

b.

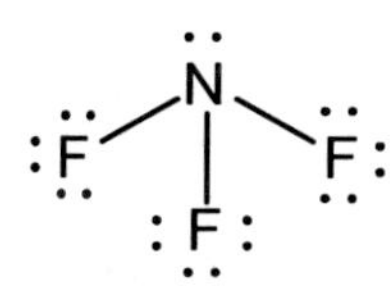

trigonal pyramid sp^3
<109.5° polar

The angles in NF_3 should be slightly less than 109.5° because the lone pair requires more space than the bonding pairs.

c.

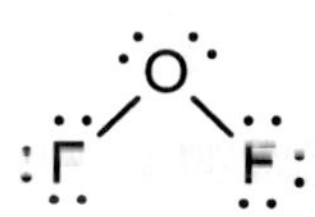

V-shaped sp^3
<109°.5 polar

d.

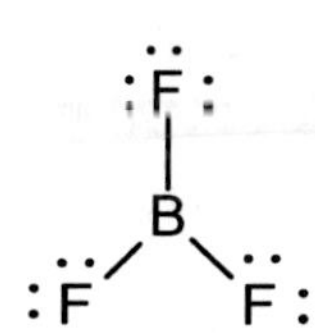

trigonal planar sp^2
120° nonpolar

e.

linear sp
180° nonpolar

f.

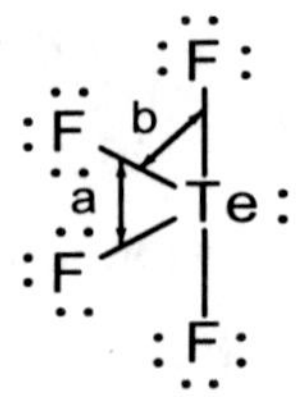

see-saw dsp^3
a) ≈120°, b) ≈90° polar

g.

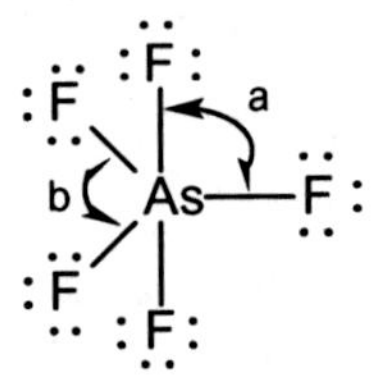

trigonal bipyramid dsp^3
a) 90°, b) 120° nonpolar

h.

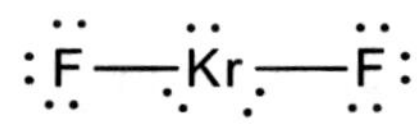

linear dsp^3
180° nonpolar

i.

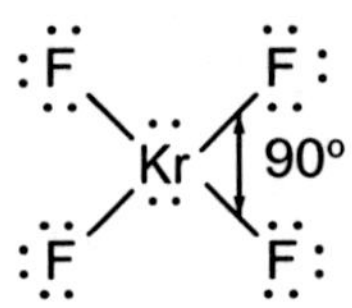

square planar d^2sp^3
90° nonpolar

j.

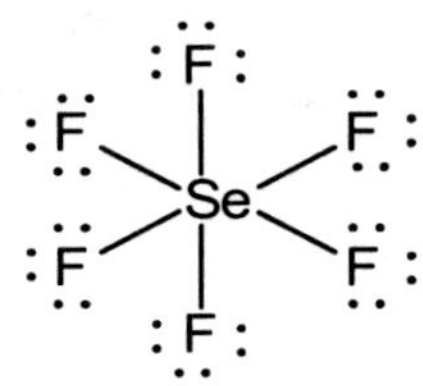

octahedral d^2sp^3
90° nonpolar

k.

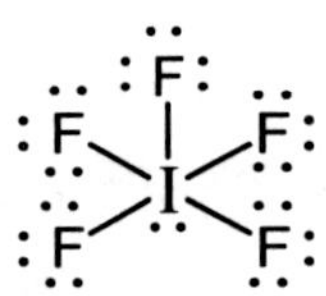

square pyramid d^2sp^3
≈90° polar

l.

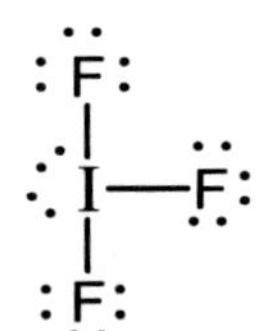

T-shaped dsp^3
≈90° polar

30. a.

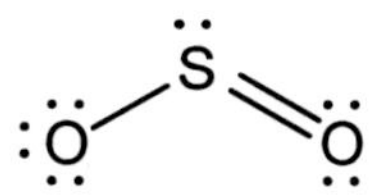

V-shaped, sp^2, 120°

Only one resonance form is shown. Resonance does not change the position of the atoms. We can predict the geometry and hybridization from any one of the resonance structures.

b.

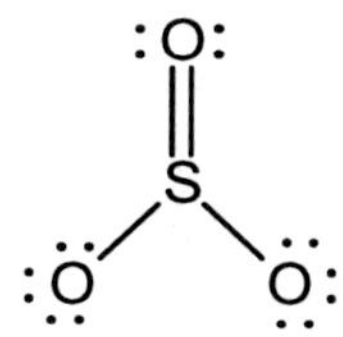

trigonal planar, 120°, sp^2
(plus two other resonance structures)

c.

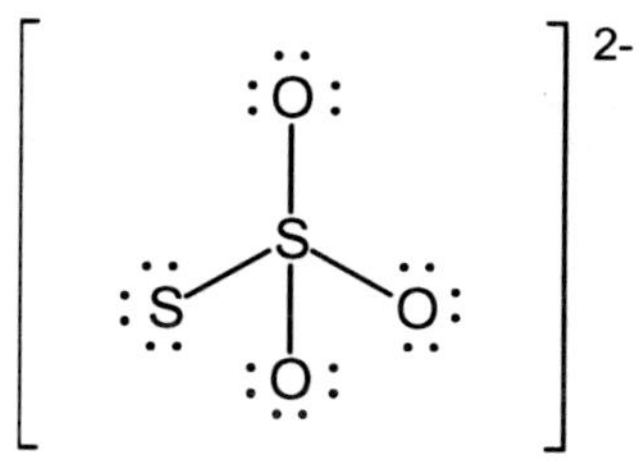

tetrahedral, 109.5°, sp^3

d.

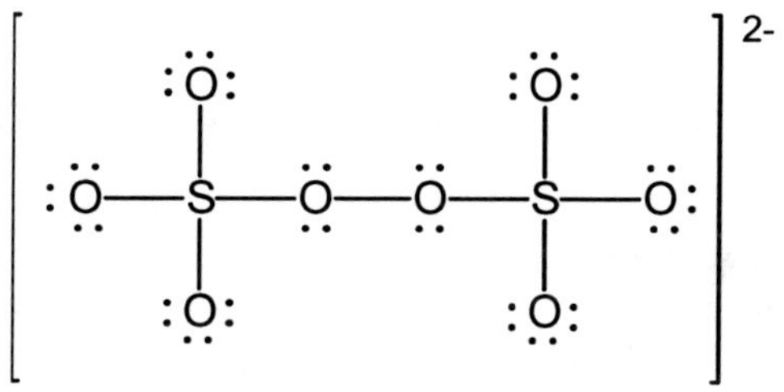

tetrahedral geometry about each S, 109.5°, sp^3 hybrids; V-shaped arrangement about peroxide O's, ≈109.5°, sp^3

e.

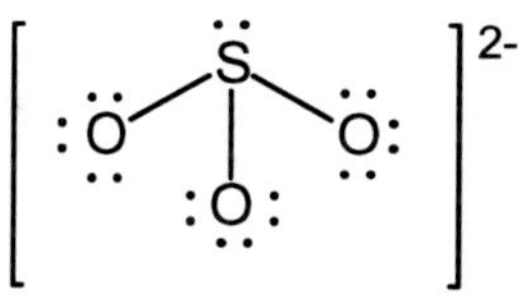

trigonal pyramid, <109.5°, sp^3

f.

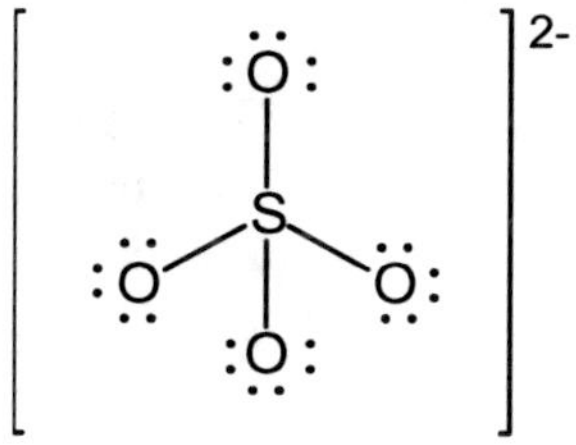

tetrahedral, 109.5°, sp^3

g.

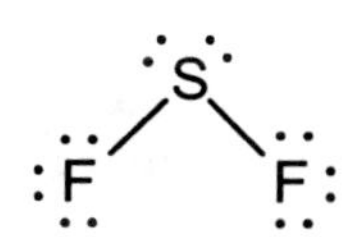

V-shaped, <109.5°, sp^3

h.

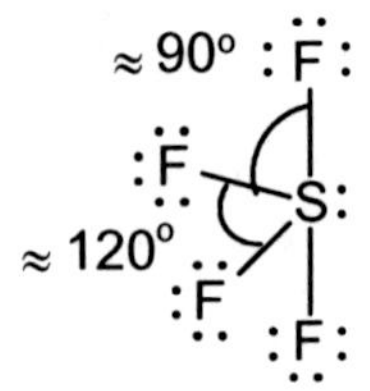

see-saw, ≈90° and ≈120°, dsp^3

i.

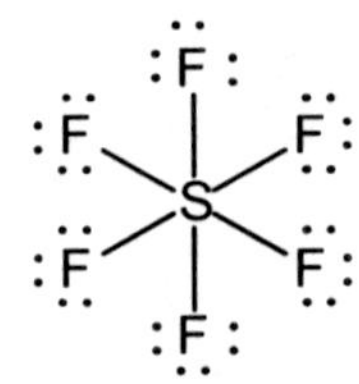

octahedral, 90°, d^2sp^3

j.

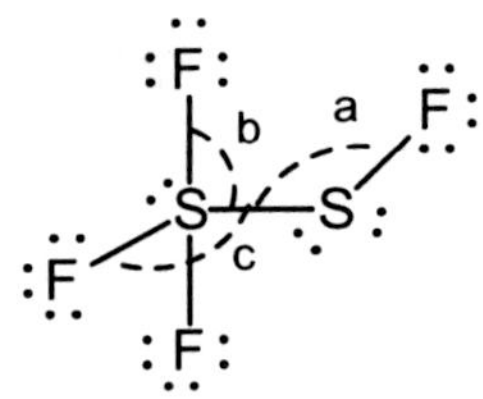

a) ≈109.5° b) ≈90° c) ≈120°
see-saw about S atom with one lone pair (dsp^3);
bent about S atom with two lone pairs (sp^3)

k.

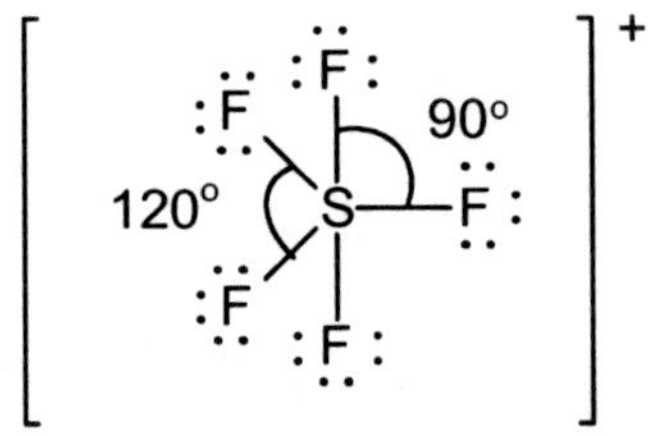

trigonal bipyramid,
90° and 120°, dsp^3

31.

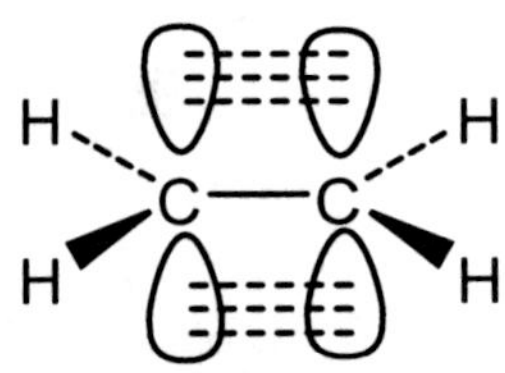

For the p orbitals to properly line up to form the π bond, all six atoms are forced into the same plane. If the atoms are not in the same plane, then the π bond could not form since the p orbitals would no longer be parallel to each other.

32. No, the CH_2 planes are mutually perpendicular to each other. The center C atom is sp hybridized and is involved in two π bonds. The p orbitals used to form each π bond must be perpendicular to each other. This forces the two CH_2 planes to be perpendicular.

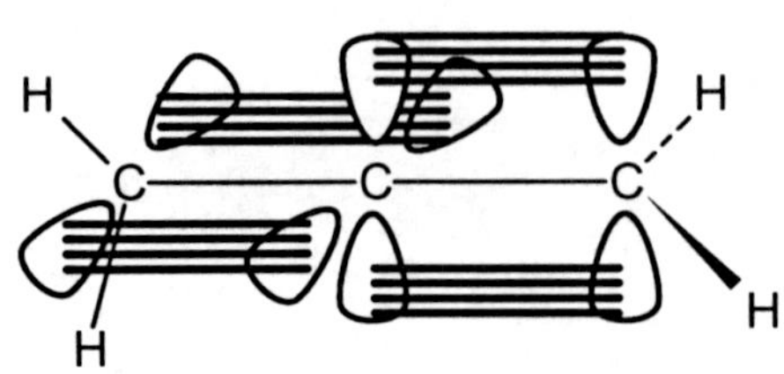

33. a. There are 33 σ and 9 π bonds. Single bonds always are σ bonds, double bonds always consist of 1 σ and 1 π bond, and triple bonds always consist of 1 σ and 2 π bonds. The 9 π bonds come from the 9 double bonds in the indigo molecule.

b. All carbon atoms are sp^2 hybridized because all have a trigonal planar arrangement of electron pairs.

34. The two nitrogen atoms in urea both have a tetrahedral arrangement of electron pairs, so both of these atoms are sp^3 hybridized. The carbon atom has a trigonal planar arrangement of electron pairs, so C is sp^2 hybridized. O is also sp^2 hybridized because it also has a trigonal planar arrangement of electron pairs.

Each of the four N–H sigma bonds are formed from overlap of an sp^3 hybrid orbital from nitrogen with a 1s orbital from hydrogen. Each of the two N–C sigma bonds are formed from an sp^3 hybrid orbital from N with an sp^2 hybrid orbital from carbon. The double bond between carbon and oxygen consists of one σ and one π bond. The σ bond in the double bond is formed from overlap of a carbon sp^2 hybrid orbital with an oxygen sp^2 hybrid orbital. The π bond in the double bond is formed from overlap of the unhybridized p atomic orbitals. Carbon and oxygen each have one unhybridized p atomic orbital, and they are assumed to be parallel to each other. When two parallel p atomic orbitals overlap side to side, a π bond results.

35. To complete the Lewis structures, just add lone pairs of electrons to satisfy the octet rule for the atoms with fewer than eight electrons.

Biacetyl ($C_4H_6O_2$) has 4(4) + 6(1) + 2(6) = 34 valence electrons.

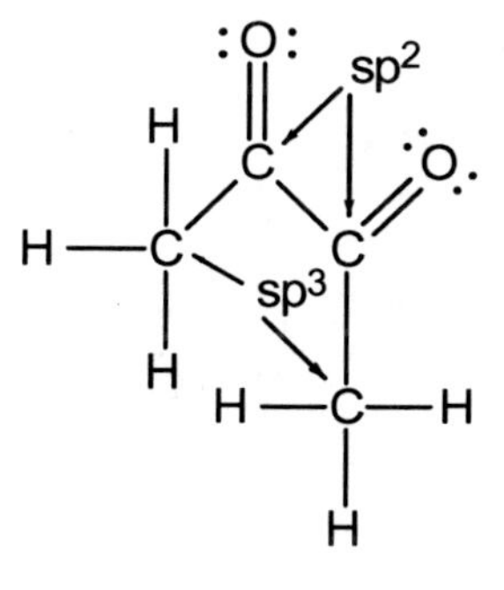

All CCO angles are 120°. The six atoms are not forced to lie in the same plane because of free rotation about the carbon-carbon single (sigma) bonds. There are 11 σ and 2 π bonds in biacetyl.

Acetoin ($C_4H_8O_2$) has 4(4) + 8(1) + 2(6) = 36 valence electrons.

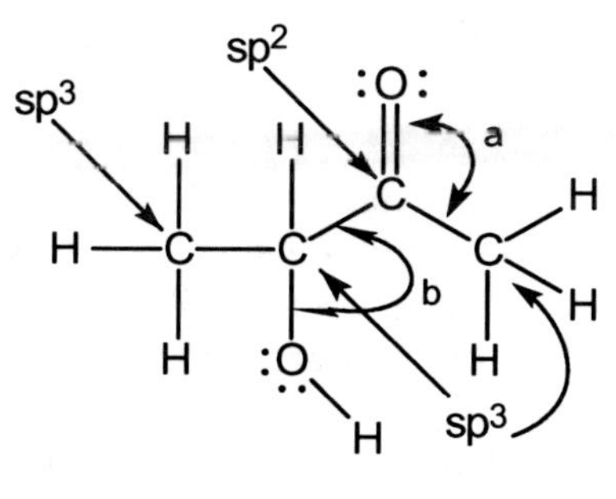

The carbon with the doubly bonded O is sp^2 hybridized. The other three C atoms are sp^3 hybridized. Angle a = 120° and angle b = 109.5°. There are 13 σ and 1 π bonds in acetoin.

Note: All single bonds are σ bonds, all double bonds are one σ and one π bond, and all triple bonds are one σ and two π bonds.

36. Acrylonitrile: C_3H_3N has 3(4) + 3(1) + 5 = 20 valence electrons.

a) 120°

b) 120°

c) 180°

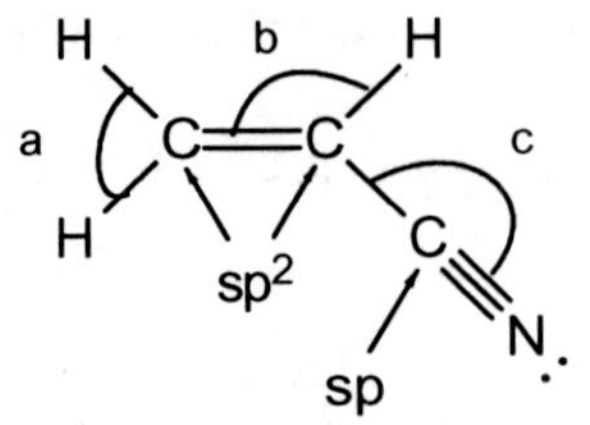

6 σ and 3 π bonds

All atoms of acrylonitrile must lie in the same plane. The π bond in the double bond dictates that the C and H atoms are all in the same plane, and the triple bond dictates that N is in the same plane with the other atoms.

Methyl methacrylate ($C_5H_8O_2$) has 5(4) + 8(1) + 2(6) = 40 valence electrons.

d) 120°

e) 120°

f) ≈109.5°

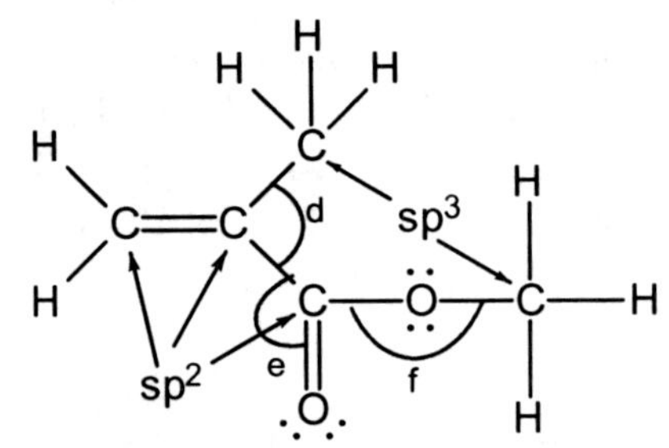

14 σ and 2 π bonds

37. a. Add lone pairs to complete octets for each O and N.

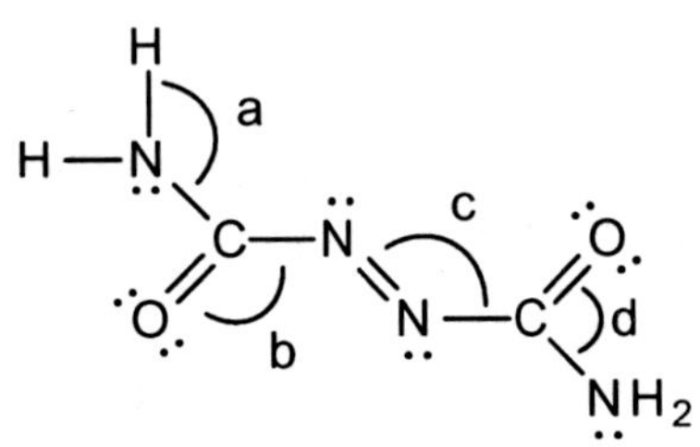

Azodicarbonamide

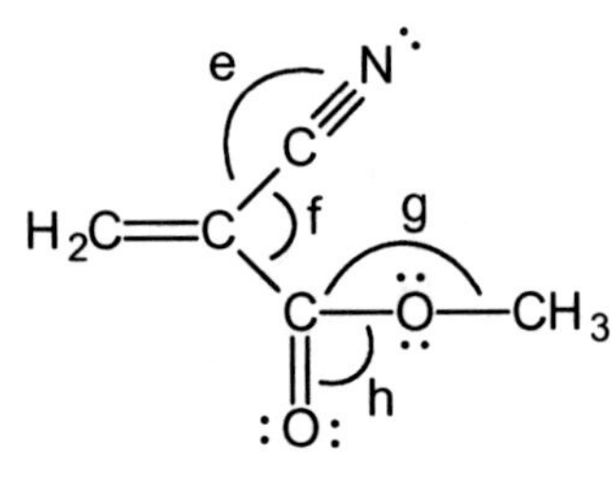

methyl cyanoacrylate

Note: NH_2, CH_2 (H_2C), and CH_3 are shorthand for nitrogen or carbon atoms singly bonded to hydrogen atoms.

b. In azodicarbonamide, the two carbon atoms are sp^2 hybridized, the two nitrogen atoms with hydrogens attached are sp^3 hybridized, and the other two nitrogens are sp^2 hybridized. In methyl cyanoacrylate, the CH_3 carbon is sp^3 hybridized, the carbon with the triple bond is sp hybridized, and the other three carbons are sp^2 hybridized.

c. Azodicarbonamide contains three π bonds and methyl cyanoacrylate contains four π bonds.

d. a) ≈109.5° b) 120° c) ≈120° d) 120° e) 180°

f) 120° g) ≈109.5° h) 120°

38. a. Piperine and capsaicin are molecules classified as organic compounds, i.e., compounds based on carbon. The majority of Lewis structures for organic compounds have all atoms with zero formal charge. Therefore, carbon atoms in organic compounds will usually form four bonds, nitrogen atoms will form three bonds and complete the octet with one lone pair of electrons, and oxygen atoms will form two bonds and complete the octet with two lone pairs of electrons. Using these guidelines, the Lewis structures are:

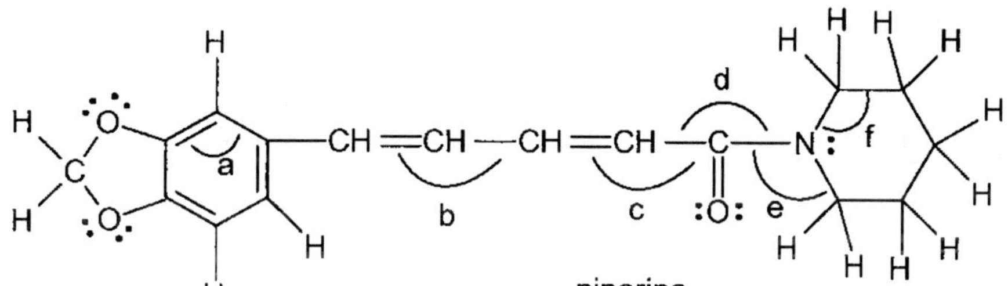

piperine

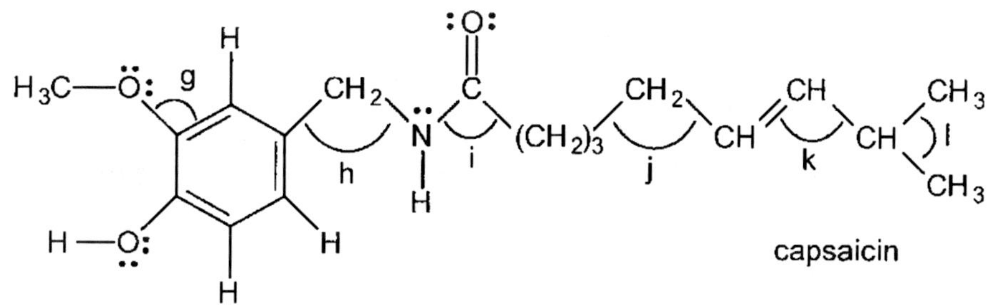

capsaicin

Note: The ring structures are all shorthand notation for rings of carbon atoms. In piperine the first ring contains six carbon atoms and the second ring contains five carbon atoms (plus nitrogen). Also notice that CH_3, CH_2, and CH are shorthand for carbon atoms singly bonded to hydrogen atoms.

b. piperine: 0 sp, 11 sp^2 and 6 sp^3 carbons; capsaicin: 0 sp, 9 sp^2, and 9 sp^3 carbons

c. The nitrogens are sp^3 hybridized in each molecule.

d. a) 120° b) 120° c) 120°

d) 120° e) ≈109.5° f) 109.5°

g) 120° h) 109.5° i) 120°

j) 109.5° k) 120° l) 109.5°

39. To complete the Lewis structure, just add lone pairs of electrons to satisfy the octet rule for the atoms that have fewer than eight electrons.

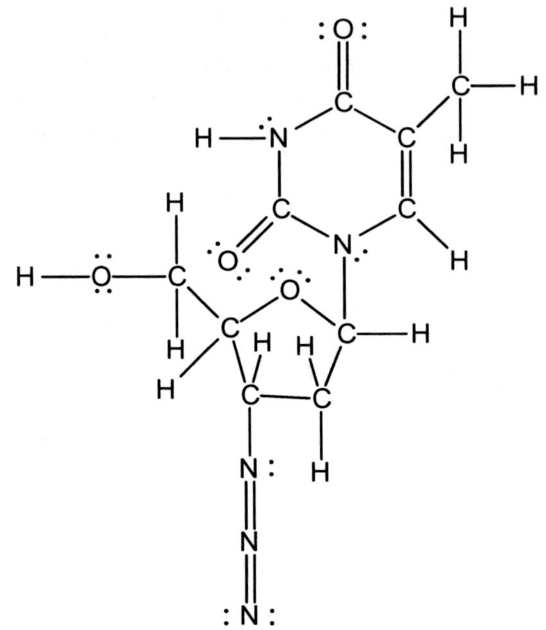

a. 6 b. 4 c. The center N in –N=N=N group

d. 33 σ e. 5 π bonds f. 180°

g. ≈109.5° h. sp^3

40. a. To complete the Lewis structure, add two lone pairs to each sulfur atom.

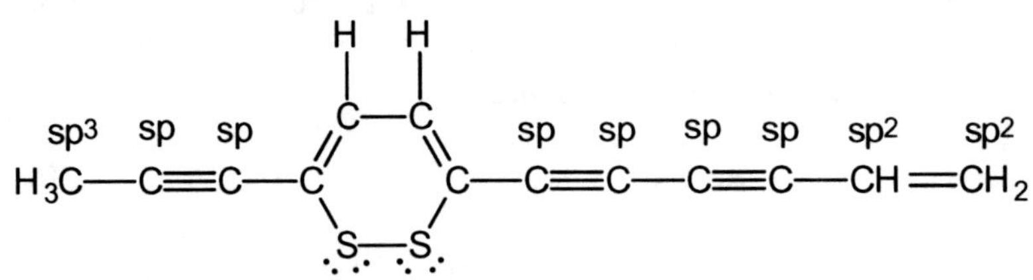

b. See the Lewis structure. The four carbon atoms in the ring are all sp^2 hybridized, and the two sulfur atoms are sp^3 hybridized.

c. 23 σ and 9 π bonds. *Note*: CH_3 (H_3C), CH_2, and CH are shorthand for carbon atoms singly bonded to hydrogen atoms.

The Molecular Orbital Model

41. a. The bonding molecular orbital is on the right and the antibonding molecular orbital is on the left. The bonding MO has the greatest electron probability between the nuclei, while the antibonding MO has greatest electron probability outside the space between the two nuclei.

b. The bonding MO is lower in energy. Because the electrons in the bonding MO have the greatest probability of lying between the two nuclei, these electrons are attracted to two different nuclei, resulting in a lower energy.

42. a.

When p orbitals are combined head-to-head and the phases are the same sign (the orbital lobes have the same sign), a sigma bonding molecular orbital is formed.

b.

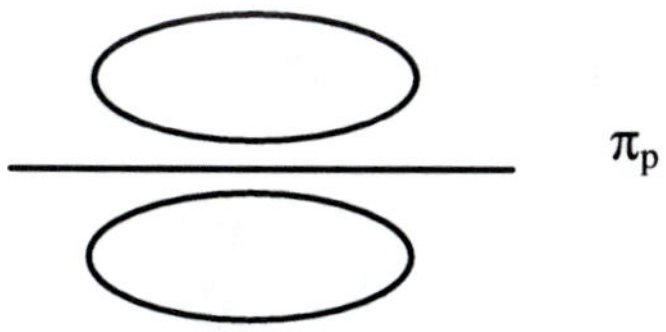

When parallel p orbitals are combined in-phase (the signs match up), a pi bonding molecular orbital is formed.

c.

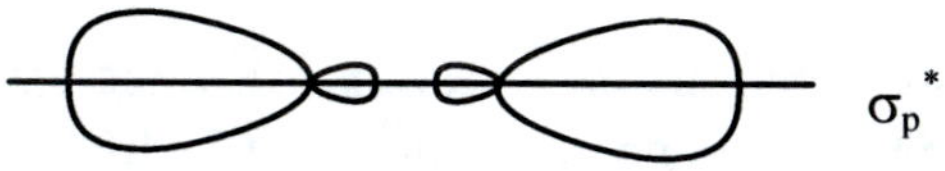

When p orbitals are combined head-to-head and the phases are opposite (the orbital lobes have opposite signs), a sigma antibonding molecular orbital is formed.

d.

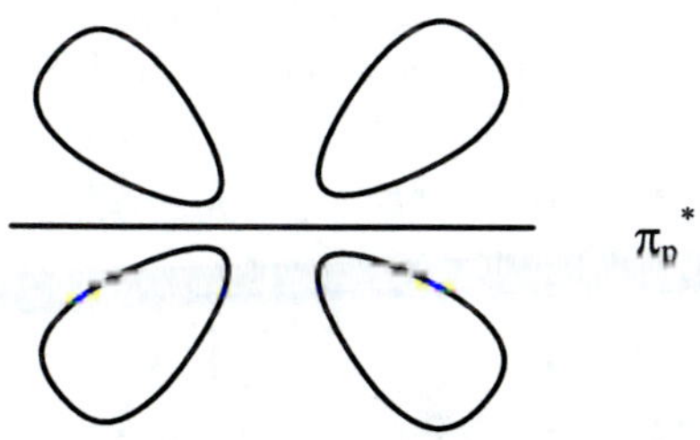

When parallel p orbitals are combined out-of-phase (the orbital lobes have opposite signs), a pi antibonding molecular orbital is formed.

43. If we calculate a nonzero bond order for a molecule, then we predict that it can exist (is stable).

a. H_2^+: $(\sigma_{1s})^1$ B.O. = bond order = (1–0)/2 = 1/2, stable

H_2: $(\sigma_{1s})^2$ B.O. = (2–0)/2 = 1, stable

H_2^-: $(\sigma_{1s})^2(\sigma_{1s}*)^1$ B.O. = (2–1)/2 = 1/2, stable

H_2^{2-}: $(\sigma_{1s})^2(\sigma_{1s}*)^2$ B.O. = (2–2)/2 = 0, not stable

b. He_2^{2+}: $(\sigma_{1s})^2$ B.O. = (2–0)/2 = 1, stable

He_2^+: $(\sigma_{1s})^2(\sigma_{1s}*)^1$ B.O. = (2–1)/2 = 1/2, stable

He_2: $(\sigma_{1s})^2(\sigma_{1s}*)^2$ B.O. = (2–2)/2 = 0, not stable

44. a. N_2^{2-}: $(\sigma_{2s})^2(\sigma_{2s}*)^2(\pi_{2p})^4(\sigma_{2p})^2(\pi_{2p}*)^2$ B.O. = bond order = (8–4)/2 = 2, stable

O_2^{2-}: $(\sigma_{2s})^2(\sigma_{2s}*)^2(\sigma_{2p})^2(\pi_{2p})^4(\pi_{2p}*)^4$ B.O. = (8–6)/2 = 1, stable

F_2^{2-}: $(\sigma_{2s})^2(\sigma_{2s}*)^2(\sigma_{2p})^2(\pi_{2p})^4(\pi_{2p}*)^4(\sigma_{2p}*)^2$ B.O. = (8–8)/2 = 0, not stable

b. Be_2: $(\sigma_{2s})^2(\sigma_{2s}*)^2$ B.O. = (2–2)/2 = 0, not stable

B_2: $(\sigma_{2s})^2(\sigma_{2s}*)^2(\pi_{2p})^2$ B.O. = (4–2)/2 = 1, stable

Ne_2: $(\sigma_{2s})^2(\sigma_{2s}*)^2(\sigma_{2p})^2(\pi_{2p})^4(\pi_{2p}*)^4(\sigma_{2p}*)^2$ B.O. = (8–8)/2 = 0, not stable

45. The electron configurations are:

a. Li_2: $(\sigma_{2s})^2$ B.O. = (2–0)/2 = 1, diamagnetic (0 unpaired e^-)

b. C_2: $(\sigma_{2s})^2(\sigma_{2s}*)^2(\pi_{2p})^4$ B.O. = (6–2)/2 = 2, diamagnetic (0 unpaired e^-)

c. S_2: $(\sigma_{3s})^2(\sigma_{3s}*)^2(\sigma_{3p})^2(\pi_{3p})^4(\pi_{3p}*)^2$ B.O. = (8–4)/2 = 2, paramagnetic (2 unpaired e^-)

46. There are 14 valence electrons in the MO electron configuration. Also, the valence shell is n = 3. Some possibilities from row 3 having 14 valence electrons are Cl_2, SCl^-, S_2^{2-}, and Ar_2^{2+}.

47. O_2: $(\sigma_{2s})^2(\sigma_{2s}*)^2(\sigma_{2p})^2(\pi_{2p})^4(\pi_{2p}*)^2$ B.O. = bond order = (8 – 4)/2 = 2

N_2: $(\sigma_{2s})^2(\sigma_{2s}*)^2(\pi_{2p})^4(\sigma_{2p})^2$ B.O. = (8 – 2)/2 = 3

In O_2, an antibonding electron is removed, which will increase the bond order to 2.5 [= (8–3)/2]. The bond order increases as an electron is removed, so the bond strengthens. In N_2, a bonding electron is removed, which decreases the bond order to 2.5 = [(7 – 2)/2]. So the bond strength weakens as an electron is removed from N_2.

48. The electron configurations are:

F_2^+: $(\sigma_{2s})^2(\sigma_{2s}*)^2(\sigma_{2p})^2(\pi_{2p})^4(\pi_{2p}*)^3$ B.O. = (8–5)/2 = 1.5; 1 unpaired e^-

F_2: $(\sigma_{2s})^2(\sigma_{2s}*)^2(\sigma_{2p})^2(\pi_{2p})^4(\pi_{2p}*)^4$ B.O. = (8–6)/2 = 1; 0 unpaired e^-

F_2^-: $(\sigma_{2s})^2(\sigma_{2s}*)^2(\sigma_{2p})^2(\pi_{2p})^4(\pi_{2p}*)^4(\sigma_{2p}*)^1$ B.O. = (8–7)/2 = 0.5; 1 unpaired e^-

From the calculated bond orders, the order of bond lengths should be $F_2^+ < F_2 < F_2^-$.

49. For CO, we will assume the same orbital ordering as that for N_2.

CO: $(\sigma_{2s})^2(\sigma_{2s}*)^2(\pi_{2p})^4(\sigma_{2p})^2$ B.O. = (8 – 2)/2 = 3; 0 unpaired electrons

O_2: $(\sigma_{2s})^2(\sigma_{2s}*)^2(\sigma_{2p})^2(\pi_{2p})^4(\pi_{2p}*)^2$ B.O. = (8 – 4)/2 = 2; 2 unpaired electrons

The most obvious differences are that CO has a larger bond order than O_2 (3 versus 2) and that CO is diamagnetic, whereas O_2 is paramagnetic.

50. Considering only the 12 valence electrons in O_2, the MO models would be:

O_2 ground state

Arrangement of electrons consistent with the Lewis structure (double bond and no unpaired electrons).

It takes energy to pair electrons in the same orbital. Thus the structure with no unpaired electrons is at a higher energy; it is an excited state.

51. The electron configurations are (assuming the same orbital order as that for N_2):

a. CO: $(\sigma_{2s})^2(\sigma_{2s}*)^2(\pi_{2p})^4(\sigma_{2p})^2$ B.O. = (8-2)/2 = 3, diamagnetic

b. CO^+: $(\sigma_{2s})^2(\sigma_{2s}*)^2(\pi_{2p})^4(\sigma_{2p})^1$ B.O. = (7-2)/2 = 2.5, paramagnetic

c. CO^{2+}: $(\sigma_{2s})^2(\sigma_{2s}*)^2(\pi_{2p})^4$ B.O. = (6-2)/2 = 2, diamagnetic

Because bond order is directly proportional to bond energy and inversely proportional to bond length:

Shortest → longest bond length: $CO < CO^+ < CO^{2+}$

Smallest → largest bond energy: $CO^{2+} < CO^+ < CO$

52. a. CN^+: $(\sigma_{2s})^2(\sigma_{2s}*)^2(\pi_{2p})^4$ B.O. = (6−2)/2 = 2, diamagnetic

b. CN: $(\sigma_{2s})^2(\sigma_{2s}*)^2(\pi_{2p})^4(\sigma_{2p})^1$ B.O. = (7−2)/2 = 2.5, paramagnetic

c. CN^-: $(\sigma_{2s})^2(\sigma_{2s}*)^2(\pi_{2p})^4(\sigma_{2p})^2$ B.O. = 3, diamagnetic

The bond orders are CN^+, 2; CN, 2.5; CN^-, 3; because bond order is directly proportional to bond energy and inversely proportional to bond length:

Shortest → longest bond length: $CN^- < CN < CN^+$

Smallest → largest bond energy: $CN^+ < CN < CN^-$

53. a. H_2: $(\sigma_{1s})^2$

B_2: $(\sigma_{2s})^2(\sigma_{2s}*)^2(\pi_{2p})^2$

b. C_2^{2-}: $(\sigma_{2s})^2(\sigma_{2s}*)^2(\pi_{2p})^4(\sigma_{2p})^2$

OF: $(\sigma_{2s})^2(\sigma_{2s}*)^2(\sigma_{2p})^2(\pi_{2p})^4(\pi_{2p}*)^3$

The bond strength will weaken if the electron removed comes from a bonding orbital. Of the molecules listed, H_2, B_2, and C_2^{2-} would be expected to have their bond strength weaken as an electron is removed. OF has the electron removed from an antibonding orbital, so its bond strength increases.

54. a. CN: $(\sigma_{2s})^2(\sigma_{2s}*)^2(\pi_{2p})^4(\sigma_{2p})^1$

NO: $(\sigma_{2s})^2(\sigma_{2s}*)^2(\pi_{2p})^4(\sigma_{2p})^2(\pi_{2p}*)^1$

b. O_2^{2+}: $(\sigma_{2s})^2(\sigma_{2s}*)^2(\sigma_{2p})^2(\pi_{2p})^4$

N_2^{2+}: $(\sigma_{2s})^2(\sigma_{2s}*)^2(\pi_{2p})^4$

If the added electron goes into a bonding orbital, the bond order would increase, making the species more stable and more likely to form. Between CN and NO, CN would most likely form CN^- since the bond order increases (unlike NO^-, where the added electron goes into an antibonding orbital). Between O_2^{2+} and N_2^{2+}, N_2^+ would most likely form since the bond order increases (unlike O_2^{2+} going to O_2^+).

55. The two types of overlap that result in bond formation for p orbitals are in-phase side-to-side overlap (π bond) and in-phase head-to-head overlap (σ bond).

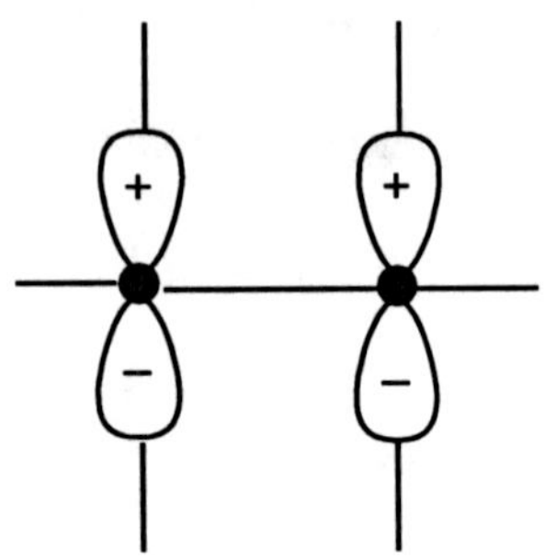

π_{2p} (in-phase; the signs match up)

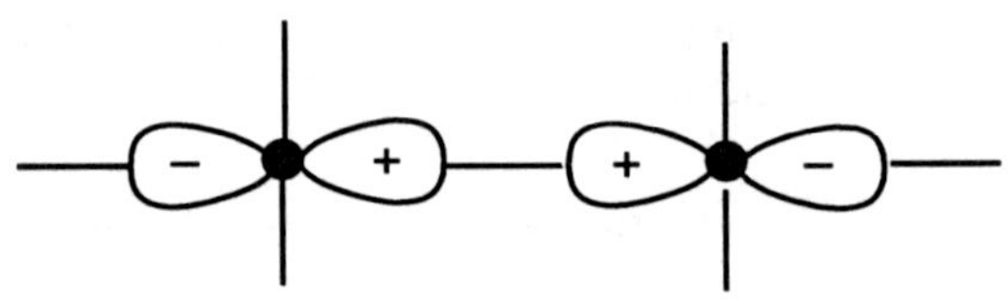

σ_{2p} (in-phase; the signs match up)

56.

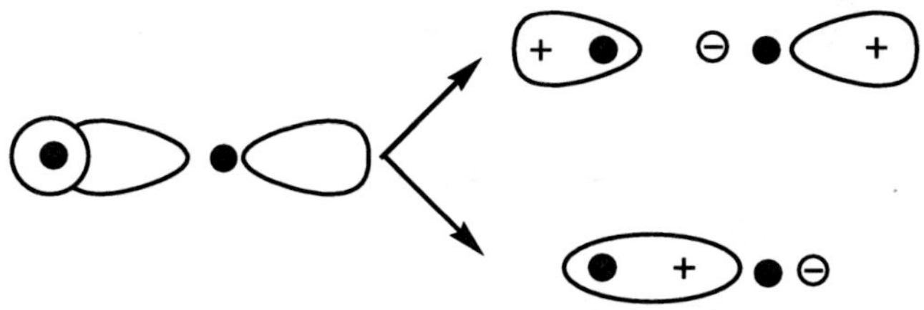

σ* (out-of-phase; the signs oppose each other)

σ (in-phase; the signs match up)

These molecular orbitals are sigma MOs because the electron density is cylindrically symmetric about the internuclear axis.

57. a. The electron density would be closer to F on average. The F atom is more electronegative than the H atom, and the 2p orbital of F is lower in energy than the 1s orbital of H.

b. The bonding MO would have more fluorine 2p character since it is closer in energy to the fluorine 2p atomic orbital.

c. The antibonding MO would place more electron density closer to H and would have a greater contribution from the higher-energy hydrogen 1s atomic orbital.

58. a. The antibonding MO will have more hydrogen 1s character because the hydrogen 1s atomic orbital is closer in energy to the antibonding MO.

b. No, the net overall overlap is zero. The p_x orbital does not have proper symmetry to overlap with a 1s orbital. The $2p_x$ and $2p_y$ orbitals are called nonbonding orbitals.

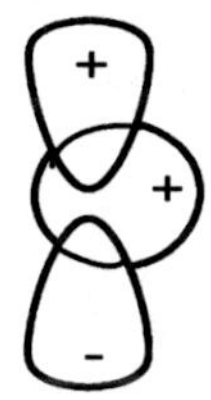

c.

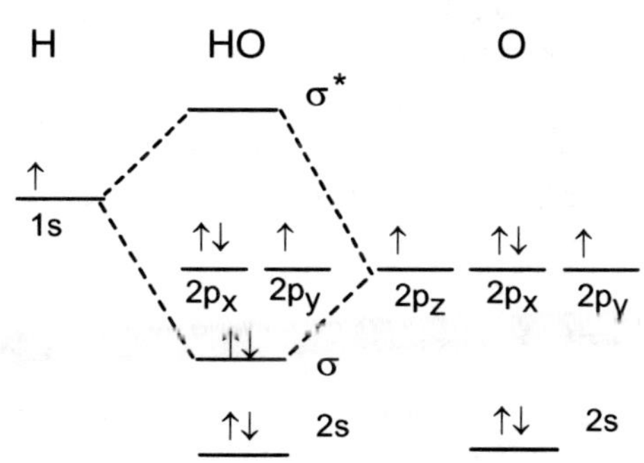

d. Bond order = (2 – 0)/2 = 1; *Note*: The 2s, $2p_x$, and $2p_y$ electrons have no effect on the bond order.

e. To form OH^+, a nonbonding electron is removed from OH. Because the number of bonding electrons and antibonding electrons is unchanged, the bond order is still equal to one.

59. C_2^{2-} has 10 valence electrons. The Lewis structure predicts sp hybridization for each carbon with two unhybridized p orbitals on each carbon.

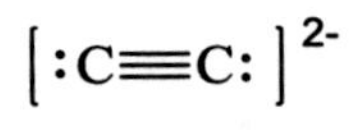

sp hybrid orbitals form the σ bond and the two unhybridized p atomic orbitals from each carbon form the two π bonds.

MO: $(\sigma_{2s})^2(\sigma_{2s}^*)^2(\pi_{2p})^4(\sigma_{2p})^2$, B.O. = (8 – 2)/2 = 3

Both give the same picture, a triple bond composed of one σ and two π bonds. Both predict the ion will be diamagnetic. Lewis structures deal well with diamagnetic (all electrons paired) species. The Lewis model cannot really predict magnetic properties.

60. Lewis structures:

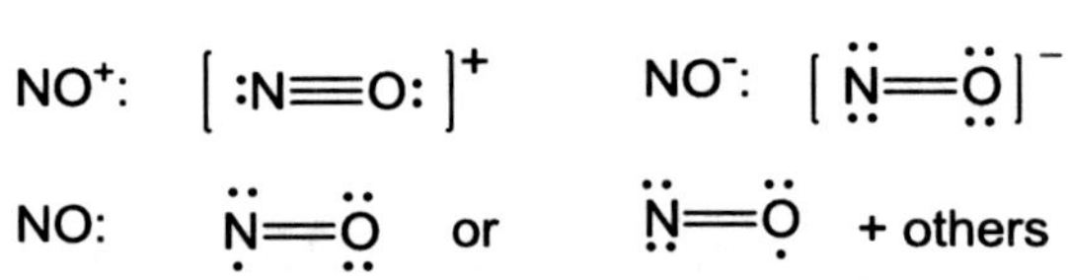

Note: Lewis structures do not handle odd numbered electron species very well.

MO model:

NO^+: $(\sigma_{2s})^2(\sigma_{2s}^*)^2(\pi_{2p})^4(\sigma_{2p})^2$, B.O. = 3, 0 unpaired e^- (diamagnetic)

NO: $(\sigma_{2s})^2(\sigma_{2s}^*)^2(\pi_{2p})^4(\sigma_{2p})^2(\pi_{2p}^*)^1$, B.O. = 2.5, 1 unpaired e^- (paramagnetic)

NO^-: $(\sigma_{2s})^2(\sigma_{2s}^*)^2(\pi_{2p})^4(\sigma_{2p})^2(\pi_{2p}^*)^2$ B.O. = 2, 2 unpaired e^- (paramagnetic)

The two models give the same results only for NO^+ (a triple bond with no unpaired electrons). Lewis structures are not adequate for NO and NO^-. The MO model gives a better representation for all three species. For NO, Lewis structures are poor for odd electron species. For NO^-, both models predict a double bond, but only the MO model correctly predicts that NO^- is paramagnetic.

61. O_3 and NO_2^- are isoelectronic, so we only need consider one of them since the same bonding ideas apply to both. The Lewis structures for O_3 are:

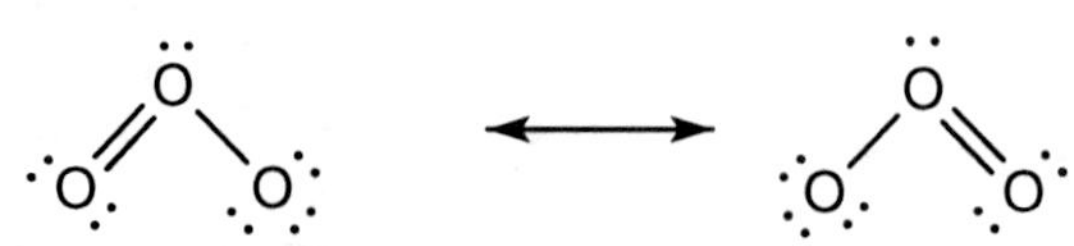

For each of the two resonance forms, the central O atom is sp^2 hybridized with one unhybridized p atomic orbital. The sp^2 hybrid orbitals are used to form the two sigma bonds to the central atom and hold the lone pair of electrons on the central O atom. The localized electron view of the π bond uses unhybridized p atomic orbitals. The π bond resonates between the two positions in the Lewis structures; the actual structure of O_3 is an average of the two resonance structures:

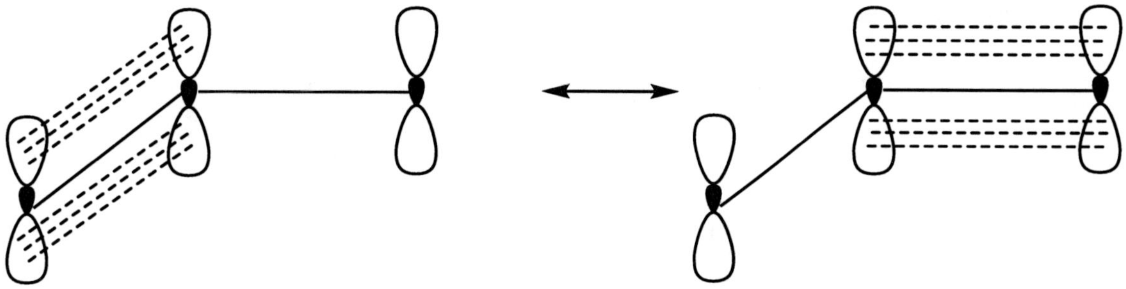

In the MO picture of the π bond, all three unhybridized p orbitals overlap at the same time, resulting in π electrons that are delocalized over the entire surface of the molecule. This is represented as:

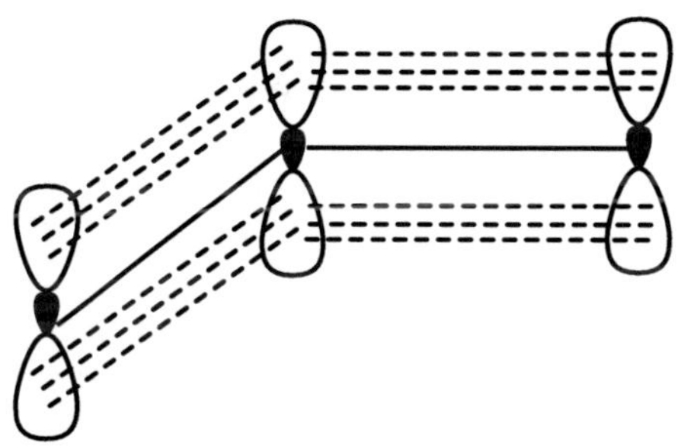

or

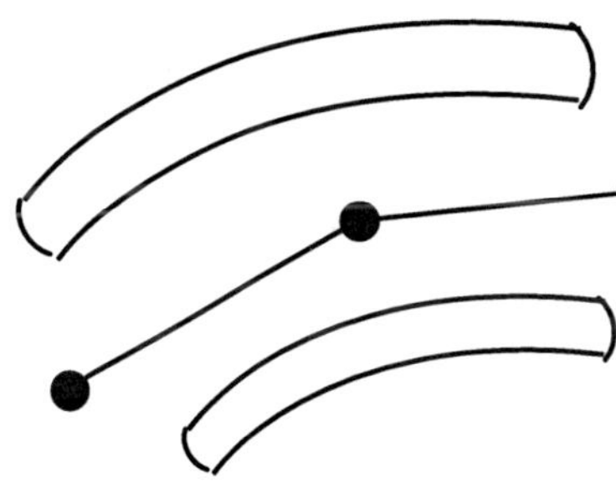

62. The Lewis structures for CO_3^{2-} are (24 e^-):

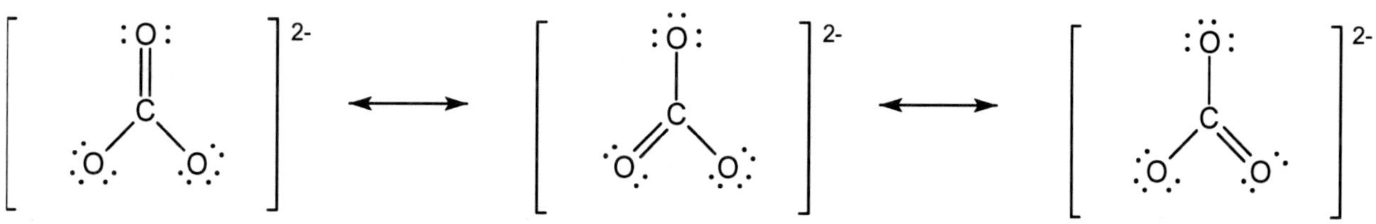

In the localized electron view, the central carbon atom is sp^2 hybridized; the sp^2 hybrid orbitals are used to form the three sigma bonds in CO_3^{2-}. The central C atom also has one unhybridized p atomic orbital that overlaps with another p atomic orbital from one of the oxygen atoms to form the π bond in each resonance structure. This localized π bond moves (resonates) from one position to another. In the molecular orbital model for CO_3^{2-}, all four atoms in CO_3^{2-} have a p atomic orbital that is perpendicular to the plane of the ion. All four of these p orbitals overlap at the same time to form a delocalized π bonding system where the π electrons can roam above and below the entire surface of the ion. The π molecular orbital system for CO_3^{2-} is analogous to that for NO_3^- which is shown in Figure 9.48 of the text.

Additional Exercises

63. a. XeO_3, $8 + 3(6) = 26\ e^-$

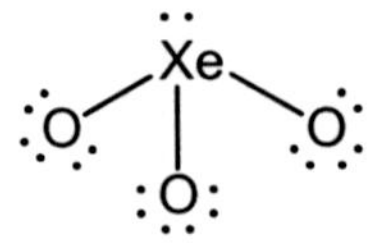

trigonal pyramid; sp^3

b. XeO_4, $8 + 4(6) = 32\ e^-$

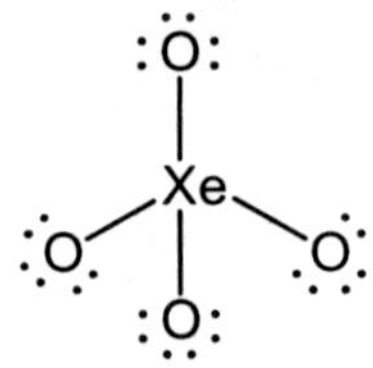

tetrahedral; sp^3

c. $XeOF_4$, $8 + 6 + 4(7) = 42\ e^-$

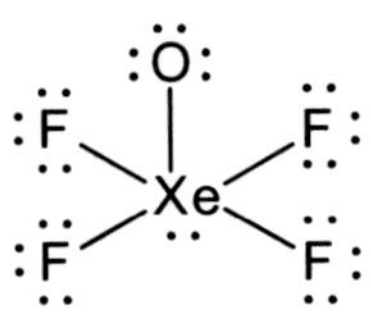

or

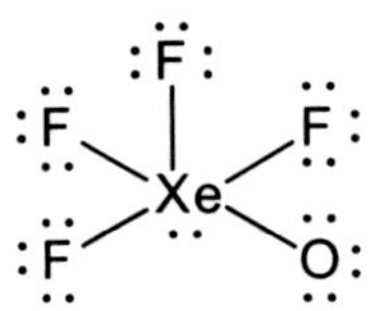

square pyramid; d^2sp^3

d. $XeOF_2$, $8 + 6 + 2(7) = 28\ e^-$

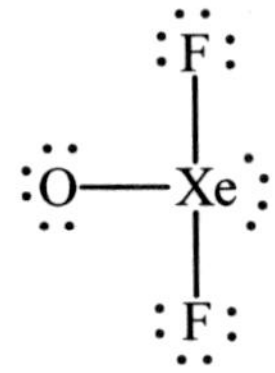

or

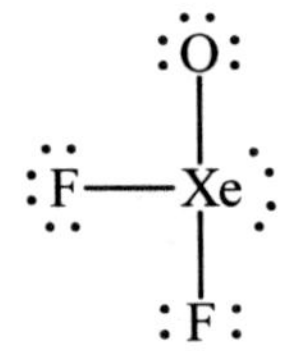

T-shaped; dsp^3

e. XeO_3F_2 has $8 + 3(6) + 2(7) = 40$ valence electrons.

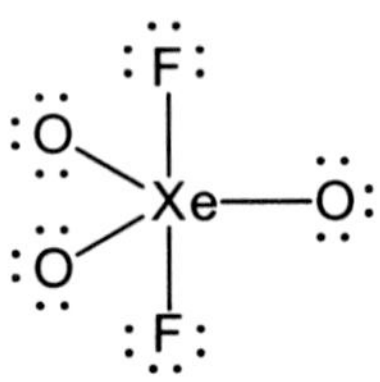

or

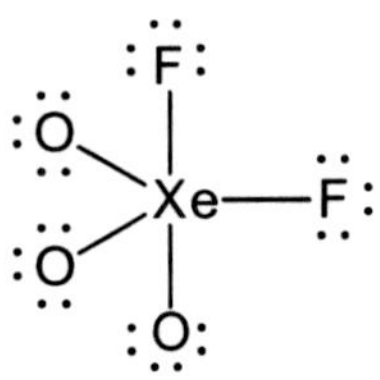

or

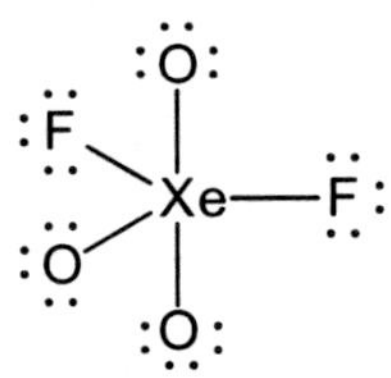

trigonal bipyramid; dsp^3

64. $FClO_2 + F^- \rightarrow F_2ClO_2^-$

$F_2ClO_2^-$, $2(7) + 7 + 2(6) + 1 = 34\ e^-$

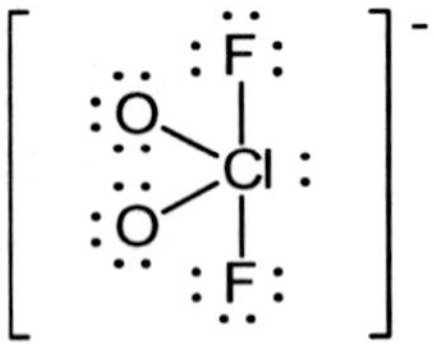

see-saw, dsp^3

$F_3ClO + F^- \rightarrow F_4ClO^-$

F_4ClO^-, $4(7) + 7 + 6 + 1 = 42\ e^-$

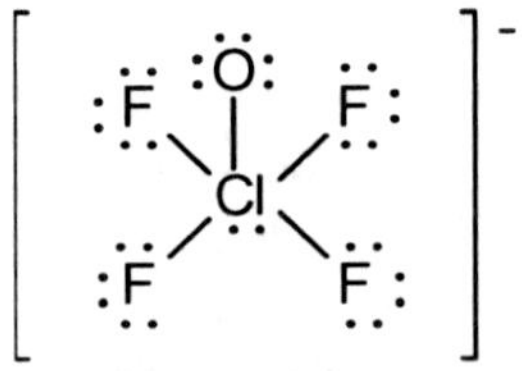

square pyramid, d^2sp^3

Note: Similar to Exercise 63c, d, and e, $F_2ClO_2^-$ has two additional Lewis structures that are possible, and F_4ClO^- has one additional Lewis structure that is possible. The predicted hybridization is unaffected.

$F_3ClO \rightarrow F^- + F_2ClO^+$

F_2ClO^+, $2(7) + 7 + 6 - 1 = 26\ e^-$

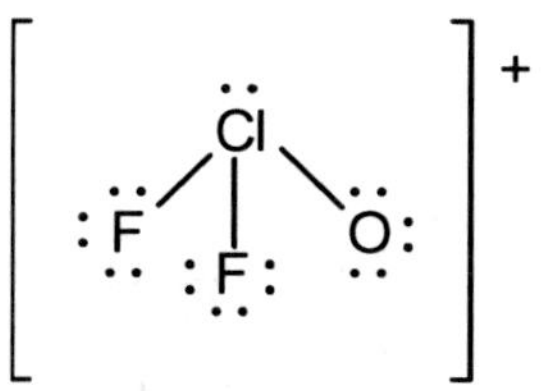

trigonal pyramid, sp^3

$F_3ClO_2 \rightarrow F^- + F_2ClO_2^+$

$F_2ClO_2^+$, $2(7) + 7 + 2(6) - 1 = 32\ e^-$

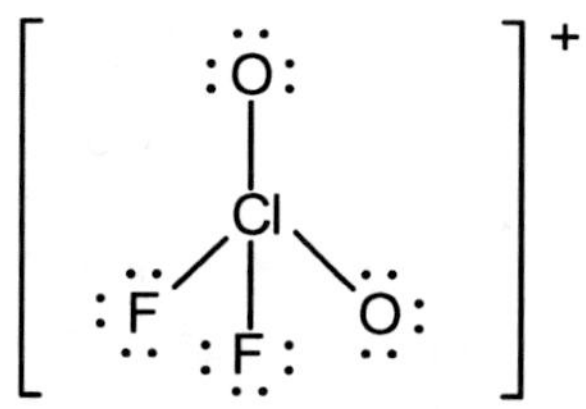

tetrahedral, sp^3

65. a. No, some atoms are in different places. Thus these are not resonance structures; they are different compounds.

b. For the first Lewis structure, all nitrogen atoms are sp^3 hybridized and all carbon atoms are sp^2 hybridized. In the second Lewis structure, all nitrogen atoms and carbon atoms are sp^2 hybridized.

c. For the reaction:

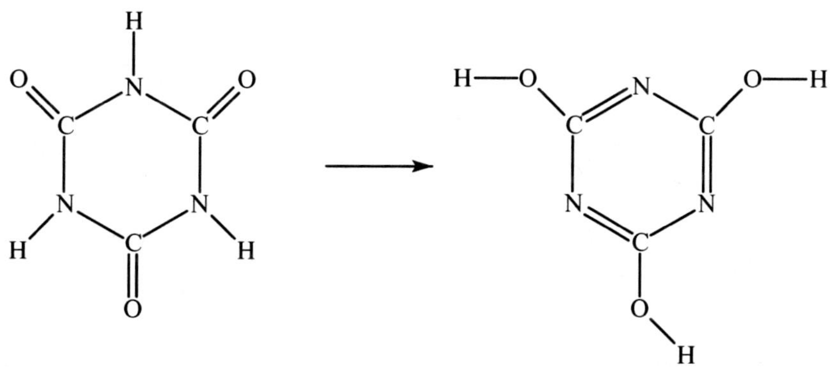

Bonds broken:	Bonds formed:
3 C=O (745 kJ/mol)	3 C=N (615 kJ/mol)
3 C–N (305 kJ/mol)	3 C–O (358 kJ/mol)
3 N–H (391 kJ/mol)	3 O–H (467 kJ/mol)

$\Delta H = 3(745) + 3(305) + 3(391) - [3(615) + 3(358) + 3(467)]$

$\Delta H = 4323\ kJ - 4320\ kJ = 3\ kJ$

The bonds are slightly stronger in the first structure with the carbon-oxygen double bonds since ΔH for the reaction is positive. However, the value of ΔH is so small that the best conclusion is that the bond strengths are comparable in the two structures.

66. a. The V-shaped (or bent) molecular structure occurs with both a trigonal planar and a tetrahedral arrangement of electron pairs. If there is a trigonal planar arrangement, the central atom is sp^2 hybridized. If there is a tetrahedral arrangement, the central atom is sp^3 hybridized.

b. The see-saw structure is a trigonal bipyramid arrangement of electron pairs which requires dsp^3 hybridization.

c. The trigonal pyramid structure occurs when a central atom has three bonded atoms and a lone pair of electrons. Whenever a central atom has four effective pairs about the central atom (exhibits a tetrahedral arrangement of electron pairs), the central atom is sp^3 hybridized.

d. A trigonal bipyramidal arrangement of electron pairs requires dsp^3 hybridization.

e. A tetrahedral arrangement of electron pairs requires sp^3 hybridization.

67. For carbon, nitrogen, and oxygen atoms to have formal charge values of zero, each C atom will form four bonds to other atoms and have no lone pairs of electrons, each N atom will form three bonds to other atoms and have one lone pair of electrons, and each O atom will form two bonds to other atoms and have two lone pairs of electrons. Following these bonding requirements gives the following two resonance structures for vitamin B_6:

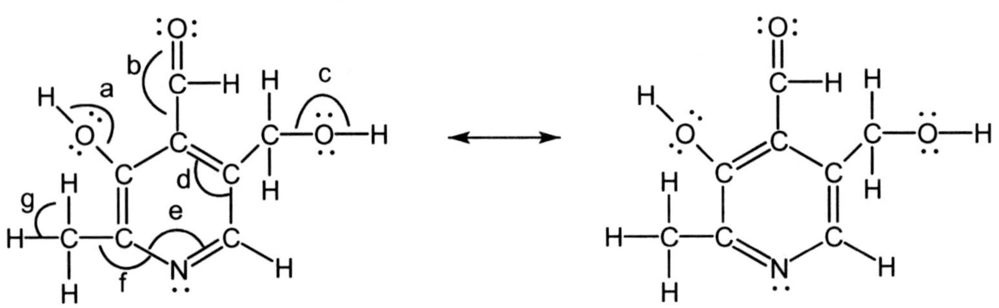

a. 21 σ bonds; 4 π bonds (The electrons in the three π bonds in the ring are delocalized.)

b. Angles a), c), and g): ≈109.5°; angles b), d), e), and f): ≈120°

c. 6 sp^2 carbons; the five carbon atoms in the ring are sp^2 hybridized, as is the carbon with the double bond to oxygen.

d. 4 sp^3 atoms; the two carbons that are not sp^2 hybridized are sp^3 hybridized, and the oxygens marked with angles a and c are sp^3 hybridized.

e. Yes, the π electrons in the ring are delocalized. The atoms in the ring are all sp^2 hybridized. This leaves a p orbital perpendicular to the plane of the ring from each atom. Overlap of all six of these p orbitals results in a π molecular orbital system where the electrons are delocalized above and below the plane of the ring (similar to benzene in Figure 9.47 of the text).

68. For carbon, nitrogen, and oxygen atoms to have formal charge values of zero, each C atom will form four bonds to other atoms and have no lone pairs of electrons, each N atom will form three bonds to other atoms and have one lone pair of electrons, and each O atom will form two bonds to other atoms and have two lone pairs of electrons. Following these bonding requirements, a Lewis structure for aspartame is:

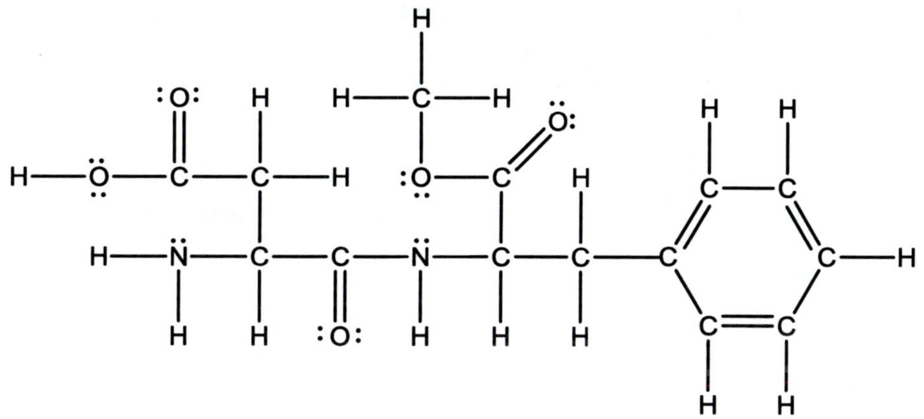

Another resonance structure could be drawn having the double bonds in the benzene ring moved over one position.

Atoms that have trigonal planar geometry of electron pairs are assumed to have sp^2 hybridization, and atoms with tetrahedral geometry of electron pairs are assumed to have sp^3 hybridization. All the N atoms have tetrahedral geometry, so they are all sp^3 hybridized (no sp^2 hybridization). The oxygens double bonded to carbon atoms are sp^2 hybridized; the other two oxygens with two single bonds are sp^3 hybridized. For the carbon atoms, the six carbon atoms in the benzene ring are sp^2 hybridized, and the three carbons double bonded to oxygen are also sp^2 hybridized (tetrahedral geometry). Answering the questions:

- 9 sp^2 hybridized C and N atoms (9 from C's and 0 from N's)
- 7 sp^3 hybridized C and O atoms (5 from C's and 2 from O's)
- 39 σ bonds and 6 π bonds (this includes the 3 π bonds in the benzene ring that are delocalized)

69.

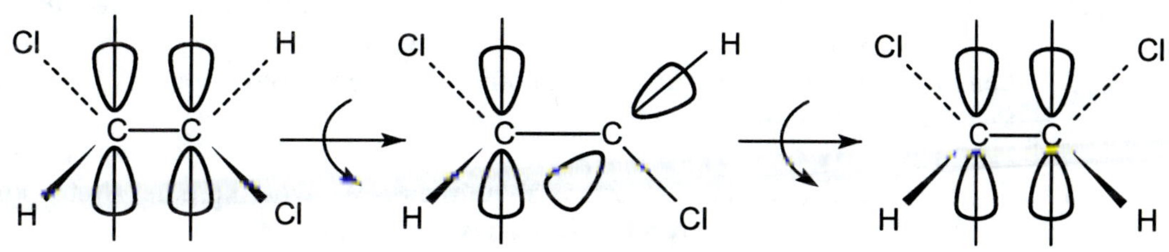

In order to rotate about the double bond, the molecule must go through an intermediate stage where the π bond is broken and the sigma bond remains intact. Bond energies are 347 kJ/mol for C–C and 614 kJ/mol for C=C. If we take the single bond as the strength of the σ bond, then the strength of the π bond is (614 – 347 =) 267 kJ/mol. In theory, 267 kJ/mol must be supplied to rotate about a carbon-carbon double bond.

70. CO, 4 + 6 = 10 e^-; CO_2, 4 + 2(6) = 16 e^-; C_3O_2, 3(4) + 2(6) = 24 e^-

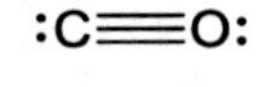 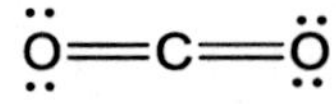 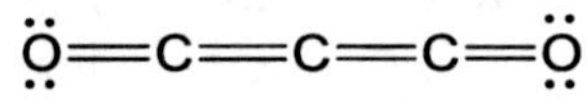

There is no molecular structure for the diatomic CO molecule. The carbon in CO is sp hybridized. CO_2 is a linear molecule, and the central carbon atom is sp hybridized. C_3O_2 is a linear molecule with all the central carbon atoms exhibiting sp hybridization.

71. a. BH_3 has 3 + 3(1) = 6 valence electrons.

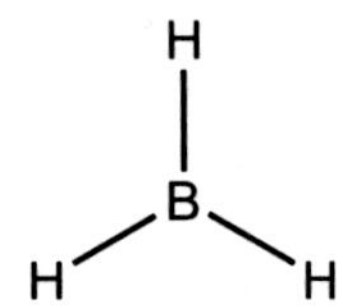

trigonal planar, nonpolar, 120°, sp^2

b. N_2F_2 has 2(5) + 2(7) = 24 valence electrons.

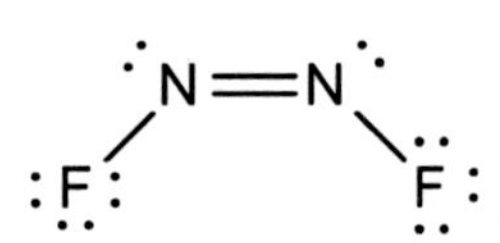

V-shaped about both N's;
≈120° about both N's;
both N atoms: sp^2

polar

Can also be:

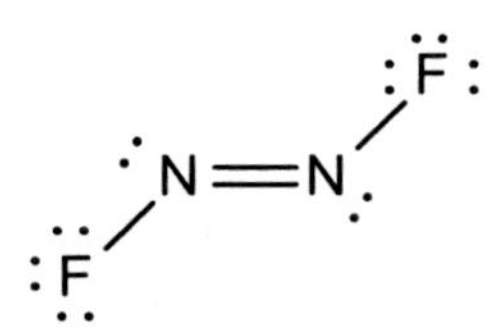

nonpolar

These are distinctly different molecules.

c. C_4H_6 has 4(4) + 6(1) = 22 valence electrons.

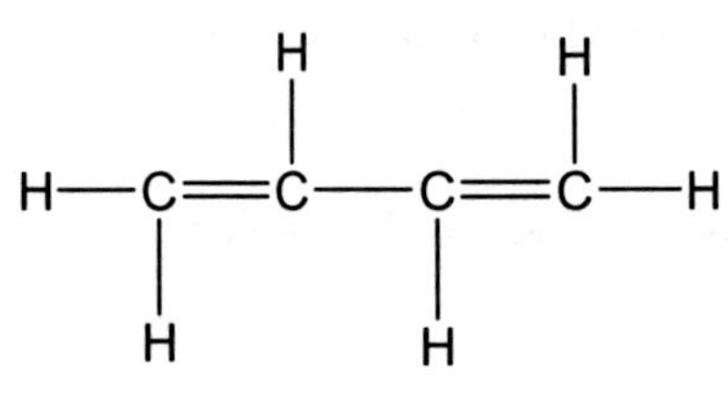

All C atoms are trigonal planar with 120° bond angles and sp^2 hybridization. Because C and H have similar electronegativity values, the C–H bonds are essentially nonpolar, so the molecule is nonpolar. All neutral compounds composed of only C and H atoms are nonpolar.

72. a. Yes, both have four sets of electrons about the P. We would predict a tetrahedral structure for both. See part d for the Lewis structures.

b. The hybridization is sp^3 for P in each structure since both structures exhibit a tetrahedral arrangement of electron pairs.

c. P has to use one of its d orbitals to form the π bond since the p orbitals are all used to form the hybrid orbitals.

d. Formal charge = number of valence electrons of an atom − [(number of lone pair electrons) + 1/2(number of shared electrons)]. The formal charges calculated for the O and P atoms are next to the atoms in the following Lewis structures.

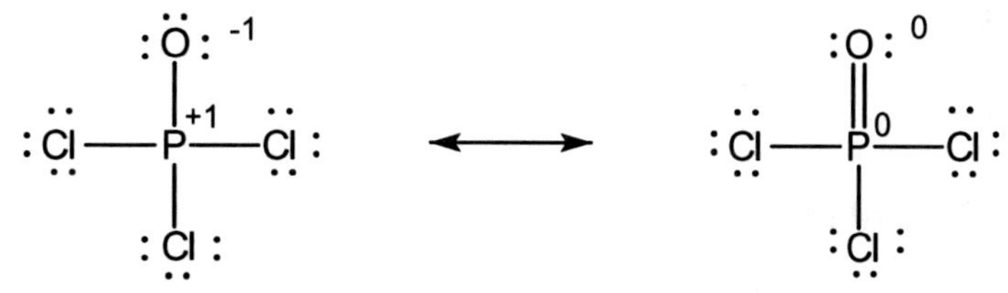

In both structures, the formal charges of the Cl atoms are all zeros. The structure with the P=O bond is favored on the basis of formal charge since it has a zero formal charge for all atoms.

73. a. The Lewis structures for NNO and NON are:

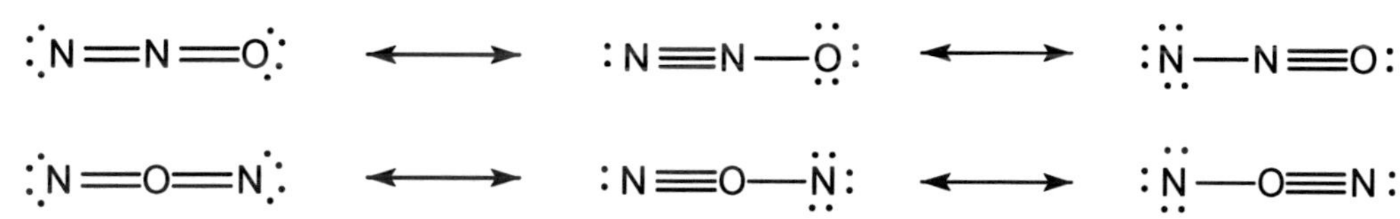

The NNO structure is correct. From the Lewis structures, we would predict both NNO and NON to be linear. However, we would predict NNO to be polar and NON to be nonpolar. Since experiments show N_2O to be polar, NNO is the correct structure.

b. Formal charge = number of valence electrons of atoms − [(number of lone pair electrons) + 1/2(number of shared electrons)].

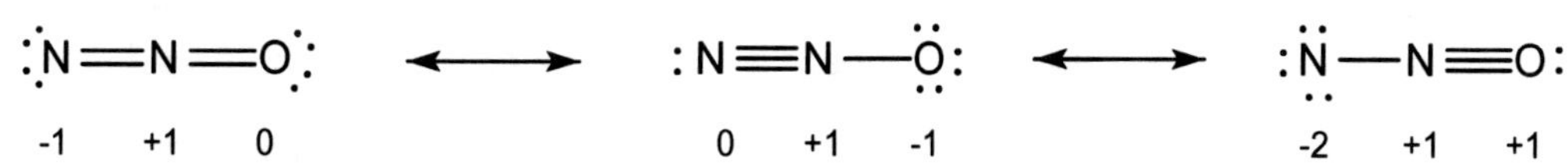

The formal charges for the atoms in the various resonance structures are below each atom. The central N is sp hybridized in all the resonance structures. We can probably ignore the third resonance structure on the basis of the relatively large formal charges as compared to the first two resonance structures.

c. The sp hybrid orbitals from the center N overlap with atomic orbitals (or appropriate hybrid orbitals) from the other two atoms to form the two sigma bonds. The remaining two unhybridized p orbitals from the center N overlap with two p orbitals from the peripheral N to form the two π bonds.

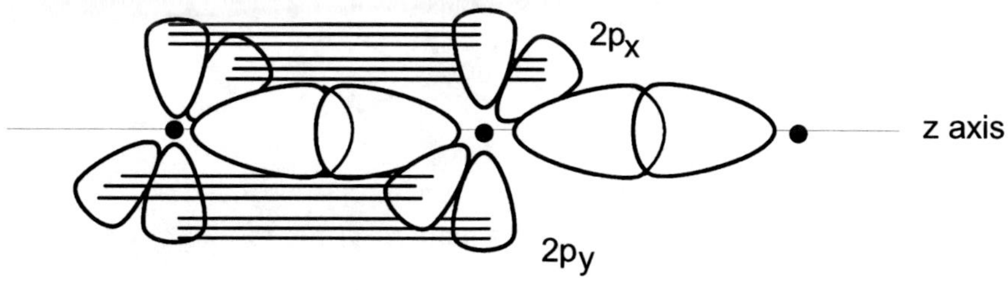

74. N_2 (ground state): $(\sigma_{2s})^2(\sigma_{2s}{*})^2(\pi_{2p})^4(\sigma_{2p})^2$; B.O. = 3; diamagnetic (0 unpaired e^-)

N_2 (1st excited state): $(\sigma_{2s})^2(\sigma_{2s}{*})^2(\pi_{2p})^4(\sigma_{2p})^1(\pi_{2p}{*})^1$

B.O. = (7 − 3)/2 = 2; paramagnetic (2 unpaired e^-)

The first excited state of N_2 should have a weaker bond and should be paramagnetic.

75. F_2: $(\sigma_{2s})^2(\sigma_{2s}{*})^2(\sigma_{2p})^2(\pi_{2p})^4(\pi_{2p}{*})^4$; F_2 should have a lower ionization energy than F. The electron removed from F_2 is in a $\pi_{2p}{*}$ antibonding molecular orbital that is higher in energy than the 2p atomic orbitals from which the electron in atomic fluorine is removed. Because the electron removed from F_2 is higher in energy than the electron removed from F, it should be easier to remove an electron from F_2 than from F.

76.

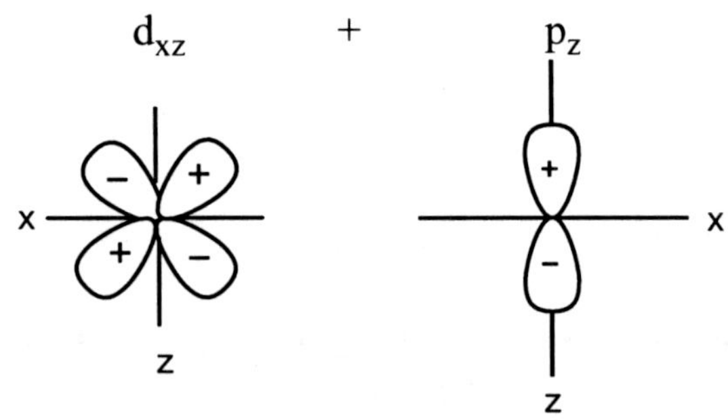

The two orbitals will overlap side to side, so when the orbitals are in phase, a π bonding molecular orbital would be expected to form.

77. Side-to-side in-phase overlap of these d orbitals would produce a π bonding molecular orbital. There would be no probability of finding an electron on the axis joining the two nuclei, which is characteristic of π MOs.

78. Molecule A has a tetrahedral arrangement of electron pairs because it is sp^3 hybridized. Molecule B has 6 electron pairs about the central atom, so it is d^2sp^3 hybridized. Molecule C has two σ and two π bonds to the central atom, so it either has two double bonds to the central atom (as in CO_2) or one triple bond and one single bond (as in HCN). Molecule C is consistent with a linear arrangement of electron pairs exhibiting sp hybridization. There are many correct possibilities for each molecule; an example of each is:

Molecule A: CH_4

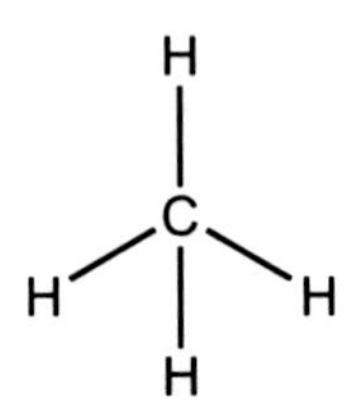

tetrahedral; 109.5°; sp^3

Molecule B: XeF_4

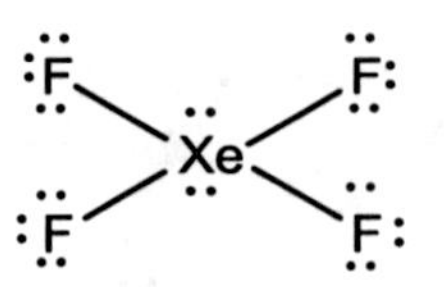

square planar; 90°; d^2sp^3

Molecule C: CO_2 or HCN

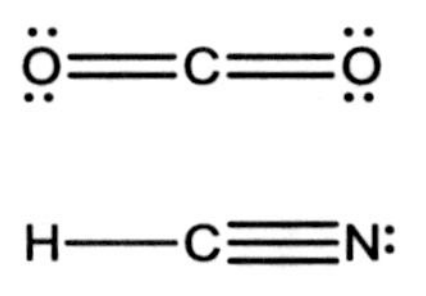

linear; 180°; sp

ChemWork Problems

The answers to the problems 79-86 (or a variation to these problems) are found in OWL. These problems are also assignable in OWL.

Challenge Problems

87. The following Lewis structure has a formal charge of zero for all of the atoms in the molecule.

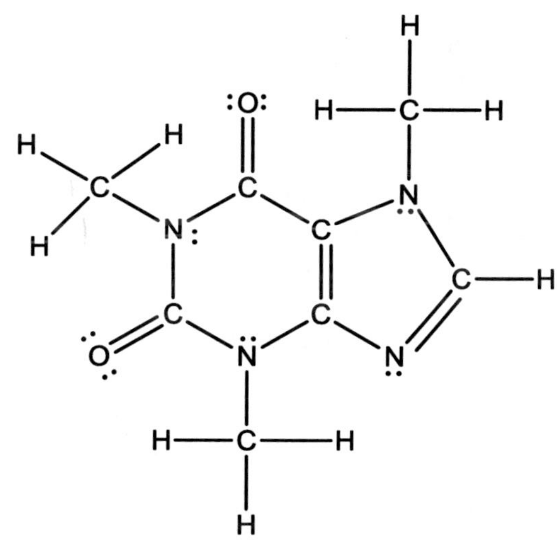

The three C atoms each bonded to three H atoms are sp^3 hybridized (tetrahedral geometry); the other five C atoms with trigonal planar geometry are sp^2 hybridized. The one N atom with the double bond is sp^2 hybridized, and the other three N atoms are sp^3 hybridized. The answers to the questions are:

- 6 total C and N atoms are sp^2 hybridized
- 6 total C and N atoms are sp^3 hybridized
- 0 C and N atoms are sp hybridized (linear geometry)
- 25 σ bonds and 4 π bonds

88. The complete Lewis structure follows on the next page. All but two of the carbon atoms are sp^3 hybridized. The two carbon atoms that contain the double bond are sp^2 hybridized (see * in the following Lewis structure).

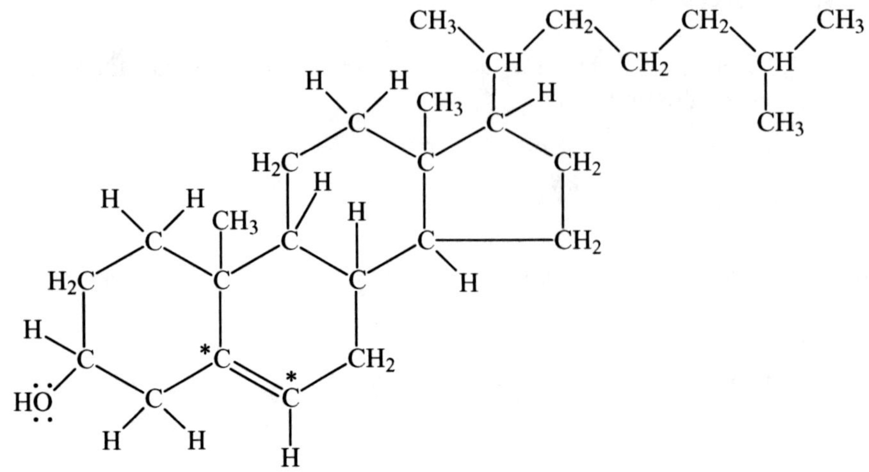

No; most of the carbons are not in the same plane since a majority of carbon atoms exhibit a tetrahedral structure (109.5° bond angles). *Note*: HO, CH, CH_2, H_2C, and CH_3 are shorthand for oxygen and carbon atoms singly bonded to hydrogen atoms.

89. a. NCN^{2-} has 5 + 4 + 5 + 2 = 16 valence electrons.

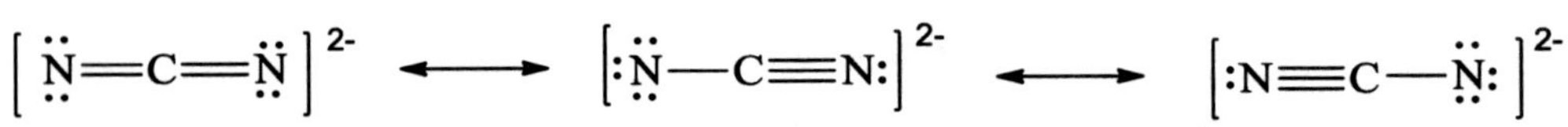

H_2NCN has 2(1) + 5 + 4 + 5 = 16 valence electrons.

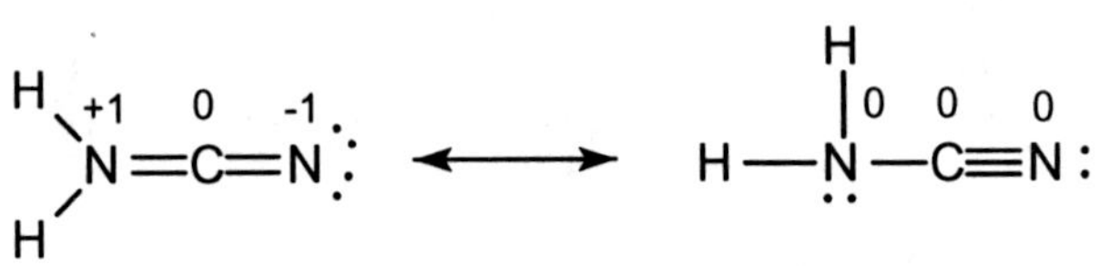

favored by formal charge

NCNC$(NH_2)_2$ has 5 + 4 + 5 + 4 + 2(5) + 4(1) = 32 valence electrons.

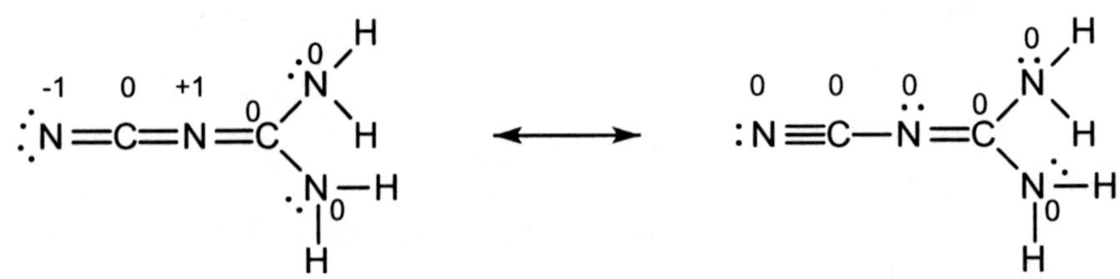

favored by formal charge

Melamine ($C_3N_6H_6$) has 3(4) + 6(5) + 6(1) = 48 valence electrons.

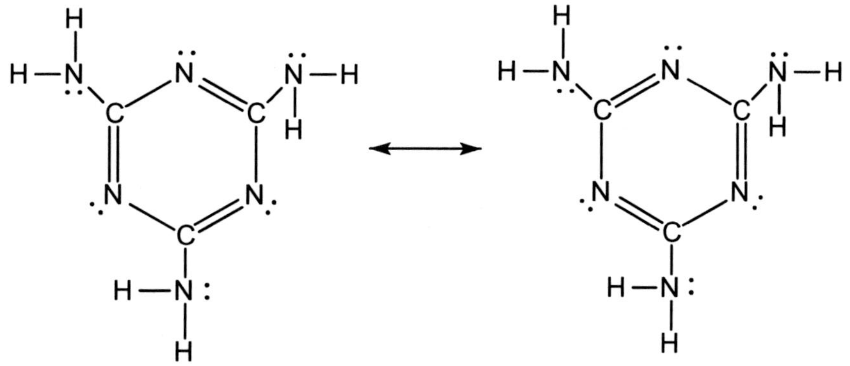

b. NCN^{2-}: C is sp hybridized. Each resonance structure predicts a different hybridization for the N atom. Depending on the resonance form, N is predicted to be sp, sp^2, or sp^3 hybridized. For the remaining compounds, we will give hybrids for the favored resonance structures as predicted from formal charge considerations.

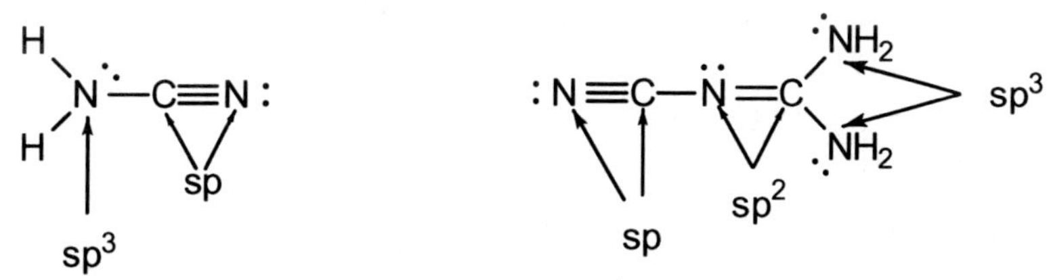

Melamine: N in NH_2 groups are all sp^3 hybridized; atoms in ring are all sp^2 hybridized.

c. NCN^{2-}: 2 σ and 2 π bonds; H_2NCN: 4 σ and 2 π bonds; dicyandiamide: 9 σ and 3 π bonds; melamine: 15 σ and 3 π bonds

d. The π-system forces the ring to be planar, just as the benzene ring is planar (see Figure 9.47 of the text).

e. The structure:

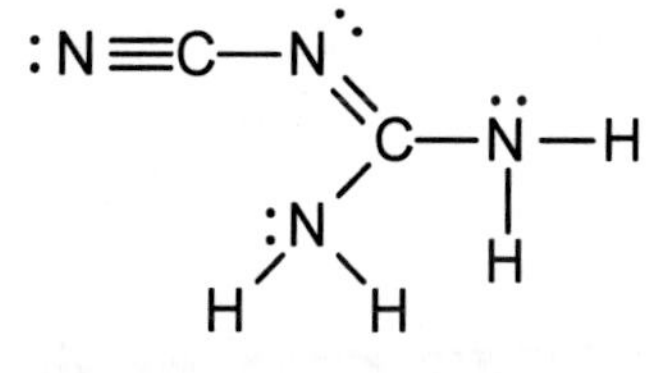

best agrees with experiments because it has three different CN bonds. This structure is also favored on the basis of formal charge.

90. One of the resonance structures for benzene is:

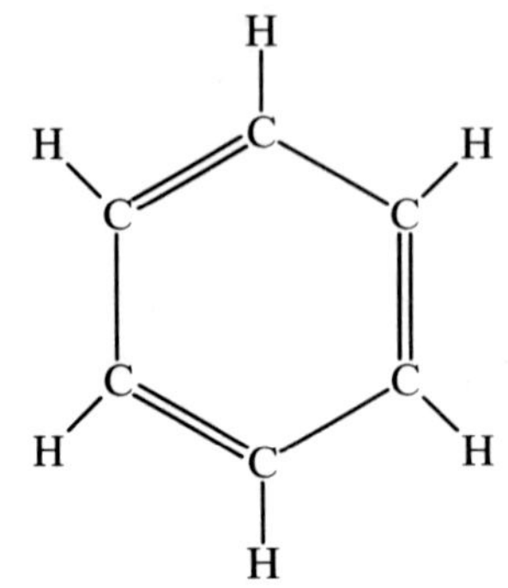

To break $C_6H_6(g)$ into C(g) and H(g) requires the breaking of 6 C–H bonds, 3 C=C bonds, and 3 C–C bonds:

$$C_6H_6(g) \rightarrow 6\ C(g) + 6\ H(g) \quad \Delta H = 6D_{C-H} + 3D_{C=C} + 3D_{C-C}$$

$$\Delta H = 6(413\ kJ) + 3(614\ kJ) + 3(347\ kJ) = 5361\ kJ$$

The question asks for ΔH_f° for $C_6H_6(g)$, which is ΔH for the reaction:

$$6\ C(s) + 3\ H_2(g) \rightarrow C_6H_6(g) \quad \Delta H = \Delta H^\circ_{f,\ C_6H_6(g)}$$

To calculate ΔH for this reaction, we will use Hess's law along with the value ΔH_f° for C(g) and the bond energy value for H_2 (D_{H_2} = 432 kJ/mol).

$6\ C(g) + 6\ H(g) \rightarrow C_6H_6(g)$	$\Delta H_1 = -5361$ kJ
$6\ C(s) \rightarrow 6\ C(g)$	$\Delta H_2 =$ 6(717 kJ)
$3\ H_2(g) \rightarrow 6\ H(g)$	$\Delta H_3 =$ 3(432 kJ)
$6\ C(s) + 3\ H_2(g) \rightarrow C_6H_6(g)$	$\Delta H = \Delta H_1 + \Delta H_2 + \Delta H_3 = 237$ kJ; $\Delta H^\circ_{f,\ C_6H_6(g)} = 237$ kJ/mol

The experimental ΔH_f° for $C_6H_6(g)$ is more stable (lower in energy) by 154 kJ than the ΔH_f° calculated from bond energies (83 − 237 = −154 kJ). This extra stability is related to benzene's ability to exhibit resonance. Two equivalent Lewis structures can be drawn for benzene. The π bonding system implied by each Lewis structure consists of three localized π bonds. This is not correct because all C–C bonds in benzene are equivalent. We say the π electrons in benzene are delocalized over the entire surface of C_6H_6 (see Section 9.5 of the text). The large discrepancy between ΔH_f° values is due to the delocalized π electrons, whose effect was not accounted for in the calculated ΔH_f° value. The extra stability associated with benzene can be called resonance stabilization. In general, molecules that exhibit resonance are usually more stable than predicted using bond energies.

91. a. $E = \frac{hc}{\lambda} = \frac{(6.626 \times 10^{-34}\ \text{J s})(2.998 \times 10^{8}\ \text{m/s})}{25 \times 10^{-9}\ \text{m}} = 7.9 \times 10^{-18}\ \text{J}$

$$7.9 \times 10^{-18}\ \text{J} \times \frac{6.022 \times 10^{23}}{\text{mol}} \times \frac{1\ \text{kJ}}{1000\ \text{J}} = 4800\ \text{kJ/mol}$$

Using ΔH values from the various reactions, 25-nm light has sufficient energy to ionize N_2 and N and to break the triple bond. Thus N_2, N_2^+, N, and N^+ will all be present, assuming excess N_2.

b. To produce atomic nitrogen but no ions, the range of energies of the light must be from 941 kJ/mol to just below 1402 kJ/mol.

$$\frac{941\ \text{kJ}}{\text{mol}} \times \frac{1\ \text{mol}}{6.022 \times 10^{23}} \times \frac{1000\ \text{J}}{1\ \text{kJ}} = 1.56 \times 10^{-18}\ \text{J/photon}$$

$$\lambda = \frac{hc}{E} = \frac{(6.6261 \times 10^{-34}\ \text{J s})(2.998 \times 10^{8}\ \text{m/s})}{1.56 \times 10^{-18}\ \text{J}} = 1.27 \times 10^{-7}\ \text{m} = 127\ \text{nm}$$

$$\frac{1402\ \text{kJ}}{\text{mol}} \times \frac{1\ \text{mol}}{6.0221 \times 10^{23}} \times \frac{1000\ \text{J}}{\text{kJ}} = 2.328 \times 10^{-18}\ \text{J/photon}$$

$$\lambda = \frac{hc}{E} = \frac{(6.6261 \times 10^{-34}\ \text{J s})(2.9979 \times 10^{8}\ \text{m/s})}{2.328 \times 10^{-18}\ \text{J}} = 8.533 \times 10^{-8}\ \text{m} = 85.33\ \text{nm}$$

Light with wavelengths in the range of $85.33\ \text{nm} < \lambda \leq 127\ \text{nm}$ will produce N but no ions.

c. N_2: $(\sigma_{2s})^2(\sigma_{2s}^*)^2(\pi_{2p})^4(\sigma_{2p})^2$; the electron removed from N_2 is in the σ_{2p} molecular orbital, which is lower in energy than the 2p atomic orbital from which the electron in atomic nitrogen is removed. Because the electron removed from N_2 is lower in energy than the electron removed from N, the ionization energy of N_2 is greater than that for N.

92. The π bonds between S atoms and between C and S atoms are not as strong. The p atomic orbitals do not overlap with each other as well as the smaller p atomic orbitals of C and O overlap.

93. O=N–Cl: The bond order of the NO bond in NOCl is 2 (a double bond).

NO: From molecular orbital theory, the bond order of this NO bond is 2.5. (See Figure 9.40 of the text.)

Both reactions apparently involve only the breaking of the N–Cl bond. However, in the reaction ONCl → NO + Cl, some energy is released in forming the stronger NO bond, lowering the value of ΔH. Therefore, the apparent N–Cl bond energy is artificially low for this reaction. The first reaction involves only the breaking of the N–Cl bond.

94. The molecular orbitals for BeH_2 are formed from the two hydrogen 1s orbitals and the 2s and one of the 2p orbitals from beryllium. One of the sigma bonding orbitals forms from overlap of the hydrogen 1s orbitals with a 2s orbital from beryllium. Assuming the z axis is the internuclear axis in the linear BeH_2 molecule, then the $2p_z$ orbital from beryllium has proper symmetry to overlap with the 1s orbitals from hydrogen; the $2p_x$ and $2p_y$ orbitals are nonbonding orbitals since they don't have proper symmetry necessary to overlap with 1s orbitals. The type of bond formed from the $2p_z$ and 1s orbitals is a sigma bond since the orbitals overlap head to head. The MO diagram for BeH_2 is:

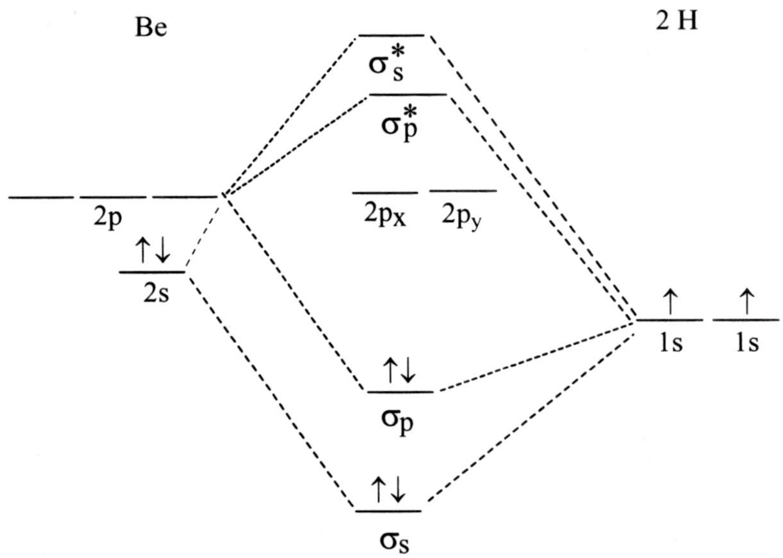

Bond order = (4 – 0)/2 = 2; the MO diagram predicts BeH_2 to be a stable species and also predicts that BeH_2 is diamagnetic. *Note*: The σ_s MO is a mixture of the two hydrogen 1s orbitals with the 2s orbital from beryllium, and the σ_p MO is a mixture of the two hydrogen 1s orbitals with the $2p_z$ orbital from beryllium. The MOs are not localized between any two atoms; instead, they extend over the entire surface of the three atoms.

95. The ground state MO electron configuration for He_2 is $(\sigma_{1s})^2(\sigma_{1s}*)^2$, giving a bond order of 0. Therefore, He_2 molecules are not predicted to be stable (and are not stable) in the lowest-energy ground state. However, in a high-energy environment, electron(s) from the anti-bonding orbitals in He_2 can be promoted into higher-energy bonding orbitals, thus giving a nonzero bond order and a "reason" to form. For example, a possible excited-state MO electron configuration for He_2 would be $(\sigma_{1s})^2(\sigma_{1s}*)^1(\sigma_{2s})^1$, giving a bond order of (3 – 1)/2 = 1. Thus excited He_2 molecules can form, but they spontaneously break apart as the electron(s) fall back to the ground state, where the bond order equals zero.

96.

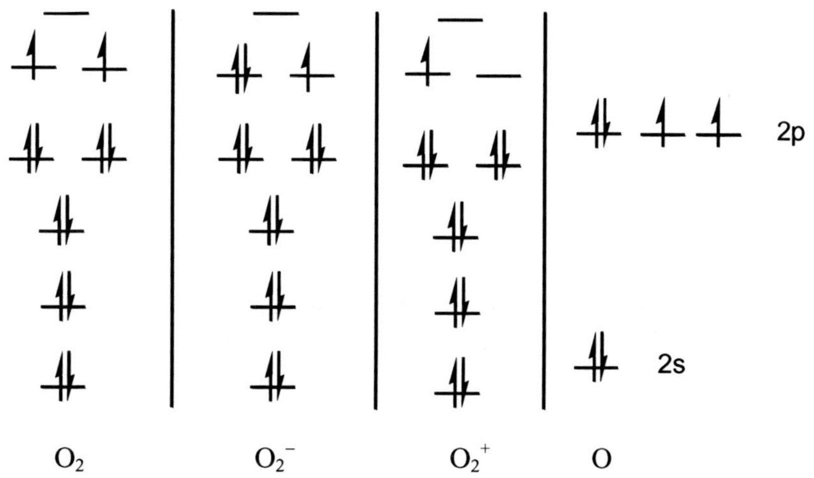

The order from lowest IE to highest IE is: $O_2^- < O_2 < O_2^+ < O$.

The electrons for O_2^-, O_2, and O_2^+ that are highest in energy are in the π^*_{2p} MOs. But for O_2^-, these electrons are paired. O_2^- should have the lowest ionization energy (its paired π^*_{2p} electron is easiest to remove). The species O_2^+ has an overall positive charge, making it harder to remove an electron from O_2^+ than from O_2. The highest-energy electrons for O (in the 2p atomic orbitals) are lower in energy than the π^*_{2p} electrons for the other species; O will have the highest ionization energy because it requires a larger quantity of energy to remove an electron from O as compared to the other species.

97. The electron configurations are:

N_2: $(\sigma_{2s})^2(\sigma_{2s}*)^2(\pi_{2p})^4(\sigma_{2p})^2$

O_2: $(\sigma_{2s})^2(\sigma_{2s}*)^2(\sigma_{2p})^2(\pi_{2p})^4(\pi_{2p}*)^2$

N_2^{2-}: $(\sigma_{2s})^2(\sigma_{2s}*)^2(\pi_{2p})^4(\sigma_{2p})^2(\pi_{2p}*)^2$

N_2^-: $(\sigma_{2s})^2(\sigma_{2s}*)^2(\pi_{2p})^4(\sigma_{2p})^2(\pi_{2p}*)^1$

O_2^+: $(\sigma_{2s})^2(\sigma_{2s}*)^2(\sigma_{2p})^2(\pi_{2p})^4(\pi_{2p}*)^1$

Note: The ordering of the σ_{2p} and π_{2p} orbitals is not important to this question.

The species with the smallest ionization energy has the electron that is easiest to remove. From the MO electron configurations, O_2, N_2^{2-}, N_2^-, and O_2^+ all contain electrons in the same higher-energy antibonding orbitals (π^*_{2p}), so they should have electrons that are easier to remove as compared to N_2, which has no π^*_{2p} electrons. To differentiate which has the easiest π^*_{2p} to remove, concentrate on the number of electrons in the orbitals attracted to the number of protons in the nucleus.

N_2^{2-} and N_2^- both have 14 protons in the two nuclei combined. Because N_2^{2-} has more electrons, one would expect N_2^{2-} to have more electron repulsions, which translates into having an easier electron to remove. Between O_2 and O_2^+, the electron in O_2 should be easier to remove. O_2 has one more electron than O_2^+, and one would expect the fewer electrons in

O_2^+ to be better attracted to the nuclei (and harder to remove). Between N_2^{2-} and O_2, both have 16 electrons; the difference is the number of protons in the nucleus. Because N_2^{2-} has two fewer protons than O_2, one would expect the N_2^{2-} to have the easiest electron to remove, which translates into the smallest ionization energy.

98. a. $F_2^-(g) \rightarrow F(g) + F^-(g)$ $\Delta H = F_2^-$ bond energy

Using Hess's law:

$F_2^-(g) \rightarrow F_2(g) + e^-$	$\Delta H = 290.$ kJ (ionization energy for F_2^-)
$F_2(g) \rightarrow 2\,F(g)$	$\Delta H = 154$ kJ (bond energy for F_2 from Table 8.4)
$F(g) + e^- \rightarrow F^-(g)$	$\Delta H = -327.8$ kJ (electron affinity for F from Table 7.7)
$F_2^-(g) \rightarrow F(g) + F^-(g)$	$\Delta H = 116$ kJ; bond energy for $F_2^- = 116$ kJ/mol

Note that F_2^- has a smaller bond energy than F_2.

b. F_2: $(\sigma_{2s})^2(\sigma_{2s}^*)^2(\sigma_{2p})^2(\pi_{2p})^4(\pi_{2p}^*)^4$ B.O. = (8 – 6)/2 = 1

F_2^-: $(\sigma_{2s})^2(\sigma_{2s}^*)^2(\sigma_{2p})^2(\pi_{2p})^4(\pi_{2p}^*)^4(\sigma_{2p}^*)^1$ B.O. = (8 – 7)/2 = 0.5

MO theory predicts that F_2 should have a stronger bond than F_2^- because F_2 has the larger bond order. As determined in part a, F_2 indeed has a stronger bond because the F_2 bond energy (154 kJ/mol) is greater than the F_2^- bond energy (116 kJ/mol).

99. a. The CO bond is polar with the negative end at the more electronegative oxygen atom. We would expect metal cations to be attracted to and bond to the oxygen end of CO on the basis of electronegativity.

b. :C≡O: FC (carbon) = 4 – 2 – 1/2(6) = –1

FC (oxygen) = 6 – 2 – 1/2(6) = +1

From formal charge, we would expect metal cations to bond to the carbon (with the negative formal charge).

c. In molecular orbital theory, only orbitals with proper symmetry overlap to form bonding orbitals. The metals that form bonds to CO are usually transition metals, all of which have outer electrons in the d orbitals. The only molecular orbitals of CO that have proper symmetry to overlap with d orbitals are the π_{2p}^* orbitals, whose shape is similar to the d orbitals. Because the antibonding molecular orbitals have more carbon character (carbon is less electronegative than oxygen), one would expect the bond to form through carbon.

100. Benzoic acid ($C_7H_6O_2$) has 7(4) + 6(1) + 2(6) = 46 valence electrons. The Lewis structure for benzoic acid is:

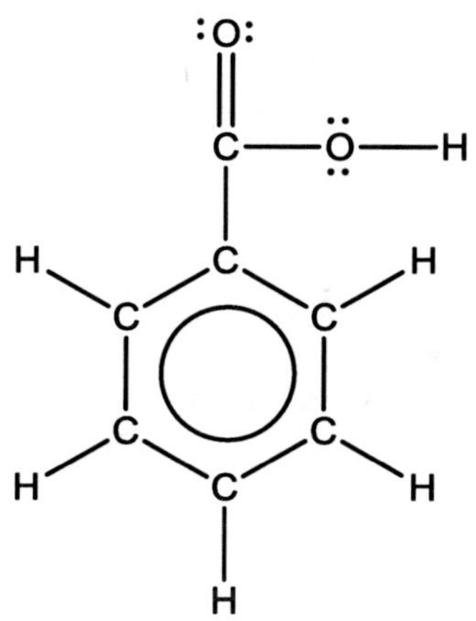

The circle in the ring indicates the delocalized π bonding in the benzene ring. The two benzene resonance Lewis structures have three alternating double bonds in the ring (see Figure 9.45).

The six carbons in the ring and the carbon bonded to the ring are all sp^2 hybridized. The five C–H sigma bonds are formed from overlap of the sp^2 hybridized carbon atoms with hydrogen 1s atomic orbitals. The seven C–C σ bonds are formed from head to head overlap of sp^2 hybrid orbitals from each carbon. The C–O single bond is formed from overlap of an sp^2 hybrid orbital on carbon with an sp^3 hybrid orbital from oxygen. The C–O σ bond in the double bond is formed from overlap of carbon sp^2 hybrid orbital with an oxygen sp^2 orbital. The π bond in the C–O double bond is formed from overlap of parallel p unhybridized atomic orbitals from C and O. The delocalized π bonding system in the ring is formed from overlap of all six unhybridized p atomic orbitals from the six carbon atoms. See Figure 9.47 for delocalized π bonding system in the benzene ring.

Integrative Problems

101. a. Li_2: $(\sigma_{2s})^2$ B.O. = (2 – 0)/2 = 1

B_2: $(\sigma_{2s})^2(\sigma_{2s}^*)^2(\pi_{2p})^2$ B.O. = (4 – 2)/2 = 1

Both have a bond order of 1.

b. B_2 has four more electrons than Li_2, so four electrons must be removed from B_2 to make it isoelectronic with Li_2. The isoelectronic ion is B_2^{4+}.

c. To form B_2^{4+}, it takes 6455 kJ of energy to remove 4 mol of electrons from 1 mol of B_2.

$$1.5 \text{ kg } B_2 \times \frac{1000 \text{g}}{1 \text{kg}} \times \frac{1 \text{mol} B_2}{21.62 \text{g} B_2} \times \frac{6455 \text{kJ}}{\text{mol} B_2} = 4.5 \times 10^5 \text{kJ}$$

102. a. HF, $1 + 7 = 8\ e^-$

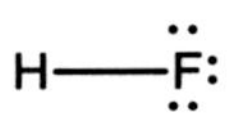

linear, sp^3 (if F is hybridized)

SbF_5, $5 + 5(7) = 40\ e^-$

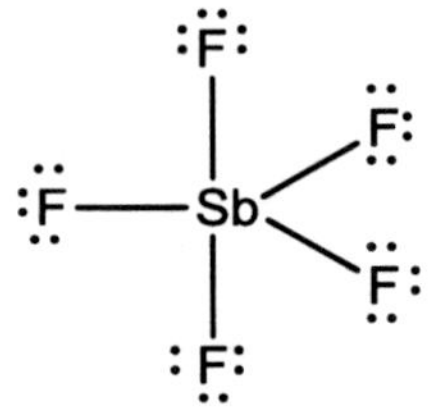

trigonal bipyramid, dsp^3

H_2F^+, $2(1) + 7 - 1 = 8\ e^-$

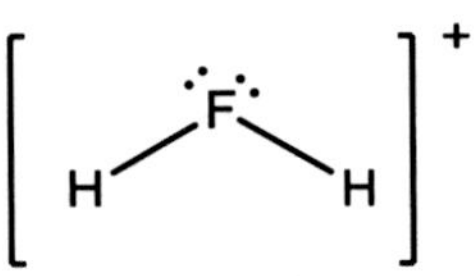

V-shaped, sp^3

SbF_6^-, $5 + 6(7) + 1 = 48\ e^-$

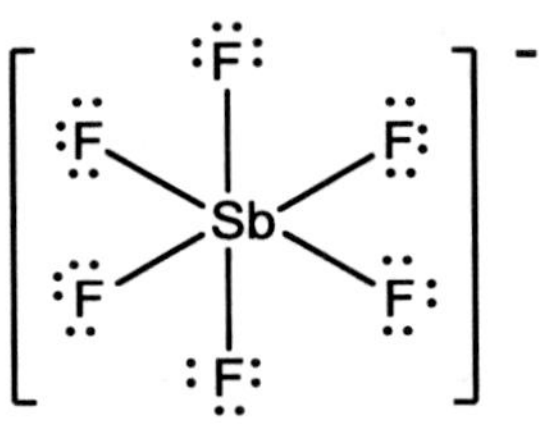

octahedral, d^2sp^3

b. $$2.93\text{ mL} \times \frac{0.975\text{g HF}}{\text{mL}} \times \frac{1\text{ mol HF}}{20.01\text{g HF}} \times \frac{1\text{ mol }[H_2F]^+[SbF_6]^-}{2\text{ mol HF}} \times \frac{256.8\text{ g}}{\text{mol}[H_2F]^+[SbF_6]^-}$$

$$= 18.3\text{ g }[H_2F]^+[SbF_6]^-$$

$$10.0\text{ mL} \times \frac{3.10\text{g }SbF_5}{\text{mL}} \times \frac{1\text{ mol }SbF_5}{216.8\text{g }SbF_5} \times \frac{1\text{ mol }[H_2F]^+[SbF_6]^-}{\text{mol }SbF_5} \times \frac{256.8\text{g}}{\text{mol }[H_2F]^+[SbF_6]^-}$$

$$= 36.7\text{ g }[H_2F]^+[SbF_6]^-$$

Because HF produces the smaller amount of product, HF is limiting and 18.3 g of $[H_2F]^+[SbF_6]^-$ can be produced.

103. Element X has 36 protons, which identifies it as Kr. Element Y has one less electron than Y^-, so the electron configuration of Y is $1s^22s^22p^5$. This is F.

KrF_3^+, $8 + 3(7) - 1 = 28\ e^-$

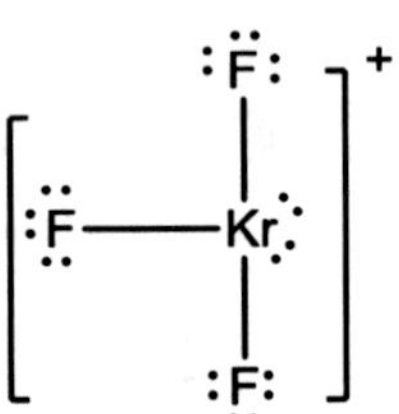

T-shaped, dsp^3

CHAPTER 10

LIQUIDS AND SOLIDS

Questions

12. Chalk is composed of the ionic compound calcium carbonate ($CaCO_3$). The electrostatic forces in ionic compounds are much stronger than the intermolecular forces in covalent compounds. Therefore, $CaCO_3$ should have a much higher boiling point than the covalent compounds found in motor oil and in H_2O. Motor oil is composed of nonpolar C–C and C–H bonds. The intermolecular forces in motor oil are therefore London dispersion forces. We generally consider these forces to be weak. However, with compounds that have large molar masses, these London dispersion forces add up significantly and can overtake the relatively strong hydrogen-bonding interactions in water.

13. Answer a is correct. Intermolecular forces are the forces between molecules that hold the substances together in the solid and liquid phases. Hydrogen bonding is a specific type of intermolecular forces. In this figure, the dotted lines represent the hydrogen bonding interactions that hold individual H_2O molecules together in the solid and liquid phases. The solid lines represent the O–H covalent bonds.

14. Hydrogen bonding occurs when hydrogen atoms are covalently bonded to highly electronegative atoms such as oxygen, nitrogen, or fluorine. Because the electronegativity difference between hydrogen and these highly electronegative atoms is relatively large, the N–H, O–H, and F–H bonds are very polar covalent bonds. This leads to strong dipole forces. Also, the small size of the hydrogen atom allows the dipoles to approach each other more closely than can occur between most polar molecules. Both of these factors make hydrogen bonding a special type of dipole interaction.

15. Atoms have an approximately spherical shape (on average). It is impossible to pack spheres together without some empty space among the spheres.

16. Critical temperature: The temperature above which a liquid cannot exist; i.e., the gas cannot be liquified by increased pressure.

    Critical pressure: The pressure that must be applied to a substance at its critical temperature to produce a liquid.

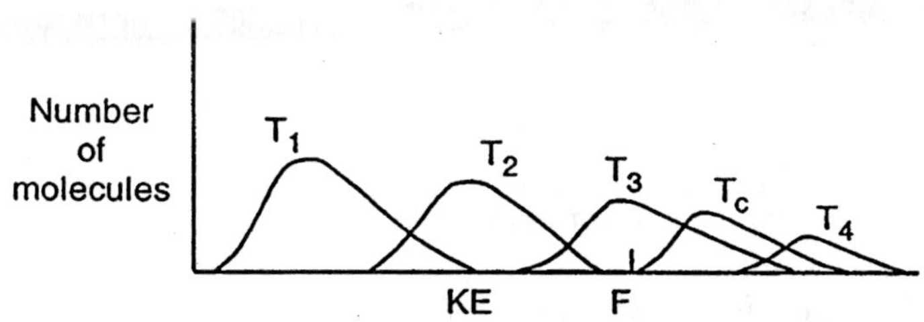

The kinetic energy distribution changes as one raises the temperature ($T_4 > T_c > T_3 > T_2 > T_1$). At the critical temperature T_c, all molecules have kinetic energies greater than the intermolecular forces F, and a liquid can't form. *Note*: The various temperature distributions shown in the plot are not to scale. The area under each temperature distribution should be equal to each other (area = total number of molecules).

17. Evaporation takes place when some molecules at the surface of a liquid have enough energy to break the intermolecular forces holding them in the liquid phase. When a liquid evaporates, the molecules that escape have high kinetic energies. The average kinetic energy of the remaining molecules is lower; thus the temperature of the liquid is lower.

18. A crystalline solid will have the simpler diffraction pattern because a regular, repeating arrangement is necessary to produce planes of atoms that will diffract the X rays in regular patterns. An amorphous solid does not have a regular repeating arrangement and will produce a complicated diffraction pattern.

19. An alloy is a substance that contains a mixture of elements and has metallic properties. In a substitutional alloy, some of the host metal atoms are replaced by other metal atoms of similar size, e.g., brass, pewter, plumber's solder. An interstitial alloy is formed when some of the interstices (holes) in the closest packed metal structure are occupied by smaller atoms, e.g., carbon steels.

20. Equilibrium: There is no change in composition; the vapor pressure is constant.

Dynamic: Two processes, vapor → liquid and liquid → vapor, are both occurring but with equal rates, so the composition of the vapor is constant.

21. a. As the strength of the intermolecular forces increase, the rate of evaporation decreases.

b. As temperature increases, the rate of evaporation increases.

c. As surface area increases, the rate of evaporation increases.

22. $H_2O(l) \rightarrow H_2O(g)$ $\Delta H° = 44$ kJ/mol; the vaporization of water is an endothermic process. In order to evaporate, water must absorb heat from the surroundings. In this case, part of the surroundings is our body. So, as water evaporates, our body supplies heat, and as a result, our body temperature can cool down. From Le Châtelier's principle, the less water vapor in the air, the more favorable the evaporation process. Thus the less humid the surroundings, the more favorably water converts into vapor, and the more heat that is lost by our bodies.

23. $C_2H_5OH(l) \rightarrow C_2H_5OH(g)$ is an endothermic process. Heat is absorbed when liquid ethanol vaporizes; the internal heat from the body provides this heat, which results in the cooling of the body.

24. The phase change $H_2O(g) \rightarrow H_2O(l)$ releases heat that can cause additional damage. Also, steam can be at a temperature greater than 100°C.

25. Sublimation will occur, allowing water to escape as $H_2O(g)$.

26. Water boils when the vapor pressure equals the external pressure. Because the external pressure is significantly lower at high altitudes, a lower temperature is required to equalize the vapor pressure of water to the external pressure. Thus food cooked in boiling water at high elevations cooks at a lower temperature, so it takes longer.

27. The strength of intermolecular forces determines relative boiling points. The types of intermolecular forces for covalent compounds are London dispersion forces, dipole forces, and hydrogen bonding. Because the three compounds are assumed to have similar molar mass and shape, the strength of the London dispersion forces will be about equal among the three compounds. One of the compounds will be nonpolar, so it only has London dispersion forces. The other two compounds will be polar, so they have additional dipole forces and will boil at a higher temperature than the nonpolar compound. One of the polar compounds will have an H covalently bonded to either N, O, or F. This gives rise to the strongest type of covalent intermolecular forces, hydrogen bonding. The compound that hydrogen bonds will have the highest boiling point, whereas the polar compound with no hydrogen bonding will boil at a temperature in the middle of the other compounds.

28. a. Both forms of carbon are network solids. In diamond, each carbon atom is surrounded by a tetrahedral arrangement of other carbon atoms to form a huge molecule. Each carbon atom is covalently bonded to four other carbon atoms.

The structure of graphite is based on layers of carbon atoms arranged in fused six-membered rings. Each carbon atom in a particular layer of graphite is surrounded by three other carbons in a trigonal planar arrangement. This requires sp^2 hybridization. Each carbon has an unhybridized p atomic orbital; all of these p orbitals in each six-membered ring overlap with each other to form a delocalized π electron system.

b. Silica is a network solid having an empirical formula of SiO_2. The silicon atoms are singly bonded to four oxygens. Each silicon atom is at the center of a tetrahedral arrangement of oxygen atoms that are shared with other silicon atoms. The structure of silica is based on a network of SiO_4 tetrahedra with shared oxygen atoms rather than discrete SiO_2 molecules.

Silicates closely resemble silica. The structure is based on interconnected SiO_4 tetrahedra. However, in contrast to silica, where the O/Si ratio is 2 : 1, silicates have O/Si ratios greater than 2 : 1 and contain silicon-oxygen anions. To form a neutral solid silicate, metal cations are needed to balance the charge. In other words, silicates are salts containing metal cations and polyatomic silicon-oxygen anions.

When silica is heated above its melting point and cooled rapidly, an amorphous (disordered) solid called glass results. Glass more closely resembles a very viscous solution than it does a crystalline solid. To affect the properties of glass, several different additives are thrown into the mixture. Some of these additives are Na_2CO_3, B_2O_3, and K_2O, with each compound serving a specific purpose relating to the properties of glass.

29. a. Both CO_2 and H_2O are molecular solids. Both have an ordered array of the individual molecules, with the molecular units occupying the lattice points. A difference within each solid lattice is the strength of the intermolecular forces. CO_2 is nonpolar and only exhibits London dispersion forces. H_2O exhibits the relatively strong hydrogen-bonding interactions. The differences in strength is evidenced by the solid-phase changes that

occur at 1 atm. $CO_2(s)$ sublimes at a relatively low temperature of −78°C. In sublimation, all of the intermolecular forces are broken. However, $H_2O(s)$ doesn't have a phase change until 0°C, and in this phase change from ice to water, only a fraction of the intermolecular forces are broken. The higher temperature and the fact that only a portion of the intermolecular forces are broken are attributed to the strength of the intermolecular forces in $H_2O(s)$ as compared to $CO_2(s)$.

Related to the intermolecular forces are the relative densities of the solid and liquid phases for these two compounds. $CO_2(s)$ is denser than $CO_2(l)$, whereas $H_2O(s)$ is less dense than $H_2O(l)$. For $CO_2(s)$ and for most solids, the molecules pack together as close as possible; hence solids are usually more dense than the liquid phase. H_2O is an exception to this. Water molecules are particularly well suited for hydrogen bonding interaction with each other because each molecule has two polar O–H bonds and two lone pairs on the oxygen. This can lead to the association of four hydrogen atoms with each oxygen atom: two by covalent bonds and two by dipoles. To keep this arrangement (which maximizes the hydrogen-bonding interactions), the $H_2O(s)$ molecules occupy positions that create empty space in the lattice. This translates into a smaller density for $H_2O(s)$ as compared to $H_2O(l)$.

b. Both NaCl and CsCl are ionic compounds with the anions at the lattice points of the unit cells and the cations occupying the empty spaces created by anions (called holes). In NaCl, the Cl^- anions occupy the lattice points of a face-centered unit cell, with the Na^+ cations occupying the octahedral holes. Octahedral holes are the empty spaces created by six Cl^- ions. CsCl has the Cl^- ions at the lattice points of a simple cubic unit cell, with the Cs^+ cations occupying the middle of the cube.

30. Because silicon carbide is made from Group 4A elements, and because it is extremely hard, one would expect SiC to form a covalent network structure similar to diamond.

31. If TiO_2 conducts electricity as a liquid, then it is an ionic solid; if not, then TiO_2 is a network solid.

32. The interparticle forces in ionic solids (the ionic bonds) are much stronger than the interparticle forces in molecular solids (dipole forces, London forces, etc.). The difference in intermolecular forces is most clearly shown in the huge difference in melting points between ionic and molecular solids. Table salt and ordinary sugar are both crystalline solids at room temperature that look very similar to each other. However, sugar can be melted easily in a saucepan during the making of candy, whereas the full heat of a stove will not melt salt. When a substance melts, some interparticle forces must be broken. Ionic solids (salt) require a much larger amount of energy to break the interparticle forces as compared to the relatively weak forces in molecular solids (sugar).

33. The mathematical equation that relates the vapor pressure of a substance to temperature is:

$$\ln P_{vap} = -\frac{\Delta H_{vap}}{R}\left(\frac{1}{T}\right) + C$$

$$y \qquad\quad m \qquad x \quad + b$$

This equation is in the form of the straight-line equation ($y = mx + b$) If one plots ln P_{vap} versus 1/T with temperature in Kelvin, the slope (m) of the straight line is $-\Delta H_{vap}/R$. Because ΔH_{vap} is always positive, the slope of the straight line will be negative.

34. The typical phase diagram for a substance shows three phases and has a positive-sloping solid-liquid equilibrium line (water is atypical). A sketch of the phase diagram for I_2 would look like this:

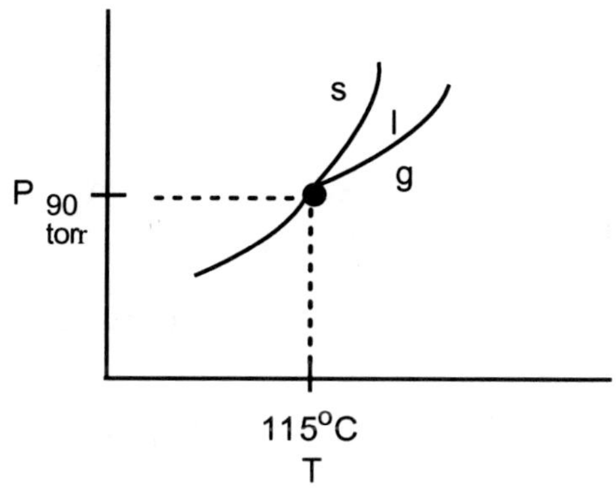

Statements a and e are true. For statement a, the liquid phase is always more dense than the gaseous phase (gases are mostly empty space). For statement e, because the triple point is at 90 torr, the liquid phase cannot exist at any pressure less than 90 torr, no matter what the temperature. For statements b, c, and d, examine the phase diagram to prove to yourself that they are false.

Exercises

Intermolecular Forces and Physical Properties

35. Ionic compounds have ionic forces. Covalent compounds all have London dispersion (LD) forces, whereas polar covalent compounds have dipole forces and/or hydrogen bonding forces. For hydrogen-bonding (H-bonding) forces, the covalent compound must have either a N–H, O–H, or F–H bond in the molecule.

a. LD only b. dipole, LD c. H-bonding, LD

d. ionic e. LD only (CH_4 is a nonpolar covalent compound.)

f. dipole, LD g. ionic

36. a. ionic b. dipole, LD (LD = London dispersion) c. LD only

d. LD only (For all practical purposes, a C – H bond can be considered as a nonpolar bond.)

e. ionic f. LD only g. H-bonding, LD

37. a. OCS; OCS is polar and has dipole-dipole forces in addition to London dispersion (LD) forces. All polar molecules have dipole forces. CO_2 is nonpolar and only has LD forces. To predict polarity, draw the Lewis structure and deduce whether the individual bond dipoles cancel.

b. SeO_2; both SeO_2 and SO_2 are polar compounds, so they both have dipole forces as well as LD forces. However, SeO_2 is a larger molecule, so it would have stronger LD forces.

c. $H_2NCH_2CH_2NH_2$; more extensive hydrogen bonding (H-bonding) is possible because two NH_2 groups are present.

d. H_2CO; H_2CO is polar, whereas CH_3CH_3 is nonpolar. H_2CO has dipole forces in addition to LD forces. CH_3CH_3 only has LD forces.

e. CH_3OH; CH_3OH can form relatively strong H-bonding interactions, unlike H_2CO.

38. Ar exists as individual atoms that are held together in the condensed phases by London dispersion forces. The molecule that will have a boiling point closest to Ar will be a nonpolar substance with about the same molar mass as Ar (39.95 g/mol); this same size nonpolar substance will have about equivalent strength of London dispersion forces. Of the choices, only Cl_2 (70.90 g/mol) and F_2 (38.00 g/mol) are nonpolar. Because F_2 has a molar mass closest to that of Ar, one would expect the boiling point of F_2 to be close to that of Ar.

39. a. Neopentane is more compact than n-pentane. There is less surface-area contact among neopentane molecules. This leads to weaker London dispersion (LD) forces and a lower boiling point.

b. HF is capable of H-bonding; HCl is not.

c. LiCl is ionic, and HCl is a molecular solid with only dipole forces and LD forces. Ionic forces are much stronger than the forces for molecular solids.

d. n-Hexane is a larger molecule, so it has stronger LD forces.

40. The electrostatic potential diagrams indicate that ethanol and acetone are polar substances, and that propane is a nonpolar substance. Ethanol, with the O–H covalent bond, will exhibit relatively strong hydrogen bonding intermolecular forces in addition to London dispersion forces. The polar acetone will exhibit dipole forces in addition to London dispersion forces, and the nonpolar propane will only exhibit London dispersion forces. Because all three compounds have about the same molar mass, the relative strengths of the London dispersion forces should be about the same. Therefore, ethanol (with the H-bonding capacity) should have the highest boiling point, with polar acetone having the next highest boiling point, and the nonpolar propane, with the weakest intermolecular forces, will have the lowest boiling point.

41. Boiling points and freezing points are assumed directly related to the strength of the intermolecular forces, whereas vapor pressure is inversely related to the strength of the intermolecular forces.

a. HBr; HBr is polar, whereas Kr and Cl_2 are nonpolar. HBr has dipole forces unlike Kr and Cl_2. So HBr has the stronger intermolecular forces and the highest boiling point.

b. NaCl; the ionic forces in NaCl are much stronger than the intermolecular forces for molecular substances, so NaCl has the highest melting point.

c. I_2; all are nonpolar, so the largest molecule (I_2) will have the strongest LD (London Dispersion) forces and the lowest vapor pressure.

d. N_2; nonpolar and smallest, so it has the weakest intermolecular forces.

e. CH_4; smallest, nonpolar molecule, so it has the weakest LD forces.

f. HF; HF can form relatively strong H-bonding interactions, unlike the others.

g. $CH_3CH_2CH_2OH$; H-bonding, unlike the others, so it has strongest intermolecular forces.

42. a. CBr_4; largest of these nonpolar molecules, so it has the strongest LD (London Dispersion) forces.

b. F_2; ionic forces in LiF are much stronger than the molecular forces in F_2 and HCl. HCl has dipole forces, whereas the nonpolar F_2 does not exhibit these. So F_2 has the weakest intermolecular forces and the lowest freezing point.

c. CH_3CH_2OH; can form H-bonding interactions, unlike the other covalent compounds.

d. H_2O_2; the H–O–O–H structure has twice the number of H-bonding sites as compared to HF, so H_2O_2 has the stronger H-bonding interactions and the greatest viscosity.

e. H_2CO; H_2CO is polar, so it has dipole forces, unlike the other nonpolar covalent compounds, so H_2CO will have the highest enthalpy of vaporization.

f. I_2; I_2 has only LD forces, whereas CsBr and CaO have much stronger ionic forces. I_2 has the weakest intermolecular forces, so it has smallest ΔH_{fusion}.

Properties of Liquids

43. The attraction of H_2O for glass is stronger than the H_2O–H_2O attraction. The miniscus is concave to increase the area of contact between glass and H_2O. The Hg–Hg attraction is greater than the Hg–glass attraction. The miniscus is convex to minimize the Hg–glass contact.

44. Water is a polar substance, and wax is a nonpolar substance; they are not attracted to each other. A molecule at the surface of a drop of water is subject to attractions only by water molecules below it and to each side. The effect of this uneven pull on the surface water molecules tends to draw them into the body of the water and causes the droplet to assume the shape that has the minimum surface area, a sphere.

45. The structure of H_2O_2 is H–O–O–H, which produces greater hydrogen bonding than in water. Thus the intermolecular forces are stronger in H_2O_2 than in H_2O, resulting in a higher normal boiling point for H_2O_2 and a lower vapor pressure.

46. CO_2 is a gas at room temperature. As melting point and boiling point increase, the strength of the intermolecular forces also increases. Therefore, the strength of forces is $CO_2 < CS_2 < CSe_2$. From a structural standpoint, this is expected. All three are linear, nonpolar molecules. Thus only London dispersion forces are present. Because the molecules increase in size from $CO_2 < CS_2 < CSe_2$, the strength of the intermolecular forces will increase in the same order.

Structures and Properties of Solids

47. $n\lambda = 2d \sin\theta,\ d = \dfrac{n\lambda}{2\sin\theta} = \dfrac{1 \times 154\,\text{pm}}{2 \times \sin 14.22°} = 313\text{ pm} = 3.13 \times 10^{-10}\text{ m}$

48. $d = \dfrac{n\lambda}{2\sin\theta} = \dfrac{2 \times 154\,\text{pm}}{2 \times \sin 22.20°} = 408\text{ pm} = 4.08 \times 10^{-10}\text{ m}$

49. $\lambda = \dfrac{2d\sin\theta}{n} = \dfrac{2 \times 1.36 \times 10^{-10}\text{ m} \times \sin 15.0°}{1} = 7.04 \times 10^{-11}\text{ m} = 0.704\text{ Å} = 70.4\text{ pm}$

50. $n\lambda = 2d \sin\theta,\ d = \dfrac{n\lambda}{2\sin\theta} = \dfrac{1 \times 2.63\text{ Å}}{2 \times \sin 15.55°} = 4.91\text{ Å} = 4.91 \times 10^{-10}\text{ m} = 491\text{ pm}$

$$\sin\theta = \frac{n\lambda}{2d} = \frac{2 \times 2.63\text{ Å}}{2 \times 4.91\text{ Å}} = 0.536,\ \theta = 32.4°$$

51. A cubic closest packed structure has a face-centered cubic unit cell. In a face-centered cubic unit, there are:

$$8\text{ corners} \times \frac{1/8\text{ atom}}{\text{corner}} + 6\text{ faces} \times \frac{1/2\text{ atom}}{\text{face}} = 4\text{ atoms}$$

The atoms in a face-centered cubic unit cell touch along the face diagonal of the cubic unit cell. Using the Pythagorean formula, where l = length of the face diagonal and r = radius of the atom:

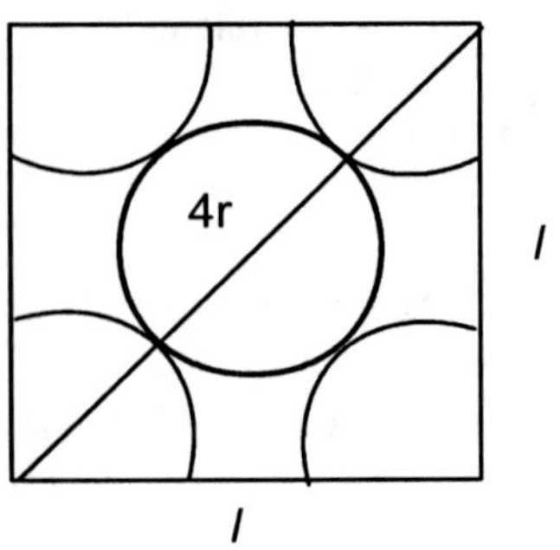

$$l^2 + l^2 = (4r)^2$$

$$2l^2 = 16r^2$$

$$l = r\sqrt{8}$$

$l = r\sqrt{8} = 197 \times 10^{-12}\text{ m} \times \sqrt{8} = 5.57 \times 10^{-10}\text{ m} = 5.57 \times 10^{-8}\text{ cm}$

Volume of a unit cell $= l^3 = (5.57 \times 10^{-8}\text{ cm})^3 = 1.73 \times 10^{-22}\text{ cm}^3$

$$\text{Mass of a unit cell} = 4 \text{ Ca atoms} \times \frac{1 \text{ mol Ca}}{6.022 \times 10^{23} \text{ atoms}} \times \frac{40.08 \text{ g Ca}}{\text{mol Ca}} = 2.662 \times 10^{-22} \text{ g Ca}$$

$$\text{Density} = \frac{\text{mass}}{\text{volume}} = \frac{2.662 \times 10^{-22} \text{ g}}{1.73 \times 10^{-22} \text{ cm}^3} = 1.54 \text{ g/cm}^3$$

52. There are four Ni atoms in each unit cell. For a unit cell:

$$\text{density} = \frac{\text{mass}}{\text{volume}} = 6.84 \text{ g/cm}^3 = \frac{4 \text{ Ni atoms} \times \frac{1 \text{ mol Ni}}{6.022 \times 10^{23} \text{ atoms}} \times \frac{58.69 \text{ g Ni}}{\text{mol Ni}}}{l^3}$$

Solving: $l = 3.85 \times 10^{-8}$ cm = cube edge length

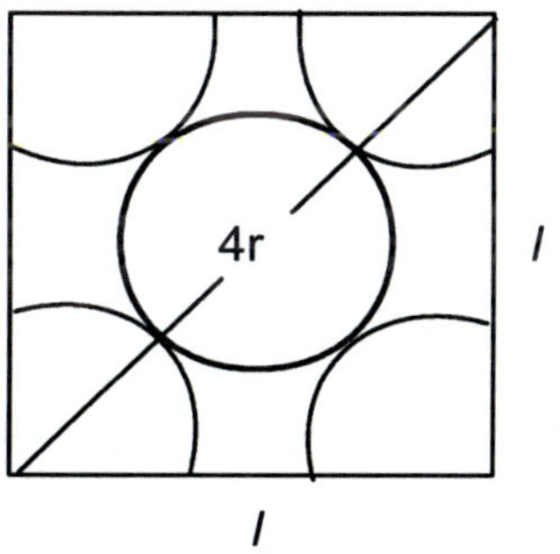

For a face-centered cube:

$$(4r)^2 = l^2 + l^2 = 2l^2$$
$$r\sqrt{8} = l, \; r = l/\sqrt{8}$$
$$r = 3.85 \times 10^{-8} \text{ cm}/\sqrt{8}$$
$$r = 1.36 \times 10^{-8} \text{ cm} = 136 \text{ pm}$$

53. The unit cell for cubic closest packing is the face-centered unit cell. The volume of a unit cell is:

$$V = l^3 = (492 \times 10^{-10} \text{ cm})^3 = 1.19 \times 10^{-22} \text{ cm}^3$$

There are four Pb atoms in the unit cell, as is the case for all face-centered cubic unit cells.

The mass of atoms in a unit cell is:

$$\text{mass} = 4 \text{ Pb atoms} \times \frac{1 \text{ mol Pb}}{6.022 \times 10^{23} \text{ atoms}} \times \frac{207.2 \text{ g Pb}}{\text{mol Pb}} = 1.38 \times 10^{-21} \text{ g}$$

$$\text{Density} = \frac{\text{mass}}{\text{volume}} = \frac{1.38 \times 10^{-21} \text{ g}}{1.19 \times 10^{-22} \text{ cm}^3} = 11.6 \text{ g/cm}^3$$

From Exercise 51, the relationship between the cube edge length l and the radius r of an atom in a face-centered unit cell is $l = r\sqrt{8}$.

$$r = \frac{l}{\sqrt{8}} = \frac{492 \text{ pm}}{\sqrt{8}} = 174 \text{ pm} = 1.74 \times 10^{-10} \text{ m}$$

54. The volume of a unit cell is:

$$V = l^3 = (383.3 \times 10^{-10}\ \text{cm})^3 = 5.631 \times 10^{-23}\ \text{cm}^3$$

There are four Ir atoms in the unit cell, as is the case for all face-centered cubic unit cells. The mass of atoms in a unit cell is:

$$\text{mass} = 4\ \text{Ir atoms} \times \frac{1\ \text{mol Ir}}{6.022 \times 10^{23}\ \text{atoms}} \times \frac{192.2\ \text{g Ir}}{\text{mol Ir}} = 1.277 \times 10^{-21}\ \text{g}$$

$$\text{Density} = \frac{\text{mass}}{\text{volume}} = \frac{1.277 \times 10^{-21}\ \text{g}}{5.631 \times 10^{-23}\ \text{cm}^3} = 22.68\ \text{g/cm}^3$$

55. A face-centered cubic unit cell contains four atoms. For a unit cell:

$$\text{mass of X} = \text{volume} \times \text{density} = (4.09 \times 10^{-8}\ \text{cm})^3 \times 10.5\ \text{g/cm}^3 = 7.18 \times 10^{-22}\ \text{g}$$

$$\text{mol X} = 4\ \text{atoms X} \times \frac{1\ \text{mol X}}{6.022 \times 10^{23}\ \text{atoms}} = 6.642 \times 10^{-24}\ \text{mol X}$$

$$\text{Molar mass} = \frac{7.18 \times 10^{-22}\ \text{g X}}{6.642 \times 10^{-24}\ \text{mol X}} = 108\ \text{g/mol};$$ the metal is silver (Ag).

56. For a face-centered unit cell, the radius r of an atom is related to the length of a cube edge l by the equation $l = r\sqrt{8}$ (see Exercise 51).

$$\text{Radius} = r = l/\sqrt{8} = 392 \times 10^{-12}\ \text{m}/\sqrt{8} = 1.39 \times 10^{-10}\ \text{m} = 1.39 \times 10^{-8}\ \text{cm}$$

The volume of a unit cell is l^3, so the mass of the unknown metal (X) in a unit cell is:

$$\text{volume} \times \text{density} = (3.92 \times 10^{-8}\ \text{cm})^3 \times \frac{21.45\ \text{g X}}{\text{cm}^3} = 1.29 \times 10^{-21}\ \text{g X}$$

Because each face-centered unit cell contains four atoms of X:

$$\text{mol X in unit cell} = 4\ \text{atoms X} \times \frac{1\ \text{mol X}}{6.022 \times 10^{23}\ \text{atoms X}} = 6.642 \times 10^{-24}\ \text{mol X}$$

Therefore, each unit cell contains 1.29×10^{-21} g X, which is equal to 6.642×10^{-24} mol X. The molar mass of X is:

$$\frac{1.29 \times 10^{-21}\ \text{g X}}{6.642 \times 10^{-24}\ \text{mol X}} = 194\ \text{g/mol}$$

The atomic mass would be 194 u. From the periodic table, the best choice for the metal is platinum.

57. For a body-centered unit cell, 8 corners $\times \frac{1/8 \text{ Ti}}{\text{corner}}$ + Ti at body center = 2 Ti atoms.

All body-centered unit cells have two atoms per unit cell. For a unit cell where l = cube edge length:

$$\text{density} = 4.50 \text{ g/cm}^3 = \frac{2 \text{ atoms Ti} \times \frac{1 \text{ mol Ti}}{6.022 \times 10^{23} \text{ atoms}} \times \frac{47.88 \text{ g Ti}}{\text{mol Ti}}}{l^3}$$

Solving: l = edge length of unit cell = 3.28×10^{-8} cm = 328 pm

Assume Ti atoms just touch along the body diagonal of the cube, so body diagonal = 4 × radius of atoms = 4r.

The triangle we need to solve is:

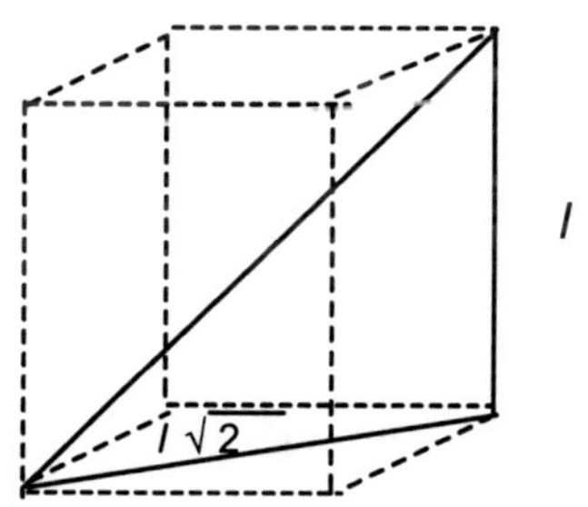

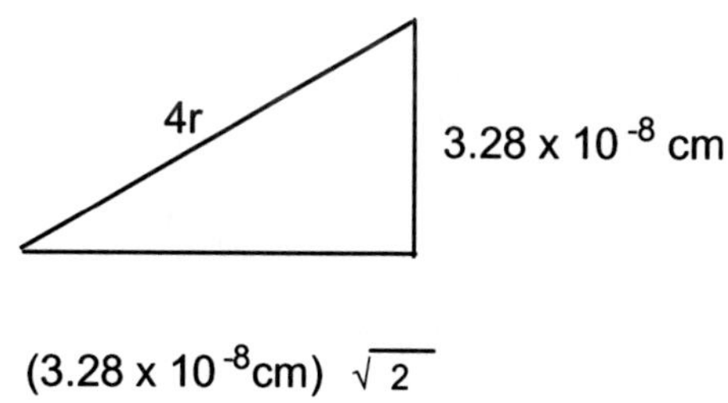

$$(4r)^2 = (3.28 \times 10^{-8} \text{ cm})^2 + [(3.28 \times 10^{-8} \text{ cm})\sqrt{2}\,]^2, \; r = 1.42 \times 10^{-8} \text{ cm} = 142 \text{ pm}$$

For a body-centered unit cell (bcc), the radius of the atom is related to the cube edge length by: $4r = l\sqrt{3}$ or $l = 4r/\sqrt{3}$.

58. From Exercise 57:

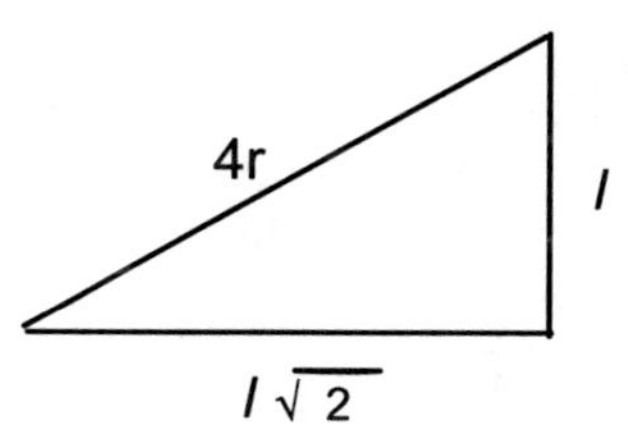

$16r^2 = l^2 + 2l^2$

$l = 4r/\sqrt{3} = (2.309)r$

$l = 2.309(222 \text{ pm}) = 513 \text{ pm} = 5.13 \times 10^{-8} \text{ cm}$

In a body-centered cubit unit cell, there are two atoms per unit cell. For a unit cell:

$$\text{density} = \frac{\text{mass}}{\text{volume}} = \frac{2 \text{ atoms Ba} \times \frac{1 \text{ mol Ba}}{6.022 \times 10^{23} \text{ atoms}} \times \frac{137.3 \text{ g Ba}}{\text{mol Ba}}}{(5.13 \times 10^{-8} \text{ cm})^3} = \frac{3.38 \text{ g}}{\text{cm}^3}$$

59. If gold has a face-centered cubic structure, then there are four atoms per unit cell, and from Exercise 51:

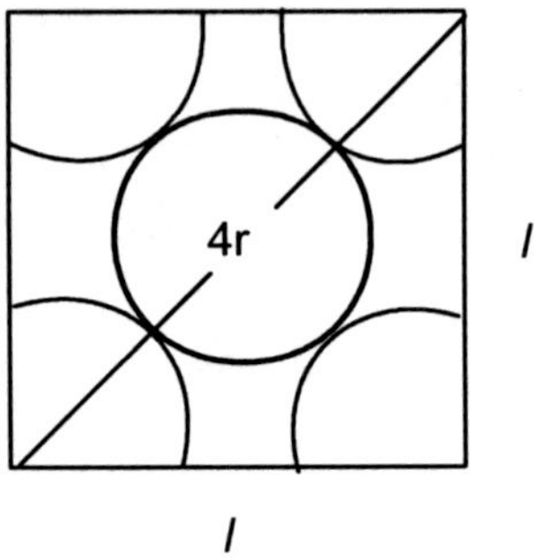

$$2l^2 = 16r^2$$

$$l = r\sqrt{8} = (144 \text{ pm})\sqrt{8} = 407 \text{ pm}$$

$$l = 407 \times 10^{-12} \text{ m} = 4.07 \times 10^{-8} \text{ cm}$$

$$\text{Density} = \frac{4 \text{ atoms Au} \times \dfrac{1 \text{ mol Au}}{6.022 \times 10^{23} \text{ atoms}} \times \dfrac{197.0 \text{ g Au}}{\text{mol Au}}}{(4.07 \times 10^{-8} \text{ cm})^3} = 19.4 \text{ g/cm}^3$$

If gold has a body-centered cubic structure, then there are two atoms per unit cell, and from Exercise 57:

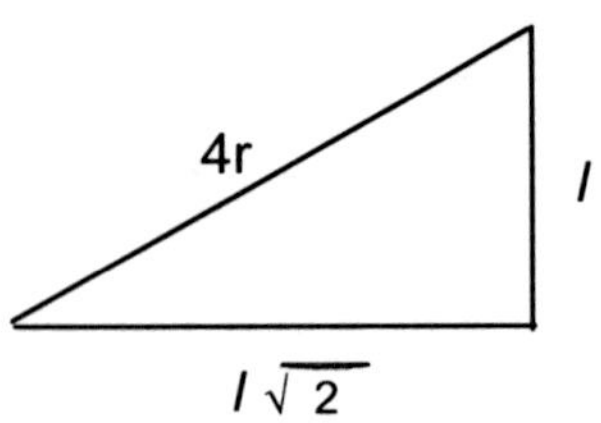

$$16r^2 = l^2 + 2l^2$$

$$l = 4r/\sqrt{3} = 333 \text{ pm} = 333 \times 10^{-12} \text{ m}$$

$$l = 333 \times 10^{-10} \text{ cm} = 3.33 \times 10^{-8} \text{ cm}$$

$$\text{Density} = \frac{2 \text{ atoms Au} \times \dfrac{1 \text{ mol Au}}{6.022 \times 10^{23} \text{ atoms}} \times \dfrac{197.0 \text{ g Au}}{\text{mol Au}}}{(3.33 \times 10^{-8} \text{ cm})^3} = 17.7 \text{ g/cm}^3$$

The measured density of gold is consistent with a face-centered cubic unit cell.

60. If face-centered cubic:

$$l = r\sqrt{8} = (137 \text{ pm})\sqrt{8} = 387 \text{ pm} = 3.87 \times 10^{-8} \text{ cm}$$

$$\text{density} = \frac{4 \text{ atoms W} \times \dfrac{1 \text{ mol}}{6.022 \times 10^{23} \text{ atoms}} \times \dfrac{183.9 \text{ g W}}{\text{mol}}}{(3.87 \times 10^{-8} \text{ cm})^3} = 21.1 \text{ g/cm}^3$$

If body-centered cubic:

$$l = \frac{4r}{\sqrt{3}} = \frac{4 \times 137 \text{ pm}}{\sqrt{3}} = 316 \text{ pm} = 3.16 \times 10^{-8} \text{ cm}$$

$$\text{density} = \frac{2\text{ atoms W} \times \dfrac{1\text{ mol}}{6.022 \times 10^{23}\text{ atoms}} \times \dfrac{183.9\text{ g W}}{\text{mol}}}{(3.16 \times 10^{-8}\text{ cm})^3} = 19.4\text{ g/cm}^3$$

The measured density of tungsten is consistent with a body-centered unit cell.

61. In a face-centered unit cell (a cubic closest packed structure), the atoms touch along the face diagonal:

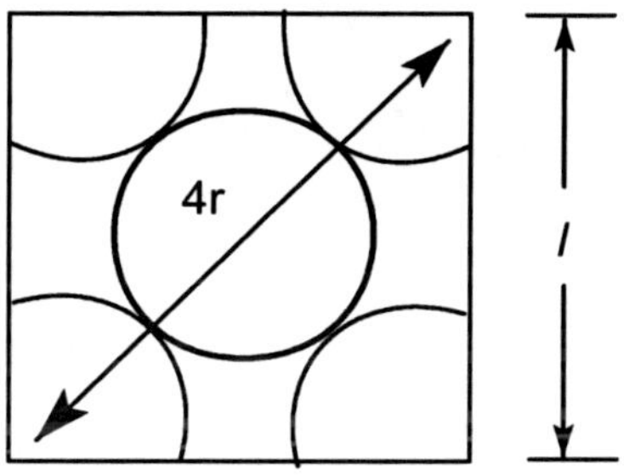

$$(4r)^2 = l^2 + l^2$$

$$l = r\sqrt{8}$$

$$V_{cube} = l^3 = (r\sqrt{8})^3 = (22.63)r^3$$

There are four atoms in a face-centered cubic cell (see Exercise 51). Each atom has a volume of $(4/3)\pi r^3$ = volume of a sphere.

$$V_{atoms} = 4 \times \frac{4}{3}\pi r^3 = (16.76)r^3$$

So $\dfrac{V_{atoms}}{V_{cube}} = \dfrac{(16.76)r^3}{(22.63)r^3} = 0.7406$, or 74.06% of the volume of each unit cell is occupied by atoms.

In a simple cubic unit cell, the atoms touch along the cube edge l:

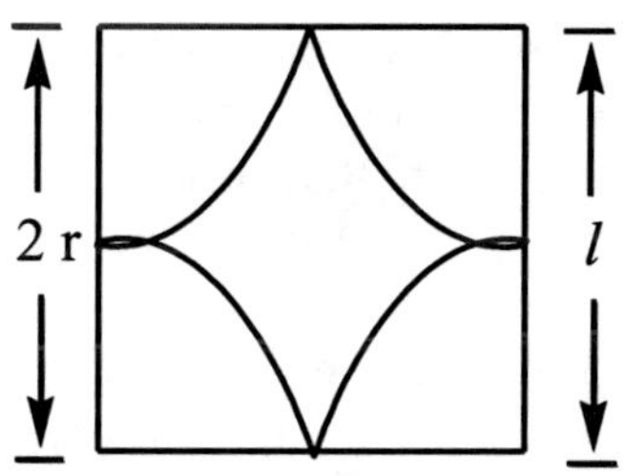

$$2(\text{radius}) = 2r = l$$

$$V_{cube} = l^3 = (2r)^3 = 8r^3$$

There is one atom per simple cubic cell (8 corner atoms × 1/8 atom per corner = 1 atom/unit cell). Each atom has an assumed volume of $(4/3)\pi r^3$ = volume of a sphere.

$$V_{atom} = \frac{4}{3}\pi r^3 = (4.189)r^3$$

So $\frac{V_{atom}}{V_{cube}} = \frac{(4.189)r^3}{8r^3}$ = 0.5236, or 52.36% of the volume of each unit cell is occupied by atoms.

A cubic closest packed structure (face-centered cubic unit cell) packs the atoms much more efficiently than a simple cubic structure.

62. From Exercise 57, a body-centered unit cell contains two net atoms, and the length of a cube edge l is related to the radius of the atom r by the equation $l = 4r/\sqrt{3}$.

Volume of unit cell = $l^3 = (4r/\sqrt{3})^3 = (12.32)r^3$

Volume of atoms in unit cell = $2 \times \frac{4}{3}\pi r^3 = (8.378)r^3$

So $\frac{V_{atoms}}{V_{cube}} = \frac{(8.378)r^3}{(12.32)r^3}$ = 0.6800 = 68.00% occupied

To determine the radius of the Fe atoms, we need to determine the cube edge length l.

$$\text{Volume of unit cell} = \left(2\ \text{Fe atoms} \times \frac{1\ \text{mol Fe}}{6.022 \times 10^{23}\ \text{atoms}} \times \frac{55.85\ \text{g Fe}}{\text{mol Fe}}\right) \times \frac{1\ \text{cm}^3}{7.86\ \text{g}}$$

$$= 2.36 \times 10^{-23}\ \text{cm}^3$$

Volume = $l^3 = 2.36 \times 10^{-23}$ cm^3, $l = 2.87 \times 10^{-8}$ cm

$l = 4r/\sqrt{3}$, $r = l\sqrt{3}/4 = 2.87 \times 10^{-8}\ \text{cm} \times \sqrt{3}/4 = 1.24 \times 10^{-8}\ \text{cm} = 124\ \text{pm}$

63. Doping silicon with phosphorus produces an n-type semiconductor. The phosphorus adds electrons at energies near the conduction band of silicon. Electrons do not need as much energy to move from filled to unfilled energy levels, so conduction increases. Doping silicon with gallium produces a p-type semiconductor. Because gallium has fewer valence electrons than silicon, holes (unfilled energy levels) at energies in the previously filled molecular orbitals are created, which induces greater electron movement (greater conductivity).

64. A rectifier is a device that produces a current that flows in one direction from an alternating current that flows in both directions. In a p-n junction, a p-type and an n-type semiconductor are connected. The natural flow of electrons in a p-n junction is for the excess electrons in the n-type semiconductor to move to the empty energy levels (holes) of the p-type semiconductor. Only when an external electric potential is connected so that electrons flow in this natural direction will the current flow easily (forward bias). If the external electric potential is connected in reverse of the natural flow of electrons, no current flows through the system (reverse bias). A p-n junction only transmits a current under forward bias, thus converting the alternating current to direct current.

65. In has fewer valence electrons than Se. Thus Se doped with In would be a p-type semiconductor.

66. To make a p-type semiconductor, we need to dope the material with atoms that have fewer valence electrons. The average number of valence electrons is four when 50-50 mixtures of Group 3A and Group 5A elements are considered. We could dope with more of the Group 3A element or with atoms of Zn or Cd. Cadmium is the most common impurity used to produce p-type GaAs semiconductors. To make an n-type GaAs semiconductor, dope with an excess Group 5A element or dope with a Group 6A element such as sulfur.

67. $E_{gap} = 2.5\text{ eV} \times 1.6 \times 10^{-19}\text{ J/eV} = 4.0 \times 10^{-19}\text{ J}$; we want $E_{gap} = E_{light} = hc/\lambda$, so:

$$\lambda = \frac{hc}{E} = \frac{(6.63 \times 10^{-34}\text{ J s})(3.00 \times 10^{8}\text{ m/s})}{4.0 \times 10^{-19}\text{ J}} = 5.0 \times 10^{-7}\text{ m} = 5.0 \times 10^{2}\text{ nm}$$

68. $$E = \frac{hc}{\lambda} = \frac{(6.626 \times 10^{-34}\text{ J s})(2.998 \times 10^{8}\text{ m/s})}{730. \times 10^{-9}\text{ m}} = 2.72 \times 10^{-19}\text{ J} = \text{energy of band gap}$$

69. Sodium chloride structure: $8\text{ corners} \times \frac{1/8\ Cl^-}{\text{corner}} + 6\text{ faces} \times \frac{1/2\ Cl^-}{\text{face}} = 4\ Cl^-$ ions

$12\text{ edges} \times \frac{1/4\ Na^+}{\text{edge}} + 1\ Na^+$ at body center $= 4\ Na^+$ ions; NaCl is the formula.

Cesium chloride structure: $1\ Cs^+$ ion at body center; $8\text{ corners} \times \frac{1/8\ Cl^-}{\text{corner}} = 1\ Cl^-$ ion

CsCl is the formula.

Zinc sulfide structure: There are four Zn^{2+} ions inside the cube.

$8\text{ corners} \times \frac{1/8\ S^{2-}}{\text{corner}} + 6\text{ faces} \times \frac{1/2\ S^{2-}}{\text{face}} = 4\ S^{2-}$ ions; ZnS is the formula.

Titanium oxide structure: $8\text{ corners} \times \frac{1/8\ Ti^{4+}}{\text{corner}} + 1\ Ti^{4+}$ at body center $= 2\ Ti^{4+}$ ions

$4\text{ faces} \times \frac{1/2\ O^{2-}}{\text{face}} + 2\ O^{2-}$ inside cube $= 4\ O^{2-}$ ions; TiO_2 is the formula.

70. Both As ions are inside the unit cell. $8\text{ corners} \times \frac{1/8\ \text{Ni}}{\text{corner}} + 4\text{ edges} \times \frac{1/4\ \text{Ni}}{\text{edge}} = 2$ Ni ions

The unit cell contains 2 ions of Ni and 2 ions of As, which gives a formula of NiAs. We would expect As to form 3– charged ions when in ionic compounds, so Ni exists as 3+ charged ions in this compound.

71. There is one octahedral hole per closest packed anion in a closest packed structure. If one-half of the octahedral holes are filled, then there is a 2 : 1 ratio of fluoride ions to cobalt ions in the crystal. The formula is CoF_2, which is composed of Co^{2+} and F^- ions.

72. There are two tetrahedral holes per closest packed anion. Let f = fraction of tetrahedral holes filled by the cations.

Na_2O: Cation-to-anion ratio $= \frac{2}{1} = \frac{2f}{1}$, f = 1; all the tetrahedral holes are filled by Na^+ cations.

CdS: Cation-to-anion ratio $= \frac{1}{1} = \frac{2f}{1}$, $f = \frac{1}{2}$; one-half the tetrahedral holes are filled by Cd^{2+} cations.

ZrI_4: Cation-to-anion ratio $= \frac{1}{4} = \frac{2f}{1}$, $f = \frac{1}{8}$; one-eighth the tetrahedral holes are filled by Zr^{4+} cations.

73. In a cubic closest packed array of anions, there are twice the number of tetrahedral holes as anions present, and an equal number of octahedral holes as anions present. A cubic closest packed array of sulfide ions will have four S^{2-} ions, eight tetrahedral holes, and four octahedral holes. In this structure we have 1/8(8) = 1 Zn^{2+} ion and 1/2(4) = 2 Al^{3+} ions present, along with the 4 S^{2-} ions. The formula is $ZnAl_2S_4$.

74. A repeating pattern in the two-dimensional structure is:

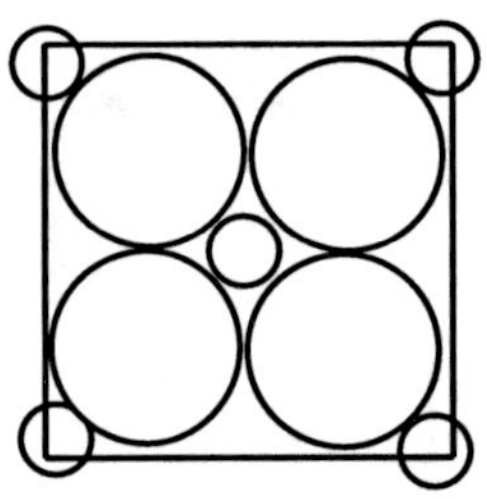

Assuming the anions A are the larger circles, there are four anions completely in this repeating square. The corner cations (smaller circles) are shared by four different repeating squares. Therefore, there is one cation in the middle of the square plus 1/4 (4) = 1 net cation from the corners. Each repeating square has two cations and four anions. The empirical formula is MA_2.

75. 8 F^- ions at corners × 1/8 F^-/corner = 1 F^- ion per unit cell; Because there is one cubic hole per cubic unit cell, there is a 2 : 1 ratio of F^- ions to metal ions in the crystal if only half of the body centers are filled with the metal ions. The formula is MF_2, where M^{2+} is the metal ion.

76. Mn ions at 8 corners: 8(1/8) = 1 Mn ion; F ions at 12 edges: 12(1/4) = 3 F ions; the formula is MnF_3. Assuming fluoride is −1 charged, then the charge on Mn is +3.

77. From Figure 10.35, MgO has the NaCl structure containing 4 Mg^{2+} ions and 4 O^{2-} ions per face-centered unit cell.

$$4 \text{ MgO formula units} \times \frac{1 \text{ mol MgO}}{6.022 \times 10^{23} \text{ atoms}} \times \frac{40.31 \text{ g MgO}}{1 \text{ mol MgO}} = 2.678 \times 10^{-22} \text{ g MgO}$$

$$\text{Volume of unit cell} = 2.678 \times 10^{-22}\text{ g MgO} \times \frac{1\text{ cm}^3}{3.58\text{ g}} = 7.48 \times 10^{-23}\text{ cm}^3$$

Volume of unit cell = l^3, l = cube edge length; $l = (7.48 \times 10^{-23}\text{ cm}^3)^{1/3} = 4.21 \times 10^{-8}$ cm

For a face-centered unit cell, the O^{2-} ions touch along the face diagonal:

$$\sqrt{2}\,l = 4r_{O^{2-}},\quad r_{O^{2-}} = \frac{\sqrt{2} \times 4.21 \times 10^{-8}\text{ cm}}{4} = 1.49 \times 10^{-8}\text{ cm}$$

The cube edge length goes through two radii of the O^{2-} anions and the diameter of the Mg^{2+} cation, so:

$$l = 2r_{O^{2-}} + 2r_{Mg^{2+}},\quad 4.21 \times 10^{-8}\text{ cm} = 2(1.49 \times 10^{-8}\text{ cm}) + 2r_{Mg^{2+}},\quad r_{Mg^{2+}} = 6.15 \times 10^{-9}\text{ cm}$$

78. Cl^-

K^+

Assuming K^+ and Cl^- just touch along the cube edge l:

$$l = 2(314\text{ pm}) = 628\text{ pm} = 6.28 \times 10^{-8}\text{ cm}$$

$$\text{Volume of unit cell} = l^3 = (6.28 \times 10^{-8}\text{ cm})^3$$

The unit cell contains four K^+ and four Cl^- ions. For a unit cell:

$$\text{density} = \frac{4\text{ KCl formula units} \times \dfrac{1\text{ mol KCl}}{6.022 \times 10^{23}\text{ formula units}} \times \dfrac{74.55\text{ g KCl}}{\text{mol KCl}}}{(6.28 \times 10^{-8}\text{ cm})^3}$$

$$= 2.00\text{ g/cm}^3$$

79. CsCl is a simple cubic array of Cl^- ions with Cs^+ in the middle of each unit cell. There is one Cs^+ and one Cl^- ion in each unit cell. Cs^+ and Cl^- ions touch along the body diagonal.

Body diagonal = $2r_{Cs^+} + 2r_{Cl^-} = \sqrt{3}\,l$, l = length of cube edge

In each unit cell:

$$\text{mass} = 1\text{ CsCl formula unit} \times \frac{1\text{ mol CsCl}}{6.022 \times 10^{23}\text{ formula units}} \times \frac{168.4\text{ g CsCl}}{\text{mol CsCl}}$$

$$= 2.796 \times 10^{-22}\text{ g}$$

$$\text{volume} = l^3 = 2.796 \times 10^{-22}\text{ g CsCl} \times \frac{1\text{ cm}^3}{3.97\text{ g CsCl}} = 7.04 \times 10^{-23}\text{ cm}^3$$

$l^3 = 7.04 \times 10^{-23}\text{ cm}^3$, $l = 4.13 \times 10^{-8}$ cm = 413 pm = length of cube edge

$$2r_{Cs^+} + 2r_{Cl^-} = \sqrt{3}\, l = \sqrt{3}(413\,\text{pm}) = 715\text{ pm}$$

The distance between ion centers = $r_{Cs^+} + r_{Cl^-}$ = 715 pm/2 = 358 pm

From ionic radius: r_{Cs^+} = 169 pm and r_{Cl^-} = 181 pm; $r_{Cs^+} + r_{Cl^-}$ = 169 + 181 = 350. pm

The distance calculated from the density is 8 pm (2.3%) greater than that calculated from tables of ionic radii.

80. a. The NaCl unit cell has a face-centered cubic arrangement of the anions with cations in the octahedral holes. There are four NaCl formula units per unit cell, and since there is a 1 : 1 ratio of cations to anions in MnO, then there would be four MnO formula units per unit cell, assuming an NaCl-type structure. The CsCl unit cell has a simple cubic structure of anions with the cations in the cubic holes. There is one CsCl formula unit per unit cell, so there would be one MnO formula unit per unit cell if a CsCl structure is observed.

$$\frac{\text{Formula units of MnO}}{\text{Unit cell}} = (4.47 \times 10^{-8}\text{ cm})^3 \times \frac{5.28\text{ g MnO}}{\text{cm}^3} \times \frac{1\text{ mol MnO}}{70.94\text{ g MnO}}$$

$$\times \frac{6.022 \times 10^{23}\text{ formula units MnO}}{\text{mol MnO}_4} = 4.00\text{ formula units MnO}$$

From the calculation, MnO crystallizes in the NaCl type structure.

b. From the NaCl structure and assuming the ions touch each other, then l = cube edge length = $2r_{Mn^{2+}} + 2r_{O^{2-}}$.

$$l = 4.47 \times 10^{-8}\text{ cm} = 2r_{Mn^{2+}} + 2(1.40 \times 10^{-8}\text{ cm}),\quad r_{Mn^{2+}} = 8.35 \times 10^{-8}\text{ cm} = 84\text{ pm}$$

81. a. CO_2: molecular b. SiO_2: network c. Si: atomic, network

d. CH_4: molecular e. Ru: atomic, metallic f. I_2: molecular

g. KBr: ionic h. H_2O: molecular i. NaOH: ionic

j. U: atomic, metallic k. $CaCO_3$: ionic l. PH_3: molecular

82. a. diamond: atomic, network b. PH_3: molecular c. H_2: molecular

d. Mg: atomic, metallic e. KCl: ionic f. quartz: network

g. NH_4NO_3: ionic h. SF_2: molecular i. Ar: atomic, group 8A

j. Cu: atomic, metallic k. $C_6H_{12}O_6$: molecular

83. a. The unit cell consists of Ni at the cube corners and Ti at the body center or Ti at the cube corners and Ni at the body center.

b. $8 \times 1/8 = 1$ atom from corners + 1 atom at body center; empirical formula = NiTi

c. Both have a coordination number of 8 (both are surrounded by 8 atoms).

84. Al: $8 \text{ corners} \times \dfrac{1/8\text{ Al}}{\text{corner}} = 1\text{ Al}$; Ni: $6 \text{ face centers} \times \dfrac{1/2\text{ Ni}}{\text{face center}} = 3\text{ Ni}$

The composition of the specific phase of the superalloy is $AlNi_3$.

85. Structure 1 (on left) | Structure 2 (on right)

Structure 1 (on left):

$8 \text{ corners} \times \dfrac{1/8\text{ Ca}}{\text{corner}} = 1\text{ Ca atom}$

$6 \text{ faces} \times \dfrac{1/2\text{ O}}{\text{face}} = 3\text{ O atoms}$

1 Ti at body center. Formula = $CaTiO_3$

Structure 2 (on right):

$8 \text{ corners} \times \dfrac{1/8\text{ Ti}}{\text{corner}} = 1\text{ Ti atom}$

$12 \text{ edges} \times \dfrac{1/4\text{ O}}{\text{corner}} = 3\text{ O atoms}$

1 Ca at body center. Formula = $CaTiO_3$

In the extended lattice of both structures, each Ti atom is surrounded by six O atoms.

86. With a cubic closest packed array of oxygen ions, we have 4 O^{2-} ions per unit cell. We need to balance the total −8 charge of the anions with a +8 charge from the Al^{3+} and Mg^{2+} cations. The only combination of ions that gives a +8 charge is 2 Al^{3+} ions and 1 Mg^{2+} ion. The formula is Al_2MgO_4.

There are an equal number of octahedral holes as anions (4) in a cubic closest packed array and twice the number of tetrahedral holes as anions in a cubic closest packed array. For the stoichiometry to work out, we need 2 Al^{3+} and 1 Mg^{2+} per unit cell. Hence one-half of the octahedral holes are filled with Al^{3+} ions, and one-eighth of the tetrahedral holes are filled with Mg^{2+} ions.

87. a. Y: 1 Y in center; Ba: 2 Ba in center

Cu: $8 \text{ corners} \times \dfrac{1/8\text{ Cu}}{\text{corner}} = 1\text{ Cu}$, $8 \text{ edges} \times \dfrac{1/4\text{ Cu}}{\text{edge}} = 2\text{ Cu}$, total = 3 Cu atoms

O: $20 \text{ edges} \times \dfrac{1/4\text{ O}}{\text{edge}} = 5\text{ oxygen}$, $8 \text{ faces} \times \dfrac{1/2\text{ O}}{\text{face}} = 4\text{ oxygen}$, total = 9 O atoms

Formula: $YBa_2Cu_3O_9$

b. The structure of this superconductor material follows the second perovskite structure described in Exercise 85. The $YBa_2Cu_3O_9$ structure is three of these cubic perovskite unit cells stacked on top of each other. The oxygen atoms are in the same places, Cu takes the place of Ti, two of the calcium atoms are replaced by two barium atoms, and one Ca is replaced by Y.

c. Y, Ba, and Cu are the same. Some oxygen atoms are missing.

$$12 \text{ edges} \times \frac{1/4 \text{ O}}{\text{edge}} = 3 \text{ O}, \; 8 \text{ faces} \times \frac{1/2 \text{ O}}{\text{face}} = 4 \text{ O}, \; \text{total} = 7 \text{ O atoms}$$

Superconductor formula is $YBa_2Cu_3O_7$.

88. a. Structure (a):

Ba: 2 Ba inside unit cell; Tl: $8 \text{ corners} \times \frac{1/8 \text{ Tl}}{\text{corner}} = 1 \text{ Tl}$

Cu: $4 \text{ edges} \times \frac{1/4 \text{ Cu}}{\text{edge}} = 1 \text{ Cu}$

O: $6 \text{ faces} \times \frac{1/2 \text{ O}}{\text{face}} + 8 \text{ edges} \times \frac{1/4 \text{ O}}{\text{edge}} = 5 \text{ O}$; Formula = $TlBa_2CuO_5$.

Structure (b):

Tl and Ba are the same as in structure (a).

Ca: 1 Ca inside unit cell; Cu: $8 \text{ edges} \times \frac{1/4 \text{ Cu}}{\text{edge}} = 2 \text{ Cu}$

O: $10 \text{ faces} \times \frac{1/2 \text{ O}}{\text{face}} + 8 \text{ edges} \times \frac{1/4 \text{ O}}{\text{edge}} = 7 \text{ O}$; Formula = $TlBa_2CaCu_2O_7$.

Structure (c):

Tl and Ba are the same, and two Ca are located inside the unit cell.

Cu: $12 \text{ edges} \times \frac{1/4 \text{ Cu}}{\text{edge}} = 3 \text{ Cu}$; O: $14 \text{ faces} \times \frac{1/2 \text{ O}}{\text{face}} + 8 \text{ edges} \times \frac{1/4 \text{ O}}{\text{edge}} = 9 \text{ O}$

Formula = $TlBa_2Ca_2Cu_3O_9$.

Structure (d): Following similar calculations, formula = $TlBa_2Ca_3Cu_4O_{11}$.

b. Structure (a) has one planar sheet of Cu and O atoms, and the number increases by one for each of the remaining structures. The order of superconductivity temperature from lowest to highest temperature is (a) < (b) < (c) < (d).

c. $TlBa_2CuO_5$: $3 + 2(2) + x + 5(-2) = 0$, $x = +3$

Only Cu^{3+} is present in each formula unit.

$TlBa_2CaCu_2O_7$: $3 + 2(2) + 2 + 2(x) + 7(-2) = 0$, $x = +5/2$

Each formula unit contains 1 Cu^{2+} and 1 Cu^{3+}.

$TlBa_2Ca_2Cu_3O_9$: $3 + 2(2) + 2(2) + 3(x) + 9(-2) = 0, \; x = +7/3$

Each formula unit contains 2 Cu^{2+} and 1 Cu^{3+}.

$TlBa_2Ca_3Cu_4O_{11}$: $3 + 2(2) + 3(2) + 4(x) + 11(-2) = 0, \; x = +9/4$

Each formula unit contains 3 Cu^{2+} and 1 Cu^{3+}.

d. This superconductor material achieves variable copper oxidation states by varying the numbers of Ca, Cu, and O in each unit cell. The mixtures of copper oxidation states are discussed in part c. The superconductor material in Exercise 87 achieves variable copper oxidation states by omitting oxygen at various sites in the lattice.

Phase Changes and Phase Diagrams

89. If we graph ln P_{vap} versus 1/T with temperature in Kelvin, the slope of the resulting straight line will be $-\Delta H_{vap}/R$.

P_{vap}	ln P_{vap}	T (Li)	1/T	T (Mg)	1/T
1 torr	0	1023 K	9.775×10^{-4} K^{-1}	893 K	11.2×10^{-4} K^{-1}
10.	2.3	1163	8.598×10^{-4}	1013	9.872×10^{-4}
100.	4.61	1353	7.391×10^{-4}	1173	8.525×10^{-4}
400.	5.99	1513	6.609×10^{-4}	1313	7.616×10^{-4}
760.	6.63	1583	6.317×10^{-4}	1383	7.231×10^{-4}

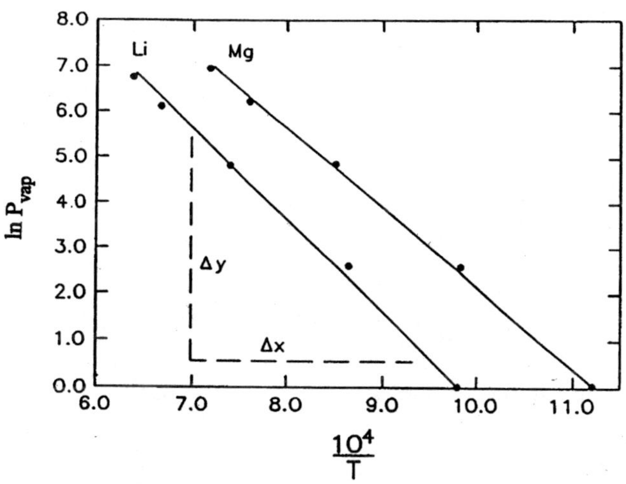

For Li:

We get the slope by taking two points (x, y) that are on the line we draw. For a line, slope = $\Delta y/\Delta x$, or we can determine the straight-line equation using a calculator. The general straight-line equation is $y = mx + b$, where m = slope and $b = y$ intercept.

The equation of the Li line is: ln $P_{vap} = -1.90 \times 10^4(1/T) + 18.6$, slope $= -1.90 \times 10^4$ K

Slope = $-\Delta H_{vap}/R$, ΔH_{vap} = –slope × R = 1.90×10^4 K × 8.3145 J/K•mol

$\Delta H_{vap} = 1.58 \times 10^5$ J/mol = 158 kJ/mol

For Mg:

The equation of the line is: ln $P_{vap} = -1.67 \times 10^4(1/T) + 18.7$, slope = -1.67×10^4 K

ΔH_{vap} = –slope × R = 1.67×10^4 K × 8.3145 J/K•mol

$\Delta H_{vap} = 1.39 \times 10^5$ J/mol = 139 kJ/mol

The bonding is stronger in Li because ΔH_{vap} is larger for Li.

90. We graph ln P_{vap} vs 1/T. The slope of the line equals $-\Delta H_{vap}/R$.

T(K)	10^3/T (K^{-1})	P_{vap} (torr)	ln P_{vap}
273	3.66	14.4	2.67
283	3.53	26.6	3.28
293	3.41	47.9	3.87
303	3.30	81.3	4.40
313	3.19	133	4.89
323	3.10	208	5.34
353	2.83	670.	6.51

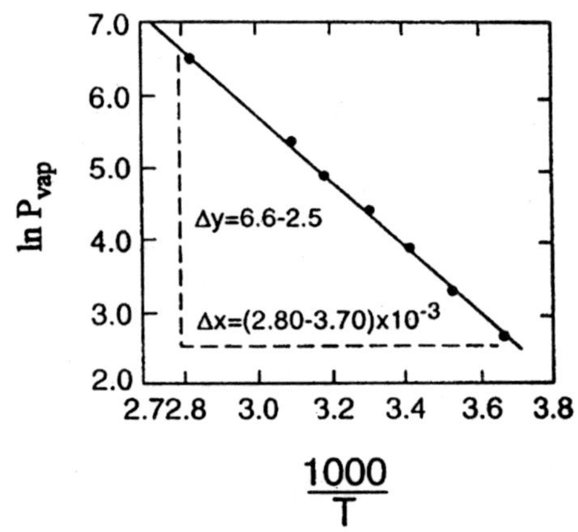

$$\text{Slope} = \frac{6.6 - 2.5}{(2.80 \times 10^{-3} - 3.70 \times 10^{-3})\ K^{-1}} = -4600\ K$$

$$-4600\ K = \frac{-\Delta H_{vap}}{R} = \frac{-\Delta H_{vap}}{8.3145 J/K \bullet mol},\ \Delta H_{vap} = 38{,}000\ J/mol = 38\ kJ/mol$$

To determine the normal boiling point, we can use the following formula:

$$\ln\left(\frac{P_1}{P_2}\right) = \frac{\Delta H_{vap}}{R}\left(\frac{1}{T_2} - \frac{1}{T_1}\right)$$

At the normal boiling point, the vapor pressure equals 1.00 atm or 760. torr. At 273 K, the vapor pressure is 14.4. torr (from data in the problem).

$$\ln\left(\frac{14.4}{760.}\right) = \frac{38{,}000 J/mol}{8.3145 J/K \bullet mol}\left(\frac{1}{T_2} - \frac{1}{273K}\right),\ -3.97 = 4.6 \times 10^3(1/T_2 - 3.66 \times 10^{-3})$$

$-8.6 \times 10^{-4} + 3.66 \times 10^{-3} = 1/T_2 = 2.80 \times 10^{-3}$, T_2 = 357 K = normal boiling point

91. At 100.°C (373 K), the vapor pressure of H_2O is 1.00 atm = 760. torr. For water, ΔH_{vap} = 40.7 kJ/mol.

$$\ln\left(\frac{P_1}{P_2}\right) = \frac{\Delta H_{vap}}{R}\left(\frac{1}{T_2} - \frac{1}{T_1}\right) \text{ or } \ln\left(\frac{P_2}{P_1}\right) = \frac{\Delta H_{vap}}{R}\left(\frac{1}{T_1} - \frac{1}{T_2}\right)$$

$$\ln\left(\frac{520.\text{ torr}}{760.\text{ torr}}\right) = \frac{40.7 \times 10^3 \text{ J/mol}}{8.3145 \text{J/K} \bullet \text{mol}}\left(\frac{1}{373\text{K}} - \frac{1}{T_2}\right),\ -7.75 \times 10^{-5} = \left(\frac{1}{373\text{K}} - \frac{1}{T_2}\right)$$

$$-7.75 \times 10^{-5} = 2.68 \times 10^{-3} - \frac{1}{T_2},\ \frac{1}{T_2} = 2.76 \times 10^{-3},\ T_2 = \frac{1}{2.76 \times 10^{-3}} = 362 \text{ K or } 89°\text{C}$$

92. At 100.°C (373 K), the vapor pressure of H_2O is 1.00 atm. For water, ΔH_{vap} = 40.7 kJ/mol.

$$\ln\left(\frac{P_1}{P_2}\right) = \frac{\Delta H_{vap}}{R}\left(\frac{1}{T_2} - \frac{1}{T_1}\right) \text{ or } \ln\left(\frac{P_2}{P_1}\right) = \frac{\Delta H_{vap}}{R}\left(\frac{1}{T_1} - \frac{1}{T_2}\right)$$

$$\ln\left(\frac{P_2}{1.00\text{atm}}\right) = \frac{40.7 \times 10^3 \text{ J/mol}}{8.3145 \text{J/K} \bullet \text{mol}}\left(\frac{1}{373\text{K}} - \frac{1}{388\text{K}}\right),\ \ln P_2 = 0.51,\ P_2 = e^{0.51} = 1.7 \text{ atm}$$

$$\ln\left(\frac{3.50}{1.00}\right) = \frac{40.7 \times 10^3 \text{ J/mol}}{8.3145 \text{J/K} \bullet \text{mol}}\left(\frac{1}{373\text{K}} - \frac{1}{T_2}\right),\ 2.56 \times 10^{-4} = \left(\frac{1}{373\text{K}} - \frac{1}{T_2}\right)$$

$$2.56 \times 10^{-4} = 2.68 \times 10^{-3} - \frac{1}{T_2},\ \frac{1}{T_2} = 2.42 \times 10^{-3},\ T_2 = \frac{1}{2.42 \times 10^{-3}} = 413 \text{ K or } 140.°\text{C}$$

93. $$\ln\left(\frac{P_1}{P_2}\right) = \frac{\Delta H_{vap}}{R}\left(\frac{1}{T_2} - \frac{1}{T_1}\right),\ \ln\left(\frac{836\text{torr}}{213\text{torr}}\right) = \frac{\Delta H_{vap}}{8.3145 \text{J/K} \bullet \text{mol}}\left(\frac{1}{313\text{K}} - \frac{1}{353\text{K}}\right)$$

Solving: $\Delta H_{vap} = 3.1 \times 10^4$ J/mol; for the normal boiling point, P = 1.00 atm = 760. torr.

$$\ln\left(\frac{760.\text{ torr}}{213\text{torr}}\right) = \frac{3.1 \times 10^4 \text{ J/mol}}{8.3145 \text{J/K} \bullet \text{mol}}\left(\frac{1}{313\text{K}} - \frac{1}{T_1}\right),\ \frac{1}{313} - \frac{1}{T_1} = 3.4 \times 10^{-4}$$

T_1 = 350. K = 77°C; the normal boiling point of CCl_4 is 77°C.

94. $$\ln\left(\frac{P_1}{P_2}\right) = \frac{\Delta H_{vap}}{R}\left(\frac{1}{T_2} - \frac{1}{T_1}\right),\ \ln\left(\frac{760\text{torr}}{400.\text{torr}}\right) = \frac{\Delta H_{vap}}{8.3145 \text{J/K} \bullet \text{mol}}\left(\frac{1}{291.1\text{K}} - \frac{1}{307.8\text{K}}\right)$$

Solving: $\Delta H_{vap} = 2.83 \times 10^4$ J/mol = 28.3 kJ/mol

95.

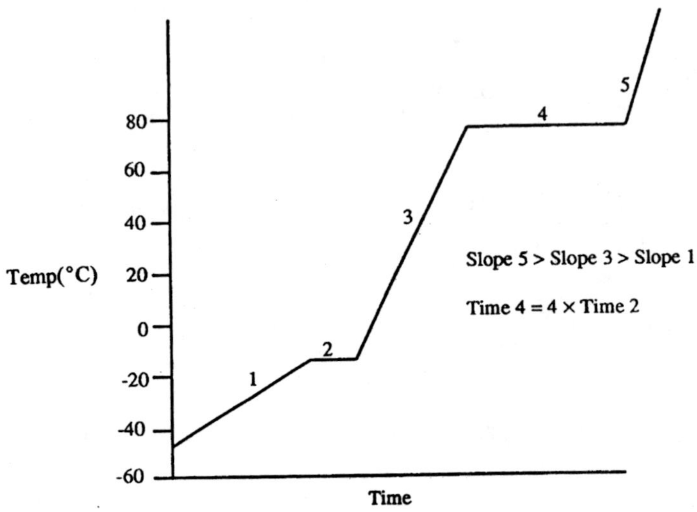

96. a. The plateau at the lowest temperature signifies the melting/freezing of the substance. Hence the freezing point is 20°C.

b. The higher temperature plateau signifies the boiling/condensation of the substance. The temperature of this plateau is 120°C.

c. $X(s) \rightarrow X(l) \quad \Delta H = \Delta H_{fusion}$; $X(l) \rightarrow X(g) \quad \Delta H = \Delta H_{vaporization}$

The heat of fusion and the heat of vaporization terms refer to enthalpy changes for the specific phase changes illustrated in the equations above. In a heating curve, energy is applied at a steady rate. So the longer, higher temperature plateau has a larger enthalpy change associated with it as compared to the shorter plateau. The higher temperature plateau occurs when a liquid is converting to a gas, so the heat of vaporization is greater than the heat of fusion. This is always the case because significantly more intermolecular forces are broken when a substance boils than when a substance melts.

97. a. Many more intermolecular forces must be broken to convert a liquid to a gas as compared with converting a solid to a liquid. Because more intermolecular forces must be broken, much more energy is required to vaporize a liquid than is required to melt a solid. Therefore, ΔH_{vap} is much larger than $\Delta H_{fus.}$

b. $1.00 \text{ g Na} \times \dfrac{1 \text{ mol Na}}{22.99 \text{ g}} \times \dfrac{2.60 \text{ kJ}}{\text{mol Na}} = 0.113 \text{ kJ} = 113 \text{ J}$ to melt 1.00 g Na

c. $1.00 \text{ g Na} \times \dfrac{1 \text{ mol Na}}{22.99 \text{ g}} \times \dfrac{97.0 \text{ kJ}}{\text{mol Na}} = 4.22 \text{ kJ} = 4220 \text{ J}$ to vaporize 1.00 g Na

d. This is the reverse process of that described in part c, so the energy change is the same quantity but opposite in sign. Therefore, $q = -4220$ J; i.e., 4220 of heat will be released.

98. Melt: $8.25 \text{ g } C_6H_6 \times \dfrac{1 \text{ mol } C_6H_6}{78.11 \text{ g}} \times \dfrac{9.92 \text{ kJ}}{\text{mol } C_6H_6} = 1.05 \text{ kJ}$

Vaporize: $8.25 \text{ g } C_6H_6 \times \dfrac{1 \text{ mol } C_6H_6}{78.11 \text{ g}} \times \dfrac{30.7 \text{ kJ}}{\text{mol } C_6H_6} = 3.24 \text{ kJ}$

As is typical, the energy required to vaporize a certain quantity of substance is much larger than the energy required to melt the same quantity of substance. A lot more intermolecular forces must be broken to vaporize a substance as compared to melting a substance.

99. To calculate q_{total}, break up the heating process into five steps.

$H_2O(s, -20.°C) \rightarrow H_2O(s, 0°C)$, $\Delta T = 20.°C$; let s_{ice} = specific heat capacity of ice:

$$q_1 = s_{ice} \times m \times \Delta T = \frac{2.03 \text{ J}}{\text{g °C}} \times 5.00 \times 10^2 \text{ g} \times 20.°C = 2.0 \times 10^4 \text{ J} = 20. \text{ kJ}$$

$H_2O(s, 0°C) \rightarrow H_2O(l, 0°C)$, $q_2 = 5.00 \times 10^2 \text{ g } H_2O \times \dfrac{1 \text{ mol}}{18.02 \text{ g}} \times \dfrac{6.02 \text{ kJ}}{\text{mol}} = 167 \text{ kJ}$

$H_2O(l, 0°C) \rightarrow H_2O(l, 100.°C)$, $q_3 = \dfrac{4.2 \text{ J}}{\text{g °C}} \times 5.00 \times 10^2 \text{ g} \times 100.°C = 2.1 \times 10^5 \text{J} = 210 \text{ kJ}$

$H_2O(l, 100.°C) \rightarrow H_2O(g, 100.°C)$, $q_4 = 5.00 \times 10^2 \text{ g} \times \dfrac{1 \text{ mol}}{18.02 \text{ g}} \times \dfrac{40.7 \text{ kJ}}{\text{mol}} = 1130 \text{ kJ}$

$H_2O(g, 100.°C) \rightarrow H_2O(g, 250.°C)$, $q_5 = \dfrac{2.0 \text{ J}}{\text{g °C}} \times 5.00 \times 10^2 \text{ g} \times 150.°C = 1.5 \times 10^5 \text{ J}$

$= 150 \text{ kJ}$

$q_{total} = q_1 + q_2 + q_3 + q_4 + q_5 = 20. + 167 + 210 + 1130 + 150 = 1680 \text{ kJ}$

100. $H_2O(g, 125°C) \rightarrow H_2O(g, 100.°C)$, $q_1 = 2.0 \text{ J/g•°C} \times 75.0 \text{ g} \times (-25°C) = -3800 \text{ J} = -3.8 \text{ kJ}$

$H_2O(g, 100.°C) \rightarrow H_2O(l, 100.°C)$, $q_2 = 75.0 \text{ g} \times \dfrac{1 \text{ mol}}{18.02 \text{ g}} \times \dfrac{-40.7 \text{ kJ}}{\text{mol}} = -169 \text{ kJ}$

$H_2O(l, 100.°C) \rightarrow H_2O(l, 0°C)$, $q_3 = 4.2 \text{ J/g•°C} \times 75.0 \text{ g} \times (-100.°C) = -32{,}000 \text{ J} = -32 \text{ kJ}$

To convert $H_2O(g)$ at 125°C to $H_2O(l)$ at 0°C requires (−3.8 kJ − 169 kJ − 32 kJ =) −205 kJ of heat removed. To convert from $H_2O(l)$ at 0°C to $H_2O(s)$ at 0°C requires:

$$q_4 = 75.0 \text{ g} \times \frac{1 \text{ mol}}{18.02 \text{ g}} \times \frac{-6.02 \text{ kJ}}{\text{mol}} = -25.1 \text{ kJ}$$

This amount of energy puts us over the −215 kJ limit (−205 kJ − 25.1 kJ = −230. kJ). Therefore, a mixture of $H_2O(s)$ and $H_2O(l)$ will be present at 0°C when 215 kJ of heat is removed from the gas sample.

101. Total mass H_2O = 18 cubes × $\frac{30.0\text{ g}}{\text{cube}}$ = 540. g; 540. g H_2O × $\frac{1\text{ mol }H_2O}{18.02\text{ g}}$ = 30.0 mol H_2O

Heat removed to produce ice at −5.0°C:

$$\left(\frac{4.18\text{ J}}{\text{g °C}} \times 540.\text{ g} \times 22.0\text{°C}\right) + \left(\frac{6.02 \times 10^3\text{ J}}{\text{mol}} \times 30.0\text{ mol}\right) + \left(\frac{2.03\text{ J}}{\text{g °C}} \times 540.\text{ g} \times 5.0\text{°C}\right)$$

$$= 4.97 \times 10^4\text{ J} + 1.81 \times 10^5\text{ J} + 5.5 \times 10^3\text{ J} = 2.36 \times 10^5\text{ J}$$

$2.36 \times 10^5\text{ J} \times \frac{1\text{ g }CF_2Cl_2}{158\text{ J}} = 1.49 \times 10^3$ g CF_2Cl_2 must be vaporized.

102. Heat released = 0.250 g Na × $\frac{1\text{ mol}}{22.99\text{ g}} \times \frac{368\text{ kJ}}{2\text{ mol}}$ = 2.00 kJ

To melt 50.0 g of ice requires: 50.0 g ice × $\frac{1\text{ mol }H_2O}{18.02\text{ g}} \times \frac{6.02\text{ kJ}}{\text{mol}}$ = 16.7 kJ

The reaction doesn't release enough heat to melt all of the ice. The temperature will remain at 0°C.

103. A: solid B: liquid C: vapor

D: solid + vapor E: solid + liquid + vapor

F: liquid + vapor G: liquid + vapor H: vapor

triple point: E critical point: G

Normal freezing point: Temperature at which solid-liquid line is at 1.0 atm (see following plot).

Normal boiling point: Temperature at which liquid-vapor line is at 1.0 atm (see following plot).

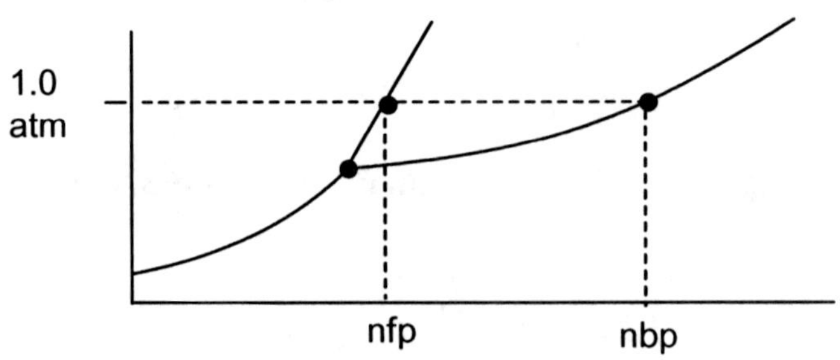

Because the solid-liquid line equilibrium has a positive slope, the solid phase is denser than the liquid phase.

104. a. 3

b. Triple point at 95.31°C: rhombic, monoclinic, gas

Triple point at 115.18°C: monoclinic, liquid, gas

Triple point at 153°C: rhombic, monoclinic, liquid

c. From the phase diagram, the monoclinic solid phase is stable at T = 100.°C and P = 1 atm.

d. Normal melting point = 115.21°C; normal boiling point = 444.6°C; the normal melting and boiling points occur at P = 1.0 atm.

e. Rhombic is the densest phase because the rhombic-monoclinic equilibrium line has a positive slope, and because the solid-liquid equilibrium lines also have positive slopes.

f. No; $P = 1.0 \times 10^{-5}$ atm is at a pressure somewhere between the 95.31 and 115.18°C triple points. At this pressure, the rhombic and gas phases are never in equilibrium with each other, so rhombic sulfur cannot sublime at $P = 1.0 \times 10^{-5}$ atm. However, monoclinic sulfur can sublime at this pressure.

g. From the phase diagram, we would start off with gaseous sulfur. At 100.°C and $\sim 1 \times 10^{-5}$ atm, S(g) would convert to the solid monoclinic form of sulfur. Finally at 100.°C and some large pressure less than 1420 atm, S(s, monoclinic) would convert to the solid rhombic form of sulfur. Summarizing, the phase changes are S(g) $\rightarrow$ S(monoclinic) $\rightarrow$ S(rhombic).

105. a. two

b. Higher-pressure triple point: graphite, diamond and liquid; lower-pressure triple point at $\sim 10^7$ Pa: graphite, liquid and vapor

c. It is converted to diamond (the more dense solid form).

d. Diamond is more dense, which is why graphite can be converted to diamond by applying pressure.

106. The following sketch of the Br_2 phase diagram is not to scale. Because the triple point of Br_2 is at a temperature below the freezing point of Br_2, the slope of the solid-liquid line is positive.

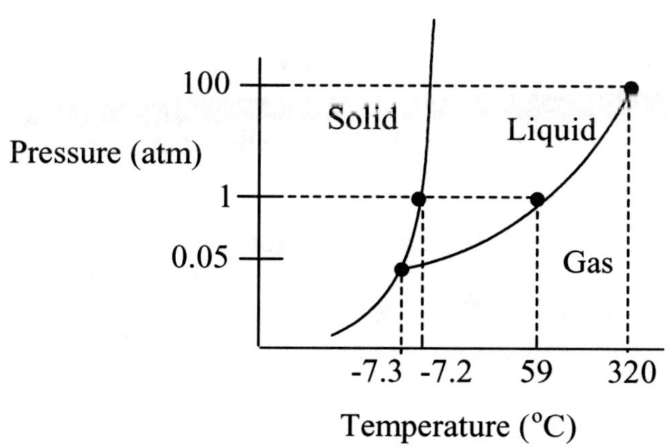

The positive slopes of all the lines indicate that $Br_2(s)$ is more dense than $Br_2(l)$, which is more dense than $Br_2(g)$. At room temperature (~22°C) and 1 atm, $Br_2(l)$ is the stable phase. $Br_2(l)$ cannot exist at a temperature below the triple-point temperature of −7.3°C or at a temperature above the critical-point temperature of 320°C. The phase changes that occur as temperature is increased at 0.10 atm are solid → liquid → gas.

107. Because the density of the liquid phase is greater than the density of the solid phase, the slope of the solid-liquid boundary line is negative (as in H_2O). With a negative slope, the melting points increase with a decrease in pressure, so the normal melting point of X should be greater than 225°C.

108.

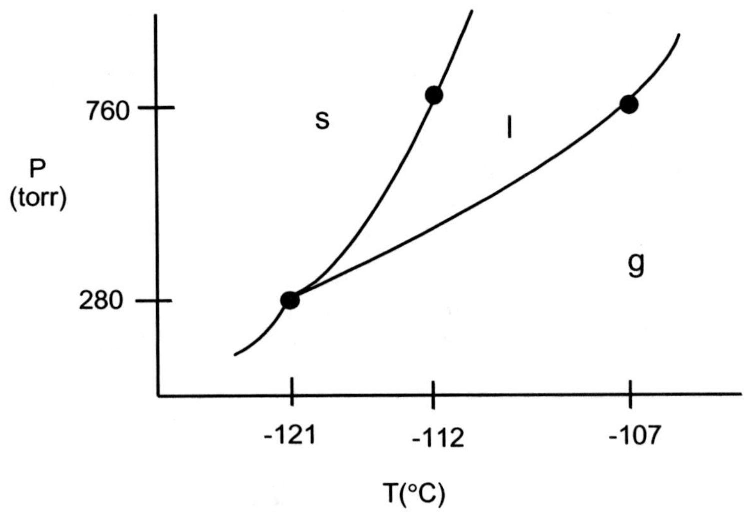

From the three points given, the slope of the solid-liquid boundary line is positive, so Xe(s) is more dense than Xe(l). Also, the positive slope of this line tells us that the melting point of Xe increases as pressure increases. The same direct relationship exists for the boiling point of Xe because the liquid-gas boundary line also has a positive slope.

Additional Exercises

109. As the physical properties indicate, the intermolecular forces are slightly stronger in D_2O than in H_2O.

110. CH_3CO_2H: H-bonding + dipole forces + London dispersion (LD) forces

CH_2ClCO_2H: H-bonding + larger electronegative atom replacing H (greater dipole) + LD forces

$CH_3CO_2CH_3$: Dipole forces (no H-bonding) + LD forces

From the intermolecular forces listed above, we predict $CH_3CO_2CH_3$ to have the weakest intermolecular forces and CH_2ClCO_2H to have the strongest. The boiling points are consistent with this view.

111. At any temperature, the plot tells us that substance A has a higher vapor pressure than substance B, with substance C having the lowest vapor pressure. Therefore, the substance with the weakest intermolecular forces is A, and the substance with the strongest intermolecular forces is C.

NH_3 can form hydrogen-bonding interactions, whereas the others cannot. Substance C is NH_3. The other two are nonpolar compounds with only London dispersion forces. Because CH_4 is smaller than SiH_4, CH_4 will have weaker LD forces and is substance A. Therefore, substance B is SiH_4.

112. As the electronegativity of the atoms covalently bonded to H increases, the strength of the hydrogen-bonding interaction increases.

$$\underset{\text{weakest}}{\text{N} \cdots \text{H–N}} < \text{N} \cdots \text{H–O} < \text{O} \cdots \text{H–O} < \text{O} \cdots \text{H–F} < \underset{\text{strongest}}{\text{F} \cdots \text{H–F}}$$

113. $8 \text{ corners} \times \dfrac{1/8 \text{ Xe}}{\text{corner}} + 1 \text{ Xe inside cell} = 2 \text{ Xe}; \; 8 \text{ edges} \times \dfrac{1/4 \text{ F}}{\text{edge}} + 2 \text{ F inside cell} = 4 \text{ F}$

The empirical formula is XeF_2.

114. One B atom and one N atom together have the same number of electrons as two C atoms. The description of physical properties sounds a lot like the properties of graphite and diamond, the two solid forms of carbon. The two forms of BN have structures similar to graphite and diamond.

115. B_2H_6: This compound contains only nonmetals, so it is probably a molecular solid with covalent bonding. The low boiling point confirms this.

SiO_2: This is the empirical formula for quartz, which is a network solid.

CsI: This is a metal bonded to a nonmetal, which generally form ionic solids. The electrical conductivity in aqueous solution confirms this.

W: Tungsten is a metallic solid as the conductivity data confirm.

116. Ar is cubic closest packed. There are four Ar atoms per unit cell, and with a face-centered unit cell, the atoms touch along the face diagonal. Let l = length of cube edge.

Face diagonal = $4r = l\sqrt{2}$, $l = 4(190. \text{ pm})/\sqrt{2} = 537 \text{ pm} = 5.37 \times 10^{-8} \text{ cm}$

$$\text{Density} = \frac{\text{mass}}{\text{volume}} = \frac{4 \text{ atoms} \times \dfrac{1 \text{ mol}}{6.022 \times 10^{23} \text{ atoms}} \times \dfrac{39.95 \text{ g}}{\text{mol}}}{(5.37 \times 10^{-8} \text{ cm})^3} = 1.71 \text{ g/cm}^3$$

117. $24.7 \text{ g } C_6H_6 \times \dfrac{1 \text{ mol}}{78.11 \text{ g}} = 0.316 \text{ mol } C_6H_6$

$$P_{C_6H_6} = \frac{nRT}{V} = \frac{0.316\text{mol} \times \frac{0.08206\text{L atm}}{\text{K mol}} \times 293.2\text{ K}}{100.0\text{ L}} = 0.0760\text{ atm, or } 57.8\text{ torr}$$

118. In order to set up an equation, we need to know what phase exists at the final temperature. To heat 20.0 g of ice from −10.0 to 0.0°C requires:

$$q = \frac{2.03\text{ J}}{\text{g °C}} \times 20.0\text{ g} \times 10.0\text{°C} = 406\text{ J}$$

To convert ice to water at 0.0°C requires:

$$q = 20.0\text{ g} \times \frac{1\text{ mol}}{18.02} \times \frac{6.02\text{ kJ}}{\text{mol}} = 6.68\text{ kJ} = 6680\text{ J}$$

To chill 100.0 g of water from 80.0 to 0.0°C requires:

$$q = \frac{4.18\text{ J}}{\text{g °C}} \times 100.0\text{ g} \times 80.0\text{°C} = 33{,}400\text{ J of heat removed}$$

From the heat values above, the liquid phase exists once the final temperature is reached (a lot more heat is lost when the 100.0 g of water is cooled to 0.0°C than the heat required to convert the ice into water). To calculate the final temperature, we will equate the heat gain by the ice to the heat loss by the water. We will keep all quantities positive in order to avoid sign errors. The heat gain by the ice will be the 406 J required to convert the ice to 0.0°C plus the 6680 J required to convert the ice at 0.0°C into water at 0.0°C plus the heat required to raise the temperature from 0.0°C to the final temperature.

$$\text{Heat gain by ice} = 406\text{ J} + 6680\text{ J} + \frac{4.18\text{ J}}{\text{g °C}} \times 20.0\text{ g} \times (T_f - 0.0\text{°C}) = 7.09 \times 10^3 + (83.6)T_f$$

$$\text{Heat loss by water} = \frac{4.18\text{ J}}{\text{g °C}} \times 100.0\text{ g} \times (80.0\text{°C} - T_f) = 3.34 \times 10^4 - 418T_f$$

Solving for the final temperature:

$$7.09 \times 10^3 + (83.6)T_f = 3.34 \times 10^4 - 418T_f,\ \ 502T_f = 2.63 \times 10^4,\ \ T_f = 52.4\text{°C}$$

119. $1.00\text{ lb} \times \frac{454\text{g}}{\text{lb}} = 454\text{ g } H_2O$; a change of 1.00°F is equal to a change of 5/9°C.

The amount of heat in J in 1 Btu is $\frac{4.18\text{ J}}{\text{g °C}} \times 454\text{g} \times \frac{5}{9}\text{°C} = 1.05 \times 10^3\text{ J} = 1.05\text{ kJ}$.

It takes 40.7 kJ to vaporize 1 mol H_2O (ΔH_{vap}). Combining these:

$$\frac{1.00 \times 10^4\text{ Bu}}{\text{h}} \times \frac{1.05\text{kJ}}{\text{Btu}} \times \frac{1\text{mol}H_2O}{40.7\text{ kJ}} = 258\text{ mol/h; or:}$$

$$\frac{258\text{mol}}{\text{h}} \times \frac{18.02\text{ g }H_2O}{\text{mol}} = 4650\text{ g/h} = 4.65\text{ kg/h}$$

120. The critical temperature is the temperature above which the vapor cannot be liquefied no matter what pressure is applied. Since N_2 has a critical temperature below room temperature (~22°C), it cannot be liquefied at room temperature. NH_3, with a critical temperature above room temperature, can be liquefied at room temperature.

ChemWork Problems

The answers to the problems 121-130 (or a variation to these problem) are found in OWL. These problems are also assignable in OWL.

Challenge Problems

131. $\Delta H = q_p = 30.79$ kJ; $\Delta E = q_p + w$, $w = -P\Delta V$

$$w = -P\Delta V = -1.00 \text{ atm}(28.90 \text{ L}) = -28.9 \text{ L atm} \times \frac{101.3 \text{ J}}{\text{L atm}} = -2930 \text{ J}$$

$$\Delta E = 30.79 \text{ kJ} + (-2.93 \text{ kJ}) = 27.86 \text{ kJ}$$

132. $XeCl_2F_2$, $8 + 2(7) + 2(7) = 36\ e^-$

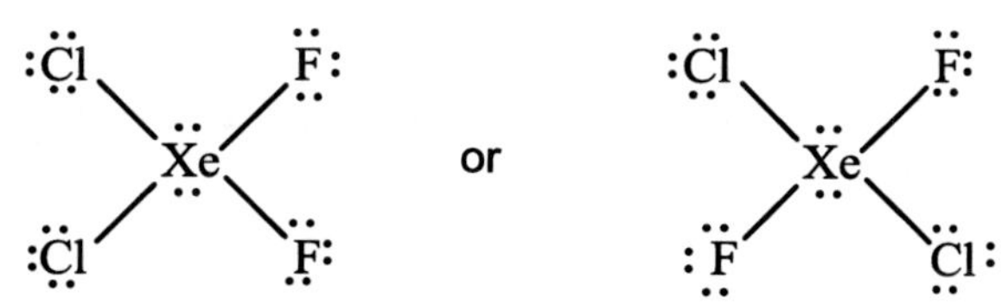

polar (Bond dipoles do not cancel each other.) or nonpolar (Bond dipoles cancel each other.)

These are two possible square planar molecular structures for $XeCl_2F_2$. One structure has the Cl atoms 90° apart; the other has the Cl atoms 180° apart. The structure with the Cl atoms 90° apart is polar; the other structure is nonpolar. The polar structure will have additional dipole forces, so it has the stronger intermolecular forces and is the liquid. The gas form of $XeCl_2F_2$ is the nonpolar form having the Cl atoms 180° apart.

133. A single hydrogen bond in H_2O has a strength of 21 kJ/mol. Each H_2O molecule forms two H bonds. Thus it should take 42 kJ/mol of energy to break all of the H bonds in water. Consider the phase transitions:

$$\text{Solid} \xrightarrow{6.0 \text{ kJ}} \text{liquid} \xrightarrow{40.7 \text{ kJ}} \text{vapor} \qquad \Delta H_{sub} = \Delta H_{fus} + \Delta H_{vap}$$

$\Delta H_{sub} = 6.0$ kJ/mol + 40.7 kJ/mol = 46.7 kJ/mol; it takes a total of 46.7 kJ/mol to convert solid H_2O to vapor. This would be the amount of energy necessary to disrupt all of the intermolecular forces in ice. Thus $(42 \div 46.7) \times 100 = 90.\%$ of the attraction in ice can be attributed to H bonding.

134. $1\text{ gal} \times \frac{3785\text{mL}}{\text{gal}} \times \frac{0.998\text{g}}{\text{mL}} \times \frac{1\text{mol}\,H_2O}{18.02\text{ g}} = 210.\text{ mol }H_2O$

From Table 10.8, the vapor pressure of H_2O at 25°C is 23.756 torr. The water will evaporate until this partial pressure is reached.

$$V = \frac{nRT}{P} = \frac{210.\text{ mol} \times \frac{0.08206\text{L atm}}{\text{K mol}} \times 298\text{K}}{23.756\text{torr} \times \frac{1\text{ atm}}{760\text{torr}}} = 1.64 \times 10^5\text{ L}$$

Dimension of cube = $(1.64 \times 10^5\text{ L} \times 1\text{ dm}^3/\text{L})^{1/3} = 54.7\text{ dm}$

$54.7\text{ dm} \times \frac{1\text{ m}}{10\text{dm}} \times \frac{1.094\text{ yards}}{\text{m}} \times \frac{3\text{ ft}}{\text{yard}} = 18.0\text{ ft}$

The cube has dimensions of 18.0 ft × 18.0 ft × 18.0 ft.

135. The structures of the two C_2H_6O compounds (20 valence e^-) are:

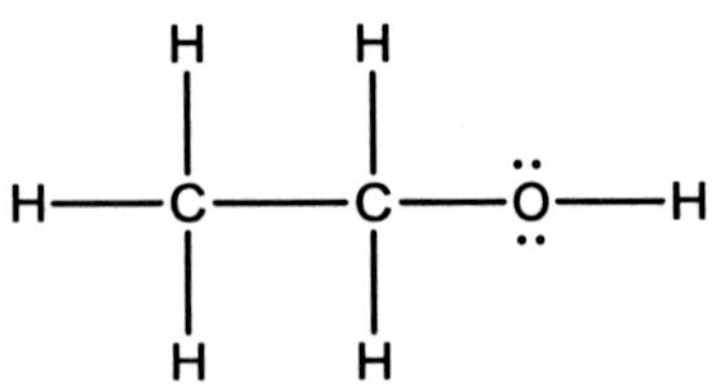

exhibits relatively strong hydrogen bonding

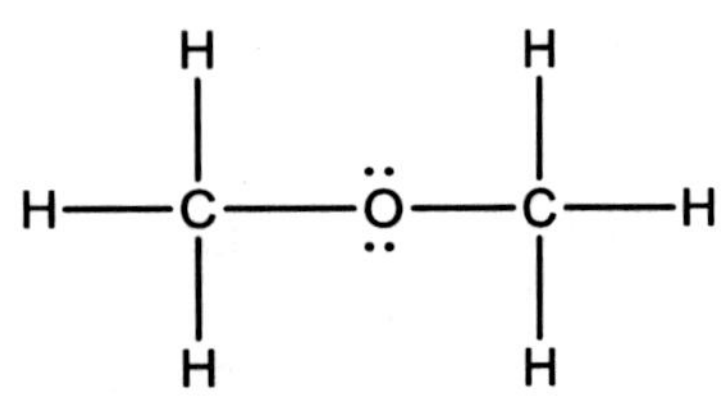

does not exhibit hydrogen bonding

The liquid will have the stronger intermolecular forces. Therefore, the first compound (ethanol) with hydrogen bonding is the liquid and the second compound (dimethyl ether) with the weaker intermolecular forces is the gas.

136. Benzene

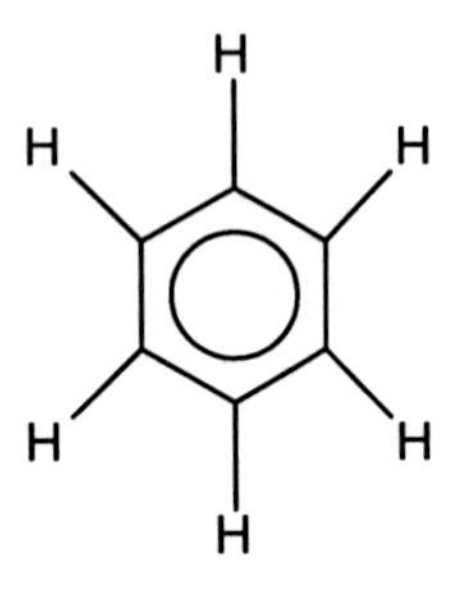

LD forces only

Naphthalene

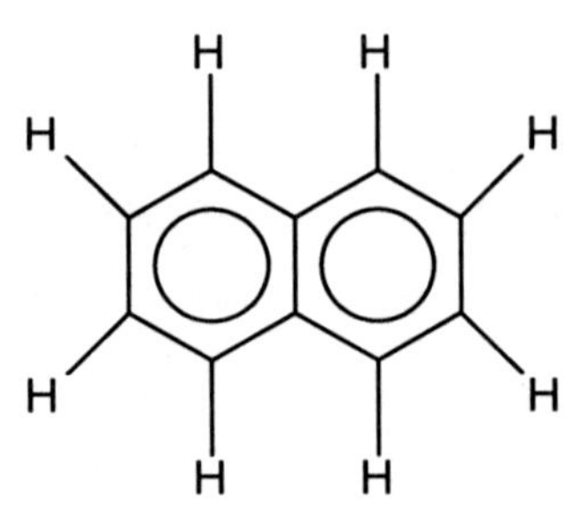

LD forces only

Note: London dispersion (LD) forces in molecules such as benzene and naphthalene are fairly large. The molecules are flat, and there is efficient surface-area contact between molecules. Large surface-area contact leads to stronger London dispersion forces.

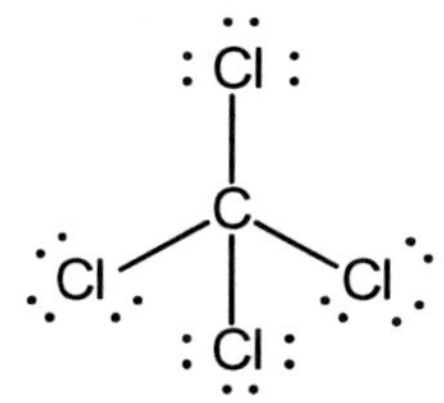

Carbon tetrachloride (CCl_4) has polar bonds but is a nonpolar molecule. CCl_4 only has LD forces.

In terms of size and shape, the order of the strength of LD forces is: $CCl_4 < C_6H_6 < C_{10}H_8$

The strengths of the LD forces are proportional to size and are related to shape. Although the size of CCl_4 is fairly large, the overall spherical shape gives rise to relatively weak LD forces as compared to flat molecules such as benzene and naphthalene. The physical properties given in the problem are consistent with the order listed above. Each of the physical properties will increase with an increase in intermolecular forces.

Acetone

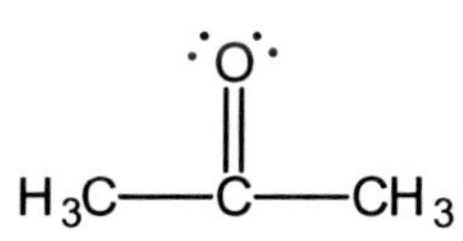

LD, dipole

Acetic acid

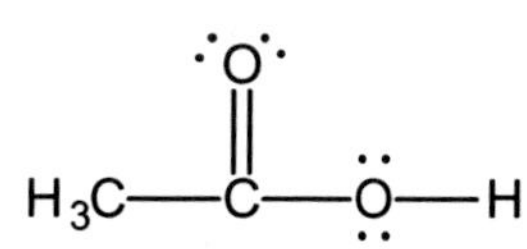

LD, dipole, H-bonding

Benzoic acid

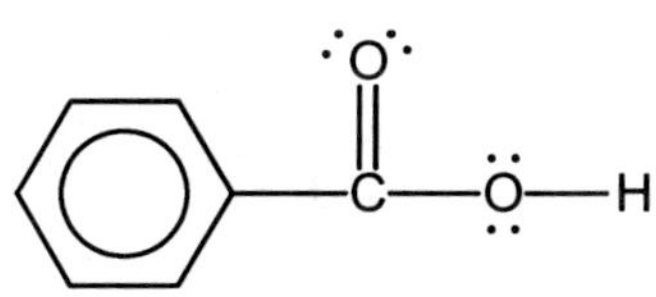

LD, dipole, H-bonding

We would predict the strength of interparticle forces of the last three molecules to be:

acetone	<	acetic acid	<	benzoic acid
polar		H-bonding		H-bonding, but large LD forces because of greater size and shape

This ordering is consistent with the values given for boiling point, melting point, and ΔH_{vap}.

The overall order of the strengths of intermolecular forces based on physical properties are:

acetone < CCl_4 < C_6H_6 < acetic acid < naphthalene < benzoic acid

The order seems reasonable except for acetone and naphthalene. Because acetone is polar, we would not expect it to boil at the lowest temperature. However, in terms of size and shape, acetone is the smallest molecule, and the LD forces in acetone must be very small compared to the other molecules. Naphthalene must have very strong LD forces because of its size and flat shape.

137. $NaCl$, $MgCl_2$, NaF, MgF_2, and AlF_3 all have very high melting points indicative of strong intermolecular forces. They are all ionic solids. $SiCl_4$, SiF_4, F_2, Cl_2, PF_5, and SF_6 are nonpolar covalent molecules. Only London dispersion (LD) forces are present. PCl_3 and SCl_2 are polar molecules. LD forces and dipole forces are present. In these eight molecular substances, the intermolecular forces are weak and the melting points low. $AlCl_3$ doesn't seem to fit in as well. From the melting point, there are much stronger forces present than in the nonmetal halides, but they aren't as strong as we would expect for an ionic solid. $AlCl_3$ illustrates a gradual transition from ionic to covalent bonding, from an ionic solid to discrete molecules.

138. Total charge of all iron ions present in a formula unit is +2 to balance the −2 charge from the one O atom. The sum of iron ions in a formula unit is 0.950. Let x = fraction Fe^{2+} ions in a formula unit and y = fraction of Fe^{3+} ions present in a formula unit.

Setting up two equations: $x + y = 0.950$ and $2x + 3y = 2.000$

Solving: $2x + 3(0.950 - x) = 2.000$, $x = 0.85$ and $y = 0.10$

$$\frac{0.10}{0.95} = 0.11 = \text{fraction of iron as } Fe^{3+} \text{ ions}$$

If all Fe^{2+}, then 1.000 Fe^{2+} ion/O^{2-} ion; 1.000 − 0.950 = 0.050 = vacant sites. 5.0% of the Fe^{2+} sites are vacant.

139. Assuming 100.00 g:

$$28.31 \text{ g O} \times \frac{1 \text{ mol}}{16.00 \text{ g}} = 1.769 \text{ mol O}; \quad 71.69 \text{ g Ti} \times \frac{1 \text{ mol}}{47.88 \text{ g}} = 1.497 \text{ mol Ti}$$

$$\frac{1.769}{1.497} = 1.182; \quad \frac{1.497}{1.769} = 0.8462; \text{ the formula is } TiO_{1.182} \text{ or } Ti_{0.8462}O.$$

For $Ti_{0.8462}O$, let $x = Ti^{2+}$ per mol O^{2-} and $y = Ti^{3+}$ per mol O^{2-}. Setting up two equations and solving:

$$x + y = 0.8462 \text{ (mass balance) and } 2x + 3y = 2 \text{ (charge balance)}$$

$$2x + 3(0.8462 - x) = 2, \; x = 0.539 \text{ mol } Ti^{2+}/\text{mol } O^{2-} \text{ and } y = 0.307 \text{ mol } Ti^{3+}/\text{mol } O^{2-}$$

$$\frac{0.539}{0.8462} \times 100 = 63.7\%$$ of the titanium ions are Ti^{2+} and 36.3% are Ti^{3+} (a 1.75 : 1 ion ratio).

140. First, we need to get the empirical formula of spinel. Assume 100.0 g of spinel:

$$37.9 \text{ g Al} \times \frac{1 \text{ mol Al}}{26.98 \text{ g Al}} = 1.40 \text{ mol Al}$$

$$17.1 \text{ g Mg} \times \frac{1 \text{ mol Mg}}{24.31 \text{ g Mg}} = 0.703 \text{ mol Mg}$$

The mole ratios are 2 : 1 : 4.

$$45.0 \text{ g O} \times \frac{1 \text{ mol O}}{16.00 \text{ g O}} = 2.81 \text{ mol O}$$

Empirical formula = Al_2MgO_4

Assume each unit cell contains an integral value (x) of Al_2MgO_4 formula units. Each Al_2MgO_4 formula unit has a mass of 24.31 + 2(26.98) + 4(16.00) = 142.27 g/mol.

$$\text{Density} = \frac{x \text{ formula units} \times \dfrac{1 \text{ mol}}{6.022 \times 10^{23} \text{ formula units}} \times \dfrac{142.27 \text{ g}}{\text{mol}}}{(8.09 \times 10^{-8} \text{ cm})^3} = \frac{3.57 \text{ g}}{\text{cm}^3}$$

Solving: $x = 8.00$

Each unit cell has 8 formula units of Al_2MgO_4 or 16 Al, 8 Mg, and 32 O ions.

141. $$\frac{\text{Density}_{Mn}}{\text{Density}_{Cu}} = \frac{\text{mass}_{Mn} \times \text{volume}_{Cu}}{\text{volume}_{Mn} \times \text{mass}_{Cu}} = \frac{\text{mass}_{Mn}}{\text{mass}_{Cu}} \times \frac{\text{volume}_{Cu}}{\text{volume}_{Mn}}$$

The type of cubic cell formed is not important; only that Cu and Mn crystallize in the same type of cubic unit cell is important. Each cubic unit cell has a specific relationship between the cube edge length l and the radius r. In all cases $l \propto$ r. Therefore, $V \propto l^3 \propto r^3$. For the mass ratio, we can use the molar masses of Mn and Cu since each unit cell must contain the same number of Mn and Cu atoms. Solving:

$$\frac{\text{density}_{Mn}}{\text{density}_{Cu}} = \frac{\text{mass}_{Mn}}{\text{mass}_{Cu}} \times \frac{\text{volume}_{Cu}}{\text{volume}_{Mn}} = \frac{54.94 \text{g/mol}}{63.55 \text{g/mol}} \times \frac{(r_{Cu})^3}{(1.056 r_{Cu})^3}$$

$$\frac{\text{density}_{Mn}}{\text{density}_{Cu}} = 0.8645 \times \left(\frac{1}{1.056}\right)^3 = 0.7341$$

$$\text{density}_{Mn} = 0.7341 \times \text{density}_{Cu} = 0.7341 \times 8.96 \text{ g/cm}^3 = 6.58 \text{ g/cm}^3$$

142. a. The arrangement of the layers is:

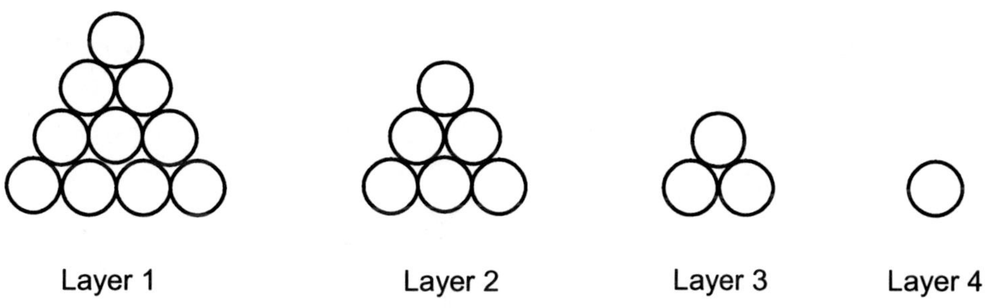

A total of 20 cannon balls will be needed.

b. The layering alternates *abcabc*, which is cubic closest packing.

c. tetrahedron

143. 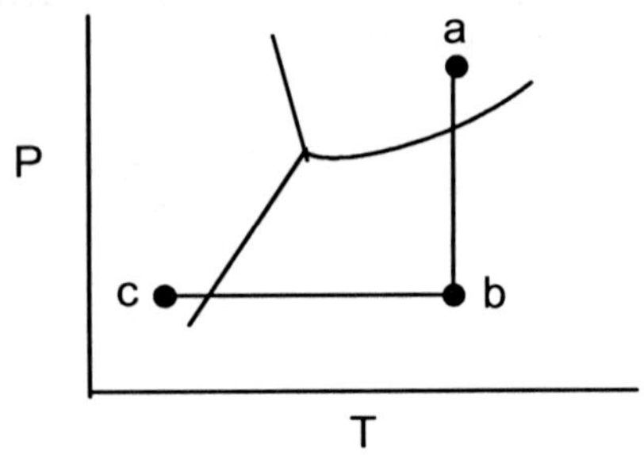

As P is lowered, we go from a to b on the phase diagram. The water boils. The boiling of water is endothermic, and the water is cooled (b → c), forming some ice. If the pump is left on, the ice will sublime until none is left. This is the basis of freeze drying.

144. w = –PΔV; assume a constant P of 1.00 atm.

$$V_{373} = \frac{nRT}{P} = \frac{1.00\text{mol} \times \frac{0.08206\text{L atm}}{\text{K mol}} \times 373\text{K}}{1.00\text{atm}} = 30.6 \text{ L for 1 mol of water vapor}$$

Because the density of $H_2O(l)$ is 1.00 g/cm^3, 1.00 mol of $H_2O(l)$ occupies 18.0 cm^3 or 0.0180 L.

$$w = -1.00 \text{ atm}(30.6 \text{ L} - 0.0180 \text{ L}) = -30.6 \text{ L atm}$$

$$w = -30.6 \text{ L atm} \times \frac{101.3\text{J}}{\text{L atm}} = -3.10 \times 10^3 \text{ J} = -3.10 \text{ kJ}$$

$$\Delta E = q + w = 40.7 \text{ kJ} - 3.10 \text{ kJ} = 37.6 \text{ kJ}$$

$\frac{37.6}{40.7} \times 100 = 92.4\%$ of the energy goes to increase the internal energy of the water.

The remainder of the energy (7.6%) goes to do work against the atmosphere.

145. For a cube: (body diagonal)^2 = (face diagonal)^2 + (cube edge length)^2

In a simple cubic structure, the atoms touch on cube edge, so the cube edge = 2r, where r = radius of sphere.

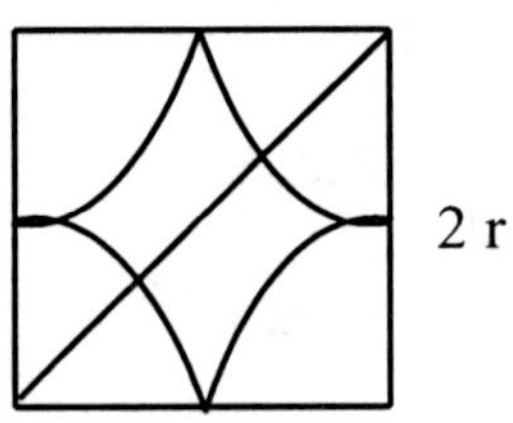

$$\text{Face diagonal} = \sqrt{(2r)^2 + (2r)^2} = \sqrt{4r^2 + 4r^2} = r\sqrt{8} = 2\sqrt{2}\ r$$

$$\text{Body diagonal} = \sqrt{(2\sqrt{2}\,r)^2 + (2r)^2} = \sqrt{12r^2} = 2\sqrt{3}\ r$$

The diameter of the hole = body diagonal – 2(radius of atoms at corners).

Diameter = $2\sqrt{3}\,r - 2r$; thus the radius of the hole is: $\dfrac{2\sqrt{3}\,r - 2r}{2} = (\sqrt{3} - 1)r$

The volume of the hole is $\dfrac{4}{3}\pi\left[(\sqrt{3} - 1)r\right]^3$.

146. Using the ionic radii values given in the question, let's calculate the density of the two structures.

Normal pressure: Rb^+ and Cl^- touch along cube edge (form NaCl structure).

Cube edge = l = 2(148 + 181) = 658 pm = 6.58×10^{-8} cm; there are four RbCl units per unit cell.

$$\text{Density} = d = \frac{4(85.47) + 4(35.45)}{6.022 \times 10^{23}(6.58 \times 10^{-8})^3} = 2.82 \text{ g/cm}^3$$

High pressure: Rb^+ and Cl^- touch along body diagonal (form CsCl structure).

$2r_- + 2r_+$ = 658 pm = body diagonal = $l\sqrt{3}$, l = 658 pm/$\sqrt{3}$ = 380. pm

Each unit cell contains 1 RbCl unit: $d = \dfrac{85.47 + 35.45}{6.022 \times 10^{23}(3.80 \times 10^{-8})^3} = 3.66 \text{ g/cm}^3$

The high-pressure form has the higher density. The density ratio is 3.66/2.82 = 1.30. We would expect this because the effect of pressure is to push things closer together and thus increase density.

Integrative Problems

147. Molar mass of XY = $\dfrac{19.0 \text{ g}}{0.132 \text{ mol}}$ = 144 g/mol

X: [Kr] $5s^2 4d^{10}$; this is cadmium, Cd.

Molar mass Y = 144 – 112.4 = 32 g/mol; Y is sulfur, S.

The semiconductor is CdS. The dopant has the electron configuration of bromine, Br. Because Br has one more valence electron than S, doping with Br will produce an n-type semiconductor.

148. Assuming 100.00 g of MO_2:

$$23.72 \text{ g O} \times \frac{1 \text{ mol O}}{16.00 \text{ g O}} = 1.483 \text{ mol O}$$

$$1.483 \text{ mol O} \times \frac{1 \text{ mol M}}{2 \text{ mol O}} = 0.7415 \text{ mol M}$$

$$100.00 \text{ g} - 23.72 \text{ g} = 76.28 \text{ g M}; \text{ molar mass M} = \frac{76.28 \text{ g}}{0.7415 \text{ mol}} = 102.9 \text{ g/mol}$$

From the periodic table, element M is rhodium (Rh).

The unit cell for cubic closest packing is face-centered cubic (4 atoms/unit cell). The atoms for a face-centered cubic unit cell are assumed to touch along the face diagonal of the cube, so the face diagonal = 4r. The distance between the centers of touching Rh atoms will be the distance of 2r, where r = radius of Rh atom.

Face diagonal = $\sqrt{2}\ l$, where l = cube edge.

Face diagonal = 4r = 2 × 269.0 × 10^{-12} m = 5.380 × 10^{-10} m

$$\sqrt{2}\, l = 4r = 5.38 \times 10^{-10} \text{ m}, \; l = \frac{5.38 \times 10^{-10} \text{ m}}{\sqrt{2}} = 3.804 \times 10^{-10} \text{ m} = 3.804 \times 10^{-8} \text{ cm}$$

$$\text{Density} = \frac{4 \text{ atoms Rh} \times \dfrac{1 \text{ mol Rh}}{6.0221 \times 10^{23} \text{ atoms}} \times \dfrac{102.9 \text{ g Rh}}{\text{mol Rh}}}{(3.804 \times 10^{-8} \text{ cm})^3} = 12.42 \text{ g/cm}^3$$

149. $$\ln\left(\frac{P_1}{P_2}\right) = \frac{\Delta H_{vap}}{R}\left(\frac{1}{T_2} - \frac{1}{T_1}\right); \; \Delta H_{vap} = \frac{296 \text{ J}}{\text{g}} \times \frac{200.6 \text{ g}}{\text{mol}} = 5.94 \times 10^4 \text{ J/mol Hg}$$

$$\ln\left(\frac{2.56 \times 10^{-3} \text{ torr}}{P_2}\right) = \frac{5.94 \times 10^4 \text{ J/mol}}{8.3145 \text{ J/K} \bullet \text{mol}}\left(\frac{1}{573 \text{ K}} - \frac{1}{298.2 \text{ K}}\right)$$

$$\ln\left(\frac{2.56 \times 10^{-3} \text{ torr}}{P_2}\right) = -11.5, \; P_2 = 2.56 \times 10^{-3} \text{ torr}/e^{-11.5} = 253 \text{ torr}$$

$$n = \frac{PV}{RT} = \frac{\left(253 \text{ torr} \times \dfrac{1 \text{ atm}}{760 \text{ torr}}\right) \times 15.0 \text{ L}}{\dfrac{0.08206 \text{ L atm}}{\text{K mol}} \times 573 \text{ K}} = 0.106 \text{ mol Hg}$$

$$0.106 \text{ mol Hg} \times \frac{6.022 \times 10^{23} \text{ atoms Hg}}{\text{mol Hg}} = 6.38 \times 10^{22} \text{ atoms Hg}$$

Marathon Problem

150. q = s × m × ΔT; heat loss by metal = heat gain by calorimeter. The change in temperature for the calorimeter is ΔT = 25.2°C ±0.2°C – 25.0°C ±0.2°C. Including the error limits, ΔT can range from 0.0 to 0.4°C. Because the temperature change can be 0.0°C, there is no way that the calculated heat capacity has any meaning.

The density experiment is also not conclusive.

$$d = \frac{4\text{ g}}{0.42\text{ cm}^3} = 10\text{ g/cm}^3 \text{ (1 significant figure)}$$

$$d_{high} = \frac{5\text{ g}}{0.40\text{ cm}^3} = 12.5\text{ g/cm}^3 = 10\text{ g/cm}^3 \text{ to 1 sig fig}$$

$$d_{low} = \frac{3\text{ g}}{0.44\text{ cm}^3} = 7\text{ g/cm}^3$$

From Table 1.5, the density of copper is 8.96 g/cm^3. The results from this experiment cannot be used to distinguish between a density of 8.96 g/cm^3 and 9.2 g/cm^3.

The crystal structure determination is more conclusive. Assuming the metal is copper:

$$\text{volume of unit cell} = (600.\text{ pm})^3 \left(\frac{1 \times 10^{-10}\text{ cm}}{1\text{ pm}}\right)^3 = 2.16 \times 10^{-22}\text{ cm}^3$$

$$\text{Cu mass in unit cell} = 4\text{ atoms} \times \frac{1\text{ mol Cu}}{6.022 \times 10^{23}\text{ atoms}} \times \frac{63.55\text{ g Cu}}{\text{mol Cu}} = 4.221 \times 10^{-22}\text{ g Cu}$$

$$d = \frac{\text{mass}}{\text{volume}} = \frac{4.221 \times 10^{-22}\text{ g}}{2.16 \times 10^{-22}\text{ cm}^3} = 1.95\text{ g/cm}^3$$

Because the density of Cu is 8.96 g/cm^3, then one can assume this metal is not copper. If the metal is not Cu, then it must be kryptonite (as the question reads). Because we don't know the molar mass of kryptonite, we cannot confirm that the calculated density would be close to 9.2 g/cm^3.

To improve the heat capacity experiment, a more precise balance is a must and a more precise temperature reading is needed. Also, a larger piece of the metal should be used so that ΔT of the calorimeter has more significant figures. For the density experiment, we would need a more precise balance and a more precise way to determine the volume. Again, a larger piece of metal would help in order to ensure more significant figures in the volume.

CHAPTER 11

PROPERTIES OF SOLUTIONS

Solution Review

11. $$\frac{585\text{ g } C_3H_7OH \times \dfrac{1\text{ mol } C_3H_7OH}{60.09\text{ g } C_3H_7OH}}{1.00\text{ L}} = 9.74\ M$$

12. $$0.250\text{ L} \times \frac{0.100\text{ mol}}{\text{L}} \times \frac{134.00\text{ g}}{\text{mol}} = 3.35\text{ g } Na_2C_2O_4$$

13. $$1.00\text{ L} \times \frac{0.040\text{ mol HCl}}{\text{L}} = 0.040\text{ mol HCl};\ 0.040\text{ mol HCl} \times \frac{1\text{ L}}{0.25\text{ mol HCl}} = 0.16\text{ L}$$

$$= 160\text{ mL}$$

14. $$1.28\text{ g } CaCl_2 \times \frac{1\text{ mol } CaCl_2}{110.98\text{ g } CaCl_2} \times \frac{1\text{ L}}{0.580\text{ mol } CaCl_2} \times \frac{1000\text{ mL}}{\text{L}} = 19.9\text{ mL}$$

15. $$\text{Mol } Na_2CO_3 = 0.0700\text{ L} \times \frac{3.0\text{ mol } Na_2CO_3}{\text{L}} = 0.21\text{ mol } Na_2CO_3$$

$Na_2CO_3(s) \rightarrow 2\ Na^+(aq) + CO_3^{2-}(aq)$; mol Na^+ = 2(0.21) = 0.42 mol

$$\text{Mol } NaHCO_3 = 0.0300\text{ L} \times \frac{1.0\text{ mol } NaHCO_3}{\text{L}} = 0.030\text{ mol } NaHCO_3$$

$NaHCO_3(s) \rightarrow Na^+(aq) + HCO_3^-(aq)$; mol Na^+ = 0.030 mol

$$M_{Na^+} = \frac{\text{total mol } Na^+}{\text{total volume}} = \frac{0.42\text{ mol} + 0.030\text{ mol}}{0.0700\text{ L} + 0.030\text{ L}} = \frac{0.45\text{ mol}}{0.1000\text{ L}} = 4.5\ M\ Na^+$$

16. a. $HNO_3(l) \rightarrow H^+(aq) + NO_3^-(aq)$

b. $Na_2SO_4(s) \rightarrow 2\ Na^+(aq) + SO_4^{2-}(aq)$

c. $Al(NO_3)_3(s) \rightarrow Al^{3+}(aq) + 3\ NO_3^-(aq)$

d. $SrBr_2(s) \rightarrow Sr^{2+}(aq) + 2\ Br^-(aq)$

e. $KClO_4(s) \rightarrow K^+(aq) + ClO_4^-(aq)$

f. $NH_4Br(s) \rightarrow NH_4^+(aq) + Br^-(aq)$

g. $NH_4NO_3(s) \rightarrow NH_4^+(aq) + NO_3^-(aq)$

h. $CuSO_4(s) \rightarrow Cu^{2+}(aq) + SO_4^{2-}(aq)$

i. $NaOH(s) \rightarrow Na^+(aq) + OH^-(aq)$

Questions

17. As the temperature increases, the gas molecules will have a greater average kinetic energy. A greater fraction of the gas molecules in solution will have a kinetic energy greater than the attractive forces between the gas molecules and the solvent molecules. More gas molecules are able to escape to the vapor phase, and the solubility of the gas decreases.

18. Henry's law is obeyed most accurately for dilute solutions of gases that do not dissociate in or react with the solvent. NH_3 is a weak base and reacts with water by the following reaction:

$$NH_3(aq) + H_2O(l) \rightarrow NH_4^+(aq) + OH^-(aq)$$

O_2 will bind to hemoglobin in the blood. Due to these reactions in the solvent, $NH_3(g)$ in water and $O_2(g)$ in blood do not follow Henry's law.

19. Because the solute is volatile, both the water and solute will transfer back and forth between the two beakers. The volume in each beaker will become constant when the concentrations of solute in the beakers are equal to each other. Because the solute is less volatile than water, one would expect there to be a larger net transfer of water molecules into the right beaker than the net transfer of solute molecules into the left beaker. This results in a larger solution volume in the right beaker when equilibrium is reached, i.e., when the solute concentration is identical in each beaker.

20. Solutions of A and B have vapor pressures less than ideal (see Figure 11.13 of the text), so this plot shows negative deviations from Rault's law. Negative deviations occur when the intermolecular forces are stronger in solution than in pure solvent and solute. This results in an exothermic enthalpy of solution. The only statement that is false is e. A substance boils when the vapor pressure equals the external pressure. Because $\chi_B = 0.6$ has a lower vapor pressure at the temperature of the plot than either pure A or pure B, one would expect this solution to require the highest temperature in order for the vapor pressure to reach the external pressure. Therefore, the solution with $\chi_B = 0.6$ will have a higher boiling point than either pure A or pure B. (Note that because $P°_B > P°_A$, B is more volatile than A, and B will have a lower boiling point temperature than A).

21. No, the solution is not ideal. For an ideal solution, the strengths of intermolecular forces in solution are the same as in pure solute and pure solvent. This results in $\Delta H_{soln} = 0$ for an ideal solution. ΔH_{soln} for methanol-water is not zero. Because $\Delta H_{soln} < 0$ (heat is released), this solution shows a negative deviation from Raoult's law.

22. The micelles form so that the ionic ends of the detergent molecules, the SO_4^- ends, are exposed to the polar water molecules on the outside, whereas the nonpolar hydrocarbon chains from the detergent molecules are hidden from the water by pointing toward the inside of the micelle. Dirt, which is basically nonpolar, is stabilized in the nonpolar interior of the micelle and is washed away. See the illustration on the following page.

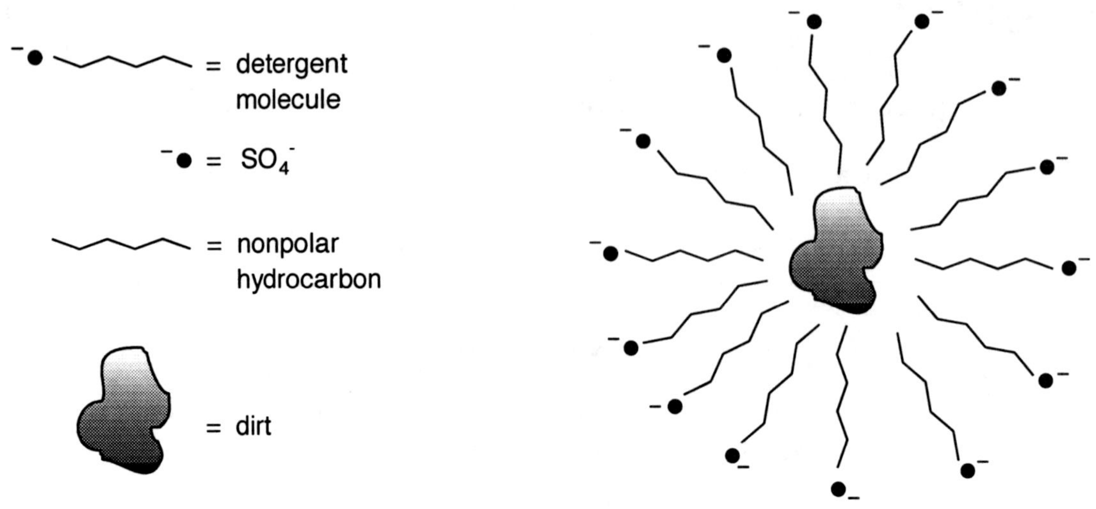

23. Normality is the number of equivalents per liter of solution. For an acid or a base, an equivalent is the mass of acid or base that can furnish 1 mole of protons (if an acid) or accept 1 mole of protons (if a base). A proton is an H^+ ion. Molarity is defined as the moles of solute per liter of solution. When the number of equivalents equals the number of moles of solute, then normality = molarity. This is true for acids which only have one acidic proton in them and for bases that accept only one proton per formula unit. Examples of acids where equivalents = moles solute are HCl, HNO_3, HF, and $HC_2H_3O_2$. Examples of bases where equivalents = moles solute are NaOH, KOH, and NH_3. When equivalents ≠ moles solute, then normality ≠ molarity. This is true for acids that donate more than one proton (H_2SO_4, H_3PO_4, H_2CO_3, etc.) and for bases that react with more than one proton per formula unit [$Ca(OH)_2$, $Ba(OH)_2$, $Sr(OH)_2$, etc.].

24. It is true that the sodium chloride lattice must be broken in order to dissolve in water, but a lot of energy is released when the water molecules hydrate the Na^+ and Cl^- ions. These two processes have relatively large values for the amount of energy associated with them, but they are opposite in sign. The end result is they basically cancel each other out resulting in a $\Delta H_{soln} \approx 0$. So energy is not the reason why ionic solids like NaCl are so soluble in water. The answer lies in nature's tendency toward the higher probability of the mixed state. Processes, in general, are favored that result in an increase in disorder because the disordered state is the easiest (most probable) state to achieve. The tendency of processes to increase disorder will be discussed in Chapter 17 when entropy, S, is introduced.

25. Only statement b is true. A substance freezes when the vapor pressure of the liquid and solid are the same. When a solute is added to water, the vapor pressure of the solution at 0°C is less than the vapor pressure of the solid, and the net result is for any ice present to convert to liquid in order to try to equalize the vapor pressures (which never can occur at 0°C). A lower temperature is needed to equalize the vapor pressure of water and ice, hence, the freezing point is depressed.

For statement a, the vapor pressure of a solution is directly related to the mole fraction of solvent (not solute) by Raoult's law. For statement c, colligative properties depend on the number of solute particles present and not on the identity of the solute. For statement d, the boiling point of water is increased because the sugar solute decreases the vapor pressure of the water; a higher temperature is required for the vapor pressure of the solution to equal the external pressure so boiling can occur.

26. This is true if the solute will dissolve in camphor. Camphor has the largest K_b and K_f constants. This means that camphor shows the largest change in boiling point and melting point as a solute is added. The larger the change in ΔT, the more precise the measurement and the more precise the calculated molar mass. However, if the solute won't dissolve in camphor, then camphor is no good and another solvent must be chosen which will dissolve the solute.

27. Isotonic solutions are those which have identical osmotic pressures. Crenation and hemolysis refer to phenomena that occur when red blood cells are bathed in solutions having a mismatch in osmotic pressures inside and outside the cell. When red blood cells are in a solution having a higher osmotic pressure than that of the cells, the cells shrivel as there is a net transfer of water out of the cells. This is called crenation. Hemolysis occurs when the red blood cells are bathed in a solution having lower osmotic pressure than that inside the cell. Here, the cells rupture as there is a net transfer of water to into the red blood cells.

28. Ion pairing is a phenomenon that occurs in solution when oppositely charged ions aggregate and behave as a single particle. For example, when NaCl is dissolved in water, one would expect sodium chloride to exist as separate hydrated Na^+ ions and Cl^- ions. A few ions, however, stay together as NaCl and behave as just one particle. Ion pairing increases in a solution as the ion concentration increases (as the molality increases).

Exercises

Solution Composition

29. Because the density of water is 1.00 g/mL, 100.0 mL of water has a mass of 100. g.

$$\text{Density} = \frac{\text{mass}}{\text{volume}} = \frac{10.0\text{ g H}_3\text{PO}_4 + 100.\text{ g H}_2\text{O}}{104\text{ mL}} = 1.06\text{ g/mL} = 1.06\text{ g/cm}^3$$

$$\text{Mol H}_3\text{PO}_4 = 10.0\text{ g} \times \frac{1\text{ mol}}{97.99\text{ g}} = 0.102\text{ mol H}_3\text{PO}_4$$

$$\text{Mol H}_2\text{O} = 100.\text{ g} \times \frac{1\text{ mol}}{18.02\text{ g}} = 5.55\text{ mol H}_2\text{O}$$

$$\text{Mole fraction of H}_3\text{PO}_4 = \frac{0.102\text{ mol H}_3\text{PO}_4}{(0.102 + 5.55)\text{ mol}} = 0.0180$$

$$\chi_{\text{H}_2\text{O}} = 1.000 - 0.0180 = 0.9820$$

$$\text{Molarity} = \frac{0.102\,\text{mol}\,H_3PO_4}{0.104\,\text{L}} = 0.981\ \text{mol/L}$$

$$\text{Molality} = \frac{0.102\,\text{mol}\,H_3PO_4}{0.100\,\text{kg}} = 1.02\ \text{mol/kg}$$

30. $$\text{Molality} = \frac{40.0\,\text{g}\,C_2H_6O_2}{60.0\,\text{g}\,H_2O} \times \frac{1000\,\text{g}}{\text{kg}} \times \frac{1\,\text{mol}\,C_2H_6O_2}{62.07\,\text{g}} = 10.7\ \text{mol/kg}$$

$$\text{Molarity} = \frac{40.0\,\text{g}\,C_2H_6O_2}{100.0\,\text{g solution}} \times \frac{1.05\,\text{g}}{\text{cm}^3} \times \frac{1000\,\text{cm}^3}{\text{L}} \times \frac{1\,\text{mol}\,C_2H_6O_2}{62.07\,\text{g}} = 6.77\ \text{mol/L}$$

$$40.0\ \text{g}\ C_2H_6O_2 \times \frac{1\,\text{mol}}{62.07\,\text{g}} = 0.644\ \text{mol}\ C_2H_6O_2;\quad 60.0\ \text{g}\ H_2O \times \frac{1\,\text{mol}}{18.02\,\text{g}} = 3.33\ \text{mol}\ H_2O$$

$$\chi_{EG} = \frac{0.644}{3.33 + 0.644} = 0.162 = \text{mole fraction ethylene glycol } (C_2H_6O_2)$$

31. Hydrochloric acid (HCl):

$$\text{molarity} = \frac{38\,\text{g HCl}}{100.\,\text{g soln}} \times \frac{1.19\,\text{g soln}}{\text{cm}^3\ \text{soln}} \times \frac{1000\,\text{cm}^3}{\text{L}} \times \frac{1\,\text{mol HCl}}{36.5\,\text{g}} = 12\ \text{mol/L}$$

$$\text{molality} = \frac{38\,\text{g HCl}}{62\,\text{g solvent}} \times \frac{1000\,\text{g}}{\text{kg}} \times \frac{1\,\text{mol HCl}}{36.5\,\text{g}} = 17\ \text{mol/kg}$$

$$38\ \text{g HCl} \times \frac{1\,\text{mol}}{36.5\,\text{g}} = 1.0\ \text{mol HCl};\quad 62\ \text{g}\ H_2O \times \frac{1\,\text{mol}}{18.0\,\text{g}} = 3.4\ \text{mol}\ H_2O$$

$$\text{mole fraction of HCl} = \chi_{HCl} = \frac{1.0}{3.4 + 1.0} = 0.23$$

Nitric acid (HNO_3):

$$\frac{70.\,\text{g}\,HNO_3}{100.\ \text{g soln}} \times \frac{1.42\,\text{g soln}}{\text{cm}^3\ \text{soln}} \times \frac{1000\,\text{cm}^3}{\text{L}} \times \frac{1\,\text{mol}\,HNO_3}{63.0\,\text{g}} = 16\ \text{mol/L}$$

$$\frac{70.\,\text{g}\,HNO_3}{30.\,\text{g solvent}} \times \frac{1000\,\text{g}}{\text{kg}} \times \frac{1\,\text{mol}\,HNO_3}{63.0\,\text{g}} = 37\ \text{mol/kg}$$

$$70.\ \text{g}\ HNO_3 \times \frac{1\,\text{mol}}{63.0\,\text{g}} = 1.1\ \text{mol}\ HNO_3;\quad 30.\ \text{g}\ H_2O \times \frac{1\,\text{mol}}{18.0\,\text{g}} = 1.7\ \text{mol}\ H_2O$$

$$\chi_{HNO_3} = \frac{1.1}{1.7 + 1.1} = 0.39$$

Sulfuric acid (H_2SO_4):

$$\frac{95\text{ g }H_2SO_4}{100.\text{ g soln}} \times \frac{1.84\text{ g soln}}{\text{cm}^3\text{ soln}} \times \frac{1000\text{cm}^3}{\text{L}} \times \frac{1\text{ mol }H_2SO_4}{98.1\text{ g }H_2SO_4} = 18\text{ mol/L}$$

$$\frac{95\text{ g }H_2SO_4}{5\text{ g }H_2O} \times \frac{1000\text{g}}{\text{kg}} \times \frac{1\text{ mol}}{98.1\text{ g}} = 194\text{ mol/kg} \approx 200\text{ mol/kg}$$

$$95\text{ g }H_2SO_4 \times \frac{1\text{ mol}}{98.1\text{ g}} = 0.97\text{ mol }H_2SO_4;\ \ 5\text{ g }H_2O \times \frac{1\text{ mol}}{18.0\text{ g}} = 0.3\text{ mol }H_2O$$

$$\chi_{H_2SO_4} = \frac{0.97}{0.97 + 0.3} = 0.76$$

Acetic acid (CH_3CO_2H):

$$\frac{99\text{ g }CH_3CO_2H}{100.\text{ g soln}} \times \frac{1.05\text{ g soln}}{\text{cm}^3\text{ soln}} \times \frac{1000\text{cm}^3}{\text{L}} \times \frac{1\text{ mol}}{60.05\text{ g}} = 17\text{ mol/L}$$

$$\frac{99\text{ g }CH_3CO_2H}{1\text{ g }H_2O} \times \frac{1000\text{g}}{\text{kg}} \times \frac{1\text{ mol}}{60.05\text{ g}} = 1600\text{ mol/kg} \approx 2000\text{ mol/kg}$$

$$99\text{ g }CH_3CO_2H \times \frac{1\text{ mol}}{60.05\text{ g}} = 1.6\text{ mol }CH_3CO_2H;\ \ 1\text{ g }H_2O \times \frac{1\text{ mol}}{18.0\text{ g}} = 0.06\text{ mol }H_2O$$

$$\chi_{CH_3CO_2H} = \frac{1.6}{1.6 + 0.06} = 0.96$$

Ammonia (NH_3):

$$\frac{28\text{ g }NH_3}{100.\text{ g soln}} \times \frac{0.90\text{ g}}{\text{cm}^3} \times \frac{1000\text{cm}^3}{\text{L}} \times \frac{1\text{ mol}}{17.0\text{ g}} = 15\text{ mol/L}$$

$$\frac{28\text{ g }NH_3}{72\text{ g }H_2O} \times \frac{1000\text{g}}{\text{kg}} \times \frac{1\text{ mol}}{17.0\text{ g}} = 23\text{ mol/kg}$$

$$28\text{ g }NH_3 \times \frac{1\text{ mol}}{17.0\text{ g}} = 1.6\text{ mol }NH_3;\ \ 72\text{ g }H_2O \times \frac{1\text{ mol}}{18.0\text{ g}} = 4.0\text{ mol }H_2O$$

$$\chi_{NH_3} = \frac{1.6}{4.0 + 1.6} = 0.29$$

32. a. If we use 100. mL (100. g) of H_2O, we need:

$$0.100\text{ kg }H_2O \times \frac{2.0\text{ mol KCl}}{\text{kg}} \times \frac{74.55\text{ g}}{\text{mol KCl}} = 14.9\text{ g} = 15\text{ g KCl}$$

Dissolve 15 g KCl in 100. mL H_2O to prepare a 2.0 *m* KCl solution. This will give us slightly more than 100 mL, but this will be the easiest way to make the solution. Because we don't know the density of the solution, we can't calculate the molarity and use a volumetric flask to make exactly 100 mL of solution.

b. If we took 15 g NaOH and 85 g H_2O, the volume probably would be less than 100 mL. To make sure we have enough solution, let's use 100. mL H_2O (100. g H_2O). Let x = mass of NaCl.

$$\text{Mass \%} = 15 = \frac{x}{100. + x} \times 100, \quad 1500 + 15x = (100.)x, \quad x = 17.6 \text{ g} \approx 18 \text{ g}$$

Dissolve 18 g NaOH in 100. mL H_2O to make a 15% NaOH solution by mass.

c. In a fashion similar to part b, let's use 100. mL CH_3OH. Let x = mass of NaOH.

$$100. \text{ mL } CH_3OH \times \frac{0.79 \text{ g}}{\text{mL}} = 79 \text{ g } CH_3OH$$

$$\text{Mass \%} = 25 = \frac{x}{79 + x} \times 100, \quad 25(79) + 25x = (100.)x, \quad x = 26.3 \text{ g} \approx 26 \text{ g}$$

Dissolve 26 g NaOH in 100. mL CH_3OH.

d. To make sure we have enough solution, let's use 100. mL (100. g) of H_2O. Let x = mol $C_6H_{12}O_6$.

$$100. \text{ g } H_2O \times \frac{1 \text{ mol } H_2O}{18.02 \text{ g}} = 5.55 \text{ mol } H_2O$$

$$\chi_{C_6H_{12}O_6} = 0.10 = \frac{x}{x + 5.55}, \quad (0.10)x + 0.56 = x, \quad x = 0.62 \text{ mol } C_6H_{12}O_6$$

$$0.62 \text{ mol } C_6H_{12}O_6 \times \frac{180.2 \text{ g}}{\text{mol}} = 110 \text{ g } C_6H_{12}O_6$$

Dissolve 110 g $C_6H_{12}O_6$ in 100. mL of H_2O to prepare a solution with $\chi_{C_6H_{12}O_6} = 0.10$.

33. $$25 \text{ mL } C_5H_{12} \times \frac{0.63 \text{ g}}{\text{mL}} = 16 \text{ g } C_5H_{12}; \; 25 \text{ mL} \times \frac{0.63 \text{ g}}{\text{mL}} \times \frac{1 \text{ mol}}{72.15 \text{ g}} = 0.22 \text{ mol } C_5H_{12}$$

$$45 \text{ mL } C_6H_{14} \times \frac{0.66 \text{ g}}{\text{mL}} = 30. \text{ g } C_6H_{14}; \; 45 \text{ mL} \times \frac{0.66 \text{ g}}{\text{mL}} \times \frac{1 \text{ mol}}{86.17 \text{ g}} = 0.34 \text{ mol } C_6H_{14}$$

$$\text{Mass \% pentane} = \frac{\text{mass pentane}}{\text{total mass}} \times 100 = \frac{16 \text{ g}}{16 \text{ g} + 30. \text{ g}} \times 100 = 35\%$$

$$\chi_{pentane} = \frac{\text{mol pentane}}{\text{total mol}} = \frac{0.22\text{ mol}}{0.22\text{ mol} + 0.34\text{ mol}} = 0.39$$

$$\text{Molality} = \frac{\text{mol pentane}}{\text{kg hexane}} = \frac{0.22\text{ mol}}{0.030\text{ kg}} = 7.3\text{ mol/kg}$$

$$\text{Molarity} = \frac{\text{mol pentane}}{\text{L solution}} = \frac{0.22\text{ mol}}{25\text{ mL} + 45\text{ mL}} \times \frac{1000\text{ mL}}{1\text{ L}} = 3.1\text{ mol/L}$$

34. $50.0\text{ mL toluene} \times \frac{0.867\text{ g}}{\text{mL}} = 43.4\text{ g toluene};\ 125\text{ mL benzene} \times \frac{0.874\text{ g}}{\text{mL}} = 109\text{ g benzene}$

$$\text{Mass \% toluene} = \frac{\text{mass of toluene}}{\text{total mass}} \times 100 = \frac{43.4\text{ g}}{43.4\text{ g} + 109\text{ g}} \times 100 = 28.5\%$$

$$\text{Molarity} = \frac{43.4\text{ g toluene}}{175\text{ mL soln}} \times \frac{1000\text{ mL}}{\text{L}} \times \frac{1\text{ mol toluene}}{92.13\text{ g toluene}} = 2.69\text{ mol/L}$$

$$\text{Molality} = \frac{43.4\text{ g toluene}}{109\text{ g benzene}} \times \frac{1000\text{ g}}{\text{kg}} \times \frac{1\text{ mol toluene}}{92.13\text{ g toluene}} = 4.32\text{ mol/kg}$$

$$43.4\text{ g toluene} \times \frac{1\text{ mol}}{92.13\text{ g}} = 0.471\text{ mol toluene}$$

$$109\text{ g benzene} \times \frac{1\text{ mol benzene}}{78.11\text{ g benzene}} = 1.40\text{ mol benzene};\quad \chi_{toluene} = \frac{0.471}{0.471 + 1.40} = 0.252$$

35. If we have 100.0 mL of wine:

$$12.5\text{ mL } C_2H_5OH \times \frac{0.789\text{ g}}{\text{mL}} = 9.86\text{ g } C_2H_5OH \text{ and } 87.5\text{ mL } H_2O \times \frac{1.00\text{ g}}{\text{mL}} = 87.5\text{ g } H_2O$$

$$\text{Mass \% ethanol} = \frac{9.86\text{ g}}{87.5\text{ g} + 9.86\text{ g}} \times 100 = 10.1\%\text{ by mass}$$

$$\text{Molality} = \frac{9.86\text{ g } C_2H_5OH}{0.0875\text{ kg } H_2O} \times \frac{1\text{ mol}}{46.07\text{ g}} = 2.45\text{ mol/kg}$$

36. $\frac{1.00\text{ mol acetone}}{1.00\text{ kg ethanol}} = 1.00\text{ molal};\ 1.00 \times 10^3\text{ g } C_2H_5OH \times \frac{1\text{ mol}}{46.07\text{ g}} = 21.7\text{ mol } C_2H_5OH$

$$\chi_{acetone} = \frac{1.00}{1.00 + 21.7} = 0.0441$$

$$1\text{ mol } CH_3COCH_3 \times \frac{58.08\text{ g } CH_3COCH_3}{\text{mol } CH_3COCH_3} \times \frac{1\text{ mL}}{0.788\text{ g}} = 73.7\text{ mL } CH_3COCH_3$$

$$1.00 \times 10^3 \text{ g ethanol} \times \frac{1\text{ mL}}{0.789\text{ g}} = 1270\text{ mL};\ \text{total volume} = 1270 + 73.7 = 1340\text{ mL}$$

$$\text{Molarity} = \frac{1.00\text{ mol}}{1.34\text{ L}} = 0.746\ M$$

37. If we have 1.00 L of solution:

$$1.37\text{ mol citric acid} \times \frac{192.12\text{ g}}{\text{mol}} = 263\text{ g citric acid } (H_3C_6H_5O_7)$$

$$1.00 \times 10^3\text{ mL solution} \times \frac{1.10\text{ g}}{\text{mL}} = 1.10 \times 10^3\text{ g solution}$$

$$\text{Mass \% of citric acid} = \frac{263\text{g}}{1.10 \times 10^3\text{ g}} \times 100 = 23.9\%$$

In 1.00 L of solution, we have 263 g citric acid and ($1.10 \times 10^3 - 263$) = 840 g of H_2O.

$$\text{Molality} = \frac{1.37\text{ mol citric acid}}{0.84\text{ kg } H_2O} = 1.6\text{ mol/kg}$$

$$840\text{ g } H_2O \times \frac{1\text{ mol}}{18.02\text{ g}} = 47\text{ mol } H_2O;\ \chi_{\text{citric acid}} = \frac{1.37}{47 + 1.37} = 0.028$$

Because citric acid is a triprotic acid, the number of protons citric acid can provide is three times the molarity. Therefore, normality = 3 × molarity:

normality = $3 \times 1.37\ M = 4.11\ N$

38. When expressing concentration in terms of normality, equivalents per liter are determined. For acid-base reactions, equivalents are equal to the moles of H^+ an acid can donate or the moles of H^+ a base can accept. For monoprotic acids like HCl, the equivalents of H^+ furnished equals the moles of acid present. Diprotic acids like H_2SO_4 furnish two equivalents of H^+ per mole of acid, whereas triprotic acids like H_3PO_4 furnish three equivalents of H^+ per mole of acid. For the bases in this problem, the equivalents of H^+ accepted equals the number of OH^- anions present in the formula ($H^+ + OH^- \rightarrow H_2O$). Finally, the equivalent mass of a substance is the mass of acid or base that can furnish or accept 1 mole of protons (H^+ ions).

a. $\text{Normality} = \frac{0.250\text{ mol HCl}}{\text{L}} \times \frac{1\text{ equivalent}}{\text{mol HCl}} = \frac{0.250\text{ equivalents}}{\text{L}}$

Equivalent mass = molar mass of HCl = 36.46 g

b. $\text{Normality} = \frac{0.105\text{ mol } H_2SO_4}{\text{L}} \times \frac{2\text{ equivalents}}{\text{mol } H_2SO_4} = \frac{0.210\text{ equivalents}}{\text{L}}$

Equivalent mass = 1/2(molar mass of H_2SO_4) = 1/2(98.09) = 49.05 g

c. $\text{Normality} = \dfrac{5.3 \times 10^{-2}\ \text{mol}\,H_3PO_4}{\text{L}} \times \dfrac{3\ \text{equivalents}}{\text{mol}\,H_3PO_4} = \dfrac{0.16\,\text{equivalents}}{\text{L}}$

Equivalent mass = 1/3(molar mass of H_3PO_4) = 1/3(97.09) = 32.66 g

d. $\text{Normality} = \dfrac{0.134\,\text{mol}\,\text{NaOH}}{\text{L}} \times \dfrac{1\ \text{equivalent}}{\text{mol}\,\text{NaOH}} = \dfrac{0.134\,\text{equivalents}}{\text{L}}$

Equivalent mass = molar mass of NaOH = 40.00 g

e. $\text{Normality} = \dfrac{0.00521\,\text{mol}\,\text{Ca(OH)}_2}{\text{L}} \times \dfrac{2\ \text{equivalents}}{\text{mol}\,\text{Ca(OH)}_2} = \dfrac{0.0104\,\text{equivalents}}{\text{L}}$

Equivalent mass = 1/2[molar mass of Ca(OH)_2] = 1/2(74.10) = 37.05 g

Energetics of Solutions and Solubility

39. Using Hess's law:

$$\begin{array}{ll} \text{NaI(s)} \rightarrow \text{Na}^+(g) + \text{I}^-(g) & \Delta H = -\Delta H_{LE} = -(-686\ \text{kJ/mol}) \\ \text{Na}^+(g) + \text{I}^-(g) \rightarrow \text{Na}^+(aq) + \text{I}^-(aq) & \Delta H = \Delta H_{hyd} = -694\ \text{kJ/mol} \\ \hline \text{NaI(s)} \rightarrow \text{Na}^+(aq) + \text{I}^-(aq) & \Delta H_{soln} = -8\ \text{kJ/mol} \end{array}$$

ΔH_{soln} refers to the heat released or gained when a solute dissolves in a solvent. Here, an ionic compound dissolves in water.

40. a.

$$\begin{array}{ll} \text{CaCl}_2(s) \rightarrow \text{Ca}^{2+}(g) + 2\ \text{Cl}^-(g) & \Delta H = -\Delta H_{LE} = -(-2247\ \text{kJ}) \\ \text{Ca}^{2+}(g) + 2\ \text{Cl}^-(g) \rightarrow \text{Ca}^{2+}(aq) + 2\ \text{Cl}^-(aq) & \Delta H = \Delta H_{hyd} \\ \hline \text{CaCl}_2(s) \rightarrow \text{Ca}^{2+}(aq) + 2\ \text{Cl}^-(aq) & \Delta H_{soln} = -46\ \text{kJ} \end{array}$$

$-46\ \text{kJ} = 2247\ \text{kJ} + \Delta H_{hyd},\ \Delta H_{hyd} -2293\ \text{kJ}$

$$\begin{array}{ll} \text{CaI}_2(s) \rightarrow \text{Ca}^{2+}(g) + 2\ \text{I}^-(g) & \Delta H = -\Delta H_{LE} = -(-2059\ \text{kJ}) \\ \text{Ca}^{2+}(g) + 2\ \text{I}^-(g) \rightarrow \text{Ca}^{2+}(aq) + 2\ \text{I}^-(aq) & \Delta H = \Delta H_{hyd} \\ \hline \text{CaI}_2(s) \rightarrow \text{Ca}^{2+}(aq) + 2\ \text{I}^-(aq) & \Delta H_{soln} = -104\ \text{kJ} \end{array}$$

$-104\ \text{kJ} = 2059\ \text{kJ} + \Delta H_{hyd},\ \Delta H_{hyd} = -2163\ \text{kJ}$

b. The enthalpy of hydration for $CaCl_2$ is more exothermic than for CaI_2. Any differences must be due to differences in hydration between Cl^- and I^-. Thus the chloride ion is more strongly hydrated than the iodide ion.

41. Both $Al(OH)_3$ and NaOH are ionic compounds. Since the lattice energy is proportional to the charge of the ions, the lattice energy of aluminum hydroxide is greater than that of sodium hydroxide. The attraction of water molecules for Al^{3+} and OH^- cannot overcome the larger

lattice energy, and $Al(OH)_3$ is insoluble. For NaOH, the favorable hydration energy is large enough to overcome the smaller lattice energy, and NaOH is soluble.

42. The dissolving of an ionic solute in water can be thought of as taking place in two steps. The first step, called the lattice-energy term, refers to breaking apart the ionic compound into gaseous ions. This step, as indicated in the problem, requires a lot of energy and is unfavorable. The second step, called the hydration-energy term, refers to the energy released when the separated gaseous ions are stabilized as water molecules surround the ions. Because the interactions between water molecules and ions are strong, a lot of energy is released when ions are hydrated. Thus the dissolution process for ionic compounds can be thought of as consisting of an unfavorable and a favorable energy term. These two processes basically cancel each other out, so when ionic solids dissolve in water, the heat released or gained is minimal, and the temperature change is minimal.

43. Water is a polar solvent and dissolves polar solutes and ionic solutes. Carbon tetrachloride (CCl_4) is a nonpolar solvent and dissolves nonpolar solutes (like dissolves like). To predict the polarity of the following molecules, draw the correct Lewis structure and then determine if the individual bond dipoles cancel or not. If the bond dipoles are arranged in such a manner that they cancel each other out, then the molecule is nonpolar. If the bond dipoles do not cancel each other out, then the molecule is polar.

a. KrF_2, $8 + 2(7) = 22\ e^-$

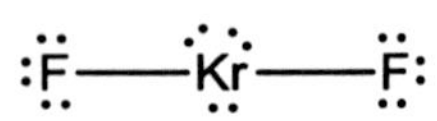

nonpolar; soluble in CCl_4

b. SF_2, $6 + 2(7) = 20\ e^-$

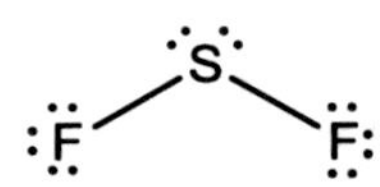

polar; soluble in H_2O

c. SO_2, $6 + 2(6) = 18\ e^-$

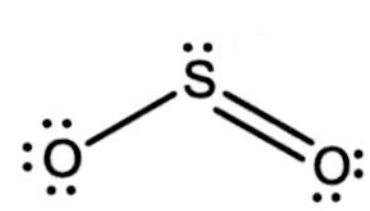

+ 1 more

polar; soluble in H_2O

d. CO_2, $4 + 2(6) = 16\ e^-$

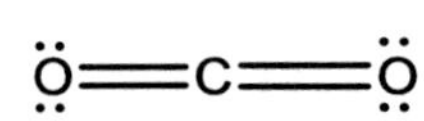

nonpolar; soluble in CCl_4

e. MgF_2 is an ionic compound so it is soluble in water.

f. CH_2O, $4 + 2(1) + 6 = 12\ e^-$

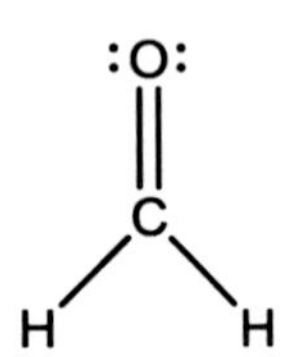

polar; soluble in H_2O

g. C_2H_4, $2(4) + 4(1) = 12\ e^-$

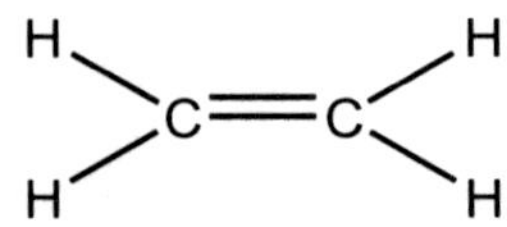

nonpolar (like all compounds made up of only carbon and hydrogen); soluble in CCl_4

44. Water is a polar solvent and dissolves polar solutes and ionic solutes. Hexane (C_6H_{14}) is a nonpolar solvent and dissolves nonpolar solutes (like dissolves like).

a. Water; $Cu(NO_3)_2$ is an ionic compound.

b. C_6H_{14}; CS_2 is a nonpolar molecule.

c. Water; CH_3OH is polar.

d. C_6H_{14}; the long nonpolar hydrocarbon chain favors a nonpolar solvent (the molecule is mostly nonpolar).

e. Water; HCl is polar.

f. C_6H_{14}; C_6H_6 is nonpolar.

45. Water exhibits H-bonding in the pure state and is classified as a polar solvent. Water will dissolve other polar solutes and ionic solutes.

a. NH_3; NH_3 is capable of H-bonding, unlike PH_3.

b. CH_3CN; CH_3CN is polar, while CH_3CH_3 is nonpolar.

c. CH_3CO_2H; CH_3CO_2H is capable of H-bonding, unlike the other compound.

46. For ionic compounds, as the charge of the ions increases and/or the size of the ions decreases, the attraction to water (hydration) increases.

a. Mg^{2+}; smaller size, higher charge

b. Be^{2+}; smaller size

c. Fe^{3+}; smaller size, higher charge

d. F^-; smaller size

e. Cl^-; smaller size

f. SO_4^{2-}; higher charge

47. As the length of the hydrocarbon chain increases, the solubility decreases. The –OH end of the alcohols can hydrogen-bond with water. The hydrocarbon chain, however, is basically nonpolar and interacts poorly with water. As the hydrocarbon chain gets longer, a greater portion of the molecule cannot interact with the water molecules, and the solubility decreases; i.e., the effect of the –OH group decreases as the alcohols get larger.

48. The main intermolecular forces are:

hexane (C_6H_{14}): London dispersion; chloroform ($CHCl_3$): dipole-dipole, London dispersion; methanol (CH_3OH): H-bonding; and H_2O: H-bonding (two places)

There is a gradual change in the nature of the intermolecular forces (weaker to stronger). Each preceding solvent is miscible in its predecessor because there is not a great change in the strengths of the intermolecular forces from one solvent to the next.

49. $C = kP$, $\dfrac{8.21 \times 10^{-4}\ \text{mol}}{\text{L}} = k \times 0.790\ \text{atm}$, $k = 1.04 \times 10^{-3}$ mol/L•atm

$C = kP,\ C = \dfrac{1.04 \times 10^{-4}\ \text{mol}}{\text{L atm}} \times 1.10\ \text{atm} = 1.14 \times 10^{-3}\ \text{mol/L}$

50. $C = kP = \dfrac{1.3 \times 10^{-3}\ \text{mol}}{\text{L atm}} \times 120\ \text{torr} \times \dfrac{1\ \text{atm}}{760\ \text{torr}} = 2.1 \times 10^{-4}\ \text{mol/L}$

Vapor Pressures of Solutions

51. $\text{Mol } C_3H_8O_3 = 164\ \text{g} \times \dfrac{1\ \text{mol}}{92.09\ \text{g}} = 1.78\ \text{mol } C_3H_8O_3$

$\text{Mol } H_2O = 338\ \text{mL} \times \dfrac{0.992\ \text{g}}{\text{mL}} \times \dfrac{1\ \text{mol}}{18.02\ \text{g}} = 18.6\ \text{mol } H_2O$

$P_{soln} = \chi_{H_2O} P^{o}_{H_2O} = \dfrac{18.6\ \text{mol}}{(1.78 + 18.6)\ \text{mol}} \times 54.74\ \text{torr} = 0.913 \times 54.74\ \text{torr} = 50.0\ \text{torr}$

52. $P_{soln} = \chi_{C_2H_5OH} P^{o}_{C_2H_5OH};\ \chi_{C_2H_5OH} = \dfrac{\text{moles of } C_2H_5OH}{\text{total moles in solution}}$

$53.6\ \text{g } C_3H_8O_3 \times \dfrac{1\ \text{mol } C_3H_8O_3}{92.09\ \text{g}} = 0.582\ \text{mol } C_3H_8O_3$

$133.7\ \text{g } C_2H_5OH \times \dfrac{1\ \text{mol } C_2H_5OH}{46.07\ \text{g}} = 2.90\ \text{mol } C_2H_5OH;$ total mol = 0.582 + 2.90

= 3.48 mol

$113\ \text{torr} = \dfrac{2.90\ \text{mol}}{3.48\ \text{mol}} \times P^{o}_{C_2H_5OH},\ P^{o}_{C_2H_5OH} = 136\ \text{torr}$

53. The normal boiling point of a substance is the boiling point at 1 atm pressure. So for this problem, P° = 760. torr at 34.5°C (the normal boiling point of diethyl ether).

$P = \chi P°;\ 698\ \text{torr} = \chi(760.\ \text{torr}),\ \chi = 0.918$ = mole fraction of diethyl ether

54. $P_B = \chi_B P^{o}_B,\ \chi_B = P_B / P^{o}_B = 0.900\ \text{atm}/0.930\ \text{atm} = 0.968$

$0.968 = \dfrac{\text{mol benzene}}{\text{total mol}};$ mol benzene $= 78.11\ \text{g } C_6H_6 \times \dfrac{1\ \text{mol}}{78.11\ \text{g}} = 1.000\ \text{mol}$

Let x = mol solute, then: $\chi_B = 0.968 = \dfrac{1.000\ \text{mol}}{1.000 + x},\quad 0.968 + (0.968)x = 1.000,\ x = 0.033\ \text{mol}$

$\text{Molar mass} = \dfrac{10.0\ \text{g}}{0.033\ \text{mol}} = 303\ \text{g/mol} \approx 3.0 \times 10^2\ \text{g/mol}$

55. $25.8 \text{ g } CH_4N_2O \times \frac{1 \text{ mol}}{60.06 \text{ g}} = 0.430 \text{ mol}$; $275 \text{ g } H_2O \times \frac{1 \text{ mol}}{18.02 \text{ g}} = 15.3 \text{ mol}$

$$\chi_{H_2O} = \frac{15.3}{15.3 + 0.430} = 0.973; \quad P_{soln} = \chi_{H_2O}P^{o}_{H_2O} = 0.973(23.8 \text{ torr}) = 23.2 \text{ torr at } 25°C$$

$P_{soln} = 0.973(71.9 \text{ torr}) = 70.0 \text{ torr at } 45°C$

56. $19.6 \text{ torr} = \chi_{H_2O}(23.8 \text{ torr})$, $\chi_{H_2O} = 0.824$; $\chi_{solute} = 1.000 - 0.824 = 0.176$

0.176 is the mol fraction of all the solute particles present. Because NaCl dissolves to produce two ions in solution (Na^+ and Cl^-), 0.176 is the mole fraction of Na^+ and Cl^- ions present (assuming complete dissociation of NaCl).

At 45°C, $P_{H_2O} = 0.824(71.9 \text{ torr}) = 59.2 \text{ torr}$

57. a. $25 \text{ mL } C_5H_{12} \times \frac{0.63 \text{ g}}{\text{mL}} \times \frac{1 \text{ mol}}{72.15 \text{ g}} = 0.22 \text{ mol } C_5H_{12}$

$45 \text{ mL } C_6H_{14} \times \frac{0.66 \text{ g}}{\text{mL}} \times \frac{1 \text{ mol}}{86.17 \text{ g}} = 0.34 \text{ mol } C_6H_{14}$; total mol = 0.22 + 0.34 = 0.56 mol

$$\chi^{L}_{pen} = \frac{\text{mol pentane in solution}}{\text{total mol in solution}} = \frac{0.22 \text{ mol}}{0.56 \text{ mol}} = 0.39, \quad \chi^{L}_{hex} = 1.00 - 0.39 = 0.61$$

$P_{pen} = \chi^{L}_{pen}P^{o}_{pen} = 0.39(511 \text{ torr}) = 2.0 \times 10^2 \text{ torr}$; $P_{hex} = 0.61(150. \text{ torr}) = 92 \text{ torr}$

$P_{total} = P_{pen} + P_{hex} = 2.0 \times 10^2 + 92 = 292 \text{ torr} = 290 \text{ torr}$

b. From Chapter 5 on gases, the partial pressure of a gas is proportional to the number of moles of gas present (at constant volume and temperature). For the vapor phase:

$$\chi^{V}_{pen} = \frac{\text{mol pentane in vapor}}{\text{total mol vapor}} = \frac{P_{pen}}{P_{total}} = \frac{2.0 \times 10^2 \text{ torr}}{290 \text{ torr}} = 0.69$$

Note: In the *Solutions Guide*, we added V or L superscripts to the mole fraction symbol to emphasize for which value we are solving. If the L or V is omitted, then the liquid phase is assumed.

58. $P_{total} = P_{CH_2Cl_2} + P_{CH_2Br_2}$; $P = \chi^{L}P^{o}$; $\chi^{L}_{CH_2Cl_2} = \frac{0.0300 \text{ mol } CH_2Cl_2}{0.0800 \text{ mol total}} = 0.375$

$P_{total} = 0.375(133 \text{ torr}) + (1.000 - 0.375)(11.4 \text{ torr}) = 49.9 + 7.13 = 57.0 \text{ torr}$

In the vapor: $\chi^{V}_{CH_2Cl_2} = \frac{P_{CH_2Cl_2}}{P_{total}} = \frac{49.9 \text{ torr}}{57.0 \text{ torr}} = 0.875$; $\chi^{V}_{CH_2Br_2} = 1.000 - 0.875 = 0.125$

Note: In the *Solutions Guide*, we added V or L superscripts to the mole fraction symbol to emphasize for which value we are solving. If the L or V is omitted, then the liquid phase is assumed.

59. $P_{total} = P_{meth} + P_{prop}$, $174 \text{ torr} = \chi^L_{meth}(303 \text{ torr}) + \chi^L_{prop}(44.6 \text{ torr})$; $\chi^L_{prop} = 1.000 - \chi^L_{meth}$

$$174 = 303\chi^L_{meth} + (1.000 - \chi^L_{meth})44.6 \text{ torr}, \ \frac{129}{258} = \chi^L_{meth} = 0.500$$

$$\chi^L_{prop} = 1.000 - 0.500 = 0.500$$

60. $P_{tol} = \chi^L_{tol}P^o_{tol}$, $P_{pen} = \chi^L_{ben}P^o_{ben}$; for the vapor, $\chi^V_A = P_A/P_{total}$. Because the mole fractions of benzene and toluene are equal in the vapor phase, $P_{tol} = P_{ben}$.

$$\chi^L_{tol}P^o_{tol} = \chi^L_{ben}P^o_{ben} = (1.00 - \chi^L_{tol})P^o_{ben}, \ \chi^L_{tol}(28 \text{ torr}) = (1.00 - \chi^L_{tol})95 \text{ torr}$$

$$123\chi^L_{tol} = 95, \ \chi^L_{tol} = 0.77; \ \ \chi^L_{ben} = 1.00 - 0.77 = 0.23$$

61. Compared to H_2O, solution d (methanol-water) will have the highest vapor pressure since methanol is more volatile than water ($P^o_{H_2O}$ = 23.8 torr at 25°C). Both solution b (glucose-water) and solution c (NaCl-water) will have a lower vapor pressure than water by Raoult's law. NaCl dissolves to give Na^+ ions and Cl^- ions; glucose is a nonelectrolyte. Because there are more solute particles in solution c, the vapor pressure of solution c will be the lowest.

62. Solution d (methanol-water); methanol is more volatile than water, which will increase the total vapor pressure to a value greater than the vapor pressure of pure water at this temperature.

63. The first diagram shows positive deviation from Raoult's law. This occurs when the solute-solvent interactions are weaker than the interactions in pure solvent and pure solute. The second diagram illustrates negative deviation from Raoult's law. This occurs when the solute-solvent interactions are stronger than the interactions in pure solvent and pure solute. The third diagram illustrates an ideal solution with no deviation from Raoult's law. This occurs when the solute-solvent interactions are about equal to the pure solvent and pure solute interactions.

a. These two molecules are named acetone (CH_3COCH_3) and water. As discussed in section 11.4 on nonideal solutions, acetone-water solutions exhibit negative deviations from Raoult's law. Acetone and water have the ability to hydrogen bond with each other, which gives the solution stronger intermolecular forces as compared to the pure states of both solute and solvent. In the pure state, acetone cannot H–bond with itself. So the middle diagram illustrating negative deviations from Raoult's law is the correct choice for acetone-water solutions.

b. These two molecules are named ethanol (CH_3CH_2OH) and water. Ethanol-water solutions show positive deviations from Raoult's law. Both substances can hydrogen bond in the pure state, and they can continue this in solution. However, the solute-solvent interactions are somewhat weaker for ethanol-water solutions due to the significant nonpolar part of ethanol (CH_3–CH_2 is the nonpolar part of ethanol). This nonpolar part of ethanol weakens the intermolecular forces in solution. So the first diagram illustrating positive deviations from Raoult's law is the correct choice for ethanol-water solutions.

c. These two molecules are named heptane (C_7H_{16}) and hexane (C_6H_{14}). Heptane and hexane are very similar nonpolar substances; both are composed entirely of nonpolar C–C bonds and relatively nonpolar C–H bonds, and both have a similar size and shape. Solutions of heptane and hexane should be ideal. So the third diagram illustrating no deviation from Raoult's law is the correct choice for heptane-hexane solutions.

d. These two molecules are named heptane (C_7H_{16}) and water. The interactions between the nonpolar heptane molecules and the polar water molecules will certainly be weaker in solution as compared to the pure solvent and pure solute interactions. This results in positive deviations from Raoult's law (the first diagram).

64. a. An ideal solution would have a vapor pressure at any mole fraction of H_2O between that of pure propanol and pure water (between 74.0 and 71.9 torr). The vapor pressures of the various solutions are not between these limits, so water and propanol do not form ideal solutions.

b. From the data, the vapor pressures of the various solutions are greater than if the solutions behaved ideally (positive deviation from Raoult's law). This occurs when the intermolecular forces in solution are weaker than the intermolecular forces in pure solvent and pure solute. This gives rise to endothermic (positive) ΔH_{soln} values.

c. The interactions between propanol and water molecules are weaker than between the pure substances because the solutions exhibit a positive deviation from Raoult's law.

d. At $\chi_{H_2O} = 0.54$, the vapor pressure is highest as compared to the other solutions. Because a solution boils when the vapor pressure of the solution equals the external pressure, the $\chi_{H_2O} = 0.54$ solution should have the lowest normal boiling point; this solution will have a vapor pressure equal to 1 atm at a lower temperature as compared to the other solutions.

Colligative Properties

65. $$\text{Molality} = m = \frac{\text{mol solute}}{\text{kg solvent}} = \frac{27.0\text{g } N_2H_4CO}{150.0\text{g } H_2O} \times \frac{1000\text{g}}{\text{kg}} \times \frac{1\text{ mol } N_2H_4CO}{60.06\text{g } N_2H_4CO} = 3.00 \text{ molal}$$

$$\Delta T_b = K_b m = \frac{0.51°C}{\text{molal}} \times 3.00 \text{ molal} = 1.5°C$$

The boiling point is raised from 100.0 to 101.5°C (assuming P = 1 atm).

66. $$\Delta T_b = 77.85°C - 76.50°C = 1.35°C;\quad m = \frac{\Delta T_b}{K_b} = \frac{1.35°C}{5.03\,°C\text{ kg/mol}} = 0.268 \text{ mol/kg}$$

$$\text{Mol biomolecule} = 0.0150 \text{ kg solvent} \times \frac{0.268 \text{ mol hydrocarbon}}{\text{kg solvent}} = 4.02 \times 10^{-3} \text{ mol}$$

From the problem, 2.00 g biomolecule was used that must contain 4.02×10^{-3} mol biomolecule. The molar mass of the biomolecule is:

$$\frac{2.00\text{ g}}{4.02 \times 10^{-3}\text{ mol}} = 498\text{ g/mol}$$

67. $\Delta T_f = K_f m,\ \Delta T_f = 1.50°C = \frac{1.86\,°C}{\text{molal}} \times m,\ m = 0.806$ mol/kg

$$0.200\text{ kg } H_2O \times \frac{0.806\text{ mol } C_3H_8O_3}{\text{kg } H_2O} \times \frac{92.09\text{ g } C_3H_8O_3}{\text{mol } C_3H_8O_3} = 14.8\text{ g } C_3H_8O_3$$

68. $\Delta T_f = 25.50°C - 24.59°C = 0.91°C = K_f m,\ m = \frac{0.91\,°C}{9.1\,°C/\text{molal}} = 0.10$ mol/kg

$$\text{Mass } H_2O = 0.0100\text{ kg t-butanol} \times \frac{0.10\text{ mol } H_2O}{\text{kg t-butanol}} \times \frac{18.02\text{ g } H_2O}{\text{mol } H_2O} = 0.018\text{ g } H_2O$$

69. $\text{Molality} = m = \frac{50.0\text{ g } C_2H_6O_2}{50.0\text{ g } H_2O} \times \frac{1000\text{ g}}{\text{kg}} \times \frac{1\text{ mol}}{62.07\text{ g}} = 16.1$ mol/kg

$\Delta T_f = K_f m = 1.86°C/\text{molal} \times 16.1\text{ molal} = 29.9°C;\ T_f = 0.0°C - 29.9°C = -29.9°C$

$\Delta T_b = K_b m = 0.51°C/\text{molal} \times 16.1\text{ molal} = 8.2°C;\ T_b = 100.0°C + 8.2°C = 108.2°C$

70. $m = \frac{\Delta T_f}{K_f} = \frac{25.0°C}{1.86\,°C\text{ kg/mol}} = 13.4$ mol $C_2H_6O_2$/kg

Because the density of water is 1.00 g/cm^3, the moles of $C_2H_6O_2$ needed are:

$$15.0\text{ L } H_2O \times \frac{1.00\text{ kg } H_2O}{\text{L } H_2O} \times \frac{13.4\text{ mol } C_2H_6O_2}{\text{kg } H_2O} = 201\text{ mol } C_2H_6O_2$$

$$\text{Volume } C_2H_6O_2 = 201\text{ mol } C_2H_6O_2 \times \frac{62.07\text{ g}}{\text{mol } C_2H_6O_2} \times \frac{1\text{ cm}^3}{1.11\text{ g}} = 11{,}200\text{ cm}^3 = 11.2\text{ L}$$

$$\Delta T_b = K_b m = \frac{0.51°C}{\text{molal}} \times 13.4\text{ molal} = 6.8°C;\quad T_b = 100.0°C + 6.8°C = 106.8°C$$

71. $\Delta T_f = K_f m,\ m = \frac{\Delta T_f}{K_f} = \frac{2.63°C}{40.\,°C\text{ kg/mol}} = \frac{6.6 \times 10^{-2}\text{ mol reserpine}}{\text{kg solvent}}$

The moles of reserpine present is:

$$0.0250\text{ kg solvent} \times \frac{6.6 \times 10^{-2}\text{ mol reserpine}}{\text{kg solvent}} = 1.7 \times 10^{-3}\text{ mol reserpine}$$

From the problem, 1.00 g reserpine was used, which must contain 1.7×10^{-3} mol reserpine. The molar mass of reserpine is:

$$\frac{1.00\text{g}}{1.7 \times 10^{-3}\text{ mol}} = 590 \text{ g/mol (610 g/mol if no rounding of numbers)}$$

72. $$m = \frac{\Delta T_b}{K_b} = \frac{0.55°\text{C}}{1.71\,°\text{C kg/mol}} = 0.32 \text{ mol/kg}$$

$$\text{Mol hydrocarbon} = 0.095 \text{ kg solvent} \times \frac{0.32 \text{ mol hydrocarbon}}{\text{kg solvent}} = 0.030 \text{ mol hydrocarbon}$$

From the problem, 3.75 g hydrocarbon was used, which must contain 0.030 mol hydrocarbon. The molar mass of the hydrocarbon is:

$$\frac{3.75\text{g}}{0.030\text{mol}} = 130 \text{ g/mol (120 g/mol if no rounding of numbers)}$$

73. a. $$M = \frac{1.0\text{g protein}}{\text{L}} \times \frac{1\text{ mol}}{9.0 \times 10^4\text{ g}} = 1.1 \times 10^{-5}\text{mol/L};\ \pi = MRT$$

$$\text{At 298 K: } \pi = \frac{1.1 \times 10^{-5}\text{ mol}}{\text{L}} \times \frac{0.08206\text{L atm}}{\text{K mol}} \times 298\text{ K} \times \frac{760\text{ torr}}{\text{atm}},\ \pi = 0.20\text{ torr}$$

Because d = 1.0 g/cm³, 1.0 L solution has a mass of 1.0 kg. Because only 1.0 g of protein is present per liter of solution, 1.0 kg of H_2O is present to the correct number of significant figures, and molality equals molarity.

$$\Delta T_f = K_f m = \frac{1.86°\text{C}}{\text{molal}} \times 1.1 \times 10^{-5}\text{molal} = 2.0 \times 10^{-5}\,°\text{C}$$

b. Osmotic pressure is better for determining the molar mass of large molecules. A temperature change of 10^{-5} °C is very difficult to measure. A change in height of a column of mercury by 0.2 mm (0.2 torr) is not as hard to measure precisely.

74. $$m = \frac{\Delta T}{K_f} = \frac{0.406°\text{C}}{1.86\,°\text{C/molal}} = 0.218 \text{ mol/kg}$$

$\pi = MRT$, where M = mol/L; we must assume that molarity = molality so that we can calculate the osmotic pressure. This is a reasonable assumption for dilute solutions when 1.00 kg of water ≈ 1.00 L of solution. Assuming NaCl exists as Na^+ and Cl^- ions in solution, a 0.218 m solution corresponds to 6.37 g NaCl dissolved in 1.00 kg of water. The volume of solution may be a little larger than 1.00 L but not by much (to three sig. figs.). The assumption that molarity = molality will be good here.

$$\pi = (0.218\ M)(0.08206 \text{ L atm/K} \cdot \text{mol})(298 \text{ K}) = 5.33 \text{ atm}$$

75. $$M = \frac{\pi}{RT} = \frac{0.745\,\text{torr} \times \dfrac{1\,\text{atm}}{760\,\text{torr}}}{\dfrac{0.08206\,\text{L atm}}{\text{K mol}} \times 300.\,\text{K}} = 3.98 \times 10^{-5}\ \text{mol/L}$$

$$1.00\ \text{L} \times \frac{3.98 \times 10^{-5}\ \text{mol}}{\text{L}} = 3.98 \times 10^{-5}\ \text{mol catalase}$$

$$\text{Molar mass} = \frac{10.00\,\text{g}}{3.98 \times 10^{-5}\ \text{mol}} = 2.51 \times 10^{5}\ \text{g/mol}$$

76. $$\pi = MRT,\ \pi = 18.6\ \text{torr} \times \frac{1\,\text{atm}}{760\,\text{torr}} = M \times \frac{0.08206\,\text{L atm}}{\text{K mol}} \times 298\ \text{K},\ M = 1.00 \times 10^{-3}\ \text{mol/L}$$

$$\text{Mol protein} = 0.0020\ \text{L} \times \frac{1.00 \times 10^{-3}\ \text{mol protein}}{\text{L}} = 2.0 \times 10^{-6}\ \text{mol protein}$$

$$\text{Molar mass} = \frac{0.15\,\text{g}}{2.0 \times 10^{-6}\ \text{mol}} = 7.5 \times 10^{4}\ \text{g/mol}$$

77. $$\pi = MRT,\ M = \frac{\pi}{RT} = \frac{15\,\text{atm}}{\dfrac{0.08206\,\text{L atm}}{\text{K mol}} \times 295\,\text{K}} = 0.62\ M$$

$$\frac{0.62\,\text{mol}}{\text{L}} \times \frac{342.30\,\text{g}}{\text{mol}\,C_{12}H_{22}O_{11}} = 212\ \text{g/L} \approx 210\ \text{g/L}$$

Dissolve 210 g of sucrose in some water and dilute to 1.0 L in a volumetric flask. To get 0.62 ±0.01 mol/L, we need 212 ±3 g sucrose.

78. $$M = \frac{\pi}{RT} = \frac{15\,\text{atm}}{\dfrac{0.08206\,\text{L atm}}{\text{K mol}} \times 295\,\text{K}} = 0.62\ M\ \text{solute particles}$$

This represents the total molarity of the solute particles. NaCl is a soluble ionic compound that breaks up into two ions, Na^+ and Cl^-. Therefore, the concentration of NaCl needed is 0.62/2 = 0.31 *M*; this NaCl concentration will produce a 0.62 *M* solute particle solution assuming complete dissociation.

$$1.0\ \text{L} \times \frac{0.31\,\text{mol NaCl}}{\text{L}} \times \frac{58.44\ \text{g NaCl}}{\text{mol NaCl}} = 18.1 \approx 18\ \text{g NaCl}$$

Dissolve 18 g of NaCl in some water and dilute to 1.0 L in a volumetric flask. To get 0.31 ±0.01 mol/L, we need 18.1 g ±0.6 g NaCl in 1.00 L solution.

Properties of Electrolyte Solutions

79. $Na_3PO_4(s) \rightarrow 3\ Na^+(aq) + PO_4^{3-}(aq)$, i = 4.0; $CaBr_2(s) \rightarrow Ca^{2+}(aq) + 2\ Br^-(aq)$, i = 3.0

$KCl(s) \rightarrow K^+(aq) + Cl^-(aq)$, i = 2.0

The effective particle concentrations of the solutions are (assuming complete dissociation):

4.0(0.010 molal) = 0.040 molal for the Na_3PO_4 solution; 3.0(0.020 molal) = 0.060 molal for the $CaBr_2$ solution; 2.0(0.020 molal) = 0.040 molal for the KCl solution; slightly greater than 0.020 molal for the HF solution because HF only partially dissociates in water (it is a weak acid).

a. The 0.010 *m* Na_3PO_4 solution and the 0.020 *m* KCl solution both have effective particle concentrations of 0.040 *m* (assuming complete dissociation), so both of these solutions should have the same boiling point as the 0.040 *m* $C_6H_{12}O_6$ solution (a nonelectrolyte).

b. $P = \chi P°$; as the solute concentration decreases, the solvent's vapor pressure increases because χ increases. Therefore, the 0.020 *m* HF solution will have the highest vapor pressure because it has the smallest effective particle concentration.

c. $\Delta T = K_f m$; the 0.020 *m* $CaBr_2$ solution has the largest effective particle concentration, so it will have the largest freezing point depression (largest ΔT).

80. The solutions of $C_{12}H_{22}O_{11}$, NaCl, and $CaCl_2$ will all have lower freezing points, higher boiling points, and higher osmotic pressures than pure water. The solution with the largest particle concentration will have the lowest freezing point, the highest boiling point, and the highest osmotic pressure. The $CaCl_2$ solution will have the largest effective particle concentration because it produces three ions per mole of compound.

a. pure water b. $CaCl_2$ solution c. $CaCl_2$ solution

d. pure water e. $CaCl_2$ solution

81. a. $m = \dfrac{5.0\text{ g NaCl}}{0.025\text{kg}} \times \dfrac{1\text{ mol}}{58.44\text{ g}} = 3.4$ molal; $NaCl(aq) \rightarrow Na^+(aq) + Cl^-(aq)$, i = 2.0

$\Delta T_f = iK_f m = 2.0 \times 1.86°C/\text{molal} \times 3.4\text{ molal} = 13°C$; $T_f = -13°C$

$\Delta T_b = iK_b m = 2.0 \times 0.51°C/\text{molal} \times 3.4\text{ molal} = 3.5°C$; $T_b = 103.5°C$

b. $m = \dfrac{2.0\text{ g Al(NO}_3)_3}{0.015\text{kg}} \times \dfrac{1\text{ mol}}{213.01\text{ g}} = 0.63\text{ mol/kg}$

$Al(NO_3)_3(aq) \rightarrow Al^{3+}(aq) + 3\ NO_3^-(aq)$, i = 4.0

$\Delta T_f = iK_f m = 4.0 \times 1.86°C/\text{molal} \times 0.63\text{ molal} = 4.7°C$; $T_f = -4.7°C$

$\Delta T_b = iK_b m = 4.0 \times 0.51°C/\text{molal} \times 0.63\text{ molal} = 1.3°C$; $T_b = 101.3°C$

82. $NaCl(s) \rightarrow Na^+(aq) + Cl^-(aq)$, i = 2.0

$$\pi = iMRT = 2.0 \times \frac{0.10\,\text{mol}}{\text{L}} \times \frac{0.08206\,\text{L atm}}{\text{K mol}} \times 293\text{ K} = 4.8\text{ atm}$$

A pressure greater than 4.8 atm should be applied to ensure purification by reverse osmosis.

83. There are six cations and six anions in the illustration which indicates six solute formula units initially. There are a total of 10 solute particles in solution (a combined ion pair counts as one solute particle). So the value for the van't Hoff factor is:

$$i = \frac{\text{moles of particles in solution}}{\text{moles of solute dissolved}} = \frac{10}{6} = 1.67$$

84. From Table 11.6, $MgSO_4$ has an observed i value of 1.3 while the observed i value for NaCl is 1.9. Both salts have an expected i value of 2. The expected i value for a salt is determined by assuming 100% of the salt breaks up into separate cations and anions. The $MgSO_4$ solution is furthest from the expected i value because it forms the most combined ion pairs in solution. So the figure on the left with the most combined ion pairs represents the $MgSO_4$ solution. The figure on the right represents the NaCl solution. When NaCl is in solution, it has very few combined ion pairs and, hence, has a van't Hoff factor very close to the expected i value.

85. a. $MgCl_2(s) \rightarrow Mg^{2+}(aq) + 2\,Cl^-(aq)$, i = 3.0 mol ions/mol solute

$\Delta T_f = iK_fm = 3.0 \times 1.86$ °C/molal × 0.050 molal = 0.28°C

Assuming water freezes at 0.00°C, the freezing point would be –0.28°C.

$\Delta T_b = iK_bm = 3.0 \times 0.51$ °C/molal × 0.050 molal = 0.077°C; T_b = 100.077°C (Assuming water boils at 100.000°C.)

b. $FeCl_3(s) \rightarrow Fe^{3+}(aq) + 3\,Cl^-(aq)$, i = 4.0 mol ions/mol solute

$\Delta T_f = iK_fm = 4.0 \times 1.86$ °C/molal × 0.050 molal = 0.37°C; T_f = –0.37°C

$\Delta T_b = iK_bm = 4.0 \times 0.51$ °C/molal × 0.050 molal = 0.10°C; T_b = 100.10°C

86. a. $MgCl_2$, i (observed) = 2.7

$\Delta T_f = iK_fm = 2.7 \times 1.86$ °C/molal × 0.050 molal = 0.25°C; T_f = –0.25°C

$\Delta T_b = iK_bm = 2.7 \times 0.51$ °C/molal × 0.050 molal = 0.069°C; T_b = 100.069°C

b. $FeCl_3$, i (observed) = 3.4

$\Delta T_f = iK_fm = 3.4 \times 1.86$ °C/molal × 0.050 molal = 0.32°C; T_f = –0.32°C

$\Delta T_b = iK_bm = 3.4 \times 0.51$°C/molal × 0.050 molal = 0.087°C; T_b = 100.087°C

87. $\Delta T_f = iK_f m,\ i = \dfrac{\Delta T_f}{K_f m} = \dfrac{0.110°C}{1.86°C/\text{molal} \times 0.0225\text{molal}} = 2.63$ for 0.0225 m $CaCl_2$

$i = \dfrac{0.440}{1.86 \times 0.0910} = 2.60$ for 0.0910 m $CaCl_2$; $i = \dfrac{1.330}{1.86 \times 0.278} = 2.57$ for 0.278 m $CaCl_2$

$i_{ave} = (2.63 + 2.60 + 2.57)/3 = 2.60$

Note that i is less than the ideal value of 3.0 for $CaCl_2$. This is due to ion pairing in solution. Also note that as molality increases, i decreases. More ion pairing appears to occur as the solute concentration increases.

88. For $CaCl_2$: $i = \dfrac{\Delta T_f}{K_f m} = \dfrac{0.440°C}{1.86\ °C/\text{molal} \times 0.091\text{molal}} = 2.6$

Percent $CaCl_2$ ionized $= \dfrac{2.6 - 1.0}{3.0 - 1.0} \times 100 = 80.\%$; 20.% ion association occurs.

For CsCl: $i = \dfrac{\Delta T_f}{K_f m} = \dfrac{0.320°C}{1.86\ °C/\text{molal} \times 0.091\text{molal}} = 1.9$

Percent CsCl ionized $= \dfrac{1.9 - 1.0}{2.0 - 1.0} \times 100 = 90.\%$; 10% ion association occurs.

The ion association is greater in the $CaCl_2$ solution.

89. a. $T_C = 5(T_F - 32)/9 = 5(-29 - 32)/9 = -34°C$

Assuming the solubility of $CaCl_2$ is temperature independent, the molality of a saturated $CaCl_2$ solution is:

$$\frac{74.5\text{ g }CaCl_2}{100.0\text{ g }H_2O} \times \frac{1000\text{g}}{\text{kg}} \times \frac{1\text{ mol}CaCl_2}{110.98\text{ g }CaCl_2} = \frac{6.71\text{ mol}CaCl_2}{\text{kg }H_2O}$$

$\Delta T_f = iK_f m = 3.00 \times 1.86\ °C\ \text{kg/mol} \times 6.71\ \text{mol/kg} = 37.4°C$

Assuming i = 3.00, a saturated solution of $CaCl_2$ can lower the freezing point of water to −37.4°C. Assuming these conditions, a saturated $CaCl_2$ solution should melt ice at −34°C (−29°F).

b. From Exercise 87, $i \approx 2.6$; $\Delta T_f = iK_f m = 2.6 \times 1.86 \times 6.71 = 32°C$; $T_f = -32°C$.

Assuming i = 2.6, a saturated $CaCl_2$ solution will not melt ice at −34°C (−29°F).

90. $\pi = iMRT,\ M = \dfrac{\pi}{iRT} = \dfrac{2.50\text{ atm}}{2.00 \times \dfrac{0.08206\text{L atm}}{\text{K mol}} \times 298\text{K}} = 5.11 \times 10^{-2}$ mol/L

$$\text{Molar mass of compound} = \frac{0.500\text{ g}}{0.1000\text{L} \times \dfrac{5.11 \times 10^{-2}\text{ mol}}{\text{L}}} = 97.8\text{ g/mol}$$

Additional Exercises

91.

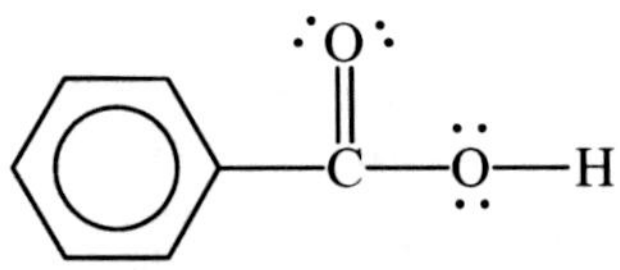

Benzoic acid is capable of hydrogen-bonding, but a significant part of benzoic acid is the nonpolar benzene ring. In benzene, a hydrogen-bonded dimer forms.

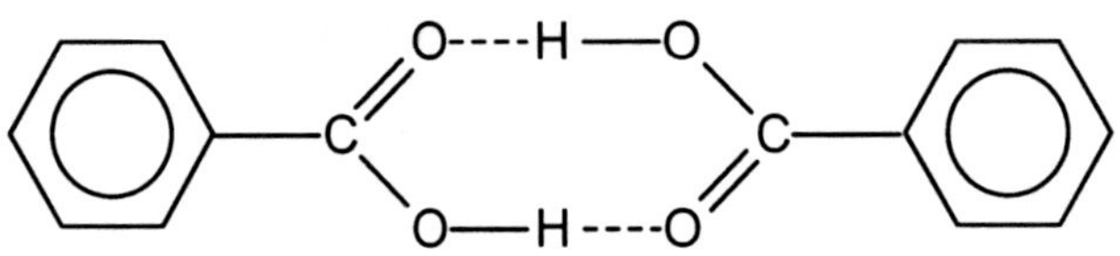

The dimer is relatively nonpolar and thus more soluble in benzene than in water.

Benzoic acid would be more soluble in a basic solution because of the reaction $C_6H_5CO_2H$ + OH^- $\rightarrow$ $C_6H_5CO_2^-$ + H_2O. By removing the acidic proton from benzoic acid, an anion forms, and like all anions, the species becomes more soluble in water.

92. Water is a polar solvent because the H_2O molecule exhibits a dipole moment, that is, H_2O is a molecule which has a partial negative charged end and a partial positive charged end. The electrostatic potential diagram for H_2O illustrates this with colors. The partial negative end of the dipole moment is the red end and the partial positive end is around the blue end.

Because water is a polar solvent, it will dissolve other polar covalent compounds. The electrostatic potential diagram for NH_3 illustrates that NH_3 has a dipole moment (the red end is the partial neative end and the light blue end is the partial positive end). Because NH_3 has a dipole moment (is polar), it will be soluble in water. However, the electrostatic potential diagram for CH_4 doesn't have one specific negative (red) end and has four blue regions arranged symmetrically about the molecule. CH_4 does not have a dipole moment (is nonpolar) and will not be soluble in water.

93. a. $NH_4NO_3(s) \rightarrow NH_4^+(aq) + NO_3^-(aq) \quad \Delta H_{soln} = ?$

Heat gain by dissolution process = heat loss by solution; we will keep all quantities positive in order to avoid sign errors. Because the temperature of the water decreased, the dissolution of NH_4NO_3 is endothermic (ΔH is positive). Mass of solution = 1.60 + 75.0 = 76.6 g.

$$\text{Heat loss by solution} = \frac{4.18\text{ J}}{°\text{C g}} \times 76.6\text{ g} \times (25.00°\text{C} - 23.34°\text{C}) = 532\text{ J}$$

$$\Delta H_{soln} = \frac{532\,J}{1.60\,g\,NH_4NO_3} \times \frac{80.05\,g\,NH_4NO_3}{mol\,NH_4NO_3} = 2.66 \times 10^4\ J/mol = 26.6\ kJ/mol$$

b. We will use Hess's law to solve for the lattice energy. The lattice-energy equation is:

$$NH_4^+(g) + NO_3^-(g) \rightarrow NH_4NO_3(s) \quad \Delta H = \text{lattice energy}$$

$NH_4^+(g) + NO_3^-(g) \rightarrow NH_4^+(aq) + NO_3^-(aq)$	$\Delta H = \Delta H_{hyd} = -630.\ kJ/mol$
$NH_4^+(aq) + NO_3^-(aq) \rightarrow NH_4NO_3(s)$	$\Delta H = -\Delta H_{soln} = -26.6\ kJ/mol$
$NH_4^+(g) + NO_3^-(g) \rightarrow NH_4NO_3(s)$	$\Delta H = \Delta H_{hyd} - \Delta H_{soln} = -657\ kJ/mol$

94. $$750.\ mL\ \text{grape juice} \times \frac{12\,mL\,C_2H_5OH}{100.\,mL\ \text{juice}} \times \frac{0.79\,g\,C_2H_5OH}{mL} \times \frac{1\,mol\,C_2H_5OH}{46.07\,g}$$

$$\times \frac{2\,mol\,CO_2}{2\,mol\,C_2H_5OH} = 1.54\ mol\ CO_2 \quad \text{(carry extra significant figure)}$$

1.54 mol CO_2 = total mol CO_2 = mol $CO_2(g)$ + mol $CO_2(aq)$ = $n_g + n_{aq}$

$$P_{CO_2} = \frac{n_gRT}{V} = \frac{n_g\left(\frac{0.08206\,L\,atm}{mol\,K}\right)(298\,K)}{75 \times 10^{-3}\,L} = 326n_g$$

$$P_{CO_2} = \frac{C}{k} = \frac{\frac{n_{aq}}{0.750\,L}}{\frac{3.1 \times 10^{-2}\,mol}{L\,atm}} = (43.0)n_{aq}$$

$P_{CO_2} = 326n_g = (43.0)n_{aq}$, and from above, $n_{aq} = 1.54 - n_g$; solving:

$326n_g = 43.0(1.54 - n_g)$, $369n_g = 66.2$, $n_g = 0.18$ mol

$P_{CO_2} = 326(0.18) = 59$ atm in gas phase;

$$C = kP_{CO_2} = \frac{3.1 \times 10^{-2}\,mol}{L\,atm} \times 59\ atm = \frac{1.8\,mol\,CO_2}{L} \text{ (in wine)}$$

95. a. Water boils when the vapor pressure equals the pressure above the water. In an open pan, $P_{atm} \approx 1.0$ atm. In a pressure cooker, $P_{inside} > 1.0$ atm, and water boils at a higher temperature. The higher the cooking temperature, the faster is the cooking time.

b. Salt dissolves in water, forming a solution with a melting point lower than that of pure water ($\Delta T_f = K_f m$). This happens in water on the surface of ice. If it is not too cold, the ice melts. This won't work if the ambient temperature is lower than the depressed freezing point of the salt solution.

c. When water freezes from a solution, it freezes as pure water, leaving behind a more concentrated salt solution. Therefore, the melt of frozen sea ice is pure water.

d. In the CO_2 phase diagram in Chapter 10, the triple point is above 1 atm, so $CO_2(g)$ is the stable phase at 1 atm and room temperature. $CO_2(l)$ can't exist at normal atmospheric pressures. Therefore, dry ice sublimes instead of boils. In a fire extinguisher, P > 1 atm, and $CO_2(l)$ can exist. When CO_2 is released from the fire extinguisher, $CO_2(g)$ forms, as predicted from the phase diagram.

e. Adding a solute to a solvent increases the boiling point and decreases the freezing point of the solvent. Thus the solvent is a liquid over a wider range of temperatures when a solute is dissolved.

96. A 92 proof ethanol solution is 46% C_2H_5OH by volume. Assuming 100.0 mL of solution:

$$\text{mol ethanol} = 46 \text{ mL } C_2H_5OH \times \frac{0.79\text{ g}}{\text{mL}} \times \frac{1\text{ mol } C_2H_5OH}{46.07\text{ g}} = 0.79 \text{ mol } C_2H_5OH$$

$$\text{molarity} = \frac{0.79\text{ mol}}{0.1000\text{L}} = 7.9\ M \text{ ethanol}$$

97. Because partial pressures are proportional to the moles of gas present, then:

$$\chi^{V}_{CS_2} = P_{CS_2}/P_{total}$$

$$P_{CS_2} = \chi^{V}_{CS_2} P_{total} = 0.855(263 \text{ torr}) = 225 \text{ torr}$$

$$P_{CS_2} = \chi^{L}_{CS_2} P^{o}_{CS_2},\ \chi^{L}_{CS_2} = \frac{P_{CS_2}}{P^{o}_{CS_2}} = \frac{225\text{ torr}}{375\text{ torr}} = 0.600$$

98. ΔH_1 and ΔH_2 refer to the breaking of intermolecular forces in pure solute and in pure solvent. ΔH_3 refers to the formation of the intermolecular forces in solution between the solute and solvent. ΔH_{soln} is the sum $\Delta H_1 + \Delta H_2 + \Delta H_3$.

a. The electrostatic potential diagram illustrates that acetone (CH_3COCH_3), like water, is a polar substance (each has a red end indicating the partial negative end of the dipole moment and a blue end indicating the partial positive end). For a polar solute in a polar solvent, ΔH_1 and ΔH_2 will be large and positive, while ΔH_3 will be a large negative value. As discussed in section 17.4 on nonideal solutions, acetone-water solutions exhibit negative deviations from Raoult's law. Acetone and water have the ability to hydrogen bond with each other, which gives the solution stronger intermolecular forces as compared to the pure states of both solute and solvent. In the pure state, acetone cannot H–bond with itself. Because acetone and water show negative deviations from Raoult's law, one would expect ΔH_{soln} to be slightly negative. Here ΔH_3 will be more than negative enough to overcome the large positive value from the ΔH_1 and ΔH_2 terms combined.

b. These two molecules are named ethanol (CH_3CH_2OH) and water. Ethanol-water solutions show positive deviations from Raoult's law. Both substances can hydrogen bond in the pure state, and they can continue this in solution. However, the solute-solvent interactions are somewhat weaker for ethanol-water solutions due to the small nonpolar part of ethanol (CH_3CH_2 is the nonpolar part of ethanol). This nonpolar part of ethanol slightly weakens the intermolecular forces in solution. So as in part a, when a polar solute and polar solvent are present, ΔH_1 and ΔH_2 are large and positive, while ΔH_3 is large and negative. For positive deviations from Raoult's law, the interactions in solution are weaker than the interactions in pure solute and pure solvent. Here, ΔH_{soln} will be slightly positive because the ΔH_3 term is not negative enough to overcome the large, positive ΔH_1 and ΔH_2 terms combined.

c. As the electrostatic potential diagrams indicate, both heptane (C_7H_{16}) and hexane (C_6H_{14}) are nonpolar substances. For a nonpolar solute dissolved in a nonpolar solvent, ΔH_1 and ΔH_2 are small and positive, while the ΔH_3 term is small and negative. These three terms have small values due to the relatively weak London dispersion forces that are broken and formed for solutions consisting of a nonpolar solute in a nonpolar solvent. Because ΔH_1, ΔH_2, and ΔH_3 are all small values, the ΔH_{soln} value will be small. Here, heptane and hexane would form an ideal solution because the relative strengths of the London dispersion forces are about equal in pure solute and pure solvent as compared to those LD forces in solution. For ideal solutions, $\Delta H_{soln} = 0$.

d. This combination represents a nonpolar solute in a polar solvent. ΔH_1 will be small due to the relative weak London dispersion forces which are broken when the solute (C_7H_{16}) expands. ΔH_2 will be large and positive because of the relatively strong hydrogen bonding interactions that must be broken when the polar solvent (water) is expanded. And finally, the ΔH_3 term will be small because the nonpolar solute and polar solvent do not interact with each other. The end result is that ΔH_{soln} is a large positive value.

99. $50.0 \text{ g } CH_3COCH_3 \times \dfrac{1 \text{ mol}}{58.08 \text{ g}} = 0.861 \text{ mol acetone}$

$$50.0 \text{ g } CH_3OH \times \frac{1 \text{ mol}}{32.04 \text{ g}} = 1.56 \text{ mol methanol}$$

$$\chi^L_{acetone} = \frac{0.861}{0.861 + 1.56} = 0.356; \ \chi^L_{methanol} = 1.000 - \chi^L_{acetone} = 0.644$$

$$P_{total} = P_{methanol} + P_{acetone} = 0.644(143 \text{ torr}) + 0.356(271 \text{ torr}) = 92.1 \text{ torr} + 96.5 \text{ torr} = 188.6 \text{ torr}$$

Because partial pressures are proportional to the moles of gas present, in the vapor phase:

$$\chi^V_{acetone} = \frac{P_{acetone}}{P_{total}} = \frac{96.5 \text{ torr}}{188.6 \text{ torr}} = 0.512; \ \chi^V_{methanol} = 1.000 - 0.512 = 0.488$$

The actual vapor pressure of the solution (161 torr) is less than the calculated pressure assuming ideal behavior (188.6 torr). Therefore, the solution exhibits negative deviations from Raoult's law. This occurs when the solute-solvent interactions are stronger than in pure solute and pure solvent.

100. $\pi = MRT = \frac{0.1\text{ mol}}{\text{L}} \times \frac{0.08206\text{ L atm}}{\text{K mol}} \times 298\text{ K} = 2.45\text{ atm} \approx 2\text{ atm}$

$\pi = 2\text{ atm} \times \frac{760\text{ mm Hg}}{\text{atm}} \approx 2000\text{ mm} \approx 2\text{ m}$

The osmotic pressure would support a mercury column of approximately 2 m. The height of a fluid column in a tree will be higher because Hg is more dense than the fluid in a tree. If we assume the fluid in a tree is mostly H_2O, then the fluid has a density of 1.0 g/cm^3. The density of Hg is 13.6 g/cm^3.

Height of fluid ≈ 2 m × 13.6 ≈ 30 m

101. $\Delta T_f = K_f m,\ m = \frac{\Delta T_f}{K_f} = \frac{0.300°\text{C}}{5.12°\text{C kg/mol}} = \frac{5.86 \times 10^{-2}\text{ mol thyroxine}}{\text{kg benzene}}$

The moles of thyroxine present are:

$0.0100\text{ kg benzene} \times \frac{5.86 \times 10^{-2}\text{ mol thyroxine}}{\text{kg benzene}} = 5.86 \times 10^{-4}\text{ mol thyroxine}$

From the problem, 0.455 g thyroxine was used; this must contain 5.86×10^{-4} mol thyroxine. The molar mass of the thyroxine is:

$\text{molar mass} = \frac{0.455\text{ g}}{5.86 \times 10^{-4}\text{ mol}} = 776\text{ g/mol}$

102. $\pi = MRT,\ M = \frac{\pi}{RT} = \frac{8.00\text{ atm}}{0.08206\text{ L atm/K}\bullet\text{mol} \times 298\text{ K}} = 0.327\text{ mol/L}$

103. Out of 100.00 g, there are:

$31.57\text{ g C} \times \frac{1\text{ mol C}}{12.01\text{ g}} = 2.629\text{ mol C};\ \frac{2.629}{2.629} = 1.000$

$5.30\text{ g H} \times \frac{1\text{ mol H}}{1.008\text{ g}} = 5.26\text{ mol H};\ \frac{5.26}{2.629} = 2.00$

$63.13\text{ g O} \times \frac{1\text{ mol O}}{16.00\text{ g}} = 3.946\text{ mol O};\ \frac{3.946}{2.629} = 1.501$

Empirical formula: $C_2H_4O_3$; use the freezing-point data to determine the molar mass.

$m = \frac{\Delta T_f}{K_f} = \frac{5.20°\text{C}}{1.86°\text{C/molal}} = 2.80\text{ molal}$

$\text{Mol solute} = 0.0250\text{ kg} \times \frac{2.80\text{ mol solute}}{\text{kg}} = 0.0700\text{ mol solute}$

$$\text{Molar mass} = \frac{10.56\text{ g}}{0.0700\text{mol}} = 151\text{ g/mol}$$

The empirical formula mass of $C_2H_4O_3$ = 76.05 g/mol. Because the molar mass is about twice the empirical mass, the molecular formula is $C_4H_8O_6$, which has a molar mass of 152.10 g/mol.

Note: We use the experimental molar mass to determine the molecular formula. Knowing this, we calculate the molar mass precisely from the molecular for
mula using the atomic masses in the periodic table.

104. a. As discussed in Figure 11.18 of the text, the water would migrate from left to right (to the side with the solution). Initially, the level of liquid in the left arm would go down, and the level in the right arm would go up. At some point the rate of solvent (H_2O) transfer will be the same in both directions, and the levels of the liquids in the two arms will stabilize. The height difference between the two arms is a measure of the osmotic pressure of the solution.

b. Initially, H_2O molecules will have a net migration into the solution side. However, the solute can now migrate into the H_2O side. Because solute and solvent transfer are both possible, the levels of the liquids will be equal once the rates of solute and solvent transfer are equal in both directions. At this point the concentration of solute will be equal in both chambers, and the levels of liquid will be equal.

105. If ideal, NaCl dissociates completely, and i = 2.00. $\Delta T_f = iK_f m$; assuming water freezes at 0.00°C:

$$1.28°C = 2 \times 1.86°C\text{ kg/mol} \times m,\quad m = 0.344\text{ mol NaCl/kg } H_2O$$

Assume an amount of solution that contains 1.00 kg of water (solvent).

$$0.344\text{ mol NaCl} \times \frac{58.44\text{ g}}{\text{mol}} = 20.1\text{ g NaCl}$$

$$\text{Mass \% NaCl} = \frac{20.1\text{ g}}{1.00 \times 10^3\text{ g} + 20.1\text{ g}} \times 100 = 1.97\%$$

106. The main factor for stabilization seems to be electrostatic repulsion. The center of a colloid particle is surrounded by a layer of same charged ions, with oppositely charged ions forming another charged layer on the outside. Overall, there are equal numbers of charged and oppositely charged ions, so the colloidal particles are electrically neutral. However, since the outer layers are the same charge, the particles repel each other and do not easily aggregate for precipitation to occur.

Heating increases the velocities of the colloidal particles. This causes the particles to collide with enough energy to break the ion barriers, allowing the colloids to aggregate and eventually precipitate out. Adding an electrolyte neutralizes the adsorbed ion layers, which allows colloidal particles to aggregate and then precipitate out.

107. $\Delta T = K_f m,\ m = \dfrac{\Delta T}{K_f} = \dfrac{2.79°C}{1.86\ °C/molal} = 1.50$ molal

a. $\Delta T = K_b m$, $\Delta T = (0.51°C/molal)(1.50\ molal) = 0.77°C$, $T_b = 100.77°C$

b. $P_{soln} = \chi_{water} P^o_{water}$, $\chi_{water} = \dfrac{mol\,H_2O}{mol\,H_2O + mol\,solute}$

Assuming 1.00 kg of water, we have 1.50 mol solute, and:

$$mol\ H_2O = 1.00 \times 10^3\ g\ H_2O \times \frac{1\ mol\ H_2O}{18.02\ g\ H_2O} = 55.5\ mol\ H_2O$$

$$\chi_{water} = \frac{55.5\ mol}{1.50 + 55.5} = 0.974;\quad P_{soln} = (0.974)(23.76\ mm\ Hg) = 23.1\ mm\ Hg$$

c. We assumed ideal behavior in solution formation, we assumed the solute was nonvolatile, and we assumed i = 1 (no ions formed).

108. a. The average values for each ion are:

300. mg Na^+, 15.7 mg K^+, 5.45 mg Ca^{2+}, 388 mg Cl^-, and 246 mg lactate ($C_3H_5O_3^-$)

Note: Because we can precisely weigh to ±0.1 mg on an analytical balance, we'll carry extra significant figures and calculate results to ±0.1 mg.

The only source of lactate is $NaC_3H_5O_3$.

$$246\ mg\ C_3H_5O_3^- \times \frac{112.06\ mg\ NaC_3H_5O_3}{89.07\ mg\ C_3H_5O_3^-} = 309.5\ mg\ sodium\ lactate$$

The only source of Ca^{2+} is $CaCl_2•2H_2O$.

$$5.45\ mg\ Ca^{2+} \times \frac{147.01\ mg\ CaCl_2 \bullet 2H_2O}{40.08\ mg\ Ca^{2+}} = 19.99\ or\ 20.0\ mg\ CaCl_2•2H_2O$$

The only source of K^+ is KCl.

$$15.7\ mg\ K^+ \times \frac{74.55\ mg\ KCl}{39.10\ mg\ K^+} = 29.9\ mg\ KCl$$

From what we have used already, let's calculate the mass of Na^+ added.

309.5 mg sodium lactate – 246.0 mg lactate = 63.5 mg Na^+

Thus we need to add an additional 236.5 mg Na^+ to get the desired 300. mg.

$$236.5\ mg\ Na^+ \times \frac{58.44\ mg\ NaCl}{22.99\ mg\ Na^+} = 601.2\ mg\ NaCl$$

Now let's check the mass of Cl^- added:

$$20.0 \text{ mg } CaCl_2{\bullet}2H_2O \times \frac{70.90 \text{ mg } Cl^-}{147.01 \text{ mg } CaCl_2 \bullet 2H_2O} = 9.6 \text{ mg } Cl^-$$

$$\begin{aligned} 20.0 \text{ mg } CaCl_2{\bullet}2H_2O &= 9.6 \text{ mg } Cl^- \\ 29.9 \text{ mg KCl} - 15.7 \text{ mg } K^+ &= 14.2 \text{ mg } Cl^- \\ 601.2 \text{ mg NaCl} - 236.5 \text{ mg } Na^+ &= 364.7 \text{ mg } Cl^- \\ \hline \text{Total } Cl^- &= 388.5 \text{ mg } Cl^- \end{aligned}$$

This is the quantity of Cl^- we want (the average amount of Cl^-).

An analytical balance can weigh to the nearest 0.1 mg. We would use 309.5 mg sodium lactate, 20.0 mg $CaCl_2{\bullet}2H_2O$, 29.9 mg KCl, and 601.2 mg NaCl.

b. To get the range of osmotic pressure, we need to calculate the molar concentration of each ion at its minimum and maximum values. At minimum concentrations, we have:

$$\frac{285 \text{ mg } Na^+}{100. \text{ mL}} \times \frac{1 \text{ mmol}}{22.99 \text{ mg}} = 0.124\ M; \quad \frac{14.1 \text{ mg } K^+}{100. \text{ mL}} \times \frac{1 \text{ mmol}}{39.10 \text{ mg}} = 0.00361\ M$$

$$\frac{4.9 \text{ mg } Ca^{2+}}{100. \text{ mL}} \times \frac{1 \text{ mmol}}{40.08 \text{ mg}} = 0.0012\ M; \quad \frac{368 \text{ mg } Cl^-}{100. \text{ mL}} \times \frac{1 \text{ mmol}}{35.45 \text{ mg}} = 0.104\ M$$

$$\frac{231 \text{ mg } C_3H_5O_3^-}{100. \text{ mL}} \times \frac{1 \text{ mmol}}{89.07 \text{ mg}} = 0.0259\ M$$ (*Note*: Molarity = mol/L = mmol/mL.)

Total = 0.124 + 0.00361 + 0.0012 + 0.104 + 0.0259 = 0.259 M

$$\pi = MRT = \frac{0.259 \text{ mol}}{\text{L}} \times \frac{0.08206 \text{ L atm}}{\text{K mol}} \times 310. \text{ K} = 6.59 \text{ atm}$$

Similarly, at maximum concentrations, the concentration for each ion is:

Na^+: 0.137 M; K^+: 0.00442 M; Ca^{2+}: 0.0015 M; Cl^-: 0.115 M; $C_3H_5O_3^-$: 0.0293 M

The total concentration of all ions is 0.287 M.

$$\pi = \frac{0.287 \text{ mol}}{\text{L}} \times \frac{0.08206 \text{ L atm}}{\text{K mol}} \times 310. \text{ K} = 7.30 \text{ atm}$$

Osmotic pressure ranges from 6.59 atm to 7.30 atm.

109. Mass of H_2O = 160. mL × $\frac{0.995 \text{ g}}{\text{mL}}$ = 159 g = 0.159 kg

Mol NaDTZ = 0.159 kg × $\frac{0.378 \text{ mol}}{\text{kg}}$ = 0.0601 mol

$$\text{Molar mass of NaDTZ} = \frac{38.4\text{ g}}{0.0601\text{mol}} = 639\text{ g/mol}$$

$$P_{soln} = \chi_{H_2O}P^o_{H_2O};\quad \text{mol } H_2O = 159\text{ g} \times \frac{1\text{ mol}}{18.02\text{ g}} = 8.82\text{ mol}$$

Sodium diatrizoate is a salt because there is a metal (sodium) in the compound. From the short-hand notation for sodium diatrizoate, NaDTZ, we can assume this salt breaks up into Na^+ and DTZ^- ions. So the moles of solute particles are 2(0.0601) = 0.120 mol solute particles.

$$\chi_{H_2O} = \frac{8.82\text{ mol}}{0.120\text{ mol} + 8.82\text{ mol}} = 0.987;\quad P_{soln} = 0.987 \times 34.1\text{ torr} = 33.7\text{ torr}$$

ChemWork Problems

The answers to the problems 110-117 (or a variation to these problem) are found in OWL. These problems are also assignable in OWL.

Challenge Problems

118. For the second vapor collected, $\chi^V_{B,2} = 0.714$ and $\chi^V_{T,2} = 0.286$. Let $\chi^L_{B,2}$ = mole fraction of benzene in the second solution and $\chi^L_{T,2}$ = mole fraction of toluene in the second solution.

$$\chi^L_{B,2} + \chi^L_{T,2} = 1.000$$

$$\chi^V_{B,2} = 0.714 = \frac{P_B}{P_{total}} = \frac{P_B}{P_B + P_T} = \frac{\chi^L_{B,2}(750.0\text{ torr})}{\chi^L_{B,2}(750.0\text{ torr}) + (1.000 - \chi^L_{B,2})(300.0\text{ torr})}$$

Solving: $\chi^L_{B,2} = 0.500 = \chi^L_{T,2}$

This second solution came from the vapor collected from the first (initial) solution, so, $\chi^V_{B,1} = \chi^V_{T,1} = 0.500$. Let $\chi^L_{B,1}$ = mole fraction benzene in the first solution and $\chi^L_{T,1}$ = mole fraction of toluene in first solution. $\chi^L_{B,1} + \chi^L_{T,1} = 1.000$.

$$\chi^V_{B,1} = 0.500 = \frac{P_B}{P_{total}} = \frac{P_B}{P_B + P_T} = \frac{\chi^L_{B,1}(750.0\text{ torr})}{\chi^L_{B,1}(750.0\text{ torr}) + (1.000 - \chi^L_{B,1})(300.0\text{ torr})}$$

Solving: $\chi^L_{B,1} = 0.286$

The original solution had $\chi_B = 0.286$ and $\chi_T = 0.714$.

119. For 30.% A by moles in the vapor, $30. = \dfrac{P_A}{P_A + P_B} \times 100$:

$$0.30 = \frac{\chi_A x}{\chi_A x + \chi_B y},\quad 0.30 = \frac{\chi_A x}{\chi_A x + (1.00 - \chi_A)y}$$

$$\chi_A x = 0.30(\chi_A x) + 0.30y - 0.30(\chi_A y),\quad \chi_A x - (0.30)\chi_A x + (0.30)\chi_A y = 0.30y$$

$$\chi_A(x - 0.30x + 0.30y) = 0.30y,\quad \chi_A = \frac{0.30y}{0.70x + 0.30y};\quad \chi_B = 1.00 - \chi_A$$

Similarly, if vapor above is 50.% A: $\chi_A = \dfrac{y}{x + y};\quad \chi_B = 1.00 - \dfrac{y}{x + y}$

If vapor above is 80.% A: $\chi_A = \dfrac{0.80y}{0.20x + 0.80y};\quad \chi_B = 1.00 - \chi_A$

If the liquid solution is 30.% A by moles, $\chi_A = 0.30$.

Thus $\chi_A^V = \dfrac{P_A}{P_A + P_B} = \dfrac{0.30x}{0.30x + 0.70y}$ and $\chi_B^V = 1.00 - \dfrac{0.30x}{0.30x + 0.70y}$

If solution is 50.% A: $\chi_A^V = \dfrac{x}{x + y}$ and $\chi_B^V = 1.00 - \chi_A^V$

If solution is 80.% A: $\chi_A^V = \dfrac{0.80x}{0.80x + 0.20y}$ and $\chi_B^V = 1.00 - \chi_A^V$

120. a. Freezing-point depression is determined using molality for the concentration units, whereas molarity units are used to determine osmotic pressure. We need to assume that the molality of the solution equals the molarity of the solution.

b. $\text{Molarity} = \dfrac{\text{moles solvent}}{\text{liters solution}};\quad \text{molality} = \dfrac{\text{moles solvent}}{\text{kg solvent}}$

When the liters of solution equal the kilograms of solvent present for a solution, then molarity equals molality. This occurs for an aqueous solution when the density of the solution is equal to the density of water, 1.00 g/cm^3. The density of a solution is close to 1.00 g/cm^3 when not a lot of solute is dissolved in solution. Therefore, molarity and molality values are close to each other only for dilute solutions.

c. $\Delta T = K_f m,\quad m = \dfrac{\Delta T}{K_f} = \dfrac{0.621°C}{1.86\ °C\ \text{kg/mol}} = 0.334\ \text{mol/kg}$

Assuming 0.334 mol/kg = 0.334 mol/L:

$$\pi = MRT = \frac{0.334\ \text{mol}}{\text{L}} \times \frac{0.08206\ \text{L atm}}{\text{K mol}} \times 298\ \text{K} = 8.17\ \text{atm}$$

d. $m = \dfrac{\Delta T}{K_b} = \dfrac{2.0^\circ C}{0.51^\circ C\ kg/mol} = 3.92\ mol/kg$

This solution is much more concentrated than the isotonic solution in part c. Here, water will leave the plant cells in order to try to equilibrate the ion concentration both inside and outside the cell. Because there is such a large concentration discrepancy, all the water will leave the plant cells, causing them to shrivel and die.

121. $m = \dfrac{\Delta T_f}{K_f} = \dfrac{0.426^\circ C}{1.86\ ^\circ C/molal} = 0.229\ molal$

Assuming a solution density = 1.00 g/mL, then 1.00 L contains 0.229 mol solute.

$NaCl \rightarrow Na^+ + Cl^-$ i = 2; so: 2(mol NaCl) + mol $C_{12}H_{22}O_{11}$ = 0.229 mol

Mass NaCl + mass $C_{12}H_{22}O_{11}$ = 20.0 g

$2n_{NaCl} + n_{C_{12}H_{22}O_{11}} = 0.229$ and $58.44(n_{NaCl}) + 342.3(n_{C_{12}H_{22}O_{11}}) = 20.0$

Solving: $n_{C_{12}H_{22}O_{11}} = 0.0425$ mol = 14.5 g and $n_{NaCl} = 0.0932$ mol = 5.45 g

Mass % $C_{12}H_{22}O_{11} = \dfrac{14.5\ g}{20.0\ g} \times 100 = 72.5\ \%$ and 27.5% NaCl by mass

$\chi_{C_{12}H_{22}O_{11}} = \dfrac{0.0425 mol}{0.0425 mol + 0.0932 mol} = 0.313$

122. a. $\pi = iMRT,\ iM = \dfrac{\pi}{RT} = \dfrac{7.83 atm}{0.08206 L\ atm/K \cdot mol \times 298 K} = 0.320\ mol/L$

Assuming 1.000 L of solution:

total mol solute particles = mol Na^+ + mol Cl^- + mol NaCl = 0.320 mol

mass solution = 1000. mL $\times \dfrac{1.071 g}{mL}$ = 1071 g solution

mass NaCl in solution = 0.0100 × 1071 g = 10.7 g NaCl

mol NaCl added to solution = 10.7 g $\times \dfrac{1\ mol}{58.44\ g}$ = 0.183 mol NaCl

Some of this NaCl dissociates into Na^+ and Cl^- (two moles of ions per mole of NaCl), and some remains undissociated. Let x = mol undissociated NaCl = mol ion pairs.

Mol solute particles = 0.320 mol = $2(0.183 - x) + x$

$0.320 = 0.366 - x,\ x = 0.046$ mol ion pairs

Fraction of ion pairs = $\dfrac{0.046}{0.183} = 0.25$, or 25%

b. $\Delta T = K_f m$, where K_f = 1.86 °C kg/mol; from part a, 1.000 L of solution contains 0.320 mol of solute particles. To calculate the molality of the solution, we need the kilograms of solvent present in 1.000 L of solution.

Mass of 1.000 L solution = 1071 g; mass of NaCl = 10.7 g

Mass of solvent in 1.000 L solution = 1071 g − 10.7 g = 1060. g

$$\Delta T = 1.86\ °\text{C kg/mol} \times \frac{0.320\,\text{mol}}{1.060\,\text{kg}} = 0.562°\text{C}$$

Assuming water freezes at 0.000°C, then T_f = −0.562°C.

123. $\chi^V_{pen} = 0.15 = \frac{P_{pen}}{P_{total}}$; $P_{pen} = \chi^L_{pen} P^o_{pen}$; $P_{total} = P_{pen} + P_{hex} = \chi^L_{pen}(511) + \chi^L_{hex}(150.)$

Because $\chi^L_{hex} = 1.000 - \chi^L_{pen}$: $P_{total} = \chi^L_{pen}(511) + (1.000 - \chi^L_{pen})(150.) = 150. + 361\chi^L_{pen}$

$$\chi^V_{pen} = \frac{P_{pen}}{P_{total}},\ 0.15 = \frac{\chi^L_{pen}(511)}{150. + 361\chi^L_{pen}},\ 0.15(150. + 361\chi^L_{pen}) = 511\chi^L_{pen}$$

$$23 + 54\chi^L_{pen} = 511\chi^L_{pen},\ \chi^L_{pen} = \frac{23}{457} = 0.050$$

124. a. $m = \frac{\Delta T_f}{K_f} = \frac{1.32°\text{C}}{5.12°\text{C kg/mol}} = 0.258$ mol/kg

$$\text{Mol unknown} = 0.01560\ \text{kg} \times \frac{0.258\,\text{mol unknown}}{\text{kg}} = 4.02 \times 10^{-3}\ \text{mol}$$

$$\text{Molar mass of unknown} = \frac{1.22\,\text{g}}{4.02 \times 10^{-3}\ \text{mol}} = 303\ \text{g/mol}$$

$$\text{Uncertainty in temperature} = \frac{0.04}{1.32} \times 100 = 3\%$$

A 3% uncertainty in 303 g/mol = 9 g/mol.

So molar mass = 303 ±9 g/mol.

b. No, codeine could not be eliminated since its molar mass is in the possible range including the uncertainty.

d. We would like the uncertainty to be ±1 g/mol. We need the freezing-point depression to be about 10 times what it was in this problem. Two possibilities are:

(1) make the solution 10 times more concentrated (may be solubility problem)

(2) use a solvent with a larger K_f value, e.g., camphor

125. $\Delta T_f = 5.51 - 2.81 = 2.70°C$; $m = \frac{\Delta T_f}{K_f} = \frac{2.70°C}{5.12°C/\text{molal}} = 0.527$ molal

Let x = mass of naphthalene (molar mass = 128.2 g/mol). Then $1.60 - x$ = mass of anthracene (molar mass = 178.2 g/mol).

$$\frac{x}{128.2} = \text{moles naphthalene and } \frac{1.60 - x}{178.2} = \text{moles anthracene}$$

$$\frac{0.527 \text{ mol solute}}{\text{kg solvent}} = \frac{\frac{x}{128.2} + \frac{1.60 - x}{178.2}}{0.0200 \text{ kg solvent}},\ 1.05 \times 10^{-2} = \frac{(178.2)x + 1.60(128.2) - (128.2)x}{128.2(178.2)}$$

$(50.0)x + 205 = 240.$, $(50.0)x = 240. - 205$, $(50.0)x = 35$, $x = 0.70$ g naphthalene

So the mixture is:

$$\frac{0.70 \text{ g}}{1.60 \text{ g}} \times 100 = 44\% \text{ naphthalene by mass and } 56\% \text{ anthracene by mass}$$

126. $$iM = \frac{\pi}{RT} = \frac{0.3950 \text{ atm}}{\frac{0.08206 \text{ L atm}}{\text{K mol}} \times 298.2 \text{ K}} = 0.01614 \text{ mol/L} = \text{total ion concentration}$$

$0.01614 \text{ mol/L} = M_{Mg^{2+}} + M_{Na^+} + M_{Cl^-}$; $M_{Cl^-} = 2M_{Mg^{2+}} + M_{Na^+}$ (charge balance)

Combining: $0.01614 = 3M_{Mg^{2+}} + 2M_{Na^+}$

Let x = mass $MgCl_2$ and y = mass NaCl; then $x + y = 0.5000$ g.

$M_{Mg^{2+}} = \frac{x}{95.21}$ and $M_{Na^+} = \frac{y}{58.44}$ (Because V = 1.000 L.)

Total ion concentration $= \frac{3x}{95.21} + \frac{2y}{58.44} = 0.01614$ mol/L

Rearranging: $3x + (3.258)y = 1.537$

Solving by simultaneous equations:

$$\begin{array}{rcl} 3x + (3.258)y & = & 1.537 \\ -3(x + y) & = & -3(0.5000) \\ \hline (0.258)y & = & 0.037,\ y = 0.14 \text{ g NaCl} \end{array}$$

Mass $MgCl_2$ = 0.5000 g – 0.14 g = 0.36 g; mass % $MgCl_2 = \frac{0.36 \text{ g}}{0.5000 \text{ g}} \times 100 = 72\%$

127. $HCO_2H \rightarrow H^+ + HCO_2^-$; only 4.2% of HCO_2H ionizes. The amount of H^+ or HCO_2^- produced is $0.042 \times 0.10\ M = 0.0042\ M$.

The amount of HCO_2H remaining in solution after ionization is $0.10\ M - 0.0042\ M = 0.10\ M$.

The total molarity of species present $= M_{HCO_2H} + M_{H^+} + M_{HCO_2^-}$

$$= 0.10 + 0.0042 + 0.0042 = 0.11\ M$$

Assuming $0.11\ M = 0.11$ molal, and assuming ample significant figures in the freezing point and boiling point of water at P = 1 atm:

$$\Delta T = K_f m = 1.86°C/\text{molal} \times 0.11\ \text{molal} = 0.20°C;\ \text{freezing point} = -0.20°C$$

$$\Delta T = K_b m = 0.51°C/\text{molal} \times 0.11\ \text{molal} = 0.056°C;\ \text{boiling point} = 100.056°C$$

128. Let χ_A^L = mole fraction A in solution, so $1.000 - \chi_A^L = \chi_B^L$. From the problem, $\chi_A^V = 2\chi_A^L$.

$$\chi_A^V = \frac{P_A}{P_{total}} = \frac{\chi_A^L(350.0\ \text{torr})}{\chi_A^L(350.0\ \text{torr}) + (1.000 - \chi_A^L)(100.0\ \text{torr})}$$

$$\chi_A^V = 2\chi_A^L = \frac{(350.0)\chi_A^L}{(250.0)\chi_A^L + 100.0},\quad (250.0)\chi_A^L = 75.0,\quad \chi_A^L = 0.300$$

The mole fraction of A in solution is 0.300.

129. a. Assuming $MgCO_3(s)$ does not dissociate, the solute concentration in water is:

$$\frac{560\,\mu g\ MgCO_3(s)}{mL} = \frac{560\ mg}{L} = \frac{560 \times 10^{-3}\ g}{L} \times \frac{1\ mol\ MgCO_3}{84.32\ g}$$

$$= 6.6 \times 10^{-3}\ mol\ MgCO_3/L$$

An applied pressure of 8.0 atm will purify water up to a solute concentration of:

$$M = \frac{\pi}{RT} = \frac{8.0\ atm}{0.08206\ L\ atm/K\ mol \times 300.\ K} = \frac{0.32\ mol}{L}$$

When the concentration of $MgCO_3(s)$ reaches 0.32 mol/L, the reverse osmosis unit can no longer purify the water. Let V = volume (L) of water remaining after purifying 45 L of H_2O. When V + 45 L of water has been processed, the moles of solute particles will equal:

$$6.6 \times 10^{-3}\ mol/L \times (45\ L + V) = 0.32\ mol/L \times V$$

Solving: $0.30 = (0.32 - 0.0066) \times V,\ V = 0.96\ L$

The minimum total volume of water that must be processed is 45 L + 0.96 L = 46 L.

Note: If $MgCO_3$ does dissociate into Mg^{2+} and CO_3^{2-} ions, then the solute concentration increases to $1.3 \times 10^{-2}\ M$, and at least 47 L of water must be processed.

b. No; a reverse osmosis system that applies 8.0 atm can only purify water with a solute concentration of less than 0.32 mol/L. Salt water has a solute concentration of 2(0.60 *M*) = 1.2 mol/L ions. The solute concentration of salt water is much too high for this reverse osmosis unit to work.

Integrative Problems

130. $$10.0\text{ mL blood} \times \frac{1\text{ L}}{1000\text{mL}} \times \frac{10\text{dL}}{1\text{ L}} \times \frac{1.0\text{mg}\,C_4H_7N_3O}{1\text{ dL blood}} \times \frac{1\text{ g}}{1000\text{mg}} \times \frac{1\text{ mol}C_4H_7N_3O}{113.13\text{g}} = 8.8 \times 10^{-7}\text{ mol }C_4H_7N_3O$$

$$\text{Mass of blood} = 10.0\text{ mL} \times \frac{1.025\text{g}}{\text{mL}} = 10.3\text{ g}$$

$$\text{Molality} = \frac{8.8 \times 10^{-7}\text{ mol}}{0.0103\text{kg}} = 8.5 \times 10^{-5}\text{ mol/kg}$$

$$\pi = M\text{RT},\quad M = \frac{8.8 \times 10^{-7}\text{ mol}}{0.0100\text{L}} = 8.8 \times 10^{-5}\text{ mol/L}$$

$$\pi = 8.8 \times 10^{-5}\text{ mol/L} \times \frac{0.08206\text{L atm}}{\text{K mol}} \times 298\text{ K} = 2.2 \times 10^{-3}\text{ atm}$$

131. $$\Delta T = imK_f,\quad i = \frac{\Delta T}{mK_f} = \frac{2.79^\circ\text{C}}{\frac{0.250\text{mol}}{0.500\text{kg}} \times \frac{1.86\ ^\circ\text{C kg}}{\text{mol}}} = 3.00$$

We have three ions in solutions, and we have twice as many anions as cations. Therefore, the formula of Q is MCl_2. Assuming 100.00 g of compound:

$$38.68\text{ g Cl} \times \frac{1\text{ mol Cl}}{35.45\text{ g}} = 1.091\text{ mol Cl}$$

$$\text{mol M} = 1.091\text{ mol Cl} \times \frac{1\text{ mol M}}{2\text{ mol Cl}} = 0.5455\text{ mol M}$$

$$\text{Molar mass of M} = \frac{61.32\text{ g M}}{0.5455\text{mol M}} = 112.4\text{ g/mol};\quad \text{M is Cd, so Q} = CdCl_2.$$

132. $$14.2\text{ mg }CO_2 \times \frac{12.01\text{mg C}}{44.01\text{mg }CO_2} = 3.88\text{ mg C};\quad \%\text{ C} = \frac{3.88\text{ mg}}{4.80\text{ mg}} \times 100 = 80.8\%\text{ C}$$

$$1.65\text{ mg }H_2O \times \frac{2.016\text{mg H}}{18.02\text{ mg }H_2O} = 0.185\text{ mg H};\quad \%\text{ H} = \frac{0.185\text{mg}}{4.80\text{ mg}} \times 100 = 3.85\%\text{ H}$$

$$\text{Mass \% O} = 100.00 - (80.8 + 3.85) = 15.4\%\text{ O}$$

Out of 100.00 g:

$$80.8 \text{ g C} \times \frac{1 \text{ mol}}{12.01 \text{ g}} = 6.73 \text{ mol C}; \quad \frac{6.73}{0.963} = 6.99 \approx 7$$

$$3.85 \text{ g H} \times \frac{1 \text{ mol}}{1.008 \text{ g}} = 3.82 \text{ mol H}; \quad \frac{3.82}{0.963} = 3.97 \approx 4$$

$$15.4 \text{ g O} \times \frac{1 \text{ mol}}{16.00 \text{ g}} = 0.963 \text{ mol O}; \quad \frac{0.963}{0.963} = 1.00$$

Therefore, the empirical formula is C_7H_4O.

$$\Delta T_f = K_f m, \ m = \frac{\Delta T_f}{K_f} = \frac{22.3\ ^\circ\text{C}}{40.\ ^\circ\text{C/molal}} = 0.56 \text{ molal}$$

$$\text{Mol anthraquinone} = 0.0114 \text{ kg camphor} \times \frac{0.56 \text{ mol anthraquinone}}{\text{kg camphor}} = 6.4 \times 10^{-3} \text{ mol}$$

$$\text{Molar mass} = \frac{1.32 \text{ g}}{6.4 \times 10^{-3} \text{ mol}} = 210 \text{ g/mol}$$

The empirical mass of C_7H_4O is 7(12) + 4(1) + 16 ≈ 104 g/mol. Because the molar mass is twice the empirical mass, the molecular formula is $C_{14}H_8O_2$.

CHAPTER 12

CHEMICAL KINETICS

Questions

10. a. Activation energy and ΔE are independent of each other. Activation energy depends on the path reactants to take to convert to products. The overall energy change ΔE only depends on the initial and final energy states of the reactants and products. ΔE is path-independent.

    b. The rate law can only be determined from experiment, not from the overall balanced reaction.

    c. Most reactions occur by a series of steps. The rate of the reaction is determined by the rate of the slowest step in the mechanism.

11. In a unimolecular reaction, a single reactant molecule decomposes to products. In a bimolecular reaction, two molecules collide to give products. The probability of the simultaneous collision of three molecules with enough energy and the proper orientation is very small, making termolecular steps very unlikely.

12. a. $T_2 > T_1$; as temperature increases, the distribution of collision energies shifts to the right. That is, as temperature increases, there are fewer collision energies with small energies and more collisions with large energies.

    b. As temperature increases, more of the collisions have the required activation energy necessary to convert reactants into products. Hence, the rate of the reaction increases with increasing temperature.

13. All of these choices would affect the rate of the reaction, but only b and c affect the rate by affecting the value of the rate constant k. The value of the rate constant depends on temperature. The value of the rate constant also depends on the activation energy. A catalyst will change the value of k because the activation energy changes. Increasing the concentration (partial pressure) of either O_2 or NO does not affect the value of k, but it does increase the rate of the reaction because both concentrations appear in the rate law.

14. One experimental method to determine rate laws is the method of initial rates. Several experiments are carried out using different initial concentrations of reactants, and the initial rate is determined for each experiment. The results are then compared to see how the initial rate depends on the initial concentrations. This allows the orders in the rate law to be determined. The value of the rate constant is determined from the experiments once the orders are known.

The second experimental method utilizes the fact that the integrated rate laws can be put in the form of a straight-line equation. Concentration versus time data are collected for a reactant as a reaction is run. These data are then manipulated and plotted to see which manipulation gives a straight line. From the straight-line plot we get the order of the reactant, and the slope of the line is mathematically related to k, the rate constant.

15. The average rate decreases with time because the reverse reaction occurs more frequently as the concentration of products increase. Initially, with no products present, the rate of the forward reaction is at its fastest, but as time goes on, the rate gets slower and slower since products are converting back into reactants. The instantaneous rate will also decrease with time. The only rate that is constant is the initial rate. This is the instantaneous rate taken at t ≈ 0. At this time, the amount of products is insignificant, and the rate of the reaction only depends on the rate of the forward reaction.

16. The most common method to experimentally determine the differential rate law is the method of initial rates. Once the differential rate law is determined experimentally, the integrated rate law can be derived. However, sometimes it is more convenient and more accurate to collect concentration versus time data for a reactant. When this is the case, then we do "proof" plots to determine the integrated rate law. Once the integrated rate law is determined, the differential rate law can be determined. Either experimental procedure allows determination of both the integrated and the differential rate law; and which rate law is determined by experiment and which is derived is usually decided by which data are easiest and most accurately collected.

17. $$\frac{\text{Rate}_2}{\text{Rate}_1} = \frac{\text{k}[\text{A}]_2^x}{\text{k}[\text{A}]_1^x} = \left(\frac{[\text{A}]_2}{[\text{A}]_1}\right)^x$$

The rate doubles as the concentration quadruples:

$$2 = (4)^x, \quad x = 1/2$$

The order is 1/2 (the square root of the concentration of reactant).

For a reactant that has an order of −1 and the reactant concentration is doubled:

$$\frac{\text{Rate}_2}{\text{Rate}_1} = (2)^{-1} = \frac{1}{2}$$

The rate will decrease by a factor of 1/2 when the reactant concentration is doubled for a −1 order reaction. Negative orders are seen for substances that hinder or slow down a reaction.

18. Some energy must be added to get the reaction started, that is, to overcome the activation energy barrier. Chemically what happens is:

$$\text{Energy} + H_2 \rightarrow 2\ H$$

The hydrogen atoms initiate a chain reaction that proceeds very rapidly. Collisions of H_2 and O_2 molecules at room temperature do not have sufficient kinetic energy to form hydrogen atoms and initiate the reaction.

19. Two reasons are:

(1) The collision must involve enough energy to produce the reaction; that is, the collision energy must be equal to or exceed the activation energy.

(2) The relative orientation of the reactants when they collide must allow formation of any new bonds necessary to produce products.

20. a. The blue plot is the catalyzed pathway. The catalyzed pathway has the lower activation. This is why the catalyzed pathway is faster.

b. ΔE_1 represents the activation energy for the uncatalyzed pathway.

c. ΔE_2 represents the energy difference between the reactants and products. Note that ΔE_2 is the same for both the catalyzed and the uncatalyzed pathways. It is the activation energy that is different for a catalyzed pathway versus an uncatalyzed pathway.

d. Because the products have a higher total energy as compared to reactants, this is an endothermic reaction.

21. Enzymes are very efficient catalysts. As is true for all catalysts, enzymes speed up a reaction by providing an alternative pathway for reactants to convert to products. This alternative pathway has a smaller activation energy and hence, a faster rate. Also true is that catalysts are not used up in the overall chemical reaction. Once an enzyme comes in contact with the correct reagent, the chemical reaction quickly occurs, and the enzyme is then free to catalyze another reaction. Because of the efficiency of the reaction step, only a relatively small amount of enzyme is needed to catalyze a specific reaction, no matter how complex the reaction.

22. The slope of the ln k versus 1/T plot (with temperature in Kelvin) is equal to $-E_a/R$. Because E_a for the catalyzed reaction will be smaller than E_a for the uncatalyzed reaction, the slope of the catalyzed plot should be less negative.

Exercises

Reaction Rates

23. The coefficients in the balanced reaction relate the rate of disappearance of reactants to the rate of production of products. From the balanced reaction, the rate of production of P_4 will be 1/4 the rate of disappearance of PH_3, and the rate of production of H_2 will be 6/4 the rate of disappearance of PH_3. By convention, all rates are given as positive values.

$$\text{Rate} = \frac{-\Delta[PH_3]}{\Delta t} = \frac{-(-0.048 \text{mol}/2.0\text{L})}{\text{s}} = 2.4 \times 10^{-3} \text{ mol/L•s}$$

$$\frac{\Delta[P_4]}{\Delta t} = -\frac{1}{4}\frac{\Delta[PH_3]}{\Delta t} = 2.4 \times 10^{-3}/4 = 6.0 \times 10^{-4} \text{ mol/L•s}$$

$$\frac{\Delta[H_2]}{\Delta t} = -\frac{6}{4}\frac{\Delta[PH_3]}{\Delta t} = 6(2.4 \times 10^{-3})/4 = 3.6 \times 10^{-3} \text{ mol/L•s}$$

24. Using the coefficients in the balanced equation to relate the rates:

$$\frac{\Delta[H_2]}{\Delta t} = 3\frac{\Delta[N_2]}{\Delta t} \text{ and } \frac{\Delta[NH_3]}{\Delta t} = -2\frac{\Delta[N_2]}{\Delta t}$$

$$\text{So}: -\frac{1}{3}\frac{\Delta[H_2]}{\Delta t} = \frac{1}{2}\frac{\Delta[NH_3]}{\Delta t} \text{ or } \frac{\Delta[NH_3]}{\Delta t} = -\frac{2}{3}\frac{\Delta[H_2]}{\Delta t}$$

Ammonia is produced at a rate equal to 2/3 of the rate of consumption of hydrogen.

25. a. $\text{Average rate} = \frac{-\Delta[H_2O_2]}{\Delta t} = \frac{-(0.500\,M - 1.000\,M)}{(2.16 \times 10^4 \text{ s} - 0)} = 2.31 \times 10^{-5} \text{ mol/L•s}$

From the coefficients in the balanced equation:

$$\frac{\Delta[O_2]}{\Delta t} = -\frac{1}{2}\frac{\Delta[H_2O_2]}{\Delta t} = 1.16 \times 10^{-5} \text{ mol/L•s}$$

b. $\frac{-\Delta[H_2O_2]}{\Delta t} = \frac{-(0.250 - 0.500)\,M}{(4.32 \times 10^4 - 2.16 \times 10^4)\text{ s}} = 1.16 \times 10^{-5} \text{ mol/L•s}$

$$\frac{\Delta[O_2]}{\Delta t} = 1/2\,(1.16 \times 10^{-5}) = 5.80 \times 10^{-6} \text{ mol/L•s}$$

Notice that as time goes on in a reaction, the average rate decreases.

26. 0.0120/0.0080 = 1.5; reactant B is used up 1.5 times faster than reactant A. This corresponds to a 3 to 2 mole ratio between B and A in the balanced equation. 0.0160/0.0080 = 2; product C is produced twice as fast as reactant A is used up, so the coefficient for C is twice the coefficient for A. A possible balanced equation is $2A + 3B \rightarrow 4C$.

27. a. The units for rate are always mol/L•s.

b. Rate = k; k must have units of mol/L•s.

c. $\text{Rate} = k[A], \quad \frac{\text{mol}}{\text{L s}} = k\left(\frac{\text{mol}}{\text{L}}\right)$

k must have units of s^{-1}.

d. $\text{Rate} = k[A]^2, \quad \frac{\text{mol}}{\text{L s}} = k\left(\frac{\text{mol}}{\text{L}}\right)^2$

k must have units of L/mol•s.

e. L^2/mol^2•s

28. $\text{Rate} = k[Cl]^{1/2}[CHCl_3], \quad \frac{\text{mol}}{\text{L s}} = k\left(\frac{\text{mol}}{\text{L}}\right)^{1/2}\left(\frac{\text{mol}}{\text{L}}\right)$; k must have units of $L^{1/2}/mol^{1/2}$•s.

Rate Laws from Experimental Data: Initial Rates Method

29. a. In the first two experiments, [NO] is held constant and $[Cl_2]$ is doubled. The rate also doubled. Thus the reaction is first order with respect to Cl_2. Or mathematically, Rate = $k[NO]^x[Cl_2]^y$.

$$\frac{0.36}{0.18} = \frac{k(0.10)^x(0.20)^y}{k(0.10)^x(0.10)^y} = \frac{(0.20)^y}{(0.10)^y},\ 2.0 = 2.0^y,\ y = 1$$

We can get the dependence on NO from the second and third experiments. Here, as the NO concentration doubles (Cl_2 concentration is constant), the rate increases by a factor of four. Thus the reaction is second order with respect to NO. Or mathematically:

$$\frac{1.45}{0.36} = \frac{k(0.20)^x(0.20)}{k(0.10)^x(0.20)} = \frac{(0.20)^x}{(0.10)^x},\ 4.0 = 2.0^x,\ x = 2;\ \text{so Rate} = k[NO]^2[Cl_2].$$

Try to examine experiments where only one concentration changes at a time. The more variables that change, the harder it is to determine the orders. Also, these types of problems can usually be solved by inspection. In general, we will solve using a mathematical approach, but keep in mind that you probably can solve for the orders by simple inspection of the data.

b. The rate constant k can be determined from the experiments. From experiment 1:

$$\frac{0.18\ \text{mol}}{\text{L min}} = k\left(\frac{0.10\ \text{mol}}{\text{L}}\right)^2\left(\frac{0.10\ \text{mol}}{\text{L}}\right),\ k = 180\ \text{L}^2/\text{mol}^2\bullet\text{min}$$

From the other experiments:

$k = 180\ \text{L}^2/\text{mol}^2\bullet\text{min}$ (second exp.); $k = 180\ \text{L}^2/\text{mol}^2\bullet\text{min}$ (third exp.)

The average rate constant is $k_{mean} = 1.8 \times 10^2\ \text{L}^2/\text{mol}^2\bullet\text{min}$.

30. a. Rate = $k[I^-]^x[S_2O_8^{2-}]^y$; $\frac{12.5 \times 10^{-6}}{6.25 \times 10^{-6}} = \frac{k(0.080)^x(0.040)^y}{k(0.040)^x(0.040)^y},\ 2.00 = 2.0^x,\ x = 1$

$$\frac{12.5 \times 10^{-6}}{6.25 \times 10^{-6}} = \frac{k(0.080)(0.040)^y}{k(0.080)(0.020)^y},\ 2.00 = 2.0^y,\ y = 1;\ \text{Rate} = k[I^-][S_2O_8^{2-}]$$

b. For the first experiment:

$$\frac{12.5 \times 10^{-6}\ \text{mol}}{\text{L s}} = k\left(\frac{0.080\ \text{mol}}{\text{L}}\right)\left(\frac{0.040\ \text{mol}}{\text{L}}\right),\ k = 3.9 \times 10^{-3}\ \text{L/mol}\bullet\text{s}$$

Each of the other experiments also gives $k = 3.9 \times 10^{-3}$ L/mol•s, so $k_{mean} = 3.9 \times 10^{-3}$ L/mol•s.

31. a. Rate = $k[NOCl]^n$; using experiments two and three:

$$\frac{2.66 \times 10^4}{6.64 \times 10^3} = \frac{k(2.0 \times 10^{16})^n}{k(1.0 \times 10^{16})^n}, \quad 4.01 = 2.0^n, \; n = 2; \; \text{Rate} = k[NOCl]^2$$

b. $$\frac{5.98 \times 10^4 \text{ molecules}}{\text{cm}^3 \text{ s}} = k\left(\frac{3.0 \times 10^{16} \text{ molecules}}{\text{cm}^3}\right)^2, \quad k = 6.6 \times 10^{-29} \text{ cm}^3/\text{molecules•s}$$

The other three experiments give (6.7, 6.6, and 6.6) × 10^{-29} cm^3/molecules•s, respectively. The mean value for k is 6.6 × 10^{-29} cm^3/molecules•s.

c. $$\frac{6.6 \times 10^{-29} \text{ cm}^3}{\text{moleculess}} \times \frac{1\text{ L}}{1000\text{cm}^3} \times \frac{6.022 \times 10^{23} \text{ molecules}}{\text{mol}} = \frac{4.0 \times 10^{-8} \text{ L}}{\text{mols}}$$

32. Rate = $k[N_2O_5]^x$; the rate laws for the first two experiments are:

$2.26 \times 10^{-3} = k(0.190)^x$ and $8.90 \times 10^{-4} = k(0.0750)^x$

Dividing the two rate laws: $2.54 = \dfrac{(0.190)^x}{(0.0750)^x} = (2.53)^x$, $x = 1$; Rate = $k[N_2O_5]$

$$k = \frac{\text{Rate}}{[N_2O_5]} = \frac{8.90 \times 10^{-4} \text{ mol/L} \bullet \text{s}}{0.0750\text{mol/L}} = 1.19 \times 10^{-2} \text{ s}^{-1}$$

The other experiments give similar values for k. $k_{mean} = 1.19 \times 10^{-2}$ s^{-1}

33. a. Rate = $k[I^-]^x[OCl^-]^y$; $\dfrac{7.91 \times 10^{-2}}{3.95 \times 10^{-2}} = \dfrac{k(0.12)^x(0.18)^y}{k(0.060)^x(0.18)^y} = 2.0^x$, $2.00 = 2.0^x$, $x = 1$

$$\frac{3.95 \times 10^{-2}}{9.88 \times 10^{-3}} = \frac{k(0.060)(0.18)^y}{k(0.030)(0.090)^y}, \quad 4.00 = 2.0 \times 2.0^y, \; 2.0 = 2.0^y, \; y = 1$$

Rate = $k[I^-][OCl^-]$

b. From the first experiment: $\dfrac{7.91 \times 10^{-2} \text{ mol}}{\text{L s}} = k\left(\dfrac{0.12\text{ mol}}{\text{L}}\right)\left(\dfrac{0.18\text{ mol}}{\text{L}}\right)$, k = 3.7 L/mol•s

All four experiments give the same value of k to two significant figures.

c. $$\text{Rate} = \frac{3.7\text{L}}{\text{mols}} \times \frac{0.15\text{mol}}{\text{L}} \times \frac{0.15\text{mol}}{\text{L}} = 0.083\text{mol/L} \bullet \text{s}$$

34. Rate = $k[NO]^x[O_2]^y$; comparing the first two experiments, $[O_2]$ is unchanged, [NO] is tripled, and the rate increases by a factor of nine. Therefore, the reaction is second order in NO (3^2 = 9). The order of O_2 is more difficult to determine. Comparing the second and third experiments:

$$\frac{3.13 \times 10^{17}}{1.80 \times 10^{17}} = \frac{k(2.50 \times 10^{18})^2(2.50 \times 10^{18})^y}{k(3.00 \times 10^{18})^2(1.00 \times 10^{18})^y}$$

$1.74 = 0.694(2.50)^y$, $2.51 = 2.50^y$, $y = 1$

Rate = $k[NO]^2[O_2]$; from experiment 1:

$$2.00 \times 10^{16} \text{ molecules/cm}^3\text{•s} = k(1.00 \times 10^{18} \text{ molecules/cm}^3)^2 \times (1.00 \times 10^{18} \text{ molecules/cm}^3)$$

$k = 2.00 \times 10^{-38}$ cm^6/molecules^2•s = k_{mean}

$$\text{Rate} = \frac{2.00 \times 10^{-38} \text{ cm}^6}{\text{molecules}^2 \text{ s}} \times \left(\frac{6.21 \times 10^{18} \text{ molecules}}{\text{cm}^3}\right)^2 \times \frac{7.36 \times 10^{18} \text{ molecules}}{\text{cm}^3}$$

Rate = 5.68×10^{18} molecules/cm^3•s

35. a. Rate = $k[Hb]^x[CO]^y$

Comparing the first two experiments, [CO] is unchanged, [Hb] doubles, and the rate doubles. Therefore, $x = 1$, and the reaction is first order in Hb. Comparing the second and third experiments, [Hb] is unchanged, [CO] triples, and the rate triples. Therefore, $y = 1$, and the reaction is first order in CO.

b. Rate = k[Hb][CO]

c. From the first experiment:

0.619 μmol/L•s = k(2.21 μmol/L)(1.00 μmol/L), k = 0.280 L/μmol•s

The second and third experiments give similar k values, so k_{mean} = 0.280 L/μmol•s

d. $$\text{Rate} = k[\text{Hb}][\text{CO}] = \frac{0.280 \text{ L}}{\mu\text{mol s}} \times \frac{3.36\,\mu\text{mol}}{\text{L}} \times \frac{2.40\,\mu\text{mol}}{\text{L}} = 2.26\ \mu\text{mol/L•s}$$

36. a. Rate = $k[ClO_2]^x[OH^-]^y$; from the first two experiments:

$2.30 \times 10^{-1} = k(0.100)^x(0.100)^y$ and $5.75 \times 10^{-2} = k(0.0500)^x(0.100)^y$

Dividing the two rate laws: $4.00 = \frac{(0.100)^x}{(0.0500)^x} = 2.00^x$, $x = 2$

Comparing the second and third experiments:

$2.30 \times 10^{-1} = k(0.100)(0.100)^y$ and $1.15 \times 10^{-1} = k(0.100)(0.0500)^y$

Dividing: $2.00 = \frac{(0.100)^y}{(0.050)^y} = 2.0^y$, $y = 1$

The rate law is Rate = $k[ClO_2]^2[OH^-]$.

2.30×10^{-1} mol/L•s = $k(0.100 \text{ mol/L})^2(0.100 \text{ mol/L})$, $k = 2.30 \times 10^2 \text{ L}^2/\text{mol}^2\text{•s} = k_{mean}$

b. $\text{Rate} = \dfrac{2.30 \times 10^2 \text{ L}^2}{\text{mol}^2 \text{ s}} \times \left(\dfrac{0.175 \text{mol}}{\text{L}}\right)^2 \times \dfrac{0.0844 \text{mol}}{\text{L}} = 0.594 \text{ mol/L•s}$

Integrated Rate Laws

37. The first assumption to make is that the reaction is first order. For a first order reaction, a graph of $\ln[H_2O_2]$ versus time will yield a straight line. If this plot is not linear, then the reaction is not first order, and we make another assumption.

Time (s)	$[H_2O_2]$ (mol/L)	$\ln[H_2O_2]$
0	1.00	0.000
120.	0.91	−0.094
300.	0.78	−0.25
600.	0.59	−0.53
1200.	0.37	−0.99
1800.	0.22	−1.51
2400.	0.13	−2.04
3000.	0.082	−2.50
3600.	0.050	−3.00

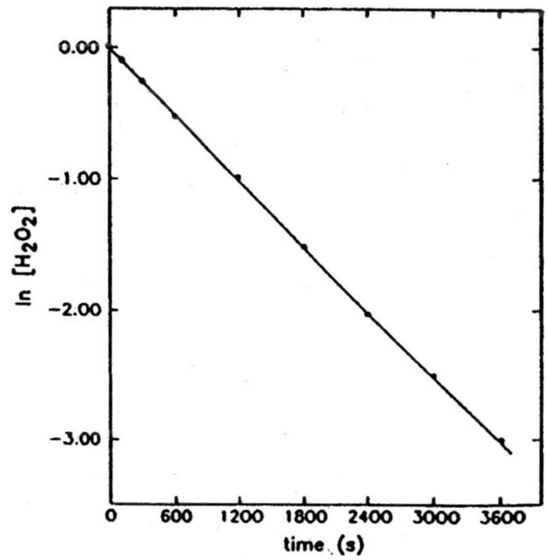

Note: We carried extra significant figures in some of the natural log values in order to reduce round-off error. For the plots, we will do this most of the time when the natural log function is involved.

The plot of $\ln[H_2O_2]$ versus time is linear. Thus the reaction is first order. The rate law and integrated rate law are Rate = $k[H_2O_2]$ and $\ln[H_2O_2] = -kt + \ln[H_2O_2]_0$.

We determine the rate constant k by determining the slope of the $\ln[H_2O_2]$ versus time plot (slope = −k). Using two points on the curve gives:

$$\text{slope} = -k = \frac{\Delta y}{\Delta x} = \frac{0-(3.00)}{0-3600.} = -8.3 \times 10^{-4} \text{ s}^{-1}, \quad k = 8.3 \times 10^{-4} \text{ s}^{-1}$$

To determine $[H_2O_2]$ at 4000. s, use the integrated rate law, where $[H_2O_2]_0$ = 1.00 *M*.

$$\ln[H_2O_2] = -kt + \ln[H_2O_2]_0 \quad \text{or} \quad \ln\left(\frac{[H_2O_2]}{[H_2O_2]_0}\right) = -kt$$

$$\ln\left(\frac{[H_2O_2]}{1.00}\right) = -8.3 \times 10^{-4}\ s^{-1} \times 4000.\ s, \quad \ln[H_2O_2] = -3.3, \quad [H_2O_2] = e^{-3.3} = 0.037\ M$$

38. a. Because the ln[A] versus time plot was linear, the reaction is first order in A. The slope of the ln[A] versus time plot equals –k. Therefore, the rate law, the integrated rate law, and the rate constant value are:

$$\text{Rate} = k[A]; \quad \ln[A] = -kt + \ln[A]_0; \quad k = 2.97 \times 10^{-2}\ \text{min}^{-1}$$

b. The half-life expression for a first order rate law is:

$$t_{1/2} = \frac{\ln 2}{k} = \frac{0.6931}{k}, \quad t_{1/2} = \frac{0.6931}{2.97 \times 10^{-2}\ \text{min}^{-1}} = 23.3\ \text{min}$$

c. $2.50 \times 10^{-3}\ M$ is 1/8 of the original amount of A present initially, so the reaction is 87.5% complete. When a first-order reaction is 87.5% complete (or 12.5% remains), then the reaction has gone through 3 half-lives:

$$100\% \xrightarrow[t_{1/2}]{} 50.0\% \xrightarrow[t_{1/2}]{} 25.0\% \xrightarrow[t_{1/2}]{} 12.5\%; \quad t = 3 \times t_{1/2} = 3 \times 23.3\ \text{min} = 69.9\ \text{min}$$

Or we can use the integrated rate law:

$$\ln\left(\frac{[A]}{[A]_0}\right) = -kt, \quad \ln\left(\frac{2.50 \times 10^{-3}\ M}{2.00 \times 10^{-2}\ M}\right) = -(2.97 \times 10^{-2}\ \text{min}^{-1})t$$

$$t = \frac{\ln(0.125)}{-2.97 \times 10^{-2}\ \text{min}^{-1}} = 70.0\ \text{min}$$

39. Assume the reaction is first order and see if the plot of $\ln[NO_2]$ versus time is linear. If this isn't linear, try the second-order plot of $1/[NO_2]$ versus time because second-order reactions are the next most common after first-order reactions. The data and plots follow.

Time (s)	$[NO_2]$ (M)	$\ln[NO_2]$	$1/[NO_2]$ (M^{-1})
0	0.500	–0.693	2.00
1.20×10^3	0.444	–0.812	2.25
3.00×10^3	0.381	–0.965	2.62
4.50×10^3	0.340	–1.079	2.94
9.00×10^3	0.250	–1.386	4.00
1.80×10^4	0.174	–1.749	5.75

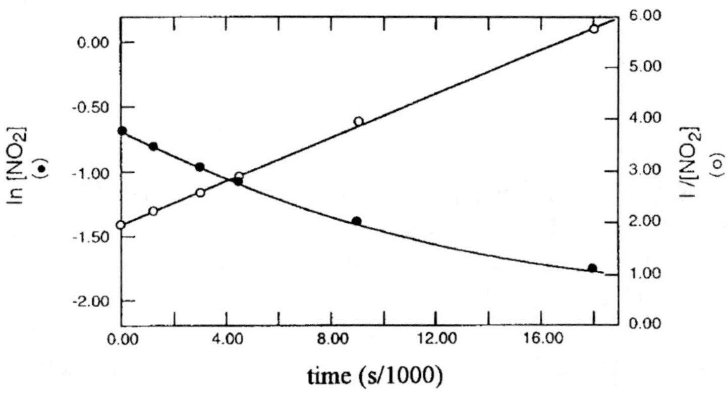

The plot of $1/[NO_2]$ versus time is linear. The reaction is second order in NO_2.

The rate law and integrated rate law are: Rate = $k[NO_2]^2$ and $\dfrac{1}{[NO_2]} = kt + \dfrac{1}{[NO_2]_0}$

The slope of the plot $1/[NO_2]$ vs. t gives the value of k. Using a couple of points on the plot:

$$\text{slope} = k = \frac{\Delta y}{\Delta x} = \frac{(5.75 - 2.00)\ M^{-1}}{(1.80 \times 10^4 - 0)\ \text{s}} = 2.08 \times 10^{-4}\ \text{L/mol•s}$$

To determine $[NO_2]$ at 2.70×10^4 s, use the integrated rate law, where $1/[NO_2]_o = 1/0.500\ M = 2.00\ M^{-1}$.

$$\frac{1}{[NO_2]} = kt + \frac{1}{[NO_2]_0},\quad \frac{1}{[NO_2]} = \frac{2.08 \times 10^{-4}\ \text{L}}{\text{mol s}} \times 2.70 \times 10^4\ \text{s} + 2.00\ M^{-1}$$

$$\frac{1}{[NO_2]} = 7.62,\quad [NO_2] = 0.131\ M$$

40. a. Because the $1/[A]$ versus time plot was linear, the reaction is second order in A. The slope of the $1/[A]$ versus time plot equals the rate constant k. Therefore, the rate law, the integrated rate law, and the rate constant value are:

$$\text{Rate} = k[A]^2;\quad \frac{1}{[A]} = kt + \frac{1}{[A]_0};\quad k = 3.60 \times 10^{-2}\ \text{L mol}^{-1}\ \text{s}^{-1}$$

b. The half-life expression for a second-order reaction is: $t_{1/2} = \dfrac{1}{k[A]_0}$

For this reaction: $$t_{1/2} = \frac{1}{3.60 \times 10^{-2}\ \text{L/mol•s} \times 2.80 \times 10^{-3}\ \text{mol/L}} = 9.92 \times 10^3\ \text{s}$$

Note: We could have used the integrated rate law to solve for $t_{1/2}$, where $[A] = (2.80 \times 10^{-3}/2)$ mol/L.

c. Because the half-life for a second-order reaction depends on concentration, we must use the integrated rate law to solve.

$$\frac{1}{[A]} = kt + \frac{1}{[A]_0}, \quad \frac{1}{7.00 \times 10^{-4}\ M} = \frac{3.60 \times 10^{-2}\ L}{mols} \times t + \frac{1}{2.80 \times 10^{-3}\ M}$$

$1.43 \times 10^3 - 357 = (3.60 \times 10^{-2})t$, $t = 2.98 \times 10^4$ s

41. a. Because the $[C_2H_5OH]$ versus time plot was linear, the reaction is zero order in C_2H_5OH. The slope of the $[C_2H_5OH]$ versus time plot equals -k. Therefore, the rate law, the integrated rate law, and the rate constant value are: Rate = $k[C_2H_5OH]^0 = k$; $[C_2H_5OH] = -kt + [C_2H_5OH]_0$; $k = 4.00 \times 10^{-5}$ mol/L•s

b. The half-life expression for a zero-order reaction is $t_{1/2} = [A]_0/2k$.

$$t_{1/2} = \frac{[C_2H_5OH]_0}{2k} = \frac{1.25 \times 10^{-2}\ mol/L}{2 \times 4.00 \times 10^{-5}\ mol/L \bullet s} = 156\ s$$

Note: We could have used the integrated rate law to solve for $t_{1/2}$, where $[C_2H_5OH] = (1.25 \times 10^{-2}/2)$ mol/L.

c. $[C_2H_5OH] = -kt + [C_2H_5OH]_0$, 0 mol/L $= -(4.00 \times 10^{-5}$ mol/L•s$)t + 1.25 \times 10^{-2}$ mol/L

$$t = \frac{1.25 \times 10^{-2}\ mol/L}{4.00 \times 10^{-5}\ mol/L \bullet s} = 313\ s$$

42. From the data, the pressure of C_2H_5OH decreases at a constant rate of 13 torr for every 100. s. Because the rate of disappearance of C_2H_5OH is not dependent on concentration, the reaction is zero order in C_2H_5OH.

$$k = \frac{13\ torr}{100.\ s} \times \frac{1\ atm}{760\ torr} = 1.7 \times 10^{-4}\ atm/s$$

The rate law and integrated rate law are:

$$\text{Rate} = k = 1.7 \times 10^{-4}\ atm/s;\ P_{C_2H_5OH} = -kt + 250.\ torr\left(\frac{1\ atm}{760\ torr}\right) = -kt + 0.329\ atm$$

At 900. s: $P_{C_2H_5OH} = -1.7 \times 10^{-4}$ atm/s $\times$ 900. s + 0.329 atm = 0.176 atm = 0.18 atm = 130 torr

43. The first assumption to make is that the reaction is first order. For a first-order reaction, a graph of $\ln[C_4H_6]$ versus t should yield a straight line. If this isn't linear, then try the second-order plot of $1/[C_4H_6]$ versus t. The data and the plots follow:

Time	195	604	1246	2180	6210 s
$[C_4H_6]$	1.6×10^{-2}	1.5×10^{-2}	1.3×10^{-2}	1.1×10^{-2}	$0.68 \times 10^{-2}\ M$
$\ln[C_4H_6]$	−4.14	−4.20	−4.34	−4.51	−4.99
$1/[C_4H_6]$	62.5	66.7	76.9	90.9	$147\ M^{-1}$

Note: To reduce round-off error, we carried extra significant figures in the data points.

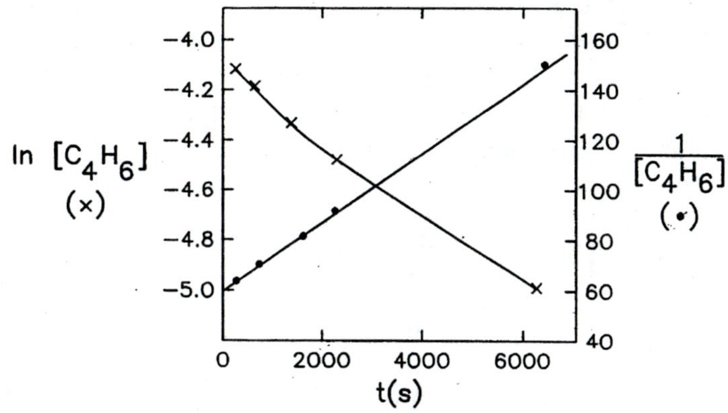

The natural log plot is not linear, so the reaction is not first order. Because the second-order plot of $1/[C_4H_6]$ versus t is linear, we can conclude that the reaction is second order in butadiene. The rate law is:

Rate = $k[C_4H_6]^2$

For a second-order reaction, the integrated rate law is $\dfrac{1}{[C_4H_6]} = kt + \dfrac{1}{[C_4H_6]_0}$.

The slope of the straight line equals the value of the rate constant. Using the points on the line at 1000. and 6000. s:

$$k = \text{slope} = \frac{144\ \text{L/mol} - 73\ \text{L/mol}}{6000.\text{s} - 1000.\text{s}} = 1.4 \times 10^{-2}\ \text{L/mol•s}$$

44. a. First, assume the reaction to be first order with respect to O. Hence a graph of ln[O] versus t would be linear if the reaction is first order.

t (s)	[O] (atoms/cm^3)	ln[O]
0	5.0×10^9	22.33
$10. \times 10^{-3}$	1.9×10^9	21.37
$20. \times 10^{-3}$	6.8×10^8	20.34
$30. \times 10^{-3}$	2.5×10^8	19.34

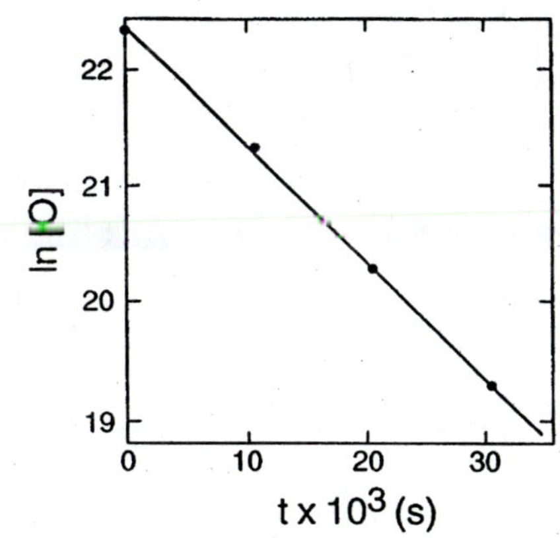

Because the graph is linear, we can conclude the reaction is first order with respect to O.

b. The overall rate law is Rate = k[NO_2][O].

Because NO_2 was in excess, its concentration is constant. Thus, for this experiment, the rate law is Rate = k′[O], where k′ = k[NO_2]. In a typical first-order plot, the slope equals –k. For this experiment, the slope equals –k′ = –k[NO_2]. From the graph:

$$\text{slope} = \frac{19.34 - 22.23}{(30. \times 10^{-3} - 0)\,\text{s}} = -1.0 \times 10^2\ \text{s}^{-1},\ \ k' = -\text{slope} = 1.0 \times 10^2\ \text{s}^{-1}$$

To determine k, the actual rate constant:

$$k' = k[\text{NO}_2],\ \ 1.0 \times 10^2\ \text{s}^{-1} = k(1.0 \times 10^{13}\ \text{molecules/cm}^3)$$

$$k = 1.0 \times 10^{-11}\ \text{cm}^3/\text{molecules}\bullet\text{s}$$

45. Because the 1/[A] versus time plot is linear with a positive slope, the reaction is second order with respect to A. The *y* intercept in the plot will equal 1/$[A]_0$. Extending the plot, the *y* intercept will be about 10, so 1/10 = 0.1 *M* = $[A]_0$.

46. a. The slope of the 1/[A] versus time plot in Exercise 45 will equal k.

$$\text{Slope} = k = \frac{(60 - 20)\,\text{L/mol}}{(5 - 1)\,\text{s}} = 10\ \text{L/mol}\bullet\text{s}$$

$$\frac{1}{[A]} = kt + \frac{1}{[A]_0} = \frac{10\,\text{L}}{\text{mol s}} \times 9\,\text{s} + \frac{1}{0.1\,M} = 100,\ \ [A] = 0.01\ M$$

b. For a second-order reaction, the half-life does depend on concentration: $t_{1/2} = \frac{1}{k[A]_0}$

$$\text{First half-life: } t_{1/2} = \frac{1}{\frac{10\,\text{L}}{\text{mol s}} \times \frac{0.1\,\text{mol}}{\text{L}}} = 1\ \text{s}$$

Second half-life ($[A]_0$ is now 0.05 *M*): $t_{1/2}$ = 1/(10 × 0.05) = 2 s

Third half-life ($[A]_0$ is now 0.025 *M*): $t_{1/2}$ = 1/(10 × 0.025) = 4 s

47. a. [A] = –kt + $[A]_0$; if k = 5.0 × 10^{-2} mol/L•s and $[A]_0$ = 1.00 × 10^{-3} *M*, then:

$$[A] = -(5.0 \times 10^{-2}\ \text{mol/L}\bullet\text{s})t + 1.00 \times 10^{-3}\ \text{mol/L}$$

b. $\frac{[A]_0}{2} = -(5.0 \times 10^{-2})t_{1/2} + [A]_0$ because at t = $t_{1/2}$, [A] = $[A]_0$/2.

$$-0.50[A]_0 = -(5.0 \times 10^{-2})t_{1/2},\ \ t_{1/2} = \frac{0.50(1.00 \times 10^{-3})}{5.0 \times 10^{-2}} = 1.0 \times 10^{-2}\ \text{s}$$

Note: We could have used the $t_{1/2}$ expression to solve ($t_{1/2} = \frac{[A]_0}{2k}$).

c. $[A] = -kt + [A]_0 = -(5.0 \times 10^{-2}\text{ mol/L•s})(5.0 \times 10^{-3}\text{ s}) + 1.00 \times 10^{-3}\text{ mol/L}$

$[A] = 7.5 \times 10^{-4}\text{ mol/L}$

$[A]_{reacted} = 1.00 \times 10^{-3}\text{ mol/L} - 7.5 \times 10^{-4}\text{ mol/L} = 2.5 \times 10^{-4}\text{ mol/L}$

$[B]_{produced} = [A]_{reacted} = 2.5 \times 10^{-4}\ M$

48. a. The integrated rate law for this zero-order reaction is $[HI] = -kt + [HI]_0$.

$$[HI] = -kt + [HI]_0,\ [HI] = -\left(\frac{1.20 \times 10^{-4}\text{ mol}}{\text{L s}}\right) \times \left(25\text{ min} \times \frac{60\text{ s}}{\text{min}}\right) + \frac{0.250\text{mol}}{\text{L}}$$

$[HI] = -0.18\text{ mol/L} + 0.250\text{ mol/L} = 0.07\ M$

b. $[HI] = 0 = -kt + [HI]_0,\ kt = [HI]_0,\ t = \dfrac{[HI]_0}{k}$

$$t = \frac{0.250\text{mol/L}}{1.20 \times 10^{-4}\text{ mol/L}\bullet\text{s}} = 2080\text{ s} = 34.7\text{ min}$$

49. If $[A]_0 = 100.0$, then after 65 s, 45.0% of A has reacted, or $[A] = 55.0$. For first order reactions:

$$\ln\left(\frac{[A]}{[A]_0}\right) = -kt,\ \ln\left(\frac{55.0}{100.0}\right) = -k(65\text{ s}),\ k = 9.2 \times 10^{-3}\text{ s}^{-1}$$

$$t_{1/2} = \frac{\ln 2}{k} = \frac{0.693}{9.2 \times 10^{-3}\text{ s}^{-1}} = 75\text{ s}$$

50. a. When a reaction is 75.0% complete (25.0% of reactant remains), this represents two half-lives (100% → 50% → 25%). The first-order half-life expression is $t_{1/2}$ = (ln 2)/k. Because there is no concentration dependence for a first-order half-life, 320. s = two half-lives, $t_{1/2}$ = 320./2 = 160. s. This is both the first half-life, the second half-life, etc.

b. $t_{1/2} = \dfrac{\ln 2}{k},\ k = \dfrac{\ln 2}{t_{1/2}} = \dfrac{\ln 2}{160.\text{ s}} = 4.33 \times 10^{-3}\text{ s}^{-1}$

At 90.0% complete, 10.0% of the original amount of the reactant remains, so $[A] = 0.100[A]_0$.

$$\ln\left(\frac{[A]}{[A]_0}\right) = -kt,\ \ln\frac{0.100[A]_0}{[A]_0} = -(4.33 \times 10^{-3}\text{ s}^{-1})t,\ t = \frac{\ln(0.100)}{-4.33 \times 10^{-3}\text{ s}^{-1}} = 532\text{ s}$$

51. For a first-order reaction, the integrated rate law is $\ln([A]/[A]_0) = -kt$. Solving for k:

$$\ln\left(\frac{0.250\text{mol/L}}{1.00\text{mol/L}}\right) = -k \times 120.\text{ s},\ k = 0.0116\text{ s}^{-1}$$

$$\ln\left(\frac{0.350\text{mol/L}}{2.00\text{mol/L}}\right) = -0.0116\text{ s}^{-1} \times t,\ t = 150.\text{ s}$$

52. $$\ln\left(\frac{[A]}{[A]_0}\right) = -kt;\ k = \frac{\ln 2}{t_{1/2}} = \frac{0.693}{56.0\text{ days}} = 0.0124\text{d}^{-1}$$

$$\ln\left(\frac{1.41 \times 10^{-7}\text{ mol/L}}{8.75 \times 10^{-5}\text{ mol/L}}\right) = -(0.0124\text{ d}^{-1})t,\ t = 519\text{ days}$$

53. Comparing experiments 1 and 2, as the concentration of AB is doubled, the initial rate increases by a factor of 4. The reaction is second order in AB.

Rate = $k[AB]^2$, 3.20×10^{-3} mol/L•s = $k(0.200\ M)^2$

$k = 8.00 \times 10^{-2}$ mol/L•s = k_{mean}

For a second order reaction:

$$t_{1/2} = \frac{1}{k[AB]_0} = \frac{1}{8.00 \times 10^{-2}\text{ L/mol}\cdot\text{s} \times 1.00\text{mol/L}} = 12.5\text{ s}$$

54. a. The integrated rate law for a second order reaction is $1/[A] = kt + 1/[A]_0$, and the half-life expression is $t_{1/2} = 1/k[A]_0$. We could use either to solve for $t_{1/2}$. Using the integrated rate law:

$$\frac{1}{(0.900/2)\text{mol/L}} = k \times 2.00\text{ s} + \frac{1}{0.900\text{mol/L}},\ k = \frac{1.11\text{L/mol}}{2.00\text{s}} = 0.555\text{ L/mol}\cdot\text{s}$$

b. $$\frac{1}{0.100\text{mol/L}} = 0.555\text{ L/mol}\cdot\text{s} \times t + \frac{1}{0.900\text{mol/L}},\ t = \frac{8.9\text{L/mol}}{0.555\text{L/mol}\cdot\text{s}} = 16\text{ s}$$

55. Successive half-lives double as concentration is decreased by one-half. This is consistent with second-order reactions, so assume the reaction is second order in A.

$$t_{1/2} = \frac{1}{k[A]_0},\ k = \frac{1}{t_{1/2}[A]_0} = \frac{1}{10.0\text{ min}(0.10\ M)} = 1.0\text{ L/mol}\cdot\text{min}$$

a. $$\frac{1}{[A]} = kt + \frac{1}{[A]_0} = \frac{1.0\text{ L}}{\text{mol min}} \times 80.0\text{ min} + \frac{1}{0.10\ M} = 90.\ M^{-1},\ [A] = 1.1 \times 10^{-2}\ M$$

b. 30.0 min = 2 half-lives, so 25% of original A is remaining.

$[A] = 0.25(0.10\ M) = 0.025\ M$

56. The consecutive half-life values of 24 hours, then 12 hours, show a direct relationship with concentration; as the concentration decreases, the half-life decreases. Assuming the drug reaction is either zero, first, or second order, only a zero order reaction shows this direct relationship between half-life and concentration. Therefore, assume the reaction is zero order in the drug.

$$t_{1/2} = \frac{[A]_0}{2k},\ k = \frac{[A]_0}{2t_{1/2}} = \frac{2.0 \times 10^{-3}\ \text{mol/L}}{2(24\text{h})} = 4.2 \times 10^{-5}\ \text{mol/L•h}$$

57. Because $[V]_0 >> [AV]_0$, the concentration of V is essentially constant in this experiment. We have a pseudo-first-order reaction in AV:

$$\text{Rate} = k[AV][V] = k'[AV],\ \text{where } k' = k[V]_0$$

The slope of the ln[AV] versus time plot is equal to –k′.

$$k' = -\text{slope} = 0.32\ s^{-1};\ k = \frac{k'}{[V]_0} = \frac{0.32\ s^{-1}}{0.20\ \text{mol/L}} = 1.6\ \text{L/mol•s}$$

58. Because $[B]_0 >> [A]_0$, the B concentration is essentially constant during this experiment, so rate = k′[A] where $k' = k[B]^2$. For this experiment, the reaction is a pseudo-first-order reaction in A.

a. $$\ln\left(\frac{[A]}{[A]_0}\right) = -k't,\ \ln\left(\frac{3.8 \times 10^{-3}\ M}{1.0 \times 10^{-2}\ M}\right) = -k' \times 8.0\ s,\ k' = 0.12\ s^{-1}$$

For the reaction: $k' = k[B]^2$, $k = 0.12\ s^{-1}/(3.0\ \text{mol/L})^2 = 1.3 \times 10^{-2}\ L^2\ mol^{-2}\ s^{-1}$

b. $$t_{1/2} = \frac{\ln 2}{k'} = \frac{0.693}{0.12\ s^{-1}} = 5.8\ s$$

c. $$\ln\left(\frac{[A]}{1.0 \times 10^{-2}\ M}\right) = -0.12\ s^{-1} \times 13.0\ s,\ \frac{[A]}{1.0 \times 10^{-2}} = e^{-0.12(13.0)} = 0.21$$

$[A] = 2.1 \times 10^{-3} M$

d. $[A]_{reacted} = 0.010\ M - 0.0021\ M = 0.008\ M$

$$[C]_{reacted} = 0.008\ M \times \frac{2\ \text{mol C}}{1\ \text{mol A}} = 0.016\ M \approx 0.02\ M$$

$[C]_{remaining} = 2.0\ M - 0.02\ M = 2.0\ M$; as expected, the concentration of C basically remains constant during this experiment since $[C]_0 >> [A]_0$.

Reaction Mechanisms

59. For elementary reactions, the rate law can be written using the coefficients in the balanced equation to determine orders.

a. Rate = $k[CH_3NC]$ b. Rate = $k[O_3][NO]$

c. Rate = $k[O_3]$ d. Rate = $k[O_3][O]$

60. From experiment (Exercise 37), we know the rate law is Rate = $k[H_2O_2]$. A mechanism consists of a series of elementary reactions where the rate law for each step can be determined using the coefficients in the balanced equation for each respective step. For a plausible mechanism, the rate law derived from a mechanism must agree with the rate law determined from experiment. To derive the rate law from the mechanism, the rate of the reaction is assumed to equal the rate of the slowest step in the mechanism.

This mechanism will agree with the experimentally determined rate law only if step 1 is the slow step (called the rate-determining step). If step 1 is slow, then Rate = $k[H_2O]_2$ which agrees with experiment.

Another important property of a mechanism is that the sum of all steps must give the overall balanced equation. Summing all steps gives:

$$H_2O_2 \rightarrow 2\ OH$$
$$H_2O_2 + OH \rightarrow H_2O + HO_2$$
$$HO_2 + OH \rightarrow H_2O + O_2$$
$$\overline{2\ H_2O_2 \rightarrow 2\ H_2O + O_2}$$

61. A mechanism consists of a series of elementary reactions in which the rate law for each step can be determined using the coefficients in the balanced equations. For a plausible mechanism, the rate law derived from a mechanism must agree with the rate law determined from experiment. To derive the rate law from the mechanism, the rate of the reaction is assumed to equal the rate of the slowest step in the mechanism.

Because step 1 is the rate-determining step, the rate law for this mechanism is Rate = $k[C_4H_9Br]$. To get the overall reaction, we sum all the individual steps of the mechanism. Summing all steps gives:

$$C_4H_9Br \rightarrow C_4H_9^+ + Br^-$$
$$C_4H_9^+ + H_2O \rightarrow C_4H_9OH_2^+$$
$$C_4H_9OH_2^+ + H_2O \rightarrow C_4H_9OH + H_3O^+$$
$$\overline{C_4H_9Br + 2\ H_2O \rightarrow C_4H_9OH + Br^- + H_3O^+}$$

Intermediates in a mechanism are species that are neither reactants nor products but that are formed and consumed during the reaction sequence. The intermediates for this mechanism are $C_4H_9^+$ and $C_4H_9OH_2^+$.

62. Because the rate of the slowest elementary step equals the rate of a reaction:

$$\text{Rate} = \text{rate of step 1} = k[NO_2]^2$$

The sum of all steps in a plausible mechanism must give the overall balanced reaction. Summing all steps gives:

$$NO_2 + NO_2 \rightarrow NO_3 + NO$$
$$NO_3 + CO \rightarrow NO_2 + CO_2$$
$$\overline{NO_2 + CO \rightarrow NO + CO_2}$$

Temperature Dependence of Rate Constants and the Collision Model

63. In the following plot, R = reactants, P = products, E_a = activation energy, and RC = reaction coordinate, which is the same as reaction progress. Note for this reaction that ΔE is positive because the products are at a higher energy than the reactants.

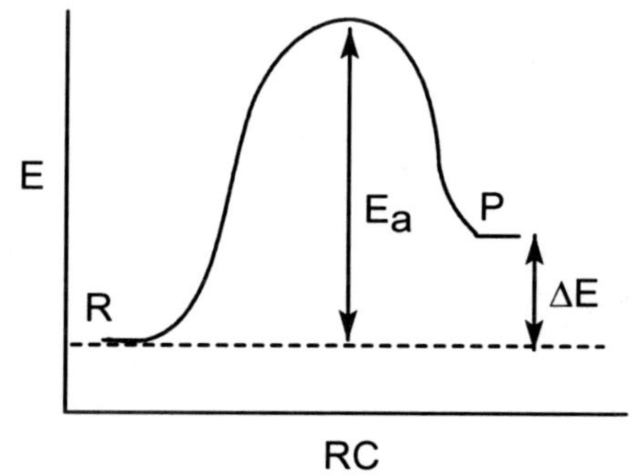

64. When ΔE is positive, the products are at a higher energy relative to reactants, and when ΔE is negative, the products are at a lower energy relative to reactants. RC = reaction coordinate, which is the same as reaction progress.

a.

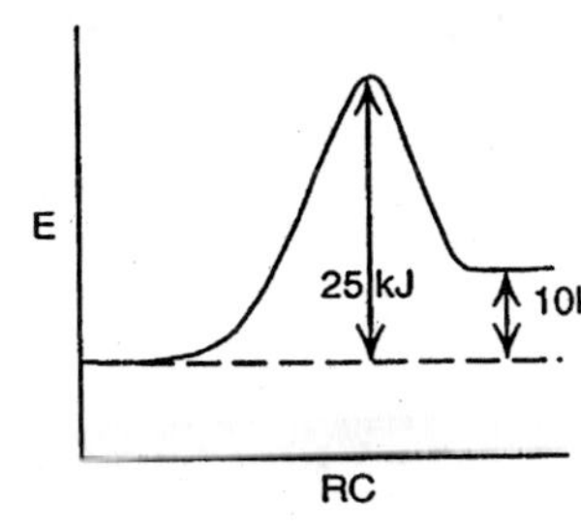

b.

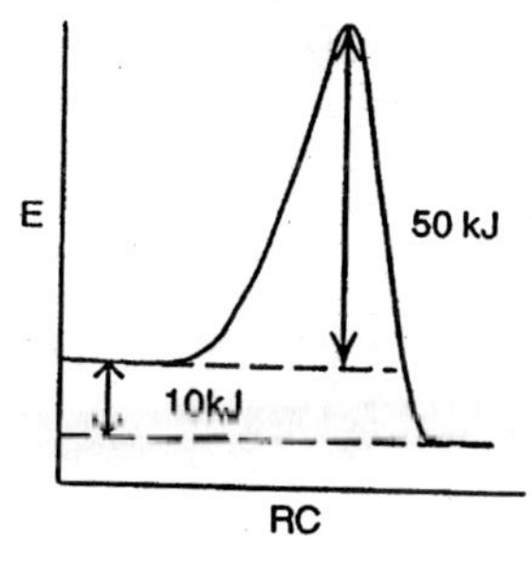

c.

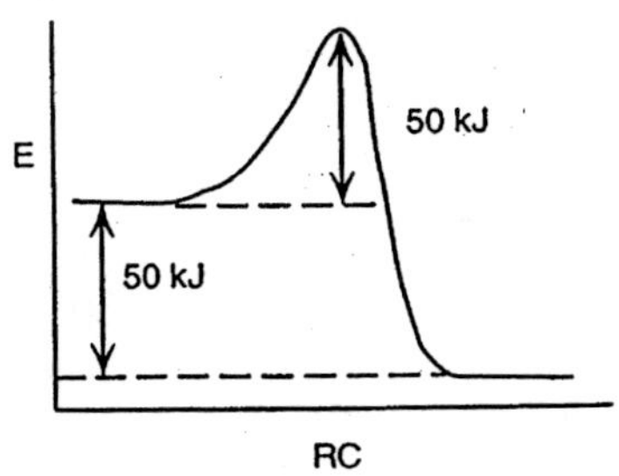

65.

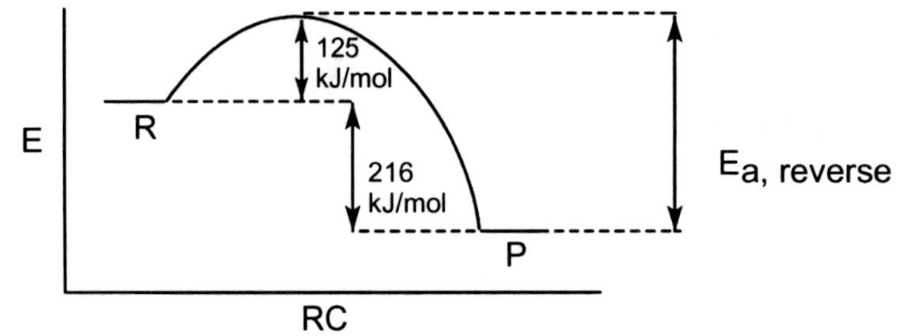

The activation energy for the reverse reaction is:

$E_{a,\ reverse}$ = 216 kJ/mol + 125 kJ/mol = 341 kJ/mol

66.

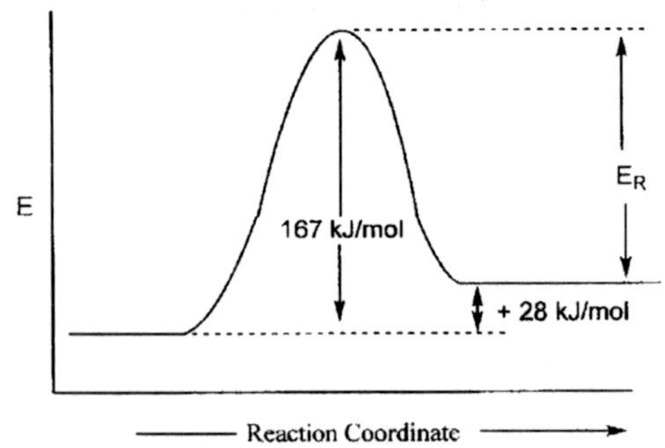

The activation energy for the reverse reaction is E_R in the diagram.

$E_R = 167 - 28 = 139$ kJ/mol

67. The Arrhenius equation is $k = A\ exp(-E_a/RT)$ or, in logarithmic form, $\ln k = -E_a/RT + \ln A$. Hence a graph of ln k versus 1/T should yield a straight line with a slope equal to $-E_a/R$ since the logarithmic form of the Arrhenius equation is in the form of a straight-line equation, $y = mx + b$. *Note*: We carried extra significant figures in the following ln k values in order to reduce round off error.

T (K)	1/T (K^{-1})	k (s^{-1})	ln k
338	2.96×10^{-3}	4.9×10^{-3}	−5.32
318	3.14×10^{-3}	5.0×10^{-4}	−7.60
298	3.36×10^{-3}	3.5×10^{-5}	−10.26

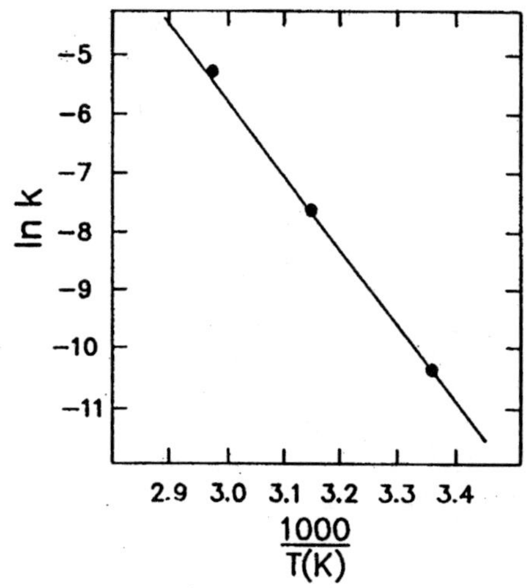

$$\text{Slope} = \frac{-10.76 - (-5.85)}{3.40 \times 10^{-3} - 3.00 \times 10^{-3}} = -1.2 \times 10^4 \text{ K} = -E_a/R$$

$$E_a = -\text{slope} \times R = 1.2 \times 10^4 \text{ K} \times \frac{8.3145 \text{J}}{\text{K mol}},\ E_a = 1.0 \times 10^5 \text{ J/mol} = 1.0 \times 10^2 \text{ kJ/mol}$$

68. From the Arrhenius equation in logarithmic form ($\ln k = -E_a/RT + \ln A$), a graph of ln k versus 1/T should yield a straight line with a slope equal to $-E_a/R$ and a *y* intercept equal to ln A.

a. $\text{Slope} = -E_a/R,\ E_a = 1.10 \times 10^4 \text{ K} \times \frac{8.3145 \text{J}}{\text{K mol}} = 9.15 \times 10^4 \text{ J/mol} = 91.5 \text{ kJ/mol}$

b. The units for A are the same as the units for k (s^{-1}).

y intercept = ln A, $A = e^{33.5} = 3.54 \times 10^{14}\ s^{-1}$

c. $\ln k = -E_a/RT + \ln A$ or $k = A \exp(-E_a/RT)$

$$k = 3.54 \times 10^{14}\ s^{-1} \times \exp\left(\frac{-9.15 \times 10^4 \text{ J/mol}}{8.3145 \text{J/K} \bullet \text{mol} \times 298 \text{K}}\right) = 3.24 \times 10^{-2}\ s^{-1}$$

69. $k = A \exp(-E_a/RT)$ or $\ln k = \frac{-E_a}{RT} + \ln A$ (the Arrhenius equation)

For two conditions: $\ln\left(\frac{k_2}{k_1}\right) = \frac{E_a}{R}\left(\frac{1}{T_1} - \frac{1}{T_2}\right)$ (Assuming A is temperature independent.)

Let $k_1 = 3.52 \times 10^{-7}$ L/mol•s, $T_1 = 555$ K; $k_2 = ?$, $T_2 = 645$ K; $E_a = 186 \times 10^3$ J/mol

$$\ln\left(\frac{k_2}{3.52 \times 10^{-7}}\right) = \frac{1.86 \times 10^5 \text{ J/mol}}{8.3145\text{J/K} \bullet \text{mol}}\left(\frac{1}{555\text{K}} - \frac{1}{645\text{K}}\right) = 5.6$$

$$\frac{k_2}{3.52 \times 10^{-7}} = e^{5.6} = 270, \ k_2 = 270(3.52 \times 10^{-7}) = 9.5 \times 10^{-5} \text{ L/mol•s}$$

70. For two conditions: $\ln\left(\frac{k_2}{k_1}\right) = \frac{E_a}{R}\left(\frac{1}{T_1} - \frac{1}{T_2}\right)$ (Assuming A is temperature independent.)

$$\ln\left(\frac{8.1 \times 10^{-2} \text{ s}^{-1}}{4.6 \times 10^{-2} \text{ s}^{-1}}\right) = \frac{E_a}{8.3145\text{J/K} \bullet \text{mol}}\left(\frac{1}{273\text{K}} - \frac{1}{293\text{K}}\right)$$

$$0.57 = \frac{E_a}{8.3145}(2.5 \times 10^{-4}), \ E_a = 1.9 \times 10^4 \text{ J/mol} = 19 \text{ kJ/mol}$$

71. $\ln\left(\frac{k_2}{k_1}\right) = \frac{E_a}{R}\left(\frac{1}{T_1} - \frac{1}{T_2}\right)$; $\frac{k_2}{k_1} = 7.00$, $T_1 = 295$ K, $E_a = 54.0 \times 10^3$ J/mol

$$\ln(7.00) = \frac{5.4 \times 10^4 \text{ J/mol}}{8.3145\text{J/K} \bullet \text{mol}}\left(\frac{1}{295\text{K}} - \frac{1}{T_2}\right), \ \frac{1}{295\text{K}} - \frac{1}{T_2} = 3.00 \times 10^{-4}$$

$$\frac{1}{T_2} = 3.09 \times 10^{-3}, \ T_2 = 324 \text{ K} = 51°\text{C}$$

72. $\ln\left(\frac{k_2}{k_1}\right) = \frac{E_a}{R}\left(\frac{1}{T_1} - \frac{1}{T_2}\right)$; because the rate doubles, $k_2 = 2k_1$.

$$\ln(2.00) = \frac{E_a}{8.3145\text{J/K} \bullet \text{mol}}\left(\frac{1}{298\text{K}} - \frac{1}{308\text{K}}\right), \ E_a = 5.3 \times 10^4 \text{ J/mol} = 53 \text{ kJ/mol}$$

73. $H_3O^+(aq) + OH^-(aq) \rightarrow 2\ H_2O(l)$ should have the faster rate. H_3O^+ and OH^- will be electrostatically attracted to each other; Ce^{4+} and Hg_2^{2+} will repel each other. The activation energy for the Ce^{4+} and Hg_2^{2+} reaction should be a larger quantity, making it the slower reaction.

74. Carbon cannot form the fifth bond necessary for the transition state because of the small atomic size of carbon and because carbon doesn't have low-energy d orbitals available to expand the octet.

Catalysts

75. a. NO is the catalyst. NO is present in the first step of the mechanism on the reactant side, but it is not a reactant. NO is regenerated in the second step and does not appear in overall balanced equation.

b. NO_2 is an intermediate. Intermediates also never appear in the overall balanced equation. In a mechanism, intermediates always appear first on the product side, whereas catalysts always appear first on the reactant side.

c. $k = A\ \exp(-E_a/RT)$; $\dfrac{k_{cat}}{k_{un}} = \dfrac{A\ \exp[-E_a(cat)/RT]}{A\ \exp[-E\,(un)/RT]} = \exp\left[\dfrac{E_a(un) - E_a(cat)}{RT}\right]$

$$\frac{k_{cat}}{k_{un}} = \exp\left(\frac{2100\text{J/mol}}{8.3145\text{J/K}\bullet\text{mol} \times 298\text{K}}\right) = e^{0.85} = 2.3$$

The catalyzed reaction is approximately 2.3 times faster than the uncatalyzed reaction at 25°C.

76. The mechanism for the chlorine catalyzed destruction of ozone is:

$$\begin{array}{ll} O_3 + Cl \rightarrow O_2 + ClO & \text{(slow)} \\ ClO + O \rightarrow O_2 + Cl & \text{(fast)} \\ \hline O_3 + O \rightarrow 2\,O_2 & \end{array}$$

Because the chlorine atom-catalyzed reaction has a lower activation energy, the Cl-catalyzed rate is faster. Hence Cl is a more effective catalyst. Using the activation energy, we can estimate the efficiency that Cl atoms destroy ozone compared to NO molecules (see Exercise 75c).

At 25°C: $\dfrac{k_{Cl}}{k_{NO}} = \exp\left[\dfrac{-E_a(Cl)}{RT} + \dfrac{E_a(NO)}{RT}\right] = \exp\left[\dfrac{(-2100 + 11{,}900)\text{J/mol}}{(8.3145\times\ 298)\text{J/mol}}\right] = e^{3.96} = 52$

At 25°C, the Cl-catalyzed reaction is roughly 52 times faster than the NO-catalyzed reaction, assuming the frequency factor A is the same for each reaction.

77. The reaction at the surface of the catalyst is assumed to follow the steps:

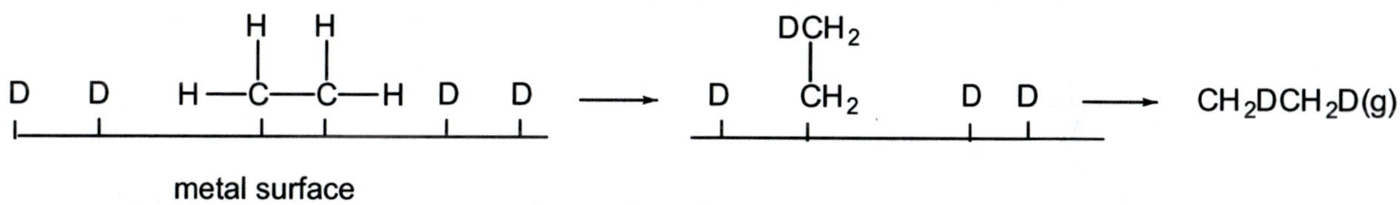

Thus $CH_2D–CH_2D$ should be the product. If the mechanism is possible, then the reaction must be:

$$C_2H_4 + D_2 \rightarrow CH_2DCH_2D$$

If we got this product, then we could conclude that this is a possible mechanism. If we got some other product, for example, CH_3CHD_2, then we would conclude that the mechanism is wrong. Even though this mechanism correctly predicts the products of the reaction, we cannot say conclusively that this is the correct mechanism; we might be able to conceive of other mechanisms that would give the same products as our proposed one.

78. a. W because it has a lower activation energy than the Os catalyst.

b. $k_w = A_w \exp[-E_a(W)/RT]$; $k_{uncat} = A_{uncat} \exp[-E_a(uncat)/RT]$; assume $A_w = A_{uncat}$.

$$\frac{k_W}{k_{uncat}} = \exp\left[\frac{-E_a(W)}{RT} + \frac{E_a(uncat)}{RT}\right]$$

$$\frac{k_w}{k_{uncat}} = \exp\left[\frac{-163{,}000 \text{J/mol} + 335{,}000 \text{J/mol}}{8.3145 \text{J/K} \bullet \text{mol} \times 298 \text{K}}\right] = 1.41 \times 10^{30}$$

The W-catalyzed reaction is approximately 10^{30} times faster than the uncatalyzed reaction.

c. Because $[H_2]$ is in the denominator of the rate law, the presence of H_2 decreases the rate of the reaction. For the decomposition to occur, NH_3 molecules must be adsorbed on the surface of the catalyst. If H_2 is also adsorbed on the catalyst surface, then there are fewer sites for NH_3 molecules to be adsorbed, and the rate decreases.

79. The rate depends on the number of reactant molecules adsorbed on the surface of the catalyst. This quantity is proportional to the concentration of reactant. However, when all the catalyst surface sites are occupied, the rate becomes independent of the concentration of reactant.

80. At high [S], the enzyme is completely saturated with substrate. Once the enzyme is completely saturated, the rate of decomposition of ES can no longer increase, and the overall rate remains constant.

81. Assuming the catalyzed and uncatalyzed reactions have the same form and orders, and because concentrations are assumed equal, the rates will be equal when the k values are equal.

$k = A \exp(-E_a/RT)$; $k_{cat} = k_{un}$ when $E_{a,cat}/RT_{cat} = E_{a,un}/RT_{un}$.

$$\frac{4.20 \times 10^4 \text{ J/mol}}{8.3145 \text{J/K} \bullet \text{mol} \times 293 \text{K}} = \frac{7.00 \times 10^4 \text{ J/mol}}{8.3145 \text{J/K} \bullet \text{mol} \times T_{un}}, \; T_{un} = 488 \text{ K} = 215°\text{C}$$

82. $$\text{Rate} = \frac{-\Delta[A]}{\Delta t} = k[A]^x$$

Assuming the catalyzed and uncatalyzed reaction have the same form and orders, and because concentrations are assumed equal, rate $\propto 1/\Delta t$, where $\Delta t = \Delta$time.

$$\frac{\text{Rate}_{cat}}{\text{Rate}_{un}} = \frac{\Delta t_{un}}{\Delta t_{cat}} = \frac{2400 \text{yr}}{\Delta t_{cat}} \text{ and } \frac{\text{rate}_{cat}}{\text{rate}_{un}} = \frac{k_{cat}}{k_{un}}$$

$$\frac{\text{Rate}_{cat}}{\text{Rate}_{un}} = \frac{k_{cat}}{k_{un}} = \frac{A \exp[-E_a(cat)/RT]}{A \exp[-E_a(un)/RT]} = \exp\left[\frac{-E_a(cat) + E_a(un)}{RT}\right]$$

$$\frac{k_{cat}}{k_{un}} = \exp\left(\frac{-5.90 \times 10^4 \text{ J/mol} + 1.84 \times 10^5 \text{ J/mol}}{8.3145\text{J/K} \bullet \text{mol} \times 600.\text{K}}\right) = 7.62 \times 10^{10}$$

$$\frac{\Delta t_{un}}{\Delta t_{cat}} = \frac{rate_{cat}}{rate_{un}} = \frac{k_{cat}}{k_{un}},\ \frac{2400\text{yr}}{\Delta t_{cat}} = 7.62 \times 10^{10},\ \Delta t_{cat} = 3.15 \times 10^{-8} \text{ yr} \approx 1 \text{ s}$$

Additional Exercises

83. Box a has 8 NO_2 molecules. Box b has 4 NO_2 molecules, and box c has 2 NO_2 molecules. Box b represents what is present after the first half-life of the reaction, and box c represents what is present after the second half-life.

a. For first order kinetics, $t_{1/2} = 0.693/k$; the half-life for a first order reaction is concentration independent. Therefore, the time for box c, the time it takes to go through two half-lives, will be 10 + 10 = 20 minutes.

b. For second order kinetics, $t_{1/2} = 1/k[A]_0$; the half-life for a second order reaction is inversely proportional to the initial concentration. So if the first half-life is 10 minutes, the second half-life will be 20 minutes. For a second order reaction, the time for box c will be 10 + 20 = 30 minutes.

c. For zero order kinetics, $t_{1/2} = [A]_0/2k$; the half-life for a zero order reaction is directly related to the initial concentration. So if this reaction was zero order, then the second half-life would decrease from 10 min to 5 min. The time for box c will be 10 + 5 = 15 minutes if the reaction is zero order.

84. Rate = $k[H_2SeO_3]^x[H^+]^y[I^-]^z$; comparing the first and second experiments:

$$\frac{3.33 \times 10^{-7}}{1.66 \times 10^{-7}} = \frac{k(2.0 \times 10^{-4})^x(2.0 \times 10^{-2})^y(2.0 \times 10^{-2})^z}{k(1.0 \times 10^{-4})^x(2.0 \times 10^{-2})^y(2.0 \times 10^{-2})^z},\ 2.01 = 2.0^x,\ x = 1$$

Comparing the first and fourth experiments:

$$\frac{6.66 \times 10^{-7}}{1.66 \times 10^{-7}} = \frac{k(1.0 \times 10^{-4})(4.0 \times 10^{-2})^y(2.0 \times 10^{-2})^z}{k(1.0 \times 10^{-4})(2.0 \times 10^{-2})^y(2.0 \times 10^{-2})^z},\ 4.01 = 2.0^y,\ y = 2$$

Comparing the first and sixth experiments:

$$\frac{13.2 \times 10^{-7}}{1.66 \times 10^{-7}} = \frac{k(1.0 \times 10^{-4})(2.0 \times 10^{-2})^2(4.0 \times 10^{-2})^z}{k(1.0 \times 10^{-4})(2.0 \times 10^{-2})^2(2.0 \times 10^{-2})^z}$$

$$7.95 = 2.0^z,\ \log(7.95) = z\log(2.0),\ z = \frac{\log(7.95)}{\log(2.0)} = 2.99 \approx 3$$

Rate = $k[H_2SeO_3][H^+]^2[I^-]^3$

Experiment 1:

$$\frac{1.66 \times 10^{-7}\ \text{mol}}{\text{L s}} = k\left(\frac{1.0 \times 10^{-4}\ \text{mol}}{\text{L}}\right)\left(\frac{2.0 \times 10^{-2}\ \text{mol}}{\text{L}}\right)^2\left(\frac{2.0 \times 10^{-2}\ \text{mol}}{\text{L}}\right)^3$$

$$k = 5.19 \times 10^5\ \text{L}^5/\text{mol}^5\text{•s} = 5.2 \times 10^5\ \text{L}^5/\text{mol}^5\text{•s} = k_{mean}$$

85. The integrated rate law for each reaction is:

$$\ln[A] = -4.50 \times 10^{-4}\ \text{s}^{-1}(t) + \ln[A]_0 \ \text{ and }\ \ln[B] = -3.70 \times 10^{-3}\ \text{s}^{-1}(t) + \ln[B]_0$$

Subtracting the second equation from the first equation ($\ln[A]_0 = \ln[B]_0$):

$$\ln[A] - \ln[B] = -4.50 \times 10^{-4}(t) + 3.70 \times 10^{-3}(t),\quad \ln\left(\frac{[A]}{[B]}\right) = 3.25 \times 10^{-3}(t)$$

When [A] = 4.00 [B], $\ln(4.00) = 3.25 \times 10^{-3}(t)$, t = 427 s.

86. The pressure of a gas is directly proportional to concentration. Therefore, we can use the pressure data to solve the problem because Rate $= -\Delta[SO_2Cl_2]/\Delta t \propto -\Delta P_{SO_2Cl_2}/\Delta t$.

Assuming a first order equation, the data and plot follow.

Time (hour)	0.00	1.00	2.00	4.00	8.00	16.00
$P_{SO_2Cl_2}$ (atm)	4.93	4.26	3.52	2.53	1.30	0.34
$\ln P_{SO_2Cl_2}$	1.595	1.449	1.258	0.928	0.262	−1.08

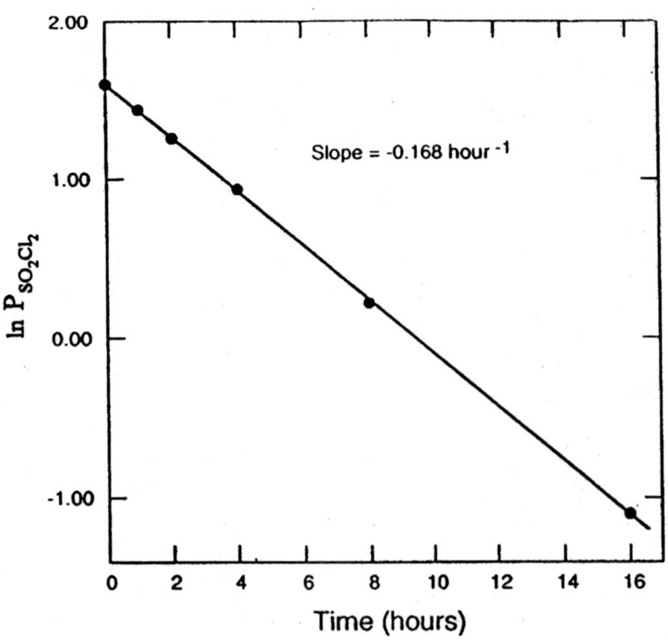

Because the $\ln P_{SO_2Cl_2}$ versus time plot is linear, the reaction is first order in SO_2Cl_2.

a. Slope of ln(P) versus t plot: $-0.168\ \text{hour}^{-1} = -k$, $k = 0.168\ \text{hour}^{-1} = 4.67 \times 10^{-5}\ \text{s}^{-1}$

Because concentration units don't appear in first-order rate constants, this value of k determined from the pressure data will be the same as if concentration data in molarity units were used.

b. $t_{1/2} = \dfrac{\ln 2}{k} = \dfrac{0.6931}{k} = \dfrac{0.6931}{0.168\text{h}^{-1}} = 4.13$ hour

c. $\ln\left(\dfrac{P_{SO_2Cl_2}}{P_0}\right) = -kt = -0.168\ \text{h}^{-1}(20.0\ \text{h}) = -3.36$, $\left(\dfrac{P_{SO_2Cl_2}}{P_0}\right) = e^{-3.36} = 3.47 \times 10^{-2}$

Fraction remaining = 0.0347 = 3.47%

87. From 338 K data, a plot of $\ln[N_2O_5]$ versus t is linear, and the slope $= -4.86 \times 10^{-3}$ (plot not included). This tells us the reaction is first order in N_2O_5 with $k = 4.86 \times 10^{-3}$ at 338 K. From 318 K data, the slope of $\ln[N_2O_5]$ versus t plot is equal to -4.98×10^{-4}, so $k = 4.98 \times 10^{-4}$ at 318 K. We now have two values of k at two temperatures, so we can solve for E_a.

$$\ln\left(\frac{k_2}{k_1}\right) = \frac{E_a}{R}\left(\frac{1}{T_1} - \frac{1}{T_2}\right),\ \ln\left(\frac{4.86 \times 10^{-3}}{4.98 \times 10^{-4}}\right) = \frac{E_a}{8.3145\text{J/K} \bullet \text{mol}}\left(\frac{1}{318\text{K}} - \frac{1}{338\text{K}}\right)$$

$E_a = 1.0 \times 10^5$ J/mol $= 1.0 \times 10^2$ kJ/mol

88. The Arrhenius equation is $k = A\exp(-E_a/RT)$ or, in logarithmic form, $\ln k = -E_a/RT + \ln A$. Hence a graph of ln k versus 1/T should yield a straight line with a slope equal to $-E_a/R$ since the logarithmic form of the Arrhenius equation is in the form of a straight-line equation, $y = mx + b$. *Note*: We carried one extra significant figure in the following ln k values in order to reduce round-off error.

T (K)	1/T (K^{-1})	k (L/mol•s)	ln k
195	5.13×10^{-3}	1.08×10^9	20.80
230.	4.35×10^{-3}	2.95×10^9	21.81
260.	3.85×10^{-3}	5.42×10^9	22.41
298	3.36×10^{-3}	12.0×10^9	23.21
369	2.71×10^{-3}	35.5×10^9	24.29

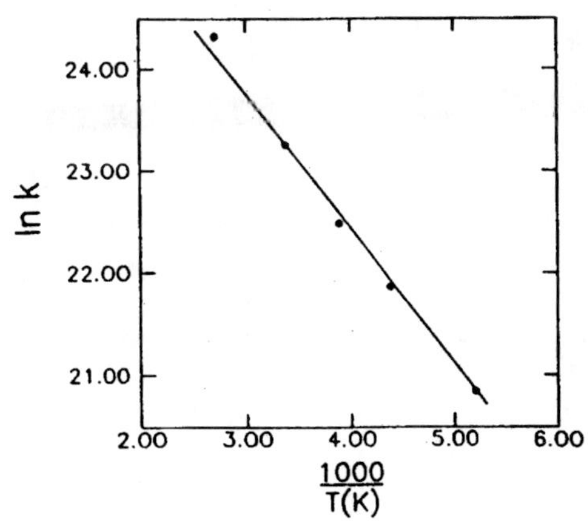

Using a couple of points on the plot:

$$\text{slope} = \frac{20.95 - 23.65}{5.00 \times 10^{-3} - 3.00 \times 10^{-3}} = \frac{-2.70}{2.00 \times 10^{-3}} = -1.35 \times 10^3\ \text{K} = \frac{-E_a}{R}$$

$E_a = 1.35 \times 10^3\ \text{K} \times 8.3145\ \text{J/K•mol} = 1.12 \times 10^4\ \text{J/mol} = 11.2\ \text{kJ/mol}$

89. To determine the rate of reaction, we need to calculate the value of the rate constant k. The activation energy data can be manipulated to determine k.

$$k = Ae^{-E_a/RT} = 0.850\ \text{s}^{-1} \times \exp\left(\frac{-26.2 \times 10^3\ \text{J/mol}}{8.3145\text{J/K} \bullet \text{mol} \times 310.2\text{K}}\right) = 3.29 \times 10^{-5}\ \text{s}^{-1}$$

Rate = k[acetycholine receptor-toxin complex]

$$\text{Rate} = 3.29 \times 10^{-5}\ \text{s}^{-1}\left(\frac{0.200\,\text{mol}}{\text{L}}\right) = 6.58 \times 10^{-6}\ \text{mol/L•s}$$

90. Rate = $k[\text{DNA}]^x[\text{CH}_3\text{I}]^y$; comparing the second and third experiments:

$$\frac{1.28 \times 10^{-3}}{6.40 \times 10^{-4}} = \frac{k(0.200)^x(0.200)^y}{k(0.100)^x(0.200)^y},\ 2.00 = 2.00^x,\ x = 1$$

Comparing the first and second experiments:

$$\frac{6.40 \times 10^{-4}}{3.20 \times 10^{-4}} = \frac{k(0.100)(0.200)^y}{k(0.100)(0.100)^y},\ 2.00 = 2.00^y,\ y = 1$$

The rate law is Rate = $k[\text{DNA}][\text{CH}_3\text{I}]$.

Mechanism I is possible because the derived rate law from the mechanism (Rate = $k[\text{DNA}][\text{CH}_3\text{I}]$) agrees with the experimentally determined rate law. The derived rate law for Mechanism II will equal the rate of the slowest step. This is step 1 in the mechanism giving a derived rate law that is Rate = $k[\text{CH}_3\text{I}]$. Because this rate law does not agree with experiment, Mechanism II would not be a possible mechanism for the reaction.

91. a. If the interval between flashes is 16.3 s, then the rate is:

1 flash/16.3 s = $6.13 \times 10^{-2}\ \text{s}^{-1}$ = k

Interval	k	T
16.3 s	$6.13 \times 10^{-2}\ \text{s}^{-1}$	21.0°C (294.2 K)
13.0 s	$7.69 \times 10^{-2}\ \text{s}^{-1}$	27.8°C (301.0 K)

$\ln\left(\frac{k_2}{k_1}\right) = \frac{E_a}{R}\left(\frac{1}{T_1} - \frac{1}{T_2}\right)$; solving using above data: $E_a = 2.5 \times 10^4$ J/mol = 25 kJ/mol

b. $\ln\left(\frac{k}{6.13 \times 10^{-2}}\right) = \frac{2.5 \times 10^4 \text{ J/mol}}{8.3145 \text{J/K} \bullet \text{mol}}\left(\frac{1}{294.2\text{K}} - \frac{1}{303.2\text{K}}\right) = 0.30$

$k = e^{0.30} \times (6.13 \times 10^{-2}) = 8.3 \times 10^{-2}\ s^{-1}$; interval = 1/k = 12 seconds.

c.

T	Interval	54-2(Intervals)
21.0°C	16.3 s	21°C
27.8°C	13.0 s	28°C
30.0°C	12 s	30°C

This rule of thumb gives excellent agreement to two significant figures.

92. $k = A\exp(-E_a/RT)$; $\frac{k_{cat}}{k_{uncat}} = \frac{A_{cat}\exp(-E_{a,\,cat}/RT)}{A_{uncat}\exp(-E_{a,\,uncat}/RT)} = \exp\left(\frac{-E_{a,\,cat} + E_{a,\,uncat}}{RT}\right)$

$$2.50 \times 10^3 = \frac{k_{cat}}{k_{uncat}} = \exp\left(\frac{-E_{a,\,cat} + 5.00 \times 10^4 \text{ J/mol}}{8.3145 \text{ J/K} \bullet \text{mol} \times 310.\text{K}}\right)$$

$\ln(2.50 \times 10^3) \times 2.58 \times 10^3 \text{ J/mol} = -E_{a,\,cat} + 5.00 \times 10^4 \text{ J/mol}$

$E_{a,\,cat} = 5.00 \times 10^4 \text{ J/mol} - 2.02 \times 10^4 \text{ J/mol} = 2.98 \times 10^4 \text{ J/mol} = 29.8 \text{ kJ/mol}$

93. a. Because $[A]_0 << [B]_0$ or $[C]_0$, the B and C concentrations remain constant at 1.00 *M* for this experiment. Thus Rate = $k[A]^2[B][C] = k'[A]^2$, where $k' = k[B][C]$.

For this pseudo-second-order reaction:

$$\frac{1}{[A]} = k't + \frac{1}{[A]_0},\quad \frac{1}{3.26 \times 10^{-5}\ M} = k'(3.00 \text{ min}) + \frac{1}{1.00 \times 10^{-4}\ M}$$

$k' = 6890$ L/mol•min = 115 L/mol•s

$$k' = k[B][C],\ k = \frac{k'}{[B][C]},\quad k = \frac{115 \text{L/mol} \bullet \text{s}}{(1.00M)(1.00M)} = 115 \text{ L}^3/\text{mol}^3 \bullet \text{s}$$

b. For this pseudo-second-order reaction:

$$\text{Rate} = k'[A]^2,\ t_{1/2} = \frac{1}{k'[A]_0} = \frac{1}{115 \text{L/mol} \bullet \text{s}(1.00 \times 10^{-4} \text{ mol/L})} = 87.0 \text{ s}$$

c. $\frac{1}{[A]} = k't + \frac{1}{[A]_0} = 115 \text{ L/mol} \bullet \text{s} \times 600.\text{ s} + \frac{1}{1.00 \times 10^{-4} \text{ mol/L}} = 7.90 \times 10^4 \text{ L/mol}$

$[A] = 1/7.90 \times 10^4 \text{ L/mol} = 1.27 \times 10^{-5} \text{ mol/L}$

From the stoichiometry in the balanced reaction, 1 mol of B reacts with every 3 mol of A.

Amount A reacted = $1.00 \times 10^{-4}\ M - 1.27 \times 10^{-5}\ M = 8.7 \times 10^{-5}\ M$

Amount B reacted = $8.7 \times 10^{-5}\ \text{mol/L} \times \dfrac{1\ \text{mol B}}{3\ \text{mol A}} = 2.9 \times 10^{-5}\ M$

$[B] = 1.00\ M - 2.9 \times 10^{-5}\ M = 1.00\ M$

As we mentioned in part a, the concentration of B (and C) remains constant because the A concentration is so small compared to the B (or C) concentration.

ChemWork Problems

The answers to the problems 94-101 (or a variation to these problem) are found in OWL. These problems are also assignable in OWL.

Challenge Problems

102. $\dfrac{-d[A]}{dt} = k[A]^3, \displaystyle\int_{[A]_0}^{[A]_t} \dfrac{d[A]}{[A]^3} = -\int_0^t k\,dt$

$$\int x^n\,dx = \frac{x^{n+1}}{n+1};\ \text{so: } -\frac{1}{2[A]^2}\Bigg|_{[A]_0}^{[A]_t} = -kt,\ -\frac{1}{2[A]_t^2} + \frac{1}{2[A]_0^2} = -kt$$

For the half-life equation, $[A]_t = 1/2[A]_0$:

$$-\frac{1}{2\left(\frac{1}{2}[A]_0\right)^2} + \frac{1}{2[A]_0^2} = -kt_{1/2},\ -\frac{4}{2[A]_0^2} + \frac{1}{2[A]_0^2} = -kt_{1/2}$$

$$-\frac{3}{2[A]_0^2} = -kt_{1/2},\ t_{1/2} = \frac{3}{2[A]_0^2 k}$$

The first half-life is $t_{1/2}$ = 40. s and corresponds to going from $[A]_0$ to 1/2 $[A]_0$. The second half-life corresponds to going from 1/2 $[A]_0$ to 1/4 $[A]_0$.

$$\text{First half-life} = \frac{3}{2[A]_0^2 k};\ \text{second half-life} = \frac{3}{2\left(\frac{1}{2}[A]_0\right)^2 k} = \frac{6}{[A]_0^2 k}$$

$$\frac{\text{First half-life}}{\text{Second half-life}} = \frac{\dfrac{3}{2[A]_0^2\,k}}{\dfrac{6}{[A]_0^2\,k}} = 3/12 = 1/4$$

Because the first half-life is 40. s, the second half-life will be four times this, or 160 s.

103. Rate = $k[I^-]^x[OCl^-]^y[OH^-]^z$; comparing the first and second experiments:

$$\frac{18.7 \times 10^{-3}}{9.4 \times 10^{-3}} = \frac{k(0.0026)^x(0.012)^y(0.10)^z}{k(0.0013)^x(0.012)^y(0.10)^z},\ 2.0 = 2.0^x,\ x = 1$$

Comparing the first and third experiments:

$$\frac{9.4 \times 10^{-3}}{4.7 \times 10^{-3}} = \frac{k(0.0013)(0.012)^y(0.10)^z}{k(0.0013)(0.0060)^y(0.10)^z},\ 2.0 = 2.0^y,\ y = 1$$

Comparing the first and sixth experiments:

$$\frac{4.8 \times 10^{-3}}{9.4 \times 10^{-3}} = \frac{k(0.0013)(0.012)(0.20)^z}{k(0.0013)(0.012)(0.10)^z},\ 1/2 = 2.0^z,\ z = -1$$

Rate = $\dfrac{k[I^-][OCl^-]}{[OH^-]}$; the presence of OH^- decreases the rate of the reaction.

For the first experiment:

$$\frac{9.4 \times 10^{-3}\ \text{mol}}{\text{L s}} = k\frac{(0.0013\text{mol/L})(0.012\text{mol/L})}{(0.10\text{mol/L})},\ k = 60.3\ s^{-1} = 60.\ s^{-1}$$

For all experiments, $k_{mean} = 60.\ s^{-1}$.

104. For second order kinetics: $\dfrac{1}{[A]} - \dfrac{1}{[A]_0} = kt$ and $t_{1/2} = \dfrac{1}{k[A]_0}$

a. $\dfrac{1}{[A]} = (0.250\ \text{L/mol•s})t + \dfrac{1}{[A]_0}$, $\dfrac{1}{[A]} = 0.250 \times 180.\ \text{s} + \dfrac{1}{1.00 \times 10^{-2}\ M}$

$\dfrac{1}{[A]} = 145\ M^{-1}$, $[A] = 6.90 \times 10^{-3}\ M$

Amount of A that reacted = 0.0100 – 0.00690 = 0.0031 M.

$[A_2] = \dfrac{1}{2}(3.1 \times 10^{-3}\ M) = 1.6 \times 10^{-3}\ M$

b. After 3.00 minutes (180. s): [A] = 3.00[B], $6.90 \times 10^{-3} M = 3.00[B]$

$[B] = 2.30 \times 10^{-3} M$

$$\frac{1}{[B]} = k_2 t + \frac{1}{[B]_0}, \quad \frac{1}{2.30 \times 10^{-3}\ M} = k_2(180.s) + \frac{1}{2.50 \times 10^{-2}\ M},$$

$k_2 = 2.19$ L/mol•s

c. $$t_{1/2} = \frac{1}{k[A]_0} = \frac{1}{0.250 \text{L/mol}\cdot\text{s} \times 1.00 \times 10^{-2} \text{ mol/L}} = 4.00 \times 10^2 \text{ s}$$

105. a. We check for first-order dependence by graphing ln[concentration] versus time for each set of data. The rate dependence on NO is determined from the first set of data because the ozone concentration is relatively large compared to the NO concentration, so $[O_3]$ is effectively constant.

Time (ms)	[NO] (molecules/cm^3)	ln[NO]
0	6.0×10^8	20.21
100.	5.0×10^8	20.03
500.	2.4×10^8	19.30
700.	1.7×10^8	18.95
1000.	9.9×10^7	18.41

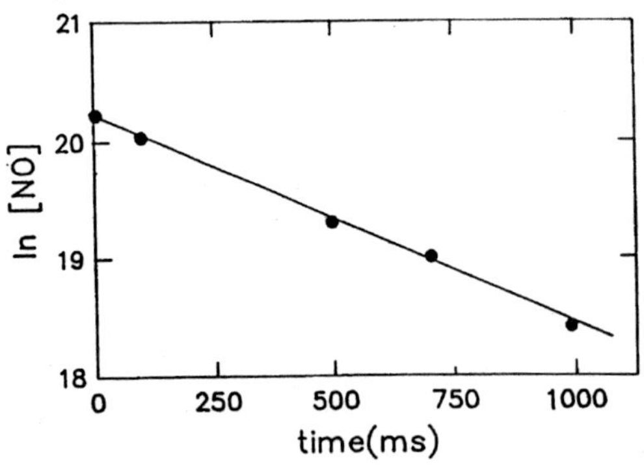

Because ln[NO] versus t is linear, the reaction is first order with respect to NO.

We follow the same procedure for ozone using the second set of data. The data and plot are:

Time (ms)	$[O_3]$ (molecules/cm^3)	$\ln[O_3]$
0	1.0×10^{10}	23.03
50.	8.4×10^9	22.85
100.	7.0×10^9	22.67
200.	4.9×10^9	22.31
300.	3.4×10^9	21.95

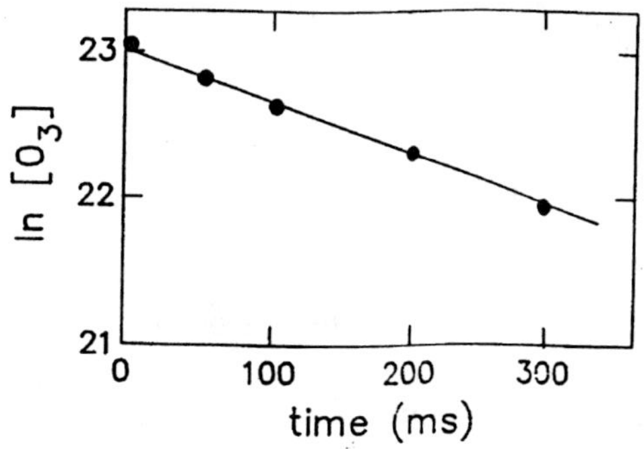

The plot of $\ln[O_3]$ versus t is linear. Hence the reaction is first order with respect to ozone.

b. Rate = $k[NO][O_3]$ is the overall rate law.

c. For NO experiment, Rate = $k'[NO]$ and $k' = -$ (slope from graph of $\ln[NO]$ versus t).

$$k' = -\text{slope} = -\frac{18.41 - 20.21}{(1000. - 0) \times 10^{-3}\ \text{s}} = 1.8\ \text{s}^{-1}$$

For ozone experiment, Rate = $k''[O_3]$ and $k'' = -$ (slope from $\ln[O_3]$ versus t plot).

$$k'' = -\text{slope} = -\frac{(21.95 - 23.03)}{(300. - 0) \times 10^{-3}\ \text{s}} = 3.6\ \text{s}^{-1}$$

d. From the NO experiment, Rate = $k[NO][O_3] = k'[NO]$ where $k' = k[O_3]$.

$k' = 1.8\ \text{s}^{-1} = k(1.0 \times 10^{14}\ \text{molecules/cm}^3)$, $k = 1.8 \times 10^{-14}\ \text{cm}^3/\text{molecules}\bullet\text{s}$

We can check this from the ozone data. Rate = $k''[O_3] = k[NO][O_3]$, where $k'' = k[NO]$.

$k'' = 3.6\ \text{s}^{-1} = k(2.0 \times 10^{14}\ \text{molecules/cm}^3)$, $k = 1.8 \times 10^{-14}\ \text{cm}^3/\text{molecules}\bullet\text{s}$

Both values of k agree.

106. On the energy profile to the right, R = reactants, P = products, E_a = activation energy, ΔE = overall energy change for the reaction, I = intermediate, and RC = reaction coordinate, which is the same as reaction progress.

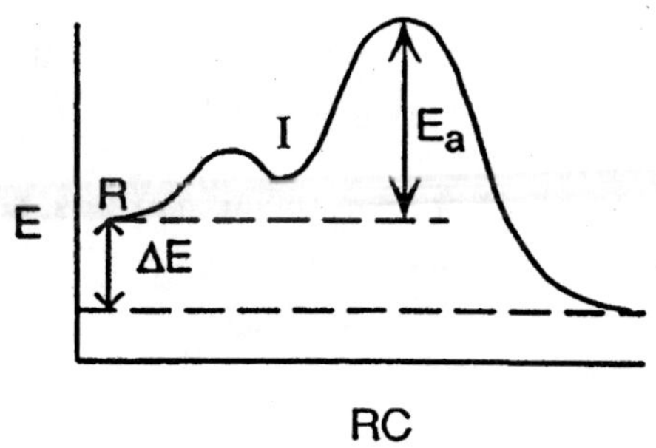

a-d. See plot to the right.

e. This is a two-step reaction since an intermediate plateau appears between the reactant and the products. This plateau represents the energy of the intermediate. The general reaction mechanism for this reaction is:

$$\begin{array}{l} R \rightarrow I \\ \underline{I \rightarrow P} \\ R \rightarrow P \end{array}$$

In a mechanism, the rate of the slowest step determines the rate of the reaction. The activation energy for the slowest step will be the largest energy barrier that the reaction must overcome. Since the second hump in the diagram is at the highest energy, the second step has the largest activation energy and will be the rate-determining step (the slow step).

107. $\ln\left(\frac{k_2}{k_1}\right) = \frac{E_a}{R}\left(\frac{1}{T_1} - \frac{1}{T_2}\right)$; assuming $\frac{rate_2}{rate_1} = \frac{k_2}{k_1} = 40.0$:

$$\ln(40.0) = \frac{E_a}{8.3145 \text{J/K} \bullet \text{mol}}\left(\frac{1}{308\text{K}} - \frac{1}{328\text{K}}\right), \quad E_a = 1.55 \times 10^5 \text{ J/mol} = 155 \text{ kJ/mol}$$

(carrying an extra sig. fig.)

Note that the activation energy is close to the F_2 bond energy. Therefore, the rate-determining step probably involves breaking the F_2 bond.

$H_2(g) + F_2(g) \rightarrow 2\ HF(g)$; for every 2 moles of HF produced, only 1 mole of the reactant is used up. Therefore, to convert the data to $P_{reactant}$ versus time, $P_{reactant} = 1.00 \text{ atm} - (1/2)P_{HF}$.

$P_{reactant}$	Time
1.000 atm	0 min
0.850 atm	30.0 min
0.700 atm	65.8 min
0.550 atm	110.4 min
0.400 atm	169.1 min
0.250 atm	255.9 min

The plot of ln $P_{reactant}$ versus time (plot not included) is linear with negative slope, so the reaction is first order with respect to the limiting reagent.

For the reactant in excess, because the values of the rate constant are the same for both experiments, one can conclude that the reaction is zero order in the excess reactant.

a. For a three-step reaction with the first step limiting, the energy-level diagram could be:

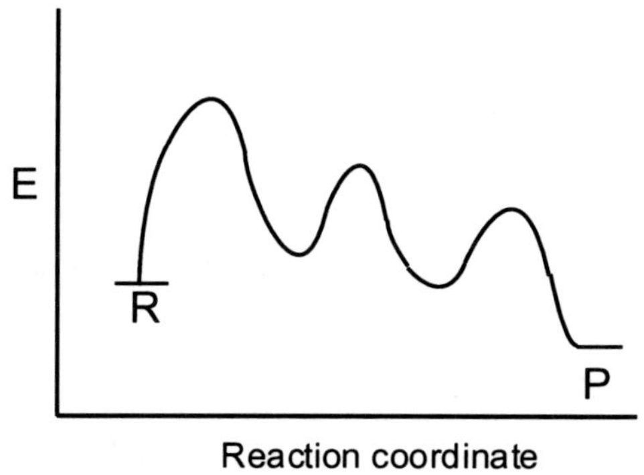

Note that the heights of the second and third humps must be lower than the first-step activation energy. However, the height of the third hump could be higher than the second hump. One cannot determine this absolutely from the information in the problem.

b. We know the reaction has a slow first step, and the calculated activation energy indicates that the rate-determining step involves breaking the F_2 bond. The reaction is also first order in one of the reactants and zero order in the other reactant. All this points to F_2 being the limiting reagent. The reaction is first order in F_2, and the rate-determining step in the mechanism is $F_2 \rightarrow 2\ F$. Possible second and third steps to complete the mechanism follow.

$F_2 \rightarrow 2\ F$	slow
$F + H_2 \rightarrow HF + H$	fast
$H + F \rightarrow HF$	fast
$F_2 + H_2 \rightarrow 2\ HF$	

c. F_2 was the limiting reactant.

108. We need the value of k at 500. K; $\ln\left(\frac{k_2}{k_1}\right) = \frac{E_a}{R}\left(\frac{1}{T_1} - \frac{1}{T_2}\right)$

$$\ln\left(\frac{k_2}{2.3 \times 10^{-12}\ \text{L/mol}\bullet\text{s}}\right) = \frac{1.11 \times 10^5\ \text{J/mol}}{8.3145\text{J/K}\bullet\text{mol}}\left(\frac{1}{273\text{K}} - \frac{1}{500\text{K}}\right) = 22.2$$

$$\frac{k_2}{2.3 \times 10^{-12}} = e^{22.2},\ k_2 = 1.0 \times 10^{-2}\ \text{L/mol}\bullet\text{s}$$

Because the decomposition reaction is an elementary reaction, the rate law can be written using the coefficients in the balanced equation. For this reaction, Rate = $k[NO_2]^2$. To solve for the time, we must use the integrated rate law for second-order kinetics. The major problem now is converting units so they match. Rearranging the ideal gas law gives n/V = P/RT. Substituting P/RT for concentration units in the second-order integrated rate equation:

$$\frac{1}{[NO_2]} = kt + \frac{1}{[NO_2]_0}, \quad \frac{1}{P/RT} = kt + \frac{1}{P_0/RT}, \quad \frac{RT}{P} - \frac{RT}{P_0} = kt, \quad t = \frac{RT}{k}\left(\frac{P_0 - P}{P \times P_0}\right)$$

$$t = \frac{(0.08206 \text{L atm/K} \bullet \text{mol})(500 \text{K})}{1.0 \times 10^{-2} \text{ L/mol} \bullet \text{s}} \times \left(\frac{2.5 \text{atm} - 1.5 \text{atm}}{1.5 \text{atm} \times 2.5 \text{atm}}\right) = 1.1 \times 10^3 \text{ s}$$

109. a. [B] >> [A], so [B] can be considered constant over the experiments. This gives us a pseudo-order rate law equation.

b. Note that in each case, the half-life doubles as times increases (in experiment 1, the first half-life is 40. s, and the second half-life is 80. s; in experiment 2, the first half-life is 20. s, and the second half-life is 40. s). This occurs only for a second-order reaction, so the reaction is second order in [A]. Between experiment 1 and experiment 2, we double [B], and the reaction rate doubles, thus it is first order in [B]. The overall rate law equation is rate = $k[A]^2[B]$.

Using $t_{1/2} = \frac{1}{k[A]_0}$, we get $k = \frac{1}{(40.\text{ s})(10.0 \times 10^{-2} \text{ mol/L})} = 0.25$ L/mol •s; but this is actually k′, where Rate = $k'[A]^2$ and $k' = k[B]$.

$$k = \frac{k'}{[B]} = \frac{0.25 \text{L/mol} \bullet \text{s}}{5.0 \text{mol/L}} = 0.050 \text{ L}^2/\text{mol}^2 \bullet \text{s}$$

110. a. Rate = $k[A]^x[B]^y$; looking at the data in experiment 2, notice that the concentration of A is cut in half every 10. s. Only first-order reactions have a half-life that is independent of concentration. The reaction is first order in A. In the data for experiment 1, notice that the half-life is 40. s. This indicates that in going from experiment 1 to experiment 2, where the B concentration doubled, the rate of reaction increased by a factor of four. This tells us that the reaction is second order in B.

Rate = $k[A][B]^2$

b. This reaction in each experiment is pseudo-first order in [A] because the concentration of B is so large, it is basically constant.

Rate = $k[B]^2[A] = k'[A]$, where $k' = k[B]^2$

For a first-order reaction, the integrated rate law is:

$$\ln\left(\frac{[A]}{[A]_0}\right) = -k't$$

Use any set of data you want to calculate kN. For example, in experiment 1, from 0 to 20. s the concentration of A decreased from 0.010 *M* to 0.0071 *M*:

$$\ln\left(\frac{0.0071}{0.010}\right) = -k'(20.\text{ s}), \quad k' = 1.7 \times 10^{-2} \text{ s}^{-1}$$

$k' = k[B]^2,\ 1.7 \times 10^{-2}\ s^{-1} = k(10.0\ mol/L)^2$

$k = 1.7 \times 10^{-4}\ L^2/mol^2{\bullet}s$

We get similar values for k using other data from either experiment 1 or experiment 2.

c. $\ln\left(\frac{[A]}{0.010M}\right) = -k't = -(1.7 \times 10^{-2}\ L^2/mol^2{\bullet}s) \times 30.\ s,\quad [A] = 6.0 \times 10^{-3}\ M$

111. Rate = $k[A]^x[B]^y[C]^z$; during the course of experiment 1, [A] and [C] are essentially constant, and Rate = $k'[B]^y$, where $k' = k[A]_0^x[C]_0^z$.

[B] (M)	Time (s)	ln[B]	1/[B] (M^{-1})
1.0×10^{-3}	0	−6.91	1.0×10^3
2.7×10^{-4}	1.0×10^5	−8.22	3.7×10^3
1.6×10^{-4}	2.0×10^5	−8.74	6.3×10^3
1.1×10^{-4}	3.0×10^5	−9.12	9.1×10^3
8.5×10^{-5}	4.0×10^5	−9.37	12×10^3
6.9×10^{-5}	5.0×10^5	−9.58	14×10^3
5.8×10^{-5}	6.0×10^5	−9.76	17×10^3

A plot of 1/[B] versus t is linear (plot not included), so the reaction is second order in B, and the integrated rate equation is:

$1/[B] = (2.7 \times 10^{-2}\ L/mol{\bullet}s)t + 1.0 \times 10^3\ L/mol;\ k' = 2.7 \times 10^{-2}\ L/mol{\bullet}s$

For experiment 2, [B] and [C] are essentially constant, and Rate = $k''[A]^x$, where $k'' = k[B]_0^y[C]_0^z = k[B]_0^2[C]_0^z$.

[A] (M)	Time (s)	ln[A]	1/[A] (M^{-1})
1.0×10^{-2}	0	−4.61	1.0×10^2
8.9×10^{-3}	1.0	−4.95	140
5.5×10^{-3}	5.0	−5.20	180
3.8×10^{-3}	8.0	−5.57	260
2.9×10^{-3}	10.0	−5.84	340
2.0×10^{-3}	13.0	−6.21	5.0×10^2

A plot of ln[A] versus t is linear, so the reaction is first order in A, and the integrated rate law is:

$\ln[A] = -(0.123\ s^{-1})t - 4.61;\ k'' = 0.123\ s^{-1}$

Note: We will carry an extra significant figure in k″.

Experiment 3: [A] and [B] are constant; Rate = $k'''[C]^z$

The plot of [C] versus t is linear. Thus $z = 0$.

The overall rate law is Rate = $k[A][B]^2$.

From Experiment 1 (to determine k):

$$k' = 2.7 \times 10^{-2}\ \text{L/mol•s} = k[A]_0^x[C]_0^z = k[A]_0 = k(2.0\ M),\ \ k = 1.4 \times 10^{-2}\ \text{L}^2/\text{mol}^2\text{•s}$$

From Experiment 2: $k'' = 0.123\ \text{s}^{-1} = k[B]_0^2,\ \ k = \dfrac{0.123\ \text{s}^{-1}}{(3.0\ M)^2} = 1.4 \times 10^{-2}\ \text{L}^2/\text{mol}^2\text{•s}$

Thus Rate = $k[A][B]^2$ and $k = 1.4 \times 10^{-2}\ \text{L}^2/\text{mol}^2\text{•s}$.

112. a. Rate = $(k_1 + k_2[H^+])[I^-]^m[H_2O_2]^n$

In all the experiments, the concentration of H_2O_2 is small compared to the concentrations of I^- and H^+. Therefore, the concentrations of I^- and H^+ are effectively constant, and the rate law reduces to:

Rate = $k_{obs}[H_2O_2]^n$, where $k_{obs} = (k_1 + k_2[H^+])[I^-]^m$

Because all plots of $\ln[H_2O_2]$ versus time are linear, the reaction is first order with respect to H_2O_2 ($n = 1$). The slopes of the $\ln[H_2O_2]$ versus time plots equal $-k_{obs}$, which equals $-(k_1 + k_2[H^+])[I^-]^m$. To determine the order of I^-, compare the slopes of two experiments in which I^- changes and H^+ is constant. Comparing the first two experiments:

$$\frac{\text{slope (exp. 2)}}{\text{slope (exp. 1)}} = \frac{-0.360}{-0.120} = \frac{-[k_1 + k_2(0.0400M)](0.3000M)^m}{-[k_1 + k_2(0.0400M)](0.1000M)^m}$$

$$3.00 = \left(\frac{0.3000M}{0.1000M}\right)^m = (3.000)^m,\ \ m = 1$$

The reaction is also first order with respect to I^-.

b. The slope equation has two unknowns, k_1 and k_2. To solve for k_1 and k_2, we must have two equations. We need to take one of the first set of three experiments and one of the second set of three experiments to generate the two equations in k_1 and k_2.

Experiment 1: Slope = $-(k_1 + k_2[H^+])[I^-]$

$$-0.120\ \text{min}^{-1} = -[k_1 + k_2(0.0400\ M)](0.1000\ M) \text{ or } 1.20 = k_1 + k_2(0.0400)$$

Experiment 4:

$$-0.0760\ \text{min}^{-1} = -[k_1 + k_2(0.0200\ M)](0.0750\ M) \text{ or } 1.01 = k_1 + k_2(0.0200)$$

Subtracting 4 from 1:

$$\begin{array}{rl} 1.20 = & k_1 + k_2(0.0400) \\ -1.01 = & -k_1 - k_2(0.0200) \\ \hline 0.19 = & \quad k_2(0.0200),\ k_2 = 9.5\ L^2/mol^2{\bullet}min \end{array}$$

$1.20 = k_1 + 9.5(0.0400),\ k_1 = 0.82\ L/mol{\bullet}min$

c. There are two pathways, one involving H^+ with Rate = $k_2[H^+][I^-][H_2O_2]$ and another not involving H^+ with Rate = $k_1[I^-][H_2O_2]$. The overall rate of reaction depends on which of these two pathways dominates, and this depends on the H^+ concentration.

Integrative Problems

113. $8.75\ h \times \dfrac{3600s}{h} = 3.15 \times 10^4\ s;\ k = \dfrac{\ln 2}{t_{1/2}} = \dfrac{\ln 2}{3.15 \times 10^4\ s} = 2.20 \times 10^{-5}\ s^{-1}$

The partial pressure of a gas is directly related to the concentration in mol/L. So, instead of using mol/L as the concentration units in the integrated first-order rate law, we can use partial pressures of SO_2Cl_2.

$$\ln\left(\frac{P}{P_0}\right) = -kt,\ \ln\left(\frac{P}{791 torr}\right) = -(2.20 \times 10^{-5}\ s^{-1}) \times 12.5\ h \times \frac{3600s}{h}$$

$$P_{SO_2Cl_2} = 294\ torr \times \frac{1\ atm}{760\ torr} = 0.387\ atm$$

$$n = \frac{PV}{RT} = \frac{0.387 atm \times 1.25 L}{\frac{0.08206 L\ atm}{K\ mol} \times 593 K} = 9.94 \times 10^{-3}\ mol\ SO_2Cl_2$$

$$9.94 \times 10^{-3}\ mol \times \frac{6.022 \times 10^{23}\ molecules}{mol} = 5.99 \times 10^{21}\ molecules\ SO_2Cl_2$$

114. $k = \dfrac{\ln 2}{t_{1/2}} = \dfrac{\ln 2}{667 s} = 1.04 \times 10^{-3}\ s^{-1}$

$$[In^+]_0 = \frac{2.38 g\ InCl \times \frac{1 mol\ InCl}{150.3 g} \times \frac{1 mol\ In^+}{mol\ InCl}}{0.500 L} = 0.0317\ mol/L$$

$$\ln\left(\frac{[In^+]}{[In^+]_0}\right) = -kt,\ \ln\left(\frac{[In^+]}{0.0317 M}\right) = -(1.04 \times 10^{-3}\ s^{-1}) \times 1.25\ h \times \frac{3600s}{h}$$

$[In^+] = 2.94 \times 10^{-4}$ mol/L

The balanced equation for the reaction is: $3\ In^+(aq) \rightarrow 2\ In(s) + In^{3+}(aq)$

$$\text{Mol In}^+ \text{ reacted} = 0.500\ L \times \frac{0.0317\ mol}{L} - 0.500\ L \times \frac{2.94 \times 10^{-4}\ mol}{L} = 1.57 \times 10^{-2}\ mol\ In^+$$

$$1.57 \times 10^{-2}\ mol\ In^+ \times \frac{2\ mol\ In}{3\ mol\ In^+} \times \frac{114.8\ g\ In}{mol\ In} = 1.20\ g\ In$$

115. $\ln\left(\frac{k_2}{k_1}\right) = \frac{E_a}{R}\left(\frac{1}{T_1} - \frac{1}{T_2}\right);\ \ln\left(\frac{1.7 \times 10^{-2}\ s^{-1}}{7.2 \times 10^{-4}\ s^{-1}}\right) = \frac{E_a}{8.3145 J/K \cdot mol}\left(\frac{1}{660.K} - \frac{1}{720.K}\right)$

$E_a = 2.1 \times 10^5$ J/mol

For k at 325°C (598 K):

$$\ln\left(\frac{1.7 \times 10^{-2}\ s^{-1}}{k}\right) = \frac{2.1 \times 10^5\ J/mol}{8.3145 J/K \cdot mol}\left(\frac{1}{598 K} - \frac{1}{720.K}\right),\ k = 1.3 \times 10^{-5}\ s^{-1}$$

For three half-lives, we go from 100% → 50% → 25% → 12.5%. After three half-lives, 12.5% of the original amount of C_2H_5I remains. Partial pressures are directly related to gas concentrations in mol/L:

$P_{C_2H_5I}$ = 894 torr × 0.125 = 112 torr after 3 half-lives

Marathon Problem

116. a. Rate = $k[CH_3X]^x[Y]^y$; for experiment 1, [Y] is in large excess, so its concentration will be constant. Rate = $k'[CH_3X]^x$, where $k' = k(3.0\ M)^y$.

A plot (not included) of $\ln[CH_3X]$ versus t is linear ($x = 1$). The integrated rate law is:

$\ln[CH_3X] = -(0.93)t - 3.99;\ k' = 0.93\ h^{-1}$

For experiment 2, [Y] is again constant, with Rate = $k''[CH_3X]^x$, where $k'' = k(4.5\ M)^y$. The natural log plot is linear again with an integrated rate law:

$\ln[CH_3X] = -(0.93)t - 5.40;\ k'' = 0.93\ h^{-1}$

Dividing the rate-constant values: $\frac{k'}{k''} = \frac{0.93}{0.93} = \frac{k(3.0)^y}{k(4.5)^y},\ 1.0 = (0.67)^y,\ y = 0$

The reaction is first order in CH_3X and zero order in Y. The overall rate law is:

Rate = $k[CH_3X]$, where $k = 0.93\ h^{-1}$ at 25°C

b. $t_{1/2} = (\ln 2)/k = 0.6931/(7.88 \times 10^8 \text{ h}^{-1}) = 8.80 \times 10^{-10}$ hour

c. $\ln\left(\frac{k_2}{k_1}\right) = \frac{E_a}{R}\left(\frac{1}{T_1} - \frac{1}{T_2}\right)$, $\ln\left(\frac{7.88 \times 10^8}{0.93}\right) = \frac{E_a}{8.3145 \text{J/K} \bullet \text{mol}}\left(\frac{1}{298\text{K}} - \frac{1}{358\text{K}}\right)$

$E_a = 3.0 \times 10^5$ J/mol $= 3.0 \times 10^2$ kJ/mol

d. From part a, the reaction is first order in CH_3X and zero order in Y. From part c, the activation energy is close to the C-X bond energy. A plausible mechanism that explains the results in parts a and c is:

$$CH_3X \rightarrow CH_3 + X \quad \text{(slow)}$$
$$CH_3 + Y \rightarrow CH_3Y \quad \text{(fast)}$$
$$\overline{CH_3X + Y \rightarrow CH_3Y + X}$$

Note: This is a plausible mechanism because the derived rate law is the same as the experimental rate law (and the sum of the steps gives the overall balanced equation).

CHAPTER 13

CHEMICAL EQUILIBRIUM

Questions

10. a. This experiment starts with only H_2 and N_2, and no NH_3 present. From the initial mixture diagram, there is three times as many H_2 as N_2 molecules. So the green line, at the highest initial concentration is the H_2 plot, the blue line is the N_2 plot, and the red line, which has an initial concentration of zero, is the NH_3 plot.

    b. $N_2(g) + 3H_2(g) \rightleftharpoons 2NH_3(g)$; when a reaction starts with only reactants present initially, the reactant concentrations decrease with time while the product concentrations increase with time. This is seen in the various plots. Also notice that the H_2 concentration initially decreases more rapidly as compared to the initial decrease in N_2 concentration. This is due to the stoichiometry in the balanced equation, which dictates that for every 1 molecule of N_2 that reacts, 3 molecules of H_2 must also react. One would expect the NH_3 plot to initially increase faster than the N_2 plot decreases (due to the 2 : 1 mole ratio in the balanced equation), and for the H_2 plot to initially decrease faster than the NH_3 plot increases (due to the 3 : 2 mole ratio). This is seen in the various plots.

    c. Equilibrium is reached when the rate of the forward reaction equals the rate of the reverse reaction. At this time, there is no net change in any of the reactant and product concentrations; so the various plots indicate equilibrium has been reached when their concentrations no longer change with time (when the plots reach a plateau).

11. No, equilibrium is a dynamic process. Both reactions:

    $H_2O + CO \rightarrow H_2 + CO_2$ and $H_2 + CO_2 \rightarrow H_2O + CO$

    are occurring at equal rates. Thus ^{14}C atoms will be distributed between CO and CO_2.

12. No, it doesn't matter from which direction the equilibrium position is reached (as long as temperature is constant). Both experiments will give the same equilibrium position because both experiments started with stoichiometric amounts of reactants or products.

13. A K value much greater than one ($K >> 1$) indicates there are relatively large concentrations of product gases/solutes as compared with the concentrations of reactant gases/solutes at equilibrium. A reaction with a very large K value is a good source of products.

14. A K value much less than one ($K << 1$) indicates that there are relatively large concentrations of reactant gases/solutes as compared with the concentrations of product gases/solutes at equilibrium. A reaction with a very small K value is a very poor source of products.

15. $H_2O(g) + CO(g) \rightleftharpoons H_2(g) + CO_2(g)$ $K = \frac{[H_2][CO_2]}{[H_2O][CO]} = 2.0$

K is a unitless number because there are an equal number of moles of product gases as moles of reactant gases in the balanced equation. Therefore, we can use units of molecules per liter instead of moles per liter to determine K.

We need to start somewhere, so let's assume 3 molecules of CO react. If 3 molecules of CO react, then 3 molecules of H_2O must react, and 3 molecules each of H_2 and CO_2 are formed. We would have 6 – 3 = 3 molecules of CO, 8 – 3 = 5 molecules of H_2O, 0 + 3 = 3 molecules of H_2, and 0 + 3 = 3 molecules of CO_2 present. This will be an equilibrium mixture if K = 2.0:

$$K = \frac{\left(\frac{3\text{ molecules } H_2}{L}\right)\left(\frac{3\text{ molecules } CO_2}{L}\right)}{\left(\frac{5\text{ molecules } H_2O}{L}\right)\left(\frac{3\text{ molecules } CO}{L}\right)} = \frac{3}{5}$$

Because this mixture does not give a value of K = 2.0, this is not an equilibrium mixture. Let's try 4 molecules of CO reacting to reach equilibrium.

Molecules CO remaining = 6 – 4 = 2 molecules of CO

Molecules H_2O remaining = 8 – 4 = 4 molecules of H_2O

Molecules H_2 present = 0 + 4 = 4 molecules of H_2

Molecules CO_2 present = 0 + 4 = 4 molecules of CO_2

$$K = \frac{\left(\frac{4\text{ molecules } H_2}{L}\right)\left(\frac{4\text{ molecules } CO_2}{L}\right)}{\left(\frac{4\text{ molecules } H_2O}{L}\right)\left(\frac{2\text{ molecules } CO}{L}\right)} = 2.0$$

Because K = 2.0 for this reaction mixture, we are at equilibrium.

16. When equilibrium is reached, there is no net change in the amount of reactants and products present because the rates of the forward and reverse reactions are equal to each other. The first diagram has 4 A_2B molecules, 2 A_2 molecules, and 1 B_2 molecule present. The second diagram has 2 A_2B molecules, 4 A_2 molecules, and 2 B_2 molecules. Therefore, the first diagram cannot represent equilibrium because there was a net change in reactants and products. Is the second diagram the equilibrium mixture? That depends on whether there is a net change between reactants and products when going from the second diagram to the third diagram. The third diagram contains the same numbers and types of molecules as the second diagram, so the second diagram is the first illustration that represents equilibrium.

The reaction container initially contained only A_2B. From the first diagram, 2 A_2 molecules and 1 B_2 molecule are present (along with 4 A_2B molecules). From the balanced reaction, these 2 A_2 molecules and 1 B_2 molecule were formed when 2 A_2B molecules decomposed. Therefore, the initial number of A_2B molecules present equals 4 + 2 = 6 molecules A_2B.

17. K and K_p are equilibrium constants, as determined by the law of mass action. For K, concentration units of mol/L are used, and for K_p, partial pressures in units of atm are used (generally). Q is called the reaction quotient. Q has the exact same form as K or K_p, but instead of equilibrium concentrations, initial concentrations are used to calculate the Q value. The use of Q is when it is compared with the K value. When Q = K (or when $Q_p = K_p$), the reaction is at equilibrium. When Q ≠ K, the reaction is not at equilibrium, and one can deduce the net change that must occur for the system to get to equilibrium.

18. $H_2(g) + I_2(g) \rightarrow 2\ HI(g) \qquad K = \frac{[HI]^2}{[H_2][I_2]}$

$H_2(g) + I_2(s) \rightarrow 2\ HI(g) \qquad K = \frac{[HI]^2}{[H_2]}$ (Solids are not included in K expressions.)

Some property differences are:

(1) the reactions have different K expressions.

(2) for the first reaction, $K = K_p$ (since $\Delta n = 0$), and for the second reaction, $K \neq K_p$ (since $\Delta n \neq 0$).

(3) a change in the container volume will have no effect on the equilibrium for reaction 1, whereas a volume change will affect the equilibrium for reaction 2 (shifts the reaction left or right depending on whether the volume is decreased or increased).

19. We always try to make good assumptions that simplify the math. In some problems we can set up the problem so that the net change x that must occur to reach equilibrium is a small number. This comes in handy when you have expressions like $0.12 - x$ or $0.727 + 2x$, etc. When x is small, we can assume that it makes little difference when subtracted from or added to some relatively big number. When this is the case, $0.12 - x \approx 0.12$ and $0.727 + 2x \approx 0.727$, etc. If the assumption holds by the 5% rule, the assumption is assumed valid. The 5% rule refers to x (or $2x$ or $3x$, etc.) that is assumed small compared to some number. If x (or $2x$ or $3x$, etc.) is less than 5% of the number the assumption was made against, then the assumption will be assumed valid. If the 5% rule fails to work, one can use a math procedure called the method of successive approximations to solve the quadratic or cubic equation. Of course, one could always solve the quadratic or cubic equation exactly. This is generally a last resort (and is usually not necessary).

20. Only statement e is correct. Addition of a catalyst has no effect on the equilibrium position; the reaction just reaches equilibrium more quickly. Statement a is false for reactants that are either solids or liquids (adding more of these has no effect on the equilibrium). Statement b is false always. If temperature remains constant, then the value of K is constant. Statement c is false for exothermic reactions where an increase in temperature decreases the value of K. For statement d, only reactions that have more reactant gases than product gases will shift left with an increase in container volume. If the moles of gas are equal, or if there are more moles of product gases than reactant gases, the reaction will not shift left with an increase in volume.

The Equilibrium Constant

21. a. $K = \frac{[NO]^2}{[N_2][O_2]}$ b. $K = \frac{[NO_2]^2}{[N_2O_4]}$

c. $K = \frac{[SiCl_4][H_2]^2}{[SiH_4][Cl_2]^2}$ d. $K = \frac{[PCl_3]^2[Br_2]^3}{[PBr_3]^2[Cl_2]^3}$

22. a. $K_p = \frac{P_{NO}^2}{P_{N_2} \times P_{O_2}}$ b. $K_p = \frac{P_{NO_2}^2}{P_{N_2O_4}}$

c. $K_p = \frac{P_{SiCl_4} \times P_{H_2}^2}{P_{SiH_4} \times P_{Cl_2}^2}$ d. $K_p = \frac{P_{PCl_3}^2 \times P_{Br_2}^3}{P_{PBr_3}^2 \times P_{Cl_2}^3}$

23. $K = 1.3 \times 10^{-2} = \frac{[NH_3]^2}{[N_2][H_2]^3}$ for $N_2(g) + 3\ H_2(g) \rightleftharpoons 2\ NH_3(g)$.

When a reaction is reversed, then $K_{new} = 1/K_{original}$. When a reaction is multiplied through by a value of n, then $K_{new} = (K_{original})^n$.

a. $1/2\ N_2(g) + 3/2\ H_2(g) \rightleftharpoons NH_3(g)$ $K' = \frac{[NH_3]^2}{[N_2]^{1/2}[H_2]^{3/2}} = K^{1/2} = (1.3 \times 10^{-2})^{1/2} = 0.11$

b. $2\ NH_3(g) \rightleftharpoons N_2(g) + 3\ H_2(g)$ $K'' = \frac{[N_2][H_2]^3}{[NH_3]^2} = \frac{1}{K} = \frac{1}{1.3 \times 10^{-2}} = 77$

c. $NH_3(g) \rightleftharpoons 1/2\ N_2(g) + 3/2\ H_2(g)$ $K''' = \frac{[N_2]^{1/2}[H_2]^{3/2}}{[NH_3]} = \left(\frac{1}{K}\right)^{1/2} = \left(\frac{1}{1.3 \times 10^{-2}}\right)^{1/2}$

$= 8.8$

d. $2\ N_2(g) + 6\ H_2(g) \rightleftharpoons 4\ NH_3(g)$ $K'''' = \frac{[NH_3]^4}{[N_2]^2[H_2]^6} = (K)^2 = (1.3 \times 10^{-2})^2 = 1.7 \times 10^{-4}$

24. $H_2(g) + Br_2(g) \rightleftharpoons 2\ HBr(g)$ $K_p = \frac{P_{HBr}^2}{(P_{H_2})(P_{Br_2})} = 3.5 \times 10^4$

a. $HBr \rightleftharpoons 1/2\ H_2 + 1/2\ Br_2$ $K_p' = \frac{(P_{H_2})^{1/2}(P_{Br_2})^{1/2}}{P_{HBr}} = \left(\frac{1}{K_p}\right)^{1/2} = \left(\frac{1}{3.5 \times 10^4}\right)^{1/2}$

$= 5.3 \times 10^{-3}$

b. $2\ HBr \rightleftharpoons H_2 + Br_2$ $K_p'' = \frac{(P_{H_2})(P_{Br_2})}{P_{HBr}^2} = \frac{1}{K_p} = \frac{1}{3.5 \times 10^4} = 2.9 \times 10^{-5}$

c. $1/2\ H_2 + 1/2\ Br_2 \rightleftharpoons HBr \quad K_p''' = \frac{P_{HBr}}{(P_{H_2})^{1/2}(P_{Br_2})^{1/2}} = (K_p)^{1/2} = 190$

25. $2\ NO(g) + 2\ H_2(g) \rightleftharpoons N_2(g) + 2\ H_2O(g) \quad K = \frac{[N_2][H_2O]^2}{[NO]^2[H_2]^2}$

$$K = \frac{(5.3 \times 10^{-2})(2.9 \times 10^{-3})^2}{(8.1 \times 10^{-3})^2(4.1 \times 10^{-5})^2} = 4.0 \times 10^6$$

26. $K = \frac{[NO]^2}{[N_2][O_2]} = \frac{(4.7 \times 10^{-4}\ M)^2}{(0.041M)(0.0078M)} = 6.9 \times 10^{-4}$

27. $[NO] = \frac{4.5 \times 10^{-3}\ \text{mol}}{3.0\ \text{L}} = 1.5 \times 10^{-3} M;\ [Cl_2] = \frac{2.4\ \text{mol}}{3.0\ \text{L}} = 0.80\ M$

$$[NOCl] = \frac{1.0\ \text{mol}}{3.0\ \text{L}} = 0.33\ M; \quad K = \frac{[NO]^2[Cl_2]}{[NOCl]^2} = \frac{(1.5 \times 10^{-3})^2(0.80)}{(0.33)^2} = 1.7 \times 10^{-5}$$

28. $[N_2O] = \frac{2.00 \times 10^{-2}\ \text{mol}}{2.00\ \text{L}};\ [N_2] = \frac{2.80 \times 10^{-4}\ \text{mol}}{2.00\ \text{L}};\ [O_2] = \frac{2.50 \times 10^{-5}\ \text{mol}}{2.00\ \text{L}}$

$$K = \frac{[N_2O]^2}{[N_2]^2[O_2]} = \frac{\left(\frac{2.00 \times 10^{-2}}{2.00}\right)^2}{\left(\frac{2.80 \times 10^{-4}}{2.00}\right)^2\left(\frac{2.50 \times 10^{-5}}{2.00}\right)} = \frac{(1.00 \times 10^{-2})^2}{(1.40 \times 10^{-4})^2(1.25 \times 10^{-5})}$$

$$= 4.08 \times 10^8$$

If the given concentrations represent equilibrium concentrations, then they should give a value of $K = 4.08 \times 10^8$.

$$\frac{(0.200)^2}{(2.00 \times 10^{-4})^2(0.00245)} = 4.08 \times 10^8$$

Because the given concentrations when plugged into the equilibrium constant expression give a value equal to K (4.08×10^8), this set of concentrations is a system at equilibrium.

29. $K_p = \frac{P_{NO}^2 \times P_{O_2}}{P_{NO_2}^2} = \frac{(6.5 \times 10^{-5})^2(4.5 \times 10^{-5})}{(0.55)^2} = 6.3 \times 10^{-13}$

30. $K_p = \frac{P_{NH_3}^2}{P_{N_2} \times P_{H_2}^3} = \frac{(3.1 \times 10^{-2})^2}{(0.85)(3.1 \times 10^{-3})^3} = 3.8 \times 10^4$

$$\frac{(0.0167)^2}{(0.525)(0.00761)^3} = 1.21 \times 10^3$$

When the given partial pressures in atmospheres are plugged into the K_p expression, the value does not equal the K_p value of 3.8×10^4. Therefore, one can conclude that the given set of partial pressures does not represent a system at equilibrium.

31. $K_p = K(RT)^{\Delta n}$, where Δn = sum of gaseous product coefficients − sum of gaseous reactant coefficients. For this reaction, $\Delta n = 3 - 1 = 2$.

$$K = \frac{[CO][H_2]^2}{[CH_3OH]} = \frac{(0.24)(1.1)^2}{(0.15)} = 1.9$$

$$K_p = K(RT)^2 = 1.9(0.08206 \text{ L atm/K•mol} \times 600.\text{ K})^2 = 4.6 \times 10^3$$

32. $K_p = K(RT)^{\Delta n}$, $K = \dfrac{K_p}{(RT)^{\Delta n}}$; $\Delta n = 2 - 3 = -1$

$$K = \frac{0.25}{(0.08206 \text{L atm/Kmol} \times 1100\text{K})^{-1}} = 23$$

33. Solids and liquids do not appear in equilibrium expressions. Only gases and dissolved solutes appear in equilibrium expressions.

a. $K = \dfrac{[H_2O]}{[NH_3]^2[CO_2]}$; $K_p = \dfrac{P_{H_2O}}{P_{NH_3}^2 \times P_{CO_2}}$

b. $K = [N_2][Br_2]^3$; $K_p = P_{N_2} \times P_{Br_2}^3$

c. $K = [O_2]^3$; $K_p = P_{O_2}^3$

d. $K = \dfrac{[H_2O]}{[H_2]}$; $K_p = \dfrac{P_{H_2O}}{P_{H_2}}$

34. a. $K_p = \dfrac{1}{(P_{O_2})^{3/2}}$

b. $K_p = \dfrac{1}{P_{CO_2}}$

c. $K_p = \dfrac{P_{CO} \times P_{H_2}}{P_{H_2O}}$

d. $K_p = \dfrac{P_{O_2}^3}{P_{H_2O}^2}$

35. $K_p = K(RT)^{\Delta n}$, where Δn equals the difference in the sum of the coefficients between gaseous products and gaseous reactants (Δn = mol gaseous products − mol gaseous reactants). When $\Delta n = 0$, then $K_p = K$. In Exercise 33, only reaction d has $\Delta n = 0$, so only reaction d has K_p = K.

36. $K_p = K$ when $\Delta n = 0$. In Exercise 34, none of the reactions have $K_p = K$ because none of the reactions have $\Delta n = 0$. The values of Δn for the various reactions are −1.5, −1, 1, and 1, respectively.

37. Because solids do not appear in the equilibrium constant expression, $K = 1/[O_2]^3$.

$$[O_2] = \frac{1.0 \times 10^{-3}\ \text{mol}}{2.0\ \text{L}};\ K = \frac{1}{[O_2]^3} = \frac{1}{\left(\frac{1.0 \times 10^{-3}}{2.0}\right)^3} = \frac{1}{(5.0 \times 10^{-4})^3} = 8.0 \times 10^9$$

38. $K_p = \frac{P_{H_2}^4}{P_{H_2O}^4}$; $P_{total} = P_{H_2O} + P_{H_2}$, 36.3 torr = 15.0 torr + P_{H_2}, P_{H_2} = 21.3 torr

Because 1 atm = 760 torr: $K_p = \frac{\left(21.3\ \text{torr} \times \frac{1\ \text{atm}}{760\ \text{torr}}\right)^4}{\left(15.0\ \text{torr} \times \frac{1\ \text{atm}}{760\ \text{torr}}\right)^4} = 4.07$

Note: Solids and pure liquids are not included in K expressions.

Equilibrium Calculations

39. $H_2O(g) + Cl_2O(g) \rightarrow 2\ HOCl(g)$ $\quad K = \frac{[HOCl]^2}{[H_2O][Cl_2O]} = 0.0900$

Use the reaction quotient Q to determine which way the reaction shifts to reach equilibrium. For the reaction quotient, initial concentrations given in a problem are used to calculate the value for Q. If Q < K, then the reaction shifts right to reach equilibrium. If Q > K, then the reaction shifts left to reach equilibrium. If Q = K, then the reaction does not shift in either direction because the reaction is already at equilibrium.

a. $$Q = \frac{[HOCl]_0^2}{[H_2O]_0[Cl_2O]_0} = \frac{\left(\frac{1.0\ \text{mol}}{1.0\ \text{L}}\right)^2}{\left(\frac{0.10\ \text{mol}}{1.0\ \text{L}}\right)\left(\frac{0.10\ \text{mol}}{1.0\ \text{L}}\right)} = 1.0 \times 10^2$$

Q > K, so the reaction shifts left to produce more reactants to reach equilibrium.

b. $$Q = \frac{\left(\frac{0.084\ \text{mol}}{2.0\ \text{L}}\right)^2}{\left(\frac{0.98\ \text{mol}}{2.0\ \text{L}}\right)\left(\frac{0.080\ \text{mol}}{2.0\ \text{L}}\right)} = 0.090 = K;\ \text{at equilibrium}$$

c. $$Q = \frac{\left(\frac{0.25\text{ mol}}{3.0\text{ L}}\right)^2}{\left(\frac{0.56\text{ mol}}{3.0\text{ L}}\right)\left(\frac{0.0010\text{mol}}{3.0\text{ L}}\right)} = 110$$

Q > K, so the reaction shifts to the left to reach equilibrium.

40. As in Exercise 39, determine Q for each reaction, compare this value to K_p (= 0.0900), and then determine which direction the reaction shifts to reach equilibrium. Note that for this reaction, K = K_p because $\Delta n = 0$.

a. $$Q = \frac{P_{HOCl}^2}{P_{H_2O} \times P_{Cl_2O}} = \frac{(1.00\text{ atm})^2}{(1.00\text{ atm})(1.00\text{ atm})} = 1.00$$

$Q > K_p$, so the reaction shifts left to reach equilibrium.

b. $$Q = \frac{(21.0\text{ torr})^2}{(200.\text{ torr})(49.8\text{ torr})} = 4.43 \times 10^{-2} < K_p$$

The reaction shifts right to reach equilibrium. *Note*: Because Q and K_p are unitless, we can use any pressure units when determining Q without changing the numerical value.

c. $$Q = \frac{(20.0\text{ torr})^2}{(296\text{ torr})(15.0\text{ torr})} = 0.0901 \approx K_p;\ \text{at equilibrium}$$

41. $CaCO_3(s) \rightleftharpoons CaO(s) + CO_2(g) \quad K_p = P_{CO_2} = 1.04$

a. $Q = P_{CO_2}$; we only need the partial pressure of CO_2 to determine Q because solids do not appear in equilibrium expressions (or Q expressions). At this temperature, all CO_2 will be in the gas phase. Q = 2.55, so $Q > K_p$; the reaction will shift to the left to reach equilibrium; the mass of CaO will decrease.

b. $Q = 1.04 = K_p$, so the reaction is at equilibrium; mass of CaO will not change.

c. $Q = 1.04 = K_p$, so the reaction is at equilibrium; mass of CaO will not change.

d. $Q = 0.211 < K_p$; the reaction will shift to the right to reach equilibrium; mass of CaO will increase.

42. $CH_3CO_2H + C_2H_5OH \rightleftharpoons CH_3CO_2C_2H_5 + H_2O \quad K = \frac{[CH_3CO_2C_2H_5][H_2O]}{[CH_3CO_2H][C_2H_5OH]} = 2.2$

a. $Q = \frac{(0.22)(0.10)}{(0.010)(0.010)} = 220 > K$; reaction will shift left to reach equilibrium, so the concentration of water will decrease.

b. $Q = \frac{(0.22)(0.0020)}{(0.0020)(0.10)} = 2.2 = K$; reaction is at equilibrium, so the concentration of water will remain the same.

c. $Q = \frac{(0.88)(0.12)}{(0.044)(6.0)} = 0.40 < K$; because Q < K, the concentration of water will increase because the reaction shifts right to reach equilibrium.

d. $Q = \frac{(4.4)(4.4)}{(0.88)(10.0)} = 2.2 = K$; at equilibrium, so the water concentration is unchanged.

e. $K = 2.2 = \frac{(2.0)[H_2O]}{(0.10)(5.0)}$, $[H_2O] = 0.55\ M$

f. Water is a product of the reaction, but it is not the solvent. Thus the concentration of water must be included in the equilibrium expression because it is a solute in the reaction. When water is the solvent, then it is not included in the equilibrium expression.

43. $K = \frac{[H_2]^2[O_2]}{[H_2O]^2}$, $2.4 \times 10^{-3} = \frac{(1.9 \times 10^{-2})^2[O_2]}{(0.11)^2}$, $[O_2] = 0.080\ M$

Moles of $O_2 = 2.0\ L \times \frac{0.080\ mol\ O_2}{L} = 0.16\ mol\ O_2$

44. $K_P = \frac{P_{NOBr}^2}{P_{NO}^2 \times P_{Br_2}}$, $109 = \frac{(0.0768)^2}{P_{NO}^2 \times 0.0159}$, $P_{NO} = 0.0583$ atm

45. $SO_2(g) + NO_2(g) \rightleftharpoons SO_3(g) + NO(g)$ $K = \frac{[SO_3][NO]}{[SO_2][NO_2]}$

To determine K, we must calculate the equilibrium concentrations. The initial concentrations are:

$$[SO_3]_0 = [NO]_0 = 0;\ \ [SO_2]_0 = [NO_2]_0 = \frac{2.00\ mol}{1.00\ L} = 2.00\ M$$

Next, we determine the change required to reach equilibrium. At equilibrium, [NO] = 1.30 mol/1.00 L = 1.30 *M*. Because there was zero NO present initially, 1.30 *M* of SO_2 and 1.30 *M* NO_2 must have reacted to produce 1.30 *M* NO as well as 1.30 *M* SO_3, all required by the balanced reaction. The equilibrium concentration for each substance is the sum of the initial concentration plus the change in concentration necessary to reach equilibrium. The equilibrium concentrations are:

$$[SO_3] = [NO] = 0 + 1.30\ M = 1.30\ M;\ \ [SO_2] = [NO_2] = 2.00\ M - 1.30\ M = 0.70\ M$$

We now use these equilibrium concentrations to calculate K:

$$K = \frac{[SO_3][NO]}{[SO_2][NO_2]} = \frac{(1.30)(1.30)}{(0.70)(0.70)} = 3.4$$

46. $S_8(g) \rightleftharpoons 4\ S_2(g) \quad K_p = \dfrac{P_{S_2}^4}{P_{S_8}}$

Initially: P_{S_8} = 1.00 atm and P_{S_2} = 0 atm

Change: Because 0.25 atm of S_8 remain at equilibrium, 1.00 atm – 0.25 atm = 0.75 atm of S_8 must have reacted in order to reach equilibrium. Because there is a 4 : 1 mole ratio between S_2 and S_8 (from the balanced reaction), 4(0.75 atm) = 3.0 atm of S_2 must have been produced when the reaction went to equilibrium (moles and pressure are directly related at constant T and V).

Equilibrium: P_{S_8} = 0.25 atm, P_{S_2} = 0 + 3.0 atm = 3.0 atm; solving for K_p:

$$K_p = \frac{(3.0)^4}{0.25} = 3.2 \times 10^2$$

47. When solving equilibrium problems, a common method to summarize all the information in the problem is to set up a table. We commonly call this table an ICE table because it summarizes *i*nitial concentrations, *c*hanges that must occur to reach equilibrium, and *e*quilibrium concentrations (the sum of the initial and change columns). For the change column, we will generally use the variable *x*, which will be defined as the amount of reactant (or product) that must react to reach equilibrium. In this problem, the reaction must shift right to reach equilibrium because there are no products present initially. Therefore, *x* is defined as the amount of reactant SO_3 that reacts to reach equilibrium, and we use the coefficients in the balanced equation to relate the net change in SO_3 to the net change in SO_2 and O_2. The general ICE table for this problem is:

	$2\ SO_3(g)$	$\rightleftharpoons$	$2\ SO_2(g)$	+	$O_2(g)$	$K = \dfrac{[SO_2]^2[O_2]}{[SO_3]^2}$
Initial	12.0 mol/3.0 L		0		0	
	Let *x* mol/L of SO_3 react to reach equilibrium.					
Change	$-x$	→	$+x$		$+x/2$	
Equil.	$4.0 - x$		x		$x/2$	

From the problem, we are told that the equilibrium SO_2 concentration is 3.0 mol/3.0 L = 1.0 *M* ($[SO_2]_e$ = 1.0 *M*). From the ICE table setup, $[SO_2]_e = x$, so x = 1.0. Solving for the other equilibrium concentrations: $[SO_3]_e = 4.0 - x = 4.0 - 1.0 = 3.0\ M$; $[O_2] = x/2 = 1.0/2 = 0.50\ M$.

$$K = \frac{[SO_2]^2[O_2]}{[SO_3]^2} = \frac{(1.0)^2(0.50)}{(3.0)^2} = 0.056$$

Alternate method: Fractions in the change column can be avoided (if you want) be defining *x* differently. If we were to let 2*x* mol/L of SO_3 react to reach equilibrium, then the ICE table setup is:

	$2\ SO_3(g)$	$\rightleftharpoons$	$2\ SO_2(g)$	+	$O_2(g)$	$K = \frac{[SO_2]^2[O_2]}{[SO_3]^2}$
Initial	4.0 *M*		0		0	
	Let 2*x* mol/L of SO_3 react to reach equilibrium.					
Change	$-2x$	→	$+2x$		$+x$	
Equil.	$4.0 - 2x$		$2x$		x	

Solving: $2x = [SO_2]_e = 1.0\ M$, $x = 0.50\ M$; $[SO_3]_e = 4.0 - 2(0.50) = 3.0\ M$; $[O_2]_e = x = 0.50\ M$

These are exactly the same equilibrium concentrations as solved for previously, thus K will be the same (as it must be). The moral of the story is to define *x* in a manner that is most comfortable for you. Your final answer is independent of how you define *x* initially.

48. When solving equilibrium problems, a common method to summarize all the information in the problem is to set up a table. We commonly call this table the ICE table because it summarizes *i*nitial concentrations, *c*hanges that must occur to reach equilibrium, and *e*quilibrium concentrations (the sum of the initial and change columns). For the change column, we will generally use the variable *x*, which will be defined as the amount of reactant (or product) that must react to reach equilibrium. In this problem, the reaction must shift right since there are no products present initially. The general ICE table for this problem is:

	$2\ NO_2(g)$	$\rightleftharpoons$	$2\ NO(g)$	+	$O_2(g)$	$K = \frac{[NO]^2[O_2]}{[NO_2]^2}$
Initial	8.0 mol/1.0 L		0		0	
	Let *x* mol/L of NO_2 react to reach equilibrium					
Change	$-x$	→	$+x$		$+x/2$	
Equil.	$8.0 - x$		x		$x/2$	

Note that we must use the coefficients in the balanced equation to determine the amount of products produced when *x* mol/L of NO_2 reacts to reach equilibrium. In the problem, we are told that $[NO]_e = 2.0\ M$. From the set-up, $[NO]_e = x = 2.0\ M$. Solving for the other concentrations: $[NO]_e = 8.0 - x = 8.0 - 2.0 = 6.0\ M$; $[O_2]_e = x/2 = 2.0/2 = 1.0\ M$. Calculating K:

$$K = \frac{[NO]^2[O_2]}{[NO_2]^2} = \frac{(2.0\ M)^2\ (1.0\ M)}{(6.0\ M)^2} = 0.11\ \text{mol/L}$$

Alternate method: Fractions in the change column can be avoided (if you want) by defining *x* differently. If we were to let 2*x* mol/L of NO_2 react to reach equilibrium, then the ICE table set-up is:

	$2\ NO_2(g)$	$\rightleftharpoons$	$2\ NO(g)$	+	$O_2(g)$	$K = \frac{[NO]^2[O_2]}{[NO_2]^2}$
	8.0 *M*		0		0	
	Let 2*x* mol/L of NO_2 react to reach equilibrium					
Change	$-2x$	→	$+2x$		$+x$	
Equil.	$8.0 - 2x$		$2x$		x	

Solving: $2x = [NO]_e = 2.0\ M,\ x = 1.0\ M;\ [NO_2]_e = 8.0 - 2(1.0) = 6.0\ M;\ [O_2]_e = x = 1.0\ M$

These are exactly the same equilibrium concentrations as solved for previously; thus K will be the same (as it must be). The moral of the story is to define x in a manner that is most comfortable for you. Your final answer is independent of how you define x initially.

49.

	$3\ H_2(g)$	+	$N_2(g)$	$\rightleftharpoons$	$2\ NH_3(g)$
Initial	$[H_2]_0$		$[N_2]_0$		0
	x mol/L of N_2 reacts to reach equilibrium				
Change	$-3x$		$-x$	$\rightarrow$	$+2x$
Equil	$[H_2]_0 - 3x$		$[N_2]_0 - x$		$2x$

From the problem:

$$[NH_3]_e = 4.0\ M = 2x,\ x = 2.0\ M;\ [H_2]_e = 5.0\ M = [H_2]_0 - 3x;\ [N_2]_e = 8.0\ M = [N_2]_0 - x$$

$$5.0\ M = [H_2]_0 - 3(2.0\ M),\ [H_2]_0 = 11.0\ M;\ 8.0\ M = [N_2]_0 - 2.0\ M,\ [N_2]_0 = 10.0\ M$$

50. $N_2(g) + 3\ H_2(g) \rightleftharpoons 2\ NH_3(g)$; with only reactants present initially, the net change that must occur to reach equilibrium is a conversion of reactants into products. At constant volume and temperature, $n \propto P$. Thus, if x atm of N_2 reacts to reach equilibrium, then $3x$ atm of H_2 must also react to form $2x$ atm of NH_3 (from the balanced equation). Let's summarize the problem in a table that lists what is present initially, what change in terms of x that occurs to reach equilibrium, and what is present at equilibrium (initial + change). This table is typically called an ICE table for *i*nitial, *c*hange, and *e*quilibrium.

$$K_p = \frac{P_{NH_3}^2}{P_{N_2} \times P_{H_2}^3}$$

	$N_2(g)$	+	$3\ H_2(g)$	$\rightleftharpoons$	$2\ NH_3(g)$
Initial	1.00 atm		2.00 atm		0
	x atm of N_2 reacts to reach equilibrium				
Change	$-x$		$-3x$	$\rightarrow$	$+2x$
Equil.	$1.00 - x$		$2.00 - 3x$		$2x$

From the setup: $P_{total} = 2.00\text{ atm} = P_{N_2} + P_{H_2} + P_{NH_3}$

$$2.00\text{ atm} = (1.00 - x) + (2.00 - 3x) + 2x = 3.00 - 2x,\quad x = 0.500$$

$$P_{H_2} = 2.00 - 3x = 2.00 - 3(0.500) = 0.50\text{ atm}$$

$$K_p = \frac{(2x)^2}{(1.00-x)(2.00-3x)^3} = \frac{[2(0.500)]^2}{(1.00-0.500)[2.00-3(0.500)]^3} = \frac{(1.00)^2}{(0.50)(0.50)^3} = 16$$

51. Q = 1.00, which is less than K. The reaction shifts to the right to reach equilibrium. Summarizing the equilibrium problem in a table:

	$SO_2(g)$ +	$NO_2(g)$ ⇌	$SO_3(g)$ +	$NO(g)$ K = 3.75
Initial	0.800 *M*	0.800 *M*	0.800 *M*	0.800 *M*
	x mol/L of SO_2 reacts to reach equilibrium			
Change	$-x$	$-x$ →	$+x$	$+x$
Equil.	$0.800 - x$	$0.800 - x$	$0.800 + x$	$0.800 + x$

Plug the equilibrium concentrations into the equilibrium constant expression:

$K = \frac{[SO_3][NO]}{[SO_2][NO_2]}$, $3.75 = \frac{(0.800 + x)^2}{(0.800 - x)^2}$; take the square root of both sides and solve for *x*:

$\frac{0.800 + x}{0.800 - x} = 1.94$, $0.800 + x = 1.55 - (1.94)x$, $(2.94)x = 0.75$, $x = 0.26\ M$

The equilibrium concentrations are:

$[SO_3] = [NO] = 0.800 + x = 0.800 + 0.26 = 1.06\ M$; $[SO_2] = [NO_2] = 0.800 - x = 0.54\ M$

52. Q = 1.00, which is less than K. Reaction shifts right to reach equilibrium.

	$H_2(g)$ +	$I_2(g)$ ⇌	$2\ HI(g)$ $K = \frac{[HI]^2}{[H_2][I_2]} = 100.$
Initial	1.00 *M*	1.00 *M*	1.00 *M*
	x mol/L of H_2 reacts to reach equilibrium		
Change	$-x$	$-x$ →	$+2x$
Equil.	$1.00 - x$	$1.00 - x$	$1.00 + 2x$

$K = 100. = \frac{(1.00 + 2x)^2}{(1.00 - x)^2}$; taking the square root of both sides:

$10.0 = \frac{1.00 + 2x}{1.00 - x}$, $10.0 - (10.0)x = 1.00 + 2x$, $(12.0)x = 9.0$, $x = 0.75\ M$

$[H_2] = [I_2] = 1.00 - 0.75 = 0.25\ M$; $[HI] = 1.00 + 2(0.75) = 2.50\ M$

53. Because only reactants are present initially, the reaction must proceed to the right to reach equilibrium. Summarizing the problem in a table:

	$N_2(g)$ +	$O_2(g)$ ⇌	$2\ NO(g)$ $K_p = 0.050$
Initial	0.80 atm	0.20 atm	0
	x atm of N_2 reacts to reach equilibrium		
Change	$-x$	$-x$ →	$+2x$
Equil.	$0.80 - x$	$0.20 - x$	$2x$

$K_p = 0.050 = \frac{P_{NO}^2}{P_{N_2} \times P_{O_2}} = \frac{(2x)^2}{(0.80 - x)(0.20 - x)}$, $0.050[0.16 - (1.00)x + x^2] = 4x^2$

$4x^2 = 8.0 \times 10^{-3} - (0.050)x + (0.050)x^2$, $(3.95)x^2 + (0.050)x - 8.0 \times 10^{-3} = 0$

Solving using the quadratic formula (see Appendix 1 of the text):

$$x = \frac{-b \pm (b^2 - 4ac)^{1/2}}{2a} = \frac{-0.050 \pm [(0.050)]^2 - 4(3.95)(-8.0 \times 10^{-3})]^{1/2}}{2(3.95)}$$

$x = 3.9 \times 10^{-2}$ atm or $x = -5.2 \times 10^{-2}$ atm; only $x = 3.9 \times 10^{-2}$ atm makes sense (x cannot be negative), so the equilibrium NO concentration is:

$$P_{NO} = 2x = 2(3.9 \times 10^{-2} \text{ atm}) = 7.8 \times 10^{-2} \text{ atm}$$

54. $H_2O(g) + Cl_2O(g) \rightleftharpoons 2\ HOCl(g)$ $\qquad K = 0.090 = \dfrac{[HOCl]^2}{[H_2O][Cl_2O]}$

a. The initial concentrations of H_2O and Cl_2O are:

$$\frac{1.0 \text{ g } H_2O}{1.0 \text{ L}} \times \frac{1 \text{ mol}}{18.02 \text{ g}} = 5.5 \times 10^{-2} \text{ mol/L}; \frac{2.0 \text{ g } Cl_2O}{1.0 \text{ L}} \times \frac{1 \text{ mol}}{86.90 \text{ g}} = 2.3 \times 10^{-2} \text{ mol/L}$$

	$H_2O(g)$	+	$Cl_2O(g)$	$\rightleftharpoons$	$2\ HOCl(g)$
Initial	$5.5 \times 10^{-2}\ M$		$2.3 \times 10^{-2}\ M$		0
	x mol/L of H_2O reacts to reach equilibrium				
Change	$-x$		$-x$	$\rightarrow$	$+2x$
Equil.	$5.5 \times 10^{-2} - x$		$2.3 \times 10^{-2} - x$		$2x$

$$K = 0.090 = \frac{(2x)^2}{(5.5 \times 10^{-2} - x)(2.3 \times 10^{-2} - x)}$$

$1.14 \times 10^{-4} - (7.02 \times 10^{-3})x + (0.090)x^2 = 4x^2$

$(3.91)x^2 + (7.02 \times 10^{-3})x - 1.14 \times 10^{-4} = 0$ (We carried extra significant figures.)

Solving using the quadratic formula:

$$\frac{-7.02 \times 10^{-3} \pm (4.93 \times 10^{-5} + 1.78 \times 10^{-3})^{1/2}}{7.82} = 4.6 \times 10^{-3} \text{ or } -6.4 \times 10^{-3}$$

A negative answer makes no physical sense; we can't have less than nothing. Thus $x = 4.6 \times 10^{-3}\ M$.

$[HOCl] = 2x = 9.2 \times 10^{-3}\ M$; $[Cl_2O] = 2.3 \times 10^{-2} - x = 0.023 - 0.0046 = 1.8 \times 10^{-2}\ M$

$[H_2O] = 5.5 \times 10^{-2} - x = 0.055 - 0.0046 = 5.0 \times 10^{-2}\ M$

b. $H_2O(g) + Cl_2O(g) \rightleftharpoons 2\ HOCl(g)$

	$H_2O(g)$	$Cl_2O(g)$		$2\ HOCl(g)$
Initial	0	0		1.0 mol/2.0 L = 0.50 M
	$2x$ mol/L of HOCl reacts to reach equilibrium			
Change	$+x$	$+x$	←	$-2x$
Equil.	x	x		$0.50 - 2x$

$$K = 0.090 = \frac{[HOCl]^2}{[H_2O][Cl_2O]} = \frac{(0.50 - 2x)^2}{x^2}$$

The expression is a perfect square, so we can take the square root of each side:

$$0.30 = \frac{0.50 - 2x}{x},\ (0.30)x = 0.50 - 2x,\ (2.30)x = 0.50$$

$x = 0.217$ (We carried extra significant figures.)

$x = [H_2O] = [Cl_2O] = 0.217 = 0.22\ M$; $[HOCl] = 0.50 - 2x = 0.50 - 0.434 = 0.07\ M$

55. $2\ SO_2(g) + O_2(g) \rightleftharpoons 2\ SO_3(g) \quad K_p = 0.25$

	$2\ SO_2(g)$	$O_2(g)$		$2\ SO_3(g)$
Initial	0.50 atm	0.50 atm		0
	$2x$ atm of SO_2 reacts to reach equilibrium			
Change	$-2x$	$-x$	→	$+2x$
Equil.	$0.50 - 2x$	$0.50 - x$		$2x$

$$K_p = 0.25 = \frac{P_{SO_3}^2}{P_{SO_2}^2 \times P_{O_2}} = \frac{(2x)^2}{(0.50 - 2x)^2(0.50 - x)}$$

This will give a cubic equation. Graphing calculators can be used to solve this expression. If you don't have a graphing calculator, an alternative method for solving a cubic equation is to use the method of successive approximations (see Appendix 1 of the text). The first step is to guess a value for x. Because the value of K is small (K < 1), not much of the forward reaction will occur to reach equilibrium. This tells us that x is small. Let's guess that x = 0.050 atm. Now we take this estimated value for x and substitute it into the equation everywhere that x appears except for one. For equilibrium problems, we will substitute the estimated value for x into the denominator and then solve for the numerator value of x. We continue this process until the estimated value of x and the calculated value of x converge on the same number. This is the same answer we would get if we were to solve the cubic equation exactly. Applying the method of successive approximations and carrying extra significant figures:

$$\frac{4x^2}{[0.50 - 2(0.050)]^2[0.50 - (0.050)]} = \frac{4x^2}{(0.40)^2(0.45)} = 0.25,\ x = 0.067$$

$$\frac{4x^2}{[0.50 - 2(0.067)]^2[0.50 - (0.067)]} = \frac{4x^2}{(0.366)^2(0.433)} = 0.25,\ x = 0.060$$

$$\frac{4x^2}{(0.38)^2(0.44)} = 0.25, \ x = 0.063; \quad \frac{4x^2}{(0.374)^2(0.437)} = 0.25, \quad x = 0.062$$

The next trial gives the same value for $x = 0.062$ atm. We are done except for determining the equilibrium concentrations. They are:

$P_{SO_2} = 0.50 - 2x = 0.50 - 2(0.062) = 0.376 = 0.38$ atm

$P_{O_2} = 0.50 - x = 0.438 = 0.44$ atm; $P_{SO_3} = 2x = 0.124 = 0.12$ atm

56. a. The reaction must proceed to products to reach equilibrium because no product is present initially. Summarizing the problem in a table where x atm of N_2O_4 reacts to reach equilibrium:

	$N_2O_4(g)$	$\rightleftharpoons$	$2\ NO_2(g)$	$K_p = 0.25$
Initial	4.5 atm		0	
Change	$-x$	$\rightarrow$	$+2x$	
Equil.	$4.5 - x$		$2x$	

$$K_p = \frac{P^2_{NO_2}}{P_{N_2O_4}} = \frac{(2x)^2}{4.5 - x} = 0.25, \ 4x^2 = 1.125 - (0.25)x, \ 4x^2 + (0.25)x - 1.125 = 0$$

We carried extra significant figures in this expression (as will be typical when we solve an expression using the quadratic formula). Solving using the quadratic formula (Appendix 1 of text):

$$x = \frac{-0.25 \pm [(0.25)^2 - 4(4)(-1.125)]^{1/2}}{2(4)} = \frac{-0.25 \pm 4.25}{8}, \ x = 0.50 \text{ (Other value is negative.)}$$

$P_{NO_2} = 2x = 1.0$ atm; $P_{N_2O_4} = 4.5 - x = 4.0$ atm

b. The reaction must shift to reactants (shifts left) to reach equilibrium.

	$N_2O_4(g)$	$\rightleftharpoons$	$2\ NO_2(g)$
Initial	0		9.0 atm
Change	$+x$	$\leftarrow$	$-2x$
Equil.	x		$9.0 - 2x$

$$K_p = \frac{(9.0 - 2x)^2}{x} = 0.25, \ 4x^2 - (36.25)x + 81 = 0 \ \text{(carrying extra significant figures)}$$

$$\text{Solving: } x = \frac{-(-36.25) \pm [(-36.25)^2 - 4(4)(81)]^{1/2}}{2(4)}, \ x = 4.0 \text{ atm}$$

The other value, 5.1, is impossible. $P_{N_2O_4} = x = 4.0$ atm; $P_{NO_2} = 9.0 - 2x = 1.0$ atm

c. No, we get the same equilibrium position starting with either pure N_2O_4 or pure NO_2 in stoichiometric amounts.

57. a. The reaction must proceed to products to reach equilibrium because only reactants are present initially. Summarizing the problem in a table:

	$2\ NOCl(g)$	$\rightleftharpoons$	$2\ NO(g)$	+	$Cl_2(g)$	$K = 1.6 \times 10^{-5}$
Initial	$\frac{2.0\ \text{mol}}{2.0\ \text{L}} = 1.0\ M$		0		0	
	$2x$ mol/L of NOCl reacts to reach equilibrium					
Change	$-2x$	$\rightarrow$	$+2x$		$+x$	
Equil.	$1.0 - 2x$		$2x$		x	

$$K = 1.6 \times 10^{-5} = \frac{[NO]^2[Cl_2]}{[NOCl]^2} = \frac{(2x)^2(x)}{(1.0 - 2x)^2}$$

If we assume that $1.0 - 2x \approx 1.0$ (from the small size of K, we know that the product concentrations will be small), then:

$$1.6 \times 10^{-5} = \frac{4x^3}{1.0^2}, \quad x = 1.6 \times 10^{-2}; \text{ now we must check the assumption.}$$

$$1.0 - 2x = 1.0 - 2(0.016) = 0.97 = 1.0 \text{ (to proper significant figures)}$$

Our error is about 3%; that is, $2x$ is 3.2% of 1.0 *M*. Generally, if the error we introduce by making simplifying assumptions is less than 5%, we go no further; the assumption is said to be valid. We call this the 5% rule. Solving for the equilibrium concentrations:

$$[NO] = 2x = 0.032\ M; \quad [Cl_2] = x = 0.016\ M; \quad [NOCl] = 1.0 - 2x = 0.97\ M \approx 1.0\ M$$

Note: If we were to solve this cubic equation exactly (a longer process), we get $x = 0.016$. This is the exact same answer we determined by making a simplifying assumption. We saved time and energy. Whenever K is a very small value ($K << 1$), always make the assumption that x is small. If the assumption introduces an error of less than 5%, then the answer you calculated making the assumption will be considered the correct answer.

b.

	$2\ NOCl(g)$	$\rightleftharpoons$	$2\ NO(g)$	+	$Cl_2(g)$
Initial	$1.0\ M$		$1.0\ M$		0
	$2x$ mol/L of NOCl reacts to reach equilibrium				
Change	$-2x$	$\rightarrow$	$+2x$		$+x$
Equil.	$1.0 - 2x$		$1.0 + 2x$		x

$$1.6 \times 10^{-5} = \frac{(1.0 + 2x)^2(x)}{(1.0 - 2x)^2} = \frac{(1.0)^2(x)}{(1.0)^2} \quad \text{(assuming } 2x << 1.0\text{)}$$

$x = 1.6 \times 10^{-5}$; assumptions are great ($2x$ is 3.2×10^{-3}% of 1.0).

$[Cl_2] = 1.6 \times 10^{-5}\ M$ and $[NOCl] = [NO] = 1.0\ M$

c. $2\ NOCl(g) \rightleftharpoons 2\ NO(g) + Cl_2(g)$

	$2\ NOCl(g)$	$\rightleftharpoons$	$2\ NO(g)$	$+\ Cl_2(g)$
Initial	$2.0\ M$		0	$1.0\ M$
	$2x$ mol/L of NOCl reacts to reach equilibrium			
Change	$-2x$	$\rightarrow$	$+2x$	$+x$
Equil.	$2.0 - 2x$		$2x$	$1.0 + x$

$$1.6 \times 10^{-5} = \frac{(2x)^2(1.0 + x)}{(2.0 - 2x)^2} = \frac{4x^2}{4.0} \quad \text{(assuming } x << 1.0\text{)}$$

Solving: $x = 4.0 \times 10^{-3}$; assumptions good (x is 0.4% of 1.0 and $2x$ is 0.4% of 2.0).

$[Cl_2] = 1.0 + x = 1.0\ M$; $[NO] = 2(4.0 \times 10^{-3}) = 8.0 \times 10^{-3}\ M$; $[NOCl] = 2.0\ M$

58. $N_2O_4(g) \rightleftharpoons 2\ NO_2(g)$ $\quad K = \dfrac{[NO_2]^2}{[N_2O_4]} = 4.0 \times 10^{-7}$

	$N_2O_4(g)$	$\rightleftharpoons$	$2\ NO_2(g)$
Initial	1.0 mol/10.0 L		0
	x mol/L of N_2O_4 reacts to reach equilibrium		
Change	$-x$	$\rightarrow$	$+2x$
Equil.	$0.10 - x$		$2x$

$$K = \frac{[NO_2]^2}{[N_2O_4]} = \frac{(2x)^2}{0.10 - x} = 4.0 \times 10^{-7}$$; because K has a small value, assume that x is small compared to 0.10, so that $0.10 - x \approx 0.10$. Solving:

$$4.0 \times 10^{-7} \approx \frac{4x^2}{0.10},\ 4x^2 = 4.0 \times 10^{-8},\ x = 1.0 \times 10^{-4}\ M$$

Checking the assumption by the 5% rule: $\dfrac{x}{0.10} \times 100 = \dfrac{1.0 \times 10^{-4}}{0.10} \times 100 = 0.10\%$

Because this number is less than 5%, we will say that the assumption is valid.

$[N_2O_4] = 0.10 - 1.0 \times 10^{-4} = 0.10\ M$; $[NO_2] = 2x = 2(1.0 \times 10^{-4}) = 2.0 \times 10^{-4}\ M$

59. $2\ CO_2(g) \rightleftharpoons 2\ CO(g) + O_2(g)$ $\quad K = \dfrac{[CO]^2[O_2]}{[CO_2]^2} = 2.0 \times 10^{-6}$

	$2\ CO_2(g)$	$\rightleftharpoons$	$2\ CO(g)$	$+\ O_2(g)$
Initial	2.0 mol/5.0 L		0	0
	$2x$ mol/L of CO_2 reacts to reach equilibrium			
Change	$-2x$	$\rightarrow$	$+2x$	$+x$
Equil.	$0.40 - 2x$		$2x$	x

$$K = 2.0 \times 10^{-6} = \frac{[CO]^2[O_2]}{[CO_2]^2} = \frac{(2x)^2(x)}{(0.40 - 2x)^2}$$; assuming $2x << 0.40$:

$$2.0 \times 10^{-6} \approx \frac{4x^3}{(0.40)^2},\ 2.0 \times 10^{-6} = \frac{4x^3}{0.16},\ x = 4.3 \times 10^{-3}\ M$$

Checking assumption: $\frac{2(4.3 \times 10^{-3})}{0.40} \times 100 = 2.2\%$; assumption is valid by the 5% rule.

$[CO_2] = 0.40 - 2x = 0.40 - 2(4.3 \times 10^{-3}) = 0.39\ M$

$[CO] = 2x = 2(4.3 \times 10^{-3}) = 8.6 \times 10^{-3}\ M$; $[O_2] = x = 4.3 \times 10^{-3}\ M$

60. $COCl_2(g) \rightleftharpoons CO(g) + Cl_2(g)$ $K_p = \frac{P_{CO} \times P_{Cl_2}}{P_{COCl_2}} = 6.8 \times 10^{-9}$

	$COCl_2(g)$		$CO(g)$	$Cl_2(g)$
Initial	1.0 atm		0	0
	x atm of $COCl_2$ reacts to reach equilibrium			
Change	$-x$	$\rightarrow$	$+x$	$+x$
Equil.	$1.0 - x$		x	x

$6.8 \times 10^{-9} = \frac{P_{CO} \times P_{Cl_2}}{P_{COCl_2}} = \frac{x^2}{1.0 - x} \approx \frac{x^2}{1.0}$ (Assuming $1.0 - x \approx 1.0$.)

$x = 8.2 \times 10^{-5}$ atm; assumption is good (x is $8.2 \times 10^{-3}\%$ of 1.0).

$P_{COCl_2} = 1.0 - x = 1.0 - 8.2 \times 10^{-5} = 1.0$ atm; $P_{CO} = P_{Cl_2} = x = 8.2 \times 10^{-5}$ atm

61. This is a typical equilibrium problem except that the reaction contains a solid. Whenever solids and liquids are present, we basically ignore them in the equilibrium problem.

$$NH_4OCONH_2(s) \rightleftharpoons 2\ NH_3(g) + CO_2(g) \qquad K_p = 2.9 \times 10^{-3}$$

	$NH_4OCONH_2(s)$		$2\ NH_3(g)$	$CO_2(g)$
Initial			0	0
	Some NH_4OCONH_2 decomposes to produce $2x$ atm of NH_3 and x atm of CO_2.			
Change		$\rightarrow$	$+2x$	$+x$
Equil.			$2x$	x

$K_p = 2.9 \times 10^{-3} = P_{NH_3}^2 \times P_{CO_2} = (2x)^2(x) = 4x^3$

$x = \left(\frac{2.9 \times 10^{-3}}{4}\right)^{1/3} = 9.0 \times 10^{-2}$ atm; $P_{NH_3} = 2x = 0.18$ atm; $P_{CO_2} = x = 9.0 \times 10^{-2}$ atm

$P_{total} = P_{NH_3} + P_{CO_2} = 0.18$ atm $+ 0.090$ atm $= 0.27$ atm

62. $NH_4Cl(s) \rightleftharpoons NH_3(g) + HCl(g)$ $K_p = P_{NH_3} \times P_{HCl}$

For this system to reach equilibrium, some of the $NH_4Cl(s)$ decomposes to form equal moles of $NH_3(g)$ and $HCl(g)$ at equilibrium. Because moles of HCl produced = moles of NH_3 produced, the partial pressures of each gas must be equal to each other.

At equilibrium: $P_{total} = P_{NH_3} + P_{HCl}$ and $P_{NH_3} = P_{HCl}$

$P_{total} = 4.4$ atm $= 2P_{NH_3}$, 2.2 atm $= P_{NH_3} = P_{HCl}$; $K_p = (2.2)(2.2) = 4.8$

Le Châtelier's Principle

63. a. No effect; adding more of a pure solid or pure liquid has no effect on the equilibrium position.

b. Shifts left; HF(g) will be removed by reaction with the glass. As HF(g) is removed, the reaction will shift left to produce more HF(g).

c. Shifts right; as $H_2O(g)$ is removed, the reaction will shift right to produce more $H_2O(g)$.

64. When the volume of a reaction container is increased, the reaction itself will want to increase its own volume by shifting to the side of the reaction that contains the most molecules of gas. When the molecules of gas are equal on both sides of the reaction, then the reaction will remain at equilibrium no matter what happens to the volume of the container.

a. Reaction shifts left (to reactants) because the reactants contain 4 molecules of gas compared with 2 molecules of gas on the product side.

b. Reaction shifts right (to products) because there are more product molecules of gas (2) than reactant molecules (1).

c. No change because there are equal reactant and product molecules of gas.

d. Reaction shifts right.

e. Reaction shifts right to produce more $CO_2(g)$. One can ignore the solids and only concentrate on the gases because gases occupy a relatively huge volume compared with solids. We make the same assumption when liquids are present (only worry about the gas molecules).

65. a. Right b. Right c. No effect; He(g) is neither a reactant nor a product.

d. Left; because the reaction is exothermic, heat is a product:

$$CO(g) + H_2O(g) \rightarrow H_2(g) + CO_2(g) + \text{heat}$$

Increasing T will add heat. The equilibrium shifts to the left to use up the added heat.

e. No effect; because the moles of gaseous reactants equals the moles of gaseous products (2 mol versus 2 mol), a change in volume will have no effect on the equilibrium.

66. a. The moles of SO_3 will increase because the reaction will shift left to use up the added $O_2(g)$.

b. Increase; because there are fewer reactant gas molecules than product gas molecules, the reaction shifts left with a decrease in volume.

c. No effect; the partial pressures of sulfur trioxide, sulfur dioxide, and oxygen are unchanged, so the reaction is still at equilibrium.

d. Increase; heat + 2 $SO_3 \rightleftharpoons$ 2 SO_2 + O_2; decreasing T will remove heat, shifting this endothermic reaction to the left to add heat.

e. Decrease

67. a. Left b. Right c. Left

d. No effect; the reactant and product concentrations/partial pressures are unchanged.

e. No effect; because there are equal numbers of product and reactant gas molecules, a change in volume has no effect on this equilibrium position.

f. Right; a decrease in temperature will shift the equilibrium to the right because heat is a product in this reaction (as is true in all exothermic reactions).

68. a. Shift to left

b. Shift to right; because the reaction is endothermic (heat is a reactant), an increase in temperature will shift the equilibrium to the right.

c. No effect; the reactant and product concentrations/partial pressures are unchanged.

d. Shift to right

e. Shift to right; because there are more gaseous product molecules than gaseous reactant molecules, the equilibrium will shift right with an increase in volume.

69. An endothermic reaction, where heat is a reactant, will shift right to products with an increase in temperature. The amount of $NH_3(g)$ will increase as the reaction shifts right, so the smell of ammonia will increase.

70. As temperature increases, the value of K decreases. This is consistent with an exothermic reaction. In an exothermic reaction, heat is a product, and an increase in temperature shifts the equilibrium to the reactant side (as well as lowering the value of K).

Additional Exercises

71.

$$O(g) + NO(g) \rightleftharpoons NO_2(g) \qquad K = 1/(6.8 \times 10^{-49}) = 1.5 \times 10^{48}$$
$$NO_2(g) + O_2(g) \rightleftharpoons NO(g) + O_3(g) \qquad K = 1/(5.8 \times 10^{-34}) = 1.7 \times 10^{33}$$

$$O_2(g) + O(g) \rightleftharpoons O_3(g) \qquad K = (1.5 \times 10^{48})(1.7 \times 10^{33}) = 2.6 \times 10^{81}$$

72. a.

$$Na_2O(s) \rightleftharpoons 2\,Na(l) + 1/2\,O_2(g) \qquad K_1$$
$$2\,Na(l) + O_2(g) \rightleftharpoons Na_2O_2(s) \qquad 1/K_3$$

$$Na_2O(s) + 1/2\,O_2(g) \rightleftharpoons Na_2O_2(s) \qquad K = (K_1)(1/K_3)$$

$$K = (K_1)(1/K_3) = \frac{2 \times 10^{-25}}{5 \times 10^{-29}} = 4 \times 10^3$$

b.

$NaO(g) \rightleftharpoons Na(l) + 1/2\ O_2(g)$	K_2
$Na_2O(s) \rightleftharpoons 2\ Na(l) + 1/2\ O_2(g)$	K_1
$2\ Na(l) + O_2(g) \rightleftharpoons Na_2O_2(s)$	$1/K_3$
$NaO(g) + Na_2O(s) \rightleftharpoons Na_2O_2(s) + Na(l)$	$K = K_2(K_1)(1/K_3) = 8 \times 10^{-2}$

c.

$2\ NaO(g) \rightleftharpoons 2\ Na(l) + O_2(g)$	$(K_2)^2$
$2\ Na(l) + O_2(g) \rightleftharpoons Na_2O_2(s)$	$1/K_3$
$2\ NaO(g) \rightleftharpoons Na_2O_2(s)$	$K = (K_2)^2(1/K_3) = 8 \times 10^{18}$

73. $5.63\text{ g } C_5H_6O_3 \times \frac{1\text{ mol } C_5H_6O_3}{114.10\text{ g}} = 0.0493\text{ mol } C_5H_6O_3$ initially

Total moles of gas at equilibrium $= n_{total} = \frac{P_{total}V}{RT} = \frac{1.63\text{ atm} \times 2.50\text{ L}}{\frac{0.08206\text{ L atm}}{\text{K mol}} \times 473\text{ K}} = 0.105\text{ mol}$

	$C_5H_6O_3(g)$	$\rightleftharpoons$	$C_2H_6(g)$	+	$3\ CO(g)$
Initial	0.0493 mol		0		0
	Let x mol $C_5H_6O_3$ react to reach equilibrium.				
Change	$-x$	$\rightarrow$	$+x$		$+3x$
Equil.	$0.0493 - x$		x		$3x$

0.105 mol total $= 0.0493 - x + x + 3x = 0.0493 + 3x$, $x = 0.0186$ mol

$$K = \frac{[C_2H_6][CO]^3}{[C_5H_6O_3]} = \frac{\left[\frac{0.0186\text{ mol } C_2H_6}{2.50\text{ L}}\right]\left[\frac{3(0.0186)\text{ mol CO}}{2.50\text{ L}}\right]^3}{\left[\frac{(0.0493 - 0.0186)\text{ mol } C_5H_6O_3}{2.50\text{ L}}\right]} = 6.74 \times 10^{-6}$$

74. a. $N_2(g) + O_2(g) \rightleftharpoons 2\ NO(g)$ $K_p = 1 \times 10^{-31} = \frac{P_{NO}^2}{P_{N_2} \times P_{O_2}} = \frac{P_{NO}^2}{(0.8)(0.2)}$

Solving: $P_{NO} = 1 \times 10^{-16}$ atm

In 1.0 cm^3 of air: $n_{NO} = \frac{PV}{RT} = \frac{(1 \times 10^{-16}\text{ atm})(1.0 \times 10^{-3}\text{ L})}{\left(\frac{0.08206\text{ L atm}}{\text{K mol}}\right)(298\text{ K})} = 4 \times 10^{-21}$ mol NO

$$\frac{4 \times 10^{-21}\text{ mol NO}}{\text{cm}^3} \times \frac{6.02 \times 10^{23}\text{ molecules}}{\text{mol NO}} = \frac{2 \times 10^3\text{ molecules NO}}{\text{cm}^3}$$

b. There is more NO in the atmosphere than we would expect from the value of K. The answer must lie in the rates of the reaction. At 25°C, the rates of both reactions:

$$N_2 + O_2 \rightarrow 2\ NO \quad \text{and} \quad 2\ NO \rightarrow N_2 + O_2$$

are so slow that they are essentially zero. Very strong bonds must be broken; the activation energy is very high. Therefore, the reaction essentially doesn't occur at low temperatures. Nitric oxide, however, can be produced in high-energy or high-temperature environments because the production of NO is endothermic. In nature, some NO is produced by lightning, and the primary manmade source is automobiles. At these high temperatures, K will increase, and the rates of the reaction will also increase, resulting in a higher production of NO. Once the NO gets into a more normal temperature environment, it doesn't go back to N_2 and O_2 because of the slow rate.

75. a.

	$2\ AsH_3(g)$	$\rightleftharpoons$	$2\ As(s)$	+	$3\ H_2(g)$
Initial	392.0 torr				0
Equil.	$392.0 - 2x$				$3x$

Using Dalton's law of partial pressure:

$$P_{total} = 488.0\ \text{torr} = P_{AsH_3} + P_{H_2} = 392.0 - 2x + 3x,\ x = 96.0\ \text{torr}$$

$$P_{H_2} = 3x = 3(96.0) = 288\ \text{torr} \times \frac{1\ \text{atm}}{760\ \text{torr}} = 0.379\ \text{atm}$$

b. $$P_{AsH_3} = 392.0 - 2(96.0) = 200.0\ \text{torr} \times \frac{1\ \text{atm}}{760\ \text{torr}} = 0.2632\ \text{atm}$$

$$K_p = \frac{(P_{H_2})^3}{(P_{AsH_3})^2} = \frac{(0.379)^3}{(0.2632)^2} = 0.786$$

76.

	$FeSCN^{2+}(aq)$	$\rightleftharpoons$	$Fe^{3+}(aq)$	+	$SCN^-(aq)$	$K = 9.1 \times 10^{-4}$
Initial	2.0 M		0		0	
	x mol/L of $FeSCN^{2+}$ reacts to reach equilibrium					
Change	$-x$	$\rightarrow$	$+x$		$+x$	
Equil.	$2.0 - x$		x		x	

$$9.1 \times 10^{-4} = \frac{[Fe^{3+}][SCN^-]}{[FeSCN^{2+}]} = \frac{x^2}{2.0 - x} = \frac{x^2}{2.0} \quad (\text{assuming } 2.0 - x \approx 2.0)$$

$x = 4.3 \times 10^{-2}\ M$; assumption good by the 5% rule (x is 2.2% of 2.0).

$[FeSCN^{2+}] = 2.0 - x = 2.0 - 4.3 \times 10^{-2} = 2.0\ M$; $[Fe^{3+}] = [SCN^-] = x = 4.3 \times 10^{-2}\ M$

77. There is a little trick we can use to solve this problem without having to solve a quadratic equation. Because K is very large (K >> 1), the reaction will have mostly products at equilibrium. So we will let the reaction go to completion (with Fe^{3+} limiting), and then solve an equilibrium problem to determine the molarity of reactants present at equilibrium (see the following set- up).

	$Fe^{3+}(aq)$	+ $SCN^-(aq)$	$\rightleftharpoons$	$FeSCN^{2+}(aq)$	$K = 1.1 \times 10^3$
Before	0.020 *M*	0.10 *M*		0	
	Let 0.020 mol/L Fe^{3+} react completely (K is large; products dominate).				
Change	−0.020	−0.020	→	+0.020	React completely
After	0	0.08		0.020	New initial
	x mol/L $FeSCN^{2+}$ reacts to reach equilibrium				
Change	$+x$	$+x$	←	$-x$	
Equil.	x	$0.08 + x$		$0.020 - x$	

$$K = 1.1 \times 10^3 = \frac{[FeSCN^{2+}]}{[Fe^{3+}][SCN^-]} = \frac{0.020 - x}{(x)(0.08 + x)} \approx \frac{0.020}{(0.08)x}$$

$x = 2 \times 10^{-4}$ *M*; x is 1% of 0.020. Assumptions are good by the 5% rule.

$x = [Fe^{3+}] = 2 \times 10^{-4}$ *M*; $[SCN^-] = 0.08 + 2 \times 10^{-4} = 0.08$ *M*

$[FeSCN^{2+}] = 0.020 - 2 \times 10^{-4} = 0.020$ *M*

Note: At equilibrium, we do indeed have mostly products present. Our assumption to first let the reaction go to completion is good.

78. a. $$P_{PCl_5} = \frac{n_{PCl_5}RT}{V} = \frac{\frac{2.450\text{g } PCl_5}{208.22\text{g/mol}} \times \frac{0.08206\text{L atm}}{\text{K mol}} \times 600.\text{K}}{0.500\text{L}} = 1.16 \text{ atm}$$

b.

	$PCl_5(g)$	$\rightleftharpoons$	$PCl_3(g)$	+ $Cl_2(g)$
Initial	1.16 atm		0	0
	x atm of PCl_5 reacts to reach equilibrium			
Change	$-x$	→	$+x$	$+x$
Equil.	$1.16 - x$		x	x

$$K_p = \frac{P_{PCl_3} \times P_{Cl_2}}{P_{PCl_5}} = 11.5$$

$$K_p = \frac{x^2}{1.16 - x} = 11.5, \quad x^2 + (11.5)x - 13.3 = 0$$

Using the quadratic formula: $x = 1.06$ atm

$P_{PCl_5} = 1.16 - 1.06 = 0.10$ atm

c. $P_{PCl_3} = P_{Cl_2} = 1.06$ atm; $P_{PCl_5} = 0.10$ atm

$P_{total} = P_{PCl_5} + P_{PCl_3} + P_{Cl_2} = 0.10 + 1.06 + 1.06 = 2.22$ atm

d. Percent dissociation $= \dfrac{x}{1.16} \times 100 = \dfrac{1.06}{1.16} \times 100 = 91.4\%$

79. $SO_2Cl_2(g) \rightleftharpoons Cl_2(g) + SO_2(g)$

	$SO_2Cl_2(g)$		$Cl_2(g)$	$SO_2(g)$	
Initial	P_0		0	0	P_0 = initial pressure of SO_2Cl_2
Change	$-x$	$\rightarrow$	$+x$	$+x$	
Equil.	$P_0 - x$		x	x	

$P_{total} = 0.900 \text{ atm} = P_0 - x + x + x = P_0 + x$

$\dfrac{x}{P_0} \times 100 = 12.5,\ P_0 = (8.00)x$

Solving: $0.900 = P_0 + x = (9.00)x,\ x = 0.100$ atm

$x = 0.100 \text{ atm} = P_{Cl_2} = P_{SO_2}$; $P_0 - x = 0.800 - 0.100 = 0.700 \text{ atm} = P_{SO_2Cl_2}$

$$K_p = \frac{P_{Cl_2} \times P_{SO_2}}{P_{SO_2Cl_2}} = \frac{(0.100)^2}{0.700} = 1.43 \times 10^{-2}$$

80. $K = \dfrac{[HF]^2}{[H_2][F_2]} = \dfrac{(0.400M)^2}{(0.0500M)(0.0100M)} = 320.$; 0.200 mol F_2/5.00 L = 0.0400 M F_2 added

After F_2 has been added, the concentrations of species present are [HF] = 0.400 M, $[H_2] = [F_2]$ = 0.0500 M. $Q = (0.400)^2/(0.0500)^2 = 64.0$; because Q < K, the reaction will shift right to reestablish equilibrium.

	$H_2(g)$	+	$F_2(g)$	$\rightleftharpoons$	2 HF(g)
Initial	0.0500 M		0.0500 M		0.400 M
	x mol/L of F_2 reacts to reach equilibrium				
Change	$-x$		$-x$	$\rightarrow$	$+2x$
Equil.	$0.0500 - x$		$0.0500 - x$		$0.400 + 2x$

$K = 320. = \dfrac{(0.400 + 2x)^2}{(0.0500 - x)^2}$; taking the square root of each side:

$17.9 = \dfrac{0.400 + 2x}{0.0500 - x}$, $0.895 - (17.9)x = 0.400 + 2x$, $(19.9)x = 0.495$, $x = 0.0249$ mol/L

[HF] = 0.400 + 2(0.0249) = 0.450 M; $[H_2] = [F_2] = 0.0500 - 0.0249 = 0.0251$ M

81. $CoCl_2(s) + 6\ H_2O(g) \rightleftharpoons CoCl_2{\bullet}6H_2O(s)$; if rain is imminent, there would be a lot of water vapor in the air. Because water vapor is a reactant gas, the reaction would shift to the right and would take on the color of $CoCl_2{\bullet}6H_2O$, pink.

82. a. Doubling the volume will decrease all concentrations by a factor of one-half.

$$Q = \frac{\frac{1}{2}[FeSCN^{2+}]_{eq}}{\left(\frac{1}{2}[Fe^{3+}]_{eq}\right)\left(\frac{1}{2}[SCN^-]_{eq}\right)} = 2K,\ Q > K$$

The reaction will shift to the left to reestablish equilibrium.

b. Adding Ag^+ will remove SCN^- through the formation of AgSCN(s). The reaction will shift to the left to produce more SCN^-.

c. Removing Fe^{3+} as $Fe(OH)_3(s)$ will shift the reaction to the left to produce more Fe^{3+}.

d. Reaction will shift to the right as Fe^{3+} is added.

83. $H^+ + OH^- \rightarrow H_2O$; sodium hydroxide (NaOH) will react with the H^+ on the product side of the reaction. This effectively removes H^+ from the equilibrium, which will shift the reaction to the right to produce more H^+ and CrO_4^{2-}. Because more CrO_4^{2-} is produced, the solution turns yellow.

84. $N_2(g) + 3\ H_2(g) \rightleftharpoons 2\ NH_3(g) + \text{heat}$

a. This reaction is exothermic, so an increase in temperature will decrease the value of K (see Table 13.3 of text.) This has the effect of lowering the amount of $NH_3(g)$ produced at equilibrium. The temperature increase, therefore, must be for kinetics reasons. As temperature increases, the reaction reaches equilibrium much faster. At low temperatures, this reaction is very slow, too slow to be of any use.

b. As $NH_3(g)$ is removed, the reaction shifts right to produce more $NH_3(g)$.

c. A catalyst has no effect on the equilibrium position. The purpose of a catalyst is to speed up a reaction so it reaches equilibrium more quickly.

d. When the pressure of reactants and products is high, the reaction shifts to the side that has fewer gas molecules. Since the product side contains two molecules of gas as compared to four molecules of gas on the reactant side, then the reaction shifts right to products at high pressures of reactants and products.

85. $PCl_5(g) \rightleftharpoons PCl_3(g) + Cl_2(g) \quad K = \dfrac{[PCl_3][Cl_2]}{[PCl_5]} = 4.5 \times 10^{-3}$

At equilibrium, $[PCl_5] = 2[PCl_3]$.

Substituting: $4.5 \times 10^{-3} = \dfrac{[PCl_3][Cl_2]}{2[PCl_3]}$, $[Cl_2] = 2(4.5 \times 10^{-3}) = 9.0 \times 10^{-3}\ M$

86. $CaCO_3(s) \rightleftharpoons CaO(s) + CO_2(g) \quad K_p = 1.16 = P_{CO_2}$

Some of the 20.0 g of $CaCO_3$ will react to reach equilibrium. The amount that reacts is the quantity of $CaCO_3$ required to produce a CO_2 pressure of 1.16 atm (from the K_p expression).

$$n_{CO_2} = \frac{P_{CO_2}V}{RT} = \frac{1.16\text{ atm} \times 10.0\text{ L}}{\dfrac{0.08206\text{ L atm}}{\text{K mol}} \times 1073\text{K}} = 0.132\text{ mol } CO_2$$

Mass $CaCO_3$ reacted $= 0.132\text{ mol } CO_2 \times \dfrac{1\text{ mol } CaCO_3}{\text{mol } CO_2} \times \dfrac{100.09\text{ g}}{\text{mol } CaCO_3} = 13.2\text{ g } CaCO_3$

Mass percent of $CaCO_3$ reacted $= \dfrac{13.2\text{ g}}{20.0\text{ g}} \times 100 = 66.0\%$

87. alpha-glucose $\rightleftharpoons$ beta-glucose $\quad K = \dfrac{[\text{beta-glucose}]}{[\text{alpha-glucose}]}$

From the problem, [alpha–glucose] = 2[beta-glucose], so:

$$K = \frac{[\text{beta-glucose}]}{2[\text{beta-glucose}]} = \frac{1}{2} = 0.50$$

88. peptide(aq) + $H_2O(l) \rightleftharpoons$ acid group(aq) + amine group(aq) $\quad K = 3.1 \times 10^{-5}$

	peptide		acid group	amine group
Initial	$\dfrac{1.0\text{ mol}}{1.0\text{ L}} = 1.0\ M$		0	0
	x mol/L peptide reacts to reach equilibrium.			
Change	$-x$	$\rightarrow$	$+x$	$+x$
Equil.	$1.0 - x$		x	x

Note: Because water is not included in the K expression, the amount of water present initially and the amount of water that reacts are not needed to solve this problem.

$K = 3.1 \times 10^{-5} = \dfrac{x(x)}{1.0 - x}$, $3.1 \times 10^{-5} \approx \dfrac{x^2}{1.0}$ (assuming $1.0 - x \approx 1.0$)

$x = \sqrt{3.1 \times 10^{-5}} = 5.6 \times 10^{-3}\ M$; assumption good (0.56% error).

[peptide] $= 1.0 - x = 1.0 - 5.6 \times 10^{-3} = 1.0\ M$

[acid group] = [amine group] $= x = 5.6 \times 10^{-3}\ M$

89. $HCO_3^-(aq) \rightleftharpoons H^+(aq) + CO_3^{2-}(aq) \qquad K = \frac{[H^+][CO_3^{2-}]}{[HCO_3^-]} = 5.6 \times 10^{-11}$

	HCO_3^-		H^+	CO_3^{2-}
Initial	0.16 mol/1.0 L		0	0
	x mol/L HCO_3^- reacts to reach equilibrium			
Change	$-x$	→	$+x$	$+x$
Equil.	$0.16 - x$		x	x

$$5.6 \times 10^{-11} = \frac{x(x)}{0.16 - x} \approx \frac{x^2}{0.16} \quad \text{(assuming } x << 0.16\text{)}$$

$x = \sqrt{5.6 \times 10^{-11}(0.16)} = 3.0 \times 10^{-6}\ M$; assumption good ($8 \times 10^{-3}$% error).

$[CO_3^{2-}] = x = 3.0 \times 10^{-6}\ M$

90. $CH_3OH(aq) \rightleftharpoons H_2CO(aq) + H_2(aq) \qquad K = 3.7 \times 10^{-10} = \frac{[H_2CO][H_2]}{[CH_3OH]}$

	CH_3OH		H_2CO	H_2
Initial	1.24 M		0	0
	x mol/L CH_3OH reacts to reach equilibrium			
Change	$-x$	→	$+x$	$+x$
Equil.	$1.24 - x$		x	x

$$3.7 \times 10^{-10} = \frac{x(x)}{1.24 - x} \approx \frac{x^2}{1.24} \quad \text{(assuming } x << 1.24\text{)}$$

$x = 2.1 \times 10^{-5}\ M$; assumption good ($1.7 \times 10^{-3}$% error).

$[H_2CO] = [H_2] = x = 2.1 \times 10^{-5}\ M$; $[CH_3OH] = 1.24 - 2.1 \times 10^{-5} = 1.24\ M$

As formaldehyde is removed from the equilibrium by forming some other substance, the equilibrium shifts right to produce more formaldehyde. Hence the concentration of methanol (a reactant) decreases as formaldehyde (a product) reacts to form formic acid.

ChemWork Problems

The answers to the problems 91-98 (or a variation to these problem) are found in OWL. These problems are also assignable in OWL.

Challenge Problems

99. P_0 (for O_2) $= n_{O_2}RT/V = (6.400\ g \times 0.08206 \times 684\ K)/(32.00\ g/mol \times 2.50\ L) = 4.49$ atm

	$CH_4(g)$ +	$2\ O_2(g)$	→	$CO_2(g)$ +	$2\ H_2O(g)$
Change	$-x$	$-2x$	→	$+x$	$+2x$

	$CH_4(g)$ +	$3/2\ O_2(g)$	→	$CO(g)$ +	$2\ H_2O(g)$
Change	$-y$	$-3/2\ y$	→	$+y$	$+2y$

Amount of O_2 reacted = 4.49 atm – 0.326 atm = 4.16 atm O_2

$2x + 3/2\,y = 4.16$ atm O_2 and $2x + 2y = 4.45$ atm H_2O

Solving using simultaneous equations:

$$\begin{array}{rcl} 2x + 2y &=& 4.45 \\ -2x - (3/2)y &=& -4.16 \\ \hline (0.50)y &=& 0.29, \quad y = 0.58 \text{ atm} = P_{CO} \end{array}$$

$$2x + 2(0.58) = 4.45, \quad x = \frac{4.45 - 1.16}{2} = 1.65 \text{ atm} = P_{CO_2}$$

100. $4.72 \text{ g } CH_3OH \times \frac{1 \text{ mol}}{32.04 \text{ g}} = 0.147 \text{ mol } CH_3OH$ initially

Graham's law of effusion: $\frac{\text{Rate}_1}{\text{Rate}_2} = \sqrt{\frac{M_2}{M_1}}$

$$\frac{\text{Rate}_{H_2}}{\text{Rate}_{CH_3OH}} = \sqrt{\frac{M_{CH_3OH}}{M_{H_2}}} = \sqrt{\frac{32.04}{2.016}} = 3.987$$

The effused mixture has 33.0 times as much H_2 as CH_3OH. When the effusion rate ratio is multiplied by the equilibrium mole ratio of H_2 to CH_3OH, the effused mixture will have 33.0 times as much H_2 as CH_3OH. Let n_{H_2} and n_{CH_3OH} equal the equilibrium moles of H_2 and CH_3OH, respectively.

$$33.0 = 3.987 \times \frac{n_{H_2}}{n_{CH_3OH}}, \quad \frac{n_{H_2}}{n_{CH_3OH}} = 8.28$$

	$CH_3OH(g)$	$\rightleftharpoons$	$CO(g)$	$+ 2\,H_2(g)$
Initial	0.147 mol		0	0
Change	$-x$	$\rightarrow$	$+x$	$+2x$
Equil.	$0.147 - x$		x	$2x$

From the ICE table, $8.28 = \frac{n_{H_2}}{n_{CH_3OH}} = \frac{2x}{0.147 - x}$

Solving: $x = 0.118$ mol

$$K = \frac{[CO][H_2]^2}{[CH_3OH]} = \frac{\left(\frac{0.118 \text{ mol}}{1.00 \text{ L}}\right)\left(\frac{2(0.118 \text{ mol})}{1.00 \text{ L}}\right)^2}{\frac{(0.147 - 0.118) \text{ mol}}{1.00 \text{ L}}} = 0.23$$

101. There is a little trick we can use to solve this problem in order to avoid solving a cubic equation. Because K for this reaction is very small (K << 1), the reaction will contain mostly reactants at equilibrium (the equilibrium position lies far to the left). We will let the products react to completion by the reverse reaction, and then we will solve the forward equilibrium problem to determine the equilibrium concentrations. Summarizing these steps in a table:

	$2\ NOCl(g)$	$\rightleftharpoons$	$2\ NO(g)$	+	$Cl_2(g)$	$K = 1.6 \times 10^{-5}$
Before	0		2.0 *M*		1.0 *M*	
	Let 1.0 mol/L Cl_2 react completely.					(K is small, reactants dominate.)
Change	+2.0	←	−2.0		−1.0	React completely
After	2.0		0		0	New initial conditions
	$2x$ mol/L of NOCl reacts to reach equilibrium					
Change	$-2x$	→	$+2x$		$+x$	
Equil.	$2.0 - 2x$		$2x$		x	

$$K = 1.6 \times 10^{-5} = \frac{(2x)^2(x)}{(2.0-2x)^2} \approx \frac{4x^3}{2.0^2} \qquad \text{(assuming } 2.0 - 2x \approx 2.0\text{)}$$

$x^3 = 1.6 \times 10^{-5}$, $x = 2.5 \times 10^{-2}$ *M*; assumption good by the 5% rule ($2x$ is 2.5% of 2.0).

[NOCl] = 2.0 − 0.050 = 1.95 *M* = 2.0 *M*; [NO] = 0.050 *M*; [Cl_2] = 0.025 *M*

Note: If we do not break this problem into two parts (a stoichiometric part and an equilibrium part), then we are faced with solving a cubic equation. The setup would be:

	$2\ NOCl$	$\rightleftharpoons$	$2\ NO$	+	Cl_2
Initial	0		2.0 *M*		1.0 *M*
Change	$+2y$	←	$-2y$		$-y$
Equil.	$2y$		$2.0 - 2y$		$1.0 - y$

$$1.6 \times 10^{-5} = \frac{(2.0-2y)^2(1.0-y)}{(2y)^2}$$

If we say that y is small to simplify the problem, then:

$$1.6 \times 10^{-5} = \frac{2.0^2}{4y^2};\ \text{we get } y = 250. \text{ This is impossible!}$$

To solve this equation, we cannot make any simplifying assumptions; we have to solve the cubic equation exactly.

102. a.

	$2\ NO(g)$	+	$Br_2(g)$	$\rightleftharpoons$	$2\ NOBr(g)$
Initial	98.4 torr		41.3 torr		0
	$2x$ torr of NO reacts to reach equilibrium				
Change	$-2x$		$-x$	→	$+2x$
Equil.	$98.4 - 2x$		$41.3 - x$		$2x$

$P_{total} = P_{NO} + P_{Br_2} + P_{NOBr} = (98.4 - 2x) + (41.3 - x) + 2x = 139.7 - x$

$P_{total} = 110.5 = 139.7 - x,\ x = 29.2$ torr; $P_{NO} = 98.4 - 2(29.2) = 40.0$ torr $= 0.0526$ atm

$P_{Br_2} = 41.3 - 29.2 = 12.1$ torr $= 0.0159$ atm; $P_{NOBr} = 2(29.2) = 58.4$ torr $= 0.0768$ atm

$$K_p = \frac{P_{NOBr}^2}{P_{NO}^2 \times P_{Br_2}} = \frac{(0.0768\text{atm})^2}{(0.0526\text{atm})^2(0.0159\text{atm})} = 134$$

b. $2\ NO(g) + Br_2(g) \rightleftharpoons 2\ NOBr(g)$

	2 NO(g)	+ Br₂(g)	⇌	2 NOBr(g)
Initial	0.30 atm	0.30 atm		0
	$2x$ atm of NO reacts to reach equilibrium			
Change	$-2x$	$-x$	→	$+2x$
Equil.	$0.30 - 2x$	$0.30 - x$		$2x$

This would yield a cubic equation, which can be difficult to solve unless you have a graphing calculator. Because K_p is pretty large, let's approach equilibrium in two steps: Assume the reaction goes to completion, and then solve the back-equilibrium problem.

	2 NO	+ Br_2	⇌	2 NOBr	
Before	0.30 atm	0.30 atm		0	
	Let 0.30 atm NO react completely.				
Change	−0.30	−0.15	→	+0.30	React completely
After	0	0.15		0.30	New initial
	$2y$ atm of NOBr reacts to reach equilibrium				
Change	$+2y$	$+y$	←	$-2y$	
Equil.	$2y$	$0.15 + y$		$0.30 - 2y$	

$$K_p = 134 = \frac{(0.30-2y)^2}{(2y)^2(0.15+y)},\quad \frac{(0.30-2y)^2}{(0.15+y)} = 134 \times 4y^2 = 536y^2$$

If $y << 0.15$: $\dfrac{(0.30)^2}{0.15} \approx 536y^2$, then $y = 0.034$; assumptions are poor (y is 23% of 0.15).

Use 0.034 as an approximation for y, and solve by successive approximations (see Appendix 1 in the text):

$$\frac{(0.30-0.068)^2}{0.15+0.034} = 536y^2,\ y = 0.023;\quad \frac{(0.30-0.046)^2}{0.15+0.023} = 536y^2,\ y = 0.026$$

$$\frac{(0.30-0.052)^2}{0.15+0.026} = 536y^2,\ y = 0.026 \text{ atm}$$ (We have converged on the correct answer.)

So: $P_{NO} = 2y = 0.052$ atm; $P_{Br_2} = 0.15 + y = 0.18$ atm; $P_{NOBr} = 0.30 - 2y = 0.25$ atm

103. $N_2(g) + 3\,H_2(g) \rightleftharpoons 2\,NH_3(g) \qquad K_p = 5.3 \times 10^5$

	$N_2(g)$	$3\,H_2(g)$		$2\,NH_3(g)$	
Initial	0	0		P_0	P_0 = initial pressure of NH_3
	$2x$ atm of NH_3 reacts to reach equilibrium				
Change	$+x$	$+3x$	$\leftarrow$	$-2x$	
Equil.	x	$3x$		$P_0 - 2x$	

From problem, $P_0 - 2x = \dfrac{P_0}{2.00}$, so $P_0 = (4.00)x$

$$K_p = \frac{[(4.00)x - 2x]^2}{(x)(3x)^3} = \frac{[(2.00)x]^2}{(x)(3x)^3} = \frac{(4.00)x^2}{27x^4} = \frac{4.00}{27x^2} = 5.3 \times 10^5,\ x = 5.3 \times 10^{-4}\text{ atm}$$

$P_0 = (4.00)x = 4.00(5.3 \times 10^{-4}\text{ atm}) = 2.1 \times 10^{-3}$ atm

104. $P_4(g) \rightleftharpoons 2\,P_2(g) \quad K_p = 0.100 = \dfrac{P_{P_2}^2}{P_{P_4}}$; $P_{P_4} + P_{P_2} = P_{total} = 1.00$ atm, $P_{P_4} = 1.00\text{ atm} - P_{P_2}$

Let $y = P_{P_2}$ at equilibrium, then $K_p = \dfrac{y^2}{1.00 - y} = 0.100$

Solving: $y = 0.270$ atm $= P_{P_2}$; $P_{P_4} = 1.00 - 0.270 = 0.73$ atm

To solve for the fraction dissociated, we need the initial pressure of P_4 (mol $\propto$ pressure).

	$P_4(g)$	$\rightleftharpoons$	$2\,P_2(g)$	
Initial	P_0		0	P_0 = initial pressure of P_4 in atm.
	x atm of P_4 reacts to reach equilibrium			
Change	$-x$	$\rightarrow$	$+2x$	
Equil.	$P_0 - x$		$2x$	

$P_{total} = P_0 - x + 2x = 1.00\text{ atm} = P_0 + x$

Solving: 0.270 atm $= P_{P_2} = 2x$, $x = 0.135$ atm; $P_0 = 1.00 - 0.135 = 0.87$ atm

Fraction dissociated $= \dfrac{x}{P_0} = \dfrac{0.135}{0.87} = 0.16$, or 16% of P_4 is dissociated to reach equilibrium.

105. $N_2O_4(g) \rightleftharpoons 2\,NO_2(g) \qquad K_p = \dfrac{P_{NO_2}^2}{P_{N_2O_4}} = \dfrac{(1.20)^2}{0.34} = 4.2$

Doubling the volume decreases each partial pressure by a factor of 2 (P = nRT/V).

$P_{NO_2} = 0.600$ atm and $P_{N_2O_4} = 0.17$ atm are the new partial pressures.

$Q = \dfrac{(0.600)^2}{0.17} = 2.1$, so $Q < K$; equilibrium will shift to the right.

	$N_2O_4(g)$	$\rightleftharpoons$	$2\ NO_2(g)$
Initial	0.17 atm		0.600 atm
	x atm of N_2O_4 reacts to reach equilibrium		
Change	$-x$	$\rightarrow$	$+2x$
Equil.	$0.17 - x$		$0.600 + 2x$

$K_p = 4.2 = \dfrac{(0.600 + 2x)^2}{(0.17 - x)}$, $4x^2 + (6.6)x - 0.354 = 0$ (carrying extra significant figures)

Solving using the quadratic formula: $x = 0.052$

$P_{NO_2} = 0.600 + 2(0.052) = 0.704$ atm; $P_{N_2O_4} = 0.17 - 0.052 = 0.12$ atm

106. a. $2\ NaHCO_3(s) \rightleftharpoons Na_2CO_3(s) + CO_2(g) + H_2O(g)$ $K_p = 0.25$

	$2\ NaHCO_3(s)$	$\rightleftharpoons$	$Na_2CO_3(s)$	$CO_2(g)$	$H_2O(g)$
Initial				0	0
	Let some $NaHCO_3(s)$ decompose to form x atm each of $CO_2(g)$ and $H_2O(g)$ at equilibrium.				
Change		$\rightarrow$		$+x$	$+x$
Equil.				x	x

$K_p = 0.25 = P_{CO_2} \times P_{H_2O}$, $0.25 = x^2$, $x = P_{CO_2} = P_{H_2O} = 0.50$ atm

b. $n_{CO_2} = \dfrac{P_{CO_2}V}{RT} = \dfrac{(0.50\text{atm})(1.00\text{L})}{(0.08206\text{L atm/K}\bullet\text{mol})(398\text{K})} = 1.5 \times 10^{-2}$ mol CO_2

Mass of Na_2CO_3 produced:

$$1.5 \times 10^{-2} \text{ mol } CO_2 \times \frac{1 \text{ mol } Na_2CO_3}{\text{mol } CO_2} \times \frac{105.99 \text{ g } Na_2CO_3}{\text{mol } Na_2CO_3} = 1.6 \text{ g } Na_2CO_3$$

Mass of $NaHCO_3$ reacted:

$$1.5 \times 10^{-2} \text{ mol } CO_2 \times \frac{2 \text{ mol } NaHCO_3}{1 \text{ mol } CO_2} \times \frac{84.01 \text{ g } NaHCO_3}{\text{mol}} = 2.5 \text{ g } NaHCO_3$$

Mass of $NaHCO_3$ remaining = 10.0 – 2.5 = 7.5 g

c. $10.0 \text{ g } NaHCO_3 \times \dfrac{1 \text{ mol } NaHCO_3}{84.01 \text{ g } NaHCO_3} \times \dfrac{1 \text{ mol } CO_2}{2 \text{ mol } NaHCO_3} = 5.95 \times 10^{-2}$ mol CO_2

When all of the $NaHCO_3$ has just been consumed, we will have 5.95×10^{-2} mol CO_2 gas at a pressure of 0.50 atm (from a).

$$V = \frac{nRT}{P} = \frac{(5.95 \times 10^{-2} \text{ mol})(0.08206 \text{ L atm/K}\bullet\text{mol})(398\text{K})}{(0.50\text{atm})} = 3.9 \text{ L}$$

107. a. $N_2(g) + 3\ H_2(g) \rightleftharpoons 2\ NH_3(g)$; because the temperature is constant, the value of K will be the same for both container volumes. Since we now the volume in the final mixture, let's calculate K using this mixture. In this final mixture, 2 N_2 molecules, 2 H_2 molecules, and 6 NH_3 molecules are present in a 1.0 L container. Using units of molecules/L for concentrations:

$$K = \frac{[NH_3]^2}{[N_2][H_2]^3} = \frac{\left(\frac{6\ NH_3\ molecules}{1.00\ L}\right)^2}{\left(\frac{2\ N_2\ molecules}{1.00\ L}\right)\left(\frac{2\ H_2\ molecules}{1.00\ L}\right)^3} = 2.25\frac{L^2}{molecules^2}$$

For the K value in typical mol/L units for the concentrations:

$$K = 2.25\frac{L^2}{molecules^2} \times \left(\frac{6.022\times10^{23}\ molecules}{mol}\right)^2 = 8.16\times10^{47}\frac{L^2}{mol^2}$$

$$= 8.16 \times 10^{47}$$

b. Because temperature is constant, the initial mixture at the larger volume must also have $K = 2.25\ L^2/molecules^2$. In the initial mixture, there are 2 NH_3 molecules, 4 N_2 molecules, and 8 H_2 molecules in some unknown volume, V.

$$K = 2.25 = \frac{\left(\frac{2\ NH_3\ molecules}{V}\right)^2}{\left(\frac{4\ N_2\ molecules}{V}\right)\left(\frac{8\ H_2\ molecules}{V}\right)^3} = \frac{4V^2}{4(512)} = \frac{V^2}{512}$$

$V = \sqrt{2.25(512)} = 33.9$ L; the volume of the initial container would be 33.9 L.

108. a. If the volume is increased, equilibrium will shift to the right, so the mole percent of N_2O_5 decomposed will be greater than 0.50%.

b.

	$2\ N_2O_5(g)$	$\rightleftharpoons$	$4\ NO_2(g)$	+	$O_2(g)$
Initial	1.000 atm		0		0
Change	−0.0050	→	+0.010		+0.0025
Equil.	0.995		0.010		0.0025

$$K_p = \frac{(0.010)^4(0.0025)}{(0.995)^2} = 2.5 \times 10^{-11}$$

The new volume is 10.0 times the old volume. Therefore, the initial partial pressure of N_2O_5 will decrease by a factor of 10.0.

$$P_{N_2O_5} = 1.00\ atm \times \frac{1.00}{10.0} = 0.100\ atm$$

	$2\ N_2O_5$	$\rightleftharpoons$	$4\ NO_2$	+	O_2
Initial	0.100 atm		0		0
Change	$-2x$	→	$+4x$		$+x$
Equil.	$0.100 - 2x$		$4x$		x

$$2.5 \times 10^{-11} = \frac{(4x)^4(x)}{(0.100 - 2x)^2} \approx \frac{(4x)^4(x)}{(0.100)^2}, \quad 2x = 2.0 \times 10^{-3}\ \text{atm} = P_{N_2O_5}\ \text{decomposed}$$

$$\frac{2.0 \times 10^{-3}}{0.100} \times 100 = 2.0\%\ N_2O_5\ \text{decomposed (moles and P are directly related)}$$

109.

	$SO_3(g)$	$\rightleftharpoons$	$SO_2(g)$	+	$1/2\ O_2(g)$	
Initial	P_0		0		0	P_0 = initial pressure of SO_3
Change	$-x$	→	$+x$		$+x/2$	
Equil.	$P_0 - x$		x		$x/2$	

Average molar mass of the mixture is:

$$\text{average molar mass} = \frac{dRT}{P} = \frac{(1.60\ \text{g/L})(0.08206\ \text{L atm/K} \cdot \text{mol})(873\ \text{K})}{1.80\ \text{atm}}$$

$$= 63.7\ \text{g/mol}$$

The average molar mass is determined by:

$$\text{average molar mass} = \frac{n_{SO_3}(80.07\ \text{g/mol}) + n_{SO_2}(64.07\ \text{g/mol}) + n_{O_2}(32.00\ \text{g/mol})}{n_{total}}$$

Because χ_A = mol fraction of component A = $n_A/n_{total} = P_A/P_{total}$:

$$63.7\ \text{g/mol} = = \frac{P_{SO_3}(80.07) + P_{SO_2}(64.07) + P_{O_2}(32.00)}{P_{total}}$$

$P_{total} = P_0 - x + x + x/2 = P_0 + x/2 = 1.80$ atm, $P_0 = 1.80 - x/2$

$$63.7 = \frac{(P_0 - x)(80.07) + x(64.07) + \frac{x}{2}(32.00)}{1.80}$$

$$63.7 = \frac{(1.80 - 3/2\,x)(80.07) + x(64.07) + \frac{x}{2}(32.00)}{1.80}$$

$115 = 144 - (120.1)x + (64.07)x + (16.00)x$, $(40.0)x = 29$, $x = 0.73$ atm

$P_{SO_3} = P_0 - x = 1.80 - (3/2)x = 0.71$ atm; $P_{SO_2} = 0.73$ atm; $P_{O_2} = x/2 = 0.37$ atm

$$K_p = \frac{P_{SO_2} \times P_{O_2}^{1/2}}{P_{SO_3}} = \frac{(0.73)(0.37)^{1/2}}{(0.71)} = 0.63$$

110. The first reaction produces equal amounts of SO_3 and SO_2. Using the second reaction, calculate the SO_3, SO_2 and O_2 partial pressures at equilibrium.

	$SO_3(g)$	$\rightleftharpoons$	$SO_2(g)$	+	$1/2\ O_2(g)$
Initial	P_0		P_0		0
Change	$-x$	$\rightarrow$	$+x$		$+x/2$
Equil.	$P_0 - x$		$P_0 + x$		$x/2$

P_0 = initial pressure of SO_3 and SO_2 after first reaction occurs.

$P_{total} = P_0 - x + P_0 + x + x/2 = 2P_0 + x/2 = 0.836$ atm

$P_{O_2} = x/2 = 0.0275$ atm, $x = 0.0550$ atm

$2P_0 + x/2 = 0.836$ atm; $2P_0 = 0.836 - 0.0275 = 0.809$ atm, $P_0 = 0.405$ atm

$P_{SO_3} = P_0 - x = 0.405 - 0.0550 = 0.350$ atm; $P_{SO_2} = P_0 + x = 0.405 + 0.0550 = 0.460$ atm

For the reaction $2\ FeSO_4(s) \rightleftharpoons Fe_2O_3(s) + SO_3(g) + SO_2(g)$:

$$K_p = P_{SO_2} \times P_{SO_3} = (0.460)(0.350) = 0.161$$

For the reaction $SO_3(g) \rightleftharpoons SO_2(g) + 1/2\ O_2(g)$:

$$K_p = \frac{P_{SO_2} \times P_{O_2}^{1/2}}{P_{SO_3}} = \frac{(0.460)(0.0275)^{1/2}}{0.350} = 0.218$$

111. When exactly 100 O_2 molecules are initially present at 5000 K and 1.000 atm:

	$O_2(g)$	$\rightleftharpoons$	$2\ O(g)$
Initial	100		0
Change	-83		$+166$
Equil.	17		166

Mole fraction O = $\chi_O = \frac{166}{183} = 0.9071$ and $\chi_{O_2} = 0.0929$; $P_{O_2} = \chi_{O_2} P_{total}$ and $P_O = \chi_O P_{total}$

Because initially $P_{total} = 1.000$ atm, $P_{O_2} = 0.0929$ atm and $P_O = 0.9071$ atm.

$$K_p = \frac{P_O^2}{P_{O_2}} = \frac{(0.9071)^2}{0.0929} = 8.86 \text{ atm}$$

At 95.0% O_2 dissociated, let x = initial partial pressure of O_2 and y = amount (atm) of O_2 that dissociates to reach equilibrium.

	O_2	$\rightleftharpoons$	2 O
Initial	x		0
Change	$-y$	→	$+2y$
Equil.	$x - y$		$2y$

$\frac{(2y)^2}{x-y} = 8.86$; $\frac{y}{x} \times 100 = 95.0$; we have two equations and two unknowns.

Solving: $x = 0.123$ atm and $y = 0.117$ atm; $P_{total} = (x - y) + 2y = 0.240$ atm

112. a. $N_2O_4(g) \rightleftharpoons 2\ NO_2(g)$

	$N_2O_4(g)$	$\rightleftharpoons$	$2\ NO_2(g)$
Initial	x		0
Change	$-(0.16)x$	→	$+(0.32)x$
Equil.	$(0.84)x$		$(0.32)x$

$(0.84)x + (0.32)x = 1.5$ atm, $x = 1.3$ atm; $K_p = \frac{(0.42)^2}{1.1} = 0.16$

b. $N_2O_4 \rightleftharpoons 2\ NO_2$; $x + y = 1.0$ atm; $\frac{y^2}{x} = K_p = 0.16$

Equil. x y

We have 2 equations and 2 unknowns. Solving:

$x = 0.67$ atm $(= P_{N_2O_4})$ and $y = 0.33$ atm $(= P_{NO_2})$

c. $N_2O_4 \rightleftharpoons 2\ NO_2$

	N_2O_4	$2\ NO_2$	
Initial	P_0	0	P_0 = initial pressure of N_2O_4
Change	$-x$	$+2x$	
Equil.	0.67 atm	0.33 atm	

$2x = 0.33$, $x = 0.165$ (using extra sig. figs.)

$P_0 - x = 0.67$, $P_0 = 0.67 + 0.165 = 0.84$ atm; $\frac{0.165}{0.84} \times 100 = 20.\%$ dissociated

113. a. Because the density (mass/volume) decreased while the mass remained constant (mass is conserved in a chemical reaction), the volume must have increased as reactants were converted to products. The volume increased because the number of moles of gas increased ($V \propto n$ at constant T and P).

$$\frac{\text{Density(initial)}}{\text{Density(equil.)}} = \frac{4.495\text{g/L}}{4.086\text{g/L}} = 1.100 = \frac{V_{equil}}{V_{initial}} = \frac{n_{equil}}{n_{initial}}$$

Assuming an initial volume of 1.000 L:

$$4.495\text{ g NOBr} \times \frac{1\text{ mol NOBr}}{109.91\text{ g}} = 0.04090\text{ mol NOBr initially}$$

	$2\ NOBr(g)$	⇌	$2\ NO(g)$	+ $Br_2(g)$
Initial	0.04090 mol		0	0
Change	$-2x$	→	$+2x$	$+x$
Equil.	$0.04090 - 2x$		$2x$	x

$$\frac{n_{equil.}}{n_{initial}} = \frac{0.04090 - 2x + 2x + x}{0.04090} = 1.100; \quad \text{solving: } x = 0.00409\text{ mol}$$

If the initial volume is 1.000 L, then the equilibrium volume will be 1.110(1.000 L) = 1.110 L. Solving for the equilibrium concentrations:

$$[NOBr] = \frac{0.03272\text{mol}}{1.100\text{L}} = 0.02975M; \quad [NO] = \frac{0.00818\text{mol}}{1.100\text{L}} = 0.00744M$$

$$[Br_2] = \frac{0.00409\text{mol}}{1.100\text{L}} = 0.00372M$$

$$K = \frac{(0.00744)^2(0.00372)}{(0.02975)^2} = 2.33 \times 10^{-4}$$

b. The argon gas will increase the volume of the container. This is because the container is a constant-pressure system, and if the number of moles increases at constant T and P, the volume must increase. An increase in volume will dilute the concentrations of all gaseous reactants and gaseous products. Because there are more moles of product gases versus reactant gases (3 mol versus 2 mol), the dilution will decrease the numerator of K more than the denominator will decrease. This causes Q < K and the reaction shifts right to get back to equilibrium.

Because temperature was unchanged, the value of K will not change. K is a constant as long as temperature is constant.

114. $CCl_4(g) \rightleftharpoons C(s) + 2\ Cl_2(g) \quad K_p = 0.76$

	$CCl_4(g)$		$2\ Cl_2(g)$	
Initial	P_0		0	P_0 = initial pressure of CCl_4
Change	$-x$	→	$+2x$	
Equil.	$P_0 - x$		$2x$	

$P_{total} = P_0 - x + 2x = P_0 + x = 1.20$ atm

$$K_p = \frac{(2x)^2}{P_0 - x} = 0.76, \quad 4x^2 = (0.76)P_0 - (0.76)x, \quad P_0 = \frac{4x^2 + (0.76)x}{0.76}$$

Substituting into $P_0 + x = 1.20$:

$$\frac{4x^2}{0.76} + x + x = 1.20 \text{ atm}, \quad (5.3)x^2 + 2x - 1.20 = 0; \text{ solving using the quadratic formula:}$$

$$x = \frac{-2 \pm (4 + 25.4)^{1/2}}{2(5.3)} = 0.32 \text{ atm}; \quad P_0 + 0.32 = 1.20, \ P_0 = 0.88 \text{ atm}$$

Integrative Problems

115. $NH_3(g) + H_2S(g) \rightleftharpoons NH_4HS(s) \qquad K = 400. = \dfrac{1}{[NH_3][H_2S]}$

	$NH_3(g)$	$H_2S(g)$
Initial	$\dfrac{2.00 \text{ mol}}{5.00 \text{ L}}$	$\dfrac{2.00 \text{ mol}}{5.00 \text{ L}}$
	x mol/L of NH_3 reacts to reach equilibrium	
Change	$-x$	$-x$
Equil.	$0.400 - x$	$0.400 - x$

$$K = 400. = \frac{1}{(0.400 - x)(0.400 - x)}, \quad 0.400 - x = \left(\frac{1}{400.}\right)^{1/2} = 0.0500, \quad x = 0.350 \ M$$

$$\text{Moles } NH_4HS(s) \text{ produced} = 5.00 \text{ L} \times \frac{0.350 \text{ mol } NH_3}{\text{L}} \times \frac{1 \text{ mol } NH_4HS}{\text{mol } NH_3} = 1.75 \text{ mol}$$

Total moles $NH_4HS(s)$ = 2.00 mol initially + 1.75 mol produced = 3.75 mol total

$$3.75 \text{ mol } NH_4HS \times \frac{51.12 \text{ g } NH_4HS}{\text{mol } NH_4HS} = 192 \text{ g } NH_4HS$$

$[H_2S]_e = 0.400 \ M - x = 0.400 \ M - 0.350 \ M = 0.050 \ M \ H_2S$

$$P_{H_2S} = \frac{n_{H_2S}RT}{V} = \frac{n_{H_2S}}{V} \times RT = \frac{0.050 \text{ mol}}{\text{L}} \times \frac{0.08206 \text{ L atm}}{\text{K mol}} \times 308.2 \text{ K} = 1.3 \text{ atm}$$

116. See the hint for Exercise 71.

$$\begin{array}{ll} 2\,C(g) \rightleftharpoons 2\,A(g) + 2\,B(g) & K_1 = (1/3.50)^2 = 8.16 \times 10^{-2} \\ 2\,A(g) + D(g) \rightleftharpoons C(g) & K_2 = 7.10 \\ \hline C(g) + D(g) \rightleftharpoons 2\,B(g) & K = K_1 \times K_2 = 0.579 \end{array}$$

$K_p = K(RT)^{\Delta n}$, $\Delta n = 2 - (1 + 1) = 0$; because $\Delta n = 0$, $K_p = K = 0.579$.

	C(g)	+	D(g)	⇌	2 B(g)
Initial	1.50 atm		1.50 atm		0
Equil.	$1.50 - x$		$1.50 - x$		$2x$

$$0.579 = K = \frac{(2x)^2}{(1.50-x)(1.50-x)} = \frac{(2x)^2}{(1.50-x)^2}$$

$$\frac{2x}{1.50-x} = (0.579)^{1/2} = 0.761,\ x = 0.413 \text{ atm}$$

P_B (at equilibrium) = $2x$ = 2(0.413) = 0.826 atm

$P_{total} = P_C + P_D + P_B = 2(1.50 - 0.413) + 0.826 = 3.00$ atm

$$P_B = \chi_B P_{total},\ \chi_B = \frac{P_B}{P_{total}} = \frac{0.826 \text{ atm}}{3.00 \text{ atm}} = 0.275$$

117. Initial moles VCl_4 = 6.6834 g VCl_4 × 1 mol VCl_4/192.74 g VCl_4 = 3.4676×10^{-2} mol VCl_4

$$\text{Total molality of solute particles} = im = \frac{\Delta T}{K_f} = \frac{5.97\,^\circ\text{C}}{29.8\,^\circ\text{C kg/mol}} = 0.200 \text{ mol/kg}$$

Because we have 0.1000 kg CCl_4, the total moles of solute particles present is:

0.200 mol/kg(0.1000 kg) = 0.0200 mol

$$2\ VCl_4 \rightleftharpoons V_2Cl_8 \qquad K = \frac{[V_2Cl_8]}{[VCl_4]^2}$$

	$2\ VCl_4$	⇌	V_2Cl_8
Initial	3.4676×10^{-2} mol		0
	$2x$ mol VCl_4 reacts to reach equilibrium		
Equil.	$3.4676 \times 10^{-2} - 2x$		x

Total moles solute particles = 0.0200 mol = mol VCl_4 + mol V_2Cl_8 = $3.4676 \times 10^{-2} - 2x + x$

$0.0200 = 3.4676 \times 10^{-2} - x,\ x = 0.0147$ mol

At equilibrium, we have 0.0147 mol V_2Cl_8 and 0.0200 – 0.0147 = 0.0053 mol VCl_4. To determine the equilibrium constant, we need the total volume of solution in order to calculate equilibrium concentrations. The total mass of solution is 100.0 g + 6.6834 g = 106.7 g.

Total volume = 106.7 g × 1 cm^3/1.696 g = 62.91 cm^3 = 0.06291 L

The equilibrium concentrations are:

$$[V_2Cl_8] = \frac{0.0147\text{ mol}}{0.06291\text{ L}} = 0.234\text{ mol/L};\ [VCl_4] = \frac{0.0053\text{ mol}}{0.06291\text{ L}} = 0.084\text{ mol/L}$$

$$K = \frac{[V_2Cl_8]}{[VCl_4]^2} = \frac{0.234}{(0.084)^2} = 33$$

118. Assuming 100.00 g naphthalene:

$$93.71\text{ g C} \times \frac{1\text{ mol C}}{12.01\text{ g}} = 7.803\text{ mol C}$$

$$6.29\text{ g H} \times \frac{1\text{ mol H}}{1.008\text{ g}} = 6.24\text{ mol H};\quad \frac{7.803}{6.24} = 1.25$$

Empirical formula = $(C_{1.25}H)_{\times 4} = C_5H_4$; molar mass = $\frac{32.8\text{ g}}{0.256\text{ mol}} = 128$ g/mol

Because the empirical mass (64.08 g/mol) is one-half of 128, the molecular formula is $C_{10}H_8$.

	$C_{10}H_8(s)$	$\rightleftharpoons$	$C_{10}H_8(g)$	$K = 4.29 \times 10^{-6} = [C_{10}H_8]$
Initial			0	
	Let some $C_{10}H_8(s)$ sublime to form x mol/L of $C_{10}H_8(g)$ at equilibrium.			
Equil.			x	

$K = 4.29 \times 10^{-6} = [C_{10}H_8] = x$

Mol $C_{10}H_8$ sublimed = 5.00 L × 4.29×10^{-6} mol/L = 2.15×10^{-5} mol $C_{10}H_8$ sublimed

Mol $C_{10}H_8$ initially = 3.00 g × $\frac{1\text{ mol }C_{10}H_8}{128.16\text{ g}}$ = 2.34×10^{-2} mol $C_{10}H_8$ initially

Percent $C_{10}H_8$ sublimed = $\frac{2.15 \times 10^{-5}\text{ mol}}{2.34 \times 10^{-2}\text{ mol}} \times 100 = 0.0919\%$

Marathon Problem

119. $\frac{2.00\text{ g}}{165\text{ g/mol}} = 0.0121$ mol XY (initially)

(0.350)(0.0121 mol) = 4.24×10^{-3} mol XY dissociated

	XY	$\rightleftharpoons$	X	+	Y
Initial	0.0121 mol		0		0
Change	–0.00424	→	+0.00424		+0.00424
Equil.	0.0079 mol		0.00424 mol		0.00424 mol

Total moles of gas = 0.0079 + 0.00424 + 0.00424 = 0.0164 mol

For an ideal gas at constant P and T, V ∝ n. So: $\dfrac{V_{final}}{V_{initial}} = \dfrac{n_{final}}{n_{initial}} = \dfrac{0.0164\text{mol}}{0.0121\text{mol}} = 1.36$

$$V_{initial} = \frac{nRT}{P} = \frac{(0.0121\text{mol})(0.008206\text{L atm/Kmol})(298\text{K})}{0.967\text{atm}} = 0.306 \text{ L}$$

$V_{final} = 0.306 \text{ L}(1.36) = 0.416 \text{ L}$

Because mass is conserved in a chemical reaction:

$$\text{density (final)} = \frac{\text{mass}}{\text{volume}} = \frac{2.00\,\text{g}}{0.416\text{L}} = 4.81\,\text{g/L}$$

$$K = \frac{[X][Y]}{[XY]} = \frac{\left(\dfrac{0.00424\text{mol}}{0.416\text{L}}\right)^2}{\left(\dfrac{0.0079\text{mol}}{0.416\text{L}}\right)} = 5.5 \times 10^{-3}$$

CHAPTER 14

ACIDS AND BASES

Questions

19. Acids are proton (H^+) donors, and bases are proton acceptors.

    HCO_3^- as an acid: $HCO_3^-(aq) + H_2O(l) \rightleftharpoons CO_3^{2-}(aq) + H_3O^+(aq)$

    HCO_3^- as a base: $HCO_3^-(aq) + H_2O(l) \rightleftharpoons H_2CO_3(aq) + OH^-(aq)$

    $H_2PO_4^-$ as an acid: $H_2PO_4^- + H_2O(l) \rightleftharpoons HPO_4^{2-}(aq) + H_3O^+(aq)$

    $H_2PO_4^-$ as a base: $H_2PO_4^- + H_2O(l) \rightleftharpoons H_3PO_4(aq) + OH^-(aq)$

20. Acidic solutions (at 25°C) have an $[H^+] > 1.0 \times 10^{-7}$ *M*, which gives a pH < 7.0. Because $[H^+][OH^-] = 1.0 \times 10^{-14}$ and pH + pOH = 14.00 for an aqueous solution at 25°C, an acidic solution must also have $[OH^-] < 1.0 \times 10^{-7}$ *M* and pOH > 7.00. From these relationships, the solutions in parts a, b, and d are acidic. The solution in part c will have a pH > 7.0 (pH = 14.00 – 4.51 = 9.49) and is therefore not acidic (solution is basic).

21. Basic solutions (at 25°C) have an $[OH^-] > 1.0 \times 10^{-7}$ *M*, which gives a pOH < 7.0. Because $[H^+][OH^-] = 1.0 \times 10^{-14}$ and pH + pOH = 14.00 for any aqueous solution at 25°C, a basic solution must also have $[H^+] < 1.0 \times 10^{-7}$ *M* and pH > 7.00. From these relationships, the solutions in parts b, c, and d are basic solutions. The solution in part a will have a pH < 7.0 (pH = 14.00 – 11.21 = 2.79) and is therefore not basic (solution is acidic).

22. When a strong acid (HX) is added to water, the reaction $HX + H_2O \rightarrow H_3O^+ + X^-$ essentially goes to completion. All strong acids in water are completely converted into H_3O^+ and X^-. Thus no acid stronger than H_3O^+ will remain undissociated in water. Similarly, when a strong base (B) is added to water, the reaction $B + H_2O \rightarrow BH^+ + OH^-$ essentially goes to completion. All bases stronger than OH^- are completely converted into OH^- and BH^+. Even though there are acids and bases stronger than H_3O^+ and OH^-, in water these acids and bases are completely converted into H_3O^+ and OH^-.

23. 10.78 (4 S.F.); 6.78 (3 S.F.); 0.78 (2 S.F.); a pH value is a logarithm. The numbers to the left of the decimal point identify the power of 10 to which $[H^+]$ is expressed in scientific notation, for example, 10^{-11}, 10^{-7}, 10^{-1}. The number of decimal places in a pH value identifies the number of significant figures in $[H^+]$. In all three pH values, the $[H^+]$ should be expressed only to two significant figures because these pH values have only two decimal places.

24. A Lewis acid must have an empty orbital to accept an electron pair, and a Lewis base must have an unshared pair of electrons.

25.

NH_3	+	NH_3	$\rightleftharpoons$	NH_2^-	+	NH_4^+
Acid		Base		Conjugate Base		Conjugate Acid

One of the NH_3 molecules acts as a base and accepts a proton to form NH_4^+. The other NH_3 molecule acts as an acid and donates a proton to form NH_2^-. NH_4^+ is the conjugate acid of the NH_3 base. In the reverse reaction, NH_4^+ donates a proton. NH_2^- is the conjugate base of the NH_3 acid. In the reverse reaction, NH_2^- accepts a proton. Conjugate acid-base pairs only differ by a H^+ in the formula.

26. a. The first equation is for the reaction of some generic acid, HA, with H_2O.

HA	+	H_2O	$\rightleftharpoons$	H_3O^+	+	A^-
Acid		Base		Conjugate Acid of H_2O		Conjugate Base of HA

HA is the proton donor (the acid) and H_2O is the proton acceptor (the base). In the reverse reaction, H_3O^+ is the proton donor (the acid) and A^- is the proton acceptor (the base).

The second equation is for some generic base, B, with some generic acid, HX. Note that B has three hydrogens bonded to it.

B	+	HX	$\rightleftharpoons$	BH^+	+	X^-
Base		Acid		Conjugate Acid of B		Conjugate Base of HX

B is the proton acceptor (the base) and HX is the proton donor (the acid). When B accepts a proton, the central atom goes from having 3 bonded hydrogens to 4 bonded hydrogens. In the reverse reaction, BH^+ is the proton donor (the acid) and X^- is the proton acceptor (the base).

b. Arrhenius acids produce H^+ in solution. So HA in the first equation is an Arrhenius acid. However, in the second equation, H^+ is not a product, so HX is not an Arrhenius acid. Both HA in the first equation and HX in the second equation are proton donors, so both are considered Brønsted-Lowry acids. All Brønsted-Lowry acids are Lewis acids, that is, all Brønsted-Lowry acids are electron pair acceptors. So both HA and HX are Lewis acids. It is the proton (H^+) that accepts the lone pair

For the bases in the two equations, H_2O and B, none of them produce OH^- in their equations, so none of them are Arrhenius bases. Both H_2O and B accept protons, so both are Brønsted-Lowry bases. As with Brønsted-Lowry acids, all Brønsted-Lowry bases are also electron-pair donors, so both H_2O and B are Lewis bases. The oxygen in H_2O and the central atom in B will always have a lone pair to donate to the Lewis acid (the proton).

27. a. These are solutuions of strong acids like HCl, HBr, HI, HNO_3, H_2SO_4, and $HClO_4$. So 0.10 *M* solutions of any of the acids would be examples of a strong electrolyte solution that is very acidic.

b. These are solutions containing salts of the conjugate acids of the bases in Table 14.3. These conjugate acids are all weak acids, and they are cations with a 1+ charge. NH_4Cl, $CH_3NH_3NO_3$, and $C_2H_5NH_3Br$ are three examples of this type of slightly acidic salts. Note that the anions used to form these salts are conjugate bases of strong acids; this is so because they have no acidic or basic properties in water (with the exception of HSO_4^-, which has weak acid properties).

c. These are solutuions of strong bases like LiOH, NaOH, KOH, RbOH, CsOH, $Ca(OH)_2$, $Sr(OH)_2$, and $Ba(OH)_2$. All of these strong bases are strong electrolytes.

d. These are solutions containing salts of the conjugate bases of the neutrally charged weak acids in Table 14.2. These conjugate bases are all weak bases, and they are anions with a 1– charge. Three examples of this type of slightly basic salts are $NaClO_2$, $KC_2H_3O_2$, and CaF_2. The cations used to form these salts are Li^+, Na^+, K^+, Rb^+, Cs^+, Ca^{2+}, Sr^{2+}, and Ba^{2+} because these cations have no acidic or basic properties in water. Notice that these are the cations of the strong bases you should memorize.

e. There are two ways to make a neutral salt solutuions. The easiest way is to combine a conjugate base of a strong acid (except for HSO_4^-) with one of the cations from a strong base. These ions have no acidic/basic properties in water, so salts of these ions are neutral. Three examples are NaCl, KNO_3, and SrI_2. Another type of strong electrolyte that can produce neutral solutions are salts that contain an ion with weak acid properties combined with an ion of opposite charge having weak base properties. If the K_a for the weak acid ion is equal to the K_b for the weak base ion, then the salt will produce a neutral solution. The most common example of this type of salt is ammonium acetate ($NH_4C_2H_3O_2$). For this salt, K_a for NH_4^+ = K_b for $C_2H_3O_2^-$ = 5.6×10^{-10}. This salt at any concentration produces a neutral solution.

28. $K_a \times K_b = K_w$, $-\log(K_a \times K_b) = -\log K_w$

$-\log K_a - \log K_b = -\log K_w$, $pK_a + pK_b = pK_w = 14.00$ ($K_w = 1.0 \times 10^{-14}$ at 25°C)

29. a. $H_2O(l) + H_2O(l) \rightleftharpoons H_3O^+(aq) + OH^-(aq)$ or

$H_2O(l) \rightleftharpoons H^+(aq) + OH^-(aq) \quad K = K_w = [H^+][OH^-]$

b. $HF(aq) + H_2O(l) \rightleftharpoons F^-(aq) + H_3O^+(aq)$ or

$$HF(aq) \rightleftharpoons H^+(aq) + F^-(aq) \quad K = K_a = \frac{[H^+][F^-]}{[HF]}$$

c. $$C_5H_5N(aq) + H_2O(l) \rightleftharpoons C_5H_5NH^+(aq) + OH^-(aq) \quad K = K_b = \frac{[C_5H_5NH^+][OH^-]}{[C_5H_5N]}$$

30. Only statement a is true (assuming the species is not amphoteric). You cannot add a base to water and get an acidic pH (pH < 7.0). For statement b, you can have negative pH values; this just indicates an $[H^+] > 1.0$ *M*. For statement c, a dilute solution of a strong acid can have a higher pH than a more concentrated weak acid solution. For statement d, the $Ba(OH)_2$ solution will have an $[OH^-]$ twice of the same concentration of KOH, but this does not correspond to a pOH value twice that of the same concentration of KOH (prove it to yourselves).

31. a. This expression holds true for solutions of strong acids having a concentration greater than 1.0×10^{-6} *M*. 0.10 *M* HCl, 7.8 *M* HNO_3, and 3.6×10^{-4} *M* $HClO_4$ are examples where this expression holds true.

b. This expression holds true for solutions of weak acids where the two normal assumptions hold. The two assumptions are that water does not contribute enough H^+ to solution to make a difference, and that the acid is less than 5% dissociated in water (from the assumption that *x* is small compared to some number). This expression will generally hold true for solutions of weak acids having a K_a value less than 1×10^{-4}, as long as there is a significant amount of weak acid present. Three example solutions are 1.5 *M* $HC_2H_3O_2$, 0.10 *M* HOCl, and 0.72 *M* HCN.

c. This expression holds true for strong bases that donate 2 OH^- ions per formula unit. As long as the concentration of the base is above 5×10^{-7} *M*, this expression will hold true. Three examples are 5.0×10^{-3} *M* $Ca(OH)_2$, 2.1×10^{-4} *M* $Sr(OH)_2$, and 9.1×10^{-5} *M* $Ba(OH)_2$.

d. This expression holds true for solutions of weak bases where the two normal assumptions hold. The assumptions are that the OH^- contribution from water is negligible and that and that the base is less than 5% ionized in water (for the 5% rule to hold). For the 5% rule to hold, you generally need bases with $K_b < 1 \times 10^{-4}$, and concentrations of weak base greater than 0.10 *M*. Three examples are 0.10 *M* NH_3, 0.54 *M* $C_6H_5NH_2$, and 1.1 *M* C_5H_5N.

32. H_2CO_3 is a weak acid with $K_{a_1} = 4.3 \times 10^{-7}$ and $K_{a_2} = 5.6 \times 10^{-11}$. The $[H^+]$ concentration in solution will be determined from the K_{a_1} reaction because $K_{a_1} >> K_{a_2}$. Because $K_{a_1} << 1$, the $[H^+] < 0.10$ *M*; only a small percentage of the 0.10 *M* H_2CO_3 will dissociate into HCO_3^- and H^+. So statement a best describes the 0.10 *M* H_2CO_3 solution. H_2SO_4 is a strong acid as well as a very good weak acid ($K_{a_1} >> 1$, $K_{a_2} = 1.2 \times 10^{-2}$). All the 0.10 *M* H_2SO_4 solution will dissociate into 0.10 *M* H^+ and 0.10 *M* HSO_4^-. However, because HSO_4^- is a good weak acid due to the relatively large K_a value, some of the 0.10 *M* HSO_4^- will dissociate into some more H^+ and SO_4^{2-}. Therefore, the $[H^+]$ will be greater than 0.10 *M* but will not reach 0.20 *M* because only some of 0.10 *M* HSO_4^- will dissociate. Statement c is best for a 0.10 *M* H_2SO_4 solution.

33. One reason HF is a weak acid is that the H–F bond is unusually strong and is difficult to break. This contributes significantly to the reluctance of the HF molecules to dissociate in water.

34. a. Sulfur reacts with oxygen to produce SO_2 and SO_3. These sulfur oxides both react with water to produce H_2SO_3 and H_2SO_4, respectively. Acid rain can result when sulfur emissions are not controlled. Note that, in general, nonmetal oxides react with water to produce acidic solutions.

b. CaO reacts with water to produce $Ca(OH)_2$, a strong base. A gardener mixes lime (CaO) into soil in order to raise the pH of the soil. The effect of adding lime is to add $Ca(OH)_2$. Note that, in general, metal oxides react with water to produce basic solutions.

Exercises

Nature of Acids and Bases

35. a. $HClO_4(aq) + H_2O(l) \rightarrow H_3O^+(aq) + ClO_4^-(aq)$. Only the forward reaction is indicated because $HClO_4$ is a strong acid and is basically 100% dissociated in water. For acids, the dissociation reaction is commonly written without water as a reactant. The common abbreviation for this reaction is $HClO_4(aq) \rightarrow H^+(aq) + ClO_4^-(aq)$. This reaction is also called the K_a reaction because the equilibrium constant for this reaction is designated as K_a.

b. Propanoic acid is a weak acid, so it is only partially dissociated in water. The dissociation reaction is:

$$CH_3CH_2CO_2H(aq) + H_2O(l) \rightleftharpoons H_3O^+(aq) + CH_3CH_2CO_2^-(aq) \quad \text{or}$$

$$CH_3CH_2CO_2H(aq) \rightleftharpoons H^+(aq) + CH_3CH_2CO_2^-(aq).$$

c. NH_4^+ is a weak acid. Similar to propanoic acid, the dissociation reaction is:

$$NH_4^+(aq) + H_2O(l) \rightleftharpoons H_3O^+(aq) + NH_3(aq) \quad \text{or} \quad NH_4^+(aq) \rightleftharpoons H^+(aq) + NH_3(aq)$$

36. The dissociation reaction (the K_a reaction) of an acid in water commonly omits water as a reactant. We will follow this practice. All dissociation reactions produce H^+ and the conjugate base of the acid that is dissociated.

a. $HCN(aq) \rightleftharpoons H^+(aq) + CN^-(aq)$ $\qquad K_a = \dfrac{[H^+][CN^-]}{[HCN]}$

b. $HOC_6H_5(aq) \rightleftharpoons H^+(aq) + OC_6H_5^-(aq)$ $\qquad K_a = \dfrac{[H^+][OC_6H_5^-]}{[HOC_6H_5]}$

c. $C_6H_5NH_3^+(aq) \rightleftharpoons H^+(aq) + C_6H_5NH_2(aq)$ $\qquad K_a = \dfrac{[H^+][C_6H_5NH_2]}{[C_6H_5NH_3^+]}$

37. An acid is a proton (H^+) donor, and a base is a proton acceptor. A conjugate acid-base pair differs by only a proton (H^+).

	Acid	Base	Conjugate Base of Acid	Conjugate Acid of Base
a.	H_2CO_3	H_2O	HCO_3^-	H_3O^+
b.	$C_5H_5NH^+$	H_2O	C_5H_5N	H_3O^+
c.	$C_5H_5NH^+$	HCO_3^-	C_5H_5N	H_2CO_3

38.

	Acid	Base	Conjugate Base of Acid	Conjugate Acid of Base
a.	$Al(H_2O)_6^{3+}$	H_2O	$Al(H_2O)_5(OH)^{2+}$	H_3O^+
b.	$HONH_3^+$	H_2O	$HONH_2$	H_3O^+
c.	HOCl	$C_6H_5NH_2$	OCl^-	$C_6H_5NH_3^+$

39. Strong acids have a $K_a >> 1$, and weak acids have $K_a < 1$. Table 14.2 in the text lists some K_a values for weak acids. K_a values for strong acids are hard to determine, so they are not listed in the text. However, there are only a few common strong acids so, if you memorize the strong acids, then all other acids will be weak acids. The strong acids to memorize are HCl, HBr, HI, HNO_3, $HClO_4$, and H_2SO_4.

a. $HClO_4$ is a strong acid.

b. HOCl is a weak acid ($K_a = 3.5 \times 10^{-8}$).

c. H_2SO_4 is a strong acid.

d. H_2SO_3 is a weak diprotic acid because the K_{a1} and K_{a2} values are much less than 1.

40. The beaker on the left represents a strong acid in solution; the acid HA is 100% dissociated into the H^+ and A^- ions. The beaker on the right represents a weak acid in solution; only a little bit of the acid HB dissociates into ions, so the acid exists mostly as undissociated HB molecules in water.

a. HNO_2: weak acid beaker

b. HNO_3: strong acid beaker

c. HCl: strong acid beaker

d. HF: weak acid beaker

e. $HC_2H_3O_2$: weak acid beaker

41. The K_a value is directly related to acid strength. As K_a increases, acid strength increases. For water, use K_w when comparing the acid strength of water to other species. The K_a values are:

$HClO_4$: strong acid ($K_a >> 1$); $HClO_2$: $K_a = 1.2 \times 10^{-2}$

NH_4^+: $K_a = 5.6 \times 10^{-10}$; H_2O: $K_a = K_w = 1.0 \times 10^{-14}$

From the K_a values, the ordering is $HClO_4 > HClO_2 > NH_4^+ > H_2O$.

42. Except for water, these are the conjugate bases of the acids in the previous exercise. In general, the weaker the acid, the stronger is the conjugate base. ClO_4^- is the conjugate base of a strong acid; it is a terrible base (worse than water). The ordering is $NH_3 > ClO_2^- > H_2O > ClO_4^-$.

43. a. HCl is a strong acid, and water is a very weak acid with $K_a = K_w = 1.0 \times 10^{-14}$. HCl is a much stronger acid than H_2O.

b. H_2O, $K_a = K_w = 1.0 \times 10^{-14}$; HNO_2, $K_a = 4.0 \times 10^{-4}$; HNO_2 is a stronger acid than H_2O because K_a for $HNO_2 > K_w$ for H_2O.

c. HOC_6H_5, $K_a = 1.6 \times 10^{-10}$; HCN, $K_a = 6.2 \times 10^{-10}$; HCN is a slightly stronger acid than HOC_6H_5 because K_a for HCN > K_a for HOC_6H_5.

44. a. H_2O; the conjugate bases of strong acids are extremely weak bases ($K_b < 1 \times 10^{-14}$).

b. NO_2^-; the conjugate bases of weak acids are weak bases ($1 \times 10^{-14} < K_b < 1$).

c. $OC_6H_5^-$; for a conjugate acid-base pair, $K_a \times K_b = K_w$. From this relationship, the stronger the acid, the weaker is the conjugate base (K_b decreases as K_a increases). Because HCN is a stronger acid than HOC_6H_5 (K_a for HCN > K_a for HOC_6H_5), $OC_6H_5^-$ will be a stronger base than CN^-.

Autoionization of Water and the pH Scale

45. At 25°C, the relationship $[H^+][OH^-] = K_w = 1.0 \times 10^{-14}$ always holds for aqueous solutions. When $[H^+]$ is greater than 1.0×10^{-7} *M*, the solution is acidic; when $[H^+]$ is less than 1.0×10^{-7} *M*, the solution is basic; when $[H^+] = 1.0 \times 10^{-7}$ *M*, the solution is neutral. In terms of $[OH^-]$, an acidic solution has $[OH^-] < 1.0 \times 10^{-7}$ *M*, a basic solution has $[OH^-] > 1.0 \times 10^{-7}$ *M*, and a neutral solution has $[OH^-] = 1.0 \times 10^{-7}$ *M*.

a. $[OH^-] = \dfrac{K_w}{[H^+]} = \dfrac{1.0 \times 10^{-14}}{1.0 \times 10^{-7}} = 1.0 \times 10^{-7}\ M$; the solution is neutral.

b. $[OH^-] = \dfrac{1.0 \times 10^{-14}}{8.3 \times 10^{-16}} = 12\ M$; the solution is basic.

c. $[OH^-] = \dfrac{1.0 \times 10^{-14}}{12} = 8.3 \times 10^{-16}\ M$; the solution is acidic.

d. $[OH^-] = \dfrac{1.0 \times 10^{-14}}{5.4 \times 10^{-5}} = 1.9 \times 10^{-10}\ M$; the solution is acidic.

46. a. $[H^+] = \dfrac{K_w}{[OH^-]} = \dfrac{1.0 \times 10^{-14}}{1.5} = 6.7 \times 10^{-15}\ M$; basic

b. $[H^+] = \dfrac{1.0 \times 10^{-14}}{3.6 \times 10^{-15}} = 2.8\ M$; acidic

c. $[H^+] = \dfrac{1.0 \times 10^{-14}}{1.0 \times 10^{-7}} = 1.0 \times 10^{-7}\ M$; neutral

d. $[H^+] = \dfrac{1.0 \times 10^{-14}}{7.3 \times 10^{-4}} = 1.4 \times 10^{-11}\ M$; basic

47. a. Because the value of the equilibrium constant increases as the temperature increases, the reaction is endothermic. In endothermic reactions, heat is a reactant, so an increase in temperature (heat) shifts the reaction to produce more products and increases K in the process.

b. $H_2O(l) \rightleftharpoons H^+(aq) + OH^-(aq)$ $K_w = 5.47 \times 10^{-14} = [H^+][OH^-]$ at 50.°C

In pure water $[H^+] = [OH^-]$, so $5.47 \times 10^{-14} = [H^+]^2$, $[H^+] = 2.34 \times 10^{-7}\ M = [OH^-]$

48. a. $H_2O(l) \rightleftharpoons H^+(aq) + OH^-(aq)$ $K_w = 2.92 \times 10^{-14} = [H^+][OH^-]$

In pure water: $[H^+] = [OH^-]$, $2.92 \times 10^{-14} = [H^+]^2$, $[H^+] = 1.71 \times 10^{-7}\ M = [OH^-]$

b. $pH = -\log[H^+] = -\log(1.71 \times 10^{-7}) = 6.767$

c. $[H^+] = K_w/[OH^-] = (2.92 \times 10^{-14})/0.10 = 2.9 \times 10^{-13}\ M$; $pH = -\log(2.9 \times 10^{-13}) = 12.54$

49. $pH = -\log[H^+]$; $pOH = -\log[OH^-]$; at 25°C, $pH + pOH = 14.00$; for Exercise 45:

a. $pH = -\log[H^+] = -\log(1.0 \times 10^{-7}) = 7.00$; $pOH = 14.00 - pH = 14.00 - 7.00 = 7.00$

b. $pH = -\log(8.3 \times 10^{-16}) = 15.08$; $pOH = 14.00 - 15.08 = -1.08$

c. $pH = -\log(12) = -1.08$; $pOH = 14.00 - (-1.08) = 15.08$

d. $pH = -\log(5.4 \times 10^{-5}) = 4.27$; $pOH = 14.00 - 4.27 = 9.73$

Note that pH is less than zero when $[H^+]$ is greater than 1.0 M (an extremely acidic solution). For Exercise 46:

a. $pOH = -\log[OH^-] = -\log(1.5) = -0.18$; $pH = 14.00 - pOH = 14.00 - (-0.18) = 14.18$

b. $pOH = -\log(3.6 \times 10^{-15}) = 14.44$; $pH = 14.00 - 14.44 = -0.44$

c. $pOH = -\log(1.0 \times 10^{-7}) = 7.00$; $pH = 14.00 - 7.00 = 7.00$

d. $pOH = -\log(7.3 \times 10^{-4}) = 3.14$; $pH = 14.00 - 3.14 = 10.86$

Note that pH is greater than 14.00 when $[OH^-]$ is greater than 1.0 M (an extremely basic solution).

50. a. $[H^+] = 10^{-pH}$, $[H^+] = 10^{-7.40} = 4.0 \times 10^{-8}\ M$

$pOH = 14.00 - pH = 14.00 - 7.40 = 6.60$; $[OH^-] = 10^{-pOH} = 10^{-6.60} = 2.5 \times 10^{-7}\ M$

or $[OH^-] = \dfrac{K_w}{[H^+]} = \dfrac{1.0 \times 10^{-14}}{4.0 \times 10^{-8}} = 2.5 \times 10^{-7}\ M$; this solution is basic since pH > 7.00.

b. $[H^+] = 10^{-15.3} = 5 \times 10^{-16}\ M$; $pOH = 14.00 - 15.3 = -1.3$; $[OH^-] = 10^{-(-1.3)} = 20\ M$; basic

c. $[H^+] = 10^{-(-1.0)} = 10\ M$; $pOH = 14.0 - (-1.0) = 15.0$; $[OH^-] = 10^{-15.0} = 1 \times 10^{-15}\ M$; acidic

d. $[H^+] = 10^{-3.20} = 6.3 \times 10^{-4}\ M$; $pOH = 14.00 - 3.20 = 10.80$; $[OH^-] = 10^{-10.80} = 1.6 \times 10^{-11}\ M$; acidic

e. $[OH^-] = 10^{-5.0} = 1 \times 10^{-5}\ M$; $pH = 14.0 - pOH = 14.0 - 5.0 = 9.0$; $[H^+] = 10^{-9.0} = 1 \times 10^{-9}\ M$; basic

f. $[OH^-] = 10^{-9.60} = 2.5 \times 10^{-10}\ M$; $pH = 14.00 - 9.60 = 4.40$; $[H^+] = 10^{-4.40} = 4.0 \times 10^{-5}\ M$; acidic

51. a. $pOH = 14.00 - 6.88 = 7.12$; $[H^+] = 10^{-6.88} = 1.3 \times 10^{-7}\ M$

$[OH^-] = 10^{-7.12} = 7.6 \times 10^{-8}\ M$; acidic

b. $[H^+] = \dfrac{1.0 \times 10^{-14}}{8.4 \times 10^{-14}} = 0.12\ M$; $pH = -\log(0.12) = 0.92$

$pOH = 14.00 - 0.92 = 13.08$; acidic

c. $pH = 14.00 - 3.11 = 10.89$; $[H^+] = 10^{-10.89} = 1.3 \times 10^{-11}\ M$

$[OH^-] = 10^{-3.11} = 7.8 \times 10^{-4}\ M$; basic

d. $pH = -\log(1.0 \times 10^{-7}) = 7.00$; $pOH = 14.00 - 7.00 = 7.00$

$[OH^-] = 10^{-7.00} = 1.0 \times 10^{-7}\ M$; neutral

52. a. $pOH = 14.00 - 9.63 = 4.37$; $[H^+] = 10^{-9.63} = 2.3 \times 10^{-10}\ M$

$[OH^-] = 10^{-4.37} = 4.3 \times 10^{-5}\ M$; basic

b. $[H^+] = \dfrac{1.0 \times 10^{-14}}{3.9 \times 10^{-6}} = 2.6 \times 10^{-9}\ M$; pH = $-\log(2.6 \times 10^{-9}) = 8.59$

pOH = 14.00 – 8.59 = 5.41; basic

c. pH = –log(0.027) = 1.57; pOH = 14.00 – 1.57 = 12.43

$[OH^-] = 10^{-12.43} = 3.7 \times 10^{-13} M$; acidic

d. pH = 14.00 – 1.22 = 12.78; $[H^+] = 10^{-12.78} = 1.7 \times 10^{-13} M$

$[OH^-] = 10^{-1.22} = 0.060\ M$; basic

53. pOH = 14.0 – pH = 14.0 – 2.1 = 11.9; $[H^+] = 10^{-pH} = 10^{-2.1} = 8 \times 10^{-3} M$ (1 sig. fig.)

$[OH^-] = \dfrac{K_w}{[H^+]} = \dfrac{1.0 \times 10^{-14}}{8 \times 10^{-3}} = 1 \times 10^{-12} M$ or $[OH^-] = 10^{-pOH} = 10^{-11.9} = 1 \times 10^{-12} M$

The sample of gastric juice is acidic because the pH is less than 7.00 at 25°C.

54. pH = 14.00 – pOH = 14.00 – 5.74 = 8.26; $[H^+] = 10^{-pH} = 10^{-8.26} = 5.5 \times 10^{-9} M$

$[OH^-] = \dfrac{K_w}{[H^+]} = \dfrac{1.0 \times 10^{-14}}{5.5 \times 10^{-9}} = 1.8 \times 10^{-6} M$ or $[OH^-] = 10^{-pOH} = 10^{-5.74} = 1.8 \times 10^{-6} M$

The solution of baking soda is basic because the pH is greater than 7.00 at 25°C.

Solutions of Acids

55. All the acids in this problem are strong acids that are always assumed to completely dissociate in water. The general dissociation reaction for a strong acid is $HA(aq) \rightarrow H^+(aq) + A^-(aq)$, where A^- is the conjugate base of the strong acid HA. For 0.250 *M* solutions of these strong acids, 0.250 *M* H^+ and 0.250 *M* A^- are present when the acids completely dissociate. The amount of H^+ donated from water will be insignificant in this problem since H_2O is a very weak acid.

a. Major species present after dissociation = H^+, ClO_4^-, and H_2O;

pH = $-\log[H^+] = -\log(0.250) = 0.602$

b. Major species = H^+, NO_3^-, and H_2O; pH = 0.602

56. Both are strong acids, which are assumed to completely dissociate in water.

$0.0500\ L \times 0.050\ mol/L = 2.5 \times 10^{-3}$ mol HBr = 2.5×10^{-3} mol H^+ + 2.5×10^{-3} mol Br^-

$0.1500\ L \times 0.10\ mol/L = 1.5 \times 10^{-2}$ mol HI = 1.5×10^{-2} mol H^+ + 1.5×10^{-2} mol I^-

$$[H^+] = \frac{(2.5 \times 10^{-3} + 1.5 \times 10^{-2})\text{ mol}}{0.2000\text{L}} = 0.088\ M; \quad pH = -\log(0.088) = 1.06$$

57. Strong acids are assumed to completely dissociate in water; for example; $HCl(aq) + H_2O(l) \rightarrow H_3O^+(aq) + Cl^-(aq)$ or $HCl(aq) \rightarrow H^+(aq) + Cl^-(aq)$.

a. A 0.10 *M* HCl solution gives 0.10 *M* H^+ and 0.10 *M* Cl^- because HCl completely dissociates. The amount of H^+ from H_2O will be insignificant.

$pH = -\log[H^+] = -\log(0.10) = 1.00$

b. 5.0 *M* H^+ is produced when 5.0 *M* $HClO_4$ completely dissociates. The amount of H^+ from H_2O will be insignificant. $pH = -\log(5.0) = -0.70$ (Negative pH values just indicate very concentrated acid solutions.)

c. 1.0×10^{-11} *M* H^+ is produced when 1.0×10^{-11} *M* HI completely dissociates. If you take the negative log of 1.0×10^{-11}, this gives pH = 11.00. This is impossible! We dissolved an acid in water and got a basic pH. What we must consider in this problem is that water by itself donates 1.0×10^{-7} *M* H^+. We can normally ignore the small amount of H^+ from H_2O except when we have a very dilute solution of an acid (as in the case here). Therefore, the pH is that of neutral water (pH = 7.00) because the amount of HI present is insignificant.

58. $HNO_3(aq) \rightarrow H^+(aq) + NO_3^-(aq)$; HNO_3 is a strong acid, which means it is assumed to completely dissociate in water. The initial concentration of HNO_3 will equal the $[H^+]$ donated by the strong acid.

a. $pH = -\log[H^+] = -\log(2.0 \times 10^{-2}) = 1.70$

b. $pH = -\log(4.0) = -0.60$

c. Because the concentration of HNO_3 is so dilute, the pH will be that of neutral water (pH = 7.00). In this problem, water is the major H^+ producer present. Whenever the strong acid has a concentration less than 1.0×10^{-7} *M*, the $[H^+]$ contribution from water must be considered.

59. $[H^+] = 10^{-pH} = 10^{-2.50} = 3.2 \times 10^{-3}$ *M*. Because HI is a strong acid, a 3.2×10^{-3} *M* HI solution will produce 3.2×10^{-3} *M* H^+, giving a pH = 2.50.

60. $[H^+] = 10^{-pH} = 10^{-4.25} = 5.6 \times 10^{-5}$ *M*. Because HBr is a strong acid, a 5.6×10^{-5} *M* HBr solution is necessary to produce a pH = 4.25 solution.

61. HCl is a strong acid. $[H^+] = 10^{-1.50} = 3.16 \times 10^{-2}$ *M* (carrying one extra sig. fig.)

$$M_1V_1 = M_2V_2, \quad V_1 = \frac{M_2V_2}{M_1} = \frac{3.16 \times 10^{-2}\text{ mol/L} \times 1.6\text{L}}{12\text{ mol/L}} = 4.2 \times 10^{-3}\text{ L}$$

Add 4.2 mL of 12 *M* HCl to water with mixing; add enough water to make 1600 mL of solution. The resulting solution will have $[H^+] = 3.2 \times 10^{-2}$ *M* and pH = 1.50.

62. $50.0 \text{ mL conc. HCl soln} \times \frac{1.19\text{ g}}{\text{mL}} \times \frac{38\text{ g HCl}}{100\text{ g conc.HCl soln}} \times \frac{1\text{ mol HCl}}{36.5\text{ g}} = 0.62 \text{ mol HCl}$

$$20.0 \text{ mL conc. } HNO_3 \text{ soln} \times \frac{1.42\text{ g}}{\text{mL}} \times \frac{70.\text{ g } HNO_3}{100\text{ g soln}} \times \frac{1\text{ mol } HNO_3}{63.0\text{ g } HNO_3} = 0.32 \text{ mol } HNO_3$$

$HCl(aq) \rightarrow H^+(aq) + Cl^-(aq)$ and $HNO_3(aq) \rightarrow H^+(aq) + NO_3^-(aq)$ (Both are strong acids.)

So we will have 0.62 + 0.32 = 0.94 mol of H^+ in the final solution.

$$[H^+] = \frac{0.94\text{ mol}}{1.00\text{ L}} = 0.94\ M;\ \text{pH} = -\log[H^+] = -\log(0.94) = 0.027 = 0.03$$

$$[OH^-] = \frac{K_w}{[H^+]} = \frac{1.0 \times 10^{-14}}{0.94} = 1.1 \times 10^{-14}\ M$$

63. a. HNO_2 ($K_a = 4.0 \times 10^{-4}$) and H_2O ($K_a = K_w = 1.0 \times 10^{-14}$) are the major species. HNO_2 is a much stronger acid than H_2O, so it is the major source of H^+. However, HNO_2 is a weak acid ($K_a < 1$), so it only partially dissociates in water. We must solve an equilibrium problem to determine $[H^+]$. In the Solutions Guide, we will summarize the *i*nitial, *c*hange, and *e*quilibrium concentrations into one table called the ICE table. Solving the weak acid problem:

	HNO_2	$\rightleftharpoons$	H^+	+	NO_2^-
Initial	0.250 M		~0		0
	x mol/L HNO_2 dissociates to reach equilibrium				
Change	$-x$	$\rightarrow$	$+x$		$+x$
Equil.	$0.250 - x$		x		x

$$K_a = \frac{[H^+][NO_2^-]}{[HNO_2]} = 4.0 \times 10^{-4} = \frac{x^2}{0.250 - x};\ \text{if we assume } x << 0.250\text{, then:}$$

$$4.0 \times 10^{-4} \approx \frac{x^2}{0.250},\quad x = \sqrt{4.0 \times 10^{-4}(0.250)} = 0.010\ M$$

We must check the assumption: $\frac{x}{0.250} \times 100 = \frac{0.010}{0.250} \times 100 = 4.0\%$

All the assumptions are good. The H^+ contribution from water ($1 \times 10^{-7}\ M$) is negligible, and x is small compared to 0.250 (percent error = 4.0%). If the percent error is less than 5% for an assumption, we will consider it a valid assumption (called the 5% rule). Finishing the problem:

$$x = 0.010\ M = [H^+];\ \text{pH} = -\log(0.010) = 2.00$$

b. CH_3CO_2H ($K_a = 1.8 \times 10^{-5}$) and H_2O ($K_a = K_w = 1.0 \times 10^{-14}$) are the major species. CH_3CO_2H is the major source of H^+. Solving the weak acid problem:

	CH_3CO_2H	$\rightleftharpoons$	H^+	+	$CH_3CO_2^-$
Initial	0.250 *M*		~0		0
	x mol/L CH_3CO_2H dissociates to reach equilibrium				
Change	$-x$	$\rightarrow$	$+x$		$+x$
Equil.	$0.250 - x$		x		x

$$K_a = \frac{[H^+][CH_3CO_2^-]}{[CH_3CO_2H]},\ 1.8 \times 10^{-5} = \frac{x^2}{0.250 - x} \approx \frac{x^2}{0.250} \quad \text{(assuming } x << 0.250\text{)}$$

$x = 2.1 \times 10^{-3}$ *M*; checking assumption: $\frac{2.1 \times 10^{-3}}{0.250} \times 100 = 0.84\%$. Assumptions good.

$[H^+] = x = 2.1 \times 10^{-3}$ *M*; $pH = -\log(2.1 \times 10^{-3}) = 2.68$

64. a. HOC_6H_5 ($K_a = 1.6 \times 10^{-10}$) and H_2O ($K_a = K_w = 1.0 \times 10^{-14}$) are the major species. The major equilibrium is the dissociation of HOC_6H_5. Solving the weak acid problem:

	HOC_6H_5	$\rightleftharpoons$	H^+	+	$OC_6H_5^-$
Initial	0.250 *M*		~0		0
	x mol/L HOC_6H_5 dissociates to reach equilibrium				
Change	$-x$	$\rightarrow$	$+x$		$+x$
Equil.	$0.250 - x$		x		x

$$K_a = 1.6 \times 10^{-10} = \frac{[H^+][OC_6H_5^-]}{[HOC_6H_5]} = \frac{x^2}{0.250 - x} \approx \frac{x^2}{0.250} \quad \text{(assuming } x << 0.250\text{)}$$

$x = [H^+] = 6.3 \times 10^{-6}$ *M*; checking assumption: x is $2.5 \times 10^{-3}\%$ of 0.250, so assumption is valid by the 5% rule.

$pH = -\log(6.3 \times 10^{-6}) = 5.20$

b. HCN ($K_a = 6.2 \times 10^{-10}$) and H_2O are the major species. HCN is the major source of H^+.

	HCN	$\rightleftharpoons$	H^+	+	CN^-
Initial	0.250 *M*		~0		0
	x mol/L HCN dissociates to reach equilibrium				
Change	$-x$	$\rightarrow$	$+x$		$+x$
Equil.	$0.250 - x$		x		x

$$K_a = 6.2 \times 10^{-10} = \frac{[H^+][CN^-]}{[HCN]} = \frac{x^2}{0.250 - x} \approx \frac{x^2}{0.250} \quad \text{(assuming } x << 0.250\text{)}$$

$x = [H^+] = 1.2 \times 10^{-5}$ *M*; checking assumption: x is $4.8 \times 10^{-3}\%$ of 0.250.

Assumptions good. $pH = -\log(1.2 \times 10^{-5}) = 4.92$

65. This is a weak acid in water. Solving the weak acid problem:

	HF	$\rightleftharpoons$	H^+	+	F^-	$K_a = 7.2 \times 10^{-4}$
Initial	0.020 *M*		~0		0	
	x mol/L HF dissociates to reach equilibrium					
Change	$-x$	$\rightarrow$	$+x$		$+x$	
Equil.	$0.020 - x$		x		x	

$$K_a = 7.2 \times 10^{-4} = \frac{[H^+][F^-]}{[HF]} = \frac{x^2}{0.020 - x} \approx \frac{x^2}{0.020} \quad \text{(assuming } x << 0.020\text{)}$$

$x = [H^+] = 3.8 \times 10^{-3}$ *M*; check assumptions:

$$\frac{x}{0.020} \times 100 = \frac{3.8 \times 10^{-3}}{0.020} \times 100 = 19\%$$

The assumption $x << 0.020$ is not good (x is more than 5% of 0.020). We must solve $x^2/(0.020 - x) = 7.2 \times 10^{-4}$ exactly by using either the quadratic formula or the method of successive approximations (see Appendix 1 of the text). Using successive approximations, we let 0.016 *M* be a new approximation for [HF]. That is, in the denominator try $x = 0.0038$ (the value of x we calculated making the normal assumption) so that $0.020 - 0.0038 = 0.016$; then solve for a new value of x in the numerator.

$$\frac{x^2}{0.020 - x} \approx \frac{x^2}{0.016} = 7.2 \times 10^{-4}, \; x = 3.4 \times 10^{-3}$$

We use this new value of x to further refine our estimate of [HF], that is, $0.020 - x = 0.020 - 0.0034 = 0.0166$ (carrying an extra sig. fig.).

$$\frac{x^2}{0.020 - x} \approx \frac{x^2}{0.0166} = 7.2 \times 10^{-4}, \; x = 3.5 \times 10^{-3}$$

We repeat until we get a self-consistent answer. This would be the same answer we would get solving exactly using the quadratic equation. In this case it is, $x = 3.5 \times 10^{-3}$. Thus:

$[H^+] = [F^-] = x = 3.5 \times 10^{-3}$ *M*; $[OH^-] = K_w/[H^+] = 2.9 \times 10^{-12}$ *M*

$[HF] = 0.020 - x = 0.020 - 0.0035 = 0.017$ *M*; pH = 2.46

Note: When the 5% assumption fails, use whichever method you are most comfortable with to solve exactly. The method of successive approximations is probably fastest when the percent error is less than ~25% (unless you have a graphing calculator).

66.

	$HClO_2$	$\rightleftharpoons$	H^+	+	ClO_2^-	$K_a = 1.2 \times 10^{-2}$
Initial	0.22 *M*		~0		0	
	x mol/L $HClO_2$ dissociates to reach equilibrium					
Change	$-x$	$\rightarrow$	$+x$		$+x$	
Equil.	$0.22 - x$		x		x	

$$K_a = 1.2 \times 10^{-2} = \frac{[H^+][ClO_2^-]}{[HClO_2]} = \frac{x^2}{0.22 - x} \approx \frac{x^2}{0.22}, \ x = 5.1 \times 10^{-2}$$

The assumption that x is small is not good (x is 23% of 0.22). Using the method of successive approximations and carrying extra significant figures:

$$\frac{x^2}{0.22 - 0.051} \approx \frac{x^2}{0.169} = 1.2 \times 10^{-2}, \ x = 4.5 \times 10^{-2}$$

$$\frac{x^2}{0.175} = 1.2 \times 10^{-2}, \ x = 4.6 \times 10^{-2} \text{ (consistent answer)}$$

$[H^+] = [ClO_2^-] = x = 4.6 \times 10^{-2}\ M$; percent dissociation $= \dfrac{4.6 \times 10^{-2}}{0.22} \times 100 = 21\%$

67. $HC_3H_5O_2$ ($K_a = 1.3 \times 10^{-5}$) and H_2O ($K_a = K_w = 1.0 \times 10^{-14}$) are the major species present. $HC_3H_5O_2$ will be the dominant producer of H^+ because $HC_3H_5O_2$ is a stronger acid than H_2O. Solving the weak acid problem:

	$HC_3H_5O_2$	$\rightleftharpoons$	H^+	+	$C_3H_5O_2^-$
Initial	0.100 M		~0		0
	x mol/L $HC_3H_5O_2$ dissociates to reach equilibrium				
Change	$-x$	$\rightarrow$	$+x$		$+x$
Equil.	$0.100 - x$		x		x

$$K_a = 1.3 \times 10^{-5} = \frac{[H^+][C_3H_5O_2^-]}{[HC_3H_5O_2]} = \frac{x^2}{0.100 - x} \approx \frac{x^2}{0.100}$$

$x = [H^+] = 1.1 \times 10^{-3}\ M$; pH $= -\log(1.1 \times 10^{-3}) = 2.96$

Assumption follows the 5% rule (x is 1.1% of 0.100).

$[H^+] = [C_3H_5O_2^-] = 1.1 \times 10^{-3}\ M$; $[OH^-] = K_w/[H^+] = 9.1 \times 10^{-12}\ M$

$[HC_3H_5O_2] = 0.100 - 1.1 \times 10^{-3} = 0.099\ M$

$$\text{Percent dissociation} = \frac{[H^+]}{[HC_3H_5O_2]_0} \times 100 = \frac{1.1 \times 10^{-3}}{0.100} \times 100 = 1.1\%$$

68. This is a weak acid in water. We must solve a weak acid problem. Let HBz = $C_6H_5CO_2H$.

$$0.56 \text{ g HBz} \times \frac{1 \text{ mol HBz}}{122.1 \text{ g}} = 4.6 \times 10^{-3} \text{ mol}; \ [HBz]_0 = 4.6 \times 10^{-3}\ M$$

$$HBz \rightleftharpoons H^+ + Bz^-$$

	HBz		H^+	Bz^-
Initial	$4.6 \times 10^{-3}\ M$		~0	0
	x mol/L HBz dissociates to reach equilibrium			
Change	$-x$	→	$+x$	$+x$
Equil.	$4.6 \times 10^{-3} - x$		x	x

$$K_a = 6.4 \times 10^{-5} = \frac{[H^+][Bz^-]}{[HBz]} = \frac{x^2}{(4.6 \times 10^{-3} - x)} \approx \frac{x^2}{4.6 \times 10^{-3}}$$

$x = [H^+] = 5.4 \times 10^{-4}$; check assumptions: $\frac{x}{4.6 \times 10^{-3}} \times 100 = \frac{5.4 \times 10^{-4}}{4.6 \times 10^{-3}} \times 100 = 12\%$

Assumption is not good (x is 12% of 4.6×10^{-3}). When assumption(s) fail, we must solve exactly using the quadratic formula or the method of successive approximations (see Appendix 1 of text). Using successive approximations:

$$\frac{x^2}{(4.6 \times 10^{-3}) - (5.4 \times 10^{-4})} = 6.4 \times 10^{-5},\ x = 5.1 \times 10^{-4}$$

$$\frac{x^2}{(4.6 \times 10^{-3}) - (5.1 \times 10^{-4})} = 6.4 \times 10^{-5},\ x = 5.1 \times 10^{-4}\ M \quad \text{(consistent answer)}$$

Thus: $x = [H^+] = [Bz^-] = [C_6H_5CO_2^-] = 5.1 \times 10^{-4}\ M$

$[HBz] = [C_6H_5CO_2H] = 4.6 \times 10^{-3} - x = 4.1 \times 10^{-3}\ M$

$pH = -\log(5.1 \times 10^{-4}) = 3.29$; $pOH = 14.00 - pH = 10.71$; $[OH^-] = 10^{-10.71} = 1.9 \times 10^{-11}\ M$

69. Major species: $HC_2H_2ClO_2$ ($K_a = 1.35 \times 10^{-3}$) and H_2O; major source of H^+: $HC_2H_2ClO_2$

$$HC_2H_2ClO_2 \rightleftharpoons H^+ + C_2H_2ClO_2^-$$

	$HC_2H_2ClO_2$		H^+	$C_2H_2ClO_2^-$
Initial	$0.10\ M$		~0	0
	x mol/L $HC_2H_2ClO_2$ dissociates to reach equilibrium			
Change	$-x$	→	$+x$	$+x$
Equil.	$0.10 - x$		x	x

$$K_a = 1.35 \times 10^{-3} = \frac{x^2}{0.10 - x} \approx \frac{x^2}{0.10},\ x = 1.2 \times 10^{-2}\ M$$

Checking the assumptions finds that x is 12% of 0.10, which fails the 5% rule. We must solve $1.35 \times 10^{-3} = x^2/(0.10 - x)$ exactly using either the method of successive approximations or the quadratic equation. Using either method gives $x = [H^+] = 1.1 \times 10^{-2}\ M$.

$pH = -\log[H^+] = -\log(1.1 \times 10^{-2}) = 1.96.$

70. $$[HC_9H_7O_4] = \frac{2\text{ tablets} \times \frac{0.325\text{ g } HC_9H_7O_4}{\text{tablet}} \times \frac{1\text{ mol } HC_9H_7O_4}{180.15\text{ g}}}{0.237\text{ L}} = 0.0152\ M$$

	$HC_9H_7O_4$	$\rightleftharpoons$	H^+	+	$C_9H_7O_4^-$
Initial	0.0152 *M*		~0		0
	x mol/L $HC_9H_7O_4$ dissociates to reach equilibrium				
Change	$-x$	→	$+x$		$+x$
Equil.	$0.0152 - x$		x		x

$$K_a = 3.3 \times 10^{-4} = \frac{[H^+][C_9H_7O_4^-]}{[HC_9H_7O_4]} = \frac{x^2}{0.0152 - x} \approx \frac{x^2}{0.0152},\ x = 2.2 \times 10^{-3}\ M$$

Assumption that $0.0152 - x \approx 0.0152$ fails the 5% rule: $\frac{2.2 \times 10^{-3}}{0.0152} \times 100 = 14\%$

Using successive approximations or the quadratic equation gives an exact answer of $x = 2.1 \times 10^{-3}\ M$.

$[H^+] = x = 2.1 \times 10^{-3}\ M$; pH = $-\log(2.1 \times 10^{-3}) = 2.68$

71. HF and HOC_6H_5 are both weak acids with K_a values of 7.2×10^{-4} and 1.6×10^{-10}, respectively. Since the K_a value for HF is much greater than the K_a value for HOC_6H_5, HF will be the dominant producer of H^+ (we can ignore the amount of H^+ produced from HOC_6H_5 because it will be insignificant).

	HF	$\rightleftharpoons$	H^+	+	F^-
Initial	1.0 *M*		~0		0
	x mol/L HF dissociates to reach equilibrium				
Change	$-x$	→	$+x$		$+x$
Equil.	$1.0 - x$		x		x

$$K_a = 7.2 \times 10^{-4} = \frac{[H^+][F^-]}{[HF]} = \frac{x^2}{1.0 - x} \approx \frac{x^2}{1.0}$$

$x = [H^+] = 2.7 \times 10^{-2}\ M$; pH = $-\log(2.7 \times 10^{-2}) = 1.57$; assumptions good.

Solving for $[OC_6H_5^-]$ using $HOC_6H_5 \rightleftharpoons H^+ + OC_6H_5^-$ equilibrium:

$$K_a = 1.6 \times 10^{-10} = \frac{[H^+][OC_6H_5^-]}{[HOC_6H_5]} = \frac{(2.7 \times 10^{-2})[OC_6H_5^-]}{1.0},\ [OC_6H_5^-] = 5.9 \times 10^{-9}\ M$$

Note that this answer indicates that only $5.9 \times 10^{-9}\ M$ HOC_6H_5 dissociates, which confirms that HF is truly the only significant producer of H^+ in this solution.

72. a. The initial concentrations are halved since equal volumes of the two solutions are mixed.

	$HC_2H_3O_2$	$\rightleftharpoons$	H^+	+	$C_2H_3O_2^-$
Initial	0.100 *M*		5.00×10^{-4} *M*		0
Equil.	$0.100 - x$		$5.00 \times 10^{-4} + x$		x

$$K_a = 1.8 \times 10^{-5} = \frac{x(5.00 \times 10^{-4} + x)}{0.100 - x} \approx \frac{x(5.00 \times 10^{-4})}{0.100}$$

$x = 3.6 \times 10^{-3}$; assumption is horrible. Using the quadratic formula:

$$x^2 + (5.18 \times 10^{-4})x - 1.8 \times 10^{-6} = 0$$

$x = 1.1 \times 10^{-3}$ *M*; $[H^+] = 5.00 \times 10^{-4} + x = 1.6 \times 10^{-3}$ *M*; pH = 2.80

b. $x = [C_2H_3O_2^-] = 1.1 \times 10^{-3}$ *M*

73. In all parts of this problem, acetic acid ($HC_2H_3O_2$) is the best weak acid present. We must solve a weak acid problem.

a.

	$HC_2H_3O_2$	$\rightleftharpoons$	H^+	+	$C_2H_3O_2^-$
Initial	0.50 *M*		~0		0
	x mol/L $HC_2H_3O_2$ dissociates to reach equilibrium				
Change	$-x$	$\rightarrow$	$+x$		$+x$
Equil.	$0.50 - x$		x		x

$$K_a = 1.8 \times 10^{-5} = \frac{[H^+][C_2H_3O_2^-]}{[HC_2H_3O_2]} = \frac{x^2}{0.50 - x} \approx \frac{x^2}{0.50}$$

$x = [H^+] = [C_2H_3O_2^-] = 3.0 \times 10^{-3}$ *M*; assumptions good.

$$\text{Percent dissociation} = \frac{[H^+]}{[HC_2H_3O_2]_0} \times 100 = \frac{3.0 \times 10^{-3}}{0.50} \times 100 = 0.60\%$$

b. The setup for solutions b and c are similar to solution a except that the final equation is different because the new concentration of $HC_2H_3O_2$ is different.

$$K_a = 1.8 \times 10^{-5} = \frac{x^2}{0.050 - x} \approx \frac{x^2}{0.050}$$

$x = [H^+] = [C_2H_3O_2^-] = 9.5 \times 10^{-4}$ *M*; assumptions good.

$$\text{Percent dissociation} = \frac{9.5 \times 10^{-4}}{0.050} \times 100 = 1.9\%$$

c. $$K_a = 1.8 \times 10^{-5} = \frac{x^2}{0.0050 - x} \approx \frac{x^2}{0.0050}$$

$x = [H^+] = [C_2H_3O_2^-] = 3.0 \times 10^{-4}$ *M*; check assumptions.

Assumption that x is negligible is borderline (6.0% error). We should solve exactly. Using the method of successive approximations (see Appendix 1 of the text):

$$1.8 \times 10^{-5} = \frac{x^2}{0.0050 - (3.0 \times 10^{-4})} = \frac{x^2}{0.0047},\ x = 2.9 \times 10^{-4}$$

Next trial also gives $x = 2.9 \times 10^{-4}$.

$$\text{Percent dissociation} = \frac{2.9 \times 10^{-4}}{5.0 \times 10^{-3}} \times 100 = 5.8\%$$

d. As we dilute a solution, all concentrations are decreased. Dilution will shift the equilibrium to the side with the greater number of particles. For example, suppose we double the volume of an equilibrium mixture of a weak acid by adding water; then:

$$Q = \frac{\left(\frac{[H^+]_{eq}}{2}\right)\left(\frac{[X^-]_{eq}}{2}\right)}{\left(\frac{[HX]_{eq}}{2}\right)} = \frac{1}{2} K_a$$

$Q < K_a$, so the equilibrium shifts to the right or toward a greater percent dissociation.

e. $[H^+]$ depends on the initial concentration of weak acid and on how much weak acid dissociates. For solutions a-c, the initial concentration of acid decreases more rapidly than the percent dissociation increases. Thus $[H^+]$ decreases.

74. a. HNO_3 is a strong acid; it is assumed 100% dissociated in solution.

b.

	HNO_2	$\rightleftharpoons$	H^+	+	NO_2^-	$K_a = 4.0 \times 10^{-4}$
Initial	0.20 M		~0		0	
	x mol/L HNO_2 dissociates to reach equilibrium					
Change	$-x$	→	$+x$		$+x$	
Equil.	$0.20 - x$		x		x	

$$K_a = 4.0 \times 10^{-4} = \frac{[H^+][NO_2^-]}{[HNO_2]} = \frac{x^2}{0.20 - x} \approx \frac{x^2}{0.20}$$

$x = [H^+] = [NO_2^-] = 8.9 \times 10^{-3}\,M$; assumptions good.

$$\text{Percent dissociation} = \frac{[H^+]}{[HNO_2]_0} \times 100 = \frac{8.9 \times 10^{-3}}{0.20} \times 100 = 4.5\%$$

c.

	HOC_6H_5	$\rightleftharpoons$	H^+	+	$OC_6H_5^-$	$K_a = 1.6 \times 10^{-10}$
Initial	0.20 M		~0		0	
	x mol/L HOC_6H_5 dissociates to reach equilibrium					
Change	$-x$	→	$+x$		$+x$	
Equil.	$0.20 - x$		x		x	

$$K_a = 1.6 \times 10^{-10} = \frac{[H^+][OC_6H_5^-]}{[HOC_6H_5]} = \frac{x^2}{0.20 - x} \approx \frac{x^2}{0.20}$$

$x = [H^+] = [OC_6H_5^-] = 5.7 \times 10^{-6}\,M$; assumptions good.

$$\text{Percent dissociation} = \frac{5.7 \times 10^{-6}}{0.20} \times 100 = 2.9 \times 10^{-3}\%$$

d. For the same initial concentration, the percent dissociation increases as the strength of the acid increases (as K_a increases).

75. Let HA symbolize the weak acid. Set up the problem like a typical weak acid equilibrium problem.

	HA	⇌	H^+	+	A^-
Initial	0.15 *M*		~0		0
	x mol/L HA dissociates to reach equilibrium				
Change	$-x$	→	$+x$		$+x$
Equil.	$0.15 - x$		x		x

If the acid is 3.0% dissociated, then $x = [H^+]$ is 3.0% of 0.15: $x = 0.030 \times (0.15\,M) = 4.5 \times 10^{-3}\,M$. Now that we know the value of x, we can solve for K_a.

$$K_a = \frac{[H^+][A^-]}{[HA]} = \frac{x^2}{0.15 - x} = \frac{(4.5 \times 10^{-3})^2}{0.15 - (4.5 \times 10^{-3})} = 1.4 \times 10^{-4}$$

76.

	HX	⇌	H^+	+	X^-	
Initial	I		~0		0	where $I = [HX]_0$
	x mol/L HX dissociates to reach equilibrium					
Change	$-x$	→	$+x$		$+x$	
Equil.	$I - x$		x		x	

From the problem, $x = 0.25(I)$ and $I - x = 0.30\,M$.

$I - 0.25(I) = 0.30\,M$, $I = 0.40\,M$ and $x = 0.25(0.40\,M) = 0.10\,M$

$$K_a = \frac{[H^+][X^-]}{[HX]} = \frac{x^2}{I - x} = \frac{(0.10)^2}{0.30} = 0.033$$

77. $HClO_4$ is a strong acid with $[H^+] = 0.040\,M$. This equals the $[H^+]$ in the trichloroacetic acid solution. Set up the problem using the K_a equilibrium reaction for CCl_3CO_2H.

	CCl_3CO_2H	⇌	H^+	+	$CCl_3CO_2^-$
Initial	0.050 *M*		~0		0
Equil.	$0.050 - x$		x		x

$$K_a = \frac{[H^+][CCl_3CO_2^-]}{[CCl_3CO_2H]} = \frac{x^2}{0.050 - x}; \text{ from the problem, } x = [H^+] = 4.0 \times 10^{-2}\ M$$

$$K_a = \frac{(4.0 \times 10^{-2})^2}{0.050 - (4.0 \times 10^{-2})} = 0.16$$

78. Set up the problem using the K_a equilibrium reaction for HOBr.

	HOBr	⇌	H^+	+	OBr^-
Initial	0.063 *M*		~0		0
	x mol/L HOBr dissociates to reach equilibrium				
Change	$-x$	→	$+x$		$+x$
Equil.	$0.063 - x$		x		x

$$K_a = \frac{[H^+][OBr^-]}{[HOBr]} = \frac{x^2}{0.063 - x}; \text{ from pH} = 4.95\text{: } x = [H^+] = 10^{-pH} = 10^{-4.95} = 1.1 \times 10^{-5}\ M$$

$$K_a = \frac{(1.1 \times 10^{-5})^2}{0.063 - 1.1 \times 10^{-5}} = 1.9 \times 10^{-9}$$

79. Major species: HCOOH and H_2O; major source of H^+: HCOOH

	HCOOH	⇌	H^+	+	$HCOO^-$	
Initial	C		~0		0	where $C = [HCOOH]_0$
	x mol/L HCOOH dissociates to reach equilibrium					
Change	$-x$	→	$+x$		$+x$	
Equil.	$C - x$		x		x	

$$K_a = 1.8 \times 10^{-4} = \frac{[H^+][HCOO^-]}{[HCOOH]} = \frac{x^2}{C - x}, \text{ where } x = [H^+]$$

$$1.8 \times 10^{-4} = \frac{[H^+]^2}{C - [H^+]}; \text{ because pH} = 2.70\text{: } [H^+] = 10^{-2.70} = 2.0 \times 10^{-3}\ M$$

$$1.8 \times 10^{-4} = \frac{(2.0 \times 10^{-3})^2}{C - (2.0 \times 10^{-3})},\ C - (2.0 \times 10^{-3}) = \frac{4.0 \times 10^{-6}}{1.8 \times 10^{-4}},\ C = 2.4 \times 10^{-2}\ M$$

A 0.024 *M* formic acid solution will have pH = 2.70.

80. Major species: $HC_2H_3O_2$ (acetic acid) and H_2O; major source of H^+: $HC_2H_3O_2$

	$HC_2H_3O_2$	⇌	H^+	+	$C_2H_3O_2^-$	
Initial	C		~0		0	where $C = [HC_2H_3O_2]_0$
	x mol/L $HC_2H_3O_2$ dissociates to reach equilibrium					
Change	$-x$	→	$+x$		$+x$	
Equil.	$C - x$		x		x	

$$K_a = 1.8 \times 10^{-5} = \frac{[H^+][C_2H_3O_2^-]}{[HC_2H_3O_2]} = \frac{x^2}{C - x}, \text{ where } x = [H^+]$$

$$1.8 \times 10^{-5} = \frac{[H^+]^2}{C - [H^+]}; \text{ from pH} = 3.0\text{: } [H^+] = 10^{-3.0} = 1 \times 10^{-3}\ M$$

$$1.8 \times 10^{-5} = \frac{(1 \times 10^{-3})^2}{C - (1 \times 10^{-3})},\ C - (1 \times 10^{-3}) = \frac{1 \times 10^{-6}}{1.8 \times 10^{-5}},\ C = 5.7 \times 10^{-2} \approx 6 \times 10^{-2}\ M$$

A 6×10^{-2} M acetic acid solution will have pH = 3.0.

81. $[HA]_0 = \frac{1.0\text{ mol}}{2.0\text{ L}} = 0.50$ mol/L; solve using the K_a equilibrium reaction.

	HA	⇌	H^+	+	A^-
Initial	0.50 M		~0		0
Equil.	$0.50 - x$		x		x

$$K_a = \frac{[H^+][A^-]}{[HA]} = \frac{x^2}{0.50 - x}; \text{ in this problem, } [HA] = 0.45\ M \text{ so:}$$

$$[HA] = 0.45\ M = 0.50\ M - x,\ x = 0.05\ M$$

$$K_a = \frac{(0.05)^2}{0.45} = 6 \times 10^{-3}$$

82. Let HSac = saccharin and I = $[HSac]_0$.

	HSac	⇌	H^+	+	Sac^-	$K_a = 10^{-11.70} = 2.0 \times 10^{-12}$
Initial	I		~0		0	
Equil.	$I - x$		x		x	

$$K_a = 2.0 \times 10^{-12} = \frac{x^2}{I - x};\ x = [H^+] = 10^{-5.75} = 1.8 \times 10^{-6}\ M$$

$$2.0 \times 10^{-12} = \frac{(1.8 \times 10^{-6})^2}{I - (1.8 \times 10^{-6})},\ I = 1.6\ M = [HSac]_0$$

$$100.0\text{ g } HC_7H_4NSO_3 \times \frac{1\text{ mol}}{183.19\text{ g}} \times \frac{1\text{ L}}{1.6\text{ mol}} \times \frac{1000\text{ mL}}{\text{L}} = 340\text{ mL}$$

Solutions of Bases

83. All K_b reactions refer to the base reacting with water to produce the conjugate acid of the base and OH^-.

a. $NH_3(aq) + H_2O(l) \rightleftharpoons NH_4^+(aq) + OH^-(aq)$ $\qquad K_b = \dfrac{[NH_4^+][OH^-]}{[NH_3]}$

b. $C_5H_5N(aq) + H_2O(l) \rightleftharpoons C_5H_5NH^+(aq) + OH^-(aq)$ $\qquad K_b = \dfrac{[C_5H_5NH^+][OH^-]}{[C_5H_5N]}$

84. a. $C_6H_5NH_2(aq) + H_2O(l) \rightleftharpoons C_6H_5NH_3^+(aq) + OH^-(aq)$ $\qquad K_b = \dfrac{[C_6H_5NH_3^+][OH^-]}{[C_6H_5NH_2]}$

b. $(CH_3)_2NH(aq) + H_2O(l) \rightleftharpoons (CH_3)_2NH_2^+(aq) + OH^-(aq)$ $\qquad K_b = \dfrac{[(CH_3)_2NH_2^+][OH^-]}{[(CH_3)_2NH]}$

85. NO_3^-: Because HNO_3 is a strong acid, NO_3^- is a terrible base ($K_b << K_w$). All conjugate bases of strong acids have no base strength.

H_2O: $K_b = K_w = 1.0 \times 10^{-14}$; NH_3: $K_b = 1.8 \times 10^{-5}$; C_5H_5N: $K_b = 1.7 \times 10^{-9}$

Base strength = $NH_3 > C_5H_5N > H_2O > NO_3^-$ (As K_b increases, base strength increases.)

86. Excluding water, these are the conjugate acids of the bases in the preceding exercise. In general, the stronger the base, the weaker is the conjugate acid. *Note*: Even though NH_4^+ and $C_5H_5NH^+$ are conjugate acids of weak bases, they are still weak acids with K_a values between K_w and 1. Prove this to yourself by calculating the K_a values for NH_4^+ and $C_5H_5NH^+$ ($K_a = K_w/K_b$).

Acid strength = $HNO_3 > C_5H_5NH^+ > NH_4^+ > H_2O$

87. a. $C_6H_5NH_2$ b. $C_6H_5NH_2$ c. OH^- d. CH_3NH_2

The base with the largest K_b value is the strongest base ($K_{b,\,C_6H_5NH_2} = 3.8 \times 10^{-10}$, $K_{b,\,CH_3NH_2} = 4.4 \times 10^{-4}$). OH^- is the strongest base possible in water.

88. a. $HClO_4$ (a strong acid) b. $C_6H_5NH_3^+$ c. $C_6H_5NH_3^+$

The acid with the largest K_a value is the strongest acid. To calculate K_a values for $C_6H_5NH_3^+$ and $CH_3NH_3^+$, use $K_a = K_w/K_b$, where K_b refers to the bases $C_6H_5NH_2$ or CH_3NH_2.

89. $NaOH(aq) \rightarrow Na^+(aq) + OH^-(aq)$; NaOH is a strong base that completely dissociates into Na^+ and OH^-. The initial concentration of NaOH will equal the concentration of OH^- donated by NaOH.

a. $[OH^-] = 0.10\ M$; $pOH = -\log[OH^-] = -\log(0.10) = 1.00$

$pH = 14.00 - pOH = 14.00 - 1.00 = 13.00$

Note that H_2O is also present, but the amount of OH^- produced by H_2O will be insignificant compared to the 0.10 M OH^- produced from the NaOH.

b. The $[OH^-]$ concentration donated by the NaOH is 1.0×10^{-10} *M*. Water by itself donates 1.0×10^{-7} *M*. In this exercise, water is the major OH^- contributor, and $[OH^-] = 1.0 \times 10^{-7}$ *M*.

$pOH = -\log(1.0 \times 10^{-7}) = 7.00$; $pH = 14.00 - 7.00 = 7.00$

c. $[OH^-] = 2.0$ *M*; $pOH = -\log(2.0) = -0.30$; $pH = 14.00 - (-0.30) = 14.30$

90. a. $Ca(OH)_2 \rightarrow Ca^{2+} + 2\ OH^-$; $Ca(OH)_2$ is a strong base and dissociates completely.

$[OH^-] = 2(0.00040) = 8.0 \times 10^{-4}$ *M*; $pOH = -\log[OH^-] = 3.10$

$pH = 14.00 - pOH = 10.90$

b. $\frac{25\text{ g KOH}}{\text{L}} \times \frac{1\text{ mol KOH}}{56.11\text{ g KOH}} = 0.45\text{ mol KOH/L}$

KOH is a strong base, so $[OH^-] = 0.45$ *M*; $pOH = -\log(0.45) = 0.35$; $pH = 13.65$

c. $\frac{150.0\text{ g NaOH}}{\text{L}} \times \frac{1\text{ mol}}{40.00\text{ g}} = 3.750$ *M*; NaOH is a strong base, so $[OH^-] = 3.750$ *M*.

$pOH = -\log(3.750) = -0.5740$ and $pH = 14.0000 - (-0.5740) = 14.5740$

Although we are justified in calculating the answer to four decimal places, in reality, the pH can only be measured to ±0.01 pH units.

91. a. Major species: K^+, OH^-, H_2O (KOH is a strong base.)

$[OH^-] = 0.015$ *M*, $pOH = -\log(0.015) = 1.82$; $pH = 14.00 - pOH = 12.18$

b. Major species: Ba^{2+}, OH^-, H_2O; $Ba(OH)_2(aq) \rightarrow Ba^{2+}(aq) + 2\ OH^-(aq)$; because each mole of the strong base $Ba(OH)_2$ dissolves in water to produce two mol OH^-, $[OH^-] = 2(0.015\ M) = 0.030$ *M*.

$pOH = -\log(0.030) = 1.52$; $pH = 14.00 - 1.52 = 12.48$

92. a. Major species: Na^+, Li^+, OH^-, H_2O (NaOH and LiOH are both strong bases.)

$[OH^-] = 0.050 + 0.050 = 0.100$ *M*; $pOH = 1.000$; $pH = 13.000$

b. Major species: Ca^{2+}, Rb^+, OH^-, H_2O; Both $Ca(OH)_2$ and RbOH are strong bases, and $Ca(OH)_2$ donates 2 mol OH^- per mol $Ca(OH)_2$.

$[OH^-] = 2(0.0010) + 0.020 = 0.022$ *M*; $pOH = -\log(0.022) = 1.66$; $pH = 12.34$

93. $pOH = 14.00 - 11.56 = 2.44$; $[OH^-] = [KOH] = 10^{-2.44} = 3.6 \times 10^{-3}$ *M*

$$0.8000 \text{ L} \times \frac{3.6 \times 10^{-3} \text{ mol KOH}}{\text{L}} \times \frac{56.11 \text{ g KOH}}{\text{mol KOH}} = 0.16 \text{ g KOH}$$

94. pH = 10.50; pOH = 14.00 – 10.50 = 3.50; $[OH^-] = 10^{-3.50} = 3.2 \times 10^{-4}\ M$

$Sr(OH)_2(aq) \rightarrow Sr^{2+}(aq) + 2\ OH^-(aq)$; $Sr(OH)_2$ donates 2 mol OH^- per mol $Sr(OH)_2$.

$$[Sr(OH)_2] = \frac{3.2 \times 10^{-4} \text{ mol OH}^-}{\text{L}} \times \frac{1 \text{ mol Sr(OH)}_2}{2 \text{ mol OH}^-} = 1.6 \times 10^{-4}\ M\ Sr(OH)_2$$

A 1.6×10^{-4} *M* $Sr(OH)_2$ solution will produce a pH = 10.50 solution.

95. NH_3 is a weak base with $K_b = 1.8 \times 10^{-5}$. The major species present will be NH_3 and H_2O ($K_b = K_w = 1.0 \times 10^{-14}$). Because NH_3 has a much larger K_b value than H_2O, NH_3 is the stronger base present and will be the major producer of OH^-. To determine the amount of OH^- produced from NH_3, we must perform an equilibrium calculation using the K_b reaction for NH_3.

	$NH_3(aq)$ + $H_2O(l)$	$\rightleftharpoons$	$NH_4^+(aq)$	+	$OH^-(aq)$
Initial	0.150 *M*		0		~0
	x mol/L NH_3 reacts with H_2O to reach equilibrium				
Change	$-x$	$\rightarrow$	$+x$		$+x$
Equil.	$0.150 - x$		x		x

$$K_b = 1.8 \times 10^{-5} = \frac{[NH_4^+][OH^-]}{[NH_3]} = \frac{x^2}{0.150 - x} \approx \frac{x^2}{0.150} \quad \text{(assuming } x \ll 0.150\text{)}$$

$x = [OH^-] = 1.6 \times 10^{-3}\ M$; check assumptions: x is 1.1% of 0.150, so the assumption $0.150 - x \approx 0.150$ is valid by the 5% rule. Also, the contribution of OH^- from water will be insignificant (which will usually be the case). Finishing the problem:

$$pOH = -\log[OH^-] = -\log(1.6 \times 10^{-3}\ M) = 2.80;\ pH = 14.00 - pOH = 14.00 - 2.80 = 11.20.$$

96. Major species: H_2NNH_2 ($K_b = 3.0 \times 10^{-6}$) and H_2O ($K_b = K_w = 1.0 \times 10^{-14}$); the weak base H_2NNH_2 will dominate OH^- production. We must perform a weak base equilibrium calculation.

	H_2NNH_2 + H_2O	$\rightleftharpoons$	$H_2NNH_3^+$	+	OH^-	$K_b = 3.0 \times 10^{-6}$
Initial	2.0 *M*		0		~0	
	x mol/L H_2NNH_2 reacts with H_2O to reach equilibrium					
Change	$-x$	$\rightarrow$	$+x$		$+x$	
Equil.	$2.0 - x$		x		x	

$$K_b = 3.0 \times 10^{-6} = \frac{[H_2NNH_3^+][OH^-]}{[H_2NNH_2]} = \frac{x^2}{2.0 - x} \approx \frac{x^2}{2.0} \quad \text{(assuming } x \ll 2.0\text{)}$$

$x = [OH^-] = 2.4 \times 10^{-3} M$; pOH = 2.62; pH = 11.38; assumptions good (x is 0.12% of 2.0).

$[H_2NNH_3^+] = 2.4 \times 10^{-3} M$; $[H_2NNH_2] = 2.0\ M$; $[H^+] = 10^{-11.38} = 4.2 \times 10^{-12} M$

97. These are solutions of weak bases in water. In each case we must solve an equilibrium weak base problem.

a. $(C_2H_5)_3N + H_2O \rightleftharpoons (C_2H_5)_3NH^+ + OH^-$ $K_b = 4.0 \times 10^{-4}$

	$(C_2H_5)_3N$		$(C_2H_5)_3NH^+$	OH^-
Initial	0.20 M		0	~0
	x mol/L of $(C_2H_5)_3N$ reacts with H_2O to reach equilibrium			
Change	$-x$	$\rightarrow$	$+x$	$+x$
Equil.	$0.20 - x$		x	x

$$K_b = 4.0 \times 10^{-4} = \frac{[(C_2H_5)_3NH^+][OH^-]}{[(C_2H_5)_3N]} = \frac{x^2}{0.20-x} \approx \frac{x^2}{0.20},\quad x = [OH^-] = 8.9 \times 10^{-3} M$$

Assumptions good (x is 4.5% of 0.20). $[OH^-] = 8.9 \times 10^{-3}\ M$

$$[H^+] = \frac{K_w}{[OH^-]} = \frac{1.0 \times 10^{-14}}{8.9 \times 10^{-3}} = 1.1 \times 10^{-12} M;\ pH = 11.96$$

b. $HONH_2 + H_2O \rightleftharpoons HONH_3^+ + OH^-$ $K_b = 1.1 \times 10^{-8}$

	$HONH_2$	$HONH_3^+$	OH^-
Initial	0.20 M	0	~0
Equil.	$0.20 - x$	x	x

$$K_b = 1.1 \times 10^{-8} = \frac{x^2}{0.20-x} \approx \frac{x^2}{0.20},\quad x = [OH^-] = 4.7 \times 10^{-5} M;\ \text{assumptions good.}$$

$[H^+] = 2.1 \times 10^{-10} M$; pH = 9.68

98. These are solutions of weak bases in water.

a. $C_6H_5NH_2 + H_2O \rightleftharpoons C_6H_5NH_3^+ + OH^-$ $K_b = 3.8 \times 10^{-10}$

	$C_6H_5NH_2$		$C_6H_5NH_3^+$	OH^-
Initial	0.40 M		0	~0
	x mol/L of $C_6H_5NH_2$ reacts with H_2O to reach equilibrium			
Change	$-x$	$\rightarrow$	$+x$	$+x$
Equil.	$0.40 - x$		x	x

$$3.8 \times 10^{-10} = \frac{x^2}{0.40-x} \approx \frac{x^2}{0.40},\quad x = [OH^-] = 1.2 \times 10^{-5} M;\ \text{assumptions good.}$$

$[H^+] = K_w/[OH^-] = 8.3 \times 10^{-10} M$; pH = 9.08

b. $CH_3NH_2 + H_2O \rightleftharpoons CH_3NH_3^+ + OH^-$ $K_b = 4.38 \times 10^{-4}$

	CH_3NH_2	$CH_3NH_3^+$	OH^-
Initial	0.40 M	0	~0
Equil.	$0.40 - x$	x	x

$$K_b = 4.38 \times 10^{-4} = \frac{x^2}{0.40 - x} \approx \frac{x^2}{0.40}, \ x = 1.3 \times 10^{-2}\ M; \text{ assumptions good.}$$

$[OH^-] = 1.3 \times 10^{-2}\ M$; $[H^+] = K_w/[OH^-] = 7.7 \times 10^{-13}\ M$; pH = 12.11

99. This is a solution of a weak base in water. We must solve the weak base equilibrium problem.

$C_2H_5NH_2 + H_2O \rightleftharpoons C_2H_5NH_3^+ + OH^-$ $K_b = 5.6 \times 10^{-4}$

	$C_2H_5NH_2$		$C_2H_5NH_3^+$	OH^-
Initial	0.20 M		0	~0
	x mol/L $C_2H_5NH_2$ reacts with H_2O to reach equilibrium			
Change	$-x$	→	$+x$	$+x$
Equil.	$0.20 - x$		x	x

$$K_b = \frac{[C_2H_5NH_3^+][OH^-]}{[C_2H_5NH_2]} = \frac{x^2}{0.20 - x} \approx \frac{x^2}{0.20} \text{ (assuming } x << 0.20)$$

$$x = 1.1 \times 10^{-2}; \text{ checking assumption: } \frac{1.1 \times 10^{-2}}{0.20} \times 100 = 5.5\%$$

The assumption fails the 5% rule. We must solve exactly using either the quadratic equation or the method of successive approximations (see Appendix 1 of the text). Using successive approximations and carrying extra significant figures:

$$\frac{x^2}{0.20 - 0.011} = \frac{x^2}{0.189} = 5.6 \times 10^{-4}, \ x = 1.0 \times 10^{-2}\ M \text{ (consistent answer)}$$

$$x = [OH^-] = 1.0 \times 10^{-2}\ M; \ [H^+] = \frac{K_w}{[OH^-]} = \frac{1.0 \times 10^{-14}}{1.0 \times 10^{-2}} = 1.0 \times 10^{-12}\ M; \text{ pH} = 12.00$$

100. $(C_2H_5)_2NH + H_2O \rightleftharpoons (C_2H_5)_2NH_2^+ + OH^-$ $K_b = 1.3 \times 10^{-3}$

	$(C_2H_5)_2NH$		$(C_2H_5)_2NH_2^+$	OH^-
Initial	0.050 M		0	~0
	x mol/L $(C_2H_5)_2NH$ reacts with H_2O to reach equilibrium			
Change	$-x$	→	$+x$	$+x$
Equil.	$0.050 - x$		x	x

$$K_b = 1.3 \times 10^{-3} = \frac{[(C_2H_5)_2NH_2^+][OH^-]}{[(C_2H_5)_2NH]} = \frac{x^2}{0.050 - x} \approx \frac{x^2}{0.050}$$

$x = 8.1 \times 10^{-3}$; assumption is bad (x is 16% of 0.20).

Using successive approximations:

$$1.3 \times 10^{-3} = \frac{x^2}{0.050 - 0.081}, \quad x = 7.4 \times 10^{-3}$$

$$1.3 \times 10^{-3} = \frac{x^2}{0.050 - 0.074}, \quad x = 7.4 \times 10^{-3} \text{ (consistent answer)}$$

$[OH^-] = x = 7.4 \times 10^{-3} M$; $[H^+] = K_w/[OH^-] = 1.4 \times 10^{-12} M$; pH = 11.85

101. To solve for percent ionization, we first solve the weak base equilibrium problem.

a. $NH_3 + H_2O \rightleftharpoons NH_4^+ + OH^-$ $\quad K_b = 1.8 \times 10^{-5}$

	NH_3		NH_4^+	OH^-
Initial	0.10 *M*		0	~0
Equil.	$0.10 - x$		x	x

$$K_b = 1.8 \times 10^{-5} = \frac{x^2}{0.10 - x} \approx \frac{x^2}{0.10}, \; x = [OH^-] = 1.3 \times 10^{-3} M; \text{ assumptions good.}$$

$$\text{Percent ionization} = \frac{x}{[NH_3]_0} \times 100 = \frac{1.3 \times 10^{-3} M}{0.10 M} \times 100 = 1.3\%$$

b. $NH_3 + H_2O \rightleftharpoons NH_4^+ + OH^-$

	NH_3		NH_4^+	OH^-
Initial	0.010 *M*		0	~0
Equil.	$0.010 - x$		x	x

$$1.8 \times 10^{-5} = \frac{x^2}{0.010 - x} \approx \frac{x^2}{0.010}, \; x = [OH^-] = 4.2 \times 10^{-4} M; \text{ assumptions good.}$$

$$\text{Percent ionization} = \frac{4.2 \times 10^{-4}}{0.010} \times 100 = 4.2\%$$

Note: For the same base, the percent ionization increases as the initial concentration of base decreases.

c. $CH_3NH_2 + H_2O \rightleftharpoons CH_3NH_3^+ + OH^-$ $\quad K_b = 4.38 \times 10^{-4}$

	CH_3NH_2		$CH_3NH_3^+$	OH^-
Initial	0.10 *M*		0	~0
Equil.	$0.10 - x$		x	x

$4.38 \times 10^{-4} = \frac{x^2}{0.10 - x} \approx \frac{x^2}{0.10}$, $x = 6.6 \times 10^{-3}$; assumption fails the 5% rule (x is 6.6% of 0.10). Using successive approximations and carrying extra significant figures:

$$\frac{x^2}{0.10 - 0.0066} = \frac{x^2}{0.093} = 4.38 \times 10^{-4}, \; x = 6.4 \times 10^{-3} \quad \text{(consistent answer)}$$

$$\text{Percent ionization} = \frac{6.4 \times 10^{-3}}{0.10} \times 100 = 6.4\%$$

102. $C_5H_5N + H_2O \rightleftharpoons C_5H_5N^+ + OH^-$ $K_b = 1.7 \times 10^{-9}$

	C_5H_5N	$C_5H_5N^+$	OH^-
Initial	0.10 *M*	0	~0
Equil.	$0.10 - x$	x	x

$K_b = 1.7 \times 10^{-9} = \frac{x^2}{0.10 - x} \approx \frac{x^2}{0.10}$, $x = [C_5H_5N] = 1.3 \times 10^{-5}$ *M*; assumptions good.

Percent C_5H_5N ionized $= \frac{1.3 \times 10^{-5}\ M}{0.10\ M} \times 100 = 1.3 \times 10^{-2}\%$

103. Using the K_b reaction to solve where PT = p-toluidine ($CH_3C_6H_4NH_2$):

$$PT + H_2O \rightleftharpoons PTH^+ + OH^-$$

	PT		PTH^+	OH^-
Initial	0.016 *M*		0	~0
	x mol/L of PT reacts with H_2O to reach equilibrium			
Change	$-x$	→	$+x$	$+x$
Equil.	$0.016 - x$		x	x

$K_b = \frac{[PTH^+][OH^-]}{[PT]} = \frac{x^2}{0.016 - x}$

Because pH = 8.60: pOH = 14.00 − 8.60 = 5.40 and $[OH^-] = x = 10^{-5.40} = 4.0 \times 10^{-6}$ *M*

$K_b = \frac{(4.0 \times 10^{-6})^2}{0.016 - (4.0 \times 10^{-6})} = 1.0 \times 10^{-9}$

104. $HONH_2 + H_2O \rightleftharpoons HONH_3^+ + OH^-$ $K_b = 1.1 \times 10^{-8}$

	$HONH_2$	$HONH_3^+$	OH^-	
Initial	I	0	~0	$I = [HONH_2]_0$
Equil.	$I - x$	x	x	

$K_b = 1.1 \times 10^{-8} = \frac{x^2}{I - x}$

From problem, pH = 10.00, so pOH = 4.00 and $x = [OH^-] = 1.0 \times 10^{-4}$ *M*.

$1.1 \times 10^{-8} = \frac{(1.0 \times 10^{-4})^2}{I - (1.0 \times 10^{-4})}$, I = 0.91 *M*

Mass $HONH_2 = 0.2500\ L \times \frac{0.91\ mol\ HONH_2}{L} \times \frac{33.03\ g\ HONH_2}{mol\ HONH_2} = 7.5\ g\ HONH_2$

Polyprotic Acids

105. $H_2SO_3(aq) \rightleftharpoons HSO_3^-(aq) + H^+(aq)$ $K_{a_1} = \dfrac{[HSO_3^-][H^+]}{[H_2SO_3]}$

$HSO_3^-(aq) \rightleftharpoons SO_3^{2-}(aq) + H^+(aq)$ $K_{a_2} = \dfrac{[SO_3^{2-}][H^+]}{[HSO_3^-]}$

106. $H_3C_6H_5O_7(aq) \rightleftharpoons H_2C_6H_5O_7^-(aq) + H^+(aq)$ $K_{a_1} = \dfrac{[H_2C_6H_5O_7^-][H^+]}{[H_3C_6H_5O_7]}$

$H_2C_6H_5O_7^-(aq) \rightleftharpoons HC_6H_5O_7^{2-}(aq) + H^+(aq)$ $K_{a_2} = \dfrac{[HC_6H_5O_7^{2-}][H^+]}{[H_2C_6H_5O_7^-]}$

$HC_6H_5O_7^{2-}(aq) \rightleftharpoons C_6H_5O_7^{3-}(aq) + H^+(aq)$ $K_{a_3} = \dfrac{[C_6H_5O_7^{3-}][H^+]}{[HC_6H_5O_7^{2-}]}$

107. For $H_2C_6H_6O_6$. $K_{a_1} = 7.9 \times 10^{-5}$ and $K_{a_2} = 1.6 \times 10^{-12}$. Because $K_{a_1} >> K_{a_2}$, the amount of H^+ produced by the K_{a_2} reaction will be negligible.

$$[H_2C_6H_6O_6]_0 = \frac{0.500\,g \times \dfrac{1\,mol\,H_2C_6H_6O_6}{176.12\,g}}{0.2000\,L} = 0.0142\,M$$

	$H_2C_6H_6O_6(aq)$	$\rightleftharpoons$	$HC_6H_6O_6^-(aq)$	+	$H^+(aq)$	$K_{a_1} = 7.9 \times 10^{-5}$
Initial	$0.0142\,M$		0		~0	
Equil.	$0.0142 - x$		x		x	

$$K_{a_1} = 7.9 \times 10^{-5} = \frac{x^2}{0.0142 - x} \approx \frac{x^2}{0.0142},\ x = 1.1 \times 10^{-3};\ \text{assumption fails the 5\% rule.}$$

Solving by the method of successive approximations:

$$7.9 \times 10^{-5} = \frac{x^2}{0.0142 - 1.1 \times 10^{-3}},\ x = 1.0 \times 10^{-3}\,M\ \text{(consistent answer)}$$

Because H^+ produced by the K_{a_2} reaction will be negligible, $[H^+] = 1.0 \times 10^{-3}$ and pH = 3.00.

108. The reactions are:

$H_3AsO_4 \rightleftharpoons H^+ + H_2AsO_4^-$ $K_{a_1} = 5.5 \times 10^{-3}$

$$H_2AsO_4^- \rightleftharpoons H^+ + HAsO_4^{2-} \quad K_{a_2} = 1.7 \times 10^{-7}$$

$$HAsO_4^{2-} \rightleftharpoons H^+ + AsO_4^{3-} \quad K_{a_3} = 5.1 \times 10^{-12}$$

We will deal with the reactions in order of importance, beginning with the largest K_a, K_{a_1}.

	H_3AsO_4	$\rightleftharpoons$	H^+	+	$H_2AsO_4^-$	$K_{a_1} = 5.5 \times 10^{-3} = \dfrac{[H^+][H_2AsO_4^-]}{[H_3AsO_4]}$
Initial	0.20 M		~0		0	
Equil.	$0.20 - x$		x		x	

$5.5 \times 10^{-3} = \dfrac{x^2}{0.20 - x} \approx \dfrac{x^2}{0.20}$, $x = 3.3 \times 10^{-2}$ M; assumption fails the 5% rule.

Solving by the method of successive approximations:

$5.5 \times 10^{-3} = x^2/(0.20 - 0.033)$, $x = 3.03 \times 10^{-2}$ (carrying an extra significant figure)

$5.5 \times 10^{-3} = x^2/(0.20 - 0.0303)$, $x = 3.06 \times 10^{-2}$

$5.5 \times 10^{-3} = x^2/(0.20 - 0.0306)$, $x = 3.05 \times 10^{-2}$ (consistent answer)

$[H^+] = [H_2AsO_4^-] = 3.05 \times 10^{-2} = 3.1 \times 10^{-2}$ M; $[H_3AsO_4] = 0.20 - 0.031 = 0.17$ M

Because $K_{a_2} = \dfrac{[H^+][HAsO_4^{2-}]}{[H_2AsO_4^-]} = 1.7 \times 10^{-7}$ is much smaller than the K_{a_1} value, very little of $H_2AsO_4^-$ (and $HAsO_4^{2-}$) dissociates as compared to H_3AsO_4. Therefore, $[H^+]$ and $[H_2AsO_4^-]$ will not change significantly by the K_{a_2} reaction. Using the previously calculated concentrations of H^+ and $H_2AsO_4^-$ to calculate the concentration of $HAsO_4^{2-}$:

$1.7 \times 10^{-7} = \dfrac{(3.1 \times 10^{-2})[HAsO_4^{2-}]}{3.1 \times 10^{-2}}$, $[HAsO_4^{2-}] = 1.7 \times 10^{-7}$ M

The assumption that the K_{a_2} reaction does not change $[H^+]$ and $[H_2AsO_4^-]$ is good. We repeat the process using K_{a_3} to get $[AsO_4^{3-}]$.

$$K_{a_3} = 5.1 \times 10^{-12} = \frac{[H^+][AsO_4^{3-}]}{[HAsO_4^{2-}]} = \frac{(3.1 \times 10^{-2})[AsO_4^{3-}]}{1.7 \times 10^{-7}}$$

$[AsO_4^{3-}] = 2.8 \times 10^{-17}$; assumption good.

So in 0.20 M analytical concentration of H_3AsO_4:

$[H_3AsO_4] = 0.17$ M; $[H^+] = [H_2AsO_4^-] = 3.1 \times 10^{-2}$ M; $[HAsO_4^{2-}] = 1.7 \times 10^{-7}$ M

$[AsO_4^{3-}] = 2.8 \times 10^{-17}$ M; $[OH^-] = K_w/[H^+] = 3.2 \times 10^{-13}$ M

109. Because K_{a_2} for H_2S is so small, we can ignore the H^+ contribution from the K_{a_2} reaction.

	H_2S	$\rightleftharpoons$	H^+		HS^-	$K_{a_1} = 1.0 \times 10^{-7}$
Initial	0.10 M		~0		0	
Equil.	$0.10 - x$		x		x	

$$K_{a_1} = 1.0 \times 10^{-7} = \frac{x^2}{0.10 - x} \approx \frac{x^2}{0.10}, \quad x = [H^+] = 1.0 \times 10^{-4}; \text{ assumptions good.}$$

$$pH = -\log(1.0 \times 10^{-4}) = 4.00$$

Use the K_{a_2} reaction to determine $[S^{2-}]$.

	HS^-	$\rightleftharpoons$	H^+	+	S^{2-}
Initial	1.0×10^{-4} M		1.0×10^{-4} M		0
Equil.	$1.0 \times 10^{-4} - x$		$1.0 \times 10^{-4} + x$		x

$$K_{a_2} = 1.0 \times 10^{-19} = \frac{(1.0 \times 10^{-4} + x)x}{(1.0 \times 10^{-4} - x)} \approx \frac{(1.0 \times 10^{-4})x}{1.0 \times 10^{-4}}$$

$x = [S^{2-}] = 1.0 \times 10^{-19}$ M; assumptions good.

110. The relevant reactions are:

$$H_2CO_3 \rightleftharpoons H^+ + HCO_3^- \quad K_{a_1} = 4.3 \times 10^{-7}; \quad HCO_3^- \rightleftharpoons H^+ + CO_3^{2-} \quad K_{a_2} = 5.6 \times 10^{-11}$$

Initially, we deal only with the first reaction (since $K_{a_1} >> K_{a_2}$), and then let those results control values of concentrations in the second reaction.

	H_2CO_3	$\rightleftharpoons$	H^+	+	HCO_3^-
Initial	0.010 M		~0		0
Equil.	$0.010 - x$		x		x

$$K_{a_1} = 4.3 \times 10^{-7} = \frac{[H^+][HCO_3^-]}{[H_2CO_3]} = \frac{x^2}{0.010 - x} \approx \frac{x^2}{0.010}$$

$x = 6.6 \times 10^{-5}$ $M = [H^+] = [HCO_3^-]$; assumptions good.

	HCO_3^-	$\rightleftharpoons$	H^+	+	CO_3^{2-}
Initial	6.6×10^{-5} M		6.6×10^{-5} M		0
Equil.	$6.6 \times 10^{-5} - y$		$6.6 \times 10^{-5} + y$		y

If y is small, then $[H^+] = [HCO_3^-]$, and $K_{a_2} = 5.6 \times 10^{-11} = \dfrac{[H^+][CO_3^{2-}]}{[HCO_3^-]} \approx y$.

$y = [CO_3^{2-}] = 5.6 \times 10^{-11}\ M;$ assumptions good.

The amount of H^+ from the second dissociation is $5.6 \times 10^{-11}\ M$ or:

$$\frac{5.6 \times 10^{-11}}{6.6 \times 10^{-5}} \times 100 = 8.5 \times 10^{-5}\%$$

This result justifies our treating the equilibria separately. If the second dissociation contributed a significant amount of H^+, we would have to treat both equilibria simultaneously.

The reaction that occurs when acid is added to a solution of HCO_3^- is:

$$HCO_3^-(aq) + H^+(aq) \rightarrow H_2CO_3(aq) \rightarrow H_2O(l) + CO_2(g)$$

The bubbles are $CO_2(g)$ and are formed by the breakdown of unstable H_2CO_3 molecules. We should write $H_2O(l) + CO_2(aq)$ or $CO_2(aq)$ for what we call carbonic acid. It is for convenience, however, that we write $H_2CO_3(aq)$.

111. The dominant H^+ producer is the strong acid H_2SO_4. A 2.0 M H_2SO_4 solution produces 2.0 M HSO_4^- and 2.0 M H^+. However, HSO_4^- is a weak acid that could also add H^+ to the solution.

	HSO_4^-	$\rightleftharpoons$	H^+	+	SO_4^{2-}
Initial	2.0 M		2.0 M		0
	x mol/L HSO_4^- dissociates to reach equilibrium				
Change	$-x$	$\rightarrow$	$+x$		$+x$
Equil.	$2.0 - x$		$2.0 + x$		x

$$K_{a_2} = 1.2 \times 10^{-2} = \frac{[H^+][SO_4^{2-}]}{[HSO_4^-]} = \frac{(2.0 + x)x}{2.0 - x} \approx \frac{2.0(x)}{2.0},\quad x = 1.2 \times 10^{-2}\ M$$

Because x is 0.60% of 2.0, the assumption is valid by the 5% rule. The amount of additional H^+ from HSO_4^- is $1.2 \times 10^{-2}\ M$. The total amount of H^+ present is:

$$[H^+] = 2.0 + (1.2 \times 10^{-2}) = 2.0\ M;\ \text{pH} = -\log(2.0) = -0.30$$

Note: In this problem, H^+ from HSO_4^- could have been ignored. However, this is not usually the case in more dilute solutions of H_2SO_4.

112. For H_2SO_4, the first dissociation occurs to completion. The hydrogen sulfate ion (HSO_4^-) is a weak acid with $K_{a_2} = 1.2 \times 10^{-2}$. We will consider this equilibrium for additional H^+ production:

	HSO_4^-	$\rightleftharpoons$	H^+	+	SO_4^{2-}
Initial	0.0050 M		0.0050 M		0
	x mol/L HSO_4^- dissociates to reach equilibrium				
Change	$-x$	$\rightarrow$	$+x$		$+x$
Equil.	$0.0050 - x$		$0.0050 + x$		x

$$K_{a_2} = 0.012 = \frac{(0.0050 + x)x}{0.0050 - x} \approx x, \ x = 0.012; \text{ assumption is horrible (240\% error).}$$

Using the quadratic formula:

$$6.0 \times 10^{-5} - (0.012)x = x^2 + (0.0050)x, \ x^2 + (0.017)x - 6.0 \times 10^{-5} = 0$$

$$x = \frac{-0.017 \pm (2.9 \times 10^{-4} + 2.4 \times 10^{-4})^{1/2}}{2} = \frac{-0.017 \pm 0.023}{2}, \ x = 3.0 \times 10^{-3}\,M$$

$$[H^+] = 0.0050 + x = 0.0050 + 0.0030 = 0.0080\,M; \ pH = 2.10$$

Note: We had to consider both H_2SO_4 and HSO_4^- for H^+ production in this problem.

Acid-Base Properties of Salts

113. One difficult aspect of acid-base chemistry is recognizing what types of species are present in solution, that is, whether a species is a strong acid, strong base, weak acid, weak base, or a neutral species. Below are some ideas and generalizations to keep in mind that will help in recognizing types of species present.

a. Memorize the following strong acids: HCl, HBr, HI, HNO_3, $HClO_4$, and H_2SO_4

b. Memorize the following strong bases: LiOH, NaOH, KOH, RbOH, CsOH, $Ca(OH)_2$, $Sr(OH)_2$, and $Ba(OH)_2$

c. Weak acids have a K_a value of less than 1 but greater than K_w. Some weak acids are listed in Table 14.2 of the text. Weak bases have a K_b value of less than 1 but greater than K_w. Some weak bases are listed in Table 14.3 of the text.

d. Conjugate bases of weak acids are weak bases; that is, all have a K_b value of less than 1 but greater than K_w. Some examples of these are the conjugate bases of the weak acids listed in Table 14.2 of the text.

e. Conjugate acids of weak bases are weak acids; that is, all have a K_a value of less than 1 but greater than K_w. Some examples of these are the conjugate acids of the weak bases listed in Table 14.3 of the text.

f. Alkali metal ions (Li^+, Na^+, K^+, Rb^+, Cs^+) and heavier alkaline earth metal ions (Ca^{2+}, Sr^{2+}, Ba^{2+}) have no acidic or basic properties in water.

g. All conjugate bases of strong acids (Cl^-, Br^-, I^-, NO_3^-, ClO_4^-, HSO_4^-) have no basic properties in water ($K_b << K_w$), and only HSO_4^- has any acidic properties in water.

Let's apply these ideas to this problem to see what type of species are present. The letters in parenthesis is(are) the generalization(s) above that identifies the species.

KOH: Strong base (b)

KNO_3: Neutral; K^+ and NO_3^- have no acidic/basic properties (f and g).

KCN: CN^- is a weak base, $K_b = K_w/K_{a,\,HCN} = 1.0 \times 10^{-14}/6.2 \times 10^{-10} = 1.6 \times 10^{-5}$ (c and d). Ignore K^+ (f).

NH_4Cl: NH_4^+ is a weak acid, $K_a = 5.6 \times 10^{-10}$ (c and e). Ignore Cl^- (g).

HCl: Strong acid (a)

The most acidic solution will be the strong acid solution, with the weak acid solution less acidic. The most basic solution will be the strong base solution, with the weak base solution less basic. The KNO_3 solution will be neutral at pH = 7.00.

Most acidic → most basic: HCl > NH_4Cl > KNO_3 > KCN > KOH

114. See Exercise 113 for some generalizations on acid-base properties of salts. The letters in parenthesis is(are) the generalization(s) listed in Exercise 113 that identifies the species.

$CaBr_2$: Neutral; Ca^{2+} and Br^- have no acidic/basic properties (f and g).

KNO_2: NO_2^- is a weak base, $K_b = K_w/K_{a,\ HNO_2} = (1.0 \times 10^{-14})/(4.0 \times 10^{-4}) = 2.5 \times 10^{-11}$ (c and d). Ignore K^+ (f).

$HClO_4$: Strong acid (a)

HNO_2: Weak acid, $K_a = 4.0 \times 10^{-4}$ (c)

$HONH_3ClO_4$: $HONH_3^+$ is a weak acid, $K_a = K_w/K_{b,\ HONH_2} = (1.0 \times 10^{-14})/(1.1 \times 10^{-8}) = 9.1 \times 10^{-7}$ (c and e). Ignore ClO_4^- (g). Note that HNO_2 has a larger K_a value than $HONH_3^+$, so HNO_2 is a stronger weak acid than $HONH_3^+$.

Using the information above (identity and the K_a or K_b values), the ordering is:

Most acidic → most basic: $HClO_4$ > HNO_2 > $HONH_3ClO_4$ > $CaBr_2$ > KNO_2

115. From the K_a values, acetic acid is a stronger acid than hypochlorous acid. Conversely, the conjugate base of acetic acid, $C_2H_3O_2^-$, will be a weaker base than the conjugate base of hypochlorous acid, OCl^-. Thus the hypochlorite ion, OCl^-, is a stronger base than the acetate ion, $C_2H_3O_2^-$. In general, the stronger the acid, the weaker the conjugate base. This statement comes from the relationship $K_w = K_a \times K_b$, which holds for all conjugate acid-base pairs.

116. Because NH_3 is a weaker base (smaller K_b value) than CH_3NH_2, the conjugate acid of NH_3 will be a stronger acid than the conjugate acid of CH_3NH_2. Thus NH_4^+ is a stronger acid than $CH_3NH_3^+$.

117. a. KCl is a soluble ionic compound that dissolves in water to produce $K^+(aq)$ and $Cl^-(aq)$. K^+ (like the other alkali metal cations) has no acidic or basic properties. Cl^- is the conjugate base of the strong acid HCl. Cl^- has no basic (or acidic) properties. Therefore, a solution of KCl will be neutral because neither of the ions has any acidic or basic properties. The 1.0 *M* KCl solution has $[H^+] = [OH^-] = 1.0 \times 10^{-7}$ *M* and pH = pOH = 7.00.

b. $KC_2H_3O_2$ is also a soluble ionic compound that dissolves in water to produce $K^+(aq)$ and $C_2H_3O_2^-(aq)$. The difference between the KCl solution and the $KC_2H_3O_2$ solution is that $C_2H_3O_2^-$ does have basic properties in water, unlike Cl^-. $C_2H_3O_2^-$ is the conjugate base of the weak acid $HC_2H_3O_2$, and as is true for all conjugate bases of weak acids, $C_2H_3O_2^-$

is a weak base in water. We must solve an equilibrium problem in order to determine the amount of OH^- this weak base produces in water.

$$C_2H_3O_2^- + H_2O \rightleftharpoons HC_2H_3O_2 + OH^- \quad K_b = \frac{K_w}{K_{a,\,C_2H_3O_2}} = \frac{1.0 \times 10^{-14}}{1.8 \times 10^{-5}}$$

	$C_2H_3O_2^-$		$HC_2H_3O_2$	OH^-	
Initial	1.0 M		0	~0	$K_b = 5.6 \times 10^{-10}$
	x mol/L of $C_2H_3O_2^-$ reacts with H_2O to reach equilibrium				
Change	$-x$	$\rightarrow$	$+x$	$+x$	
Equil.	$1.0 - x$		x	x	

$$K_b = 5.6 \times 10^{-10} = \frac{[HC_2H_3O_2][OH^-]}{[C_2H_3O_2^-]}, \quad 5.6 \times 10^{-10} = \frac{x^2}{1.0 - x} \approx \frac{x^2}{1.0}$$

$x = [OH^-] = 2.4 \times 10^{-5}$ M ; assumptions good

pOH = 4.62; pH = 14.00 – 4.62 = 9.38; $[H^+] = 10^{-9.38} = 4.2 \times 10^{-10}$ M

118. $C_2H_5NH_3Cl \rightarrow C_2H_5NH_3^+ + Cl^-$; $C_2H_5NH_3^+$ is the conjugate acid of the weak base $C_2H_5NH_2$ ($K_b = 5.6 \times 10^{-4}$). As is true for all conjugate acids of weak bases, $C_2H_5NH_3^+$ is a weak acid. Cl^- has no basic (or acidic) properties. Ignore Cl^-. Solving the weak acid problem:

$$C_2H_5NH_3^+ \rightleftharpoons C_2H_5NH_2 + H^+ \quad K_a = K_w/5.6 \times 10^{-4} = 1.8 \times 10^{-11}$$

	$C_2H_5NH_3^+$		$C_2H_5NH_2$	H^+
Initial	0.25 M		0	~0
	x mol/L $C_2H_5NH_3^+$ dissociates to reach equilibrium			
Change	$-x$	$\rightarrow$	$+x$	$+x$
Equil.	$0.25 - x$		x	x

$$K_a = 1.8 \times 10^{-11} = \frac{[C_2H_5NH_2][H^+]}{[C_2H_5NH_3^+]} = \frac{x^2}{0.25 - x} \approx \frac{x^2}{0.25} \quad \text{(assuming } x \ll 0.25\text{)}$$

$x = [H^+] = 2.1 \times 10^{-6}$ M; pH = 5.68; assumptions good.

$[C_2H_5NH_2] = [H^+] = 2.1 \times 10^{-6}$ M; $[C_2H_5NH_3^+] = 0.25$ M; $[Cl^-] = 0.25$ M

$[OH^-] = K_w/[H^+] = 4.8 \times 10^{-9}$ M

119. a. $CH_3NH_3Cl \rightarrow CH_3NH_3^+ + Cl^-$: $CH_3NH_3^+$ is a weak acid. Cl^- is the conjugate base of a strong acid. Cl^- has no basic (or acidic) properties.

$$CH_3NH_3^+ \rightleftharpoons CH_3NH_2 + H^+ \quad K_a = \frac{[CH_3NH_2][H^+]}{[CH_3NH_3^+]} = \frac{K_w}{K_b} = \frac{1.00 \times 10^{-14}}{4.38 \times 10^{-4}}$$

$$= 2.28 \times 10^{-11}$$

$$CH_3NH_3^+ \rightleftharpoons CH_3NH_2 + H^+$$

	$CH_3NH_3^+$	$\rightleftharpoons$	CH_3NH_2	+	H^+
Initial	0.10 *M*		0		~0
	x mol/L $CH_3NH_3^+$ dissociates to reach equilibrium				
Change	$-x$	→	$+x$		$+x$
Equil.	$0.10 - x$		x		x

$$K_a = 2.28 \times 10^{-11} = \frac{x^2}{0.10 - x} \approx \frac{x^2}{0.10} \quad \text{(assuming } x << 0.10\text{)}$$

$x = [H^+] = 1.5 \times 10^{-6}$ *M*; pH = 5.82; assumptions good.

b. $NaCN \rightarrow Na^+ + CN^-$: CN^- is a weak base. Na^+ has no acidic (or basic) properties.

$$CN^- + H_2O \rightleftharpoons HCN + OH^- \qquad K_b = \frac{K_w}{K_a} = \frac{1.0 \times 10^{-14}}{6.2 \times 10^{-10}}$$

	CN^- + H_2O	$\rightleftharpoons$	HCN	+	OH^-	
Initial	0.050 *M*		0		~0	$K_b = 1.6 \times 10^{-5}$
	x mol/L CN^- reacts with H_2O to reach equilibrium					
Change	$-x$	→	$+x$		$+x$	
Equil.	$0.050 - x$		x		x	

$$K_b = 1.6 \times 10^{-5} = \frac{[HCN][OH^-]}{[CN^-]} = \frac{x^2}{0.050 - x} \approx \frac{x^2}{0.050}$$

$x = [OH^-] = 8.9 \times 10^{-4}$ *M*; pOH = 3.05; pH = 10.95; assumptions good.

120. a. $KNO_2 \rightarrow K^+ + NO_2^-$: NO_2^- is a weak base. Ignore K^+.

$$NO_2^- + H_2O \rightleftharpoons HNO_2 + OH^- \qquad K_b = \frac{K_w}{K_a} = \frac{1.0 \times 10^{-14}}{4.0 \times 10^{-4}} = 2.5 \times 10^{-11}$$

	NO_2^-	HNO_2	OH^-
Initial	0.12 *M*	0	~0
Equil.	$0.12 - x$	x	x

$$K_b = 2.5 \times 10^{-11} = \frac{[OH^-][HNO_2]}{[NO_2^-]} = \frac{x^2}{0.12 - x} \approx \frac{x^2}{0.12}$$

$x = [OH^-] = 1.7 \times 10^{-6}$ *M*; pOH = 5.77; pH = 8.23; assumptions good.

b. $NaOCl \rightarrow Na^+ + OCl^-$: OCl^- is a weak base. Ignore Na^+.

$$OCl^- + H_2O \rightleftharpoons HOCl + OH^- \qquad K_b = \frac{K_w}{K_a} = \frac{1.0 \times 10^{-14}}{3.5 \times 10^{-8}} = 2.9 \times 10^{-7}$$

	OCl^-	HOCl	OH^-
Initial	0.45 *M*	0	~0
Equil.	$0.45 - x$	x	x

$$K_b = 2.9 \times 10^{-7} = \frac{[HOCl][OH^-]}{[OCl^-]} = \frac{x^2}{0.45 - x} \approx \frac{x^2}{0.45}$$

$x = [OH^-] = 3.6 \times 10^{-4}$ *M*; pOH = 3.44; pH = 10.56; assumptions good.

c. $NH_4ClO_4 \rightarrow NH_4^+ + ClO_4^-$: NH_4^+ is a weak acid. ClO_4^- is the conjugate base of a strong acid. ClO_4^- has no basic (or acidic) properties.

$$NH_4^+ \rightleftharpoons NH_3 + H^+ \qquad K_a = \frac{K_w}{K_b} = \frac{1.0 \times 10^{-14}}{1.8 \times 10^{-5}} = 5.6 \times 10^{-10}$$

	NH_4^+	NH_3	H^+
Initial	0.40 *M*	0	~0
Equil.	$0.40 - x$	x	x

$$K_a = 5.6 \times 10^{-10} = \frac{[NH_3][H^+]}{[NH_4^+]} = \frac{x^2}{0.40 - x} \approx \frac{x^2}{0.40}$$

$x = [H^+] = 1.5 \times 10^{-5}$ *M*; pH = 4.82; assumptions good.

121. $NaN_3 \rightarrow Na^+ + N_3^-$; azide ($N_3^-$) is a weak base because it is the conjugate base of a weak acid. All conjugate bases of weak acids are weak bases ($K_w < K_b < 1$). Ignore Na^+.

$$N_3^- + H_2O \rightleftharpoons HN_3 + OH^- \qquad K_b = \frac{K_w}{K_a} = \frac{1.0 \times 10^{-14}}{1.9 \times 10^{-5}} = 5.3 \times 10^{-10}$$

	N_3^-		HN_3	OH^-
Initial	0.010 *M*		0	~0
	x mol/L of N_3^- reacts with H_2O to reach equilibrium			
Change	$-x$	$\rightarrow$	$+x$	$+x$
Equil.	$0.010 - x$		x	x

$$K_b = \frac{[HN_3][OH^-]}{[N_3^-]},\quad 5.3 \times 10^{-10} = \frac{x^2}{0.010 - x} \approx \frac{x^2}{0.010} \quad \text{(assuming } x \ll 0.010\text{)}$$

$x = [OH^-] = 2.3 \times 10^{-6}$ *M*; $[H^+] = \dfrac{1.0 \times 10^{-14}}{2.3 \times 10^{-6}} = 4.3 \times 10^{-9}$ *M*; assumptions good.

$[HN_3] = [OH^-] = 2.3 \times 10^{-6}$ *M*; $[Na^+] = 0.010$ *M*; $[N_3^-] = 0.010 - 2.3 \times 10^{-6} = 0.010$ *M*

122. $$\frac{30.0\,\text{mg papH}^+\text{Cl}^-}{\text{mL soln}} \times \frac{1000\,\text{mL}}{\text{L}} \times \frac{1\,\text{g}}{1000\,\text{mg}} \times \frac{1\,\text{mol papH}^+\text{Cl}^-}{378.85\,\text{g}} \times \frac{1\,\text{mol papH}^+}{\text{mol papH}^+\text{Cl}^-} = 0.0792\ M$$

$$\text{papH}^+ \rightleftharpoons \text{pap} + H^+ \qquad K_a = \frac{K_w}{K_{b,\,pap}} = \frac{2.1 \times 10^{-14}}{8.33 \times 10^{-9}} = 2.5 \times 10^{-6}$$

	papH$^+$	pap	H^+
Initial	0.0792 *M*	0	~0
Equil.	$0.0792 - x$	x	x

$$K_a = 2.5 \times 10^{-6} = \frac{x^2}{0.0792 - x} \approx \frac{x^2}{0.0792},\quad x = [H^+] = 4.4 \times 10^{-4}\ M$$

pH = $-\log(4.4 \times 10^{-4}) = 3.36$; assumptions good.

123. All these salts contain Na^+, which has no acidic/basic properties, and a conjugate base of a weak acid (except for NaCl, where Cl^- is a neutral species). All conjugate bases of weak acids are weak bases since K_b values for these species are between K_w and 1. To identify the species, we will use the data given to determine the K_b value for the weak conjugate base. From the K_b value and data in Table 14.2 of the text, we can identify the conjugate base present by calculating the K_a value for the weak acid. We will use A^- as an abbreviation for the weak conjugate base.

	A^- + H_2O	$\rightleftharpoons$	HA	+	OH^-
Initial	0.100 mol/1.00 L		0		~0
	x mol/L A^- reacts with H_2O to reach equilibrium				
Change	$-x$	→	$+x$		$+x$
Equil.	$0.100 - x$		x		x

$$K_b = \frac{[HA][OH^-]}{[A^-]} = \frac{x^2}{0.100 - x}; \text{ from the problem, pH} = 8.07:$$

$$pOH = 14.00 - 8.07 = 5.93;\ [OH^-] = x = 10^{-5.93} = 1.2 \times 10^{-6}\ M$$

$$K_b = \frac{(1.2 \times 10^{-6})^2}{0.100 - (1.2 \times 10^{-6})} = 1.4 \times 10^{-11} = K_b \text{ value for the conjugate base of a weak acid.}$$

The K_a value for the weak acid equals K_w/K_b: $K_a = \dfrac{1.0 \times 10^{-14}}{1.4 \times 10^{-11}} = 7.1 \times 10^{-4}$

From Table 14.2 of the text, this K_a value is closest to HF. Therefore, the unknown salt is NaF.

124. $BHCl \rightarrow BH^+ + Cl^-$; Cl^- is the conjugate base of the strong acid HCl, so Cl^- has no acidic/basic properties. BH^+ is a weak acid because it is the conjugate acid of a weak base B. Determining the K_a value for BH^+:

	BH^+	$\rightleftharpoons$	B	+	H^+
Initial	0.10 M		0		~0
	x mol/L BH^+ dissociates to reach equilibrium				
Change	$-x$	→	$+x$		$+x$
Equil.	$0.10 - x$		x		x

$$K_a = \frac{[B][H^+]}{[BH^+]} = \frac{x^2}{0.10 - x}; \text{ from the problem, pH} = 5.82:$$

$$[H^+] = x = 10^{-5.82} = 1.5 \times 10^{-6}\ M;\ K_a = \frac{(1.5 \times 10^{-6})^2}{0.10 - (1.5 \times 10^{-6})} = 2.3 \times 10^{-11}$$

K_b for the base B = $K_w/K_a = (1.0 \times 10^{-14})/(2.3 \times 10^{-11}) = 4.3 \times 10^{-4}$.

From Table 14.3 of the text, this K_b value is closest to CH_3NH_2, so the unknown salt is CH_3NH_3Cl.

125. B^- is a weak base. Use the weak base data to determine K_b for B^-.

	B^- + H_2O	⇌	HB	+	OH^-
Initial	0.050 *M*		0		~0
Equil.	0.050 – *x*		*x*		*x*

From pH = 9.00: pOH = 5.00, $[OH^-] = 10^{-5.00} = 1.0 \times 10^{-5}\ M = x$.

$$K_b = \frac{[HB][OH^-]}{[B^-]} = \frac{x^2}{0.050 - x} = \frac{(1.0 \times 10^{-5})^2}{0.050 - (1.0 \times 10^{-5})} = 2.0 \times 10^{-9}$$

Because B^- is a weak base, HB will be a weak acid. Solve the weak acid problem.

	HB	⇌	H^+	+	B^-
Initial	0.010 *M*		~0		0
Equil.	0.010 – *x*		*x*		*x*

$$K_a = \frac{K_w}{K_b} = \frac{1.0 \times 10^{-14}}{2.0 \times 10^{-9}},\ 5.0 \times 10^{-6} = \frac{x^2}{0.010 - x} \approx \frac{x^2}{0.010}$$

$x = [H^+] = 2.2 \times 10^{-4}\ M$; pH = 3.66; assumptions good.

126. From the pH, $C_7H_4ClO_2^-$ is a weak base. Use the weak base data to determine K_b for $C_7H_4ClO_2^-$ (which we will abbreviate as CB^-).

	CB^- + H_2O	⇌	HCB	+	OH^-
Initial	0.20 *M*		0		~0
Equil.	0.20 – *x*		*x*		*x*

Because pH = 8.65, pOH = 5.35 and $[OH^-] = 10^{-5.35} = 4.5 \times 10^{-6} = x$.

$$K_b = \frac{[HCB][OH^-]}{[CB^-]} = \frac{x^2}{0.20 - x} = \frac{(4.5 \times 10^{-6})^2}{0.20 - (4.5 \times 10^{-6})} = 1.0 \times 10^{-10}$$

Because CB^- is a weak base, HCB, chlorobenzoic acid, is a weak acid. Solving the weak acid problem:

	HCB	⇌	H^+	+	CB^-
Initial	0.20 *M*		~0		0
Equil.	0.20 – *x*		*x*		*x*

$$K_a = \frac{K_w}{K_b} = \frac{1.0 \times 10^{-14}}{1.0 \times 10^{-10}},\ 1.0 \times 10^{-4} = \frac{x^2}{0.20 - x} \approx \frac{x^2}{0.20}$$

$x = [H^+] = 4.5 \times 10^{-3}\ M$; pH = 2.35; assumptions good.

127. Major species present: $Al(H_2O)_6^{3+}$ ($K_a = 1.4 \times 10^{-5}$), NO_3^- (neutral), and H_2O ($K_w = 1.0 \times 10^{-14}$); $Al(H_2O)_6^{3+}$ is a stronger acid than water, so it will be the dominant H^+ producer.

$$Al(H_2O)_6^{3+} \rightleftharpoons Al(H_2O)_5(OH)^{2+} + H^+$$

	$Al(H_2O)_6^{3+}$		$Al(H_2O)_5(OH)^{2+}$	H^+
Initial	0.050 *M*		0	~0
	x mol/L $Al(H_2O)_6^{3+}$ dissociates to reach equilibrium			
Change	$-x$	→	$+x$	$+x$
Equil.	$0.050 - x$		x	x

$$K_a = 1.4 \times 10^{-5} = \frac{[Al(H_2O)_5(OH)^{2+}][H^+]}{[Al(H_2O)_6^{3+}]} = \frac{x^2}{0.050 - x} \approx \frac{x^2}{0.050}$$

$x = 8.4 \times 10^{-4}\ M = [H^+]$; pH = $-\log(8.4 \times 10^{-4}) = 3.08$; assumptions good.

128. Major species: $Co(H_2O)_6^{3+}$ ($K_a = 1.0 \times 10^{-5}$), Cl^- (neutral), and H_2O ($K_w = 1.0 \times 10^{-14}$); $Co(H_2O)_6^{3+}$ will determine the pH because it is a stronger acid than water. Solving the weak acid problem in the usual manner:

$$Co(H_2O)_6^{3+} \rightleftharpoons Co(H_2O)_5(OH)^{2+} + H^+ \quad K_a = 1.0 \times 10^{-5}$$

	$Co(H_2O)_6^{3+}$	$Co(H_2O)_5(OH)^{2+}$	H^+
Initial	0.10 *M*	0	~0
Equil.	$0.10 - x$	x	x

$$K_a = 1.0 \times 10^{-5} = \frac{x^2}{0.10 - x} \approx \frac{x^2}{0.10}, \quad x = [H^+] = 1.0 \times 10^{-3}\ M$$

pH = $-\log(1.0 \times 10^{-3}) = 3.00$; assumptions good.

129. Reference Table 14.6 of the text and the solution to Exercise 113 for some generalizations on acid-base properties of salts.

a. $NaNO_3 \rightarrow Na^+ + NO_3^-$ neutral; neither species has any acidic/basic properties.

b. $NaNO_2 \rightarrow Na^+ + NO_2^-$ basic; NO_2^- is a weak base, and Na^+ has no effect on pH.

$$NO_2^- + H_2O \rightleftharpoons HNO_2 + OH^- \quad K_b = \frac{K_w}{K_{a,\,HNO_2}} = \frac{1.0 \times 10^{-14}}{4.0 \times 10^{-4}} = 2.5 \times 10^{-11}$$

c. $C_5H_5NHClO_4 \rightarrow C_5H_5NH^+ + ClO_4^-$ acidic; $C_5H_5NH^+$ is a weak acid, and ClO_4^- has no effect on pH.

$$C_5H_5NH^+ \rightleftharpoons H^+ + C_5H_5N \quad K_a = \frac{K_w}{K_{b,\,C_5H_5N}} = \frac{1.0 \times 10^{-14}}{1.7 \times 10^{-9}} = 5.9 \times 10^{-6}$$

d. $NH_4NO_2 \rightarrow NH_4^+ + NO_2^-$ acidic; NH_4^+ is a weak acid ($K_a = 5.6 \times 10^{-10}$), and NO_2^- is a weak base ($K_b = 2.5 \times 10^{-11}$). Because $K_{a,\,NH_4^+} > K_{b,\,NO_2^-}$, the solution is acidic.

$$NH_4^+ \rightleftharpoons H^+ + NH_3 \quad K_a = 5.6 \times 10^{-10}; \quad NO_2^- + H_2O \rightleftharpoons HNO_2 + OH^- \quad K_b = 2.5 \times 10^{-11}$$

e. $KOCl \rightarrow K^+ + OCl^-$ basic; OCl^- is a weak base, and K^+ has no effect on pH.

$$OCl^- + H_2O \rightleftharpoons HOCl + OH^- \quad K_b = \frac{K_w}{K_{a,\,HOCl}} = \frac{1.0 \times 10^{-14}}{3.5 \times 10^{-8}} = 2.9 \times 10^{-7}$$

f. $NH_4OCl \rightarrow NH_4^+ + OCl^-$ basic; NH_4^+ is a weak acid, and OCl^- is a weak base. Because $K_{b,\,OCl^-} > K_{a,\,NH_4^+}$, the solution is basic.

$$NH_4^+ \rightleftharpoons NH_3 + H^+ \quad K_a = 5.6 \times 10^{-10}; \quad OCl^- + H_2O \rightleftharpoons HOCl + OH^- \quad K_b = 2.9 \times 10^{-7}$$

130. a. $Sr(NO_3)_2 \rightarrow Sr^{2+} + 2\ NO_3^-$ neutral; Sr^{2+} and NO_3^- have no effect on pH.

b. $NH_4C_2H_3O_2 \rightarrow NH_4^+ + C_2H_3O_2^-$ neutral; NH_4^+ is a weak acid, and $C_2H_3O_2^-$ is a weak base. Because $K_{a,\,NH_4^+} = K_{b,\,C_2H_3O_2^-}$, the solution will be neutral (pH = 7.00).

$$NH_4^+ \rightleftharpoons NH_3 + H^+ \quad K_a = \frac{K_w}{K_{b,\,NH_3}} = \frac{1.0 \times 10^{-14}}{1.8 \times 10^{-5}} = 5.6 \times 10^{-10}$$

$$C_2H_3O_2^- + H_2O \rightleftharpoons HC_2H_3O_2 + OH^- \quad K_b = \frac{K_w}{K_{a,\,HC_2H_3O_2}} = \frac{1.0 \times 10^{-14}}{1.8 \times 10^{-5}} = 5.6 \times 10^{-10}$$

c. $CH_3NH_3Cl \rightarrow CH_3NH_3^+ + Cl^-$ acidic; $CH_3NH_3^+$ is a weak acid, and Cl^- has no effect on pH. Because only a weak acid is present, the solution will be acidic.

$$CH_3NH_3^+ \rightleftharpoons H^+ + CH_3NH_2 \quad K_a = \frac{K_w}{K_{b,\,CH_3NH_2}} = \frac{1.00 \times 10^{-14}}{4.38 \times 10^{-4}} = 2.28 \times 10^{-11}$$

d. $C_6H_5NH_3ClO_2 \rightarrow C_6H_5NH_3^+ + ClO_2^-$ acidic; $C_6H_5NH_3^+$ is a weak acid, and ClO_2^- is a very weak base. Because $K_{a,\,C_6H_5NH_3^+} > K_{b,\,ClO_2^-}$, the solution is acidic.

$$C_6H_5NH_3^+ \rightleftharpoons H^+ + C_6H_5NH_2 \quad K_a = \frac{K_w}{K_{b,\,C_6H_5NH_2}} = \frac{1.0 \times 10^{-14}}{3.8 \times 10^{-10}} = 2.6 \times 10^{-5}$$

$$ClO_2^- + H_2O \rightleftharpoons HClO_2 + OH^- \quad K_b = \frac{K_w}{K_{a,\,HClO_2}} = \frac{1.0 \times 10^{-14}}{1.2 \times 10^{-2}} = 8.3 \times 10^{-13}$$

e. $NH_4F \rightarrow NH_4^+ + F^-$ acidic; NH_4^+ is a weak acid, and F^- is a weak base. Because $K_{a,\,NH_4^+} > K_{b,\,F^-}$, the solution is acidic.

$$NH_4^+ \rightleftharpoons H^+ + NH_3 \quad K_a = 5.6 \times 10^{-10}; \quad F^- + H_2O \rightleftharpoons HF + OH^- \quad K_b = 1.4 \times 10^{-11}$$

f. $CH_3NH_3CN \rightarrow CH_3NH_3^+ + CN^-$ basic; $CH_3NH_3^+$ is a weak acid, and CN^- is a weak base. Because $K_{b,\,CN^-} > K_{a,\,CH_3NH_3^+}$, the solution is basic.

$$CH_3NH_3^+ \rightleftharpoons H^+ + CH_3NH_2 \quad K_a = 2.28 \times 10^{-11}$$

$$CN^- + H_2O \rightleftharpoons HCN + OH^- \quad K_b = \frac{K_w}{K_{a,\,HCN}} = \frac{1.0 \times 10^{-14}}{6.2 \times 10^{-10}} = 1.6 \times 10^{-5}$$

Relationships Between Structure and Strengths of Acids and Bases

131. a. $HIO_3 < HBrO_3$; as the electronegativity of the central atom increases, acid strength increases.

b. $HNO_2 < HNO_3$; as the number of oxygen atoms attached to the central nitrogen atom increases, acid strength increases.

c. $HOI < HOCl$; same reasoning as in a.

d. $H_3PO_3 < H_3PO_4$; same reasoning as in b.

132. a. $BrO_3^- < IO_3^-$; these are the conjugate bases of the acids in Exercise 131a. Since $HBrO_3$ is the stronger acid, the conjugate base of $HBrO_3$ (BrO_3^-) will be the weaker base. IO_3^- will be the stronger base because HIO_3 is the weaker acid.

b. $NO_3^- < NO_2^-$; these are the conjugate bases of the acids in Exercise 131b. Conjugate base strength is inversely related to acid strength.

c. $OCl^- < OI^-$; these are the conjugate bases of the acids in Exercise 131c.

133. a. $H_2O < H_2S < H_2Se$; as the strength of the H–X bond decreases, acid strength increases.

b. $CH_3CO_2H < FCH_2CO_2H < F_2CHCO_2H < F_3CCO_2H$; as the electronegativity of neighboring atoms increases, acid strength increases.

c. $NH_4^+ < HONH_3^+$; same reason as in b.

d. $NH_4^+ < PH_4^+$; same reason as in a.

134. In general, the stronger the acid, the weaker is the conjugate base.

a. $SeH^- < SH^- < OH^-$; these are the conjugate bases of the acids in Exercise 133a. The ordering of the base strength is the opposite of the acids.

b. $PH_3 < NH_3$ (See Exercise 133d.)

c. $HONH_2 < NH_3$ (See Exercise133c.)

135. In general, metal oxides form basic solutions when dissolved in water, and nonmetal oxides form acidic solutions in water.

a. Basic; $CaO(s) + H_2O(l) \rightarrow Ca(OH)_2(aq)$; $Ca(OH)_2$ is a strong base.

b. Acidic; $SO_2(g) + H_2O(l) \rightarrow H_2SO_3(aq)$; H_2SO_3 is a weak diprotic acid.

c. Acidic; $Cl_2O(g) + H_2O(l) \rightarrow 2\ HOCl(aq)$; HOCl is a weak acid.

136. a. Basic; $Li_2O(s) + H_2O(l) \rightarrow 2\ LiOH(aq)$; LiOH is a strong base.

b. Acidic; $CO_2(g) + H_2O(l) \rightarrow H_2CO_3(aq)$; H_2CO_3 is a weak diprotic acid.

c. Basic; $SrO(s) + H_2O(l) \rightarrow Sr(OH)_2(aq)$; $Sr(OH)_2$ is a strong base.

Lewis Acids and Bases

137. A Lewis base is an electron pair donor, and a Lewis acid is an electron pair acceptor.

a. $B(OH)_3$, acid; H_2O, base b. Ag^+, acid; NH_3, base c. BF_3, acid; F^-, base

138. a. Fe^{3+}, acid; H_2O, base b. H_2O, acid; CN^-, base c. HgI_2, acid; I^-, base

139. $Al(OH)_3(s) + 3\ H^+(aq) \rightarrow Al^{3+}(aq) + 3\ H_2O(l)$ (Brønsted-Lowry base, H^+ acceptor)

$Al(OH)_3(s) + OH^-(aq) \rightarrow Al(OH)_4^-(aq)$ (Lewis acid, electron pair acceptor)

140. $Zn(OH)_2(s) + 2\ H^+(aq) \rightarrow Zn^{2+}(aq) + 2\ H_2O(l)$ (Brønsted-Lowry base)

$Zn(OH)_2(s) + 2\ OH^-(aq) \rightarrow Zn(OH)_4^{2-}(aq)$ (Lewis acid)

141. Fe^{3+} should be the stronger Lewis acid. Fe^{3+} is smaller and has a greater positive charge. Because of this, Fe^{3+} will be more strongly attracted to lone pairs of electrons as compared to Fe^{2+}.

142. The Lewis structures for the reactants and products are:

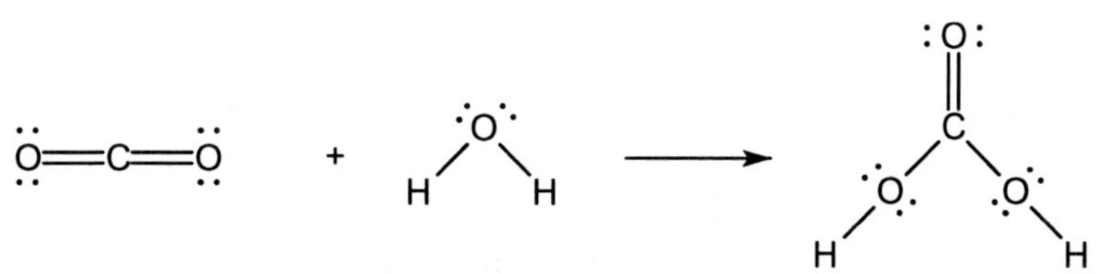

In this reaction, H_2O donates a pair of electrons to carbon in CO_2, which is followed by a proton shift to form H_2CO_3. H_2O is the Lewis base, and CO_2 is the Lewis acid.

Additional Exercises

143. At pH = 2.000, $[H^+] = 10^{-2.000} = 1.00 \times 10^{-2}\ M$

At pH = 4.000, $[H^+] = 10^{-4.000} = 1.00 \times 10^{-4}\ M$

$$\text{Moles } H^+ \text{ present} = 0.0100\ \text{L} \times \frac{0.0100\ \text{mol}\ H^+}{\text{L}} = 1.00 \times 10^{-4}\ \text{mol}\ H^+$$

Let V = total volume of solution at pH = 4.000:

$$1.00 \times 10^{-4}\ \text{mol/L} = \frac{1.00 \times 10^{-4}\ \text{mol}\ H^+}{V},\quad V = 1.00\ \text{L}$$

Volume of water added = 1.00 L – 0.0100 L = 0.99 L = 990 mL

144. Conjugate acid-base pairs differ by an H^+ in the formula. Pairs in parts a, c, and d are conjugate acid-base pairs. For part b, HSO_4^- is the conjugate base of H_2SO_4. In addition, HSO_4^- is the conjugate acid of SO_4^{2-}.

145. The light bulb is bright because a strong electrolyte is present; that is, a solute is present that dissolves to produce a lot of ions in solution. The pH meter value of 4.6 indicates that a weak acid is present. (If a strong acid were present, the pH would be close to zero.) Of the possible substances, only HCl (strong acid), NaOH (strong base), and NH_4Cl are strong electrolytes. Of these three substances, only NH_4Cl contains a weak acid (the HCl solution would have a pH close to zero, and the NaOH solution would have a pH close to 14.0). NH_4Cl dissociates into NH_4^+ and Cl^- ions when dissolved in water. Cl^- is the conjugate base of a strong acid, so it has no basic (or acidic properties) in water. NH_4^+, however, is the conjugate acid of the weak base NH_3, so NH_4^+ is a weak acid and would produce a solution with a pH = 4.6 when the concentration is ~1.0 *M*. NH_4Cl is the solute.

146. $CO_2(aq) + H_2O(l) \rightleftharpoons H_2CO_3(aq) \qquad K = \dfrac{[H_2CO_3]}{[CO_2]}$

During exercise: $[H_2CO_3] = 26.3$ m*M* and $[CO_2] = 1.63$ m*M*, so: $K = \dfrac{26.3\text{ m}M}{1.63\text{ m}M} = 16.1$

At rest: $K = 16.1 = \dfrac{24.9\text{ m}M}{[CO_2]}$, $[CO_2] = 1.55$ m*M*

147. a. In the lungs there is a lot of O_2, and the equilibrium favors $Hb(O_2)_4$. In the cells there is a lower concentration of O_2, and the equilibrium favors HbH_4^{4+}.

b. CO_2 is a weak acid, $CO_2 + H_2O \rightleftharpoons HCO_3^- + H^+$. Removing CO_2 essentially decreases H^+, which causes the hemoglobin reaction to shift right. $Hb(O_2)_4$ is then favored, and O_2 is not released by hemoglobin in the cells. Breathing into a paper bag increases CO_2 in the blood, thus increasing $[H^+]$, which shifts the hemoglobin reaction left.

c. CO_2 builds up in the blood, and it becomes too acidic, driving the hemoglobin equilibrium to the left. Hemoglobin can't bind O_2 as strongly in the lungs. Bicarbonate ion acts as a base in water and neutralizes the excess acidity.

148. $CaO(s) + H_2O(l) \rightarrow Ca(OH)_2(aq)$; $Ca(OH)_2(aq) \rightarrow Ca^{2+}(aq) + 2\ OH^-(aq)$

$$[OH^-] = \frac{0.25\text{ g CaO} \times \dfrac{1\text{ mol CaO}}{56.08\text{ g}} \times \dfrac{1\text{ mol Ca(OH)}_2}{1\text{ mol CaO}} \times \dfrac{2\text{ mol OH}^-}{\text{mol Ca(OH)}_2}}{1.5\text{ L}} = 5.9 \times 10^{-3}\ M$$

$pOH = -\log(5.9 \times 10^{-3}) = 2.23$, $pH = 14.00 - 2.23 = 11.77$

149.

	HBz	$\rightleftharpoons$	H^+	+	Bz^-
Initial	C		~0		0
	x mol/L HBz dissociates to reach equilibrium				
Change	$-x$	$\rightarrow$	$+x$		$+x$
Equil.	$C - x$		x		x

$HBz = C_6H_5CO_2H$

$C = [HBz]_0$ = concentration of HBz that dissolves to give saturated solution.

$$K_a = \frac{[H^+][Bz^-]}{[HBz]} = 6.4 \times 10^{-5} = \frac{x^2}{C - x}\text{, where } x = [H^+]$$

$$6.4 \times 10^{-5} = \frac{[H^+]^2}{C - [H^+]};\ pH = 2.80;\ [H^+] = 10^{-2.80} = 1.6 \times 10^{-3}\,M$$

$$C - (1.6 \times 10^{-3}) = \frac{(1.6 \times 10^{-3})^2}{6.4 \times 10^{-5}} = 4.0 \times 10^{-2}$$

$$C = (4.0 \times 10^{-2}) + (1.6 \times 10^{-3}) = 4.2 \times 10^{-2}\,M$$

The molar solubility of $C_6H_5CO_2H$ is 4.2×10^{-2} mol/L.

150. $[H^+]_0 = (1.0 \times 10^{-2}) + (1.0 \times 10^{-2}) = 2.0 \times 10^{-2}\,M$ from strong acids HCl and H_2SO_4.

HSO_4^- is a good weak acid ($K_a = 0.012$). However, HCN is a poor weak acid ($K_a = 6.2 \times 10^{-10}$) and can be ignored. Calculating the H^+ contribution from HSO_4^-:

	HSO_4^-	$\rightleftharpoons$	H^+	+	SO_4^{2-}	$K_a = 0.012$
Initial	0.010 *M*		0.020 *M*		0	
Equil.	$0.010 - x$		$0.020 + x$		x	

$$K_a = \frac{x(0.020 + x)}{0.010 - x},\ 0.012 \approx \frac{x(0.020)}{0.010},\ x = 0.0060;\ \text{assumption poor (60\% error).}$$

Using the quadratic formula: $x^2 + (0.032)x - 1.2 \times 10^{-4} = 0$, $x = 3.4 \times 10^{-3}\,M$

$[H^+] = 0.020 + x = 0.020 + (3.4 \times 10^{-3}) = 0.023\,M$; pH = 1.64

151. For this problem we will abbreviate $CH_2{=}CHCO_2H$ as Hacr and $CH_2{=}CHCO_2^-$ as acr^-.

a. Solving the weak acid problem:

	Hacr	$\rightleftharpoons$	H^+	+	acr^-	$K_a = 5.6 \times 10^{-5}$
Initial	0.10 *M*		~0		0	
Equil.	$0.10 - x$		x		x	

$$\frac{x^2}{0.10 - x} = 5.6 \times 10^{-5} \approx \frac{x^2}{0.10},\ x = [H^+] = 2.4 \times 10^{-3}\,M;\ pH = 2.62;\ \text{assumptions good.}$$

b. $$\text{Percent dissociation} = \frac{[H^+]}{[Hacr]_0} \times 100 = \frac{2.4 \times 10^{-3}}{0.10} \times 100 = 2.4\%$$

c. acr^- is a weak base and the major source of OH^- in this solution.

	acr^-	+ H_2O	$\rightleftharpoons$	Hacr	+	OH^-	$K_b = \frac{K_w}{K_a} = \frac{1.0 \times 10^{-14}}{5.6 \times 10^{-5}}$
Initial	0.050 *M*			0		~0	$K_b = 1.8 \times 10^{-10}$
Equil.	$0.050 - x$			x		x	

$$K_b = \frac{[Hacr][OH^-]}{[acr^-]}, \quad 1.8 \times 10^{-10} = \frac{x^2}{0.050 - x} \approx \frac{x^2}{0.050}$$

$x = [OH^-] = 3.0 \times 10^{-6}\ M$; pOH = 5.52; pH = 8.48; assumptions good.

152. In deciding whether a substance is an acid or a base, strong or weak, you should keep in mind a couple of ideas:

(1) There are only a few common strong acids and strong bases, all of which should be memorized. Common strong acids = HCl, HBr, HI, HNO_3, $HClO_4$, and H_2SO_4. Common strong bases = LiOH, NaOH, KOH, RbOH, CsOH, $Ca(OH)_2$, $Sr(OH)_2$, and $Ba(OH)_2$.

(2) All other acids and bases are weak and will have K_a and K_b values of less than 1 but greater than K_w (1.0×10^{-14}). Reference Table 14.2 for K_a values for some weak acids and Table 14.3 for K_b values for some weak bases. There are too many weak acids and weak bases to memorize them all. Therefore, use the tables of K_a and K_b values to help you identify weak acids and weak bases. Appendix 5 contains more complete tables of K_a and K_b values.

a. weak acid ($K_a = 4.0 \times 10^{-4}$)
b. strong acid
c. weak base ($K_b = 4.38 \times 10^{-4}$)
d. strong base
e. weak base ($K_b = 1.8 \times 10^{-5}$)
f. weak acid ($K_a = 7.2 \times 10^{-4}$)
g. weak acid ($K_a = 1.8 \times 10^{-4}$)
h. strong base
i. strong acid

153. a. HA is a weak acid. Most of the acid is present as HA molecules; only one set of H^+ and A^- ions is present. In a strong acid, all of the acid would be dissociated into H^+ and A^- ions.

b. This picture is the result of 1 out of 10 HA molecules dissociating.

Percent dissociation = $\frac{1}{10} \times 100 = 10\%$ (an exact number)

	HA	⇌	H^+	+	A^-	$K_a = \frac{[H^+][A^-]}{[HA]}$
Initial	0.20 *M*		~0		0	
	x mol/L HA dissociates to reach equilibrium					
Change	$-x$	→	$+x$		$+x$	
Equil.	$0.20 - x$		x		x	

$[H^+] = [A^-] = x = 0.10 \times 0.20\ M = 0.020\ M$; $[HA] = 0.20 - 0.020 = 0.18\ M$

$$K_a = \frac{(0.020)^2}{0.18} = 2.2 \times 10^{-3}$$

154. $\frac{1.0 \text{ g quinine}}{1.9000 \text{L}} \times \frac{1 \text{ mol quinine}}{324.4 \text{ g quinine}} = 1.6 \times 10^{-3}\ M$ quinine; let Q = quinine = $C_{20}H_{24}N_2O_2$.

$$Q + H_2O \rightleftharpoons QH^+ + OH^- \quad K_b = 10^{-5.1} = 8 \times 10^{-6}$$

	Q		QH^+	OH^-
Initial	$1.6 \times 10^{-3}\ M$		0	~0
	x mol/L quinine reacts with H_2O to reach equilibrium			
Change	$-x$	→	$+x$	$+x$
Equil.	$1.6 \times 10^{-3} - x$		x	x

$$K_b = 8 \times 10^{-6} = \frac{[QH^+][OH^-]}{[Q]} = \frac{x^2}{(1.6 \times 10^{-3} - x)} \approx \frac{x^2}{1.6 \times 10^{-3}}$$

$x = 1 \times 10^{-4}$; assumption fails 5% rule (x is 6% of 0.0016). Using successive approximations:

$$\frac{x^2}{(1.6 \times 10^{-3} - 1 \times 10^{-4})} = 8 \times 10^{-6},\ x = 1 \times 10^{-4}\ M \text{ (consistent answer)}$$

$x = [OH^-] = 1 \times 10^{-4}\ M$; pOH = 4.0; pH = 10.0

155. Let cod = codeine, $C_{18}H_{21}NO_3$; using the K_b reaction to solve:

$$\text{cod} + H_2O \rightleftharpoons \text{codH}^+ + OH^-$$

	cod		codH^+	OH^-
Initial	$1.7 \times 10^{-3}\ M$		0	~0
	x mol/L codeine reacts with H_2O to reach equilibrium			
Change	$-x$	→	$+x$	$+x$
Equil.	$1.7 \times 10^{-3} - x$		x	x

$K_b = \frac{x^2}{1.7 \times 10^{-3} - x}$; pH = 9.59; pOH = 14.00 − 9.59 = 4.41

$[OH^-] = x = 10^{-4.41} = 3.9 \times 10^{-5}\ M$; $K_b = \frac{(3.9 \times 10^{-5})^2}{1.7 \times 10^{-3} - (3.9 \times 10^{-5})} = 9.2 \times 10^{-7}$

156. Codeine = $C_{18}H_{21}NO_3$; codeine sulfate = $C_{36}H_{44}N_2O_{10}S$

The formula for codeine sulfate works out to $(\text{codeineH}^+)_2SO_4^{2-}$, where $\text{codeineH}^+ = HC_{18}H_{21}NO_3^+$. Two codeine molecules are protonated by H_2SO_4, forming the conjugate acid of codeine. The SO_4^{2-} then acts as the counter ion to give a neutral compound. Codeine sulfate is an ionic compound that is more soluble in water than codeine, allowing more of the drug into the bloodstream.

157. a.

$$Fe(H_2O)_6^{3+} + H_2O \rightleftharpoons Fe(H_2O)_5(OH)^{2+} + H_3O^+$$

	$Fe(H_2O)_6^{3+}$	$Fe(H_2O)_5(OH)^{2+}$	H_3O^+
Initial	$0.10\ M$	0	~0
Equil.	$0.10 - x$	x	x

$$K_a = \frac{[Fe(H_2O)_5(OH)^{2+}][H_3O^+]}{[Fe(H_2O)_6^{3+}]},\ 6.0 \times 10^{-3} = \frac{x^2}{0.10 - x} \approx \frac{x^2}{0.10}$$

$x = 2.4 \times 10^{-2}$; assumption is poor (x is 24% of 0.10). Using successive approximations:

$$\frac{x^2}{0.10 - 0.024} = 6.0 \times 10^{-3},\ x = 0.021$$

$$\frac{x^2}{0.10 - 0.021} = 6.0 \times 10^{-3},\ x = 0.022;\quad \frac{x^2}{0.10 - 0.022} = 6.0 \times 10^{-3},\ x = 0.022$$

$x = [H^+] = 0.022\ M$; pH = 1.66

b. Because of the lower charge, Fe^{2+}(aq) will not be as strong an acid as Fe^{3+}(aq). A solution of iron(II) nitrate will be less acidic (have a higher pH) than a solution with the same concentration of iron(III) nitrate.

158. One difficult aspect of acid-base chemistry is recognizing what types of species are present in solution, that is, whether a species is a strong acid, strong base, weak acid, weak base, or a neutral species. Below are some ideas and generalizations to keep in mind that will help in recognizing types of species present.

a. Memorize the following strong acids: HCl, HBr, HI, HNO_3, $HClO_4$, and H_2SO_4

b. Memorize the following strong bases: LiOH, NaOH, KOH, RbOH, CsOH, $Ca(OH)_2$, $Sr(OH)_2$, and $Ba(OH)_2$

c. Weak acids have a K_a value of less than 1 but greater than K_w. Some weak acids are listed in Table 14.2 of the text. Weak bases have a K_b value of less than 1 but greater than K_w. Some weak bases are listed in Table 14.3 of the text.

d. Conjugate bases of weak acids are weak bases; that is, all have a K_b value of less than 1 but greater than K_w. Some examples of these are the conjugate bases of the weak acids listed in Table 14.2 of the text.

e. Conjugate acids of weak bases are weak acids; that is, all have a K_a value of less than 1 but greater than K_w. Some examples of these are the conjugate acids of the weak bases listed in Table 14.3 of the text.

f. Alkali metal ions (Li^+, Na^+, K^+, Rb^+, Cs^+) and some alkaline earth metal ions (Ca^{2+}, Sr^{2+}, Ba^{2+}) have no acidic or basic properties in water.

g. Conjugate bases of strong acids (Cl^-, Br^-, I^-, NO_3^-, ClO_4^-, HSO_4^-) have no basic properties in water ($K_b << K_w$), and only HSO_4^- has any acidic properties in water.

Let's apply these ideas to this problem to see what types of species are present.

a. HI: Strong acid; HF: weak acid ($K_a = 7.2 \times 10^{-4}$)

NaF: F^- is the conjugate base of the weak acid HF, so F^- is a weak base. The K_b value for $F^- = K_w/K_{a,\,HF} = 1.4 \times 10^{-11}$. Na^+ has no acidic or basic properties.

NaI: Neutral (pH = 7.0); Na^+ and I^- have no acidic/basic properties.

In order of increasing pH, we place the compounds from most acidic (lowest pH) to most basic (highest pH). Increasing pH: HI < HF < NaI < NaF.

b. NH_4Br: NH_4^+ is a weak acid ($K_a = 5.6 \times 10^{-10}$), and Br^- is a neutral species.

HBr: Strong acid

KBr: Neutral; K^+ and Br^- have no acidic/basic properties.

NH_3: Weak base, $K_b = 1.8 \times 10^{-5}$

Increasing pH: HBr < NH_4Br < KBr < NH_3
Most acidic ... Most basic

c. $C_6H_5NH_3NO_3$: $C_6H_5NH_3^+$ is a weak acid ($K_a = K_w/K_{b,C_6H_5NH_2} = 1.0 \times 10^{-14}/3.8 \times 10^{-10} = 2.6 \times 10^{-5}$), and NO_3^- is a neutral species.

$NaNO_3$: Neutral; Na^+ and NO_3^- have no acidic/basic properties.

NaOH: Strong base

HOC_6H_5: Weak acid ($K_a = 1.6 \times 10^{-10}$)

KOC_6H_5: $OC_6H_5^-$ is a weak base ($K_b = K_w/K_{a,\,HOC_6H_5} = 6.3 \times 10^{-5}$), and K^+ is a neutral species.

$C_6H_5NH_2$: Weak base ($K_b = 3.8 \times 10^{-10}$)

HNO_3: Strong acid

This is a little more difficult than the previous parts of this problem because two weak acids and two weak bases are present. Between the weak acids, $C_6H_5NH_3^+$ is a stronger weak acid than HOC_6H_5 since the K_a value for $C_6H_5NH_3^+$ is larger than the K_a value for HOC_6H_5. Between the two weak bases, because the K_b value for $OC_6H_5^-$ is larger than the K_b value for $C_6H_5NH_2$, $OC_6H_5^-$ is a stronger weak base than $C_6H_5NH_2$.

Increasing pH: HNO_3 < $C_6H_5NH_3NO_3$ < HOC_6H_5 < $NaNO_3$ < $C_6H_5NH_2$ < KOC_6H_5 < NaOH
Most acidic ... Most basic

159. The solution is acidic from $HSO_4^- \rightleftharpoons H^+ + SO_4^{2-}$. Solving the weak acid problem:

	HSO_4^-	$\rightleftharpoons$	H^+	+	SO_4^{2-}	$K_a = 1.2 \times 10^{-2}$
Initial	0.10 *M*		~0		0	
Equil.	$0.10 - x$		x		x	

$$1.2 \times 10^{-2} = \frac{[H^+][SO_4^{2-}]}{[HSO_4^-]} = \frac{x^2}{0.10 - x} \approx \frac{x^2}{0.10},\quad x = 0.035$$

Assumption is not good (x is 35% of 0.10). Using successive approximations:

$$\frac{x^2}{0.10 - x} = \frac{x^2}{0.10 - 0.035} = 1.2 \times 10^{-2},\ x = 0.028$$

$$\frac{x^2}{0.10 - 0.028} = 1.2 \times 10^{-2},\ x = 0.029;\quad \frac{x^2}{0.10 - 0.029} = 1.2 \times 10^{-2},\ x = 0.029$$

$x = [H^+] = 0.029\ M$; pH = 1.54

160. a. $NH_3 + H_3O^+ \rightleftharpoons NH_4^+ + H_2O$

$$K_{eq} = \frac{[NH_4^+]}{[NH_3][H^+]} = \frac{1}{K_a \text{ for } NH_4^+} = \frac{K_b \text{ for } NH_3}{K_w} = \frac{1.8 \times 10^{-5}}{1.0 \times 10^{-14}} = 1.8 \times 10^9$$

b. $NO_2^- + H_3O^+ \rightleftharpoons HNO_2 + H_2O \quad K_{eq} = \frac{[HNO_2]}{[NO_2^-][H^+]} = \frac{1}{K_a \text{ for } HNO_2} = \frac{1}{4.0 \times 10^{-4}}$

$= 2.5 \times 10^3$

c. $NH_4^+ + OH^- \rightleftharpoons NH_3 + H_2O \quad K_{eq} = \frac{1}{K_b \text{ for } NH_3} = \frac{1}{1.8 \times 10^{-5}} = 5.6 \times 10^4$

d. $HNO_2 + OH^- \rightleftharpoons H_2O + NO_2^-$

$$K_{eq} = \frac{[NO_2^-]}{[HNO_2][OH^-]} \times \frac{[H^+]}{[H^+]} = \frac{K_a \text{ for } HNO_2}{K_w} = \frac{4.0 \times 10^{-4}}{1.0 \times 10^{-14}} = 4.0 \times 10^{10}$$

161. a. H_2SO_3 b. $HClO_3$ c. H_3PO_3

NaOH and KOH are soluble ionic compounds composed of Na^+ and K^+ cations and OH^- anions. All soluble ionic compounds dissolve to form the ions from which they are formed. In oxyacids, the compounds are all covalent compounds in which electrons are shared to form bonds (unlike ionic compounds). When these compounds are dissolved in water, the covalent bond between oxygen and hydrogen breaks to form H^+ ions.

ChemWork Problems

The answers to the problems 162-169 (or a variation to these problems) are found in OWL. These problems are also assignable in OWL.

Challenge Problems

170. The pH of this solution is not 8.00 because water will donate a significant amount of H^+ from the autoionization of water. You can't add an acid to water and get a basic pH. The pertinent equations are:

$H_2O \rightleftharpoons H^+ + OH^- \quad K_w = [H^+][OH^-] = 1.0 \times 10^{-14}$

$HCl \rightarrow H^+ + Cl^-$ K_a is very large, so we assume that only the forward reaction occurs.

In any solution, the overall net positive charge must equal the overall net negative charge (called the charge balance). For this problem:

[positive charge] = [negative charge], so $[H^+] = [OH^-] + [Cl^-]$

From K_w, $[OH^-] = K_w/[H^+]$, and from 1.0×10^{-8} *M* HCl, $[Cl^-] = 1.0 \times 10^{-8}$ *M*. Substituting into the charge balance equation:

$$[H^+] = \frac{1.0 \times 10^{-14}}{[H^+]} + 1.0 \times 10^{-8},\ [H^+]^2 - (1.0 \times 10^{-8})[H^+] - 1.0 \times 10^{-14} = 0$$

Using the quadratic formula to solve:

$$[H^+] = \frac{-(-1.0 \times 10^{-8}) \pm [(-1.0 \times 10^{-8})^2 - 4(1)(-1.0 \times 10^{-14})]^{1/2}}{2(1)},\ [H^+] = 1.1 \times 10^{-7}\ M$$

$$pH = -\log(1.1 \times 10^{-7}) = 6.96$$

171. Because this is a very dilute solution of NaOH, we must worry about the amount of OH^- donated from the autoionization of water.

$$NaOH \rightarrow Na^+ + OH^-$$

$$H_2O \rightleftharpoons H^+ + OH^- \quad K_w = [H^+][OH^-] = 1.0 \times 10^{-14}$$

This solution, like all solutions, must be charge balanced; that is, [positive charge] = [negative charge]. For this problem, the charge balance equation is:

$$[Na^+] + [H^+] = [OH^-],\ \text{where } [Na^+] = 1.0 \times 10^{-7}\ M \text{ and } [H^+] = \frac{K_w}{[OH^-]}$$

Substituting into the charge balance equation:

$$1.0 \times 10^{-7} + \frac{1.0 \times 10^{-14}}{[OH^-]} = [OH^-],\ [OH^-]^2 - (1.0 \times 10^{-7})[OH^-] - 1.0 \times 10^{-14} = 0$$

Using the quadratic formula to solve:

$$[OH^-] = \frac{-(-1.0 \times 10^{-7}) \pm [(-1.0 \times 10^{-7})^2 - 4(1)(-1.0 \times 10^{-14})]^{1/2}}{2(1)}$$

$[OH^-] = 1.6 \times 10^{-7}$ *M*; $pOH = -\log(1.6 \times 10^{-7}) = 6.80$; pH = 7.20

172. $Ca(OH)_2\,(s) \rightarrow Ca^{2+}(aq) + 2\ OH^-(aq)$

This is a very dilute solution of $Ca(OH)_2$, so we can't ignore the OH^- contribution from H_2O. From the dissociation of $Ca(OH)_2$ alone, $2[Ca^{2+}] = [OH^-]$. Including the H_2O autoionization into H^+ and OH^-, the overall charge balance is:

$$2[Ca^{2+}] + [H^+] = [OH^-]$$

$2(3.0 \times 10^{-7}\ M) + K_w/[OH^-] = [OH^-],\ \ [OH^-]^2 = (6.0 \times 10^{-7})[OH^-] + K_w$

$[OH^-]^2 - (6.0 \times 10^{-7})[OH^-] - 1.0 \times 10^{-14} = 0$; using quadratic formula: $[OH^-] = 6.2 \times 10^{-7}\ M$

173. $HA \rightleftharpoons H^+ + A^-$ $\quad K_a = 1.00 \times 10^{-6}$

	HA	H^+	A^-	
Initial	C	~0	0	$C = [HA]_0$, for pH = 4.000,
Equil.	$C - 1.00 \times 10^{-4}$	1.00×10^{-4}	1.00×10^{-4}	$x = [H^+] = 1.00 \times 10^{-4}\ M$

$$K_a = \frac{(1.00 \times 10^{-4})^2}{(C - 1.00 \times 10^{-4})} = 1.00 \times 10^{-6};\ \text{solving: } C = 0.0101\ M$$

The solution initially contains 50.0×10^{-3} L × 0.0101 mol/L = 5.05×10^{-4} mol HA. We then dilute to a total volume V in liters. The resulting pH = 5.000, so $[H^+] = 1.00 \times 10^{-5}$. In the typical weak acid problem, $x = [H^+]$, so:

	HA	$\rightleftharpoons$	H^+	+	A^-
Initial	5.05×10^{-4} mol/V		~0		0
Equil.	$(5.05 \times 10^{-4}/V) - (1.00 \times 10^{-5})$		1.00×10^{-5}		1.00×10^{-5}

$$K_a = \frac{(1.00 \times 10^{-5})^2}{(5.05 \times 10^{-4}/V) - (1.00 \times 10^{-5})} = 1.00 \times 10^{-6}$$

$1.00 \times 10^{-4} = (5.05 \times 10^{-4}/V) - 1.00 \times 10^{-5}$

V = 4.59 L; 50.0 mL are present initially, so we need to add 4540 mL of water.

174. $HBrO \rightleftharpoons H^+ + BrO^-$ $\quad K_a = 2 \times 10^{-9}$

	HBrO		H^+	BrO^-
Initial	$1.0 \times 10^{-6}\ M$		~0	0
	x mol/L HBrO dissociates to reach equilibrium			
Change	$-x$	$\rightarrow$	$+x$	$+x$
Equil.	$1.0 \times 10^{-6} - x$		x	x

$$K_a = 2 \times 10^{-9} = \frac{x^2}{(1.0 \times 10^{-6} - x)} \approx \frac{x^2}{1.0 \times 10^{-6}},\ x = [H^+] = 4 \times 10^{-8}\ M;\ \text{pH} = 7.4$$

Let's check the assumptions. This answer is impossible! We can't add a small amount of an acid to water and get a basic solution. The highest possible pH for an acid in water is 7.0. In the correct solution we would have to take into account the autoionization of water.

175. Major species present are H_2O, $C_5H_5NH^+$ [$K_a = K_w/K_{b,C_5H_5N} = (1.0 \times 10^{-14})/(1.7 \times 10^{-9}) = 5.9 \times 10^{-6}$], and F^- [$K_b = K_w/K_{a,HF} = (1.0 \times 10^{-14})/(7.2 \times 10^{-4}) = 1.4 \times 10^{-11}$]. The reaction to consider is the best acid present ($C_5H_5NH^+$) reacting with the best base present (F^-). Let's solve by first setting up an ICE table.

	$C_5H_5NH^+(aq)$	+	$F^-(aq)$	$\rightleftharpoons$	$C_5H_5N(aq)$	+	$HF(aq)$
Initial	0.200 *M*		0.200 *M*		0		0
Change	$-x$		$-x$	$\rightarrow$	$+x$		$+x$
Equil.	$0.200 - x$		$0.200 - x$		x		x

$$K = K_{a, C_5H_5NH^+} \times \frac{1}{K_{a, HF}} = 5.9 \times 10^{-6} \times \frac{1}{7.2 \times 10^{-4}} = 8.2 \times 10^{-3}$$

$$K = \frac{[C_5H_5N][HF]}{[C_5H_5NH^+][F^-]},\ 8.2 \times 10^{-3} = \frac{x^2}{(0.200-x)^2};\ \text{taking the square root of both sides:}$$

$$0.091 = \frac{x}{0.200-x},\quad x = 0.018 - (0.091)x,\ x = 0.016\ M$$

From the setup to the problem, $x = [C_5H_5N] = [HF] = 0.016\ M$, and $0.200 - x = 0.200 - 0.016 = 0.184\ M = [C_5H_5NH^+] = [F^-]$. To solve for the $[H^+]$, we can use either the K_a equilibrium for $C_5H_5NH^+$ or the K_a equilibrium for HF. Using $C_5H_5NH^+$ data:

$$K_{a, C_5H_5NH^+} = 5.9 \times 10^{-6} = \frac{[C_5H_5N][H^+]}{[C_5H_5NH^+]} = \frac{(0.016)[H^+]}{0.184},\ [H^+] = 6.8 \times 10^{-5}\ M$$

$pH = -\log(6.8 \times 10^{-5}) = 4.17$

As one would expect, because the K_a for the weak acid is larger than the K_b for the weak base, a solution of this salt should be acidic.

176. Major species: NH_4^+, OCl^-, and H_2O; K_a for $NH_4^+ = (1.0 \times 10^{-14})/(1.8 \times 10^{-5}) = 5.6 \times 10^{-10}$ and K_b for $OCl^- = (1.0 \times 10^{-14})/(3.5 \times 10^{-8}) = 2.9 \times 10^{-7}$.

Because OCl^- is a better base than NH_4^+ is an acid, the solution will be basic. The dominant equilibrium is the best acid (NH_4^+) reacting with the best base (OCl^-) present.

	NH_4^+	+	OCl^-	$\rightleftharpoons$	NH_3	+	HOCl
Initial	0.50 *M*		0.50 *M*		0		0
Change	$-x$		$-x$	$\rightarrow$	$+x$		$+x$
Equil.	$0.50 - x$		$0.50 - x$		x		x

$$K = K_{a, NH_4^+} \times \frac{1}{K_{a, HOCl}} = (5.6 \times 10^{-10})/(3.5 \times 10^{-8}) = 0.016$$

$$K = 0.016 = \frac{[NH_3][HOCl]}{[NH_4^+][OCl^-]} = \frac{x(x)}{(0.50-x)(0.50-x)}$$

$$\frac{x^2}{(0.50-x)^2} = 0.016,\ \frac{x}{0.50-x} = (0.016)^{1/2} = 0.13,\ x = 0.058\ M$$

To solve for the H^+, use any pertinent K_a or K_b value. Using K_a for NH_4^+:

$$K_{a,\,NH_4^+} = 5.6 \times 10^{-10} = \frac{[NH_3][H^+]}{[NH_4^+]} = \frac{(0.058)[H^+]}{0.50 - 0.058},\ [H^+] = 4.3 \times 10^{-9}\ M,\ pH = 8.37$$

177. Because NH_3 is so concentrated, we need to calculate the OH^- contribution from the weak base NH_3.

	NH_3 + H_2O	⇌	NH_4^+	+	OH^-	$K_b = 1.8 \times 10^{-5}$
Initial	15.0 *M*		0		0.0100 *M*	(Assume no volume change.)
Equil.	$15.0 - x$		x		$0.0100 + x$	

$$K_b = 1.8 \times 10^{-5} = \frac{x(0.0100 + x)}{15.0 - x} \approx \frac{x(0.0100)}{15.0},\ x = 0.027;$$ assumption is horrible (x is 270% of 0.0100).

Using the quadratic formula:

$$(1.8 \times 10^{-5})(15.0 - x) = (0.0100)x + x^2,\ x^2 + (0.0100)x - 2.7 \times 10^{-4} = 0$$

$$x = 1.2 \times 10^{-2}\ M,\ [OH^-] = (1.2 \times 10^{-2}) + 0.0100 = 0.022\ M$$

178. For 0.0010% dissociation: $[NH_4^+] = 1.0 \times 10^{-5}(0.050) = 5.0 \times 10^{-7}\ M$

$$NH_3 + H_2O \rightleftharpoons NH_4^+ + OH^- \quad K_b = \frac{(5.0 \times 10^{-7})[OH^-]}{0.050 - 5.0 \times 10^{-7}} = 1.8 \times 10^{-5}$$

Solving: $[OH^-] = 1.8\ M$; assuming no volume change:

$$1.0\ L \times \frac{1.8\ mol\ NaOH}{L} \times \frac{40.00\ g\ NaOH}{mol\ NaOH} = 72\ g\ of\ NaOH$$

179. $$1.000\ L \times \frac{1.00 \times 10^{-4}\ mol\ HA}{L} = 1.00 \times 10^{-4}\ mol\ HA$$

25.0% dissociation gives:

$$\text{moles } H^+ = 0.250 \times (1.00 \times 10^{-4}) = 2.50 \times 10^{-5}\ mol$$

$$\text{moles } A^- = 0.250 \times (1.00 \times 10^{-4}) = 2.50 \times 10^{-5}\ mol$$

$$\text{moles HA} = 0.750 \times (1.00 \times 10^{-4}) = 7.50 \times 10^{-5}\ mol$$

$$1.00 \times 10^{-4} = K_a = \frac{[H^+][A^-]}{[HA]} = \frac{\left(\frac{2.50 \times 10^{-5}}{V}\right)\left(\frac{2.50 \times 10^{-5}}{V}\right)}{\left(\frac{7.50 \times 10^{-5}}{V}\right)}$$

$$1.00 \times 10^{-4} = \frac{(2.50 \times 10^{-5})^2}{(7.50 \times 10^{-5})(V)},\quad V = \frac{(2.50 \times 10^{-5})^2}{(1.00 \times 10^{-4})(7.50 \times 10^{-5})} = 0.0833\text{ L} = 83.3\text{ mL}$$

The volume goes from 1000. mL to 83.3 mL, so 917 mL of water evaporated.

180. $HC_2H_3O_2 \rightleftharpoons H^+ + C_2H_3O_2^-$ $K_a = 1.8 \times 10^{-5}$

	$HC_2H_3O_2$	H^+	$C_2H_3O_2^-$
Initial	1.00 M	~0	0
Equil.	$1.00 - x$	x	x

$$1.8 \times 10^{-5} = \frac{x^2}{1.00 - x} \approx \frac{x^2}{1.00},\quad x = [H^+] = 4.24 \times 10^{-3}\,M \text{ (using one extra sig. fig.)}$$

$pH = -\log(4.24 \times 10^{-3}) = 2.37$; assumptions good.

We want to double the pH to 2(2.37) = 4.74 by addition of the strong base NaOH. As is true with all strong bases, they are great at accepting protons. In fact, they are so good that we can assume they accept protons 100% of the time. The best acid present will react the strong base. This is $HC_2H_3O_2$. The initial reaction that occurs when the strong base is added is:

$$HC_2H_3O_2 + OH^- \rightarrow C_2H_3O_2^- + H_2O$$

Note that this reaction has the net effect of converting $HC_2H_3O_2$ into its conjugate base, $C_2H_3O_2^-$.

For a pH = 4.74, let's calculate the ratio of $[C_2H_3O_2^-]/[HC_2H_3O_2]$ necessary to achieve this pH.

$$HC_2H_3O_2 \rightleftharpoons H^+ + C_2H_3O_2^- \quad K_a = \frac{[H^+][C_2H_3O_2^-]}{[HC_2H_3O_2]}$$

When pH = 4.74, $[H^+] = 10^{-4.74} = 1.8 \times 10^{-5}$.

$$K_a = 1.8 \times 10^{-5} = \frac{(1.8 \times 10^{-5})[C_2H_3O_2^-]}{[HC_2H_3O_2]},\quad \frac{[C_2H_3O_2^-]}{[HC_2H_3O_2]} = 1.0$$

For a solution having pH = 4.74, we need to have equal concentrations (equal moles) of $C_2H_3O_2^-$ and $HC_2H_3O_2$. Therefore, we need to add an amount of NaOH that will convert one-half of the $HC_2H_3O_2$ into $C_2H_3O_2^-$. This amount is 0.50 M NaOH.

	$HC_2H_3O_2$	+ OH^-	$\rightarrow$	$C_2H_3O_2^-$ + H_2O
Before	1.00 M	0.50 M		0
Change	−0.50	−0.50	$\rightarrow$	+0.50
After completion	0.50 M	0		0.50 M

From the preceding stoichiometry problem, adding enough NaOH(s) to produce a 0.50 M OH^- solution will convert one-half the $HC_2H_3O_2$ into $C_2H_3O_2^-$; this results in a solution with pH = 4.74.

$$\text{Mass NaOH} = 1.00\text{ L} \times \frac{0.50\text{ mol NaOH}}{\text{L}} \times \frac{40.00\text{ g NaOH}}{\text{mol}} = 20.\text{ g NaOH}$$

181. PO_4^{3-} is the conjugate base of HPO_4^{2-}. The K_a value for HPO_4^{2-} is $K_{a_3} = 4.8 \times 10^{-13}$.

$$PO_4^{3-}(aq) + H_2O(l) \rightleftharpoons HPO_4^{2-}(aq) + OH^-(aq) \quad K_b = \frac{K_w}{K_{a_3}} = \frac{1.0 \times 10^{-14}}{4.8 \times 10^{-13}} = 0.021$$

HPO_4^{2-} is the conjugate base of $H_2PO_4^-$ ($K_{a_2} = 6.2 \times 10^{-8}$).

$$HPO_4^{2-} + H_2O \rightleftharpoons H_2PO_4^- + OH^- \quad K_b = \frac{K_w}{K_{a_1}} = \frac{1.0 \times 10^{-14}}{6.2 \times 10^{-8}} = 1.6 \times 10^{-7}$$

$H_2PO_4^-$ is the conjugate base of H_3PO_4 ($K_{a_1} = 7.5 \times 10^{-3}$).

$$H_2PO_4^- + H_2O \rightleftharpoons H_3PO_4 + OH^- \quad K_b = \frac{K_w}{K_{a_1}} = \frac{1.0 \times 10^{-14}}{7.5 \times 10^{-3}} = 1.3 \times 10^{-12}$$

From the K_b values, PO_4^{3-} is the strongest base. This is expected because PO_4^{3-} is the conjugate base of the weakest acid (HPO_4^{2-}).

182. Major species: Na^+, PO_4^{3-} (a weak base), H_2O; From the K_b values calculated in Exercise 181, the dominant producer of OH^- is the K_b reaction for PO_4^{3-}. We can ignore the contribution of OH^- from the K_b reactions for HPO_4^{2-} and $H_2PO_4^-$. From Exercise 181, K_b for PO_4^{3-} = 0.021.

	PO_4^{3-}	+	H_2O	$\rightleftharpoons$	HPO_4^{2-}	+	OH^-	$K_b = 0.021$
Initial	0.10 M				0		~0	
Equil.	$0.10 - x$				x		x	

$K_b = 0.021 = \dfrac{x^2}{0.10 - x}$; because K_b is so large, the 5% assumption will not hold. Solving using the quadratic equation:

$$x^2 + (0.021)x - 0.0021 = 0,\ \ x = [OH^-] = 3.7 \times 10^{-2}\ M,\ \ \text{pOH} = 1.43,\ \ \text{pH} = 12.57$$

183. a. $NH_4(HCO_3) \rightarrow NH_4^+ + HCO_3^-$

$$K_{a,\,NH_4^+} = \frac{1.0 \times 10^{-14}}{1.8 \times 10^{-5}} = 5.6 \times 10^{-10};\ \ K_{b,\,HCO_3^-} = \frac{K_w}{K_{a_1}} = \frac{1.0 \times 10^{-14}}{4.3 \times 10^{-7}} = 2.3 \times 10^{-8}$$

The solution is basic because HCO_3^- is a stronger base than NH_4^+ is as an acid. The acidic properties of HCO_3^- were ignored because K_{a_2} is very small (4.8×10^{-11}).

b. $NaH_2PO_4 \rightarrow Na^+ + H_2PO_4^-$; ignore Na^+.

$$K_{a_2, H_2PO_4^-} = 6.2 \times 10^{-8};\ K_{b, H_2PO_4^-} = \frac{K_w}{K_{a_1}} = \frac{1.0 \times 10^{-14}}{7.5 \times 10^{-3}} = 1.3 \times 10^{-12}$$

Solution is acidic because $K_a > K_b$.

c. $Na_2HPO_4 \rightarrow 2\ Na^+ + HPO_4^{2-}$; ignore Na^+.

$$K_{a_3, HPO_4^{2-}} = 4.8 \times 10^{-13};\ K_{b, HPO_4^{2-}} = \frac{K_w}{K_{a_2}} = \frac{1.0 \times 10^{-14}}{6.2 \times 10^{-8}} = 1.6 \times 10^{-7}$$

Solution is basic because $K_b > K_a$.

d. $NH_4(H_2PO_4) \rightarrow NH_4^+ + H_2PO_4^-$

NH_4^+ is a weak acid, and $H_2PO_4^-$ is also acidic (see part b). Solution with both ions present will be acidic.

e. $NH_4(HCO_2) \rightarrow NH_4^+ + HCO_2^-$; from Appendix 5, $K_{a, HCO_2H} = 1.8 \times 10^{-4}$.

$$K_{a, NH_4^+} = 5.6 \times 10^{-10};\ K_{b, HCO_2^-} = \frac{K_w}{K_a} = \frac{1.0 \times 10^{-14}}{1.8 \times 10^{-4}} = 5.6 \times 10^{-11}$$

Solution is acidic because NH_4^+ is a stronger acid than HCO_2^- is a base.

184. a. $HCO_3^- + HCO_3^- \rightleftharpoons H_2CO_3 + CO_3^{2-}$

$$K_{eq} = \frac{[H_2CO_3][CO_3^{2-}]}{[HCO_3^-][HCO_3^-]} \times \frac{[H^+]}{[H^+]} = \frac{K_{a_2}}{K_{a_1}} = \frac{5.6 \times 10^{-11}}{4.3 \times 10^{-7}} = 1.3 \times 10^{-4}$$

b. $[H_2CO_3] = [CO_3^{2-}]$ since the reaction in part a is the principal equilibrium reaction.

c. $H_2CO_3 \rightleftharpoons 2\ H^+ + CO_3^{2-}$ $\quad K_{eq} = \frac{[H^+]^2[CO_3^{2-}]}{[H_2CO_3]} = K_{a_1} \times K_{a_2}$

Because $[H_2CO_3] = [CO_3^{2-}]$ from part b, $[H^+]^2 = K_{a_1} \times K_{a_2}$.

$[H^+] = (K_{a_1} \times K_{a_2})^{1/2}$, or taking the –log of both sides: $pH = \frac{pK_{a_1} + pK_{a_2}}{2}$

d. $[H^+] = [(4.3 \times 10^{-7}) \times (5.6 \times 10^{-11})]^{1/2}$, $[H^+] = 4.9 \times 10^{-9}\ M$; pH = 8.31

185. $$\text{Molality} = m = \frac{0.100\text{ g} \times \frac{1\text{ mol}}{100.0\text{ g}}}{0.5000\text{ kg}} = 2.00 \times 10^{-3}\text{ mol/kg} \approx 2.00 \times 10^{-3}\text{ mol/L}$$

$\Delta T_f = iK_f m$, $0.0056°C = i(1.86°C/\text{molal})(2.00 \times 10^{-3}\ \text{molal})$, i = 1.5

If i = 1.0, the percent dissociation of the acid = 0%, and if i = 2.0, the percent dissociation of the acid = 100%. Because i = 1.5, the weak acid is 50.% dissociated.

$$HA \rightleftharpoons H^+ + A^- \qquad K_a = \frac{[H^+][A^-]}{[HA]}$$

Because the weak acid is 50.% dissociated:

$$[H^+] = [A^-] = [HA]_0 \times 0.50 = 2.00 \times 10^{-3}\,M \times 0.50 = 1.0 \times 10^{-3}\,M$$

$$[HA] = [HA]_0 - \text{amount HA reacted} = 2.00 \times 10^{-3}\,M - 1.0 \times 10^{-3}\,M = 1.0 \times 10^{-3}\,M$$

$$K_a = \frac{[H^+][A^-]}{[HA]} = \frac{(1.0 \times 10^{-3})(1.0 \times 10^{-3})}{1.0 \times 10^{-3}} = 1.0 \times 10^{-3}$$

186. a. Assuming no ion association between $SO_4^{2-}(aq)$ and $Fe^{3+}(aq)$, then i = 5 for $Fe_2(SO_4)_3$.

$$\pi = iMRT = 5(0.0500 \text{ mol/L})(0.08206 \text{ L atm/K•mol})(298 \text{ K}) = 6.11 \text{ atm}$$

b. $Fe_2(SO_4)_3(aq) \rightarrow 2\ Fe^{3+}(aq) + 3\ SO_4^{2-}(aq)$

Under ideal circumstances, 2/5 of π calculated above results from Fe^{3+} and 3/5 results from SO_4^{2-}. The contribution to π from SO_4^{2-} is 3/5 × 6.11 atm = 3.67 atm. Because SO_4^{2-} is assumed unchanged in solution, the SO_4^{2-} contribution in the actual solution will also be 3.67 atm. The contribution to the actual osmotic pressure from the $Fe(H_2O)_6^{3+}$ dissociation reaction is 6.73 – 3.67 = 3.06 atm.

The initial concentration of $Fe(H_2O)_6^{2+}$ is 2(0.0500) = 0.100 M. The set up for the weak acid problem is:

$$Fe(H_2O)_6^{3+} \rightleftharpoons H^+ + Fe(OH)(H_2O)_5^{2+} \qquad K_a = \frac{[H^+][Fe(OH)(H_2O)_5^{2+}]}{[Fe(H_2O)_6^{3+}]}$$

	$Fe(H_2O)_6^{3+}$	H^+	$Fe(OH)(H_2O)_5^{2+}$
Initial	0.100 M	~0	0
	x mol/L of $Fe(H_2O)_6^{3+}$ reacts to reach equilibrium		
Equil.	$0.100 - x$	x	x

$$\pi = iMRT;\ \text{total ion concentration} = iM = \frac{\pi}{RT} = \frac{3.06\text{atm}}{0.8206\text{L atm/K} \bullet \text{mol}(298)} = 0.125\,M$$

$$0.125\,M = 0.100 - x + x + x = 0.100 + x,\ x = 0.025\,M$$

$$K_a = \frac{[H^+][Fe(OH)(H_2O)_5^{2+}]}{[Fe(H_2O)_6^{3+}]} = \frac{x^2}{0.100 - x} = \frac{(0.025)^2}{(0.100 - 0.025)} = \frac{(0.025)^2}{0.075}$$

$$K_a = 8.3 \times 10^{-3}$$

Integrative Problems

187. $$[\text{IO}^-] = \frac{2.14\text{ g NaIO} \times \frac{1\text{ mol NaIO}}{165.89\text{ g}} \times \frac{1\text{ mol IO}^-}{\text{mol NaIO}}}{1.25\text{ L}} = 1.03 \times 10^{-2}\ M\ \text{IO}^-$$

$$\text{IO}^- + \text{H}_2\text{O} \rightleftharpoons \text{HIO} + \text{OH}^- \qquad \text{K}_b = \frac{[\text{HIO}][\text{OH}^-]}{[\text{IO}^-]}$$

	IO^-	HIO	OH^-
Initial	$1.03 \times 10^{-2}\ M$	0	~0
Equil.	$1.03 \times 10^{-2} - x$	x	x

$\text{K}_b = \dfrac{x^2}{1.03 \times 10^{-2} - x}$; from the problem, pOH = 14.00 – 11.32 = 2.68.

$[\text{OH}^-] = 10^{-2.68} = 2.1 \times 10^{-3}\ M = x$; $\text{K}_b = \dfrac{(2.1 \times 10^{-3})^2}{1.03 \times 10^{-2} - 2.1 \times 10^{-3}} = 5.4 \times 10^{-4}$

188. If NaOCN is limiting:

$$10.0\text{ g NaOCN} \times \frac{1\text{ mol NaOCN}}{65.01\text{ g NaOCN}} \times \frac{2\text{ mol HNCO}}{2\text{ mol NaOCN}} = 0.154\text{ mol HNCO}$$

If $H_2C_2O_4$ is limiting:

$$10.0\text{ g H}_2\text{C}_2\text{O}_4 \times \frac{1\text{ mol H}_2\text{C}_2\text{O}_4}{90.04\text{ g H}_2\text{C}_2\text{O}_4} \times \frac{2\text{ mol HNCO}}{\text{mol H}_2\text{C}_2\text{O}_4} = 0.222\text{ mol HNCO}$$

Because NaOCN produces the smaller amount of product, NaOCN is limiting and 0.154 mol HNCO can be produced.

$$\text{HNCO} \rightleftharpoons \text{H}^+ + \text{NCO}^- \qquad \text{K}_a = 1.2 \times 10^{-4}$$

	HNCO	H^+	NCO^-
Initial	0.154 mol/0.100 L	~0	0
Equil.	$1.54 - x$	x	x

$\text{K}_a = 1.2 \times 10^{-4} = \dfrac{x^2}{1.54 - x} \approx \dfrac{x^2}{1.54}$, $x = [\text{H}^+] = 1.4 \times 10^{-2}\ M$

pH = –log(1.4×10^{-2}) = 1.85; assumptions good.

189. $$\text{Molar mass} = \frac{\text{dRT}}{\text{P}} = \frac{5.11\text{ g/L} \times \frac{0.08206\text{ L atm}}{\text{K mol}} \times 298\text{ K}}{1.00\text{ atm}} = 125\text{ g/mol}$$

$[\text{HA}]_0 = \dfrac{1.50\text{ g} \times \frac{1\text{ mol}}{125\text{ g}}}{0.100\text{ L}} = 0.120\ M$; pH = 1.80, $[\text{H}^+] = 10^{-1.80} = 1.6 \times 10^{-2}\ M$

	HA	$\rightleftharpoons$	H^+	+	A^-	
Initial	0.120 *M*		~0		0	
Equil.	$0.120 - x$		x		x	where $x = [H^+] = 1.6 \times 10^{-2}$ *M*

$$K_a = \frac{[H^+][A^-]}{[HA]} = \frac{(1.6 \times 10^{-2})^2}{0.120 - 0.016} = 2.5 \times 10^{-3}$$

Marathon Problems

190. Let's abbreviate HCO_2H with HA and $CH_3CH_2CO_2H$ with HB. Fom the problem, we have:

0.0500 *M* HCO_2H (HA), $K_a = 1.77 \times 10^{-4}$; 0.150 *M* $CH_3CH_2CO_2H$ (HB), $K_a = 1.34 \times 10^{-5}$

Because two comparable weak acids are present, each will contribute to the total pH. In order to solve for the pH, we need to determine some relationships that must hold true in this solution. Two equations that must hold true are the K_a expressions for the two acids:

$$\frac{[H^+][A^-]}{[HA]} = 1.77 \times 10^{-4}; \quad \frac{[H^+][B^-]}{[HB]} = 1.34 \times 10^{-5}$$

Another relationship that must hold true recognizes that all of the HA originally dissolved (0.0500 *M*) must be present as either A^- or HA at equilibrium. So:

$$0.0500 = [HA] + [A^-]$$

A similar relationship holds for HB: $0.150 = [HB] + [B^-]$

The last relationship we need recognizes that in any solution, the net positive charge must equal the net negative charge. This is called the charge balance equation. This equation is:

$$[H^+] = [A^-] + [B^-] + [OH^-] = [A^-] + [B^-] + K_w/[H^+]$$

We now have five equations and five unknowns ([HA], [HB], $[A^-]$, $[B^-]$, and $[H^+]$). We should be able to manipulate the equations to solve for $[H^+]$. Solving:

$$[H^+] = [A^-] + [B^-] + K_w/[H^+]; \quad [H^+]^2 = [H^+][A^-] + [H^+][B^-] + K_w$$

$$[H^+][A^-] = (1.77 \times 10^{-4})[HA] = (1.77 \times 10^{-4})\,(0.0500 - [A^-])$$

If $[A^-] << 0.0500$, then $[H^+][A^-] \approx (1.77 \times 10^{-4})\,(0.0500) = 8.85 \times 10^{-6}$.

Similarly, assume $[H^+][B^-] \approx (1.34 \times 10^{-5})(0.150) = 2.01 \times 10^{-6}$.

$$[H^+]^2 = 8.85 \times 10^{-6} + 2.01 \times 10^{-6} + 1.00 \times 10^{-14}, \quad [H^+] = 3.30 \times 10^{-3} \text{ mol/L}$$

Checking assumptions: $[H^+][A^-] \approx 8.85 \times 10^{-6}$, $[A^-] \approx \dfrac{8.85 \times 10^{-6}}{3.30 \times 10^{-3}} \approx 2.68 \times 10^{-3}$

We assumed $0.0500 - [A^-] \approx 0.0500$. This assumption is borderline (2.68×10^{-3} is 5.4% of 0.0500). The HB assumption is good (0.4% error).

Using successive approximations to refine the $[H^+][A^-]$ value gives $[H^+][A^-] = 8.39 \times 10^{-6}$, which gives:

$$[H^+] = 3.22 \times 10^{-3}\ M,\ \ pH = -\log(3.22 \times 10^{-3}) = 2.492$$

Note: If we treat each acid separately:

H^+ from HA $= 2.9 \times 10^{-3}\ M$
H^+ from HB $= 1.4 \times 10^{-3}\ M$
$4.3 \times 10^{-3}\ M = [H^+]_{total}$

This assumes the acids did not suppress each other's ionization. They do, and we expect the $[H^+]$ to be less than $4.3 \times 10^{-3}\ M$. We get such an answer ($[H^+] = 3.22 \times 10^{-3}\ M$).

191. a. Strongest acid from group I = HCl; weakest base (smallest K_b) from group II = $NaNO_2$

0.20 *M* HCl + 0.20 *M* $NaNO_2$; major species = H^+, Cl^-, Na^+, NO_2^-, and H_2O; let the H^+ react to completion with the NO_2^-; then solve the back equilibrium problem.

	H^+	+ NO_2^-	$\rightarrow$ HNO_2	
Before	0.10 *M*	0.10 *M*	0	(Molarities are halved due to dilution.)
After	0	0	0.10 *M*	

	HNO_2	$\rightleftharpoons$	H^+	+ NO_2^-	$K_a = 4.0 \times 10^{-4}$
Initial	0.10 *M*		0	0	
Change	$-x$	$\rightarrow$	$+x$	$+x$	
Equil.	$0.10 - x$		x	x	

$$\frac{x^2}{0.10 - x} = 4.0 \times 10^{-4};\ \text{solving: } x = [H^+] = 6.1 \times 10^{-3}\ M;\ \ pH = 2.21$$

b. Weakest acid from group I = $(C_2H_5)_3NHCl$; best base from group II = KOI; the dominant equilibrium will be the best base reacting with the best acid.

	OI^-	+ $(C_2H_5)_3NH^+$	$\rightleftharpoons$	HOI	+ $(C_2H_5)_3N$
Initial	0.10 *M*	0.10 *M*		0	0
Equil.	$0.10 - x$	$0.10 - x$		x	x

$$K = \frac{K_{a,\ (C_2H_5)_3NH^+}}{K_{a,\ HOI}} = \frac{1.0 \times 10^{-14}}{4.0 \times 10^{-4}} \times \frac{1}{2.0 \times 10^{-11}} = 1.25 \text{ (carrying extra sig. fig.)}$$

$$\frac{x^2}{(0.10-x)^2} = 1.25, \quad \frac{x}{0.10-x} = 1.12, \quad x = 0.053\ M$$

So [HOI] = 0.053 M and $[OI^-] = 0.10 - x = 0.047\ M$; using the K_a equilibrium constant for HOI to solve for $[H^+]$:

$$2.0 \times 10^{-11} = \frac{[H^+](0.047)}{(0.053)}, \quad [H^+] = 2.3 \times 10^{-11}\ M; \quad pH = 10.64$$

c. K_a for $(C_2H_5)_3NH^+ = \dfrac{1.0 \times 10^{-14}}{4.0 \times 10^{-4}} = 2.5 \times 10^{-11}$

K_b for $NO_2^- = \dfrac{1.0 \times 10^{-14}}{4.0 \times 10^{-4}} = 2.5 \times 10^{-11}$

Because $K_a = K_b$, mixing $(C_2H_5)_3NHCl$ with $NaNO_2$ will result in a solution with pH = 7.00.

CHAPTER 15

ACID-BASE EQUILIBRIA

Questions

9. When an acid dissociates, ions are produced. The common ion effect is observed when one of the product ions in a particular equilibrium is added from an outside source. For a weak acid dissociating to its conjugate base and H^+, the common ion would be the conjugate base; this would be added by dissolving a soluble salt of the conjugate base into the acid solution. The presence of the conjugate base from an outside source shifts the equilibrium to the left so less acid dissociates.

10. $pH = pK_a + \log\frac{[base]}{[acid]}$; when [acid] > [base], then $\frac{[base]}{[acid]} < 1$ and $\log\left(\frac{[base]}{[acid]}\right) < 0$.

From the Henderson-Hasselbalch equation, if the log term is negative, then pH < pK_a. When one has more acid than base in a buffer, the pH will be on the acidic side of the pK_a value; that is, the pH is at a value lower than the pK_a value. When one has more base than acid in a buffer ([conjugate base] > [weak acid]), then the log term in the Henderson-Hasselbalch equation is positive, resulting in pH > pK_a. When one has more base than acid in a buffer, the pH is on the basic side of the pK_a value; that is, the pH is at a value greater than the pK_a value. The other scenario you can run across in a buffer is when [acid] = [base]. Here, the log term is equal to zero, and pH = pK_a.

11. The more weak acid and conjugate base present, the more H^+ and/or OH^- that can be absorbed by the buffer without significant pH change. When the concentrations of weak acid and conjugate base are equal (so that pH = pK_a), the buffer system is equally efficient at absorbing either H^+ or OH^-. If the buffer is overloaded with weak acid or with conjugate base, then the buffer is not equally efficient at absorbing either H^+ or OH^-.

12. a. The red plot is the pH curve for the strong acid and the blue plot is the pH curve for the weak acid. The pH at the equivalence point is 7.00 for the strong acid-strong base titration, while the pH is greater than 7.00 if a weak acid is titrated. Another point one could look at is the initial point. Because both acids have the same concentration, the strong acid curve will be at the lowest initial pH. Actually, any point at any volume up to the equivalence point for the strong acid plot will have a lower pH than the weak acid plot (assuming equal concentrations and volumes). Another difference would be the pH at the halfway point to equivalence. For the weak acid titration, the pH of solution equals the pK_a value for the weak acid at the halfway point to equivalence; this is not the case when a strong acid is titrated.

    b. A buffer is a solution that resists pH change. From this definition, both titrations have regions where the pH doesn't change much on addition of strong base, so both could be labeled to have buffer regions. However, we don't normally include strong acids as a component of buffer solutions. Strong acids certainly can absorb added OH^- by reacting with it to form water. But when more strong acid is added, the H^+ concentration increases steadily; there is nothing present in a strong acid solution to react with added H^+.

This is not the case in the weak acid-strong base titration. After some OH^- has been added, some weak acid is converted into its conjugate base. We now have a typical buffer solution because there are significant amounts of weak acid and conjugate base present at the same time. The buffer region extends from a little past the initial point in the titration up to just a little before the equivalence point. This entire region is a buffer region because both the weak acid and conjugate base are present in significant quantities in this region.

c. True; $HA + OH^- \rightarrow A^- + H_2O$; both reactions have the same neutralization reaction. In both cases, the equivalence point is reached when enough OH^- has been added to exactly react with the acid present initially. Because all acid concentrations and volumes are the same, we have equal moles of each acid which requires the same moles of OH^- to reach the equivalence point. Therefore, each acid requires the same volume of 0.10 *M* NaOH to reach the equivalence point.

d. False; the pH for the strong acid-strong base titration will be 7.00 at the equivalence point. The pH for the weak acid-strong base titration will be greater than 7.00 at the equivalence point. In both titrations, the major species present at the equivalence points are Na^+, H_2O, and the conjugate base of the acid titrated. Because the conjugate base of a strong acid has no basic characteristics, pH = 7.00 at the equivalence point. However, the conjugate base of a weak acid is a weak base. A weak base is present at the equivalence point of a weak acid-strong base titration, so the pH is basic (pH > 7.0).

13. a. Let's call the acid HA, which is a weak acid. When HA is present in the beakers, it exists in the undissociated form, making it a weak acid. A strong acid would exist as separate H^+ and A^- ions.

b. Beaker a contains 4 HA molecules and 2 A^- ions, beaker b contains 6 A^- ions, beaker c contains 6 HA molecules, beaker d contains 6 A^- and 6 OH^- ions, and beaker e contains 3 HA molecules and 3 A^- ions. $HA + OH^- \rightarrow A^- + H_2O$; this is the neutralization reaction that occurs when OH^- is added. We start off the titration with a beaker full of weak acid (beaker c). When some OH^- is added, we convert some weak acid HA into its conjugate base A^- (beaker a). At the halfway point to equivalence, we have converted exactly one-half of the initial amount of acid present into its conjugate base (beaker e). We finally reach the equivalence point when we have added just enough OH^- to convert all of the acid present initially into its conjugate base (beaker b). Past the equivalence point, we have added an excess of OH^-, so we have excess OH^- present as well as the conjugate base of the acid produced from the neutralization reaction (beaker d). The order of the beakers from start to finish is:

beaker c → beaker a → beaker e → beaker b → beaker d

c. $pH = pK_a$ when a buffer solution is present that has equal concentrations of the weak acid and conjugate base. This is beaker e.

d. The equivalence point is when just enough OH^- has been added to exactly react with all of the acid present initially. This is beaker b.

e. Past the equivalence point, the pH is dictated by the concentration of excess OH^- added from the strong base. We can ignore the amount of hydroxide added by the weak conjugate base that is also present. This is beaker d.

14. Titration i is a strong acid titrated by a strong base. The pH is very acidic until just before the equivalence point; at the equivalence point, pH = 7.00, and past the equivalence the pH is very basic. Titration ii is a strong base titrated by a strong acid. Here, the pH is very basic until just before the equivalence point; at the equivalence point, pH = 7.00, and past the equivalence point the pH is very acidic. Titration iii is a weak base titrated by a strong acid. The pH starts out basic because a weak base is present. However, the pH will not be as basic as in titration ii, where a strong base is titrated. The pH drops as HCl is added; then at the halfway point to equivalence, pH = pK_a. Because $K_b = 4.4 \times 10^{-4}$ for CH_3NH_2, $CH_3NH_3^+$ has $K_a = K_w/K_b = 2.3 \times 10^{-11}$ and $pK_a = 10.64$. So, at the halfway point to equivalence for this weak base-strong acid titration, pH = 10.64. The pH continues to drop as HCl is added; then at the equivalence point the pH is acidic (pH < 7.00) because the only important major species present is a weak acid (the conjugate acid of the weak base). Past the equivalence point the pH becomes more acidic as excess HCl is added. Titration iv is a weak acid titrated by a strong base. The pH starts off acidic, but not nearly as acidic as the strong acid titration (i). The pH increases as NaOH is added; then at the halfway point to equivalence, pH = pK_a for HF = $-\log(7.2 \times 10^{-4}) = 3.14$. The pH continues to increase past the halfway point; then at the equivalence point the pH is basic (pH > 7.0) because the only important major species present is a weak base (the conjugate base of the weak acid). Past the equivalence point the pH becomes more basic as excess NaOH is added.

a. All require the same volume of titrant to reach the equivalence point. At the equivalence point for all these titrations, moles acid = moles base ($M_AV_A = M_BV_B$). Because all the molarities and volumes are the same in the titrations, the volume of titrant will be the same (50.0 mL titrant added to reach equivalence point).

b. Increasing initial pH: i < iv < iii < ii; the strong acid titration has the lowest pH, the weak acid titration is next, followed by the weak base titration, with the strong base titration having the highest pH.

c. i < iv < iii < ii; the strong acid titration has the lowest pH at the halfway point to equivalence, and the strong base titration has the highest halfway point pH. For the weak acid titration, pH = pK_a = 3.14, and for the weak base titration, pH = pK_a = 10.64.

d. Equivalence point pH: iii < ii = i < iv; the strong-by-strong titrations have pH = 7.00 at the equivalence point. The weak base titration has an acidic pH at the equivalence point, and a weak acid titration has a basic equivalence point pH.
The only different answer when the weak acid and weak base are changed would be for part c. This is for the halfway point to equivalence, where pH = pK_a.

$$HOC_6H_5;\ K_a = 1.6 \times 10^{-10},\ pK_a = -\log(1.6 \times 10^{-10}) = 9.80$$

$$C_5H_5NH^+,\ K_a = \frac{K_w}{K_{b,\,C_5H_5N}} = \frac{1.0 \times 10^{-14}}{1.7 \times 10^{-9}} = 5.9 \times 10^{-6},\ pK_a = 5.23$$

From the pK_a values, the correct ordering at the halfway point to equivalence would be i < iii < iv < ii. Note that for the weak base-strong acid titration using C_5H_5N, the pH is acidic at the halfway point to equivalence, whereas the weak acid-strong base titration using HOC_6H_5 is basic at the halfway point to equivalence. This is fine; this will always happen when the weak base titrated has a $K_b < 1 \times 10^{-7}$ (so K_a of the conjugate acid is greater than 1×10^{-7}) and when the weak acid titrated has a $K_a < 1 \times 10^{-7}$ (so K_b of the conjugate base is greater than 1×10^{-7}).

15. The three key points to emphasize in your sketch are the initial pH, the pH at the halfway point to equivalence, and the pH at the equivalence point. For all the weak bases titrated, pH = pK_a at the halfway point to equivalence (50.0 mL HCl added) because [weak base] = [conjugate acid] at this point. Here, the weak base with $K_b = 1 \times 10^{-5}$ has a conjugate acid with $K_a = 1 \times 10^{-9}$, so pH = 9.0 at the halfway point. The weak base with $K_b = 1 \times 10^{-10}$ has a pH = 4.0 at the halfway point to equivalence. For the initial pH, the strong base has the highest pH (most basic), whereas the weakest base has the lowest pH (least basic). At the equivalence point (100.0 mL HCl added), the strong base titration has pH = 7.0. The weak bases titrated have acidic pH's because the conjugate acids of the weak bases titrated are the major species present. The weakest base has the strongest conjugate acid so its pH will be lowest (most acidic) at the equivalence point.

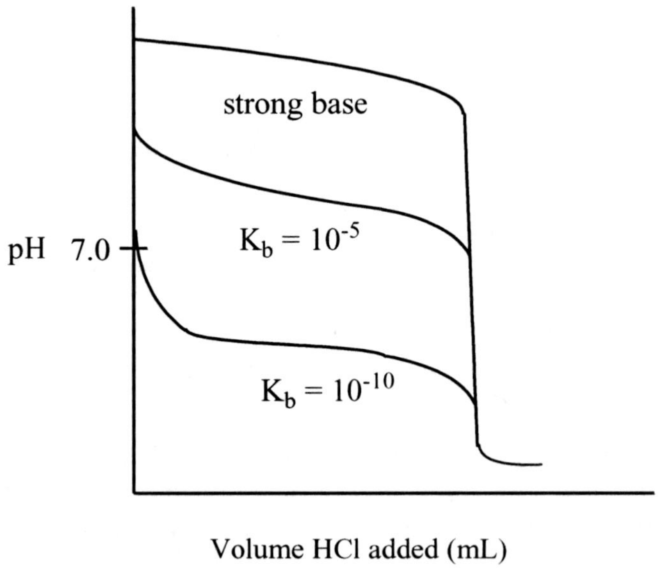

16. $HIn \rightleftharpoons H^+ + In^-$ $\quad K_a = \frac{[H^+][In^-]}{[HIn]}$

Indicators are weak acids themselves. The special property they have is that the acid form of the indicator (HIn) has one distinct color, whereas the conjugate base form (In^-) has a different distinct color. Which form dominates and thus determines the color of the solution is determined by the pH. An indicator is chosen in order to match the pH of the color change at about the pH of the equivalence point.

Exercises

Buffers

17. Only the third (lower) beaker represents a buffer solution. A weak acid and its conjugate base must both be present in large quantities in order to have a buffer solution. This is only the case in the third beaker. The first beaker represents a beaker full of strong acid which is 100% dissociated. The second beaker represents a weak acid solution. In a weak acid solution, only a small fraction of the acid is dissociated. In this representation, 1/10 of the weak acid has dissociated. The only B^- present in this beaker is from the dissociation of the weak acid. A buffer solution has B^- added from another source.

18. A buffer solution is a solution containing a weak acid plus its conjugate base or a weak base plus its conjugate acid. Solution c contains a weak acid (HOCl) plus its conjugate base (OCl^-), so it is a buffer. Solution e is also a buffer solution. It contains a weak base (H_2NNH_2) plus its conjugate acid ($H_2NNH_3^+$).

    Solution a contains a strong acid (HBr) and a weak acid (HOBr). Solution b contains a strong acid ($HClO_4$) and a strong base (RbOH). Solution d contains a strong base (KOH) and a weak base ($HONH_2$).

19. When strong acid or strong base is added to a bicarbonate-carbonate mixture, the strong acid(base) is neutralized. The reaction goes to completion, resulting in the strong acid(base) being replaced with a weak acid(base), resulting in a new buffer solution. The reactions are:

$$H^+(aq) + CO_3^{2-}(aq) \rightarrow HCO_3^-(aq); \quad OH^- + HCO_3^-(aq) \rightarrow CO_3^{2-}(aq) + H_2O(l)$$

20. Similar to the HCO_3^-/CO_3^{2-} buffer discussed in Exercise 19, the $HONH_3^+/HONH_2$ buffer absorbs added OH^- and H^+ in the same fashion.

$$HONH_2(aq) + H^+(aq) \rightarrow HONH_3^+(aq)$$

$$HONH_3^+(aq) + OH^-(aq) \rightarrow HONH_2(aq) + H_2O(l)$$

21. a. This is a weak acid problem. Let $HC_3H_5O_2$ = HOPr and $C_3H_5O_2^-$ = OPr^-.

	HOPr(aq)	⇌	H^+(aq)	+	OPr^-(aq)	$K_a = 1.3 \times 10^{-5}$
Initial	0.100 *M*		~0		0	
	x mol/L HOPr dissociates to reach equilibrium					
Change	$-x$	→	$+x$		$+x$	
Equil.	$0.100 - x$		x		x	

$$K_a = 1.3 \times 10^{-5} = \frac{[H^+][OPr^-]}{[HOPr]} = \frac{x^2}{0.100 - x} \approx \frac{x^2}{0.100}$$

$x = [H^+] = 1.1 \times 10^{-3}\ M$; pH = 2.96; assumptions good by the 5% rule.

b. This is a weak base problem.

	$OPr^-(aq)$ + $H_2O(l)$	$\rightleftharpoons$	$HOPr(aq)$	+ $OH^-(aq)$	$K_b = \frac{K_w}{K_a} = 7.7 \times 10^{-10}$
Initial	0.100 *M*		0	~0	
	x mol/L OPr^- reacts with H_2O to reach equilibrium				
Change	$-x$	→	$+x$	$+x$	
Equil.	$0.100 - x$		x	x	

$$K_b = 7.7 \times 10^{-10} = \frac{[HOPr][OH^-]}{[OPr^-]} = \frac{x^2}{0.100 - x} \approx \frac{x^2}{0.100}$$

$x = [OH^-] = 8.8 \times 10^{-6}$ *M*; pOH = 5.06; pH = 8.94; assumptions good.

c. Pure H_2O, $[H^+] = [OH^-] = 1.0 \times 10^{-7}$ *M*; pH = 7.00

d. This solution contains a weak acid and its conjugate base. This is a buffer solution. We will solve for the pH through the weak acid equilibrium reaction.

	$HOPr(aq)$	$\rightleftharpoons$	$H^+(aq)$	+ $OPr^-(aq)$	$K_a = 1.3 \times 10^{-5}$
Initial	0.100 *M*		~0	0.100 *M*	
	x mol/L HOPr dissociates to reach equilibrium				
Change	$-x$	→	$+x$	$+x$	
Equil.	$0.100 - x$		x	$0.100 + x$	

$$1.3 \times 10^{-5} = \frac{(0.100 + x)(x)}{0.100 - x} \approx \frac{(0.100)(x)}{0.100} = x = [H^+]$$

$[H^+] = 1.3 \times 10^{-5}$ *M*; pH = 4.89; assumptions good.

Alternately, we can use the Henderson-Hasselbalch equation to calculate the pH of buffer solutions.

$$pH = pK_a + \log\frac{[base]}{[acid]} = pK_a + \log\left(\frac{0.100}{0.100}\right) = pK_a = -\log(1.3 \times 10^{-5}) = 4.89$$

The Henderson-Hasselbalch equation will be valid when an assumption of the type $0.1 + x \approx 0.1$ that we just made in this problem is valid. From a practical standpoint, this will almost always be true for useful buffer solutions. If the assumption is not valid, the solution will have such a low buffering capacity it will not be of any use to control the pH. *Note*: The Henderson-Hasselbalch equation can only be used to solve for the pH of buffer solutions.

22. a. Weak base problem:

	$HONH_2$ + H_2O	$\rightleftharpoons$	$HONH_3^+$	+ OH^-	$K_b = 1.1 \times 10^{-8}$
Initial	0.100 *M*		0	~0	
	x mol/L $HONH_2$ reacts with H_2O to reach equilibrium				
Change	$-x$	→	$+x$	$+x$	
Equil.	$0.100 - x$		x	x	

$$K_b = 1.1 \times 10^{-8} = \frac{x^2}{0.100 - x} \approx \frac{x^2}{0.100}$$

$x = [OH^-] = 3.3 \times 10^{-5}\ M$; pOH = 4.48; pH = 9.52; assumptions good.

b. Weak acid problem (Cl^- has no acidic/basic properties);

	$HONH_3^+$	$\rightleftharpoons$	$HONH_2$	+	H^+
Initial	0.100 *M*		0		~0
	x mol/L $HONH_3^+$ dissociates to reach equilibrium				
Change	$-x$	→	$+x$		$+x$
Equil.	$0.100 - x$		x		x

$$K_a = \frac{K_w}{K_b} = 9.1 \times 10^{-7} = \frac{[HONH_2][H^+]}{[HONH_3^+]} = \frac{x^2}{0.100 - x} \approx \frac{x^2}{0.100}$$

$x = [H^+] = 3.0 \times 10^{-4}\ M$; pH = 3.52; assumptions good.

c. Pure H_2O, pH = 7.00

d. Buffer solution where $pK_a = -\log(9.1 \times 10^{-7}) = 6.04$. Using the Henderson-Hasselbalch equation:

$$pH = pK_a + \log\frac{[base]}{[acid]} = 6.04 + \log\frac{[HONH_2]}{[HONH_3^+]} = 6.04 + \log\frac{(0.100)}{(0.100)} = 6.04$$

23. 0.100 *M* $HC_3H_5O_2$: Percent dissociation $= \frac{[H^+]}{[HC_3H_5O_2]_0} \times 100 = \frac{1.1 \times 10^{-3}\ M}{0.100\ M} \times 100$

$= 1.1\%$

0.100 *M* $HC_3H_5O_2$ + 0.100 *M* $NaC_3H_5O_2$: % dissociation $= \frac{1.3 \times 10^{-5}}{0.100} \times 100 = 1.3 \times 10^{-2}\ \%$

The percent dissociation of the acid decreases from 1.1% to 1.3×10^{-2} % (a factor of 85) when $C_3H_5O_2^-$ is present. This is known as the common ion effect. The presence of the conjugate base of the weak acid inhibits the acid dissociation reaction.

24. 0.100 *M* $HONH_2$: Percent ionization $\frac{[OH^-]}{[HONH_2]_0} \times 100 = \frac{3.3 \times 10^{-5}\ M}{0.100\ M} \times 100$

$= 3.3 \times 10^{-2}\ \%$

0.100 *M* $HONH_2$ + 0.100 *M* $HONH_3^+$: % ionization $= \frac{1.1 \times 10^{-8}}{0.100} \times 100 = 1.1 \times 10^{-5}\%$

The percent ionization decreases by a factor of 3000. The presence of the conjugate acid of the weak base inhibits the weak base reaction with water. This is known as the common ion effect.

25. a. We have a weak acid (HOPr = $HC_3H_5O_2$) and a strong acid (HCl) present. The amount of H^+ donated by the weak acid will be negligible. To prove it, consider the weak acid equilibrium reaction:

	$HOPr$	$\rightleftharpoons$	H^+	+	OPr^-	$K_a = 1.3 \times 10^{-5}$
Initial	0.100 *M*		0.020 *M*		0	
	x mol/L HOPr dissociates to reach equilibrium					
Change	$-x$	→	$+x$		$+x$	
Equil.	$0.100 - x$		$0.020 + x$		x	

$[H^+] = 0.020 + x \approx 0.020$ *M*; pH = 1.70; assumption good ($x = 6.5 \times 10^{-5}$ is << 0.020).

Note: The H^+ contribution from the weak acid HOPr was negligible. The pH of the solution can be determined by only considering the amount of strong acid present.

b. Added H^+ reacts completely with the best base present, OPr^-.

	OPr^-	+	H^+	→	$HOPr$	
Before	0.100 *M*		0.020 *M*		0	
Change	−0.020		−0.020	→	+0.020	Reacts completely
After	0.080		0		0.020 *M*	

After reaction, a weak acid, HOPr , and its conjugate base, OPr^-, are present. This is a buffer solution. Using the Henderson-Hasselbalch equation where $pK_a = -\log(1.3 \times 10^{-5}) = 4.89$:

$$pH = pK_a + \log\frac{[\text{base}]}{[\text{acid}]} = 4.89 + \log\frac{(0.080)}{(0.020)} = 5.49; \text{ assumptions good.}$$

c. This is a strong acid problem. $[H^+] = 0.020$ *M*; pH = 1.70

d. Added H^+ reacts completely with the best base present, OPr^-.

	OPr^-	+	H^+	→	$HOPr$	
Before	0.100 *M*		0.020 *M*		0.100 *M*	
Change	−0.020		−0.020	→	+0.020	Reacts completely
After	0.080		0		0.120	

A buffer solution results (weak acid + conjugate base). Using the Henderson-Hasselbalch equation:

$$pH = pK_a + \log\frac{[\text{base}]}{[\text{acid}]} = 4.89 + \log\frac{(0.080)}{(0.120)} = 4.71; \text{ assumptions good.}$$

26. a. Added H^+ reacts completely with $HONH_2$ (the best base present) to form $HONH_3^+$.

	$HONH_2$	+	H^+	→	$HONH_3^+$	
Before	0.100 *M*		0.020 *M*		0	
Change	−0.020		−0.020	→	+0.020	Reacts completely
After	0.080		0		0.020	

After this reaction, a buffer solution exists; that is, a weak acid ($HONH_3^+$) and its conjugate base ($HONH_2$) are present at the same time. Using the Henderson-Hasselbalch equation to solve for the pH where $pK_a = -\log(K_w/K_b) = 6.04$:

$$pH = pK_a + \log\frac{[\text{base}]}{[\text{acid}]} = 6.04 + \log\frac{(0.080)}{(0.020)} = 6.04 + 0.60 = 6.64$$

b. We have a weak acid and a strong acid present at the same time. The H^+ contribution from the weak acid, $HONH_3^+$, will be negligible. So we have to consider only the H^+ from HCl. $[H^+] = 0.020\ M$; pH = 1.70

c. This is a strong acid in water. $[H^+] = 0.020\ M$; pH = 1.70

d. Major species: H_2O, Cl^-, $HONH_2$, $HONH_3^+$, H^+

H^+ will react completely with $HONH_2$, the best base present.

	$HONH_2$	+	H^+	→	$HONH_3^+$	
Before	0.100 *M*		0.020 *M*		0.100 *M*	
Change	–0.020		–0.020	→	+0.020	Reacts completely
After	0.080		0		0.120	

A buffer solution results after reaction. Using the Henderson-Hasselbalch equation:

$$pH = 6.04 + \log\frac{[HONH_2]}{[HONH_3^+]} = 6.04 + \log\frac{(0.080)}{(0.120)} = 6.04 - 0.18 = 5.86$$

27. a. OH^- will react completely with the best acid present, HOPr.

	HOPr	+	OH^-	→	OPr^-	+ H_2O	
Before	0.100 *M*		0.020 *M*		0		
Change	–0.020		–0.020	→	+0.020		Reacts completely
After	0.080		0		0.020		

A buffer solution results after the reaction. Using the Henderson-Hasselbalch equation:

$$pH = pK_a + \log\frac{[\text{base}]}{[\text{acid}]} = 4.89 + \log\frac{(0.020)}{(0.080)} = 4.29;\ \text{assumptions good.}$$

b. We have a weak base and a strong base present at the same time. The amount of OH^- added by the weak base will be negligible. To prove it, let's consider the weak base equilibrium:

	OPr^-	+ H_2O	⇌	HOPr	+	OH^-	$K_b = 7.7 \times 10^{-10}$
Initial	0.100 *M*			0		0.020 *M*	
	x mol/L OPr^- reacts with H_2O to reach equilibrium						
Change	$-x$		→	$+x$		$+x$	
Equil.	$0.100 - x$			x		$0.020 + x$	

$[OH^-] = 0.020 + x \approx 0.020\ M$; pOH = 1.70; pH = 12.30; assumption good.

Note: The OH^- contribution from the weak base OPr^- was negligible ($x = 3.9 \times 10^{-9}\ M$ as compared to 0.020 *M* OH^- from the strong base). The pH can be determined by only considering the amount of strong base present.

c. This is a strong base in water. $[OH^-] = 0.020\ M$; pOH = 1.70; pH = 12.30

d. OH^- will react completely with HOPr, the best acid present.

	HOPr	+	OH^-	→	OPr^-	+	H_2O	
Before	0.100 *M*		0.020 *M*		0.100 *M*			
Change	–0.020		–0.020	→	+0.020			Reacts completely
After	0.080		0		0.120			

Using the Henderson-Hasselbalch equation to solve for the pH of the resulting buffer solution:

$$pH = pK_a + \log\frac{[base]}{[acid]} = 4.89 + \log\frac{(0.120)}{(0.080)} = 5.07;\ \text{assumptions good.}$$

28. a. We have a weak base and a strong base present at the same time. The OH^- contribution from the weak base, $HONH_2$, will be negligible. Consider only the added strong base as the primary source of OH^-.

$[OH^-] = 0.020\ M$; pOH = 1.70; pH = 12.30

b. Added strong base will react to completion with the best acid present, $HONH_3^+$.

	OH^-	+	$HONH_3^+$	→	$HONH_2$	+	H_2O	
Before	0.020 *M*		0.100 *M*		0			
Change	–0.020		–0.020	→	+0.020			Reacts completely
After	0		0.080		0.020			

The resulting solution is a buffer (a weak acid and its conjugate base). Using the Henderson-Hasselbalch equation:

$$pH = 6.04 + \log\frac{(0.020)}{(0.080)} = 6.04 - 0.60 = 5.44$$

c. This is a strong base in water. $[OH^-] = 0.020\ M$; pOH = 1.70; pH = 12.30

d. Major species: H_2O, Cl^-, Na^+, $HONH_2$, $HONH_3^+$, OH^-; again, the added strong base reacts completely with the best acid present, $HONH_3^+$.

	$HONH_3^+$	+	OH^-	→	$HONH_2$	+	H_2O	
Before	0.100 *M*		0.020 *M*		0.100 *M*			
Change	–0.020		–0.020	→	+0.020			Reacts completely
After	0.080		0		0.120			

A buffer solution results. Using the Henderson-Hasselbalch equation:

$$pH = 6.04 + \log\frac{[HONH_2]}{[HONH_3^+]} = 6.04 + \log\frac{(0.120)}{(0.080)} = 6.04 + 0.18 = 6.22$$

29. Consider all the results to Exercises 21, 25, and 27:

Solution	Initial pH	After Added H^+	After Added OH^-
a	2.96	1.70	4.29
b	8.94	5.49	12.30
c	7.00	1.70	12.30
d	4.89	4.71	5.07

The solution in Exercise 21d is a buffer; it contains both a weak acid ($HC_3H_5O_2$) and a weak base ($C_3H_5O_2^-$). Solution d shows the greatest resistance to changes in pH when either a strong acid or a strong base is added, which is the primary property of buffers.

30. Consider all of the results to Exercises 22, 26, and 28.

Solution	Initial pH	After Added H^+	After Added OH^-
a	9.52	6.64	12.30
b	3.52	1.70	5.44
c	7.00	1.70	12.30
d	6.04	5.86	6.22

The solution in Exercise 22d is a buffer; it shows the greatest resistance to a change in pH when strong acid or base is added. The solution in Exercise 22d contains a weak acid ($HONH_3^+$) and a weak base ($HONH_2$), which constitutes a buffer solution.

31. Major species: HNO_2, NO_2^- and Na^+. Na^+ has no acidic or basic properties. One appropriate equilibrium reaction you can use is the K_a reaction of HNO_2, which contains both HNO_2 and NO_2^-. However, you could also use the K_b reaction for NO_2^- and come up with the same answer. Solving the equilibrium problem (called a buffer problem):

	HNO_2	$\rightleftharpoons$	NO_2^-	+	H^+
Initial	1.00 M		1.00 M		~0
	x mol/L HNO_2 dissociates to reach equilibrium				
Change	$-x$	$\rightarrow$	$+x$		$+x$
Equil.	$1.00 - x$		$1.00 + x$		x

$$K_a = 4.0 \times 10^{-4} = \frac{[NO_2^-][H^+]}{[HNO_2]} = \frac{(1.00 + x)(x)}{1.00 - x} \approx \frac{(1.00)(x)}{1.00} \quad \text{(assuming } x << 1.00\text{)}$$

$x = 4.0 \times 10^{-4}\ M = [H^+]$; assumptions good ($x$ is 4.0×10^{-2} % of 1.00).

$$pH = -\log(4.0 \times 10^{-4}) = 3.40$$

Note: We would get the same answer using the Henderson-Hasselbalch equation. Use whichever method you prefer.

32. Major species: HF, F^-, K^+, and H_2O. K^+ has no acidic or basic properties. This is a solution containing a weak acid and its conjugate base. This is a buffer solution. One appropriate equilibrium reaction you can use is the K_a reaction of HF, which contains both HF and F^-. However, you could also use the K_b reaction for F^- and come up with the same answer. Alternately, you could use the Henderson-Hasselblach equation to solve for the pH. For this problem, we will use the K_a reaction and set up an ICE table to solve for the pH.

	HF	$\rightleftharpoons$	F^-	+	H^+
Initial	0.60 *M*		1.00 *M*		~0
	x mol/L HF dissociates to reach equilibrium				
Change	$-x$	$\rightarrow$	$+x$		$+x$
Equil.	$0.60 - x$		$1.00 + x$		x

$$K_a = 7.2 \times 10^{-4} = \frac{[F^-][H^+]}{[HF]} = \frac{(1.00 + x)(x)}{0.60 - x} \approx \frac{(1.00)(x)}{0.60} \quad \text{(assuming } x << 0.60\text{)}$$

$x = [H^+] = 0.60 \times (7.2 \times 10^{-4}) = 4.3 \times 10^{-4}$ *M*; assumptions good.

$pH = -\log(4.3 \times 10^{-4}) = 3.37$

33. Major species after NaOH added: HNO_2, NO_2^-, Na^+, and OH^-. The OH^- from the strong base will react with the best acid present (HNO_2). Any reaction involving a strong base is assumed to go to completion. Because all species present are in the same volume of solution, we can use molarity units to do the stoichiometry part of the problem (instead of moles). The stoichiometry problem is:

	OH^-	+	HNO_2	$\rightarrow$	NO_2^- + H_2O	
Before	0.10 mol/1.00 L		1.00 *M*		1.00 *M*	
Change	-0.10 *M*		-0.10 *M*	$\rightarrow$	$+0.10$ *M*	Reacts completely
After	0		0.90		1.10	

After all the OH^- reacts, we are left with a solution containing a weak acid (HNO_2) and its conjugate base (NO_2^-). This is what we call a buffer problem. We will solve this buffer problem using the K_a equilibrium reaction. One could also use the K_b equilibrium reaction or use the Henderson-Hasselbalch equation to solve for the pH.

	HNO_2	$\rightleftharpoons$	NO_2^-	+	H^+
Initial	0.90 *M*		1.10 *M*		~0
	x mol/L HNO_2 dissociates to reach equilibrium				
Change	$-x$	$\rightarrow$	$+x$		$+x$
Equil.	$0.90 - x$		$1.10 + x$		x

$$K_a = 4.0 \times 10^{-4} = \frac{(1.10 + x)(x)}{0.90 - x} \approx \frac{(1.10)(x)}{0.90}, \quad x = [H^+] = 3.3 \times 10^{-4}\ M;\ pH = 3.48;$$

assumptions good.

Note: The added NaOH to this buffer solution changes the pH only from 3.40 to 3.48. If the NaOH were added to 1.0 L of pure water, the pH would change from 7.00 to 13.00.

Major species after HCl added: HNO_2, NO_2^-, H^+, Na^+, Cl^-; the added H^+ from the strong acid will react completely with the best base present (NO_2^-).

	H^+	+	NO_2^-	$\rightarrow$	HNO_2	
Before	$\frac{0.20\text{ mol}}{1.00\text{ L}}$		1.00 *M*		1.00 *M*	
Change	–0.20 *M*		–0.20 *M*	$\rightarrow$	+0.20 *M*	Reacts completely
After	0		0.80		1.20	

After all the H^+ has reacted, we have a buffer solution (a solution containing a weak acid and its conjugate base). Solving the buffer problem:

	HNO_2	$\rightleftharpoons$	NO_2^-	+	H^+
Initial	1.20 *M*		0.80 *M*		0
Equil.	$1.20 - x$		$0.80 + x$		$+x$

$$K_a = 4.0 \times 10^{-4} = \frac{(0.80 + x)(x)}{1.20 - x} \approx \frac{(0.80)(x)}{1.20},\ x = [H^+] = 6.0 \times 10^{-4}\ M;\ pH = 3.22;$$

assumptions good.

Note: The added HCl to this buffer solution changes the pH only from 3.40 to 3.22. If the HCl were added to 1.0 L of pure water, the pH would change from 7.00 to 0.70.

34. Major species after NaOH added: HF, F^-, K^+, Na^+, OH^-, and H_2O. The OH^- from the strong base will react with the best acid present (HF). Any reaction involving a strong base is assumed to go to completion. Because all species present are in the same volume of solution, we can use molarity units to do the stoichiometry part of the problem (instead of moles). The stoichiometry problem is:

	OH^-	+	HF	$\rightarrow$	F^-	+ H_2O	
Before	0.10 mol/1.00 L		0.60 *M*		1.00 *M*		
Change	–0.10 *M*		–0.10 *M*	$\rightarrow$	+0.10 *M*		Reacts completely
After	0		0.50		1.10		

After all the OH^- reacts, we are left with a solution containing a weak acid (HF) and its conjugate base (F^-). This is what we call a buffer problem. We will solve this buffer problem using the K_a equilibrium reaction. One could also use the K_b equilibrium reaction or use the Henderson-Hasselbalch equation to solve for the pH.

	HF	$\rightleftharpoons$	F^-	+	H^+
Initial	0.50 *M*		1.10 *M*		~0
	x mol/L HF dissociates to reach equilibrium				
Change	$-x$	$\rightarrow$	$+x$		$+x$
Equil.	$0.50 - x$		$1.10 + x$		x

$$K_a = 7.2 \times 10^{-4} = \frac{(1.10 + x)(x)}{0.50 - x} \approx \frac{(1.10)(x)}{0.50},\ x = [H^+] = 3.3 \times 10^{-4}\ M;\ pH = 3.48;$$

assumptions good.

Note: The added NaOH to this buffer solution changes the pH only from 3.37 to 3.48. If the NaOH were added to 1.0 L of pure water, the pH would change from 7.00 to 13.00.

Major species after HCl added: HF, F^-, H^+, K^+, Cl^-, and H_2O; the added H^+ from the strong acid will react completely with the best base present (F^-).

	H^+	+	F^-	$\rightarrow$	HF	
Before	$\frac{0.20 \text{ mol}}{1.00 \text{ L}}$		1.00 *M*		0.60 *M*	
Change	–0.20 *M*		–0.20 *M*	$\rightarrow$	+0.20 *M*	Reacts completely
After	0		0.80		0.80	

After all the H^+ has reacted, we have a buffer solution (a solution containing a weak acid and its conjugate base). Solving the buffer problem:

	HF	$\rightleftharpoons$	F^-	+	H^+
Initial	0.80 *M*		0.80 *M*		0
Equil.	$0.80 - x$		$0.80 + x$		x

$$K_a = 7.2 \times 10^{-4} = \frac{(0.80 + x)(x)}{0.80 - x} \approx \frac{(0.80)(x)}{0.80}, \ x = [H^+] = 7.2 \times 10^{-4}\ M;\ pH = 3.14;$$

assumptions good.

Note: The added HCl to this buffer solution changes the pH only from 3.37 to 3.14. If the HCl were added to 1.0 L of pure water, the pH would change from 7.00 to 0.70.

35. a.

	$HC_2H_3O_2$	$\rightleftharpoons$	H^+	+	$C_2H_3O_2^-$	$K_a = 1.8 \times 10^{-5}$
Initial	0.10 *M*		~0		0.25 *M*	
	x mol/L $HC_2H_3O_2$ dissociates to reach equilibrium					
Change	$-x$	$\rightarrow$	$+x$		$+x$	
Equil.	$0.10 - x$		x		$0.25 + x$	

$$1.8 \times 10^{-5} = \frac{x(0.25 + x)}{(0.10 - x)} \approx \frac{x(0.25)}{0.10} \quad \text{(assuming } 0.25 + x \approx 0.25 \text{ and } 0.10 - x \approx 0.10\text{)}$$

$x = [H^+] = 7.2 \times 10^{-6}\ M$; pH = 5.14; assumptions good by the 5% rule.

Alternatively, we can use the Henderson-Hasselbalch equation:

$$pH = pK_a + \log\frac{[\text{base}]}{[\text{acid}]}, \text{ where } pK_a = -\log(1.8 \times 10^{-5}) = 4.74$$

$$pH = 4.74 + \log\frac{(0.25)}{(0.10)} = 4.74 + 0.40 = 5.14$$

The Henderson-Hasselbalch equation will be valid when assumptions of the type, $0.10 - x \approx 0.10$, that we just made are valid. From a practical standpoint, this will almost always be true for useful buffer solutions. *Note*: The Henderson-Hasselbalch equation can only be used to solve for the pH of buffer solutions.

b. $pH = 4.74 + \log\frac{(0.10)}{(0.25)} = 4.74 + (-0.40) = 4.34$

c. $pH = 4.74 + \log\frac{(0.20)}{(0.080)} = 4.74 + 0.40 = 5.14$

d. $pH = 4.74 + \log\frac{(0.080)}{(0.20)} = 4.74 + (-0.40) = 4.34$

36. We will use the Henderson-Hasselbalch equation to solve for the pH of these buffer solutions.

a. $pH = pK_a + \log\frac{[base]}{[acid]}$; $[base] = [C_2H_5NH_2] = 0.50\ M$; $[acid] = [C_2H_5NH_3^+] = 0.25\ M$

$$K_a = \frac{K_w}{K_b} = \frac{1.0 \times 10^{-14}}{5.6 \times 10^{-4}} = 1.8 \times 10^{-11}$$

$$pH = -\log(1.8 \times 10^{-11}) + \log\frac{(0.50\ M)}{(0.25\ M)} = 10.74 + 0.30 = 11.04$$

b. $pH = 10.74 + \log\frac{(0.25\ M)}{(0.50\ M)} = 10.74 + (-0.30) = 10.44$

c. $pH = 10.74 + \log\frac{(0.50\ M)}{(0.50\ M)} = 10.74 + 0 = 10.74$

37. $$[HC_7H_5O_2] = \frac{21.5\text{ g } HC_7H_5O_2 \times \frac{1\text{ mol } HC_7H_5O_2}{122.12\text{ g}}}{0.2000\text{L}} = 0.880\ M$$

$$[C_7H_5O_2^-] = \frac{37.7\text{ g } NaC_7H_5O_2 \times \frac{1\text{ mol } NaC_7H_5O_2}{144.10\text{ g}} \times \frac{1\text{ mol } C_7H_5O_2^-}{\text{mol } NaC_7H_5O_2}}{0.2000\text{L}} = 1.31\ M$$

We have a buffer solution since we have both a weak acid and its conjugate base present at the same time. One can use the K_a reaction or the K_b reaction to solve. We will use the K_a reaction for the acid component of the buffer.

	$HC_7H_5O_2$	$\rightleftharpoons$ H^+	+ $C_7H_5O_2^-$
Initial	0.880 M	~0	1.31 M
	x mol/L of $HC_7H_5O_2$ dissociates to reach equilibrium		
Change	$-x$	$\rightarrow$ $+x$	$+x$
Equil.	$0.880 - x$	x	$1.31 + x$

$$K_a = 6.4 \times 10^{-5} = \frac{x(1.31 + x)}{0.880 - x} \approx \frac{x(1.31)}{0.880},\ x = [H^+] = 4.3 \times 10^{-5}\ M$$

$pH = -\log(4.3 \times 10^{-5}) = 4.37$; assumptions good.

Alternatively, we can use the Henderson-Hasselbalch equation to calculate the pH of buffer solutions.

$$pH = pK_a + \log\frac{[\text{base}]}{[\text{acid}]} = pK_a + \log\frac{[C_7H_5O_2^-]}{[HC_7H_5O_2]}$$

$$pH = -\log(6.4 \times 10^{-5}) + \log\left(\frac{1.31}{0.880}\right) = 4.19 + 0.173 = 4.36$$

Within round-off error, this is the same answer we calculated solving the equilibrium problem using the K_a reaction.

The Henderson-Hasselbalch equation will be valid when an assumption of the type $1.31 + x \approx 1.31$ that we just made in this problem is valid. From a practical standpoint, this will almost always be true for useful buffer solutions. If the assumption is not valid, the solution will have such a low buffering capacity that it will be of no use to control the pH. *Note*: The Henderson-Hasselbalch equation can <u>only</u> be used to solve for the pH of buffer solutions.

38. $50.0 \text{ g } NH_4Cl \times \frac{1 \text{ mol } NH_4Cl}{53.49 \text{ g } NH_4Cl} = 0.935 \text{ mol } NH_4Cl$ added to 1.00 L; $[NH_4^+] = 0.935\ M$

Using the Henderson Hasselbalch equation to solve for the pH of this buffer solution:

$$pH = pK_a + \log\frac{[NH_3]}{[NH_4^+]} = -\log(5.6 \times 10^{-10}) + \log\left(\frac{0.75}{0.935}\right) = 9.25 - 0.096 = 9.15$$

39. $[H^+] \text{ added} = \frac{0.010 \text{ mol}}{0.2500 \text{ L}} = 0.040\ M$; the added H^+ reacts completely with NH_3 to form NH_4^+.

a.

	NH_3	+	H^+	$\rightarrow$	NH_4^+	
Before	0.050 *M*		0.040 *M*		0.15 *M*	
Change	–0.040		–0.040	$\rightarrow$	+0.040	Reacts completely
After	0.010		0		0.19	

A buffer solution still exists after H^+ reacts completely. Using the Henderson-Hasselbalch equation:

$$pH = pK_a + \log\frac{[NH_3]}{[NH_4^+]} = -\log(5.6 \times 10^{-10}) + \log\left(\frac{0.010}{0.19}\right) = 9.25 + (-1.28) = 7.97$$

b.

	NH_3	+	H^+	$\rightarrow$	NH_4^+	
Before	0.50 *M*		0.040 *M*		1.50 *M*	
Change	–0.040		–0.040	$\rightarrow$	+0.040	Reacts completely
After	0.46		0		1.54	

A buffer solution still exists. $pH = pK_a + \log\frac{[NH_3]}{[NH_4^+]}$, $9.25 + \log\left(\frac{0.46}{1.54}\right) = 8.73$

The two buffers differ in their capacity and not their initial pH (both buffers had an initial pH = 8.77). Solution b has the greatest capacity since it has the largest concentrations of weak acid and conjugate base. Buffers with greater capacities will be able to absorb more added H^+ or OH^-.

40. a. pK_b for $C_6H_5NH_2 = -\log(3.8 \times 10^{-10}) = 9.42$; pK_a for $C_6H_5NH_3^+ = 14.00 - 9.42 = 4.58$

$$pH = pK_a + \log\frac{[C_6H_5NH_2]}{[C_6H_5NH_3^+]},\ 4.20 = 4.58 + \log\frac{0.50\,M}{[C_6H_5NH_3^+]}$$

$$-0.38 = \log\frac{0.50\,M}{[C_6H_5NH_3^+]},\ [C_6H_5NH_3^+] = [C_6H_5NH_3Cl] = 1.2\,M$$

b. $$4.0\text{ g NaOH} \times \frac{1\text{ mol NaOH}}{40.00\text{ g}} \times \frac{1\text{ mol OH}^-}{\text{mol NaOH}} = 0.10\text{ mol OH}^-;\ [OH^-] = \frac{0.10\text{ mol}}{1.0\text{ L}} = 0.10\,M$$

	$C_6H_5NH_3^+$	+	OH^-	$\rightarrow$	$C_6H_5NH_2$	+	H_2O	
Before	1.2 M		0.10 M		0.50 M			
Change	−0.10		−0.10	$\rightarrow$	+0.10			Reacts completely
After	1.1		0		0.60			

A buffer solution exists. $pH = 4.58 + \log\left(\frac{0.60}{1.1}\right) = 4.32$

41. $$pH = pK_a + \log\frac{[C_2H_3O_2^-]}{[HC_2H_3O_2]};\ pK_a = -\log(1.8 \times 10^{-5}) = 4.74$$

Because the buffer components, $C_2H_3O_2^-$ and $HC_2H_3O_2$, are both in the same volume of solution, the concentration ratio of $[C_2H_3O_2^-] : [HC_2H_3O_2]$ will equal the mole ratio of mol $C_2H_3O_2^-$ to mol $HC_2H_3O_2$.

$$5.00 = 4.74 + \log\frac{\text{mol } C_2H_3O_2^-}{\text{mol } HC_2H_3O_2};\ \text{mol } HC_2H_3O_2 = 0.5000\text{ L} \times \frac{0.200\text{ mol}}{\text{L}} = 0.100\text{ mol}$$

$$0.26 = \log\frac{\text{mol } C_2H_3O_2^-}{0.100\text{ mol}},\ \frac{\text{mol } C_2H_3O_2^-}{0.100} = 10^{0.26} = 1.8,\ \text{mol } C_2H_3O_2^- = 0.18\text{ mol}$$

$$\text{Mass } NaC_2H_3O_2 = 0.18\text{ mol } NaC_2H_3O_2 \times \frac{82.03\text{ g}}{\text{mol}} = 15\text{ g } NaC_2H_3O_2$$

42. $$pH = pK_a + \log\frac{[NO_2^-]}{[HNO_2]},\ 3.55 = -\log(4.0 \times 10^{-4}) + \log\frac{[NO_2^-]}{[HNO_2]}$$

$$3.55 = 3.40 + \log\frac{[NO_2^-]}{[HNO_2]},\ \frac{[NO_2^-]}{[HNO_2]} = 10^{0.15} = 1.4 = \frac{\text{mol } NO_2^-}{\text{mol } HNO_2}$$

Let x = volume (L) of HNO_2 solution needed; then $1.00 - x$ = volume of $NaNO_2$ solution needed to form this buffer solution.

$$\frac{\text{Mol } NO_2^-}{\text{Mol } HNO_2} = 1.4 = \frac{(1.00 - x) \times \dfrac{0.50 \text{ mol } NaNO_2}{L}}{x \times \dfrac{0.50 \text{ mol } HNO_2}{L}} = \frac{0.50 - (0.50)x}{(0.50)x}$$

$(0.70)x = 0.50 - (0.50)x$, $(1.20)x = 0.50$, $x = 0.42$ L

We need 0.42 L of 0.50 M HNO_2 and 1.00 – 0.42 = 0.58 L of 0.50 M $NaNO_2$ to form a pH = 3.55 buffer solution.

43. $C_5H_5NH^+ \rightleftharpoons H^+ + C_5H_5N \quad K_a = \dfrac{K_w}{K_b} = \dfrac{1.0 \times 10^{-14}}{1.7 \times 10^{-9}} = 5.9 \times 10^{-6}$

$pK_a = -\log(5.9 \times 10^{-6}) = 5.23$

We will use the Henderson-Hasselbalch equation to calculate the concentration ratio necessary for each buffer.

$$pH = pK_a + \log\frac{[\text{base}]}{[\text{acid}]}, \quad pH = 5.23 + \log\frac{[C_5H_5N]}{[C_5H_5NH^+]}$$

a. $4.50 = 5.23 + \log\dfrac{[C_5H_5N]}{[C_5H_5NH^+]}, \quad \dfrac{[C_5H_5N]}{[C_5H_5NH^+]} = 10^{-0.73} = 0.19$

b. $5.00 = 5.23 + \log\dfrac{[C_5H_5N]}{[C_5H_5NH^+]}, \quad \dfrac{[C_5H_5N]}{[C_5H_5NH^+]} = 10^{-0.23} = 0.59$

c. $5.23 = 5.23 + \log\dfrac{[C_5H_5N]}{[C_5H_5NH^+]}, \quad \dfrac{[C_5H_5N]}{[C_5H_5NH^+]} = 10^{0.0} = 1.0$

d. $5.50 = 5.23 + \log\dfrac{[C_5H_5N]}{[C_5H_5NH^+]}, \quad \dfrac{[C_5H_5N]}{[C_5H_5NH^+]} = 10^{0.27} = 1.9$

44. $NH_4^+ \rightleftharpoons H^+ + NH_3 \quad K_a = K_w/K_b = 5.6 \times 10^{-10}$; $pK_a = -\log(5.6 \times 10^{-10}) = 9.25$; we will use the Henderson-Hasselbalch equation to calculate the concentration ratio necessary for each buffer.

$$pH = pK_a + \log\frac{[\text{base}]}{[\text{acid}]}, \quad pH = 9.25 + \log\frac{[NH_3]}{[NH_4^+]}$$

a. $9.00 = 9.25 + \log\dfrac{[NH_3]}{[NH_4^+]}, \quad \dfrac{[NH_3]}{[NH_4^+]} = 10^{-0.25} = 0.56$

b. $8.80 = 9.25 + \log\dfrac{[NH_3]}{[NH_4^+]}, \quad \dfrac{[NH_3]}{[NH_4^+]} = 10^{-0.45} = 0.35$

c. $10.00 = 9.25 + \log\frac{[NH_3]}{[NH_4^+]}$, $\frac{[NH_3]}{[NH_4^+]} = 10^{0.75} = 5.6$

d. $9.60 = 9.25 + \log\frac{[NH_3]}{[NH_4^+]}$, $\frac{[NH_3]}{[NH_4^+]} = 10^{0.35} = 2.2$

45. $pH = pK_a + \log\frac{[HCO_3^-]}{[H_2CO_3]}$, $7.40 = -\log(4.3 \times 10^{-7}) + \log\frac{[HCO_3^-]}{0.0012}$

$\log\frac{[HCO_3^-]}{0.0012} = 7.40 - 6.37 = 1.03$, $\frac{[HCO_3^-]}{0.0012} = 10^{1.03}$, $[HCO_3^-] = 1.3 \times 10^{-2}\ M$

46. At pH = 7.40: $7.40 = -\log(4.3 \times 10^{-7}) + \log\frac{[HCO_3^-]}{[H_2CO_3]}$

$\log\frac{[HCO_3^-]}{[H_2CO_3]} = 7.40 - 6.37 = 1.03$, $\frac{[HCO_3^-]}{[H_2CO_3]} = 10^{1.03}$, $\frac{[H_2CO_3]}{[HCO_3^-]} = 10^{-1.03} = 0.093$

At pH = 7.35: $\log\frac{[HCO_3^-]}{[H_2CO_3]} = 7.35 - 6.37 = 0.98$, $\frac{[HCO_3^-]}{[H_2CO_3]} = 10^{0.98}$

$\frac{[H_2CO_3]}{[HCO_3^-]} = 10^{-0.98} = 0.10$

The $[H_2CO_3]$: $[HCO_3^-]$ concentration ratio must increase from 0.093 to 0.10 in order for the onset of acidosis to occur.

47. A best buffer has large and equal quantities of weak acid and conjugate base. Because [acid] = [base] for a best buffer, $pH = pK_a + \log\frac{[base]}{[acid]} = pK_a + 0 = pK_a$ ($pH \approx pK_a$ for a best buffer).

The best acid choice for a pH = 7.00 buffer would be the weak acid with a pK_a close to 7.0 or $K_a \approx 1 \times 10^{-7}$. HOCl is the best choice in Table 14.2 ($K_a = 3.5 \times 10^{-8}$; $pK_a = 7.46$). To make this buffer, we need to calculate the [base] : [acid] ratio.

$$7.00 = 7.46 + \log\frac{[base]}{[acid]},\quad \frac{[OCl^-]}{[HOCl]} = 10^{-0.46} = 0.35$$

Any OCl^-/HOCl buffer in a concentration ratio of 0.35 : 1 will have a pH = 7.00. One possibility is [NaOCl] = 0.35 *M* and [HOCl] = 1.0 *M*.

48. For a pH = 5.00 buffer, we want an acid with a pK_a close to 5.00. For a conjugate acid-base pair, $14.00 = pK_a + pK_b$. So, for a pH = 5.00 buffer, we want the base to have a pK_b close to (14.0 – 5.0 =) 9.0 or a K_b close to 1×10^{-9}. The best choice in Table 14.3 is pyridine (C_5H_5N) with $K_b = 1.7 \times 10^{-9}$.

$$pH = pK_a + \log\frac{[\text{base}]}{[\text{acid}]};\quad K_a = \frac{K_w}{K_b} = \frac{1.0 \times 10^{-14}}{1.7 \times 10^{-9}} = 5.9 \times 10^{-6}$$

$$5.00 = -\log(5.9 \times 10^{-6}) + \log\frac{[\text{base}]}{[\text{acid}]},\quad \frac{[C_5H_5N]}{[C_5H_5NH^+]} = 10^{-0.23} = 0.59$$

There are many possibilities to make this buffer. One possibility is a solution of $[C_5H_5N]$ = 0.59 *M* and $[C_5H_5NHCl]$ = 1.0 *M*. The pH of this solution will be 5.00 because the base to acid concentration ratio is 0.59 : 1.

49. K_a for $H_2NNH_3^+ = K_w/K_{b,\,H_2NNH_2} = 1.0 \times 10^{-14}/3.0 \times 10^{-6} = 3.3 \times 10^{-9}$

$$pH = pK_a + \log\frac{[H_2NNH_2]}{[H_2NNH_3^+]} = -\log(3.3 \times 10^{-9}) + \log\left(\frac{0.40}{0.80}\right) = 8.48 + (-0.30) = 8.18$$

pH = pK_a for a buffer when [acid] = [base]. Here, the acid ($H_2NNH_3^+$) concentration needs to decrease, while the base (H_2NNH_2) concentration needs to increase in order for $[H_2NNH_3^+]$ = $[H_2NNH_2]$. Both of these changes are accomplished by adding a strong base (like NaOH) to the original buffer. The added OH^- from the strong base converts the acid component of the buffer into the conjugate base. Here, the reaction is $H_2NNH_3^+ + OH^- \rightarrow H_2NNH_2 + H_2O$. Because a strong base is reacting, the reaction is assumed to go to completion. The following set-up determines the number of moles of OH^-(x) that must be added so that mol $H_2NNH_3^+$ = mol H_2NNH_2 . When mol acid = mol base in a buffer, then [acid] = [base] and pH = pK_a.

	$H_2NNH_3^+$ +	OH^-	→	H_2NNH_2 + H_2O	
Before	1.0 L × 0.80 mol/L	x		1.0 L × 0.40 mol/L	
Change	$-x$	$-x$	→	$+x$	Reacts completely
After	$0.80 - x$	0		$0.40 + x$	

We want mol $H_2NNH_3^+$ = mol H_2NNH_2. Therefore:

$$0.80 - x = 0.40 + x,\quad 2x = 0.40,\quad x = 0.20 \text{ mol } OH^-$$

When 0.20 mol OH^- is added to the initial buffer, mol $H_2NNH_3^+$ is decreased to 0.60 mol, while mol H_2NNH_2 is increased to 0.60 mol. Therefore, 0.20 mol of NaOH must be added to the initial buffer solution in order to produce a solution where pH = pK_a.

50. $$pH = pK_a + \log\frac{[OCl^-]}{[HOCl]} = -\log(3.5 \times 10^{-8}) + \log\left(\frac{0.90}{0.20}\right) = 7.46 + 0.65 = 8.11$$

pH = pK_a when [HOCl] = $[OCl^-]$ (or when mol HOCl = mol OCl^-). Here, the moles of the base component of the buffer must decrease, while the moles of the acid component of the buffer must increase in order to achieve a solution where pH = pK_a. Both of these changes occur when a strong acid (like HCl) is added. Let x = mol H^+ added from the strong acid HCl.

	H^+	+	OCl^-	$\rightarrow$	$HOCl$	
Before	x		$1.0\ L \times 0.90\ mol/L$		$1.0\ L \times 0.20\ mol/L$	
Change	$-x$		$-x$	$\rightarrow$	$+x$	Reacts completely
After	0		$0.90 - x$		$0.20 + x$	

We want mol HOCl = mol OCl^-. Therefore:

$$0.90 - x = 0.20 + x,\ 2x = 0.70,\ x = 0.35 \text{ mol } H^+$$

When 0.35 mol H^+ is added, mol OCl^- is decreased to 0.55 mol, while the mol HOCl is increased to 0.55 mol Therefore, 0.35 mol of HCl must be added to the original buffer solution in order to produce a solution where pH = pK_a.

51. The reaction $OH^- + CH_3NH_3^+ \rightarrow CH_3NH_2 + H_2O$ goes to completion for solutions a, c, and d (no reaction occurs between the species in solution b because both species are bases). After the OH^- reacts completely, there must be both $CH_3NH_3^+$ and CH_3NH_2 in solution for it to be a buffer. The important components of each solution (after the OH^- reacts completely) is(are):

a. 0.05 *M* CH_3NH_2 (no $CH_3NH_3^+$ remains, no buffer)

b. 0.05 *M* OH^- and 0.1 *M* CH_3NH_2 (two bases present, no buffer)

c. 0.05 *M* OH^- and 0.05 *M* CH_3NH_2 (too much OH^- added, no $CH_3NH_3^+$ remains, no buffer)

d. 0.05 *M* CH_3NH_2 and 0.05 *M* $CH_3NH_3^+$ (a buffer solution results)

Only the combination in mixture d results in a buffer. Note that the concentrations are halved from the initial values. This is so because equal volumes of two solutions were added together, which halves the concentrations.

52. a. No; a solution of a strong acid (HNO_3) and its conjugate base (NO_3^-) is not generally considered a buffer solution.

b. No; two acids are present (HNO_3 and HF), so it is not a buffer solution.

c. Yes; H^+ reacts completely with F^-. Since equal volumes are mixed, the initial concentrations in the mixture are 0.10 *M* HNO_3 and 0.20 *M* NaF.

	H^+	+	F^-	$\rightarrow$	HF	
Before	0.10 *M*		0.20 *M*		0	
Change	−0.10		−0.10	$\rightarrow$	+0.10	Reacts completely
After	0		0.10		0.10	

After H^+ reacts completely, a buffer solution results; that is, a weak acid (HF) and its conjugate base (F^-) are both present in solution in large quantities.

d. No; a strong acid (HNO_3) and a strong base (NaOH) do not form buffer solutions. They will neutralize each other to form H_2O.

53. Added OH^- converts $HC_2H_3O_2$ into $C_2H_3O_2^-$: $HC_2H_3O_2 + OH^- \rightarrow C_2H_3O_2^- + H_2O$

From this reaction, the moles of $C_2H_3O_2^-$ produced equal the moles of OH^- added. Also, the total concentration of acetic acid plus acetate ion must equal 2.0 *M* (assuming no volume change on addition of NaOH). Summarizing for each solution:

$[C_2H_3O_2^-] + [HC_2H_3O] = 2.0\ M$ and $[C_2H_3O_2^-]$ produced $= [OH^-]$ added

a. $pH = pK_a + \log\frac{[C_2H_3O_2^-]}{[HC_2H_3O_2]}$; for $pH = pK_a$, $\log\frac{[C_2H_3O_2^-]}{[HC_2H_3O_2]} = 0$

Therefore, $\frac{[C_2H_3O_2^-]}{[HC_2H_3O_2]} = 1.0$ and $[C_2H_3O_2^-] = [HC_2H_3O_2]$.

Because $[C_2H_3O_2^-] + [HC_2H_3O_2] = 2.0\ M$:

$[C_2H_3O_2^-] = [HC_2H_3O_2] = 1.0\ M = [OH^-]$ added

To produce a 1.0 *M* $C_2H_3O_2^-$ solution, we need to add 1.0 mol of NaOH to 1.0 L of the 2.0 *M* $HC_2H_3O_2$ solution. The resulting solution will have $pH = pK_a = 4.74$.

b. $4.00 = 4.74 + \log\frac{[C_2H_3O_2^-]}{[HC_2H_3O_2]}, \frac{[C_2H_3O_2^-]}{[HC_2H_3O_2]} = 10^{-0.74} = 0.18$

$[C_2H_3O_2^-] = 0.18[HC_2H_3O_2]$ or $[HC_2H_3O_2] = 5.6[C_2H_3O_2^-]$

Because $[C_2H_3O_2^-] + [HC_2H_3O_2] = 2.0\ M$:

$[C_2H_3O_2^-] + 5.6[C_2H_3O_2^-] = 2.0\ M$, $[C_2H_3O_2^-] = \frac{2.0}{6.6} = 0.30\ M = [OH^-]$ added

We need to add 0.30 mol of NaOH to 1.0 L of 2.0 *M* $HC_2H_3O_2$ solution to produce 0.30 *M* $C_2H_3O_2^-$. The resulting solution will have pH = 4.00.

c. $5.00 = 4.74 + \log\frac{[C_2H_3O_2^-]}{[HC_2H_3O_2]}, \frac{[C_2H_3O_2^-]}{[HC_2H_3O_2]} = 10^{0.26} = 1.8$

$1.8[HC_2H_3O_2] = [C_2H_3O_2^-]$ or $[HC_2H_3O_2] = 0.56[C_2H_3O_2^-]$

$1.56[C_2H_3O_2^-] = 2.0\ M$, $[C_2H_3O_2^-] = 1.3\ M = [OH^-]$ added

We need to add 1.3 mol of NaOH to 1.0 L of 2.0 *M* $HC_2H_3O_2$ to produce a solution with pH = 5.00.

54. When H^+ is added, it converts $C_2H_3O_2^-$ into $HC_2H_3O_2$: $C_2H_3O_2^- + H^+ \rightarrow HC_2H_3O_2$. From this reaction, the moles of $HC_2H_3O_2$ produced must equal the moles of H^+ added and the total concentration of acetate ion + acetic acid must equal 1.0 *M* (assuming no volume change). Summarizing for each solution:

$[C_2H_3O_2^-] + [HC_2H_3O_2] = 1.0\ M$ and $[HC_2H_3O_2] = [H^+]$ added

a. $pH = pK_a + \log\frac{[C_2H_3O_2^-]}{[HC_2H_3O_2]}$; for $pH = pK_a$, $[C_2H_3O_2^-] = [HC_2H_3O_2]$.

For this to be true, $[C_2H_3O_2^-] = [HC_2H_3O_2] = 0.50\ M = [H^+]$ added, which means that 0.50 mol of HCl must be added to 1.0 L of the initial solution to produce a solution with $pH = pK_a$.

b. $4.20 = 4.74 + \log\frac{[C_2H_3O_2^-]}{[HC_2H_3O_2]}$, $\frac{[C_2H_3O_2^-]}{[HC_2H_3O_2]} = 10^{-0.54} = 0.29$

$[C_2H_3O_2^-] = 0.29[HC_2H_3O_2]$; $0.29[HC_2H_3O_2] + [HC_2H_3O_2] = 1.0\ M$

$[HC_2H_3O_2] = 0.78\ M = [H^+]$ added

0.78 mol of HCl must be added to produce a solution with pH = 4.20.

c. $5.00 = 4.74 + \log\frac{[C_2H_3O_2^-]}{[HC_2H_3O_2]}$, $\frac{[C_2H_3O_2^-]}{[HC_2H_3O_2]} = 10^{0.26} = 1.8$

$[C_2H_3O_2^-] = 1.8[HC_2H_3O_2]$; $1.8[HC_2H_3O_2] + [HC_2H_3O_2] = 1.0\ M$

$[HC_2H_3O_2] = 0.36\ M = [H^+]$ added

0.36 mol of HCl must be added to produce a solution with pH = 5.00.

Acid-Base Titrations

55.

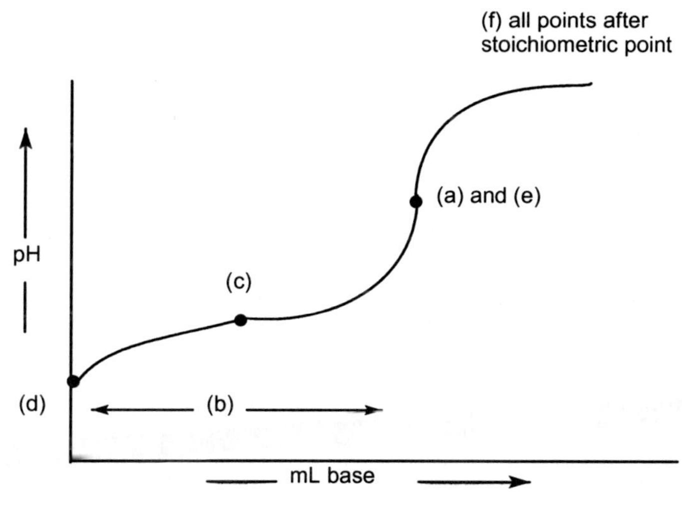

$HA + OH^- \rightarrow A^- + H_2O$; added OH^- from the strong base converts the weak acid HA into its conjugate base A^-. Initially, before any OH^- is added (point d), HA is the dominant species present. After OH^- is added, both HA and A^- are present, and a buffer solution results (region b). At the equivalence point (points a and e), exactly enough OH^- has been added to convert all the weak acid HA into its conjugate base A^-. Past the equivalence point (region f), excess OH^- is present. For the answer to part b, we included almost the entire buffer region. The maximum buffer region (or the region which is the best buffer solution) is around the halfway point to equivalence (point c). At this point, enough OH^- has been added to convert exactly one-half of the weak acid present initially into its conjugate base, so [HA] = [A^-] and pH = pK_a. A "best" buffer has about equal concentrations of weak acid and conjugate base present.

56.

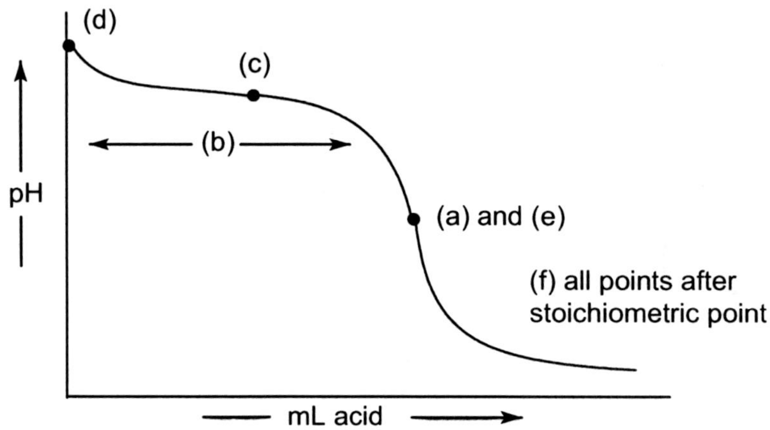

$B + H^+ \rightarrow BH^+$; added H^+ from the strong acid converts the weak base B into its conjugate acid BH^+. Initially, before any H^+ is added (point d), B is the dominant species present. After H^+ is added, both B and BH^+ are present, and a buffered solution results (region b). At the equivalence point (points a and e), exactly enough H^+ has been added to convert all the weak base present initially into its conjugate acid BH^+. Past the equivalence point (region f), excess H^+ is present. For the answer to b, we included almost the entire buffer region. The maximum buffer region is around the halfway point to equivalence (point c), where [B] = [BH^+]. Here, pH = pK_a, which is a characteristic of a best buffer.

57. This is a strong acid ($HClO_4$) titrated by a strong base (KOH). Added OH^- from the strong base will react completely with the H^+ present from the strong acid to produce H_2O.

a. Only strong acid present. [H^+] = 0.200 *M*; pH = 0.699

b. mmol OH^- added = $10.0\text{ mL} \times \dfrac{0.100\text{ mmol OH}^-}{\text{mL}} = 1.00\text{ mmol OH}^-$

mmol H^+ present = $40.0\text{ mL} \times \dfrac{0.200\text{ mmol H}^+}{\text{mL}} = 8.0\text{ mmol H}^+$

Note: The units millimoles are usually easier numbers to work with. The units for molarity are moles per liter but are also equal to millimoles per milliliter.

	H^+	+	OH^-	$\rightarrow$ H_2O	
Before	8.00 mmol		1.00 mmol		
Change	−1.00 mmol		−1.00 mmol		Reacts completely
After	7.00 mmol		0		

The excess H^+ determines the pH. $[H^+]_{excess} = \dfrac{7.00 \text{ mmol } H^+}{40.0 \text{ mL} + 10.0 \text{ mL}} = 0.140\ M$

pH = –log(0.140) = 0.854

c. mmol OH^- added = 40.0 mL × 0.100 M = 4.00 mmol OH^-

	H^+	+	OH^-	→	H_2O
Before	8.00 mmol		4.00 mmol		
After	4.00 mmol		0		

$$[H^+]_{excess} = \frac{4.00 \text{ mmol}}{(40.0 + 40.0) \text{ mL}} = 0.0500\ M; \ \text{pH} = 1.301$$

d. mmol OH^- added = 80.0 mL × 0.100 M = 8.00 mmol OH^-; this is the equivalence point because we have added just enough OH^- to react with all the acid present. For a strong acid-strong base titration, pH = 7.00 at the equivalence point because only neutral species are present (K^+, ClO_4^-, H_2O).

e. mmol OH^- added = 100.0 mL × 0.100 M = 10.0 mmol OH^-

	H^+	+	OH^-	→	H_2O
Before	8.00 mmol		10.0 mmol		
After	0		2.0 mmol		

Past the equivalence point, the pH is determined by the excess OH^- present.

$$[OH^-]_{excess} = \frac{2.0 \text{ mmol}}{(40.0 + 100.0) \text{ mL}} = 0.014\ M; \ \text{pOH} = 1.85; \ \text{pH} = 12.15$$

58. This is a strong base, $Ba(OH)_2$, titrated by a strong acid, HCl. The added strong acid will neutralize the OH^- from the strong base. As is always the case when a strong acid and/or strong base reacts, the reaction is assumed to go to completion.

a. Only a strong base is present, but it breaks up into 2 moles of OH^- ions for every mole of $Ba(OH)_2$. $[OH^-] = 2 \times 0.100\ M = 0.200\ M$; pOH = 0.699; pH = 13.301

b. mmol OH^- present = 80.0 mL × $\dfrac{0.100 \text{ mmol } Ba(OH)_2}{\text{mL}} \times \dfrac{2 \text{ mmol } OH^-}{\text{mmol } Ba(OH)_2}$

= 16.0 mmol OH^-

mmol H^+ added = 20.0 mL × $\dfrac{0.400 \text{ mmol } H^+}{\text{mL}}$ = 8.00 mmol H^+

	OH^-	+	H^+	→	H_2O	
Before	16.0 mmol		8.00 mmol			
Change	–8.00 mmol		–8.00 mmol			Reacts completely
After	8.0 mmol		0			

$$[OH^-]_{excess} = \frac{8.0 \text{ mmol} OH^-}{80.0 \text{ mL} + 20.0 \text{ mL}} = 0.080\ M;\ \ pOH = 1.10;\ \ pH = 12.90$$

c. mmol H^+ added = 30.0 mL × 0.400 M = 12.0 mmol H^+

	OH^-	+	H^+	→ H_2O
Before	16.0 mmol		12.0 mmol	
After	4.0 mmol		0	

$$[OH^-]_{excess} = \frac{4.0 \text{ mmol} OH^-}{(80.0 + 30.0) \text{ mL}} = 0.036\ M;\ \ pOH = 1.44;\ \ pH = 12.56$$

d. mmol H^+ added = 40.0 mL × 0.400 M = 16.0 mmol H^+; this is the equivalence point. Because the H^+ will exactly neutralize the OH^- from the strong base, all we have in solution is Ba^{2+}, Cl^-, and H_2O. All are neutral species, so pH = 7.00.

e. mmol H^+ added = 80.0 mL × 0.400 M = 32.0 mmol H^+

	OH^-	+	H^+	→ H_2O
Before	16.0 mmol		32.0 mmol	
After	0		16.0 mmol	

$$[H^+]_{excess} = \frac{16.0 \text{ mmol} H^+}{(80.0 + 80.0) \text{ mL}} = 0.100\ M;\ \ pH = 1.000$$

59. This is a weak acid ($HC_2H_3O_2$) titrated by a strong base (KOH).

a. Only weak acid is present. Solving the weak acid problem:

	$HC_2H_3O_2$	⇌	H^+	+	$C_2H_3O_2^-$
Initial	0.200 M		~0		0
	x mol/L $HC_2H_3O_2$ dissociates to reach equilibrium				
Change	$-x$	→	$+x$		$+x$
Equil.	$0.200 - x$		x		x

$$K_a = 1.8 \times 10^{-5} = \frac{x^2}{0.200 - x} \approx \frac{x^2}{0.200},\ \ x = [H^+] = 1.9 \times 10^{-3}\ M$$

pH = 2.72; assumptions good.

b. The added OH^- will react completely with the best acid present, $HC_2H_3O_2$.

$$\text{mmol } HC_2H_3O_2 \text{ present} = 100.0 \text{ mL} \times \frac{0.200 \text{ mmol } HC_2H_3O_2}{\text{mL}} = 20.0 \text{ mmol } HC_2H_3O_2$$

$$\text{mmol } OH^- \text{ added} = 50.0 \text{ mL} \times \frac{0.100 \text{ mmol } OH^-}{\text{mL}} = 5.00 \text{ mmol } OH^-$$

	$HC_2H_3O_2$	+	OH^-	→	$C_2H_3O_2^-$ + H_2O	
Before	20.0 mmol		5.00 mmol		0	
Change	–5.00 mmol		–5.00 mmol	→	+5.00 mmol	Reacts Completely
After	15.0 mmol		0		5.00 mmol	

After reaction of all of the strong base, we have a buffer solution containing a weak acid ($HC_2H_3O_2$) and its conjugate base ($C_2H_3O_2^-$). We will use the Henderson-Hasselbalch equation to solve for the pH.

$$pH = pK_a + \log\frac{[C_2H_3O_2^-]}{[HC_2H_3O_2]} = -\log(1.8 \times 10^{-5}) + \log\left(\frac{5.00\text{mmol}/V_T}{15.0\text{mmol}/V_T}\right),\ \text{where } V_T = \text{total volume}$$

$$pH = 4.74 + \log\left(\frac{5.00}{15.0}\right) = 4.74 + (-0.477) = 4.26$$

Note that the total volume cancels in the Henderson-Hasselbalch equation. For the [base]/[acid] term, the mole ratio equals the concentration ratio because the components of the buffer are always in the same volume of solution.

c. mmol OH^- added = 100.0 mL × (0.100 mmol OH^-/mL) = 10.0 mmol OH^-; the same amount (20.0 mmol) of $HC_2H_3O_2$ is present as before (it doesn't change). As before, let the OH^- react to completion, then see what is remaining in solution after this reaction.

	$HC_2H_3O_2$	+	OH^-	→	$C_2H_3O_2^-$ + H_2O
Before	20.0 mmol		10.0 mmol		0
After	10.0 mmol		0		10.0 mmol

A buffer solution results after reaction. Because $[C_2H_3O_2^-] = [HC_2H_3O_2] = 10.0$ mmol/total volume, $pH = pK_a$. This is always true at the halfway point to equivalence for a weak acid-strong base titration, $pH = pK_a$.

$$pH = -\log(1.8 \times 10^{-5}) = 4.74$$

d. mmol OH^- added = 150.0 mL × 0.100 M = 15.0 mmol OH^-. Added OH^- reacts completely with the weak acid.

	$HC_2H_3O_2$	+	OH^-	→	$C_2H_3O_2^-$ + H_2O
Before	20.0 mmol		15.0 mmol		0
After	5.0 mmol		0		15.0 mmol

We have a buffer solution after all the OH^- reacts to completion. Using the Henderson-Hasselbalch equation:

$$pH = 4.74 + \log\frac{[C_2H_3O_2^-]}{[HC_2H_3O_2]} = 4.74 + \log\left(\frac{15.0\text{ mmol}}{5.0\text{ mmol}}\right)$$

$$pH = 4.74 + 0.48 = 5.22$$

e. mmol OH^- added = 200.00 mL × 0.100 M = 20.0 mmol OH^-; as before, let the added OH^- react to completion with the weak acid; then see what is in solution after this reaction.

	$HC_2H_3O_2$	+	OH^-	→	$C_2H_3O_2^-$ + H_2O
Before	20.0 mmol		20.0 mmol		0
After	0		0		20.0 mmol

This is the equivalence point. Enough OH^- has been added to exactly neutralize all the weak acid present initially. All that remains that affects the pH at the equivalence point is the conjugate base of the weak acid ($C_2H_3O_2^-$). This is a weak base equilibrium problem.

	$C_2H_3O_2^-$ + H_2O ⇌	$HC_2H_3O_2$ +	OH^-	$K_b = \frac{K_w}{K_b} = \frac{1.0 \times 10^{-14}}{1.8 \times 10^{-5}}$
Initial	20.0 mmol/300.0 mL	0	0	$K_b = 5.6 \times 10^{-9}$
	x mol/L $C_2H_3O_2^-$ reacts with H_2O to reach equilibrium			
Change	$-x$ →	$+x$	$+x$	
Equil.	$0.0667 - x$	x	x	

$$K_b = 5.6 \times 10^{-10} = \frac{x^2}{0.0667 - x} \approx \frac{x^2}{0.0667}, \quad x = [OH^-] = 6.1 \times 10^{-6}\ M$$

pOH = 5.21; pH = 8.79; assumptions good.

f. mmol OH^- added = 250.0 mL × 0.100 M = 25.0 mmol OH^-

	$HC_2H_3O_2$	+	OH^-	→	$C_2H_3O_2^-$ + H_2O
Before	20.0 mmol		25.0 mmol		0
After	0		5.0 mmol		20.0 mmol

After the titration reaction, we have a solution containing excess OH^- and a weak base $C_2H_3O_2^-$. When a strong base and a weak base are both present, assume that the amount of OH^- added from the weak base will be minimal; that is, the pH past the equivalence point is determined by the amount of excess strong base.

$$[OH^-]_{excess} = \frac{5.0\ \text{mmol}}{100.0\ \text{mL} + 250.0\ \text{mL}} = 0.014\ M; \quad \text{pOH} = 1.85; \quad \text{pH} = 12.15$$

60. This is a weak base (H_2NNH_2) titrated by a strong acid (HNO_3). To calculate the pH at the various points, let the strong acid react completely with the weak base present; then see what is in solution.

a. Only a weak base is present. Solve the weak base equilibrium problem.

	H_2NNH_2 + H_2O ⇌	$H_2NNH_3^+$ +	OH^-
Initial	0.100 M	0	~0
Equil.	0.100 - x	x	x

$$K_b = 3.0 \times 10^{-6} = \frac{x^2}{0.100 - x} \approx \frac{x^2}{0.100},\ x = [OH^-] = 5.5 \times 10^{-4}\ M$$

pOH = 3.26; pH = 10.74; assumptions good.

b. $\text{mmol } H_2NNH_2 \text{ present} = 100.0 \text{ mL} \times \frac{0.100 \text{ mmol } H_2NNH_2}{\text{mL}} = 10.0 \text{ mmol } H_2NNH_2$

$\text{mmol } H^+ \text{ added} = 20.0 \text{ mL} \times \frac{0.200 \text{ mmol } H^+}{\text{mL}} = 4.00 \text{ mmol } H^+$

	H_2NNH_2	+	H^+	→	$H_2NNH_3^+$	
Before	10.0 mmol		4.00 mmol		0	
Change	–4.00 mmol		–4.00 mmol	→	+4.00 mmol	Reacts completely
After	6.0 mmol		0		4.00 mmol	

A buffer solution results after the titration reaction. Solving using the Henderson-Hasselbalch equation:

$$pH = pK_a + \log\frac{[\text{base}]}{[\text{acid}]};\quad K_a = \frac{K_w}{K_b} = \frac{1.0 \times 10^{-14}}{3.0 \times 10^{-6}} = 3.3 \times 10^{-9}$$

$$pH = -\log(3.3 \times 10^{-9}) + \log\left(\frac{6.0 \text{ mmol}/V_T}{4.00 \text{ mmol}/V_T}\right)$$, where V_T = total volume, which cancels.

$pH = 8.48 + \log(1.5) = 8.48 + 0.18 = 8.66$

c. $\text{mmol } H^+ \text{ added} = 25.0 \text{ mL} \times 0.200\ M = 5.00 \text{ mmol } H^+$

	H_2NNH_2	+	H^+	→	$H_2NNH_3^+$
Before	10.0 mmol		5.00 mmol		0
After	5.0 mmol		0		5.00 mmol

This is the halfway point to equivalence, where $[H_2NNH_3^+] = [H_2NNH_2]$. At this point, $pH = pK_a$ (which is characteristic of the halfway point for any weak base-strong acid titration).

$pH = -\log(3.3 \times 10^{-9}) = 8.48$

d. $\text{mmol } H^+ \text{ added} = 40.0 \text{ mL} \times 0.200\ M = 8.00 \text{ mmol } H^+$

	H_2NNH_2	+	H^+	→	$H_2NNH_3^+$
Before	10.0 mmol		8.00 mmol		0
After	2.0 mmol		0		8.00 mmol

A buffer solution results.

$$pH = pK_a + \log\frac{[\text{base}]}{[\text{acid}]} = 8.48 + \log\left(\frac{2.0\ \text{mmol}/V_T}{8.00\ \text{mmol}/V_T}\right) = 8.48 + (-0.60) = 7.88$$

e. mmol H^+ added = 50.0 mL × 0.200 M = 10.0 mmol H^+

	H_2NNH_2	+	H^+	→	$H_2NNH_3^+$
Before	10.0 mmol		10.0 mmol		0
After	0		0		10.0 mmol

As is always the case in a weak base-strong acid titration, the pH at the equivalence point is acidic because only a weak acid ($H_2NNH_3^+$) is present. Solving the weak acid equilibrium problem:

	$H_2NNH_3^+$	⇌	H^+	+	H_2NNH_2
Initial	10.0 mmol/150.0 mL		0		0
Equil.	$0.0667 - x$		x		x

$$K_a = 3.3 \times 10^{-9} = \frac{x^2}{0.0667 - x} \approx \frac{x^2}{0.0667},\quad x = [H^+] = 1.5 \times 10^{-5}\ M$$

pH = 4.82; assumptions good.

f. mmol H^+ added = 100.0 mL × 0.200 M = 20.0 mmol H^+

	H_2NNH_2	+	H^+	→	$H_2NNH_3^+$
Before	10.0 mmol		20.0 mmol		0
After	0		10.0 mmol		10.0 mmol

Two acids are present past the equivalence point, but the excess H^+ will determine the pH of the solution since $H_2NNH_3^+$ is a weak acid.

$$[H^+]_{excess} = \frac{10.0\ \text{mmol}}{100.0\ \text{mL} + 100.0\ \text{mL}} = 0.0500\ M;\ \text{pH} = 1.301$$

61. We will do sample calculations for the various parts of the titration. All results are summarized in Table 15.1 at the end of Exercise 64.

At the beginning of the titration, only the weak acid $HC_3H_5O_3$ is present. Let HLac = $HC_3H_5O_3$ and Lac^- = $C_3H_5O_3^-$.

	HLac	⇌	H^+	+	Lac^-	$K_a = 10^{-3.86} = 1.4 \times 10^{-4}$
Initial	0.100 M		~0		0	
	x mol/L HLac dissociates to reach equilibrium					
Change	$-x$	→	$+x$		$+x$	
Equil.	$0.100 - x$		x		x	

$$1.4 \times 10^{-4} = \frac{x^2}{0.100 - x} \approx \frac{x^2}{0.100},\ x = [H^+] = 3.7 \times 10^{-3}\ M;\ pH = 2.43;\ \text{assumptions good.}$$

Up to the stoichiometric point, we calculate the pH using the Henderson-Hasselbalch equation. This is the buffer region. For example, at 4.0 mL of NaOH added:

$$\text{initial mmol HLac present} = 25.0\text{ mL} \times \frac{0.100\text{ mmol}}{\text{mL}} = 2.50\text{ mmol HLac}$$

$$\text{mmol } OH^- \text{ added} = 4.0\text{ mL} \times \frac{0.100\text{ mmol}}{\text{mL}} = 0.40\text{ mmol } OH^-$$

Note: The units millimoles are usually easier numbers to work with. The units for molarity are moles per liter but are also equal to millimoles per milliliter.

The 0.40 mmol of added OH^- converts 0.40 mmol HLac to 0.40 mmol Lac^- according to the equation:

$HLac + OH^- \rightarrow Lac^- + H_2O$ Reacts completely since a strong base is added.

mmol HLac remaining = 2.50 – 0.40 = 2.10 mmol; mmol Lac^- produced = 0.40 mmol

We have a buffer solution. Using the Henderson-Hasselbalch equation where $pK_a = 3.86$:

$$pH = pK_a + \log\frac{[Lac^-]}{[HLac]} = 3.86 + \log\frac{(0.40)}{(2.10)}$$ (Total volume cancels, so we can use use the ratio of moles or millimoles.)

$pH = 3.86 - 0.72 = 3.14$

Other points in the buffer region are calculated in a similar fashion. Perform a stoichiometry problem first, followed by a buffer problem. The buffer region includes all points up to and including 24.9 mL OH^- added.

At the stoichiometric point (25.0 mL OH^- added), we have added enough OH^- to convert all of the HLac (2.50 mmol) into its conjugate base (Lac^-). All that is present is a weak base. To determine the pH, we perform a weak base calculation.

$$[Lac^-]_0 = \frac{2.50\text{ mmol}}{25.0\text{ mL} + 25.0\text{ mL}} = 0.0500\ M$$

	Lac^- + H_2O	$\rightleftharpoons$	HLac	+	OH^-	$K_b = \frac{1.0 \times 10^{-14}}{1.4 \times 10^{-4}} = 7.1 \times 10^{-11}$
Initial	0.0500 M		0		0	
	x mol/L Lac reacts with H_2O to reach equilibrium					
Change	$-x$	$\rightarrow$	$+x$		$+x$	
Equil.	$0.0500 - x$		x		x	

$$K_b = \frac{x^2}{0.0500 - x} \approx \frac{x^2}{0.0500} = 7.1 \times 10^{-11}$$

$x = [OH^-] = 1.9 \times 10^{-6}\ M$; pOH = 5.72; pH = 8.28; assumptions good.

Past the stoichiometric point, we have added more than 2.50 mmol of NaOH. The pH will be determined by the excess OH^- ion present. An example of this calculation follows.

At 25.1 mL: OH^- added $= 25.1 \text{ mL} \times \dfrac{0.100 \text{ mmol}}{\text{mL}} = 2.51 \text{ mmol } OH^-$

2.50 mmol OH^- neutralizes all the weak acid present. The remainder is excess OH^-.

Excess $OH^- = 2.51 - 2.50 = 0.01 \text{ mmol } OH^-$

$$[OH^-]_{excess} = \frac{0.01 \text{ mmol}}{(25.0 + 25.1) \text{ mL}} = 2 \times 10^{-4} M; \;\; pOH = 3.7; \;\; pH = 10.3$$

All results are listed in Table 15.1 at the end of the solution to Exercise 64.

62. Results for all points are summarized in Table 15.1 at the end of the solution to Exercise 64. At the beginning of the titration, we have a weak acid problem:

$HOPr = HC_3H_5O_2$
$OPr^- = C_3H_5O_2^-$

	HOPr	⇌	H^+	+	OPr^-
Initial	0.100 *M*		~0		0
	x mol/L HOPr acid dissociates to reach equilibrium				
Change	$-x$	→	$+x$		$+x$
Equil.	$0.100 - x$		x		x

$$K_a = \frac{[H^+][OPr^-]}{[HOPr]} = 1.3 \times 10^{-5} = \frac{x^2}{0.100 - x} \approx \frac{x^2}{0.100}$$

$x = [H^+] = 1.1 \times 10^{-3} M$; pH = 2.96; assumptions good.

The buffer region is from 4.0 to 24.9 mL of OH^- added. We will do a sample calculation at 24.0 mL OH^- added.

Initial mmol HOPr present $= 25.0 \text{ mL} \times \dfrac{0.100 \text{ mmol}}{\text{mL}} = 2.50 \text{ mmol HOPr}$

mmol OH^- added $= 24.0 \text{ mL} \times \dfrac{0.100 \text{ mmol}}{\text{mL}} = 2.40 \text{ mmol } OH^-$

The added strong base converts HOPr into OPr^-.

	HOPr	+	OH^-	→	OPr^-	+	H_2O	
Before	2.50 mmol		2.40 mmol		0			
Change	−2.40		−2.40	→	+2.40			Reacts completely
After	0.10 mmol		0		2.40 mmol			

A buffer solution results. Using the Henderson-Hasselbalch equation where $pK_a = -\log(1.3 \times 10^{-5}) = 4.89$:

$$\text{pH} = \text{pK}_a + \log\frac{[\text{base}]}{[\text{acid}]} = 4.89 + \log\frac{[\text{OPr}^-]}{[\text{HOPr}]}$$

$$\text{pH} = 4.89 + \log\left(\frac{2.40}{0.10}\right) = 4.89 + 1.38 = 6.27$$ (Volume cancels, so we can use the millimole ratio in the log term.)

All points in the buffer region 4.0 mL to 24.9 mL are calculated this way. See Table 15.1 at the end of Exercise 64 for all the results.

At the stoichiometric point (25.0 mL NaOH added), only a weak base (OPr^-) is present:

	OPr^- + H_2O	$\rightleftharpoons$	OH^-	+ $HOPr$
Initial	$\frac{2.50\text{ mmol}}{50.0\text{ mL}} = 0.0500\ M$		0	0
	x mol/L OPr^- reacts with H_2O to reach equilibrium			
Change	$-x$	→	$+x$	$+x$
Equil.	$0.0500 - x$		x	x

$$K_b = \frac{[\text{OH}^-][\text{HOPr}]}{[\text{OPr}^-]} = \frac{K_w}{K_a} = 7.7 \times 10^{-10} = \frac{x^2}{0.0500 - x} \approx \frac{x^2}{0.0500}$$

$x = 6.2 \times 10^{-6}\ M = [OH^-]$, pOH = 5.21, pH = 8.79; assumptions good.

Beyond the stoichiometric point, the pH is determined by the excess strong base added. The results are the same as those in Exercise 61 (see Table 15.1).

For example, at 26.0 mL NaOH added:

$$[\text{OH}^-] = \frac{2.60\text{ mmol} - 2.50\text{ mmol}}{(25.0 + 26.0)\text{ mL}} = 2.0 \times 10^{-3}\ M;\ \text{pOH} = 2.70;\ \text{pH} = 11.30$$

63. At beginning of the titration, only the weak base NH_3 is present. As always, solve for the pH using the K_b reaction for NH_3.

	NH_3 + H_2O	$\rightleftharpoons$	NH_4^+	+ OH^-	$K_b = 1.8 \times 10^{-5}$
Initial	$0.100\ M$		0	~0	
Equil.	$0.100 - x$		x	x	

$$K_b = \frac{x^2}{0.100 - x} \approx \frac{x^2}{0.100} = 1.8 \times 10^{-5}$$

$x = [OH^-] = 1.3 \times 10^{-3}\ M$; pOH = 2.89; pH = 11.11; assumptions good.

In the buffer region (4.0 – 24.9 mL), we can use the Henderson-Hasselbalch equation:

$$K_a = \frac{1.0 \times 10^{-14}}{1.8 \times 10^{-5}} = 5.6 \times 10^{-10};\ \text{pK}_a = 9.25;\ \text{pH} = 9.25 + \log\frac{[\text{NH}_3]}{[\text{NH}_4^+]}$$

We must determine the amounts of NH_3 and NH_4^+ present after the added H^+ reacts completely with the NH_3. For example, after 8.0 mL HCl added:

$$\text{initial mmol } NH_3 \text{ present} = 25.0 \text{ mL} \times \frac{0.100 \text{ mmol}}{\text{mL}} = 2.50 \text{ mmol } NH_3$$

$$\text{mmol } H^+ \text{ added} = 8.0 \text{ mL} \times \frac{0.100 \text{ mmol}}{\text{mL}} = 0.80 \text{ mmol } H^+$$

Added H^+ reacts with NH_3 to completion: $NH_3 + H^+ \rightarrow NH_4^+$

mmol NH_3 remaining = 2.50 – 0.80 = 1.70 mmol; mmol NH_4^+ produced = 0.80 mmol

$pH = 9.25 + \log\frac{1.70}{0.80} = 9.58$ (Mole ratios can be used since the total volume cancels.)

Other points in the buffer region are calculated in similar fashion. Results are summarized in Table 15.1 at the end of Exercise 64.

At the stoichiometric point (25.0 mL H^+ added), just enough HCl has been added to convert all the weak base (NH_3) into its conjugate acid (NH_4^+). Perform a weak acid calculation.

$[NH_4^+]_0$ = 2.50 mmol/50.0 mL = 0.0500 M

	NH_4^+	$\rightleftharpoons$	H^+	+	NH_3	$K_a = 5.6 \times 10^{-10}$
Initial	0.0500 M		0		0	
Equil.	0.0500 - x		x		x	

$5.6 \times 10^{-10} = \frac{x^2}{0.0500 - x} \approx \frac{x^2}{0.0500}$, $x = [H^+] = 5.3 \times 10^{-6}$ M; pH = 5.28; assumptions good.

Beyond the stoichiometric point, the pH is determined by the excess H^+. For example, at 28.0 mL of H^+ added:

$$H^+ \text{ added} = 28.0 \text{ mL} \times \frac{0.100 \text{ mmol}}{\text{mL}} = 2.80 \text{ mmol } H^+$$

Excess H^+ = 2.80 mmol – 2.50 mmol = 0.30 mmol excess H^+

$$[H^+]_{excess} = \frac{0.30 \text{ mmol}}{(25.0 + 28.0) \text{ mL}} = 5.7 \times 10^{-3} M; \text{ pH} = 2.24$$

All results are summarized in Table 15.1 at the end of Exercise 64.

64. Initially, a weak base problem:

	py	+	H_2O	$\rightleftharpoons$	Hpy^+	+	OH^-	py is pyridine.
Initial	0.100 M				0		~0	
Equil.	0.100 – x				x		x	

$$K_b = \frac{[Hpy^+][OH^-]}{[py]} = \frac{x^2}{0.100 - x} \approx \frac{x^2}{0.100} \approx 1.7 \times 10^{-9}$$

$x = [OH^-] = 1.3 \times 10^{-5}$ *M*; pOH = 4.89; pH = 9.11; assumptions good.

Buffer region (4.0 – 24.5 mL): Added H^+ reacts completely with py: py + $H^+ \rightarrow Hpy^+$. Determine the moles (or millimoles) of py and Hpy^+ after reaction, then use the Henderson-Hasselbalch equation to solve for the pH.

$$K_a = \frac{K_w}{K_b} = \frac{1.0 \times 10^{-14}}{1.7 \times 10^{-9}} = 5.9 \times 10^{-6};\ pK_a = 5.23;\ pH = 5.23 + \log\frac{[py]}{[Hpy^+]}$$

Results in the buffer region are summarized in Table 15.1, which follows this problem. See Exercise 63 for a similar sample calculation.

At the stoichiometric point (25.0 mL H^+ added), this is a weak acid problem since just enough H^+ has been added to convert all the weak base into its conjugate acid. The initial concentration of $[Hpy^+]$ = 0.0500 *M*.

	Hpy^+	$\rightleftharpoons$	py	+	H^+	$K_a = 5.9 \times 10^{-6}$
Initial	0.0500 *M*		0		0	
Equil.	0.0500 – x		x		x	

$$5.9 \times 10^{-6} = \frac{x^2}{0.0500 - x} \approx \frac{x^2}{0.0500},\ x = [H^+] = 5.4 \times 10^{-4}\ M;\ pH = 3.27;\ \text{assumptions good.}$$

Beyond the equivalence point, the pH determination is made by calculating the concentration of excess H^+. See Exercise 63 for an example. All results are summarized in Table 15.1.

Table 15.1 Summary of pH Results for Exercises 61-64 (Plot on next page)

Titrant mL	Exercise 61	Exercise 62	Exercise 63	Exercise 64
0.0	2.43	2.96	11.11	9.11
4.0	3.14	4.17	9.97	5.95
8.0	3.53	4.56	9.58	5.56
12.5	3.86	4.89	9.25	5.23
20.0	4.46	5.49	8.65	4.63
24.0	5.24	6.27	7.87	3.85
24.5	5.6	6.6	7.6	3.5
24.9	6.3	7.3	6.9	–
25.0	8.28	8.79	5.28	3.27
25.1	10.3	10.3	3.7	–
26.0	11.30	11.30	2.71	2.71
28.0	11.75	11.75	2.24	2.24
30.0	11.96	11.96	2.04	2.04

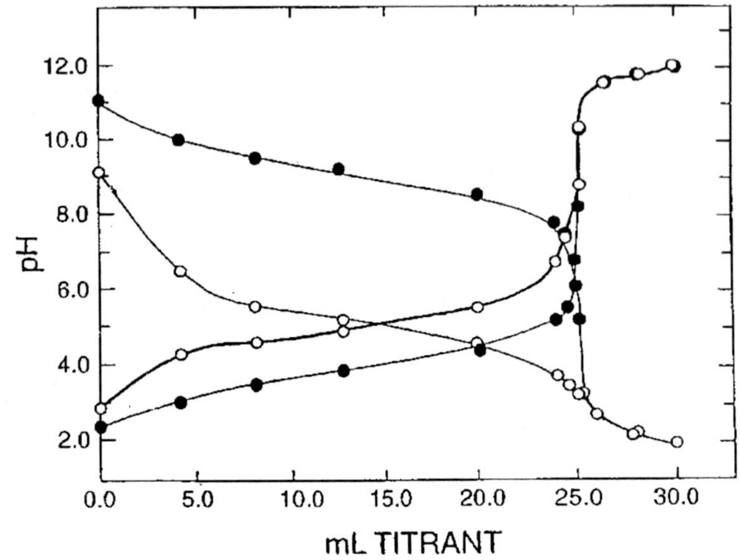

65. a. This is a weak acid-strong base titration. At the halfway point to equivalence, [weak acid] = [conjugate base], so pH = pK_a (always for a weak acid-strong base titration).

$pH = -\log(6.4 \times 10^{-5}) = 4.19$

mmol $HC_7H_5O_2$ present = 100.0 mL × 0.10 M = 10. mmol $HC_7H_5O_2$. For the equivalence point, 10. mmol of OH^- must be added. The volume of OH^- added to reach the equivalence point is:

$$10.\text{ mmol OH}^- \times \frac{1\text{ mL}}{0.10\text{ mmol OH}^-} = 1.0 \times 10^2\text{ mL OH}^-$$

At the equivalence point, 10. mmol of $HC_7H_5O_2$ is neutralized by 10. mmol of OH^- to produce 10. mmol of $C_7H_5O_2^-$. This is a weak base. The total volume of the solution is 100.0 mL + 1.0 × 10^2 mL = 2.0 × 10^2 mL. Solving the weak base equilibrium problem:

$$C_7H_5O_2^- + H_2O \rightleftharpoons HC_7H_5O_2 + OH^- \quad K_b = \frac{1.0 \times 10^{-14}}{6.4 \times 10^{-5}} = 1.6 \times 10^{-10}$$

	$C_7H_5O_2^-$	$HC_7H_5O_2$	OH^-
Initial	10. mmol/2.0 × 10^2 mL	0	0
Equil.	$0.050 - x$	x	x

$$K_b = 1.6 \times 10^{-10} = \frac{x^2}{0.050 - x} \approx \frac{x^2}{0.050}, \; x = [OH^-] = 2.8 \times 10^{-6}\,M$$

pOH = 5.55; pH = 8.45; assumptions good.

b. At the halfway point to equivalence for a weak base-strong acid titration, pH = pK_a because [weak base] = [conjugate acid].

$$K_a = \frac{K_w}{K_b} = \frac{1.0 \times 10^{-14}}{5.6 \times 10^{-4}} = 1.8 \times 10^{-11}; \; pH = pK_a = -\log(1.8 \times 10^{-11}) = 10.74$$

For the equivalence point (mmol acid added = mmol base present):

mmol $C_2H_5NH_2$ present = 100.0 mL × 0.10 M = 10. mmol $C_2H_5NH_2$

$$\text{mL H}^+ \text{ added} = 10.\text{ mmol H}^+ \times \frac{1\text{ mL}}{0.20\text{ mmol H}^+} = 50.\text{ mL H}^+$$

The strong acid added completely converts the weak base into its conjugate acid. Therefore, at the equivalence point, $[C_2H_5NH_3^+]_0$ = 10. mmol/(100.0 + 50.) mL = 0.067 M. Solving the weak acid equilibrium problem:

	$C_2H_5NH_3^+$	$\rightleftharpoons$	H^+	+	$C_2H_5NH_2$
Initial	0.067 M		0		0
Equil.	$0.067 - x$		x		x

$$K_a = 1.8 \times 10^{-11} = \frac{x^2}{0.067 - x} \approx \frac{x^2}{0.067},\ x = [H^+] = 1.1 \times 10^{-6}\ M$$

pH = 5.96; assumptions good.

c. In a strong acid-strong base titration, the halfway point has no special significance other than that exactly one-half of the original amount of acid present has been neutralized.

mmol H^+ present = 100.0 mL × 0.50 M = 50. mmol H^+

$$\text{mL OH}^- \text{ added} = 25\text{ mmol OH}^- \times \frac{1\text{ mL}}{0.25\text{ mmol}} = 1.0 \times 10^2\text{ mL OH}^-$$

	H^+	+	OH^-	$\rightarrow$	H_2O
Before	50. mmol		25 mmol		
After	25 mmol		0		

$$[H^+]_{excess} = \frac{25\text{ mmol}}{(100.0 + 1.0 \times 10^2)\text{ mL}} = 0.13\ M;\ \text{pH} = 0.89$$

At the equivalence point of a strong acid-strong base titration, only neutral species are present (Na^+, Cl^-, and H_2O), so the pH = 7.00.

66. 50.0 mL × 1.0 M = 50. mmol CH_3NH_2 present initially; $CH_3NH_2 + H^+ \rightarrow CH_3NH_3^+$

a. 50.0 mL × 0.50 M = 25. mmol HCl added. The added H^+ will convert one-half of the CH_3NH_2 into $CH_3NH_3^+$. This is the halfway point to equivalence, where $[CH_3NH_2] = [CH_3NH_3^+]$.

$$\text{pH} = \text{p}K_a + \log\frac{[CH_3NH_2]}{[CH_3NH_3^+]} = \text{p}K_a;\ K_a = \frac{1.0 \times 10^{-14}}{4.4 \times 10^{-4}} = 2.3 \times 10^{-11}$$

$$\text{pH} = \text{p}K_a = -\log(2.3 \times 10^{-11}) = 10.64$$

b. It will take 100. mL of HCl solution to reach the stoichiometric (equivalence) point. Here the added H^+ will convert all of the CH_3NH_2 into its conjugate acid, $CH_3NH_3^+$.

$$[CH_3NH_3^+]_0 = \frac{50.\text{ mmol}}{150.\text{ mL}} = 0.33\ M$$

	$CH_3NH_3^+$	$\rightleftharpoons$	H^+	+	CH_3NH_2	$K_a = \frac{K_w}{K_b} = 2.3 \times 10^{-11}$
Initial	0.33 *M*		0		0	
Equil.	$0.33 - x$		x		x	

$2.3 \times 10^{-11} = \frac{x^2}{0.33 - x} \approx \frac{x^2}{0.33}$, $x = [H^+] = 2.8 \times 10^{-6}\ M$; pH = 5.55; assumptions good.

67. $75.0\text{ mL} \times \frac{0.10\text{ mmol}}{\text{mL}} = 7.5$ mmol HA; $30.0\text{ mL} \times \frac{0.10\text{ mmol}}{\text{mL}} = 3.0$ mmol OH^- added

The added strong base reacts to completion with the weak acid to form the conjugate base of the weak acid and H_2O.

	HA	+	OH^-	→	A^-	+	H_2O
Before	7.5 mmol		3.0 mmol		0		
After	4.5 mmol		0		3.0 mmol		

A buffer results after the OH^- reacts to completion. Using the Henderson-Hasselbalch equation:

$$pH = pK_a + \log\frac{[A^-]}{[HA]},\quad 5.50 = pK_a + \log\left(\frac{3.0\text{ mmol}/105.0\text{ mL}}{4.5\text{ mmol}/105.0\text{ mL}}\right)$$

$pK_a = 5.50 - \log(3.0/4.5) = 5.50 - (-0.18) = 5.68$; $K_a = 10^{-5.68} = 2.1 \times 10^{-6}$

68. Mol H^+ added = 0.0400 L × 0.100 mol/L = 0.00400 mol H^+

The added strong acid reacts to completion with the weak base to form the conjugate acid of the weak base and H_2O. Let B = weak base:

	B	+	H^+	→	BH^+
Before	0.0100 mol		0.00400 mol		0
After	0.0060		0		0.0400 mol

After the H^+ reacts to completion, we have a buffer solution. Using the Henderson-Hasselbalch equation:

$$pH = pK_a + \log\frac{[\text{base}]}{[\text{acid}]},\quad 8.00 = pK_a + \log\frac{(0.0060/V_T)}{(0.00400/V_T)}$$, where V_T = total volume of solution

$$pK_a = 8.00 - \log\frac{(0.0060)}{(0.00400)} = 8.00 - 0.18,\ pK_a = 7.82$$

For a conjugate acid-base pair, $pK_a + pK_b = 14,00$, so:

$$pK_b = 14.00 - 7.82 = 6.18;\ K_b = 10^{-6.18} = 6.6 \times 10^{-7}$$

Indicators

69. $HIn \rightleftharpoons In^- + H^+ \quad K_a = \frac{[In^-][H^+]}{[HIn]} = 1.0 \times 10^{-9}$

a. In a very acid solution, the HIn form dominates, so the solution will be yellow.

b. The color change occurs when the concentration of the more dominant form is approximately ten times as great as the less dominant form of the indicator.

$\frac{[HIn]}{[In^-]} = \frac{10}{1};\ K_a = 1.0 \times 10^{-9} = \left(\frac{1}{10}\right)[H^+],\ [H^+] = 1 \times 10^{-8}\ M;\ pH = 8.0$ at color change

c. This is way past the equivalence point (100.0 mL OH^- added), so the solution is very basic, and the In^- form of the indicator dominates. The solution will be blue.

70. The color of the indicator will change over the approximate range of $pH = pK_a \pm 1 = 5.3 \pm 1$. Therefore, the useful pH range of methyl red where it changes color would be about 4.3 (red) to 6.3 (yellow). Note that at $pH < 4.3$, the HIn form of the indicator dominates, and the color of the solution is the color of HIn (red). At $pH > 6.3$, the In^- form of the indicator dominates, and the color of the solution is the color of In^- (yellow). In titrating a weak acid with base, we start off with an acidic solution with $pH < 4.3$, so the color would change from red to reddish orange at $pH \approx 4.3$. In titrating a weak base with acid, the color change would be from yellow to yellowish orange at $pH \approx 6.3$. Only a weak base-strong acid titration would have an acidic pH at the equivalence point, so only in this type of titration would the color change of methyl red indicate the approximate endpoint.

71. At the equivalence point, P^{2-} is the major species. P^{2-} is a weak base in water because it is the conjugate base of a weak acid.

	P^{2-}	+	H_2O	$\rightleftharpoons$	HP^-	+	OH^-
Initial	$\frac{0.5\ g}{0.1\ L} \times \frac{1\ mol}{204.2\ g} = 0.024\ M$				0		~0 (carry extra sig. fig.)
Equil.	$0.024 - x$				x		x

$$K_b = \frac{[HP^-][OH^-]}{P^{2-}} = \frac{K_w}{K_a} = \frac{1.0 \times 10^{-14}}{10^{-5.51}},\ 3.2 \times 10^{-9} = \frac{x^2}{0.024 - x} \approx \frac{x^2}{0.024}$$

$x = [OH^-] = 8.8 \times 10^{-6}\ M$; pOH = 5.1; pH = 8.9; assumptions good.

Phenolphthalein would be the best indicator for this titration because it changes color at around $pH \approx 9$ (from acid color to base color).

72. $HIn \rightleftharpoons In^- + H^+ \quad K_a = \dfrac{[In^-][H^+]}{[HIn]} = 10^{-3.00} = 1.0 \times 10^{-3}$

At 7.00% conversion of HIn into In^-, $[In^-]/[HIn] = 7.00/93.00$.

$$K_a = 1.0 \times 10^{-3} = \frac{[In^-]}{[HIn]} \times [H^+] = \frac{7.00}{93.00} \times [H^+],\ [H^+] = 1.3 \times 10^{-2}\ M,\ pH = 1.89$$

The color of the base form will start to show when the pH is increased to 1.89.

73. When choosing an indicator, we want the color change of the indicator to occur approximately at the pH of the equivalence point. Because the pH generally changes very rapidly at the equivalence point, we don't have to be exact. This is especially true for strong acid-strong base titrations. The following are some indicators where the color change occurs at about the pH of the equivalence point:

Exercise	pH at Eq. Pt.	Indicator
57	7.00	bromthymol blue or phenol red
59	8.79	o-cresolphthalein or phenolphthalein

74.

Exercise	pH at Eq. Pt.	Indicator
58	7.00	bromthymol blue or phenol red
60	4.82	bromcresol green

75.

Exercise	pH at Eq. Pt.	Indicator
61	8.28	o-cresolphthalein or phenolphthalein
63	5.28	bromcresol green

76.

Exercise	pH at Eq. Pt.	Indicator
62	8.79	o-cresolphthalein or phenolphthalein
64	3.27	2,4-dinitrophenol

In the titration in Exercise 64, it will be very difficult to mark the equivalence point. The pH break at the equivalence point is too small.

77. $pH > 5$ for bromcresol green to be blue. $pH < 8$ for thymol blue to be yellow. The pH is between 5 and 8.

78. The pH will be less than about 0.5 because crystal violet is yellow at a pH less than about 0.5. The methyl orange result only tells us that the pH is less than about 3.5.

79. a. yellow b. green (Both yellow and blue forms are present.) c. yellow d. blue

80. a. yellow b. yellow

c. green (Both yellow and blue forms are present.) d. colorless

Additional Exercises

81. $NH_3 + H_2O \rightleftharpoons NH_4^+ + OH^-$ $K_b = \dfrac{[NH_4^+][OH^-]}{[NH_3]}$; taking the –log of the K_b expression:

$$-\log K_b = -\log[OH^-] - \log\frac{[NH_4^+]}{[NH_3]}, \quad -\log[OH^-] = -\log K_b + \log\frac{[NH_4^+]}{[NH_3]}$$

$$pOH = pK_b + \log\frac{[NH_4^+]}{[NH_3]} \text{ or } pOH = pK_b + \log\frac{[acid]}{[base]}$$

82. a. $pH = pK_a = -\log(6.4 \times 10^{-5}) = 4.19$ since $[HBz] = [Bz^-]$, where $HBz = C_6H_5CO_2H$ and $[Bz^-] = C_6H_5CO_2^-$.

b. $[Bz^-]$ will increase to 0.120 *M* and [HBz] will decrease to 0.080 *M* after OH^- reacts completely with HBz. The Henderson-Hasselbalch equation is derived from the K_a dissociation reaction.

$$pH = pK_a + \log\frac{[Bz^-]}{[HBz]}, \quad pH = 4.19 + \log\frac{(0.120)}{(0.080)} = 4.37; \text{ assumptions good.}$$

c.

	Bz^-	+	H_2O	$\rightleftharpoons$	HBz	+	OH^-
Initial	0.120 *M*				0.080 *M*		0
Equil.	$0.120 - x$				$0.080 + x$		x

$$K_b = \frac{K_w}{K_a} = \frac{1.0 \times 10^{-14}}{6.4 \times 10^{-5}} = \frac{(0.080 + x)(x)}{(0.120 - x)} \approx \frac{(0.080)(x)}{0.120}$$

$x = [OH^-] = 2.34 \times 10^{-10}$ *M* (carrying extra sig. fig.); assumptions good.

$pOH = 9.63$; $pH = 4.37$

d. We get the same answer. Both equilibria involve the two major species, benzoic acid and benzoate anion. Both equilibria must hold true. K_b is related to K_a by K_w and $[OH^-]$ is related to $[H^+]$ by K_w, so all constants are interrelated.

83. a. The optimum pH for a buffer is when $pH = pK_a$. At this pH a buffer will have equal neutralization capacity for both added acid and base. As shown next, because the pK_a for $TRISH^+$ is 8.1, the optimal buffer pH is about 8.1.

$K_b = 1.19 \times 10^{-6}$; $K_a = K_w/K_b = 8.40 \times 10^{-9}$; $pK_a = -\log(8.40 \times 10^{-9}) = 8.076$

b. $$pH = pK_a + \log\frac{[TRIS]}{[TRISH^+]}, \quad 7.00 = 8.076 + \log\frac{[TRIS]}{[TRISH^+]}$$

$$\frac{[TRIS]}{[TRISH^+]} = 10^{-1.08} = 0.083 \text{ (at pH = 7.00)}$$

$$9.00 = 8.076 + \log\frac{[\text{TRIS}]}{[\text{TRISH}^+]},\quad \frac{[\text{TRIS}]}{[\text{TRISH}^+]} = 10^{0.92} = 8.3 \quad (\text{at pH} = 9.00)$$

c. $\dfrac{50.0\text{ g TRIS}}{2.0\text{ L}} \times \dfrac{1\text{ mol}}{121.14\text{ g}} = 0.206\ M = 0.21\ M = [\text{TRIS}]$

$$\frac{65.0\text{ g TRISHCl}}{2.0\text{ L}} \times \frac{1\text{ mol}}{157.60\text{ g}} = 0.206\ M = 0.21\ M = [\text{TRISHCl}] = [\text{TRISH}^+]$$

$$\text{pH} = \text{p}K_a + \log\frac{[\text{TRIS}]}{[\text{TRISH}^+]} = 8.076 + \log\frac{(0.21)}{(0.21)} = 8.08$$

The amount of H^+ added from HCl is: $(0.50 \times 10^{-3}\text{ L}) \times 12\text{ mol/L} = 6.0 \times 10^{-3}\text{ mol H}^+$

The H^+ from HCl will convert TRIS into TRISH^+. The reaction is:

	TRIS	+	H^+	→	TRISH^+	
Before	0.21 M		$\dfrac{6.0 \times 10^{-3}}{0.2005} = 0.030\ M$		0.21 M	
Change	−0.030		−0.030	→	+0.030	Reacts completely
After	0.18		0		0.24	

Now use the Henderson-Hasselbalch equation to solve this buffer problem.

$$\text{pH} = 8.076 + \log\left(\frac{0.18}{0.24}\right) = 7.95$$

84. $\text{pH} = \text{p}K_a + \log\dfrac{[C_2H_3O_2^-]}{[HC_2H_3O_2]},\quad 4.00 = -\log(1.8 \times 10^{-5}) + \log\dfrac{[C_2H_3O_2^-]}{[HC_2H_3O_2]}$

$\dfrac{[C_2H_3O_2^-]}{[HC_2H_3O_2]} = 0.18$; this is also equal to the mole ratio between $C_2H_3O_2^-$ and $HC_2H_3O_2$.

Let x = volume of 1.00 M $HC_2H_3O_2$ and y = volume of 1.00 M $NaC_2H_3O_2$

$x + y = 1.00$ L, $x = 1.00 - y$

$x(1.00\text{ mol/L}) = \text{mol } HC_2H_3O_2$; $y(1.00\text{ mol/L}) = \text{mol } NaC_2H_3O_2 = \text{mol } C_2H_3O_2^-$

Thus: $\dfrac{y}{x} = 0.18$ or $\dfrac{y}{1.00 - y} = 0.18$; solving: $y = 0.15$ L, so $x = 1.00 - 0.15 = 0.85$ L.

We need 850 mL of 1.00 M $HC_2H_3O_2$ and 150 mL of 1.00 M $NaC_2H_3O_2$ to produce a buffer solution at pH = 4.00.

85. A best buffer is when pH ≈ $\text{p}K_a$; these solutions have about equal concentrations of weak acid and conjugate base. Therefore, choose combinations that yield a buffer where pH ≈ $\text{p}K_a$; that is, look for acids whose $\text{p}K_a$ is closest to the pH.

a. Potassium fluoride + HCl will yield a buffer consisting of HF (pK_a = 3.14) and F^-.

b. Benzoic acid + NaOH will yield a buffer consisting of benzoic acid (pK_a = 4.19) and benzoate anion.

c. Sodium acetate + acetic acid (pK_a = 4.74) is the best choice for pH = 5.0 buffer since acetic acid has a pK_a value closest to 5.0.

d. HOCl and NaOH: This is the best choice to produce a conjugate acid-base pair with pH = 7.0. This mixture would yield a buffer consisting of HOCl (pK_a = 7.46) and OCl^-. Actually, the best choice for a pH = 7.0 buffer is an equimolar mixture of ammonium chloride and sodium acetate. NH_4^+ is a weak acid ($K_a = 5.6 \times 10^{-10}$), and $C_2H_3O_2^-$ is a weak base ($K_b = 5.6 \times 10^{-10}$). A mixture of the two will give a buffer at pH = 7.0 because the weak acid and weak base are the same strengths (K_a for NH_4^+ = K_b for $C_2H_3O_2^-$). $NH_4C_2H_3O_2$ is commercially available, and its solutions are used for pH = 7.0 buffers.

e. Ammonium chloride + NaOH will yield a buffer consisting of NH_4^+ (pK_a = 9.26) and NH_3.

86. Let's abbreviate the carboxylic acid group in alanine as RCOOH and the amino group in alanine as RNH_2. The K_a reaction for the carboxylic acid group is:

$$RCOOH \rightleftharpoons RCOO^- + H^+ \qquad K_a = 4.5 \times 10^{-7}$$

From Le Chatelier's principle, if we have a very acidic solution, a lot of H^+ is present. This drives the K_a reaction to the left, and the dominant form of the carboxylic acid group will be RCOOH (an overall neutral charge). If we have a very basic solution, the excess OH^- will remove H^+ from solution. As H^+ is removed, the K_a reaction shifts right, and the dominant form of the carboxylic acid group will be $RCOO^-$ (an overall 1– charged ion).

The K_b reaction for the amino group is:

$$RNH_2 + H_2O \rightleftharpoons RNH_3^+ + OH^-$$

If we have a very acidic solution, the excess protons present will remove OH^- from solution, and the dominant form of the amino group will be RNH_3^+ (an overall 1+ charged ion). If we have a very basic solution, a lot of OH^- is present, and the dominant form of the amino group will be RNH_2 (an overall neutral charge).

In alanine, both an RCOOH group and an RNH_2 group are present. The dominant form of alanine in a very acidic solution will be the form with the protons attached to the two groups that have acid-base properties. This form of alanine is:

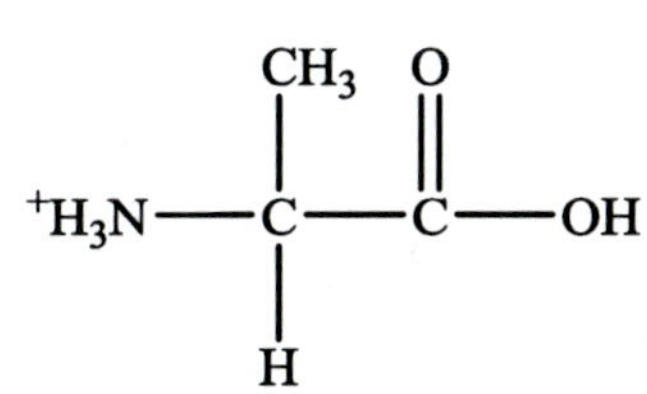

which has an overall charge of 1+. The dominant form of alanine in a very basic solution will be in the form with the protons removed from the two groups that have acid-base properties. This form of alanine is:

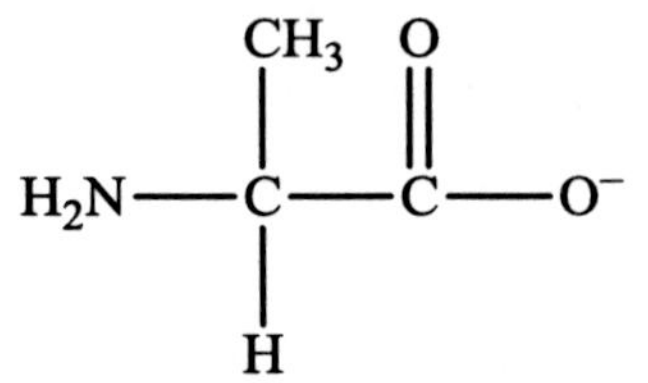

which has an overall charge of 1–.

87. a. $pH = pK_a + \log\frac{[\text{base}]}{[\text{acid}]}$, $7.15 = -\log(6.2 \times 10^{-8}) + \log\frac{[HPO_4^{2-}]}{[H_2PO_4^-]}$

$$7.15 = 7.21 + \log\frac{[HPO_4^{2-}]}{[H_2PO_4^-]},\ \frac{[HPO_4^{2-}]}{[H_2PO_4^-]} = 10^{-0.06} = 0.9,\ \frac{[H_2PO_4^-]}{[HPO_4^{2-}]} = \frac{1}{0.9} = 1.1 \approx 1$$

b. A best buffer has approximately equal concentrations of weak acid and conjugate base, so $pH \approx pK_a$ for a best buffer. The pK_a value for a $H_3PO_4/H_2PO_4^-$ buffer is $-\log(7.5 \times 10^{-3}) = 2.12$. A pH of 7.15 is too high for a $H_3PO_4/H_2PO_4^-$ buffer to be effective. At this high of pH, there would be so little H_3PO_4 present that we could hardly consider it a buffer; this solution would not be effective in resisting pH changes, especially when a strong base is added.

88. At pH = 0.00, $[H^+] = 10^{-0.00} = 1.0\ M$. We begin with 1.0 L × 2.0 mol/L OH^- = 2.0 mol OH^-. We will need 2.0 mol HCl to neutralize the OH^-, plus an additional 1.0 mol excess H^+ to reduce the pH to 0.00. We need 3.0 mol HCl total assuming 1.0 L of solution.

89. a. $HC_2H_3O_2 + OH^- \rightleftharpoons C_2H_3O_2^- + H_2O$

$$K_{eq} = \frac{[C_2H_3O_2^-]}{[HC_2H_3O_2][OH^-]} \times \frac{[H^+]}{[H^+]} = \frac{K_{a,\,HC_2H_3O_2}}{K_w} = \frac{1.8 \times 10^{-5}}{1.0 \times 10^{-14}} = 1.8 \times 10^9$$

Note: This reaction is the reverse of the K_b reaction for $C_2H_3O_2^-$.

b. $C_2H_3O_2^- + H^+ \rightleftharpoons HC_2H_3O_2$ $K_{eq} = \frac{[HC_2H_3O_2]}{[H^+][C_2H_3O_2^-]} = \frac{1}{K_{a,\,HC_2H_3O_2}} = 5.6 \times 10^4$

c. $HCl + NaOH \rightarrow NaCl + H_2O$

Net ionic equation is $H^+ + OH^- \rightleftharpoons H_2O$; $K_{eq} = \frac{1}{K_w} = 1.0 \times 10^{14}$

90. a. Because all acids are the same initial concentration, the pH curve with the highest pH at 0 mL of NaOH added will correspond to the titration of the weakest acid. This is pH curve f.

b. The pH curve with the lowest pH at 0 mL of NaOH added will correspond to the titration of the strongest acid. This is pH curve a.

The best point to look at to differentiate a strong acid from a weak acid titration (if initial concentrations are not known) is the equivalence point pH. If the pH = 7.00, the acid titrated is a strong acid; if the pH is greater than 7.00, the acid titrated is a weak acid.

c. For a weak acid-strong base titration, the pH at the halfway point to equivalence is equal to the pK_a value. The pH curve, which represents the titration of an acid with $K_a = 1.0 \times 10^{-6}$, will have a pH = $-\log(1 \times 10^{-6}) = 6.0$ at the halfway point. The equivalence point, from the plots, occurs at 50 mL NaOH added, so the halfway point is 25 mL. Plot d has a pH ≈ 6.0 at 25 mL of NaOH added, so the acid titrated in this pH curve (plot d) has $K_a \approx 1 \times 10^{-6}$.

91. In the final solution: $[H^+] = 10^{-2.15} = 7.1 \times 10^{-3}$ *M*

Beginning mmol HCl = 500.0 mL × 0.200 mmol/mL = 100. mmol HCl

Amount of HCl that reacts with NaOH = 1.50×10^{-2} mmol/mL × V

$$\frac{7.1 \times 10^{-3}\ \text{mmol}}{\text{mL}} = \frac{\text{final mmol H}^+}{\text{total volume}} = \frac{100. - (0.0150)\text{V}}{500.0 + \text{V}}$$

$3.6 + (7.1 \times 10^{-3})\text{V} = 100. - (1.50 \times 10^{-2})\text{V}$, $(2.21 \times 10^{-2})\text{V} = 100. - 3.6$

$\text{V} = 4.36 \times 10^3$ mL = 4.36 L = 4.4 L NaOH

92. For a titration of a strong acid with a strong base, the added OH^- reacts completely with the H^+ present. To determine the pH, we calculate the concentration of excess H^+ or OH^- after the neutralization reaction, and then calculate the pH.

0 mL: $[H^+] = 0.100$ *M* from HNO_3; pH = 1.000

4.0 mL: Initial mmol H^+ present = 25.0 mL × $\frac{0.100\ \text{mmol H}^+}{\text{mL}}$ = 2.50 mmol H^+

mmol OH^- added = 4.0 mL × $\frac{0.100\ \text{mmol OH}^-}{\text{mL}}$ = 0.40 mmol OH^-

0.40 mmol OH^- reacts completely with 0.40 mmol H^+: $OH^- + H^+ \rightarrow H_2O$

$$[H^+]_{excess} = \frac{(2.50 - 0.40)\ \text{mmol}}{(25.0 + 4.0)\ \text{mL}} = 7.24 \times 10^{-2}\ M;\ \text{pH} = 1.140$$

We follow the same procedure for the remaining calculations.

8.0 mL: $$[H^+]_{excess} = \frac{(2.50 - 0.80)\ \text{mmol}}{33.0\ \text{mL}} = 5.15 \times 10^{-2}\ M;\ \text{pH} = 1.288$$

12.5 mL: $[H^+]_{excess} = \dfrac{(2.50 - 1.25)\text{ mmol}}{37.5\text{ mL}} = 3.33 \times 10^{-2}\,M$; pH = 1.478

20.0 mL: $[H^+]_{excess} = \dfrac{(2.50 - 2.00)\text{ mmol}}{45.0\text{ mL}} = 1.1 \times 10^{-2}\,M$; pH = 1.96

24.0 mL: $[H^+]_{excess} = \dfrac{(2.50 - 2.40)\text{ mmol}}{49.0\text{ mL}} = 2.0 \times 10^{-3}\,M$; pH = 2.70

24.5 mL: $[H^+]_{excess} = \dfrac{(2.50 - 2.45)\text{ mmol}}{49.5\text{ mL}} = 1 \times 10^{-3}\,M$; pH = 3.0

24.9 mL: $[H^+]_{excess} = \dfrac{(2.50 - 2.49)\text{ mmol}}{49.9\text{ mL}} = 2 \times 10^{-4}\,M$; pH = 3.7

25.0 mL: Equivalence point; we have a neutral solution because there is no excess H^+ or OH^- remaining after the neutralization reaction. pH = 7.00

25.1 mL: Base in excess; $[OH^-]_{excess} = \dfrac{(2.51 - 2.50)\text{ mmol}}{50.1\text{ mL}} = 2 \times 10^{-4}\,M$; pOH = 3.7

pH = 14.00 − 3.7 = 10.3

26.0 mL: $[OH^-]_{excess} = \dfrac{(2.60 - 2.50)\text{ mmol}}{51.0\text{ mL}} = 2.0 \times 10^{-3}\,M$; pOH = 2.70; pH = 11.30

28.0 mL: $[OH^-]_{excess} = \dfrac{(2.80 - 2.50)\text{ mmol}}{53.0\text{ mL}} = 5.7 \times 10^{-3}\,M$; pOH = 2.24; pH = 11.76

30.0 mL: $[OH^-]_{excess} = \dfrac{(3.00 - 2.50)\text{ mmol}}{55.0\text{ mL}} = 9.1 \times 10^{-3}\,M$; pOH = 2.04; pH = 11.96

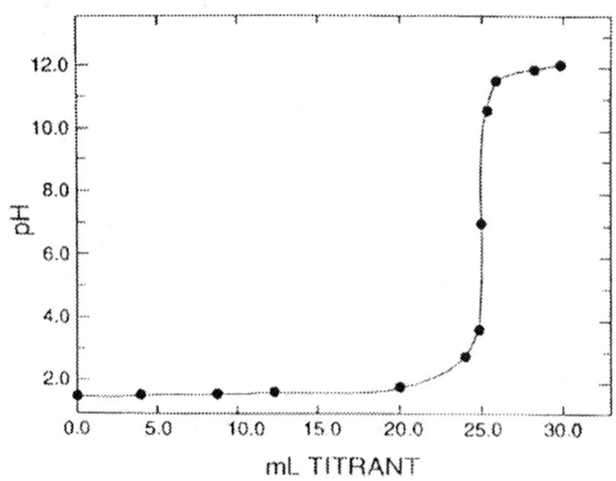

93. $HC_2H_3O_2 \rightleftharpoons H^+ + C_2H_3O_2^-$; let C_0 = initial concentration of $HC_2H_3O_2$

From normal weak acid setup: $K_a = 1.8 \times 10^{-5} = \frac{[H^+][C_2H_3O_2^-]}{[HC_2H_3O_2]} = \frac{[H^+]^2}{C_0 - [H^+]}$

$$[H^+] = 10^{-2.68} = 2.1 \times 10^{-3}\ M;\ 1.8 \times 10^{-5} = \frac{(2.1 \times 10^{-3})^2}{C_0 - (2.1 \times 10^{-3})},\ C_0 = 0.25\ M$$

25.0 mL × 0.25 mmol/mL = 6.3 mmol $HC_2H_3O_2$

Need 6.3 mmol KOH = V_{KOH} × 0.0975 mmol/mL, V_{KOH} = 65 mL

94. Mol acid = $0.210\text{ g} \times \frac{1\text{ mol}}{192\text{ g}} = 0.00109$ mol

Mol OH^- added = $0.0305\text{ L} \times \frac{0.108\text{ mol NaOH}}{\text{L}} \times \frac{1\text{ mol OH}^-}{\text{mol NaOH}} = 0.00329\text{ mol OH}^-$

$$\frac{\text{Mol OH}^-}{\text{Mol acid}} = \frac{0.00329}{0.00109} = 3.02$$

The acid is triprotic (H_3A) because 3 mol of OH^- are required to react with 1 mol of the acid; that is, the acid must have 3 mol H^+ in the formula to react with 3 mol of OH^-.

95. $HA + OH^- \rightarrow A^- + H_2O$, where HA = acetylsalicylic acid (assuming it is a monoprotic acid).

mmol HA present = $27.36\text{ mL OH}^- \times \frac{0.5106\text{ mmol OH}^-}{\text{mL OH}^-} \times \frac{1\text{ mmol HA}}{\text{mmol OH}^-} = 13.97$ mmol HA

$$\text{Molar mass of HA} = \frac{2.51\text{ g HA}}{13.97 \times 10^{-3}\text{ mol HA}} = 180.\text{ g/mol}$$

To determine the K_a value, use the pH data. After complete neutralization of acetylsalicylic acid by OH^-, we have 13.97 mmol of A^- produced from the neutralization reaction. A^- will react completely with the added H^+ and re-form acetylsalicylic acid HA.

mmol H^+ added = $13.68\text{ mL} \times \frac{0.5106\text{ mmol H}^+}{\text{mL}} = 6.985\text{ mmol H}^+$

	A^-	+	H^+	→	HA	
Before	13.97 mmol		6.985 mmol		0	
Change	−6.985		−6.985	→	+6.985	Reacts completely
After	6.985 mmol		0		6.985 mmol	

We have back titrated this solution to the halfway point to equivalence, where pH = pK_a (assuming HA is a weak acid). This is true because after H^+ reacts completely, equal millimoles of HA and A^- are present, which only occurs at the halfway point to equivalence. Assuming acetylsalicylic acid is a weak monoprotic acid, then pH = pK_a = 3.48. $K_a = 10^{-3.48}$ $= 3.3 \times 10^{-4}$.

96. NaOH added = 50.0 mL × $\frac{0.500\,\text{mmol}}{\text{mL}}$ = 25.0 mmol NaOH

NaOH left unreacted = 31.92 mL HCl × $\frac{0.289\,\text{mmol}}{\text{mL}} \times \frac{1\,\text{mmol NaOH}}{\text{mmol HCl}}$ = 9.22 mmol NaOH

NaOH reacted with aspirin = 25.0 – 9.22 = 15.8 mmol NaOH

15.8 mmol NaOH × $\frac{1\,\text{mmol aspirin}}{2\,\text{mmol NaOH}} \times \frac{180.2\,\text{mg}}{\text{mmol}}$ = 1420 mg = 1.42 g aspirin

Purity = $\frac{1.42\,\text{g}}{1.427\,\text{g}} \times 100 = 99.5\%$

Here, a strong base is titrated by a strong acid. The equivalence point will be at pH = 7.0. Bromthymol blue would be the best indicator since it changes color at pH ≈ 7 (from base color to acid color), although phenolphthalein is commonly used for the indicator. See Fig. 15.8 of the text.

97. 50.0 mL × 0.100 *M* = 5.00 mmol NaOH initially

At pH = 10.50, pOH = 3.50, $[OH^-] = 10^{-3.50} = 3.2 \times 10^{-4}$ *M*

mmol OH^- remaining = 3.2×10^{-4} mmol/mL × 73.75 mL = 2.4×10^{-2} mmol

mmol OH^- that reacted = 5.00 – 0.024 = 4.98 mmol

Because the weak acid is monoprotic, 23.75 mL of the weak acid solution contains 4.98 mmol HA.

$[HA]_0 = \frac{4.98\,\text{mmol}}{23.75\,\text{mL}} = 0.210\,M$

98. $HA + OH^- \rightarrow A^- + H_2O$; it takes 25.0 mL of 0.100 *M* NaOH to reach the equivalence point, where mmol HA = mmol OH^- = 25.0 mL(0.100 *M*) = 2.50 mmol. At the equivalence point, some HCl is added. The H^+ from the strong acid reacts to completion with the best base present, A^-.

	H^+	+	A^-	→	HA
Before	13.0 mL × 0.100 *M*		2.5 mmol		0
Change	–1.3 mmol		–1.3 mmol		+1.3 mmol
After	0		1.2 mmol		1.3 mmol

A buffer solution is present after the H^+ has reacted completely.

$\text{pH} = \text{p}K_a + \log\frac{[A^-]}{[HA]}$, $4.70 = \text{p}K_a + \log\left(\frac{1.2\,\text{mmol}/V_T}{1.3\,\text{mmol}/V_T}\right)$, where V_T = total volume

Because the log term will be negative [log(1.2/1.3) = –0.035)], the $\text{p}K_a$ value of the acid must be greater than 4.70 ($\text{p}K_a$ = 4.70 + 0.035 = 4.74).

99. At equivalence point: 16.00 mL × 0.125 mmol/mL = 2.00 mmol OH^- added; there must be 2.00 mmol HX present initially.

$HX + OH^- \rightarrow X^- + H_2O$ (neutralization rection)

2.00 mL NaOH added = 2.00 mL × 0.125 mmol/mL = 0.250 mmol OH^-; 0.250 mmol of OH^- added will convert 0.250 mmol HX into 0.250 mmol X^-. Remaining HX = 2.00 − 0.250 = 1.75 mmol HX; this is a buffer solution where $[H^+] = 10^{-6.912} = 1.22 \times 10^{-7}$ *M*. Because total volume cancels:

$$K_a = \frac{[H^+][X^-]}{[HX]} = \frac{1.22 \times 10^{-7}(0.250/V_T)}{1.75/V_T} = \frac{1.22 \times 10^{-7}(0.250)}{1.75} = 1.74 \times 10^{-8}$$

Note: We could solve for K_a using the Henderson-Hasselbalch equation.

ChemWork Problems

The answers to the problems 100 -107 (or a variation to these problems) are found in OWL. These problems are also assignable in OWL.

Challenge Problems

108. At 4.0 mL NaOH added: $\left|\frac{\Delta pH}{\Delta mL}\right| = \left|\frac{2.43 - 3.14}{0 - 4.0}\right| = 0.18$

The other points are calculated in a similar fashion. The results are summarized and plotted below. As can be seen from the plot, the advantage of this approach is that it is much easier to accurately determine the location of the equivalence point.

mL	pH	|ΔpH/ΔmL|
0	2.43	–
4.0	3.14	0.18
8.0	3.53	0.098
12.5	3.86	0.073
20.0	4.46	0.080
24.0	5.24	0.20
24.5	5.6	0.7
24.9	6.3	2
25.0	8.28	20
25.1	10.3	20
26.0	11.30	1
28.0	11.75	0.23
30.0	11.96	0.11

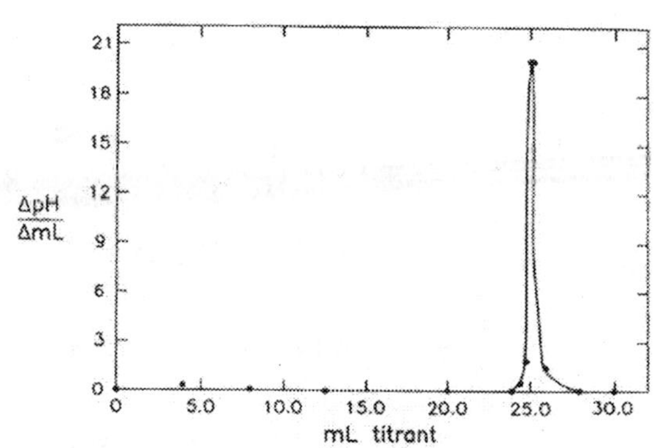

109. mmol $HC_3H_5O_2$ present initially = 45.0 mL × $\frac{0.750\,\text{mmol}}{\text{mL}}$ = 33.8 mmol $HC_3H_5O_2$

mmol $C_3H_5O_2^-$ present initially = 55.0 mL × $\frac{0.700\,\text{mmol}}{\text{mL}}$ = 38.5 mmol $C_3H_5O_2^-$

The initial pH of the buffer is:

$$pH = pK_a + \log\frac{[C_3H_5O_2^-]}{[HC_3H_5O_2]} = -\log(1.3 \times 10^{-5}) + \log\frac{\dfrac{38.5\text{ mmol}}{100.0\text{ mL}}}{\dfrac{33.8\text{ mmol}}{100.0\text{ mL}}} = 4.89 + \log\frac{38.5}{33.8} = 4.95$$

Note: Because the buffer components are in the same volume of solution, we can use the mole (or millimole) ratio in the Henderson-Hasselbalch equation to solve for pH instead of using the concentration ratio of $[C_3H_5O_2^-]$: $[HC_3H_5O_2]$.

When NaOH is added, the pH will increase, and the added OH^- will convert $HC_3H_5O_2$ into $C_3H_5O_2^-$. The pH after addition of OH^- increases by 2.5%, so the resulting pH is:

$$4.95 + 0.025(4.95) = 5.07$$

At this pH, a buffer solution still exists, and the millimole ratio between $C_3H_5O_2^-$ and $HC_3H_5O_2$ is:

$$pH = pK_a + \log\frac{\text{mmol}C_3H_5O_2^-}{\text{mmol}HC_3H_5O_2},\quad 5.07 = 4.89 + \log\frac{\text{mmol}C_3H_5O_2^-}{\text{mmol}HC_3H_5O_2}$$

$$\frac{\text{mmol}C_3H_5O_2^-}{\text{mmol}HC_3H_5O_2} = 10^{0.18} = 1.5$$

Let x = mmol OH^- added to increase pH to 5.07. Because OH^- will essentially react to completion with $HC_3H_5O_2$, the setup to the problem using millimoles is:

	$HC_3H_5O_2$	+	OH^-	→	$C_3H_5O_2^-$	
Before	33.8 mmol		x mmol		38.5 mmol	
Change	$-x$		$-x$	→	$+x$	Reacts completely
After	$33.8 - x$		0		$38.5 + x$	

$$\frac{\text{mmol}C_3H_5O_2^-}{\text{mmol}HC_3H_5O_2} = 1.5 = \frac{38.5 + x}{33.8 - x},\quad 1.5(33.8 - x) = 38.5 + x,\quad x = 4.9\text{ mmol }OH^-\text{ added}$$

The volume of NaOH necessary to raise the pH by 2.5% is:

$$4.9\text{ mmol NaOH} \times \frac{1\text{ mL}}{0.10\text{ mmol NaOH}} = 49\text{ mL}$$

49 mL of 0.10 *M* NaOH must be added to increase the pH by 2.5%.

110. $0.400\ \text{mol/L} \times V_{NH_3}$ = mol NH_3 = mol NH_4^+ after reaction with HCl at the equivalence point.

At the equivalence point: $[NH_4^+]_0 = \frac{\text{mol}\ NH_4^+}{\text{total volume}} = \frac{0.400 \times V_{NH_3}}{1.50 \times V_{NH_3}} = 0.267\ M$

	NH_4^+	$\rightleftharpoons$	H^+	+	NH_3
Initial	0.267 M		0		0
Equil.	$0.267 - x$		x		x

$$K_a = \frac{K_w}{K_b} = \frac{1.0 \times 10^{-14}}{1.8 \times 10^{-5}},\ 5.6 \times 10^{-10} = \frac{x^2}{0.267 - x} \approx \frac{x^2}{0.267}$$

$x = [H^+] = 1.2 \times 10^{-5}\ M$; pH = 4.92; assumption good.

111. For HOCl, $K_a = 3.5 \times 10^{-8}$ and $pK_a = -\log(3.5 \times 10^{-8}) = 7.46$. This will be a buffer solution because the pH is close to the pK_a value.

$$pH = pK_a + \log\frac{[OCl^-]}{[HOCl]},\ 8.00 = 7.46 + \log\frac{[OCl^-]}{[HOCl]},\ \frac{[OCl^-]}{[HOCl]} = 10^{0.54} = 3.5$$

1.00 L × 0.0500 M = 0.0500 mol HOCl initially. Added OH^- converts HOCl into OCl^-. The total moles of OCl^- and HOCl must equal 0.0500 mol. Solving where n = moles:

$n_{OCl^-} + n_{HOCl} = 0.0500$ and $n_{OCl^-} = (3.5)n_{HOCl}$

$(4.5)n_{HOCl} = 0.0500$, $n_{HOCl} = 0.011$ mol; $n_{OCl^-} = 0.039$ mol

Need to add 0.039 mol NaOH to produce 0.039 mol OCl^-.

0.039 mol = V × 0.0100 M, V = 3.9 L NaOH

Note: Normal buffer assumptions hold.

112. 50.0 mL × 0.100 M = 5.00 mmol H_2SO_4; 30.0 mL × 0.100 M = 3.00 mmol HOCl

25.0 mL × 0.200 M = 5.00 mmol NaOH; 10.0 mL × 0.150 M = 1.50 mmol KOH

25.0 mL × 0.100 M = 2.50 mmol $Ba(OH)_2$ = 5.00 mmol OH^-; we've added 11.50 mmol OH^- total.

Let OH^- react completely with the best acid present (H_2SO_4).

10.00 mmol OH^- + 5.00 mmol H_2SO_4 → 0.00 mmol H_2O + 5.00 mmol SO_4^{2-}

OH^- still remains after reacting completely with H_2SO_4. OH^- will then react with the next best acid (HOCl). The remaining 1.50 mmol OH^- will convert 1.50 mmol HOCl into 1.50 mmol OCl^-, resulting in a solution with 1.50 mmol OCl^- and (3.00 – 1.50 =) 1.50 mmol

HOCl. The major species at this point are HOCl, OCl^-, SO_4^{2-}, and H_2O plus cations that don't affect pH. SO_4^{2-} is an extremely weak base ($K_b = 8.3 \times 10^{-13}$). We have a buffer solution composed of HOCl and OCl^-. Because [HOCl] = [OCl^-]:

$[H^+] = K_a = 3.5 \times 10^{-8}$ *M*; pH = 7.46; assumptions good.

113. The first titration plot (from 0 – 100.0 mL) corresponds to the titration of H_2A by OH^-. The reaction is $H_2A + OH^- \rightarrow HA^- + H_2O$. After all the H_2A has been reacted, the second titration (from 100.0 – 200.0 mL) corresponds to the titration of HA^- by OH^-. The reaction is $HA^- + OH^- \rightarrow A^{2-} + H_2O$.

a. At 100.0 mL of NaOH, just enough OH^- has been added to react completely with all of the H_2A present (mol OH^- added = mol H_2A present initially). From the balanced equation, the mol of HA^- produced will equal the mol of H_2A present initially. Because mol of HA^- present at 100.0 mL OH^- added equals the mol of H_2A present initially, exactly 100.0 mL more of NaOH must be added to react with all of the HA^-. The volume of NaOH added to reach the second equivalence point equals 100.0 mL + 100.0 mL = 200.0 mL.

b. $H_2A + OH^- \rightarrow HA^- + H_2O$ is the reaction occurring from 0 – 100.0 mL NaOH added.

$HA^- + OH^- \rightarrow A^{2-} + H_2O$ is the reaction occurring from 100.0 – 200.0 mL NaOH added.

i. No reaction has taken place, so H_2A and H_2O are the major species.

ii. Adding OH^- converts H_2A into HA^-. The major species between 0 and 100.0 mL NaOH added are H_2A, HA^-, H_2O, and Na^+.

iii. At 100.0 mL NaOH added, mol of OH^- = mol H_2A, so all of the H_2A present initially has been converted into HA^-. The major species are HA^-, H_2O, and Na^+.

iv. Between 100.0 and 200.0 mL NaOH added, the OH^- converts HA^- into A^{2-}. The major species are HA^-, A^{2-}, H_2O, and Na^+.

v. At the second equivalence point (200.0 mL), just enough OH^- has been added to convert all of the HA^- into A^{2-}. The major species are A^{2-}, H_2O, and Na^+.

vi. Past 200.0 mL NaOH added, excess OH^- is present. The major species are OH^-, A^{2-}, H_2O, and Na^+.

c. 50.0 mL of NaOH added corresponds to the first halfway point to equivalence. Exactly one-half of the H_2A present initially has been converted into its conjugate base HA^-, so [H_2A] = [HA^-] in this buffer solution.

$$H_2A \rightleftharpoons HA^- + H^+ \qquad K_{a_1} = \frac{[HA^-][H^+]}{[H_2A]}$$

When [HA^-] = [H_2A], then $K_{a_1} = [H^+]$ or pK_{a_1} = pH.

Here, pH = 4.0, so $pK_{a_1} = 4.0$ and $K_{a_1} = 10^{-4.0} = 1 \times 10^{-4}$.

150.0 mL of NaOH added correspond to the second halfway point to equivalence, where $[HA^-] = [A^{2-}]$ in this buffer solution.

$$HA^- \rightleftharpoons A^{2-} + H^+ \quad K_{a_2} = \frac{[A^{2-}][H^+]}{[HA^-]}$$

When $[A^{2-}] = [HA^-]$, then $K_{a_2} = [H^+]$ or $pK_{a_2} = pH$.

Here, pH = 8.0, so $pK_{a_2} = 8.0$ and $K_{a_2} = 10^{-8.0} = 1 \times 10^{-8}$.

114. We will see only the first stoichiometric point in the titration of salicylic acid because K_{a_2} is so small. For adipic acid, the K_a values are fairly close to each other. Both protons will be titrated almost simultaneously, giving us only one break. The stoichiometric points will occur when 1 mol of OH^- is added per mole of salicylic acid present and when 2 mol of OH^- is added per mole of adipic acid present. Thus the 25.00-mL volume corresponded to the titration of salicylic acid, and the 50.00-mL volume corresponded to the titration of adipic acid.

115. a. Na^+ is present in all solutions. The added H^+ from HCl reacts completely with CO_3^{2-} to convert it into HCO_3^- (points A-C). After all of the CO_3^{2-} is reacted (after point C, the first equivalence point), H^+ then reacts completely with the next best base present, HCO_3^- (points C-E). Point E represents the second equivalence point. The major species present at the various points after H^+ reacts completely follow.

A. CO_3^{2-}, H_2O, Na^+

B. CO_3^{2-}, HCO_3^-, H_2O, Cl^-, Na^+

C. HCO_3^-, H_2O, Cl^-, Na^+

D. HCO_3^-, CO_2 (H_2CO_3), H_2O, Cl^-, Na^+

E. CO_2 (H_2CO_3), H_2O, Cl^-, Na^+

F. H^+ (excess), CO_2 (H_2CO_3), H_2O, Cl^-. Na^+

b. $H_2CO_3 \rightleftharpoons HCO_3^- + H^+ \quad K_{a_1} = 4.3 \times 10^{-7}$

$HCO_3^- \rightleftharpoons CO_3^{2-} + H^+ \quad K_{a_2} = 5.6 \times 10^{-11}$

The first titration reaction occurring between points A and C is:

$$H^+ + CO_3^{2-} \rightarrow HCO_3^-$$

At point B, enough H^+ has been added to convert one-half of the CO_3^{2-} into its conjugate acid. At this halfway point to equivalence, $[CO_3^{2-}] = [HCO_3^-]$. For this buffer solution,

$$pH = pK_{a_2} = -\log(5.6 \times 10^{-11}) = 10.25$$

The second titration reaction occurring between points C and E is:

$$H^+ + HCO_3^- \rightarrow H_2CO_3$$

Point D is the second halfway point to equivalence, where $[HCO_3^-] = [H_2CO_3]$. Here, $pH = pK_{a_1} = -\log(4.3 \times 10^{-7}) = 6.37$.

116. a. V_1 corresponds to the titration reaction of $CO_3^{2-} + H^+ \rightarrow HCO_3^-$; V_2 corresponds to the titration reaction of $HCO_3^- + H^+ \rightarrow H_2CO_3$.

Here, there are two sources of HCO_3^-: $NaHCO_3$ and the titration of Na_2CO_3, so $V_2 > V_1$.

b. V_1 corresponds to two titration reactions: $OH^- + H^+ \rightarrow H_2O$ and $CO_3^{2-} + H^+ \rightarrow HCO_3^-$.

V_2 corresponds to just one titration reaction: $HCO_3^- + H^+ \rightarrow H_2CO_3$.

Here, $V_1 > V_2$ due to the presence of OH^- , which is titrated in the V_1 region.

c. 0.100 mmol HCl/mL × 18.9 mL = 1.89 mmol H^+; Because the first stoichiometric point only involves the titration of Na_2CO_3 by H^+, 1.89 mmol of CO_3^{2-} has been converted into HCO_3^-. The sample contains 1.89 mmol Na_2CO_3 × 105.99 mg/mmol = 2.00×10^2 mg = 0.200 g Na_2CO_3.

The second stoichiometric point involves the titration of HCO_3^- by H^+.

$$\frac{0.100 \text{ mmol} H^+}{\text{mL}} \times 36.7 \text{ mL} = 3.67 \text{ mmol } H^+ = 3.67 \text{ mmol } HCO_3^-$$

1.89 mmol $NaHCO_3$ came from the first stoichiometric point of the Na_2CO_3 titration.

3.67 − 1.89 = 1.78 mmol HCO_3^- came from $NaHCO_3$ in the original mixture.

1.78 mmol $NaHCO_3$ × 84.01 mg $NaHCO_3$/mmol = 1.50×10^2 mg $NaHCO_3$ = 0.150 g $NaHCO_3$

$$\text{Mass \% } Na_2CO_3 = \frac{0.200 \text{ g}}{(0.200 + 0.150) \text{ g}} \times 100 = 57.1\% \ Na_2CO_3$$

$$\text{Mass \% } NaHCO_3 = \frac{0.150 \text{ g}}{0.350 \text{ g}} \times 100 = 42.9\% \ NaHCO_3$$

117. An indicator changes color at pH ≈ pK_a ±1.The results from each indicator tells us something about the pH. The conclusions are summarized below:

Results from	pH
bromphenol blue	≥ ~5.0
bromcresol purple	≤ ~5.0
bromcresol green *	pH ≈ pK_a ≈ 4.8
alizarin	≤ ~5.5

*For bromcresol green, the resultant color is green.
This is a combination of the extremes (yellow and blue).
This occurs when pH ≈ pK_a of the indicator.

From the indicator results, the pH of the solution is about 5.0. We solve for K_a by setting up the typical weak acid problem.

	HX	⇌	H^+	+	X^-
Initial	1.0 *M*		~0		0
Equil.	$1.0 - x$		x		x

$$K_a = \frac{[H^+][X^-]}{[HX]} = \frac{x^2}{1.0 - x}\text{; because pH} \approx 5.0,\ [H^+] = x \approx 1 \times 10^{-5} M.$$

$$K_a \approx \frac{(1 \times 10^{-5})^2}{1.0 - 1 \times 10^{-5}} \approx 1 \times 10^{-10}$$

118. Phenolphthalein will change color at pH ≈ 9. Phenolphthalein will mark the second end point of the titration. Therefore, we have titrated both protons on malonic acid.

$H_2Mal + 2\ OH^- \rightarrow 2\ H_2O + Mal^{2-}$ where H_2Mal = malonic acid

$$31.50\,\text{mL} \times \frac{0.0984\,\text{mmol NaOH}}{\text{mL}} \times \frac{1\,\text{mmol}\,H_2Mal}{2\,\text{mol NaOH}} = 1.55\ \text{mmol}\ H_2Mal$$

$$[H_2Mal] = \frac{1.55\,\text{mmol}}{25.00\,\text{mL}} = 0.0620\ M$$

Integrative Problems

119. $$\text{pH} = \text{p}K_a + \log\frac{[C_7H_4O_2F^-]}{[C_7H_5O_2F]} = 2.90 + \log\left[\frac{(55.0\text{mL} \times 0.472M)/130.0\text{mL}}{(75.0\text{mL} \times 0.275M)/130.0\text{mL}}\right]$$

$$\text{pH} = 2.90 + \log\left(\frac{26.0}{20.6}\right) = 2.90 + 0.101 = 3.00$$

120. a. 1.00 L × 0.100 mol/L = 0.100 mol HCl added to reach stoichiometric point.

The 10.00-g sample must have contained 0.100 mol of NaA. $\dfrac{10.00\,\text{g}}{0.100\,\text{mol}} = 100.\ \text{g/mol}$

b. 500.0 mL of HCl added represents the halfway point to equivalence. Thus pH = $\text{p}K_a$ = 5.00 and $K_a = 1.0 \times 10^{-5}$. At the equivalence point, enough H^+ has been added to convert all the A^- present initially into HA. The concentration of HA at the equivalence point is:

$$[HA]_0 = \frac{0.100\,\text{mol}}{1.10\,\text{L}} = 0.0909\ M$$

	HA	⇌	H^+	+	A^-	$K_a = 1.0 \times 10^{-5}$
Initial	0.0909 *M*		0		0	
Equil.	0.0909 - *x*		*x*		*x*	

$$K_a = 1.0 \times 10^{-5} = \frac{x^2}{0.0909 - x} \approx \frac{x^2}{0.0909}$$

$x = 9.5 \times 10^{-4}\ M = [H^+]$; pH = 3.02; assumptions good.

121. The added OH^- from the strong base reacts to completion with the best acid present, HF. To determine the pH, see what is in solution after the OH^- reacts to completion.

$$OH^- \text{ added} = 38.7 \text{ g soln} \times \frac{1.50 \text{ g NaOH}}{100.0 \text{ g soln}} \times \frac{1 \text{ mol NaOH}}{40.00 \text{ g}} \times \frac{1 \text{ mol } OH^-}{\text{mol NaOH}} = 0.0145 \text{ mol } OH^-$$

For the 0.174 *m* HF solution, if we had exactly 1 kg of H_2O, then the solution would contain 0.174 mol HF.

$$0.174 \text{ mol HF} \times \frac{20.01 \text{ g}}{\text{mol HF}} = 3.48 \text{ g HF}$$

Mass of solution = 1000.00 g H_2O + 3.48 g HF = 1003.48 g

$$\text{Volume of solution} = 1003.48 \text{ g} \times \frac{1 \text{ mL}}{1.10 \text{ g}} = 912 \text{ mL}$$

$$\text{Mol HF} = 250. \text{ mL} \times \frac{0.174 \text{ mol HF}}{912 \text{ mL}} = 4.77 \times 10^{-2} \text{ mol HF}$$

	OH^-	+	HF	→	F^-	+	H_2O
Before	0.0145 mol		0.0477 mol		0		
Change	−0.0145		−0.0145		+0.0145		
After	0		0.0332 mol		0.0145 mol		

After reaction, a buffer solution results containing HF, a weak acid, and F^-, its conjugate base. Let V_T = total volume of solution.

$$pH = pK_a + \log\frac{[F^-]}{[HF]} = -\log(7.2 \times 10^{-4}) + \log\left(\frac{0.0145/V_T}{0.0332/V_T}\right)$$

$$pH = 3.14 + \log\left(\frac{0.0145}{0.0332}\right) = 3.14 + (-0.360),\ pH = 2.78$$

Marathon Problem

122. Major species: PO_4^{3-}, OH^-, H^+, CN^-, Na^+, K^+, Cl^-, H_2O

PO_4^{3-}	OH^-	H^+	CN^-
5.00 mmol	5.00 mmol	15.0 mmol	7.50 mmol

PO_4^{3-} and CN^- are weak bases.

$$PO_4^{3-} + H_2O \rightleftharpoons HPO_4^{2-} + OH^- \quad K_b = K_w/K_{a_3} = 2.1 \times 10^{-2}$$

$$CN^- + H_2O \rightleftharpoons HCN + OH^- \quad K_b = K_w/K_a = 1.6 \times 10^{-5}$$

One of the keys to this problem is to recognize that pK_{a_2} for H_3PO_4 = 7.21 [$-\log(6.2 \times 10^{-8})$ = 7.21]. The K_{a_2} reaction for H_3PO_4 is:

$$H_2PO_4^- \rightleftharpoons HPO_4^{2-} + H^+ \quad K_{a_2} = 6.2 \times 10^{-8};\ pK_{a_2} = 7.21$$

The pH of the final solution will equal 7.21 when we have a buffer solution where $[H_2PO_4^-]$ = $[HPO_4^{2-}]$. To solve this problem, we need to determine the quantity of HNO_3 that must be added so that the final solution contains equal moles of $H_2PO_4^-$ and HPO_4^{2-}.

To start the problem, let's see what is in solution after we let the best acid and best base react in a series of reactions. In each of the following reactions, something strong is reacting, so we assume the reactions will go to completion. The first reaction to run to completion is the strong acid reacting with the strong base: $H^+ + OH^- \rightarrow H_2O$.

	H^+ +	OH^-	$\rightarrow H_2O$
Before	15.0 mmol	5.00 mmol	
After	10.0 mmol	0	

After all of the strong base is neutralized, the next best base present is PO_4^{3-}.

	H^+ +	PO_4^{3-}	$\rightarrow$	HPO_4^{2-}
Before	10.0 mmol	5.00 mmol		0
After	5.0 mmol	0		5.00 mmol

The next best base present is CN^-.

	H^+ +	CN^-	$\rightarrow$	HCN
Before	5.0 mmol	7.50 mmol		0
After	0	2.5 mmol		5.0 mmol

We need to add 2.5 mmol H^+ to convert all the CN^- into HCN; then all that remains is 5.00 mmol HPO_4^{2-} and 7.5 mmol HCN (a very weak acid with $K_a = 6.2 \times 10^{-10}$). From here, we would need to add another 2.5 mmol H^+ in order to convert one-half of the HPO_4^{2-} present into its conjugate acid so that $[HPO_4^{2-}] = [H_2PO_4^-]$ and pH = pK_{a_2} = 7.21. Adding 5.0 mmol H^+ to the original solution:

	H^+	+	CN^-	$\rightarrow$	HCN
Before	5.0 mmol		2.5 mmol		5.0 mmol
After	2.5 mmol		0		7.5 mmol

	H^+	+	HPO_4^{2-}	$\rightarrow$	$H_2PO_4^-$
Before	2.5 mmol		5.00 mmol		0
After	0		2.5 mmol		2.5 mmol

After 5.0 mmol H^+ (HNO_3) has been added to the original mixture, we have a final solution containing equal moles of HPO_4^{2-} and $H_2PO_4^-$ so that pH = pK_{a_2} = 7.21. Note that HCN, with $K_a = 6.2 \times 10^{-10}$, is too weak of an acid to interfere with the $H_2PO_4^-$/ HPO_4^{2-} buffer.

$$\text{Volume } HNO_3 = 5.0 \text{ mmol } HNO_3 \times \frac{1\text{ mL}}{0.100\text{ mmol } HNO_3} = 50.\text{ mL } HNO_3$$

CHAPTER 16

SOLUBILITY AND COMPLEX ION EQUILIBRIA

Questions

8. $MX(s) \rightleftharpoons M^{n+}(aq) + X^{n-}(aq)$ $K_{sp} = [M^{n+}][X^{n-}]$; the K_{sp} reaction always refers to a solid breaking up into its ions. The representations all show 1 : 1 salts, i.e., the formula of the solid contains 1 cation for every 1 anion (either +1 and −1, or +2 and −2, or +3 and −3). The solution with the largest number of ions (largest $[M^{n+}]$ and $[X^{n-}]$) will have the largest K_{sp} value. From the representations, the second beaker has the largest number of ions present, so this salt has the largest K_{sp} value. Conversely, the third beaker, with the fewest number of hydrated ions, will have the smallest K_{sp} value.

9. K_{sp} values can only be directly compared to determine relative solubilities when the salts produce the same number of ions (have the same stoichiometry). Here, Ag_2S and CuS do not produce the same number of ions when they dissolve, so each has a different mathematical relationship between the K_{sp} value and the molar solubility. To determine which salt has the larger molar solubility, you must do the actual calculations and compare the two molar solubility values.

10. The solubility product constant (K_{sp}) is an equilibrium constant that has only one value for a given solid at a given temperature. Solubility, on the other hand, can have many values for a given solid at a given temperature. In pure water, the solubility is some value, yet the solubility is another value if a common ion is present. And the actual solubility when a common ion is present varies according to the concentration of the common ion. However, in all cases the product of the ion concentrations must satisfy the K_{sp} expression and give that one unique K_{sp} value at that particular temperature.

11. i. This is the result when you have a salt that breaks up into two ions. Examples of these salts include AgCl, $SrSO_4$, $BaCrO_4$, and $ZnCO_3$.

    ii. This is the result when you have a salt that breaks up into three ions, either two cations and one anion or one cation and two anions. Some examples are SrF_2, Hg_2I_2, and Ag_2SO_4.

    iii. This is the result when you have a salt that breaks up into four ions, either three cations and one anion (Ag_3PO_4) or one cation and three anions (ignoring the hydroxides, there are no examples of this type of salt in Table 16.1).

    iv. This is the result when you have a salt that breaks up into five ions, either three cations and two anions [$Sr_3(PO_4)_2$] or two cations and three anions (no examples of this type of salt are in Table 16.1).

12. The obvious choice is that the metal ion reacts with PO_4^{3-} and forms an insoluble phosphate salt. The other possibility is due to the weak base properties of PO_4^{3-} (PO_4^{3-} is the conjugate base of the weak acid HPO_4^{2-}, so it is a weak base). Because PO_4^{3-} is a weak base in water, OH^- ions are present at a fairly large concentration. Hence the other potential precipitate is the metal ion reacting with OH^- to form an insoluble hydroxide salt.

13. For the K_{sp} reaction of a salt dissolving into its respective ions, a common ion would be one of the ions in the salt added from an outside source. When a common ion (a product in the K_{sp} reaction) is present, the K_{sp} equilibrium shifts to the left, resulting in less of the salt dissolving into its ions (solubility decreases).

14. S^{2-} is a very basic anion and reacts significantly with H^+ to form HS^- ($S^{2-} + H^+ \rightleftharpoons HS^-$). Thus, the actual concentration of S^{2-} in solution depends on the amount of H^+ present. In basic solutions, little H^+ is present, which shifts the above reaction to the left. In basic solutions, the S^{2-} concentration is relatively high. So, in basic solutions, a wider range of sulfide salts will precipitate. However, in acidic solutions, added H^+ shifts the equilibrium to the right resulting in a lower S^{2-} concentration. In acidic solutions, only the least soluble sulfide salts will precipitate out of solution.

15. Some people would automatically think that an increase in temperature would increase the solubility of a salt. This is not always the case as some salts show a decrease in solubility as temperature increases. The two major methods used to increase solubility of a salt both involve removing one of the ions in the salt by reaction. If the salt has an ion with basic properties, adding H^+ will increase the solubility of the salt because the added H^+ will react with the basic ion, thus removing it from solution. More salt dissolves in order to to make up for the lost ion. Some examples of salts with basic ions are AgF, $CaCO_3$, and $Al(OH)_3$. The other way to remove an ion is to form a complex ion. For example, the Ag^+ ion in silver salts forms the complex ion $Ag(NH_3)_2^+$ as ammonia is added. Silver salts increase their solubility as NH_3 is added because the Ag^+ ion is removed through complex ion formation.

16. Because the formation constants are usually very large numbers, the stepwise reactions can be assumed to essentially go to completion. Thus an equilibrium mixture of a metal ion and a specific ligand will mostly contain the final complex ion in the stepwise formation reactions.

17. In 2.0 *M* NH_3, the soluble complex ion $Ag(NH_3)_2^+$ forms, which increases the solubility of AgCl(s). The reaction is $AgCl(s) + 2\ NH_3 \rightleftharpoons Ag(NH_3)_2^+ + Cl^-$. In 2.0 *M* NH_4NO_3, NH_3 is only formed by the dissociation of the weak acid NH_4^+. There is not enough NH_3 produced by this reaction to dissolve AgCl(s) by the formation of the complex ion.

18. Unlike AgCl(s), $PbCl_2(s)$ shows a significant increase in solubility with an increase in temperature. Hence add NaCl to the solution containing the metal ion to form the chloride salt precipitate, and then heat the solution. If the precipitate dissolves, then $PbCl_2$ is present, and the metal ion is Pb^{2+}. If the precipitate does not dissolve with an increase in temperature, then AgCl is the precipitate, and Ag^+ is the metal ion present.

Exercises

Solubility Equilibria

19. a. $AgC_2H_3O_2(s) \rightleftharpoons Ag^+(aq) + C_2H_3O_2^-(aq)$ $K_{sp} = [Ag^+][C_2H_3O_2^-]$

b. $Al(OH)_3(s) \rightleftharpoons Al^{3+}(aq) + 3\ OH^-(aq)$ $K_{sp} = [Al^{3+}][OH^-]^3$

c. $Ca_3(PO_4)_2(s) \rightleftharpoons 3\ Ca^{2+}(aq) + 2\ PO_4^{3-}(aq)$ $K_{sp} = [Ca^{2+}]^3[PO_4^{3-}]^2$

20. a. $Ag_2CO_3(s) \rightleftharpoons 2\ Ag^+(aq) + CO_3^{2-}(aq)$ $K_{sp} = [Ag^+]^2[CO_3^{2-}]$

b. $Ce(IO_3)_3(s) \rightleftharpoons Ce^{3+}(aq) + 3\ IO_3^-(aq)$ $K_{sp} = [Ce^{3+}][IO_3^-]^3$

c. $BaF_2(s) \rightleftharpoons Ba^{2+}(aq) + 2\ F^-(aq)$ $K_{sp} = [Ba^{2+}][F^-]^2$

21. In our setup, s = solubility of the ionic solid in mol/L. This is defined as the maximum amount of a salt that can dissolve. Because solids do not appear in the K_{sp} expression, we do not need to worry about their initial and equilibrium amounts.

a.

	$CaC_2O_4(s)$	$\rightleftharpoons$	$Ca^{2+}(aq)$	+	$C_2O_4^{2-}(aq)$
Initial			0		0
	s mol/L of $CaC_2O_4(s)$ dissolves to reach equilibrium				
Change	$-s$	$\rightarrow$	$+s$		$+s$
Equil.			s		s

From the problem, $s = 4.8 \times 10^{-5}$ mol/L.

$K_{sp} = [Ca^{2+}][C_2O_4^{2-}] = (s)(s) = s^2$, $K_{sp} = (4.8 \times 10^{-5})^2 = 2.3 \times 10^{-9}$

b.

	$BiI_3(s)$	$\rightleftharpoons$	$Bi^{3+}(aq)$	+	$3\ I^-(aq)$
Initial			0		0
	s mol/L of $BiI_3(s)$ dissolves to reach equilibrium				
Change	$-s$	$\rightarrow$	$+s$		$+3s$
Equil.			s		$3s$

$K_{sp} = [Bi^{3+}][I^-]^3 = (s)(3s)^3 = 27\ s^4$, $K_{sp} = 27(1.32 \times 10^{-5})^4 = 8.20 \times 10^{-19}$

22. a.

	$Pb_3(PO_4)_2(s)$	$\rightleftharpoons$	$3\ Pb^{2+}(aq)$	+	$2\ PO_4^{3-}(aq)$
Initial			0		0
	s mol/L of $Pb_3(PO_4)_2(s)$ dissolves to reach equilibrium = molar solubility				
Change	$-s$	$\rightarrow$	$+3s$		$+2s$
Equil.			$3s$		$2s$

$K_{sp} = [Pb^{2+}]^3[PO_4^{3-}]^2 = (3s)^3(2s)^2 = 108\ s^5$, $K_{sp} = 108(6.2 \times 10^{-12})^5 = 9.9 \times 10^{-55}$

b.

	$Li_2CO_3(s)$	$\rightleftharpoons$	$2\ Li^+(aq)$	+	$CO_3^{2-}(aq)$
Initial	s = solubility (mol/L)		0		0
Equil.			$2s$		s

$K_{sp} = [Li^+]^2[CO_3^{2-}] = (2s)^2(s) = 4s^3,\ K_{sp} = 4(7.4 \times 10^{-2})^3 = 1.6 \times 10^{-3}$

23. $$\text{Solubility} = s = \frac{0.14\text{ g Ni(OH)}_2}{\text{L}} \times \frac{1\text{ mol Ni(OH)}_2}{92.71\text{ g Ni(OH)}_2} = 1.5 \times 10^{-3}\text{ mol/L}$$

	$Ni(OH)_2(s)$	$\rightleftharpoons$	$Ni^{2+}(aq)$	+	$2\ OH^-(aq)$
Initial			0		1.0×10^{-7} *M* (from water)
	s mol/L of $Ni(OH)_2$ dissolves to reach equailibrium				
Change	$-s$	→	$+s$		$+2s$
Equil.			s		$1.0 \times 10^{-7} + 2s$

From the calculated molar solubility, $1.0 \times 10^{-7} + 2s \approx 2s$.

$K_{sp} = [Ni^{2+}][OH^-]^2 = s(2s)^2 = 4s^3,\ K_{sp} = 4(1.5 \times 10^{-3})^3 = 1.4 \times 10^{-8}$

24.

	$M_2X_3(s)$	$\rightleftharpoons$	$2\ M^{3+}(aq)$	+	$3\ X^{2-}(aq)$	$K_{sp} = [M^{3+}]^2[X^{2-}]^3$
Initial	s = solubility (mol/L)		0		0	
	s mol/L of $M_2X_3(s)$ dissolves to reach equilibrium					
Change	$-s$	→	$+2s$		$+3s$	
Equil.			$2s$		$3s$	

$$K_{sp} = (2s)^2(3s)^3 = 108s^5;\quad s = \frac{3.60 \times 10^{-7}\text{ g}}{\text{L}} \times \frac{1\text{ mol } M_2X_3}{288\text{ g}} = 1.25 \times 10^{-9}\text{ mol/L}$$

$K_{sp} = 108(1.25 \times 10^{-9})^5 = 3.30 \times 10^{-43}$

25.

	$PbBr_2(s)$	$\rightleftharpoons$	$Pb^{2+}(aq)$	+	$2\ Br^-(aq)$
Initial			0		0
	s mol/L of $PbBr_2(s)$ dissolves to reach equilibrium				
Change	$-s$	→	$+s$		$+2s$
Equil.			s		$2s$

From the problem, $s = [Pb^{2+}] = 2.14 \times 10^{-2}$ *M*. So:

$K_{sp} = [Pb^{2+}][Br^-]^2 = s(2s)^2 = 4s^3,\ K_{sp} = 4(2.14 \times 10^{-2})^3 = 3.92 \times 10^{-5}$

26.

	$Ag_2C_2O_4(s)$	$\rightleftharpoons$	$2\ Ag^+(aq)$	+	$C_2O_4^{2-}(aq)$
Initial	s = solubility (mol/L)		0		0
Equil.			$2s$		s

From problem, $[Ag^+] = 2s = 2.2 \times 10^{-4}\ M$, $s = 1.1 \times 10^{-4}\ M$

$K_{sp} = [Ag^+]^2[C_2O_4^{2-}] = (2s)^2(s) = 4s^3 = 4(1.1 \times 10^{-4})^3 = 5.3 \times 10^{-12}$

27. In our setup, s = solubility in mol/L. Because solids do not appear in the K_{sp} expression, we do not need to worry about their initial or equilibrium amounts.

a.

	$Ag_3PO_4(s)$	$\rightleftharpoons$	$3\ Ag^+(aq)$	+	$PO_4^{3-}(aq)$
Initial			0		0
	s mol/L of $Ag_3PO_4(s)$ dissolves to reach equilibrium				
Change	$-s$	$\rightarrow$	$+3s$		$+s$
Equil.			$3s$		s

$K_{sp} = 1.8 \times 10^{-18} = [Ag^+]^3[PO_4^{3-}] = (3s)^3(s) = 27s^4$

$27s^4 = 1.8 \times 10^{-18}$, $s = (6.7 \times 10^{-20})^{1/4} = 1.6 \times 10^{-5}$ mol/L = molar solubility

b.

	$CaCO_3(s)$	$\rightleftharpoons$	$Ca^{2+}(aq)$	+	$CO_3^{2-}(aq)$
Initial	s = solubility (mol/L)		0		0
Equil.			s		s

$K_{sp} = 8.7 \times 10^{-9} = [Ca^{2+}][CO_3^{2-}] = s^2$, $s = 9.3 \times 10^{-5}$ mol/L

c.

	$Hg_2Cl_2(s)$	$\rightleftharpoons$	$Hg_2^{2+}(aq)$	+	$2\ Cl^-(aq)$
Initial	s = solubility (mol/L)		0		0
Equil.			s		$2s$

$K_{sp} = 1.1 \times 10^{-18} = [Hg_2^{2+}][Cl^-]^2 = (s)(2s)^2 = 4s^3$, $s = 6.5 \times 10^{-7}$ mol/L

28. a.

	$PbI_2(s)$	$\rightleftharpoons$	$Pb^{2+}(aq)$	+	$2\ I^-(aq)$
Initial	s = solubility (mol/L)		0		0
Equil.			s		$2s$

$K_{sp} = 1.4 \times 10^{-8} = [Pb^{2+}][I^-]^2 = s(2s)^2 = 4s^3$

$s = (1.4 \times 10^{-8}/4)^{1/3} = 1.5 \times 10^{-3}$ mol/L = molar solubility

b.

	$CdCO_3(s)$	$\rightleftharpoons$	$Cd^{2+}(aq)$	+	$CO_3^{2-}(aq)$
Initial	s = solubility (mol/L)		0		0
Equil.			s		s

$K_{sp} = 5.2 \times 10^{-12} = [Cd^{2+}][CO_3^{2-}] = s^2$, $s = 2.3 \times 10^{-6}$ mol/L

c.	$Sr_3(PO_4)_2(s)$	$\rightleftharpoons$	$3\ Sr^{2+}(aq)$	+	$2\ PO_4^{3-}(aq)$
Initial	s = solubility (mol/L)		0		0
Equil.			$3s$		$2s$

$K_{sp} = 1 \times 10^{-31} = [Sr^{2+}]^3[PO_4^{3-}]^2 = (3s)^3(2s)^2 = 108s^5,\ s = 2 \times 10^{-7}$ mol/L

29. KBT dissolves to form the potassium ion (K^+) and the bitartrate ion (abbreviated as BT^-).

	KBT(s)	$\rightleftharpoons$	$K^+(aq)$	+	$BT^-(aq)$	$K_{sp} = 3.8 \times 10^{-4}$
Initial	s = solubility (mol/L)		0		0	
Equil.			s		s	

$3.8 \times 10^{-4} = [K^+][BT^-] = s(s) = s^2,\ s = 1.9 \times 10^{-2}$ mol/L

$$0.2500\text{L} \times \frac{1.9 \times 10^{-2}\ \text{mol KBT}}{\text{L}} \times \frac{188.2\ \text{g KBT}}{\text{mol}} = 0.89\ \text{g KBT}$$

30.

	$BaSO_4(s)$	$\rightleftharpoons$	$Ba^{2+}(aq)$	+	$SO_4^{2-}(aq)$	$K_{sp} = 1.5 \times 10^{-9}$
Initial	s = solubility (mol/L)		0		0	
Equil.			s		s	

$1.5 \times 10^{-9} = [Ba^{2+}][SO_4^{2-}] = s^2,\ s = 3.9 \times 10^{-5}$ mol/L

$$0.1000\text{L} \times \frac{3.9 \times 10^{-5}\ \text{mol BaSO}_4}{\text{L}} \times \frac{233.4\ \text{g BaSO}_4}{\text{mol}} = 9.1 \times 10^{-4}\ \text{g BaSO}$$

31.

	$Mg(OH)_2(s)$	$\rightleftharpoons$	$Mg^{2+}(aq)$	+	$2\ OH^-(aq)$	
Initial	s = solubility (mol/L)		0		$1.0 \times 10^{-7}\ M$	(from water)
Equil.			s		$1.0 \times 10^{-7} + 2s$	

$K_{sp} = [Mg^{2+}][OH^-]^2 = s(1.0 \times 10^{-7} + 2s)^2$; assume that $1.0 \times 10^{-7} + 2s \approx 2s$, then:

$K_{sp} = 8.9 \times 10^{-12} = s(2s)^2 = 4s^3,\ s = 1.3 \times 10^{-4}$ mol/L

Assumption is good (1.0×10^{-7} is 0.04% of $2s$). Molar solubility = 1.3×10^{-4} mol/L

32.

	$Cd(OH)_2(s)$	$\rightleftharpoons$	$Cd^{2+}(aq)$	+	$2\ OH^-(aq)$	$K_{sp} = 5.9 \times 10^{-15}$
Initial	s = solubility (mol/L)		0		$1.0 \times 10^{-7}\ M$	
Equil.			s		$1.0 \times 10^{-7} + 2s$	

$K_{sp} = [Cd^{2+}][OH^-]^2 = s(1.0 \times 10^{-7} + 2s)^2$; assume that $1.0 \times 10^{-7} + 2s \approx 2s$, then:

$K_{sp} = 5.9 \times 10^{-15} = s(2s)^2 = 4s^3, \quad s = 1.1 \times 10^{-5}$ mol/L

Assumption is good (1.0×10^{-7} is 0.4% of $2s$). Molar solubility = 1.1×10^{-5} mol/L

33. Let s = solubility of $Al(OH)_3$ in mol/L. *Note*: Because solids do not appear in the K_{sp} expression, we do not need to worry about their initial or equilibrium amounts.

	$Al(OH)_3(s)$	$\rightleftharpoons$	$Al^{3+}(aq)$	+	$3\ OH^-(aq)$
Initial			0		1.0×10^{-7} *M* (from water)
	s mol/L of $Al(OH)_3(s)$ dissolves to reach equilibrium = molar solubility				
Change	$-s$	$\rightarrow$	$+s$		$+3s$
Equil.			s		$1.0 \times 10^{-7} + 3s$

$K_{sp} = 2 \times 10^{-32} = [Al^{3+}][OH^-]^3 = (s)(1.0 \times 10^{-7} + 3s)^3 \approx s(1.0 \times 10^{-7})^3$

$s = \dfrac{2 \times 10^{-32}}{1.0 \times 10^{-21}} = 2 \times 10^{-11}$ mol/L; assumption good ($1.0 \times 10^{-7} + 3s \approx 1.0 \times 10^{-7}$).

34. Let s = solubility of $Co(OH)_3$ in mol/L.

	$Co(OH)_3(s)$	$\rightleftharpoons$	$Co^{3+}(aq)$	+	$3\ OH^-(aq)$
Initial			0		1.0×10^{-7} *M* (from water)
	s mol/L of $Co(OH)_3(s)$ dissolves to reach equilibrium = molar solubility				
Change	$-s$	$\rightarrow$	$+s$		$+3s$
Equil.			s		$1.0 \times 10^{-7} + 3s$

$K_{sp} = 2.5 \times 10^{-43} = [Co^{3+}][OH^-]^3 = (s)(1.0 \times 10^{-7} + 3s)^3 \approx s(1.0 \times 10^{-7})^3$

$s = \dfrac{2.5 \times 10^{-43}}{1.0 \times 10^{-21}} = 2.5 \times 10^{-22}$ mol/L; assumption good ($1.0 \times 10^{-7} + 3s \approx 1.0 \times 10^{-7}$).

35. a. Because both solids dissolve to produce three ions in solution, we can compare values of K_{sp} to determine relative solubility. Because the K_{sp} for CaF_2 is the smallest, $CaF_2(s)$ has the smallest molar solubility.

b. We must calculate molar solubilities because each salt yields a different number of ions when it dissolves.

	$Ca_3(PO_4)_2(s)$	$\rightleftharpoons$	$3\ Ca^{2+}(aq)$	+	$2\ PO_4^{3-}(aq)$	$K_{sp} = 1.3 \times 10^{-32}$
Initial	s = solubility (mol/L)		0		0	
Equil.			$3s$		$2s$	

$K_{sp} = [Ca^{2+}]^3[PO_4^{3-}]^2 = (3s)^3(2s)^2 = 108s^5, \quad s = (1.3 \times 10^{-32}/108)^{1/5} = 1.6 \times 10^{-7}$ mol/L

	$FePO_4(s)$	$\rightleftharpoons$	$Fe^{3+}(aq)$	+	$PO_4^{3-}(aq)$	$K_{sp} = 1.0 \times 10^{-22}$
Initial	s = solubility (mol/L)		0		0	
Equil.			s		s	

$K_{sp} = [Fe^{3+}][PO_4^{3-}] = s^2,\ s = \sqrt{1.0 \times 10^{-22}} = 1.0 \times 10^{-11}$ mol/L

$FePO_4$ has the smallest molar solubility.

36. a.

	$FeC_2O_4(s)$	$\rightleftharpoons$	$Fe^{2+}(aq)$	+	$C_2O_4^{2-}(aq)$
Equil.	s = solubility (mol/L)		s		s

$K_{sp} = 2.1 \times 10^{-7} = [Fe^{2+}][C_2O_4^{2-}] = s^2,\ s = 4.6 \times 10^{-4}$ mol/L

	$Cu(IO_4)_2(s)$	$\rightleftharpoons$	$Cu^{2+}(aq)$	+	$2\ IO_4^-(aq)$
Equil.			s		$2s$

$K_{sp} = 1.4 \times 10^{-7} = [Cu^{2+}][IO_4^-]^2 = s(2s)^2 = 4s^3,\ s = (1.4 \times 10^{-7}/4)^{1/3} = 3.3 \times 10^{-3}$ mol/L

By comparing calculated molar solubilities, $FeC_2O_4(s)$ is less soluble (in mol/L).

b. Each salt produces three ions in solution, so we can compare K_{sp} values to determine relative molar solubilities. Therefore, $Mn(OH)_2(s)$ will be less soluble (in mol/L) because it has a smaller K_{sp} value.

37. a.

	$Fe(OH)_3(s)$	$\rightleftharpoons$	$Fe^{3+}(aq)$	+	$3\ OH^-(aq)$
Initial			0		$1 \times 10^{-7}\ M$ (from water)
	s mol/L of $Fe(OH)_3(s)$ dissolves to reach equilibrium = molar solubility				
Change	$-s$	$\rightarrow$	$+s$		$+3s$
Equil.			s		$1 \times 10^{-7} + 3s$

$K_{sp} = 4 \times 10^{-38} = [Fe^{3+}][OH^-]^3 = (s)(1 \times 10^{-7} + 3s)^3 \approx s(1 \times 10^{-7})^3$

$s = 4 \times 10^{-17}$ mol/L; assumption good ($3s << 1 \times 10^{-7}$)

b.

	$Fe(OH)_3(s)$	$\rightleftharpoons$	$Fe^{3+}(aq)$	+	$3\ OH^-(aq)$	pH = 5.0, $[OH^-] = 1 \times 10^{-9}\ M$
Initial			0		$1 \times 10^{-9}\ M$	(buffered)
	s mol/L dissolves to reach equilibrium					
Change	$-s$	$\rightarrow$	$+s$			(assume no pH change in buffer)
Equil.			s		1×10^{-9}	

$K_{sp} = 4 \times 10^{-38} = [Fe^{3+}][OH^-]^3 = (s)(1 \times 10^{-9})^3,\ s = 4 \times 10^{-11}$ mol/L = molar solubility

c.

	$Fe(OH)_3(s)$	$\rightleftharpoons$	$Fe^{3+}(aq)$	+	$3\ OH^-(aq)$	pH = 11.0, $[OH^-] = 1 \times 10^{-3}\ M$
Initial			0		$0.001\ M$	(buffered)
	s mol/L dissolves to reach equilibrium					
Change	$-s$	$\rightarrow$	$+s$			(assume no pH change)
Equil.			s		0.001	

$K_{sp} = 4 \times 10^{-38} = [Fe^{3+}][OH^-]^3 = (s)(0.001)^3$, $s = 4 \times 10^{-29}$ mol/L = molar solubility

Note: As $[OH^-]$ increases, solubility decreases. This is the common ion effect.

38.

	$Co(OH)_2(s)$	$\rightleftharpoons$	$Co^{2+}(aq)$ +	$2\ OH^-(aq)$	pH = 11.00, $[OH^-] = 1.0 \times 10^{-3}\ M$
Initial	s = solubility (mol/L)		0	1.0×10^{-3}	(buffered)
Equil.			s	1.0×10^{-3}	(assume no pH change)

$K_{sp} = 2.5 \times 10^{-16} = [Co^{2+}][OH^-]^2 = s(1.0 \times 10^{-3})^2$, $s = 2.5 \times 10^{-10}$ mol/L

39. a.

	$Ag_2SO_4(s)$	$\rightleftharpoons$	$2\ Ag^+(aq)$ +	$SO_4^{2-}(aq)$
Initial	s = solubility (mol/L)		0	0
Equil.			$2s$	s

$K_{sp} = 1.2 \times 10^{-5} = [Ag^+]^2[SO_4^{2-}] = (2s)^2 s = 4s^3$, $s = 1.4 \times 10^{-2}$ mol/L

b.

	$Ag_2SO_4(s)$	$\rightleftharpoons$	$2\ Ag^+(aq)$ +	$SO_4^{2-}(aq)$
Initial	s = solubility (mol/L)		0.10 M	0
Equil.			$0.10 + 2s$	s

$K_{sp} = 1.2 \times 10^{-5} = (0.10 + 2s)^2(s) \approx (0.10)^2(s)$, $s = 1.2 \times 10^{-3}$ mol/L; assumption good.

c.

	$Ag_2SO_4(s)$	$\rightleftharpoons$	$2\ Ag^+(aq)$ +	$SO_4^{2-}(aq)$
Initial	s = solubility (mol/L)		0	0.20 M
Equil			$2s$	$0.20 + s$

$1.2 \times 10^{-5} = (2s)^2(0.20 + s) \approx 4s^2(0.20)$, $s = 3.9 \times 10^{-3}$ mol/L; assumption good.

Note: Comparing the solubilities of parts b and c to that of part a illustrates that the solubility of a salt decreases when a common ion is present.

40. a.

	$PbI_2(s)$	$\rightleftharpoons$	$Pb^{2+}(aq)$ +	$2\ I^-(aq)$
Initial	s = solubility (mol/L)		0	0
Equil.			s	$2s$

$K_{sp} = 1.4 \times 10^{-8} = [Pb^{2+}][I^-]^2 = 4s^3$, $s = 1.5 \times 10^{-3}$ mol/L

b.

	$PbI_2(s)$	$\rightleftharpoons$	$Pb^{2+}(aq)$ +	$2\ I^-(aq)$
Initial	s = solubility (mol/L)		0.10 M	0
Equil.			$0.10 + s$	$2s$

$1.4 \times 10^{-8} = (0.10 + s)(2s)^2 \approx (0.10)(2s)^2 = (0.40)s^2$, $s = 1.9 \times 10^{-4}$ mol/L; assumption good.

c.

	$PbI_2(s)$	$\rightleftharpoons$	$Pb^{2+}(aq)$	+	$2\ I^-(aq)$
Initial	s = solubility (mol/L)		0		0.010 M
Equil.			s		$0.010 + 2s$

$1.4 \times 10^{-8} = (s)(0.010 + 2s)^2 \approx (s)(0.010)^2$, $s = 1.4 \times 10^{-4}$ mol/L; assumption good.

Note that in parts b and c, the presence of a common ion decreases the solubility as compared to the solubility of $PbI_2(s)$ in water.

41.

	$Ca_3(PO_4)_2(s)$	$\rightleftharpoons$	$3\ Ca^{2+}(aq)$	+	$2\ PO_4^{3-}(aq)$
Initial			0		0.20 M
	s mol/L of $Ca_3(PO_4)_2(s)$ dissolves to reach equilibrium				
Change	$-s$	$\rightarrow$	$+3s$		$+2s$
Equil.			$3s$		$0.20 + 2s$

$K_{sp} = 1.3 \times 10^{-32} = [Ca^{2+}]^3[PO_4^{3-}]^2 = (3s)^3(0.20 + 2s)^2$

Assuming $0.20 + 2s \approx 0.20$: $1.3 \times 10^{-32} = (3s)^3(0.20)^2 = 27s^3(0.040)$

s = molar solubility = 2.3×10^{-11} mol/L; assumption good.

42.

	$Pb_3(PO_4)_2(s)$	$\rightleftharpoons$	$3\ Pb^{2+}(aq)$	+	$2\ PO_4^{3-}(aq)$	$K_{sp} = 1 \times 10^{-54}$
Initial	s = solubility (mol/L)		0.10 M		0	
Equil.			$0.10 + 3s$		$2s$	

$1 \times 10^{-54} = (0.10 + 3s)^3(2s)^2 \approx (0.10)^3(2s)^2$, $s = 2 \times 10^{-26}$ mol/L; assumption good.

43.

	$Ce(IO_3)_3(s)$	$\rightleftharpoons$	$Ce^{3+}(aq)$	+	$3\ IO_3^-(aq)$
Initial	s = solubility (mol/L)		0		0.20 M
Equil.			s		$0.20 + 3s$

$K_{sp} = [Ce^{3+}][IO_3^-]^3 = s(0.20 + 3s)^3$

From the problem, $s = 4.4 \times 10^{-8}$ mol/L; solving for K_{sp}:

$K_{sp} = (4.4 \times 10^{-8})[0.20 + 3(4.4 \times 10^{-8})]^3 = 3.5 \times 10^{-10}$

44.

	$Pb(IO_3)_2(s)$	$\rightleftharpoons$	$Pb^{2+}(aq)$	+	$2\ IO_3^-(aq)$
Initial	s = solubility (mol/L)		0		0.10 M
Equil.			s		$0.10 + 2s$

$K_{sp} = [Pb^{2+}][IO_3^-]^2 = (s)(0.10 + 2s)^2$

From the problem, $s = 2.6 \times 10^{-11}$ mol/L; solving for K_{sp}:

$K_{sp} = (2.6 \times 10^{-11})[0.10 + 2(2.6 \times 10^{-11})]^2 = 2.6 \times 10^{-13}$

45. If the anion in the salt can act as a base in water, then the solubility of the salt will increase as the solution becomes more acidic. Added H^+ will react with the base, forming the conjugate acid. As the basic anion is removed, more of the salt will dissolve to replenish the basic anion. The salts with basic anions are Ag_3PO_4, $CaCO_3$, $CdCO_3$ and $Sr_3(PO_4)_2$. Hg_2Cl_2 and PbI_2 do not have any pH dependence because Cl^- and I^- are terrible bases (the conjugate bases of a strong acids).

$$Ag_3PO_4(s) + H^+(aq) \rightarrow 3\ Ag^+(aq) + HPO_4^{2-}(aq) \xrightarrow{\text{excess } H^+} 3\ Ag^+(aq) + H_3PO_4(aq)$$

$$CaCO_3(s) + H^+ \rightarrow Ca^{2+} + HCO_3^- \xrightarrow{\text{excess } H^+} Ca^{2+} + H_2CO_3\ [H_2O(l) + CO_2(g)]$$

$$CdCO_3(s) + H^+ \rightarrow Cd^{2+} + HCO_3^- \xrightarrow{\text{excess } H^+} Cd^{2+} + H_2CO_3\ [H_2O(l) + CO_2(g)]$$

$$Sr_3(PO_4)_2(s) + 2\ H^+ \rightarrow 3\ Sr^{2+} + 2\ HPO_4^{2-} \xrightarrow{\text{excess } H^+} 3\ Sr^{2+} + 2\ H_3PO_4$$

46. a. AgF b. $Pb(OH)_2$ c. $Sr(NO_2)_2$ d. $Ni(CN)_2$

All these salts have anions that are bases. The anions of the other choices are conjugate bases of strong acids. They have no basic properties in water and, therefore, do not have solubilities that depend on pH.

Precipitation Conditions

47.

	ZnS(s)	$\rightleftharpoons$	$Zn^{2+}(aq)$	+ $S^{2-}(aq)$	$K_{sp} = [Zn^{2+}][S^{2-}]$
Initial	s = solubility (mol/L)		0.050 M	0	
Equil.			$0.050 + s$	s	

$K_{sp} = 2.5 \times 10^{-22} = (0.050 + s)(s) \approx (0.050)s$, $s = 5.0 \times 10^{-21}$ mol/L; assumption good.

$$\text{Mass ZnS that dissolves} = 0.3000\ L \times \frac{5.0 \times 10^{-21}\ \text{mol ZnS}}{L} \times \frac{97.45\ \text{g ZnS}}{\text{mol}} = 1.5 \times 10^{-19}\ \text{g}$$

48. For 99% of the Mg^{2+} to be removed, we need, at equilibrium, $[Mg^{2+}] = 0.01(0.052\ M)$. Using the K_{sp} equilibrium constant, calculate the $[OH^-]$ required to reach this reduced $[Mg^{2+}]$.

$Mg(OH)_2(s) \rightleftharpoons Mg^{2+}(aq) + 2\ OH^-(aq)$ $K_{sp} = 8.9 \times 10^{-12}$

$8.9 \times 10^{-12} = [Mg^{2+}][OH^-]^2 = [0.01(0.052\ M)]\ [OH^-]^2$, $[OH^-] = 1.3 \times 10^{-4}\ M$ (extra sig. fig.)

$pOH = -\log(1.3 \times 10^{-4}) = 3.89$; $pH = 10.11$; at a pH = 10.1, 99% of the Mg^{2+} in seawater will be removed as $Mg(OH)_2(s)$.

49. The formation of $Mg(OH)_2(s)$ is the only possible precipitate. $Mg(OH)_2(s)$ will form if Q > K_{sp}.

$Mg(OH)_2(s) \rightleftharpoons Mg^{2+}(aq) + 2\ OH^-(aq) \quad K_{sp} = [Mg^{2+}][OH^-]^2 = 8.9 \times 10^{-12}$

$$[Mg^{2+}]_0 = \frac{100.0\ mL \times 4.0 \times 10^{-4}\ mmol Mg^{2+}/mL}{100.0\ mL + 100.0\ mL} = 2.0 \times 10^{-4}\ M$$

$$[OH^-]_0 = \frac{100.0\ mL \times 2.0 \times 10^{-4}\ mmol OH^-/mL}{200.0\ mL} = 1.0 \times 10^{-4}\ M$$

$$Q = [Mg^{2+}]_0[OH^-]_0^2 = (2.0 \times 10^{-4}\ M)(1.0 \times 10^{-4})^2 = 2.0 \times 10^{-12}$$

Because $Q < K_{sp}$, $Mg(OH)_2(s)$ will not precipitate, so no precipitate forms.

50. $AgCN(s) \rightleftharpoons Ag^+(aq) + CN^-(aq) \qquad K_{sp} = 2.2 \times 10^{-12}$

$Q = [Ag^+]_0[CN^-]_0 = (1.0 \times 10^{-5})(2.0 \times 10^{-6}) = 2.0 \times 10^{-11}$

Because $Q > K_{sp}$, $AgCN(s)$ will form as a precipitate.

51. $PbF_2(s) \rightleftharpoons Pb^{2+}(aq) + 2\ F^-(aq) \qquad K_{sp} = 4 \times 10^{-8}$

$$[Pb^{2+}]_0 = \frac{mmol Pb^{2+}(aq)}{total mL solution} = \frac{100.0\ mL \times \frac{1.0 \times 10^{-2}\ mmol Pb^{2+}}{mL}}{100.0\ mL + 100.0\ mL} = 5.0 \times 10^{-3}\ M$$

$$[F^-]_0 = \frac{mmol F^-}{total mL solution} = \frac{100.0\ mL \times \frac{1.0 \times 10^{-3}\ mmol F^-}{mL}}{200.0\ mL} = 5.0 \times 10^{-4}\ M$$

$$Q = [Pb^{2+}]_0[F^-]_0^2 = (5.0 \times 10^{-3})(5.0 \times 10^{-4})^2 = 1.3 \times 10^{-9}$$

Because $Q < K_{sp}$, $PbF_2(s)$ will not form as a precipitate.

52. $Ce(IO_3)_3(s) \rightleftharpoons Ce^{3+}(aq) + 3\ IO_3^-(aq) \quad K_{sp} = 3.2 \times 10^{-10}$

$Q = [Ce^{3+}]_0[IO_3^-]_0^3 = (2.0 \times 10^{-3})(1.0 \times 10^{-2})^3 = 2.0 \times 10^{-9}$

Because $Q > K_{sp}$, $Ce(IO_3)_3(s)$ will form as a precipitate.

53. The concentrations of ions are large, so Q will be greater than K_{sp}, and $BaC_2O_4(s)$ will form. To solve this problem, we will assume that the precipitation reaction goes to completion; then we will solve an equilibrium problem to get the actual ion concentrations. This makes the math reasonable.

$$100.\ mL \times \frac{0.200\ mmol K_2C_2O_4}{mL} = 20.0\ mmol\ K_2C_2O_4$$

$$150.\ mL \times \frac{0.250\ mmol BaBr_2}{mL} = 37.5\ mmol\ BaBr_2$$

	$Ba^{2+}(aq)$ +	$C_2O_4^{2-}(aq)$	→	$BaC_2O_4(s)$	$K = 1/K_{sp} >> 1$
Before	37.5 mmol	20.0 mmol		0	
Change	−20.0	−20.0	→	+20.0	Reacts completely (K is large).
After	17.5	0		20.0	

New initial concentrations (after complete precipitation) are:

$$[Ba^{2+}] = \frac{17.5\text{ mmol}}{250.\text{ mL}} = 7.00 \times 10^{-2}\ M;\ [C_2O_4^{2-}] = 0\ M$$

$$[K^+] = \frac{2(20.0\text{ mmol})}{250.\text{ mL}} = 0.160\ M;\ [Br^-] = \frac{2(37.5\text{ mmol})}{250.\text{ mL}} = 0.300\ M$$

For K^+ and Br^-, these are also the final concentrations. We can't have 0 M $C_2O_4^{2-}$. For Ba^{2+} and $C_2O_4^{2-}$, we need to perform an equilibrium calculation.

	$BaC_2O_4(s)$	⇌	$Ba^{2+}(aq)$ +	$C_2O_4^{2-}(aq)$	$K_{sp} = 2.3 \times 10^{-8}$
Initial			0.0700 M	0	
	s mol/L of $BaC_2O_4(s)$ dissolves to reach equilibrium				
Equil.			0.0700 + s	s	

$K_{sp} = 2.3 \times 10^{-8} = [Ba^{2+}][C_2O_4^{2-}] = (0.0700 + s)(s) \approx (0.0700)s$

$s = [C_2O_4^{2-}] = 3.3 \times 10^{-7}$ mol/L; $[Ba^{2+}] = 0.0700\ M$; assumption good ($s << 0.0700$).

54. $$[Ba^{2+}]_0 = \frac{75.0\text{ mL} \times \frac{0.020\text{ mmol}}{\text{mL}}}{200.\text{ mL}} = 7.5 \times 10^{-3}\ M$$

$$[SO_4^{2-}]_0 = \frac{125\text{ mL} \times \frac{0.040\text{ mmol}}{\text{mL}}}{200.\text{ mL}} = 2.5 \times 10^{-2}\ M$$

$Q = [Ba^{2+}]_0[SO_4^{2-}]_0 = (7.5 \times 10^{-3})(2.5 \times 10^{-2}) = 1.9 \times 10^{-4} > K_{sp}\ (1.5 \times 10^{-9})$

A precipitate of $BaSO_4(s)$ will form.

	$BaSO_4(s)$	⇌	Ba^{2+} +	SO_4^{2-}	
Before			0.0075 M	0.025 M	
	Let 0.0075 mol/L Ba^{2+} react with SO_4^{2-} to completion because $K_{sp} << 1$.				
Change		←	−0.0075	−0.0075	Reacts completely
After			0	0.0175	New initial (carry extra sig. fig.)
	s mol/L $BaSO_4$ dissolves to reach equilibrium				
Change	−s	→	+s	+s	
Equil.			s	0.0175 + s	

$K_{sp} = 1.5 \times 10^{-9} = [Ba^{2+}][SO_4^{2-}] = (s)(0.0175 + s) \approx s(0.0175)$

$s = 8.6 \times 10^{-8}$ mol/L; $[Ba^{2+}] = 8.6 \times 10^{-8}\ M$; $[SO_4^{2-}] = 0.018\ M$; assumption good.

55. 50.0 mL × 0.00200 M = 0.100 mmol Ag^+; 50.0 mL × 0.0100 M = 0.500 mmol IO_3^-

From the small K_{sp} value, assume $AgIO_3(s)$ precipitates completely. After reaction, 0.400 mmol IO_3^- is remaining. Now, let some $AgIO_3(s)$ dissolve in solution with excess IO_3^- to reach equilibrium.

	$AgIO_3(s)$	$\rightleftharpoons$	$Ag^+(aq)$	+	$IO_3^-(aq)$
Initial			0		$\frac{0.400\text{ mmol}}{100.0\text{ mL}} = 4.00 \times 10^{-3}\ M$
	s mol/L $AgIO_3(s)$ dissolves to reach equilibrium				
Equil.			s		$4.00 \times 10^{-3} + s$

$K_{sp} = [Ag^+][IO_3^-] = 3.0 \times 10^{-8} = s(4.00 \times 10^{-3} + s) \approx s(4.00 \times 10^{-3})$

$s = 7.5 \times 10^{-6}$ mol/L = $[Ag^+]$; assumption good.

56. 50.0 mL × 0.10 M = 5.0 mmol Pb^{2+}; 50.0 mL × 1.0 M = 50. mmol Cl^-. For this solution, Q > K_{sp}, so $PbCl_2$ precipitates. Assume precipitation of $PbCl_2(s)$ is complete. 5.0 mmol Pb^{2+} requires 10. mmol of Cl^- for complete precipitation, which leaves 40. mmol Cl^- in excess. Now, let some of the $PbCl_2(s)$ redissolve to establish equilibrium

	$PbCl_2(s)$	$\rightleftharpoons$	$Pb^{2+}(aq)$	+	$2\ Cl^-(aq)$
Initial			0		$\frac{40.\text{ mmol}}{100.0\text{ mL}} = 0.40\ M$
	s mol/L of $PbCl_2(s)$ dissolves to reach equilibrium				
Equil.			s		$0.40 + 2s$

$K_{sp} = [Pb^{2+}][Cl^-]^2,\ 1.6 \times 10^{-5} = s(0.40 + 2s)^2 \approx s(0.40)^2$

$s = 1.0 \times 10^{-4}$ mol/L; assumption good.

At equilibrium: $[Pb^{2+}] = s = 1.0 \times 10^{-4}$ mol/L; $[Cl^-] = 0.40 + 2s,\ 0.40 + 2(1.0 \times 10^{-4}) = 0.40\ M$

57. $Ag_3PO_4(s) \rightleftharpoons 3\ Ag^+(aq) + PO_4^{3-}(aq)$; when Q is greater than K_{sp}, precipitation will occur. We will calculate the $[Ag^+]_0$ necessary for Q = K_{sp}. Any $[Ag^+]_0$ greater than this calculated number will cause precipitation of $Ag_3PO_4(s)$. In this problem, $[PO_4^{3-}]_0 = [Na_3PO_4]_0 = 1.0 \times 10^{-5}\ M$.

$K_{sp} = 1.8 \times 10^{-18}$; $Q = 1.8 \times 10^{-18} = [Ag^+]_0^3[PO_4^{3-}]_0 = [Ag^+]_0^3(1.0 \times 10^{-5}\ M)$

$$[Ag^+]_0 = \left(\frac{1.8 \times 10^{-18}}{1.0 \times 10^{-5}}\right)^{1/3},\quad [Ag^+]_0 = 5.6 \times 10^{-5}\ M$$

When $[Ag^+]_0 = [AgNO_3]_0$ is greater than $5.6 \times 10^{-5}\ M$, precipitation of $Ag_3PO_4(s)$ will occur.

58. $Al(OH)_3(s) \rightleftharpoons Al^{3+}(aq) + 3\ OH^-(aq) \qquad K_{sp} = 2 \times 10^{-32}$

$Q = 2 \times 10^{-32} = [Al^{3+}]_0[OH^-]_0^3 = (0.2)[OH^-]_0^3$, $[OH^-]_0 = 4.6 \times 10^{-11}$ (carrying extra sig. fig.)

pOH = $-\log(4.6 \times 10^{-11}) = 10.3$; when the pOH of the solution equals 10.3, $K_{sp} = Q$. For precipitation, we want $Q > K_{sp}$. This will occur when $[OH^-]_0 > 4.6 \times 10^{-11}$ or when pOH < 10.3. Because pH + pOH = 14.00, precipitation of $Al(OH)_3(s)$ will begin when pH > 3.7 because this corresponds to a solution with pOH < 10.3.

59. For each lead salt, we will calculate the $[Pb^{2+}]_0$ necessary for $Q = K_{sp}$. Any $[Pb^{2+}]_0$ greater than this value will cause precipitation of the salt ($Q > K_{sp}$).

$PbF_2(s) \rightleftharpoons Pb^{2+}(aq) + 2\ F^-(aq) \qquad K_{sp} = 4 \times 10^{-8};\ Q = 4 \times 10^{-8} = [Pb^{2+}]_0[F^-]_0^2$

$$[Pb^{2+}]_0 = \frac{4 \times 10^{-8}}{(1 \times 10^{-4})^2} = 4\ M$$

$PbS(s) \rightleftharpoons Pb^{2+}(aq) + S^{2-}(aq) \qquad K_{sp} = 7 \times 10^{-29};\ Q = 7 \times 10^{-29} = [Pb^{2+}]_0[S^{2-}]_0$

$$[Pb^{2+}]_0 = \frac{7 \times 10^{-29}}{1 \times 10^{-4}} = 7 \times 10^{-25}\ M$$

$Pb_3(PO_4)_2(s) \rightleftharpoons 3\ Pb^{2+}(aq) + 2\ PO_4^{3-}(aq) \quad K_{sp} = 1 \times 10^{-54}$

$$Q = 1 \times 10^{-54} = [Pb^{2+}]_0^3[PO_4^{3-}]_0^2$$

$$[Pb^{2+}]_0 = \left[\frac{1 \times 10^{-54}}{(1 \times 10^{-4})^2}\right]^{1/3} = 5 \times 10^{-16}\ M$$

From the calculated $[Pb^{2+}]_0$, the least soluble salt is PbS(s), and it will form first. $Pb_3(PO_4)_2(s)$ will form second, and $PbF_2(s)$ will form last because it requires the largest $[Pb^{2+}]_0$ in order for precipitation to occur.

60. From Table 16.1, K_{sp} for $NiCO_3 = 1.4 \times 10^{-7}$ and K_{sp} for $CuCO_3 = 2.5 \times 10^{-10}$. From the K_{sp} values, $CuCO_3$ will precipitate first since it has the smaller K_{sp} value and will be the least soluble. For $CuCO_3(s)$, precipitation begins when:

$$[CO_3^{2-}] = \frac{K_{sp,\ CuCO_3}}{[Cu^{2+}]} = \frac{2.5 \times 10^{-10}}{0.25\ M} = 1.0 \times 10^{-9}\ M\ CO_3^{2-}$$

For $NiCO_3(s)$ to precipitate:

$$[CO_3^{2-}] = \frac{K_{sp,\ NiCO_3}}{[Ni^{2+}]} = \frac{1.4 \times 10^{-7}}{0.25\ M} = 5.6 \times 10^{-7}\ M\ CO_3^{2-}$$

Determining the $[Cu^{2+}]$ when $CuCO_3(s)$ begins to precipitate:

$$[Cu^{2+}] = \frac{K_{sp,\,CuCO_3}}{[CO_3^{2-}]} = \frac{2.5 \times 10^{-10}}{5.6 \times 10^{-7} M} = 4.5 \times 10^{-4}\, M\, Cu^{2+}$$

For successful separation, 1% Cu^{2+} or less of the initial amount of Cu^{2+} (0.25 *M*) must be present before $NiCO_3(s)$ begins to precipitate. The percent of Cu^{2+} present when $NiCO_3(s)$ begins to precipitate is:

$$\frac{4.5 \times 10^{-4}\, M}{0.25\, M} \times 100 = 0.18\%\ Cu^{2+}$$

Because less than 1% of the initial amount of Cu^{2+} remains, the metals can be separated through slow addition of $Na_2CO_3(aq)$.

Complex Ion Equilibria

61. a.

$$Ni^{2+} + CN^- \rightleftharpoons NiCN^+ \qquad K_1$$
$$NiCN^+ + CN^- \rightleftharpoons Ni(CN)_2 \qquad K_2$$
$$Ni(CN)_2 + CN^- \rightleftharpoons Ni(CN)_3^- \qquad K_3$$
$$Ni(CN)_3^- + CN^- \rightleftharpoons Ni(CN)_4^{2-} \qquad K_4$$

$$Ni^{2+} + 4\,CN^- \rightleftharpoons Ni(CN)_4^{2-} \qquad K_f = K_1K_2K_3K_4$$

Note: The various K constants are included for your information. Each NH_3 adds with a corresponding K value associated with that reaction. The overall formation constant K_f for the overall reaction is equal to the product of all the stepwise K values.

b.

$$V^{3+} + C_2O_4^{2-} \rightleftharpoons VC_2O_4^+ \qquad K_1$$
$$VC_2O_4^+ + C_2O_4^{2-} \rightleftharpoons V(C_2O_4)_2^- \qquad K_2$$
$$V(C_2O_4)_2^- + C_2O_4^{2-} \rightleftharpoons V(C_2O_4)_3^{3-} \qquad K_3$$

$$V^{3+} + 3\,C_2O_4^{2-} \rightleftharpoons V(C_2O_4)_3^{3-} \qquad K_f = K_1K_2K_3$$

62. a.

$$Co^{3+} + F^- \rightleftharpoons CoF^{2+} \qquad K_1$$
$$CoF^{2+} + F^- \rightleftharpoons CoF_2^+ \qquad K_2$$
$$CoF_2^+ + F^- \rightleftharpoons CoF_3 \qquad K_3$$
$$CoF_3 + F^- \rightleftharpoons CoF_4^- \qquad K_4$$
$$CoF_4^- + F^- \rightleftharpoons CoF_5^{2-} \qquad K_5$$
$$CoF_5^{2-} + F^- \rightleftharpoons CoF_6^{3-} \qquad K_6$$

$$Co^{3+} + 6\,F^- \rightleftharpoons CoF_6^{3-} \qquad K_f = K_1K_2K_3K_4K_5K_6$$

b.

$$Zn^{2+} + NH_3 \rightleftharpoons ZnNH_3^{2+} \qquad K_1$$
$$ZnNH_3^{2+} + NH_3 \rightleftharpoons Zn(NH_3)_2^{2+} \qquad K_2$$
$$Zn(NH_3)_2^{2+} + NH_3 \rightleftharpoons Zn(NH_3)_3^{2+} \qquad K_3$$
$$Zn(NH_3)_3^{2+} + NH_3 \rightleftharpoons Zn(NH_3)_4^{2+} \qquad K_4$$

$$Zn^{2+} + 4\ NH_3 \rightleftharpoons Zn(NH_3)_4^{2+} \qquad K_f = K_1K_2K_3K_4$$

63. $Fe^{3+}(aq) + 6\ CN^-(aq) \rightleftharpoons Fe(CN)_6^{3-}$ $\qquad K = \dfrac{[Fe(CN)_6^{3-}]}{[Fe^{3+}][CN^-]^6}$

$$K = \frac{1.5\times10^{-3}}{(8.5\times10^{-40})(0.11)^6} = 1.0 \times 10^{42}$$

64. $Cu^{2+}(aq) + 4\ NH_3(aq) \rightleftharpoons Cu(NH_3)_4^{2+}(aq)$ $\qquad K = \dfrac{[Cu(NH_3)_4^{2+}]}{[Cu^{2+}][NH_3]^4}$

$$K = \frac{1.0 \times 10^{-3}}{(1.8 \times 10^{-17})(1.5)^4} = 1.1 \times 10^{13}$$

65. $Hg^{2+}(aq) + 2\ I^-(aq) \rightleftharpoons HgI_2(s)$; $\quad HgI_2(s) + 2\ I^-(aq) \rightleftharpoons HgI_4^{2-}(aq)$

orange ppt $\qquad$ soluble complex ion

66. $Ag^+(aq) + Cl^-(aq) \rightleftharpoons AgCl(s)$, white ppt.; $AgCl(s) + 2\ NH_3(aq) \rightleftharpoons Ag(NH_3)_2^+(aq) + Cl^-(aq)$

$Ag(NH_3)_2^+(aq) + Br^-(aq) \rightleftharpoons AgBr(s) + 2\ NH_3(aq)$, pale yellow ppt.

$AgBr(s) + 2\ S_2O_3^{2-}(aq) \rightleftharpoons Ag(S_2O_3)_2^{3-}(aq) + Br^-(aq)$

$Ag(S_2O_3)_2^{3-}(aq) + I^-(aq) \rightleftharpoons AgI(s) + 2\ S_2O_3^{2-}(aq)$, yellow ppt.

The least soluble salt (smallest K_{sp} value) must be AgI because it forms in the presence of Cl^- and Br^-. The most soluble salt (largest K_{sp} value) must be AgCl since it forms initially but never re-forms. The order of K_{sp} values is K_{sp} (AgCl) > K_{sp} (AgBr) > K_{sp} (AgI).

67. The formation constant for HgI_4^{2-} is an extremely large number. Because of this, we will let the Hg^{2+} and I^- ions present initially react to completion and then solve an equilibrium problem to determine the Hg^{2+} concentration.

	$Hg^{2+}(aq)$	+ $4\ I^-(aq)$	$\rightleftharpoons$	$HgI_4^{2-}(aq)$	$K = 1.0 \times 10^{30}$
Before	0.010 *M*	0.78 *M*		0	
Change	−0.010	−0.040	→	+0.010	Reacts completely (K is large).
After	0	0.74		0.010	New initial
	x mol/L HgI_4^{2-} dissociates to reach equilibrium				
Change	+*x*	+4*x*	←	−*x*	
Equil.	*x*	0.74 + 4*x*		0.010 − *x*	

$$K = 1.0 \times 10^{30} = \frac{[HgI_4^{2-}]}{[Hg^{2+}][I^-]^4} = \frac{(0.010 - x)}{(x)(0.74 + 4x)^4}; \text{ making normal assumptions:}$$

$$1.0 \times 10^{30} = \frac{(0.010)}{(x)(0.74)^4}, \quad x = [Hg^{2+}] = 3.3 \times 10^{-32}\ M; \quad \text{assumptions good.}$$

Note: 3.3×10^{-32} mol/L corresponds to one Hg^{2+} ion per 5×10^7 L. It is very reasonable to approach this problem in two steps. The reaction does essentially go to completion.

68. $Ni^{2+}(aq) + 6\ NH_3(aq) \rightleftharpoons Ni(NH_3)_6^{2+}(aq) \qquad K = 5.5 \times 10^8$

	Ni^{2+}	NH_3		$Ni(NH_3)_6^{2+}$
Initial	0	3.0 M		0.10 mol/0.50 L = 0.20 M
	x mol/L $Ni(NH_3)_6^{2+}$ dissociates to reach equilibrium			
Change	$+x$	$+6x$	←	$-x$
Equil.	x	$3.0 + 6x$		$0.20 - x$

$$K = 5.5 \times 10^8 = = \frac{[Ni(NH_3)_6^{2+}]}{[Ni^{2+}][NH_3]^6} = \frac{(0.20 - x)}{(x)(3.0 + 6x)^6}, \quad 5.5 \times 10^8 \approx \frac{(0.20)}{(x)(3.0)^6}$$

$x = [Ni^{2+}] = 5.0 \times 10^{-13}\ M$; $[Ni(NH_3)_6^{2+}] = 0.20\ M - x = 0.20\ M$; assumptions good.

69. $[X^-]_0 = 5.00\ M$ and $[Cu^+]_0 = 1.0 \times 10^{-3}\ M$ because equal volumes of each reagent are mixed.

Because the K values are much greater than 1, assume the reaction goes completely to CuX_3^{2-}, and then solve an equilibrium problem.

$Cu^+(aq) + 3\ X^-(aq) \rightleftharpoons CuX_3^{2-}(aq) \qquad K = K_1 \times K_2 \times K_3 = 1.0 \times 10^9$

	Cu^+	X^-	CuX_3^{2-}	
Before	$1.0 \times 10^{-3}\ M$	5.00 M	0	
After	0	$5.00 - 3(10^{-3}) \approx 5.00$	1.0×10^{-3}	Reacts completely
Equil.	x	$5.00 + 3x$	$1.0 \times 10^{-3} - x$	

$$K = \frac{(1.0 \times 10^{-3} - x)}{(x)(5.00 + 3x)^3} = 1.0 \times 10^9 \approx \frac{1.0 \times 10^{-3}}{(x)(5.00)^3}, \quad x = [Cu^+] = 8.0 \times 10^{-15}\ M$$

Assumptions good.

$[CuX_3^{2-}] = 1.0 \times 10^{-3} - 8.0 \times 10^{-15} = 1.0 \times 10^{-3}\ M$

$$K_3 = \frac{[CuX_3^{2-}]}{[CuX_2^-][X^-]} = 1.0 \times 10^3 = \frac{(1.0 \times 10^{-3})}{[CuX_2^-](5.00)}, \quad [CuX_2^-] = 2.0 \times 10^{-7}\ M$$

Summarizing:

$[CuX_3^{2-}] = 1.0 \times 10^{-3}\ M$ (answer a)

$[CuX_2^-] = 2.0 \times 10^{-7}\ M$ (answer b)

$[Cu^{2+}] = 8.0 \times 10^{-15}\ M$ (answer c)

70. $[Be^{2+}]_0 = 5.0 \times 10^{-5}$ M and $[F^-]_0 = 4.0$ M because equal volumes of each reagent are mixed, so all concentrations given in the problem are diluted by a factor of one-half.

Because the K values are large, assume all reactions go to completion, and then solve an equilibrium problem.

	$Be^{2+}(aq)$	+ $4\ F^-(aq)$	$\rightleftharpoons$ $BeF_4^{2-}(aq)$	$K = K_1K_2K_3K_4 = 7.5 \times 10^{12}$
Before	5.0×10^{-5} M	4.0 M	0	
After	0	4.0 M	5.0×10^{-5} M	
Equil.	x	$4.0 + 4x$	$5.0 \times 10^{-5} - x$	

$$K = 7.5 \times 10^{12} = \frac{[BeF_4^{2-}]}{[Be^{2+}][F^-]^4} = \frac{5.0 \times 10^{-5} - x}{x(4.0 + 4x)^4} \approx \frac{5.0 \times 10^{-5}}{x(4.0)^4}$$

$x = [Be^{2+}] = 2.6 \times 10^{-20}$ M; assumptions good. $[F^-] = 4.0$ M; $[BeF_4^{2-}] = 5.0 \times 10^{-5}$ M

Now use the stepwise K values to determine the other concentrtations.

$$K_1 = 7.9 \times 10^4 = \frac{[BeF^+]}{[Be^{2+}][F^-]} = \frac{[BeF^+]}{(2.6 \times 10^{-20})(4.0)},\ [BeF^+] = 8.2 \times 10^{-15}\ M$$

$$K_2 = 5.8 \times 10^3 = \frac{[BeF_2]}{[BeF^+][F^-]} = \frac{[BeF_2]}{(8.2 \times 10^{-15})(4.0)},\ [BeF_2] = 1.9 \times 10^{-10}\ M$$

$$K_3 = 6.1 \times 10^2 = \frac{[BeF_3^-]}{[BeF_2][F^-]} = \frac{[BeF_3^-]}{(1.9 \times 10^{-10})(4.0)},\ [BeF_3^-] = 4.6 \times 10^{-7}\ M$$

71. a.

	$AgI(s)$	$\rightleftharpoons$ $Ag^+(aq)$	+ $I^-(aq)$	$K_{sp} = [Ag^+][I^-] = 1.5 \times 10^{-16}$
Initial	s = solubility (mol/L)	0	0	
Equil.		s	s	

$K_{sp} = 1.5 \times 10^{-16} = s^2$, $s = 1.2 \times 10^{-8}$ mol/L

b.

$$AgI(s) \rightleftharpoons Ag^+ + I^- \qquad K_{sp} = 1.5 \times 10^{-16}$$
$$Ag^+ + 2\ NH_3 \rightleftharpoons Ag(NH_3)_2^+ \qquad K_f = 1.7 \times 10^7$$

$$AgI(s) + 2\ NH_3(aq) \rightleftharpoons Ag(NH_3)_2^+(aq) + I^-(aq) \qquad K = K_{sp} \times K_f = 2.6 \times 10^{-9}$$

	$AgI(s)$ + $2\ NH_3$	$\rightleftharpoons$ $Ag(NH_3)_2^+$	+ I^-
Initial	3.0 M	0	0
	s mol/L of AgBr(s) dissolves to reach equilibrium = molar solubility		
Equil.	$3.0 - 2s$	s	s

$$K = \frac{[Ag(NH_3)_2^+][I^-]}{[NH_3]^2} = \frac{s^2}{(3.0 - 2s)^2} = 2.6 \times 10^{-9} \approx \frac{s^2}{(3.0)^2},\ s = 1.5 \times 10^{-4}\ \text{mol/L}$$

Assumption good.

c. The presence of NH_3 increases the solubility of AgI. Added NH_3 removes Ag^+ from solution by forming the complex ion, $Ag(NH_3)_2^+$. As Ag^+ is removed, more AgI(s) will dissolve to replenish the Ag^+ concentration.

72.

$$AgBr(s) \rightleftharpoons Ag^+ + Br^- \qquad K_{sp} = 5.0 \times 10^{-13}$$
$$Ag^+ + 2\,S_2O_3^{2-} \rightleftharpoons Ag(S_2O_3)_2^{3-} \qquad K_f = 2.9 \times 10^{13}$$

$$AgBr(s) + 2\,S_2O_3^{2-} \rightleftharpoons Ag(S_2O_3)_2^{3-} + Br^- \qquad K = K_{sp} \times K_f = 14.5 \text{ (Carry extra sig. figs.)}$$

	$AgBr(s)$ +	$2\,S_2O_3^{2-}(aq)$ $\rightleftharpoons$	$Ag(S_2O_3)_2^{3-}(aq)$ +	$Br^-aq)$
Initial		0.500 M	0	0
	s mol/L AgBr(s) dissolves to reach equilibrium			
Change	$-s$	$-2s$ →	$+s$	$+s$
Equil.		$0.500 - 2s$	s	s

$$K = \frac{s^2}{(0.500 - 2s)^2} = 14.5\text{; taking the square root of both sides:}$$

$$\frac{s}{0.500 - 2s} = 3.81,\ s = 1.91 - (7.62)s,\ s = 0.222 \text{ mol/L}$$

$$1.00 \text{ L} \times \frac{0.222 \text{ mol AgBr}}{\text{L}} \times \frac{187.8 \text{ g AgBr}}{\text{mol AgBr}} = 41.7 \text{ g AgBr} = 42 \text{ g AgBr}$$

73.

$$AgCl(s) \rightleftharpoons Ag^+ + Cl^- \qquad K_{sp} = 1.6 \times 10^{-10}$$
$$Ag^+ + 2\,NH_3 \rightleftharpoons Ag(NH_3)_2^+ \qquad K_f = 1.7 \times 10^7$$

$$AgCl(s) + 2\,NH_3(aq) \rightleftharpoons Ag(NH_3)_2^+(aq) + Cl^-(aq) \qquad K = K_{sp} \times K_f = 2.7 \times 10^{-3}$$

	$AgCl(s) + 2\,NH_3$ $\rightleftharpoons$	$Ag(NH_3)_2^+$ +	Cl^-
Initial	1.0 M	0	0
	s mol/L of AgCl(s) dissolves to reach equilibrium = molar solubility		
Equil.	$1.0 - 2s$	s	s

$$K = 2.7 \times 10^{-3} = \frac{[Ag(NH_3)_2^+][Cl^-]}{[NH_3]^2} = \frac{s^2}{(1.0 - 2s)^2}\text{; taking the square root:}$$

$$\frac{s}{1.0 - 2s} = (2.7 \times 10^{-3})^{1/2} = 5.2 \times 10^{-2},\ s = 4.7 \times 10^{-2} \text{ mol/L}$$

In pure water, the solubility of AgCl(s) is $(1.6 \times 10^{-10})^{1/2} = 1.3 \times 10^{-5}$ mol/L. Notice how the presence of NH_3 increases the solubility of AgCl(s) by over a factor of 3500.

74. a.

	$CuCl(s)$ $\rightleftharpoons$	$Cu^+(aq)$ +	$Cl^-(aq)$
Initial	s = solubility (mol/L)	0	0
Equil.		s	s

$K_{sp} = 1.2 \times 10^{-6} = [Cu^+][Cl^-] = s^2, \; s = 1.1 \times 10^{-3}$ mol/L

b. Cu^+ forms the complex ion $CuCl_2^-$ in the presence of Cl^-. We will consider both the K_{sp} reaction and the complex ion reaction at the same time.

$CuCl(s) \rightleftharpoons Cu^+(aq) + Cl^-(aq)$	$K_{sp} = 1.2 \times 10^{-6}$
$Cu^+(aq) + 2\,Cl^-(aq) \rightleftharpoons CuCl_2^-(aq)$	$K_f = 8.7 \times 10^4$
$CuCl(s) + Cl^-(aq) \rightleftharpoons CuCl_2^-(aq)$	$K = K_{sp} \times K_f = 0.10$

	$CuCl(s)$ + Cl^-	$\rightleftharpoons$ $CuCl_2^-$	
Initial	0.10 *M*	0	
Equil.	$0.10 - s$	s	where s = solubility of CuCl(s) in mol/L

$$K = 0.10 = \frac{[CuCl_2^-]}{[Cl^-]} = \frac{s}{0.10 - s}, \quad 1.0 \times 10^{-2} \,!\, 0.10\,s = s, \quad s = 9.1 \times 10^{-3} \text{ mol/L}$$

75. Test tube 1: Added Cl^- reacts with Ag^+ to form a silver chloride precipitate. The net ionic equation is $Ag^+(aq) + Cl^-(aq) \rightarrow AgCl(s)$. Test tube 2: Added NH_3 reacts with Ag^+ ions to form a soluble complex ion, $Ag(NH_3)_2^+$. As this complex ion forms, Ag^+ is removed from the solution, which causes the AgCl(s) to dissolve. When enough NH_3 is added, all the silver chloride precipitate will dissolve. The equation is $AgCl(s) + 2\,NH_3(aq) \rightarrow Ag(NH_3)_2^+(aq) + Cl^-(aq)$. Test tube 3: Added H^+ reacts with the weak base, NH_3, to form NH_4^+. As NH_3 is removed from the $Ag(NH_3)_2^+$ complex ion, Ag^+ ions are released to solution and can then react with Cl^- to re-form AgCl(s). The equations are $Ag(NH_3)_2^+(aq) + 2\,H^+(aq) \rightarrow Ag^+(aq) + 2\,NH_4^+(aq)$ and $Ag^+(aq) + Cl^-(aq) \rightarrow AgCl(s)$.

76. In NH_3, Cu^{2+} forms the soluble complex ion $Cu(NH_3)_4^{2+}$. This increases the solubility of $Cu(OH)_2(s)$ because added NH_3 removes Cu^{2+} from the equilibrium causing more $Cu(OH)_2(s)$ to dissolve. In HNO_3, H^+ removes OH^- from the K_{sp} equilibrium causing more $Cu(OH)_2(s)$ to dissolve. Any salt with basic anions will be more soluble in an acid solution. $AgC_2H_3O_2(s)$ will be more soluble in either NH_3 or HNO_3. This is because Ag^+ forms the complex ion $Ag(NH_3)_2^+$, and $C_2H_3O_2^-$ is a weak base, so it will react with added H^+. AgCl(s) will be more soluble only in NH_3 due to $Ag(NH_3)_2^+$ formation. In acid, Cl^- is a horrible base, so it doesn't react with added H^+. AgCl(s) will not be more soluble in HNO_3.

Additional Exercises

77. Mol Ag^+ added $= 0.200 \text{ L} \times \dfrac{0.24 \text{ mol } AgNO_3}{\text{L}} \times \dfrac{1 \text{ mol } Ag^+}{\text{mol } AgNO_3} = 0.048$ mol Ag^+

The added Ag^+ will react with the halogen ions to form a precipitate. Because the K_{sp} values are small, we can assume these precipitation reactions go to completion. The order of precipitation will be AgI(s) first (the least soluble compound since K_{sp} is the smallest), followed by AgBr(s), with AgCl(s) forming last [AgCl(s) is the most soluble compound listed since it has the largest K_{sp}].

Let the Ag^+ react with I^- to completion.

	$Ag^+(aq)$ +	$I^-(aq)$ →	$AgI(s)$	$K = 1/K_{sp} >> 1$
Before	0.048 mol	0.018 mol	0	
Change	−0.018	−0.018	+0.018	I^- is limiting.
After	0.030 mol	0	0.018 mol	

Let the Ag^+ remaining react next with Br^- to completion.

	$Ag^+(aq)$ +	$Br^-(aq)$ →	$AgBr(s)$	$K = 1/K_{sp} >> 1$
Before	0.030 mol	0.018 mol	0	
Change	−0.018	−0.018	+0.018	Br^- is limiting.
After	0.012 mol	0	0.018 mol	

Finally, let the remaining Ag^+ react with Cl^- to completion.

	$Ag^+(aq)$ +	$Cl^-(aq)$ →	$AgCl(s)$	$K = 1/K_{sp} >> 1$
Before	0.012 mol	0.018 mol	0	
Change	−0.012	−0.012	+0.012	Ag^+ is limiting.
After	0	0.006 mol	0.012 mol	

Some of the AgCl will redissolve to produce some Ag^+ ions; we can't have $[Ag^+] = 0$ *M*. Calculating how much AgCl(s) redissolves:

	$AgCl(s)$ ⇌	$Ag^+(aq)$ +	$Cl^-(aq)$	$K_{sp} = 1.6 \times 10^{-10}$
Initial	s = solubility (mol/L)	0	0.006 mol/0.200 L = 0.03 *M*	
	s mol/L of AgCl dissolves to reach equilibrium			
Change	$-s$ →	$+s$	$+s$	
Equil.		s	$0.03 + s$	

$K_{sp} = 1.6 \times 10^{-10} = [Ag^+][Cl^-] = s(0.03 + s) \approx (0.03)s$

$s = 5 \times 10^{-9}$ mol/L; the assumption that $0.03 + s \approx 0.03$ is good.

Mol AgCl present = 0.012 mol − 5×10^{-9} mol = 0.012 mol

Mass AgCl present = $0.012 \text{ mol AgCl} \times \frac{143.4 \text{ g}}{\text{mol AgCl}} = 1.7 \text{ g AgCl}$

$[Ag^+] = s = 5 \times 10^{-9}$ mol/L

78. $AgX(s) \rightleftharpoons Ag^+(aq) + X^-(aq) \quad K_{sp} = [Ag^+][X^-]$

$AgY(s) \rightleftharpoons Ag^+(aq) + Y^-(aq) \quad K_{sp} = [Ag^+][Y^-]$

For conjugate acid-base pairs, the weaker the acid, the stronger is the conjugate base. Because HX is a stronger acid (has a larger K_a value) than HY, Y^- will be a stronger base than X^-. In acidic solution, Y^- will have a greater affinity for the H^+ ions. Therefore, AgY(s) will be more soluble in acidic solution because more Y^- will be removed through reaction with H^+, which will cause more AgY(s) to dissolve.

79.

	$Ca_5(PO_4)_3OH(s)$	$\rightleftharpoons$	$5\ Ca^{2+}$	+	$3\ PO_4^{3-}$	+	OH^-
Initial	s = solubility (mol/L)		0		0		1.0×10^{-7} from water
Equil.			$5s$		$3s$		$s + 1.0 \times 10^{-7} \approx s$

$K_{sp} = 6.8 \times 10^{-37} = [Ca^{2+}]^5[PO_4^{3-}]^3[OH^-] = (5s)^5(3s)^3(s)$

$6.8 \times 10^{-37} = (3125)(27)s^9$, $s = 2.7 \times 10^{-5}$ mol/L; assumption is good.

The solubility of hydroxyapatite will increase as the solution gets more acidic because both phosphate and hydroxide can react with H^+.

	$Ca_5(PO_4)_3F(s)$	$\rightleftharpoons$	$5\ Ca^{2+}(aq)$	+	$3\ PO_4^{3-}(aq)$	+	$F^-(aq)$
Initial	s = solubility (mol/L)		0		0		0
Equil.			$5s$		$3s$		s

$K_{sp} = 1 \times 10^{-60} = (5s)^5(3s)^3(s) = (3125)(27)s^9$, $s = 6 \times 10^{-8}$ mol/L

The hydroxyapatite in tooth enamel is converted to the less soluble fluorapatite by fluoride-treated water. The less soluble fluorapatite is more difficult to remove, making teeth less susceptible to decay.

80. $\dfrac{1\ \text{mg}\ F^-}{L} \times \dfrac{1\ \text{g}}{1000\ \text{mg}} \times \dfrac{1\ \text{mol}\ F^-}{19.00\ \text{g}\ F^-} = 5.3 \times 10^{-5}\ M\ F^- = 5 \times 10^{-5}\ M\ F^-$

$CaF_2(s) \rightleftharpoons Ca^{2+}(aq) + 2\ F^-(aq)$ $K_{sp} = [Ca^{2+}][F^-]^2 = 4.0 \times 10^{-11}$; precipitation will occur when $Q > K_{sp}$. Let's calculate $[Ca^{2+}]$ so that $Q = K_{sp}$.

$Q = 4.0 \times 10^{-11} = [Ca^{2+}]_0[F^-]_0^2 = [Ca^{2+}]_0(5 \times 10^{-5})^2$, $[Ca^{2+}]_0 = 2 \times 10^{-2}\ M$

$CaF_2(s)$ will precipitate when $[Ca^{2+}]_0 > 2 \times 10^{-2}\ M$. Therefore, hard water should have a calcium ion concentration of less than $2 \times 10^{-2}\ M$ in order to avoid $CaF_2(s)$ formation.

81. $CaF_2(s) \rightleftharpoons Ca^{2+}(aq) + 2\ F^-(aq)$ $K_{sp} = [Ca^{2+}][F^-]^2$

We need to determine the F^- concentration present in a 1.0 M HF solution. Solving the weak acid equilibrium problem:

	$HF(aq)$	$\rightleftharpoons$	$H^+(aq)$	+	$F^-(aq)$	$K_a = \dfrac{[H^+][F^-]}{[HF]}$
Initial	1.0 M		~0		0	
Equil.	$1.0 - x$		x		x	

$K_a = 7.2 \times 10^{-4} = \frac{x(x)}{1.0 - x} \approx \frac{x^2}{1.0}$, $x = [F^-] = 2.7 \times 10^{-2}$ M; assumption good.

Next, calculate the Ca^{2+} concentration necessary for $Q = K_{sp,\, CaF_2}$.

$Q = [Ca^{2+}]_0[F^-]_0^2$, $4.0 \times 10^{-11} = [Ca^{2+}]_0(2.7 \times 10^{-2})^2$, $[Ca^{2+}]_0 = 5.5 \times 10^{-8}$ mol/L

$$\text{Mass } Ca(NO_3)_2 = 1.0 \text{ L} \times \frac{5.5 \times 10^{-8} \text{ mol } Ca^{2+}}{\text{L}} \times \frac{1 \text{ mol } Ca(NO_3)_2}{\text{mol } Ca^{2+}} \times \frac{164.10 \text{ g } Ca(NO_3)_2}{\text{mol}}$$

Mass $Ca(NO_3)_2 = 9.0 \times 10^{-6}$ g $Ca(NO_3)_2$

For precipitation of $CaF_2(s)$ to occur, we need $Q > K_{sp}$. When 9.0×10^{-6} g $Ca(NO_3)_2$ has been added to 1.0 L of solution, $Q = K_{sp}$. So precipitation of $CaF_2(s)$ will begin to occur when just over 9.0×10^{-6} g $Ca(NO_3)_2$ has been added.

82. | | $Mn(OH)_2(s)$ | $\rightleftharpoons$ | $Mn^{2+}(aq)$ | + | $2\,OH^-(aq)$ | $K_{sp} = [Mn^{2+}][OH^-]^2$ |
|---|---|---|---|---|---|---|
| Initial | s = solubility (mol/L) | | 0 | | 1.0×10^{-7} M (from water) | |
| Equil. | | | s | | $1.0 \times 10^{-7} + 2s$ | |

$K_{sp} = 2.0 \times 10^{-13} = s(1.0 \times 10^{-7} + 2s) \approx s(2s)^2 = 4s^3$, $s = 3.7 \times 10^{-5}$ mol/L; assumption good.

$$1.3 \text{ L} \times \frac{3.7 \times 10^{-5} \text{ mol } Mn(OH)_2}{\text{L}} \times \frac{88.96 \text{ g } Mn(OH)_2}{\text{mol}} = 4.3 \times 10^{-3} \text{ g } Mn(OH)_2$$

83. $$s = \text{solubility} = \frac{0.24 \text{ g } PbI_2 \times \frac{1 \text{ mol } PbI_2}{461.0 \text{ g}}}{0.2000 \text{ L}} = 2.6 \times 10^{-3}\ M$$

	$PbI_2(s)$	$\rightleftharpoons$	$Pb^{2+}(aq)$	+	$2\,I^-(aq)$	$K_{sp} = [Pb^{2+}][I^-]^2$
Initial	s = solubility (mol/L)		0		0	
Equil.			s		$2s$	

$K_{sp} = s(2s)^2 = 4s^3$, $K_{sp} = 4(2.6 \times 10^{-3})^3 = 7.0 \times 10^{-8}$

84. | | $C_7H_5BiO_4(s) + H_2O(l)$ | $\rightleftharpoons$ | $C_7H_4O_3^{2-}(aq)$ | + | $Bi^{3+}(aq)$ | + | $OH^-(aq)$ |
|---|---|---|---|---|---|---|---|
| Initial | s = solubility (mol/L) | | 0 | | 0 | | 1.0×10^{-7} M (from water) |
| Equil. | | | s | | s | | $1.0 \times 10^{-7} + s$ |

$K = [C_7H_4O_3^{2-}][Bi^{3+}][OH^-] = s(s)(1.0 \times 10^{-7} + s)$; from the problem, $s = 3.2 \times 10^{-19}$ mol/L:

$$K = (3.2 \times 10^{-19})^2(1.0 \times 10^{-7} + 3.2 \times 10^{-19}) = 1.0 \times 10^{-44}$$

85.

$Mn^{2+} + C_2O_4^{2-} \rightleftharpoons MnC_2O_4$	$K_1 = 7.9 \times 10^3$
$MnC_2O_4 + C_2O_4^{2-} \rightleftharpoons Mn(C_2O_4)_2^{2-}$	$K_2 = 7.9 \times 10^1$
$Mn^{2+}(aq) + 2\ C_2O_4^{2-}(aq) \rightleftharpoons Mn(C_2O_4)_2^{2-}(aq)$	$K_f = K_1K_2 = 6.2 \times 10^5$

86.

	Cr^{3+}	+ H_2EDTA^{2-}	$\rightleftharpoons$	$CrEDTA^-$	+ $2\ H^+$	
Before	0.0010 *M*	0.050 *M*		0	1.0×10^{-6} *M*	(Buffer)
Change	−0.0010	−0.0010	→	+0.0010	No change	Reacts completely
After	0	0.049		0.0010	1.0×10^{-6}	New initial
	x mol/L $CrEDTA^-$ dissociates to reach equilibrium					
Change	+*x*	+*x*	←	−*x*		
Equil.	*x*	0.049 + *x*		0.0010 − *x*	1.0×10^{-6}	(Buffer)

$$K_f = 1.0 \times 10^{23} = \frac{[CrEDTA^-][H^+]^2}{[Cr^{3+}][H_2EDTA^{2-}]} = \frac{(0.0010-x)(1.0 \times 10^{-6})^2}{(x)(0.049+x)}$$

$$1.0 \times 10^{23} \approx \frac{(0.0010)(1.0 \times 10^{-12})}{x(0.049)},\ x = [Cr^{3+}] = 2.0 \times 10^{-37}\ M;\ \text{assumptions good.}$$

87. a.

	$Pb(OH)_2(s)$	$\rightleftharpoons$	Pb^{2+}	+ $2\ OH^-$
Initial	*s* = solubility (mol/L)		0	1.0×10^{-7} *M* from water
Equil.			*s*	$1.0 \times 10^{-7} + 2s$

$$K_{sp} = 1.2 \times 10^{-15} = [Pb^{2+}][OH^-]^2 = s(1.0 \times 10^{-7} + 2s)^2 \approx s(2s^2) = 4s^3$$

$s = [Pb^{2+}] = 6.7 \times 10^{-6}\ M$; assumption is good by the 5% rule.

b.

	$Pb(OH)_2(s)$	$\rightleftharpoons$	Pb^{2+}	+ $2\ OH^-$	
Initial			0	0.10 *M*	pH = 13.00, $[OH^-]$ = 0.10 *M*
	s mol/L $Pb(OH)_2(s)$ dissolves to reach equilibrium				
Equil.			*s*	0.10	(Buffered solution)

$1.2 \times 10^{-15} = (s)(0.10)^2,\ s = [Pb^{2+}] = 1.2 \times 10^{-13}\ M$

c. We need to calculate the Pb^{2+} concentration in equilibrium with $EDTA^{4-}$. Since K is large for the formation of $PbEDTA^{2-}$, let the reaction go to completion and then solve an equilibrium problem to get the Pb^{2+} concentration.

	Pb^{2+}	+ $EDTA^{4-}$	$\rightleftharpoons$	$PbEDTA^{2-}$	$K = 1.1 \times 10^{18}$
Before	0.010 *M*	0.050 *M*		0	
	0.010 mol/L Pb^{2+} reacts completely (large K)				
Change	−0.010	−0.010	→	+0.010	Reacts completely
After	0	0.040		0.010	New initial
	x mol/L $PbEDTA^{2-}$ dissociates to reach equilibrium				
Equil.	*x*	0.040 + *x*		0.010 − *x*	

$$1.1 \times 10^{18} = \frac{(0.010 - x)}{(x)(0.040 + x)} \approx \frac{0.010}{x(0.040)},\ x = [Pb^{2+}] = 2.3 \times 10^{-19}\ M;\ \text{assumptions good.}$$

Now calculate the solubility quotient for $Pb(OH)_2$ to see if precipitation occurs. The concentration of OH^- is 0.10 M since we have a solution buffered at pH = 13.00.

$$Q = [Pb^{2+}]_0[OH^-]_0^2 = (2.3 \times 10^{-19})(0.10)^2 = 2.3 \times 10^{-21} < K_{sp}\ (1.2 \times 10^{-15})$$

$Pb(OH)_2(s)$ will not form since Q is less than K_{sp}.

88. $1.0\text{ mL} \times \dfrac{1.0\text{ mmol}}{\text{mL}} = 1.0\text{ mmol } Cd^{2+}$ added to the ammonia solution

Thus $[Cd^{2+}]_0 = 1.0 \times 10^{-3}$ mol/L. We will first calculate the equilibrium Cd^{2+} concentration using the complex ion equilibrium and then determine if this Cd^{2+} concentration is large enough to cause precipitation of $Cd(OH)_2(s)$.

	Cd^{2+}	+ $4\ NH_3$	$\rightleftharpoons$	$Cd(NH_3)_4^{2+}$	$K_f = 1.0 \times 10^7$
Before	$1.0 \times 10^{-3}\ M$	$5.0\ M$		0	
Change	-1.0×10^{-3}	-4.0×10^{-3}	$\rightarrow$	$+1.0 \times 10^{-3}$	Reacts completely
After	0	$4.996 \approx 5.0$		1.0×10^{-3}	New initial
	x mol/L $Cd(NH_3)_4^{2+}$ dissociates to reach equilibrium				
Change	$+x$	$+4x$	$\leftarrow$	$-x$	
Equil.	x	$5.0 + 4x$		$0.0010 - x$	

$$K_f = 1.0 \times 10^7 = \frac{(0.010 - x)}{(x)(5.0 + 4x)^4} \approx \frac{(0.010)}{(x)(5.0)^4}$$

$x = [Cd^{2+}] = 1.6 \times 10^{-13}\ M$; assumptions good. This is the maximum $[Cd^{2+}]$ possible. Now we will determine if $Cd(OH)_2(s)$ forms at this concentration of Cd^{2+}. In 5.0 M NH_3 we can calculate the pH:

	NH_3	+ H_2O	$\rightleftharpoons$	NH_4^+	+ OH^-	$K_b = 1.8 \times 10^{-5}$
Initial	$5.0\ M$			0	~0	
Equil.	$5.0 - y$			y	y	

$$K_b = 1.8 \times 10^{-5} = \frac{[NH_4^+][OH^-]}{[NH_3]} = \frac{y^2}{5.0 - y} \approx \frac{y^2}{5.0},\ y = [OH^-] = 9.5 \times 10^{-3}\ M;\ \text{assumptions good.}$$

We now calculate the value of the solubility quotient, Q:

$$Q = [Cd^{2+}][OH^-]^2 = (1.6 \times 10^{-13})(9.5 \times 10^{-3})^2$$

$Q = 1.4 \times 10^{-17} < K_{sp}\ (5.9 \times 10^{-15})$; therefore, no precipitate forms.

89. a.

$$Cu(OH)_2 \rightleftharpoons Cu^{2+} + 2\ OH^- \qquad K_{sp} = 1.6 \times 10^{-19}$$
$$Cu^{2+} + 4\ NH_3 \rightleftharpoons Cu(NH_3)_4^{2+} \qquad K_f = 1.0 \times 10^{13}$$

$$Cu(OH)_2(s) + 4\ NH_3(aq) \rightleftharpoons Cu(NH_3)_4^{2+}(aq) + 2\ OH^-(aq) \qquad K = K_{sp}K_f = 1.6 \times 10^{-6}$$

b. $Cu(OH)_2(s) + 4\,NH_3 \rightleftharpoons Cu(NH_3)_4^{2+} + 2\,OH^-$ $K = 1.6 \times 10^{-6}$

		NH_3	$Cu(NH_3)_4^{2+}$	OH^-
Initial		5.0 *M*	0	0.0095 *M*
	s mol/L $Cu(OH)_2$ dissolves to reach equilibrium			
Equil.		$5.0 - 4s$	s	$0.0095 + 2s$

$$K = 1.6 \times 10^{-6} = \frac{[Cu(NH_3)_4^{2+}][OH^-]^2}{[NH_3]^4} = \frac{s(0.0095 + 2s)^2}{(5.0 - 4s)^4}$$

If *s* is small: $1.6 \times 10^{-6} = \dfrac{s(0.0095)^2}{(5.0)^4}$, $s = 11.$ mol/L

Assumptions are horrible. We will solve the problem by successive approximations.

$$s_{calc} = \frac{1.6 \times 10^{-6}\,(5.0 - 4s_{guess})^4}{(0.0095 + 2s_{guess})^2}$$; the results from six trials are:

s_{guess}: 0.10, 0.050, 0.060, 0.055, 0.056

s_{calc}: 1.6×10^{-2}, 0.071, 0.049, 0.058, 0.056

Thus the solubility of $Cu(OH)_2$ is 0.056 mol/L in 5.0 *M* NH_3.

90. a.

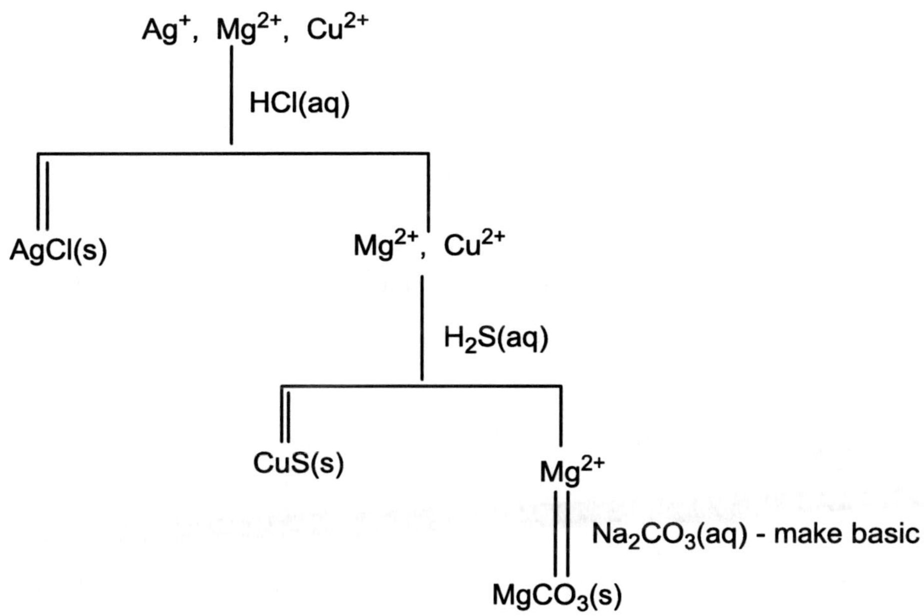

b.

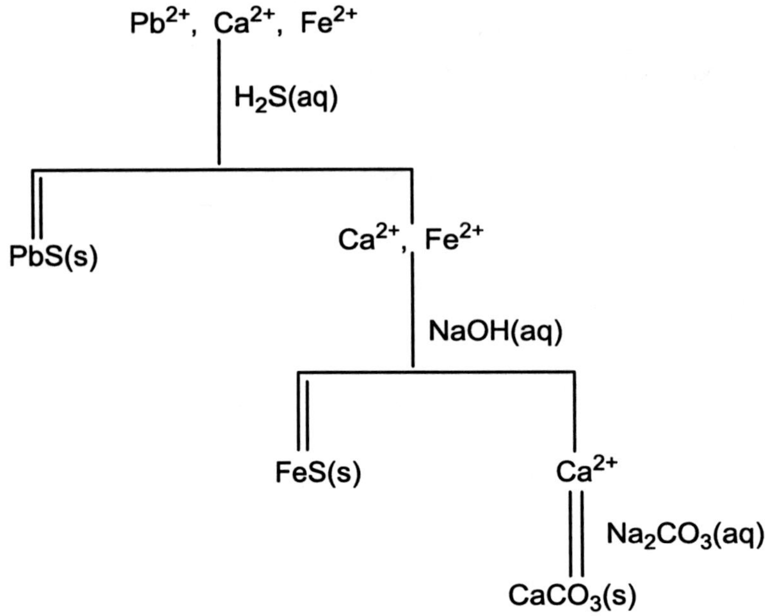

c.

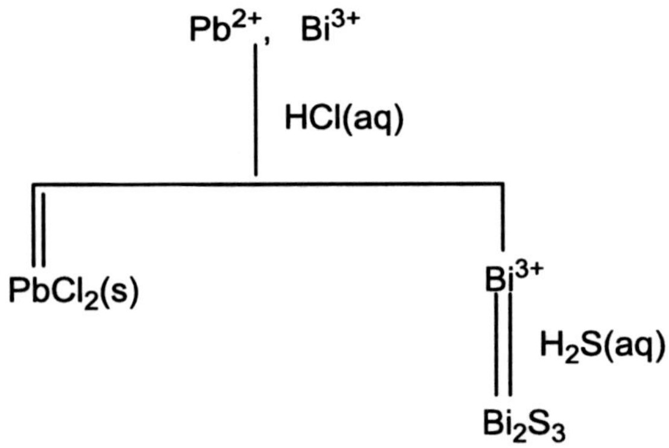

91.

	$Ba(OH)_2(s)$	$\rightleftharpoons$	$Ba^{2+}(aq)$ +	$2\ OH^-(aq)$	$K_{sp} = [Ba^{2+}][OH^-]^2 = 5.0 \times 10^{-3}$
Initial	s = solubility (mol/L)		0	~0	
Equil.			s	$2s$	

$K_{sp} = 5.0 \times 10^{-3} = s(2s)^2 = 4s^3$, $s = 0.11$ mol/L; assumption good.

$[OH^-] = 2s = 2(0.11) = 0.22$ mol/L; pOH = 0.66, pH = 13.34

	$Sr(OH)_2(s)$	$\rightleftharpoons$	$Sr^{2+}(aq)$ +	$2\ OH^-(aq)$	$K_{sp} = [Sr^{2+}][OH^-]^2 = 3.2 \times 10^{-4}$
Equil.			s	$2s$	

$K_{sp} = 3.2 \times 10^{-4} = 4s^3$, $s = 0.043$ mol/L; asssumption good.

$[OH^-] = 2(0.043) = 0.086\ M$; pOH = 1.07, pH = 12.93

	$Ca(OH)_2(s)$	$\rightleftharpoons$	$Ca^{2+}(aq)$	+	$2\ OH^-(aq)$	$K_{sp} = [Ca^{2+}][OH^-]^2 = 1.3 \times 10^{-6}$
Equil.			s		$2s$	

$K_{sp} = 1.3 \times 10^{-6} = 4s^3$, $s = 6.9 \times 10^{-3}$ mol/L; assumption good.

$[OH^-] = 2(6.9 \times 10^{-3}) = 1.4 \times 10^{-2}$ mol/L; pOH = 1.85, pH = 12.15

92. $K_{sp} = 6.4 \times 10^{-9} = [Mg^{2+}][F^-]^2$, $6.4 \times 10^{-9} = (0.00375 - y)(0.0625 - 2y)^2$

This is a cubic equation. No simplifying assumptions can be made since y is relatively large. Solving cubic equations is difficult unless you have a graphing calculator. However, if you don't have a graphing calculator, one way to solve this problem is to make the simplifying assumption to run the precipitation reaction to completion. This assumption is made because of the very small value for K, indicating that the ion concentrations are very small. Once this assumption is made, the problem becomes much easier to solve.

ChemWork Problems

The answers to the problems 93-98 (or a variation to these problems) are found in OWL. These problems are also assignable in OWL.

Challenge Problems

99. a.

$$CuBr(s) \rightleftharpoons Cu^+ + Br^- \qquad K_{sp} = 1.0 \times 10^{-5}$$

$$Cu+ + 3\ CN^- \rightleftharpoons Cu(CN)_3^{2-} \qquad K_f = 1.0 \times 10^{11}$$

$$CuBr(s) + 3\ CN^- \rightleftharpoons Cu(CN)_3^{2-} + Br^- \qquad K = 1.0 \times 10^6$$

Because K is large, assume that enough CuBr(s) dissolves to completely use up the 1.0 M CN^-; then solve the back equilibrium problem to determine the equilibrium concentrations.

	$CuBr(s)$	+ $3\ CN^-$	$\rightleftharpoons$	$Cu(CN)_3^{2-}$	+ Br^-
Before	x	1.0 M		0	0
	x mol/L of CuBr(s) dissolves to react completely with 1.0 M CN^-				
Change	$-x$	$-3x$	$\rightarrow$	$+x$	$+x$
After	0	$1.0 - 3x$		x	x

For reaction to go to completion, $1.0 - 3x = 0$ and $x = 0.33$ mol/L. Now solve the back-equilibrium problem.

	$CuBr(s) + 3\ CN^-$	$\rightleftharpoons$	$Cu(CN)_3^{2-}$	$+\ Br^-$
Initial	0		0.33 M	0.33 M
	Let y mol/L of $Cu(CN)_3^{2-}$ react to reach equilibrium.			
Change	$+3y$	$\leftarrow$	$-y$	$-y$
Equil.	3y		$0.33 - y$	$0.33 - y$

$$K = 1.0 \times 10^6 = \frac{(0.33 - y)^2}{(3y)^3} \approx \frac{(0.33)^2}{27y^3},\ y = 1.6 \times 10^{-3}\ M;\ \text{assumption good.}$$

Of the initial 1.0 M CN^-, only $3(1.6 \times 10^{-3}) = 4.8 \times 10^{-3}$ M is present at equilibrium. Indeed, enough CuBr(s) did dissolve to essentially remove the initial 1.0 M CN^-. This amount, 0.33 mol/L, is the solubility of CuBr(s) in 1.0 M NaCN.

b. $[Br^-] = 0.33 - y = 0.33 - 1.6 \times 10^{-3} = 0.33\ M$

c. $[CN^-] = 3y = 3(1.6 \times 10^{-3}) = 4.8 \times 10^{-3}\ M$

100.

$$Ag^+ + NH_3 \rightleftharpoons AgNH_3^+ \qquad K_1 = 2.1 \times 10^3$$
$$AgNH_3^+ + NH_3 \rightleftharpoons Ag(NH_3)_2^+ \qquad K_2 = 8.2 \times 10^3$$
$$Ag^+ + 2\ NH_3 \rightleftharpoons Ag(NH_3)_2^+ \qquad K = K_1K_2 = 1.7 \times 10^7$$

The initial concentrations are halved because equal volumes of the two solutions are mixed. Let the reaction go to completion since K is large; then solve an equilibrium problem.

	$Ag^+(aq)$	+	$2\ NH_3(aq)$	$\rightleftharpoons$	$Ag(NH_3)_2^+(aq)$
Before	0.20 M		2.0 M		0
After	0		1.6		0.20
Equil.	x		$1.6 + 2x$		$0.20 - x$

$$K = 1.7 \times 10^7 = \frac{[Ag(NH_3)_2^+]}{[Ag^+][NH_3]^2} = \frac{0.20 - x}{x(1.6 + 2x)^2} \approx \frac{0.20}{x(1.6)^2},\ x = 4.6 \times 10^{-9}\ M;$$

assumptions good.

$[Ag^+] = x = 4.6 \times 10^{-9}\ M$; $[NH_3] = 1.6\ M$; $[Ag(NH_3)_2^+] = 0.20\ M$

Use either the K_1 or K_2 equilibrium expression to calculate $[AgNH_3^+]$.

$$AgNH_3^+ + NH_3 \rightleftharpoons Ag(NH_3)_2^+ \qquad K_2 = 8.2 \times 10^3$$

$$8.2 \times 10^3 = \frac{[Ag(NH_3)_2^+]}{[AgNH_3^+][NH_3]} = \frac{0.20}{[AgNH_3^+](1.6)},\ [AgNH_3^+] = 1.5 \times 10^{-5}\ M$$

101. a.

	$AgBr(s)$	$\rightleftharpoons$	$Ag^+(aq)$	+	$Br^-(aq)$	$K_{sp} = [Ag^+][Br^-] = 5.0 \times 10^{-13}$
Initial	s = solubility (mol/L)		0		0	
Equil.			s		s	

$K_{sp} = 5.0 \times 10^{-13} = s^2$, $s = 7.1 \times 10^{-7}$ mol/L

b.

$$AgBr(s) \rightleftharpoons Ag^+ + Br^- \qquad K_{sp} = 5.0 \times 10^{-13}$$
$$Ag^+ + 2\,NH_3 \rightleftharpoons Ag(NH_3)_2^+ \qquad K_f = 1.7 \times 10^7$$

$$AgBr(s) + 2\,NH_3(aq) \rightleftharpoons Ag(NH_3)_2^+(aq) + Br^-(aq) \qquad K = K_{sp} \times K_f = 8.5 \times 10^{-6}$$

	AgBr(s) +	2 NH_3	⇌ $Ag(NH_3)_2^+$	+ Br^-
Initial		3.0 *M*	0	0
	s mol/L of AgBr(s) dissolves to reach equilibrium = molar solubility			
Equil.		3.0 – 2*s*	*s*	*s*

$$K = \frac{[Ag(NH_3)_2^+][Br^-]}{[NH_3]^2} = \frac{s^2}{(3.0 - 2s)^2} = 8.5 \times 10^{-6} \approx \frac{s^2}{(3.0)^2}, \; s = 8.7 \times 10^{-3}\text{ mol/L}$$

Assumption good.

c. The presence of NH_3 increases the solubility of AgBr. Added NH_3 removes Ag^+ from solution by forming the complex ion, $Ag(NH_3)_2^+$. As Ag^+ is removed, more AgBr(s) will dissolve to replenish the Ag^+ concentration.

d. $$\text{Mass AgBr} = 0.2500\text{ L} \times \frac{8.7 \times 10^{-3}\text{ mol AgBr}}{\text{L}} \times \frac{187.8\text{ g AgBr}}{\text{mol AgBr}} = 0.41\text{ g AgBr}$$

e. Added HNO_3 will have no effect on the AgBr(s) solubility in pure water. Neither H^+ nor NO_3^- react with Ag^+ or Br^- ions. Br^- is the conjugate base of the strong acid HBr, so it is a terrible base. However, added HNO_3 will reduce the solubility of AgBr(s) in the ammonia solution. NH_3 is a weak base ($K_b = 1.8 \times 10^{-5}$). Added H^+ will react with NH_3 to form NH_4^+. As NH_3 is removed, a smaller amount of the $Ag(NH_3)_2^+$ complex ion will form, resulting in a smaller amount of AgBr(s) that will dissolve.

102. $$[NH_3]_0 = \frac{3.00\,M}{2} = 1.50\,M; \; [Cu^{2+}]_0 = \frac{2.00 \times 10^{-3}\,M}{2} = 1.00 \times 10^{-3}\,M$$

Because $[NH_3]_0 >> [Cu^{2+}]_0$, and because K_1, K_2, K_3 and K_4 are all large, $Cu(NH_3)_4^{2+}$ will be the dominant copper-containing species. The net reaction will be $Cu^{2+} + 4\,NH_3 \rightarrow Cu(NH_3)_4^{2+}$. Here, $1.00 \times 10^{-3}\,M\,Cu^{2+}$ plus $4(1.00 \times 10^{-3}\,M)$ NH_3 will produce $1.00 \times 10^{-3}\,M\,Cu(NH_3)_4^{2+}$. At equilibrium:

$$[Cu(NH_3)_4^{2+}] \approx 1.00 \times 10^{-3}\,M$$

$$[NH_3] = [NH_3]_0 - [NH_3]_{reacted} = 1.50\,M - 4(1.00 \times 10^{-3}\,M) - 1.50\,M$$

Calculate $[Cu(NH_3)_3^{2+}]$ from the K_4 reaction:

$$1.55 \times 10^2 = \frac{[Cu(NH_3)_4^{2+}]}{[Cu(NH_3)_3^{2+}][NH_3]} = \frac{1.00 \times 10^{-3}}{[Cu(NH_3)_3^{2+}](1.50)}, \; [Cu(NH_3)_3^{2+}] = 4.30 \times 10^{-6}\,M$$

Calculate $[Cu(NH_3)_2^{2+}]$ from K_3 reaction:

$$1.00 \times 10^3 = \frac{[Cu(NH_3)_3^{2+}]}{[Cu(NH_3)_2^{2+}][NH_3]} = \frac{4.30 \times 10^{-6}}{[Cu(NH_3)_2^{2+}](1.50)},\ [Cu(NH_3)_2^{2+}] = 2.87 \times 10^{-9}\ M$$

Calculate $[Cu(NH_3)_3^{2+}]$ from the K_2 reaction:

$$3.88 \times 10^3 = \frac{[Cu(NH_3)_2^{2+}]}{[CuNH_3^{2+}][NH_3]} = \frac{2.87 \times 10^{-9}}{[CuNH_3^{2+}](1.50)},\ [CuNH_3^{2+}] = 4.93 \times 10^{-13}\ M$$

Calculate $[Cu^{2+}]$ from the K_1 reaction:

$$1.86 \times 10^4 = \frac{[CuNH_3^{2+}]}{[Cu^{2+}][NH_3]} = \frac{4.93 \times 10^{-13}}{[Cu^{2+}](1.50)},\ [Cu^{2+}] = 1.77 \times 10^{-17}\ M$$

The assumptions are valid. $Cu(NH_3)_4^{2+}$ is clearly the dominant copper-containing component.

103.

$$AgCN(s) \rightleftharpoons Ag^+(aq) + CN^-(aq) \qquad K_{sp} = 2.2 \times 10^{-12}$$
$$H^+(aq) + CN^-(aq) \rightleftharpoons HCN(aq) \qquad K = 1/K_{a,HCN} = 1.6 \times 10^9$$

$$AgCN(s) + H^+(aq) \rightleftharpoons Ag^+(aq) + HCN(aq) \qquad K = 2.2 \times 10^{-12}(1.6 \times 10^9) = 3.5 \times 10^{-3}$$

	AgCN(s) +	$H^+(aq)$	$\rightleftharpoons$ $Ag^+(aq)$ +	HCN(aq)
Initial		1.0 *M*	0	0
	s mol/L AgCN(s) dissolves to reach equilibrium			
Equil.		1.0 – *s*	*s*	*s*

$$3.5 \times 10^{-3} = \frac{[Ag^+][HCN]}{[H^+]} = \frac{s(s)}{1.0 - s} \approx \frac{s^2}{1.0},\ s = 5.9 \times 10^{-2}$$

Assumption fails the 5% rule (*s* is 5.9% of 1.0 *M*). Using the method of successive approximations:

$$3.5 \times 10^{-3} = \frac{s^2}{1.0 - 0.059},\ s = 5.7 \times 10^{-2}$$

$$3.5 \times 10^{-3} = \frac{s^2}{1.0 - 0.057},\ s = 5.7 \times 10^{-2} \text{ (consistent answer)}$$

The molar solubility of AgCN(s) in 1.0 *M* H^+ is 5.7×10^{-2} mol/L.

104. Solubility in pure water:

	$CaC_2O_4(s)$	$\rightleftharpoons$	$Ca^{2+}(aq)$ +	$C_2O_4^{2-}(aq)$	$K_{sp} = 2 \times 10^{-9}$
Initial	*s* = solubility (mol/L)		0	0	
Equil.			*s*	*s*	

$K_{sp} = s^2 = 2 \times 10^{-9}$, s = solubility = $4.47 \times 10^{-5} = 4 \times 10^{-5}$ mol/L

Solubility in 0.10 M H^+:

$$\begin{array}{rcll} CaC_2O_4(s) & \rightleftharpoons & Ca^{2+} + C_2O_4^{2-} & K_{sp} = 2 \times 10^{-9} \\ C_2O_4^{2-} + H^+ & \rightleftharpoons & HC_2O_4^- & K = 1/K_{a_2} = 1.6 \times 10^4 \\ HC_2O_4^- + H^+ & \rightleftharpoons & H_2C_2O_4 & K = 1/K_{a_1} = 15 \\ \hline CaC_2O_4(s) + 2\,H^+ & \rightleftharpoons & Ca^{2+} + H_2C_2O_4 & K_{overall} = 5 \times 10^{-4} \end{array}$$

	$CaC_2O_4(s)$ +	$2\,H^+$	$\rightleftharpoons$	Ca^{2+} +	$H_2C_2O_4$
Initial		0.10 M		0	0
	s mol/L of $CaC_2O_4(s)$ dissolves to reach equilibrium				
Equil.		$0.10 - 2s$		s	s

$$5 \times 10^{-4} = \frac{s^2}{(0.10 - 2s)^2}, \quad \frac{s}{0.10 - 2s} = (5 \times 10^{-4})^{1/2}, \quad s = 2 \times 10^{-3} \text{ mol/L}$$

$$\frac{\text{Solubility in } 0.10\,M\ H^+}{\text{Solubility in pure water}} = \frac{2 \times 10^{-3} \text{ mol/L}}{4 \times 10^{-5} \text{ mol/L}} = 50$$

$CaC_2O_4(s)$ is 50 times more soluble in 0.10 M H^+ than in pure water. This increase in solubility is due to the weak base properties of $C_2O_4^{2-}$.

105.
$$\begin{array}{ll} H_2S(aq) \rightleftharpoons H^+(aq) + HS^-(aq) & K_{a_1} = 1.0 \times 10^{-7} \\ HS^-(aq) \rightleftharpoons H^+(aq) + S^{2-}(aq) & K_{a_2} = 1 \times 10^{-19} \\ \hline H_2S(aq) \rightleftharpoons 2\,H^+(aq) + S^{2-}(aq) & K = K_{a_1} \times K_{a_2} = 1 \times 10^{-26} = \dfrac{[H^+]^2[S^{2-}]}{[H_2S]} \end{array}$$

Because K is very small, only a tiny fraction of the H_2S will react. At equilibrium, $[H_2S]$ = 0.10 M and $[H^+] = 1 \times 10^{-3}$.

$$[S^{2-}] = \frac{K[H_2S]}{[H^+]^2} = \frac{(1 \times 10^{-26})(0.10)}{(1 \times 10^{-3})^2} = 1 \times 10^{-21}\,M$$

$$NiS(s) \rightleftharpoons Ni^{2+}(aq) + S^{2-}(aq) \qquad K_{sp} = [Ni^{2+}][S^{2-}] = 3 \times 10^{-21}$$

Precipitation of NiS will occur when $Q > K_{sp}$. We will calculate $[Ni^{2+}]$ for $Q = K_{sp}$.

$$Q = K_{sp} = [Ni^{2+}][S^{2-}] = 3.0 \times 10^{-21}, \ [Ni^{2+}] = \frac{3.0 \times 10^{-21}}{1 \times 10^{-21}} = 3\,M = \text{maximum concentration}$$

106. We need to determine $[S^{2-}]_0$ that will cause precipitation of CuS(s) but not MnS(s).

For CuS(s):

$$CuS(s) \rightleftharpoons Cu^{2+}(aq) + S^{2-}(aq) \quad K_{sp} = [Cu^{2+}][S^{2-}] = 8.5 \times 10^{-45}$$

$$[Cu^{2+}]_0 = 1.0 \times 10^{-3}\,M, \quad \frac{K_{sp}}{[Cu^{2+}]_0} = \frac{8.5 \times 10^{-45}}{1.0 \times 10^{-3}} = 8.5 \times 10^{-42}\,M = [S^{2-}]$$

This $[S^{2-}]$ represents the concentration that we must exceed to cause precipitation of CuS because if $[S^{2-}]_0 > 8.5 \times 10^{-42}\,M$, $Q > K_{sp}$.

For MnS(s):

$$MnS(s) \rightleftharpoons Mn^{2+}(aq) + S^{2-}(aq) \qquad K_{sp} = [Mn^{2+}][S^{2-}] = 2.3 \times 10^{-13}$$

$$[Mn^{2+}]_0 = 1.0 \times 10^{-3}\,M, \quad \frac{K_{sp}}{[Mn^{2+}]} = \frac{2.3 \times 10^{-13}}{1.0 \times 10^{-3}} = 2.3 \times 10^{-10}\,M = [S^{2-}]$$

This value of $[S^{2-}]$ represents the largest concentration of sulfide that can be present without causing precipitation of MnS. That is, for this value of $[S^{2-}]$, $Q = K_{sp}$, and no precipitatation of MnS occurs. However, for any $[S^{2-}]_0 > 2.3 \times 10^{-10}\,M$, MnS(s) will form.

We must have $[S^{2-}]_0 > 8.5 \times 10^{-42}\,M$ to precipitate CuS, but $[S^{2-}]_0 < 2.3 \times 10^{-10}\,M$ to prevent precipitation of MnS.

The question asks for a pH that will precipitate CuS(s) but not MnS(s). We need to first choose an initial concentration of S^{2-} that will do this. Let's choose $[S^{2-}]_0 = 1.0 \times 10^{-10}\,M$ because this will clearly cause CuS(s) to precipitate but is still less than the $[S^{2-}]_0$ required for MnS(s) to precipitate. The problem now is to determine the pH necessary for a 0.1 M H_2S solution to have $[S^{2-}] = 1.0 \times 10^{-10}\,M$. Let's combine the K_{a_1} and K_{a_2} equations for H_2S to determine the required $[H^+]$.

$$H_2S(aq) \rightleftharpoons H^+(aq) + HS^-(aq) \qquad K_{a_1} = 1.0 \times 10^{-7}$$
$$HS^-(aq) \rightleftharpoons H^+(aq) + S^{2-}(aq) \qquad K_{a_2} = 1 \times 10^{-19}$$

$$H_2S(aq) \rightleftharpoons 2H^+(aq) + S^{2-}(aq) \qquad K = K_{a_1} \times K_{a_2} = 1.0 \times 10^{-26}$$

$$1 \times 10^{-26} = \frac{[H^+]^2[S^{2-}]}{[H_2S]} = \frac{[H^+]^2(1 \times 10^{-10})}{0.10}, \quad [H^+] = 3 \times 10^{-9}\,M$$

$pH = -\log(3 \times 10^{-9}) = 8.5$. So, if $pH = 8.5$, $[S^{2-}] = 1 \times 10^{-10}\,M$, which will cause precipitation of CuS(s) but not MnS(s).

Note: Any pH less than 8.7 would be a correct answer to this problem.

107. $Mg^{2+} + P_3O_{10}^{5-} \rightleftharpoons MgP_3O_{10}^{3-} \quad K = 4.0 \times 10^8$

$$[Mg^{2+}]_0 = \frac{50. \times 10^{-3}\text{ g}}{\text{L}} \times \frac{1\text{ mol}}{24.31\text{ g}} = 2.1 \times 10^{-3}\,M$$

$$[P_3O_{10}{}^{5-}]_0 = \frac{40.\text{ g } Na_5P_3O_{10}}{L} \times \frac{1\text{ mol}}{367.86\text{ g}} = 0.11\ M$$

Assume the reaction goes to completion because K is large. Then solve the back-equilibrium problem to determine the small amount of Mg^{2+} present.

	Mg^{2+} +	$P_3O_{10}{}^{5-}$	$\rightleftharpoons$	$MgP_3O_{10}{}^{3-}$	
Before	$2.1 \times 10^{-3}\ M$	$0.11\ M$		0	
Change	-2.1×10^{-3}	-2.1×10^{-3}	$\rightarrow$	$+2.1 \times 10^{-3}$	React completely
After	0	0.11		2.1×10^{-3}	New initial condition
	x mol/L $MgP_3O_{10}{}^{3-}$ dissociates to reach equilibrium				
Change	$+x$	$+x$	$\leftarrow$	$-x$	
Equil.	x	$0.11 + x$		$2.1 \times 10^{-3} - x$	

$$K = 4.0 \times 10^8 = \frac{[MgP_3O_{10}{}^{3-}]}{[Mg^{2+}][P_3O_{10}{}^{5-}]} = \frac{2.1 \times 10^{-3} - x}{x(0.11 + x)} \quad (\text{assume } x << 2.1 \times 10^{-3})$$

$$4.0 \times 10^8 \approx \frac{2.1 \times 10^{-3}}{x(0.11)},\quad x = [Mg^{2+}] = 4.8 \times 10^{-11}\ M;\ \text{assumptions good.}$$

108. $MX(s) \rightleftharpoons M^{n+}(aq) + X^{n-}(aq)$; $\Delta T = K_f m$, $m = \dfrac{\Delta T}{K_f} = \dfrac{0.028°C}{1.86°C/\text{molal}} = 0.015$ mol/kg

$$\frac{0.015\text{ mol}}{\text{kg}} \times \frac{1\text{ kg}}{1000\text{ g}} \times 250\text{ g} = 0.00375 \text{ mol total solute particles (carrying extra sig. fig.)}$$

0.0375 mol = mol M^{n+} + mol X^{n-}, mol M^{n+} = mol X^{n-} = 0.0375/2

Because the density of the solution is 1.0 g/mL, 250 g = 250 mL of solution.

$$[M^{n+}] = \frac{(0.00375/2)\text{ mol } M^{n+}}{0.25\text{ L}} = 7.5 \times 10^{-3}\ M,\ [X^{n-}] = \frac{(0.00375/2)\text{ mol } X^{n-}}{0.25\text{ L}} = 7.5 \times 10^{-3}\ M$$

$$K_{sp} = [M^{n+}][X^{n-}] = (7.5 \times 10^{-3})^2 = 5.6 \times 10^{-5}$$

109. a.

	$SrF_2(s)$	$\rightleftharpoons$	$Sr^{2+}(aq)$ +	$2\ F^-(aq)$
Initial			0	0
	s mol/L SrF_2 dissolves to reach equilibrium			
Equil.			s	$2s$

$[Sr^{2+}][F^-]^2 = K_{sp} = 7.9 \times 10^{-10} = 4s^3$, $s = 5.8 \times 10^{-4}$ mol/L in pure water

b. Greater, because some of the F^- would react with water:

$$F^- + H_2O \rightleftharpoons HF + OH^- \quad K_b = \frac{K_w}{K_{a,\,HF}} = 1.4 \times 10^{-11}$$

This lowers the concentration of F^-, forcing more SrF_2 to dissolve.

c. $SrF_2(s) \rightleftharpoons Sr^{2+} + 2\,F^-$ $K_{sp} = 7.9 \times 10^{-10} = [Sr^{2+}][F^-]^2$

Let s = solubility = $[Sr^{2+}]$; then $2s$ = total F^- concentration.

Since F^- is a weak base, some of the F^- is converted into HF. Therefore:

total F^- concentration = $2s = [F^-] + [HF]$

$HF \rightleftharpoons H^+ + F^-$ $K_a = 7.2 \times 10^{-4} = \dfrac{[H^+][F^-]}{[HF]} = \dfrac{1.0 \times 10^{-2}[F^-]}{[HF]}$ (since pH = 2.00 buffer)

$7.2 \times 10^{-2} = \dfrac{[F^-]}{[HF]}$, $[HF] = 14[F^-]$; solving:

$[Sr^{2+}] = s$; $2s = [F^-] + [HF] = [F^-] + 14[F^-]$, $2s = 15[F^-]$, $[F^-] = 2s/15$

$K_{sp} = 7.9 \times 10^{-10} = [Sr^{2+}][F^-]^2 = (s)\left(\dfrac{2s}{15}\right)^2$, $s = 3.5 \times 10^{-3}$ mol/L in pH = 2.00 solution

Integrative Problems

110.

	$M_3X_2(s)$	$\rightleftharpoons$	$3\,M^{2+}(aq)$	+	$2\,X^{3-}(aq)$	$K_{sp} = [M^{2+}]^3[X^{3-}]^2$
Initial	s = solubility (mol/L)		0		0	
Equil.			$3s$		$2s$	

$K_{sp} = (3s)^3(2s)^2 = 108\,s^5$; total ion concentration = $3s + 2s = 5s$

$\pi = iMRT$, iM = total ion concentration = $\dfrac{\pi}{RT} = \dfrac{2.64 \times 10^{-2}\text{ atm}}{0.08206\text{ L atm/K}\bullet\text{mol} \times 298\text{K}}$

$= 1.08 \times 10^{-3}$ mol/L

$5s = 1.08 \times 10^{-3}$ mol/L, $s = 2.16 \times 10^{-4}$ mol/L; $K_{sp} = 108s^5 = 108(2.16 \times 10^{-4})^5$

$K_{sp} = 5.08 \times 10^{-17}$

111. Major species: H^+, HSO_4^-, Ba^{2+}, NO_3^-, and H_2O; Ba^{2+} will react with the SO_4^{2-} produced from the K_a reaction for HSO_4^-.

$HSO_4^- \rightleftharpoons H^+ + SO_4^{2-}$ $K_{a_2} = 1.2 \times 10^{-2}$

$Ba^{2+} + SO_4^{2-} \rightleftharpoons BaSO_4(s)$ $K = 1/K_{sp} = 1/(1.5 \times 10^{-9}) = 6.7 \times 10^8$

$Ba^{2+} + HSO_4^- \rightleftharpoons H^+ + BaSO_4(s)$ $K_{overall} = (1.2 \times 10^{-2}) \times (6.7 \times 10^8) = 8.0 \times 10^6$

Because $K_{overall}$ is so large, the reaction essentially goes to completion. Because H_2SO_4 is a strong acid, $[HSO_4^-]_0 = [H^+]_0 = 0.10\,M$.

	Ba^{2+}	+ HSO_4^-	$\rightleftharpoons$	H^+	+ $BaSO_4(s)$	
Before	0.30 *M*	0.10 *M*		0.10 *M*		
Change	–0.10	–0.10	→	+0.10		
After	0.20	0		0.20 *M*		New initial
Change	$+x$	$+x$		$-x$		
Equil.	$0.20 + x$	x		$0.20 - x$		

$K = 8.0 \times 10^6 = \frac{0.20 - x}{(0.20 + x)x} \approx \frac{0.20}{0.20(x)}$, $x = 1.3 \times 10^{-7}$ *M*; assumptions good.

$[H^+] = 0.20 - 1.3 \times 10^{-7} = 0.20$ *M*; pH $= -\log(0.20) = 0.70$

$[Ba^{2+}] = 0.20 + 1.3 \times 10^{-7} = 0.20$ *M*

From the initial reaction essentially going to completion, 1.0 L(0.10 mol HSO_4^-/L) = 0.10 mol HSO_4^- reacted; this will produce 0.10 mol $BaSO_4(s)$. Only 1.3×10^{-7} mol of this dissolves to reach equilibrium, so 0.10 mol $BaSO_4(s)$ is produced.

$$0.10 \text{ mol BaSO}_4 \times \frac{233.4 \text{ g BaSO}_4}{\text{mol}} = 23 \text{ g BaSO}_4 \text{ produced}$$

112. M: $[Xe]6s^24f^{14}5d^{10}$; this is mercury, Hg. Because X^- has 54 electrons, X has 53 protons and is iodine, I. The identity of Q = Hg_2I_2.

$$[I^-]_0 = \frac{1.98 \text{ g NaI} \times \frac{1 \text{ mol NaI}}{149.9 \text{ g}} \times \frac{1 \text{ mol I}^-}{\text{mol NaI}}}{0.150 \text{ L}} = 0.0881 \text{ mol/L}$$

	$Hg_2I_2(s)$	$\rightleftharpoons$	$Hg_2^{2+}(aq)$	+	$2\,I^-(aq)$	$K_{sp} = 4.5 \times 10^{-29}$
Initial	s = solubility (mol/L)		0		0.0881 *M*	
Equil.			s		$0.0881 + 2s$	

$K_{sp} = 4.5 \times 10^{-29} = [Hg_2^{2+}][I^-]^2 = s(0.0881 + 2s)^2 \approx s(0.0881)^2$

$s = 5.8 \times 10^{-27}$ mol/L; assumption good.

Marathon Problem

113. a. In very acidic solutions, the reaction that occurs to increase the solubility is $Al(OH)_3(s) + 3H^+ \rightarrow Al^{3+}(aq) + 3H_2O(l)$. In very basic solutions, the reaction that occurs to increase solubility is $Al(OH)_3(s) + OH^-(aq) \rightarrow Al(OH)_4^-(aq)$.

b. $Al(OH)_3(s) \rightleftharpoons Al^{3+} + 3\,OH^-$; $Al(OH)_3(s) + OH^- \rightleftharpoons Al(OH)_4^-$

S = solubility = total Al^{3+} concentration = $[Al^{3+}] + [Al(OH)_4^-]$

$$[Al^{3+}] = \frac{K_{sp}}{[OH^-]^3} = K_{sp} \times \frac{[H^+]^3}{K_w^{\ 3}}, \text{ because } [OH^-]^3 = (K_w/[H^+])^3$$

$$\frac{[Al(OH)_4^-]}{[OH^-]} = K; \ [OH^-] = \frac{K_w}{[H^+]}; \ [Al(OH)_4^-] = K[OH^-] = \frac{KK_w}{[H^+]}$$

$$S = [Al^{3+}] + [Al(OH)_4^-] = [H^+]^3K_{sp}/K_w^{\ 3} + KK_w/[H^+]$$

c. $K_{sp} = 2 \times 10^{-32}$; $K_w = 1.0 \times 10^{-14}$; $K = 40.0$

$$S = \frac{[H^+]^3(2 \times 10^{-32})}{(1.0 \times 10^{-14})^3} + \frac{40.0(1.0 \times 10^{-14})}{[H^+]} = [H^+]^3(2 \times 10^{10}) + \frac{4.0 \times 10^{-13}}{[H^+]}$$

pH	solubility (S, mol/L)	log S
4.0	2×10^{-2}	–1.7
5.0	2×10^{-5}	–4.7
6.0	4.2×10^{-7}	–6.38
7.0	4.0×10^{-6}	–5.40
8.0	4.0×10^{-5}	–4.40
9.0	4.0×10^{-4}	–3.40
10.0	4.0×10^{-3}	–2.40
11.0	4.0×10^{-2}	–1.40
12.0	4.0×10^{-1}	–0.40

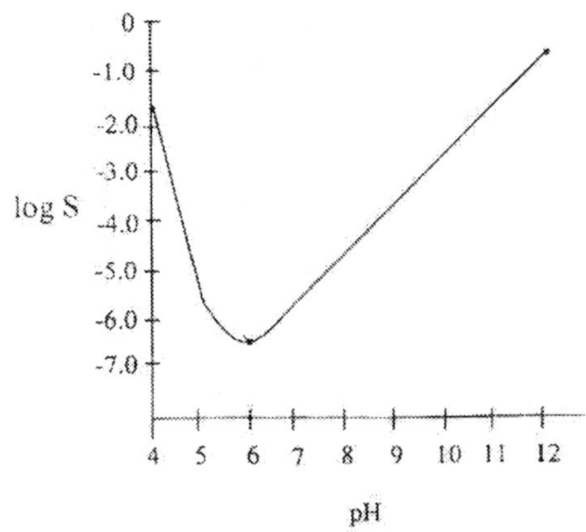

As expected, the solubility of $Al(OH)_3(s)$ is increased by very acidic solutions and by very basic solutions.

CHAPTER 17

SPONTANEITY, ENTROPY, AND FREE ENERGY

Questions

11. Living organisms need an external source of energy to carry out these processes. Green plants use the energy from sunlight to produce glucose from carbon dioxide and water by photosynthesis. In the human body, the energy released from the metabolism of glucose helps drive the synthesis of proteins. For all processes combined, ΔS_{univ} must be greater than zero (the second law).

12. Dispersion increases the entropy of the universe because the more widely something is dispersed, the greater the disorder. We must do work to overcome this disorder. In terms of the second law, it would be more advantageous to prevent contamination of the environment rather than to clean it up later. As a substance disperses, we have a much larger area that must be decontaminated.

13. As a process occurs, ΔS_{univ} will increase; ΔS_{univ} cannot decrease. Time, like ΔS_{univ}, only goes in one direction.

14. The introduction of mistakes is an effect of entropy. The purpose of redundant information is to provide a control to check the "correctness" of the transmitted information.

15. This reaction is kinetically slow but thermodynamically favorable ($\Delta G < 0$). Thermodynamics only tells us if a reaction can occur. To answer the question will it occur, one also needs to consider the kinetics (speed of reaction). The ultraviolet light provides the activation energy for this slow reaction to occur.

16. a. Thermodynamics is concerned with only the initial and final states of a reaction, and not the pathway required to get from reactants to products. It is the kinetics of a reaction that concentrates on the pathway and speed of a reaction. For these plots, the reaction with the smallest activation energy will be fastest. The activation energy is the energy reactants must have in order to overcome the energy barrier to convert to products. So the fastest reaction is reaction 5 with the smallest activation. Reactions 1, 2, and 4 all should have the same speed because they have the same activation energy. (This assumes all other kinetic factors are the same.) Reaction 3 will be the slowest since it has the largest activation energy.

    b. If the products have lower energy than the reactants, then the reaction is exothermic. Reaction 1, 2, 3, and 5 are exothermic. If the products have higher energy than the reactants, then the reaction is endothermic. Only reaction 4 is endothermic. Note that it is the thermodynamics of a reaction that dictates whether a reaction is exothermic or endothermic. The kinetics plays no factor in these designations.

c. The thermodynamics of reaction determine potential energy change. In an exothermic reaction, some of the potential energy stored in chemical bonds is converted to thermal energy (heat is released as the potential energy decreases). The exothermic reaction with the greatest loss of potential energy is the reaction with the largest energy difference (ΔE) between reactants and products. Reactions 2 and 5 both have the same largest ΔE values, so reactions 2 and 5 have the greatest change in potential energy. Reactions 1 and 3 both have the same smallest ΔE values, so reactions 1 and 3 have the smallest change in potential energy.

17. $\Delta S_{surr} = -\Delta H/T$; heat flow ($\Delta H$) into or out of the system dictates ΔS_{surr}. If heat flows into the surroundings, the random motions of the surroundings increase, and the entropy of the surroundings increases. The opposite is true when heat flows from the surroundings into the system (an endothermic reaction). Although the driving force described here really results from the change in entropy of the surroundings, it is often described in terms of energy. Nature tends to seek the lowest possible energy.

18. When solid NaCl dissolves, the Na^+ and Cl^- ions are randomly dispersed in water. The ions have access to a larger volume and a larger number of possible positions. Hence positional disorder has increased and ΔS is positive.

For a salt dissolving in water, ΔH is usually a value close to zero (see Section 11.2 of the text). Sometimes ΔH is positive and sometimes ΔH is negative for a salt dissolving in water, but the magnitude of ΔH is small. So the only prediction that can be made is that ΔH for this process will be close to zero.

19. Note that these substances are not in the solid state but are in the aqueous state; water molecules are also present. There is an apparent increase in ordering when these ions are placed in water as compared to the separated state. The hydrating water molecules must be in a highly ordered arrangement when surrounding these anions.

20. $\Delta G° = -RT \ln K = \Delta H° - T\Delta S°$; $HX(aq) \rightleftharpoons H^+(aq) + X^-(aq)$ K_a reaction; the value of K_a for HF is less than one, while the other hydrogen halide acids have $K_a > 1$. In terms of $\Delta G°$, HF must have a positive $\Delta G°_{rxn}$ value, while the other HX acids have $\Delta G°_{rxn} < 0$. The reason for the sign change in the K_a value, between HF versus HCl, HBr, and HI is entropy. ΔS for the dissociation of HF is very large and negative. There is a high degree of ordering that occurs as the water molecules associate (hydrogen bond) with the small F^- ions. The entropy of hydration strongly opposes HF dissociating in water, so much so that it overwhelms the favorable hydration energy making HF a weak acid.

21. One can determine $\Delta S°$ and $\Delta H°$ for the reaction using the standard entropies and standard enthalpies of formation in Appendix 4; then use the equation $\Delta G° = \Delta H° - T\Delta S°$. One can also use the standard free energies of formation in Appendix 4. And finally, one can use Hess's law to calculate $\Delta G°$. Here, reactions having known $\Delta G°$ values are manipulated to determine $\Delta G°$ for a different reaction.

For temperatures other than 25°C, $\Delta G°$ is estimated using the $\Delta G° = \Delta H° - T\Delta S°$ equation. The assumptions made are that the $\Delta H°$ and $\Delta S°$ values determined from Appendix 4 data are temperature-independent. We use the same $\Delta H°$ and $\Delta S°$ values as determined when T =

25°C; then we plug in the new temperature in Kelvin into the equation to estimate ΔG° at the new temperature.

22. The sign of ΔG tells us if a reaction is spontaneous or not at whatever concentrations are present (at constant T and P). The magnitude of ΔG equals w_{max}. When ΔG < 0, the magnitude tells us how much work, in theory, could be harnessed from the reaction. When ΔG > 0, the magnitude tells us the minimum amount of work that must be supplied to make the reaction occur. ΔG° gives us the same information only when the concentrations for all reactants and products are at standard conditions (1 atm for gases, 1 *M* for solute). These conditions rarely occur.

    ΔG° = –RT ln K; from this equation, one can calculate K for a reaction if ΔG° is known at that temperature. Therefore, ΔG° gives the equilibrium position for a reaction. To determine K at a temperature other than 25°C, one needs to know ΔG° at that temperature. We assume ΔH° and ΔS° are temperature-independent and use the equation ΔG° = ΔH° – TΔS° to estimate ΔG° at the different temperature. For K = 1, we want ΔG° = 0, which occurs when ΔH° = TΔS°. Again, assume ΔH° and ΔS° are temperature-independent; then solve for T (= ΔH°/ΔS°). At this temperature, K = 1 because ΔG° = 0. This only works for reactions where the signs of ΔH° and ΔS° are the same (either both positive or both negative). When the signs are opposite, K will always be greater than 1 (when ΔH° is negative and ΔS° is positive) or K will always be less than 1 (when ΔH° is positive and ΔS° is negative). When the signs of ΔH° and ΔS° are opposite, K can never equal 1.

23. The light source for the first reaction is necessary for kinetic reasons. The first reaction is just too slow to occur unless a light source is available. The kinetics of a reaction are independent of the thermodynamics of a reaction. Even though the first reaction is more favorable thermodynamically (assuming standard conditions), it is unfavorable for kinetic reasons. The second reaction has a negative ΔG° value and is a fast reaction, so the second reaction which occurs very quickly is favored both kinetically and thermodynamically. When considering if a reaction will occur, thermodynamics and kinetics must both be considered.

24. Using Le Chatelier's principle, a decrease in pressure (volume increases) will favor the side with the greater number of particles. Thus 2 I(g) will be favored at low pressure.

    Looking at ΔG: $\Delta G = \Delta G° + RT\ \ln(P_I^2/P_{I_2})$; $\ln(P_I^2/P_{I_2}) > 0$ when $P_I = P_{I_2} = 10$ atm and ΔG is positive (not spontaneous). But at $P_I = P_{I_2} = 0.10$ atm, the logarithm term is negative. If |RT ln Q| > ΔG°, then ΔG becomes negative, and the reaction is spontaneous.

Exercises

Spontaneity, Entropy, and the Second Law of Thermodynamics: Free Energy

25. a, b, and c; from our own experiences, salt water, colored water, and rust form without any outside intervention. It takes an outside energy source to clean a bedroom, so this process is not spontaneous.

26. c and d; it takes an outside energy source to build a house and to launch and keep a satellite in orbit, so these processes are not spontaneous.

27. Possible arrangements for one molecule:

1 way 1 way

Both are equally probable.

Possible arrangements for two molecules:

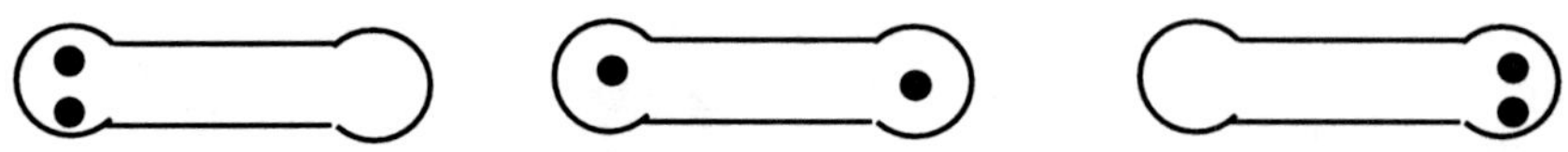

1 way 2 ways, most probable 1 way

Possible arrangement for three molecules:

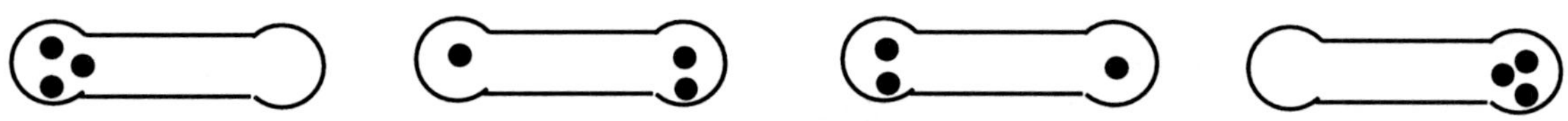

1 way 3 ways 3 ways 1 way

equally most probable

28. a. The most likely arrangement would be 3 molecules in each flask. There are 20 different ways (microstates) to achieve 3 gas molecules in each flask. Let the letters A-F represent the six molecules. The different ways to have groups of three are: ABC, ABD, ABE, ABF, ACD, ACE, ACF, ADE, ADF, AEF, BCD, BCE, BCF, BDE, BDF, BEF, CDE, CDF, CEF, and DEF.

b. With six molecules and two bulbs, there are $2^6 = 64$ possible arrangements. Because there are 20 ways to achieve 3 gas molecules in each bulb, the probability expressed as a percent of this arrangement is:

$$\frac{20}{64} \times 100 = 31.25\%$$

29. We draw all the possible arrangements of the two particles in the three levels.

2 kJ	___	___	x	___	x	xx
1 kJ	___	x	___	xx	x	___
0 kJ	xx	x	x	___	___	___
Total E =	0 kJ	1 kJ	2 kJ	2 kJ	3 kJ	4 kJ

The most likely total energy is 2 kJ.

30.

2 kJ	___	___	AB	___	___	B	A	B	A
1 kJ	___	AB	___	B	A	___	___	A	B
0 kJ	AB	___	___	A	B	A	B	___	___
E_{total} =	0 kJ	2 kJ	4 kJ	1 kJ	1 kJ	2 kJ	2 kJ	3 kJ	3 kJ

The most likely total energy is 2 kJ.

31. a. H_2 at 100°C and 0.5 atm; higher temperature and lower pressure means greater volume and hence larger positional probability.

b. N_2; N_2 at STP has the greater volume because P is smaller and T is larger.

c. $H_2O(l)$ has a larger positional probability than $H_2O(s)$.

32. Of the three phases (solid, liquid, and gas), solids are most ordered (have the smallest positional probability) and gases are most disordered (have the largest positional probability). Thus a , b, and f (melting, sublimation, and boiling) involve an increase in the entropy of the system since going from a solid to a liquid or from a solid to a gas or from a liquid to a gas increases disorder (increases positional probability). For freezing (process c), a substance goes from the more disordered liquid state to the more ordered solid state; hence, entropy decreases. Process d (mixing) involves an increase in disorder (an increase in positional probability), while separation (phase e) increases order (decreases positional probability). So, of all the processes, a, b, d, and f result in an increase in the entropy of the system.

33. a. Boiling a liquid requires heat. Hence this is an endothermic process. All endothermic processes decrease the entropy of the surroundings (ΔS_{surr} is negative).

b. This is an exothermic process. Heat is released when gas molecules slow down enough to form the solid. In exothermic processes, the entropy of the surroundings increases (ΔS_{surr} is positive).

34. a. $\Delta S_{surr} = \dfrac{-\Delta H}{T} = \dfrac{-(-2221\text{ kJ})}{298\text{ K}} = 7.45\text{ kJ/K} = 7.45 \times 10^3\text{ J/K}$

b. $\Delta S_{surr} = \dfrac{-\Delta H}{T} = \dfrac{-112\text{ kJ}}{298\text{ K}} = -0.376\text{ kJ/K} = -376\text{ J/K}$

35. $\Delta G = \Delta H - T\Delta S$; when ΔG is negative, then the process will be spontaneous.

a. $\Delta G = \Delta H - T\Delta S = 25 \times 10^3\text{ J} - (300.\text{ K})(5.0\text{ J/K}) = 24{,}000\text{ J}$; not spontaneous

b. $\Delta G = 25{,}000\text{ J} - (300.\text{ K})(100.\text{ J/K}) = -5000\text{ J}$; spontaneous

c. Without calculating ΔG, we know this reaction will be spontaneous at all temperatures. ΔH is negative and ΔS is positive ($-T\Delta S < 0$). ΔG will always be less than zero with these sign combinations for ΔH and ΔS.

d. $\Delta G = -1.0 \times 10^4\text{ J} - (200.\text{ K})(-40.\text{ J/K}) = -2000\text{ J}$; spontaneous

36. $\Delta G = \Delta H - T\Delta S$; a process is spontaneous when $\Delta G < 0$. For the following, assume ΔH and ΔS are temperature-independent.

a. When ΔH and ΔS are both negative, ΔG will be negative below a certain temperature where the favorable ΔH term dominates. When $\Delta G = 0$, then $\Delta H = T\Delta S$. Solving for this temperature:

$$T = \frac{\Delta H}{\Delta S} = \frac{-18{,}000\ \text{J}}{-60.\ \text{J/K}} = 3.0 \times 10^2\ \text{K}$$

At $T < 3.0 \times 10^2$ K, this process will be spontaneous ($\Delta G < 0$).

b. When ΔH and ΔS are both positive, ΔG will be negative above a certain temperature where the favorable ΔS term dominates.

$$T = \frac{\Delta H}{\Delta S} = \frac{18{,}000\ \text{J}}{60.\ \text{J/K}} = 3.0 \times 10^2\ \text{K}$$

At $T > 3.0 \times 10^2$ K, this process will be spontaneous ($\Delta G < 0$).

c. When ΔH is positive and ΔS is negative, this process can never be spontaneous at any temperature because ΔG can never be negative.

d. When ΔH is negative and ΔS is positive, this process is spontaneous at all temperatures because ΔG will always be negative.

37. At the boiling point, $\Delta G = 0$, so $\Delta H = T\Delta S$.

$$\Delta S = \frac{\Delta H}{T} = \frac{27.5\ \text{kJ/mol}}{(273+35)\ \text{K}} = 8.93 \times 10^{-2}\ \text{kJ/K•mol} = 89.3\ \text{J/K•mol}$$

38. At the boiling point, $\Delta G = 0$, so $\Delta H = T\Delta S$. $T = \dfrac{\Delta H}{\Delta S} = \dfrac{58.51 \times 10^3\ \text{J/mol}}{92.92\text{J/K•mol}} = 629.7\ \text{K}$

39. a. $NH_3(s) \rightarrow NH_3(l)$; $\Delta G = \Delta H - T\Delta S = 5650$ J/mol – 200. K (28.9 J/K•mol)

$\Delta G = 5650$ J/mol – 5780 J/mol = –130 J/mol

Yes, NH_3 will melt because $\Delta G < 0$ at this temperature.

b. At the melting point, $\Delta G = 0$, so $T = \dfrac{\Delta H}{\Delta S} = \dfrac{5650\ \text{J/mol}}{28.9\text{J/K•mol}} = 196$ K.

40. $C_2H_5OH(l) \rightarrow C_2H_5OH(g)$; at the boiling point, $\Delta G = 0$ and $\Delta S_{univ} = 0$. For the vaporization process, ΔS is a positive value, whereas ΔH is a negative value. To calculate ΔS_{sys}, we will determine ΔS_{surr} from ΔH and the temperature; then $\Delta S_{sys} = -\Delta S_{surr}$ for a system at equilibrium.

$$\Delta S_{surr} = \frac{-\Delta H}{T} = \frac{38.7 \times 10^3 \text{ J/mol}}{351\text{K}} = -110. \text{ J/K•mol}$$

$$\Delta S_{sys} = -\Delta S_{surr} = -(-110.) = 110. \text{ J/K•mol}$$

Chemical Reactions: Entropy Changes and Free Energy

41. a. Decrease in positional probability; ΔS° will be negative. There is only one way to achieve 12 gas molecules all in one bulb, but there are many more ways to achieve the gas molecules equally distributed in each flask.

b. Decrease in positional probability; ΔS° is negative for the liquid to solid phase change.

c. Decrease in positional probability; ΔS° is negative because the moles of gas decreased when going from reactants to products (3 moles → 0 moles). Changes in the moles of gas present as reactants are converted to products dictates predicting positional probability. The gaseous phase always has the larger positional probability associated with it.

d. Increase in positional probability; ΔS° is positive for the liquid to gas phase change. The gas phase always has the larger positional probability.

42. a. Decrease in positional probability ($\Delta n < 0$); ΔS°(−)

b. Decrease in positional probability ($\Delta n < 0$); ΔS°(−)

c. Increase in positional probability; ΔS°(+)

d. Increase in positional probability; ΔS°(+)

43. a. $C_{graphite}(s)$; diamond has a more ordered structure (has a smaller positional probability) than graphite.

b. $C_2H_5OH(g)$; the gaseous state is more disordered (has a larger positional probability) than the liquid state.

c. $CO_2(g)$; the gaseous state is more disordered (has a larger positional probability) than the solid state.

44. a. He (10 K); S = 0 at 0 K

b. N_2O; more complicated molecule, so has the larger positional probability.

c. $NH_3(l)$; the liquid state is more disordered (has a larger positional probability) than the solid state.

45. a. $2\ H_2S(g) + SO_2(g) \rightarrow 3\ S_{rhombic}(s) + 2\ H_2O(g)$; because there are more molecules of reactant gases than product molecules of gas ($\Delta n = 2 - 3 < 0$), ΔS° will be negative.

$\Delta S° = \sum n_p S°_{products} - \sum n_r S°_{reactants}$

$\Delta S° = [3 \text{ mol } S_{rhombic}(s)\ (32\ J/K{\bullet}mol) + 2 \text{ mol } H_2O(g)\ (189\ J/K{\bullet}mol)]$

$- [2 \text{ mol } H_2S(g)\ (206\ J/K{\bullet}mol) + 1 \text{ mol } SO_2(g)\ (248\ J/K{\bullet}mol)]$

$\Delta S° = 474\ J/K - 660.\ J/K = -186\ J/K$

b. $2\ SO_3(g) \rightarrow 2\ SO_2(g) + O_2(g)$; because Δn of gases is positive ($\Delta n = 3 - 2$), $\Delta S°$ will be positive.

$\Delta S = 2\ mol(248\ J/K{\bullet}mol) + 1\ mol(205\ J/K{\bullet}mol) - [2\ mol(257\ J/K{\bullet}mol)] = 187\ J/K$

c. $Fe_2O_3(s) + 3\ H_2(g) \rightarrow 2\ Fe(s) + 3\ H_2O(g)$; because Δn of gases $= 0$ ($\Delta n = 3 - 3$), we can't easily predict if $\Delta S°$ will be positive or negative.

$\Delta S = 2\ mol(27\ J/K{\bullet}mol) + 3\ mol(189\ J/K{\bullet}mol) -$

$[1\ mol(90.\ J/K{\bullet}mol) + 3\ mol\ (141\ J/K{\bullet}mol)] = 138\ J/K$

46. a. $H_2(g) + 1/2\ O_2(g) \rightarrow H_2O(l)$; since Δn of gases is negative, $\Delta S°$ will be negative.

$\Delta S° = 1 \text{ mol } H_2O(l)\ (70.\ J/K{\bullet}mol) -$

$[1 \text{ mol } H_2(g)\ (131\ J/K{\bullet}mol) + 1/2 \text{ mol } O_2(g)\ (205\ J/K{\bullet}mol)]$

$\Delta S° = 70.\ J/K - 234\ J/K = -164\ J/K$

b. $2\ CH_3OH(g) + 3\ O_2(g) \rightarrow 2\ CO_2(g) + 4\ H_2O(g)$; because Δn of gases is positive, $\Delta S°$ will be positive.

$[2\ mol\ (214\ J/K{\bullet}mol) + 4\ mol\ (189\ J/K{\bullet}mol)] -$

$[2\ mol\ (240.\ J/K{\bullet}mol) + 3\ mol\ (205\ J/K{\bullet}mol)] = 89\ J/K$

c. $HCl(g) \rightarrow H^+(aq) + Cl^-(aq)$; the gaseous state dominates predictions of $\Delta S°$. Here, the gaseous state is more disordered than the ions in solution, so $\Delta S°$ will be negative.

$\Delta S° = 1 \text{ mol } H^+(0) + 1 \text{ mol } Cl^-(57\ J/K{\bullet}mol) - 1 \text{ mol } HCl(187\ J/K{\bullet}mol) = -130.\ J/K$

47. $C_2H_2(g) + 4\ F_2(g) \rightarrow 2\ CF_4(g) + H_2(g)$; $\Delta S° = 2S°_{CF_4} + S°_{H_2} - [S°_{C_2H_2} + 4S°_{F_2}]$

$-358\ J/K = (2\ mol) S°_{CF_4} + 131\ J/K - [201\ J/K + 4(203\ J/K)]$, $S°_{CF_4} = 262\ J/K{\bullet}mol$

48. $CS_2(g) + 3\ O_2(g) \rightarrow CO_2(g) + 2\ SO_2(g)$; $\Delta S° = S°_{CO_2} + 2S°_{SO_2} - [3S°_{O_2} + S°_{CS_2}]$

$-143\ J/K = 214\ J/K + 2(248\ J/K) - 3(205\ J/K) - (1\ mol) S°_{CS_2}$, $S°_{CS_2} = 238\ J/K{\bullet}mol$

49. a. $S_{rhombic} \rightarrow S_{monoclinic}$; this phase transition is spontaneous ($\Delta G < 0$) at temperatures above 95°C. $\Delta G = \Delta H - T\Delta S$; for ΔG to be negative only above a certain temperature, then ΔH is positive and ΔS is positive (see Table 17.5 of text).

b. Because ΔS is positive, $S_{rhombic}$ is the more ordered crystalline structure (has the smaller positional probability).

50. $P_4(s,\alpha) \rightarrow P_4(s,\beta)$

a. At T < –76.9°C, this reaction is spontaneous, and the sign of ΔG is (–). At –76.9°C, ΔG = 0, and above –76.9 °C, the sign of ΔG is (+). This is consistent with ΔH (–) and ΔS (–).

b. Because the sign of ΔS is negative, the β form has the more ordered structure (has the smaller positional probability).

51. a. When a bond is formed, energy is released, so ΔH is negative. There are more reactant molecules of gas than product molecules of gas ($\Delta n < 0$), so ΔS will be negative.

b. $\Delta G = \Delta H - T\Delta S$; for this reaction to be spontaneous ($\Delta G < 0$), the favorable enthalpy term must dominate. The reaction will be spontaneous at low temperatures (at a temperature below some number), where the ΔH term dominates.

52. Because there are more product gas molecules than reactant gas molecules ($\Delta n > 0$), ΔS will be positive. From the signs of ΔH and ΔS, this reaction is spontaneous at all temperatures. It will cost money to heat the reaction mixture. Because there is no thermodynamic reason to do this, the purpose of the elevated temperature must be to increase the rate of the reaction, that is, kinetic reasons.

53. a.

	$CH_4(g)$	+ $2\ O_2(g)$	→ $CO_2(g)$	+ $2\ H_2O(g)$	
ΔH_f°	–75 kJ/mol	0	–393.5	–242	
ΔG_f°	–51 kJ/mol	0	–394	–229	Data from Appendix 4
S°	186 J/K•mol	205	214	189	

$\Delta H^\circ = \sum n_p \Delta H^\circ_{f,\ products} - \sum n_r \Delta H^\circ_{f,\ reactants}$; $\Delta S^\circ = \sum n_p S^\circ_{products} - \sum n_r S^\circ_{reactants}$

ΔH° = 2 mol(–242 kJ/mol) + 1 mol(–393.5 kJ/mol) – [1 mol(–75 kJ/mol)] = –803 kJ

ΔS° = 2 mol(189 J/K•mol) + 1 mol(214 J/K•mol)
– [1 mol(186 J/K•mol) + 2 mol(205 J/K•mol)] = –4 J/K

There are two ways to get ΔG°. We can use $\Delta G^\circ = \Delta H^\circ - T\Delta S^\circ$ (be careful of units):

$$\Delta G^\circ = \Delta H^\circ - T\Delta S^\circ = -803 \times 10^3 \text{ J} - (298 \text{ K})(-4 \text{ J/K}) = -8.018 \times 10^5 \text{ J} = -802 \text{ kJ}$$

or we can use ΔG_f° values, where $\Delta G^\circ = \sum n_p \Delta G^\circ_{f,\ products} - \sum n_r \Delta G^\circ_{f,\ reactants}$:

$\Delta G° = 2\text{ mol}(-229\text{ kJ/mol}) + 1\text{ mol}(-394\text{ kJ/mol}) - [1\text{ mol}(-51\text{ kJ/mol})]$

$\Delta G° = -801$ kJ (Answers are the same within round off error.)

b. $6\ CO_2(g) + 6\ H_2O(l) \rightarrow C_6H_{12}O_6(s) + 6\ O_2(g)$

	$6\ CO_2(g)$	$6\ H_2O(l)$	$C_6H_{12}O_6(s)$	$6\ O_2(g)$
ΔH_f°	−393.5 kJ/mol	−286	−1275	0
S°	214 J/K•mol	70.	212	205

$\Delta H° = -1275 - [6{-286}) + 6(-393.5)] = 2802$ kJ

$\Delta S° = 6(205) + 212 - [6(214) + 6(70.)] = -262$ J/K

$\Delta G° = 2802\text{ kJ} - (298\text{ K})(-0.262\text{ kJ/K}) = 2880.$ kJ

c. $P_4O_{10}(s) + 6\ H_2O(l) \rightarrow 4\ H_3PO_4(s)$

	$P_4O_{10}(s)$	$6\ H_2O(l)$	$4\ H_3PO_4(s)$
ΔH_f° (kJ/mol)	−2984	−286	−1279
S° (J/K•mol)	229	70.	110.

$\Delta H° = 4\text{ mol}(-1279\text{ kJ/mol}) - [1\text{ mol}(-2984\text{ kJ/mol}) + 6\text{ mol}(-286\text{ kJ/mol})] = -416$ kJ

$\Delta S° = 4(110.) - [229 + 6(70.)] = -209$ J/K

$\Delta G° = \Delta H° - T\Delta S° = -416\text{ kJ} - (298\text{ K})(-0.209\text{ kJ/K}) = -354$ kJ

d. $HCl(g) + NH_3(g) \rightarrow NH_4Cl(s)$

	$HCl(g)$	$NH_3(g)$	$NH_4Cl(s)$
ΔH_f° (kJ/mol)	−92	−46	−314
S° (J/K•mol)	187	193	96

$\Delta H° = -314 - [-92 - 46] = -176$ kJ; $\Delta S° = 96 - [187 + 193] = -284$ J/K

$\Delta G° = \Delta H° - T\Delta S° = -176\text{ kJ} - (298\text{ K})(-0.284\text{ kJ/K}) = -91$ kJ

54. a. $\Delta H° = 2(-46\text{ kJ}) = -92$ kJ; $\Delta S° = 2(193\text{ J/K}) - [3(131\text{ J/K}) + 192\text{ J/K}] = -199$ J/K

$\Delta G° = \Delta H° - T\Delta S° = -92\text{ kJ} - 298\text{ K}(-0.199\text{ kJ/K}) = -33$ kJ

b. $\Delta G°$ is negative, so this reaction is spontaneous at standard conditions.

c. $\Delta G° = 0$ when $T = \frac{\Delta H°}{\Delta S°} = \frac{-92\,kJ}{-0.199\,kJ/K} = 460\ K$

At T < 460 K and standard pressures (1 atm), the favorable ΔH° term dominates, and the reaction is spontaneous (ΔG° < 0).

55. $\Delta G° = -58.03\ kJ - (298\ K)(-0.1766\ kJ/K) = -5.40\ kJ$

$\Delta G° = 0 = \Delta H° - T\Delta S°,\ T = \frac{\Delta H°}{\Delta S°} = \frac{-58.03\ kJ}{-0.1766\,kJ/K} = 328.6\ K$

ΔG° is negative below 328.6 K, where the favorable ΔH° term dominates.

56. $H_2O(l) \rightarrow H_2O(g)$; ΔG° = 0 at the boiling point of water at 1 atm and 100. °C.

$\Delta H° = T\Delta S°,\ \Delta S° = \frac{\Delta H°}{T} = \frac{40.6 \times 10^3\ J/mol}{373\,K} = 109\ J/K{\bullet}mol$

At 90.°C: $\Delta G° = \Delta H° - T\Delta S° = 40.6\ kJ/mol - (363\ K)(0.109\ kJ/K{\bullet}mol) = 1.0\ kJ/mol$

As expected, ΔG° > 0 at temperatures below the boiling point of water at 1 atm (process is nonspontaneous).

At 110.°C: $\Delta G° = \Delta H° - T\Delta S° = 40.6\ kJ/mol - (383\ K)(0.109\ J/K{\bullet}mol) = -1.1\ kJ/mol$

When ΔG° < 0, the boiling of water is spontaneous at 1 atm, and T > 100. °C (as expected).

57.

$CH_4(g) \rightarrow 2\,H_2(g) + C(s)$	$\Delta G° = -(-51\ kJ)$
$2\,H_2(g) + O_2(g) \rightarrow 2\,H_2O(l)$	$\Delta G° = -474\ kJ)$
$C(s) + O_2(g) \rightarrow CO_2(g)$	$\Delta G° = -394\ kJ$
$CH_4(g) + 2\,O_2(g) \rightarrow 2\,H_2O(l) + CO_2(g)$	$\Delta G° = -817\ kJ$

58.

$6\,C(s) + 6\,O_2(g) \rightarrow 6\,CO_2(g)$	$\Delta G° = 6(-394\ kJ)$
$3\,H_2(g) + 3/2\,O_2(g) \rightarrow 3\,H_2O(l)$	$\Delta G° = 3(-237\ kJ)$
$6\,CO_2(g) + 3\,H_2O(l) \rightarrow C_6H_6(l) + 15/2\,O_2(g)$	$\Delta G° = -1/2\,(-6399\ kJ)$
$6\,C(s) + 3\,H_2(g) \rightarrow C_6H_6(l)$	$\Delta G° = 125\ kJ$

59. $\Delta G° = \sum n_p \Delta G°_{f,\,products} - \sum n_r \Delta G°_{f,\,reactants}$, $-374\ kJ = -1105\ kJ - \Delta G°_{f,\,SF_4}$

$\Delta G°_{f,\,SF_4} = -731\ kJ/mol$

60. $-5490.\ kJ = 8(-394\ kJ) + 10(-237\ kJ) - 2\Delta G°_{f,\,C_4H_{10}}$, $\Delta G°_{f,\,C_4H_{10}} = -16\ kJ/mol$

61. a. $\Delta G° = 2\ mol(0) + 3\ mol(-229\ kJ/mol) - [1\ mol(-740.\ kJ/mol) + 3\ mol(0)] = 53\ kJ$

b. Because $\Delta G°$ is positive, this reaction is not spontaneous at standard conditions and 298 K.

c. $\Delta G° = \Delta H° - T\Delta S°,\ \Delta S° = \dfrac{\Delta H° - \Delta G°}{T} = \dfrac{100.\ \text{kJ} - 53\ \text{kJ}}{298\ \text{K}} = 0.16\ \text{kJ/K}$

We need to solve for the temperature when $\Delta G° = 0$:

$$\Delta G° = 0 = \Delta H° - T\Delta S°,\ \Delta H° = T\Delta S°,\ T = \frac{\Delta H°}{\Delta S°} = \frac{100.\ \text{kJ}}{0.16\ \text{kJ/K}} = 630\ \text{K}$$

This reaction will be spontaneous ($\Delta G < 0$) at $T > 630$ K, where the favorable entropy term will dominate.

62. a. $\Delta G° = 2(-270.\ \text{kJ}) - 2(-502\ \text{kJ}) = 464\ \text{kJ}$

b. Because $\Delta G°$ is positive, this reaction is not spontaneous at standard conditions at 298 K.

c. $\Delta G° = \Delta H° - T\Delta S°,\ \Delta H° = \Delta G° + T\Delta S° = 464\ \text{kJ} + 298\ \text{K}(0.179\ \text{kJ/K}) = 517\ \text{kJ}$

We need to solve for the temperature when $\Delta G° = 0$:

$$\Delta G° = 0 = \Delta H° - T\Delta S°,\ T = \frac{\Delta H°}{\Delta S°} = \frac{517\ \text{kJ}}{0.179\ \text{kJ/K}} = 2890\ \text{K}$$

This reaction will be spontaneous at standard conditions ($\Delta G° < 0$) when $T > 2890$ K. Here the favorable entropy term will dominate.

63. $CH_4(g) + CO_2(g) \rightarrow CH_3CO_2H(l)$

$\Delta H° = -484 - [-75 + (-393.5)] = -16\ \text{kJ};\ \ \Delta S° = 160. - (186 + 214) = -240.\ \text{J/K}$

$\Delta G° = \Delta H° - T\Delta S° = -16\ \text{kJ} - (298\ \text{K})(-0.240\ \text{kJ/K}) = 56\ \text{kJ}$

At standard concentrations, this reaction is spontaneous only at temperatures below $T = \Delta H°/\Delta S° = 67$ K (where the favorable $\Delta H°$ term will dominate, giving a negative $\Delta G°$ value). This is not practical. Substances will be in condensed phases and rates will be very slow at this extremely low temperature.

$CH_3OH(g) + CO(g) \rightarrow CH_3CO_2H(l)$

$\Delta H° = -484 - [-110.5 + (-201)] = -173\ \text{kJ};\ \ \Delta S° = 160. - (198 + 240.) = -278\ \text{J/K}$

$\Delta G° = -173\ \text{kJ} - (298\ \text{K})(-0.278\ \text{kJ/K}) = -90.\ \text{kJ}$

This reaction also has a favorable enthalpy and an unfavorable entropy term. But this reaction, at standard concentrations, is spontaneous at temperatures below $T = \Delta H°/\Delta S° = 622$ K (a much higher temperature than the first reaction). So the reaction of CH_3OH and CO will be preferred at standard concentrations. It is spontaneous at high enough temperatures that the rates of reaction should be reasonable.

64. $C_2H_4(g) + H_2O(g) \rightarrow CH_3CH_2OH(l)$

$\Delta H° = -278 - (52 - 242) = -88$ kJ; $\Delta S° = 161 - (219 + 189) = -247$ J/K

When $\Delta G° = 0$, $\Delta H° = T\Delta S°$, so $T = \dfrac{\Delta H°}{\Delta S°} = \dfrac{-88 \times 10^3 \text{ J}}{-247 \text{ J/K}} = 360$ K.

Since the signs of ΔH° and ΔS° are both negative, this reaction at standard concentrations will be spontaneous at temperatures below 360 K (where the favorable ΔH° term will dominate).

$C_2H_6(g) + H_2O(g) \rightarrow CH_3CH_2OH(l) + H_2(g)$

$\Delta H° = -278 - (-84.7 - 242) = 49$ kJ; $\Delta S° = 131 + 161 - (229.5 + 189) = -127$ J/K

Both ΔH° and ΔS° have unfavorable signs. ΔG° can never be negative when ΔH° is positive and ΔS° is negative. So this reaction can never be spontaneous at standard conditions.

Thus the reaction $C_2H_4(g) + H_2O(g) \rightarrow C_2H_5OH(l)$ would be preferred at standard conditions.

Free Energy: Pressure Dependence and Equilibrium

65. $\Delta G = \Delta G° + RT \ln Q$; for this reaction: $\Delta G = \Delta G° + RT \ln \dfrac{P_{NO_2} \times P_{O_2}}{P_{NO} \times P_{O_3}}$

$\Delta G° = 1 \text{ mol}(52 \text{ kJ/mol}) + 1 \text{ mol}(0) - [1 \text{ mol}(87 \text{ kJ/mol}) + 1 \text{ mol}(163 \text{ kJ/mol})] = -198$ kJ

$$\Delta G = -198 \text{ kJ} + \frac{8.3145 \text{ J/K} \bullet \text{mol}}{1000 \text{ J/kJ}}(298 \text{K}) \ln \frac{(1.00 \times 10^{-7})(1.00 \times 10^{-3})}{(1.00 \times 10^{-6})(1.00 \times 10^{-6})}$$

$\Delta G = -198 \text{ kJ} + 9.69 \text{ kJ} = -188$ kJ

66. $\Delta G° = 3(0) + 2(-229) - [2(-34) + 1(-300.)] = -90.$ kJ

$$\Delta G = \Delta G° + RT \ln \frac{P^2_{H_2O}}{P^2_{H_2S} \times P_{SO_2}} = -90. \text{ kJ} + \frac{(8.3145)(298)}{1000} \text{ kJ} \left[\ln \frac{(0.030)^2}{(1.0 \times 10^{-4})(0.010)} \right]$$

$\Delta G = -90. \text{ kJ} + 39.7 \text{ kJ} = -50.$ kJ

67. $\Delta G = \Delta G° + RT \ln Q = \Delta G° + RT \ln \dfrac{P_{N_2O_4}}{P^2_{NO_2}}$

$\Delta G° = 1 \text{ mol}(98 \text{ kJ/mol}) - 2 \text{ mol}(52 \text{ kJ/mol}) = -6$ kJ

a. These are standard conditions, so ΔG = ΔG° because Q = 1 and ln Q = 0. Because ΔG° is negative, the forward reaction is spontaneous. The reaction shifts right to reach equilibrium.

b. $\Delta G = -6 \times 10^3 \text{ J} + 8.3145 \text{ J/K}\bullet\text{mol (298 K)} \ln\dfrac{0.50}{(0.21)^2}$

$\Delta G = -6 \times 10^3 \text{ J} + 6.0 \times 10^3 \text{ J} = 0$

Because $\Delta G = 0$, this reaction is at equilibrium (no shift).

c. $\Delta G = -6 \times 10^3 \text{ J} + 8.3145 \text{ J/K}\bullet\text{mol (298 K)} \ln\dfrac{1.6}{(0.29)^2}$

$\Delta G = -6 \times 10^3 \text{ J} + 7.3 \times 10^3 \text{ J} = 1.3 \times 10^3 \text{ J} = 1 \times 10^3 \text{ J}$

Because ΔG is positive, the reverse reaction is spontaneous, and the reaction shifts to the left to reach equilibrium.

68. a. $\Delta G = \Delta G° + RT \ln\dfrac{P_{NH_3}^2}{P_{N_2} \times P_{H_2}^2}$; $\Delta G° = 2\Delta G°_{f,\,NH_3} = 2(-17) = -34 \text{ kJ}$

$\Delta G = -34 \text{ kJ} + \dfrac{(8.3145 \text{J/K}\bullet\text{mol})(298\text{K})}{1000 \text{J/kJ}} \ln\dfrac{(50.)^2}{(200.)(200.)^3}$

$\Delta G = -34 \text{ kJ} - 33 \text{ kJ} = -67 \text{ kJ}$

b. $\Delta G = -34 \text{ kJ} + \dfrac{(8.3145 \text{J/K}\bullet\text{mol})(298\text{K})}{1000 \text{J/kJ}} \ln\dfrac{(200.)^2}{(200.)(600.)^3}$

$\Delta G = -34 \text{ kJ} - 34.4 \text{ kJ} = -68 \text{ kJ}$

69. $NO(g) + O_3(g) \rightleftharpoons NO_2(g) + O_2(g)$; $\Delta G° = \Sigma n_p \Delta G°_{f,\,\text{products}} - \Sigma n_r \Delta G°_{f,\,\text{reactants}}$

$\Delta G° = 1 \text{ mol}(52 \text{ kJ/mol}) - [1 \text{ mol}(87 \text{ kJ/mol}) + 1 \text{ mol}(163 \text{ kJ/mol})] = -198 \text{ kJ}$

$\Delta G° = -RT \ln K,\ K = \exp\dfrac{-\Delta G°}{RT} = \exp\left[\dfrac{-(-1.98 \times 10^5 \text{ J})}{8.3145 \text{J/K}\bullet\text{mol}(298\text{K})}\right] = e^{79.912} = 5.07 \times 10^{34}$

Note: When determining exponents, we will round off after the calculation is complete. This helps eliminate excessive round off error.

70. $\Delta G° = 2 \text{ mol}(-229 \text{ kJ/mol}) - [2 \text{ mol}(-34 \text{ kJ/mol}) + 1 \text{ mol}(-300. \text{ kJ/mol})] = -90. \text{ kJ}$

$K = \exp\dfrac{-\Delta G°}{RT} = \exp\left[\dfrac{-(-9.0 \times 10^4 \text{ J})}{8.3145 \text{J/K}\bullet\text{mol}(298\text{K})}\right] = e^{36.32} = 5.9 \times 10^{15}$

$\Delta G° = \Delta H° - T\Delta S°$; because there is a decrease in the number of moles of gaseous particles, $\Delta S°$ is negative. Because $\Delta G°$ is negative, $\Delta H°$ must be negative. The reaction will be spontaneous at low temperatures (the favorable $\Delta H°$ term dominates at low temperatures).

71. At 25.0°C: $\Delta G° = \Delta H° - T\Delta S° = -58.03 \times 10^3$ J/mol − (298.2 K)(−176.6 J/K•mol)

$= -5.37 \times 10^3$ J/mol

$$\Delta G° = -RT \ln K, \ \ln K = \frac{-\Delta G°}{RT} = \exp\left[\frac{-(-5.37 \times 10^3 \text{ J/mol})}{(8.3145\text{J/K} \bullet \text{mol})(298.2\text{K})}\right] = 2.166$$

$K = e^{2.166} = 8.72$

At 100.0°C: $\Delta G° = -58.03 \times 10^3$ J/mol − (373.2 K)(−176.6 J/K•mol) = 7.88×10^3 J/mol

$$\ln K = \frac{-(7.88 \times 10^3 \text{ J/mol})}{(8.3145\text{J/K} \bullet \text{mol})(373.2\text{K})} = -2.540, \ K = e^{-2.540} = 0.0789$$

Note: When determining exponents, we will round off after the calculation is complete. This helps eliminate excessive round off error.

72. a. $\Delta G° = 3(191.2) - 78.2 = 495.4$ kJ; $\Delta H° = 3(241.3) - 132.8 = 591.1$ kJ

$$\Delta S° = \frac{\Delta H° - \Delta G°}{T} = \frac{591.1\text{ kJ} - 495.4\text{ kJ}}{298\text{K}} = 0.321 \text{ kJ/K} = 321 \text{ J/K}$$

b. $$\Delta G° = -RT \ln K, \ \ln K = \frac{-\Delta G°}{RT} = \frac{-495{,}400\text{J}}{8.3145\text{J/K} \bullet \text{mol}(298\text{K})} = -199.942$$

$K = e^{-199.942} = 1.47 \times 10^{-87}$

c. Assuming $\Delta H°$ and $\Delta S°$ are temperature-independent:

$\Delta G°_{3000} = 591.1$ kJ − 3000. K(0.321 kJ/K) = −372 kJ

$$\ln K = \frac{-(-372{,}000\text{J})}{8.3145\text{J/K} \bullet \text{mol}(3000\text{K})} = 14.914, \ K = e^{14.914} = 3.00 \times 10^6$$

73. When reactions are added together, the equilibrium constants are multiplied together to determine the K value for the final reaction.

$H_2(g) + O_2(g) \rightleftharpoons H_2O_2(g)$	$K = 2.3 \times 10^6$
$H_2O(g) \rightleftharpoons H_2(g) + 1/2O_2(g)$	$K = (1.8 \times 10^{37})^{-1/2}$
$H_2O(g) + 1/2O_2(g) \rightleftharpoons H_2O_2(g)$	$K = 2.3 \times 10^6(1.8 \times 10^{37})^{-1/2} = 5.4 \times 10^{-13}$

$$\Delta G° = -RT \ln K = \frac{-8.3145\text{J}}{\text{K} \bullet \text{mol}}(600.\text{ K}) \ln(5.4 \times 10^{-13}) = 1.4 \times 10^5 \text{ J/mol} = 140 \text{ kJ/mol}$$

74. a.

	ΔH_f° (kJ/mol)	S° (J/K•mol)
$NH_3(g)$	−46	193
$O_2(g)$	0	205
$NO(g)$	90.	211
$H_2O(g)$	−242	189
$NO_2(g)$	34	240.
$HNO_3(l)$	−174	156
$H_2O(l)$	−286	70.

$4\ NH_3(g) + 5\ O_2(g) \rightarrow 4\ NO(g) + 6\ H_2O(g)$

$\Delta H^\circ = 6(-242) + 4(90.) - [4(-46)] = -908$ kJ

$\Delta S^\circ = 4(211) + 6(189) - [4(193) + 5(205)] = 181$ J/K

$\Delta G^\circ = -908$ kJ $-$ (298 K)(0.181 kJ/K) $= -962$ kJ

$$\Delta G^\circ = -RT \ln K,\ \ln K = \frac{-\Delta G^\circ}{RT} = \left[\frac{-(-962 \times 10^3\ J)}{8.3145 J/K \bullet mol \times 298 K}\right] = 388$$

$\ln K = 2.303 \log K$, $\log K = 168$, $K = 10^{168}$ (an extremely large number)

$2\ NO(g) + O_2(g) \rightarrow 2\ NO_2(g)$

$\Delta H^\circ = 2(34) - [2(90.)] = -112$ kJ; $\Delta S^\circ = 2(240.) - [2(211) + (205)] = -147$ J/K

$\Delta G^\circ = -112$ kJ $-$ (298 K)(-0.147 kJ/K) $= -68$ kJ

$$K = \exp\frac{-\Delta G^\circ}{RT} = \exp\left[\frac{-(-68{,}000 J)}{8.3145 J/K \bullet mol(298 K)}\right] = e^{27.44} = 8.3 \times 10^{11}$$

Note: When determining exponents, we will round off after the calculation is complete.

$3\ NO_2(g) + H_2O(l) \rightarrow 2\ HNO_3(l) + NO(g)$

$\Delta H^\circ = 2(-174) + (90.) - [3(34) + (-286)] = -74$ kJ

$\Delta S^\circ = 2(156) + (211) - [3(240.) + (70.)] = -267$ J/K

$\Delta G^\circ = -74$ kJ $-$ (298 K)(-0.267 kJ/K) $= 6$ kJ

$$K = \exp\frac{-\Delta G^\circ}{RT} = \exp\left[\frac{-6000 J}{8.3145 J/K \bullet mol(298 K)}\right] = e^{-2.4} = 9 \times 10^{-2}$$

b. $\Delta G^\circ = -RT \ln K$; T = 825°C = (825 + 273) K = 1098 K; we must determine ΔG° at 1098 K.

$\Delta G^\circ_{1098} = \Delta H^\circ - T\Delta S^\circ = -908$ kJ $-$ (1098 K)(0.181 kJ/K) $= -1107$ kJ

$$K = \exp\frac{-\Delta G^\circ}{RT} = \exp\left[\frac{-(-1.107 \times 10^6\ J)}{8.3145 J/K \bullet mol\ (1098K)}\right] = e^{121.258} = 4.589 \times 10^{52}$$

c. There is no thermodynamic reason for the elevated temperature because ΔH° is negative and ΔS° is positive. Thus the purpose for the high temperature must be to increase the rate of the reaction.

75. a. $\Delta G^\circ = -RT \ln K,\ K = \exp\left[\frac{-(-30{,}500 J)}{8.3145 J/K \bullet mol \times 298K}\right] = 2.22 \times 10^5$

b. $C_6H_{12}O_6(s) + 6\ O_2(g) \rightarrow 6\ CO_2(g) + 6\ H_2O(l)$

$\Delta G^\circ = 6\ mol(-394\ kJ/mol) + 6\ mol(-237\ kJ/mol) - 1\ mol(-911\ kJ/mol) = -2875\ kJ$

$\frac{2875 kJ}{mol\, glucose} \times \frac{1\, mol\, ATP}{30.5 kJ} = 94.3$ mol ATP; 94.3 molecules ATP/molecule glucose

This is an overstatement. The assumption that all the free energy goes into this reaction is false. Actually, only 38 moles of ATP are produced by metabolism of 1 mole of glucose.

76. a. $\ln K = \frac{-\Delta G^\circ}{RT} = \frac{-14{,}000 J}{(8.3145 J/K \bullet mol)(298K)} = -5.65,\quad K = e^{-5.65} = 3.5 \times 10^{-3}$

b.

Glutamic acid + $NH_3 \rightarrow$ Glutamine + H_2O	$\Delta G^\circ = 14$ kJ
ATP + $H_2O \rightarrow$ ADP + $H_2PO_4^-$	$\Delta G^\circ = -30.5$ kJ
Glutamic acid + ATP + $NH_3 \rightarrow$ Glutamine + ADP + $H_2PO_4^-$	$\Delta G^\circ = 14 - 30.5 = -17$ kJ

$\ln K = \frac{-\Delta G^\circ}{RT} = \frac{-(-17{,}000 J)}{8.3145 J/K \bullet mol(298K)} = 6.86,\quad K = e^{6.86} = 9.5 \times 10^2$

77. $K = \frac{P_{NF_3}^2}{P_{N_2} \times P_{F_2}^3} = \frac{(0.48)^2}{0.021(0.063)^3} = 4.4 \times 10^4$

$\Delta G^\circ_{800} = -RT \ln K = -8.3145$ J/K•mol (800. K) ln $(4.4 \times 10^4) = -7.1 \times 10^4$ J/mol = −71 kJ/mol

78. $2\ SO_2(g) + O_2(g) \rightarrow 2\ SO_3(g)$; $\Delta G^\circ = 2(-371\ kJ) - [2(-300.\ kJ)] = -142\ kJ$

$\Delta G^\circ = -RT \ln K,\ \ln K = \frac{-\Delta G^\circ}{RT} = \frac{-(-142.000 J)}{8.3145 J/K \bullet mol\ (298K)} = 57.311$

$K = e^{57.311} = 7.76 \times 10^{24}$

$K = 7.76 \times 10^{24} = \frac{P_{SO_3}^2}{P_{SO_2}^2 \times P_{O_2}} = \frac{(2.0)^2}{P_{SO_2}^2 \times (0.50)},\quad P_{SO_2} = 1.0 \times 10^{-12}$ atm

From the negative value of $\Delta G°$, this reaction is spontaneous at standard conditions. There are more molecules of reactant gases than product gases, so $\Delta S°$ will be negative (unfavorable). Therefore, this reaction must be exothermic ($\Delta H° < 0$). When $\Delta H°$ and $\Delta S°$ are both negative, the reaction will be spontaneous at relatively low temperatures where the favorable $\Delta H°$ term dominates.

79. The equation $\ln K = \frac{-\Delta H°}{R}\left(\frac{1}{T}\right) + \frac{\Delta S°}{R}$ is in the form of a straight line equation ($y = mx + b$). A graph of ln K versus 1/T will yield a straight line with slope = $m = -\Delta H°/R$ and a y intercept = $b = \Delta S°/R$.

From the plot:

$$\text{slope} = \frac{\Delta y}{\Delta x} = \frac{0 - 40.}{3.0 \times 10^{-3}\ K^{-1} - 0} = -1.3 \times 10^4\ K$$

$-1.3 \times 10^4\ K = -\Delta H°/R$, $\Delta H° = 1.3 \times 10^4\ K \times 8.3145\ J/K{\bullet}mol = 1.1 \times 10^5\ J/mol$

y intercept = $40. = \Delta S°/R$, $\Delta S° = 40. \times 8.3145\ J/K{\bullet}mol = 330\ J/K{\bullet}mol$

As seen here, when $\Delta H°$ is positive, the slope of the ln K versus 1/T plot is negative. When $\Delta H°$ is negative as in an exothermic process, then the slope of the ln K versus 1/T plot will be positive (slope = $-\Delta H°/R$).

80. The ln K versus 1/T plot gives a straight line with slope = $-\Delta H°/R$ and y intercept = $\Delta S°/R$.

$1.352 \times 10^4\ K = -\Delta H°/R$, $\Delta H° = -(8.3145\ J/K{\bullet}mol)(1.352 \times 10^4\ K)$

$\Delta H° = -1.124 \times 10^5\ J/mol = -112.4\ kJ/mol$

$-14.51 = \Delta S°/R$, $\Delta S° = (-14.51)(8.3145\ J/K{\bullet}mol) = -120.6\ J/K{\bullet}mol$

Note that the signs for $\Delta H°$ and $\Delta S°$ make sense. When a bond forms, $\Delta H° < 0$ and $\Delta S° < 0$.

Additional Exercises

81. From Appendix 4, $S° = 198\ J/K{\bullet}mol$ for CO(g) and $S° = 27\ J/K{\bullet}mol$ for Fe(s).

Let S_l^o = S° for $Fe(CO)_5(l)$ and S_g^o = S° for $Fe(CO)_5(g)$.

$\Delta S° = -677\ J/K = 1\ mol(S_l^o) - [1\ mol\ (27\ J/K{\bullet}mol) + 5\ mol(198\ J/K{\bullet}mol]$

$S_l^o = 340.\ J/K{\bullet}mol$

$\Delta S° = 107\ J/K = 1\ mol\ (S_g^o) - 1\ mol\ (340.\ J/K{\bullet}mol)$

S_g^o = S° for $Fe(CO)_5(g) = 447\ J/K{\bullet}mol$

82. When an ionic solid dissolves, one would expect the disorder of the system to increase, so ΔS_{sys} is positive. Because temperature increased as the solid dissolved, this is an exothermic process, and ΔS_{surr} is positive ($\Delta S_{surr} = -\Delta H/T$). Because the solid did dissolve, the dissolving process is spontaneous, so ΔS_{univ} is positive.

83. ΔS will be negative because 2 mol of gaseous reactants form 1 mol of gaseous product. For ΔG to be negative, ΔH must be negative (exothermic). For exothermic reactions, K decreases as T increases, resulting in a smaller quantity of product and a larger quantity of reactants present at equilibrium. Therefore, the ratio of the partial pressure of PCl_5 (a product) to the partial pressure of PCl_3 (a reactant) will decrease when T is raised.

84. At boiling point, ΔG = 0 so $\Delta S = \dfrac{\Delta H_{vap}}{T}$; for methane: $\Delta S = \dfrac{8.20 \times 10^3 \text{ J/mol}}{112 \text{ K}}$

$= 73.2$ J/mol•K

For hexane: $\Delta S = \dfrac{28.9 \times 10^3 \text{ J/mol}}{342 \text{ K}} = 84.5$ J/mol•K. $V_{met} = \dfrac{nRT}{P}$

$$V_{met} = \frac{1.00 \text{ mol}(0.08206 \text{ L atm/K} \bullet \text{mol})(112 \text{ K})}{1.00 \text{ atm}} = 9.19 \text{ L}; \; V_{hex} = \frac{nRT}{P} = R(342 \text{ K}) = 28.1 \text{ L}$$

Hexane has the larger molar volume at the boiling point, so hexane should have the larger entropy. As the volume of a gas increases, positional disorder increases.

85. solid I → solid II; equilibrium occurs when ΔG = 0.

$$\Delta G = \Delta H - T\Delta S, \; \Delta H = T\Delta S, \; T = \Delta H/\Delta S = \frac{-743.1 \text{ J/mol}}{-17.0 \text{ J/K} \bullet \text{mol}} = 43.7 \text{ K} = -229.5°\text{C}$$

86. a. $\Delta G° = -RT \ln K = -(8.3145 \text{ J/K•mol})(298 \text{ K}) \ln 0.090 = 6.0 \times 10^3 \text{ J/mol} = 6.0 \text{ kJ/mol}$

b. H–O–H + Cl–O–Cl → 2 H–O–Cl

On each side of the reaction there are 2 H–O bonds and 2 O–Cl bonds. Both sides have the same number and type of bonds. Thus $\Delta H \approx \Delta H° \approx 0$.

c. $\Delta G° = \Delta H° - T\Delta S°, \; \Delta S° = \dfrac{\Delta H° - \Delta G°}{T} = \dfrac{0 - 6.0 \times 10^3 \text{ J}}{298 \text{ K}} = -20.$ J/K

d. For $H_2O(g)$, $\Delta H_f° = -242$ kJ/mol and $S° = 189$ J/K•mol.

$\Delta H° = 0 = 2\Delta H°_{f, HOCl} - [1 \text{ mol}(-242 \text{ kJ/mol}) + 1 \text{ mol}(80.3 \text{ J/K/mol})]$, $\Delta H°_{f, HOCl}$

$= -81$ kJ/mol

$-20. \text{ J/K} = 2S°_{HOCl} - [1 \text{ mol}(189 \text{ J/K•mol}) + 1 \text{ mol}(266.1 \text{ J/K•mol})], S°_{HOCl} = 218$ J/K•mol

e. Assuming ΔH° and ΔS° are T-independent: $\Delta G°_{500} = 0 - (500. \text{ K})(-20. \text{ J/K}) = 1.0 \times 10^4$ J

$$\Delta G° = -RT \ln K,\ K = \exp\left(\frac{-\Delta G°}{RT}\right) = \exp\left[\frac{-1.0 \times 10^4}{(8.3145)(500.)}\right] = e^{-2.41} = 0.090$$

f. $\Delta G = \Delta G° + RT \ln \frac{P^2_{HOCl}}{P_{H_2O} \times P_{Cl_2O}}$; from part a, $\Delta G° = 6.0$ kJ/mol.

We should express all partial pressures in atm. However, we perform the pressure conversion the same number of times in the numerator and denominator, so the factors of 760 torr/atm will all cancel. Thus we can use the pressures in units of torr.

$$\Delta G = \frac{6.0\,\text{kJ/mol} + (8.3145\,\text{J/K} \bullet \text{mol})(298\,\text{K})}{1000\,\text{J/kJ}} \ln\left[\frac{(0.10)^2}{(18)(2.0)}\right] = 6.0 - 20. = -14 \text{ kJ/mol}$$

87. $Ba(NO_3)_2(s) \rightleftharpoons Ba^{2+}(aq) + 2\ NO_3^-(aq)$ $K = K_{sp}$; $\Delta G° = -561 + 2(-109) - (-797) = 18$ kJ

$$\Delta G° = -RT \ln K_{sp},\ \ln K_{sp} = \frac{-\Delta G°}{RT} = \frac{-18{,}000\,\text{J}}{8.3145\,\text{J/K} \bullet \text{mol}(298\,\text{K})} = -7.26$$

$$K_{sp} = e^{-7.26} = 7.0 \times 10^{-4}$$

88. $K^+(blood) \rightleftharpoons K^+(muscle)$ $\Delta G° = 0$; $\Delta G = RT \ln\left(\frac{[K^+]_m}{[K^+]_b}\right)$; $\Delta G = w_{max}$

$$\Delta G = \frac{8.3145\,\text{J}}{\text{K mol}}(310.\ \text{K}) \ln\left(\frac{0.15}{0.0050}\right),\ \Delta G = 8.8 \times 10^3 \text{ J/mol} = 8.8 \text{ kJ/mol}$$

At least 8.8 kJ of work must be applied to transport 1 mol K^+.

Other ions will have to be transported in order to maintain electroneutrality. Either anions must be transported into the cells, or cations (Na^+) in the cell must be transported to the blood. The latter is what happens: $[Na^+]$ in blood is greater than $[Na^+]$ in cells as a result of this pumping.

89.

$HgbO_2 \rightarrow Hgb + O_2$	$\Delta G° = -(-70 \text{ kJ})$
$Hgb + CO \rightarrow HgbCO$	$\Delta G° = -80 \text{ kJ}$
$HgbO_2 + CO \rightarrow HgbCO + O_2$	$\Delta G° = -10 \text{ kJ}$

$$\Delta G° = -RT \ln K,\ K = \exp\left(\frac{-\Delta G°}{RT}\right) = \exp\left[\frac{-(-10 \times 10^3\ \text{J})}{(8.3145\,\text{J/K} \bullet \text{mol})(298\text{K})}\right] = 60$$

90. $HF(aq) \rightleftharpoons H^+(aq) + F^-(aq)$; $\Delta G = \Delta G° + RT \ln \frac{[H^+][F^-]}{[HF]}$

$$\Delta G° = -RT \ln K = -(8.3145 \text{ J/K•mol})(298 \text{ K}) \ln(7.2 \times 10^{-4}) = 1.8 \times 10^4 \text{ J/mol}$$

a. The concentrations are all at standard conditions, so $\Delta G = \Delta G = 1.8 \times 10^4$ J/mol (Q = 1.0 and ln Q = 0). Because $\Delta G°$ is positive, the reaction shifts left to reach equilibrium.

b. $\Delta G = 1.8 \times 10^4 \text{ J/mol} + (8.3145 \text{ J/K•mol})(298 \text{ K}) \ln \dfrac{(2.7 \times 10^{-2})^2}{0.98}$

$\Delta G = 1.8 \times 10^4 \text{ J/mol} - 1.8 \times 10^4 \text{ J/mol} = 0$

$\Delta G = 0$, so the reaction is at equilibrium (no shift).

c. $\Delta G = 1.8 \times 10^4 \text{ J/mol} + 8.3145(298 \text{ K}) \ln \dfrac{(1.0 \times 10^{-5})^2}{1.0 \times 10^{-5}} = -1.1 \times 10^4$ J/mol; shifts right

d. $\Delta G = 1.8 \times 10^4 + 8.3145(298) \ln \dfrac{7.2 \times 10^{-4}(0.27)}{0.27} = 1.8 \times 10^4 - 1.8 \times 10^4 = 0$; at equilibrium

e. $\Delta G = 1.8 \times 10^4 + 8.3145(298) \ln \dfrac{1.0 \times 10^{-3}(0.67)}{0.52} = 2 \times 10^3$ J/mol; shifts left

91. ΔS is more favorable (less negative) for reaction 2 than for reaction 1, resulting in $K_2 > K_1$. In reaction 1, seven particles in solution are forming one particle in solution. In reaction 2, four particles are forming one, which results in a smaller decrease in positional probability than for reaction 1.

92. A graph of ln K versus 1/T will yield a straight line with slope equal to $-\Delta H°/R$ and *y* intercept equal to $\Delta S°/R$.

Temp (°C)	T (K)	1000/T (K^{-1})	K_w	ln K_w
0	273	3.66	1.14×10^{-15}	−34.408
25	298	3.36	1.00×10^{-14}	−32.236
35	308	3.25	2.09×10^{-14}	−31.499
40.	313	3.19	2.92×10^{-14}	−31.165
50.	323	3.10	5.47×10^{-14}	−30.537

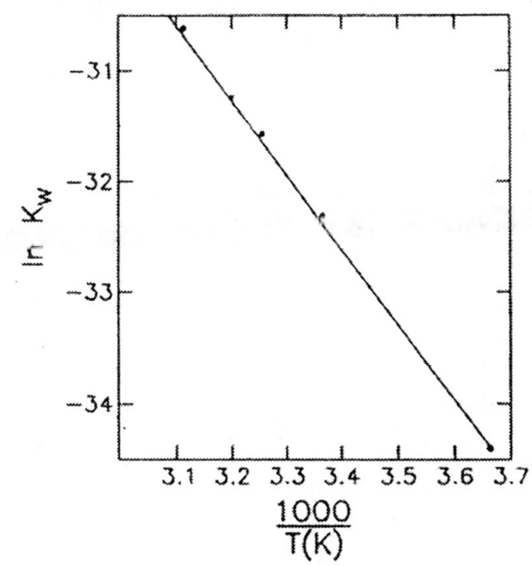

The straight-line equation (from a calculator) is $\ln K = -6.91 \times 10^3\left(\frac{1}{T}\right) - 9.09$.

$$\text{Slope} = -6.91 \times 10^3 \text{ K} = \frac{-\Delta H^\circ}{R}$$

$\Delta H^\circ = -(-6.91 \times 10^3 \text{ K} \times 8.3145 \text{ J/K•mol}) = 5.75 \times 10^4 \text{ J/mol}$

y intercept $= -9.09 = \frac{\Delta S^\circ}{R}$, $\Delta S^\circ = -9.09 \times 8.3145 \text{ J/K•mol} = -75.6 \text{ J/K•mol}$

93. $\Delta G^\circ = -RT \ln K$; when K = 1.00, $\Delta G^\circ = 0$ since ln 1.00 = 0. $\Delta G^\circ = 0 = \Delta H^\circ - T\Delta S^\circ$

$\Delta H^\circ = 3(-242 \text{ kJ}) - [-826 \text{ kJ}] = 100. \text{ kJ}$; $\Delta S^\circ = [2(27 \text{ J/K}) + 3(189 \text{ J/K})] - [90. \text{ J/K} + 3(131 \text{ J/K})] = 138 \text{ J/K}$

$$\Delta H^\circ = T\Delta S^\circ,\ T = \frac{\Delta H^\circ}{\Delta S^\circ} = \frac{100.\text{kJ}}{0.138\text{kJ/K}} = 725 \text{ K}$$

94. Equilibrium occurs when the minimum in free energy has been reached. For this reaction, the minimum in free energy is when 1/3 of A has reacted. Because equilibrium lies closer to reactants for this reaction, the equilibrium constant will be less than 1 (K < 1) which dictates that ΔG° is greater than 0 ($\Delta G^\circ > 0$; $\Delta G^\circ = -RT\ln K$).

	2 A(g)	⇌	B(g)	$K_p = \frac{P_B}{{P_A}^2}$
Initial	3.0 atm		0	
Change	$-2x$	→	$+x$	
Equil.	$3.0 - 2x$		x	

From the plot, equilibrium occurs when 1/3 of the A(g) has reacted.

$2x$ = amount A reacted to reach equilibrium = 1/3(3.0 atm) = 1.0 atm, $x = 0.50$ atm

$$K_p = \frac{x}{(3.0 - 2x)^2} = \frac{0.50}{(2.0)^2} = 0.125 = 0.13 \text{ (to 2 sig. figs.)}$$

ChemWork Problems

The answers to the problems 95-102 (or a variation to these problems) are found in OWL. These problems are also assignable in OWL.

Challenge Problems

103 a. Vessel 1: At 0°C, this system is at equilibrium, so $\Delta S_{univ} = 0$ and $\Delta S = \Delta S_{surr}$. Because the vessel is perfectly insulated, q = 0, so $\Delta S_{surr} = 0 = \Delta S_{sys}$.

b. Vessel 2: The presence of salt in water lowers the freezing point of water to a temperature below 0°C. In vessel 2, the conversion of ice into water will be spontaneous at 0°C, so $\Delta S_{univ} > 0$. Because the vessel is perfectly insulated, $\Delta S_{surr} = 0$. Therefore, ΔS_{sys} must be positive ($\Delta S > 0$) in order for ΔS_{univ} to be positive.

104. The liquid water will evaporate at first and eventually equilibrium will be reached (physical equilibrium).

- Because evaporation is an endothermic process, ΔH is positive.
- Because $H_2O(g)$ is more disordered (greater positional probability), ΔS is positive.
- The water will become cooler (the higher energy water molecules leave), thus ΔT_{water} will be negative.
- The vessel is insulated (q = 0), so $\Delta S_{surr} = 0$.
- Because the process occurs, it is spontaneous, so ΔS_{univ} is positive.

105. $3\ O_2(g) \rightleftharpoons 2\ O_3(g)$; ΔH° = 2(143 kJ) = 286 kJ; ΔG° = 2(163 kJ) = 326 kJ

$$\ln K = \frac{-\Delta G^\circ}{RT} = \frac{-326 \times 10^3\ \text{J}}{(8.3145\text{J/K} \bullet \text{mol})(298\text{K})} = -131.573,\ K = e^{-131.573} = 7.22 \times 10^{-58}$$

We need the value of K at 230. K. From Section 17.8 of the text:

$$\ln K = \frac{-\Delta G^\circ}{RT} + \frac{\Delta S^\circ}{R}$$

For two sets of K and T:

$$\ln K_1 = \frac{-\Delta H^\circ}{R}\left(\frac{1}{T_1}\right) + \frac{\Delta S}{R};\ \ln K_2 = \frac{-\Delta H^\circ}{R}\left(\frac{1}{T_2}\right) + \frac{\Delta S^\circ}{R}$$

Subtracting the first expression from the second:

$$\ln K_2 - \ln K_1 = \frac{\Delta H^\circ}{R}\left(\frac{1}{T_1} - \frac{1}{T_2}\right) \text{ or } \ln\frac{K_2}{K_1} = \frac{\Delta H^\circ}{R}\left(\frac{1}{T_1} - \frac{1}{T_2}\right)$$

Let $K_2 = 7.22 \times 10^{-58}$, $T_2 = 298$ K; $K_1 = K_{230}$, $T_1 = 230.$ K; ΔH° = 286 × 10³ J

$$\ln\frac{7.22 \times 10^{-58}}{K_{230}} = \frac{286 \times 10^3}{8.3145}\left(\frac{1}{230.} - \frac{1}{298}\right) = 34.13$$

$$\frac{7.22 \times 10^{-58}}{K_{230}} = e^{34.13} = 6.6 \times 10^{14},\ K_{230} = 1.1 \times 10^{-72}$$

$$K_{230} = 1.1 \times 10^{-72} = \frac{P_{O_3}^2}{P_{O_3}^3} = \frac{P_{O_3}^2}{(1.0 \times 10^{-3}\ \text{atm})^3},\ P_{O_3} = 3.3 \times 10^{-41}\ \text{atm}$$

The volume occupied by one molecule of ozone is:

$$V = \frac{nRT}{P} = \frac{(1/6.022\times 10^{23}\ \text{mol})(0.08206\text{L atm/K}\bullet\text{mol})(230\text{K})}{(3.3\times 10^{-41}\ \text{atm})} = 9.5\times 10^{17}\ \text{L}$$

Equilibrium is probably not maintained under these conditions. When only two ozone molecules are in a volume of 9.5×10^{17} L, the reaction is not at equilibrium. Under these conditions, Q > K, and the reaction shifts left. But with only 2 ozone molecules in this huge volume, it is extremely unlikely that they will collide with each other. At these conditions, the concentration of ozone is not large enough to maintain equilibrium.

106. Arrangement I and V: $S = k \ln W;\ W = 1;\ S = k \ln 1 = 0$

Arrangement II and IV: $W = 4;\ S = k \ln 4 = 1.38 \times 10^{-23}\ \text{J/K} \ln 4,\ S = 1.91 \times 10^{-23}\ \text{J/K}$

Arrangement III: $W = 6;\ S = k \ln 6 = 2.47 \times 10^{-23}\ \text{J/K}$

107. a. From the plot, the activation energy of the reverse reaction is $E_a + (-\Delta G°) = E_a - \Delta G°$ ($\Delta G°$ is a negative number as drawn in the diagram).

$$k_f = A\exp\left(\frac{-E_a}{RT}\right) \text{ and } k_r = A\exp\left[\frac{-(E_a - \Delta G^o)}{RT}\right],\quad \frac{k_f}{k_r} = \frac{A\exp\left(\frac{-E_a}{RT}\right)}{A\exp\left[\frac{-(E_a - \Delta G^o)}{RT}\right]}$$

If the A factors are equal: $\dfrac{k_f}{k_r} = \exp\left[\dfrac{-E_a}{RT} + \dfrac{(E_a - \Delta G^o)}{RT}\right] = \exp\left(\dfrac{-\Delta G^o}{RT}\right)$

From $\Delta G° = -RT \ln K$, $K = \exp\left(\dfrac{-\Delta G^o}{RT}\right)$; because K and $\dfrac{k_f}{k_r}$ are both equal to the same expression, $K = k_f/k_r$.

b. A catalyst will lower the activation energy for both the forward and reverse reactions (but not change $\Delta G°$). Therefore, a catalyst must increase the rate of both the forward and reverse reactions.

108. At equilibrium:

$$P_{H_2} = \frac{nRT}{V} = \frac{\left(\dfrac{1.10\times 10^{13}\ \text{molecules}}{6.022\times 10^{23}\ \text{molecules/mol}}\right)\left(\dfrac{0.08206\text{L atm}}{\text{K}\bullet\text{mol}}\right)(298\text{K})}{1.00\text{L}}$$

$$P_{H_2} = 4.47\times 10^{-10}\ \text{atm}$$

The pressure of H_2 decreased from 1.00 atm to 4.47×10^{-10} atm. Essentially all of the H_2 and Br_2 has reacted. Therefore, P_{HBr} = 2.00 atm because there is a 2 : 1 mole ratio between HBr and H_2 in the balanced equation. Because we began with equal moles of H_2 and Br_2, we will have equal moles of H_2 and Br_2 at equilibrium. Therefore, $P_{H_2} = P_{Br_2} = 4.47 \times 10^{-10}$ atm.

$$K = \frac{P^2_{HBr}}{P_{H_2} \times P_{Br_2}} = \frac{(2.00)^2}{(4.47 \times 10^{-10})^2} = 2.00 \times 10^{19}; \quad \text{assumptions good.}$$

$$\Delta G° = -RT \ln K = -(8.3145 \text{ J/K•mol})(298 \text{ K}) \ln(2.00 \times 10^{19}) = -1.10 \times 10^5 \text{ J/mol}$$

$$\Delta S° = \frac{\Delta H° - \Delta G°}{T} = \frac{-103{,}800\text{J/mol} - (-1.0 \times 10^5 \text{ J/mol})}{298\text{K}} = 20 \text{ J/K•mol}$$

109. a. $\Delta G° = G^\circ_B - G^\circ_A = 11{,}718 - 8996 = 2722$ J

$$K = \exp\left(\frac{-\Delta G°}{RT}\right) = \exp\left[\frac{-2722\text{J}}{(8.3145\text{J/K} \bullet \text{mol})(298\text{K})}\right] = 0.333$$

b. When Q = 1.00 > K, the reaction shifts left. Let x = atm of B(g), which reacts to reach equilibrium.

	A(g) ⇌	B(g)
Initial	1.00 atm	1.00 atm
Equil.	$1.00 + x$	$1.00 - x$

$$K = \frac{P_B}{P_A} = \frac{1.00 - x}{1.00 + x} = 0.333, \quad 1.00 - x = 0.333 + (0.333)x, \quad x = 0.50 \text{ atm}$$

$P_B = 1.00 - 0.50 = 0.50$ atm; $P_A = 1.00 + 0.50 = 1.50$ atm

c. $\Delta G = \Delta G° + RT \ln Q = \Delta G° + RT \ln(P_B/P_A)$

$\Delta G = 2722 \text{ J} + (8.3145)(298) \ln (0.50/1.50) = 2722 \text{ J} - 2722 \text{ J} = 0$ (carrying extra sig. figs.)

110. From Exercise 79, $\ln K = \frac{-\Delta H°}{RT} + \frac{\Delta S°}{R}$. For K at two temperatures T_1 and T_2, the equation can be manipulated to give (see Exercise 105): $\ln \frac{K_2}{K_1} = \frac{\Delta H°}{R}\left(\frac{1}{T_1} - \frac{1}{T_2}\right)$

$$\ln\left(\frac{3.25 \times 10^{-2}}{8.84}\right) = \frac{\Delta H°}{8.3145\text{J/K} \bullet \text{mol}}\left(\frac{1}{298\text{K}} - \frac{1}{348\text{K}}\right)$$

$-5.61 = (5.8 \times 10^{-5} \text{ mol/J})(\Delta H°)$, $\Delta H° = -9.7 \times 10^4$ J/mol

For K = 8.84 at T = 25°C:

$$\ln 8.84 = \frac{-(-9.7 \times 10^4 \text{ J/mol})}{(8.3145\text{J/K} \bullet \text{mol})(298\text{K})} + \frac{\Delta S°}{8.3145\text{J/K} \bullet \text{mol}}, \quad \frac{\Delta S°}{8.3145} = -37$$

$\Delta S° = -310$ J/K•mol

We get the same value for $\Delta S°$ using $K = 3.25 \times 10^{-2}$ at T = 348 K data.

$\Delta G° = -RT \ln K$; when K = 1.00, then $\Delta G° = 0$ since ln 1.00 = 0. Assuming $\Delta H°$ and $\Delta S°$ do not depend on temperature:

$$\Delta G° = 0 = \Delta H° - T\Delta S°,\ \ \Delta H° = T\Delta S°,\ \ T = \frac{\Delta H°}{\Delta S°} = \frac{-9.7 \times 10^4\ \text{J/mol}}{-310\ \text{J/K} \bullet \text{mol}} = 310\ \text{K}$$

111. $K = P_{CO_2}$; to ensure Ag_2CO_3 from decomposing, P_{CO_2} should be greater than K.

From Exercise 79, $\ln K = \frac{\Delta H°}{RT} + \frac{\Delta S°}{R}$. For two conditions of K and T, the equation is:

$$\ln\frac{K_2}{K_1} = \frac{\Delta H°}{R}\left(\frac{1}{T_1} + \frac{1}{T_2}\right)$$

Let T_1 = 25°C = 298 K, $K_1 = 6.23 \times 10^{-3}$ torr; T_2 = 110.°C = 383 K, K_2 = ?

$$\ln\frac{K_2}{6.23 \times 10^{-3}\ \text{torr}} = \frac{79.14 \times 10^3\ \text{J/mol}}{8.3145\ \text{J/K} \bullet \text{mol}}\left(\frac{1}{298\text{K}} - \frac{1}{383\text{K}}\right)$$

$$\ln\frac{K_2}{6.23 \times 10^{-3}} = 7.1,\ \ \frac{K_2}{6.23 \times 10^{-3}} = e^{7.1} = 1.2 \times 10^3,\ \ K_2 = 7.5\ \text{torr}$$

To prevent decomposition of Ag_2CO_3, the partial pressure of CO_2 should be greater than 7.5 torr.

112. From the problem, $\chi^L_{C_6H_6} = \chi^L_{CCl_4} = 0.500$. We need the pure vapor pressures ($P°$) in order to calculate the vapor pressure of the solution. Use the thermodynamic data to determine the pure vapor pressure values.

$$C_6H_6(l) \rightleftharpoons C_6H_6(g)\ \ \ K = P_{C_6H_6} = P°_{C_6H_6}\ \text{at } 25°\text{C}$$

$$\Delta G°_{rxn} = \Delta G°_{f,\,C_6H_6(g)} - \Delta G°_{f,\,C_6H_6(l)} = 129.66\ \text{kJ/mol} - 124.50\ \text{kJ/mol} = 5.16\ \text{kJ/mol}$$

$$\Delta G° = -RT \ln K,\ \ln K = \frac{-\Delta G°}{RT} = \frac{-5.16 \times 10^3\ \text{J/mol}}{(8.3145\ \text{J/K} \bullet \text{mol})(298\text{K})} = -2.08$$

$$K = P°_{C_6H_6} = e^{-2.08} = 0.125\ \text{atm}$$

For CCl_4: $\Delta G°_{rxn} = \Delta G°_{f,\,CCl_4(g)} - \Delta G°_{f,\,CCl_4(l)} = -60.59\ \text{kJ/mol} - (-65.21\ \text{kJ/mol})$
$= 4.62$ kJ/mol

$$K = P°_{CCl_4} = \exp\left(\frac{-\Delta G°}{RT}\right) = \exp\left(\frac{-4620\ \text{J/mol}}{8.3145\ \text{J/K} \bullet \text{mol} \times 298\text{K}}\right) = 0.155\ \text{atm}$$

$P_{C_6H_6} = \chi^L_{C_6H_6} P°_{C_6H_6} = 0.500(0.125\ \text{atm}) = 0.0625$ atm; $P°_{CCl_4} = 0.500(0.155\ \text{atm})$
$= 0.0775$ atm

$$\chi^{V}_{C_6H_6} = \frac{P_{C_6H_6}}{P_{tot}} = \frac{0.0625\text{atm}}{0.0625\text{atm} + 0.0775\text{atm}} = \frac{0.0625}{0.1400} = 0.446$$

$$\chi^{V}_{CCl_4} = 1.000 - 0.446 = 0.554$$

113. $NaCl(s) \rightleftharpoons Na^+(aq) + Cl^-(aq) \qquad K = K_{sp} = [Na^+][Cl^-]$

$\Delta G° = [(-262\text{ kJ}) + (-131\text{ kJ})] - (-384\text{ kJ}) = -9\text{ kJ} = -9000\text{ J}$

$$\Delta G° = -RT\ln K_{sp},\ K_{sp} = \exp\left[\frac{-(-9000\text{J})}{8.3145\text{J/K}\bullet\text{mol} \times 298\text{K}}\right] = 38 = 40$$

	$NaCl(s)$	$\rightleftharpoons$	$Na^+(aq)$	+	$Cl^-(aq)$	$K_{sp} = 40$
Initial	s = solubility (mol/L)		0		0	
Equil.			s		s	

$K_{sp} = 40 = s(s),\ s = (40)^{1/2} = 6.3 = 6\ M = [Cl^-]$

114. $\Delta S_{surr} = -\Delta H/T = -q_P/T$

q = heat loss by hot water = moles × molar heat capacity × ΔT

$$q = 1.00 \times 10^3\text{ g } H_2O \times \frac{1\text{ mol } H_2O}{18.02\text{ g}} \times \frac{75.4\text{ J}}{\text{K mol}} \times (298.2 - 363.2) = -2.72 \times 10^5\text{ J}$$

$$\Delta S_{surr} = \frac{-(-2.72\times10^5\text{ J})}{298.2\text{ K}} = 912\text{ J/K}$$

115. $HX(aq) \rightleftharpoons H^+(aq) + X^-(aq) \qquad K_a = \frac{[H^+][X^-]}{[HX]}$

	$HX(aq)$	$\rightleftharpoons$	$H^+(aq)$	+	$X^-(aq)$
Initial	0.10 M		~0		0
Equil.	$0.10 - x$		x		x

From problem, $x = [H^+] = 10^{-5.83} = 1.5 \times 10^{-6}$; $K_a = \frac{(1.5 \times 10^{-6})^2}{0.10 - 1.5 \times 10^{-6}} = 2.3 \times 10^{-11}$

$\Delta G° = -RT\ln K = -8.3145\text{ J/K}\bullet\text{mol}(298\text{ K})\ln(2.3 \times 10^{-11}) = 6.1 \times 10^4\text{ J/mol} = 61\text{ kJ/mol}$

Integrative Problems

116. Use the thermodynamic data to calculate the boiling point of the solvent.

At boiling point: $\Delta G = 0 = \Delta H - T\Delta S,\ T = \frac{\Delta H}{\Delta S} = \frac{33.90 \times 10^3\text{ J/mol}}{95.95\text{ J/K}\bullet\text{mol}} = 353.3\text{ K}$

$\Delta T = K_b m$, (355.4 K − 353.3 K) = 2.5 K kg/mol(m), $m = \dfrac{2.1}{2.5} = 0.84$ mol/kg

$$\text{Mass solvent} = 150.\text{ mL} \times \frac{0.879\text{ g}}{\text{mL}} \times \frac{1\text{ kg}}{1000\text{ g}} = 0.132\text{ kg}$$

$$\text{Mass solute} = 0.132\text{ kg solvent} \times \frac{0.84\text{ mol solute}}{\text{kg solvent}} \times \frac{142\text{ g}}{\text{mol}} = 15.7\text{ g} = 16\text{ g solute}$$

117. Because the partial pressure of C(g) decreased, the net change that occurs for this reaction to reach equilibrium is for products to convert to reactants.

	A(g)	+	2 B(g)	⇌	C(g)
Initial	0.100 atm		0.100 atm		0.100 atm
Change	$+x$		$+2x$	←	$-x$
Equil.	$0.100 + x$		$0.100 + 2x$		$0.100 - x$

From the problem, $P_C = 0.040$ atm = $0.100 - x$, $x = 0.060$ atm

The equilibrium partial pressures are: $P_A = 0.100 + x = 0.100 + 0.060 = 0.160$ atm, $P_B = 0.100 + 2(0.60) = 0.22$ atm, and $P_C = 0.040$ atm

$$K = \frac{0.040}{0.160(0.22)^2} = 5.2$$

$\Delta G° = -RT \ln K = -8.3145$ J/K•mol(298 K) ln(5.2) = -4.1×10^3 J/mol = −4.1 kJ/mol

118. $\Delta G° = \Delta H° - T\Delta S° = -28.0 \times 10^3\text{ J} - 298\text{ K}(-175\text{ J/K}) = 24{,}200\text{ J}$

$$\Delta G° = -RT \ln K,\ \ln K = \frac{-\Delta G°}{RT} = \frac{-24{,}000\text{J}}{8.3145\text{ J/K} \bullet \text{mol} \times 298\text{K}} = -9.767$$

$K = e^{-9.767} = 5.73 \times 10^{-5}$

	B	+	H_2O	⇌	BH^+	+	OH^-	$K = K_b = 5.73 \times 10^{-5}$
Initial	0.125 M				0		~0	
Change	$-x$				$+x$		$+x$	
Equil.	$0.125 - x$				x		x	

$$K_b = 5.73 \times 10^{-5} = \frac{[BH^+][OH^-]}{[B]} = \frac{x^2}{0.125 - x} \approx \frac{x^2}{0.125},\ x = [OH^-] = 2.68 \times 10^{-3}\,M$$

pH = −log(2.68×10^{-3}) = 2.572; pOH = 14.000 − 2.572 = 11.428; assumptions good

Marathon Problem

119. a. $\Delta S°$ will be negative because there is a decrease in the number of moles of gas.

b. Because $\Delta S°$ is negative, $\Delta H°$ must be negative for the reaction to be spontaneous at some temperatures. Therefore, ΔS_{surr} is positive.

c. $Ni(s) + 4\ CO(g) \rightleftharpoons Ni(CO)_4(g)$

$\Delta H° = -607 - [4(-110.5)] = -165$ kJ; $\Delta S° = 417 - [4(198) + (30.)] = -405$ J/K

d. $\Delta G° = 0 = \Delta H° - T\Delta S°,\ T = \dfrac{\Delta H°}{\Delta S°} = \dfrac{-165 \times 10^3\ J}{-405\ J/K} = 407$ K or 134°C

e. T = 50.°C + 273 = 323 K

$\Delta G°_{323} = -165\ kJ - (323\ K)(-0.405\ kJ/K) = -34\ kJ$

$$\ln K = \frac{-\Delta G°}{RT} = \frac{-(-34{,}000 J)}{8.3145\ J/K \bullet mol(323K)} = 12.66,\ K = e^{12.66} = 3.1 \times 10^5$$

f. T = 227°C + 273 = 500. K

$\Delta G°_{500} = -165\ kJ - (500.\ K)(-0.405\ kJ/K) = 38\ kJ$

$$\ln K = \frac{-38{,}000 J}{(8.3145\ J/K \bullet mol)(500.K)} = -9.14,\ K = e^{-9.14} = 1.1 \times 10^{-4}$$

g. The temperature change causes the value of the equilibrium constant to change from a large value favoring formation of $Ni(CO)_4$ to a small value favoring the decomposition of $Ni(CO)_4$ into pure Ni and CO. This is exactly what is wanted in order to purify a nickel sample.

h. $Ni(CO)_4(l) \rightleftharpoons Ni(CO)_4(g)$ $\quad K = P_{Ni(CO)_4}$

At 42°C (the boiling point): $\Delta G° = 0 = \Delta H° - T\Delta S°$

$$\Delta S° = \frac{\Delta H°}{T} = \frac{29.0 \times 10^3\ J}{315K} = 92.1\ J/K$$

At 152°C: $\Delta G°_{152} = \Delta H° - T\Delta S° = 29.0 \times 10^3\ J - 425\ K(92.1\ J/K) = -10{,}100\ J$

$$\Delta G° = -RT \ln K,\ \ln K = \frac{-(-10{,}100 J)}{8.3145\ J/K \bullet mol(425K)} = 2.858,\ K_p = e^{2.858} = 17.4$$

A maximum pressure of 17.4 atm can be attained before $Ni(CO)_4(g)$ will liquify.

CHAPTER 18

ELECTROCHEMISTRY

Review of Oxidation-Reduction Reactions

15. Oxidation: increase in oxidation number; loss of electrons

    Reduction: decrease in oxidation number; gain of electrons

16. See Table 4.2 in Chapter 4 of the text for rules for assigning oxidation numbers.

    a. H (+1), O (−2), N (+5)

    b. Cl (−1), Cu (+2)

    c. O (0)

    d. H (+1), O (−1)

    e. H(+1), O (−2), C (0)

    f. Ag (0)

    g. Pb (+2), O (−2), S (+6)

    h. O (−2), Pb (+4)

    i. Na (+1), O (−2), C (+3)

    j. O (−2), C (+4)

    k. $(NH_4)_2Ce(SO_4)_3$ contains NH_4^+ ions and SO_4^{2-} ions. Thus cerium exists as the Ce^{4+} ion.

    H (+1), N (−3), Ce (+4), S (+6), O (−2)

    l. O (−2), Cr (+3)

17. The species oxidized shows an increase in oxidation numbers and is called the reducing agent. The species reduced shows a decrease in oxidation numbers and is called the oxidizing agent. The pertinent oxidation numbers are listed by the substance oxidized and the substance reduced.

	Redox?	Ox. Agent	Red. Agent	Substance Oxidized	Substance Reduced
a.	Yes	H_2O	CH_4	CH_4 (C, −4 → +2)	H_2O (H, +1 → 0)
b.	Yes	$AgNO_3$	Cu	Cu (0 → +2)	$AgNO_3$ (Ag, +1 → 0)
c.	Yes	HCl	Zn	Zn (0 → +2)	HCl (H, +1 → 0)

    d. No; there is no change in any of the oxidation numbers.

18. a. $\underset{-3\ +1}{4\ NH_3(g)} + \underset{0}{5\ O_2(g)} \rightarrow \underset{+2\ -2}{4\ NO(g)} + \underset{+1\ -2}{6\ H_2O(g)}$ oxidation numbers

    $\underset{+2\ -2}{2\ NO(g)} + \underset{0}{O_2(g)} \rightarrow \underset{+4\ -2}{2\ NO_2(g)}$

$$\underset{+4\ -2}{3\ NO_2(g)} + \underset{+1\ -2}{H_2O(l)} \rightarrow \underset{+1\ +5\ -2}{2\ HNO_3(aq)} + \underset{+2\ -2}{NO(g)}$$

All three reactions are oxidation-reduction reactions since there is a change in oxidation numbers of some of the elements in each reaction.

b. $4\ NH_3 + 5\ O_2 \rightarrow 4\ NO + 6\ H_2O$; O_2 is the oxidizing agent and NH_3 is the reducing agent.

$2\ NO + O_2 \rightarrow 2\ NO_2$; O_2 is the oxidizing agent and NO is the reducing agent.

$3\ NO_2 + H_2O \rightarrow 2\ HNO_3 + NO$; NO_2 is both the oxidizing and reducing agent.

Questions

19. Electrochemistry is the study of the interchange of chemical and electrical energy. A redox (oxidation-reduction) reaction is a reaction in which one or more electrons are transferred. In a galvanic cell, a spontaneous redox reaction occurs that produces an electric current. In an electrolytic cell, electricity is used to force a nonspontaneous redox reaction to occur.

20. Mass balance indicates that we have the same number and type of atoms on both sides of the equation (so that mass is conserved). Similarly, net charge must also be conserved. We cannot have a buildup of charge on one side of the reaction or the other. In redox equations, electrons are used to balance the net charge between reactants and products.

21. Magnesium is an alkaline earth metal; Mg will oxidize to Mg^{2+}. The oxidation state of hydrogen in HCl is +1. To be reduced, the oxidation state of H must decrease. The obvious choice for the hydrogen product is $H_2(g)$, where hydrogen has a zero oxidation state. The balanced reaction is $Mg(s) + 2HCl(aq) \rightarrow MgCl_2(aq) + H_2(g)$. Mg goes from the 0 to the +2 oxidation state by losing two electrons. Each H atom goes from the +1 to the 0 oxidation state by gaining one electron. Since there are two H atoms in the balanced equation, then a total of two electrons are gained by the H atoms. Hence two electrons are transferred in the balanced reaction. When the electrons are transferred directly from Mg to H^+, no work is obtained. In order to harness this reaction to do useful work, we must control the flow of electrons through a wire. This is accomplished by making a galvanic cell that separates the reduction reaction from the oxidation reaction in order to control the flow of electrons through a wire to produce a voltage.

22. Galvanic cells use spontaneous redox reactions to produce a voltage. For a spontaneous redox reaction, the key is to have an overall positive E°_{cell} value when manipulating the half-reactions. For any two half-reactions, the half-reaction with the most positive reduction potential will always be the cathode reaction. The remaining half-reaction (the one with the most negative E°_{red}) will be reversed and become the anode half-reaction. This combination will always yield an overall reaction having a positive standard cell potential that can be used to run a galvanic cell.

23. An extensive property is one that depends directly on the amount of substance. The free-energy change for a reaction depends on whether 1 mole of product is produced or 2 moles of product is produced or 1 million moles of product is produced. This is not the case for cell

potentials, which do not depend on the amount of substance. The equation that relates ΔG to E is ΔG = −nFE. It is the n term that converts the intensive property E into the extensive property ΔG. n is the number of moles of electrons transferred in the balanced reaction that ΔG is associated with.

24. $$E = E^{\circ}_{cell} - \frac{0.0591}{n}\log Q$$

A concentration cell has the same anode and cathode contents; thus $E^{\circ}_{cell} = 0$ for a concentration cell. No matter which half-reaction you choose, the opposite half-reaction is occurring in the other cell. The driving force to produce a voltage is the −log Q term in the Nernst equation. Q is determined by the concentration of ions in the anode and cathode compartments. The larger the difference in concentrations, the larger is the −log Q term, and the larger is the voltage produced. Therefore, the driving force for concentration cells is the difference in ion concentrations between the cathode and anode compartments. When the ion concentrations are equal, Q = 1 and log Q = 0, and no voltage is produced.

25. A potential hazard when jump starting a car is the possibility for the electrolysis of $H_2O(l)$ to occur. When $H_2O(l)$ is electrolyzed, the products are the explosive gas mixture of $H_2(g)$ and $O_2(g)$. A spark produced during jump-starting a car could ignite any $H_2(g)$ and $O_2(g)$ produced. Grounding the jumper cable far from the battery minimizes the risk of a spark nearby the battery, where $H_2(g)$ and $O_2(g)$ could be collecting.

26. Metals corrode because they oxidize easily. Referencing Table 18.1, most metals are associated with negative standard reduction potentials. This means that the reverse reactions, the oxidation half-reactions, have positive oxidation potentials indicating that they oxidize fairly easily. Another key point is that the reduction of O_2 (which is a reactant in corrosion processes) has a more positive E°_{red} than most of the metals (for O_2, $E^{\circ}_{red} = 0.40$ V). This means that when O_2 is coupled with most metals, the reaction will be spontaneous since $E^{\circ}_{cell} > 0$, so corrosion occurs.

The noble metals (Ag, Au, and Pt) all have standard reduction potentials greater than that of O_2. Therefore, O_2 is not capable of oxidizing these metals at standard conditions.

Note: The standard reduction potential for $Pt \rightarrow Pt^{2+} + 2\ e^-$ is not in Table 18.1. As expected, its reduction potential is greater than that of O_2 ($E^{\circ}_{Pt} = 1.19$ V).

27. You need to know the identity of the metal so that you know which molar mass to use. You need to know the oxidation state of the metal ion in the salt so that the moles of electrons transferred can be determined. And finally, you need to know the amount of current and the time the current was passed through the electrolytic cell. If you know these four quantities, then the mass of metal plated out can be calculated.

28. For a galvanic cell, we want a combination of the half-reactions that give a positive overall cell potential. For these two cell compartments, the combination that gives a positive overall cell potential is:

$$\begin{array}{ll} (Ag^+ + e^- \rightarrow Ag) \times 2 & E° = 0.80\text{ V} \\ Cu \rightarrow Cu^{2+} + 2e^- & -E° = -0.34\text{ V} \\ \hline 2\,Ag^+(aq) + Cu(s) \rightarrow 2\,Ag(s) + Cu^{2+}(aq) & E^\circ_{cell} = 0.46\text{ V} \end{array}$$

a. From above, silver metal is a product when these two compartments make a galvanic cell ($E^\circ_{cell} > 0$). The silver compartment is the cathode (Ag^+ is reduced) and the copper compartment is the anode (Cu is oxidized). As is always the case, electrons flow from the anode (copper compartment) to the cathode (silver compartment).

b. The reverse of the spontaneous reaction produces copper metal [$2\,Ag(s) + Cu^{2+}(aq) \rightarrow 2\,Ag^+(aq) + Cu(s)$]. This reverse reaction is nonspontaneous ($E^\circ_{cell} = -0.46$ V), so this will be an electrolytic cell. For this nonspontaneous reaction, the copper compartment is the cathode and the silver compartment is the anode. Electron flow will be from the silver compartment to the copper compartment (from the anode to the cathode).

c. From the initial work above, $E^\circ_{cell} = 0.46$ V when these compartments form a galvanic cell.

d. From part b, $2\,Ag(s) + Cu^{2+}(aq) \rightarrow 2\,Ag^+(aq) + Cu(s)$ $E^\circ_{cell} = -0.46$ V; this is the electrolytic cell reaction which tells us that a potential greater than 0.46 V must be applied to force this nonspontaneous reaction to occur.

Balancing Oxidation-Reduction Equations

29. Use the method of half-reactions described in Section 18.1 of the text to balance these redox reactions. The first step always is to separate the reaction into the two half-reactions, and then to balance each half-reaction separately.

a. $3\,I^- \rightarrow I_3^- + 2e^-$ $ClO^- \rightarrow Cl^-$

$2e^- + 2H^+ + ClO^- \rightarrow Cl^- + H_2O$

Adding the two balanced half-reactions so electrons cancel:

$$3\,I^-(aq) + 2\,H^+(aq) + ClO^-(aq) \rightarrow I_3^-(aq) + Cl^-(aq) + H_2O(l)$$

b. $As_2O_3 \rightarrow H_3AsO_4$ $NO_3^- \rightarrow NO + 2\,H_2O$

$As_2O_3 \rightarrow 2\,H_3AsO_4$ $4\,H^+ + NO_3^- \rightarrow NO + 2\,H_2O$

Left 3 – O; right 8 – O $(3\,e^- + 4\,H^+ + NO_3^- \rightarrow NO + 2\,H_2O) \times 4$

Right hand side has 5 extra O. Balance the oxygen atoms first using H_2O, then balance H using H^+, and finally, balance charge using electrons. This gives:

$$(5\,H_2O + As_2O_3 \rightarrow 2\,H_3AsO_4 + 4\,H^+ + 4\,e^-) \times 3$$

Common factor is a transfer of 12 e^-. Add half-reactions so that electrons cancel.

$$12\ e^- + 16\ H^+ + 4\ NO_3^- \rightarrow 4\ NO + 8\ H_2O$$
$$15\ H_2O + 3\ As_2O_3 \rightarrow 6\ H_3AsO_4 + 12\ H^+ + 12\ e^-$$

$$7\ H_2O(l) + 4\ H^+(aq) + 3\ As_2O_3(s) + 4\ NO_3^-(aq) \rightarrow 4\ NO(g) + 6\ H_3AsO_4(aq)$$

c. $(2\ Br \rightarrow Br_2 + 2\ e^-) \times 5$

$$MnO_4^- \rightarrow Mn^{2+} + 4\ H_2O$$
$$(5\ e^- + 8\ H^+ + MnO_4^- \rightarrow Mn^{2+} + 4\ H_2O) \times 2$$

Common factor is a transfer of 10 e^-.

$$10\ Br^- \rightarrow 5\ Br_2 + 10\ e^-$$
$$10\ e^- + 16\ H^+ + 2\ MnO_4^- \rightarrow 2\ Mn^{2+} + 8\ H_2O$$

$$16\ H^+(aq) + 2\ MnO_4^-(aq) + 10\ Br^-(aq) \rightarrow 5\ Br_2(l) + 2\ Mn^{2+}(aq) + 8\ H_2O(l)$$

d. $CH_3OH \rightarrow CH_2O$
$(CH_3OH \rightarrow CH_2O + 2\ H^+ + 2\ e^-) \times 3$

$$Cr_2O_7^{2-} \rightarrow 2\ Cr^{3+}$$
$$14\ H^+ + Cr_2O_7^{2-} \rightarrow 2\ Cr^{3+} + 7\ H_2O$$
$$6\ e^- + 14\ H^+ + Cr_2O_7^{2-} \rightarrow 2\ Cr^{3+} + 7\ H_2O$$

Common factor is a transfer of 6 e^-.

$$3\ CH_3OH \rightarrow 3\ CH_2O + 6\ H^+ + 6\ e^-$$
$$6\ e^- + 14\ H^+ + Cr_2O_7^{2-} \rightarrow 2\ Cr^{3+} + 7\ H_2O$$

$$8\ H^+(aq) + 3\ CH_3OH(aq) + Cr_2O_7^{2-}(aq) \rightarrow 2\ Cr^{3+}(aq) + 3\ CH_2O(aq) + 7\ H_2O(l)$$

30. a. $(Cu \rightarrow Cu^{2+} + 2\ e^-) \times 3$

$$NO_3^- \rightarrow NO + 2\ H_2O$$
$$(3\ e^- + 4\ H^+ + NO_3^- \rightarrow NO + 2\ H_2O) \times 2$$

Adding the two balanced half-reactions so that electrons cancel:

$$3\ Cu \rightarrow 3\ Cu^{2+} + 6\ e^-$$
$$6\ e^- + 8\ H^+ + 2\ NO_3^- \rightarrow 2\ NO + 4\ H_2O$$

$$3\ Cu(s) + 8\ H^+(aq) + 2\ NO_3^-(aq) \rightarrow 3\ Cu^{2+}(aq) + 2\ NO(g) + 4\ H_2O(l)$$

b. $(2\ Cl^- \rightarrow Cl_2 + 2\ e^-) \times 3$

$$Cr_2O_7^{2-} \rightarrow 2\ Cr^{3+} + 7\ H_2O$$
$$6\ e^- + 14\ H^+ + Cr_2O_7^{2-} \rightarrow 2\ Cr^{3+} + 7\ H_2O$$

Add the two half-reactions with six electrons transferred:

$$6\ Cl^- \rightarrow 3\ Cl_2 + 6\ e^-$$
$$6\ e^- + 14\ H^+ + Cr_2O_7^{2-} \rightarrow 2\ Cr^{3+} + 7\ H_2O$$

$$14\ H^+(aq) + Cr_2O_7^{2-}(aq) + 6\ Cl^-(aq) \rightarrow 3\ Cl_2(g) + 2\ Cr^{3+}(aq) + 7\ H_2O(l)$$

c. $Pb \rightarrow PbSO_4$
$Pb + H_2SO_4 \rightarrow PbSO_4 + 2\ H^+$
$Pb + H_2SO_4 \rightarrow PbSO_4 + 2\ H^+ + 2\ e^-$

$$PbO_2 \rightarrow PbSO_4$$
$$PbO_2 + H_2SO_4 \rightarrow PbSO_4 + 2\ H_2O$$
$$2\ e^- + 2\ H^+ + PbO_2 + H_2SO_4 \rightarrow PbSO_4 + 2\ H_2O$$

Add the two half-reactions with two electrons transferred:

$$2\ e^- + 2\ H^+ + PbO_2 + H_2SO_4 \rightarrow PbSO_4 + 2\ H_2O$$
$$Pb + H_2SO_4 \rightarrow PbSO_4 + 2\ H^+ + 2\ e^-$$

$$Pb(s) + 2\ H_2SO_4(aq) + PbO_2(s) \rightarrow 2\ PbSO_4(s) + 2\ H_2O(l)$$

This is the reaction that occurs in an automobile lead-storage battery.

d.

$$Mn^{2+} \rightarrow MnO_4^-$$
$$(4\ H_2O + Mn^{2+} \rightarrow MnO_4^- + 8\ H^+ + 5\ e^-) \times 2$$

$$NaBiO_3 \rightarrow Bi^{3+} + Na^+$$
$$6\ H^+ + NaBiO_3 \rightarrow Bi^{3+} + Na^+ + 3\ H_2O$$
$$(2\ e^- + 6\ H^+ + NaBiO_3 \rightarrow Bi^{3+} + Na^+ + 3\ H_2O) \times 5$$

$$8\ H_2O + 2\ Mn^{2+} \rightarrow 2\ MnO_4^- + 16\ H^+ + 10\ e^-$$
$$10\ e^- + 30\ H^+ + 5\ NaBiO_3 \rightarrow 5\ Bi^{3+} + 5\ Na^+ + 15\ H_2O$$

$$8\ H_2O + 30\ H^+ + 2\ Mn^{2+} + 5\ NaBiO_3 \rightarrow 2\ MnO_4^- + 5\ Bi^{3+} + 5\ Na^+ + 15\ H_2O + 16\ H^+$$

Simplifying:

$$14\ H^+(aq) + 2\ Mn^{2+}(aq) + 5\ NaBiO_3(s) \rightarrow 2\ MnO_4^-(aq) + 5\ Bi^{3+}(aq) + 5\ Na^+(aq) + 7\ H_2O(l)$$

e.

$$H_3AsO_4 \rightarrow AsH_3$$
$$H_3AsO_4 \rightarrow AsH_3 + 4\ H_2O$$
$$8\ e^- + 8\ H^+ + H_3AsO_4 \rightarrow AsH_3 + 4\ H_2O$$

$$(Zn \rightarrow Zn^{2+} + 2\ e^-) \times 4$$

$$8\ e^- + 8\ H^+ + H_3AsO_4 \rightarrow AsH_3 + 4\ H_2O$$
$$4\ Zn \rightarrow 4\ Zn^{2+} + 8\ e^-$$

$$8\ H^+(aq) + H_3AsO_4(aq) + 4\ Zn(s) \rightarrow 4\ Zn^{2+}(aq) + AsH_3(g) + 4\ H_2O(l)$$

31. Use the same method as with acidic solutions. After the final balanced equation, convert H^+ to OH^- as described in Section 18.1 of the text. The extra step involves converting H^+ into H_2O by adding equal moles of OH^- to each side of the reaction. This converts the reaction to a basic solution while still keeping it balanced.

a.

$$Al \rightarrow Al(OH)_4^-$$
$$4\ H_2O + Al \rightarrow Al(OH)_4^- + 4\ H^+$$
$$4\ H_2O + Al \rightarrow Al(OH)_4^- + 4\ H^+ + 3\ e^-$$

$$MnO_4^- \rightarrow MnO_2$$
$$3\ e^- + 4\ H^+ + MnO_4^- \rightarrow MnO_2 + 2\ H_2O$$

$$4\ H_2O + Al \rightarrow Al(OH)_4^- + 4\ H^+ + 3\ e^-$$
$$3\ e^- + 4\ H^+ + MnO_4^- \rightarrow MnO_2 + 2\ H_2O$$

$$2\ H_2O(l) + Al(s) + MnO_4^-(aq) \rightarrow Al(OH)_4^-(aq) + MnO_2(s)$$

H^+ doesn't appear in the final balanced reaction, so we are done.

b.

$$Cl_2 \rightarrow Cl^-$$
$$2\ e^- + Cl_2 \rightarrow 2\ Cl^-$$

$$Cl_2 \rightarrow OCl^-$$
$$2\ H_2O + Cl_2 \rightarrow 2\ OCl^- + 4\ H^+ + 2\ e^-$$

$$2\ e^- + Cl_2 \rightarrow 2\ Cl^-$$
$$2\ H_2O + Cl_2 \rightarrow 2\ OCl^- + 4\ H^+ + 2\ e^-$$
$$\overline{2\ H_2O + 2\ Cl_2 \rightarrow 2\ Cl^- + 2\ OCl^- + 4\ H^+}$$

Now convert to a basic solution. Add 4 OH^- to both sides of the equation. The 4 OH^- will react with the 4 H^+ on the product side to give 4 H_2O. After this step, cancel identical species on both sides (2 H_2O). Applying these steps gives: 4 OH^- + 2 Cl_2 → 2 Cl^- + 2 OCl^- + 2 H_2O, which can be further simplified to:

$$2\ OH^-(aq) + Cl_2(g) \rightarrow Cl^-(aq) + OCl^-(aq) + H_2O(l)$$

c. $NO_2^- \rightarrow NH_3$ $\qquad$ $Al \rightarrow AlO_2^-$

$6\ e^- + 7\ H^+ + NO_2^- \rightarrow NH_3 + 2\ H_2O$ $\qquad$ $(2\ H_2O + Al \rightarrow AlO_2^- + 4\ H^+ + 3\ e^-) \times 2$

Common factor is a transfer of 6 e^-.

$$6e^- + 7\ H^+ + NO_2^- \rightarrow NH_3 + 2\ H_2O$$
$$4\ H_2O + 2\ Al \rightarrow 2\ AlO_2^- + 8\ H^+ + 6\ e^-$$
$$\overline{OH^- + 2\ H_2O + NO_2^- + 2\ Al \rightarrow NH_3 + 2\ AlO_2^- + H^+ + OH^-}$$

Reducing gives $OH^-(aq) + H_2O(l) + NO_2^-(aq) + 2\ Al(s) \rightarrow NH_3(g) + 2\ AlO_2^-(aq)$.

32. a. $Cr \rightarrow Cr(OH)_3$ $\qquad$ $CrO_4^{2-} \rightarrow Cr(OH)_3$

$3\ H_2O + Cr \rightarrow Cr(OH)_3 + 3\ H^+ + 3\ e^-$ $\qquad$ $3\ e^- + 5\ H^+ + CrO_4^{2-} \rightarrow Cr(OH)_3 + H_2O$

$$3\ H_2O + Cr \rightarrow Cr(OH)_3 + 3\ H^+ + 3\ e^-$$
$$3\ e^- + 5\ H^+ + CrO_4^{2-} \rightarrow Cr(OH)_3 + H_2O$$
$$\overline{2\ OH^- + 2\ H^+ + 2\ H_2O + Cr + CrO_4^{2-} \rightarrow 2\ Cr(OH)_3 + 2\ OH^-}$$

Two OH^- were added above to each side to convert to a basic solution. The two OH^- react with the 2 H^+ on the reactant side to produce 2 H_2O. The overall balanced equation is:

$$4\ H_2O(l) + Cr(s) + CrO_4^{2-}(aq) \rightarrow 2\ Cr(OH)_3(s) + 2\ OH^-(aq)$$

b. $S^{2-} \rightarrow S$ $\qquad$ $MnO_4^- \rightarrow MnS$

$(S^{2-} \rightarrow S + 2\ e^-) \times 5$ $\qquad$ $MnO_4^- + S^{2-} \rightarrow MnS$

$(5\ e^- + 8\ H^+ + MnO_4^- + S^{2-} \rightarrow MnS + 4\ H_2O) \times 2$

Common factor is a transfer of 10 e^-.

$$5\ S^{2-} \rightarrow 5\ S + 10\ e^-$$
$$10\ e^- + 16\ H^+ + 2\ MnO_4^- + 2\ S^{2-} \rightarrow 2\ MnS + 8\ H_2O$$
$$\overline{16\ OH^- + 16\ H^+ + 7\ S^{2-} + 2\ MnO_4^- \rightarrow 5\ S + 2\ MnS + 8\ H_2O + 16\ OH^-}$$

$$16\ H_2O + 7\ S^{2-} + 2\ MnO_4^- \rightarrow 5\ S + 2\ MnS + 8\ H_2O + 16\ OH^-$$

Reducing gives $8\ H_2O(l) + 7\ S^{2-}(aq) + 2\ MnO_4^-(aq) \rightarrow 5\ S(s) + 2\ MnS(s) + 16\ OH^-(aq)$.

c. $CN^- \rightarrow CNO^-$

$(H_2O + CN^- \rightarrow CNO^- + 2\ H^+ + 2\ e^-) \times 3$

$MnO_4^- \rightarrow MnO_2$

$(3\ e^- + 4\ H^+ + MnO_4^- \rightarrow MnO_2 + 2\ H_2O) \times 2$

Common factor is a transfer of 6 electrons.

$$3\ H_2O + 3\ CN^- \rightarrow 3\ CNO^- + 6\ H^+ + 6\ e^-$$
$$6\ e^- + 8\ H^+ + 2\ MnO_4^- \rightarrow 2\ MnO_2 + 4\ H_2O$$

$$2\ OH^- + 2\ H^+ + 3\ CN^- + 2\ MnO_4^- \rightarrow 3\ CNO^- + 2\ MnO_2 + H_2O + 2\ OH^-$$

Reducing gives:

$$H_2O(l) + 3\ CN^-(aq) + 2\ MnO_4^-(aq) \rightarrow 3\ CNO^-(aq) + 2\ MnO_2(s) + 2\ OH^-(aq)$$

33. $NaCl + H_2SO_4 + MnO_2 \rightarrow Na_2SO_4 + MnCl_2 + Cl_2 + H_2O$

We could balance this reaction by the half-reaction method, which is generally the preferred method. However, sometimes a redox reaction is not so complicated and thus balancing by inspection is a possibility. Let's try inspection here. To balance Cl^-, we need 4 NaCl:

$$4\ NaCl + H_2SO_4 + MnO_2 \rightarrow Na_2SO_4 + MnCl_2 + Cl_2 + H_2O$$

Balance the Na^+ and $SO_4^{2!}$ ions next:

$$4\ NaCl + 2\ H_2SO_4 + MnO_2 \rightarrow 2\ Na_2SO_4 + MnCl_2 + Cl_2 + H_2O$$

On the left side: 4 H and 10 O; on the right side: 8 O not counting H_2O

We need 2 H_2O on the right side to balance H and O:

$$4\ NaCl(aq) + 2\ H_2SO_4(aq) + MnO_2(s) \rightarrow 2\ Na_2SO_4(aq) + MnCl_2(aq) + Cl_2(g) + 2\ H_2O(l)$$

34. $Au + HNO_3 + HCl \rightarrow AuCl_4^- + NO$

Only deal with ions that are reacting (omit H^+): $Au + NO_3^- + Cl^- \rightarrow AuCl_4^- + NO$

The balanced half-reactions are:

$Au + 4\ Cl^- \rightarrow AuCl_4^- + 3\ e^-$ $\qquad$ $3\ e^- + 4\ H^+ + NO_3^- \rightarrow NO + 2\ H_2O$

Adding the two balanced half-reactions:

$$Au(s) + 4\ Cl^-(aq) + 4\ H^+(aq) + NO_3^-(aq) \rightarrow AuCl_4^-(aq) + NO(g) + 2\ H_2O(l)$$

Exercises

Galvanic Cells, Cell Potentials, Standard Reduction Potentials, and Free Energy

35. The reducing agent causes reduction to occur; it does this by containing the species which is oxidized. Oxidation occurs at the anode, so the reducing agent will be in the anode compartment. The oxidizing agent causes oxidation to occur; it does this by containing the species which is reduced. Reduction occurs at the cathode, so the oxidizing agent will be in the cathode compartment. Electron flow is always from the anode compartment to the cathode compartment.

36. A galvanic cell at standard conditions must have a positive overall standard cell potential ($E^{\circ}_{cell} > 0$). The only combination of the half-reactions that gives a positive cell potential is:

$$Cu^{2+} + 2e^- \rightarrow Cu \qquad E°(\text{cathode}) = 0.34 \text{ V}$$
$$Zn \rightarrow Zn^{2+} + 2e^- \qquad -E°(\text{anode}) = 0.76 \text{ V}$$

$$Cu^{2+}(aq) + Zn(s) \rightarrow Cu(s) + Zn^{2+}(aq) \qquad E^{\circ}_{cell} = 1.10 \text{ V}$$

a. The reducing agent causes reduction to occur since it always contains the species which is oxidized. Zn is oxidized in the galvanic cell, so Zn is the reducing agent. The oxidizing agent causes oxidation to occur since it always contains the species which is reduced. Cu^{2+} is reduced in the galvanic cell, so Cu^{2+} is the oxidizing agent. Electrons will flow from the zinc compartment (the anode) to the copper compartment (the cathode).

b. From the work above, $E^{\circ}_{cell} = 1.10$ V.

c. The pure metal that is a product in the spontaneous reaction is copper. So the copper electrode will increase in mass as $Cu^{2+}(aq)$ is reduced to Cu(s). The zinc electrode will decrease in mass for this galvanic cell as Zn(s) is oxidized to $Zn^{2+}(aq)$.

37. A typical galvanic cell diagram is:

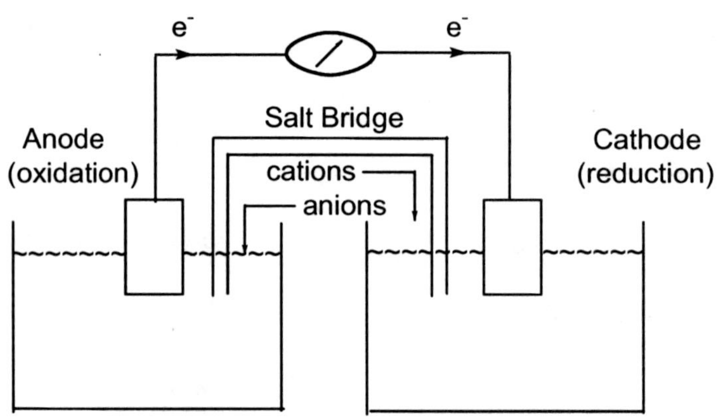

The diagram for all cells will look like this. The contents of each half-cell compartment will be identified for each reaction, with all solute concentrations at 1.0 *M* and all gases at 1.0 atm. For Exercises 37 and 38, the flow of ions through the salt bridge was not asked for in the questions. If asked, however, cations always flow into the cathode compartment, and anions always flow into the anode compartment. This is required to keep each compartment electrically neutral.

a. Table 18.1 of the text lists balanced reduction half-reactions for many substances. For this overall reaction, we need the Cl_2 to Cl^- reduction half-reaction and the Cr^{3+} to $Cr_2O_7^{2-}$ oxidation half-reaction. Manipulating these two half-reactions gives the overall balanced equation.

$$(Cl_2 + 2\,e^- \rightarrow 2\,Cl^-) \times 3$$
$$7\,H_2O + 2\,Cr^{3+} \rightarrow Cr_2O_7^{2-} + 14\,H^+ + 6\,e^-$$

$$7\,H_2O(l) + 2\,Cr^{3+}(aq) + 3\,Cl_2(g) \rightarrow Cr_2O_7^{2-}(aq) + 6\,Cl^-(aq) + 14\,H^+(aq)$$

The contents of each compartment are:

Cathode: Pt electrode; Cl_2 bubbled into solution, Cl^- in solution

Anode: Pt electrode; Cr^{3+}, H^+, and $Cr_2O_7^{2-}$ in solution

We need a nonreactive metal to use as the electrode in each case, since all the reactants and products are in solution. Pt is a common choice. Another possibility is graphite.

b.
$$Cu^{2+} + 2\,e^- \rightarrow Cu$$
$$Mg \rightarrow Mg^{2+} + 2e^-$$

$$Cu^{2+}(aq) + Mg(s) \rightarrow Cu(s) + Mg^{2+}(aq)$$

Cathode: Cu electrode; Cu^{2+} in solution; anode: Mg electrode; Mg^{2+} in solution

38. Reference the answer to Exercise 37 for a typical galvanic cell diagram. The contents of each half-cell compartment are identified below with all solute concentrations at 1.0 *M* and all gases at 1.0 atm.

a. Reference Table 18.1 for the balanced half-reactions.

$$5\,e^- + 6\,H^+ + IO_3^- \rightarrow 1/2\,I_2 + 3\,H_2O$$
$$(Fe^{2+} \rightarrow Fe^{3+} + e^-) \times 5$$

$$6\,H^+ + IO_3^- + 5\,Fe^{2+} \rightarrow 5\,Fe^{3+} + 1/2\,I_2 + 3\,H_2O$$

or $12\,H^+(aq) + 2\,IO_3^-(aq) + 10\,Fe^{2+}(aq) \rightarrow 10\,Fe^{3+}(aq) + I_2(aq) + 6\,H_2O(l)$

Cathode: Pt electrode; IO_3^-, I_2 and H_2SO_4 (H^+ source) in solution.

Note: $I_2(s)$ would make a poor electrode since it sublimes.

Anode: Pt electrode; Fe^{2+} and Fe^{3+} in solution

b.

$$(Ag^+ + e^- \rightarrow Ag) \times 2$$
$$Zn \rightarrow Zn^{2+} + 2\ e^-$$
$$\overline{Zn(s) + 2\ Ag^+(aq) \rightarrow 2\ Ag(s) + Zn^{2+}(aq)}$$

Cathode: Ag electrode; Ag^+ in solution; anode: Zn electrode; Zn^{2+} in solution

39. To determine E° for the overall cell reaction, we must add the standard reduction potential to the standard oxidation potential ($E^\circ_{cell} = E^\circ_{red} + E^\circ_{ox}$). Reference Table 18.1 for values of standard reduction potentials. Remember that $E^\circ_{ox} = -E^\circ_{red}$ and that standard potentials are <u>not</u> multiplied by the integer used to obtain the overall balanced equation.

37a. $E^\circ_{cell} = E^\circ_{Cl_2 \rightarrow Cl^-} + E^\circ_{Cr^{3+} \rightarrow Cr_2O_7^{2-}} = 1.36\ V + (-1.33\ V) = 0.03\ V$

37b. $E^\circ_{cell} = E^\circ_{Cu^{2+} \rightarrow Cu} + E^\circ_{Mg \rightarrow Mg^{2+}} = 0.34\ V + 2.37\ V = 2.71\ V$

40. 38a. $E^\circ_{cell} = E^\circ_{IO_3^- \rightarrow I_2} + E^\circ_{Fe^{2+} \rightarrow Fe^{3+}} = 1.20\ V + (-0.77\ V) = 0.43\ V$

38b. $E^\circ_{cell} = E^\circ_{Ag^+ \rightarrow Ag} + E^\circ_{Zn \rightarrow Zn^{2+}} = 0.80\ V + 0.76\ V = 1.56\ V$

41. Reference the answer to Exercise 37 for a typical galvanic cell design. The contents of each half-cell compartment are identified below with all solute concentrations at 1.0 *M* and all gases at 1.0 atm. For each pair of half-reactions, the half-reaction with the largest (most positive) standard reduction potential will be the cathode reaction, and the half-reaction with the smallest (most negative) reduction potential will be reversed to become the anode reaction. Only this combination gives a spontaneous overall reaction, i.e., a reaction with a positive overall standard cell potential. Note that in a galvanic cell as illustrated in Exercise 37 the cations in the salt bridge migrate to the cathode, and the anions migrate to the anode.

a.

$$Cl_2 + 2\ e^- \rightarrow 2\ Cl^- \qquad E^\circ = 1.36\ V$$
$$2\ Br^- \rightarrow Br_2 + 2\ e^- \qquad -E^\circ = -1.09\ V$$
$$\overline{Cl_2(g) + 2\ Br^-(aq) \rightarrow Br_2(aq) + 2\ Cl^-(aq) \qquad E^\circ_{cell} = 0.27\ V}$$

The contents of each compartment are:

Cathode: Pt electrode; $Cl_2(g)$ bubbled in, Cl^- in solution

Anode: Pt electrode; Br_2 and Br^- in solution

b.

$$(2\ e^- + 2\ H^+ + IO_4^- \rightarrow IO_3^- + H_2O) \times 5 \qquad E^\circ = 1.60\ V$$
$$(4\ H_2O + Mn^{2+} \rightarrow MnO_4^- + 8\ H^+ + 5\ e^-) \times 2 \qquad -E^\circ = -1.51\ V$$
$$\overline{10\ H^+ + 5\ IO_4^- + 8\ H_2O + 2\ Mn^{2+} \rightarrow 5\ IO_3^- + 5\ H_2O + 2\ MnO_4^- + 16\ H^+ \qquad E^\circ_{cell} = 0.09\ V}$$

This simplifies to:

$$3\ H_2O(l) + 5\ IO_4^-(aq) + 2\ Mn^{2+}(aq) \rightarrow 5\ IO_3^-\ (aq) + 2\ MnO_4^-\ (aq) + 6\ H^+(aq) \qquad E^\circ_{cell} = 0.09\ V$$

Cathode: Pt electrode; IO_4^-, IO_3^-, and H_2SO_4 (as a source of H^+) in solution

Anode: Pt electrode; Mn^{2+}, MnO_4^-, and H_2SO_4 in solution

42. Reference the answer to Exercise 37 for a typical galvanic cell design. The contents of each half-cell compartment are identified below, with all solute concentrations at 1.0 *M* and all gases at 1.0 atm.

a. $H_2O_2 + 2\ H^+ + 2\ e^- \rightarrow 2\ H_2O \qquad E^\circ = 1.78\ V$

$H_2O_2 \rightarrow O_2 + 2\ H^+ + 2\ e^- \qquad -E^\circ = -0.68\ V$

$$2\ H_2O_2(aq) \rightarrow 2\ H_2O(l) + O_2(g) \qquad E^\circ_{cell} = 1.10\ V$$

Cathode: Pt electrode; H_2O_2 and H^+ in solution

Anode: Pt electrode; $O_2(g)$ bubbled in, H_2O_2 and H^+ in solution

b. $(Fe^{3+} + 3\ e^- \rightarrow Fe) \times 2 \qquad E^\circ = -0.036\ V$

$(Mn \rightarrow Mn^{2+} + 2\ e^-) \times 3 \qquad -E^\circ = 1.18\ V$

$$2\ Fe^{3+}(aq) + 3\ Mn(s) \rightarrow 2\ Fe(s) + 3\ Mn^{2+}(aq) \qquad E^\circ_{cell} = 1.14\ V$$

Cathode: Fe electrode; Fe^{3+} in solution; anode: Mn electrode; Mn^{2+} in solution

43. In standard line notation, the anode is listed first, and the cathode is listed last. A double line separates the two compartments. By convention, the electrodes are on the ends with all solutes and gases toward the middle. A single line is used to indicate a phase change. We also included all concentrations.

37a. Pt | Cr^{3+} (1.0 *M*), $Cr_2O_7^{2-}$ (1.0 *M*), H^+ (1.0 *M*) || Cl_2 (1.0 atm) | Cl^- (1.0 *M*) | Pt

37b. Mg | Mg^{2+} (1.0 *M*) || Cu^{2+} (1.0 *M*) | Cu

41a. Pt | Br^- (1.0 *M*), Br_2 (1.0 *M*) || Cl_2 (1.0 atm) | Cl^- (1.0 *M*) | Pt

41b. Pt | Mn^{2+} (1.0 *M*), MnO_4^- (1.0 *M*), H^+ (1.0 *M*) || IO_4^- (1.0 *M*), H^+ (1.0 *M*), IO_3^- (1.0 *M*) | Pt

44. 38a. Pt | Fe^{2+} (1.0 *M*), Fe^{3+} (1.0 *M*) || IO_3^- (1.0 *M*), H^+ (1.0 *M*), I_2 (1.0 *M*) | Pt

38b. Zn | Zn^{2+} (1.0 *M*) || Ag^+ (1.0 *M*) | Ag

42a. Pt | H_2O_2 (1.0 *M*), H^+ (1.0 *M*) | O_2 (1.0 atm)|| H_2O_2 (1.0 *M*), H^+ (1.0 *M*) | Pt

42b. Mn | Mn^{2+} (1.0 *M*) || Fe^{3+} (1.0 *M*) | Fe

45. Locate the pertinent half-reactions in Table 18.1, and then figure which combination will give a positive standard cell potential. In all cases, the anode compartment contains the species with the smallest standard reduction potential. For part a, the copper compartment is the anode, and in part b, the cadmium compartment is the anode.

a.
$$Au^{3+} + 3\,e^- \rightarrow Au \qquad E° = 1.50\text{ V}$$
$$(Cu^+ \rightarrow Cu^{2+} + e^-) \times 3 \qquad -E° = -0.16\text{ V}$$
$$Au^{3+}(aq) + 3\,Cu^+(aq) \rightarrow Au(s) + 3\,Cu^{2+}(aq) \qquad E°_{cell} = 1.34\text{ V}$$

b.
$$(VO_2^+ + 2\,H^+ + e^- \rightarrow VO^{2+} + H_2O) \times 2 \qquad E° = 1.00\text{ V}$$
$$Cd \rightarrow Cd^{2+} + 2e^- \qquad -E° = 0.40\text{ V}$$
$$2\,VO_2^+(aq) + 4\,H^+(aq) + Cd(s) \rightarrow 2\,VO^{2+}(aq) + 2\,H_2O(l) + Cd^{2+}(aq) \qquad E°_{cell} = 1.40\text{ V}$$

46. a.
$$(H_2O_2 + 2\,H^+ + 2\,e^- \rightarrow 2\,H_2O) \times 3 \qquad E° = 1.78\text{ V}$$
$$2\,Cr^{3+} + 7\,H_2O \rightarrow Cr_2O_7^{2-} + 14\,H^+ + 6\,e^- \qquad -E° = -1.33\text{ V}$$
$$3\,H_2O_2(aq) + 2\,Cr^{3+}(aq) + H_2O(l) \rightarrow Cr_2O_7^{2-}(aq) + 8\,H^+(aq) \qquad E°_{cell} = 0.45\text{ V}$$

b.
$$(2\,H^+ + 2\,e^- \rightarrow H_2) \times 3 \qquad E° = 0.00\text{ V}$$
$$(Al \rightarrow Al^{3+} + 3\,e^-) \times 2 \qquad -E° = 1.66\text{ V}$$
$$6\,H^+(aq) + 2\,Al(s) \rightarrow 3\,H_2(g) + 2\,Al^{3+}(aq) \qquad E°_{cell} = 1.66\text{ V}$$

47. a.
$$(5\,e^- + 8\,H^+ + MnO_4^- \rightarrow Mn^{2+} + 4\,H_2O) \times 2 \qquad E° = 1.51\text{ V}$$
$$(2\,I^- \rightarrow I_2 + 2\,e^-) \times 5 \qquad -E° = -0.54\text{ V}$$
$$16\,H^+(aq) + 2\,MnO_4^-(aq) + 10\,I^-(aq) \rightarrow 5\,I_2(aq) + 2\,Mn^{2+}(aq) + 8\,H_2O(l) \qquad E°_{cell} = 0.97\text{ V}$$

This reaction is spontaneous at standard conditions because $E°_{cell} > 0$.

b.
$$(5\,e^- + 8\,H^+ + MnO_4^- \rightarrow Mn^{2+} + 4\,H_2O) \times 2 \qquad E° = 1.51\text{ V}$$
$$(2\,F^- \rightarrow F_2 + 2\,e^-) \times 5 \qquad -E° = -2.87\text{ V}$$
$$16\,H^+(aq) + 2\,MnO_4^-(aq) + 10\,F^-(aq) \rightarrow 5\,F_2(aq) + 2\,Mn^{2+}(aq) + 8\,H_2O(l) \qquad E°_{cell} = -1.36\text{ V}$$

This reaction is not spontaneous at standard conditions because $E°_{cell} < 0$.

48. a.
$$H_2 \rightarrow 2H^+ + 2\,e^- \qquad E° = 0.00\text{ V}$$
$$H_2 + 2\,e^- \rightarrow 2H^- \qquad -E° = -2.23\text{ V}$$
$$2H_2(g) \rightarrow 2H^+(aq) + 2\,H^-(aq) \qquad E°_{cell} = -2.23\text{ V}$$

Not spontaneous

b.

$Au^{3+} + 3\,e^- \rightarrow Au$	$E° = 1.50$ V
$(Ag \rightarrow Ag^+ + e^-) \times 3$	$-E° = -0.80$ V
$Au^{3+}(aq) + 3\,Ag(s) \rightarrow Au(s) + 3\,Ag^+(aq)$	$E^o_{cell} = 0.70$ V

Spontaneous ($E^o_{cell} > 0$)

49.

$Cl_2 + 2\,e^- \rightarrow 2\,Cl^-$	$E° = 1.36$ V
$(ClO_2^- \rightarrow ClO_2 + e^-) \times 2$	$-E° = -0.954$ V
$2\,ClO_2^-(aq) + Cl_2(g) \rightarrow 2\,ClO_2(aq) + 2\,Cl^-(aq)$	$E^o_{cell} = 0.41$ V $= 0.41$ J/C

$\Delta G° = -nFE^o_{cell} = -(2 \text{ mol } e^-)(96{,}485 \text{ C/mol } e^-)(0.41 \text{ J/C}) = -7.9 \times 10^4 \text{ J} = -79 \text{ kJ}$

50. a.

$(4\,H^+ + NO_3^- + 3\,e^- \rightarrow NO + 2\,H_2O) \times 2$	$E° = 0.96$ V
$(Mn \rightarrow Mn^{2+} + 2\,e^-) \times 3$	$-E° = 1.18$ V
$3\,Mn(s) + 8\,H^+(aq) + 2\,NO_3^-(aq) \rightarrow 2\,NO(g) + 4\,H_2O(l) + 3\,Mn^{2+}(aq)$	$E^o_{cell} = 2.14$ V

$(2\,e^- + 2\,H^+ + IO_4^- \rightarrow IO_3^- + H_2O) \times 5$	$E° = 1.60$ V
$(Mn^{2+} + 4\,H_2O \rightarrow MnO_4^- + 8\,H^+ + 5\,e^-) \times 2$	$-E° = -1.51$ V
$5\,IO_4^-(aq) + 2\,Mn^{2+}(aq) + 3\,H_2O(l) \rightarrow 5\,IO_3^-(aq) + 2\,MnO_4^-(aq) + 6\,H^+(aq)$	$E^o_{cell} = 0.09$ V

b. Nitric acid oxidation (see above for E^o_{cell}):

$\Delta G° = -nFE^o_{cell} = -(6 \text{ mol } e^-)(96{,}485 \text{ C/mol } e^-)(2.14 \text{ J/C}) = -1.24 \times 10^6 \text{ J} = -1240 \text{ kJ}$

Periodate oxidation (see above for E^o_{cell}):

$\Delta G° = -(10 \text{ mol } e^-)(96{,}485 \text{ C/mol } e^-)(0.09 \text{ J/C})(1 \text{ kJ}/1000 \text{ J}) = -90 \text{ kJ}$

51. Because the cells are at standard conditions, $w_{max} = \Delta G = \Delta G° = -nFE^o_{cell}$. See Exercise 45 for the balanced overall equations and for E^o_{cell}.

45a. $w_{max} = -(3 \text{ mol } e^-)(96{,}485 \text{ C/mol } e^-)(1.34 \text{ J/C}) = -3.88 \times 10^5 \text{ J} = -388 \text{ kJ}$

45b. $w_{max} = -(2 \text{ mol } e^-)(96{,}485 \text{ C/mol } e^-)(1.40 \text{ J/C}) = -2.70 \times 10^5 \text{ J} = -270. \text{ kJ}$

52. Because the cells are at standard conditions, $w_{max} = \Delta G = \Delta G° = -nFE^o_{cell}$. See Exercise 46 for the balanced overall equations and for E^o_{cell}.

46a. $w_{max} = -(6 \text{ mol } e^-)(96{,}485 \text{ C/mol } e^-)(0.45 \text{ J/C}) = -2.6 \times 10^5 \text{ J} = -260 \text{ kJ}$

46b. $w_{max} = -(6 \text{ mol } e^-)(96{,}485 \text{ C/mol } e^-)(1.66 \text{ J/C}) = -9.61 \times 10^5 \text{ J} = -961 \text{ kJ}$

53. $2\,H_2O + 2\,e^- \rightarrow H_2 + 2\,OH^-$

$\Delta G° = \Sigma n_p \Delta G^o_{f,\,products} - \Sigma n_r \Delta G^o_{f,\,reactants} = 2(-157) - [2(-237)] = 160. \text{ kJ}$

$$\Delta G° = -nFE°,\ E° = \frac{-\Delta G°}{nF} = \frac{-1.60 \times 10^5\ J}{(2\ mol\ e^-)(96{,}485\ C/mol\ e^-)} = -0.829\ J/C = -0.829\ V$$

The two values agree to two significant figures (–0.83 V in Table 18.1).

54. $Fe^{2+} + 2\ e^- \rightarrow Fe \quad E° = -0.44\ V = -0.44\ J/C$

$\Delta G° = -nFE° = -(2\ mol\ e^-)(96{,}485\ C/mol\ e^-)(-0.44\ J/C)(1\ kJ/1000\ J) = 85\ kJ$

$85\ kJ = 0 - [\Delta G^o_{f,Fe^{2+}} + 0],\ \Delta G^o_{f,Fe^{2+}} = -85\ kJ$

We can get $\Delta G^o_{f,Fe^{3+}}$ two ways. Consider: $Fe^{3+} + e^- \rightarrow Fe^{2+} \quad E° = 0.77\ V$

$\Delta G° = -(1\ mol\ e)(96{,}485\ C/mol\ e^-)(0.77\ J/C) = -74{,}300\ J = -74\ kJ$

$$\begin{array}{ll} Fe^{2+} \rightarrow Fe^{3+} + e^- & \Delta G° = 74\ kJ \\ Fe \rightarrow Fe^{2+} + 2\ e^- & \Delta G° = -85\ kJ \\ \hline Fe \rightarrow Fe^{3+} + 3\ e^- & \Delta G° = -11\ kJ,\quad \Delta G^o_{f,Fe^{3+}} = -11\ kJ/mol \end{array}$$

or consider: $Fe^{3+} + 3\ e^- \rightarrow Fe \quad E° = -0.036\ V$

$\Delta G° = -(3\ mol\ e^-)(96{,}485\ C/mol\ e^-)(-0.036\ J/C) = 10{,}400\ J \approx 10.\ kJ$

$10.\ kJ = 0 - [\Delta G^o_{f,Fe^{3+}} + 0],\ \Delta G^o_{f,Fe^{3+}} = -10.\ kJ/mol$; round-off error explains the 1-kJ discrepancy.

55. $\Delta G° = [6\ mol(-394\ kJ/mol) + 6\ mol(-237\ kJ/mol] - [1\ mol(-911\ kJ/mol + 6\ mol(0)]$
$= -2875\ kJ$

Carbon is oxidized in this combustion reaction. In $C_6H_{12}O_6$, H has a +1 oxidation state, and oxygen has a –2 oxidation, so $6(x) + 12(+1) + 6(-2) = 0$, x = oxidation state of C in $C_6H_{12}O_6$ = 0. In CO_2, O has an oxidation state of –2, so $y + 2(-2) = 0$, y = oxidation state of C in CO_2 = +4. Carbon goes from the 0 oxidation state in $C_6H_{12}O_6$ to the +4 oxidation state in CO_2, so each carbon atom loses 4 electrons. Because the balanced reaction has 6 mol of carbon, 6(4) = 24 mol electrons are transferred in the balanced equation.

$$\Delta G° = -nFE°,\ E° = \frac{-\Delta G°}{nF} = \frac{-(-2875 \times 10^3\ J)}{(24\ mol\ e^-)(96{,}485\ C/mol\ e^-)} = 1.24\ J/C = 1.24\ V$$

56. $CH_3OH(l) + 3/2\ O_2(g) \rightarrow CO_2(g) + 2\ H_2O(l) \quad \Delta G° = 2(-237) + (-394) - [-166] = -702\ kJ$

The balanced half-reactions are:

$H_2O + CH_3OH \rightarrow CO_2 + 6\ H^+ + 6\ e^-$ and $O_2 + 4\ H^+ + 4\ e^- \rightarrow 2\ H_2O$

For 3/2 mol O_2, 6 mol of electrons will be transferred (n = 6).

$$\Delta G° = -nFE°,\ E° = \frac{-\Delta G°}{nF} = \frac{-(-702{,}000 J)}{(6\ mol\ e^-)(96{,}485 C/mol\ e^-)} = 1.21\ J/C = 1.21\ V$$

57. Good oxidizing agents are easily reduced. Oxidizing agents are on the left side of the reduction half-reactions listed in Table 18.1. We look for the largest, most positive standard reduction potentials to correspond to the best oxidizing agents. The ordering from worst to best oxidizing agents is:

	K^+	<	H_2O	<	Cd^{2+}	<	I_2	<	$AuCl_4^-$	<	IO_3^-
E° (V)	−2.92		−0.83		−0.40		0.54		0.99		1.20

58. Good reducing agents are easily oxidized. The reducing agents are on the right side of the reduction half-reactions listed in Table 18.1. The best reducing agents have the most negative standard reduction potentials (E°) or the most positive standard oxidation potentials E°_{ox} (= −E°). The ordering from worst to best reducing agents is:

	F^-	<	H_2O	<	I_2	<	Cu^+	<	H^-	<	K
−E° (V)	−2.87		−1.23		−1.20		−0.16		2.23		2.92

59. a. $2\ H^+ + 2\ e^- \rightarrow H_2$ E° = 0.00 V; $Cu \rightarrow Cu^{2+} + 2\ e^-$ −E° = −0.34 V

E°_{cell} = −0.34 V; no, H^+ cannot oxidize Cu to Cu^{2+} at standard conditions ($E^\circ_{cell} < 0$).

b. $Fe^{3+} + e^- \rightarrow Fe^{2+}$ E° = 0.77 V; $2\ I^- \rightarrow I_2 + 2\ e^-$ −E° = -0.54 V

E°_{cell} = 0.77 − 0.54 = 0.23 V; yes, Fe^{3+} can oxidize I^- to I_2.

c. $H_2 \rightarrow 2\ H^+ + 2\ e^-$ −E° = 0.00 V; $Ag^+ + e^- \rightarrow Ag$ E° = 0.80 V

E°_{cell} = 0.80 V; yes, H_2 can reduce Ag^+ to Ag at standard conditions ($E^\circ_{cell} > 0$).

60. a. $H_2 \rightarrow 2\ H^+ + 2\ e^-$ −E° = 0.00 V; $Ni^{2+} + 2\ e^- \rightarrow Ni$ E° = −0.23 V

E°_{cell} = −0.23 V; no, H_2 cannot reduce Ni^{2+} to Ni at standard conditions ($E^\circ_{cell} < 0$).

b. $Fe^{2+} \rightarrow Fe^{3+} + e^-$ −E° = −0.77 V; $VO_2^+ + 2\ H^+ + e^- \rightarrow VO^{2+} + H_2O$ E° = 1.00 V

E°_{cell} = 1.00 − 0.77 = 0.23 V; yes, Fe^{2+} can reduce VO_2^+ at standard conditions.

c. $Fe^{2+} \rightarrow Fe^{3+} + e^-$ −E° = −0.77 V; $Cr^{3+} + e^- \rightarrow Cr^{2+}$ E° = -0.50 V

E°_{cell} = −0.50 − 0.77 = −1.27 V

No, Fe^{2+} cannot reduce Cr^{3+} to Cr^{2+} at standard conditions($E^\circ_{cell} < 0$).

61. $Cl_2 + 2\ e^- \rightarrow 2\ Cl^-$ E° = 1.36 V $Ag^+ + e^- \rightarrow Ag$ E° = 0.80 V

$Pb^{2+} + 2\ e^- \rightarrow Pb$ E° = −0.13 V $Zn^{2+} + 2\ e^- \rightarrow Zn$ E° = −0.76 V

$Na^+ + e^- \rightarrow Na$ E° = −2.71 V

a. Oxidizing agents (species reduced) are on the left side of the preceding reduction half-reactions. Of the species available, Ag^+ would be the best oxidizing agent since it has the largest E° value. Note that Cl_2 is a better oxidizing agent than Ag^+, but it is not one of the choices listed.

b. Reducing agents (species oxidized) are on the right side of the reduction half-reactions. Of the species available, Zn would be the best reducing agent since it has the largest –E° value.

c. $SO_4^{2-} + 4\ H^+ + 2\ e^- \rightarrow H_2SO_3 + H_2O$ E° = 0.20 V; SO_4^{2-} can oxidize Pb and Zn at standard conditions. When SO_4^{2-} is coupled with these reagents, E°_{cell} is positive.

d. $Al \rightarrow Al^{3+} + 3\ e^-$ –E° = 1.66 V; Al can reduce Ag^+ and Zn^{2+} at standard conditions because $E^\circ_{cell} > 0$.

62.

$Br_2 + 2\ e^- \rightarrow 2\ Br^-$	E° = 1.09 V	$La^{3+} + 3\ e^- \rightarrow La$	E° = –2.37 V
$2\ H^+ + 2\ e^- \rightarrow H_2$	E° = 0.00 V	$Ca^{2+} + 2\ e^- \rightarrow Ca$	E° = –2.76 V
$Cd^{2+} + 2\ e^- \rightarrow Cd$	E° = –0.40 V		

a. Oxidizing agents are on the left side of the preceding reduction half-reactions. Br_2 is the best oxidizing agent (largest E°).

b. Reducing agents are on the right side of the reduction half-reactions. Ca is the best reducing agent (largest –E°).

c. $MnO_4^- + 8\ H^+ + 5\ e^- \rightarrow Mn^{2+} + 4\ H_2O$ E° = 1.51 V; permanganate can oxidize Br^-, H_2, Cd, and Ca at standard conditions. When MnO_4^- is coupled with these reagents, E°_{cell} is positive. *Note*: La is not one of the choices given in the question or it would have been included.

d. $Zn \rightarrow Zn^{2+} + 2\ e^-$ –E° = 0.76 V; zinc can reduce Br_2 and H^+ beause $E^\circ_{cell} > 0$.

63. a. $2\ Br^- \rightarrow Br_2 + 2\ e^-$ –E° = –1.09 V; $2\ Cl^- \rightarrow Cl_2 + 2\ e^-$ –E° = –1.36 V; E° > 1.09 V to oxidize Br^-; E° < 1.36 V to not oxidize Cl^-; $Cr_2O_7^{2-}$, O_2, MnO_2, and IO_3^- are all possible since when all of these oxidizing agents are coupled with Br^-, $E^\circ_{cell} > 0$, and when coupled with Cl^-, $E^\circ_{cell} < 0$ (assuming standard conditions).

b. $Mn \rightarrow Mn^{2+} + 2\ e^-$ –E° = 1.18; $Ni \rightarrow Ni^{2+} + 2\ e^-$ –E° = 0.23 V; any oxidizing agent with –0.23 V > E° > –1.18 V will work. $PbSO_4$, Cd^{2+}, Fe^{2+}, Cr^{3+}, Zn^{2+}, and H_2O will be able to oxidize Mn but not Ni (assuming standard conditions).

64. a. $Cu^{2+} + 2\ e^- \rightarrow Cu$ E° = 0.34 V; $Cu^{2+} + e^- \rightarrow Cu^+$ E° = 0.16 V; to reduce Cu^{2+} to Cu but not reduce Cu^{2+} to Cu^+, the reducing agent must have a standard oxidation potential E°_{ox} = –E°) between –0.34 and –0.16 V (so E°_{cell} is positive only for the Cu^{2+} to Cu reduction). The reducing agents (species oxidized) are on the right side of the half-reactions in Table 18.1. The reagents at standard conditions that have E°_{ox} (= –E°) between –0.34 and –0.16 V are Ag (in 1.0 *M* Cl^-) and H_2SO_3.

b. $Br_2 + 2\,e^- \rightarrow 2\,Br^-$ $E° = 1.09$ V; $I_2 + 2\,e^- \rightarrow 2\,I^-$ $E° = 0.54$ V; from Table 18.1, VO^{2+}, Au (in 1.0 *M* Cl^-), NO, ClO_2^-, Hg_2^{2+}, Ag, Hg, Fe^{2+}, H_2O_2, and MnO_4^- are all capable at standard conditions of reducing Br_2 to Br^- but not reducing I_2 to I^-. When these reagents are coupled with Br_2, $E^o_{cell} > 0$, and when coupled with I_2, $E^o_{cell} < 0$.

65.

$$ClO^- + H_2O + 2\,e^- \rightarrow 2\,OH^- + Cl^- \qquad E° = 0.90 \text{ V}$$
$$2\,NH_3 + 2\,OH^- \rightarrow N_2H_4 + 2\,H_2O + 2\,e^- \qquad -E° = 0.10 \text{ V}$$

$$ClO^-(aq) + 2\,NH_3(aq) \rightarrow Cl^-(aq) + N_2H_4(aq) + H_2O(l) \qquad E^o_{cell} = 1.00 \text{ V}$$

Because E^o_{cell} is positive for this reaction, at standard conditions, ClO^- can spontaneously oxidize NH_3 to the somewhat toxic N_2H_4.

66.

$$Tl^{3+} + 2\,e^- \rightarrow Tl^+ \qquad E° = 1.25 \text{ V}$$
$$3\,I^- \rightarrow I_3^- + 2\,e^- \qquad -E° = -0.55 \text{ V}$$

$$Tl^{3+} + 3\,I^- \rightarrow Tl^+ + I_3^- \qquad E^o_{cell} = 0.70 \text{ V}$$

In solution, Tl^{3+}can oxidize I^- to I_3^-. Thus we expect TlI_3 to be thallium(I) triiodide.

The Nernst Equation

67.

$$H_2O_2 + 2\,H^+ + 2\,e^- \rightarrow 2\,H_2O \qquad E° = 1.78 \text{ V}$$
$$(Ag \rightarrow Ag^+ + e^-) \times 2 \qquad -E° = -0.80 \text{ V}$$

$$H_2O_2(aq) + 2\,H^+(aq) + 2\,Ag(s) \rightarrow 2\,H_2O(l) + 2\,Ag^+(aq) \qquad E^o_{cell} = 0.98 \text{ V}$$

a. A galvanic cell is based on spontaneous redox reactions. At standard conditions, this reaction produces a voltage of 0.98 V. Any change in concentration that increases the tendency of the forward reaction to occur will increase the cell potential. Conversely, any change in concentration that decreases the tendency of the forward reaction to occur (increases the tendency of the reverse reaction to occur) will decrease the cell potential. Using Le Chatelier's principle, increasing the reactant concentrations of H_2O_2 and H^+ from 1.0 to 2.0 *M* will drive the forward reaction further to right (will further increase the tendency of the forward reaction to occur). Therefore, E_{cell} will be greater than E^o_{cell}.

b. Here, we decreased the reactant concentration of H^+ and increased the product concentration of Ag^+ from the standard conditions. This decreases the tendency of the forward reaction to occur, which will decrease E_{cell} as compared to E^o_{cell} ($E_{cell} < E^o_{cell}$).

68. The concentrations of Fe^{2+} in the two compartments are now 0.01 and 1×10^{-7} *M*. The driving force for this reaction is to equalize the Fe^{2+} concentrations in the two compartments. This occurs if the compartment with 1×10^{-7} *M* Fe^{2+} becomes the anode (Fe will be oxidized to Fe^{2+}) and the compartment with the 0.01 *M* Fe^{2+} becomes the cathode (Fe^{2+} will be reduced to Fe). Electron flow, as always for galvanic cells, goes from the anode to the cathode, so electron flow will go from the right compartment ($[Fe^{2+}] = 1 \times 10^{-7}$ *M*) to the left compartment ($[Fe^{2+}] = 0.01$ *M*).

69. For concentration cells, the driving force for the reaction is the difference in ion concentrations between the anode and cathode. In order to equalize the ion concentrations, the anode always has the smaller ion concentration. The general setup for this concentration cell is:

Cathode:	$Ag^+(x\ M) + e^- \rightarrow Ag$	$E° = 0.80$ V
Anode:	$Ag \rightarrow Ag^+\ (y\ M) + e^-$	$-E° = -0.80$ V
	$Ag^+(\text{cathode}, x\ M) \rightarrow Ag^+\ (\text{anode}, y\ M)$	$E°_{cell} = 0.00$ V

$$E_{cell} = E°_{cell} - \frac{0.0591}{n}\log Q = \frac{-0.0591}{1}\log\frac{[Ag^+]_{anode}}{[Ag^+]_{cathode}}$$

For each concentration cell, we will calculate the cell potential using the preceding equation. Remember that the anode always has the smaller ion concentration.

a. Both compartments are at standard conditions ($[Ag^+] = 1.0\ M$), so $E_{cell} = E°_{cell} = 0$ V. No voltage is produced since no reaction occurs. Concentration cells only produce a voltage when the ion concentrations are not equal.

b. Cathode = 2.0 M Ag^+; anode = 1.0 M Ag^+; electron flow is always from the anode to the cathode, so electrons flow to the right in the diagram.

$$E_{cell} = \frac{-0.0591}{n}\log\frac{[Ag^+]_{anode}}{[Ag^+]_{cathode}} = \frac{-0.0591}{1}\log\frac{1.0}{2.0} = 0.018\text{ V}$$

c. Cathode = 1.0 M Ag^+; anode = 0.10 M Ag^+; electrons flow to the left in the diagram.

$$E_{cell} = \frac{-0.0591}{n}\log\frac{[Ag^+]_{anode}}{[Ag^+]_{cathode}} = \frac{-0.0591}{1}\log\frac{0.10}{1.0} = 0.059\text{ V}$$

d. Cathode = 1.0 M Ag^+; anode = 4.0×10^{-5} M Ag^+; electrons flow to the left in the diagram.

$$E_{cell} = \frac{-0.0591}{n}\log\frac{4.0\times10^{-5}}{1.0} = 0.26\text{ V}$$

e. The ion concentrations are the same; thus $\log([Ag^+]_{anode}/[Ag^+]_{cathode}) = \log(1.0) = 0$ and $E_{cell} = 0$. No electron flow occurs.

70. As is the case for all concentration cells, $E°_{cell} = 0$, and the smaller ion concentration is always in the anode compartment. The general Nernst equation for the Ni | $Ni^{2+}(x\ M)$ || $Ni^{2+}(y\ M)$ | Ni concentration cell is:

$$E_{cell} = E°_{cell} - \frac{0.0591}{n}\log Q = \frac{-0.0591}{2}\log\frac{[Ni^{2+}]_{anode}}{[Ni^{2+}]_{cathode}}$$

a. Both compartments are at standard conditions ($[Ni^{2+}] = 1.0\ M$), and $E_{cell} = E°_{cell} = 0$ V. No electron flow occurs.

b. Cathode = 2.0 *M* Ni^{2+}; anode = 1.0 *M* Ni^{2+}; electron flow is always from the anode to the cathode, so electrons flow to the right in the diagram.

$$E_{cell} = \frac{-0.0591}{2}\log\frac{[Ni^{2+}]_{anode}}{[Ni^{2+}]_{cathode}} = \frac{-0.0591}{2}\log\frac{1.0}{2.0} = 8.9 \times 10^{-3}\ V$$

c. Cathode = 1.0 *M* Ni^{2+}; anode = 0.10 *M* Ni^{2+}; electrons flow to the left in the diagram.

$$E_{cell} = \frac{-0.0591}{2}\log\frac{0.10}{1.0} = 0.030\ V$$

d. Cathode = 1.0 *M* Ni^{2+}; anode = 4.0×10^{-5} *M* Ni^{2+}; electrons flow to the left in the diagram.

$$E_{cell} = \frac{-0.0591}{2}\log\frac{4.0\times10^{-5}}{1.0} = 0.13\ V$$

e. Because both concentrations are equal, log(2.5/2.5) = log 1.0 = 0, and E_{cell} = 0. No electron flow occurs.

71. n = 2 for this reaction (lead goes from Pb → Pb^{2+} in $PbSO_4$).

$$E = E° - \frac{0.0591}{2}\log\frac{1}{[H^+]^2[HSO_4^-]^2} = 2.04\ V - \frac{0.0591}{2}\log\frac{1}{(4.5)^2(4.5)^2}$$

E = 2.04 V + 0.077 V = 2.12 V

72.

$Cr_2O_7^{2-} + 14\ H^+ + 6\ e^- \rightarrow 2\ Cr^{3+} + 7\ H_2O$	E° = 1.33 V
$(Al \rightarrow Al^{3+} + 3\ e^-) \times 2$	–E° = 1.66 V
$Cr_2O_7^{2-} + 14\ H^+ + 2\ Al \rightarrow 2\ Al^{3+} + 2\ Cr^{3+} + 7\ H_2O$	E^o_{cell} = 2.99 V

$$E = E° - \frac{0.0591}{n}\log Q,\quad E = 2.99 - \frac{0.0591}{6}\log\frac{[Al^{3+}]^2[Cr^{3+}]^2}{[Cr_2O_7^{2-}][H^+]^{14}}$$

$$3.01 = 2.99 - \frac{0.0591}{n}\log Q\frac{(0.30)^2(0.15)^2}{(0.55)[H^+]^{14}},\quad \frac{-6(0.02)}{0.0591} = \log\left(\frac{3.7\times10^{-3}}{[H^+]^{14}}\right)$$

$$\frac{3.7\times10^{-3}}{[H^+]^{14}} = 10^{-2.0} = 0.01,\ [H^+]^{14} = 0.37,\ [H^+] = 0.93 = 0.9\ M,\ pH = -\log(0.9) = 0.05$$

73.

$Cu^{2+} + 2\ e^- \rightarrow Cu$	E° = 0.34 V
$Zn \rightarrow Zn^{2+} + 2\ e^-$	–E° = 0.76 V
$Cu^{2+}(aq) + Zn(s) \rightarrow Zn^{2+}(aq) + Cu(s)$	E^o_{cell} = 1.10 V

Because Zn^{2+} is a product in the reaction, the Zn^{2+} concentration increases from 1.00 to 1.20 *M*. This means that the reactant concentration of Cu^{2+} must decrease from 1.00 to 0.80 *M* (from the 1 : 1 mole ratio in the balanced reaction).

$$E_{cell} = E^{o}_{cell} - \frac{0.0591}{n}\log Q = 1.10\text{ V} - \frac{0.0591}{2}\log\frac{[Zn^{2+}]}{[Cu^{2+}]}$$

$$E_{cell} = 1.10\text{ V} - \frac{0.0591}{2}\log\frac{1.20}{0.80} = 1.10\text{ V} - 0.0052\text{ V} = 1.09\text{ V}$$

74. $(Pb^{2+} + 2\,e^- \rightarrow Pb) \times 3$ $\quad E° = -0.13$ V

$(Al \rightarrow Al^{3+} + 3\,e^-) \times 2$ $\quad -E° = 1.66$ V

$3\,Pb^{2+}(aq) + 2\,Al(s) \rightarrow 3\,Pb(s) + 2\,Al^{3+}(aq)$ $\quad E^{o}_{cell} = 1.53$ V

From the balanced reaction, when the Al^{3+} has increased by 0.60 mol/L (Al^{3+} is a product in the spontaneous reaction), then the Pb^{2+} concentration has decreased by 3/2 (0.60 mol/L) = 0.90 *M*.

$$E_{cell} = 1.53\text{ V} - \frac{0.0591}{6}\log\frac{[Al^{3+}]^2}{[Pb^{2+}]^3} = 1.53 - \frac{0.0591}{6}\log\frac{(1.60)^2}{(0.10)^3}$$

$E_{cell} = 1.53\text{ V} - 0.034\text{ V} = 1.50\text{ V}$

75. See Exercises 37, 39, and 41 for balanced reactions and standard cell potentials. Balanced reactions are necessary to determine n, the moles of electrons transferred.

37a. $7\,H_2O + 2\,Cr^{3+} + 3\,Cl_2 \rightarrow Cr_2O_7^{2-} + 6\,Cl^- + 14\,H^+$ $\quad E^{o}_{cell} = 0.03\text{ V} = 0.03\text{ J/C}$

$\Delta G° = -nFE^{o}_{cell} = -(6\text{ mol e}^-)(96{,}485\text{ C/mol e}^-)(0.03\text{ J/C}) = -1.7 \times 10^4\text{ J} = -20\text{ kJ}$

$E_{cell} = E^{o}_{cell} - \frac{0.0591}{n}\log Q$; at equilibrium, $E_{cell} = 0$ and Q = K, so:

$$E^{o}_{cell} = \frac{0.0591}{n}\log K,\ \log K = \frac{nE°}{0.0591} = \frac{6(0.03)}{0.0591} = 3.05,\ K = 10^{3.05} = 1 \times 10^3$$

Note: When determining exponents, we will round off to the correct number of significant figures after the calculation is complete in order to help eliminate excessive round-off error.

37b. $\Delta G° = -(2\text{ mol e}^-)(96{,}485\text{ C/mol e}^-)(2.71\text{ J/C}) = -5.23 \times 10^5\text{ J} = -523\text{ kJ}$

$$\log K = \frac{2(2.71)}{0.0591} = 91.709,\ K = 5.12 \times 10^{91}$$

41a. $\Delta G° = -(2\text{ mol e}^-)(96{,}485\text{ C/mol}^-)(0.27\text{ J/C}) = -5.21 \times 10^4\text{ J} = -52\text{ kJ}$

$$\log K = \frac{2(0.27)}{0.0591} = 9.14,\ K = 1.4 \times 10^9$$

41b. $\Delta G° = -(10\text{ mol e}^-)(96{,}485\text{ C/mol e}^-)(0.09\text{ J/C}) = -8.7 \times 10^4\text{ J} = -90\text{ kJ}$

$$\log K = \frac{10(0.09)}{0.0591} = 15.23,\ K = 2 \times 10^{15}$$

76. $\Delta G° = -nFE°_{cell}$; $E°_{cell} = \dfrac{0.0591}{n}\log K$, $\log K = \dfrac{nE°}{0.0591}$

38a. $\Delta G° = -(10 \text{ mol e}^-)(96{,}485 \text{ C/mol e}^-)(0.43 \text{ J/C}) = -4.1 \times 10^5 \text{ J} = -410 \text{ kJ}$

$$\log K = \frac{10(0.43)}{0.0591} = 72.76,\ K = 10^{72.76} = 5.8 \times 10^{72}$$

38b. $\Delta G° = -(2 \text{ mol e}^-)(96{,}485 \text{ C/mol e}^-)(1.56 \text{ J/C}) = -3.01 \times 10^5 \text{ J} = -301 \text{ kJ}$

$$\log K = \frac{2(1.56)}{0.0591} = 52.792,\ K = 6.19 \times 10^{52}$$

42a. $\Delta G° = -(2 \text{ mol }^-)(96{,}485 \text{ C/mol e}^-)(1.10 \text{ J/C}) = -2.12 \times 10^5 \text{ J} = -212 \text{ kJ}$

$$\log K = \frac{2(1.10)}{0.0591} = 37.225,\ K = 1.68 \times 10^{37}$$

42b. $\Delta G° = -(6 \text{ mol e}^-)(96{,}485 \text{ C/mol e}^-)(1.14 \text{ J/C}) = -6.60 \times 10^5 \text{ J} = -660. \text{ kJ}$

$$\log K = \frac{6(1.14)}{0.0591} = 115.736,\ K = 5.45 \times 10^{115}$$

77. a.

$$Fe^{2+} + 2e^- \rightarrow Fe \qquad E° = -0.44 \text{ V}$$
$$Zn \rightarrow Zn^{2+} + 2e^- \qquad -E° = 0.76 \text{ V}$$

$$Fe^{2+}(aq) + Zn(s) \rightarrow Zn^{2+}(aq) + Fe(s) \qquad E°_{cell} = 0.32 \text{ V} = 0.32 \text{ J/C}$$

b. $\Delta G° = -nFE°_{cell} = -(2 \text{ mol e}^-)(96{,}485 \text{ C/mol e}^-)(0.32 \text{ J/C}) = -6.2 \times 10^4 \text{ J} = -62 \text{ kJ}$

$$E°_{cell} = \frac{0.0591}{n}\log K,\ \log K = \frac{nE°}{0.0591} = \frac{2(0.32)}{0.0591} = 10.83,\ K = 10^{10.83} = 6.8 \times 10^{10}$$

c. $E_{cell} = E°_{cell} - \dfrac{0.0591}{n}\log Q = 0.32 \text{ V} - \dfrac{0.0591}{n}\log\dfrac{[Zn^{2+}]}{[Fe^{2+}]}$

$$E_{cell} = 0.32 - \frac{0.0591}{2}\log\frac{0.10}{1.0 \times 10^{-5}} = 0.32 - 0.12 = 0.20 \text{ V}$$

78. a.

$$Au^{3+} + 3e^- \rightarrow Au \qquad E° = 1.50 \text{ V}$$
$$(Tl \rightarrow Tl^+ + e^-) \times 3 \qquad -E° = 0.34 \text{ V}$$

$$Au^{3+}(aq) + 3\,Tl(s) \rightarrow Au(s) + 3\,Tl^+(aq) \qquad E°_{cell} = 1.84 \text{ V}$$

b. $\Delta G° = -nFE°_{cell} = -(3 \text{ mol e}^-)(96{,}485 \text{ C/mol e}^-)(1.84 \text{ J/C}) = -5.33 \times 10^5 \text{ J} = -533 \text{ kJ}$

$$\log K = \frac{nE°}{0.0591} = \frac{3(1.84)}{0.0591} = 93.401,\ K = 10^{93.401} = 2.52 \times 10^{93}$$

c. At 25°C, $E_{cell} = E^{o}_{cell} - \frac{0.0591}{n} \log Q$, where $Q = \frac{[Tl^+]^3}{[Au^{3+}]}$.

$$E_{cell} = 1.84\text{ V} - \frac{0.0591}{3}\log\frac{[Tl^+]^3}{[Au^{3+}]} = 1.84 - \frac{0.0591}{3}\log\frac{(1.0\times10^{-4})^3}{1.0\times10^{-2}}$$

$E_{cell} = 1.84 - (-0.20) = 2.04$ V

79. $Cu^{2+}(aq) + H_2(g) \rightarrow 2\,H^+(aq) + Cu(s)$ $E^{o}_{cell} = 0.34\text{ V} - 0.00\text{ V} = 0.34\text{ V}$; n = 2 mol electrons

$P_{H_2} = 1.0$ atm and $[H^+] = 1.0\,M$: $E_{cell} = E^{o}_{cell} - \frac{0.0591}{n}\log\frac{1}{[Cu^{2+}]}$

a. $E_{cell} = 0.34\text{ V} - \frac{0.0591}{2}\log\frac{1}{2.5\times10^{-4}} = 0.34\text{ V} - 0.11\text{V} = 0.23\text{ V}$

b. $0.195\text{ V} = 0.34\text{ V} - \frac{0.0591}{2}\log\frac{1}{[Cu^{2+}]}$, $\log\frac{1}{[Cu^{2+}]} = 4.91$, $[Cu^{2+}] = 10^{-4.91}$
$= 1.2\times10^{-5}\,M$

Note: When determining exponents, we will carry extra significant figures.

80. $3\,Ni^{2+}(aq) + 2\,Al(s) \rightarrow 2\,Al^{3+}(aq) + 3\,Ni(s)$ $E^{o}_{cell} = -0.23 + 1.66 = 1.43$ V; n = 6 mol electrons for this reaction.

a. $E_{cell} = 1.43\text{ V} - \frac{0.0591}{6}\log\frac{[Al^{3+}]^2}{[Ni^{2+}]^3} = 1.43 - \frac{0.0591}{6}\log\frac{(7.2\times10^{-3})^2}{(1.0)^3}$

$E_{cell} = 1.43\text{ V} - (-0.042\text{ V}) = 1.47$ V

b. $1.62\text{ V} = 1.43\text{ V} - \frac{0.0591}{6}\log\frac{[Al^{3+}]^2}{(1.0)^3}$, $\log[Al^{3+}]^2 = -19.29$

$[Al^{3+}]^2 = 10^{-19.29}$, $[Al^{3+}] = 2.3\times10^{-10}\,M$

81. $Cu^{2+}(aq) + H_2(g) \rightarrow 2\,H^+(aq) + Cu(s)$ $E^{o}_{cell} = 0.34\text{ V} - 0.00\text{ V} = 0.34\text{ V}$; n = 2

$P_{H_2} = 1.0$ atm and $[H^+] = 1.0\,M$: $E_{cell} = E^{o}_{cell} - \frac{0.0591}{2}\log\frac{1}{[Cu^{2+}]}$

Use the K_{sp} expression to calculate the Cu^{2+} concentration in the cell.

$Cu(OH)_2(s) \rightleftharpoons Cu^{2+}(aq) + 2\,OH^-(aq)$ $K_{sp} = 1.6\times10^{-19} = [Cu^{2+}][OH^-]^2$

From problem, $[OH^-] = 0.10\,M$, so: $[Cu^{2+}] = \frac{1.6\times10^{-19}}{(0.10)^2} = 1.6\times10^{-17}\,M$

$$E_{cell} = E^{o}_{cell} - \frac{0.0591}{2}\log\frac{1}{[Cu^{2+}]} = 0.34\text{ V} - \frac{0.0591}{2}\log\frac{1}{1.6\times10^{-17}}$$

$E_{cell} = 0.34 - 0.50 = -0.16$ V

Because $E_{cell} < 0$, the forward reaction is not spontaneous, but the reverse reaction is spontaneous. The Cu electrode becomes the anode and $E_{cell} = 0.16$ V for the reverse reaction. The cell reaction is $2\ H^+(aq) + Cu(s) \rightarrow Cu^{2+}(aq) + H_2(g)$.

82. $3\ Ni^{2+}(aq) + 2\ Al(s) \rightarrow 2\ Al^{3+}(aq) + 3\ Ni(s)$ $E^\circ_{cell} = -0.23\ V + 1.66\ V = 1.43\ V$; n = 6

$$E_{cell} = E^\circ_{cell} - \frac{0.0591}{n}\log\frac{[Al^{3+}]^2}{[Ni^{2+}]^3},\quad 1.82\ V = 1.43\ V - \frac{0.0591}{6}\log\frac{[Al^{3+}]^2}{(1.0)^3}$$

$\log [Al^{3+}]^2 = -39.59$, $[Al^{3+}]^2 = 10^{-39.59}$, $[Al^{3+}] = 1.6 \times 10^{-20}\ M$

$Al(OH)_3(s) \rightleftharpoons Al^{3+}(aq) + 3\ OH^-(aq)$ $K_{sp} = [Al^{3+}][OH^-]^3$; from the problem, $[OH^-] = 1.0 \times 10^{-4}\ M$.

$K_{sp} = (1.6 \times 10^{-20})(1.0 \times 10^{-4})^3 = 1.6 \times 10^{-32}$

83. Cathode: $M^{2+} + 2e^- \rightarrow M(s)$ $E^\circ = -0.31$ V

Anode: $M(s) \rightarrow M^{2+} + 2e^-$ $-E^\circ = 0.31$ V

M^{2+} (cathode) $\rightarrow M^{2+}$ (anode) $E^\circ_{cell} = 0.00$ V

$$E_{cell} = 0.44\ V = 0.00\ V - \frac{0.0591}{2}\log\frac{[M^{2+}]_{anode}}{[M^{2+}]_{cathode}},\quad 0.44 = -\frac{0.0591}{2}\log\frac{[M^{2+}]_{anode}}{1.0}$$

$$\log [M^{2+}]_{anode} = -\frac{2(0.44)}{0.0591} = -14.89,\quad [M^{2+}]_{anode} = 1.3 \times 10^{-15}\ M$$

Because we started with equal numbers of moles of SO_4^{2-} and M^{2+}, $[M^{2+}] = [SO_4^{2-}]$ at equilibrium.

$K_{sp} = [M^{2+}][SO_4^{2-}] = (1.3 \times 10^{-15})^2 = 1.7 \times 10^{-30}$

84. a. Ag^+ (x *M*, anode) $\rightarrow$ Ag^+ (0.10 *M*, cathode); for the silver concentration cell, $E^\circ = 0.00$ (as is always the case for concentration cells) and n = 1.

$$E = 0.76\ V = 0.00 - \frac{0.0591}{1}\log\frac{[Ag^+]_{anode}}{[Ag^+]_{cathode}}$$

$$0.76 = -(0.0591)\log\frac{[Ag^+]_{anode}}{0.10},\quad \frac{[Ag^+]_{anode}}{0.10} = 10^{-12.86},\quad [Ag^+]_{anode} = 1.4 \times 10^{-14}\ M$$

b. $Ag^+(aq) + 2\ S_2O_3^{2-}(aq) \rightleftharpoons Ag(S_2O_3)_2^{3-}(aq)$

$$K = \frac{[Ag(S_2O_3)_2^{3-}]}{[Ag^+][S_2O_3^{2-}]^2} = \frac{1.0 \times 10^{-3}}{1.4 \times 10^{-14}(0.050)^2} = 2.9 \times 10^{13}$$

85. a. Possible reaction: $I_2(s) + 2\ Cl^-(aq) \rightarrow 2\ I^-(aq) + Cl_2(g)$ $E^\circ_{cell} = 0.54\ V - 1.36\ V = -0.82\ V$

This reaction is not spontaneous at standard conditions because $E^\circ_{cell} < 0$; no reaction occurs.

b. Possible reaction: $Cl_2(g) + 2\ I^-(aq) \rightarrow I_2(s) + 2\ Cl^-(aq)$ $E^\circ_{cell} = 0.82$ V; this reaction is spontaneous at standard conditions because $E^\circ_{cell} > 0$. The reaction will occur.

$Cl_2(g) + 2\ I^-(aq) \rightarrow I_2(s) + 2\ Cl^-(aq)$ $E^\circ_{cell} = 0.82\ V = 0.82\ J/C$

$\Delta G^\circ = -nFE^\circ_{cell} = -(2\text{ mol e}^-)(96{,}485\text{ C/mol e}^-)(0.82\text{ J/C}) = -1.6 \times 10^5\text{ J} = -160\text{ kJ}$

$$E^\circ = \frac{0.0591}{n}\log K,\ \log K = \frac{nE^\circ}{0.0591} = \frac{2(0.82)}{0.0591} = 27.75,\ K = 10^{27.75} = 5.6 \times 10^{27}$$

c. Possible reaction: $2\ Ag(s) + Cu^{2+}(aq) \rightarrow Cu(s) + 2\ Ag^+(aq)$ $E^\circ_{cell} = -0.46$ V; no reaction occurs.

d. Fe^{2+} can be oxidized or reduced. The other species present are H^+, SO_4^{2-}, H_2O, and O_2 from air. Only O_2 in the presence of H^+ has a large enough standard reduction potential to oxidize Fe^{2+} to Fe^{3+} (resulting in $E^\circ_{cell} > 0$). All other combinations, including the possible reduction of Fe^{2+}, give negative cell potentials. The spontaneous reaction is:

$$4\ Fe^{2+}(aq) + 4\ H^+(aq) + O_2(g) \rightarrow 4\ Fe^{3+}(aq) + 2\ H_2O(l) \quad E^\circ_{cell} = 1.23 - 0.77 = 0.46\ V$$

$\Delta G^\circ = -nFE^\circ_{cell} = -(4\text{ mol e}^-)(96{,}485\text{ C/mol e}^-)(0.46\text{ J/C})(1\text{ kJ/1000 J}) = -180\text{ kJ}$

$$\log K = \frac{4(0.46)}{0.0591} = 31.13,\ K = 1.3 \times 10^{31}$$

86. a.

$$\begin{array}{ll} Cu^+ + e^- \rightarrow Cu & E^\circ = 0.52\ V \\ Cu^+ \rightarrow Cu^{2+} + e^- & -E^\circ = -0.16\ V \\ \hline 2\ Cu^+(aq) \rightarrow Cu^{2+}(aq) + Cu(s) & E^\circ_{cell} = 0.36\ V;\ \text{spontaneous} \end{array}$$

$\Delta G^\circ = -nFE^\circ_{cell} = -(1\text{ mol e}^-)(96{,}485\text{ C/mol e}^-)(0.36\text{ J/C}) = -34{,}700\text{ J} = -35\text{ kJ}$

$$E^\circ_{cell} = \frac{0.0591}{n}\log K,\ \log K = \frac{nE^\circ}{0.0591} = \frac{1(0.36)}{0.0591} = 6.09,\ K = 10^{6.09} = 1.2 \times 10^6$$

b.

$$\begin{array}{ll} Fe^{2+} + 2\ e^- \rightarrow Fe & E^\circ = -0.44\ V \\ (Fe^{2+} \rightarrow Fe^{3+} + e^-) \times 2 & -E^\circ = -0.77\ V \\ \hline 3\ Fe^{2+}(aq) \rightarrow 2\ Fe^{3+}(aq) + Fe(s) & E^\circ_{cell} = -1.21\ V;\ \text{not spontaneous} \end{array}$$

c. $HClO_2 + 2\,H^+ + 2\,e^- \rightarrow HClO + H_2O$ $\quad E° = 1.65\text{ V}$

$HClO_2 + H_2O \rightarrow ClO_3^- + 3\,H^+ + 2\,e^-$ $\quad -E° = -1.21\text{ V}$

$2\,HClO_2(aq) \rightarrow ClO_3^-(aq) + H^+(aq) + HClO(aq) \quad E°_{cell} = 0.44\text{ V}$; spontaneous

$$\Delta G° = -nFE°_{cell} = -(2\text{ mol e}^-)(96{,}485\text{ C/mol e}^-)(0.44\text{ J/C}) = -84{,}900\text{ J} = -85\text{ kJ}$$

$$\log K = \frac{nE°}{0.0591} = \frac{2(0.44)}{0.0591} = 14.89,\ K = 7.8 \times 10^{14}$$

87. $(Cr^{2+} \rightarrow Cr^{3+} + e^-) \times 2$

$Co^{2+} + 2\,e^- \rightarrow Co$

$2\,Cr^{2+}(aq) + Co^{2+}(aq) \rightarrow 2\,Cr^{3+}(aq) + Co(s)$

$$E°_{cell} = \frac{0.0591}{n}\log K = \frac{0.0591}{2}\log(2.79 \times 10^7) = 0.220\text{ V}$$

$$E = E° - \frac{0.0591}{n}\log\frac{[Cr^{3+}]^2}{[Cr^{2+}]^2[Co^{2+}]} = 0.220\text{ V} - \frac{0.0591}{2}\log\frac{(2.0)^2}{(0.30)^2(0.20)} = 0.151\text{ V}$$

$$\Delta G = -nFE = -(2\text{ mol e}^-)(96{,}485\text{ C/mol e}^-)(0.151\text{ J/C}) = -2.91 \times 10^4\text{ J} = -29.1\text{ kJ}$$

88. $2\,Ag^+(aq) + Cu(s) \rightarrow Cu^{2+}(aq) + 2\,Ag(s) \quad E°_{cell} = 0.80 - 0.34 = 0.46\text{ V}$ and $n = 2$

Because $[Ag^+] = 1.0\,M$, $E_{cell} = 0.46\text{ V} - \dfrac{0.0591}{2}\log[Cu^{2+}]$.

Use the equilibrium reaction to calculate the Cu^{2+} concentration in the cell.

$$Cu^{2+}(aq) + 4\,NH_3(aq) \rightleftharpoons Cu(NH_3)_4^{2+}(aq) \quad K = \frac{[Cu(NH_3)_4^{2+}]}{[Cu^{2+}][NH_3]^4} = 1.0 \times 10^{13}$$

From the problem, $[NH_3] = 5.0\,M$ and $[Cu(NH_3)_4^{2+}] = 0.010\,M$:

$$1.0 \times 10^{13} = \frac{0.010}{[Cu^{2+}](5.0)^4},\ [Cu^{2+}] = 1.6 \times 10^{-18}\,M$$

$$E_{cell} = 0.46 - \frac{0.0591}{2}\log(1.6 \times 10^{-18}) = 0.46 - (-0.53) = 0.99\text{ V}$$

89. The K_{sp} reaction is $FeS(s) \rightleftharpoons Fe^{2+}(aq) + S^{2-}(aq)$ $\quad K = K_{sp}$. Manipulate the given equations so that when added together we get the K_{sp} reaction. Then we can use the value of $E°_{cell}$ for the reaction to determine K_{sp} (by using the equation $\log K = nE°/0.0591$).

$FeS + 2\,e^- \rightarrow Fe + S^{2-}$ $\quad E° = -1.01\text{ V}$

$Fe \rightarrow Fe^{2+} + 2\,e^-$ $\quad -E° = 0.44\text{ V}$

$Fe(s) \rightarrow Fe^{2+}(aq) + S^{2-}(aq) \quad E°_{cell} = -0.57\text{ V}$

$$\log K_{sp} = \frac{nE^o}{0.0591} = \frac{2(-0.57)}{0.0591} = -19.29,\ K_{sp} = 10^{-19.29} = 5.1 \times 10^{-20}$$

90.

$$Al^{3+} + 3\ e^- \rightarrow Al \qquad E° = -1.66\ V$$
$$Al + 6\ F^- \rightarrow AlF_6^{3-} + 3\ e^- \qquad -E° = 2.07\ V$$

$$Al^{3+}(aq) + 6\ F^-(aq) \rightarrow AlF_6^{3-}(aq) \qquad E^o_{cell} = 0.41\ V \quad K = ?$$

$$\log K = \frac{nE^o}{0.0591} = \frac{3(0.41)}{0.0591} = 20.81,\ K = 10^{20.81} = 6.5 \times 10^{20}$$

91.

$$e^- + AgI \rightarrow Ag + I^- \qquad E^o_{AgI} = ?$$
$$Ag \rightarrow Ag^+ + e^- \qquad -E° = -0.80\ V$$

$$AgI(s) \rightarrow Ag^+(aq) + I^-(aq) \qquad E^o_{cell} = E^o_{AgI} - 0.80,\ K = K_{sp} = 1.5 \times 10^{-16}$$

For this overall reaction:

$$E^o_{cell} = \frac{0.0591}{n}\log K_{sp} = \frac{0.0591}{1}\log(1.5 \times 10^{-16}) = -0.94\ V$$

$$E^o_{cell} = -0.94V = E^o_{AgI} - 0.80V,\ E^o_{AgI} = -0.94 + 0.80 = -0.14\ V$$

92.

$$CuI + e^- \rightarrow Cu + I^- \qquad E^o_{CuI} = ?$$
$$Cu \rightarrow Cu^+ + e^- \qquad -E° = -0.52\ V$$

$$CuI(s) \rightarrow Cu^+(aq) + I^-(aq) \qquad E^o_{cell} = E^o_{CuI} - 0.52\ V$$

For this overall reaction, $K = K_{sp} = 1.1 \times 10^{-12}$:

$$E^o_{cell} = \frac{0.0591}{n}\log K_{sp} = \frac{0.0591}{1}\log(1.1\times10^{-12}) = -0.71\ V$$

$$E^o_{cell} = -0.71\ V = E^o_{CuI} - 0.52,\ E^o_{CuI} = -0.19\ V$$

Electrolysis

93. a. $Al^{3+} + 3\ e^- \rightarrow Al$; 3 mol e^- are needed to produce 1 mol Al from Al^{3+}.

$$1.0 \times 10^3\ g\ Al \times \frac{1\ mol\ Al}{26.98\ g\ Al} \times \frac{3\ mol\ e^-}{mol\ Al} \times \frac{96{,}485\ C}{mol\ e^-} \times \frac{1\ s}{100.0\ C} = 1.07 \times 10^5\ s$$
$$= 30.\ hours$$

b. $$1.0\ g\ Ni \times \frac{1\ mol\ Ni}{58.69\ g\ Ni} \times \frac{2\ mol\ e^-}{mol\ Ni} \times \frac{96{,}485\ C}{mol\ e^-} \times \frac{1\ s}{100.0\ C} = 33\ s$$

c. $$5.0\ mol\ Ag \times \frac{1\ mol\ e^-}{mol\ Ag} \times \frac{96{,}485\ C}{mol\ e^-} \times \frac{1\ s}{100.0\ C} = 4.8 \times 10^3\ s = 1.3\ hours$$

94. The oxidation state of bismuth in BiO^+ is +3 because oxygen has a −2 oxidation state in this ion. Therefore, 3 moles of electrons are required to reduce the bismuth in BiO^+ to Bi(s).

$$10.0 \text{ g Bi} \times \frac{1 \text{ mol Bi}}{209.0 \text{ g Bi}} \times \frac{3 \text{ mol } e^-}{\text{mol Bi}} \times \frac{96{,}485 \text{ C}}{\text{mol } e^-} \times \frac{1 \text{ s}}{25.0 \text{ C}} = 554 \text{ s} = 9.23 \text{ min}$$

95. $15 \text{ A} = \frac{15 \text{ C}}{\text{s}} \times \frac{60 \text{ s}}{\text{min}} \times \frac{60 \text{ min}}{\text{h}} = 5.4 \times 10^4$ C of charge passed in 1 hour

a. $$5.4 \times 10^4 \text{ C} \times \frac{1 \text{ mol } e^-}{96{,}485 \text{ C}} \times \frac{1 \text{ mol Co}}{2 \text{ mol } e^-} \times \frac{58.93 \text{ g Co}}{\text{mol Co}} = 16 \text{ g Co}$$

b. $$5.4 \times 10^4 \text{ C} \times \frac{1 \text{ mol } e^-}{96{,}485 \text{ C}} \times \frac{1 \text{ mol Hf}}{4 \text{ mol } e^-} \times \frac{178.5 \text{ g Hf}}{\text{mol Hf}} = 25 \text{ g Hf}$$

c. $2 I^- \rightarrow I_2 + 2 e^-$; $$5.4 \times 10^4 \text{ C} \times \frac{1 \text{ mol } e^-}{96{,}485 \text{ C}} \times \frac{1 \text{ mol } I_2}{2 \text{ mol } e^-} \times \frac{253.8 \text{ g } I_2}{\text{mol } I_2} = 71 \text{ g } I_2$$

d. $CrO_3(l) \rightarrow Cr^{6+} + 3\, O^{2-}$; 6 mol e^- are needed to produce 1 mol Cr from molten CrO_3.

$$5.4 \times 10^4 \text{ C} \times \frac{1 \text{ mol } e^-}{96{,}485 \text{ C}} \times \frac{1 \text{ mol Cr}}{6 \text{ mol } e^-} \times \frac{52.00 \text{ g Cr}}{\text{mol Cr}} = 4.9 \text{ g Cr}$$

96. Al is in the +3 oxidation in Al_2O_3, so 3 mol e^- are needed to convert Al^{3+} into Al(s).

$$2.00 \text{ h} \times \frac{60 \text{ min}}{\text{h}} \times \frac{60 \text{ s}}{\text{min}} \times \frac{1.00 \times 10^6 \text{ C}}{\text{s}} \times \frac{1 \text{ mol } e^-}{96{,}485 \text{ C}} \times \frac{1 \text{ mol Al}}{3 \text{ mol } e^-} \times \frac{26.98 \text{ g Al}}{\text{mol Al}} = 6.71 \times 10^5 \text{ g}$$

97. $$74.1 \text{ s} \times \frac{2.00 \text{ C}}{\text{s}} \times \frac{1 \text{ mol } e^-}{96{,}485 \text{ C}} \times \frac{1 \text{ mol M}}{3 \text{ mol } e^-} = 5.12 \times 10^{-4} \text{ mol M}$$, where M = unknown metal

Molar mass $= \frac{0.107 \text{ g M}}{5.12 \times 10^{-4} \text{ mol M}} = \frac{209 \text{ g}}{\text{mol}}$; the element is bismuth.

98. Alkaline earth metals form +2 ions, so 2 mol of e^- are transferred to form the metal M.

$$\text{Moles of M} = 748 \text{ s} \times \frac{5.00 \text{ C}}{\text{s}} \times \frac{1 \text{ mol } e^-}{96{,}485 \text{ C}} \times \frac{1 \text{ mol M}}{2 \text{ mol } e^-} \times \frac{1 \text{ mol } e^-}{96{,}485 \text{ C}} = 1.94 \times 10^{-2} \text{ mol M}$$

Molar mass of M $= \frac{0.471 \text{ g M}}{1.94 \times 10^{-2} \text{ mol M}} = 24.3$ g/mol; $MgCl_2$ was electrolyzed.

99. F_2 is produced at the anode: $2 F^- \rightarrow F_2 + 2 e^-$

$$2.00 \text{ h} \times \frac{60 \text{ min}}{\text{h}} \times \frac{60 \text{ s}}{\text{min}} \times \frac{10.0 \text{ C}}{\text{s}} \times \frac{1 \text{ mol } e^-}{96{,}485 \text{ C}} = 0.746 \text{ mol } e^-$$

$0.746 \text{ mol e}^- \times \frac{1 \text{ mol} F_2}{2 \text{ mol e}^-} = 0.373 \text{ mol } F_2$; $PV = nRT$, $V = \frac{nRT}{P}$

$$\frac{(0.373 \text{ mol})(0.08206 \text{ L atm/K} \cdot \text{mol})(298 \text{ K})}{1.00 \text{ atm}} = 9.12 \text{ L } F_2$$

K is produced at the cathode: $K^+ + e^- \rightarrow K$

$$0.746 \text{ mol e}^- \times \frac{1 \text{ mol K}}{\text{mol e}^-} \times \frac{39.10 \text{ g K}}{\text{mol K}} = 29.2 \text{ g K}$$

100. The half-reactions for the electrolysis of water are:

$$(2 e^- + 2 H_2O \rightarrow H_2 + 2 OH^-) \times 2$$
$$2 H_2O \rightarrow 4 H^+ + O_2 + 4 e^-$$
$$\overline{2 H_2O(l) \rightarrow 2 H_2(g) + O_2(g)}$$

Note: $4 H^+ + 4 OH^-$ react to give $4 H_2O$, and n = 4 for this reaction.

$$15.0 \text{ min} \times \frac{60 \text{ s}}{\text{min}} \times \frac{2.50 \text{ C}}{\text{s}} \times \frac{1 \text{ mol e}^-}{96{,}485 \text{ C}} \times \frac{2 \text{ mol } H_2}{4 \text{ mol e}^-} = 1.17 \times 10^{-2} \text{ mol } H_2$$

At STP, 1 mol of an ideal gas occupies a volume of 22.42 L (see Chapter 5 of the text).

$$1.17 \times 10^{-2} \text{ mol } H_2 \times \frac{22.42 \text{ L}}{\text{mol } H_2} = 0.262 \text{ L} = 262 \text{ mL } H_2$$

$$1.17 \times 10^{-2} \text{ mol } H_2 \times \frac{1 \text{ mol } O_2}{2 \text{ mol } H_2} \times \frac{22.42 \text{ L}}{\text{mol } O_2} = 0.131 \text{ L} = 131 \text{ mL } O_2$$

101. $Al^{3+} + 3 e^- \rightarrow Al$; 3 mol e^- are needed to produce Al from Al^{3+}

$$2000 \text{ lb Al} \times \frac{453.6 \text{ g}}{\text{lb}} \times \frac{1 \text{ mol Al}}{26.98 \text{ g}} \times \frac{3 \text{ mol e}^-}{\text{mol Al}} \times \frac{96{,}485 \text{ C}}{\text{mol e}^-} = 1 \times 10^{10} \text{ C of electricity needed}$$

$$\frac{1 \times 10^{10} \text{ C}}{24 \text{ h}} \times \frac{1 \text{ h}}{60 \text{ min}} \times \frac{1 \text{ min}}{60 \text{ s}} = 1 \times 10^5 \text{ C/s} = 1 \times 10^5 \text{ A}$$

102. Barium is in the +2 oxidation state in $BaCl_2$. $Ba^{2+} + 2e^- \rightarrow Ba$

$$1.00 \times 10^6 \text{ g Ba} \times \frac{1 \text{ mol Ba}}{137.3 \text{ g}} \times \frac{2 \text{ mol e}^-}{\text{mol Ba}} \times \frac{96{,}485 \text{ C}}{\text{mol e}^-} = 1.41 \times 10^9 \text{ C of electricity needed}$$

$$\frac{1.41 \times 10^9 \text{ C}}{4.00 \text{ h}} \times \frac{1 \text{ h}}{3600 \text{ s}} = 9.79 \times 10^4 \text{ A}$$

103. $2.30 \text{ min} \times \frac{60 \text{ s}}{\text{min}} = 138 \text{ s}$; $138 \text{ s} \times \frac{2.00 \text{ C}}{\text{s}} \times \frac{1 \text{ mol e}^-}{96{,}485 \text{ C}} \times \frac{1 \text{ mol Ag}}{\text{mol e}^-} = 2.86 \times 10^{-3} \text{ mol Ag}$

$[Ag^+] = 2.86 \times 10^{-3} \text{ mol } Ag^+/0.250 \text{ L} = 1.14 \times 10^{-2}\ M$

104. $0.50\ L \times 0.010\ mol\ Pt^{4+}/L = 5.0 \times 10^{-3}\ mol\ Pt^{4+}$

To plate out 99% of the Pt^{4+}, we will produce $0.99 \times 5.0 \times 10^{-3}$ mol Pt.

$$0.99 \times 5.0 \times 10^{-3}\ mol\ Pt \times \frac{4\ mol\ e^-}{mol\ Pt} \times \frac{96{,}485\ C}{mol\ e^-} \times \frac{1\ s}{4.00\ C} \times \frac{1\ mol\ Ag}{mol\ e^-} = 480\ s$$

105. $Au^{3+} + 3\ e^- \rightarrow Au \quad E° = 1.50\ V \qquad Ni^{2+} + 2\ e^- \rightarrow Ni \quad E° = -0.23\ V$

$Ag^+ + e^- \rightarrow Ag \quad E° = 0.80\ V \qquad Cd^{2+} + 2\ e^- \rightarrow Cd \quad E° = -0.40\ V$

$2\ H_2O + 2e^- \rightarrow H_2 + 2\ OH^- \quad E° = -0.83\ V$ (Water can also be reduced.)

Au(s) will plate out first since it has the most positive reduction potential, followed by Ag(s), which is followed by Ni(s), and finally Cd(s) will plate out last since it has the most negative reduction potential of the metals listed. Water will not interfere with the plating process.

106. To begin plating out Pd:

$$E = 0.62 - \frac{0.0591}{2}\log\frac{[Cl^-]^4}{[PdCl_4^{\ 2-}]} = 0.62 - \frac{0.0591}{2}\log\frac{(1.0)^4}{0.020}$$

$E = 0.62\ V - 0.050\ V = 0.57\ V$

When 99% of Pd has plated out, $[PdCl_4^-] = \dfrac{0.020}{100} = 0.00020\ M$.

$$E = 0.62 - \frac{0.0591}{2}\log\frac{(1.0)^4}{2.0\times10^{-4}} = 0.62\ V - 0.11V = 0.51\ V$$

To begin Pt plating: $E = 0.73\ V - \dfrac{0.0591}{2}\log\dfrac{(1.0)^4}{0.020} = 0.73 - 0.050 = 0.68\ V$

When 99% of Pt plated: $E = 0.73 - \dfrac{0.0591}{2}\log\dfrac{(1.0)^4}{2.0\times10^{-4}} = 0.73 - 0.11 = 0.62\ V$

To begin Ir plating: $E = 0.77\ V - \dfrac{0.0591}{3}\log\dfrac{(1.0)^4}{0.020} = 0.77 - 0.033 = 0.74\ V$

When 99% of Ir plated: $E = 0.77 - \dfrac{0.0591}{3}\log\dfrac{(1.0)^4}{2.0 \times 10^{-4}} = 0.77 - 0.073 = 0.70\ V$

Yes, because the range of potentials for plating out each metal does not overlap, we should be able to separate the three metals. The exact potential to apply depends on the oxidation reaction. The order of plating will be Ir(s) first, followed by Pt(s), and finally, Pd(s) as the potential is gradually increased.

107. Species present: Na^+, SO_4^{2-}, and H_2O. H_2O and Na^+ can be reduced and H_2O and SO_4^{2-} can be oxidized. From the potentials, H_2O is the most easily reduced and the most easily oxidized species present. This is the case because water, of the species present, has the most positive reduction potential as well as the most positive oxidation potential. The reactions are:

Cathode: $2\ H_2O + 2\ e^- \rightarrow H_2(g) + 2\ OH^-$; anode: $2\ H_2O \rightarrow O_2(g) + 4\ H^+ + 4\ e^-$

108. a. The spoon is where Cu^{2+} is reduced to Cu, so the spoon will be the cathode. The anode will be the copper strip where Cu is oxidized to Cu^{2+}.

b. Cathode reaction: $Cu^{2+} + 2\ e^- \rightarrow Cu$; anode reaction: $Cu \rightarrow Cu^{2+} + 2\ e^-$

109. Reduction occurs at the cathode, and oxidation occurs at the anode. First, determine all the species present; then look up pertinent reduction and/or oxidation potentials in Table 18.1 for all these species. The cathode reaction will be the reaction with the most positive reduction potential, and the anode reaction will be the reaction with the most positive oxidation potential.

a. Species present: Ni^{2+} and Br^-; Ni^{2+} can be reduced to Ni, and Br^- can be oxidized to Br_2 (from Table 18.1). The reactions are:

Cathode: $Ni^{2+} + 2e^- \rightarrow Ni$ $\quad E° = -0.23$ V

Anode: $2\ Br^- \rightarrow Br_2 + 2\ e^-$ $\quad -E° = -1.09$ V

b. Species present: Al^{3+} and F^-; Al^{3+} can be reduced, and F^- can be oxidized. The reactions are:

Cathode: $Al^{3+} + 3\ e^- \rightarrow Al$ $\quad E° = -1.66$ V

Anode: $2\ F^- \rightarrow F_2 + 2\ e^-$ $\quad -E° = -2.87$ V

c. Species present: Mn^{2+} and I^-; Mn^{2+} can be reduced, and I^- can be oxidized. The reactions are:

Cathode: $Mn^{2+} + 2\ e^- \rightarrow Mn$ $\quad E° = -1.18$ V

Anode: $2\ I^- \rightarrow I_2 + 2\ e^-$ $\quad -E° = -0.54$ V

110. Reduction occurs at the cathode, and oxidation occurs at the anode. First, determine all the species present; then look up pertinent reduction and/or oxidation potentials in Table 18.1 for all these species. The cathode reaction will be the reaction with the most positive reduction potential, and the anode reaction will be the reaction with the most positive oxidation potential.

a. Species present: K^+ and F^-; K^+ can be reduced to K, and F^- can be oxidized to F_2 (from Table 18.1). The reactions are:

Cathode: $K^+ + e^- \rightarrow K$ $\quad E° = -2.92$ V

Anode: $2\ F^- \rightarrow F_2 + 2\ e^-$ $\quad -E° = -2.87$ V

b. Species present: Cu^{2+} and Cl^-; Cu^{2+} can be reduced, and Cl^- can be oxidized. The reactions are:

Cathode: $Cu^{2+} + 2\ e^- \rightarrow Cu$ $\quad E° = 0.34\ V$

Anode: $2\ Cl^- \rightarrow Cl_2 + 2\ e^-$ $\quad -E° = -1.36\ V$

c. Species present: Mg^{2+} and I^-; Mg^{2+} can be reduced, and I^- can be oxidized. The reactions are:

Cathode: $Mg^{2+} + 2\ e^- \rightarrow Mg$ $\quad E° = -2.37\ V$

Anode: $2\ I^- \rightarrow I_2 + 2\ e^-$ $\quad -E° = -0.54\ V$

111. These are all in aqueous solutions, so we must also consider the reduction and oxidation of H_2O in addition to the potential redox reactions of the ions present. For the cathode reaction, the species with the most positive reduction potential will be reduced, and for the anode reaction, the species with the most positive oxidation potential will be oxidized.

a. Species present: Ni^{2+}, Br^-, and H_2O. Possible cathode reactions are:

$Ni^{2+} + 2e^- \rightarrow Ni$ $\quad E° = -0.23\ V$

$2\ H_2O + 2\ e^- \rightarrow H_2 + 2\ OH^-$ $\quad E° = -0.83\ V$

Because it is easier to reduce Ni^{2+} than H_2O (assuming standard conditions), Ni^{2+} will be reduced by the preceding cathode reaction.

Possible anode reactions are:

$2\ Br^- \rightarrow Br_2 + 2\ e^-$ $\quad -E° = -1.09\ V$

$2\ H_2O \rightarrow O_2 + 4\ H^+ + 4\ e^-$ $\quad -E° = -1.23\ V$

Because Br^- is easier to oxidize than H_2O (assuming standard conditions), Br^- will be oxidized by the preceding anode reaction.

b. Species present: Al^{3+}, F^-, and H_2O; Al^{3+} and H_2O can be reduced. The reduction potentials are $E° = -1.66\ V$ for Al^{3+} and $E° = -0.83\ V$ for H_2O (assuming standard conditions). H_2O will be reduced at the cathode ($2\ H_2O + 2\ e^- \rightarrow H_2 + 2\ OH^-$).

F^- and H_2O can be oxidized. The oxidation potentials are $-E° = -2.87\ V$ for F^- and $-E° = -1.23\ V$ for H_2O (assuming standard conditions). From the potentials, we would predict H_2O to be oxidized at the anode ($2\ H_2O \rightarrow O_2 + 4\ H^+ + 4\ e^-$).

c. Species present: Mn^{2+}, I^-, and H_2O; Mn^{2+} and H_2O can be reduced. The possible cathode reactions are:

$Mn^{2+} + 2\ e^- \rightarrow Mn$ $\quad E° = -1.18\ V$

$2\ H_2O + 2\ e^- \rightarrow H_2 + 2\ OH^-$ $\quad E° = -0.83\ V$

Reduction of H_2O will occur at the cathode since $E^{o}_{H_2O}$ is most positive.

I^- and H_2O can be oxidized. The possible anode reactions are:

$$2\,I^- \rightarrow I_2 + 2\,e^- \qquad -E° = -0.54\text{ V}$$

$$2\,H_2O \rightarrow O_2 + 4\,H^+ + 4\,e^- \qquad -E° = -1.23\text{ V}$$

Oxidation of I^- will occur at the anode since $-E^\circ_{I^-}$ is most positive.

112. These are all in aqueous solutions, so we must also consider the reduction and oxidation of H_2O in addition to the potential redox reactions of the ions present. For the cathode reaction, the species with the most positive reduction potential will be reduced, and for the anode reaction, the species with the most positive oxidation potential will be oxidized.

a. Species present: K^+, F^-, and H_2O. Possible cathode reactions are:

$$K^+ + e^- \rightarrow K \qquad E° = -2.92\text{ V}$$

$$2\,H_2O + 2\,e^- \rightarrow H_2 + 2\,OH^- \qquad E° = -0.83\text{ V}$$

Because it is easier to reduce H_2O than K^+ (assuming standard conditions), H_2O will be reduced by the preceding cathode reaction.

Possible anode reactions are:

$$2\,F^- \rightarrow F_2 + 2\,e^- \qquad -E° = -2.87\text{ V}$$

$$2\,H_2O \rightarrow 4\,H^+ + O_2 + 4\,e^- \qquad -E° = -1.23\text{ V}$$

Because H_2O is easier to oxidize than F^- (assuming standard conditions), H_2O will be oxidized by the preceding anode reaction.

b. Species present: Cu^{2+}, Cl^-, and H_2O; Cu^{2+} and H_2O can be reduced. The reduction potentials are E° = 0.34 V for Cu^{2+} and E° = −0.83 V for H_2O (assuming standard conditions). Cu^{2+} will be reduced to Cu at the cathode ($Cu^{2+} + 2\,e^- \rightarrow Cu$).

Cl^- and H_2O can be oxidized. The oxidation potentials are −E° = −1.36 V for Cl^- and −E° = −1.23 V for H_2O (assuming standard conditions). From the potentials, we would predict H_2O to be oxidized at the anode ($2\,H_2O \rightarrow 4\,H^+ + O_2 + 4\,e^-$). *Note*: In real life, Cl^- is oxidized to Cl_2 when water is present due to a phenomenon called overvoltage (see Section 18.8 of the text). Because overvoltage is difficult to predict, we will generally ignore it.

c. Species present: Mg^{2+}, I^-, and H_2O: The only possible cathode reactions are:

$$2\,H_2O + 2\,e^- \rightarrow H_2 + 2\,OH^- \qquad E° = -0.83\text{ V}$$

$$Mg^{2+} + 2\,e^- \rightarrow Mg \qquad E° = -2.37\text{ V}$$

Reduction of H_2O will occur at the cathode since $E^\circ_{H_2O}$ is more positive.

The only possible anode reactions are:

$2\ I^- \rightarrow I_2 + 2\ e^-$ $\quad -E° = -0.54$ V

$2\ H_2O \rightarrow O_2 + 4\ H^+ + 4\ e^-$ $\quad -E° = -1.23$ V

Oxidation of I^- will occur at the anode because $-E^o_{H_2O}$ is more positive.

Additional Exercises

113. The oxidation state of gold in $Au(CN)_2^-$ is +1. Each mole of gold produced requires 1 mole of electrons gained (+1 → 0). The only oxygen containing reactant is H_2O. Each mole of oxygen goes from the −2 → 0 oxidation state as H_2O is converted into O_2. One mole of O_2 contains 2 moles of O, so 4 moles of electrons are lost when 1 mole of O_2 is formed. In order to balance the electrons, we need 4.00 moles of gold for every mole of O_2 produced or 0.250 moles of O_2 for every 1.00 mole of gold formed.

114.
$$2(6\ e^- + 14\ H^+ + Cr_2O_7^{2-} \rightarrow 2\ Cr^{3+} + 7\ H_2O)$$
$$3\ H_2O + C_2H_5OH \rightarrow 2\ CO_2 + 12\ H^+ + 12\ e^-$$

$$16\ H^+ + 2\ Cr_2O_7^{2-} + C_2H_5OH \rightarrow 4\ Cr^{3+} + 2\ CO_2 + 11\ H_2O$$

$$0.03105\text{ L}\left(\frac{0.0600\text{ mol }Cr_2O_7^{2-}}{\text{L}}\right)\left(\frac{1\text{ mol }C_2H_5OH}{2\text{ mol }Cr_2O_7^{2-}}\right)\left(\frac{46.07\text{ g}}{\text{mol }C_2H_5OH}\right) = 0.0429\text{ g }C_2H_5OH$$

$$\frac{0.0429\text{ g }C_2H_5OH}{30.0\text{ g blood}} \times 100 = 0.143\%\ C_2H_5OH$$

115. The half-reaction for the SCE is:

$Hg_2Cl_2 + 2\ e^- \rightarrow 2\ Hg + 2\ Cl^-$ $\quad E_{SCE} = 0.242$ V

For a spontaneous reaction to occur, E_{cell} must be positive. Using the standard reduction potentials in Table 18.1 and the given the SCE potential, deduce which combination will produce a positive overall cell potential.

a. $Cu^{2+} + 2\ e^- \rightarrow Cu$ $\quad E° = 0.34$ V

$E_{cell} = 0.34 - 0.242 = 0.10$ V; SCE is the anode.

b. $Fe^{3+} + e^- \rightarrow Fe^{2+}$ $\quad E° = 0.77$ V

$E_{cell} = 0.77 - 0.242 = 0.53$ V; SCE is the anode.

c. $AgCl + e^- \rightarrow Ag + Cl^-$ $\quad E° = 0.22$ V

$E_{cell} = 0.242 - 0.22 = 0.02$ V; SCE is the cathode.

d. $Al^{3+} + 3\ e^- \rightarrow Al$ $E° = -1.66\ V$

$E_{cell} = 0.242 + 1.66 = 1.90\ V$; SCE is the cathode.

e. $Ni^{2+} + 2\ e^- \rightarrow Ni$ $E° = -0.23\ V$

$E_{cell} = 0.242 + 0.23 = 0.47\ V$; SCE is the cathode.

116. The potential oxidizing agents are NO_3^- and H^+. Hydrogen ion cannot oxidize Pt under either condition. Nitrate cannot oxidize Pt unless there is Cl^- in the solution. Aqua regia has both Cl^- and NO_3^-. The overall reaction is:

$$(NO_3^- + 4\ H^+ + 3\ e^- \rightarrow NO + 2\ H_2O) \times 2 \qquad E° = 0.96\ V$$
$$(4\ Cl^- + Pt \rightarrow PtCl_4^{2-} + 2\ e^-) \times 3 \qquad -E° = -0.755\ V$$

$$12\ Cl^-(aq) + 3\ Pt(s) + 2\ NO_3^-(aq) + 8\ H^+(aq) \rightarrow 3\ PtCl_4^{2-}(aq) + 2\ NO(g) + 4\ H_2O(l) \qquad E^o_{cell} = 0.21\ V$$

117. $Ag^+(aq) + Cu(s) \rightarrow Cu^{2+}(aq) + 2\ Ag(s)$ $E^o_{cell} = 0.80 - 0.34\ V = 0.46\ V$; a galvanic cell produces a voltage as the forward reaction occurs. Any stress that increases the tendency of the forward reaction to occur will increase the cell potential, whereas a stress that decreases the tendency of the forward reaction to occur will decrease the cell potential.

a. Added Cu^{2+} (a product ion) will decrease the tendency of the forward reaction to occur, which will decrease the cell potential.

b. Added NH_3 removes Cu^{2+} in the form of $Cu(NH_3)_4^{2+}$. Because a product ion is removed, this will increase the tendency of the forward reaction to occur, which will increase the cell potential.

c. Added Cl^- removes Ag^+ in the form of AgCl(s). Because a reactant ion is removed, this will decrease the tendency of the forward reaction to occur, which will decrease the cell potential.

d. $Q_1 = \frac{[Cu^{2+}]_0}{[Ag^+]_0^2}$; as the volume of solution is doubled, each concentration is halved.

$$Q_2 = \frac{1/2\ [Cu^{2+}]_0}{(1/2\ [Ag^+]_0)^2} = \frac{2[Cu^{2+}]_0}{[Ag^+]_0^2} = 2Q_1$$

The reaction quotient is doubled because the concentrations are halved. Because reactions are spontaneous when $Q < K$, and because Q increases when the solution volume doubles, the reaction is closer to equilibrium, which will decrease the cell potential.

e. Because Ag(s) is not a reactant in this spontaneous reaction, and because solids do not appear in the reaction quotient expressions, replacing the silver electrode with a platinum electrode will have no effect on the cell potential.

118. $(Al^{3+} + 3\,e^- \rightarrow Al) \times 2$ $E° = -1.66$ V
$(M \rightarrow M^{2+} + 2\,e^-) \times 3$ $-E° = ?$

$3\,M(s) + 2\,Al^{3+}(aq) \rightarrow 2\,Al(s) + 3\,M^{2+}(aq)$ $E°_{cell} = -E° - 1.66$ V

$$\Delta G° = -nFE°_{cell},\ -411 \times 10^3\ J = -(6\,mol\,e^-)(96{,}485\,C/mol\,e^-)E°_{cell},\ E°_{cell} = 0.71\ V$$

$E°_{cell} = -E° - 1.66\ V = 0.71\ V,\ -E° = 2.37$ or $E° = -2.37$

From Table 18.1, the reduction potential for $Mg^{2+} + 2\,e^- \rightarrow Mg$ is −2.37 V, which fits the data. Hence, the metal is magnesium.

119. a. $\Delta G° = \sum n_p \Delta G°_{f,\,products} - \sum n_r \Delta G°_{f,\,reactants} = 2(-480.) + 3(86) - [3(-40.)] = -582$ kJ

From oxidation numbers, n = 6. $\Delta G° = -nFE°,\ E° = \frac{-\Delta G°}{nF} = \frac{-(-582{,}000\,J)}{6(96{,}485)\,C} = 1.01\ V$

$$\log K = \frac{nE°}{0.0591} = \frac{6(1.01)}{0.0591} = 102.538,\ K = 10^{102.538} = 3.45 \times 10^{102}$$

b. $2\,e^- + Ag_2S \rightarrow 2\,Ag + S^{2-}) \times 3$ $E°_{Ag_2S} = ?$
$(Al \rightarrow Al^{3+} + 3\,e^-) \times 2$ $-E° = 1.66$ V

$3\,Ag_2S(s) + 2\,Al(s) \rightarrow 6\,Ag(s) + 3\,S^{2-}(aq) + 2\,Al^{3+}(aq)$ $E°_{cell} = 1.01\ V = E°_{Ag_2S} + 1.66V$

$E°_{Ag_2S} = 1.01\ V - 1.66\ V = -0.65\ V$

120. $Zn \rightarrow Zn^{2+} + 2\,e^-$ $-E° = 0.76$ V; $Fe \rightarrow Fe^{2+} + 2\,e^-$ $-E° = 0.44$ V

It is easier to oxidize Zn than Fe, so the Zn would be preferentially oxidized, protecting the iron of the *Monitor's* hull.

121. Aluminum has the ability to form a durable oxide coating over its surface. Once the HCl dissolves this oxide coating, Al is exposed to H^+ and is easily oxidized to Al^{3+}, i.e., the Al foil disappears after the oxide coating is dissolved.

122. Only statement e is true. The attached metals that are more easily oxidized than iron are called sacrificial metals. For statement a, corrosion is a spontaneous process, like the ones harnessed to make galvanic cells. For statement b, corrosion of steel is the oxidation of iron coupled with the reduction of oxygen. For statement c, cars rust more easily in high-moisture areas (the humid areas) because water is a reactant in the reduction half-reaction as well as providing a medium for ion migration (a salt bridge of sorts). For statement d, salting roads adds ions to the corrosion process, which increases the conductivity of the aqueous solution and, in turn, accelerates corrosion.

123. For C_2H_5OH, H has a +1 oxidation state, and O has a −2 oxidation state. This dictates a −2 oxidation state for C. For CO_2, O has a −2 oxidation state, so carbon has a +4 oxidation state.

Six moles of electrons are transferred per mole of carbon oxidized (C goes from −2 → +4). Two moles of carbon are in the balanced reaction, so n = 12.

$w_{max} = -1320 \text{ kJ} = \Delta G = -nFE, \; -1320 \times 10^3 \text{ J} = -nFE = -(12 \text{ mol } e^-)(96{,}485 \text{ C/mol } e^-)E$

$E = 1.14 \text{ J/C} = 1.14 \text{ V}$

124. $2 H_2(g) + O_2(g) \rightarrow 2 H_2O(l)$; oxygen goes from the zero oxidation state to the −2 oxidation state in H_2O. Because 2 mol of O are in the balanced reaction, n = 4 moles of electrons transferred.

a. $E^\circ_{cell} = \dfrac{0.0591}{n}\log K = \dfrac{0.0591}{4}\log(1.28 \times 10^{83}), \; E^\circ_{cell} = 1.23 \text{ V}$

$\Delta G^\circ = -nFE^\circ_{cell} = -(4 \text{ mol } e^-)(96{,}485 \text{ C/mol } e^-)(1.23 \text{ J/C}) = -4.75 \times 10^5 \text{ J} = -475 \text{ kJ}$

b. Because the moles of gas decrease as reactants are converted into products, ΔS° will be negative (unfavorable). Because the value of ΔG° is negative, ΔH° must be negative to override the unfavorable ΔS° ($\Delta G^\circ = \Delta H^\circ - T\Delta S^\circ$).

c. $\Delta G = w_{max} = \Delta H - T\Delta S$. Because ΔS is negative, as T increases, ΔG becomes more positive (closer to zero). Therefore, w_{max} will decrease as T increases.

125.
$$\begin{array}{ll} O_2 + 2 H_2O + 4 e^- \rightarrow 4 OH^- & E^\circ = 0.40 \text{ V} \\ (H_2 + 2 OH^- \rightarrow 2 H_2O + 2 e^-) \times 2 & -E^\circ = 0.83 \text{ V} \\ \hline 2 H_2(g) + O_2(g) \rightarrow 2 H_2O(l) & E^\circ_{cell} = 1.23 \text{ V} = 1.23 \text{ J/C} \end{array}$$

Because standard conditions are assumed, $w_{max} = \Delta G^\circ$ for 2 mol H_2O produced.

$\Delta G^\circ = -nFE^\circ_{cell} = -(4 \text{ mol } e^-)(96{,}485 \text{ C/mol } e^-)(1.23 \text{ J/C}) = -475{,}000 \text{ J} = -475 \text{ kJ}$

For 1.00×10^3 g H_2O produced, w_{max} is:

$$1.00 \times 10^3 \text{ g } H_2O \times \frac{1 \text{ mol } H_2O}{18.02 \text{ g } H_2O} \times \frac{-475 \text{ kJ}}{2 \text{ mol } H_2O} = -13{,}200 \text{ kJ} = w_{max}$$

The work done can be no larger than the free energy change. The best that could happen is that all of the free energy released would go into doing work, but this does not occur in any real process because there is always waste energy in a real process. Fuel cells are more efficient in converting chemical energy into electrical energy; they are also less massive. The major disadvantage is that they are expensive. In addition, $H_2(g)$ and $O_2(g)$ are an explosive mixture if ignited; much more so than fossil fuels.

126. Cadmium goes from the zero oxidation state to the +2 oxidation state in $Cd(OH)_2$. Because 1 mol of Cd appears in the balanced reaction, n = 2 mol electrons transferred. At standard conditions:

$w_{max} = \Delta G^\circ = -nFE^\circ$

$w_{max} = -(2 \text{ mol } e^-)(96{,}485 \text{ C/mol } e^-)(1.10 \text{ J/C}) = -2.12 \times 10^5 \text{ J} = -212 \text{ kJ}$

127. $(CO + O^{2-} \rightarrow CO_2 + 2\ e^-) \times 2$

$O_2 + 4\ e^- \rightarrow 2\ O^{2-}$

$CO + O_2 \rightarrow 2\ CO_2$

$$\Delta G = -nFE, \quad E = \frac{-\Delta G^o}{nF} = \frac{-(-380 \times 10^3\ J)}{(4\ mole^-)(96{,}485\ C/mole^-)} = 0.98\ V$$

128. For a spontaneous process, $E_{cell} > 0$. In each electron transfer step, we need to couple a reduction half-reaction with an oxidation half-reaction. To determine the correct order, each step must have a positive cell potential in order to be spontaneous. The only possible order for spontaneous electron transfer is:

Step 1:

$cyt\ a(Fe^{3+}) + e^- \rightarrow cyt\ a(Fe^{2+})$ $\quad E = 0.385\ V$

$cyt\ c(Fe^{2+}) \rightarrow cyt\ c(Fe^{3+}) + e^-$ $\quad -E = -0.254\ V$

$cyt\ a(Fe^{3+}) + cyt\ c(Fe^{2+}) \rightarrow cyt\ a(Fe^{2+}) + cyt\ c((Fe^{3+})$ $\quad E_{cell} = 0.131\ V$

Step 2:

$cyt\ c(Fe^{3+}) + e^- \rightarrow cyt\ c(Fe^{2+})$ $\quad E = 0.254\ V$

$cyt\ b(Fe^{2+}) \rightarrow cyt\ b(Fe^{3+}) + e^-$ $\quad -E = -0.030\ V$

$cyt\ c(Fe^{3+}) + cyt\ b(Fe^{2+}) \rightarrow cyt\ c(Fe^{2+}) + cyt\ b((Fe^{3+})$ $\quad E_{cell} = 0.224\ V$

Step 3 would involve the reduction half-reaction of $cyt\ b(Fe^{3+}) + e^- \rightarrow cyt\ b(Fe^{2+})$ coupled with some oxidation half-reaction.

This is the only order that utilizes all three cytochromes and has each step with a positive cell potential. Therefore, electron transport through these cytochromes occurs from cytochrome a to cytochrome c to cytochrome b to some other substance and eventually to oxygen in O_2.

129. $$\frac{150. \times 10^3\ g\ C_6H_8N_2}{h} \times \frac{1\ h}{60\ min} \times \frac{1\ min}{60\ s} \times \frac{1\ mol\ C_6H_8N_2}{108.14\ g\ C_6H_8N_2} \times \frac{2\ mole^-}{mol\ C_6H_8N_2} \times \frac{96{,}485\ C}{mole^-}$$

$= 7.44 \times 10^4$ C/s, or a current of 7.44×10^4 A

130. If the metal M forms 1+ ions, then the atomic mass of M would be:

$$mol\ M = 150.\ s \times \frac{1.25\ C}{s} \times \frac{1\ mole^-}{96{,}485\ C} \times \frac{1\ mol\ M}{1\ mole^-} = 1.94 \times 10^{-3}\ mol\ M$$

$$\text{Atomic mass of M} = \frac{0.109\ g\ M}{1.94 \times 10^{-3}\ mol\ M} = 56.2\ g/mol$$

From the periodic table, the only metal with an atomic mass close to 56.2 g/mol is iron, but iron does not form stable 1+ ions. If M forms 2+ ions, then the atomic mass would be:

$$\text{mol M} = 150.\text{ s} \times \frac{1.25\text{ C}}{\text{s}} \times \frac{1\text{ mol e}^-}{96{,}485\text{ C}} \times \frac{1\text{ mol M}}{2\text{ mol e}^-} = 9.72 \times 10^{-4}\text{ mol M}$$

$$\text{Atomic mass of M} = \frac{0.109\text{ g M}}{9.72 \times 10^{-4}\text{ mol M}} = 112\text{ g/mol}$$

Cadmium has an atomic mass of 112.4 g/mol and does form stable 2+ ions. Cd^{2+} is a much more logical choice than Fe^+.

131. $15\text{ kWh} = \frac{15000\text{ J h}}{\text{s}} \times \frac{60\text{ s}}{\text{min}} \times \frac{60\text{ min}}{\text{h}} = 5.4 \times 10^7\text{ J or } 5.4 \times 10^4\text{ kJ}$ (Hall-Heroult process)

To melt 1.0 kg Al requires: $1.0 \times 10^3\text{ g Al} \times \frac{1\text{ mol Al}}{26.98\text{ g}} \times \frac{10.7\text{ kJ}}{\text{mol Al}} = 4.0 \times 10^2\text{ kJ}$

It is feasible to recycle Al by melting the metal because, in theory, it takes less than 1% of the energy required to produce the same amount of Al by the Hall-Heroult process.

132. In the electrolysis of aqueous sodium chloride, H_2O is reduced in preference to Na^+, and Cl^- is oxidized in preference to H_2O. The anode reaction is $2\,Cl^- \rightarrow Cl_2 + 2\,e^-$, and the cathode reaction is $2\,H_2O + 2\,e^- \rightarrow H_2 + 2\,OH^-$. The overall reaction is:

$$2\,H_2O(l) + 2\,Cl^-(aq) \rightarrow Cl_2(g) + H_2(g) + 2\,OH^-(aq)$$

From the 1 : 1 mol ratio between Cl_2 and H_2 in the overall balanced reaction, if 257 L of $Cl_2(g)$ is produced, then 257 L of $H_2(g)$ will also be produced because moles and volume of gas are directly proportional at constant T and P (see Chapter 5 of text).

133. $\text{Moles of e}^- = 50.0\text{ min} \times \frac{60\text{ s}}{\text{min}} \times \frac{2.50\text{ C}}{\text{s}} \times \frac{1\text{ mol e}^-}{96{,}485\text{ C}} = 7.77 \times 10^{-2}\text{ mol e}^-$

$$\text{Moles of Ru} = 2.618\text{ g Ru} \times \frac{1\text{ mol Ru}}{101.1\text{ g Ru}} = 2.590 \times 10^{-2}\text{ mol Ru}$$

$$\frac{\text{Moles of e}^-}{\text{Moles of Ru}} = \frac{7.77 \times 10^{-2}\text{ mol e}^-}{2.590 \times 10^{-2}\text{ mol Ru}} = 3.00;$$ the charge on the ruthenium ions is +3.

($Ru^{3+} + 3\,e^- \rightarrow Ru$)

ChemWork Problems

The answers to the problems 134-139 (or a variation to these problems) are found in OWL. These problems are also assignable in OWL.

Challenge Problems

140. a. HCl(aq) dissociates to $H^+(aq) + Cl^-(aq)$. For simplicity, let's use H^+ and Cl^- separately.

$$H^+ \rightarrow H_2 \qquad\qquad Fe \rightarrow HFeCl_4$$
$$(2\ H^+ + 2\ e^- \rightarrow H_2) \times 3 \qquad\qquad (H^+ + 4\ Cl^- + Fe \rightarrow HFeCl_4 + 3\ e^-) \times 2$$

$$6\ H^+ + 6\ e^- \rightarrow 3\ H_2$$
$$2\ H^+ + 8\ Cl^- + 2\ Fe \rightarrow 2\ HFeCl_4 + 6\ e^-$$
$$\overline{8\ H^+ + 8\ Cl^- + 2\ Fe \rightarrow 2\ HFeCl_4 + 3\ H_2}$$

or $8\ HCl(aq) + 2\ Fe(s) \rightarrow 2\ HFeCl_4(aq) + 3\ H_2(g)$

b.

$$IO_3^- \rightarrow I_3^- \qquad\qquad I^- \rightarrow I_3^-$$
$$3\ IO_3^- \rightarrow I_3^- \qquad\qquad (3\ I^- \rightarrow I_3^- + 2\ e^-) \times 8$$
$$3\ IO_3^- \rightarrow I_3^- + 9\ H_2O$$
$$16\ e^- + 18\ H^+ + 3\ IO_3^- \rightarrow I_3^- + 9\ H_2O$$

$$16\ e^- + 18\ H^+ + 3\ IO_3^- \rightarrow I_3^- + 9\ H_2O$$
$$24\ I^- \rightarrow 8\ I_3^- + 16\ e^-$$
$$\overline{18\ H^+ + 24\ I^- + 3\ IO_3^- \rightarrow 9\ I_3^- + 9\ H_2O}$$

Reducing: $6\ H^+(aq) + 8\ I^-(aq) + IO_3^-(aq) \rightarrow 3\ I_3^-(aq) + 3\ H_2O(l)$

c. $(Ce^{4+} + e^- \rightarrow Ce^{3+}) \times 97$

$$Cr(NCS)_6^{4-} \rightarrow Cr^{3+} + NO_3^- + CO_2 + SO_4^{2-}$$
$$54\ H_2O + Cr(NCS)_6^{4-} \rightarrow Cr^{3+} + 6\ NO_3^- + 6\ CO_2 + 6\ SO_4^{2-} + 108\ H^+$$

Charge on left $= -4$. Charge on right $= +3 + 6(-1) + 6(-2) + 108(+1) = +93$. Add 97 e^- to the product side, and then add the two balanced half-reactions with a common factor of 97 e^- transferred.

$$54\ H_2O + Cr(NCS)_6^{4-} \rightarrow Cr^{3+} + 6\ NO_3^- + 6\ CO_2 + 6\ SO_4^{2-} + 108\ H^+ + 97\ e^-$$
$$97\ e^- + 97\ Ce^{4+} \rightarrow 97\ Ce^{3+}$$

$$97\ Ce^{4+}(aq) + 54\ H_2O(l) + Cr(NCS)_6^{4-}(aq) \rightarrow 97\ Ce^{3+}(aq) + Cr^{3+}(aq) + 6\ NO_3^-(aq)$$
$$+ 6\ CO_2(g) + 6\ SO_4^{2-}(aq) + 108\ H^+(aq)$$

This is very complicated. A check of the net charge is a good check to see if the equation is balanced. Left: Charge $= 97(+4) - 4 = +384$. Right: Charge $= 97(+3) + 3 + 6(-1) + 6(-2) + 108(+1) = +384$.

d.

$$CrI_3 \rightarrow CrO_4^{2-} + IO_4^- \qquad\qquad Cl_2 \rightarrow Cl^-$$
$$(16\ H_2O + CrI_3 \rightarrow CrO_4^{2-} + 3\ IO_4^- + 32\ H^+ + 27\ e^-) \times 2 \qquad\qquad (2\ e^- + Cl_2 \rightarrow 2\ Cl^-) \times 27$$

Common factor is a transfer of 54 e^-.

$$54\ e^- + 27\ Cl_2 \rightarrow 54\ Cl^-$$
$$32\ H_2O + 2\ CrI_3 \rightarrow 2\ CrO_4^{2-} + 6\ IO_4^- + 64\ H^+ + 54\ e^-$$

$$32\ H_2O + 2\ CrI_3 + 27\ Cl_2 \rightarrow 54\ Cl^- + 2\ CrO_4^{2-} + 6\ IO_4^- + 64\ H^+$$

Add 64 OH^- to both sides and convert 64 H^+ into 64 H_2O.

$$64\ OH^- + 32\ H_2O + 2\ CrI_3 + 27\ Cl_2 \rightarrow 54\ Cl^- + 2\ CrO_4^{2-} + 6\ IO_4^- + 64\ H_2O$$

Reducing gives:

$$64\ OH^-(aq) + 2\ CrI_3(s) + 27\ Cl_2(g) \rightarrow 54\ Cl^-(aq) + 2\ CrO_4^{2-}(aq) + 6\ IO_4^-(aq) + 32\ H_2O(l)$$

e.

$$Ce^{4+} \rightarrow Ce(OH)_3$$
$$(e^- + 3\ H_2O + Ce^{4+} \rightarrow Ce(OH)_3 + 3\ H^+) \times 61$$

$$Fe(CN)_6^{4-} \rightarrow Fe(OH)_3 + CO_3^{2-} + NO_3^-$$
$$Fe(CN)_6^{4-} \rightarrow Fe(OH)_3 + 6\ CO_3^{2-} + 6\ NO_3^-$$

There are 39 extra O atoms on right. Add 39 H_2O to left; then add 75 H^+ to right to balance H^+.

$$39\ H_2O + Fe(CN)_6^{4-} \rightarrow Fe(OH)_3 + 6\ CO_3^{2-} + 6\ NO_3^- + 75\ H^+$$

Net charge = 4– Net charge = 57+

Add 61 e^- to the product side, and then add the two balanced half-reactions with a common factor of 61 e^- transferred.

$$39\ H_2O + Fe(CN)_6^{4-} \rightarrow Fe(OH)_3 + 6\ CO_3^- + 6\ NO_3^- + 75\ H^+ + 61\ e^-$$
$$61\ e^- + 183\ H_2O + 61\ Ce^{4+} \rightarrow 61\ Ce(OH)_3 + 183\ H^+$$

$$222\ H_2O + Fe(CN)_6^{4-} + 61\ Ce^{4+} \rightarrow 61\ Ce(OH)_3 + Fe(OH)_3 + 6\ CO_3^{2-} + 6\ NO_3^- + 258\ H^+$$

Adding 258 OH^- to each side and then reducing gives:

$$258\ OH^-(aq) + Fe(CN)_6^{4-}(aq) + 61\ Ce^{4+}(aq) \rightarrow 61\ Ce(OH)_3(s) + Fe(OH)_3(s) + 6\ CO_3^{2-}(aq) + 6\ NO_3^-(aq) + 36\ H_2O(l)$$

141. $\Delta G° = -nFE° = \Delta H° - T\Delta S°,\ E° = \dfrac{T\Delta S°}{nF} - \dfrac{\Delta H°}{nF}$

If we graph E° versus T we should get a straight line ($y = mx + b$). The slope of the line is equal to $\Delta S°/nF$, and the y intercept is equal to $-\Delta H°/nF$. From the preceding equation, E° will have a small temperature dependence when $\Delta S°$ is close to zero.

142. a. We can calculate ΔG° from ΔG° = ΔH° − TΔS° and then E° from ΔG° = −nFE°, or we can use the equation derived in Exercise 141. For this reaction, n = 2 (from oxidation states).

$$E^\circ_{-20} = \frac{T\Delta\Delta^\circ - \Delta H^\circ}{nF} = \frac{(253\text{K})(263.5\text{J/K}) + 315.9 \times 10^3\ \text{J}}{(2\,\text{mol e}^-)(96{,}485\text{C/mol e}^-)} = 1.98\ \text{J/C} = 1.98\ \text{V}$$

b. $$E_{-20} = E_{-20} - \frac{RT}{nF}\ln Q = 1.98\ \text{V} - \frac{RT}{nF}\ln\frac{1}{[H^+]^2[HSO_4^-]^2}$$

$$E_{-20} = 1.98\ \text{V} - \frac{(8.3145\text{J/K}\bullet\text{mol})(253\text{K})}{(2\,\text{mol e}^-)(96{,}485\text{C/mol e}^-)}\ln\frac{1}{(4.5)^2(4.5)^2} = 1.98\ \text{V} + 0.066\ \text{V}$$

$$= 2.05\ \text{V}$$

c. From Exercise 71, E = 2.12 V at 25°C. As the temperature decreases, the cell potential decreases. Also, oil becomes more viscous at lower temperatures, which adds to the difficulty of starting an engine on a cold day. The combination of these two factors results in batteries failing more often on cold days than on warm days.

143.

$$(Ag^+ + e^- \rightarrow Ag) \times 2 \qquad E^\circ = 0.80\ \text{V}$$

$$Pb \rightarrow Pb^{2+} + 2\,e^- \qquad -E^\circ = -(-0.13)$$

$$2\,Ag^+(aq) + Pb(s) \rightarrow 2\,Ag(s) + Pb^{2+}(aq) \qquad E^\circ_{cell} = 0.93\ \text{V}$$

$$E = E^\circ - \frac{0.0591}{n}\log\frac{[Pb^{2+}]}{[Ag^+]^2},\quad 0.83\ \text{V} = 0.93\ \text{V} - \frac{0.0591}{n}\log\frac{(1.8)}{[Ag^+]^2}$$

$$\log\frac{(1.8)}{[Ag^+]^2} = \frac{0.10(2)}{0.0591} = 3.4,\quad \frac{(1.8)}{[Ag^+]^2} = 10^{3.4},\ [Ag^+] = 0.027\ M$$

	$Ag_2SO_4(s)$	⇌	$2\,Ag^+(aq)$	+	$SO_4^{2-}(aq)$	$K_{sp} = [Ag^+]^2[SO_4^{2-}]$
Initial	s = solubility (mol/L)		0		0	
Equil.			$2s$		s	

From problem: $2s = 0.027\ M$, $s = 0.027/2$

$K_{sp} = (2s)^2(s) = (0.027)^2(0.027/2) = 9.8 \times 10^{-6}$

144. a. $Zn(s) + Cu^{2+}(aq) \rightarrow Zn^{2+}(aq) + Cu(s) \quad E^\circ_{cell} = 1.10\ \text{V}$

$$E_{cell} = 1.10\ \text{V} - \frac{0.0591}{2}\log\frac{[Zn^{2+}]}{[Cu^{2+}]}$$

$$E_{cell} = 1.10\ \text{V} - \frac{0.0591}{2}\log\frac{0.10}{2.50} = 1.10\ \text{V} - (-0.041\ \text{V}) = 1.14\ \text{V}$$

b. $10.0 \text{ h} \times \frac{60 \text{ min}}{\text{h}} \times \frac{60 \text{ s}}{\text{min}} \times \frac{10.0 \text{ C}}{\text{s}} \times \frac{1 \text{ mol e}^-}{96{,}485 \text{ C}} \times \frac{1 \text{ mol Cu}}{2 \text{ mol e}^-} = 1.87 \text{ mol Cu produced}$

The Cu^{2+} concentration decreases by 1.87 mol/L, and the Zn^{2+} concentration will increase by 1.87 mol/L.

$[Cu^{2+}] = 2.50 - 1.87 = 0.63\ M$; $[Zn^{2+}] = 0.10 + 1.87 = 1.97\ M$

$$E_{cell} = 1.10 \text{ V} - \frac{0.0591}{2} \log \frac{1.97}{0.63} = 1.10 \text{ V} - 0.015 \text{ V} = 1.09 \text{ V}$$

c. $1.87 \text{ mol Zn consumed} \times \frac{65.38 \text{ g Zn}}{\text{mol Zn}} = 122 \text{ g Zn}$

Mass of electrode = 200. – 122 = 78 g Zn

$$1.87 \text{ mol Cu formed} \times \frac{63.55 \text{ g Cu}}{\text{mol Cu}} = 119 \text{ g Cu}$$

Mass of electrode = 200. + 119 = 319 g Cu

d. Three things could possibly cause this battery to go dead:

(1) All the Zn is consumed.

(2) All the Cu^{2+} is consumed.

(3) Equilibrium is reached ($E_{cell} = 0$).

We began with 2.50 mol Cu^{2+} and 200. g Zn × 1 mol Zn/65.38 g Zn = 3.06 mol Zn. Because there is a 1 : 1 mole relationship between Cu^{2+} and Zn in the balanced cell equation, Cu^{2+} is the limiting reagent and will run out first. To react all the Cu^{2+} requires:

$$2.50 \text{ mol Cu}^{2+} \times \frac{2 \text{ mol e}^-}{\text{mol Cu}^{2+}} \times \frac{96{,}485 \text{ C}}{\text{mol e}^-} \times \frac{1 \text{ s}}{10.0 \text{ C}} \times \frac{1 \text{ h}}{3600 \text{ s}} = 13.4 \text{ h}$$

For equilibrium to be reached: $E = 0 = 1.10 \text{ V} - \frac{0.0591}{2} \log \frac{[Zn^{2+}]}{[Cu^{2+}]}$

$$\frac{[Zn^{2+}]}{[Cu^{2+}]} = K = 10^{2(1.10)/0.0591} = 1.68 \times 10^{37}$$

This is such a large equilibrium constant that virtually all the Cu^{2+} must react to reach equilibrium. So the battery will go dead in 13.4 hours.

145.

$2 H^+ + 2 e^- \rightarrow H_2$	$E° = 0.000$ V
$Fe \rightarrow Fe^{2+} + 2e^-$	$-E° = -(-0.440 \text{V})$
$2 H^+(aq) + Fe(s) \rightarrow H_2(g) + Fe^{3+}(aq)$	$E°_{cell} = 0.440$ V

$$E_{cell} = E^{o}_{cell} - \frac{0.0591}{n}\log Q\text{, where n = 2 and } Q = \frac{P_{H_2} \times [Fe^{3+}]}{[H^+]^2}$$

To determine K_a for the weak acid, first use the electrochemical data to determine the H^+ concentration in the half-cell containing the weak acid.

$$0.333\text{ V} = 0.440\text{ V} - \frac{0.0591}{2}\log\frac{1.00\text{ atm}(1.00 \times 10^{-3}\ M)}{[H^+]^2}$$

$$\frac{0.107(2)}{0.0591} = \log\frac{1.0 \times 10^{-3}}{[H^+]^2},\quad \frac{1.0 \times 10^{-3}}{[H^+]^2} = 10^{3.621} = 4.18 \times 10^3,\quad [H^+] = 4.89 \times 10^{-4}\ M$$

Now we can solve for the K_a value of the weak acid HA through the normal setup for a weak acid problem.

	$HA(aq)$	$\rightleftharpoons$	$H^+(aq)$	+	$A^-(aq)$	$K_a = \frac{[H^+][A^-]}{[HA]}$
Initial	1.00 *M*		~0		0	
Equil.	$1.00 - x$		x		x	

$$K_a = \frac{x^2}{1.00 - x},\text{ where } x = [H^+] = 4.89 \times 10^{-4}\ M,\ K_a = \frac{(4.89 \times 10^{-4})^2}{1.00 - 4.89 \times 10^{-4}} = 2.39 \times 10^{-7}$$

146. a. Nonreactive anions are present in each half-cell to balance the cation charges.

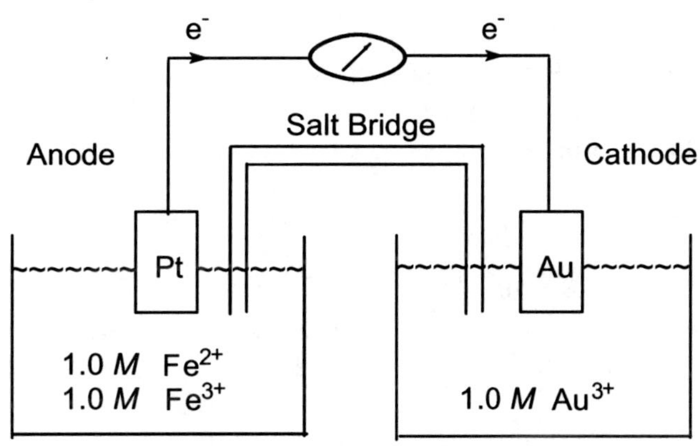

b. $Au^{3+}(aq) + 3\ Fe^{2+}(aq) \rightarrow 3\ Fe^{3+}(aq) + Au(s) \qquad E^{o}_{cell} = 1.50 - 0.77 = 0.73\text{ V}$

$$E_{cell} = E^{o}_{cell} - \frac{0.0591}{n}\log Q = 0.73\text{ V} - \frac{0.0591}{3}\log\frac{[Fe^{3+}]^3}{[Au^{3+}][Fe^{2+}]^3}$$

Because $[Fe^{3+}] = [Fe^{2+}] = 1.0\ M$: $0.31\text{ V} = 0.73\text{ V} - \frac{0.0591}{3}\log\frac{1}{[Au^{3+}]}$

$$\frac{3(-0.42)}{0.0591} = -\log\frac{1}{[Au^{3+}]},\quad \log[Au^{3+}] = -21.32,\quad [Au^{3+}] = 10^{-21.32} = 4.8 \times 10^{-22}\ M$$

$Au^{3+}(aq) + 4\ Cl^-(aq) \rightleftharpoons AuCl_4^-(aq)$; because the equilibrium Au^{3+} concentration is so small, assume $[AuCl_4^-] \approx [Au^{3+}]_0 \approx 1.0\ M$, i.e., assume K is large, so the reaction essentially goes to completion.

$$K = \frac{[AuCl_4^-]}{[Au^{3+}][Cl^-]^4} = \frac{1.0}{(4.8\times10^{-22})(0.10)^4} = 2.1 \times 10^{25}; \text{ assumption good (K is large).}$$

147. a. $E_{cell} = E_{ref} + 0.05916\ pH,\ 0.480\ V = 0.250\ V + 0.05916\ pH$

$$pH = \frac{0.480-0.250}{0.05916} = 3.888; \text{ uncertainty} = \pm1\ mV = \pm\ 0.001\ V$$

$$pH_{max} = \frac{0.481-0.250}{0.05916} = 3.905; \quad pH_{min} = \frac{0.479-0.250}{0.05916} = 3.871$$

Thus, if the uncertainty in potential is ±0.001 V, then the uncertainty in pH is ±0.017, or about ±0.02 pH units. For this measurement, $[H^+] = 10^{-3.888} = 1.29 \times 10^{-4}\ M$. For an error of +1 mV, $[H^+] = 10^{-3.905} = 1.24 \times 10^{-4}\ M$. For an error of −1 mV, $[H^+] = 10^{-3.871} = 1.35 \times 10^{-4}\ M$. So the uncertainty in $[H^+]$ is $\pm0.06 \times 10^{-4}\ M = \pm6 \times 10^{-6}\ M$.

b. From part a, we will be within ±0.02 pH units if we measure the potential to the nearest ±0.001 V (±1 mV).

148. a. $E° = 0$ (concentration cell), $E = 0 - \dfrac{0.0591}{2}\log\left(\dfrac{1.0\times10^{-4}}{1.00}\right)$, $E = 0.12\ V$

b. $Cu^{2+}(aq) + 4\ NH_3(aq) \rightleftharpoons Cu(NH_3)_4^{2+}(aq)$ $K = 1.0 \times 10^{13}$; because K is so large, this reaction lies far to the right. Assume the reaction goes to completion, then solve an equilibrium problem for the new Cu^{2+} concentration at the anode.

	$Cu^{2+}(aq)$	+ $4\ NH_3(aq)$	$\rightleftharpoons$	$Cu(NH_3)_4^{2+}(aq)$
Before	$1.0 \times 10^{-4}\ M$	$2.0\ M$		0
Change	-1.0×10^{-4}	-4.0×10^{-4}	→	$+1.0 \times 10^{-4}$
After	0	2.0		1.0×10^{-4}

Now allow the reaction to get to equilibrium.

	$Cu^{2+}(aq)$	+ $4\ NH_3(aq)$	$\rightleftharpoons$	$Cu(NH_3)_4^{2+}(aq)$
Initial	0	$2.0\ M$		$1.0 \times 10^{-4}\ M$
Change	$+x$	$+4x$	←	$-x$
Equil.	x	$2.0 + 4x$		$1.0 \times 10^{-4} - x$

$$\frac{(1.0 \times 10^{-4} - x)}{x(2.0 + 4x)^4} \approx \frac{1.0 \times 10^{-4}}{x(2.0)^4} = 1.0 \times 10^{13},\ x = [Cu^{2+}] = 6.3 \times 10^{-19}\ M;$$

assumptions good.

In the concentration cell, $[Cu^{2+}]_{anode} = 6.3 \times 10^{-19}$ *M*. Solving for the cell potential:

$$E = 0 - \frac{0.0591}{2}\log\left(\frac{6.3 \times 10^{-19}}{1.00}\right), \quad E = 0.54 \text{ V}$$

149. a.

$$\begin{array}{ll} (Ag^+ + e^- \rightarrow Ag) \times 2 & E° = \ \ 0.80 \text{ V} \\ Cu \rightarrow Cu^{2+} + 2\,e^- & -E° = -0.34 \text{ V} \\ \hline 2\,Ag^+(aq) + Cu(s) \rightarrow 2\,Ag(s) + Cu^{2+}(aq) & E°_{cell} = 0.46 \text{ V} \end{array}$$

$$E_{cell} = E°_{cell} - \frac{0.0591}{n}\log Q, \text{ where n = 2 and } Q = \frac{[Cu^{2+}]}{[Ag^+]^2}.$$

To calculate E_{cell}, we need to use the K_{sp} data to determine $[Ag^+]$.

	$AgCl(s)$	$\rightleftharpoons$	$Ag^+(aq)$	+	$Cl^-(aq)$	$K_{sp} = 1.6 \times 10^{-10} = [Ag^+][Cl^-]$
Initial	s = solubility (mol/L)		0		0	
Equil.			s		s	

$K_{sp} = 1.6 \times 10^{-10} = s^2$, $s = [Ag^+] = 1.3 \times 10^{-5}$ mol/L

$$E_{cell} = 0.46 \text{ V} - \frac{0.0591}{2}\log\frac{2.0}{(1.3 \times 10^{-5})^2} = 0.46 \text{ V} - 0.30 = 0.16 \text{ V}$$

b. $Cu^{2+}(aq) + 4\,NH_3(aq) \rightleftharpoons Cu(NH_4)_4^{2+}(aq)$ $\quad K = 1.0 \times 10^{13} = \frac{[Cu(NH_3)_4^{2+}]}{[Cu^{2+}][NH_3]^4}$

Because K is very large for the formation of $Cu(NH_3)_4^{2+}$, the forward reaction is dominant. At equilibrium, essentially all the 2.0 *M* Cu^{2+} will react to form 2.0 *M* $Cu(NH_3)_4^{2+}$. This reaction requires 8.0 *M* NH_3 to react with all the Cu^{2+} in the balanced equation. Therefore, the moles of NH_3 added to 1.0-L solution will be larger than 8.0 mol since some NH_3 must be present at equilibrium. In order to calculate how much NH_3 is present at equilibrium, we need to use the electrochemical data to determine the Cu^{2+} concentration.

$$E_{cell} = E°_{cell} - \frac{0.0591}{n}\log Q, \quad 0.52 \text{ V} = 0.46 \text{ V} - \frac{0.0591}{2}\log\frac{[Cu^{2+}]}{(1.3 \times 10^{-5})^2}$$

$$\log\frac{[Cu^{2+}]}{(1.3 \times 10^{-5})^2} = \frac{-0.06(2)}{0.0591} = -2.03, \quad \frac{[Cu^{2+}]}{(1.3 \times 10^{-5})^2} = 10^{-2.03} = 9.3 \times 10^{-3}$$

$[Cu^{2+}] = 1.6 \times 10^{-12} = 2 \times 10^{-12}$ *M*

(We carried extra significant figures in the calculation.)

Note: Our assumption that the 2.0 *M* Cu^{2+} essentially reacts to completion is excellent because only 2×10^{-12} *M* Cu^{2+} remains after this reaction. Now we can solve for the equilibrium $[NH_3]$.

$$K = 1.0 \times 10^{13} = \frac{[Cu(NH_3)_4^{2+}]}{[Cu^{2+}][NH_3]^4} = \frac{(2.0)}{(2 \times 10^{-12})[NH_3]^4}, \quad [NH_3] = 0.6\ M$$

Because 1.0 L of solution is present, 0.6 mol NH_3 remains at equilibrium. The total moles of NH_3 added is 0.6 mol plus the 8.0 mol NH_3 necessary to form 2.0 *M* $Cu(NH_3)_4^{2+}$. Therefore, 8.0 + 0.6 = 8.6 mol NH_3 was added.

150. Standard reduction potentials can only be manipulated and added together when electrons in the reduction half-reaction exactly cancel with the electrons in the oxidation half-reaction. We will solve this problem by applying the equation ΔG° = −nFE° to the half-reactions.

$$M^{3+} + 3\ e^- \rightarrow M \quad \Delta G° = -nFE° = -3(96{,}485)(0.10) = -2.9 \times 10^4\ J$$

Because M and e^- have $\Delta G_f^\circ = 0$: $-2.9 \times 10^4\ J = -\Delta G^\circ_{f,M^{3+}}$, $\Delta G^\circ_{f,M^{3+}} = 2.9 \times 10^4\ J$

$$M^{2+} + 2\ e^- \rightarrow M \quad \Delta G° = -nFE° = -2(96{,}485)(0.50) = -9.6 \times 10^4\ J$$

$$-9.6 \times 10^4\ J = -\Delta G^\circ_{f,M^{2+}}, \quad \Delta G^\circ_{f,M^{2+}} = 9.6 \times 10^4\ J$$

$$M^{3+} + e^- \rightarrow M^{2+} \qquad \Delta G° = 9.6 \times 10^4\ J - (2.9 \times 10^4\ J) = 6.7 \times 10^4\ J$$

$$E° = \frac{-\Delta G°}{nF} = \frac{-(6.7 \times 10^4)}{(1)(96{,}485)} = -0.69\ V; \quad M^{3+} + e^- \rightarrow M^{2+} \quad E° = -0.69\ V$$

151. $2\ Ag^+(aq) + Ni(s) \rightarrow Ni^{2+}(aq) + Ag(s)$; the cell is dead at equilibrium (E = 0).

$$E^\circ_{cell} = 0.80\ V + 0.23\ V = 1.03\ V$$

$$0 = 1.03\ V - \frac{0.0591}{2}\log K, \quad K = 7.18 \times 10^{34}$$

K is very large. Let the forward reaction go to completion.

	$2\ Ag^+(aq)$	+ $Ni(s)$	→	$Ni^{2+}(aq)$	+ $2\ Ag(s)$	$K = [Ni^{2+}]/[Ag^+]^2 = 7.18 \times 10^{34}$
Before	1.0 *M*			1.0 *M*		
Change	−1.0		→	+0.50		
After	0			1.5 *M*		

Now solve the back-equilibrium problem.

	$2\ Ag^+(aq) + Ni(s)$	⇌	$Ni^{2+}(aq)$	+ $2\ Ag(s)$
Initial	0		1.5 *M*	
Change	+2*x*	←	−*x*	
Equil.	2*x*		1.5 − *x*	

$$K = 7.18 \times 10^{34} = \frac{1.5 - x}{(2x)^2} \approx \frac{1.5}{(2x)^2}$$; solving, $x = 2.3 \times 10^{-18}$ *M*. Assumptions good.

$[Ag^+] = 2x = 4.6 \times 10^{-18}$ *M*; $[Ni^{2+}] = 1.5 - 2.3 \times 10^{-18} = 1.5$ *M*

152. a. $Ag_2CrO_4(s) + 2\ e^- \rightarrow 2\ Ag(s) + CrO_4^{2-}(aq)$ $E° = 0.446$ V

$Hg_2Cl_2 + 2\ e^- \rightarrow 2\ Hg + 2\ Cl^-$ $E_{SCE} = 0.242$ V

SCE will be the oxidation half-reaction with $E_{cell} = 0.446 - 0.242 = 0.204$ V.

$\Delta G = -nFE_{cell} = -2(96{,}485)(0.204)J = -3.94 \times 10^4\ J = -39.4$ kJ

b. In SCE, we assume all concentrations are constant. Therefore, only CrO_4^{2-} appears in the Q expression, and it will appear in the numerator since CrO_4^{2-} is produced in the reduction half-reaction. To calculate E_{cell} at nonstandard CrO_4^{2-} concentrations, we use the following equation.

$$E_{cell} = E^{\circ}_{cell} - \frac{0.0591}{2}\log[CrO_4^{2-}] = 0.204\ V - \frac{0.0591}{2}\log[CrO_4^{2-}]$$

c. $$E_{cell} = 0.204 - \frac{0.0591}{2}\log(1.00 \times 10^{-5}) = 0.204\ V - (-0.148\ V) = 0.352\ V$$

d. $0.504\ V = 0.204\ V - (0.0591/2)\log[CrO_4^{2-}]$

$\log[CrO_4^{2-}] = -10.152$, $[CrO_4^{2-}] = 10^{-10.152} = 7.05 \times 10^{-11}$ *M*

e. $Ag_2CrO_4 + 2\ e^- \rightarrow 2\ Ag + CrO_4^{2-}$ $E° = 0.446$ V

$(Ag \rightarrow Ag^+ + e^-) \times 2$ $-E° = -0.80$ V

$Ag_2CrO_4(s) \rightarrow 2\ Ag^+(aq) + CrO_4^-(aq)$ $E^{\circ}_{cell} = -0.35$ V; $K = K_{sp} = ?$

$$E^{\circ}_{cell} = \frac{0.0591}{n}\log K_{sp},\ \log K_{sp} = \frac{(-0.35\ V)(2)}{0.0591} = -11.84,\ K_{sp} = 10^{-11.84} = 1.4 \times 10^{-12}$$

153. $(Ag^+ + e^- \rightarrow Ag) \times 2$ $E° = 0.80$ V

$Cd \rightarrow Cd^{2+} + 2\ e^-$ $-E° = 0.40$ V

$2\ Ag^+(aq) + Cd(s) \rightarrow Cd^{2+}(aq) + 2\ Ag(s)$ $E^{\circ}_{cell} = 1.20$ V

Overall complex ion reaction:

$Ag^+(aq) + 2\ NH_3(aq) \rightarrow Ag(NH_3)_2^+(aq)$ $K = K_1K_2 = 2.1 \times 10^3(8.2 \times 10^3) = 1.7 \times 10^7$

Because K is large, we will let the reaction go to completion and then solve the back-equilibrium problem.

	$Ag^+(aq)$	+ $2\ NH_3(aq)$	$\rightleftharpoons$	$Ag(NH_3)_2^+(aq)$	$K = 1.7 \times 10^7$
Before	1.00 *M*	15.0 *M*		0	
After	0	13.0		1.00	New initial
Change	x	$+2x$	$\leftarrow$	$-x$	
Equil.	x	$13.0 + 2x$		$1.00 - x$	

$$K = \frac{[Ag(NH_3)_2^+]}{[Ag^+][NH_3]^2};\quad 1.7 \times 10^7 = \frac{1.00 - x}{x(13.0 + 2x)^2} \approx \frac{1.00}{x(13.0)^2}$$

Solving: $x = 3.5 \times 10^{-10}\ M = [Ag^+]$; assumptions good.

$$E = E° - \frac{0.0591}{2}\log\frac{[Cd^{2+}]}{[Ag^+]^2} = 1.20\ V - \frac{0.0591}{2}\log\left[\frac{1.0}{(3.5 \times 10^{-10})^2}\right]$$

$E = 1.20 - 0.56 = 0.64$ V

154. a.

$$3 \times (e^- + 2\ H^+ + NO_3^- \rightarrow NO_2 + H_2O) \qquad E° = 0.775\ V$$
$$2\ H_2O + NO \rightarrow NO_3^- + 4\ H^+ + 3\ e^- \qquad -E° = -0.957\ V$$

$$2\ H^+(aq) + 2\ NO_3^-(aq) + NO(g) \rightarrow 3\ NO_2(g) + H_2O(l) \qquad E°_{cell} = -0.182\ V,\ K = ?$$

$$\log K = \frac{E°}{0.0591} = \frac{3(-0.182)}{0.0591} = -9.239,\ K = 10^{-9.239} = 5.77 \times 10^{-10}$$

b. Let C = concentration of $HNO_3 = [H^+] = [NO_3^-]$.

$$5.77 \times 10^{-10} = \frac{P^3_{NO_2}}{P_{NO} \times [H^+]^2 \times [NO_3^-]^2} = \frac{P^3_{NO_2}}{P_{NO} \times C^4}$$

If 0.20% NO_2 by moles and P_{total} = 1.00 atm:

$$P_{NO_2} = \frac{0.20\ mol\ NO_2}{100.\ mol\ total} \times 1.00\ atm = 2.0 \times 10^{-3}\ atm;\ P_{NO} = 1.00 - 0.0020 = 1.00\ atm$$

$$5.77 \times 10^{-10} = \frac{(2.0 \times 10^{-3})^3}{(1.00)C^4},\ C = 1.9\ M\ HNO_3$$

Integrative Problems

155. a.

$$(In^+ + e^- \rightarrow In) \times 2 \qquad E° = -0.126\ V$$
$$In^+ \rightarrow In^{3+} + 2\ e^- \qquad -E° = 0.444\ V$$

$$3\ In^+(aq) \rightarrow In^{3+}(aq) + 2\ In(s) \qquad E°_{cell} = 0.318$$

$$\log K = \frac{nE°}{0.0591} = \frac{2(0.318)}{0.0591} = 10.761,\ K = 10^{10.761} = 5.77 \times 10^{10}$$

b. $\Delta G° = -nFE° = -(2 \text{ mol e}^-)(96{,}485 \text{ C/mol e}^-)(0.318 \text{ J/C}) = -6.14 \times 10^5 \text{ J} = -61.4 \text{ kJ}$

$$\Delta G^\circ_{rxn} = -61.4 \text{ kJ} = [2(0) + 1(-97.9 \text{ kJ}] - 3\,\Delta G^\circ_{f,\,In^+},\quad \Delta G^\circ_{f,\,In^+} = -12.2 \text{ kJ/mol}$$

156. $E^\circ_{cell} = 0.400 \text{ V} - 0.240 \text{ V} = 0.160 \text{ V};\ E = E° - \dfrac{0.0591}{n}\log Q$

$$0.180 = 0.160 - \frac{0.0591}{n}\log(9.32 \times 10^{-3}),\quad 0.020 = \frac{0.120}{n},\quad n = 6$$

Six moles of electrons are transferred in the overall balanced reaction. We now have to figure out how to get 6 mol e^- into the overall balanced equation. The two possibilities are to have ion charges of +1 and +6 or +2 and +3; only these two combinations yield a 6 when common multiples are determined when adding the reduction half-reaction to the oxidation half-reaction. Because N forms +2 charged ions, M must form for +3 charged ions. The overall cell reaction can now be determined.

$$(M^{3+} + 3\,e^- \rightarrow M) \times 2 \qquad E^\circ = 0.400 \text{ V}$$
$$(N \rightarrow N^{2+} + 2\,e^-) \times 3 \qquad -E^\circ = -0.240 \text{ V}$$

$$2\,M^{3+}(aq) + 3\,N(s) \rightarrow 3\,N^{2+}(aq) + 2\,M(s) \qquad E^\circ_{cell} = 0.160 \text{ V}$$

$$Q = 9.32 \times 10^{-3} = \frac{[N^{2+}]_0^{\,3}}{[M^{3+}]_0^{\,2}} = \frac{(0.10)^3}{[M^{3+}]^2},\quad [M^{3+}] = 0.33\,M$$

$$w_{max} = \Delta G = -nFE = -6(96{,}485)(0.180) = -1.04 \times 10^5 \text{ J} = -104 \text{ kJ}$$

The maximum amount of work this cell could produce is 104 kJ.

157. Chromium(III) nitrate $[Cr(NO_3)_3]$ has chromium in the +3 oxidation state.

$$1.15 \text{ g Cr} \times \frac{1 \text{ mol Cr}}{52.00 \text{ g}} \times \frac{3 \text{ mol e}^-}{\text{mol Cr}} \times \frac{96{,}485 \text{ C}}{\text{mol e}^-} = 6.40 \times 10^3 \text{ C of charge}$$

For the Os cell, 6.40×10^3 C of charge also was passed.

$$3.15 \text{ g Os} \times \frac{1 \text{ mol Os}}{190.2 \text{ g}} = 0.0166 \text{ mol Os};\ 6.40 \times 10^3 \text{ C} \times \frac{1 \text{ mol e}^-}{96{,}485 \text{ C}} = 0.0663 \text{ mol e}^-$$

$$\frac{\text{Moles of e}^-}{\text{Moles of Os}} = \frac{0.0663}{0.0166} = 3.99 \approx 4$$

This salt is composed of Os^{4+} and NO_3^- ions. The compound is $Os(NO_3)_4$, osmium(IV) nitrate.

For the third cell, identify X by determining its molar mass. Two moles of electrons are transferred when X^{2+} is reduced to X.

$$\text{Molar mass} = \frac{2.11\text{ g X}}{6.40 \times 10^3\text{ C} \times \dfrac{1\text{ mol e}^-}{96{,}485\text{ C}} \times \dfrac{1\text{ mol X}}{2\text{ mol e}^-}} = 63.6\text{ g/mol}$$

This is copper (Cu), which has an electron configuration of $[Ar]4s^1 3d^{10}$.

158. a. In a concentration cell, the anode always has the smaller ion concentration. So the cell with AgCl(s) at the bottom has the smaller $[Ag^+]$ and is the anode, while the compartment with $[Ag^+] = 1.0\ M$ is the cathode. Electron flow is from the anode to the cathode.

b. $AgCl(s) \rightleftharpoons Ag^+(aq) + Cl^-(aq)$ $K_{sp} = [Ag^+][Cl^-]$; use the concentration cell potential of 0.58 V to calculate the silver ion concentration in the anode compartment. For a concentration cell, E° = 0, and for Ag^+/Ag half-reactions, n = 1.

$$E = 0.58\text{ V} = 0 - \frac{0.0591}{1}\log\frac{[Ag^+]_{anode}}{[Ag^+]_{cathode}}$$

$$\frac{-0.58}{0.0591} = \log\frac{[Ag^+]_{anode}}{1.0},\quad [Ag^+]_{anode} = 1.5 \times 10^{-10}\ M$$

At the anode, we have $[Ag^+] = 1.5 \times 10^{-10}\ M$ and $[Cl^-] = 1.0\ M$:

$$K_{sp} = [Ag^+][Cl^-] = 1.5 \times 10^{-10}(1.0) = 1.5 \times 10^{-10}$$

Marathon Problems

159. a.

$$\begin{array}{ll} Cu^{2+} + 2e^- \rightarrow Cu & E° = 0.34\text{ V} \\ V \rightarrow V^{2+} + 2e^- & -E° = 1.20\text{ V} \\ \hline Cu^{2+}(aq) + V(s) \rightarrow Cu(s) + V^{2+}(aq) & E°_{cell} = 1.54\text{ V} \end{array}$$

$$E_{cell} = E°_{cell} - \frac{0.0591}{n}\log Q,\text{ where } n = 2 \text{ and } Q = \frac{[V^{2+}]}{[Cu^{2+}]} = \frac{[V^{2+}]}{1.00\ M}.$$

To determine E_{cell}, we must know the initial $[V^{2+}]$, which can be determined from the stoichiometric point data. At the stoichiometric point, moles H_2EDTA^{2-} added = moles V^{2+} present initially.

$$\text{Mol } V^{2+} \text{ present initially} = 0.5000\text{ L} \times \frac{0.0800\text{ mol } H_2EDTA^{2-}}{\text{L}} \times \frac{1\text{ mol } V^{2+}}{\text{mol } H_2EDTA^{2-}} = 0.0400\text{ mol } V^{2+}$$

$$[V^{2+}]_0 = \frac{0.0400\text{ mol } V^{2+}}{1.00\text{ L}} = 0.0400\ M$$

$$E_{cell} = 1.54\text{ V} - \frac{0.0591}{2}\log\frac{0.0400}{1.00} = 1.54\text{ V} - (-0.0413) = 1.58\text{ V}$$

b. Use the electrochemical data to solve for the equilibrium $[V^{2+}]$.

$$E_{cell} = E^{\circ}_{cell} - \frac{0.0591}{n}\log\frac{[V^{2+}]}{[Cu^{2+}]}, \quad 1.98\text{ V} = 1.54\text{ V} - \frac{0.0591}{2}\log\frac{[V^{2+}]}{1.00\,M}$$

$[V^{2+}] = 10^{-(0.44)(2)/0.0591} = 1.3 \times 10^{-15}\ M$

$$H_2EDTA^{2-}(aq) + V^{2+}(aq) \rightleftharpoons VEDTA^{2-}(aq) + 2\,H^+(aq) \qquad K = \frac{[VEDTA^{2-}][H^+]^2}{[H_2EDTA^{2-}][V^{2+}]}$$

In this titration reaction, equal moles of V^{2+} and H_2EDTA^{2-} are reacted at the stoichiometric point. Therefore, equal moles of both reactants must be present at equilibrium, so $[H_2EDTA^{2-}] = [V^{2+}] = 1.3 \times 10^{-15}\ M$. In addition, because $[V^{2+}]$ at equilibrium is very small compared to the initial 0.0400 *M* concentration, the reaction essentially goes to completion. The moles of $VEDTA^{2-}$ produced will equal the moles of V^{2+} reacted (= 0.0400 mol). At equilibrium, $[VEDTA^{2-}]$ = 0.0400 mol/(1.00 L + 0.5000 L) = 0.0267 *M*. Finally, because we have a buffer solution, the pH is assumed not to change, so $[H^+] = 10^{-10.00} = 1.0 \times 10^{-10}\ M$. Calculating K for the reaction:

$$K = \frac{[VEDTA^{2-}][H^+]^2}{[H_2EDTA^{2-}][V^{2+}]} = \frac{(0.0267)(1.0 \times 10^{-10})^2}{(1.3 \times 10^{-15})(1.3 \times 10^{-15})} = 1.6 \times 10^8$$

c. At the halfway point, 250.0 mL of H_2EDTA^{2-} has been added to 1.00 L of 0.0400 *M* V^{2+}. Exactly one-half the 0.0400 mol of V^{2+} present initially has been converted into $VEDTA^{2-}$. Therefore, 0.0200 mol of V^{2+} remains in 1.00 + 0.2500 = 1.25 L solution.

$$E_{cell} = 1.54\text{ V} - \frac{0.0591}{2}\log\frac{[V^{2+}]}{[Cu^{2+}]} = 1.54 - \frac{0.0591}{2}\log\frac{(0.0200/1.25)}{1.00}$$

$E_{cell} = 1.54 - (-0.0531) = 1.59\text{ V}$

160. Begin by choosing any reduction potential as 0.00 V. For example, let's assume

$B^{2+} + 2\,e^- \rightarrow B \qquad E° = 0.00\text{ V}$

From the data, when B/B^{2+} and E/E^{2+} are together as a cell, E° = 0.81 V.

$E^{2+} + 2\,e^- \rightarrow E$ must have a potential of −0.81 V or 0.81 V since E may be involved in either the reduction or the oxidation half-reaction. We will arbitrarily choose E to have a potential of −0.81 V.

Setting the reduction potential at −0.81 for E and 0.00 for B, we get the following table of potentials.

$B^{2+} + 2\,e^- \rightarrow B$	0.00 V
$E^{2+} + 2\,e^- \rightarrow E$	−0.81 V
$D^{2+} + 2\,e^- \rightarrow D$	0.19 V
$C^{2+} + 2\,e^- \rightarrow C$	−0.94 V
$A^{2+} + 2\,e^- \rightarrow A$	0.53 V

From largest to smallest:

$D^{2+} + 2\ e^- \rightarrow D$	0.19 V
$B^{2+} + 2\ e^- \rightarrow B$	0.00 V
$A^{2+} + 2\ e^- \rightarrow A$	−0.53 V
$E^{2+} + 2\ e^- \rightarrow E$	−0.81 V
$C^{2+} + 2\ e^- \rightarrow C$	−0.94 V

$A^{2+} + 2\ e^- \rightarrow A$ is in the middle. Let's call this 0.00 V. The other potentials would be:

$D^{2+} + 2\ e^- \rightarrow D$	0.72 V
$B^{2+} + 2\ e^- \rightarrow B$	0.53 V
$A^{2+} + 2\ e^- \rightarrow A$	0.00 V
$E^{2+} + 2\ e^- \rightarrow E$	−0.28 V
$C^{2+} + 2\ e^- \rightarrow C$	−0.41 V

Of course, since the reduction potential of E could have been assumed to 0.81 V instead of −0.81 V, we can also get:

$C^{2+} + 2\ e^- \rightarrow C$	0.41 V
$E^{2+} + 2\ e^- \rightarrow E$	0.28 V
$A^{2+} + 2\ e^- \rightarrow A$	0.00 V
$B^{2+} + 2\ e^- \rightarrow B$	−0.53 V
$D^{2+} + 2\ e^- \rightarrow D$	−0.72 V

One way to determine which table is correct is to add metal C to a solution with D^{2+} present, as well as to add metal D to a different solution with C^{2+} present. If metal D forms in the first solution, then the first table is correct. If metal C forms in the second solution, then the second table is correct.

CHAPTER 19

THE NUCLEUS: A CHEMIST'S VIEW

Questions

1. Characteristic frequencies of energies emitted in a nuclear reaction suggest that discrete energy levels exist in the nucleus. The extra stability of certain numbers of nucleons and the predominance of nuclei with even numbers of nucleons suggest that the nuclear structure might be described by using quantum numbers.

2. No, coal-fired power plants also pose risks. A partial list of risks is:

Coal	Nuclear
Air pollution	Radiation exposure to workers
Coal mine accidents	Disposal of wastes
Health risks to miners (black lung disease)	Meltdown
	Terrorists
	Public fear

3. Beta-particle production has the net effect of turning a neutron into a proton. Radioactive nuclei having too many neutrons typically undergo β-particle decay. Positron production has the net effect of turning a proton into a neutron. Nuclei having too many protons typically undergo positron decay.

4. a. Nothing; binding energy is related to thermodynamic stability, and is not related to kinetics. Binding energy indicates nothing about how fast or slow a specific nucleon decays.

   b. ^{56}Fe has the largest binding energy per nucleon, so it is the most stable nuclide. ^{56}Fe has the greatest mass loss per nucleon when the protons and neutrons are brought together to form the ^{56}Fe nucleus. The least stable nuclide shown, having the smallest binding energy per nucleon, is ^{2}H.

   c. Fusion refers to combining two light nuclei having relatively small binding energies per nucleon to form a heavier nucleus which has a larger binding energy per nucleon. The difference in binding energies per nucleon is related to the energy released in a fusion reaction. Nuclides to the left of ^{56}Fe can undergo fusion.

   Nuclides to the right of ^{56}Fe can undergo fission. In fission, a heavier nucleus having a relatively small binding energy per nucleon is split into two smaller nuclei having larger binding energy per nucleons. The difference in binding energies per nucleon is related to the energy released in a fission reaction.

5. The transuranium elements are the elements having more protons than uranium. They are synthesized by bombarding heavier nuclei with neutrons and positive ions in a particle accelerator.

6. All radioactive decay follows first-order kinetics. A sample is analyzed for the ^{176}Lu and ^{176}Hf content, from which the first-order rate law can be applied to determine the age of the sample. The reason ^{176}Lu decay is valuable for dating very old objects is the extremely long half-life. Substances formed a long time ago that have short half-lives have virtually no nuclei remaining. On the other hand, ^{176}Lu decay hasn't even approached one half-life when dating 5-billion-year-old objects.

7. $\Delta E = \Delta mc^2$; the key difference is the mass change when going from reactants to products. In chemical reactions, the mass change is indiscernible. In nuclear processes, the mass change is discernible. It is the conversion of this discernible mass change into energy that results in the huge energies associated with nuclear processes.

8. Effusion is the passage of a gas through a tiny orifice into an evacuated container. Graham's law of effusion says that the effusion of a gas in inversely proportional to the square root of the mass of its particle. The key to effusion, and to the gaseous diffusion process, is that they are both directly related to the velocity of the gas molecules, which is inversely related to the molar mass. The lighter $^{235}UF_6$ gas molecules have a faster average velocity than the heavier $^{238}UF_6$ gas molecules. The difference in average velocity is used in the gaseous diffusion process to enrich the ^{235}U content in natural uranium.

9. The temperatures of fusion reactions are so high that all physical containers would be destroyed. At these high temperatures, most of the electrons are stripped from the atoms. A plasma of gaseous ions is formed that can be controlled by magnetic fields.

10. The linear model postulates that damage from radiation is proportional to the dose, even at low levels of exposure. Thus any exposure is dangerous. The threshold model, on the other hand, assumes that no significant damage occurs below a certain exposure, called the threshold exposure. A recent study supported the linear model.

Exercises

Radioactive Decay and Nuclear Transformations

11. All nuclear reactions must be charge balanced and mass balanced. To charge balance, balance the sum of the atomic numbers on each side of the reaction, and to mass balance, balance the sum of the mass numbers on each side of the reaction.

a. $^{3}_{1}H \rightarrow ^{3}_{2}He + ^{0}_{-1}e$

b. $^{8}_{3}Li \rightarrow ^{8}_{4}Be + ^{0}_{-1}e$

$^{8}_{4}Be \rightarrow 2\ ^{4}_{2}He$

$^{8}_{3}Li \rightarrow 2\ ^{4}_{2}He + ^{0}_{-1}e$

c. ${}^{7}_{4}Be + {}^{0}_{-1}e \rightarrow {}^{7}_{3}Li$ d. ${}^{8}_{5}B \rightarrow {}^{8}_{4}Be + {}^{0}_{+1}e$

12. All nuclear reactions must be charge balanced and mass balanced. To charge balance, balance the sum of the atomic numbers on each side of the reaction, and to mass balance, balance the sum of the mass numbers on each side of the reaction.

a. ${}^{60}_{27}Co \rightarrow {}^{60}_{28}Ni + {}^{0}_{-1}e$ b. ${}^{97}_{43}Tc + {}^{0}_{-1}e \rightarrow {}^{97}_{42}Mo$

c. ${}^{99}_{43}Tc \rightarrow {}^{99}_{44}Ru + {}^{0}_{-1}e$ d. ${}^{239}_{94}Pu \rightarrow {}^{235}_{92}U + {}^{4}_{2}He$

13. All nuclear reactions must be charge balanced and mass balanced. To charge balance, balance the sum of the atomic numbers on each side of the reaction, and to mass balance, balance the sum of the mass numbers on each side of the reaction.

a. ${}^{238}_{92}U \rightarrow {}^{4}_{2}He + {}^{234}_{90}Th$; this is alpha-particle production.

b. ${}^{234}_{90}Th \rightarrow {}^{234}_{91}Pa + {}^{0}_{-1}e$; this is β-particle production.

14. a. ${}^{51}_{24}Cr + {}^{0}_{-1}e \rightarrow {}^{51}_{23}V$ b. ${}^{131}_{53}I \rightarrow {}^{0}_{-1}e + {}^{131}_{54}Xe$ c. ${}^{32}_{15}P \rightarrow {}^{0}_{-1}e + {}^{32}_{16}S$

15. a. ${}^{68}_{31}Ga + {}^{0}_{-1}e \rightarrow {}^{68}_{30}Zn$ b. ${}^{62}_{29}Cu \rightarrow {}^{0}_{+1}e + {}^{62}_{28}Ni$

c. ${}^{212}_{87}Fr \rightarrow {}^{4}_{2}He + {}^{208}_{85}At$ d. ${}^{129}_{51}Sb \rightarrow {}^{0}_{-1}e + {}^{129}_{52}Te$

16. a. ${}^{73}_{31}Ga \rightarrow {}^{73}_{32}Ge + {}^{0}_{-1}e$ b. ${}^{192}_{78}Pt \rightarrow {}^{188}_{76}Os + {}^{4}_{2}He$

c. ${}^{205}_{83}Bi \rightarrow {}^{205}_{82}Pb + {}^{0}_{+1}e$ d. ${}^{241}_{96}Cm + {}^{0}_{-1}e \rightarrow {}^{241}_{95}Am$

17. ${}^{235}_{92}U \rightarrow {}^{207}_{82}Pb + ?\ {}^{4}_{2}He + ?\ {}^{0}_{-1}e$

From the two possible decay processes, only alpha-particle decay changes the mass number. So the mass number change of 28 from 235 to 207 must be done in the decay series by seven alpha particles. The atomic number change of 10 from 92 to 82 is due to both alpha-particle production and beta-particle production. However, because we know that seven alpha-particles are in the complete decay process, we must have four beta-particle decays in order to balance the atomic number. The complete decay series is summarized as:

$${}^{235}_{92}U \rightarrow {}^{207}_{82}Pb + 7\ {}^{4}_{2}He + 4\ {}^{0}_{-1}e$$

18. ${}^{247}_{97}Bk \rightarrow {}^{207}_{82}Pb + ?\ {}^{4}_{2}He + {}^{0}_{-1}e$; the change in mass number (247 – 207 = 40) is due exclusively to the alpha-particles. A change in mass number of 40 requires 10 ${}^{4}_{2}He$ particles to be produced. The atomic number only changes by 97 – 82 = 15. The 10 alpha-particles change the atomic number by 20, so 5 ${}^{0}_{-1}e$ (5 beta-particles) are produced in the decay series of ${}^{247}Bk$ to ${}^{207}Pb$.

19. a. $^{241}_{95}Am \rightarrow ^{4}_{2}He + ^{237}_{93}Np$

b. $^{241}_{95}Am \rightarrow 8\,^{4}_{2}He + 4\,^{0}_{-1}e + ^{209}_{83}Bi$; the final product is $^{209}_{83}Bi$.

c. $^{241}_{95}Am \rightarrow ^{237}_{93}Np + \alpha \rightarrow ^{233}_{91}Pa + \alpha \rightarrow ^{233}_{92}U + \beta \rightarrow ^{229}_{90}Th + \alpha \rightarrow ^{225}_{88}Ra + \alpha$

$\downarrow$

$^{213}_{84}Po + \beta \leftarrow ^{213}_{83}Bi + \alpha \leftarrow ^{217}_{85}At + \alpha \leftarrow ^{221}_{87}Fr + \alpha \leftarrow ^{225}_{89}Ac + \beta$

$\downarrow$

$^{209}_{82}Pb + \alpha \rightarrow ^{209}_{83}Bi + \beta$

The intermediate radionuclides are:

$^{237}_{93}Np$, $^{233}_{91}Pa$, $^{233}_{92}U$, $^{229}_{90}Th$, $^{225}_{88}Ra$, $^{225}_{89}Ac$, $^{221}_{87}Fr$, $^{217}_{85}At$, $^{213}_{83}Bi$, $^{213}_{84}Po$, and $^{209}_{82}Pb$

20. The complete decay series is:

$^{232}_{90}Th \rightarrow ^{228}_{88}Ra + ^{4}_{2}He \rightarrow ^{228}_{89}Ac + ^{0}_{-1}e \rightarrow ^{228}_{90}Th + ^{0}_{-1}e \rightarrow ^{224}_{88}Ra + ^{4}_{2}He$

$\downarrow$

$^{0}_{-1}e + ^{212}_{84}Po \leftarrow ^{0}_{-1}e + ^{212}_{83}Bi \leftarrow ^{4}_{2}He + ^{212}_{82}Pb \leftarrow ^{4}_{2}He + ^{216}_{84}Po \leftarrow ^{220}_{86}Rn + ^{4}_{2}He$

$\downarrow$

$^{208}_{82}Pb + ^{4}_{2}He$

21. $^{53}_{26}Fe$ has too many protons. It will undergo either positron production, electron capture, and/or alpha-particle production. $^{59}_{26}Fe$ has too many neutrons and will undergo beta-particle production. (See Table 19.2 of the text.)

22. Reference Table 19.2 of the text for potential radioactive decay processes. ^{17}F and ^{18}F contain too many protons or too few neutrons. Electron capture and positron production are both possible decay mechanisms that increase the neutron to proton ratio. Alpha-particle production also increases the neutron-to-proton ratio, but it is not likely for these light nuclei. ^{21}F contains too many neutrons or too few protons. Beta-particle production lowers the neutron-to-proton ratio, so we expect ^{21}F to be a beta-emitter.

23. a. $^{249}_{98}Cf + ^{18}_{8}O \rightarrow ^{263}_{106}Sg + 4\,^{1}_{0}n$

b. $^{259}_{104}Rf$; $^{263}_{106}Sg \rightarrow ^{4}_{2}He + ^{259}_{104}Rf$

24. a. $^{240}_{95}Am + ^{4}_{2}He \rightarrow ^{243}_{97}Bk + ^{1}_{0}n$

b. $^{238}_{92}U + ^{12}_{6}C \rightarrow ^{244}_{98}Cf + 6\,^{1}_{0}n$

c. $^{249}_{98}Cf + ^{15}_{7}N \rightarrow ^{260}_{105}Db + 4\,^{1}_{0}n$

d. $^{249}_{98}Cf + ^{10}_{5}B \rightarrow ^{257}_{103}Lr + 2\,^{1}_{0}n$

Kinetics of Radioactive Decay

25. All radioactive decay follows first-order kinetics where $t_{1/2} = (\ln 2)/k$.

$$t_{1/2} = \frac{\ln 2}{k} = \frac{0.693}{1.0 \times 10^{-3}\ h^{-1}} = 690\ h$$

26. $$k = \frac{\ln 2}{t_{1/2}} = \frac{0.69315}{433 yr} \times \frac{1\ yr}{365 d} \times \frac{1\ d}{24 h} \times \frac{1\ h}{3600 s} = 5.08 \times 10^{-11}\ s^{-1}$$

$$\text{Rate} = kN = 5.08 \times 10^{-11}\ s^{-1} \times 5.00\ g \times \frac{1\ mol}{241 g} \times \frac{6.022 \times 10^{23}\ \text{nuclei}}{mol}$$

$$= 6.35 \times 10^{11}\ \text{decays/s}$$

6.35×10^{11} alpha particles are emitted each second from a 5.00-g ^{241}Am sample.

27. Kr-81 is most stable because it has the longest half-life. Kr-73 is hottest (least stable); it decays most rapidly because it has the shortest half-life.

12.5% of each isotope will remain after 3 half-lives:

$$100\% \xrightarrow[t_{1/2}]{} 50\% \xrightarrow[t_{1/2}]{} 25\% \xrightarrow[t_{1/2}]{} 12.5\%$$

For Kr-73: t = 3(27 s) = 81 s; for Kr-74: t = 3(11.5 min) = 34.5 min

For Kr-76: t = 3(14.8 h) = 44.4 h; for Kr-81: t = 3(2.1×10^5 yr) = 6.3×10^5 yr

28. a. $$k = \frac{\ln 2}{t_{1/2}} = \frac{0.6931}{12.8 d} \times \frac{1\ d}{24 h} \times \frac{1\ h}{3600 s} = 6.27 \times 10^{-7}\ s^{-1}$$

b. $$\text{Rate} = kN = 6.27 \times 10^{-7}\ s^{-1} \times \left(28.0 \times 10^{-3}\ g \times \frac{1\ mol}{64.0\ g} \times \frac{6.022 \times 10^{23}\ \text{nuclei}}{mol}\right)$$

Rate = 1.65×10^{14} decays/s

c. 25% of the ^{64}Cu will remain after 2 half-lives (100% decays to 50% after one half-life, which decays to 25% after a second half-life). Hence 2(12.8 days) = 25.6 days is the time frame for the experiment.

29. Units for N and N_0 are usually number of nuclei but can also be grams if the units are identical for both N and N_0. In this problem, m_0 = the initial mass of $^{47}Ca^{2+}$ to be ordered.

$$k = \frac{\ln 2}{t_{1/2}};\ \ln\left(\frac{N}{N_0}\right) = -kt = \frac{-(0.693)t}{t_{1/2}},\ \ln\left(\frac{5.0\ \mu g\ Ca^{2+}}{m_0}\right) = \frac{-0.693(2.0\ d)}{4.5\ d} = -0.31$$

$$\frac{5.0}{m_0} = e^{-0.31} = 0.73,\ m_0 = 6.8\ \mu g \text{ of } {}^{47}Ca^{2+} \text{ needed initially}$$

$$6.8\ \mu g\ {}^{47}Ca^{2+} \times \frac{107.0\ \mu g\ {}^{47}CaCO_3}{47.0\ \mu g\ {}^{47}Ca^{2+}} = 15\ \mu g\ {}^{47}CaCO_3 \text{ should be ordered at the minimum.}$$

30. a. $0.0100 \text{ Ci} \times \dfrac{3.7\times10^{10}\text{ decays/s}}{\text{Ci}} = 3.7\times10^{8}\text{ decays/s};\ k = \dfrac{\ln 2}{t_{1/2}}$

$$\text{Rate} = kN,\quad \frac{3.7\times10^{8}\text{ decays}}{\text{s}} = \left(\frac{0.6931}{2.87\text{ h}}\times\frac{1\text{ h}}{3600\text{s}}\right)\times N,\ N = 5.5\times10^{12}\text{ atoms of }^{38}\text{S}$$

$$5.5\times10^{12}\text{ atoms }^{38}\text{S}\times\frac{1\text{ mol }^{38}\text{S}}{6.02\times10^{23}\text{ atoms}}\times\frac{1\text{ mol Na}_2{}^{38}\text{SO}_4}{\text{mol }^{38}\text{S}} = 9.1\times10^{-12}\text{mol Na}_2{}^{38}\text{SO}_4$$

$$9.1\times10^{-12}\text{mol Na}_2{}^{38}\text{SO}_4\times\frac{148.0\text{ g Na}_2{}^{38}\text{SO}_4}{\text{mol Na}_2{}^{38}\text{SO}_4} = 1.3\times10^{-9}\text{ g} = 1.3\text{ ng Na}_2{}^{38}\text{SO}_4$$

b. 99.99% decays, 0.01% left; $\ln\left(\dfrac{0.01}{100}\right) = -kt = \dfrac{-(0.6931)t}{2.87\text{ h}}$, $t = 38.1$ hours $\approx$ 40 hours

31. $t = 68.0$ yr; $k = \dfrac{\ln 2}{t_{1/2}}$; $\ln\left(\dfrac{N}{N_0}\right) = -kt = \dfrac{-(0.6931)\,68.0\text{ yr}}{28.9\text{yr}} = -1.63,\ \left(\dfrac{N}{N_0}\right) = e^{-1.63} = 0.196$

19.6% of the ^{90}Sr remains as of July 16, 2013.

32. Assuming 2 significant figures in 1/100:

$\ln(N/N_0) = -kt$; $N = (0.010)N_0$; $t_{1/2} = (\ln 2)/k$

$$\ln(0.010) = \frac{-(\ln 2)t}{t_{1/2}} = \frac{-(0.693)t}{8.0\text{ d}},\quad t = 53\text{ days}$$

33. $k = \dfrac{\ln 2}{t_{1/2}}$; $\ln\left(\dfrac{N}{N_0}\right) = -kt = \dfrac{(\ln 2)t}{t_{1/2}}$; $\ln\left(\dfrac{N}{N_0}\right) = \dfrac{-(0.693)(48.0\text{ h})}{6.0\text{ h}} = 5.5$

$\dfrac{N}{N_0} = e^{-5.5} = 0.0041$; the fraction of ^{99}Tc that remains is 0.0041, or 0.41%.

34. $175\text{ mg Na}_3{}^{32}\text{PO}_4\times\dfrac{32.0\text{ mg }^{32}\text{P}}{165.0\text{ mg Na}_3{}^{32}\text{PO}_4} = 33.9\text{ mg }^{32}\text{P};\ k = \dfrac{\ln 2}{t_{1/2}}$

$\ln\left(\dfrac{N}{N_0}\right) = -kt = \dfrac{-(0.6931)t}{t_{1/2}},\ \ln\left(\dfrac{m}{33.9\text{ mg}}\right) = \dfrac{-0.6931(35.0\text{ d})}{14.3\text{ d}}$; carrying extra sig. figs.:

$\ln(m) = -1.696 + 3.523 = 1.827,\ m = e^{1.827} = 6.22\text{ mg }^{32}\text{P}$ remains

35. $$\ln\left(\frac{N}{N_0}\right) = -kt = \frac{-(\ln 2)t}{t_{1/2}},\quad \ln\left(\frac{1.0\text{ g}}{m_0}\right) = \frac{-0.693\left(3.0\text{ d}\times\dfrac{24\text{ h}}{\text{d}}\times\dfrac{60\text{ min}}{\text{h}}\right)}{1.0\times10^{3}\text{ min}}$$

$$\ln\left(\frac{1.0\text{ g}}{m_0}\right) = -3.0, \quad \frac{1.0}{m_0} = e^{-3.0}, \quad m_0 = 20.\text{ g } ^{82}\text{Br needed}$$

$$20.\text{ g } ^{82}\text{Br} \times \frac{1\text{ mol } ^{82}\text{Br}}{82.0\text{ g}} \times \frac{1\text{ mol Na}^{82}\text{Br}}{\text{mol } ^{82}\text{Br}} \times \frac{105.0\text{ g Na}^{82}\text{Br}}{\text{mol Na}^{82}\text{Br}} = 26\text{ g Na}^{82}\text{Br}$$

36. Assuming the current year is 2013, t = 67 yr.

$$\ln\left(\frac{N}{N_0}\right) = -kt = \frac{-(0.693)t}{t_{1/2}}, \quad \ln\left(\frac{N}{5.5}\right) = \frac{-0.693(67\text{yr})}{12.3\text{yr}}, \quad N = \frac{0.13\text{ decay events}}{\text{min} \bullet 100.\text{ g water}}$$

37. $$k = \frac{\ln 2}{t_{1/2}}; \quad \ln\left(\frac{N}{N_0}\right) = -kt = \frac{-(0.693)t}{t_{1/2}}, \quad \ln\left(\frac{N}{13.6}\right) = \frac{-0.693(15{,}000\text{ yr})}{5730\text{yr}} = -1.8$$

$$\frac{N}{13.6} = e^{-1.8} = 0.17, \quad N = 13.6 \times 0.17 = 2.3 \text{ counts per minute per g of C}$$

If we had 10. mg C, we would see:

$$10.\text{ mg} \times \frac{1\text{ g}}{1000\text{mg}} \times \frac{2.3\text{ counts}}{\text{min g}} = \frac{0.023\text{counts}}{\text{min}}$$

It would take roughly 40 min to see a single disintegration. This is too long to wait, and the background radiation would probably be much greater than the ^{14}C activity. Thus ^{14}C dating is not practical for very small samples.

38. $$\ln\left(\frac{N}{N_0}\right) = -kt = \frac{-(0.6931)t}{t_{1/2}}, \quad \ln\left(\frac{1.2}{13.6}\right) = \frac{-(0.6931)t}{5730\text{yr}}, \quad t = 2.0 \times 10^4 \text{ yr}$$

39. Assuming 1.000 g ^{238}U present in a sample, then 0.688 g ^{206}Pb is present. Because 1 mol ^{206}Pb is produced per mol ^{238}U decayed:

$$^{238}\text{U decayed} = 0.688\text{ g Pb} \times \frac{1\text{ mol Pb}}{206\text{g Pb}} \times \frac{1\text{ mol U}}{\text{mol Pb}} \times \frac{238\text{g U}}{\text{mol U}} = 0.795\text{ g } ^{238}\text{U}$$

Original mass ^{238}U present = 1.000 g + 0.795 g = 1.795 g ^{238}U

$$\ln\left(\frac{N}{N_0}\right) = -kt = \frac{-(\ln 2)t}{t_{1/2}}, \quad \ln\left(\frac{1.000\text{g}}{1.795\text{g}}\right) = \frac{-0.693(t)}{4.5 \times 10^9\text{ yr}}, \quad t = 3.8 \times 10^9 \text{ yr}$$

40. a. The decay of ^{40}K is not the sole source of ^{40}Ca.

b. Decay of ^{40}K is the sole source of ^{40}Ar and no ^{40}Ar is lost over the years.

c. $$\frac{0.95\text{ g } ^{40}\text{Ar}}{1.00\text{ g } ^{40}\text{K}} = \text{current mass ratio}$$

0.95 g of ^{40}K decayed to ^{40}Ar. 0.95 g of ^{40}K is only 10.7% of the total ^{40}K that decayed, or:

$0.107(m) = 0.95$ g, $m = 8.9$ g = total mass of ^{40}K that decayed

Mass of ^{40}K when the rock was formed was 1.00 g + 8.9 g = 9.9 g.

$$\ln\left(\frac{1.00\text{ g }^{40}K}{9.9\text{ g }^{40}K}\right) = -kt = \frac{-(\ln 2)t}{t_{1/2}} = \frac{-(0.6931)t}{1.27\times 10^{9}\text{ yr}},\quad t = 4.2\times 10^{9}\text{ years old}$$

d. If some ^{40}Ar escaped, then the measured ratio of $^{40}Ar/^{40}K$ is less than it should be. We would calculate the age of the rock to be less than it actually is.

Energy Changes in Nuclear Reactions

41. $$\Delta E = \Delta mc^2,\ \Delta m = \frac{\Delta E}{c^2} = \frac{3.9\times 10^{23}\text{ kg m}^2/\text{s}^2}{(3.00\times 10^{8}\text{ m/s})^2} = 4.3\times 10^{6}\text{ kg}$$

The sun loses 4.3×10^6 kg of mass each second. *Note*: 1 J = 1 kg m^2/s^2

42. $$\frac{1.8\times 10^{14}\text{ kJ}}{\text{s}}\times\frac{1000\text{J}}{\text{kJ}}\times\frac{3600\text{s}}{\text{h}}\times\frac{24\text{ h}}{\text{day}} = 1.6\times 10^{22}\text{ J/day}$$

$$\Delta E = \Delta mc^2,\ \Delta m = \frac{\Delta E}{c^2} = \frac{1.6\times 10^{22}\text{ J}}{(3.00\times 10^{8}\text{ m/s})^2} = 1.8\times 10^{5}\text{ kg}$$ of solar material provides 1 day of solar energy to the earth

$$1.6\times 10^{22}\text{ J}\times\frac{1\text{ kJ}}{1000\text{J}}\times\frac{1\text{ g}}{32\text{ kJ}}\times\frac{1\text{ kg}}{1000\text{g}} = 5.0\times 10^{14}\text{ kg}$$ of coal is needed to provide the same amount of energy

43. We need to determine the mass defect Δm between the mass of the nucleus and the mass of the individual parts that make up the nucleus. Once Δm is known, we can then calculate ΔE (the binding energy) using $E = mc^2$. *Note*: 1 J = 1 kg m^2/s^2.

For $^{232}_{94}Pu$ (94 e, 94 p, 138 n):

mass of ^{232}Pu nucleus $= 3.85285\times 10^{-22}$ g − mass of 94 electrons

mass of ^{232}Pu nucleus $= 3.85285\times 10^{-22}\text{ g} - 94(9.10939\times 10^{-28})\text{ g} = 3.85199\times 10^{-22}$ g

$\Delta m = 3.85199\times 10^{-22}$ g − (mass of 94 protons + mass of 138 neutrons)

$$\Delta m = 3.85199\times 10^{-22}\text{ g} - [94(1.67262\times 10^{-24}) + 138(1.67493\times 10^{-24})]\text{ g} = -3.168\times 10^{-24}\text{ g}$$

For 1 mol of nuclei: $\Delta m = -3.168\times 10^{-24}$ g/nuclei × 6.0221×10^{23} nuclei/mol $= -1.908$ g/mol

$$\Delta E = \Delta mc^2 = (-1.908\times 10^{-3}\text{ kg/mol})(2.9979\times 10^{8}\text{ m/s})^2 = -1.715\times 10^{14}\text{ J/mol}$$

For $^{231}_{91}Pa$ (91 e, 91 p, 140 n):

mass of ^{231}Pa nucleus = 3.83616×10^{-22} g − $91(9.10939 \times 10^{-28})$ g = 3.83533×10^{-22} g

$$\Delta m = 3.83533 \times 10^{-22}\text{g} - [91(1.67262 \times 10^{-24}) + 140(1.67493 \times 10^{-24})]\ \text{g}$$
$$= -3.166 \times 10^{-24}\ \text{g}$$

$$\Delta E = \Delta mc^2 = \frac{-3.166 \times 10^{-27}\ \text{kg}}{\text{nuclei}} \times \frac{6.0221 \times 10^{23}\ \text{nuclei}}{\text{mol}} \times \left(\frac{2.9979 \times 10^8\ \text{m}}{\text{s}}\right)^2$$
$$= -1.714 \times 10^{14}\ \text{J/mol}$$

44. From the text, the mass of a proton = 1.00728 u, the mass of a neutron = 1.00866 u, and the mass of an electron = 5.486×10^{-4} u.

Mass of $^{56}_{26}Fe$ nucleus = mass of atom − mass of electrons = 55.9349 − 26(0.0005486)
= 55.9206 u

$26\ ^1_1H + 30\ ^1_0n \rightarrow\ ^{56}_{26}Fe$; Δm = 55.9206 u − [26(1.00728) + 30(1.00866)] u = −0.5285 u

$$\Delta E = \Delta mc^2 = -0.5285\ \text{u} \times \frac{1.6605 \times 10^{-27}\ \text{kg}}{\text{u}} \times (2.9979 \times 10^8\ \text{m/s})^2 = -7.887 \times 10^{-11}\ \text{J}$$

$$\frac{\text{Binding energy}}{\text{Nucleon}} = \frac{7.887 \times 10^{-11}\ \text{J}}{56\ \text{nucleons}} = 1.408 \times 10^{-12}\ \text{J/nucleon}$$

45. Let m_e = mass of electron; for ^{12}C (6e, 6p, and 6n): Mass defect = Δm = [mass of ^{12}C nucleus] − [mass of 6 protons + mass of 6 neutrons]. *Note*: Atomic masses given include the mass of the electrons.

$$\Delta m = 12.00000\ \text{u} - 6m_e - [6(1.00782 - m_e) + 6(1.00866)];\ \text{mass of electrons cancel.}$$

$$\Delta m = 12.00000 - [6(1.00782) + 6(1.00866)] = -0.09888\ \text{u}$$

$$\Delta E = \Delta mc^2 = -0.09888\ \text{u} \times \frac{1.6605 \times 10^{-27}\ \text{kg}}{\text{u}} \times (2.9979 \times 10^8\ \text{m/s})^2 = -1.476 \times 10^{-11}\ \text{J}$$

$$\frac{\text{Binding energy}}{\text{Nucleon}} = \frac{1.476 \times 10^{-11}\ \text{J}}{12\ \text{nucleons}} = 1.230 \times 10^{-12}\ \text{J/nucleon}$$

For ^{235}U (92e, 92p, and 143n):

$$\Delta m = 235.0439 - 92m_e - [92(1.00782 - m_e) + 143(1.00866)] = -1.9139\ \text{u}$$

$$\Delta E = \Delta mc^2 = -1.9139 \times \frac{1.66054 \times 10^{-27}\ \text{kg}}{\text{u}} \times (2.99792 \times 10^8\ \text{m/s})^2 = -2.8563 \times 10^{-10}\ \text{J}$$

$$\frac{\text{Binding energy}}{\text{Nucleon}} = \frac{2.8563 \times 10^{-10}\ \text{J}}{235\ \text{nucleons}} = 1.2154 \times 10^{-12}\ \text{J/nucleon}$$

Because ^{56}Fe is the most stable known nucleus, the binding energy per nucleon for ^{56}Fe (1.408×10^{-12} J/nucleon) will be larger than that of ^{12}C or ^{235}U (see Figure 19.9 of the text).

46. For $^{2}_{1}H$: Mass defect = Δm = mass of $^{2}_{1}H$ nucleus – mass of proton – mass of neutron. The mass of the ^{2}H nucleus will equal the atomic mass of ^{2}H minus the mass of the electron in an ^{2}H atom. From the text, the pertinent masses are $m_e = 5.49 \times 10^{-4}$ u, $m_p = 1.00728$ u, and $m_n = 1.00866$ u.

$$\Delta m = 2.01410\ \text{u} - 0.000549\ \text{u} - (1.00728\ \text{u} + 1.00866\ \text{u}) = -2.39 \times 10^{-3}\ \text{u}$$

$$\Delta E = \Delta mc^2 = -2.39 \times 10^{-3}\ \text{u} \times \frac{1.6605 \times 10^{-27}\ \text{kg}}{\text{u}} \times (2.998 \times 10^{8}\ \text{m/s})^2 = -3.57 \times 10^{-13}\ \text{J}$$

$$\frac{\text{Binding energy}}{\text{Nucleon}} = \frac{3.57 \times 10^{-13}\ \text{J}}{2\ \text{nucleons}} = 1.79 \times 10^{-13}\ \text{J/nucleon}$$

For $^{3}_{1}H$: $\Delta m = 3.01605 - 0.000549 - [1.00728 + 2(1.00866)] = -9.10 \times 10^{-3}$ u

$$\Delta E = -9.10 \times 10^{-3}\ \text{u} \times \frac{1.6605 \times 10^{-27}\ \text{kg}}{\text{u}} \times (2.998 \times 10^{8}\ \text{m/s})^2 = -1.36 \times 10^{-12}\ \text{J}$$

$$\frac{\text{Binding energy}}{\text{Nucleon}} = \frac{1.36 \times 10^{-12}\ \text{J}}{3\ \text{nucleons}} = 4.53 \times 10^{-13}\ \text{J/nucleon}$$

47. Let m_{Li} = mass of ^{6}Li nucleus; an ^{6}Li nucleus has 3p and 3n.

$-0.03434\ \text{u} = m_{Li} - (3m_p + 3m_n) = m_{Li} - [3(1.00728\ \text{u}) + 3(1.00866\ \text{u})]$

$m_{Li} = 6.01348$ u

Mass of ^{6}Li atom = 6.01348 u + $3m_e$ = 6.01348 + 3(5.49×10^{-4} u) = 6.01513 u (includes mass of 3 e^-)

48. $$\text{Binding energy} = \frac{1.326 \times 10^{-12}\ \text{J}}{\text{nucleon}} \times 27\ \text{nucleons} = 3.580 \times 10^{-11}\ \text{J for each } ^{27}\text{Mg nucleus}$$

$$\Delta E = \Delta mc^2,\quad \Delta m = \frac{\Delta E}{c^2} = \frac{-3.580 \times 10^{-11}\ \text{J}}{(2.9979 \times 10^{8}\ \text{m/s})^2} = -3.983\ 10^{-28}\ \text{kg}$$

$$\Delta m = -3.983\ 10^{-28}\ \text{kg} \times \frac{1\ \text{u}}{1.6605 \times 10^{-27}\ \text{kg}} = -0.2399\ \text{u} = \text{mass defect}$$

Let m_{Mg} = mass of ^{27}Mg nucleus; an ^{27}Mg nucleus has 12 p and 15 n.

$-0.2399\text{ u} = m_{Mg} - (12m_p + 15m_n) = m_{Mg} - [12(1.00728\text{ u}) + 15(1.00866\text{ u})]$

$m_{Mg} = 26.9764\text{ u}$

Mass of ^{27}Mg atom = 26.9764 u + $12m_e$, $26.9764 + 12(5.49 \times 10^{-4}\text{ u}) = 26.9830\text{ u}$ (includes mass of 12 e^-)

49. $^1_1\text{H} + ^1_1\text{H} \rightarrow ^2_1\text{H} + ^{0}_{+1}\text{e}$; $\Delta m = (2.01410\text{ u} - m_e + m_e) - 2(1.00782\text{ u} - m_e)$

$\Delta m = 2.01410 - 2(1.00782) + 2(0.000549) = -4.4 \times 10^{-4}$ u for two protons reacting

When 2 mol of protons undergoes fusion, $\Delta m = -4.4 \times 10^{-4}$ g.

$\Delta E = \Delta mc^2 = -4.4 \times 10^{-7}\text{ kg} \times (3.00 \times 10^8\text{ m/s})^2 = -4.0 \times 10^{10}\text{ J}$

$$\frac{-4.0 \times 10^{10}\text{ J}}{2\text{ mol protons}} \times \frac{1\text{ mol}}{1.01\text{ g}} = -2.0 \times 10^{10}\text{ J/g of hydrogen nuclei}$$

50. $^2_1\text{H} + ^3_1\text{H} \rightarrow ^4_2\text{He} + ^1_0\text{n}$; using atomic masses, the masses of the electrons cancel when determining Δm for this nuclear reaction.

$\Delta m = [4.00260 + 1.00866 - (2.01410 + 3.01605)]\text{ u} = -1.889 \times 10^{-2}\text{ u}$

For the production of 1 mol of ^4_2He: $\Delta m = -1.889 \times 10^{-2}\text{ g} = -1.889 \times 10^{-5}\text{ kg}$

$\Delta E = \Delta mc^2 = -1.889 \times 10^{-5}\text{ kg} \times (2.9979 \times 10^8\text{ m/s})^2 = -1.698 \times 10^{12}\text{ J/mol}$

For 1 nucleus of ^4_2He $$\frac{-1.698 \times 10^{12}\text{ J}}{\text{mol}} \times \frac{1\text{ mol}}{6.0221 \times 10^{23}\text{ nuclei}} = -2.820 \times 10^{-12}\text{ J/nucleus}$$

Detection, Uses, and Health Effects of Radiation

51. The Geiger-Müller tube has a certain response time. After the gas in the tube ionizes to produce a "count," some time must elapse for the gas to return to an electrically neutral state. The response of the tube levels off because at high activities, radioactive particles are entering the tube faster than the tube can respond to them.

52. Not all of the emitted radiation enters the Geiger-Müller tube. The fraction of radiation entering the tube must be constant.

53. Water is produced in this reaction by removing an OH group from one substance and H from the other substance. There are two ways to do this:

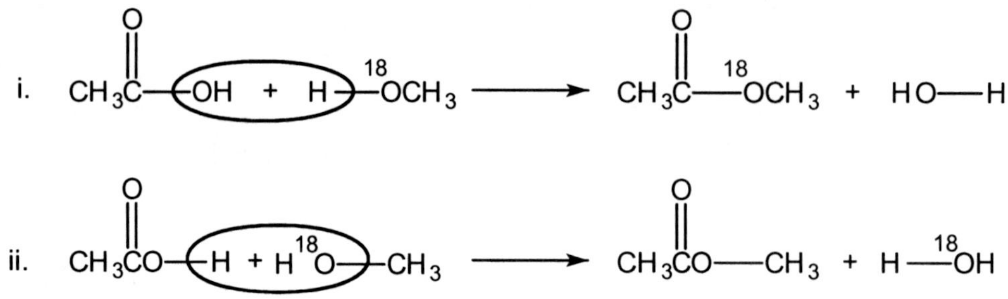

Because the water produced is not radioactive, methyl acetate forms by the first reaction in which all the oxygen-18 ends up in methyl acetate.

54. The only product in the fast-equilibrium step is assumed to be $N^{16}O^{18}O_2$, where N is the central atom. However, this is a reversible reaction where $N^{16}O^{18}O_2$ will decompose to NO and O_2. Because any two oxygen atoms can leave $N^{16}O^{18}O_2$ to form O_2, we would expect (at equilibrium) one-third of the NO present in this fast equilibrium step to be $N^{16}O$ and two-thirds to be $N^{18}O$. In the second step (the slow step), the intermediate $N^{16}O^{18}O_2$ reacts with the scrambled NO to form the NO_2 product, where N is the central atom in NO_2. Any one of the three oxygen atoms can be transferred from $N^{16}O^{18}O_2$ to NO when the NO_2 product is formed. The distribution of ^{18}O in the product can best be determined by forming a probability table.

	$N^{16}O$ (1/3)	$N^{18}O$ (2/3)
^{16}O (1/3) from $N^{16}O^{18}O_2$	$N^{16}O_2$ (1/9)	$N^{18}O^{16}O$ (2/9)
^{18}O (2/3) from $N^{16}O^{18}O_2$	$N^{16}O^{18}O$ (2/9)	$N^{18}O_2$ (4/9)

From the probability table, 1/9 of the NO_2 is $N^{16}O_2$, 4/9 of the NO_2 is $N^{18}O_2$, and 4/9 of the NO_2 is $N^{16}O^{18}O$ (2/9 + 2/9 = 4/9). *Note*: $N^{16}O^{18}O$ is the same as $N^{18}O^{16}O$. In addition, $N^{16}O^{18}O_2$ is not the only NO_3 intermediate formed; $N^{16}O_2{}^{18}O$ and $N^{18}O_3$ can also form in the fast-equilibrium first step. However, the distribution of ^{18}O in the NO_2 product is the same as calculated above, even when these other NO_3 intermediates are considered.

55. $^{235}_{92}U + ^{1}_{0}n \rightarrow ^{144}_{58}Ce + ^{90}_{38}Sr + ?\ ^{1}_{0}n + ?\ ^{0}_{-1}e$; to balance the atomic number, we need 4 beta-particles, and to balance the mass number, we need 2 neutrons.

56. $^{238}_{92}U + ^{1}_{0}n \rightarrow ^{239}_{92}U \rightarrow ^{0}_{-1}e + ^{239}_{93}Np \rightarrow ^{0}_{-1}e + ^{239}_{94}Pu$; plutonium-239 is the fissionable material in breeder reactors.

57. Release of Sr is probably more harmful. Xe is chemically unreactive. Strontium is in the same family as calcium and could be absorbed and concentrated in the body in a fashion similar to Ca. This puts the radioactive Sr in the bones; red blood cells are produced in bone marrow. Xe would not be readily incorporated into the body.

The chemical properties determine where a radioactive material may be concentrated in the body or how easily it may be excreted. The length of time of exposure and what is exposed to radiation significantly affects the health hazard. (See Exercise 58 for a specific example.)

58. (i) and (ii) mean that Pu is not a significant threat outside the body. Our skin is sufficient to keep out the alpha-particles. If Pu gets inside the body, it is easily oxidized to Pu^{4+} (iv), which is chemically similar to Fe^{3+} (iii). Thus Pu^{4+} will concentrate in tissues where Fe^{3+} is found. One of these is the bone marrow, where red blood cells are produced. Once inside the body, alpha-particles cause considerable damage.

Additional Exercises

59. The most abundant isotope is generally the most stable isotope. The periodic table predicts that the most stable isotopes for exercises a-d are ^{39}K, ^{56}Fe, ^{23}Na, and ^{204}Tl. (Reference Table 19.2 of the text for potential decay processes.)

a. Unstable; ^{45}K has too many neutrons and will undergo beta-particle production.

b. Stable

c. Unstable; ^{20}Na has too few neutrons and will most likely undergo electron capture or positron production. Alpha-particle production makes too severe of a change to be a likely decay process for the relatively light ^{20}Na nuclei. Alpha-particle production usually occurs for heavy nuclei.

d. Unstable; ^{194}Tl has too few neutrons and will undergo electron capture, positron production, and/or alpha-particle production.

60. a. Cobalt is a component of vitamin B_{12}. By monitoring the cobalt-57 decay, one can study the pathway of vitamin B_{12} in the body.

b. Calcium is present in the bones in part as $Ca_3(PO_4)_2$. Bone metabolism can be studied by monitoring the calcium-47 decay as it is taken up in bones.

c. Iron is a component of hemoglobin found in red blood cells. By monitoring the iron-59 decay, one can study red blood cell processes.

61. $$N = 180\text{ lb} \times \frac{453.6\text{ g}}{\text{lb}} \times \frac{18\text{ g C}}{100\text{ g body}} \times \frac{1.6 \times 10^{-10}\text{ g }^{14}C}{100\text{ g C}} \times \frac{1\text{ mol }^{14}C}{14\text{ g }^{14}C}$$

$$\times \frac{6.022 \times 10^{23}\text{ nuclei }^{14}C}{\text{mol }^{14}C} = 1.0 \times 10^{15}\text{ nuclei }^{14}C$$

$$\text{Rate} = kN;\ k = \frac{\ln 2}{t_{1/2}} = \frac{0.693}{5730\text{yr}} \times \frac{1\text{ yr}}{365\text{d}} \times \frac{1\text{ d}}{24\text{h}} \times \frac{1\text{ h}}{3600\text{s}} = 3.8 \times 10^{-12}\text{ s}^{-1}$$

$$\text{Rate} = kN;\ k = 3.8 \times 10^{-12}\text{ s}^{-1}(1.0 \times 10^{15}\ ^{14}\text{C nuclei}) = 3800\text{decays/s}$$

A typical 180 lb person produces 3800 beta particles each second.

62. $t_{1/2} = 5730$ yr; $k = (\ln 2)/t_{1/2}$; $\ln(N/N_0) = -kt$; $\ln\frac{15.1}{15.3} = \frac{-(\ln 2)t}{5730\,\text{yr}}$, t = 109 yr

No; from ^{14}C dating, the painting was produced during the early 1900s.

63. The third-life will be the time required for the number of nuclides to reach one-third of the original value ($N_0/3$).

$$\ln\left(\frac{N}{N_0}\right) = -kt = \frac{-(0.6931)t}{t_{1/2}},\quad \ln\left(\frac{1}{3}\right) = \frac{-(0.6931)t}{31.4\ \text{yr}},\quad t = 49.8\ \text{yr}$$

The third-life of this nuclide is 49.8 years.

64. $\ln(N/N_0) = -kt$; $k = (\ln 2)/t_{1/2}$; $N = 0.001 \times N_0$

$$\ln\left(\frac{0.001 \times N_0}{N_0}\right) = \frac{-(\ln 2)t}{24{,}100\,\text{yr}},\quad \ln(0.001) = -(2.88 \times 10^{-5})t,\quad t = 2 \times 10^5\ \text{yr} = 200{,}000\ \text{yr}$$

65. $$\ln\left(\frac{N}{N_0}\right) = -kt = \frac{-(\ln 2)t}{12.3\ \text{yr}},\quad \ln\left(\frac{0.17 \times N_0}{N_0}\right) = -(5.64 \times 10^{-2})t,\quad t = 31.4\ \text{yr}$$

It takes 31.4 years for the tritium to decay to 17% of the original amount. Hence the watch stopped fluorescing enough to be read in 1975 (1944 + 31.4).

66. $\Delta m = -2(5.486 \times 10^{-4}\ \text{u}) = -1.097 \times 10^{-3}\,\text{u}$

$$\Delta E = \Delta mc^2 = -1.097 \times 10^{-3}\,\text{u} \times \frac{1.6605 \times 10^{-27}\ \text{kg}}{\text{u}} \times (2.9979 \times 10^8\ \text{m/s})^2$$

$$= -1.637 \times 10^{-13}\,\text{J}$$

$E_{photon} = 1/2(1.637 \times 10^{-13}\,\text{J}) = 8.185 \times 10^{-14}\,\text{J} = hc/\lambda$

$$\lambda = \frac{hc}{E} = \frac{6.6261 \times 10^{-34}\ \text{J s} \times 2.9979 \times 10^8\ \text{m/s}}{8.185 \times 10^{-14}\ \text{J}} = 2.427 \times 10^{-12}\,\text{m} = 2.427 \times 10^{-3}\,\text{nm}$$

67. $$20{,}000\ \text{ton TNT} \times \frac{4 \times 10^9\ \text{J}}{\text{ton TNT}} \times \frac{1\ \text{mol}\ ^{235}\text{U}}{2 \times 10^{13}\ \text{J}} \times \frac{235\,\text{g}\ ^{235}\text{U}}{\text{mol}\ ^{235}\text{U}} = 940\ \text{g}\ ^{235}\text{U} \approx 900\ \text{g}\ ^{235}\text{U}$$

This assumes that all of the ^{235}U undergoes fission.

68. In order to sustain a nuclear chain reaction, the neutrons produced by the fission must be contained within the fissionable material so that they can go on to cause other fissions. The fissionable material must be closely packed together to ensure that neutrons are not lost to the outside. The critical mass is the mass of material in which exactly one neutron from each fission event causes another fission event so that the process sustains itself. A supercritical situation occurs when more than one neutron from each fission event causes another fission event. In this case, the process rapidly escalates and the heat build up causes a violent explosion.

69. Mass of nucleus = atomic mass – mass of electron = 2.01410 u – 0.000549 u = 2.01355 u

$$u_{rms} = \left(\frac{3\,RT}{M}\right)^{1/2} = \left(\frac{3(8.3145 J/K \bullet mol)(4 \times 10^7\ K)}{2.01355 g(1\ kg/1000 g)}\right)^{1/2} = 7 \times 10^5\ m/s$$

$$KE_{avg} = \frac{1}{2}mu^2 = \frac{1}{2}\left(2.01355u \times \frac{1.66 \times 10^{-27}\ kg}{u}\right)(7 \times 10^5\ m/s)^2 = 8 \times 10^{-16}\ J/nuclei$$

We could have used $KE_{ave} = (3/2)RT$ to determine the same average kinetic energy.

70. ${}^1_1H + {}^1_0n \rightarrow 2\ {}^1_1H + {}^1_0n + {}^{\ 1}_{-1}H$; mass ${}^{\ 1}_{-1}H$ = mass ${}^1_1H$ = 1.00728 u = mass of proton = m_p

$$\Delta m = 3m_p + m_n - (m_p + m_n) = 2m_p = 2(1.00728) = 2.01456\ u$$

$$\Delta E = \Delta mc^2 = 2.01456\ amu \times \frac{1.66056 \times 10^{-27}\ kg}{amu} \times (2.997925 \times 10^8\ m/s)^2$$

$\Delta E = 3.00660 \times 10^{-10}$ J of energy is absorbed per nuclei, or 1.81062×10^{14} J/mol nuclei.

The source of energy is the kinetic energy of the proton and the neutron in the particle accelerator.

71. All evolved oxygen in O_2 comes from water and not from carbon dioxide.

72. Sr-90 is an alkaline earth metal having chemical properties similar to calcium. Sr-90 can collect in bones, replacing some of the calcium. Once embedded inside the human body, beta-particles can do significant damage. Rn-222 is a noble gas, so one would expect Rn to be unreactive and pass through the body quickly; it does. The problem with Rn-222 is the rate at which it produces alpha-particles. With a short half-life, the few moments that Rn-222 is in the lungs, a significant number of decay events can occur; each decay event produces an alpha-particle that is very effective at causing ionization and can produce a dense trail of damage.

ChemWork Problems

The answers to the problems 73-78 (or a variation to these problems) are found in OWL. These problems are also assignable in OWL.

Challenge Problems

79. $k = \frac{\ln 2}{t_{1/2}}$; $\ln\left(\frac{N}{N_0}\right) = -kt = \frac{-(0.693)t}{t_{1/2}}$

For ^{238}U: $\ln\left(\frac{N}{N_0}\right) = \frac{-(0.693)(4.5 \times 10^9\ yr)}{4.5 \times 10^9\ yr} = -0.693,\ \frac{N}{N_0} = e^{-0.693} = 0.50$

For ^{235}U: $\ln\left(\frac{N}{N_0}\right) = \frac{-(0.693)(4.5 \times 10^9\ yr)}{7.1 \times 10^8\ yr} = -4.39,\ \frac{N}{N_0} = e^{-4.39} = 0.012$

If we have a current sample of 10,000 uranium nuclei, 9928 nuclei of ^{238}U and 72 nuclei of ^{235}U are present. Now let's calculate the initial number of nuclei that must have been present 4.5×10^9 years ago to produce these 10,000 uranium nuclei.

For ^{238}U: $\frac{N}{N_0} = 0.50,\ N_0 = \frac{N}{0.50} = \frac{9928\,\text{nuclei}}{0.50} = 2.0 \times 10^4\ ^{238}\text{U nuclei}$

For ^{235}U: $N_0 = \frac{N}{0.012} = \frac{72\,\text{nuclei}}{0.012} = 6.0 \times 10^3\ ^{235}\text{U nuclei}$

So 4.5 billion years ago, the 10,000-nuclei sample of uranium was composed of 2.0×10^4 ^{238}U nuclei and 6.0×10^3 ^{235}U nuclei. The percent composition 4.5 billion years ago would have been:

$$\frac{2.0 \times 10^4\ ^{238}\text{U nuclei}}{(6.0 \times 10^3 + 2.0 \times 10^4)\,\text{total nuclei}} \times 100 = 77\%\ ^{238}\text{U and } 23\%\ ^{235}\text{U}$$

80. Total activity injected = 86.5×10^{-3} Ci

Activity withdrawn $= \frac{3.6 \times 10^{-6}\ \text{Ci}}{2.0\,\text{mL}\ H_2O} = \frac{1.8 \times 10^{-6}\ \text{Ci}}{\text{mL}\ H_2O}$

Assuming no significant decay occurs, then the total volume of water in the body multiplied by 1.8×10^{-6} Ci/mL must equal the total activity injected.

$V \times \frac{1.8 \times 10^{-6}\ \text{Ci}}{\text{mL}\ H_2O} = 8.65 \times 10^{-2}\ \text{Ci},\ V = 4.8 \times 10^4\ \text{mL}\ H_2O$

Assuming a density of 1.0 g/mL for water, the mass percent of water in this 150-lb person is:

$$\frac{4.8 \times 10^4\ \text{mL}\ H_2O \times \frac{1.0\,\text{g}\ H_2O}{\text{mL}} \times \frac{1\,\text{lb}}{453.6\,\text{g}}}{150\,\text{lb}} \times 100 = 71\%$$

81. Assuming that the radionuclide is long-lived enough that no significant decay occurs during the time of the experiment, the total counts of radioactivity injected are:

$0.10\ \text{mL} \times \frac{5.0 \times 10^3\ \text{cpm}}{\text{mL}} = 5.0 \times 10^2\ \text{cpm}$

Assuming that the total activity is uniformly distributed only in the rat's blood, the blood volume is:

$V \times \frac{48\,\text{cpm}}{\text{mL}} = 5.0 \times 10^2\ \text{cpm},\ V = 10.4\ \text{mL} = 10.\ \text{mL}$

82. a. From Table 18.1: $2\ H_2O + 2\ e^- \rightarrow H_2 + 2\ OH^-$ $E° = -\ 0.83$ V

$$E^\circ_{cell} = E^\circ_{H_2O} - E^\circ_{Zr} = -\ 0.83\ V + 2.36\ V = 1.53\ V$$

Yes, the reduction of H_2O to H_2 by Zr is spontaneous at standard conditions because $E^\circ_{cell} > 0$.

b. $(2\ H_2O + 2\ e^- \rightarrow H_2 + 2\ OH^-) \times 2$

$Zr + 4\ OH^- \rightarrow ZrO_2•H_2O + H_2O + 4\ e^-$

$3\ H_2O(l) + Zr(s) \rightarrow 2\ H_2(g) + ZrO_2•H_2O(s)$

c. $\Delta G° = -nFE° = -(4\ \text{mol e}^-)(96{,}485\ \text{C/mol e}^-)(1.53\ \text{J/C}) = -5.90 \times 10^5\ \text{J} = -590.\ \text{kJ}$

$$E = E° - \frac{0.0591}{n}\log Q;\ \text{at equilibrium, } E = 0 \text{ and } Q = K.$$

$$E° = \frac{0.0591}{n}\log K,\ \log K = \frac{4(1.53)}{0.0591} = 104,\ K \approx 10^{104}$$

d. $$1.00 \times 10^3\ \text{kg Zr} \times \frac{1000\text{g}}{\text{kg}} \times \frac{1\ \text{mol Zr}}{91.22\ \text{g Zr}} \times \frac{2\ \text{mol}H_2}{\text{mol Zr}} = 2.19 \times 10^4\ \text{mol } H_2$$

$$2.19 \times 10^4\ \text{mol } H_2 \times \frac{2.016\ \text{g } H_2}{\text{mol} H_2} = 4.42 \times 10^4\ \text{g } H_2$$

$$V = \frac{nRT}{P} = \frac{(2.19\times 10^4\ \text{mol})(0.08206\text{L atm/mol• K})(1273\text{K})}{1.0\text{atm}} = 2.3 \times 10^6\ \text{L } H_2$$

e. Probably yes; less radioactivity overall was released by venting the H_2 than what would have been released if the H_2 had exploded inside the reactor (as happened at Chernobyl). Neither alternative is pleasant, but venting the radioactive hydrogen is the less unpleasant of the two alternatives.

83. a. ^{12}C; it takes part in the first step of the reaction but is regenerated in the last step. ^{12}C is not consumed, so it is not a reactant.

b. ^{13}N, ^{13}C, ^{14}N, ^{15}O, and ^{15}N are the intermediates.

c. $4\,^1_1H \rightarrow\ ^4_2He + 2\,^0_{+1}e$; $\Delta m = 4.00260\ \text{u} - 2\ m_e + 2\ m_e - [4(1.00782\ \text{u} - m_e)]$

$\Delta m = 4.00260 - 4(1.00782) + 4(0.000549) = -0.02648$ u for four protons reacting

For 4 mol of protons, $\Delta m = -0.02648$ g, and ΔE for the reaction is:

$$\Delta E = \Delta mc^2 = -\ 2.648 \times 10^{-5}\ \text{kg} \times (2.9979 \times 10^8\ \text{m/s})^2 = -2.380 \times 10^{12}\ \text{J}$$

For 1 mol of protons reacting: $\dfrac{-\ 2.380 \times 10^{12}\ \text{J}}{4\ \text{mol}\ ^1\text{H}} = -5.950 \times 10^{11}\ \text{J/mol}\ ^1\text{H}$

84. a. $^{238}_{92}U \rightarrow {}^{222}_{86}Rn + ?\ {}^{4}_{2}He + ?\ {}^{0}_{-1}e$; to account for the mass number change, four alpha-particles are needed. To balance the number of protons, two beta-particles are needed.

$^{222}_{86}Rn \rightarrow {}^{4}_{2}He + {}^{218}_{84}Po$; polonium-218 is produced when ^{222}Rn decays.

b. Alpha-particles cause significant ionization damage when inside a living organism. Because the half-life of ^{222}Rn is relatively short, a significant number of alpha-particles will be produced when ^{222}Rn is present (even for a short period of time) in the lungs.

c. $^{222}_{86}Rn \rightarrow {}^{4}_{2}He + {}^{218}_{84}Po$; $^{218}_{84}Po \rightarrow {}^{4}_{2}He + {}^{214}_{82}Pb$; polonium-218 is produced when radon-222 decays. ^{218}Po is a more potent alpha-particle producer since it has a much shorter half-life than ^{222}Rn. In addition, ^{218}Po is a solid, so it can get trapped in the lung tissue once it is produced. Once trapped, the alpha-particles produced from polonium-218 (with its very short half-life) can cause significant ionization damage.

d. Rate = kN; $\text{rate} = \frac{4.0\ \text{pCi}}{\text{L}} \times \frac{1 \times 10^{-12}\ \text{Ci}}{\text{pCi}} \times \frac{3.7 \times 10^{10}\ \text{decays/sec}}{\text{Ci}} = 0.15\ \text{decays/s•L}$

$$k = \frac{\ln 2}{t_{1/2}} = \frac{0.6391}{3.82\ \text{d}} \times \frac{1\ \text{d}}{24\ \text{h}} \times \frac{1\ \text{h}}{3600\text{s}} = 2.10 \times 10^{-6}\ \text{s}^{-1}$$

$$N = \frac{\text{rate}}{K} = \frac{0.15\text{decays/s} \bullet \text{L}}{2.10 \times 10^{-6}\ \text{s}^{-1}} = 7.1 \times 10^{4}\ {}^{222}\text{Rn atoms/L}$$

$$\frac{7.1 \times 10^{4}\ {}^{222}\text{Rn atoms}}{\text{L}} \times \frac{1\ \text{mol}\ {}^{222}\text{Rn}}{6.02 \times 10^{23}\ \text{atoms}} = 1.2 \times 10^{-19}\ \text{mol}\ {}^{222}\text{Rn/L}$$

85. $$\text{Moles of I}^- = \frac{33\ \text{counts}}{\text{min}} \times \frac{1\ \text{mol I} \bullet \text{min}}{5.0 \times 10^{11}\ \text{counts}} = 6.6 \times 10^{-11}\ \text{mol I}^-$$

$$[\text{I}^-] = \frac{6.6 \times 10^{-11}\ \text{mol I}^-}{0.150\text{L}} = 4.4 \times 10^{-10}\text{mol/L}$$

	$Hg_2I_2(s)$ →	$Hg_2^{2+}(aq)$	+	$2\ I^-(aq)$	$K_{sp} = [Hg_2^{2+}][I^-]^2$
Initial	s = solubility (mol/L)	0		0	
Equil.		s		$2s$	

From the problem, $2s = 4.4 \times 10^{-10}$ mol/L, $s = 2.2 \times 10^{-10}$ mol/L.

$K_{sp} = (s)(2s)^2 = (2.2 \times 10^{-10})(4.4 \times 10^{-10})^2 = 4.3 \times 10^{-29}$

86. $^{2}_{1}H + {}^{2}_{1}H \rightarrow {}^{4}_{2}He$; Q for $^{2}_{1}H = 1.6 \times 10^{-19}$ C; mass of deuterium = 2 u.

$$E = \frac{9.0 \times 10^{9}\ \text{J} \bullet \text{m/C}^2 (Q_1 Q_2)}{r} = \frac{9.0 \times 10^{9}\ \text{J} \bullet \text{m/C}^2 (1.6 \times 10^{-19}\ \text{C})^2}{2 \times 10^{-15}\ \text{m}}$$

$= 1 \times 10^{-13}$ J per alpha particle

$KE = 1/2\ mv^2$; $1 \times 10^{-13}\ J = 1/2\ (2\ u \times 1.66 \times 10^{-27}\ kg/u)v^2$, $v = 8 \times 10^6\ m/s$

From the kinetic molecular theory discussed in Chapter 5:

$$u_{rms} = \left(\frac{3RT}{M}\right)^{1/2}, \text{ where M = molar mass in kilograms} = 2 \times 10^{-3}\ \text{kg/mol for deuterium}$$

$$8 \times 10^6\ m/s = \left[\frac{3(8.3145 J/K \bullet mol)(T)}{2 \times 10^{-3}\ kg}\right]^{1/2}, \quad T = 5 \times 10^9\ K$$

Integrative Problems

87. $^{249}_{97}Bk + ^{22}_{10}Ne \rightarrow ^{267}_{107}Bh + ?$; this equation is charge balanced, but it is not mass balanced. The products are off by 4 mass units. The only possibility to account for the 4 mass units is to have 4 neutrons produced. The balanced equation is:

$$^{249}_{97}Bk + ^{22}_{10}Ne \rightarrow ^{267}_{107}Bh + 4\,^{1}_{0}n$$

$$\ln\left(\frac{N}{N_0}\right) = -kt = \frac{-(0.6931)t}{t_{1/2}}, \quad \ln\left(\frac{11}{199}\right) = \frac{-(0.6931)t}{15.0\ s}, \quad t = 62.7\ s$$

Bh: $[Rn]7s^2 5f^{14} 6d^5$ is the expected electron configuration.

88. $^{58}_{26}Fe + 2\,^{1}_{0}n \rightarrow ^{60}_{27}Co + ?$; in order to balance the equation, the missing particle has no mass and a charge of 1−; this is an electron.

An atom of $^{60}_{27}Co$ has 27 e, 27 p, and 33 n. The mass defect of the ^{60}Co nucleus is:

$$\Delta m = (59.9338 - 27m_e) - [27(1.00782 - m_e) + 33(1.00866)] = -0.5631\ u$$

$$\Delta E = \Delta mc^2 = -0.5631\ u \times \frac{1.6605 \times 10^{-27}\ kg}{u} \times (2.9979 \times 10^8\ m/s)^2 = -8.403 \times 10^{-11}\ J$$

$$\frac{\text{Binding energy}}{\text{Nucleon}} = \frac{8.403 \times 10^{-11}\ J}{60\ \text{nucleons}} = 1.401 \times 10^{-12}\ \text{J/nucleon}$$

The emitted particle was an electron, which has a mass of 9.109×10^{-31} kg. The deBroglie wavelength is:

$$\lambda = \frac{h}{mv} = \frac{6.626 \times 10^{-34}\ J\ s}{9.109 \times 10^{-31}\ kg \times (0.90 \times 2.998 \times 10^8\ m/s)} = 2.7 \times 10^{-12}\ m$$

CHAPTER 20

THE REPRESENTATIVE ELEMENTS

Questions

1. The gravity of the earth is not strong enough to keep the light H_2 molecules in the atmosphere.

2. (1) Ammonia production and (2) hydrogenation of vegetable oils.

3. The acidity decreases. Solutions of Be^{2+} are acidic, while solutions of the other M^{2+} ions are neutral.

4. Size decreases from left to right and increases going down the periodic table. So, going one element right and one element down would result in a similar size for the two elements diagonal to each other. The ionization energies will be similar for the diagonal elements since the periodic trends also oppose each other. Electron affinities are harder to predict, but atoms with similar size and ionization energy should also have similar electron affinities.

5. For Groups 1A-3A, the small sizes of H (as compared to Li), Be (as compared to Mg), and B (as compared to Al) seem to be the reason why these elements have nonmetallic properties, while others in the Groups 1A-3A are strictly metallic. The small sizes of H, Be, and B also cause these species to polarize the electron cloud in nonmetals, thus forcing a sharing of electrons when bonding occurs. For Groups 4A-6A, a major difference between the first and second members of a group is the ability to form π bonds. The smaller elements form stable π bonds, while the larger elements are not capable of good overlap between parallel p orbitals and, in turn, do not form strong π bonds. For Group 7A, the small size of F as compared to Cl is used to explain the low electron affinity of F and the weakness of the F–F bond.

6. SiC would have a covalent network structure similar to diamond.

7. Solids have stronger intermolecular forces than liquids. In order to maximize the hydrogen bonding in the solid phase, ice is forced into an open structure. This open structure is why $H_2O(s)$ is less dense than $H_2O(l)$.

8. Nitrogen fixation is the process of transforming atmospheric N_2 to other nitrogen-containing compounds. Some examples are:

$$N_2(g) + 3\ H_2(g) \rightarrow 2\ NH_3(g)$$

$$N_2(g) + O_2(g) \rightarrow 2\ NO(g)$$

$$N_2(g) + 2\ O_2(g) \rightarrow 2\ NO_2(g)$$

9. Group 1A and 2A metals are all easily oxidized. They must be produced in the absence of materials (H_2O, O_2) that are capable of oxidizing them.

10. The bonds in SnX_4 compounds have a large covalent character. SnX_4 acts as discrete molecules held together by weak London dispersion forces. SnX_2 compounds are ionic and are held in the solid state by strong ionic forces. Because the intermolecular forces are weaker for SnX_4 compounds, they are more volatile (have a lower boiling point).

Exercises

Group 1A Elements

11. a. $\Delta H° = -110.5 - [-75 + (-242)] = 207$ kJ; $\Delta S° = 198 + 3(131) - [186 + 189] = 216$ J/K

b. $\Delta G° = \Delta H° - T\Delta S°$; $\Delta G° = 0$ when $T = \dfrac{\Delta H°}{\Delta S°} = \dfrac{207 \times 10^3 \text{ J}}{216 \text{ J/K}} = 958$ K

At T > 958 K and standard pressures, the favorable ΔS° term dominates, and the reaction is spontaneous (ΔG° < 0).

12. a. $\Delta H° = 2(-46 \text{ kJ}) = -92$ kJ; $\Delta S° = 2(193 \text{ J/K}) - [3(131 \text{ J/K}) + 192 \text{ J/K}] = -199$ J/K;

$\Delta G° = \Delta H° - T\Delta S° = -92 \text{ kJ} - 298 \text{ K}(-0.199 \text{ kJ/K}) = -33$ kJ

b. Because ΔG° is negative, this reaction is spontaneous at standard conditions.

c. $\Delta G° = 0$ when $T = \dfrac{\Delta H°}{\Delta S°} = \dfrac{-92 \text{ kJ}}{-0.199 \text{ kJ/K}} = 460$ K

At T < 460 K and standard pressures, the favorable ΔH° term dominates, and the reaction is spontaneous (ΔG° < 0).

13. $4 \text{ Li(s)} + O_2\text{(g)} \rightarrow 2 \text{ Li}_2\text{O(s)}$

$2 \text{ Li(s)} + \text{S(s)} \rightarrow \text{Li}_2\text{S(s)}$; $2 \text{ Li(s)} + Cl_2\text{(g)} \rightarrow 2 \text{ LiCl(s)}$

$12 \text{ Li(s)} + P_4\text{(s)} \rightarrow 4 \text{ Li}_3\text{P(s)}$; $2 \text{ Li(s)} + H_2\text{(g)} \rightarrow 2 \text{ LiH(s)}$

$2 \text{ Li(s)} + 2 \text{ H}_2\text{O(l)} \rightarrow 2 \text{ LiOH(aq)} + H_2\text{(g)}$; $2 \text{ Li(s)} + 2 \text{ HCl(aq)} \rightarrow 2 \text{ LiCl(aq)} + H_2\text{(g)}$

14. We need another reactant beside NaCl(aq) because oxygen and hydrogen are in some of the products. The obvious choice is H_2O.

$$2 \text{ NaCl(aq)} + 2 \text{ H}_2\text{O(l)} \rightarrow Cl_2\text{(g)} + H_2\text{(g)} + 2 \text{ NaOH(aq)}$$

Note that hydrogen is reduced and chlorine is oxidized in this electrolysis process.

15. When lithium reacts with excess oxygen, Li_2O forms, which is composed of Li^+ and O^{2-} ions. This is called an oxide salt. When sodium reacts with oxygen, Na_2O_2 forms, which is composed of Na^+ and O_2^{2-} ions. This is called a peroxide salt. When potassium (or rubidium or cesium) reacts with oxygen, KO_2 forms, which is composed of K^+ and O_2^- ions. For your information, this is called a superoxide salt. So the three types of alkali metal oxides that can form differ in the oxygen anion part of the formula (O^{2-} versus O_2^{2-} versus O_2^-). Each of these anions has unique bonding arrangements and oxidation states.

16. The first illustration is an example of a covalent hydride like H_2O. Covalent hydrides are just binary covalent compounds formed between hydrogen and some other nonmetal and exist as individual molecules. The middle illustration represents interstitial (or metallic) hydrides. In interstitial hydrides, hydrogen atoms occupy the holes of a transition metal crystal. These hydrides are more like solid solutions than true compounds. The third illustration represents ionic (or salt-like) hydrides like LiH. Ionic hydrides form when hydrogen reacts with a metal from Group 1A or 2A. The metals lose electrons to form cations and the hydrogen atoms gain electrons to form the hydride anions (H^-). These are just ionic compounds formed between a metal and hydrogen.

17. The small size of the Li^+ cation results in a much greater attraction to water. The attraction to water is not so great for the other alkali metal ions. Thus lithium salts tend to absorb water.

18. Counting over in the periodic table, the next alkali metal will be element 119. It will be located under Fr. One would expect the physical properties of element 119 to follow the trends shown in Table 20.4. Element 119 should have the smallest ionization energy, the most negative standard reduction potential, the largest radius, and the smallest melting point of all the alkali metals listed in Table 20.4. It should also be radioactive like Fr.

Group 2A Elements

19. $CaCO_3(s) + H_2SO_4(aq) \rightarrow CaSO_4(aq) + H_2O(l) + CO_2(g)$

20. $2\ Sr(s) + O_2(g) \rightarrow 2\ SrO(s)$; $Sr(s) + S(s) \rightarrow SrS(s)$

$Sr(s) + Cl_2(g) \rightarrow SrCl_2(s)$; $6\ Sr(s) + P_4(s) \rightarrow 2\ Sr_3P_2(s)$

$Sr(s) + H_2(g) \rightarrow SrH_2(s)$; $Sr(s) + 2\ H_2O(l) \rightarrow Sr(OH)_2(aq) + H_2(g)$

$Sr(s) + 2\ HCl(aq) \rightarrow SrCl_2(aq) + H_2(g)$

21. $Ba^{2+} + 2\ e^- \rightarrow Ba$; $6.00\ h \times \frac{60\ min}{h} \times \frac{60\ s}{min} \times \frac{2.50 \times 10^5\ C}{s} \times \frac{1\ mol\ e^-}{96{,}485\ C} \times \frac{1\ mol\ Ba}{2\ mol\ e^-}$

$$\times \frac{137.3\ g\ Ba}{mol\ Ba} = 3.84 \times 10^6\ g\ Ba$$

22. $MgCl_2$ is composed of Mg^{2+} ions; Mg^{2+} gains 2 electrons to form magnesium metal in electrolysis.

$$1.00 \times 10^6 \text{ g Mg} \times \frac{1 \text{ mol Mg}}{24.31 \text{ g Mg}} \times \frac{2 \text{ mol e}^-}{1 \text{ mol Mg}} \times \frac{96{,}485 \text{ C}}{1 \text{ mol e}^-} \times \frac{1 \text{ s}}{5.00 \times 10^4 \text{ C}} \times \frac{1 \text{ h}}{3600 \text{s}}$$
$$= 44.1 \text{ hours}$$

23. Beryllium has a small size and a large electronegativity as compared to the other alkaline earth metals. The electronegativity of Be is so high that it does not readily give up electrons to nonmetals, as is the case for the other alkaline earth metals. Instead, Be has significant covalent character in its bonds; it prefers to share valence electrons rather than give them up to form ionic bonds.

24. The alkaline earth ions that give water the hard designation are Ca^{2+} and Mg^{2+}. These ions interfere with the action of detergents and form unwanted precipitates with soaps. Large-scale water softeners remove Ca^{2+} by precipitating out the calcium ions as $CaCO_3$. In homes, Ca^{2+} and Mg^{2+} (plus other cations) are removed by ion exchange. See Figure 20.6 for a schematic of a typical cation exchange resin.

Group 3A Elements

25. Element 113: $[Rn]7s^25f^{14}6d^{10}7p^1$; element 113 would fall below Tl in the periodic table. Like Tl, we would expect element 113 to form +1 and +3 oxidation states in its compounds.

26. Tl_2O_3, thallium(III) oxide; Tl_2O, thallium(I) oxide; $InCl_3$, indium(III) chloride; InCl, indium(I) chloride

27. $B_2H_6(g) + 3\ O_2(g) \rightarrow 2\ B(OH)_3(s)$

28. $B_2O_3(s) + 3\ Mg(s) \rightarrow 3\ MgO(s) + 2\ B(s)$

29. $2\ Ga(s) + 3\ F_2(g) \rightarrow 2\ GaF_3(s)$; $4\ Ga(s) + 3\ O_2(g) \rightarrow 2\ Ga_2O_3(s)$

$2Ga(s) + 3\ S(s) \rightarrow Ga_2S_3(s)$; $2\ Ga(s) + 6\ HCl(aq) \rightarrow 2\ GaCl_3(aq) + 3\ H_2(g)$

30. $2\ Al(s) + 2\ NaOH(aq) + 6\ H_2O(l) \rightarrow 2\ Al(OH)_4^-(aq) + 2\ Na^+(aq) + 3\ H_2(g)$

31. An amphoteric substance is one that can behave as either an acid or as a base. Al_2O_3 dissolves in both acidic and basic solutions. The reactions are:

$$Al_2O_3(s) + 6\ H^+(aq) \rightarrow 2Al^{3+}(aq) + 3\ H_2O(l)$$

$$Al_2O_3(s) + 2\ OH^-(aq) + 3\ H_2O(l) \rightarrow 2\ Al(OH)_4^-(aq)$$

32. Compounds called boranes have three-centered bonds. Three-centered bonds occur when a single H atom forms bridging bonds between two boron atoms. The bonds have two electrons bonding all three atoms together. The bond is electron-deficient and makes boranes very reactive.

Group 4A Elements

33. Compounds containing Si–Si single and multiple bonds are rare, unlike compounds of carbon. The bond strengths of the Si–Si and C–C single bonds are similar. The difference in bonding properties must be for other reasons. One reason is that silicon does not form strong π bonds, unlike carbon. Another reason is that silicon forms particularly strong sigma bonds to oxygen, resulting in compounds with Si–O bonds instead of Si–Si bonds.

34. CO_2 is a molecular substance composed of individual CO_2 molecules. SiO_2 does not exist as discreet molecules. Instead, SiO_2 is the empirical formula for quartz, which is composed of a network of SiO_4 tetrahedra with shared oxygen atoms between the various tetrahedra. The major reason for the difference in structures is that carbon has the ability to form π bonds, whereas silica does not form stable π bonds. In order to form discrete CO_2 molecules, π bonds must form. Because silicon does not form stable bonds, silicon atoms achieve a noble gas configuration by forming several Si–O single bonds. These Si–O single bonds extend in all directions, giving the network structure of quartz.

35. 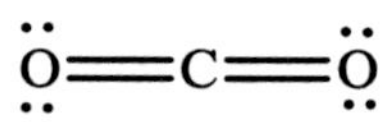

The darker green orbitals about carbon are sp hybrid orbitals. The lighter green orbitals about each oxygen are sp^2 hybrid orbitals, and the gold orbitals about all of the atoms are unhybridized p atomic orbitals. In each double bond in CO_2, one sigma and one π bond exists. The two carbon-oxygen sigma bonds are formed from overlap of sp hybrid orbitals from carbon with a sp^2 hybrid orbital from each oxygen. The two carbon-oxygen π bonds are formed from side-to-side overlap of the unhybridized p atomic orbitals from carbon with an unhybridized p atomic orbital from each oxygen. These two π bonds are oriented perpendicular to each other as illustrated in the figure.

36. The formulas for the stable oxides of carbon are CO, CO_2, and C_3O_2.

CO, $4 + 6 = 10\ e^-$; CO_2, $4 + 2(6) = 16\ e^-$; C_3O_2, $3(4) + 2(6) = 24\ e^-$

 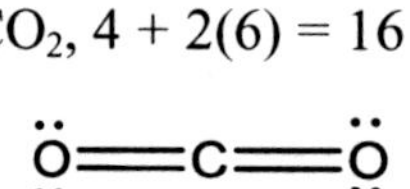 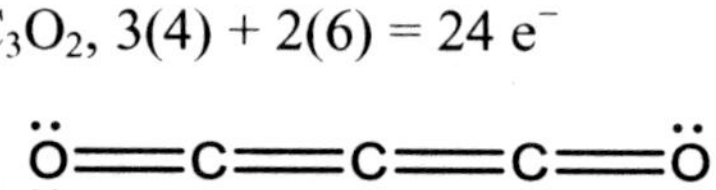

For the bonding in CO, both carbon and oxygen are sp hybridized. In the CO triple bond, 1 sigma and 2 π bonds exist. Overlap of sp hybrid orbitals from carbon and oxygen form the sigma bond. The two π bonds are formed from side-to-side overlap of unhybridized p atomic orbitals from carbon and oxygen. These π bonds are oriented perpendicular to each other.

For the bonding in C_3O_2, the carbon atoms are all sp hybridized while the terminal oxygen atoms are sp^2 hybridized. In each double bond in C_3O_2, one sigma and one π bond exists. The sigma bonds between the carbon and oxygen atoms are formed from overlap of an sp hybrid orbital from carbon with an sp^2 hybrid orbital from oxygen. The sigma bond between the carbon atoms are both formed from overlap of sp hybrid orbitals from each carbon. All four π bonds in the C_3O_2 molecule are formed from side-to-side overlap of unhybridized p atomic orbitals from the various atoms.

37. a. $SiO_2(s) + 2\ C(s) \rightarrow Si(s) + 2\ CO(g)$

b. $SiCl_4(l) + 2\ Mg(s) \rightarrow Si(s) + 2\ MgCl_2(s)$

c. $Na_2SiF_6(s) + 4\ Na(s) \rightarrow Si(s) + 6\ NaF(s)$

38. $Sn(s) + 2\ Cl_2(g) \rightarrow SnCl_4(s)$; $Sn(s) + O_2(g) \rightarrow SnO_2(s)$

$Sn(s) + 2\ HCl(aq) \rightarrow SnCl_2(aq) + H_2(g)$

39. Pb_3O_4: we assign −2 for the oxidation state of O. The sum of the oxidation states of Pb must be +8. We get this if two of the lead atoms are Pb(II) and one is Pb(IV). Therefore, the mole ratio of lead(II) to lead(IV) is 2 : 1.

40. $Sn(s) + 2F_2(g) \rightarrow SnF_4(s)$, tin(IV) fluoride; $Sn(s) + F_2(g) \rightarrow SnF_2(s)$, tin(II) fluoride

Group 5A Elements

41. NO_4^{3-}

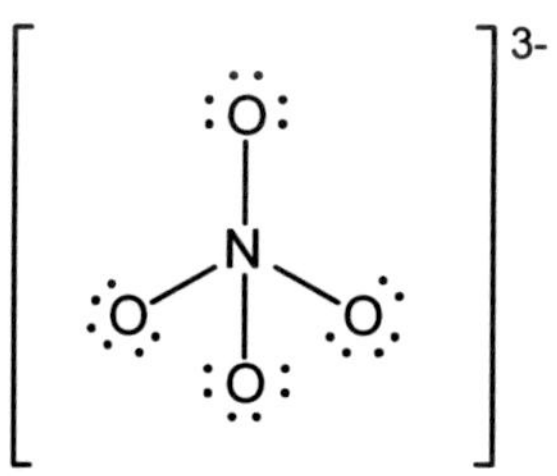

Both NO_4^{3-} and PO_4^{3-} have 32 valence electrons, so both have similar Lewis structures. From the Lewis structure for NO_4^{3-}, the central N atom has a tetrahedral arrangement of electron pairs. N is small. There is probably not enough room for all 4 oxygen atoms around N. P is larger; thus PO_4^{3-} is stable.

PO_3^-

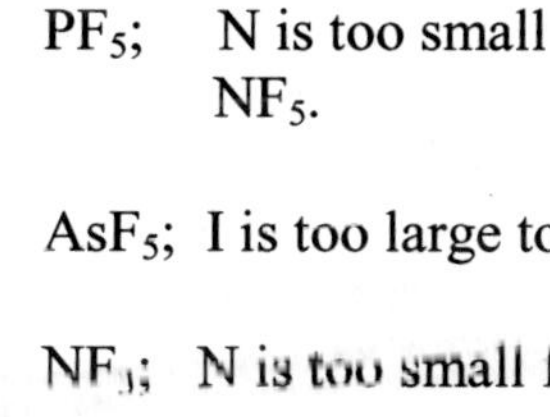

PO_3^- and NO_3^- each have 24 valence electrons, so both have similar Lewis structures. From the Lewis structure for PO_3^-, PO_3^- has a trigonal planar arrangement of electron pairs about the central P atom (two single bonds and one double bond). P=O bonds are not particularly stable, while N=O bonds are stable. Thus NO_3^- is stable.

42. a. PF_5; N is too small and doesn't have low-energy d-orbitals to expand its octet to form NF_5.

b. AsF_5; I is too large to fit 5 atoms of I around As.

c. NF_3; N is too small for three large bromine atoms to fit around it.

43. Production of bismuth:

$$2\ Bi_2S_3(s) + 9\ O_2(g) \rightarrow 2\ Bi_2O_3(s) + 6\ SO_2(g);\ 2\ Bi_2O_3(s) + 3\ C(s) \rightarrow 4\ Bi(s) + 3\ CO_2(g)$$

Production of antimony:

$$2\ Sb_2S_3(s) + 9\ O_2(g) \rightarrow 2\ Sb_2O_3(s) + 6\ SO_2(g);\ 2\ Sb_2O_3(s) + 3\ C(s) \rightarrow 4\ Sb(s) + 3\ CO_2(g)$$

44. $4\ As(s) + 3\ O_2(g) \rightarrow As_4O_6(s)$; $4\ As(s) + 5\ O_2(g) \rightarrow As_4O_{10}(s)$

$As_4O_6(s) + 6\ H_2O(l) \rightarrow 4\ H_3AsO_3(aq)$; $As_4O_{10}(s) + 6\ H_2O(l) \rightarrow 4\ H_3AsO_4(aq)$

45. NH_3, $5 + 3(1) = 8\ e^-$

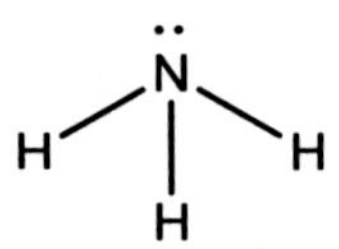

Trigonal pyramid; sp^3

$AsCl_5$, $5 + 5(7) = 40\ e^-$

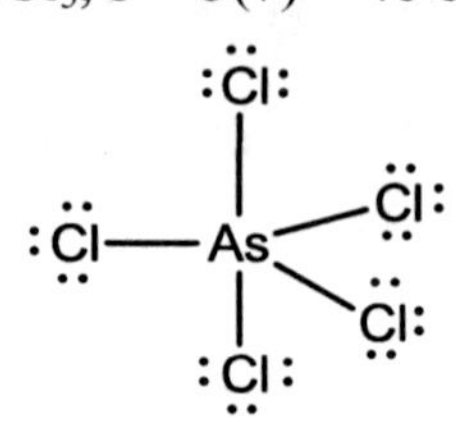

Trigonal bipyramid; dsp^3

PF_6^-, $5 + 6(7) + 1 = 48\ e^-$

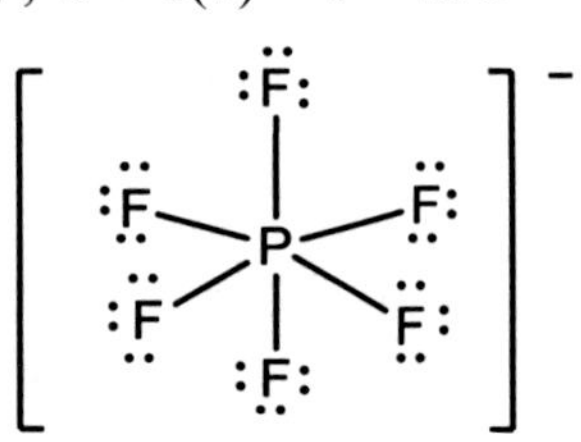

Octahedral; d^2sp^3

Nitrogen does not have low-energy d orbitals it can use to expand its octet. Both NF_5 and NCl_6^- would require nitrogen to have more than 8 valence electrons around it; this never happens.

46. MO model (where B.O. = bond order):

NO^+: $(\sigma_{2s})^2(\sigma_{2s}^*)^2(\pi_{2p})^4(\sigma_{2p})^2$; B.O = $(8 - 2)/2 = 3$, 0 unpaired e^- (diamagnetic)

NO: $(\sigma_{2s})^2(\sigma_{2s}^*)^2(\pi_{2p})^4(\sigma_{2p})^2(\pi_{2p}^*)^1$; B.O. = 2.5, 1 unpaired e^- (paramagnetic)

NO^-: $(\sigma_{2s})^2(\sigma_{2s}^*)^2(\pi_{2p})^4(\sigma_{2p})^2(\pi_{2p}^*)^2$; B.O. = 2, 2 unpaired e^- (paramagnetic)

Lewis structures: NO^+:

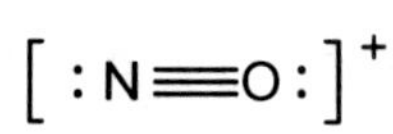

NO:

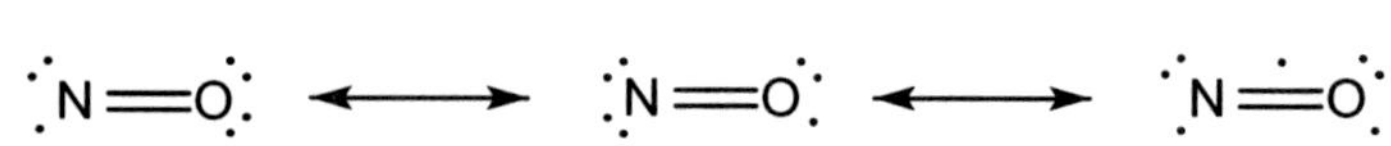

NO^-:

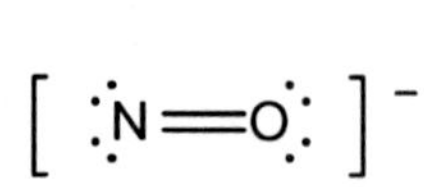

The two models give the same results only for NO^+ (a triple bond with no unpaired electrons). Lewis structures are not adequate for NO and NO^-. The MO model gives a better representation for all three species. For NO, Lewis structures are poor for odd electron species. For NO^-, both models predict a double bond, but only the MO model correctly predicts that NO^- is paramagnetic.

47. $1/2\ N_2(g) + 1/2\ O_2(g) \rightarrow NO(g)$ $\Delta G° = \Delta G^\circ_{f,\,NO} = 87$ kJ/mol; by definition, ΔG°_f for a compound equals the free energy change that would accompany the formation of 1 mol of that compound from its elements in their standard states. NO (and some other oxides of nitrogen) have weaker bonds as compared to the triple bond of N_2 and the double bond of O_2. Because of this, NO (and some other oxides of nitrogen) have higher (positive) standard free energies of formation as compared to the relatively stable N_2 and O_2 molecules.

48. $\Delta H° = 2(90.\text{ kJ}) - (0 + 0) = 180.\text{ kJ}$; $\Delta S° = 2(211\text{ J/K}) - (192 + 205) = 25\text{ J/K}$

$\Delta G° = 2(87\text{ kJ}) - (0) = 174\text{ kJ}$

At the high temperatures in automobile engines, the reaction $N_2 + O_2 \rightarrow 2\ NO$ becomes spontaneous since the favorable $\Delta S°$ term will become dominate. In the atmosphere, even though $2\ NO \rightarrow N_2 + O_2$ is spontaneous at the cooler temperatures of the atmosphere, it doesn't occur because the rate is slow. Therefore, higher concentrations of NO are present in the atmosphere as compared to what is predicted by thermodynamics.

49. The pollution provides sources of nitrogen and phosphorus nutrients so that the algae can grow. The algae consume oxygen, which decrease the dissolved oxygen levels below that required for other aquatic life to survive, and fish die.

50. For a buffer solution: $pH = pK_a + \log\frac{[\text{base}]}{[\text{acid}]}$; $pH = -\log(6.2\times10^{-8}) + \log\frac{[HPO_4^{\ 2-}]}{[H_2PO_4^{\ -}]}$

From the problem, $\frac{[H_2PO_4^{\ -}]}{[HPO_4^{\ 2-}]} = 1.1$, so: $pH = 7.21 + \log\frac{1}{1.1}$, $pH = 7.21 - 0.041 = 7.17$

51. The acidic hydrogens in the oxyacids of phosphorus all are bonded to oxygen. The hydrogens bonded directly to phosphorus are not acidic. H_3PO_4 has three oxygen-bonded hydrogens, and it is a triprotic acid. H_3PO_3 has only two of the hydrogens bonded to oxygen, and it is a diprotic acid. The third oxyacid of phosphorus, H_3PO_2, has only one of the hydrogens bonded to an oxygen; it is a monoprotic acid.

52. TSP = Na_3PO_4; $PO_4^{\ 3-}$ is the conjugate base of the weak acid $HPO_4^{\ 2-}$ ($K_a = 4.8\times10^{-13}$). All conjugate bases of weak acids are effective bases ($K_b = K_w/K_a = 1.0\times10^{-14}/4.8\times10^{-13} = 2.1\times10^{-2}$). The weak base reaction of $PO_4^{\ 3-}$ with H_2O is $PO_4^{\ 3-} + H_2O \rightleftharpoons HPO_4^{\ 2-} + OH^-$ $K_b = 2.1\times10^{-2}$.

Group 6A Elements

53. O=O–O → O=O + O

Break O–O bond: $\Delta H = \frac{146\text{kJ}}{\text{mol}} \times \frac{1\text{ mol}}{6.022\times10^{23}} = 2.42\times10^{-22}\text{kJ} = 2.42\times10^{-19}\text{J}$

A photon of light must contain at least 2.42×10^{-19} J to break one O–O bond.

$$E_{photon} = \frac{hc}{\lambda},\ \lambda = \frac{(6.626\times10^{-34}\ \text{J s})(2.998\times10^{8}\ \text{m/s})}{2.42\times10^{-19}\ \text{J}} = 8.21\times10^{-7}\ \text{m} = 821\ \text{nm}$$

54. From Figure 7.2 in the text, light from violet to green will work.

55. $H_2SeO_4(aq) + 3\ SO_2(g) \rightarrow Se(s) + 3\ SO_3(g) + H_2O(l)$

56. a. $2\ SO_2(g) + O_2(g) \rightarrow 2\ SO_3(g)$

b. $SO_3(g) + H_2O(l) \rightarrow H_2SO_4(aq)$

c. $C_{12}H_{22}O_{11}(s) + 11\ H_2SO_4(conc) \rightarrow 12\ C(s) + 11\ H_2SO_4{\bullet}H_2O(l)$

57. In the upper atmosphere, O_3 acts as a filter for ultraviolet (UV) radiation:

$$O_3 \xrightarrow{hv} O_2 + O$$

O_3 is also a powerful oxidizing agent. It irritates the lungs and eyes, and at high concentration, it is toxic. The smell of a "spring thunderstorm" is O_3 formed during lightning discharges. Toxic materials don't necessarily smell bad. For example, HCN smells like almonds.

58. Chlorine is a good oxidizing agent. Similarly, ozone is a good oxidizing agent. After chlorine reacts, residues of chloro compounds are left behind. Long-term exposure to some chloro compounds may cause cancer. Ozone would not break down and form harmful substances. The major problem with ozone is that because virtually no ozone is left behind after initial treatment, the water supply is not protected against recontamination. In contrast, for chlorination, significant residual chlorine remains after treatment, thus reducing (eliminating) the risk of recontamination.

59. O_2: $(\sigma_{2s})^2(\sigma_{2s}{*})^2(\sigma_{2p})^2(\pi_{2p})^4(\pi_{2p}{*})^2$; the MO electron configuration of O_2 has two unpaired electrons in the degenerate π antibonding (π^{*}_{2p}) orbitals. A substance with unpaired electrons is paramagnetic (see Figure 9.39).

60. SO_2, $6 + 2(6) = 18\ e^-$ SO_3, $6 + 3(6) = 24\ e^-$

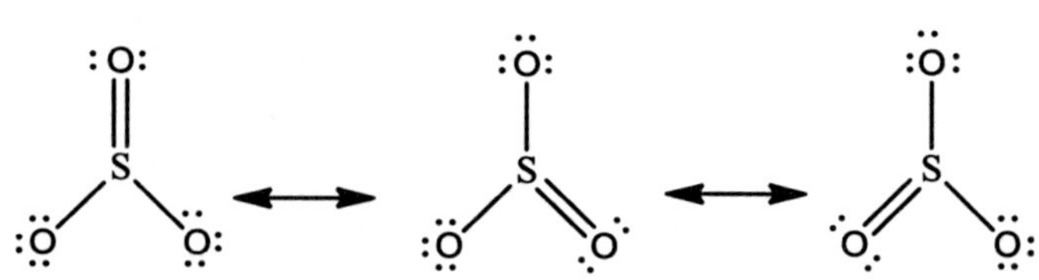

The molecular structure of SO_2 is bent with a 119° bond angle (close to the predicted 120° trigonal planar geometry). The molecular structure of SO_3 is trigonal planar with 120° bond angles. Both SO_2 and SO_3 exhibit resonance. Both sulfurs in SO_2 and SO_3 are sp^2 hybridized. To explain the equal bond lengths that occur in SO_2 and SO_3, the molecular orbital model assumes that the π electrons are delocalized over the entire surface of the molecule. The orbitals that form the delocalized π bonding system are unhybridized p atomic orbitals from the sulfurs and oxygens in each molecule. When all of the p atomic orbitals overlap together, there is a cloud of electron density above and below the entire surface of the molecule. Because the π electrons are delocalized over the entire surface of the molecule in SO_2 and SO_3, all of the S–O bonds in each molecule are equivalent.

Group 7A Elements

61. O_2F_2 has 2(6) + 2(7) = 26 valence e^-; from the following Lewis structure, each oxygen atom has a tetrahedral arrangement of electron pairs. Therefore, bond angles are ≈109.5° and each O is sp^3 hybridized.

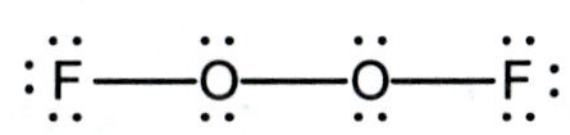

Formal charge	0	0	0	0
Oxidation state	−1	+1	+1	−1

Oxidation states are more useful. We are forced to assign +1 as the oxidation state for oxygen. Oxygen is very electronegative, and +1 is not a stable oxidation state for this element.

62. OF_2, 6 + 2(7) = 20 e^-

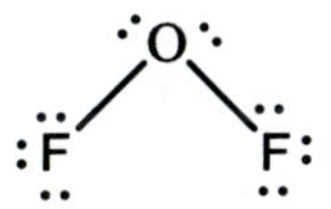

V-shaped; <109.5°; sp^3

Because fluorine is more electronegative than oxygen, each fluorine atom in OF_2 has a −1 oxidation number, which results in oxygen having a +2 oxidation number. Oxygen is very electronegative, so +2 is an extremely unstable oxidation state for this element. One would expect OF_2, with oxygen in the +2 oxidation state, to be an even stronger oxidizing agent than O_2F_2, which has oxygen in the +1 oxidation state.

63. SF_2, 6 + 2(7) = 20 e^-

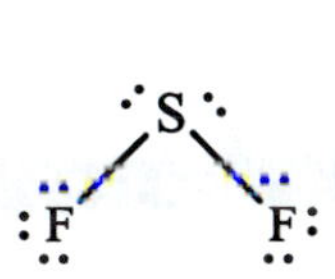

V-shaped; <109.5°

SF_4, 6 + 4(7) = 34 e^-

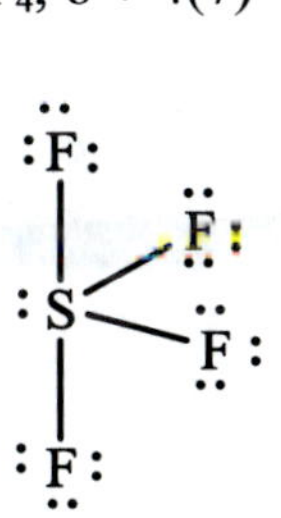

See-saw; ≈90, ≈120

SF_6, 6 + 6(7) = 48 e^-

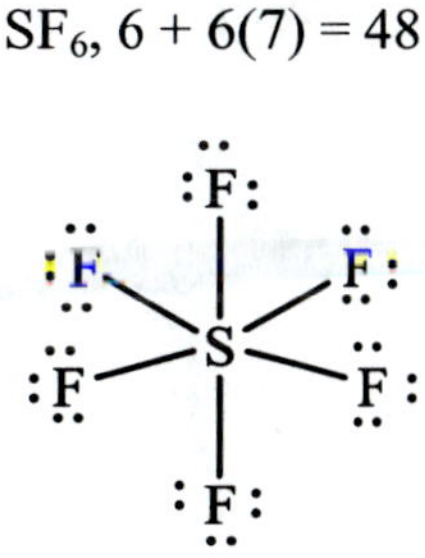

Octahedral; 90°

OF_4 would have the same Lewis structure as SF_4. In order to form OF_4, the central oxygen atom must expand its octet. O is too small and doesn't have low-energy d orbitals available to expand its octet. Therefore, OF_4 would not be a stable compound.

64. Selenium should form compounds similar to those that sulfur forms because both are group 6A nonmetals. Because sulfur forms covalent compounds with halogens having SX_2, SX_4, and SX_6 formulas, one would predict selenium and chlorine to form covalent compounds having the formulas $SeCl_2$, $SeCl_4$, and $SeCl_6$.

65. The oxyacid strength increases as the number of oxygens in the formula increase. Therefore, the order of the oxyacids from weakest to strongest acid is $HOCl < HClO_2 < HClO_3 < HClO_4$.

66. One reason is that the H–F bond is stronger than the other hydrohalides, making it more difficult to form H^+ and F^-. The main reason HF is a weak acid is entropy. When $F^-(aq)$ forms from the dissociation of HF, there is a high degree of ordering that takes place as water molecules hydrate this small ion. Entropy is considerably more unfavorable for the formation of hydrated F^- than for the formation of the other hydrated halides. The result of the more unfavorable $\Delta S°$ term is a positive $\Delta G°$ value that leads to a K_a value less than one.

Group 8A Elements

67. Xe has one more valence electron than I. Thus the isoelectric species will have I plus one extra electron substituted for Xe, giving a species with a net minus one charge.

a. IO_4^- b. IO_3^- c. IF_2^- d. IF_4^- e. IF_6^-

68. a. KrF_2, $8 + 2(7) = 22\ e^-$

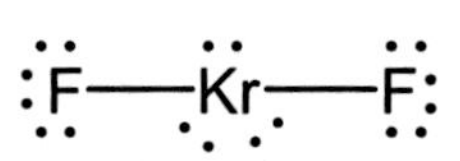

Linear; 180°; dsp^3

b. KrF_4, $8 + 4(7) = 36\ e^-$

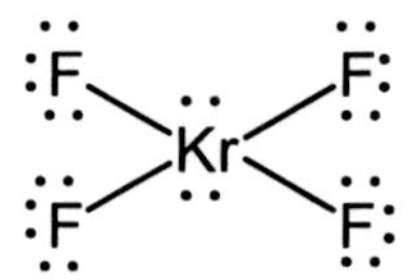

Square planar; 90°; d^2sp^3

c. XeO_2F_2, $8 + 2(6) + 2(7) = 34\ e^-$

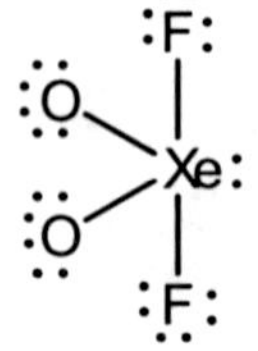

or

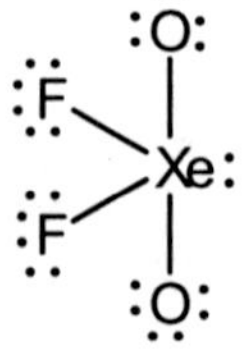

or

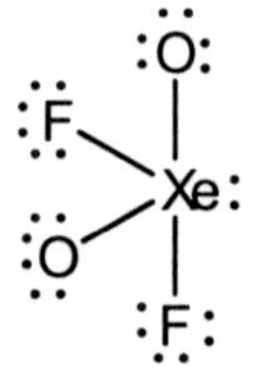

All are see-saw; $\approx 90°$ and $\approx 120°$; dsp^3

d. XeO_2F_4, 8 + 2(6) + 4(7) = 48 e^-

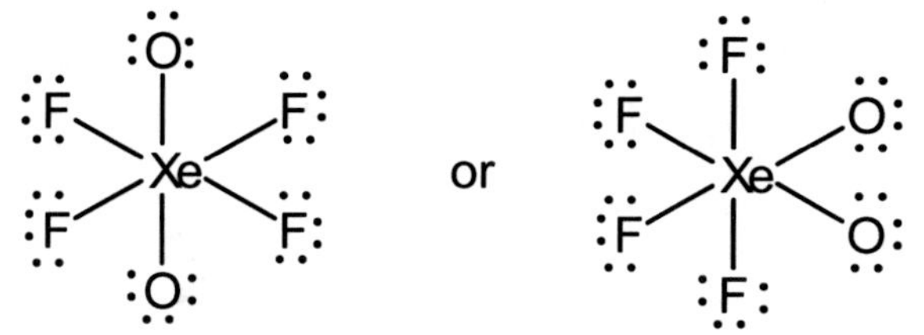

All are octahedral; 90°; d^2sp^3

69. Helium is unreactive and doesn't combine with any other elements. It is a very light gas and would easily escape the earth's gravitational pull as the planet was formed.

70. 10.0 m × 10.0 m × 10.0 m = 1.00×10^3 m^3; from Table 20.22, volume % Ar = 0.9%.

$$1.00 \times 10^3 \text{ m}^3 \times \left(\frac{10\text{ dm}}{\text{m}}\right)^3 \times \frac{1\text{ L}}{\text{dm}^3} \times \frac{0.9\text{ L Ar}}{100\text{ L air}} = 9 \times 10^3 \text{ L of Ar in the room}$$

$$PV = nRT,\ n = \frac{PV}{RT} = \frac{(1.0\text{ atm})(9 \times 10^3\text{ L})}{(0.08206\text{ L atm/K} \bullet \text{mol})(298\text{ K})} = 4 \times 10^2 \text{ mol Ar}$$

$$4 \times 10^2 \text{ mol Ar} \times \frac{39.95\text{ g}}{\text{mol}} = 2 \times 10^4 \text{ g Ar in the room}$$

$$4 \times 10^2 \text{ mol Ar} \times \frac{6.022 \times 10^{23}\text{ atoms}}{\text{mol}} = 2 \times 10^{26} \text{ atoms Ar in the room}$$

$$\text{A 2-L breath contains: } 2\text{ L air} \times \frac{0.9\text{ L Ar}}{100\text{ L air}} = 2 \times 10^{-2} \text{ L Ar}$$

$$n = \frac{PV}{RT} = \frac{(1.0\text{ atm})(2 \times 10^{-2}\text{ L})}{(0.08206\text{ L atm/K} \bullet \text{mol})(298\text{ K})} = 8 \times 10^{-4} \text{ mol Ar}$$

$$8 \times 10^{-4} \text{ mol Ar} \times \frac{6.022 \times 10^{23}\text{ atoms}}{\text{mol}} = 5 \times 10^{20} \text{ atoms of Ar in a 2-L breath}$$

Because Ar and Rn are both noble gases, both species will be relatively unreactive. However, all nuclei of Rn are radioactive, unlike most nuclei of Ar. The radioactive decay products of Rn can cause biological damage when inhaled.

71. One would expect RnF_2, RnF_4, and maybe RnF_6 to form in a fashion similar to XeF_2, XeF_4, and XeF_6. The chemistry of radon is difficult to study because radon isotopes are all radioactive. The hazards of dealing with radioactive materials are immense.

72. RnF_2, 8 + 2(7) = 22 e^- RnF_4, 8 + 4(7) = 36 e^- RnF_6, 8 + 6(7) = 50 e^-

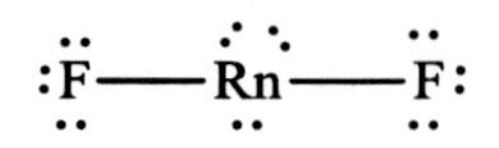

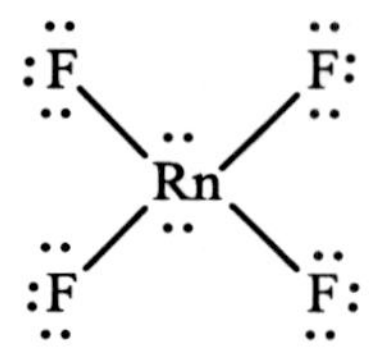

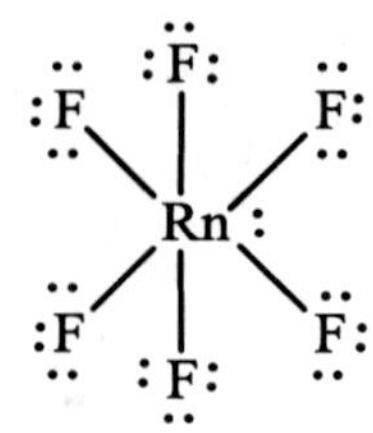

Linear; 180° Square planar; 90°

The structure for RnF_6 is difficult to predict. For six electron pairs about a central atom, the geometry is octahedral with 90° bond angles. RnF_6 has seven electron pairs about the central Rn atom, so the structure is not octahedral. We will call the molecular structure of RnF_6 a distorted octahedral structure with exact bond angles that are hard to predict.

Additional Exercises

73.

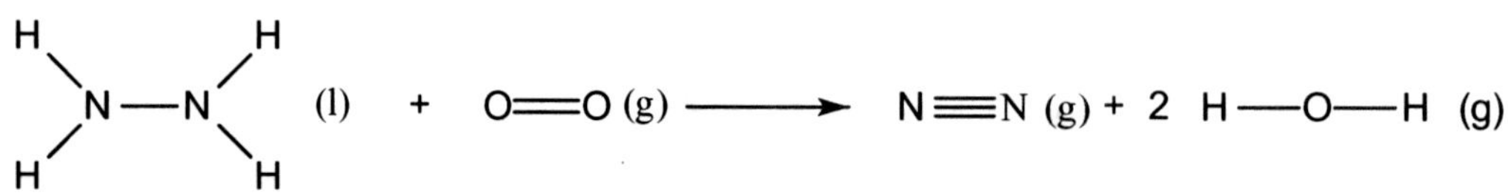

Bonds broken:

1 N–N (160. kJ/mol)

4 N–H (391 kJ/mol)

1 O=O (495 kJ/mol)

Bonds formed:

1 N≡N (941 kJ/mol)

2 × 2 O–H (467 kJ/mol)

ΔH = 160. + 4(391) + 495 – [941 + 4(467)] = 2219 kJ – 2809 kJ = –590. kJ

74. The *inert pair effect* refers to the difficulty of removing the pair of s electrons from some of the elements in the fifth and sixth periods of the periodic table. As a result, multiple oxidation states are exhibited for the heavier elements of Groups 3A and 4A. In^+, In^{3+}, Tl^+, and Tl^{3+} oxidation states are all important to the chemistry of In and Tl.

75. Ga(I): $[Ar]4s^23d^{10}$, no unpaired e^-; Ga(III): $[Ar]3d^{10}$, no unpaired e^-

Ga(II): $[Ar]4s^13d^{10}$, 1 unpaired e^-; note that the s electrons are lost before the d electrons.

If the compound contained Ga(II), it would be paramagnetic, and if the compound contained Ga(I) and Ga(III), it would be diamagnetic. This can be determined easily by measuring the mass of a sample in the presence and in the absence of a magnetic field. Paramagnetic compounds will have an apparent increase in mass in a magnetic field.

76. The π electrons are free to move in graphite, thus giving it greater conductivity (lower resistance). The electrons in graphite have the greatest mobility within sheets of carbon atoms, resulting in a lower resistance in the plane of the sheets (basal plane). Electrons in diamond are not mobile (high resistance). The structure of diamond is uniform in all directions; thus resistivity has no directional dependence in diamond.

77.

	$CaCO_3(s)$	$\rightleftharpoons$	$Ca^{2+}(aq)$	+	$CO_3^{2-}(aq)$
Initial	s = solubility (mol/L)		0		0
Equil.			s		s

$K_{sp} = 8.7 \times 10^{-9} = [Ca^{2+}][CO_3^{2-}] = s^2$, $s = 9.3 \times 10^{-5}$ mol/L

78.

	Pb^{2+}	+ H_2EDTA^{2-}	$\rightleftharpoons$	$PbEDTA^{2-}$	+ $2\,H^+$	
Before	0.0050 M	0.075 M		0	1.0×10^{-7} M	(Buffer, $[H^+]$ constant)
Change	−0.0050	−0.0050	→	+0.0050	No change	Reacts completely
After	0	0.070		0.0050	1.0×10^{-7}	New initial conditions
	x mol/L $PbEDTA^{2-}$ dissociates to reach equilibrium					
Change	$+x$	$+x$	←	$-x$		
Equil.	x	$0.070 + x$		$0.0050 - x$	1.0×10^{-7}	(Buffer)

$$K = 6.7 \times 10^{21} = \frac{[PbEDTA^{2-}][H^+]^2}{[Pb^{2+}][H_2EDTA^{2-}]} = \frac{(0.0050 - x)(1.0 \times 10^{-7})^2}{(x)(0.070 + x)}$$

$$6.7 \times 10^{21} \approx \frac{(0.0050)(1.0 \times 10^{-14})}{(x)(0.070)}, \quad x = [Pb^{2+}] = 1.1 \times 10^{-37}\,M;\ \text{assumptions good.}$$

79. a. $AgCl(s) \xrightarrow{hv} Ag(s) + Cl$; the reactive chlorine atom is trapped in the crystal. When light is removed, Cl reacts with silver atoms to re-form AgCl; i.e., the reverse reaction occurs. In pure AgCl, the Cl atoms escape, making the reverse reaction impossible.

b. Over time, chlorine is lost and the dark silver metal is permanent.

80. $AsCl_4^+$, $5 + 4(7) - 1 = 32\ e^-$ $AsCl_6^-$, $5 + 6(7) + 1 = 48\ e^-$

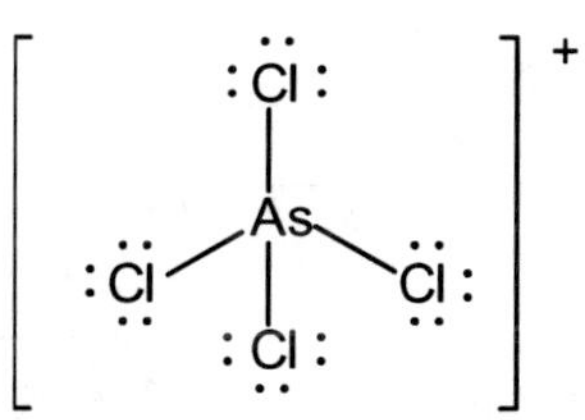

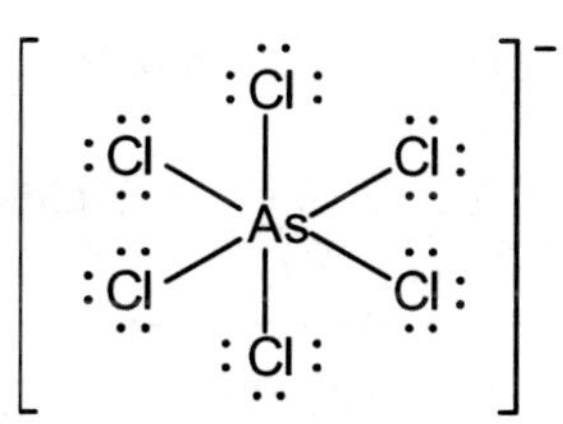

The reaction is a Lewis acid-base reaction. A chloride ion acts as a Lewis base when it is transferred from one $AsCl_5$ to another. Arsenic is the Lewis acid (electron pair acceptor).

81. Table 20.2 lists the mass percents of various elements in the human body. If we consider the mass percents through sulfur, that will cover 99.5% of the body mass, which is fine for a reasonable estimate. 150 lb × 454 g/lb = 68,000 g. We will carry an extra significant figure in some of the following calculations.

Moles of O = 0.650 × 68,000 g × 1 mol O/16.00 g O = 2760 mol

Moles of C = 0.180 × 68,000 g × 1 mol C/12.01 g C = 1020 mol

Moles of H = 0.100 × 68,000 g × 1 mol H/1.008 g H = 6750 mol

Moles of N = 0.030 × 68,000 g × 1 mol N/14.01 g N = 150 mol

Moles of Ca = 0.014 × 68,000 g × 1 mol Ca/40.08 g Ca = 24 mol

Moles of P = 0.010 × 68,000 g × 1 mol P/30.97 g P = 22 mol

Moles of Mg = 0.0050 × 68,000 g × 1 mol Mg/24.31 g Mg = 14 mol

Moles of K = 0.0034 × 68,000 g × 1 mol K/39.10 g K = 5.9 mol

Moles of S = 0.0026 × 68,000 g × 1 mol S/32.07 g S = 5.5 mol

Total moles of elements in 150-lb body = 10,750 mol atoms

$$10{,}750 \text{ mol atoms} \times \frac{6.022 \times 10^{23} \text{atoms}}{\text{mol atoms}} = 6.474 \times 10^{27} \text{ atoms} \approx 6.5 \times 10^{27} \text{ atoms}$$

82. There are medical studies that have shown an inverse relationship between the incidence of cancer and the selenium levels in soil. The foods grown in these soils and eventually digested are assumed to somehow furnish protection from cancer. Selenium is also involved in the activity of vitamin E and certain enzymes in the human body. In addition, selenium deficiency has been shown to be connected to the occurrence of congestive heart failure.

83. Strontium and calcium are both alkaline earth metals, so both have similar chemical properties. Because milk is a good source of calcium, strontium could replace some calcium in milk without much difficulty.

84. A disproportion reaction is an oxidation-reduction reaction in which one species will act as both the oxidizing agent and the reducing agent. The species reacts with itself, forming products with higher and lower oxidation states. For example, $2\ Cu^+ \rightarrow Cu + Cu^{2+}$ is a disproportion reaction.

$HClO_2$ will disproportionate at standard conditions because $E^\circ_{cell} > 0$:

$$HClO_2 + 2\ H^+ + 2\ e^- \rightarrow HClO + H_2O \qquad E^\circ = 1.65\ V$$
$$HClO_2 + H_2O \rightarrow ClO_3^- + 3\ H^+ + 2\ e^- \qquad -E^\circ = -1.21\ V$$

$$2\ HClO_2(aq) \rightarrow HClO(aq) + ClO_3^-(aq) + H^+(aq) \qquad E^\circ_{cell} = 0.44\ V$$

85. +6 oxidation state: SO_4^{2-}, SO_3, SF_6

+4 oxidation state: SO_3^{2-}, SO_2, SF_4

+2 oxidation state: SCl_2

0 oxidation state: S_8 and all other elemental forms of sulfur

−2 oxidation state: H_2S, Na_2S

86. ClF, $7 + 7 = 14\ e^-$

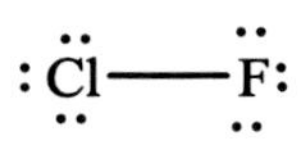

Linear; no bond angle present

ClF_3, $7 + 3(7) = 28\ e^-$

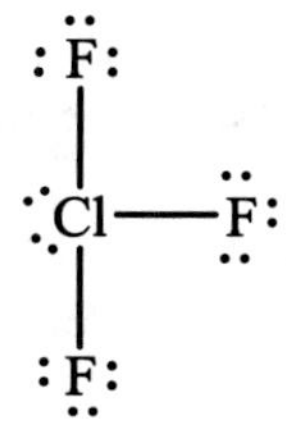

T-shaped; ≈90°

ClF_5, $7 + 5(7) = 42\ e^-$

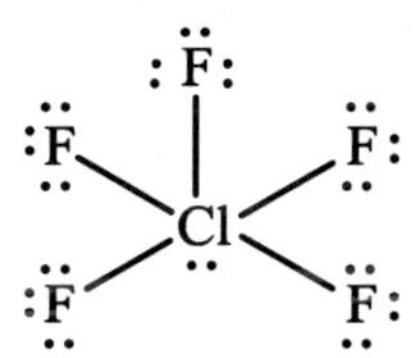

Square pyramid; ≈90°

In order to form FCl_3, F would have to expand its octet of electrons. Fluorine is too small and doesn't have low-energy d orbitals available to expand its octet. Therefore, FCl_3 would not be a stable compound.

ChemWork Problems

The answers to the problems 87-96 (or a variation to these problems) are found in OWL. These problems are also assignable in OWL.

Challenge Problems

97. The reaction is $X(s) + 2H_2O(l) \rightarrow H_2(g) + X(OH)_2(aq)$.

$$\text{Mol X} = \text{mol } H_2 = \frac{PV}{RT} = \frac{1.00\ \text{atm} \times 6.10\ \text{L}}{\frac{0.08206\ \text{L atm}}{\text{K mol}} \times 298\ \text{K}} = 0.249\ \text{mol}$$

$$\text{Molar mass X} = \frac{10.00\ \text{g X}}{0.249\ \text{mol X}} = 40.2\ \text{g/mol};\ \text{X is Ca.}$$

$Ca(s) + 2\ H_2O(l) \rightarrow H_2(g) + Ca(OH)_2(aq)$; $Ca(OH)_2$ is a strong base.

$$[OH^-] = \frac{10.00\ \text{g Ca} \times \frac{1\ \text{mol Ca}}{40.08\ \text{g}} \times \frac{1\ \text{mol Ca(OH)}_2}{\text{mol Ca}} \times \frac{2\ \text{mol OH}^-}{\text{mol Ca(OH)}_2}}{10.0\ \text{L}} = 0.0499\ M$$

pOH = –log(0.0499) = 1.302, pH = 14.000 – 1.302 = 12.698

98. White tin is stable at normal temperatures. Gray tin is stable at temperatures below 13.2°C. Thus, for the phase change Sn(gray) → Sn(white), ΔG is (−) at T > 13.2°C, and ΔG is (+) at T < 13.2°C. This is only possible if ΔH is (+) and ΔS is (+). Thus, gray tin has the more ordered structure (has the smaller positional probability).

99. $PbX_4 \rightarrow PbX_2 + X_2$; from the equation, mol PbX_4 = mol PbX_2. Let x = molar mass of the halogen. Setting up an equation where mol PbX_4 = mol PbX_2:

$$\frac{25.00\text{ g}}{207.2 + 4x} = \frac{16.12\text{ g}}{207.2 + 2x};\quad \text{solving, } x = 127.1;\ \text{the halogen is iodine, I.}$$

100. In order to form a π bond, the d and p orbitals must overlap "side to side" instead of "head to head" as in sigma bonds. A representation of the side-to-side overlap follows. For a bonding orbital to form, the phases of the lobes must match (positive to positive and negative to negative).

d + p

101. For the reaction:

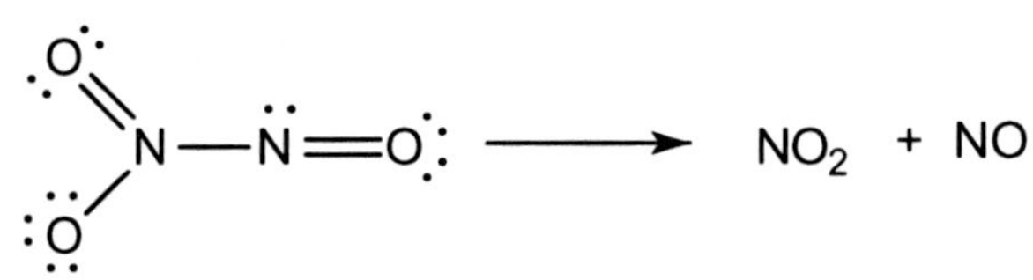

the activation energy must in some way involve breaking a nitrogen-nitrogen single bond.

For the reaction:

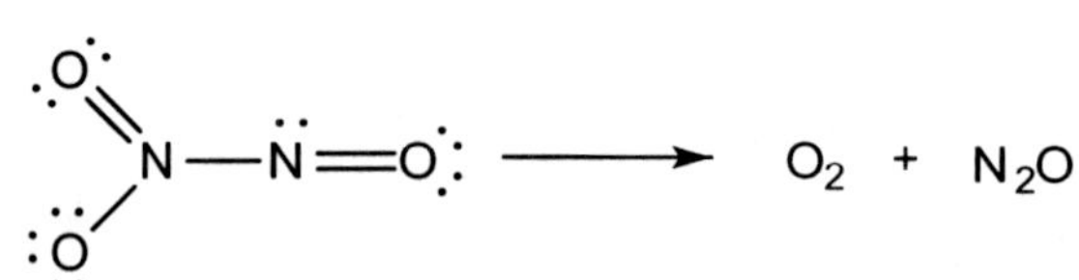

at some point nitrogen-oxygen bonds must be broken. N–N single bonds (160. kJ/mol) are weaker than N–O single bonds (201 kJ/mol). In addition, resonance structures indicate that there is more double-bond character in the N–O bonds than in the N–N bond. Thus NO_2 and NO are preferred by kinetics because of the lower activation energy.

102. a. The sum of the two steps gives the overall balanced equation.

$$O_3(g) + NO(g) \rightarrow NO_2(g) + O_2(g)$$
$$NO_2(g) + O(g) \rightarrow NO(g) + O_2(g)$$
$$O_3(g) + O(g) \rightarrow 2\,O_2(g) \quad \text{overall equation}$$

b. NO is the catalyst. NO is present in the first step of the mechanism on the reactant side, but it is not a reactant because it is regenerated in the second step and does not appear in the overall balanced equation.

c. NO_2 is an intermediate. It is produced in the first step, but is consumed in the second step. Intermediates also never appear in the overall balanced equation. In a mechanism, intermediates always appear first on the product side, while catalysts always appear first on the reactant side.

d. The rate of the slow step in a mechanism gives the rate law for the reaction. From the problem, the rate determining step (the slow step) is step 1. The derived rate law is:

$$\text{Rate} = k[O_3][NO]$$

Because NO is a catalyst and not a proposed intermediate, it can appear in the rate law.

e. The mechanism for the chlorine-catalyzed destruction of ozone is:

$$O_3(g) + Cl(g) \rightarrow O_2(g) + ClO(g) \quad \text{slow step 1}$$
$$ClO(g) + O(g) \rightarrow O_2(g) + Cl(g) \quad \text{fast step 2}$$
$$O_3(g) + O(g) \rightarrow 2\,O_2(g) \quad \text{overall equation}$$

103. $NH_3 + NH_3 \rightleftharpoons NH_4^+ + NH_2^- \qquad K = [NH_4^+][NH_2^-] = 1.8 \times 10^{-12}$

NH_3 is the solvent, so it is not included in the K expression. In a neutral solution of ammonia:

$$[NH_4^+] = [NH_2^-];\ \ 1.8 \times 10^{-12} = [NH_4^+]^2,\ \ [NH_4^+] = 1.3 \times 10^{-6}\,M = [NH_2^-]$$

We could abbreviate this autoionization as: $NH_3 \rightleftharpoons H^+ + NH_2^-$, where $[H^+] = [NH_4^+]$.

This abbreviation is synonymous with the abbreviation used for the autoionization of water ($H_2O \rightleftharpoons H^+ + OH^-$). So $pH = pNH_4^+ = -\log(1.3 \times 10^{-6}) = 5.89$.

104. The representation indicates that we have an equimolar mixture of $N_2(g)$ and $H_2(g)$ (6 molecules of each are shown). To solve the problem, let's assume a reaction between 3.00x moles of N_2 and 3.00x moles of H_2 (equimolar). The reaction going to completion is summarized in the following table. Note that H_2 is limiting.

	$N_2(g)$	+	$3\ H_2(g)$	$\rightarrow$	$2\ NH_3(g)$
Before	$3.00x$ mol		$3.00x$ mol		0
Change	$-1.00x$ mol		$-3.00x$ mol	$\rightarrow$	$+2.00x$ mol
After	$2.00x$ mol		0		$2.00x$ mol

When an equimolar mixture is reacted, the total moles of gas present decreases from $6.00x$ moles initially to $4.00x$ moles after completion.

a. The total pressure in the piston apparatus is a constant 1.00 atm. After the reaction, we have $2.00x$ moles of N_2 and $2.00x$ moles of NH_3. One-half of the moles of gas present are NH_3 molecules, so one-half of the total pressure is due to the NH_3 molecules. $P_{NH_3} =$ 0.500 atm.

b. $\chi_{NH_3} = \dfrac{\text{moles NH}_3}{\text{total moles}} = \dfrac{2.00x \text{ mol}}{(2.00x + 2.00x)\text{ mol}} = 0.500$

c. At constant P and T, volume is directly proportional to n, the moles of gas present. Because n decreased from $6.00x$ to $4.00x$ moles, the volume will decrease by the same factor.

$V_{final} = 15.0\text{ L }(4/6) = 10.0\text{ L}$

105. Let n_{SO_2} = initial moles SO_2 present. The reaction is summarized in the following table (O_2 is in excess).

	$2\ SO_2$	+	$O_2(g)$	$\rightarrow$	$2\ SO_3(g)$
Initial	n_{SO_2}		2.00 mol		0
Change	$-n_{SO_2}$		$-n_{SO_2}/2$		$+n_{SO_2}$
Final	0		$2.00 - n_{SO_2}/2$		n_{SO_2}

Density = d = mass/volume; let d_i = initial density of gas mixture and d_f = final density of gas mixture after reaction. Because mass is conserved in a chemical reaction, $mass_i = mass_f$.

$$\frac{d_f}{d_i} = \frac{mass_f/V_f}{mass_i/V_i} = \frac{V_i}{V_f}$$

At constant P and T, $V \propto n$, so $\dfrac{d_f}{d_i} = \dfrac{V_i}{V_f} = \dfrac{n_i}{n_f}$; setting up an equation:

$$\frac{d_f}{d_i} = \frac{0.8471\text{ g/L}}{0.8000\text{ g/L}} = 1.059,\ \ 1.059 = \frac{n_i}{n_f} = \frac{n_{SO_2} + 2.00}{(2.00 - n_{SO_2}/2) + n_{SO_2}} = \frac{n_{SO_2} + 2.00}{2.00 + n_{SO_2}/2}$$

Solving: $n_{SO_2} = 0.25$ mol; so, 0.25 moles of SO_3 is formed.

$$0.25\text{ mol } SO_3 \times \frac{80.07\text{ g } SO_3}{\text{mol}} = 20.\text{ g } SO_3$$

Integrative Exercises

106. 1.75×10^8 g pitchblende $\times \dfrac{1 \text{ metric ton}}{1.0 \times 10^6 \text{ g}} \times \dfrac{1.0 \text{ g Ra}}{7.0 \text{ metric tons}} \times \dfrac{1 \text{ mol Ra}}{226 \text{ g Ra}}$

$$\times \frac{6.022 \times 10^{23} \text{ atoms Ra}}{\text{mol Ra}} = 6.7 \times 10^{22} \text{ atoms Ra}$$

Radioactive decay follows first-order kinetics.

$$\ln\left(\frac{N}{N_0}\right) = -kt = \frac{-(\ln 2)t}{t_{1/2}}; \ \ln\left(\frac{N}{15.0 \text{ mg}}\right) = \frac{-0.693(100. \text{ yr})}{1.60 \times 10^3 \text{ yr}}, \ N = 14.4 \text{ mg Ra}$$

$$14.4 \times 10^{-3} \text{ g Ra} \times \frac{1 \text{ mol Ra}}{226 \text{ g Ra}} \times \frac{6.022 \times 10^{23} \text{ atoms Ra}}{\text{mol Ra}} = 3.84 \times 10^{19} \text{ atoms Ra}$$

107. a. Moles of $In(CH_3)_3 = \dfrac{PV}{RT} = \dfrac{2.00 \text{ atm} \times 2.56 \text{ L}}{0.08206 \text{ L atm/K} \bullet \text{mol} \times 900. \text{K}} = 0.0693$ mol

Moles of $PH_3 = \dfrac{PV}{RT} = \dfrac{3.00 \text{ atm} \times 1.38 \text{ L}}{0.08206 \text{ L atm/K} \bullet \text{mol} \times 900. \text{K}} = 0.0561$ mol

Because the reaction requires a 1 : 1 mole ratio between these reactants, the reactant with the small number of moles (PH_3) is limiting.

$$0.0561 \text{ mol } PH_3 \times \frac{1 \text{ mol InP}}{\text{mol } PH_3} \times \frac{145.8 \text{ g InP}}{\text{mol InP}} = 8.18 \text{ g InP}$$

The actual yield of InP is: $0.87 \times 8.18 \text{ g} = 7.1$ g InP

b. $\lambda = \dfrac{hc}{E} = \dfrac{6.626 \times 10^{-34} \text{ J s} \times 2.998 \times 10^8 \text{ m/s}}{2.03 \times 10^{-19} \text{ J}} = 9.79 \times 10^{-7} \text{ m} = 979$ nm

From the Figure 7.2 of the text, visible light has wavelengths between 4×10^{-7} and 7×10^{-7} m. Therefore, this wavelength is not visible to humans; it is in the infrared region of the electromagnetic radiation spectrum.

c. $[Kr]5s^2 4d^{10} 5p^4$ is the electron configuration for tellurium, Te. Because Te has more valence electrons than P, this would form an n-type semiconductor (n-type doping).

108. a. $-307 \text{ kJ} = (-1136 + x) - [(-254 \text{ kJ}) + 3(-96 \text{ kJ})], \ x = \Delta H^{\circ}_{f, NI_3} = 287$ kJ/mol

b. IF_2^+, $7 + 2(7) - 1 = 20\ e^-$

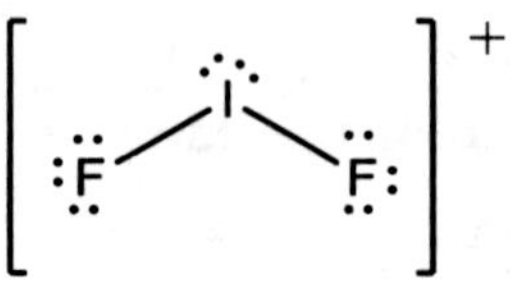

V-shaped; sp^3

BF_4^-, $3 + 4(7) + 1 = 32\ e^-$

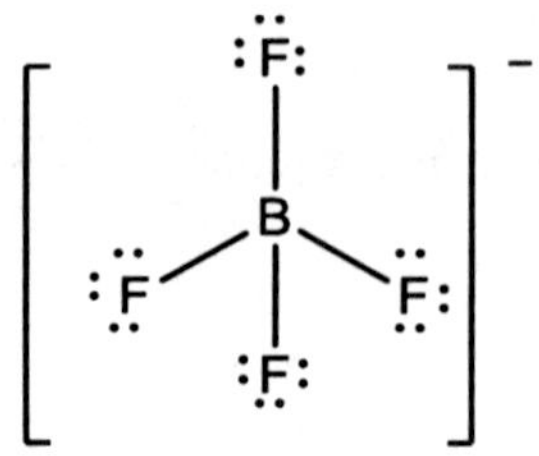

Tetrahedral; sp^3

109. a. Because the hydroxide ion has a 1– charge, Te has a +6 oxidation state.

b. Assuming Te is limiting:

$$(0.545\ \text{cm})^3 \times \frac{6.240\ \text{g}}{\text{cm}^3} \times \frac{1\ \text{mol Te}}{127.6} \times \frac{1\ \text{mol TeF}_6}{\text{mol Te}} = 7.92 \times 10^{-3}\ \text{mol TeF}_6$$

Assuming F_2 is limiting:

$$\text{Mol F}_2 = n = \frac{PV}{RT} = \frac{1.06\ \text{atm} \times 2.34\ \text{L}}{0.08206\ \text{L atm/K} \bullet \text{mol} \times 298\ \text{K}} = 0.101\ \text{mol F}_2$$

$$0.101\ \text{mol F}_2 \times \frac{1\ \text{mol TeF}_6}{3\ \text{mol F}_2} = 3.37 \times 10^{-2}\ \text{mol TeF}_6$$

Because Te produces the smaller amount of product, Te is limiting and 7.92×10^{-3} mol TeF_6 can be produced. From the first equation given in the question, the moles of TeF_6 reacted equals the moles of $Te(OH)_6$ produced. So 7.92×10^{-3} mol $Te(OH)_6$ can be produced.

$$[\text{Te(OH)}_6] = \frac{7.92 \times 10^{-3}\ \text{mol Te(OH)}_6}{0.115\ \text{L}} = 6.89 \times 10^{-2}\ M$$

Because $K_{a_1} > K_{a_2}$, the amount of protons produced by the K_{a_2} reaction will be insignificant.

$$\text{Te(OH)}_6 \rightleftharpoons \text{Te(OH)}_5\text{O}^- + \text{H}^+ \qquad K_{a_1} = 10^{-7.68} = 2.1 \times 10^{-8}$$

	$Te(OH)_6$	$Te(OH)_5O^-$	H^+
Initial	0.0689 M	0	~0
Equil.	$0.0689 - x$	x	x

$$K_{a_1} = 2.1 \times 10^{-8} = \frac{x^2}{0.0689 - x} \approx \frac{x^2}{0.0689},\quad x = [\text{H}^+] = 3.8 \times 10^{-5}\ M$$

$\text{pH} = -\log(3.8 \times 10^{-5}) = 4.42$; assumptions good.

Marathon Problems

110. The answers to the clues are:

(1) HI has the second highest boiling point; (2) HF is the weak hydrogen halide acid; (3) He was first discovered from the sun's emission spectrum; (4) of the elements in Table 20.13, Bi will have the most metallic character (metallic character increases down a group); (5) Te is a semiconductor; (6) S has both rhombic and monoclinic solid forms; (7) Cl_2 is a yellow-green gas; (8) O is the most abundant element in and near the earth's crust; (9) Se appears to furnish some form of protection against cancer; (10) Kr is the smallest noble gas that forms compounds such as KrF_2 and KrF_4 (the symbol in reverse order is rk); (11) As forms As_4 molecules; (12) N_2 is a major inert component of air, and N is often found in fertilizers and explosives.
Filling in the blank spaces with the answers to the clues, the message is "If he bites, close ranks."

111. The answer to the clues are:

(1) BeO is amphoteric; (2) From Table 20.2, N makes up about 3.0% of the human body; (3) Fr has the $7s^1$ valence electron configuration; (4) Na has the least negative E° value (the symbol in reverse is an); (5) in intracellular fluids, K^+ is the more concentrated alkali metal ion; (6) only Li forms Li_3N; (7) In is the first Group 3A element to form stable +1 and +3 ions in its compounds (the second letter of the symbol is n).

Inserting the symbols into the blanks gives Ben Franklin for the name of the American scientist.

CHAPTER 21

TRANSITION METALS AND COORDINATION CHEMISTRY

Questions

5. The lanthanide elements are located just before the 5d transition metals. The lanthanide contraction is the steady decrease in the atomic radii of the lanthanide elements when going from left to right across the periodic table. As a result of the lanthanide contraction, the sizes of the 4d and 5d elements are very similar. This leads to a greater similarity in the chemistry of the 4d and 5d elements in a given vertical group.

6. Only the Cr^{3+} ion can form four different compounds with H_2O ligands and Cl^- ions. The Cr^{2+} ion could form only three different compounds, while the Cr^{4+} ion could form five different compounds.

   The Cl^- ions that form precipitates with Ag^+ are the counterions, not the ligands in the complex ion. The four compounds and mol AgCl precipitate that would form with 1 mol of compound are:

Compound	Mol AgCl(s)
$[Cr(H_2O)_6]Cl_3$	3 mol
$[Cr(H_2O)_5Cl]Cl_2$	2 mol
$[Cr(H_2O)_4Cl_2]Cl$	1 mol
$[Cr(H_2O)_3Cl_3]$	0 mol

7.

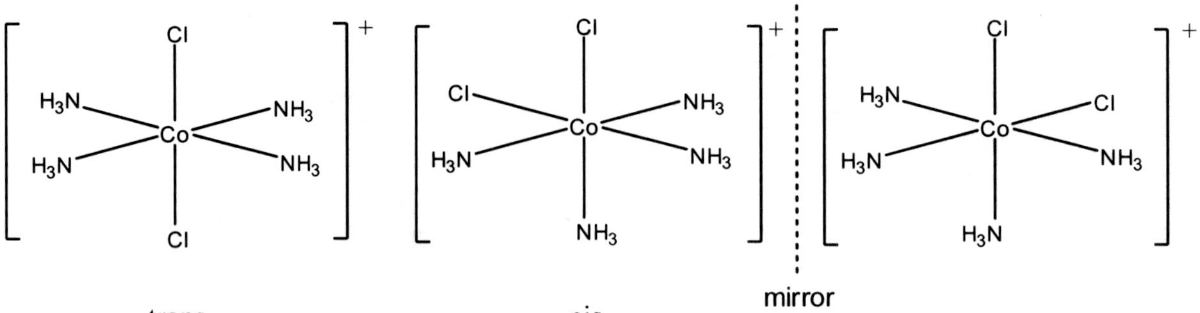

trans (mirror image is superimposable) — cis — The mirror image of the cis isomer is also superimposable.

No; both the trans and the cis forms of $Co(NH_3)_4Cl_2^+$ have mirror images that are superimposable. For the cis form, the mirror image only needs a 90° rotation to produce the original structure. Hence neither the trans nor cis form is optically active.

8. The transition metal ion must form octahedral complex ions; only with the octahedral geometry are two different arrangements of d electrons possible in the split d orbitals. These two arrangements depend on whether a weak field or strong field is present. For four unpaired electrons, the two possible weak field cases are for transition metal ions with $3d^4$ or $3d^6$ electron configurations:

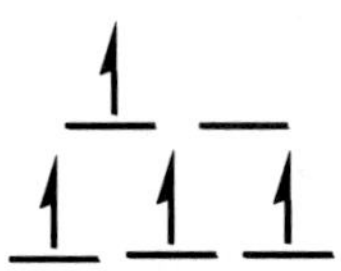
small Δ

d^4

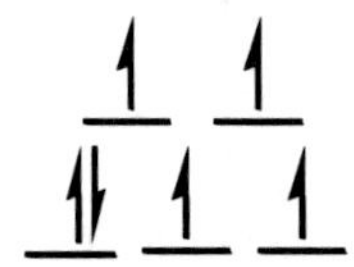
small Δ

d^6

Of these two, only d^6 ions have no unpaired electron in the strong field case.

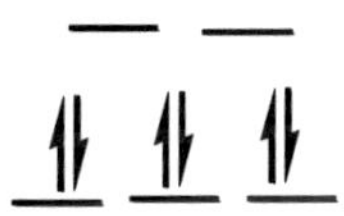
large Δ

Therefore, the transition metal ion has a $3d^6$ arrangement of electrons. Two possible metal ions that are $3d^6$ are Fe^{2+} and Co^{3+}. Thus one of these ions is present in the four coordination compounds, and each of these complex ions has a coordination number of 6.

The colors of the compounds are related to the magnitude of Δ (the d-orbital splitting value). The weak-field compounds will have the smallest Δ, so the λ of light absorbed will be longest. Using Table 21.16, the green solution (absorbs 650-nm light) and the blue solution (absorbs 600-nm light) absorb the longest-wavelength light; these solutions contain the complex ions that are the weak-field cases with four unpaired electrons. The red solution (absorbs 490-nm light) and yellow solution (absorbs 450-nm light) contain the two strong-field case complex ions because they absorb the shortest-wavelength (highest-energy) light. These complex ions are diamagnetic.

9. $Fe_2O_3(s) + 6\ H_2C_2O_4(aq) \rightarrow 2\ Fe(C_2O_4)_3^{3-}(aq) + 3\ H_2O(l) + 6\ H^+(aq)$; the oxalate anion forms a soluble complex ion with iron in rust (Fe_2O_3), which allows rust stains to be removed.

10. a.

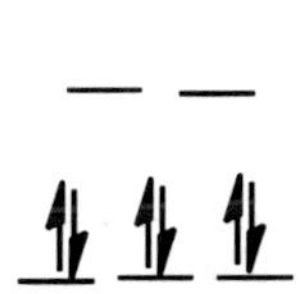
Low Spin, large Δ

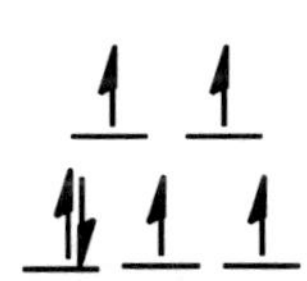
High Spin, small Δ

A d^6 octahedral crystal field diagram can either be low spin (0 upaired electrons) or high spin (4 unpaired electrons). The diagram in the question is for the low spin d^6 crystal field.

b.

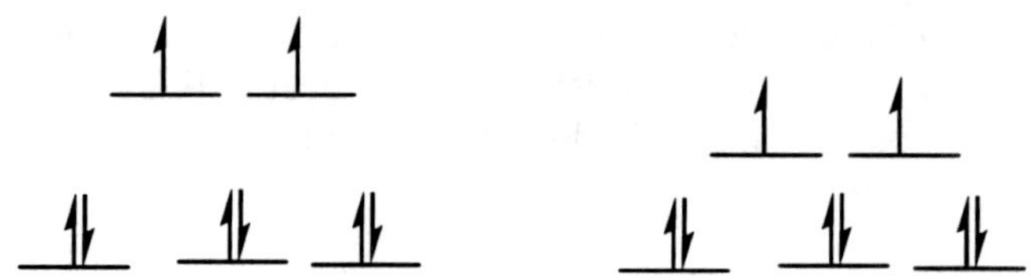

Low Spin, large Δ High Spin, small Δ

There is only one possible crystal field diagram for a d^8 species. Hence, one cannot tell from the diagram whether it is low spin or high spin; both have 2 unpaired electrons in the e_g orbitals.

c.

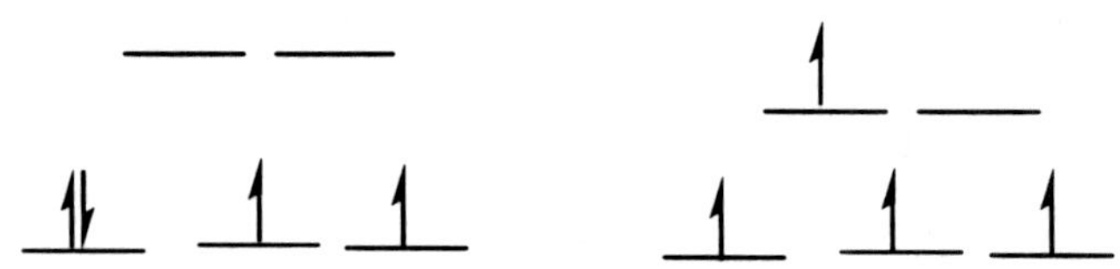

Low Spin, large Δ High Spin, small Δ

There are two possible octahedral crystal field diagrams for a d^4 species. Low spin has 2 unpaired electrons and high spin has 4 unpaired electrons. The diagram in the question is for a high spin d^4 crystal field.

11. a. $CoCl_4^{2-}$; Co^{2+}: $4s^0 3d^7$; all tetrahedral complexes are a weak field (high spin).

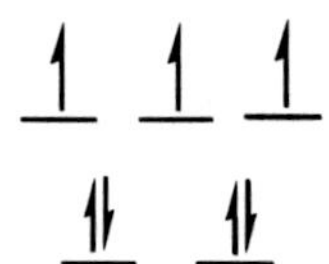

small Δ

$CoCl_4^{2-}$ is an example of a weak-field case having three unpaired electrons.

b. $Co(CN)_6^{3-}$: Co^{3+} : $4s^0 3d^6$; because CN^- is a strong-field ligand, $Co(CN)_6^{3-}$ will be a strong-field case (low-spin case).

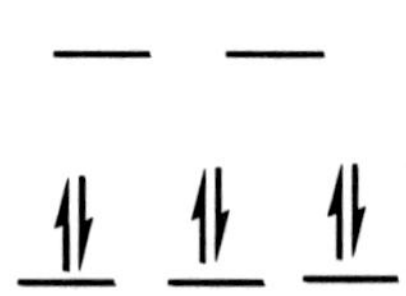

large Δ

CN^- is a strong-field ligand, so $Co(CN)_6^{3-}$ will be a low-spin case having zero unpaired electrons.

12. a. The coordination compound has the formula $[Co(H_2O)_6]Cl_2$. The complex ion is $Co(H_2O)_6^{2+}$, and the counterions are the Cl^- ions. The geometry would be octahedral, and the electron configuration of Co^{2+} is $[Ar]3d^7$.

b. The coordination compound is $Na_3[Ag(S_2O_3)_2]$.The compound consists of Na^+ counter-ions and the $Ag(S_2O_3)_2^{3-}$ complex ion. The complex ion is linear, and the electron configuration of Ag^+ is $[Kr]4d^{10}$.

c. The reactant coordination compound is $[Cu(NH_3)_4]Cl_2$. The complex ion is $Cu(NH_3)_4^{2+}$, and the counterions are Cl^- ions. The complex ion is tetrahedral (given in the question), and the electron configuration of Cu^{2+} is $[Ar]3d^9$. The product coordination compound is $[Cu(NH_3)_4]Cl$. The complex ion is $Cu(NH_3)_4^+$ with Cl^- counter ions. The complex ion is tetrahedral, and the electron configuration of Cu^+ is $[Ar]3d^{10}$.

13. From Table 21.16, the red octahedral $Co(H_2O)_6^{2+}$ complex ion absorbs blue-green light ($\lambda \approx$ 490 nm), whereas the blue tetrahedral $CoCl_4^{2-}$ complex ion absorbs orange light ($\lambda \approx 600$ nm). Because tetrahedral complexes have a d-orbital splitting much less than octahedral complexes, one would expect the tetrahedral complex to have a smaller energy difference between split d orbitals. This translates into longer-wavelength light absorbed ($E = hc/\lambda$) for tetrahedral complex ions compared to octahedral complex ions. Information from Table 21.16 confirms this.

14. Co^{2+}: $[Ar]3d^7$; the corresponding d-orbital splitting diagram for tetrahedral Co^{2+} complexes is:

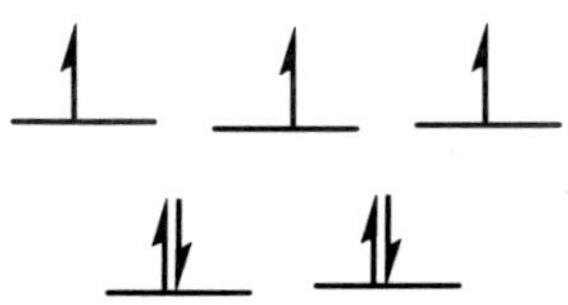

All tetrahedral complexes are high spin since the d-orbital splitting is small. Ions with two or seven d electrons should give the most stable tetrahedral complexes because they have the greatest number of electrons in the lower-energy orbitals as compared with the number of electrons in the higher-energy orbitals.

15. Linkage isomers differ in the way that the ligand bonds to the metal. SCN^- can bond through the sulfur or through the nitrogen atom. NO_2^- can bond through the nitrogen or through the oxygen atom. OCN^- can bond through the oxygen or through the nitrogen atom. N_3^-, $NH_2CH_2CH_2NH_2$, and I^- are not capable of linkage isomerism.

16. Cu^{2+}: $[Ar]3d^9$; Cu^+: $[Ar]3d^{10}$; Cu(II) is d^9 and Cu(I) is d^{10}. Color is a result of the electron transfer between split d orbitals. This cannot occur for the filled d orbitals in Cu(I). Cd^{2+}, like Cu^+, is also d^{10}. We would not expect $Cd(NH_3)_4Cl_2$ to be colored since the d orbitals are filled in this Cd^{2+} complex.

17. Sc^{3+} has no electrons in d orbitals. Ti^{3+} and V^{3+} have d electrons present. The color of transition metal complexes results from electron transfer between split d orbitals. If no d electrons are present, no electron transfer can occur, and the compounds are not colored.

18. Metals are easily oxidized by oxygen and other substances to form the metal cations. Because of this, metals are found in nature combined with nonmetals such as oxygen, sulfur, and the halogens. These compounds are called ores. To recover and use the metals, we must separate them from their ores and reduce the metal ions. Then, because most metals are unsuitable for use in the pure state, we must form alloys with the metals in order to form materials having desirable properties.

19. At high altitudes, the oxygen content of air is lower, so less oxyhemoglobin is formed, which diminishes the transport of oxygen in the blood. A serious illness called high-altitude sickness can result from the decrease of O_2 in the blood. High-altitude acclimatization is the phenomenon that occurs with time in the human body in response to the lower amounts of oxyhemoglobin in the blood. This response is to produce more hemoglobin and hence, increase the oxyhemoglobin in the blood. High-altitude acclimatization takes several weeks to take hold for people moving from lower altitudes to higher altitudes.

20. CN^- and CO form much stronger complexes with Fe^{2+} than O_2. Thus O_2 cannot be transported by hemoglobin in the presence of CN^- or CO because the binding sites prefer the toxic CN^- and CO ligands.

Exercises

Transition Metals and Coordination Compounds

21. a. Ni: $[Ar]4s^23d^8$ b. Cd: $[Kr]5s^24d^{10}$

c. Zr: $[Kr]5s^24d^2$ d. Os: $[Xe]6s^24f^{14}5d^6$

22. Transition metal ions lose the s electrons before the d electrons.

a. Ni^{2+}: $[Ar]3d^8$ b. Cd^{2+}: $[Kr]4d^{10}$

c. Zr^{3+}: $[Kr]4d^1$; Zr^{4+}: [Kr] d. Os^{2+}: $[Xe]4f^{14}5d^6$; Os^{3+}: $[Xe]4f^{14}5d^5$

23. Transition metal ions lose the s electrons before the d electrons.

a. Ti: $[Ar]4s^23d^2$ b. Re: $[Xe]6s^24f^{14}5d^5$ c. Ir: $[Xe]6s^24f^{14}5d^7$

Ti^{2+}: $[Ar]3d^2$ Re^{2+}: $[Xe]4f^{14}5d^5$ Ir^{2+}: $[Xe]4f^{14}5d^7$

Ti^{4+}: [Ar] or $[Ne]3s^23p^6$ Re^{3+}: $[Xe]4f^{14}5d^4$ Ir^{3+}: $[Xe]4f^{14}5d^6$

24. Cr and Cu are exceptions to the normal filling order of electrons.

a. Cr: $[Ar]4s^13d^5$ b. Cu: $[Ar]4s^13d^{10}$ c. V: $[Ar]4s^23d^3$

Cr^{2+}: $[Ar]3d^4$ Cu^+: $[Ar]3d^{10}$ V^{2+}: $[Ar]3d^3$

Cr^{3+}: $[Ar]3d^3$ Cu^{2+}: $[Ar]3d^9$ V^{3+}: $[Ar]3d^2$

25. a. With K^+ and CN^- ions present, iron has a 3+ charge. Fe^{3+}: $[Ar]3d^5$

b. With a Cl^- ion and neutral NH_3 molecules present, silver has a 1+ charge. Ag^+: $[Kr]4d^{10}$

c. With Br^- ions and neutral H_2O molecules present, nickel has a 2+ charge. Ni^{2+}: $[Ar]3d^8$

d. With NO_2^- ions, an I^- ion, and neutral H_2O molecules present, chromium has a 3+ charge. Cr^{3+}: $[Ar]3d^3$

26. a. With NH_4^+ ions, Cl^- ions, and neutral H_2O molecules present, iron has a 2+ charge. Fe^{2+}: $[Ar]3d^6$

b. With I^- ions and neutral NH_3 and $NH_2CH_2CH_2NH_2$ molecules present, cobalt has a 2+ charge. Co^{2+}: $[Ar]3d^7$

c. With Na^+ and F^- ions present, tantalum has a 5+ charge. Ta^{5+}: $[Xe]4f^{14}$ (expected)

d. Each platinum complex ion must have an overall charge if the two complex ions are counterions to each. Knowing that platinum forms 2+ and 4+ charged ions, we can deduce that the six coordinate complex ion has a 4+ charged platinum ion and the four coordinate complex ion has a 2+ charged ion. With I^- ions and neutral NH_3 molecules present, the two complex ions are $[Pt(NH_3)_4I_2]^{2+}$ and $[PtI_4]^{2-}$.

Pt^{2+}: $[Xe]4f^{14}5d^8$; Pt^{4+}: $[Xe]4f^{14}5d^6$

27. a. molybdenum(IV) sulfide; molybdenum(VI) oxide

b. MoS_2, +4; MoO_3, +6; $(NH_4)_2Mo_2O_7$, +6; $(NH_4)_6Mo_7O_{24}\bullet 4H_2O$, +6

28. a. 4 O ions on faces × 1/2 O/face = 2 O ions, 2 O ions inside body; total: 4 O ions

8 Ti ions on corners × 1/8 Ti/corner + 1 Ti ion/body center = 2 Ti ions

Formula of the unit cell is Ti_2O_4. The empirical formula is TiO_2.

b.
$$\overset{+4\,-2}{2\,TiO_2} + \overset{0}{3\,C} + \overset{0}{4\,Cl_2} \rightarrow \overset{+4\,-1}{2\,TiCl_4} + \overset{+4\,-2}{CO_2} + \overset{+2\,-2}{2\,CO}$$
Cl is reduced, and C is oxidized. Cl_2 is the oxidizing agent, and C is the reducing agent.

$$\overset{+4\,-1}{TiCl_4} + \overset{0}{O_2} \rightarrow \overset{+4\,-2}{TiO_2} + \overset{0}{2\,Cl_2}$$
O is reduced, and Cl is oxidized. O_2 is the oxidizing agent, and $TiCl_4$ is the reducing agent.

29. NH_3 is a weak base which produces OH^- ions in solution. The white precipitate is $Cu(OH)_2(s)$.

$$Cu^{2+}(aq) + 2\,OH^-(aq) \rightarrow Cu(OH)_2(s)$$

With excess NH_3 present, Cu^{2+} forms a soluble complex ion, $Cu(NH_3)_4^{2+}$.

$$Cu(OH)_2(s) + 4\,NH_3(aq) \rightarrow Cu(NH_3)_4^{2+}(aq) + 2\,OH^-(aq)$$

30. CN^- is a weak base, so OH^- ions are present that lead to precipitation of $Ni(OH)_2(s)$. As excess CN^- is added, the $Ni(CN)_4^{2-}$ complex ion forms. The two reactions are:

$Ni^{2+}(aq) + 2\ OH^-(aq) \rightarrow Ni(OH)_2(s)$; the precipitate is $Ni(OH)_2(s)$.

$Ni(OH)_2(s) + 4\ CN^-(aq) \rightarrow Ni(CN)_4^{2-}(aq) + 2\ OH^-(aq)$; $Ni(CN)_4^{2-}$ is a soluble species.

31. Because each compound contains an octahedral complex ion, the formulas for the compounds are $[Co(NH_3)_6]I_3$, $[Pt(NH_3)_4I_2]I_2$, $Na_2[PtI_6]$, and $[Cr(NH_3)_4I_2]I$. Note that in some cases the I^- ions are ligands bound to the transition metal ion as required for a coordination number of 6, while in other cases the I^- ions are counter ions required to balance the charge of the complex ion. The $AgNO_3$ solution will only precipitate the I^- counterions and will not precipitate the I^- ligands. Therefore, 3 moles of AgI will precipitate per mole of $[Co(NH_3)_6]I_3$, 2 moles of AgI will precipitate per mole of $[Pt(NH_3)_4I_2]I_2$, 0 moles of AgI will precipitate per mole of $Na_2[PtI_6]$, and 1 mole of AgI will precipitate per mole of $[Cr(NH_3)_4I_2]I$.

32. $BaCl_2$ gives no precipitate, so SO_4^{2-} must be in the coordination sphere ($BaSO_4$ is insoluble). A precipitate with $AgNO_3$ means the Cl^- is not in the coordination sphere. Because there are only four ammonia molecules in the coordination sphere, SO_4^{2-} must be acting as a bidentate ligand. The structure is:

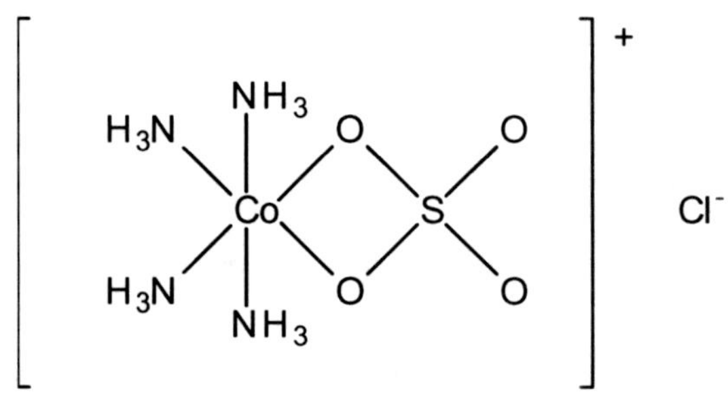

33. To determine the oxidation state of the metal, you must know the charges of the various common ligands (see Table 21.13 of the text).

a. hexacyanomanganate(II) ion

b. cis-tetraamminedichlorocobalt(III) ion

c. pentaamminechlorocobalt(II) ion

34. To determine the oxidation state of the metal, you must know the charges of the various common ligands (see Table 21.13 of the text).

a. pentaamminechlororuthenium(III) ion
b. hexacyanoferrate(II) ion
c. tris(ethylenediamine)manganese(II) ion
d. pentaamminenitrocobalt(III) ion
e. tetracyanonicklate(II) ion
f. tetraamminedichlorochromium(III) ion
g. tris(oxalato)ferrate(III) ion
h. tetraaquadithiocyanatocobalt(III) ion

35. a. hexaamminecobalt(II) chloride
b. hexaaquacobalt(III) iodide
c. potassium tetrachloroplatinate(II)
d. potassium hexachloroplatinate(II)
e. pentaamminechlorocobalt(III) chloride
f. triamminetrinitrocobalt(III)

36. a. pentaaquabromochromium(III) bromide
b. sodium hexacyanocobaltate(III)
c. bis(ethylenediamine)dinitroiron(III) chloride
d. tetraamminediiodoplatinum(IV) tetraiodoplatinate(II)

37. a. $K_2[CoCl_4]$
b. $[Pt(H_2O)(CO)_3]Br_2$
c. $Na_3[Fe(CN)_2(C_2O_4)_2]$
d. $[Cr(NH_3)_3Cl(H_2NCH_2CH_2NH_2)]I_2$

38. a. $FeCl_4^-$
b. $[Ru(NH_3)_5H_2O]^{3+}$
c. $[Cr(CO)_4(OH)_2]^+$
d. $[Pt(NH_3)Cl_3]^-$

39. a.

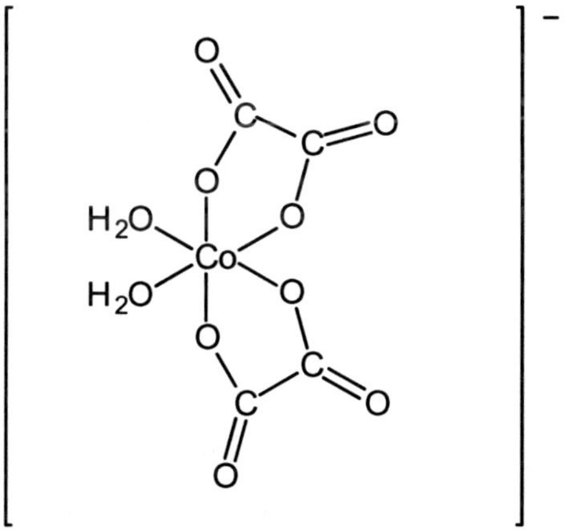

cis

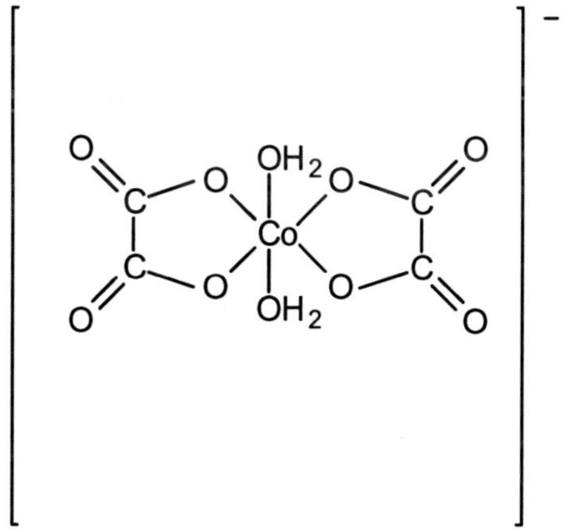

trans

Note: $C_2O_4^{2-}$ is a bidentate ligand. Bidentate ligands bond to the metal at two positions that are 90° apart from each other in octahedral complexes. Bidentate ligands do not bond to the metal at positions 180° apart.

b.

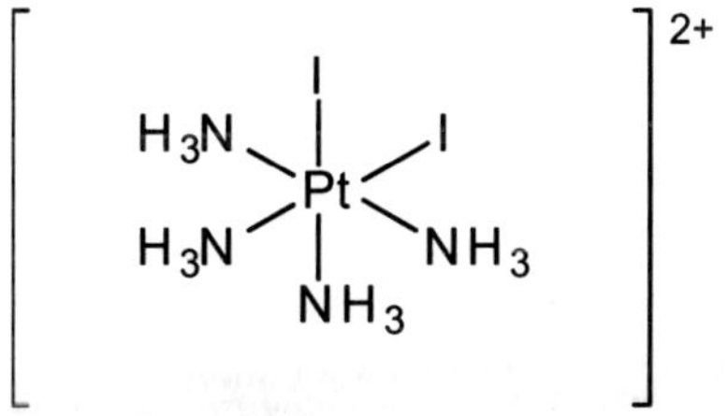

cis

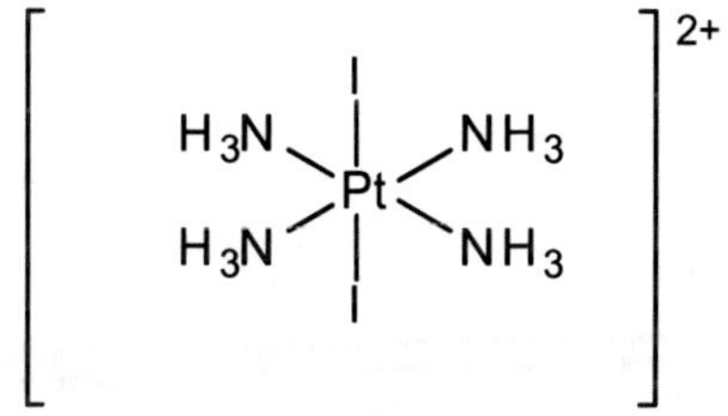

trans

c.

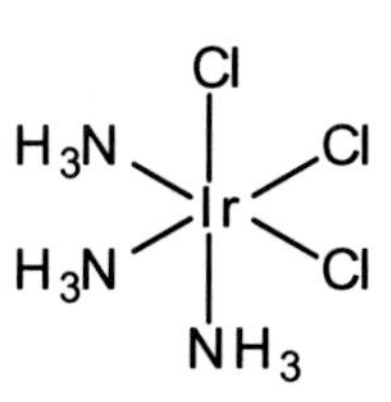

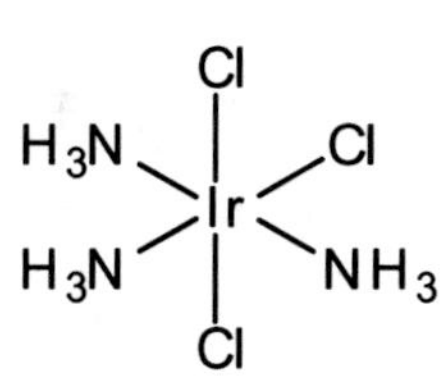

d.

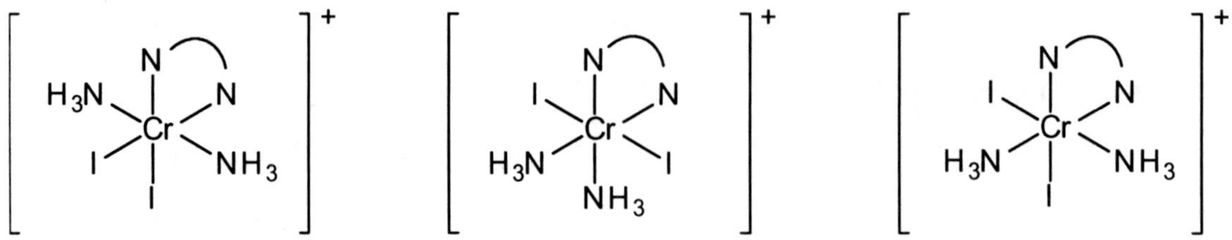

Note: en =  are abbreviations for the bidentate ligand ethylenediamine ($H_2NCH_2CH_2NH_2$).

40. a.

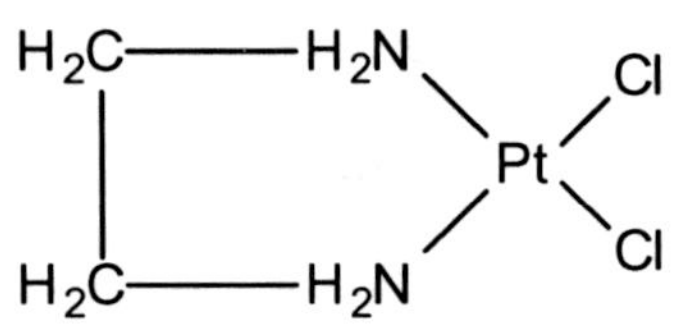

b.

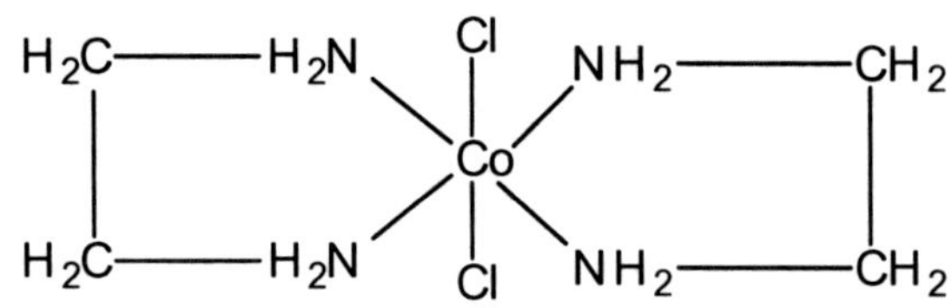

c.

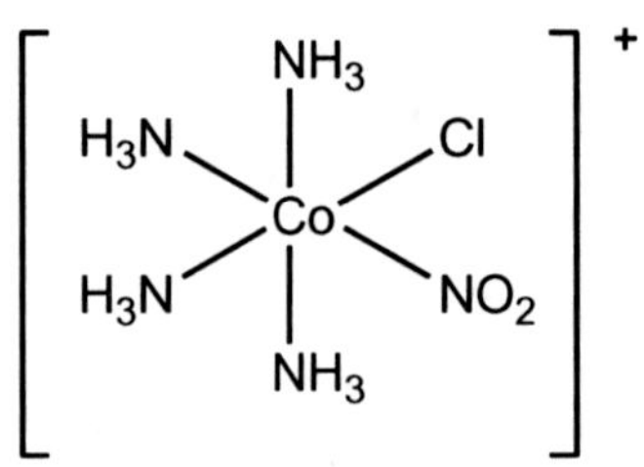

d.

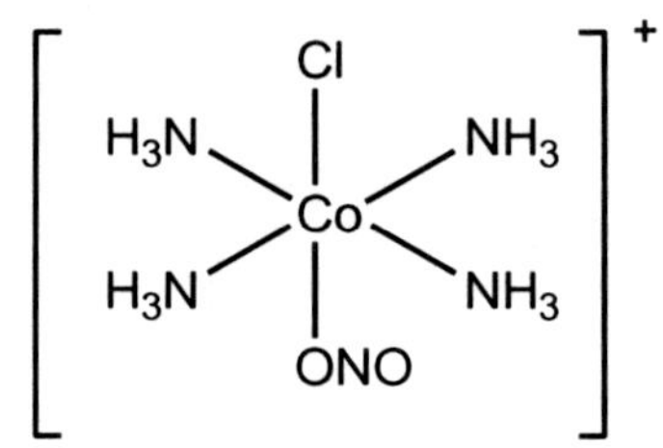

e.

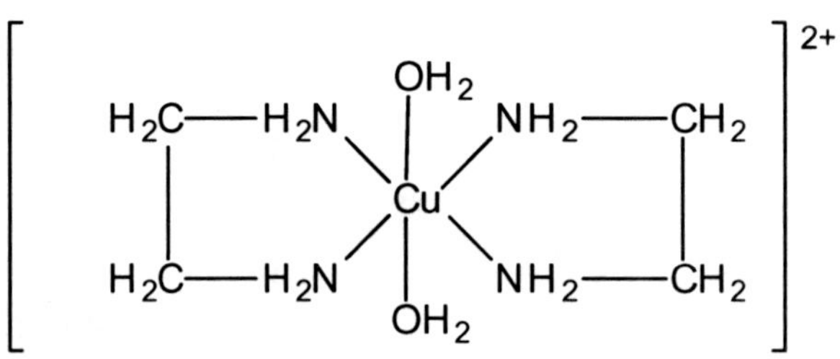

41. monodentate

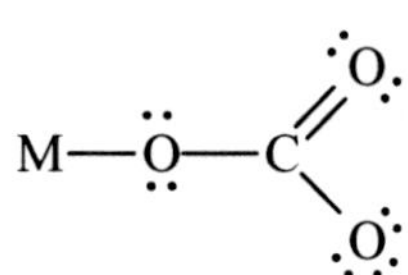

bidentate

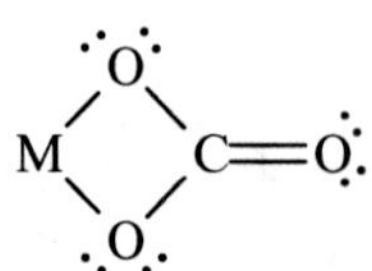

bridging

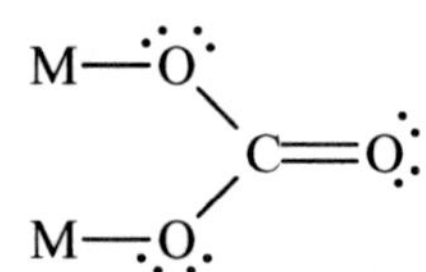

42.

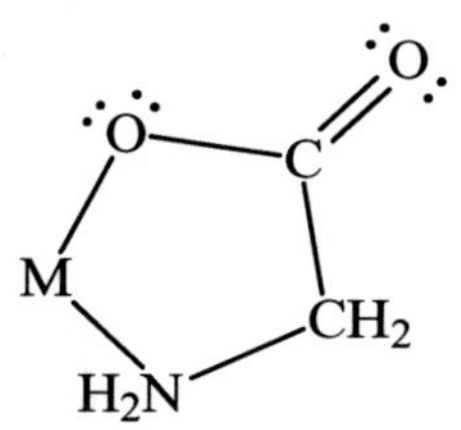

M = transition metal ion

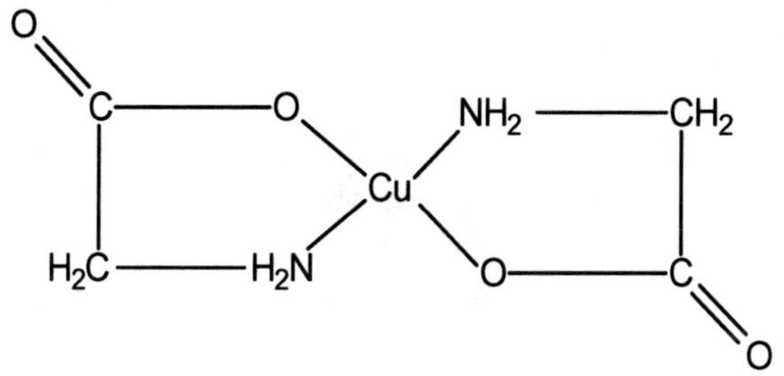

and

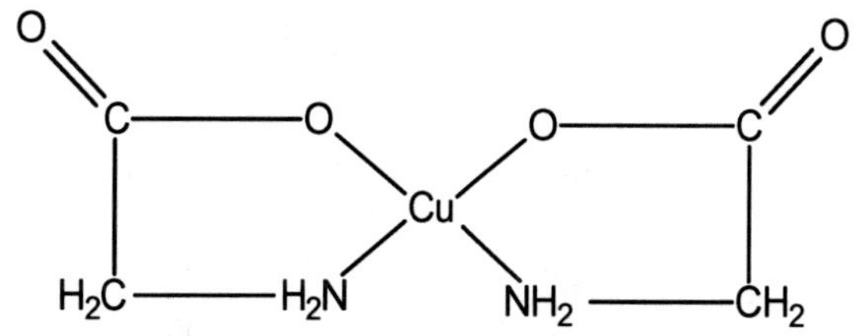

43. a. 2; forms bonds through the lone pairs on the two oxygen atoms.

b. 3; forms bonds through the lone pairs on the three nitrogen atoms.

c. 4; forms bonds through the two nitrogen atoms and the two oxygen atoms.

d. 4; forms bonds through the four nitrogen atoms.

44. M = transition metal ion

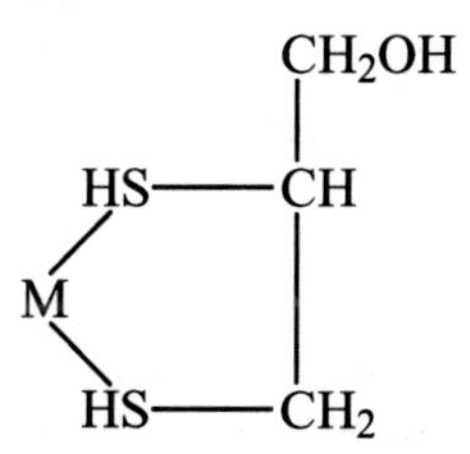

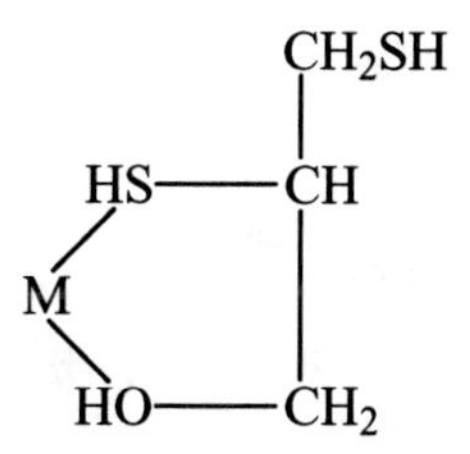

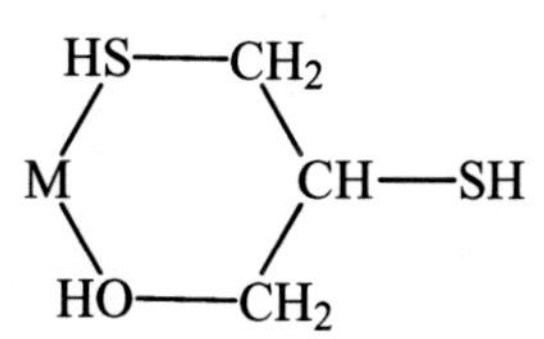

45.

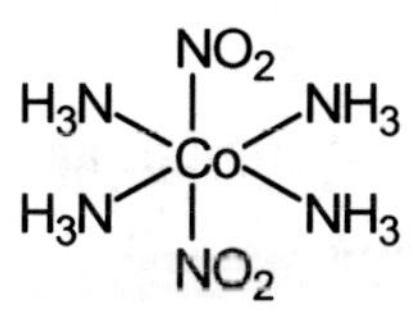

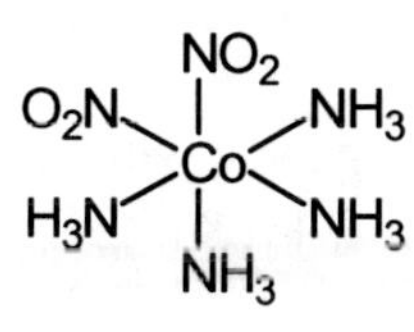

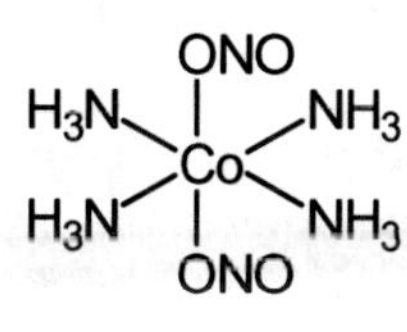

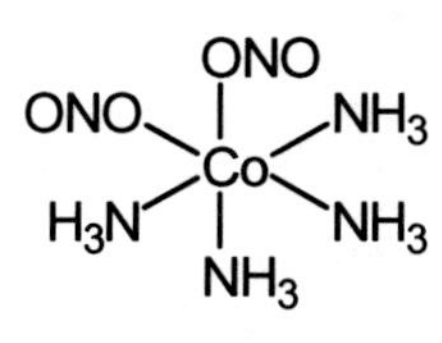

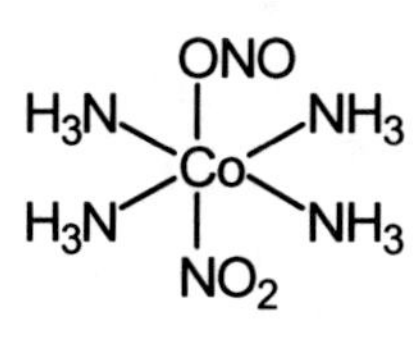

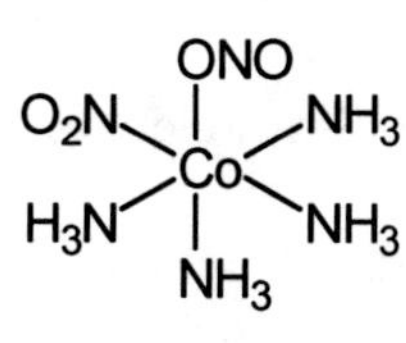

46.

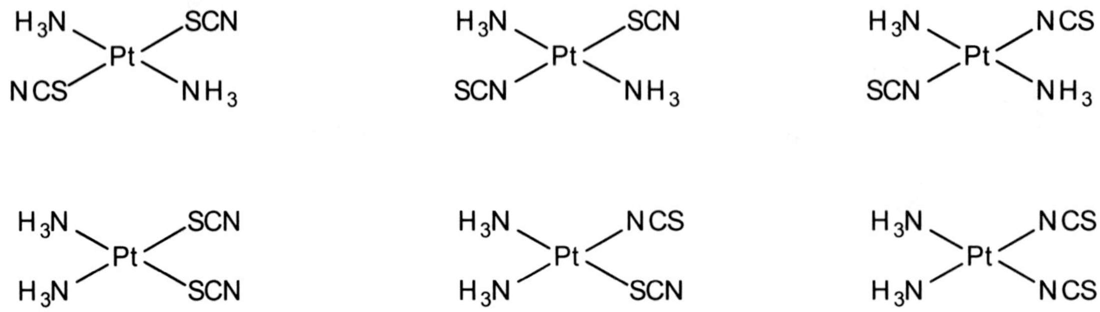

47. Similar to the molecules discussed in Figures 21.16 and 21.17 of the text, $Cr(acac)_3$ and cis-$Cr(acac)_2(H_2O)_2$ are optically active. The mirror images of these two complexes are nonsuperimposable. There is a plane of symmetry in trans-$Cr(acac)_2(H_2O)_2$, so it is not optically active. A molecule with a plane of symmetry is never optically active because the mirror images are always superimposable. A plane of symmetry is a plane through a molecule where one side reflects the other side of the molecule.

48. There are five geometrical isomers (labeled i-v). Only isomer v, where the CN^-, Br^-, and H_2O ligands are cis to each other, is optically active. The nonsuperimposable mirror image is shown for isomer v.

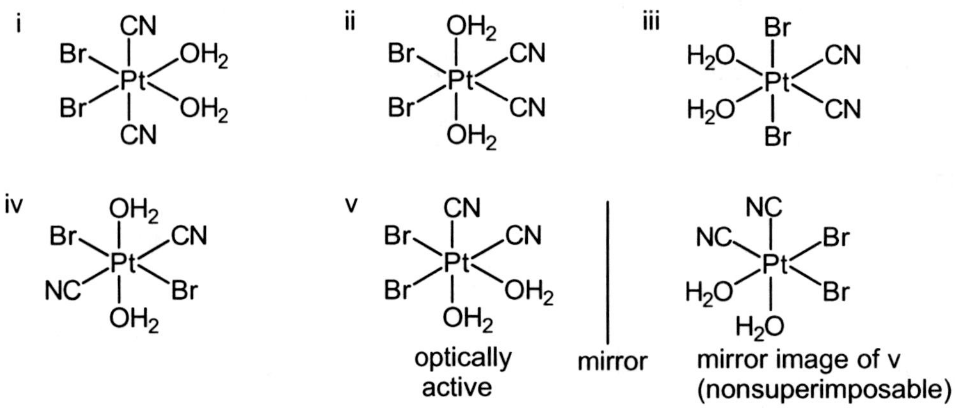

Bonding, Color, and Magnetism in Coordination Compounds

49. NH_3 and H_2O are neutral charged ligands, while chloride and bromide are 1– charged ligands. The metal ions in the three compounds are Cr^{3+}: $[Ar]d^3$, Co^{3+}: $[Ar]d^6$, and Fe^{3+}: $[Ar]d^5$.

a. With five electrons each in a different orbital, this diagram is for the weak-field $[Fe(H_2O)_6]^{3+}$ complex ion.

b. With three electrons, this diagram is for the $[Cr(NH_3)_5Cl]^{2+}$ complex ion.

c. With six electrons all paired up, this diagram is for the strong-field $[Co(NH_3)_4Br_2]^+$ complex ion.

50. The metal ions are both d^5 (Fe^{3+}: $[Ar]d^5$ and Mn^{2+}: $[Ar]d^5$). One of the diagrams (a) is for a weak-field (high-spin) d^5 complex ion while the other diagram (b) is for a strong-field (low-spin) d^5 complex ion. From the spectrochemical series, CN^- is a strong-field ligand while H_2O is in the middle of the series. Because the iron complex ion has CN^- for the ligands as well as having a higher metal ion charge (+3 vs. +2), one would expect $[Fe(CN)_6]^{3-}$ to have the strong-field diagram in b, while $[Mn(H_2O)_6]^{2+}$ would have the weak-field diagram in a.

51. a. Fe^{2+}: $[Ar]3d^6$

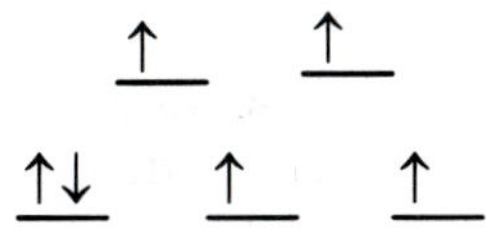

High spin, small Δ

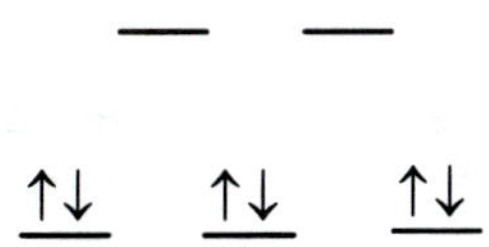

Low spin, large Δ

b. Fe^{3+}: $[Ar]3d^5$

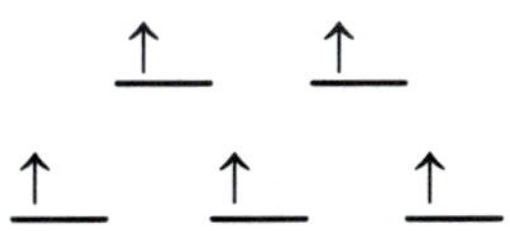

High spin, small Δ

c. Ni^{2+}: $[Ar]3d^8$

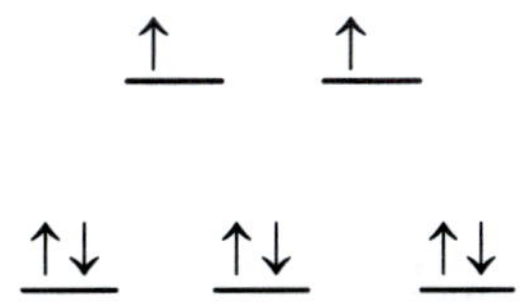

52. a. Zn^{2+}: $[Ar]3d^{10}$

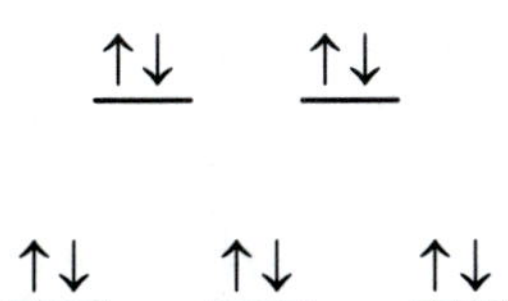

b. Co^{2+}: $[Ar]3d^7$

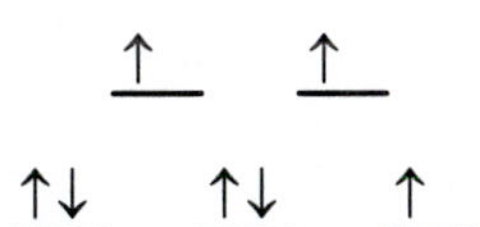

High spin, small Δ

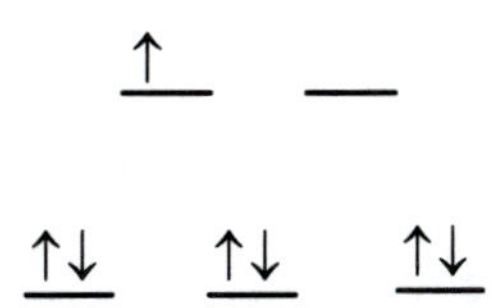

Low spin, large Δ

c. Ti^{3+}: $[Ar]3d^1$

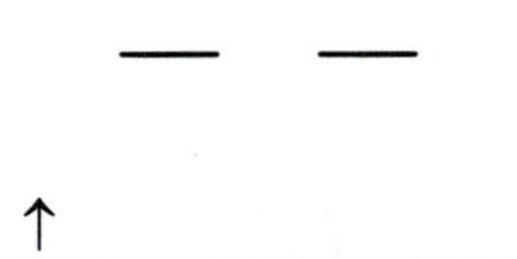

53. Because fluorine has a −1 charge as a ligand, chromium has a +2 oxidation state in CrF_6^{4-}. The electron configuration of Cr^{2+} is $[Ar]3d^4$. For four unpaired electrons, this must be a weak-field (high-spin) case where the splitting of the d-orbitals is small and the number of unpaired electrons is maximized. The crystal field diagram for this ion is:

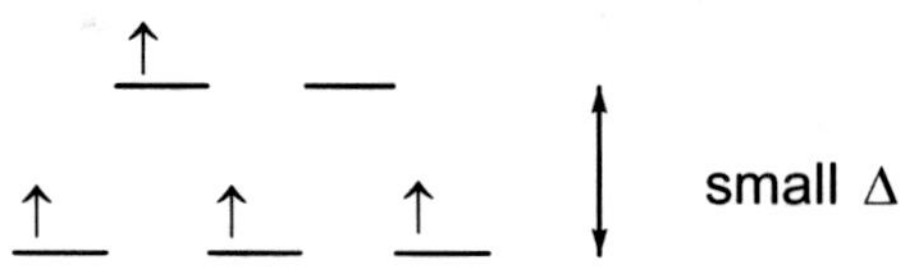

54. NH_3 and H_2O are neutral ligands, so the oxidation states of the metals are Co^{3+} and Fe^{2+}. Both have six d electrons ($[Ar]3d^6$). To explain the magnetic properties, we must have a strong-field for $Co(NH_3)_6^{3+}$ and a weak-field for $Fe(H_2O)_6^{2+}$.

Co^{3+}: $[Ar]3d^6$

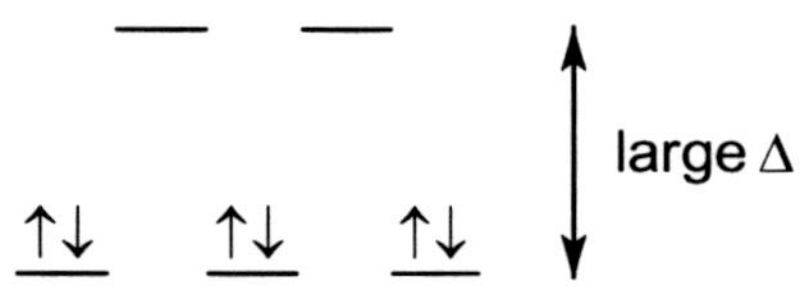

Fe^{2+}: $[Ar]3d^6$

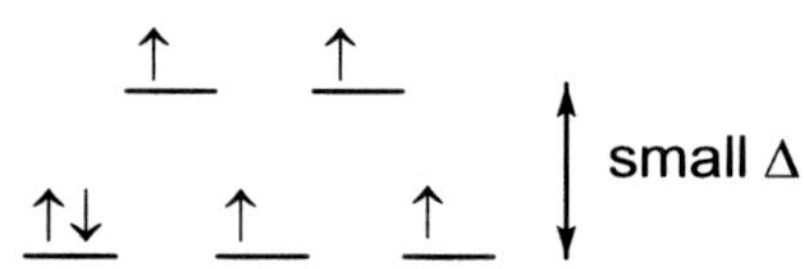

Only this splitting of d-orbitals gives a diamagnetic $Co(NH_3)_6^{3+}$ (no unpaired electrons) and a paramagnetic $Fe(H_2O)_6^{2+}$ (unpaired electrons present).

55. To determine the crystal field diagrams, you need to determine the oxidation state of the transition metal, which can only be determined if you know the charges of the ligands (see Table 21.13). The electron configurations and the crystal field diagrams follow.

a. Ru^{2+}: $[Kr]4d^6$, no unpaired e^-

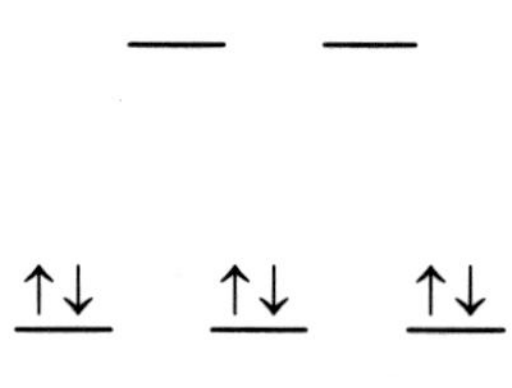

Low spin, large Δ

b. Ni^{2+}: $[Ar]3d^8$, 2 unpaired e^-

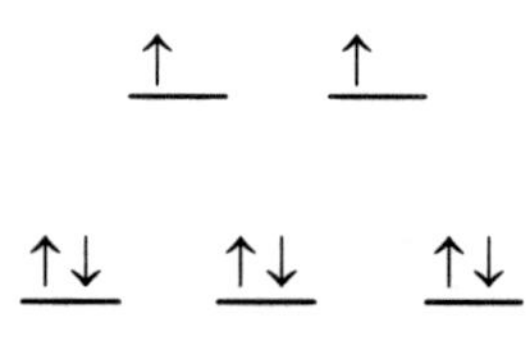

c. V^{3+}: $[Ar]3d^2$, 2 unpaired e^-

Note: Ni^{2+} must have 2 unpaired electrons, whether high-spin or low-spin, and V^{3+} must have 2 unpaired electrons, whether high-spin or low-spin.

56. In both compounds, iron is in the +3 oxidation state with an electron configuration of $[Ar]3d^5$. Fe^{3+} complexes have one unpaired electron when a strong-field case and five unpaired electrons when a weak-field case. $Fe(CN)_6^{2-}$ is a strong-field case, and $Fe(SCN)_6^{3-}$ is a weak-field case. Therefore, cyanide (CN^-) is a stronger-field ligand than thiocyanate (SCN^-).

57. All have octahedral Co^{3+} ions, so the difference in d orbital splitting and the wavelength of light absorbed only depends on the ligands. From the spectrochemical series, the order of the ligands from strongest to weakest field is $CN^- > en > H_2O > I^-$. The strongest-field ligand produces the greatest d-orbital splitting (largest Δ) and will absorb light having the smallest wavelength. The weakest-field ligand produces the smallest Δ and absorbs light having the longest wavelength. The order is:

$$Co(CN)_6^{3-} < Co(en)_3^{3+} < Co(H_2O)_6^{3+} < CoI_6^{3-}$$

shortest λ absorbed — longest λ absorbed

58. Replacement of water ligands by ammonia ligands resulted in shorter wavelengths of light being absorbed. Energy and wavelength are inversely related, so the presence of the NH_3 ligands resulted in a larger d-orbital splitting (larger Δ). Therefore, NH_3 is a stronger-field ligand than H_2O.

59. From Table 21.16 of the text, the violet complex ion absorbs yellow-green light ($\lambda \approx 570$ nm), the yellow complex ion absorbs blue light ($\lambda \approx 450$ nm), and the green complex ion absorbs red light ($\lambda \approx 650$ nm). The spectrochemical series shows that NH_3 is a stronger-field ligand than H_2O, which is a stronger-field ligand than Cl^-. Therefore, $Cr(NH_3)_6^{3+}$ will have the largest d-orbital splitting and will absorb the lowest-wavelength electromagnetic radiation ($\lambda \approx 450$ nm) since energy and wavelength are inversely related ($\lambda = hc/E$). Thus the yellow solution contains the $Cr(NH_3)_6^{3+}$ complex ion. Similarly, we would expect the $Cr(H_2O)_4Cl_2^+$ complex ion to have the smallest d-orbital splitting since it contains the weakest-field ligands. The green solution with the longest wavelength of absorbed light contains the $Cr(H_2O)_4Cl_2^+$ complex ion. This leaves the violet solution, which contains the $Cr(H_2O)_6^{3+}$ complex ion. This makes sense because we would expect $Cr(H_2O)_6^{3+}$ to absorb light of a wavelength between that of $Cr(NH_3)_6^{3+}$ and $Cr(H_2O)_4Cl_2^+$.

60. All these complex ions contain Co^{3+} bound to different ligands, so the difference in d-orbital splitting for each complex ion is due to the difference in ligands. The spectrochemical series indicates that CN^- is a stronger-field ligand than NH_3 which is a stronger-field ligand than F^-. Therefore, $Co(CN)_6^{3-}$ will have the largest d-orbital splitting and will absorb the lowest-wavelength electromagnetic radiation ($\lambda = 290$ nm) since energy and wavelength are inversely related ($\lambda = hc/E$). $Co(NH_3)_6^{3+}$ will absorb 440-nm electromagnetic radiation, while CoF_6^{3-} will absorb the longest-wavelength electromagnetic radiation ($\lambda = 770$ nm) since F^- is the weakest-field ligand present.

61. $CoBr_6^{4-}$ has an octahedral structure, and $CoBr_4^{2-}$ has a tetrahedral structure (as do most Co^{2+} complexes with four ligands). Coordination complexes absorb electromagnetic radiation (EMR) of energy equal to the energy difference between the split d-orbitals. Because the tetrahedral d-orbital splitting is less than one-half the octahedral d-orbital splitting, tetrahedral complexes will absorb lower energy EMR, which corresponds to longer-wavelength EMR ($E = hc/\lambda$). Therefore, $CoBr_6^{2-}$ will absorb EMR having a wavelength shorter than 3.4×10^{-6} m.

62. In both complexes, nickel is in the +2 oxidation state: Ni^{2+}: [Ar]$3d^8$. The differences in unpaired electrons must be due to differences in molecular structure. $NiCl_4^{2-}$ is a tetrahedral complex, and $Ni(CN)_4^{2-}$ is a square planar complex. The corresponding d-orbital splitting diagrams are:

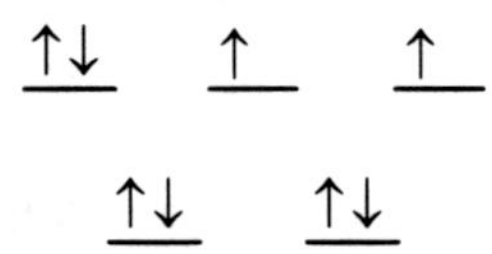

$NiCl_4^{2-}$

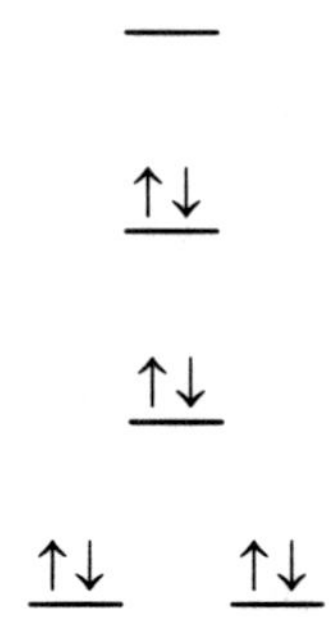

$Ni(CN)_4^{2-}$

63. Because the ligands are Cl^-, iron is in the +3 oxidation state. Fe^{3+}: [Ar]$3d^5$

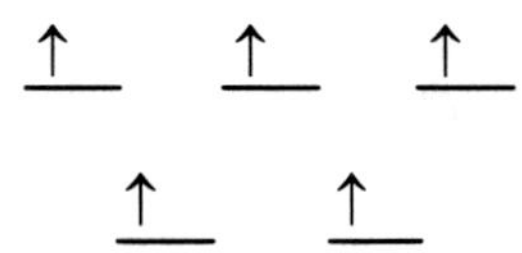

Because all tetrahedral complexes are high spin, there are 5 unpaired electrons in $FeCl_4^-$.

64. Pd is in the +2 oxidation state in $PdCl_4^{2-}$; Pd^{2+}: [Kr]$4d^8$. If $PdCl_4^{2-}$ were a tetrahedral complex, then it would have 2 unpaired electrons and would be paramagnetic (see diagram below). Instead, $PdCl_4^{2-}$ has a square planar molecular structure with the d-orbital splitting diagram also shown below. Note that all electrons are paired in the square planar diagram; this explains the diamagnetic properties of $PdCl_4^{2-}$.

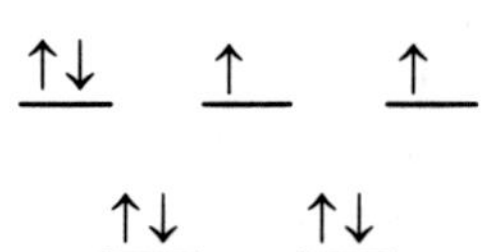

tetrahedral d^8

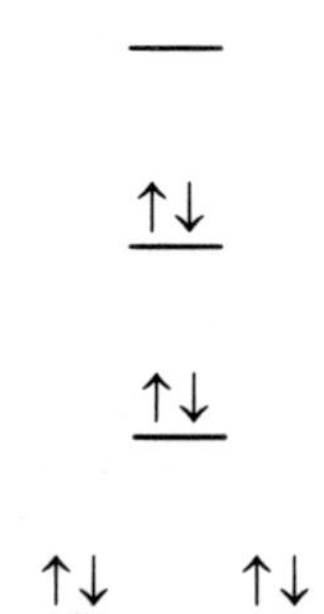

square planar d^8

Metallurgy

65. a. To avoid fractions, let's first calculate ΔH for the reaction:

$6\ FeO(s) + 6\ CO(g) \rightarrow 6\ Fe(s) + 6\ CO_2(g)$

$6\ FeO + 2\ CO_2 \rightarrow 2\ Fe_3O_4 + 2\ CO$	$\Delta H° = -2(18\ kJ)$
$2\ Fe_3O_4 + CO_2 \rightarrow 3\ Fe_2O_3 + CO$	$\Delta H° = -(-39\ kJ)$
$3\ Fe_2O_3 + 9\ CO \rightarrow 6\ Fe + 9\ CO_2$	$\Delta H° = 3(-23\ kJ)$
$6\ FeO(s) + 6\ CO(g) \rightarrow 6\ Fe(s) + 6\ CO_2(g)$	$\Delta H° = -66\ kJ$

So for: $FeO(s) + CO(g) \rightarrow Fe(s) + CO_2(g)$ $\qquad \Delta H° = \dfrac{-66\,kJ}{6} = -11\ kJ$

b. $\Delta H° = 2(-110.5\ kJ) - (-393.5\ kJ + 0) = 172.5\ kJ$

$\Delta S° = 2(198\ J/K) - (214\ J/K + 6\ J/K) = 176\ J/K$

$\Delta G° = \Delta H° - T\Delta S°,\ \Delta G° = 0 \text{ when } T = \dfrac{\Delta H°}{\Delta S°} = \dfrac{172.5\,kJ}{0.176\,kJ/K} = 980.\ K$

Due to the favorable ΔS° term, this reaction is spontaneous at T > 980. K. From Figure 21.36 of the text, this reaction takes place in the blast furnace at temperatures greater than 980. K, as required by thermodynamics.

66. a. $\Delta H° = 2(-1117) + (-393.5) - [3(-826) + (-110.5)] = -39\ kJ$

$\Delta S° = 2(146) + 214 - [3(90.) + 198] = 38\ J/K$

b. $\Delta G° = \Delta H° - T\Delta S°;\ T = 800. + 273 = 1073\ K$

$\Delta G° = -39\ kJ - 1073\ K(0.038\ kJ/K) = -39\ kJ - 41\ kJ = -80.\ kJ$

67. Fe_2O_3: iron has a +3 oxidation state; Fe_3O_4: iron has a +8/3 oxidation state. The three iron ions in Fe_3O_4 must have a total charge of +8. The only combination that works is to have two Fe^{3+} ions and one Fe^{2+} ion per formula unit. This makes sense from the other formula for magnetite, $FeO\bullet Fe_2O_3$. FeO has an Fe^{2+} ion, and Fe_2O_3 has two Fe^{3+} ions.

68. $3\ Fe + C \rightarrow Fe_3C$; $\Delta H° = 21 - [3(0) + 0] = 21\ kJ$; $\Delta S° = 108 - [3(27) + 6] = 21\ J/K$

$\Delta G° = \Delta H° - T\Delta S°$; when ΔH° and ΔS° are both positive, the reaction is spontaneous at high temperatures, where the favorable ΔS° term becomes dominant. Thus, to incorporate carbon into steel, high temperatures are needed for thermodynamic reasons but will also be beneficial for kinetic reasons (as the temperature increases, the rate of the reaction will increase). The relative amount of Fe_3C (cementite) that remains in the steel depends on the cooling process. If the steel is cooled slowly, there is time for the equilibrium to shift back to the left; small crystals of carbon form, giving a relatively ductile steel. If cooling is rapid, there is not enough time for the equilibrium to shift back to the left; Fe_3C is still present in the steel, and the steel is more brittle. Which cooling process occurs depends on the desired properties of the steel. The process of tempering fine-tunes the steel to the desired properties by repeated heating and cooling.

69. Review Section 18.1 for balancing reactions in basic solution by the half-reaction method.

$$(2\ CN^- + Ag \rightarrow Ag(CN)^{2-} + e^-) \times 4$$
$$4\ e^- + O_2 + 4\ H^+ \rightarrow 2\ H_2O$$

$$8\ CN^- + 4\ Ag + O_2 + 4\ H^+ \rightarrow 4\ Ag(CN)_2^- + 2\ H_2O$$

Adding 4 OH^- to both sides and canceling out 2 H_2O on both sides of the equation gives the balanced equation:

$$8\ CN^-(aq) + 4\ Ag(s) + O_2(g) + 2\ H_2O(l) \rightarrow 4\ Ag(CN)_2^-(aq) + 4\ OH^-(aq)$$

70. $Mn + HNO_3 \rightarrow Mn^{2+} + NO_2$

$$Mn \rightarrow Mn^{2+} + 2\ e^- \qquad HNO_3 \rightarrow NO_2$$
$$HNO_3 \rightarrow NO_2 + H_2O$$
$$(e^- + H^+ + HNO_3 \rightarrow NO_2 + H_2O) \times 2$$

$$Mn \rightarrow Mn^{2+} + 2\ e^-$$
$$2\ e^- + 2\ H^+ + 2\ HNO_3 \rightarrow 2\ NO_2 + 2\ H_2O$$

$$2\ H^+(aq) + Mn(s) + 2\ HNO_3(aq) \rightarrow Mn^{2+}(aq) + 2\ NO_2(g) + 2\ H_2O(l)$$

$Mn^{2+} + IO_4^- \rightarrow MnO_4^- + IO_3^-$

$$(4\ H_2O + Mn^{2+} \rightarrow MnO_4^- + 8\ H^+ + 5\ e^-) \times 2 \qquad (2\ e^- + 2\ H^+ + IO_4^- \rightarrow IO_3^- + H_2O) \times 5$$

$$8\ H_2O + 2\ Mn^{2+} \rightarrow 2\ MnO_4^- + 16\ H^+ + 10\ e^-$$
$$10\ e^- + 10\ H^+ + 5\ IO_4^- \rightarrow 5\ IO_3^- + 5\ H_2O$$

$$3\ H_2O(l) + 2\ Mn^{2+}(aq) + 5\ IO_4^-(aq) \rightarrow 2\ MnO_4^-(aq) + 5\ IO_3^-(aq) + 6\ H^+(aq)$$

Additional Exercises

71. $$0.112\ g\ Eu_2O_3 \times \frac{304.0\ g\ Eu}{352.0\ g\ Eu_2O_3} = 0.0967\ g\ Eu; \quad \text{mass \% Eu} = \frac{0.0967\ g}{0.286\ g} \times 100 = 33.8\%\ Eu$$

Mass % O = 100.00 – (33.8 + 40.1 + 4.71) = 21.4% O

Assuming 100.00 g of compound:

$$33.8\ g\ Eu \times \frac{1\ mol}{152.0\ g} = 0.222\ mol\ Eu; \quad 40.1\ g\ C \times \frac{1\ mol}{12.01\ g} = 3.34\ mol\ C$$

$$4.71\ g\ H \times \frac{1\ mol}{1.008\ g} = 4.67\ mol\ H; \quad 21.4\ g\ O \times \frac{1\ mol}{16.00\ g} = 1.34\ mol\ O$$

$$\frac{3.34}{0.222} = 15.0, \quad \frac{4.67}{0.222} = 21.0, \quad \frac{1.34}{0.222} = 6.04$$

The molecular formula is $EuC_{15}H_{21}O_6$. Because each $acac^-$ ligand has a formula of $C_5H_7O_2^-$, an abbreviated molecular formula is $Eu(acac)_3$.

72. The complex ion is $PtCl_4^{2-}$, which is composed of Pt^{2+} and four Cl^- ligands. Pt^{2+}: $[Xe]4f^{14}5d^8$. With square planar geometry, geometric (cis-trans) isomerism is possible. Cisplatin is the cis isomer of the compound and has the following structural formula.

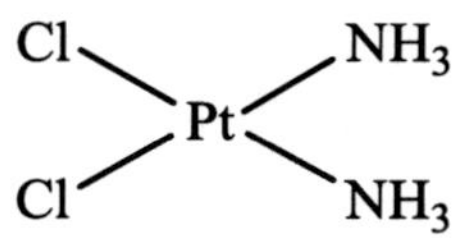

73.

$$\begin{array}{lr} (Au(CN)_2^- + e^- \rightarrow Au + 2\ CN^-) \times 2 & E° = -0.60\ V \\ Zn + 4\ CN^- \rightarrow Zn(CN)_4^{2-} + 2\ e^- & -E° = 1.26\ V \\ \hline 2\ Au(CN)_2^-(aq) + Zn(s) \rightarrow 2\ Au(s) + Zn(CN)_4^{2-}(aq) & E^\circ_{cell} = 0.66\ V \end{array}$$

$$\Delta G° = -nFE^\circ_{cell} = -(2\ mol\ e^-)(96{,}485\ C/mol\ e^-)(0.66\ J/C) = -1.3 \times 10^5\ J = -130\ kJ$$

$$E° = \frac{0.0591}{n}\log K, \ \log K = \frac{nE°}{0.0591} = \frac{2(0.66)}{0.0591} = 22.34, \ K = 10^{22.34} = 2.2 \times 10^{22}$$

Note: We carried extra significant figures to determine K.

74. a. In the following structures, we omitted the 4 NH_3 ligands coordinated to the outside cobalt atoms.

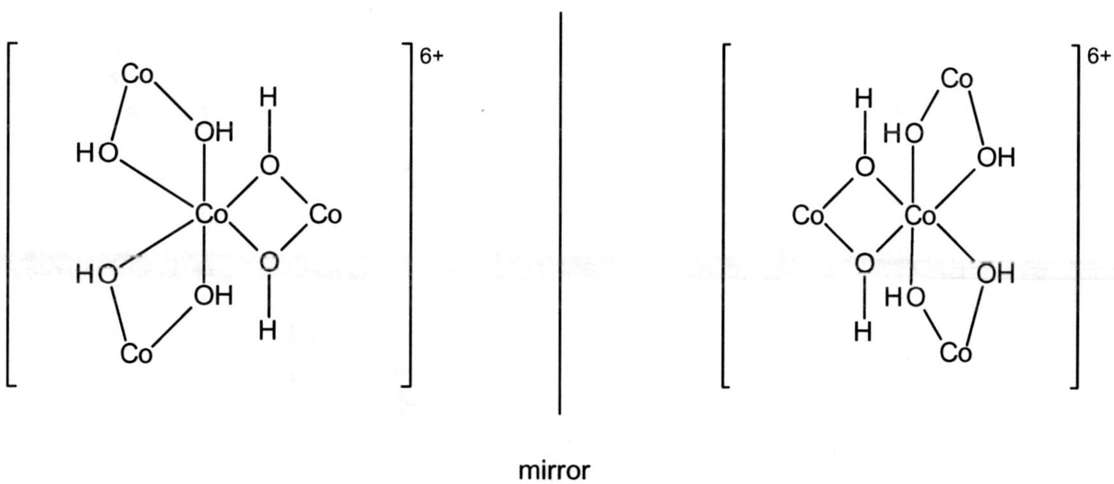

b. All are Co(III). The three "ligands" each contain 2 OH^- and 4 NH_3 groups. If each cobalt is in the +3 oxidation state, then each ligand has a +1 overall charge. The 3+ charge from the three ligands, along with the 3+ charge of the central cobalt atom, gives the overall complex a +6 charge. This is balanced by the 6– charge of the six Cl^- ions.

c. Co^{3+}: $[Ar]3d^6$; there are zero unpaired electrons if a low-spin (strong-field) case.

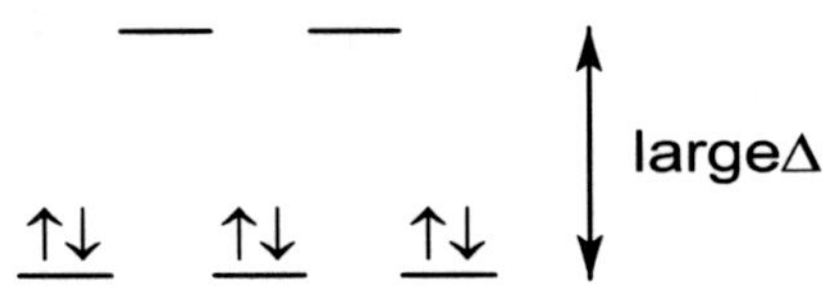

75. There are four geometrical isomers (labeled i-iv). Isomers iii and iv are optically active, and the nonsuperimposable mirror images are shown.

i.

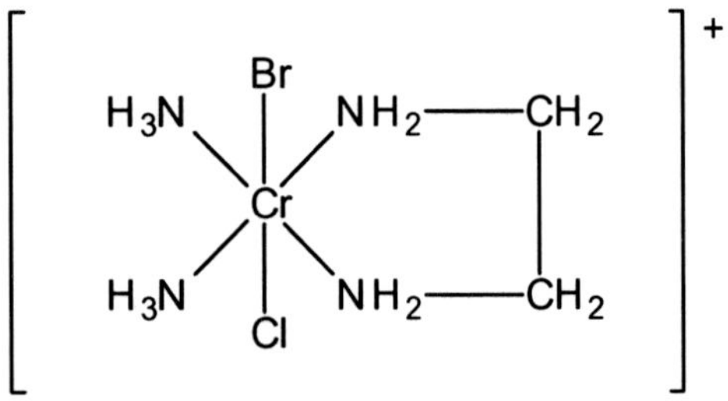

ii.

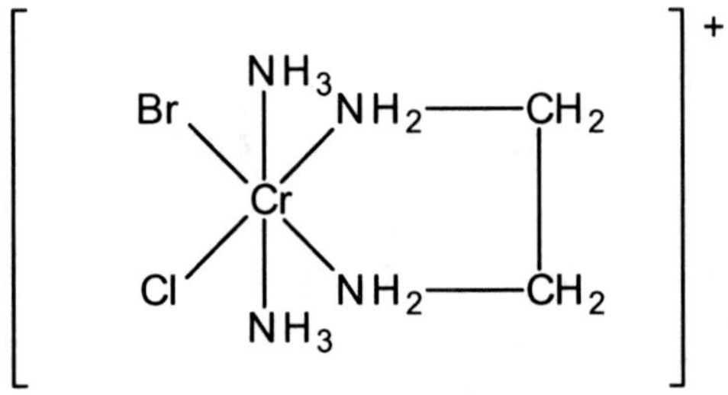

iii.

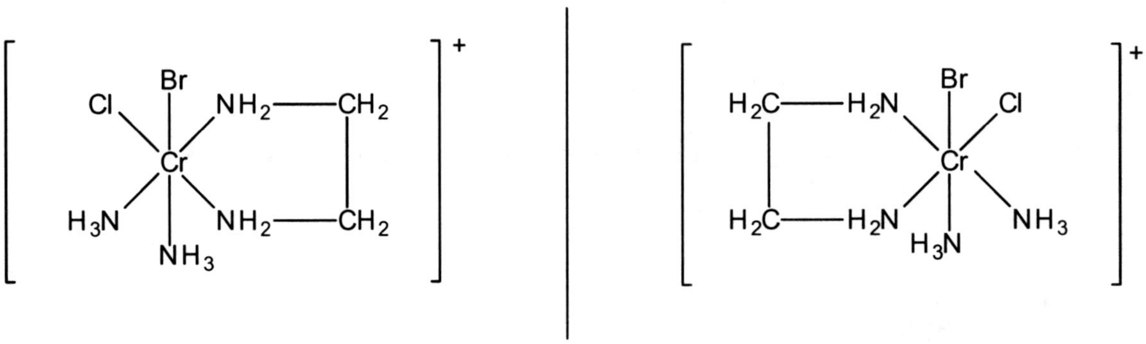

optically active mirror mirror image of iii (nonsuperimposable)

iv.

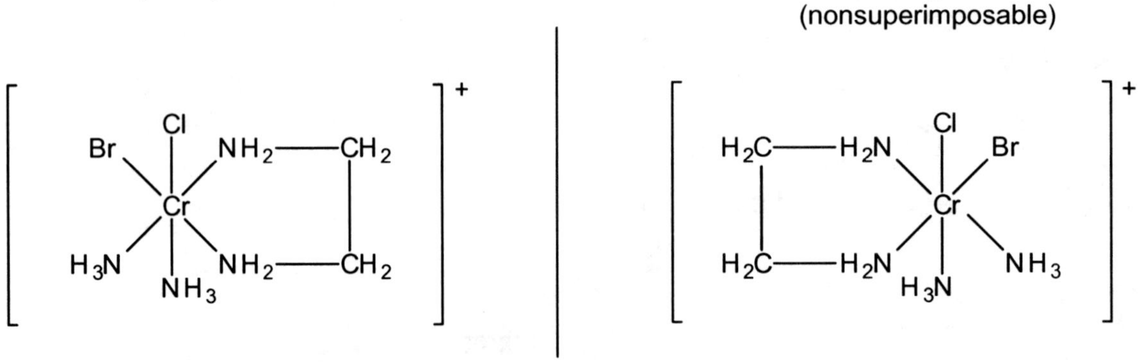

optically active mirror mirror image of iv (nonsuperimposable)

76. a. $Be(tfa)_2$ exhibits optical isomerism. A representation for the tetrahedral optical isomers is:

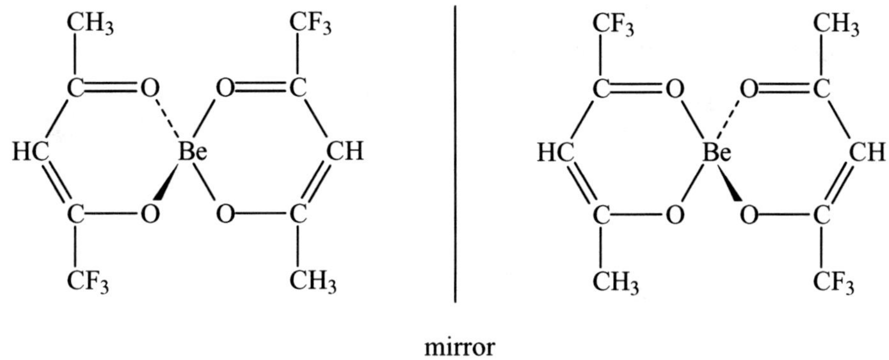

Note: The dotted line indicates a bond pointing into the plane of the paper, and the wedge indicates a bond pointing out of the plane of the paper.

b. Square planar $Cu(tfa)_2$ molecules exhibit geometric isomerism. In one geometric isomer, the CF_3 groups are cis to each other, and in the other isomer, the CF_3 groups are trans.

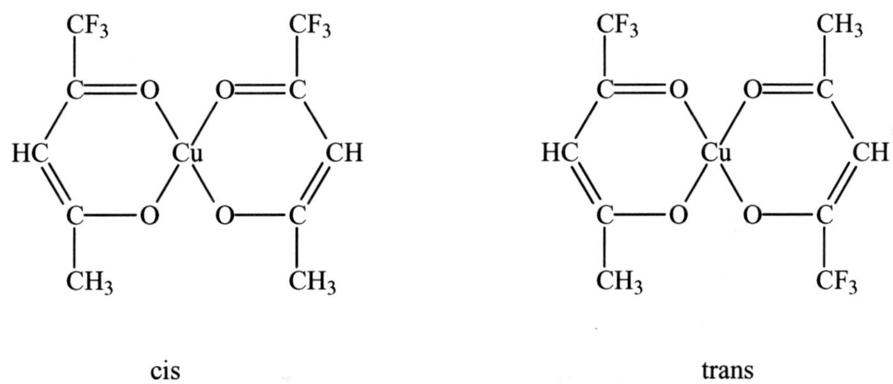

77. Octahedral Cr^{2+} complexes should be used. Cr^{2+}: $[Ar]3d^4$; high-spin (weak-field) Cr^{2+} complexes have 4 unpaired electrons, and low-spin (strong-field) Cr^{2+} complexes have 2 unpaired electrons. Ni^{2+}: $[Ar]3d^8$; octahedral Ni^{2+} complexes will always have 2 unpaired electrons, whether high or low spin. Therefore, Ni^{2+} complexes cannot be used to distinguish weak- from strong-field ligands by examining magnetic properties. Alternatively, the ligand field strengths can be measured using visible spectra. Either Cr^{2+} or Ni^{2+} complexes can be used for this method.

78. a. sodium tris(oxalato)nickelate(II) b. potassium tetrachlorocobaltate(II)

c. tetraamminecopper(II) sulfate

d. chlorobis(ethylenediamine)thiocyanatocobalt(III) chloride

79. a. $[Co(C_5H_5N)_6]Cl_3$ b. $[Cr(NH_3)_5I]I_2$ c. $[Ni(NH_2CH_2CH_2NH_2)_3]Br_2$

d. $K_2[Ni(CN)_4]$ e. $[Pt(NH_3)_4Cl_2][PtCl_4]$

80. a. $Ru(phen)_3^{2+}$ exhibits optical isomerism [similar to $Co(en)_3^{3+}$ in Figure 21.16 of the text].

b. Ru^{2+}: $[Kr]4d^6$; because there are no unpaired electrons, Ru^{2+} is a strong-field (low-spin) case.

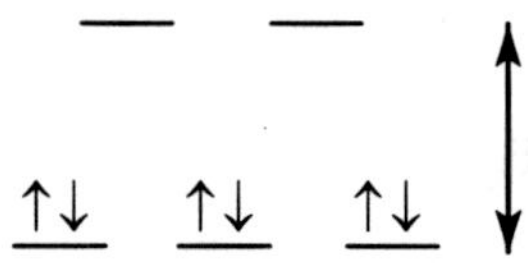

large Δ

81.

$HbO_2 \rightarrow Hb + O_2$		$\Delta G° = -(-70\text{ kJ})$
$Hb + CO \rightarrow HbCO$		$\Delta G° = -80\text{ kJ}$
$HbO_2 + CO \rightarrow HbCO + O_2$		$\Delta G° = -10\text{ kJ}$

$$\Delta G° = -RT \ln K, \quad K = \exp\left(\frac{-\Delta G°}{RT}\right) = \exp\left[\frac{-(-10 \times 10^3\text{ J})}{(8.3145\text{J/K} \bullet \text{mol})(298\text{kJ})}\right] = 60$$

82.

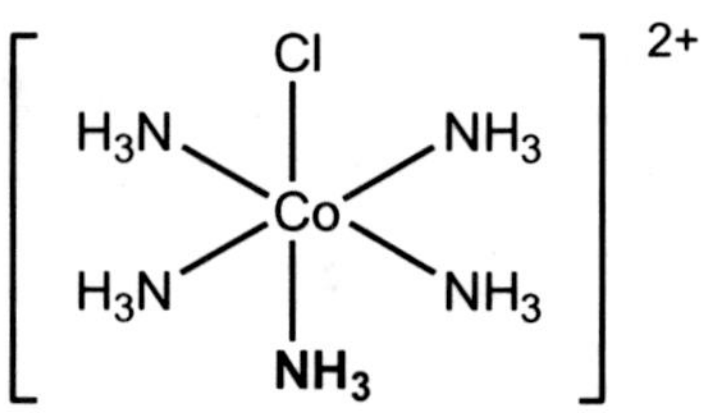

To form the trans isomer, Cl^- would replace the NH_3 ligand that is bold in the structure above. If any of the other four NH_3 molecules are replaced by Cl^-, the cis isomer results. Therefore, the expected ratio of the cis:trans isomer in the product is 4 : 1.

ChemWork Problems

The answers to the problems 83-88 (or a variation to these problems) are found in OWL. These problems are also assignable in OWL.

Challenge Problems

89. $Ni^{2+} = d^8$; if ligands A and B produced very similar crystal fields, the cis-$[NiA_2B_4]^{2+}$ complex ion would give the following octahedral crystal field diagram for a d^8 ion:

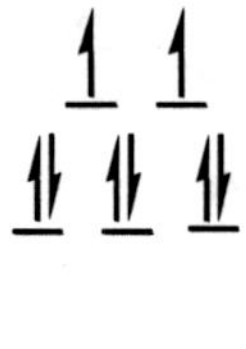

This is paramagnetic.

Because it is given that the complex ion is diamagnetic, the A and B ligands must produce different crystal fields, giving a unique d-orbital splitting diagram that would result in a diamagnetic species.

90. Cr^{3+}: $[Ar]d^3$; looking at figure 21.20 of the text, the d_{z^2} orbital will be destabilized the most when stronger field ligands are on the z-axis. The d_{xy} orbital will be destabilized the least by this arrangement of ligands. Assuming the A ligands produce a significantly stronger field than the B ligands, we will assume the d_{xz} and d_{yz} orbitals will be destabilized the second most of all the orbitals due to the lobes of these orbitals lying in the z-axis plane. This leaves the $d_{x^2-y^2}$ orbitals, whose relative position is very difficult to predict. We will assume the $d_{x^2-y^2}$ orbital is destabilized more than the d_{xy} orbital because the lobes point directly at the B ligands, but is not destabilized as much as the other orbitals which all have lobes in the z-plane. Making these assumptions, the d-orbital splitting diagram is (assuming a low-spin case):

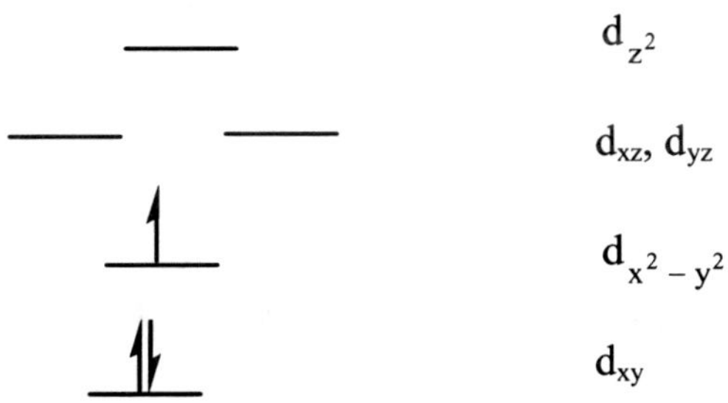

Note: It could be that the relative positions of the $d_{x^2-y^2}$ with the d_{xz} and d_{yz} orbitals are switched.

91. a. Consider the following electrochemical cell:

$$Co^{3+} + e^- \rightarrow Co^{2+} \qquad E° = 1.82 \text{ V}$$
$$Co(en)_3{}^{2+} \rightarrow Co(en)_3{}^{3+} + e^- \qquad -E° = ?$$
$$Co^{3+} + Co(en)_3{}^{2+} \rightarrow Co^{2+} + Co(en)_3{}^{3+} \qquad E°_{cell} = 1.82 - E°$$

The equilibrium constant for this overall reaction is:

$$Co^{3+} + 3\text{ en} \rightarrow Co(en)_3{}^{3+} \qquad K_1 = 2.0 \times 10^{47}$$
$$Co(en)_3{}^{2+} \rightarrow Co^{2+} + 3\text{ en} \qquad K_2 = 1/1.5 \times 10^{12}$$
$$Co^{3+} + Co(en)_3{}^{2+} \rightarrow Co(en)_3{}^{3+} + Co^{2+} \qquad K = K_1K_2 = \frac{2.0 \times 10^{47}}{1.5 \times 10^{12}} = 1.3 \times 10^{35}$$

From the Nernst equation for the overall reaction:

$$E°_{cell} = \frac{0.0591}{n}\log K = \frac{0.0591}{1}\log(1.3 \times 10^{35}), \ E°_{cell} = 2.08 \text{ V}$$

$$E°_{cell} = 1.82 - E° = 2.08 \text{ V}, \ E° = 1.82 \text{ V} - 2.08 \text{ V} = -0.26 \text{ V}$$

b. The stronger oxidizing agent will be the more easily reduced species and will have the more positive standard reduction potential. From the reduction potentials, Co^{3+} ($E° =$ 1.82 V) is a much stronger oxidizing agent than $Co(en)_3{}^{3+}$ ($E° = -0.26$ V).

c. In aqueous solution, Co^{3+} forms the hydrated transition metal complex $Co(H_2O)_6{}^{3+}$. In both complexes, $Co(H_2O)_6{}^{3+}$ and $Co(en)_3{}^{3+}$, cobalt exists as Co^{3+}, which has 6 d electrons. Assuming a strong-field case for each complex ion, the d-orbital splitting diagram for each is:

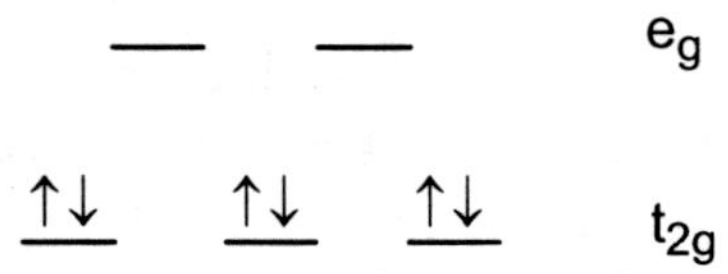

When each complex gains an electron, the electron enters a higher energy e_g orbital. Since en is a stronger-field ligand than H_2O, the d-orbital splitting is larger for $Co(en)_3{}^{3+}$, and it takes more energy to add an electron to $Co(en)_3{}^{3+}$ than to $Co(H_2O)_6{}^{3+}$. Therefore, it is more favorable for $Co(H_2O)_6{}^{3+}$ to gain an electron than for $Co(en)_3{}^{3+}$ to gain an electron.

92. $\overset{II}{(H_2O)_5Cr}–Cl–\overset{III}{Co(NH_3)_5} \rightarrow \overset{III}{(H_2O)_5Cr}–Cl–\overset{II}{Co(NH_3)_5} \rightarrow Cr(H_2O)_5Cl^{2+}$ + Co(II) complex

Yes; after the oxidation, the ligands on Cr(III) won't exchange. Since Cl^- is in the coordination sphere, it must have formed a bond to Cr(II) before the electron transfer occurred (as proposed through the formation of the intermediate).

93. No; in all three cases, six bonds are formed between Ni^{2+} and nitrogen, so ΔH values should be similar. ΔS° for formation of the complex ion is most negative for 6 NH_3 molecules reacting with a metal ion (7 independent species become 1). For penten reacting with a metal ion, 2 independent species become 1, so ΔS° is least negative of all three of the reactions. Thus the chelate effect occurs because the more bonds a chelating agent can form to the metal, the less unfavorable ΔS° becomes for the formation of the complex ion, and the larger the formation constant.

94.

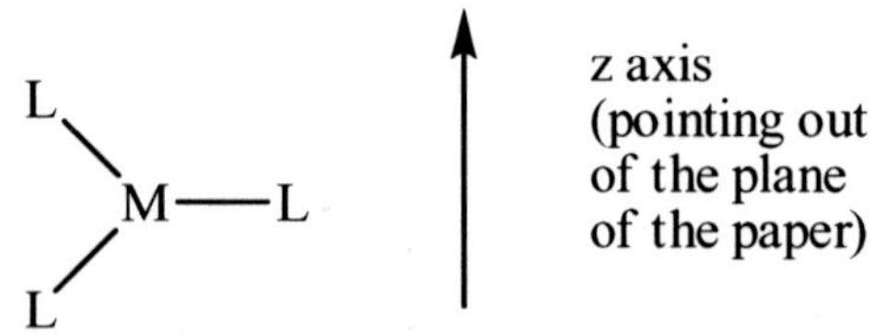

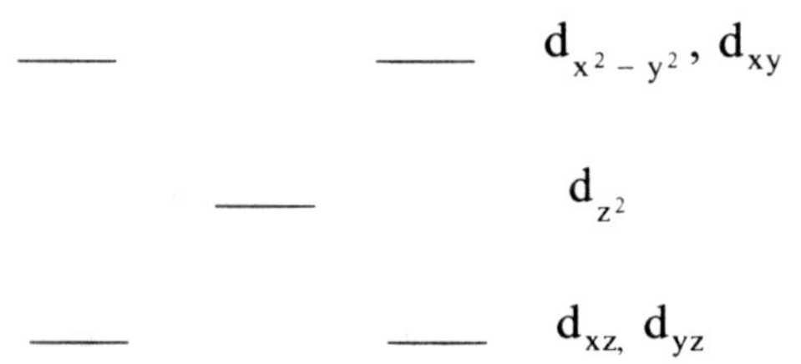

The $d_{x^2-y^2}$ and d_{xy} orbitals are in the plane of the three ligands and should be destabilized the most. The amount of destabilization should be about equal when all the possible interactions are considered. The d_{z^2} orbital has some electron density in the xy plane (the doughnut) and should be destabilized a lesser amount than the $d_{x^2-y^2}$ and d_{xy} orbitals. The d_{xz} and d_{yz} orbitals have no electron density in the plane and should be lowest in energy.

95.

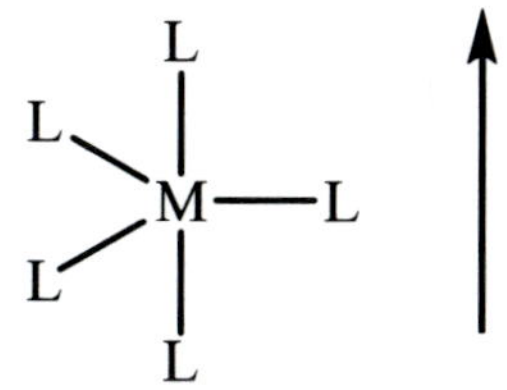

z axis (pointing out of the plane of the paper)

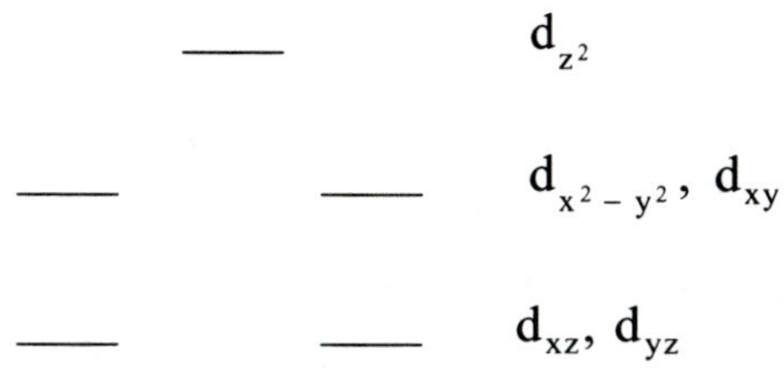

The d_{z^2} orbital will be destabilized much more than in the trigonal planar case (see Exercise 94). The d_{z^2} orbital has electron density on the z axis directed at the two axial ligands. The $d_{x^2-y^2}$ and d_{xy} orbitals are in the plane of the three trigonal planar ligands and should be destabilized a lesser amount than the d_{z^2} orbital; only a portion of the electron density in the $d_{x^2-y^2}$ and d_{xy} orbitals is directed at the ligands. The d_{xz} and d_{yz} orbitals will be destabilized the least since the electron density is directed between the ligands.

96. For a linear complex ion with ligands on the x axis, the $d_{x^2-y^2}$ orbital will be destabilized the most, with the lobes pointing directly at the ligands. The d_{yz} orbital has the fewest interactions with x-axis ligands, so it is destabilized the least. The d_{xy} and d_{xz} orbitals will have similar destabi-lization but will have more interactions with x-axis ligands than the d_{yz} orbital. Finally, the d_{z^2} orbital with the doughnut of electron density in the xy plane will probably be destabilized more than the d_{xy} and d_{xz} orbitals but will have nowhere near the amount of destabilization that occurs with the $d_{x^2-y^2}$ orbital. The only difference that would occur in the diagram if the ligands were on the y axis is the relative positions of the d_{xy}, d_{xz}, and d_{yz} orbitals. The d_{xz} will have the smallest destabilization of all these orbitals, whereas the d_{xy} and d_{yz} orbitals will be degenerate since we expect both to be destabilized equivalently from y-axis ligands. The d-orbital splitting diagrams are:

a.

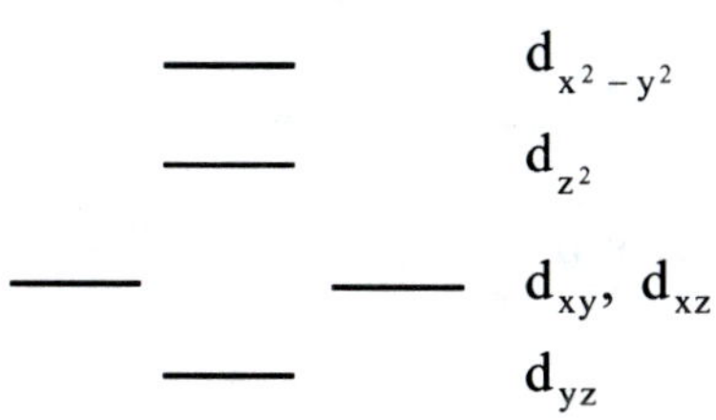

linear x-axis ligands

b.

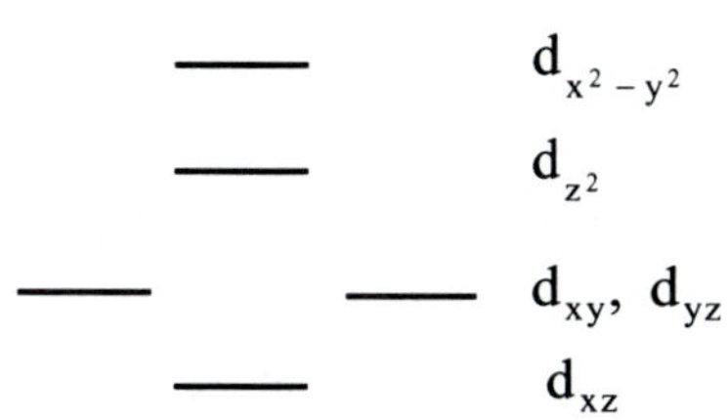

linear y-axis ligands

97. Ni^{2+}: $[Ar]3d^8$; the coordinate system for trans-$[Ni(NH_3)_2(CN)_4]^{2-}$ is shown below. Because CN^- produces a much stronger crystal field, it will dominate the d-orbital splitting. From the coordinate system, the CN^- ligands are in a square planar arrangement. Therefore, the diagram will most likely resemble the square planar diagram given in Figure 21.28. Note that the relative position of d_{z^2} orbital is hard to predict. With the NH_3 ligands on the z axis, we will assume the $d_z{}^2$ orbital is destabilized more than the d_{xy} orbital. However, this is only an assumption. It could be that the d_{xy} orbital is destabilized more.

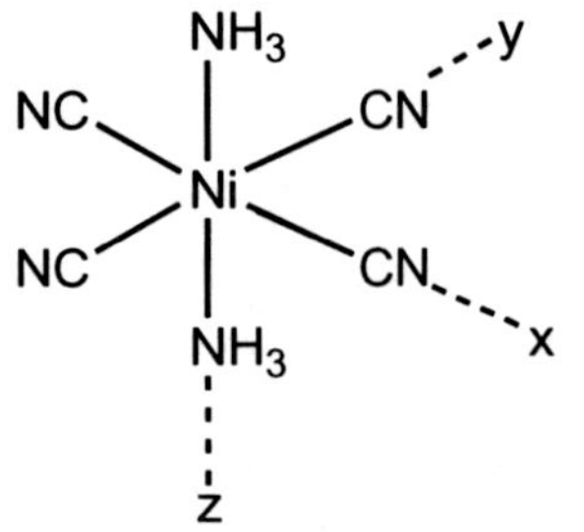

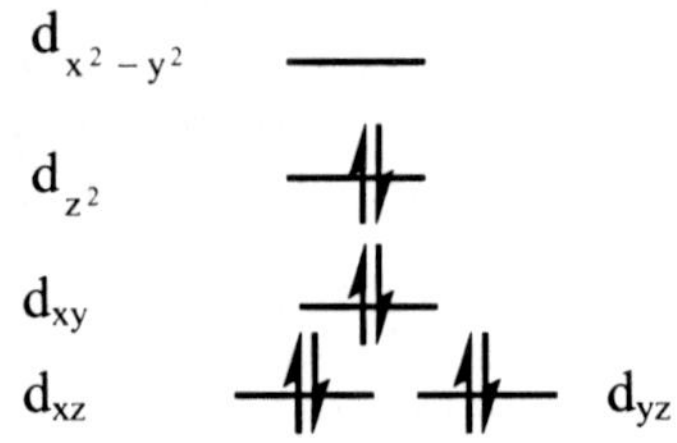

Integrative Problems

98. a. Because O is in the −2 oxidation state, iron must be in the +6 oxidation state. Fe^{6+}: $[Ar]3d^2$.

b. Using the half-reaction method of balancing redox reactions, the balanced equation is:

$$10\ H^+(aq) + 2\ FeO_4^{2-}(aq) + 2\ NH_3(aq) \rightarrow 2\ Fe^{3+}(aq) + N_2(g) + 8\ H_2O(l)$$

$$0.0250\ L \times \frac{0.243\ mol\ FeO_4^{2-}}{L} \times \frac{1\ mol\ N_2}{2\ mol\ FeO_4^{2-}} = 3.04 \times 10^{-3}\ mol\ N_2$$

(if FeO_4^{2-} limiting)

$$0.0550\ L \times \frac{1.45\ mol\ NH_3}{L} \times \frac{1\ mol\ N_2}{2\ mol\ NH_3} = 3.99 \times 10^{-2}\ mol\ N_2 \text{ (if } NH_3 \text{ limiting)}$$

Because the FeO_4^{2-} reagent produces the smaller quantity of N_2, FeO_4^{2-} is limiting and 3.04×10^{-3} mol N_2 can be produced.

$$V_{N_2} = \frac{nRT}{P} = \frac{3.04 \times 10^{-3}\ mol\ N_2 \times \frac{0.08206\ L\ atm}{K\ mol} \times 298\ K}{1.50\ atm}$$

$$V_{N_2} = 0.0496\ L = 49.6\ mL\ N_2$$

99. i. $0.0203\ g\ CrO_3 \times \frac{52.00\ g\ Cr}{100.0\ g\ CrO_3} = 0.0106\ g\ Cr$; $\%\ Cr = \frac{0.0106\ g}{0.105\ g} \times 100 = 10.1\%\ Cr$

ii. $32.93\ mL\ HCl \times \frac{0.100\ mmol\ HCl}{mL} \times \frac{1\ mmol\ NH_3}{mmol\ HCl} \times \frac{17.03\ mg\ NH_3}{mmol} = 56.1\ mg\ NH_3$

$$\%\ NH_3 = \frac{56.1\ mg}{341\ mg} \times 100 = 16.5\%\ NH_3$$

iii. 73.53% + 16.5% + 10.1% = 100.1%; the compound must be composed of only Cr, NH_3, and I.

Out of 100.00 g of compound:

$$10.1 \text{ g Cr} \times \frac{1 \text{ mol}}{52.00 \text{ g}} = 0.194 \text{ mol}; \qquad \frac{0.194}{0.194} = 1.00$$

$$16.5 \text{ g NH}_3 \times \frac{1 \text{ mol}}{17.03 \text{ g}} = 0.969 \text{ mol}; \qquad \frac{0.969}{0.194} = 4.99$$

$$73.53 \text{ g I} \times \frac{1 \text{ mol}}{126.9 \text{ g}} = 0.5794 \text{ mol}; \qquad \frac{0.5794}{0.194} = 2.99$$

$Cr(NH_3)_5I_3$ is the empirical formula. If we assume an octahedral complex ion, then compound A is made of the octahedral $[Cr(NH_3)_5I]^{2+}$ complex ion and two I^- ions as counterions; the formula is $[Cr(NH_3)_5I]I_2$. Lets check this proposed formula using the freezing-point data.

iv. $\Delta T_f = iK_f m$; for $[Cr(NH_3)_5I]I_2$, i = 3.0 (assuming complete dissociation).

$$\text{Molality} = m = \frac{0.601 \text{ g complex}}{1.000 \times 10^{-2} \text{ kg H}_2\text{O}} \times \frac{1 \text{ mol complex}}{517.9 \text{ g complex}} = 0.116 \text{ mol/kg}$$

$\Delta T_f = 3.0 \times 1.86$ °C kg/mol $\times$ 0.116 mol/kg = 0.65°C

Because ΔT_f is close to the measured value, this is consistent with the formula $[Cr(NH_3)_5I]I_2$. So our assumption of an octahedral complex ion is probably a good assumption.

100. a. $$\lambda = \frac{hc}{E} = \frac{6.626 \times 10^{-34} \text{ J s} \times 2.998 \times 10^8 \text{ m/s}}{1.75 \times 10^4 \text{ cm}^{-1} \times \frac{1.986 \times 10^{-23} \text{ J}}{\text{cm}^{-1}}} = 5.72 \times 10^{-7} \text{ m} = 572 \text{ nm}$$

b. There are three resonance structures for NCS^-. From a formal charge standpoint, the following resonance structure is best.

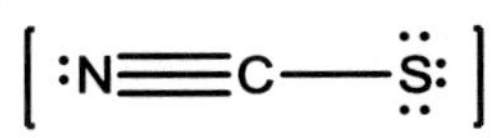

The N in this resonance structure is sp hybridized. Because the sp hybrid orbitals are 180° apart, one would expect that when the lone pair in an sp hybrid orbital on N is donated to the Cr^{3+} ion, the 180° bond angle would stay intact between Cr, N, C, and S.

Similar to $Co(en)_2Cl_2^+$ discussed in Figure 21.17 of the text, $[Co(en)_2(NCS)_2]^+$ would exhibit cis-trans isomerism (geometric isomerism), and only the cis form would exhibit optical isomerism. For $[Co(en)_2(NCS)_2]^+$, NCS^- just replaces the Cl^- ions in the isomers drawn in Figure 21.17. The trans isomer would not exhibit optical isomerism.

Marathon Problem

101. $CrCl_3 \cdot 6H_2O$ contains nine possible ligands, only six of which are used to form the octahedral complex ion. The three species not present in the complex ion will either be counterions to balance the charge of the complex ion and/or waters of hydration. The number of counterions for each compound can be determined from the silver chloride precipitate data, and the number of waters of hydration can be determined from the dehydration data. In all experiments, the ligands in the complex ion do not react.

Compound I:

$$\text{mol } CrCl_3 \cdot 6H_2O = 0.27 \text{ g} \times \frac{1 \text{ mol}}{266.5 \text{ g}} = 1.0 \times 10^{-3} \text{ mol } CrCl_3 \cdot 6H_2O$$

$$\text{mol waters of hydration} = 0.036 \text{ g } H_2O \times \frac{1 \text{ mol}}{18.02 \text{ g}} = 2.0 \times 10^{-3} \text{ mol } H_2O$$

$$\frac{\text{mol waters of hydration}}{\text{mol compound}} = \frac{2.0 \times 10^{-3} \text{ mol}}{1.0 \times 10^{-3} \text{ mol}} = 2.0$$

In compound I, two of the H_2O molecules are waters of hydration, so the other four water molecules are present in the complex ion. Therefore, the formula for compound I must be $[Cr(H_2O)_4Cl_2]Cl \cdot 2H_2O$. Two of the Cl^- ions are present as ligands in the octahedral complex ion, and one Cl^- ion is present as a counterion. The AgCl precipitate data that refer to this compound are the one that produces 1430 mg AgCl:

$$\text{mol } Cl^- \text{ from compound I} = 0.1000 \text{ L} \times \frac{0.100 \text{ mol} [Cr(H_2O)_4Cl_2]Cl \bullet 2H_2O}{\text{L}}$$

$$\times \frac{1 \text{ mol } Cl^-}{\text{mol} [Cr(H_2O)_4Cl_2]Cl \bullet 2H_2O} = 0.0100 \text{ mol } Cl^-$$

$$\text{mass AgCl produced} = 0.0100 \text{ mol } Cl^- \times \frac{1 \text{ mol AgCl}}{\text{mol } Cl^-} \times \frac{143.4 \text{ g AgCl}}{\text{mol AgCl}} = 1.43 \text{ g}$$

$$= 1430 \text{ mg AgCl}$$

Compound II:

$$\frac{\text{mol waters of hydration}}{\text{mol compound}} = \frac{0.018 \text{ g } H_2O \times \dfrac{1 \text{ mol}}{18.02 \text{ g}}}{1.0 \times 10^{-3} \text{ mol compound}} = 1.0$$

The formula for compound II must be $[Cr(H_2O)_5Cl]Cl_2 \cdot H_2O$. The 2870-mg AgCl precipitate data refer to this compound. For 0.0100 mol of compound II, 0.0200 mol Cl^- is present as counterions:

$$\text{mass AgCl produced} = 0.0200 \text{ mol } Cl^- \times \frac{1 \text{ mol AgCl}}{\text{mol } Cl^-} \times \frac{143.4 \text{ g}}{\text{mol}} = 2.87 \text{ g}$$

$$= 2870 \text{ mg AgCl}$$

Compound III:

This compound has no mass loss on dehydration, so there are no waters of hydration present. The formula for compound III must be $[Cr(H_2O)_6]Cl_3$. 0.0100 mol of this compound produces 4300 mg of AgCl(s) when treated with $AgNO_3$.

$$0.0300 \text{ mol Cl}^- \times \frac{1 \text{ mol AgCl}}{\text{mol Cl}^-} \times \frac{143.4 \text{ g AgCl}}{\text{mol AgCl}} = 4.30 \text{ g} = 4.30 \times 10^3 \text{ mg AgCl}$$

The structural formulas for the compounds are:

Compound I

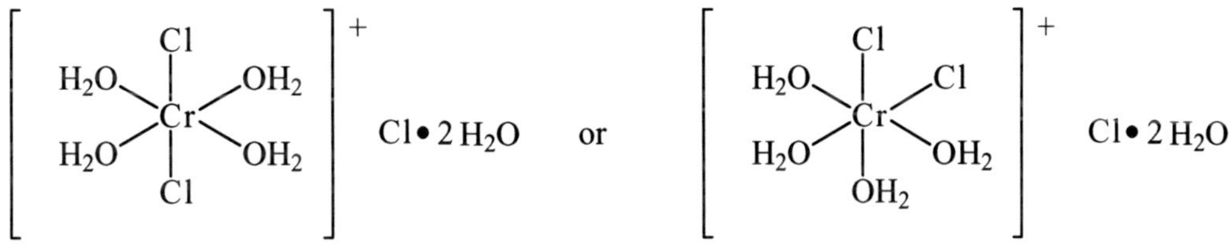

Compound II

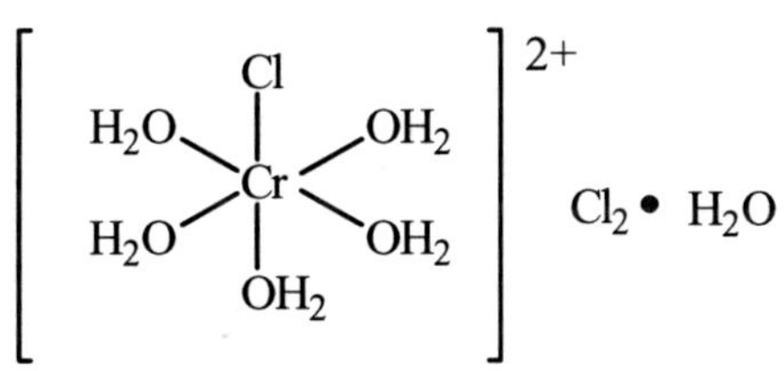

Compound III

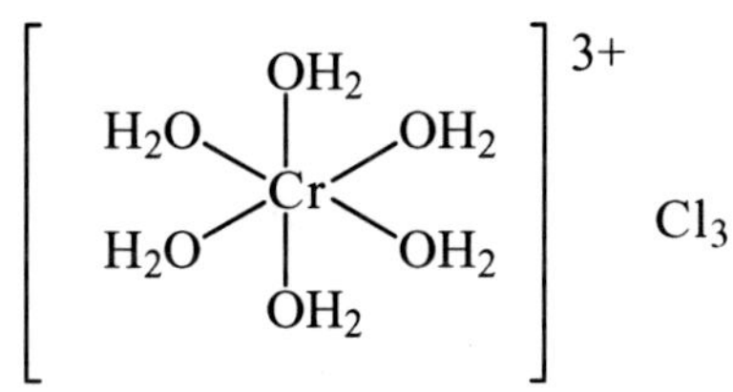

From Table 21.16 of the text, the violet compound will be the one that absorbs light with the shortest wavelength (highest energy). This should be compound III. H_2O is a stronger-field ligand than Cl^-; compound III with the most coordinated H_2O molecules will have the largest d-orbital splitting and will absorb the higher-energy light.

The magnetic properties would be the same for all three compounds. Cr^{3+} is a d^3 ion. With only three electrons present, all Cr^{3+}complexes will have three unpaired electrons, whether strong field or weak field. If $Cr^{2+}$was present with the d^4 configuration, then the magnetic properties might be different for the complexes and could be worth examining.

CHAPTER 22

ORGANIC AND BIOLOGICAL MOLECULES

Questions

1. a. 1-sec-butylpropane

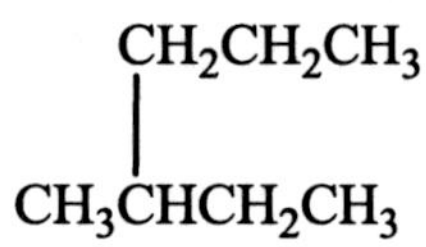

3-methylhexane is correct.

b. 4-methylhexane

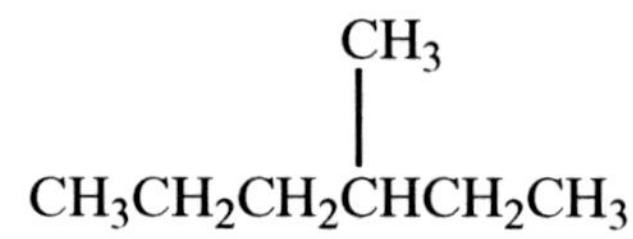

3-methylhexane is correct.

c. 2-ethylpentane

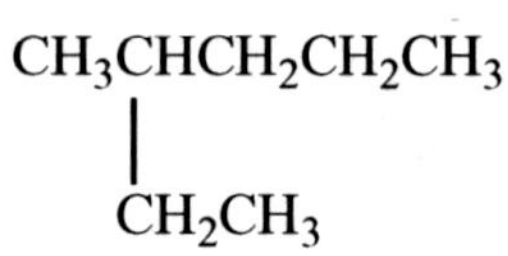

3-methylhexane is correct.

d. 1-ethyl-1-methylbutane

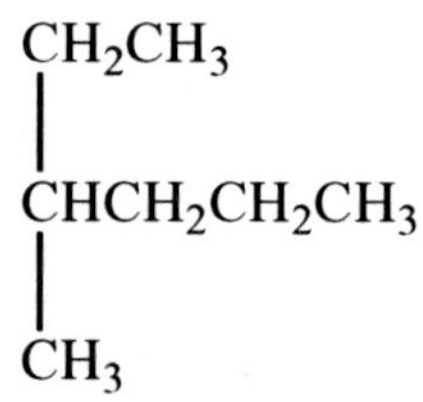

3-methylhexane is correct.

e. 3-methylhexane

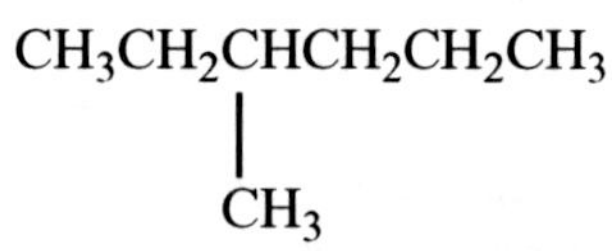

f. 4-ethylpentane

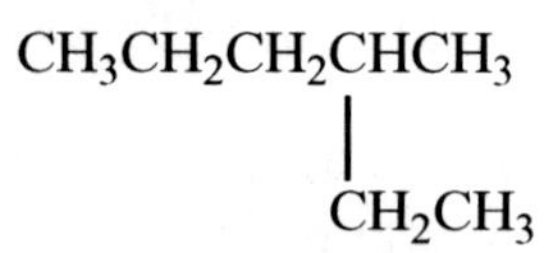

3-methylhexane is correct.

All six of these compounds are the same. They only differ from each other by rotations about one or more carbon-carbon single bonds. Only one isomer of C_7H_{16} is present in all of these names, 3-methylhexane.

2. a. C_6H_{12} can exhibit structural, geometric, and optical isomerism. Two structural isomers (of many) are:

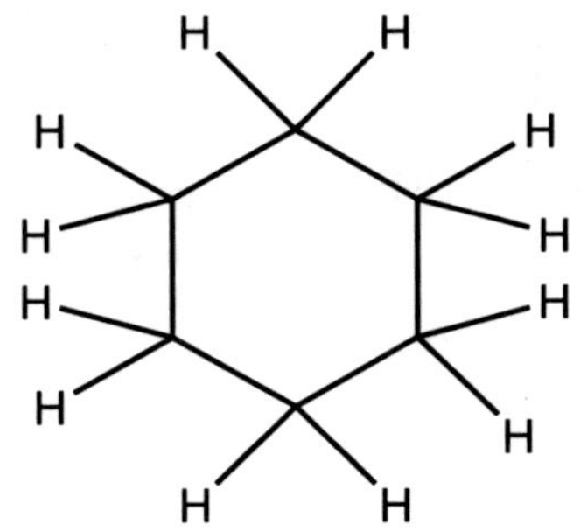

cyclohexane

$CH_2{=}CHCH_2CH_2CH_2CH_3$

1-hexene

The structural isomer 2-hexene (plus others) exhibits geometric isomerism.

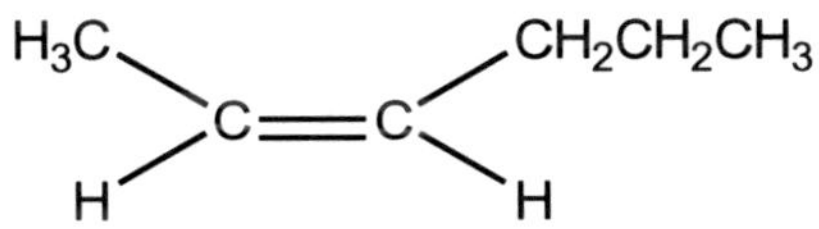

cis

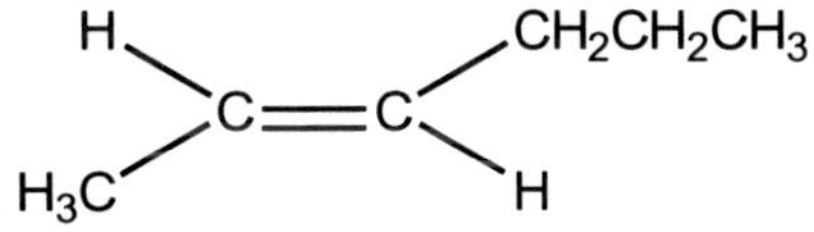

trans

The structural isomer 3-methyl-1-pentene exhibits optical isomerism (the asterisk marks the chiral carbon).

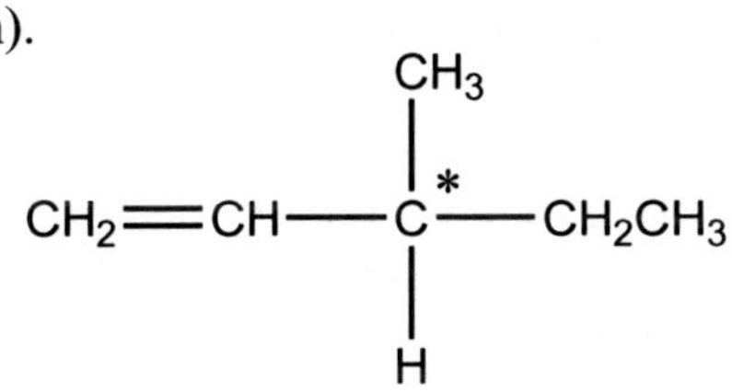

Optical isomerism is also possible with some of the cyclobutane and cyclopropane structural isomers.

b. $C_5H_{12}O$ can exhibit structural and optical isomerism. Two structural isomers (of many) are:

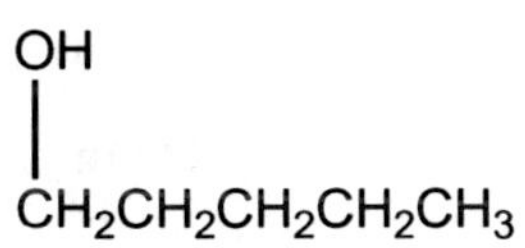

1-pentanol

$CH_3{-}O{-}CH_2CH_2CH_2CH_3$

butyl methyl ether

Two of the optically active isomers having a $C_5H_{12}O$ formula are:

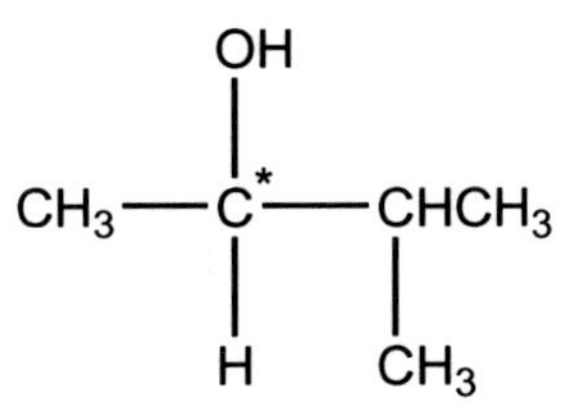

3-methyl-2-butanol

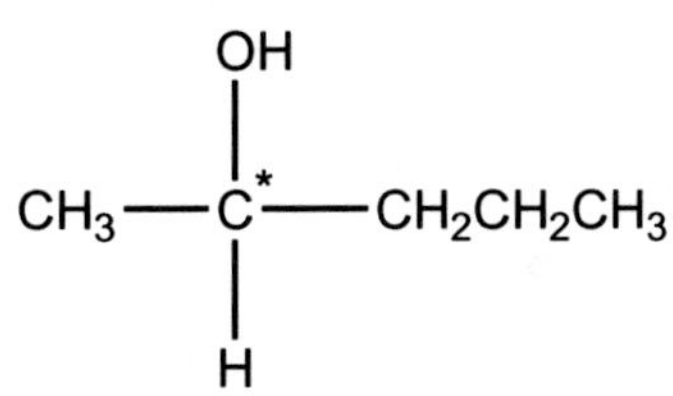

2-pentanol

No isomers of $C_5H_{12}O$ exhibit geometric isomerism because no double bonds or ring structures are possible with 12 hydrogens present.

c. We will assume the structure having the $C_6H_4Br_2$ formula is a benzene ring derivative. $C_6H_4Br_2$ exhibits structural isomerism only. Two structural isomers of $C_6H_4Br_2$ are:

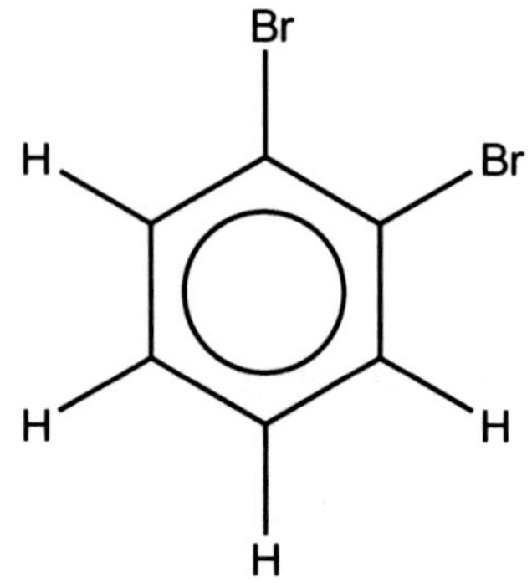

o-dibromobenzene
or 1,2-dibromobenzene

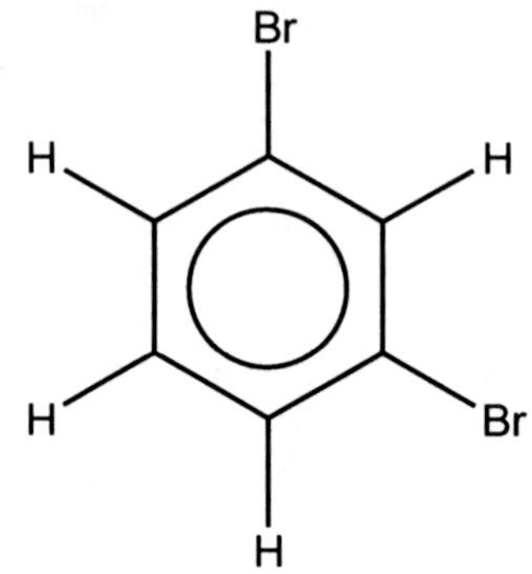

m-dibromobenzene
or 1,3-dibromobenzene

The benzene ring is planar and does not exhibit geometric isomerism. It also does not exhibit optical activity. All carbons only have three atoms bonded to them; it is impossible for benzene to be optically active.

Note: There are possible noncyclic structural isomers having the formula $C_6H_4Br_2$. These noncyclic isomers can, in theory, exhibit geometrical and optical isomerism. But they are beyond the introduction to organic chemistry given in this text.

3. a.

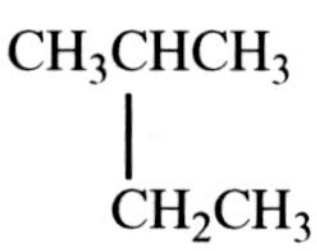

The longest chain is 4 carbons long. The correct name is 2-methylbutane.

b.

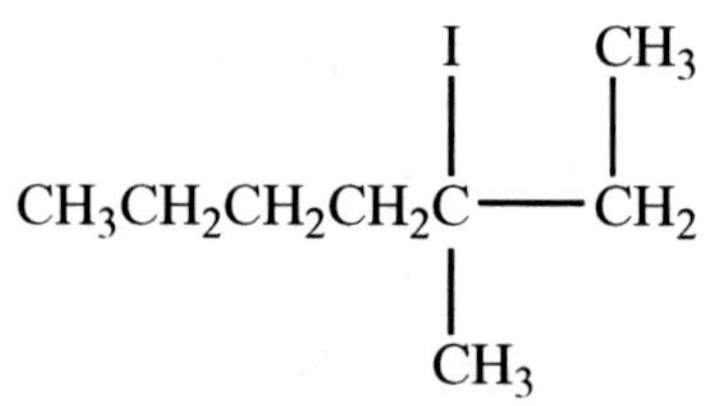

The longest chain is 7 carbons long, and we would start the numbering system at the other end for lowest possible numbers. The correct name is 3-iodo-3-methyl-heptane.

c.

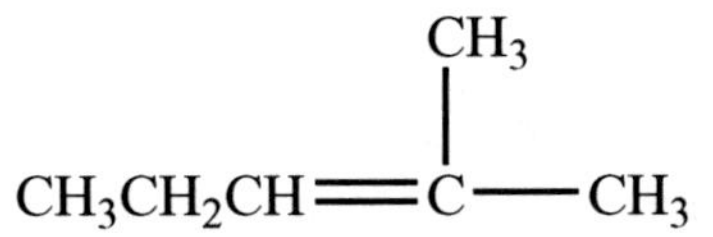

This compound cannot exhibit cis–trans isomerism since one of the double bonded carbons has the same two groups (CH_3) attached. The numbering system should also start at the other end to give the double bond the lowest possible number. 2-methyl-2-pentene is correct.

d.

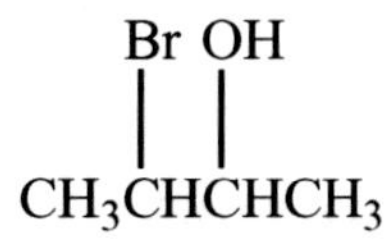

The OH functional group gets the lowest number. 3-bromo-2-butanol is correct.

4. a. 2-Chloro-2-butyne would have 5 bonds to the second carbon. Carbon never expands its octet.

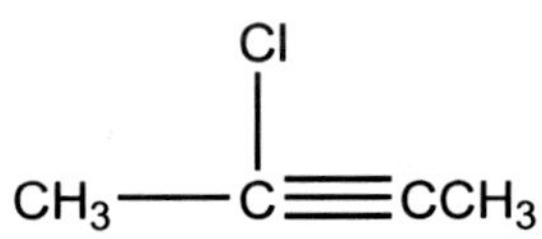

b. 2-Methyl-2-propanone would have 5 bonds to the second carbon.

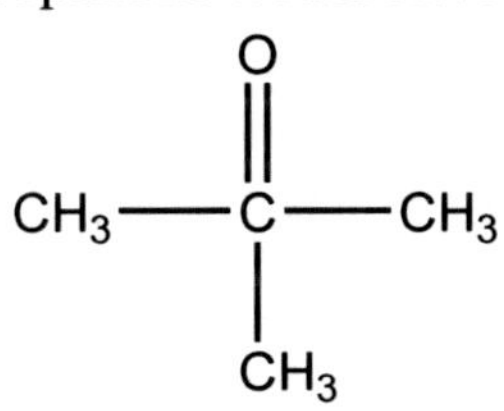

c. Carbon-1 in 1,1-dimethylbenzene would have 5 bonds.

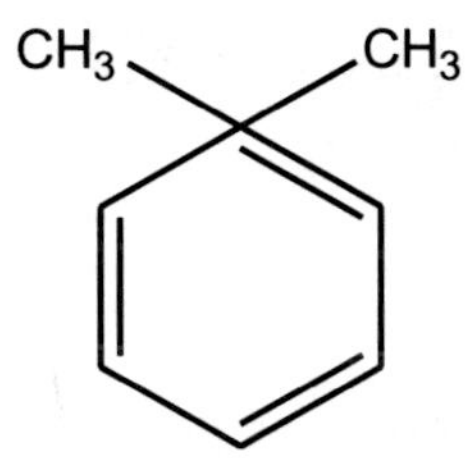

d. You cannot have an aldehyde functional group off a middle carbon in a chain. Aldehyde groups:

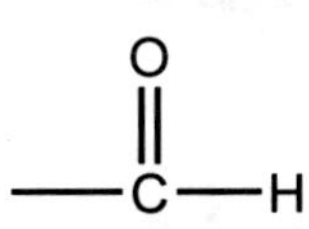

can only be at the beginning and/or the end of a chain of carbon atoms.

e. You cannot have a carboxylic acid group off a middle carbon in a chain. Carboxylic groups:

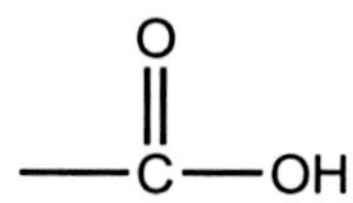

must be at the beginning and/or the end of a chain of carbon atoms.

f. In cyclobutanol, the 1 and 5 positions refer to the same carbon atom. 5,5-Dibromo-1-cyclobutanol would have five bonds to carbon-1. This is impossible; carbon never expands its octet.

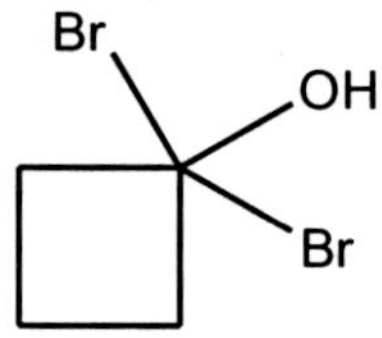

5. Hydrocarbons are nonpolar substances exhibiting only London dispersion forces. Size and shape are the two most important structural features relating to the strength of London dispersion forces. For size, the bigger the molecule (the larger the molar mass), the stronger are the London dispersion forces, and the higher is the boiling point. For shape, the more branching present in a compound, the weaker are the London dispersion forces, and the lower is the boiling point.

6. In order to hydrogen-bond, the compound must have at least one N–H, O–H or H–F covalent bond in the compound. In Table 22.4, alcohols and carboxylic acids have an O–H covalent bond, so they can hydrogen-bond. In addition, primary and secondary amines have at least one N–H covalent bond, so they can hydrogen-bond.

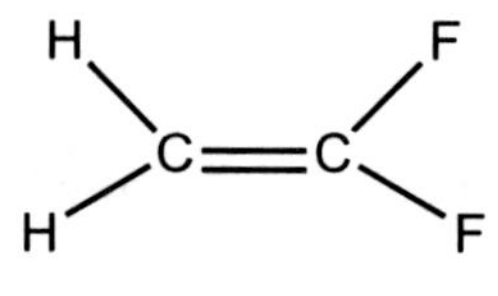

CH_2CF_2 cannot form hydrogen bonds because it has no hydrogens covalently bonded to the fluorine atoms.

7. The amide functional group is: 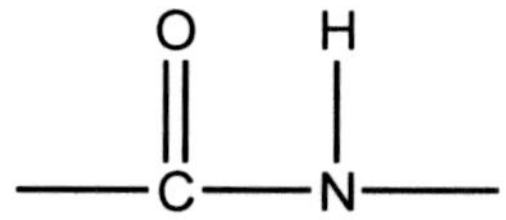

When the amine end of one amino acid reacts with the carboxylic acid end of another amino acid, the two amino acids link together by forming an amide functional group. A polypeptide has many amino acids linked together, with each linkage made by the formation of an amide functional group. Because all linkages result in the presence of the amide functional group, the resulting polymer is called a polyamide. For nylon, the monomers also link together by forming the amide functional group (the amine end of one monomer reacts with the carboxylic acid end of another monomer to give the amide functional group linkage). Hence nylon is also a polyamide.

The correct order of strength is:

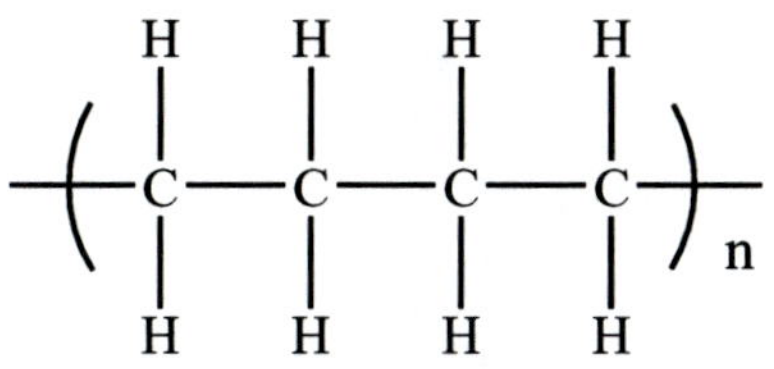

polyhydrocarbon
weakest fibers

<

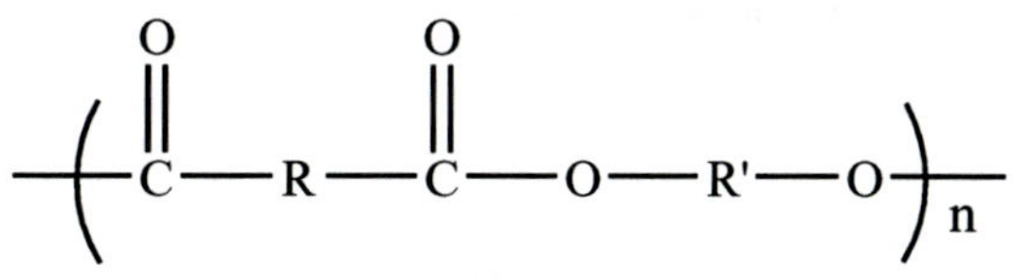

polyester

<

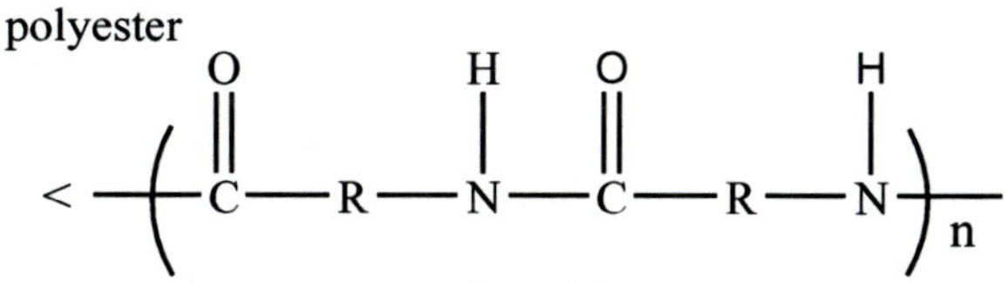

polyamide
strongest fibers

The difference in strength is related to the types of intermolecular forces present. All these types of polymers have London dispersion forces. However, the polar ester group in polyesters and the polar amide group in polyamides give rise to additional dipole forces. The polyamide has the ability to form relatively strong hydrogen-bonding interactions, hence why it would form the strongest fibers.

8. a.

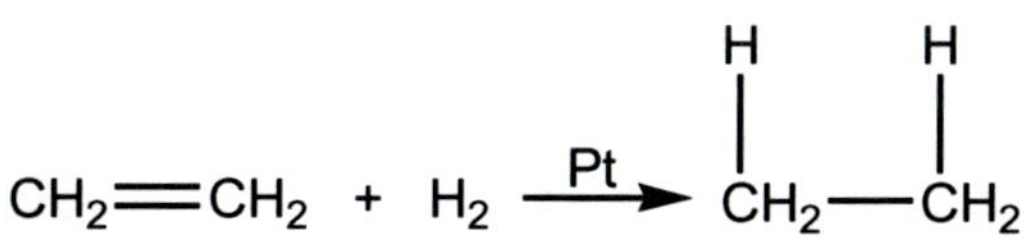

ethene ethane

b.

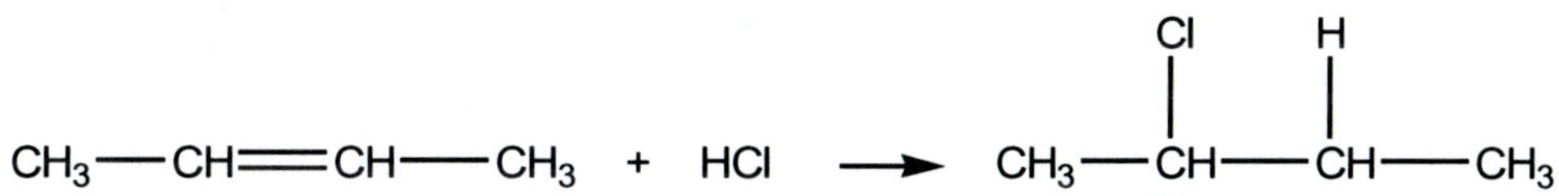

2-butene (cis or trans) 2-chlorobutane

c.

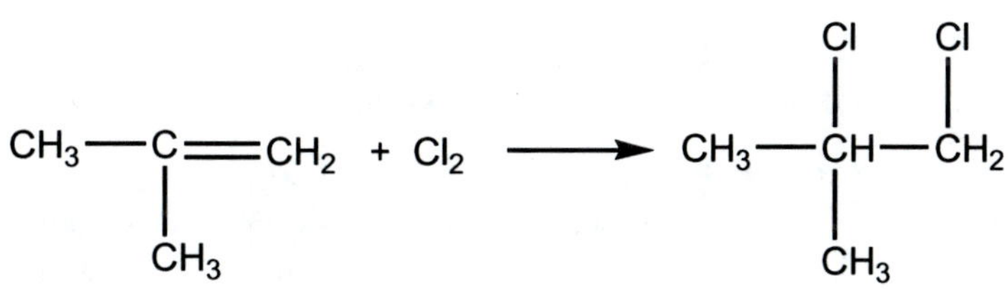

2-methyl-1-propene
(or 2-methylpropene)

1,2-dichloro-2-methylpropane

d.

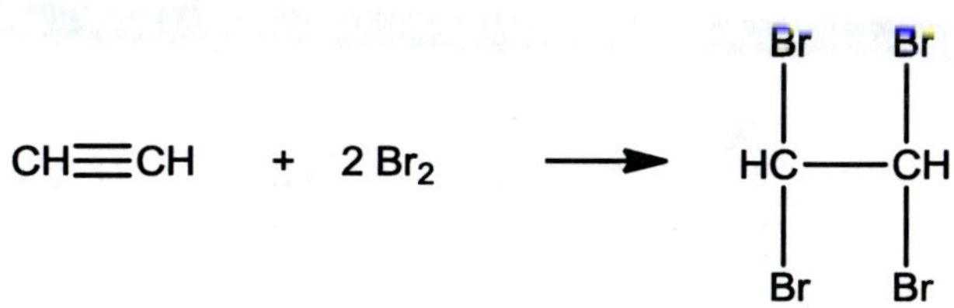

ethyne 1,1,2,2-tetrabromoethane

Another possibility would be:

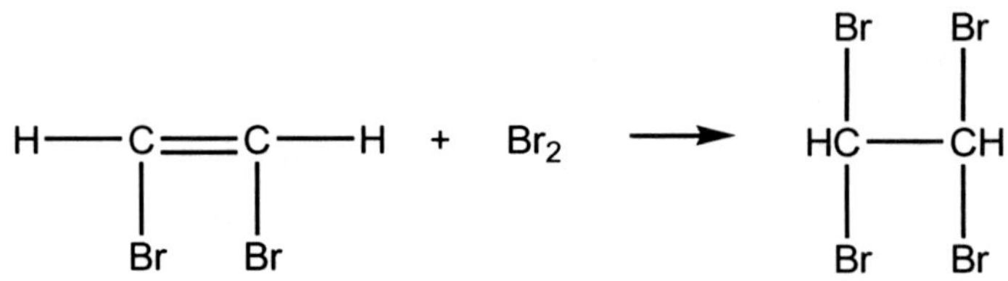

1,2-dibromoethene 1,1,2,2-tetrabromoethane

e.

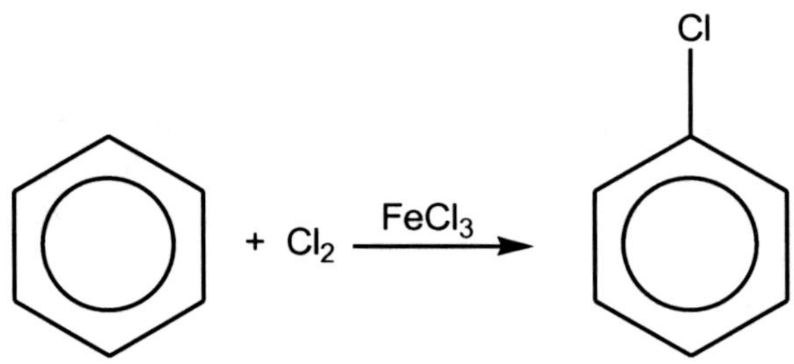

benzene chlorobenzene

f. $$CH_3CH_3 \xrightarrow[500^oC]{Cr_2O_3} CH_2{=}CH_2 + H_2$$

ethane ethene

or 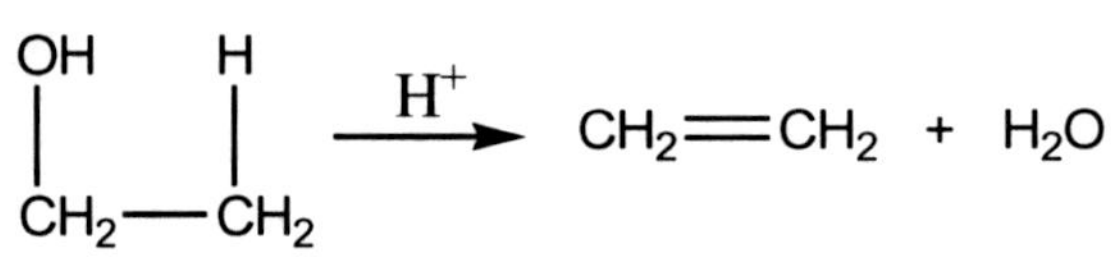

ethanol ethene

This reaction is not explicitly discussed in the text. This is the reverse of the reaction used to produce alcohols. This reaction is reversible. Which organic substance dominates is determined by LeChâtelier's principle. For example, if the alcohol is wanted, then water is removed as reactants are converted to products, driving the reaction to produce more water (and more alcohol).

9.

a. 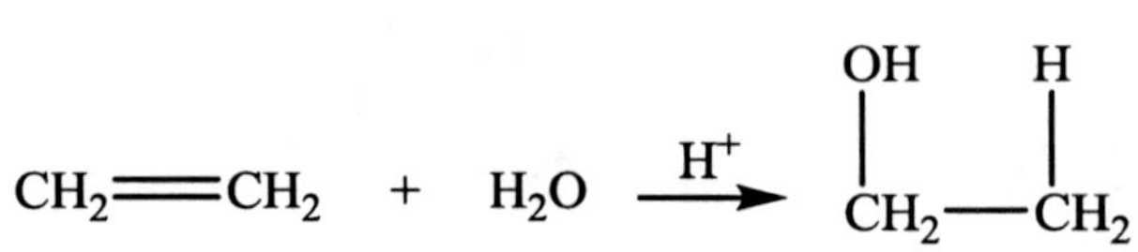 1° alcohol

b. 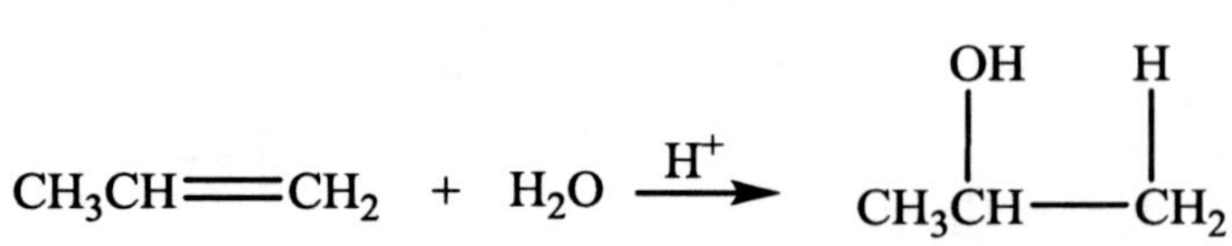 2° alcohol

major product

c. 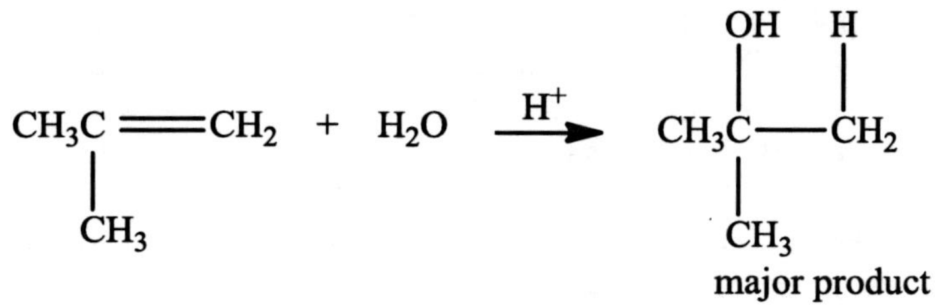 3° alcohol

d. 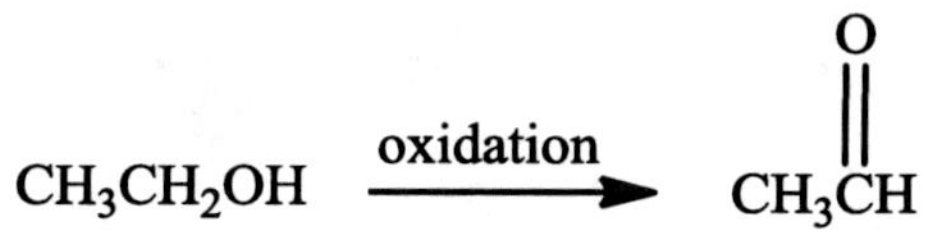 aldehyde

e. 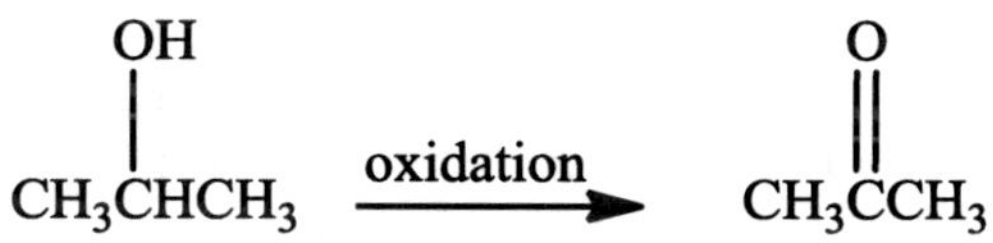 ketone

f.  carboxylic acid

or

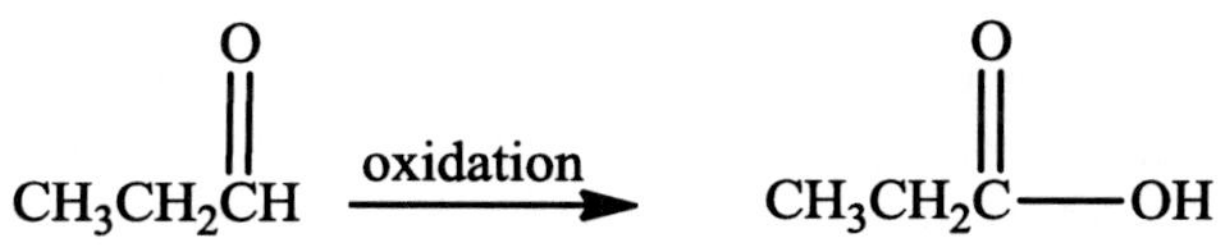

g. 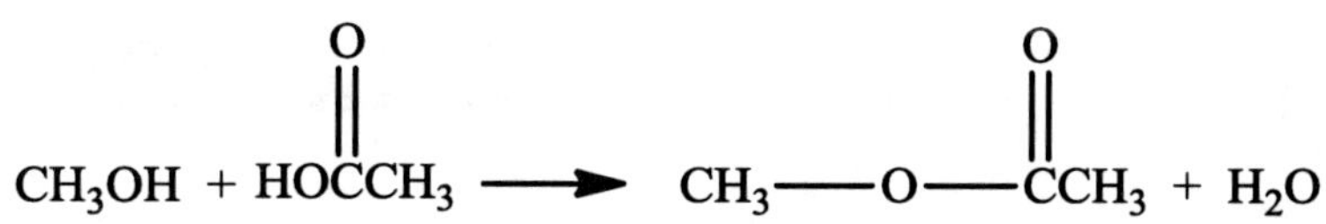 ester

10. Polystyrene is an addition polymer formed from the monomer styrene.

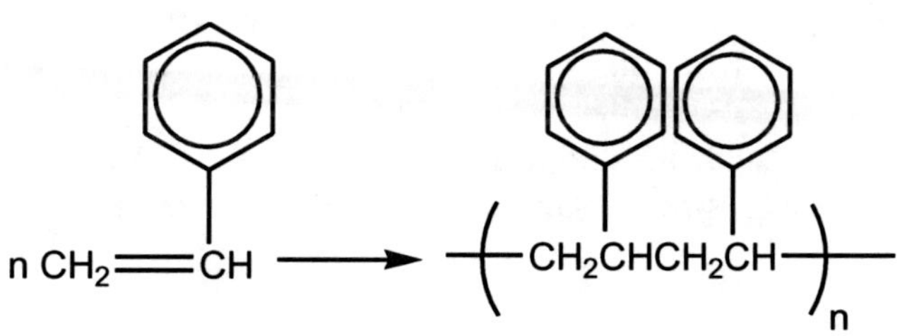

a. Syndiotactic polystyrene has all of the benzene ring side groups aligned on alternate sides of the chain. This ordered alignment of the side groups allows individual polymer chains of polystyrene to pack together efficiently, maximizing the London dispersion forces. Stronger London dispersion forces translate into stronger polymers.

b. By copolymerizing with butadiene, double bonds exist in the carbon backbone of the polymer. These double bonds can react with sulfur to form crosslinks (bonds) between individual polymer chains. The crosslinked polymer is stronger.

c. The longer the chain of polystyrene, the stronger are the London dispersion forces between polymer chains.

d. In linear (versus branched) polystyrene, chains pack together more efficiently, resulting in stronger London dispersion forces.

11. a. A polyester forms when an alcohol functional group reacts with a carboxylic acid functional group. The monomer for a homopolymer polyester must have an alcohol functional group and a carboxylic acid functional group present within the structure of the monomer.

b. A polyamide forms when an amine functional group reacts with a carboxylic acid functional group. For a copolymer polyamide, one monomer would have at least two amine functional groups present, and the other monomer would have at least two carboxylic acid functional groups present. For polymerization to occur, each monomer must have two reactive functional groups present.

c. To form an addition polymer, a carbon-carbon double bond must be present. Polyesters and polyamides are condensation polymers. To form a polyester, the monomer would need the alcohol and carboxylic acid functional groups present. To form a polyamide, the monomer would need the amine and carboxylic acid functional groups present. The two possibilities are for the monomer to have a carbon-carbon double bond, an alcohol functional group, and a carboxylic acid functional group present or to have a carbon-carbon double bond, an amine functional group, and a carboxylic acid functional group present.

12. Proteins are polymers made up of monomer units called amino acids. One of the functions of proteins is to provide structural integrity and strength for many types of tissues. In addition, proteins transport and store oxygen and nutrients, catalyze many reactions in the body, fight invasion by foreign objects, participate in the body's many regulatory systems, and transport electrons in the process of metabolizing nutrients.

Carbohydrate polymers, such as starch and cellulose, are composed of the monomer units called monosaccharides or simple sugars. Carbohydrates serve as a food source for most organisms.

Nucleic acids are polymers made up of monomer units called nucleotides. Nucleic acids store and transmit genetic information and are also responsible for the synthesis of various proteins needed by a cell to carry out its life functions.

Exercises

Hydrocarbons

13. i.

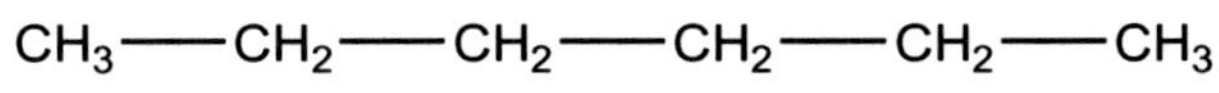

ii.

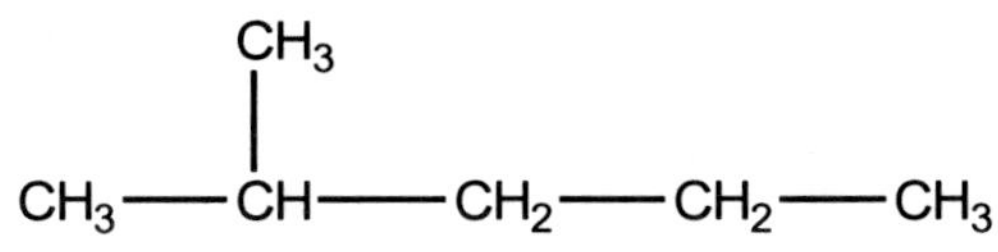

iii.

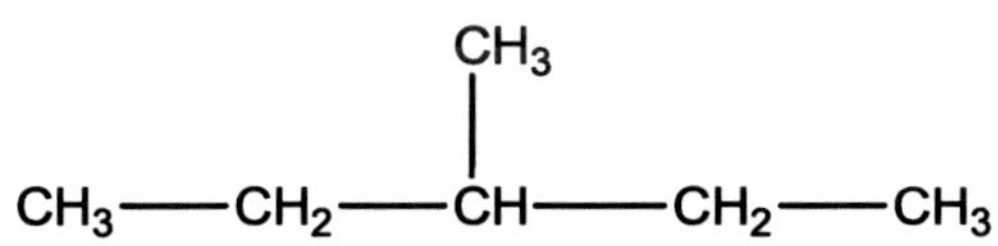

iv.

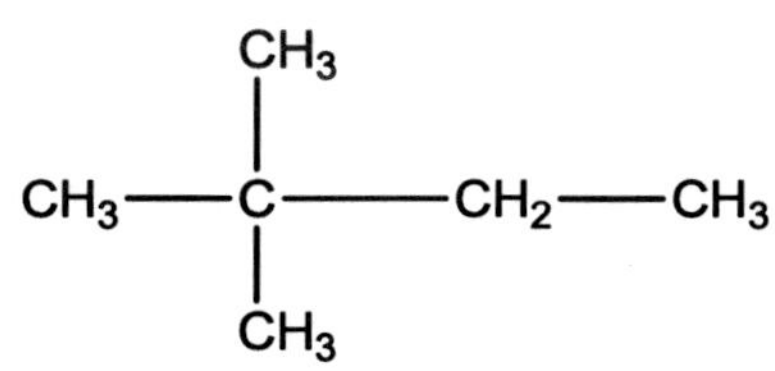

v.

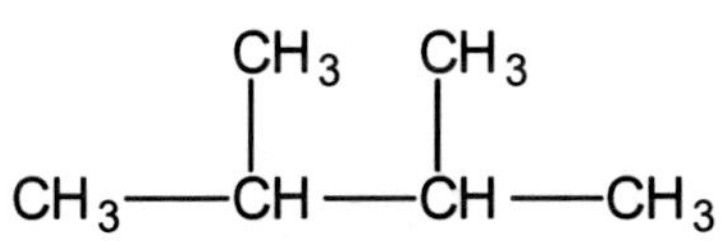

All other possibilities are identical to one of these five compounds.

14. See Exercise 13 for the structures. The names of structures i-v respectively, are hexane (or n-hexane), 2-methylpentane, 3-methylpentane, 2,2-dimethylbutane, and 2,3-dimethylbutane.

15. A difficult task in this problem is recognizing different compounds from compounds that differ by rotations about one or more C–C bonds (called conformations). The best way to distinguish different compounds from conformations is to name them. Different name = different compound; same name = same compound, so it is not an isomer but instead is a conformation.

a.

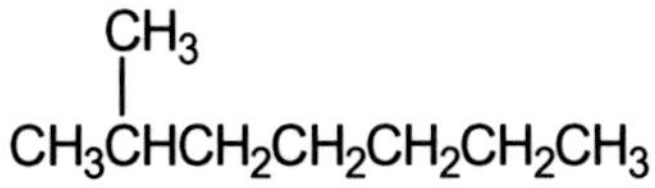

2-methylheptane

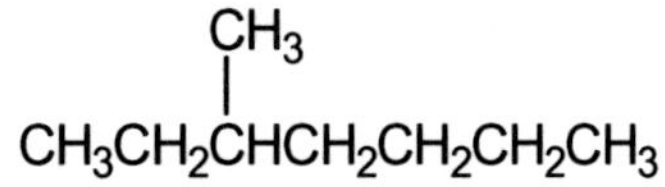

3-methylheptane

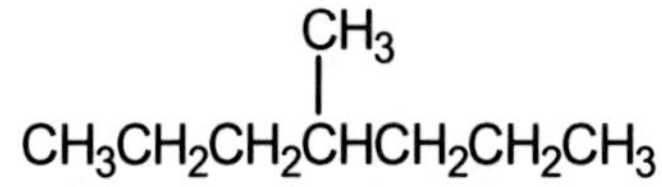

4-methylheptane

b.

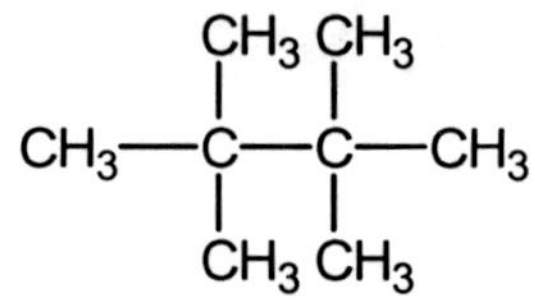

2,2,3,3-tetramethylbutane

16. a.

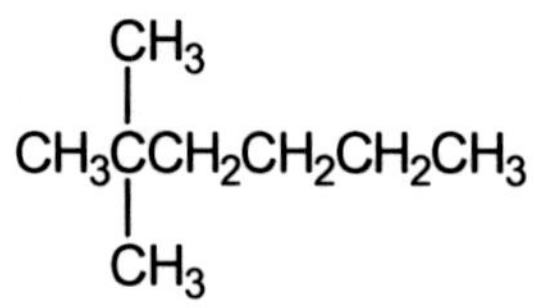

2,2-dimethylhexane

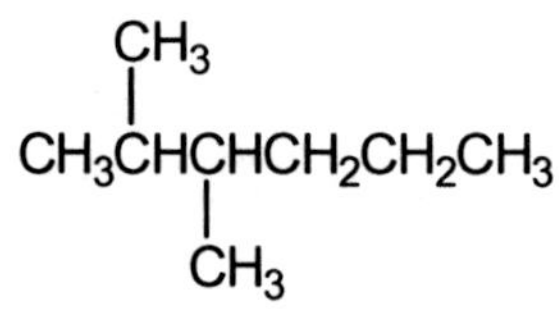

2,3-dimethylhexane

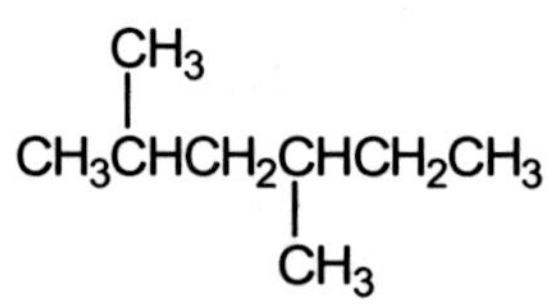

2,4-dimethylhexane

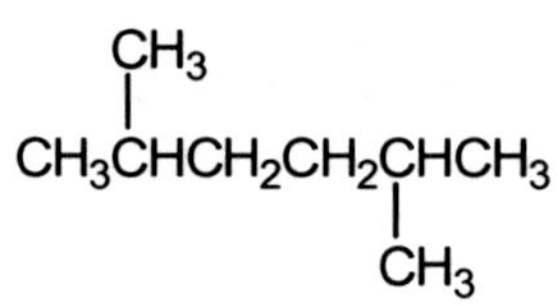

2,5-dimethylhexane

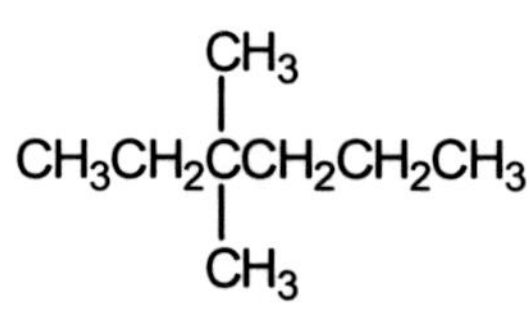

3,3-dimethylhexane

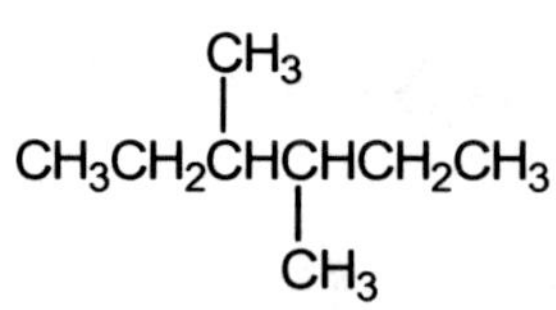

3,4-dimethylhexane

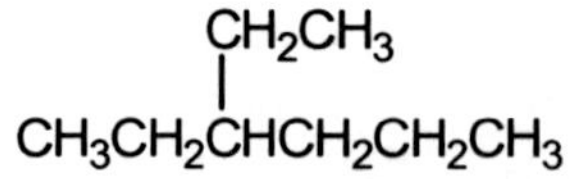

3-ethylhexane

b.

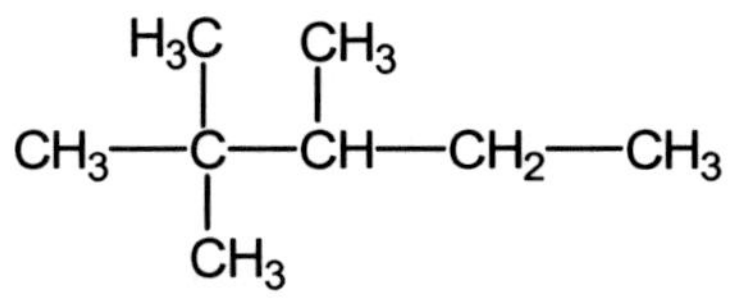

2,2,3-trimethylpentane

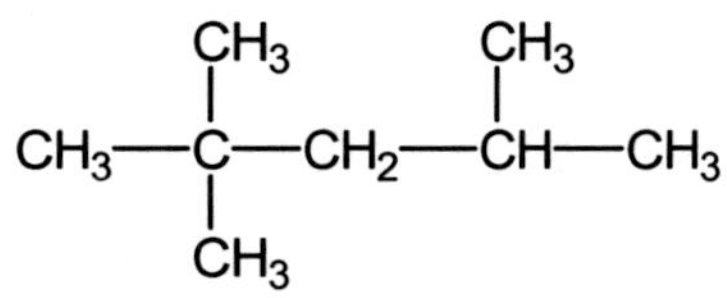

2,2,4-trimethylpentane

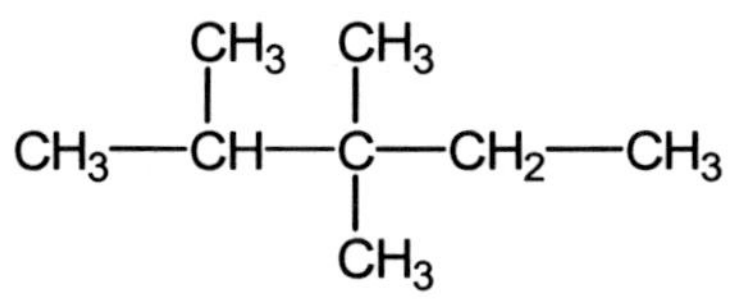

2,3,3-trimethylpentane

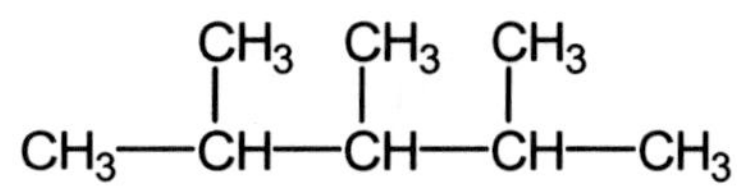

2,3,4-trimethylpentane

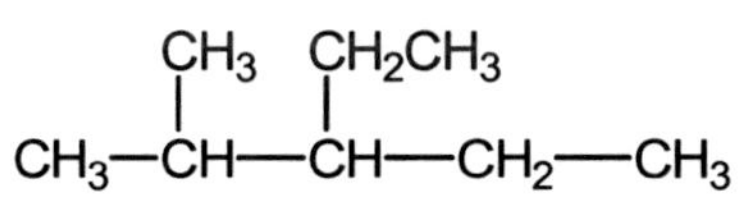

3-ethyl-2-methylpentane

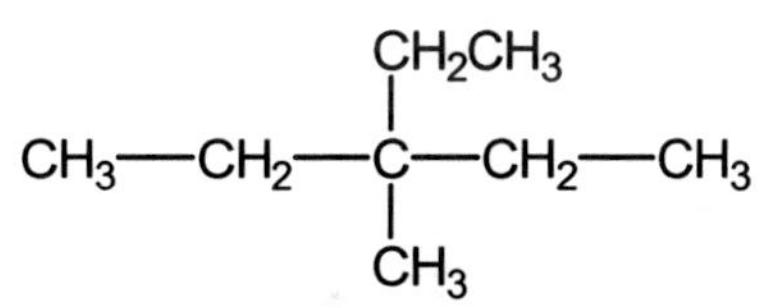

3-ethyl-3-methylpentane

17. a. 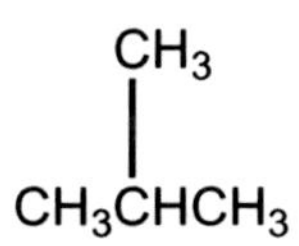

b. 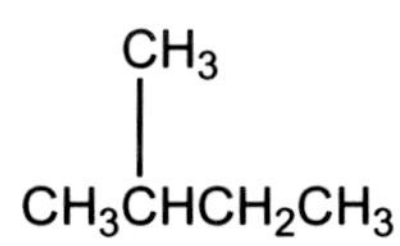

c. 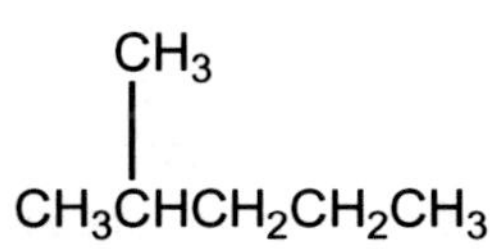

d. 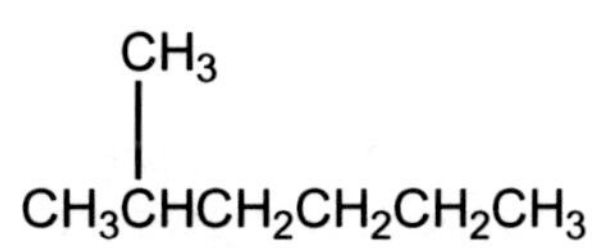

18. a. 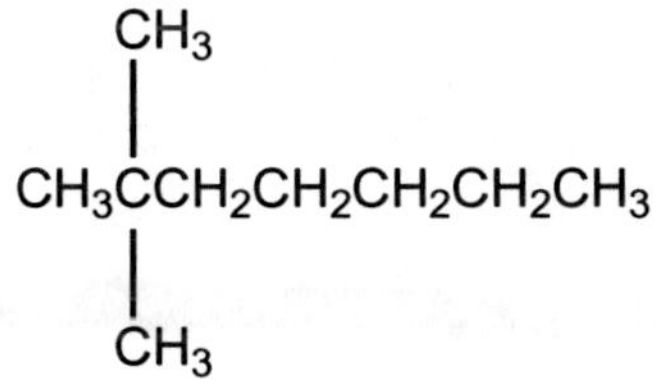

b. 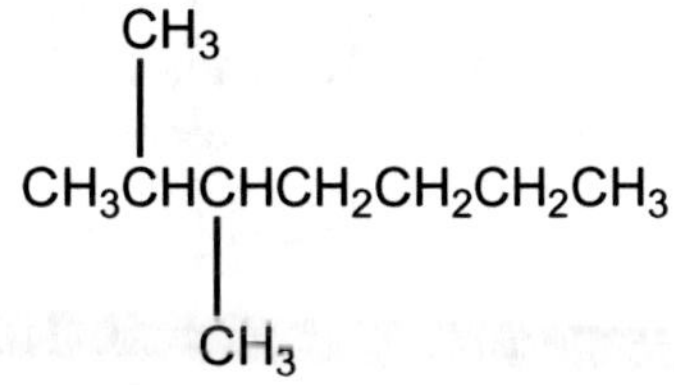

c. 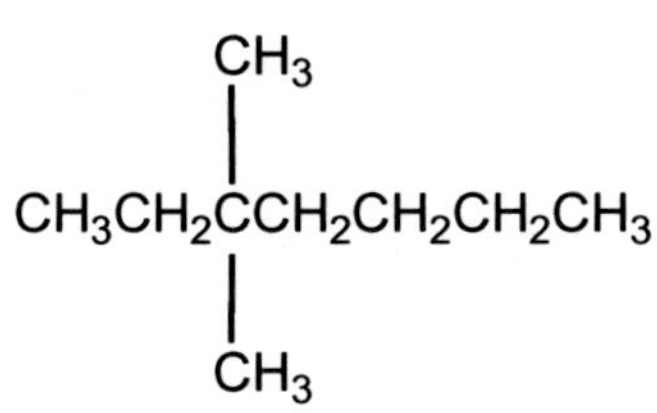

d. 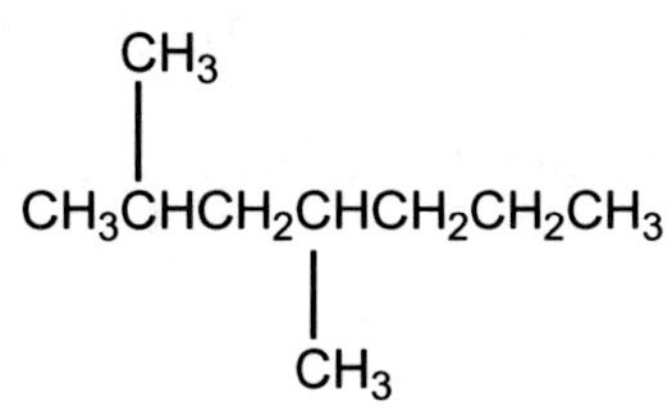

19. a.

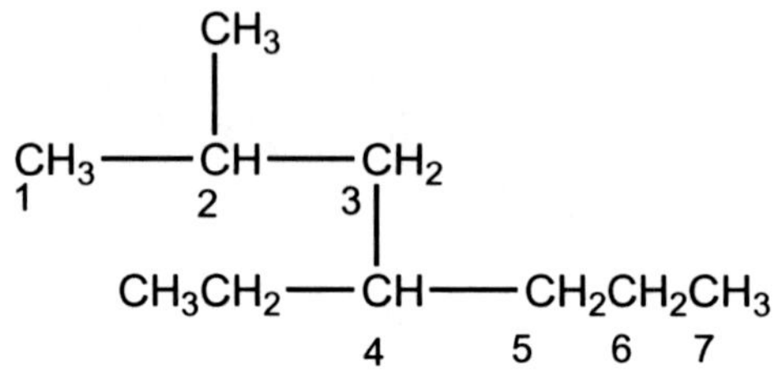

b.

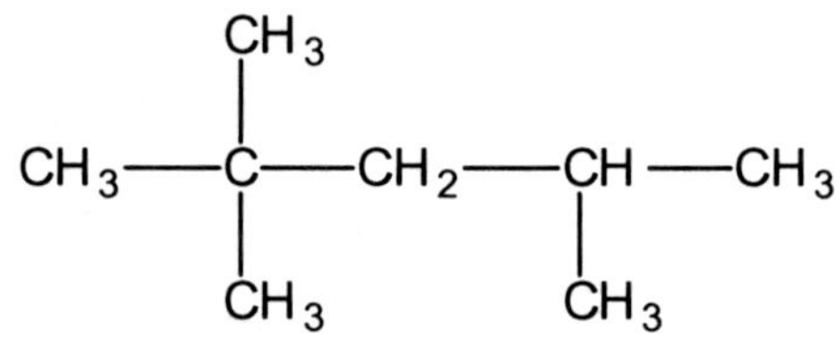

c.

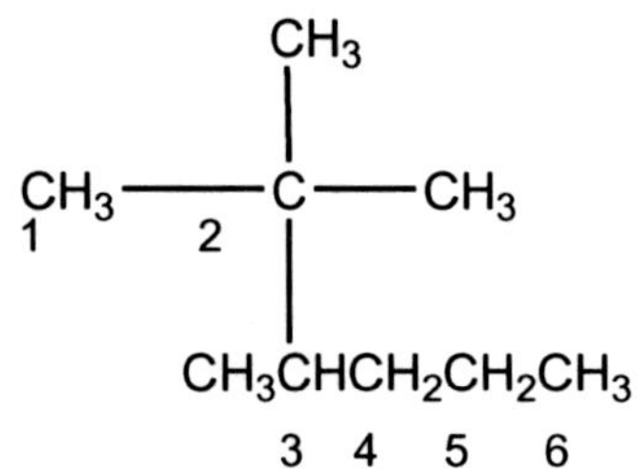

d. For 3-isobutylhexane, the longest chain is 7 carbons long. The correct name is 4-ethyl-2-methylheptane. For 2-tert-butylpentane, the longest chain is 6 carbons long. The correct name is 2,2,3-trimethylhexane.

20.

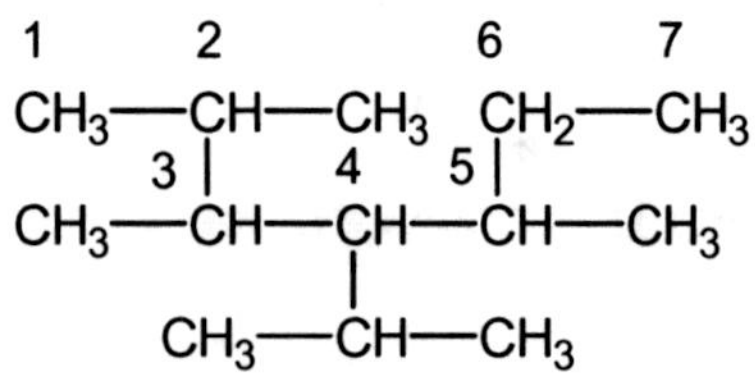

4-isopropyl-2,3,5-trimethylheptane

21. a. 2,2,4-trimethylhexane b. 5-methylnonane c. 2,2,4,4-tetramethylpentane

d. 3-ethyl-3-methyloctane

Note: For alkanes, always identify the longest carbon chain for the base name first, then number the carbons to give the lowest overall numbers for the substituent groups.

22. The hydrogen atoms in ring compounds are commonly omitted. In organic compounds, carbon atoms satisfy the octet rule of electrons by forming four bonds to other atoms. Therefore, add C-H bonds to the carbon atoms in the ring in order to give each C atom four bonds. You can also determine the formula of these cycloalkanes by using the general formula C_nH_{2n}.

a. isopropylcyclobutane; C_7H_{14}

b. 1-tert-butyl-3-methylcyclopentane; $C_{10}H_{20}$

c. 1,3-dimethyl-2-propylcyclohexane; $C_{11}H_{22}$

23. $CH_3—CH_2—CH_2—CH_3$

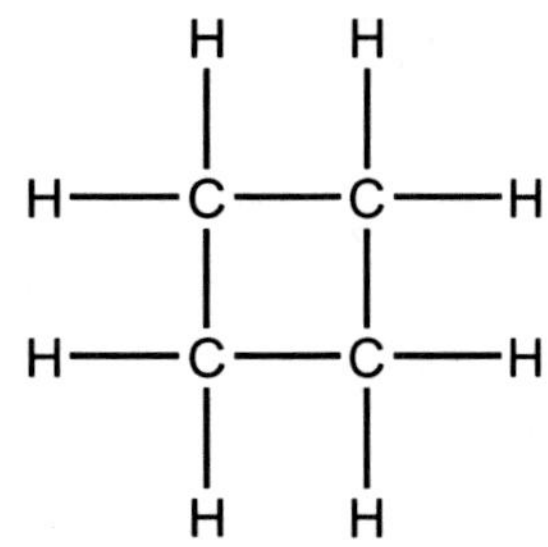

Each carbon is bonded to four other carbon and/or hydrogen atoms in a saturated hydrocarbon (only single bonds are present).

24. $CH_2=CH_2$ $HC\equiv C—CH=CH_2$

An unsaturated hydrocarbon has at least one carbon-carbon double and/or triple bond in the structure.

25. a. 1-butene b. 4-methyl-2-hexene c. 2,5-dimethyl-3-heptene

Note: The multiple bond is assigned the lowest number possible.

26. a. 2,3-dimethyl-2-butene b. 4-methyl-2-hexyne

c. 2,3-dimethyl-1-pentene

27. a. $CH_3–CH_2–CH=CH–CH_2–CH_3$ b. $CH_3–CH=CH–CH=CH–CH_2CH_3$

c.

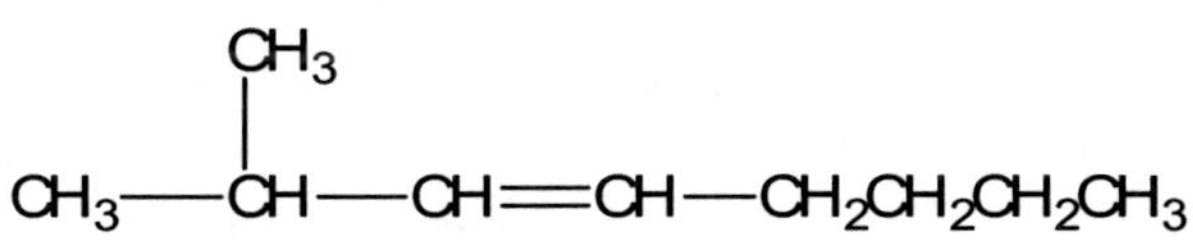

28.

a. 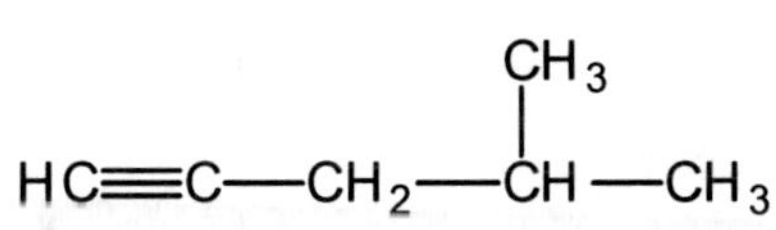

b. 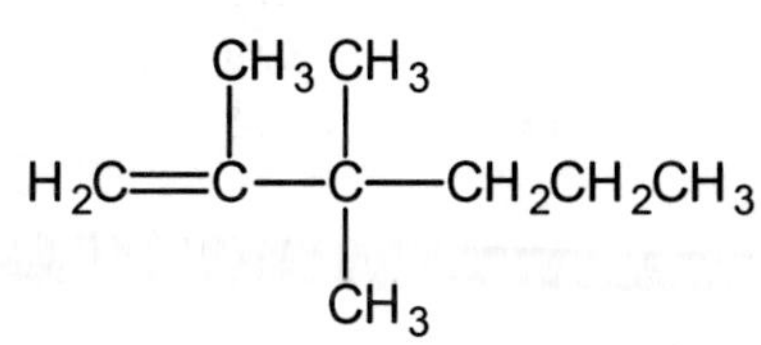

c. 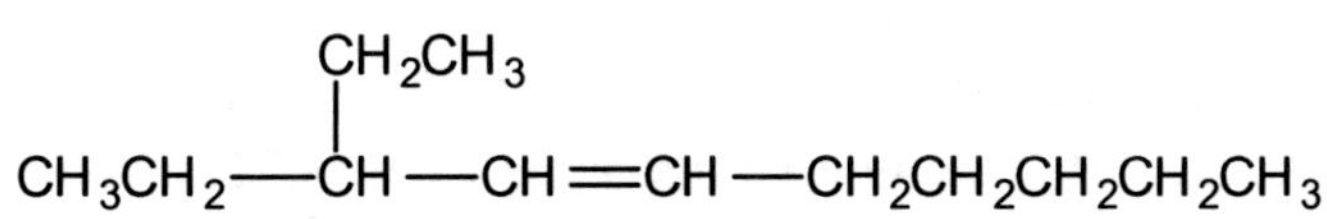

29. a.

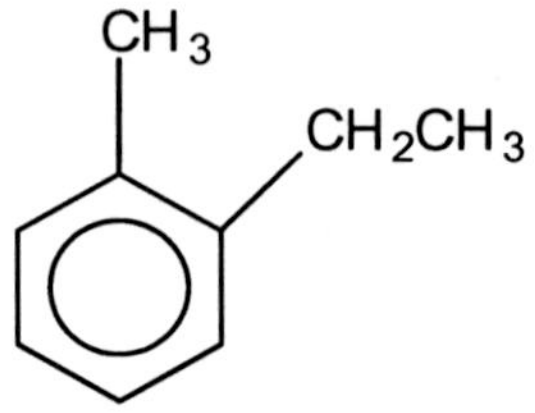

b.

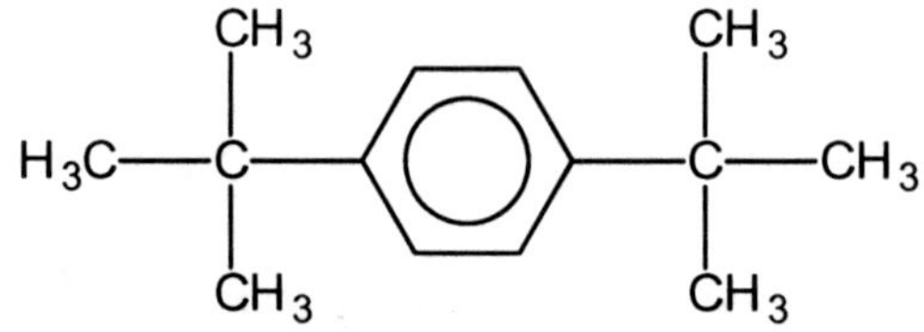

c.

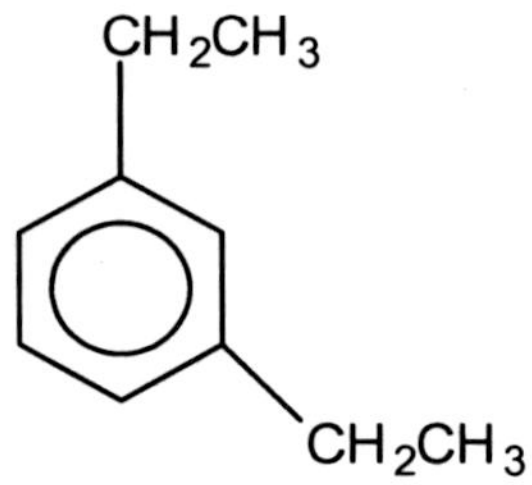

d.

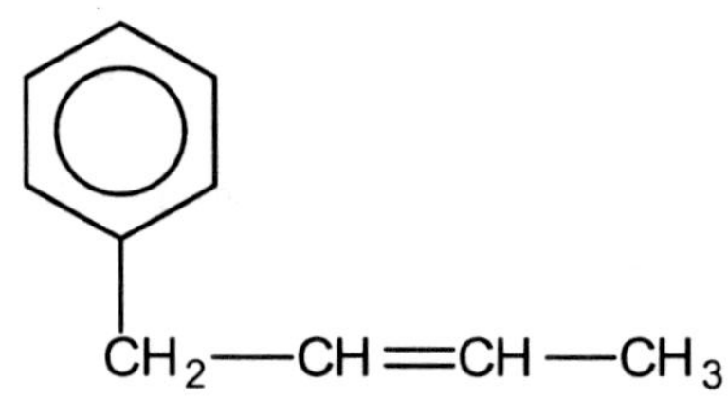

30. isopropylbenzene or 2-phenylpropane

31. a. 1,3-dichlorobutane

b. 1,1,1-trichlorobutane

c. 2,3-dichloro-2,4-dimethylhexane

d. 1,2-difluoroethane

32. a. 3-chloro-l-butene

b. 1-ethyl-3-methycyclopentene

c. 3-chloro-4-propylcyclopentene

d. 1,2,4-trimethylcyclohexane

e. 2-bromotoluene (or o-bromotoluene or 1-bromo-2-methylbenzene)

f. 1-bromo-2-methylcyclohexane

g. 4-bromo-3-methylcyclohexene

Note: If the location of the double bond is not given in the name, it is assumed to be located between C_1 and C_2. Also, when the base name can be numbered in equivalent ways, give the first substituent group the lowest number; e.g., for part f, 1-bromo-2-methylcyclohexane is preferred to 2-bromo-1-methycyclohexane.

Isomerism

33. $CH_2Cl–CH_2Cl$, 1,2-dichloroethane: In this compound, there is free rotation about the C–C single bond that doesn't lead to different compounds. $CHCl=CHCl$, 1,2-dichloroethene: This compound, however, has no free rotation about the C=C double bond. This creates the cis and trans isomers, which are different compounds.

34. a. All of these structures have the formula C_5H_8. The compounds with the same physical properties will be the compounds that are identical to each other, i.e., compounds that only differ by rotations of C–C single bonds. To recognize identical compounds, name them. The names of the compounds are:

i. trans-1,3-pentadiene ii. cis-1,3-pentadiene

iii. cis-1,3-pentadiene iv. 2-methyl-1,3-butadiene

Compounds ii and iii are identical compounds, so they would have the same physical properties.

b. Compound i is a trans isomer because the bulkiest groups off the C_3=C_4 double bond are on opposite sides of the double bond.

c. Compound iv does not have carbon atoms in a double bond that each have two different groups attached. Compound iv does not exhibit cis-trans isomerism.

35. To exhibit cis-trans isomerism, each carbon in the double bond must have two structurally different groups bonded to it. In Exercise 25, this occurs for compounds b and c. The cis isomer has the bulkiest groups on the same side of the double bond while the trans isomer has the bulkiest groups on opposite sides of the double bond. The cis and trans isomers for 25b and 25c are:

25 b.

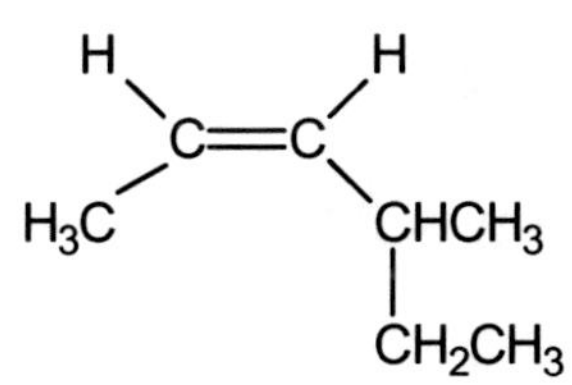

cis

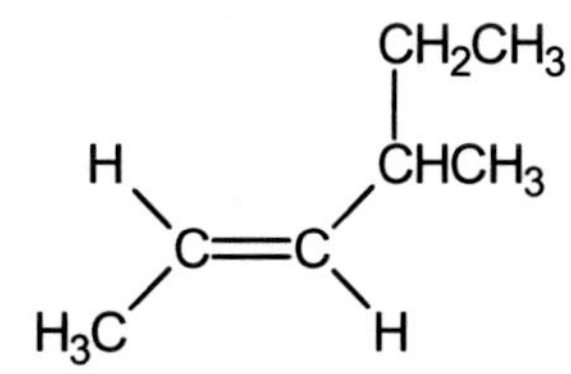

trans

25 c.

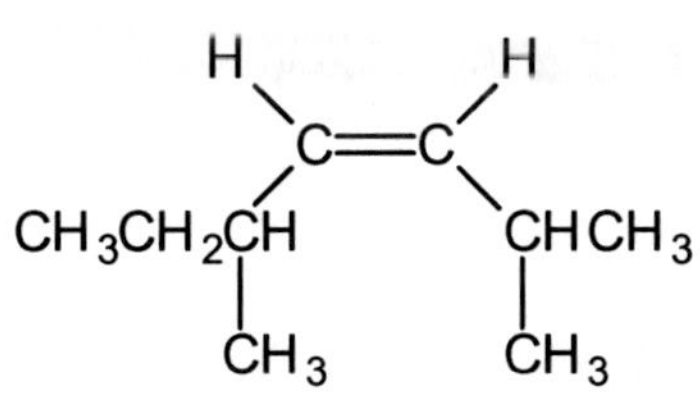

cis

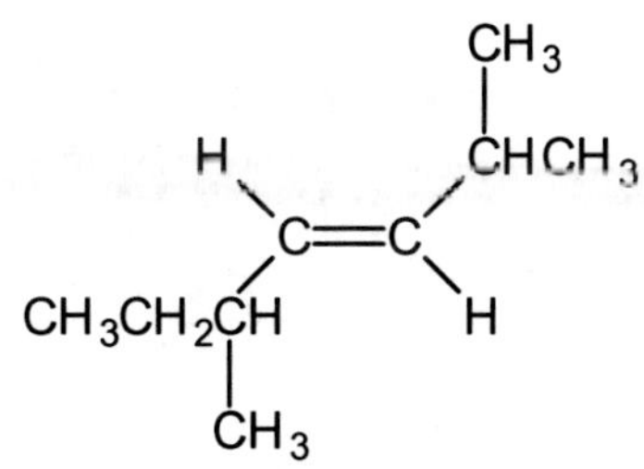

trans

Similarly, all the compounds in Exercise 27 exhibit *cis-trans* isomerism.

In compound a of Exercise 25, the first carbon in the double bond does not contain two different groups. The first carbon in the double bond contains two H atoms. To illustrate that this compound does not exhibit *cis-trans* isomerism, let's look at the potential *cis-trans* isomers.

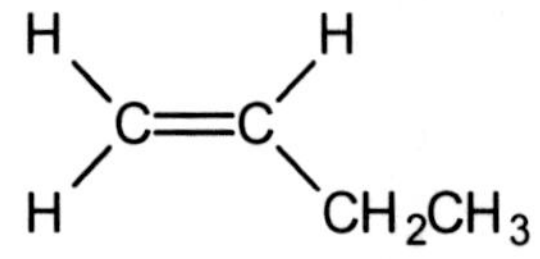

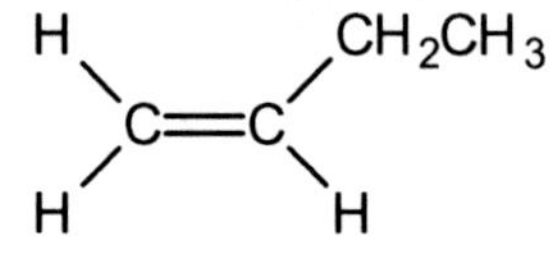

These are the same compounds; they only differ by a simple rotation of the molecule. Therefore, they are not isomers of each other but instead are the same compound.

36. In Exercise 26, none of the compounds can exhibit *cis-trans* isomerism since none of the carbons with the multiple bond have two different groups bonded to each. In Exercise 28, only 3-ethyl-4-decene can exhibit *cis-trans* isomerism since the fourth and fifth carbons each have two different groups bonded to the carbon atoms with the double bond.

37. C_5H_{10} has the general formula for alkenes, C_nH_{2n}. To distinguish the different isomers from each other, we will name them. Each isomer must have a different name.

$CH_2{=}CHCH_2CH_2CH_3$

1-pentene

$CH_3CH{=}CHCH_2CH_3$

2-pentene

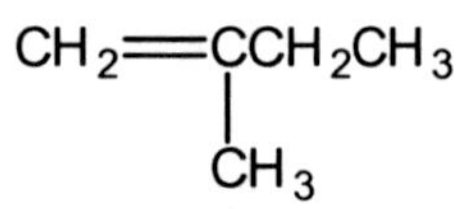

2-methyl-1-butene

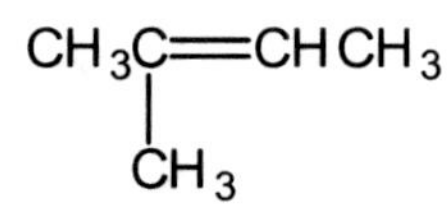

2-methyl-2-butene

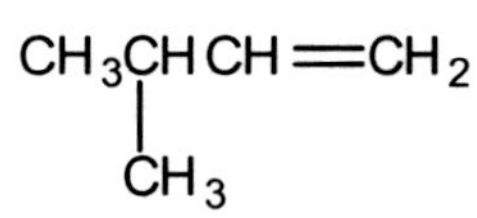

3-methyl-1-butene

38. Only 2-pentene exhibits cis-trans isomerism. The isomers are:

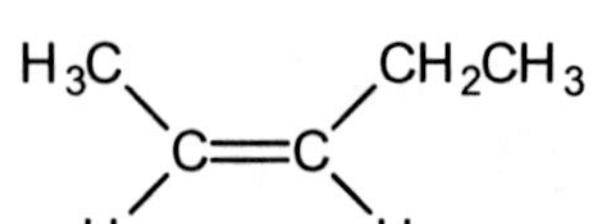

cis

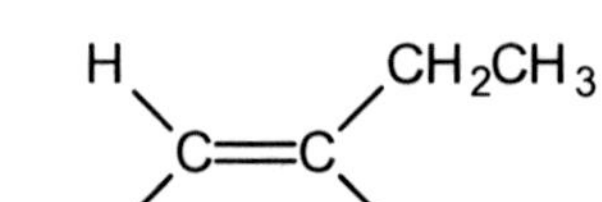

trans

The other isomers of C_5H_{10} do not contain carbons in the double bonds that each have two different groups attached.

39. To help distinguish the different isomers, we will name them.

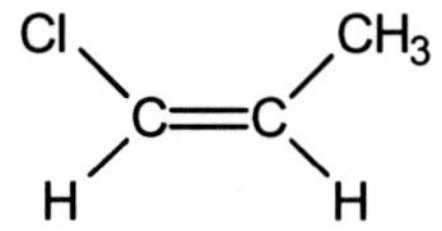

cis-1-chloro-1-propene

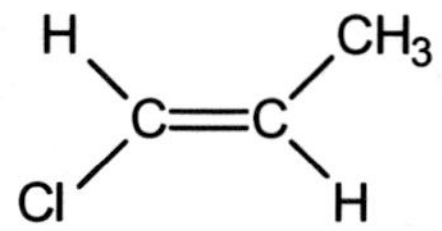

trans-1-chloro-1-propene

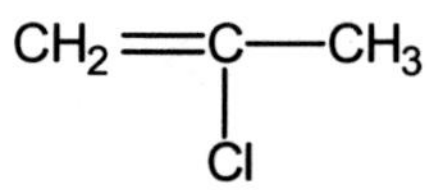

2-chloro-1-propene

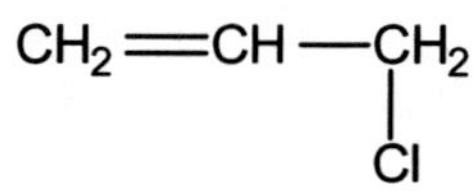

3-chloro-1-propene

chlorocyclopropane

40. $HCBrCl–CH=CH_2$

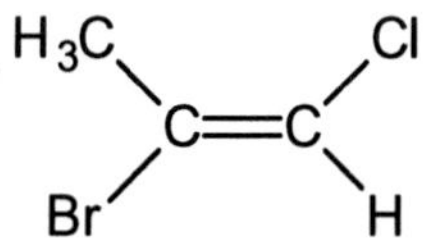

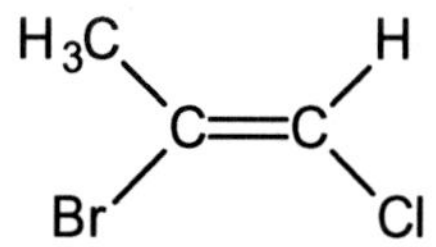

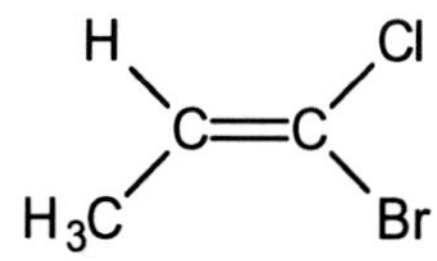

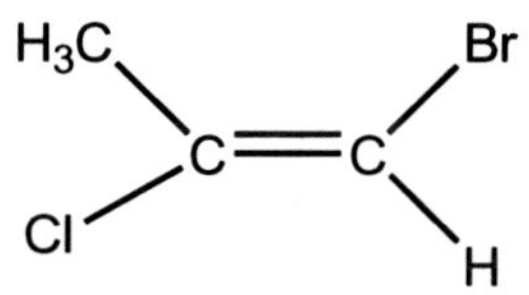

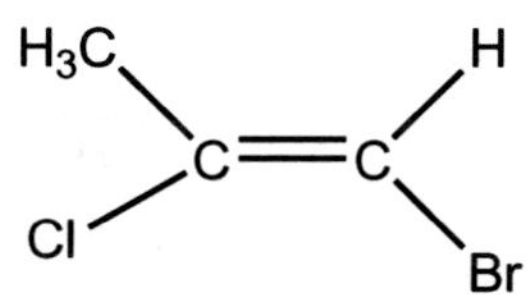

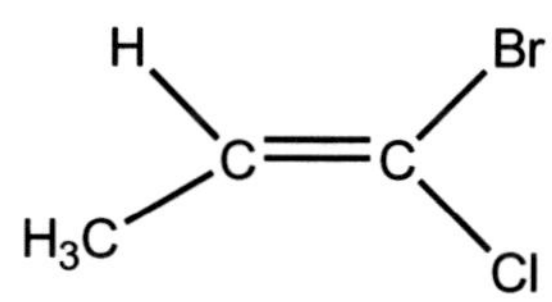

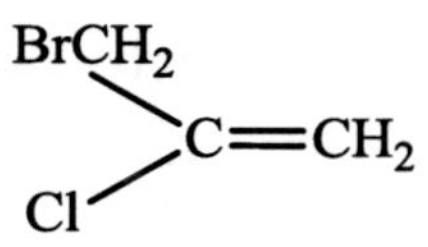

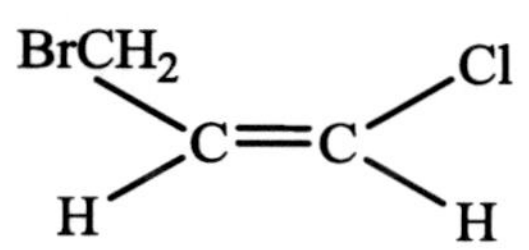

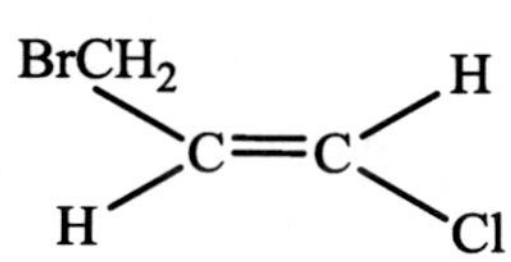

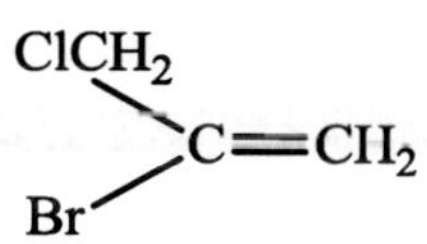

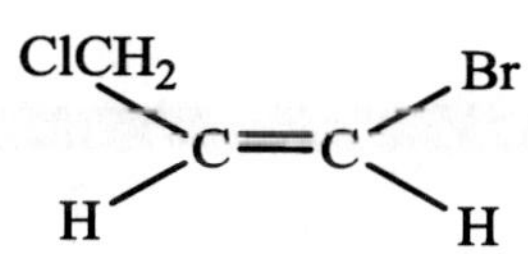

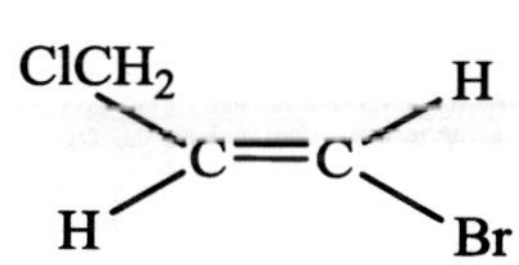

The cyclic isomers of bromochloropropene (C_3H_4BrCl) are:

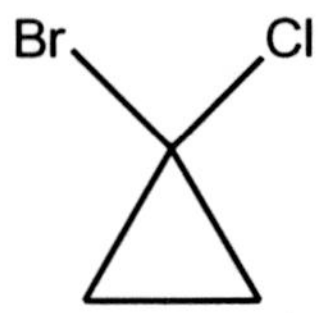

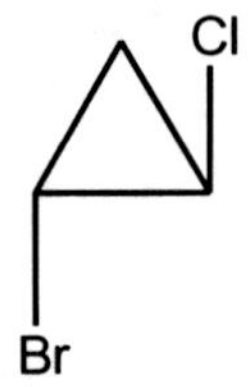

trans

cis

41.

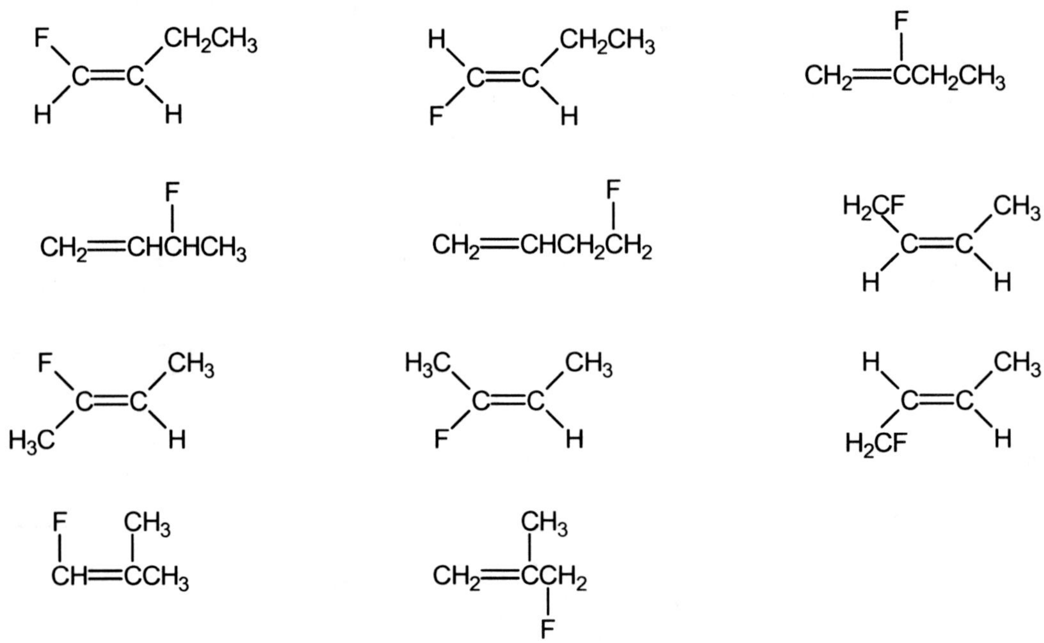

42. The *cis* isomer has the CH_3 groups on the same side of the ring. The *trans* isomer has the CH_3 groups on opposite sides of the ring.

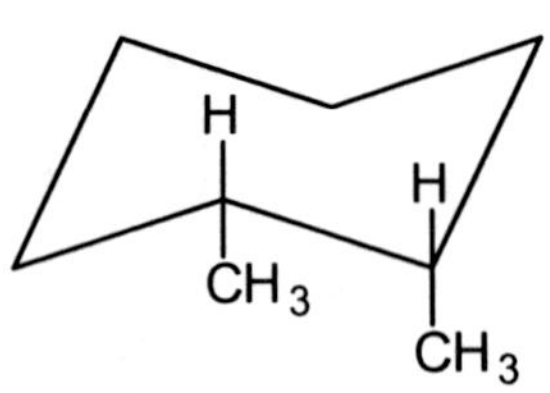

cis

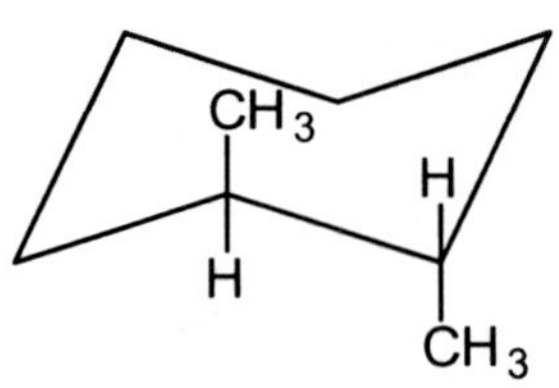

trans

The cyclic structural and geometric isomers of C_4H_7F are:

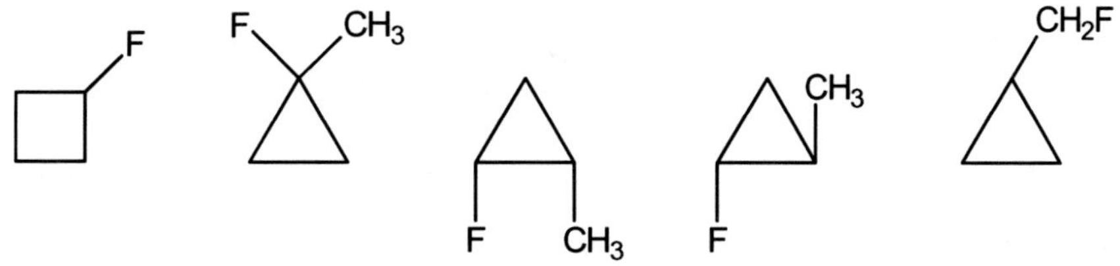

cis trans

43. a. 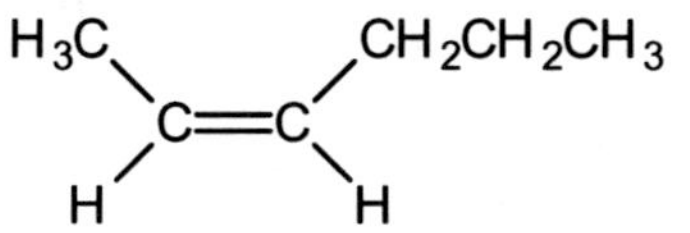 b. 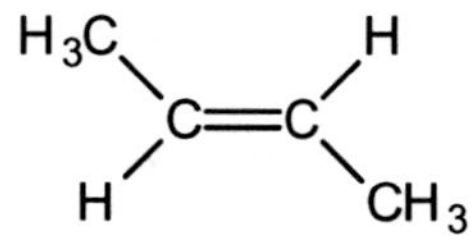 c. 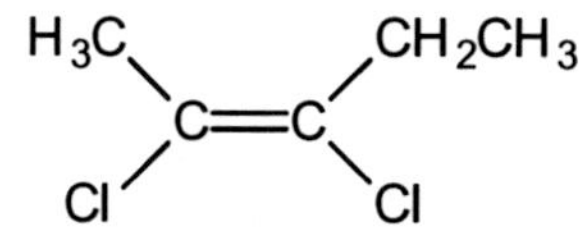

44. a. cis-1-bromo-1-propene b. cis-4-ethyl-3-methyl-3-heptene

c. trans-1,4-diiodo-2-propyl-1-pentene

Note: In general, cis-trans designations refer to the relative positions of the largest groups. In compound b, the largest group off the first carbon in the double bond is CH_2CH_3, and the largest group off the second carbon in the double bond is $CH_2CH_2CH_3$. Because their relative placement is on the same side of the double bond, this is the cis isomer.

45. a.

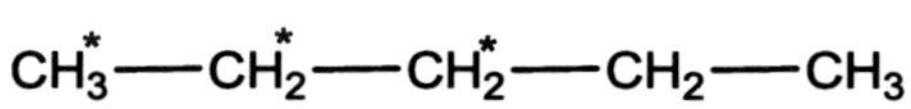

There are three different types of hydrogens in n-pentane (see asterisks). Thus there are three monochloro isomers of n-pentane (1-chloropentane, 2-chloropentane, and 3-chloropentane).

b.

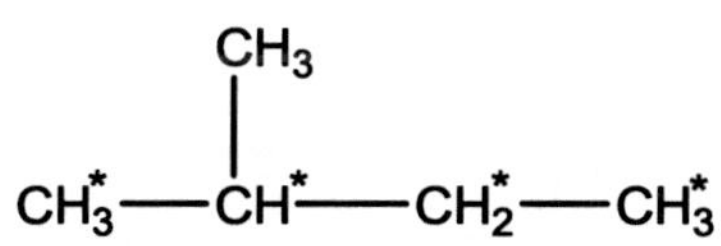

There are four different types of hydrogens in 2-methylbutane, so four monochloro isomers of 2-methylbutane are possible.

c.

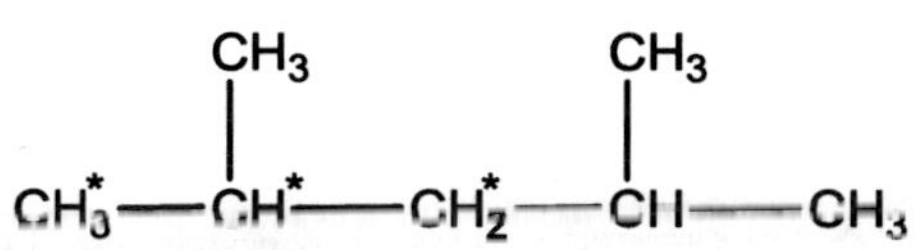

There are three different types of hydrogens, so three monochloro isomers are possible.

d.

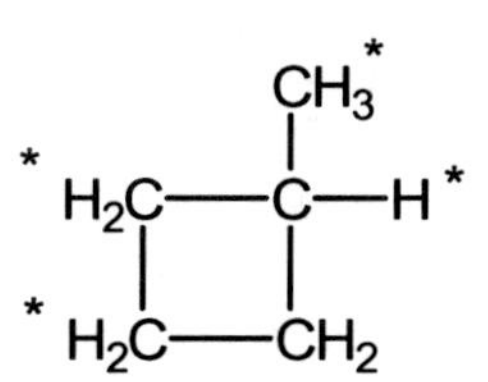

There are four different types of hydrogens, so four monochloro isomers are possible.

46. a.

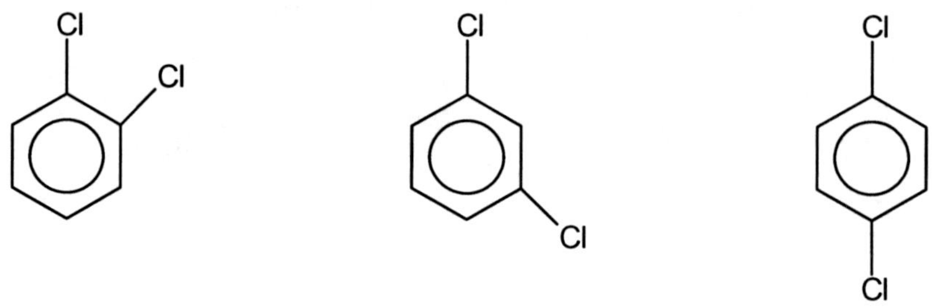

ortho meta para

b. There are three trichlorobenzenes (1,2,3-trichlorobenzene, 1,2,4-trichlorobenzene, and 1,3,5-trichlorobenzene).

c. The meta isomer will be very difficult to synthesize.

d. 1,3,5-Trichlorobenzene will be the most difficult to synthesize since all Cl groups are meta to each other in this compound.

Functional Groups

47. Reference Table 22.5 for the common functional groups.

a. ketone b. aldehyde

c. carboxylic acid d. amine

48. a.

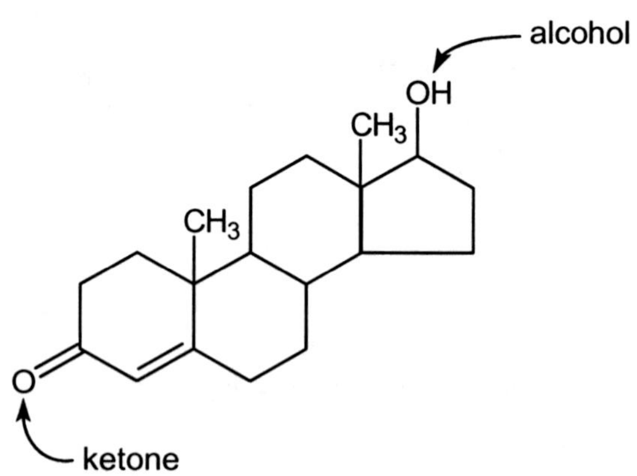

b.

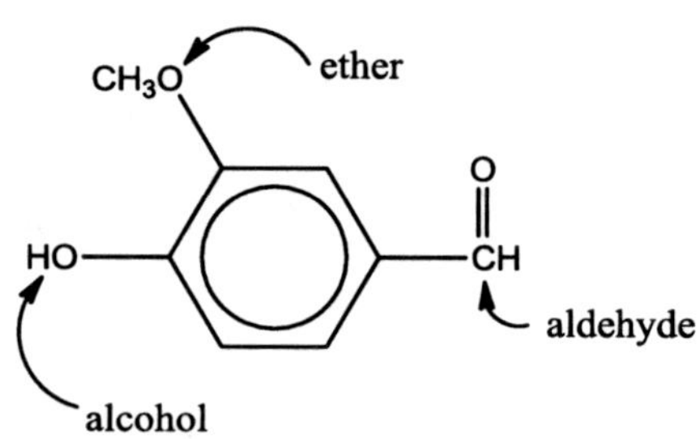

c.

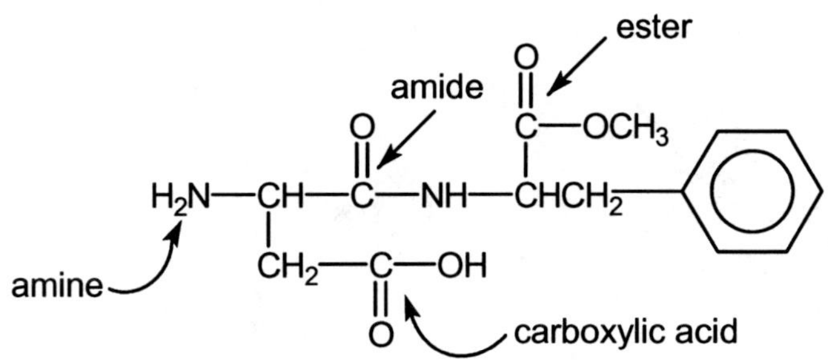

Note: The amide functional group 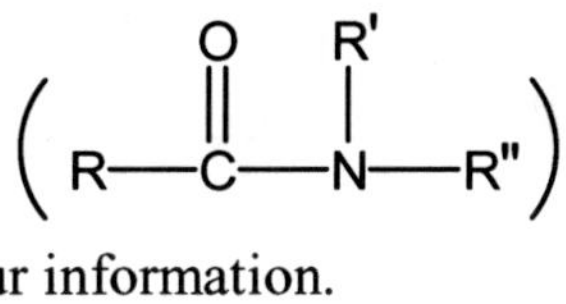 is not covered in Section 22.4 of the text. We point it out for your information.

49. a.

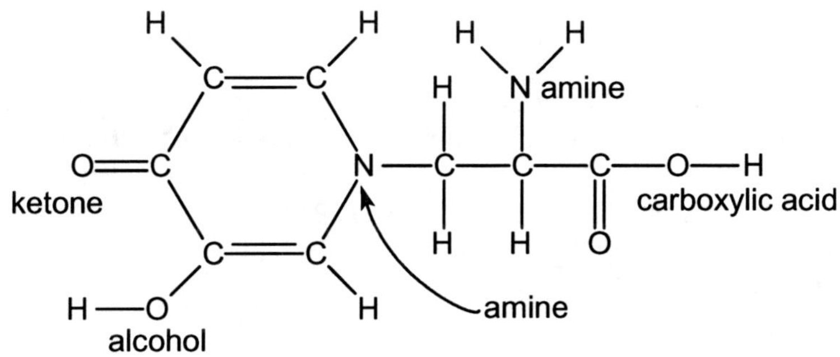

b. 5 carbons in the ring and the carbon in $-CO_2H$: sp^2; the other two carbons: sp^3

c. 24 sigma bonds; 4 pi bonds

50. Hydrogen atoms are usually omitted from ring structures. In organic compounds, the carbon atoms form four bonds. With this in mind, the following structure has the missing hydrogen atoms included in order to give each carbon atom the four bond requirement.

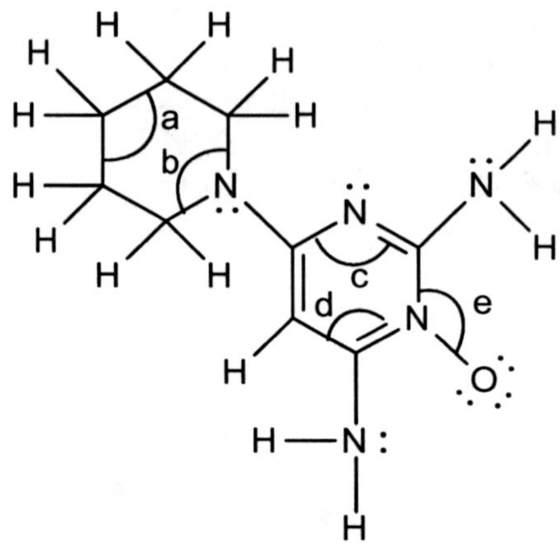

a. Minoxidil would be more soluble in acidic solution. The nitrogens with lone pairs can be protonated, forming a water soluble cation.

b. The two nitrogens in the ring with double bonds are sp^2 hybridized. The other three N's are sp^3 hybridized.

c. The five carbon atoms in the ring with one nitrogen are all sp^3 hybridized. The four carbon atoms in the other ring with double bonds are all sp^2 hybridized.

d. Angles a and b $\approx 109.5°$; angles c, d, and e $\approx 120°$

e. 31 sigma bonds

f. 3 pi bonds

51. a. 3-chloro-1-butanol; because the carbon containing the OH group is bonded to just 1 other carbon (1 R group), this is a primary alcohol.

b. 3-methyl-3-hexanol; because the carbon containing the OH group is bonded to three other carbons (3 R groups), this is a tertiary alcohol.

c. 2-methylcyclopentanol; secondary alcohol (2 R groups bonded to carbon containing the OH group). *Note*: In ring compounds, the alcohol group is assumed to be bonded to C_1, so the number designation is commonly omitted for the alcohol group.

52.

a. 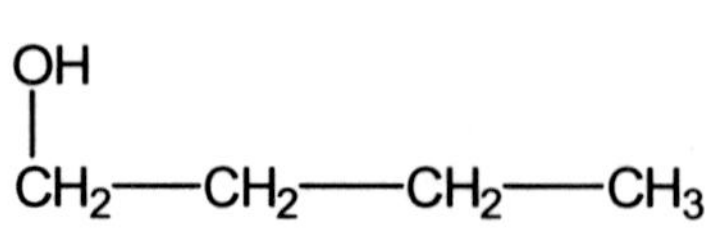 primary alcohol

b. 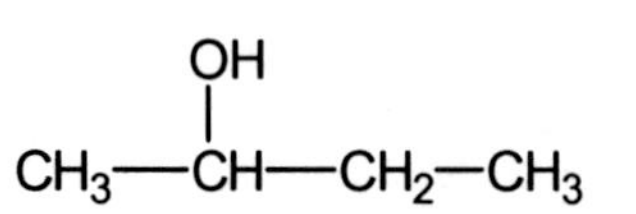 secondary alcohol

c. 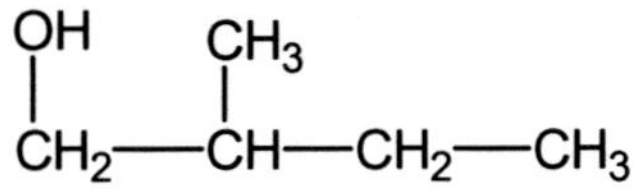 primary alcohol

d. 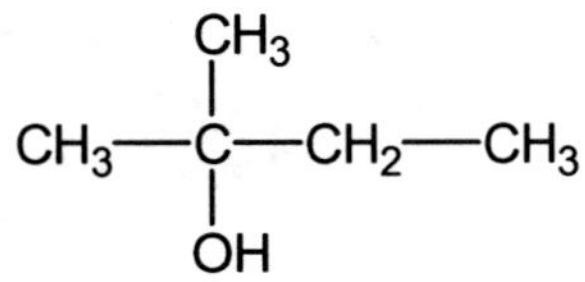 tertiary alcohol

53.

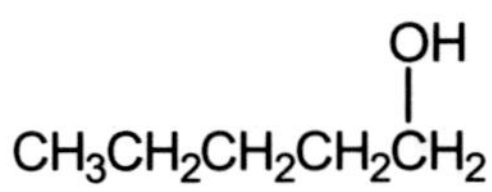

1-pentanol

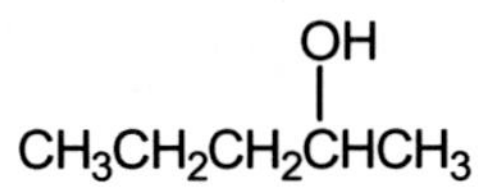

2-pentanol

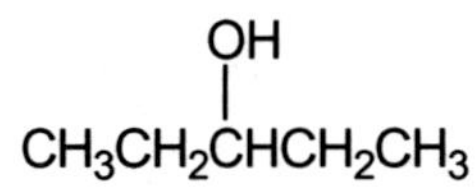

3-pentanol

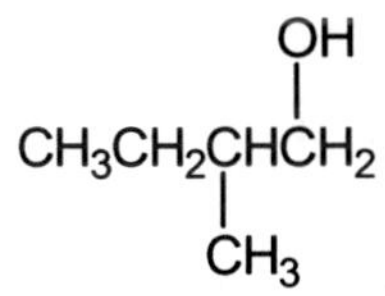

2-methyl-1-butanol

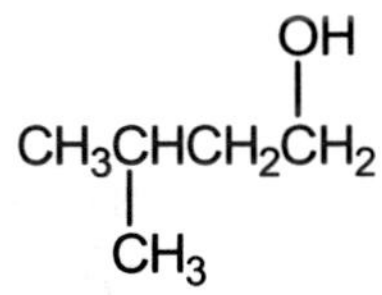

3-methyl-1-butanol

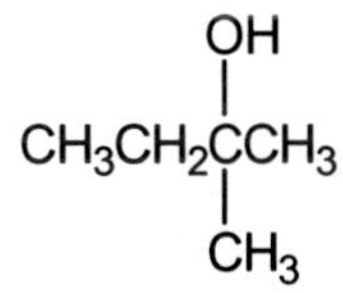

2-methyl-2-butanol

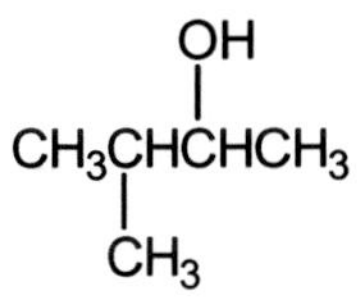

3-methyl-2-butanol

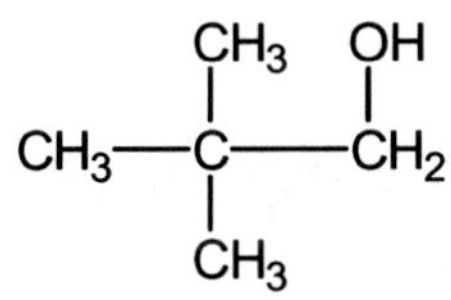

2,2-dimethyl-1-propanol

There are six isomeric ethers with formula $C_5H_{12}O$. The structures follow:

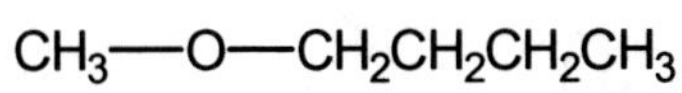

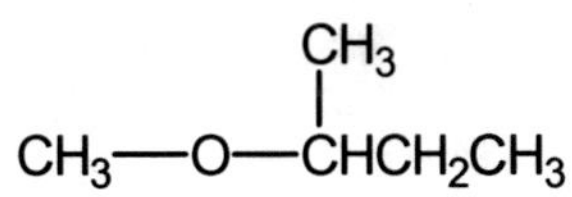

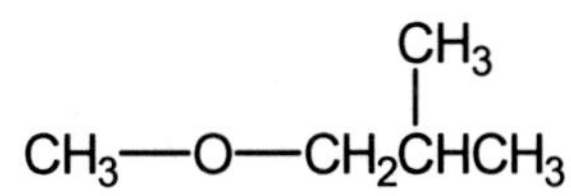

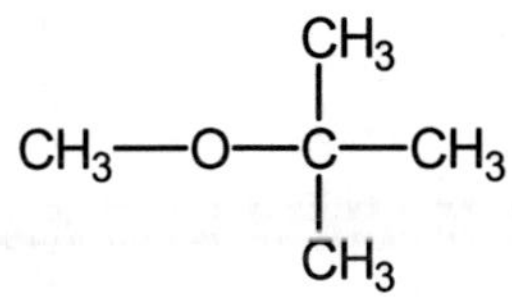

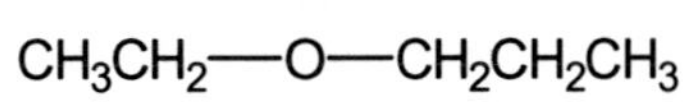

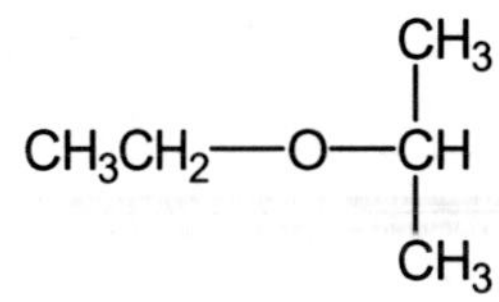

54. There are four aldehydes and three ketones with formula $C_5H_{10}O$. The structures follow:

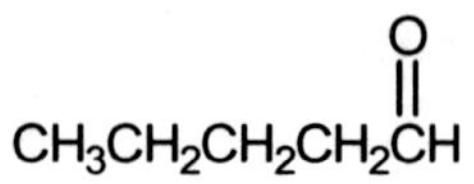

pentanal

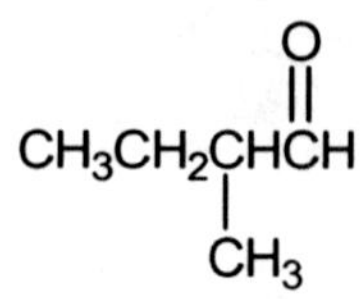

2-methylbutanal

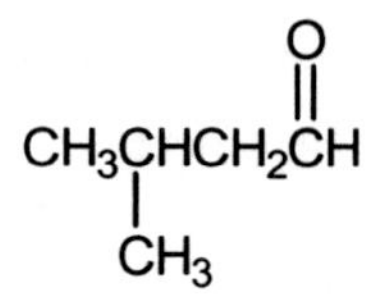

3-methylbutanal

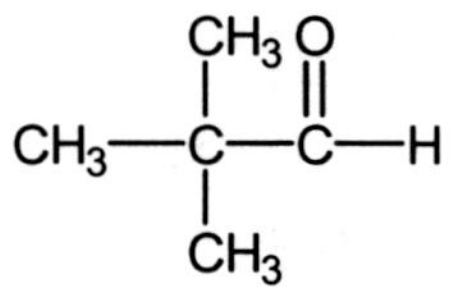

2,2-dimethylpropanal

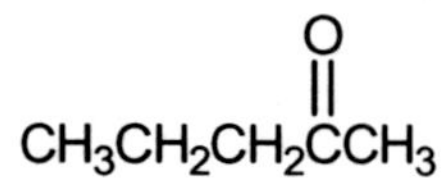

2-pentanone

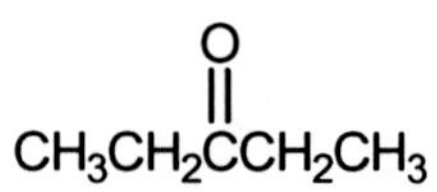

3-pentanone

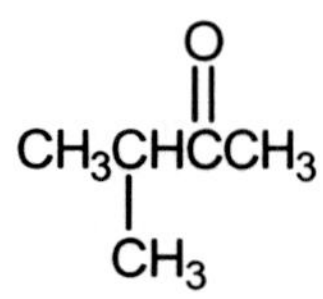

3-methyl-2-butanone

55. a. 4,5-dichloro-3-hexanone b. 2,3-dimethylpentanal

c. 3-methylbenzaldehyde or m-methylbenzaldehyde

56. a.

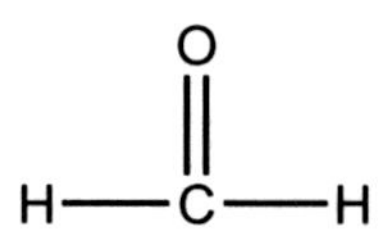

b.

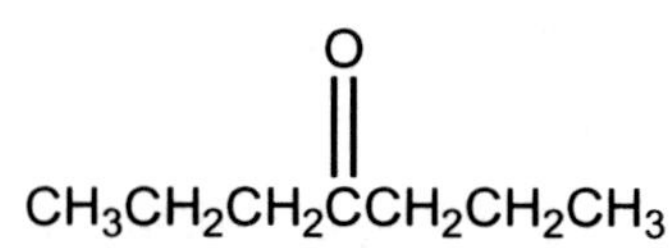

c.

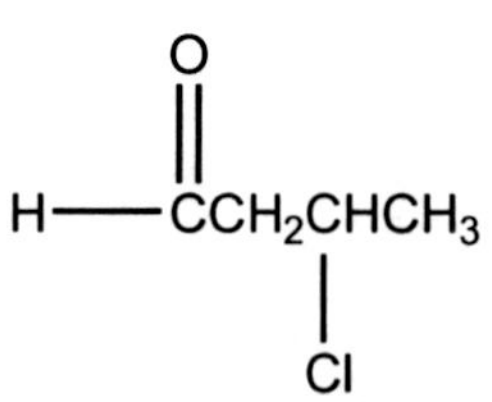

d.

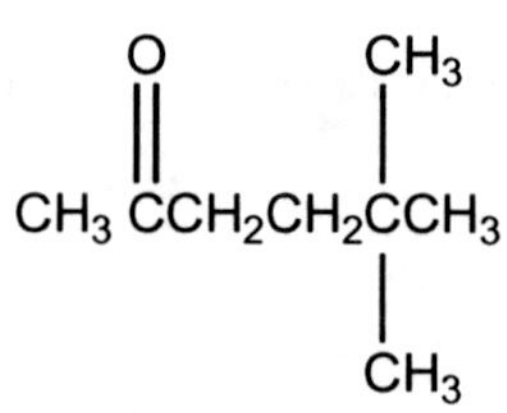

57. a. 4-chlorobenzoic acid or p-chlorobenzoic acid

b. 3-ethyl-2-methylhexanoic acid

c. methanoic acid (common name = formic acid)

58. a.

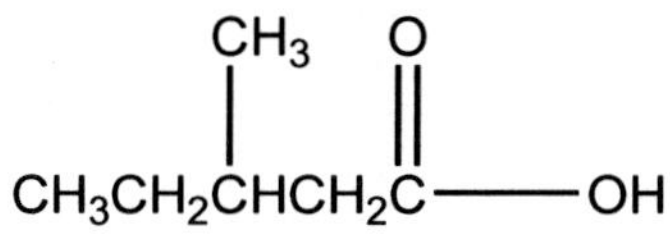

b.

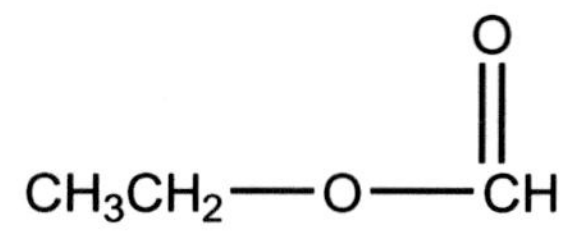

c.

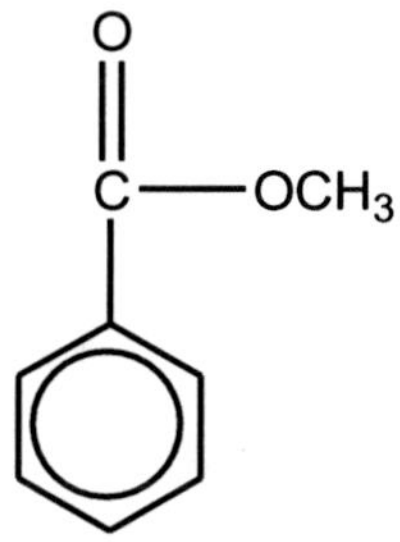

d.

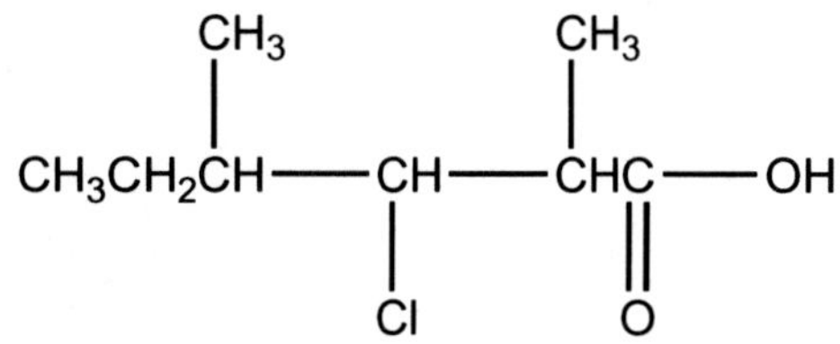

59. Only statement d is false. The other statements refer to compounds having the same formula but different attachment of atoms; they are structural isomers.

a.

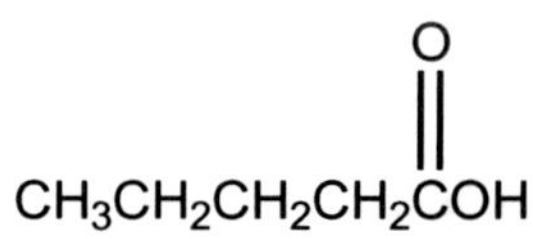

Both have a formula of $C_5H_{10}O_2$.

b.

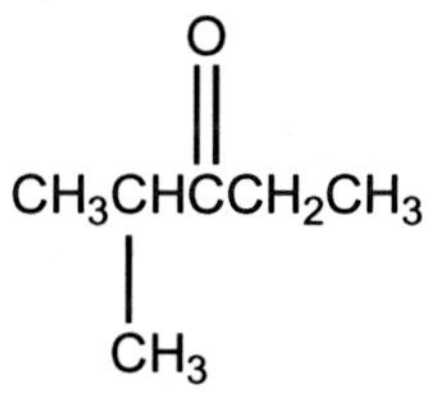

Both have a formula of $C_6H_{12}O$.

c.

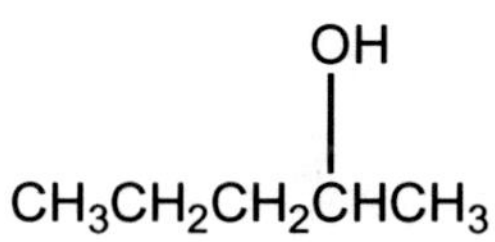

Both have a formula of $C_5H_{12}O$.

d.

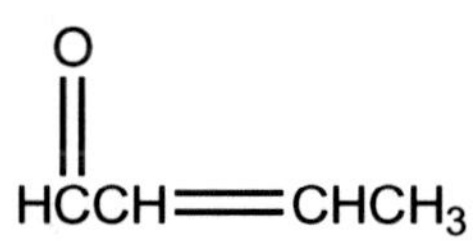

2-Butenal has a formula of C_4H_6O while the alcohol has a formula of C_4H_8O.

e.

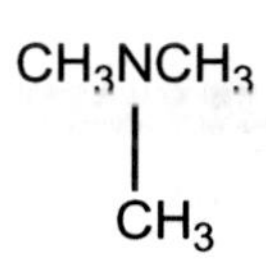

Both have a formula of C_3H_9N.

60. a. trans-2-butene: 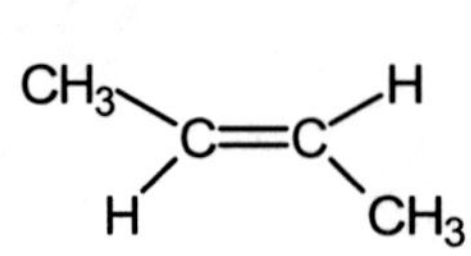 , formula = C_4H_8

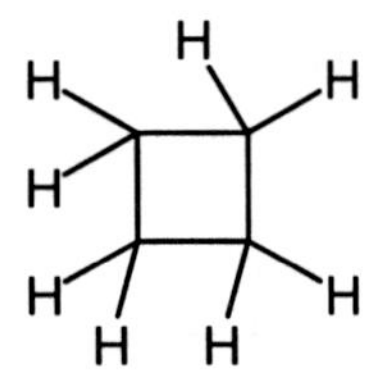 or 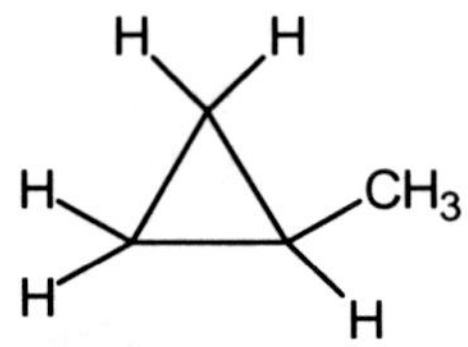

b. propanoic acid: $CH_3CH_2C(=O)—OH$, formula = $C_3H_6O_2$

$CH_3C(=O)—O—CH_3$ or $HC(=O)—O—CH_2CH_3$

c. butanal: $CH_3CH_2CH_2CH(=O)$, formula = C_4H_8O

$CH_3CH_2C(=O)CH_3$

d. butylamine: $CH_3CH_2CH_2CH_2NH_2$, formula = $C_4H_{11}N$:

A secondary amine has two R groups bonded to N.

$CH_3—N(CH_2CH_2CH_3)—H$ $CH_3—N(CH_3CHCH_3)—H$ $CH_3CH_2—N(CH_2CH_3)—H$

e. A tertiary amine has three R groups bonded to N. (See answer d for structure of butylamine.)

$CH_3—N(CH_2CH_3)—CH_3$

f. 2-methyl-2-propanol: $CH_3C(CH_3)(OH)CH_3$, formula = $C_4H_{10}O$

$CH_3—O—CH_2CH_2CH_3$ $CH_3—O—CH(CH_3)CH_3$ $CH_3CH_2—O—CH_2CH_3$

g. A secondary alcohol has two R groups attached to the carbon bonded to the OH group. (See answer f for the structure of 2-methyl-2-propanol.)

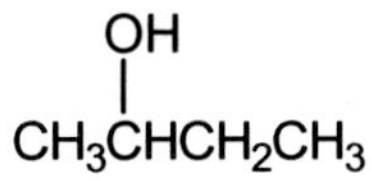

Reactions of Organic Compounds

61.

a. 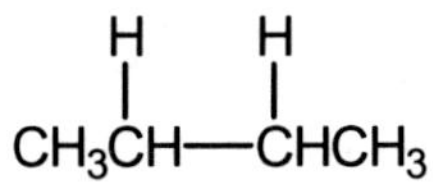

b. 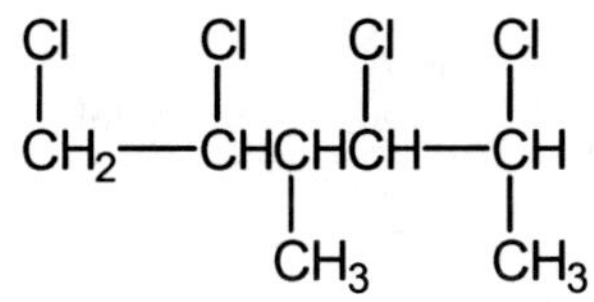

c. 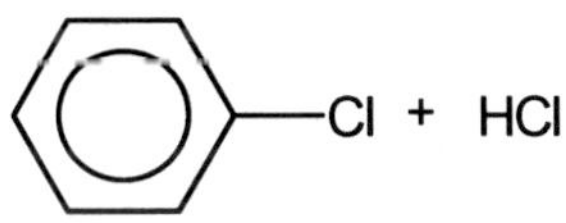

d. $C_4H_8(g) + 6\ O_2(g) \rightarrow 4\ CO_2(g) + 4\ H_2O(g)$

62. a. The two possible products for the addition of HOH to this alkene are:

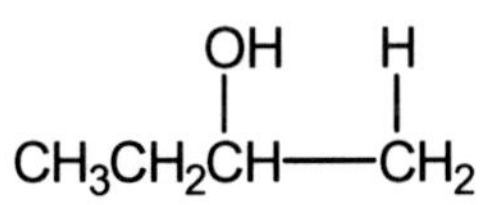

major product

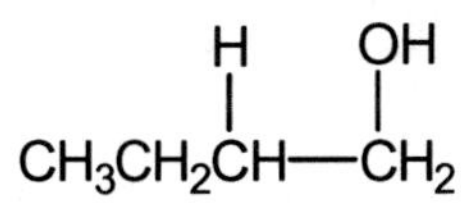

minor product

We would get both products in this reaction. Using the rule given in the problem, the first compound listed is the major product. In the reactant, the terminal carbon has more hydrogens bonded to it (2 versus 1), so H forms a bond to this carbon, and OH forms a bond to the other carbon in the double bond for the major product. We will list only the major product for the remaining parts to this problem.

b.

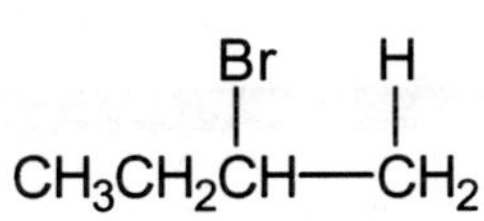

c.

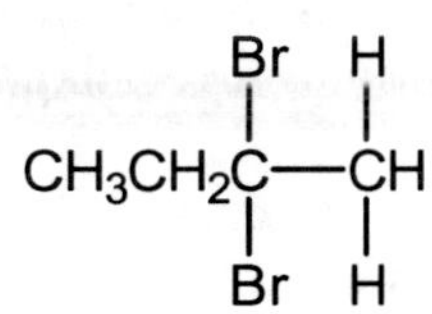

d.

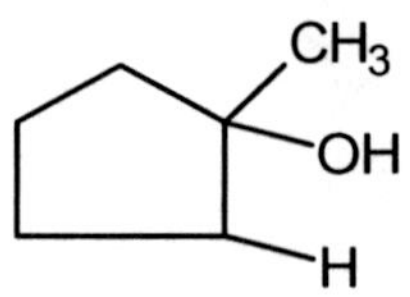

e.

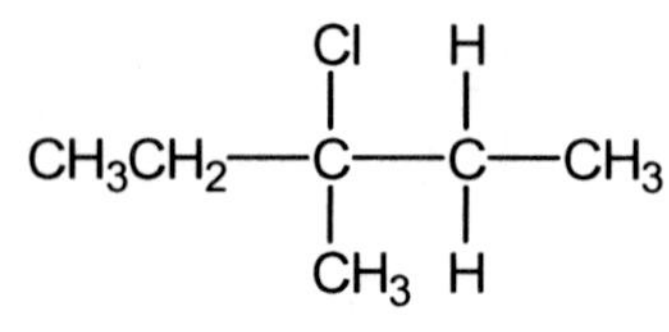

63.

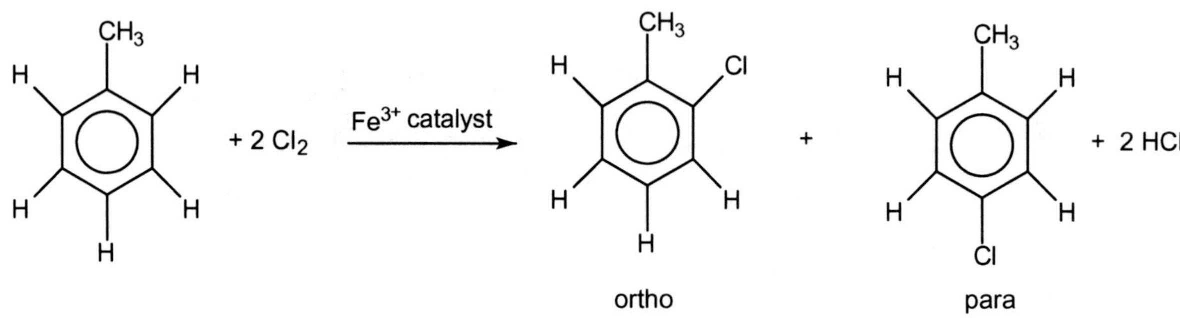

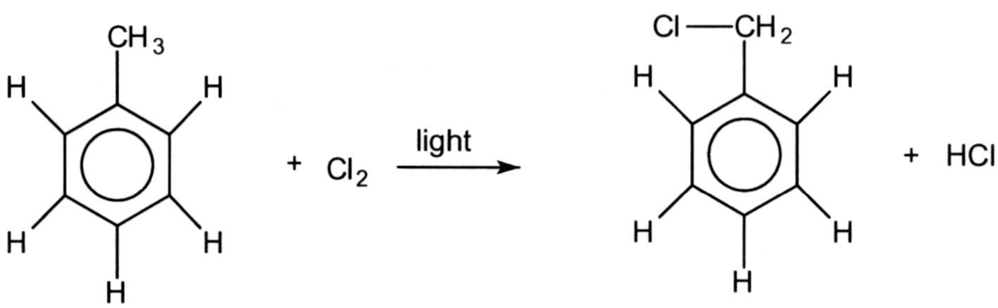

To substitute for the benzene ring hydrogens, an iron(III) catalyst must be present. Without this special iron catalyst, the benzene ring hydrogens are unreactive. To substitute for an alkane hydrogen, light must be present. For toluene, the light-catalyzed reaction substitutes a chlorine for a hydrogen in the methyl group attached to the benzene ring.

64. When $CH_2{=}CH_2$ reacts with HCl, there is only one possible product, chloroethane. When Cl_2 is reacted with CH_3CH_3 (in the presence of light), there are six possible products because any number of the six hydrogens in ethane can be substituted for by Cl. The light-catalyzed substitution reaction is very difficult to control; hence it is not a very efficient method of producing monochlorinated alkanes.

65. Primary alcohols (a, d, and f) are oxidized to aldehydes, which can be oxidized further to carboxylic acids. Secondary alcohols (b, e, and f) are oxidized to ketones, and tertiary alcohols (c and f) do not undergo this type of oxidation reaction. Note that compound f contains a primary, secondary, and tertiary alcohol. For the primary alcohols (a, d, and f), we listed both the aldehyde and the carboxylic acid as possible products.

a. 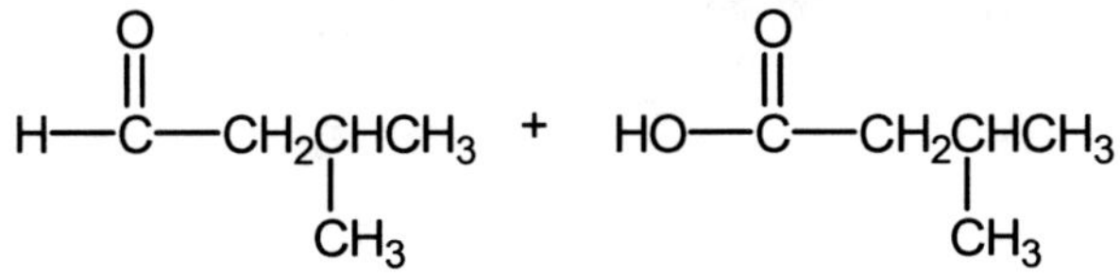

b. 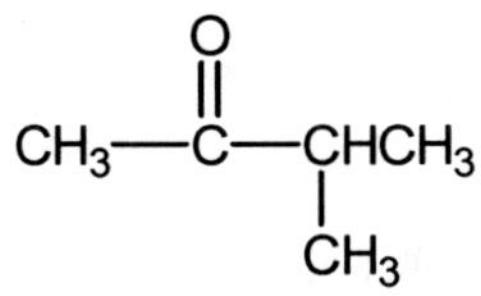

c. No reaction

d. 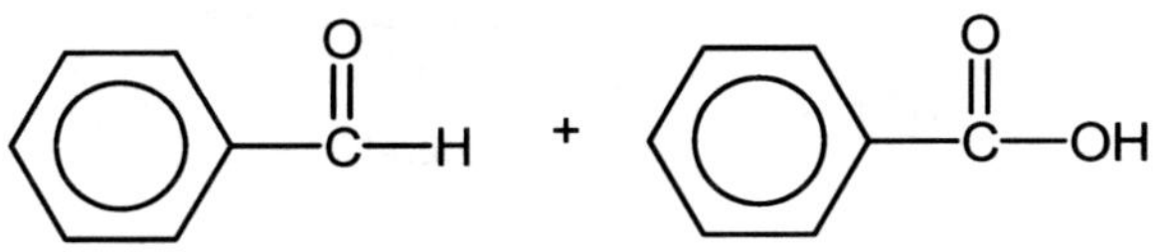

e. 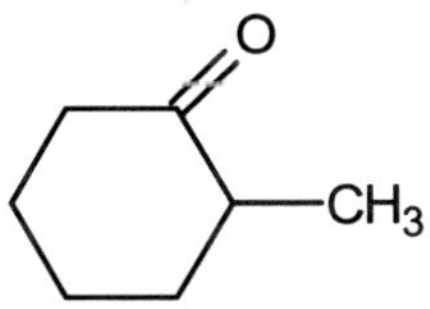

f. 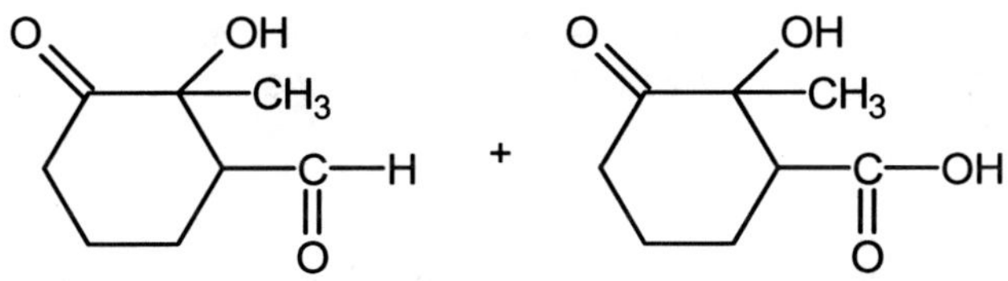

66. a.

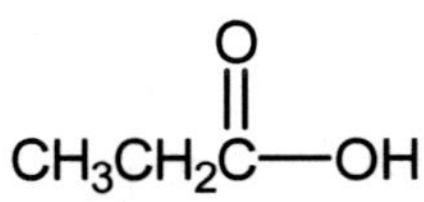

b.

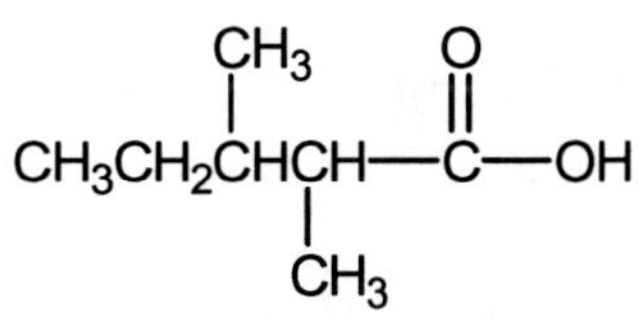

c.

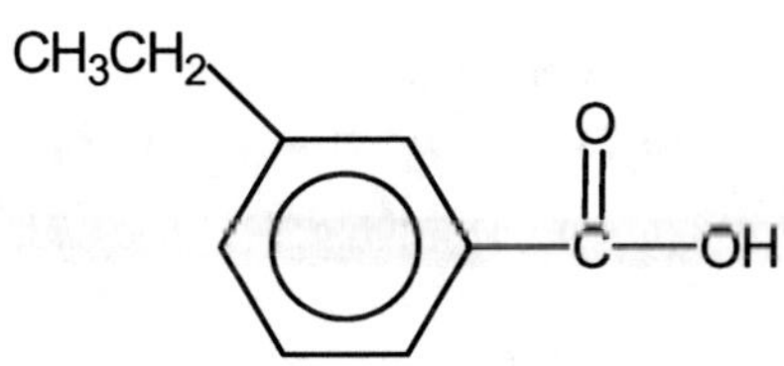

67. a. $CH_3CH{=}CH_2 + Br_2 \rightarrow CH_3CHBrCH_2Br$ (addition reaction of Br_2 with propene)

b. 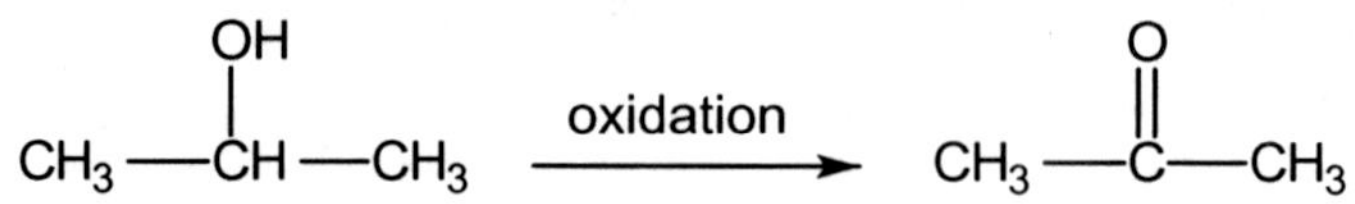

Oxidation of 2-propanol yields acetone (2-propanone).

c. 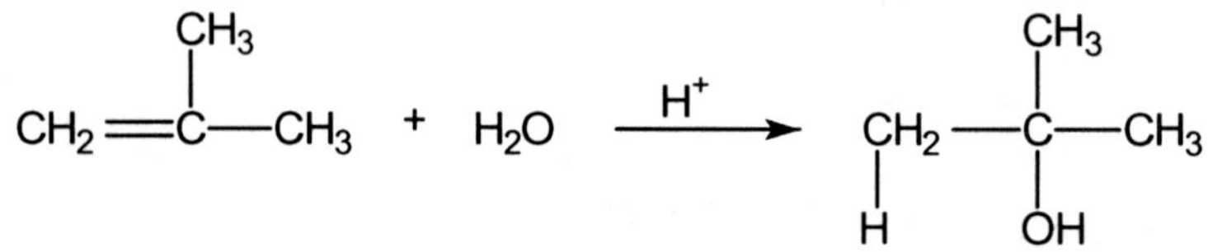

Addition of H_2O to 2-methylpropene would yield tert-butyl alcohol (2-methyl-2-propanol) as the major product.

d. 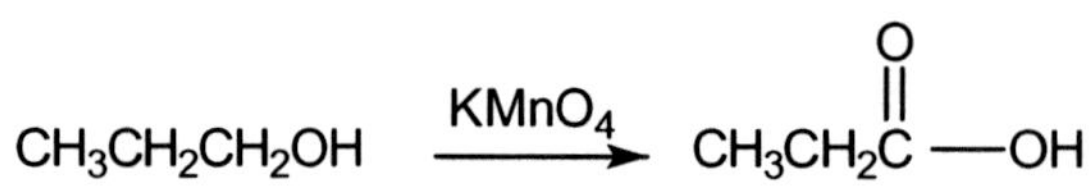

Oxidation of 1-propanol would eventually yield propanoic acid. Propanal is produced first in this reaction and is then oxidized to propanoic acid.

68. a. $CH_2{=}CHCH_2CH_3$ will react with Br_2 without any catalyst present. $CH_3CH_2CH_2CH_3$ reacts with Br_2 only when ultraviolet light is present.

b. $CH_3CH_2CH_2\overset{\overset{O}{\|}}{C}OH$ is an acid, so this compound should react positively with a base like $NaHCO_3$. The other compound is a ketone, which will not react with a base.

c. $CH_3CH_2CH_2OH$ can be oxidized with $KMnO_4$ to propanoic acid. 2-Propanone (a ketone) will not react with $KMnO_4$.

d. $CH_3CH_2NH_2$ is an amine, so it behaves as a base in water. Dissolution of some of this base in water will produce a solution with a basic pH. The ether, CH_3OCH_3, will not produce a basic pH when dissolved in water.

69. Reaction of a carboxylic acid with an alcohol can produce these esters.

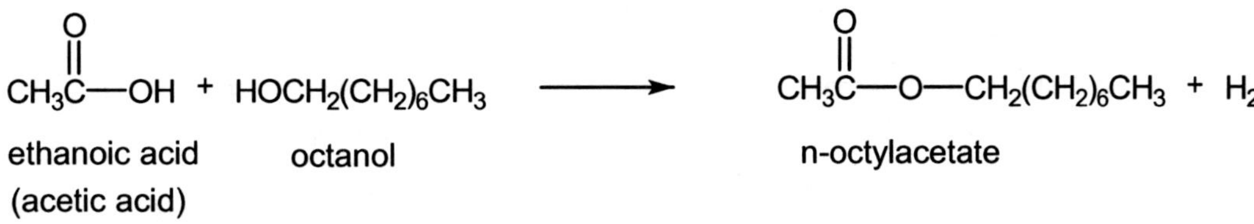

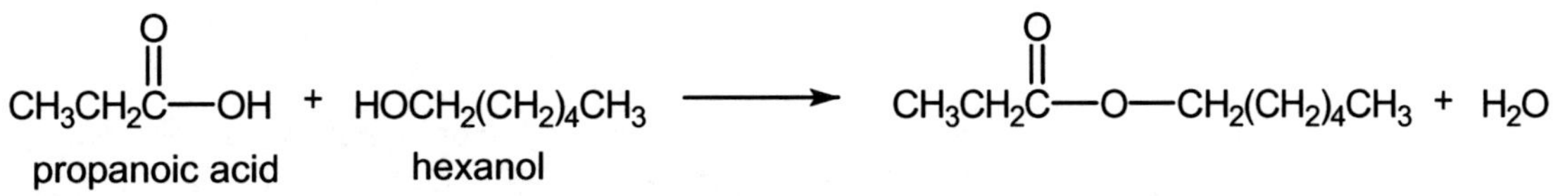

70. When an alcohol is reacted with a carboxylic acid, an ester is produced.

a.

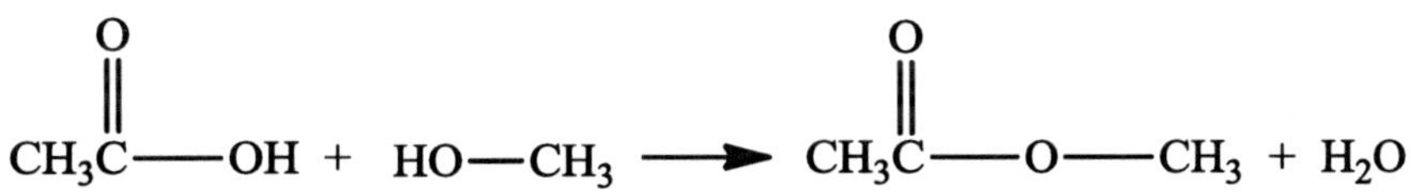

b.

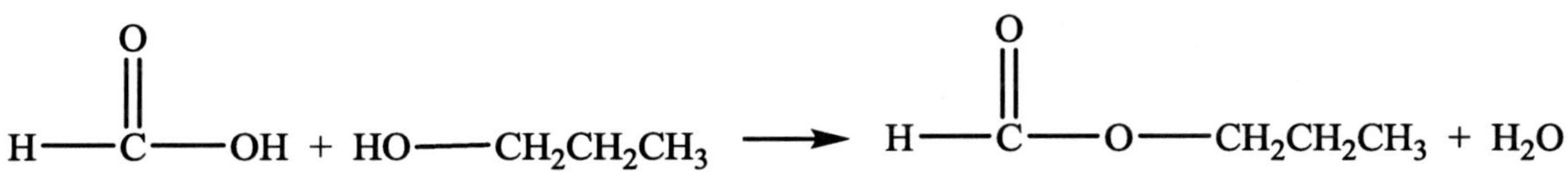

Polymers

71. The backbone of the polymer contains only carbon atoms, which indicates that Kel-F is an addition polymer. The smallest repeating unit of the polymer and the monomer used to produce this polymer are:

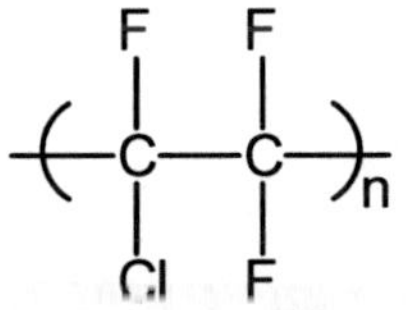

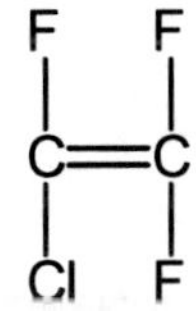

Note: Condensation polymers generally have O or N atoms in the backbone of the polymer.

72. a. repeating unit:

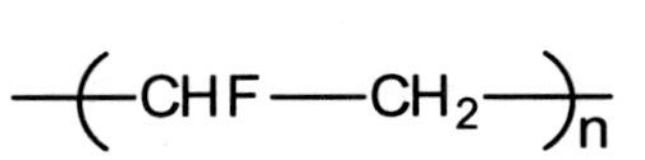

monomer: $CHF{=}CH_2$

b.

repeating unit: 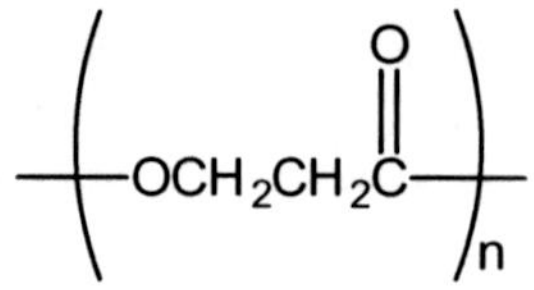 monomer: $HO–CH_2CH_2–CO_2H$

c.

repeating unit:

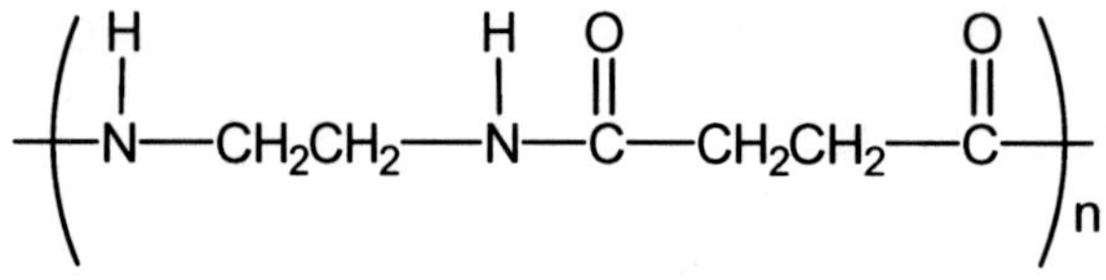

copolymer of: $H_2NCH_2CH_2NH_2$ and $HO_2CCH_2CH_2CO_2H$

d. monomer:

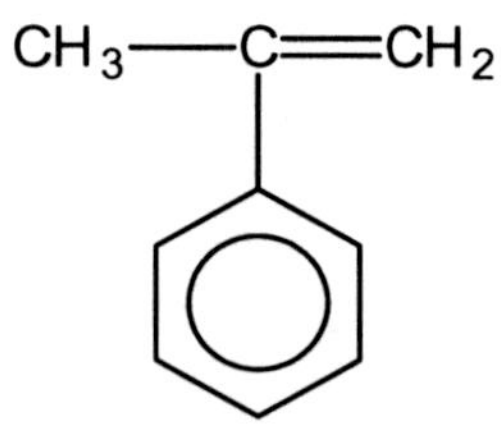

e. monomer:

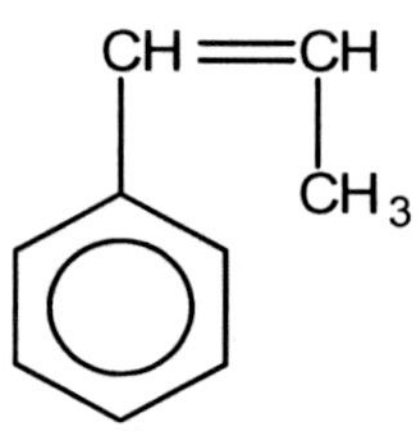

f. copolymer of:

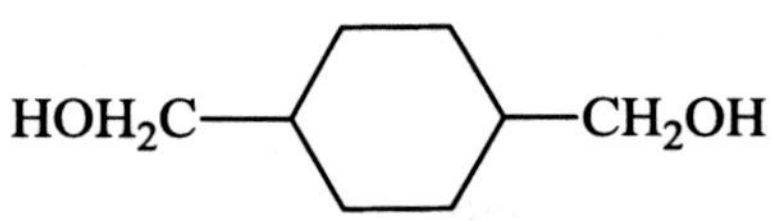 and 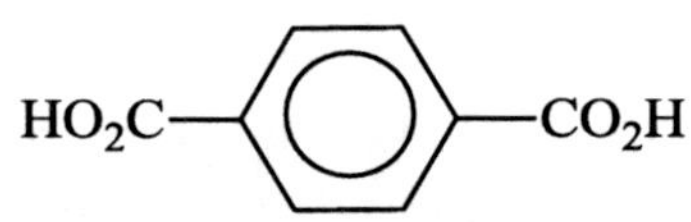

Addition polymers: a, d, and e; condensation polymers: b, c, and f; copolymer: c and f

73.

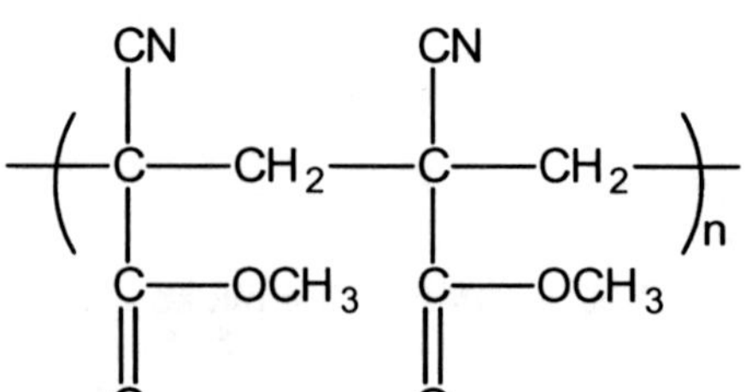

Super glue is an addition polymer formed by reaction of the C=C bond in methyl cyanoacrylate.

74. a. 2-methyl-1,3-butadiene

b.

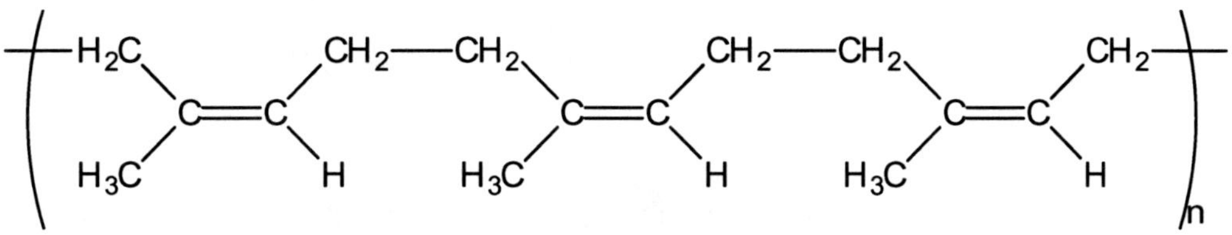

cis-polyisoprene (natural rubber)

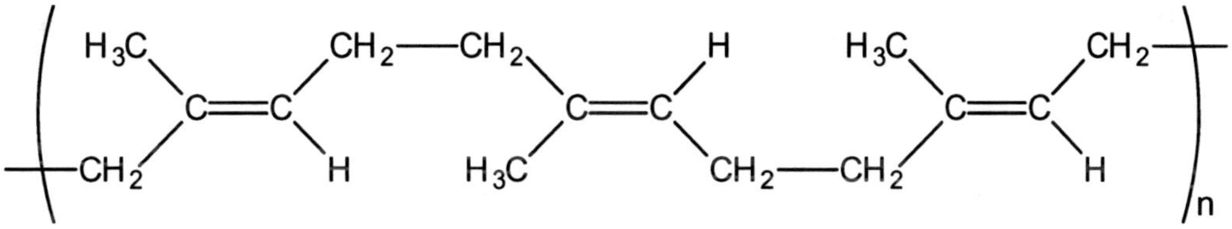

trans-polyisoprene (gutta percha)

75. H_2O is eliminated when Kevlar forms. Two repeating units of Kevlar are:

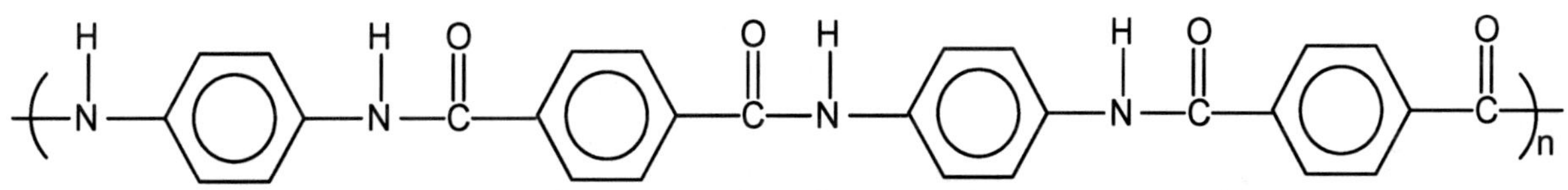

76. This condensation polymer forms by elimination of water. The ester functional group repeats, hence the term polyester.

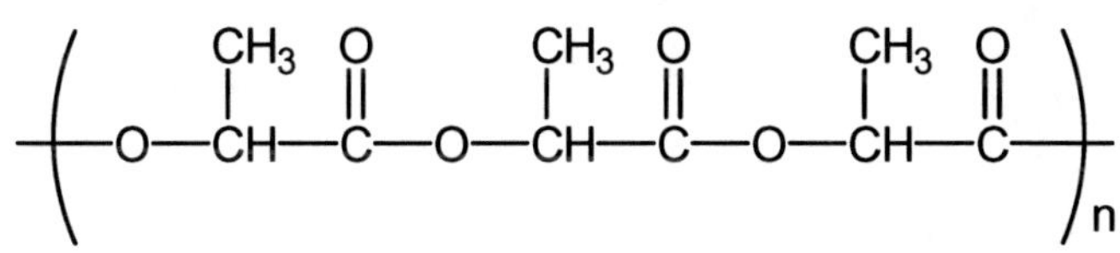

77. This is a condensation polymer, where two molecules of H_2O form when the monomers link together.

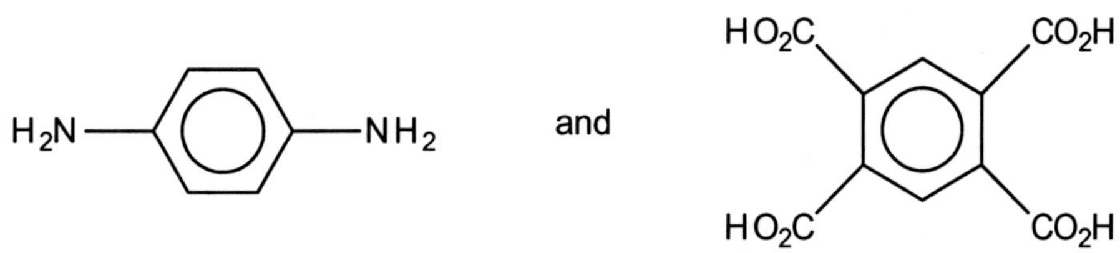

78.

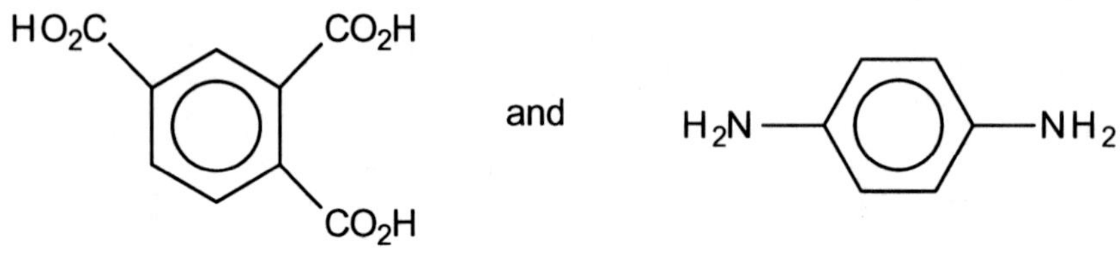

79. Divinylbenzene has two reactive double bonds that are used during formation of the polymer. The key is for the double bonds to insert themselves into two different polymer chains during the polymerization process. When this occurs, the two chains are bonded together (are cross-linked). The chains cannot move past each other because of the crosslinks, making the polymer more rigid.

80. a.

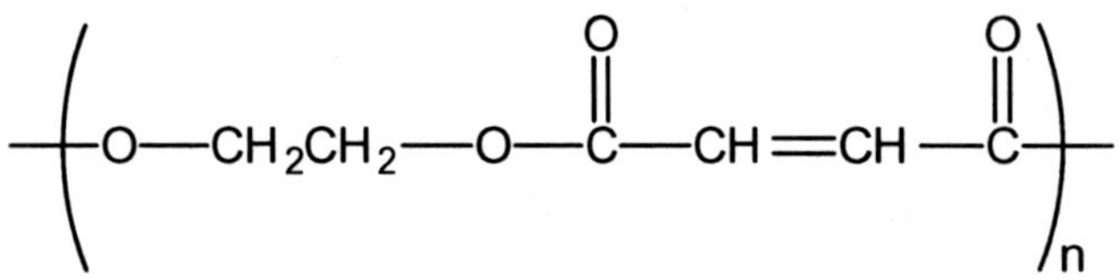

b.

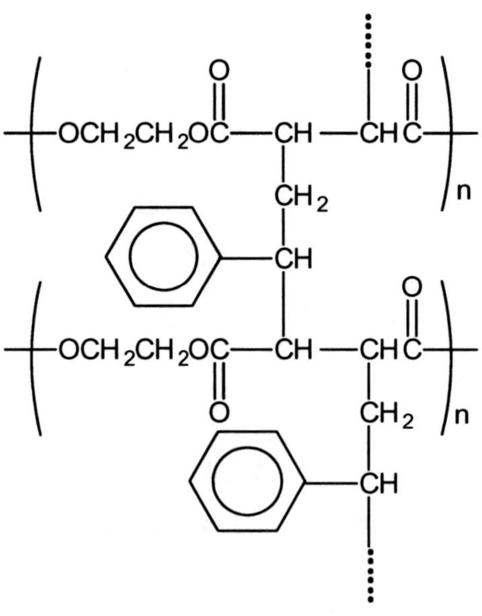

81. a. The polymer formed using 1,2-diaminoethane will exhibit relatively strong hydrogen-bonding interactions between adjacent polymer chains. Hydrogen bonding is not present in the ethylene glycol polymer (a polyester polymer forms), so the 1,2-diaminoethane polymer will be stronger.

b. The presence of rigid groups (benzene rings or multiple bonds) makes the polymer stiffer. Hence the monomer with the benzene ring will produce the more rigid polymer.

c. Polyacetylene will have a double bond in the carbon backbone of the polymer.

$$n\ HC{\equiv}CH \longrightarrow \left(HC{=}CH \right)_n$$

The presence of the double bond in polyacetylene will make polyacetylene a more rigid polymer than polyethylene. Polyethylene doesn't have C=C bonds in the backbone of the polymer (the double bonds in the monomers react to form the polymer).

82. At low temperatures, the polymer is coiled into balls. The forces between poly(lauryl methacrylate) and oil molecules will be minimal, and the effect on viscosity will be minimal. At higher temperatures, the chains of the polymer will unwind and become tangled with the oil molecules, increasing the viscosity of the oil. Thus the presence of the polymer counteracts the temperature effect, and the viscosity of the oil remains relatively constant.

Natural Polymers

83. a. Serine, tyrosine, and threonine contain the -OH functional group in the R group.

b. Aspartic acid and glutamic acid contain the -COOH functional group in the R group.

c. An amine group has a nitrogen bonded to other carbon and/or hydrogen atoms. Histidine, lysine, arginine, and tryptophan contain the amine functional group in the R group.

d. The amide functional group is:

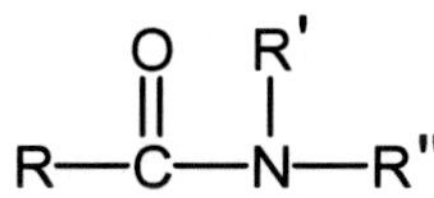

This functional group is formed when individual amino acids bond together to form the peptide linkage. Glutamine and asparagine have the amide functional group in the R group.

84. Crystalline amino acids exist as zwitterions, $^{+}H_3NCRHCOO^{-}$, held together by ionic forces. The ionic interparticle forces are strong. Before the temperature gets high enough to melt the solid, the amino acid decomposes.

85. a. Aspartic acid and phenylalanine make up aspartame.

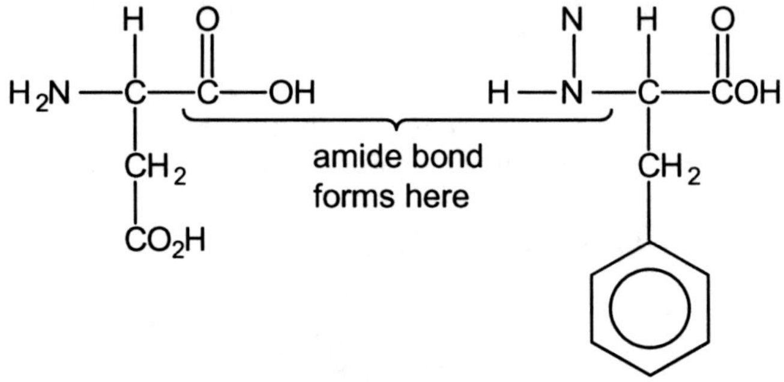

b. Aspartame contains the methyl ester of phenylalanine. This ester can hydrolyze to form methanol:

$RCO_2CH_3 + H_2O \rightleftharpoons RCO_2H + HOCH_3$

86.

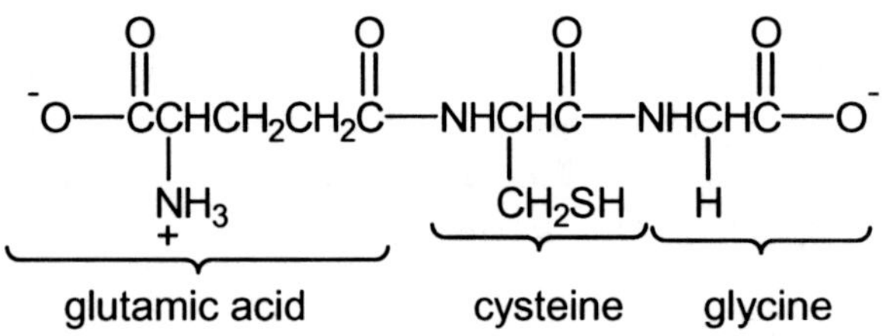

Glutamic acid, cysteine, and glycine are the three amino acids in glutathione. Glutamic acid uses the −COOH functional group in the R group to bond to cysteine instead of the carboxylic acid group bonded to the α-carbon. The cysteine-glycine bond is the typical peptide linkage.

87.

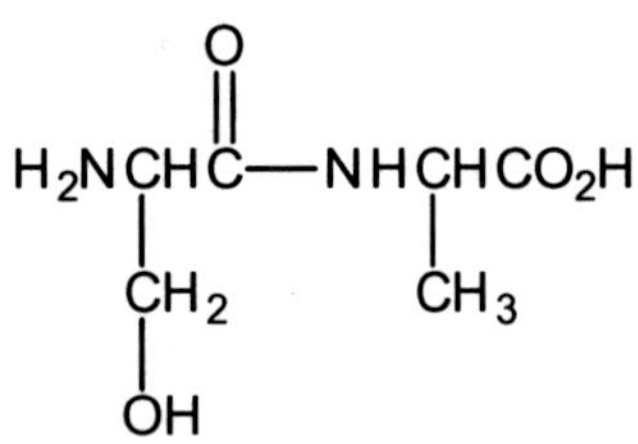

ser - ala

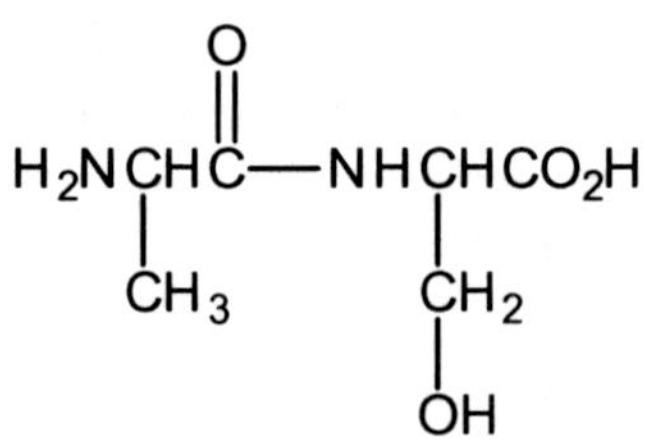

ala - ser

88.

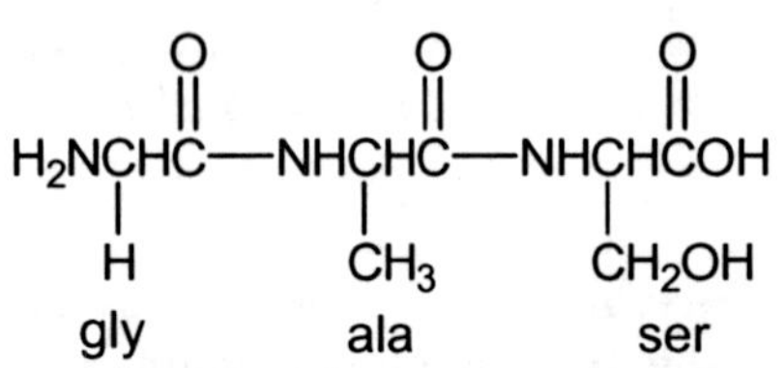

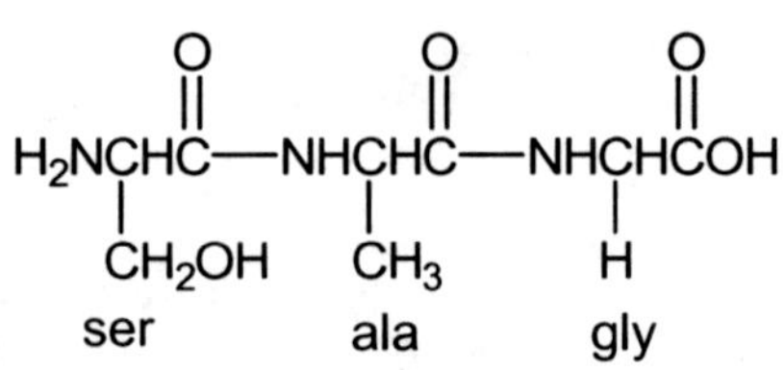

There are six possible tripeptides with gly, ala, and ser. The other four tripeptides are gly-ser-ala, ser-gly-ala, ala-gly-ser, and ala-ser-gly.

89. a. Six tetrapeptides are possible. From NH_2 to CO_2H end:

phe-phe-gly-gly, gly-gly-phe-phe, gly-phe-phe-gly,

phe-gly-gly-phe, phe-gly-phe-gly, gly-phe-gly-phe

b. Twelve tetrapeptides are possible. From NH_2 to CO_2H end:

phe-phe-gly-ala, phe-phe-ala-gly, phe-gly-phe-ala,

phe-gly-ala-phe, phe-ala-phe-gly, phe-ala-gly-phe,

gly-phe-phe-ala, gly-phe-ala-phe, gly-ala-phe-phe

ala-phe-phe-gly, ala-phe-gly-phe, ala-gly-phe-phe

90. There are 5 possibilities for the first amino acid, 4 possibilities for the second amino acid, 3 possibilities for the third amino acid, 2 possibilities for the fourth amino acid, and 1 possibility for the last amino acid. The number of possible sequences is:

$5 \times 4 \times 3 \times 2 \times 1 = 5! = 120$ different pentapeptides

91. a. Ionic: Need NH_2 on side chain of one amino acid with CO_2H on side chain of the other amino acid. The possibilities are:

NH_2 on side chain = His, Lys, or Arg; CO_2H on side chain = Asp or Glu

b. Hydrogen bonding: Need N–H or O–H bond present in side chain. The hydrogen bonding interaction occurs between the X– H bond and a carbonyl group from any amino acid.

X–H · · · · · · · O = C (carbonyl group)

Ser Asn	Any amino acid
Glu Thr	
Tyr Asp	
His Gln	
Arg Lys	

c. Covalent: Cys–Cys (forms a disulfide linkage)

d. London dispersion: All amino acids with nonpolar R groups. They are:

Gly, Ala, Pro, Phe, Ile, Trp, Met, Leu, and Val

e. Dipole-dipole: Need side chain with OH group. Tyr, Thr and Ser all could form this specific dipole-dipole force with each other since all contain an OH group in the side chain.

92. Reference Exercise 91 for a more detailed discussion of these various interactions.

a. Covalent b. Hydrogen bonding

c. Ionic d. London dispersion

93. Glutamic acid: R = $-CH_2CH_2CO_2H$; valine: R = $-CH(CH_3)_2$; a polar side chain is replaced by a nonpolar side chain. This could affect the tertiary structure of hemoglobin and the ability of hemoglobin to bind oxygen.

94. Glutamic acid: R = $-CH_2CH_2COOH$; glutamine: R = $-CH_2CH_2CONH_2$; the R groups only differ by OH versus NH_2. Both of these groups are capable of forming hydrogen-bonding interactions, so the change in intermolecular forces is minimal. Thus this change is not critical because the secondary and tertiary structures of hemoglobin should not be greatly affected.

95. See Figures 22.29 and 22.30 of the text for examples of the cyclization process.

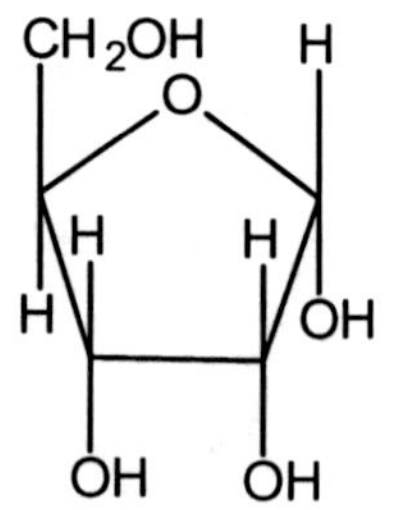

D-Ribose

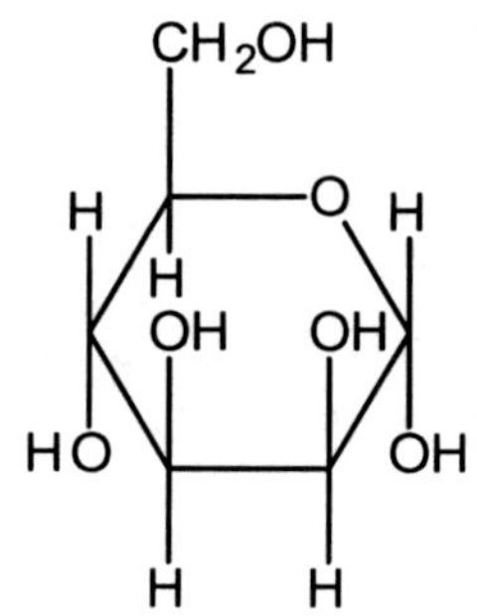

D-Mannose

96. The chiral carbon atoms are marked with asterisks. A chiral carbon atom has four different substituent groups attached.

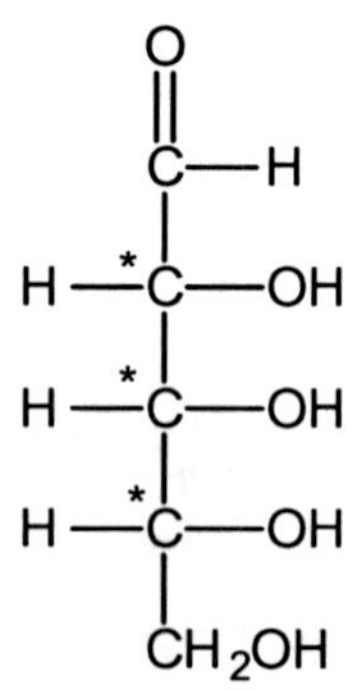

D-Ribose

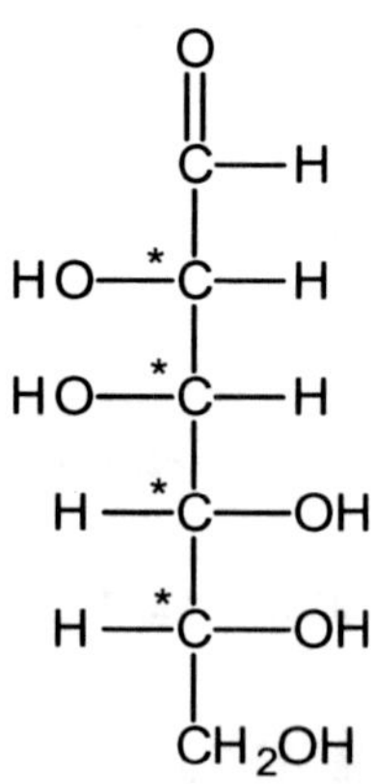

D-Mannose

97. The aldohexoses contain 6 carbons and the aldehyde functional group. Glucose, mannose, and galactose are aldohexoses. Ribose and arabinose are aldopentoses since they contain 5 carbons with the aldehyde functional group. The ketohexose (6 carbons + ketone functional group) is fructose, and the ketopentose (5 carbons + ketone functional group) is ribulose.

98. This is an example of Le Châtelier's principle at work. For the equilibrium reactions among the various forms of glucose, reference Figure 22.30 of the text. The chemical tests involve reaction of the aldehyde group found only in the open-chain structure. As the aldehyde group is reacted, the equilibrium between the cyclic forms of glucose, and the open-chain structure will shift to produce more of the open-chain structure. This process continues until either the glucose or the chemicals used in the tests run out.

99. The α and β forms of glucose differ in the orientation of a hydroxy group on one specific carbon in the cyclic forms (see Figure 22.30 of the text). Starch is a polymer composed of only α-D-glucose, and cellulose is a polymer composed of only β-D-glucose.

100. Humans do not possess the necessary enzymes to break the β-glycosidic linkages found in cellulose. Cows, however, do possess the necessary enzymes to break down cellulose into the β-D-glucose monomers and therefore can derive nutrition from cellulose.

101. A chiral carbon has four different groups attached to it. A compound with a chiral carbon is optically active. Isoleucine and threonine contain more than the one chiral carbon atom (see asterisks).

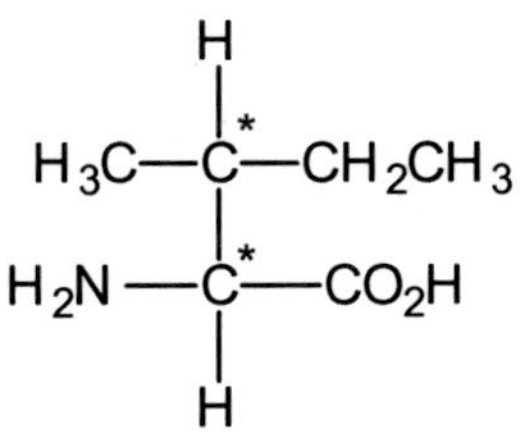

isoleucine

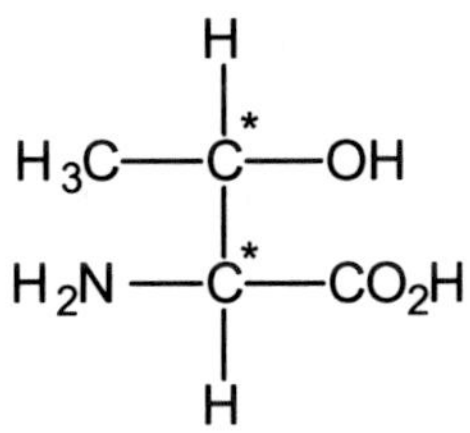

threonine

102. There is no chiral carbon atom in glycine since it contains no carbon atoms with four different groups bonded to it.

103. Only one of the isomers is optically active. The chiral carbon in this optically active isomer is marked with an asterisk.

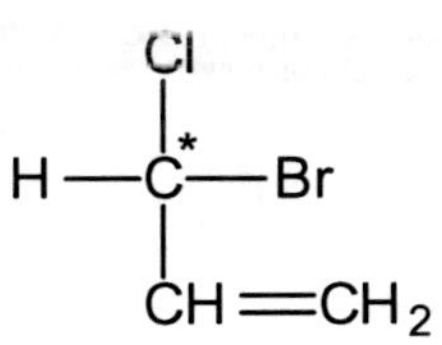

104.

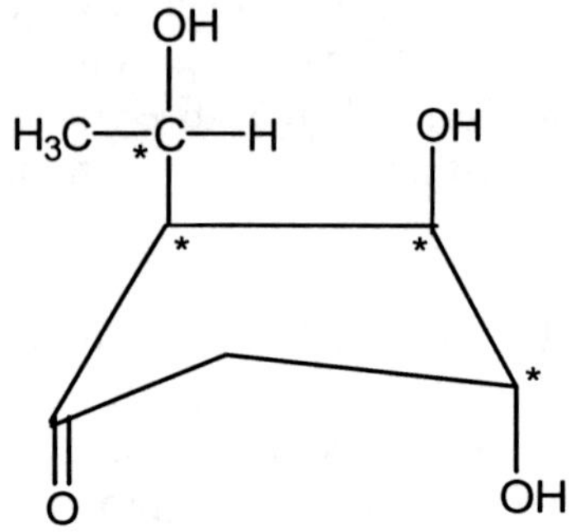

The compound has four chiral carbon atoms. The fourth group bonded to the three chiral carbon atoms in the ring is a hydrogen atom.

105. The complementary base pairs in DNA are cytosine (C) and guanine (G) and thymine (T) and adenine (A). The complementary sequence is C–C–A–G–A–T–A–T–G

106. For each letter, there are 4 choices; A, T, G, or C. Hence the total number of codons is $4 \times 4 \times 4 = 64$.

107. Uracil will hydrogen bond to adenine. The dashed lines represent the H-bonding interactions.

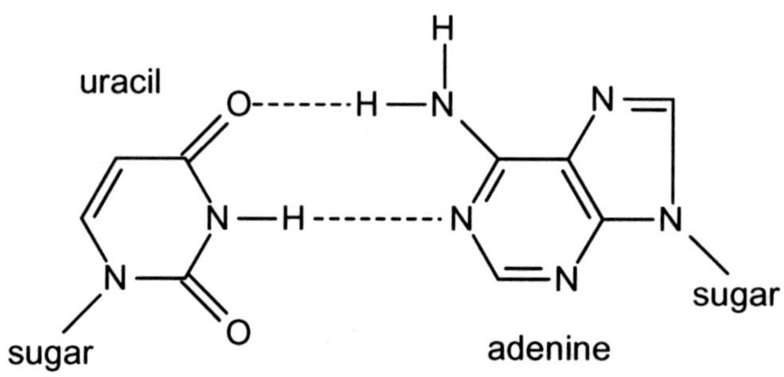

108. The tautomer could hydrogen bond to guanine, forming a G–T base pair instead of A–T.

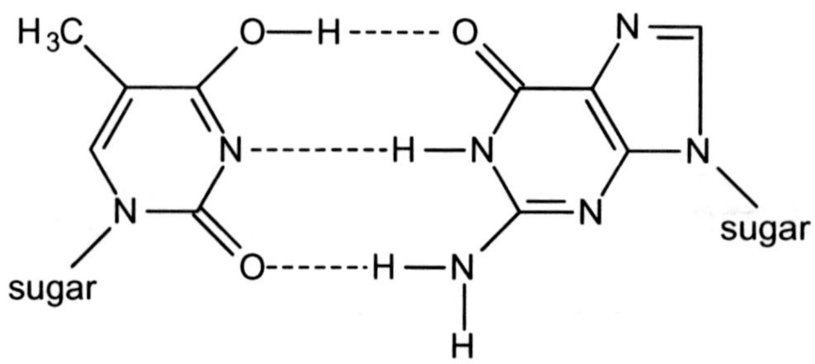

109. Base pair:

RNA	DNA
A	T
G	C
C	G
U	A

a. Glu: CTT, CTC   Val: CAA, CAG, CAT, CAC

Met: TAC   Trp: ACC

Phe: AAA, AAG   Asp: CTA, CTG

b. DNA sequence for trp-glu-phe-met:

ACC –CTT –AAA –TAC
or or
CTC AAG

c. Due to glu and phe, there is a possibility of four different DNA sequences. They are:

ACC–CTT–AAA–TAC or ACC–CTC–AAA–TAC or

ACC–CTT–AAG–TAC or ACC–CTC–AAG –TAC

d.

T—A—C—C—T—G—A—A—G

met (T—A—C) asp (C—T—G) phe (A—A—G)

e. TAC–CTA–AAG; TAC–CTA–AAA; TAC–CTG–AAA

110. In sickle cell anemia, glutamic acid is replaced by valine. DNA codons: Glu: CTT, CTC; Val: CAA, CAG, CAT, CAC; replacing the middle T with an A in the code for Glu will code for Val.

CTT → CAT or CTC → CAC
Glu Val Glu Val

Additional Exercises

111. We omitted the hydrogens for clarity. The number of hydrogens bonded to each carbon is the number necessary to form four bonds.

a.

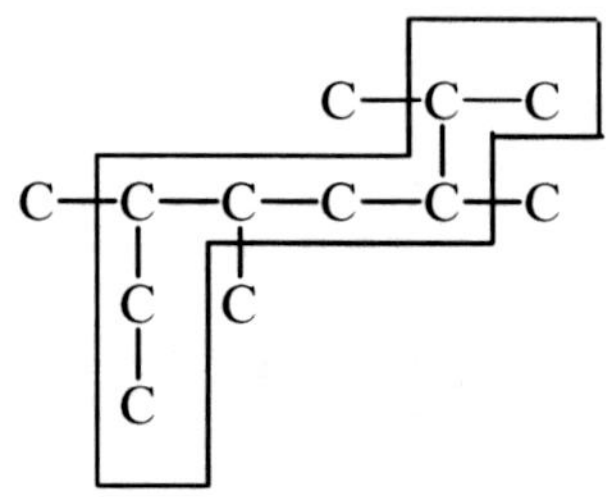

2,3,5,6-tetramethyloctane

b.

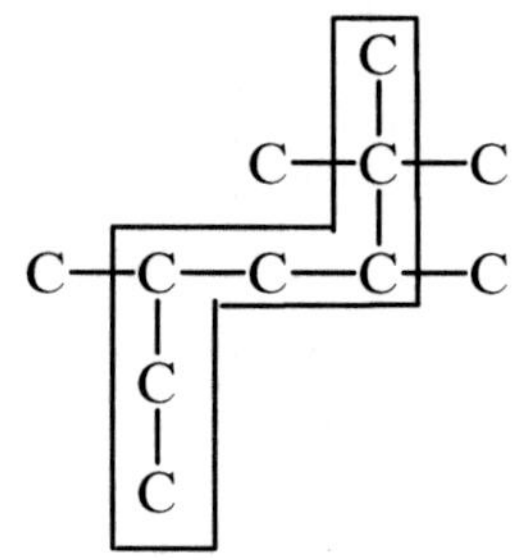

2,2,3,5-tetramethylheptane

c.

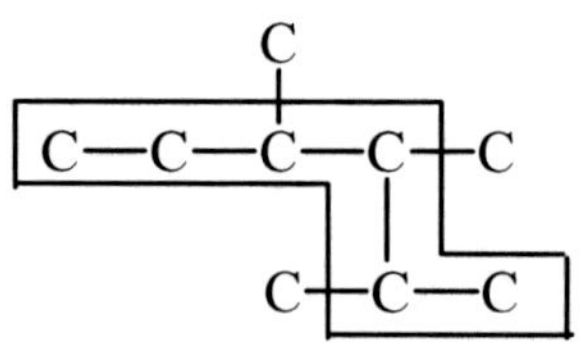

2,3,4-trimethylhexane

d.

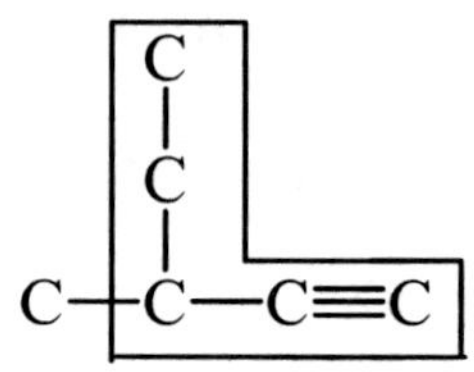

3-methyl-1-pentyne

112. a. Only one monochlorination product can form (1-chloro-2,2-dimethylpropane). The other possibilities differ from this compound by a simple rotation, so they are not different compounds.

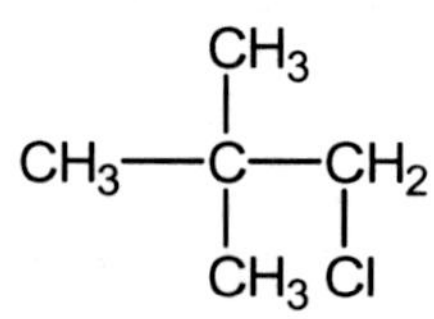

b. Three different monochlorination products are possible (ignoring cis-trans isomers).

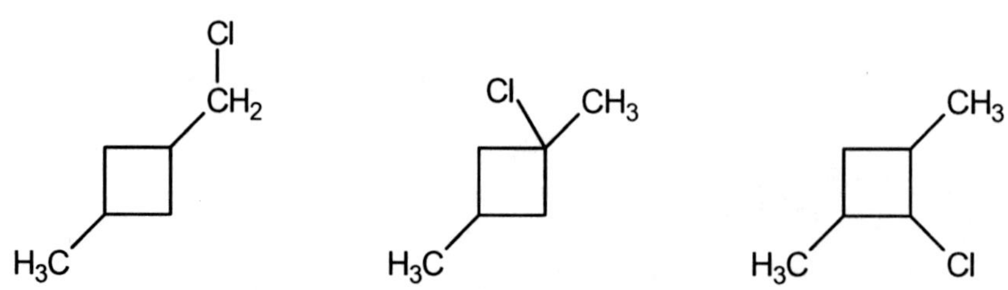

c. Two different monochlorination products are possible (the other possibilities differ by a simple rotation of one of these two compounds).

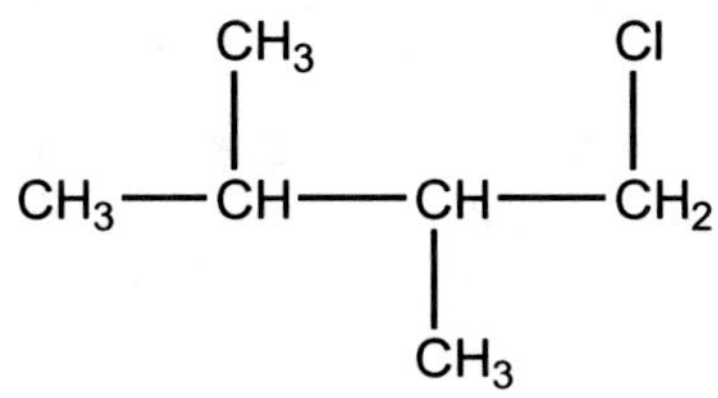

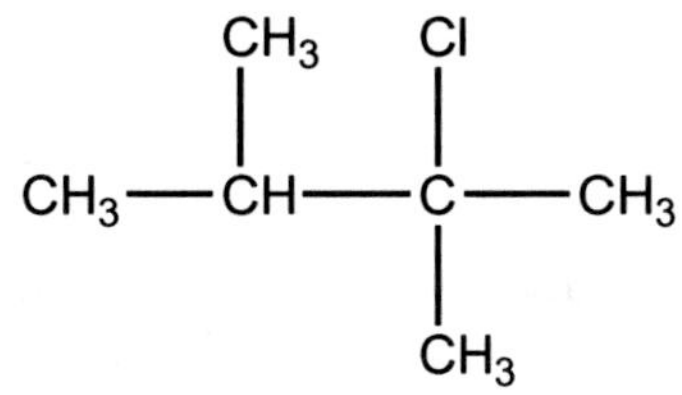

113.

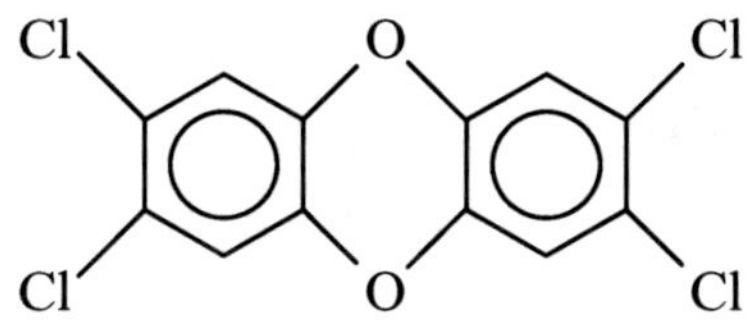

There are many possibilities for isomers. Any structure with four chlorines replacing four hydrogens in any four of the numbered positions would be an isomer; i.e., 1,2,3,4-tetrachloro-dibenzo-p-dioxin is a possible isomer.

114. We would expect compounds b and d to boil at the higher temperatures because they exhibit additional dipole forces that the nonpolar compounds in a, c, and e do not exhibit. London dispersion (LD) forces are the intermolecular forces exhibited by compounds a, c, and e. Size and shape are the two main factors that affect the strength of LD forces. Compounds a and e have a formula of C_5H_{12}, and the bigger compound c has a formula of C_6H_{14}. The smaller compounds in a and e will boil at the two lowest boiling points. Between a and e, compound a has a more elongated structure which leads to stronger LD forces; compound a boils at 36°C, and compound e boils at 9.5°C.

115. The isomers are:

$CH_3—O—CH_3$ CH_3CH_2OH

dimethyl ether, –23°C ethanol, 78.5°C

Ethanol, with its ability to form the relatively strong hydrogen-bonding interactions, boils at the higher temperature.

116. The isomers are:

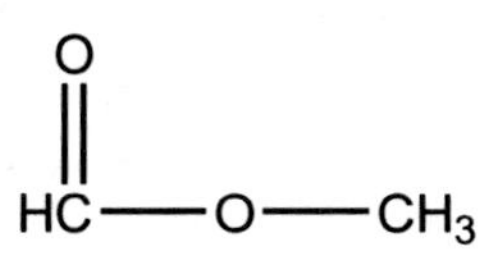

boils at lowest temperature
(no H-bonding)

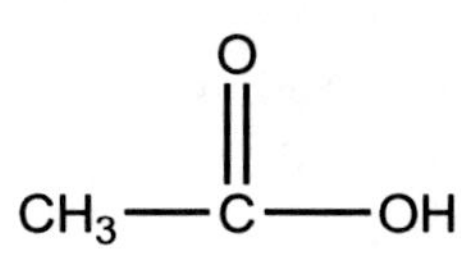

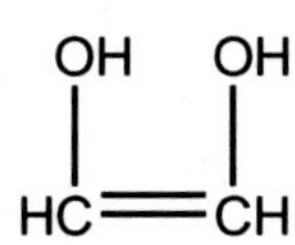

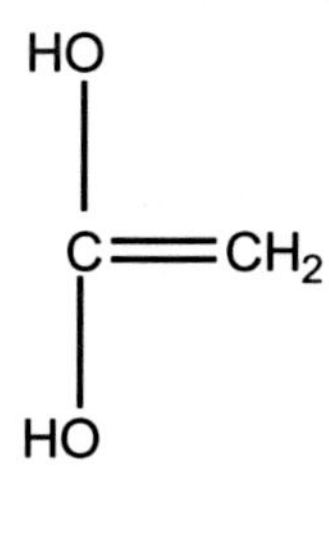

With the exception of the first isomer, the other isomers can form the relatively strong hydrogen bonding interactions. The isomers that can hydrogen bond will boil at higher temperatures.

117. Alcohols consist of two parts, the polar OH group and the nonpolar hydrocarbon chain attached to the OH group. As the length of the nonpolar hydrocarbon chain increases, the solubility of the alcohol decreases in water, a very polar solvent. In methyl alcohol (methanol), the polar OH group overrides the effect of the nonpolar CH_3 group, and methyl alcohol is soluble in water. In stearyl alcohol, the molecule consists mostly of the long nonpolar hydrocarbon chain, so it is insoluble in water.

118. $CH_3CH_2CH_2CH_2CH_2CH_2CH_2COOH + OH^- \rightarrow CH_3–(CH_2)_6–COO^- + H_2O$; octanoic acid is more soluble in 1 *M* NaOH. Added OH^- will remove the acidic proton from octanoic acid, creating a charged species. As is the case with any substance with an overall charge, solubility in water increases. When morphine is reacted with H^+, the amine group is protonated, creating a positive charge on morphine ($R_3N + H^+ \rightarrow R_3NH$). By treating morphine with HCl, an ionic compound results that is more soluble in water and in the bloodstream than the neutral covalent form of morphine.

119. The structures, the types of intermolecular forces exerted, and the boiling points for the compounds are:

$$CH_3CH_2CH_2\overset{\displaystyle O}{\overset{\|}{C}}OH \qquad\qquad CH_3CH_2CH_2CH_2CH_2OH$$

butanoic acid, 164°C — LD + dipole + H bonding

1-pentanol, 137°C — LD + H bonding

$$CH_3CH_2CH_2CH_2\overset{\displaystyle O}{\overset{\|}{C}}H \qquad\qquad CH_3CH_2CH_2CH_2CH_2CH_3$$

pentanal, 103°C — LD + dipole

n-hexane, 69°C — LD (London dispersion) only

All these compounds have about the same molar mass. Therefore, the London dispersion (LD) forces in each are about the same. The other types of forces determine the boiling-point order. Since butanoic acid and 1-pentanol both exhibit hydrogen bonding interactions, these two compounds will have the two highest boiling points. Butanoic acid has the highest boiling point since it exhibits H bonding along with dipole-dipole forces due to the polar C=O bond.

120. $85.63 \text{ g C} \times \dfrac{1 \text{ mol C}}{12.01 \text{ g C}} = 7.130 \text{ mol C}$; $14.37 \text{ g H} \times \dfrac{1 \text{ mol H}}{1.008 \text{ g H}} = 14.26 \text{ mol H}$

Because the mol H to mol C ratio is 2 : 1 (14.26/7.130 = 2.000), the empirical formula is CH_2.

The empirical formula mass ≈ 12 + 2(1) = 14. Since 4 × 14 = 56 puts the molar mass between 50 and 60, the molecular formula is C_4H_8. The isomers of C_4H_8 are:

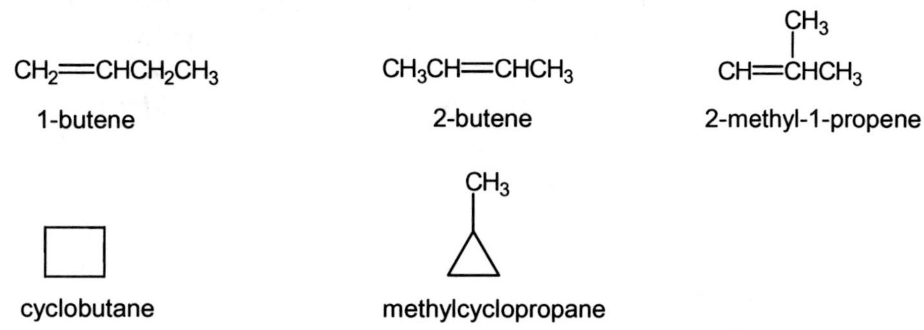

Only the alkenes will react with H_2O to produce alcohols, and only 1-butene will produce a secondary alcohol for the major product and a primary alcohol for the minor product.

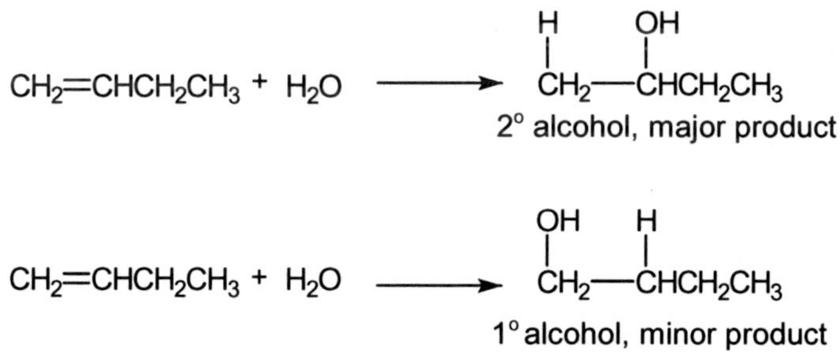

2-Butene will produce only a secondary alcohol when reacted with H_2O, and 2-methyl-1-propene will produce a tertiary alcohol as the major product and a primary alcohol as the minor product.

121. $KMnO_4$ will oxidize primary alcohols to aldehydes and then to carboxylic acids. Secondary alcohols are oxidized to ketones by $KMnO_4$. Tertiary alcohols and ethers are not oxidized by $KMnO_4$. The three isomers and their reactions with $KMnO_4$ are:

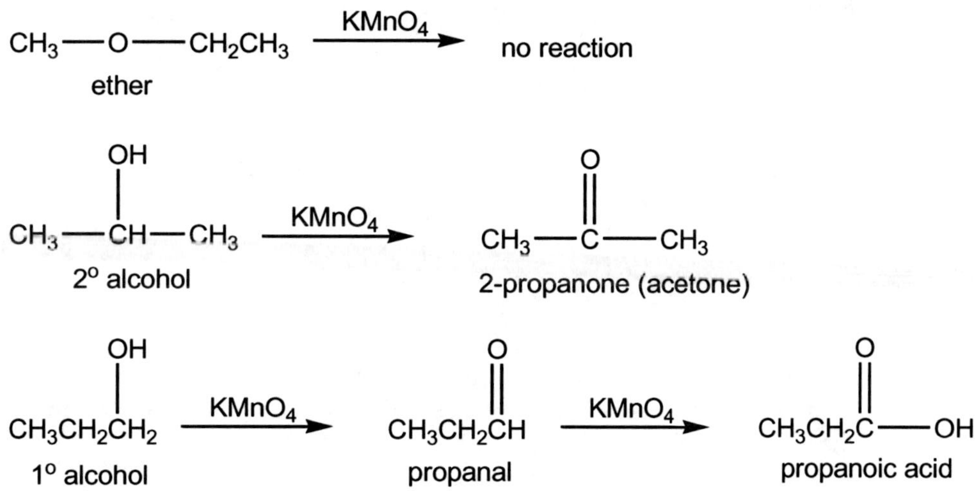

The products of the reactions with excess $KMnO_4$ are 2-propanone and propanoic acid.

122. When addition polymerization of monomers with C=C bonds occurs, the backbone of the polymer chain consists of only carbon atoms. Because the backbone contains oxygen atoms, this is not an addition polymer; it is a condensation polymer. Because the ester functional group is present, we have a polyester condensation polymer. To form an ester functional group, we need the carboxylic acid and alcohol functional groups present in the monomers. From the structure of the polymer, we have a copolymer formed by the following monomers.

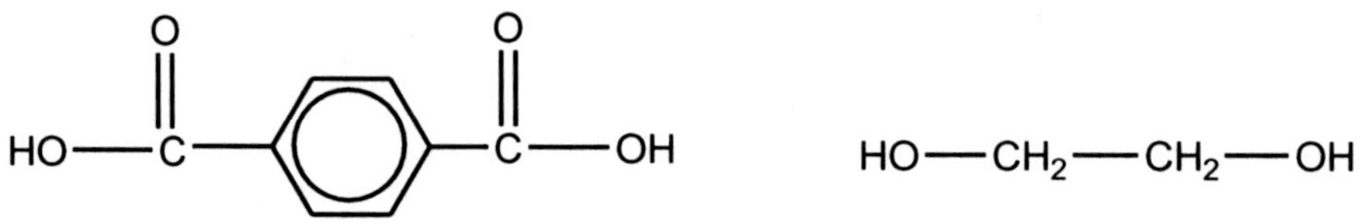

123. In nylon, hydrogen-bonding interactions occur due to the presence of N–H bonds in the polymer. For a given polymer chain length, there are more N–H groups in Nylon-46 as compared to Nylon-6. Hence Nylon-46 forms a stronger polymer compared to Nylon-6 due to the increased hydrogen-bonding interactions.

124. The monomers for nitrile are CH_2=CHCN (acrylonitrile) and CH_2=CHCH=CH_2 (butadiene). The structure of the polymer nitrile is:

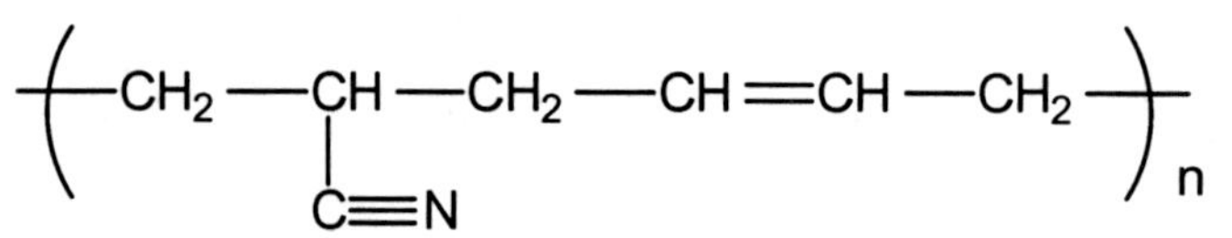

125. a.

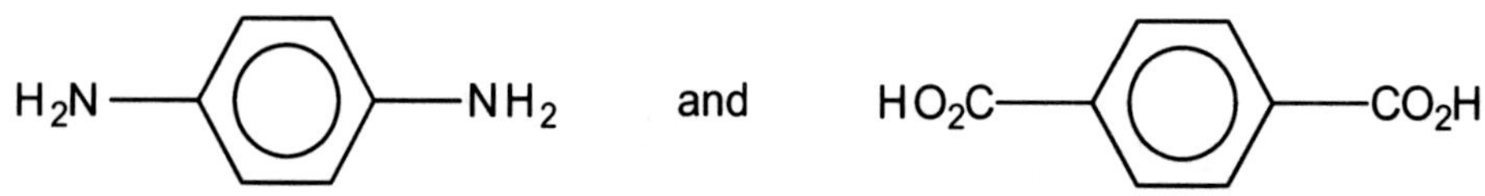

b. Repeating unit:

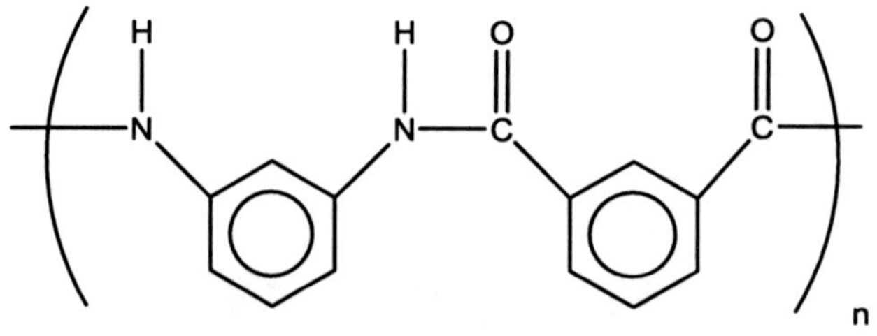

The two polymers differ in the substitution pattern on the benzene rings. The Kevlar chain is straighter, and there is more efficient hydrogen-bonding between Kevlar chains than between Nomex chains.

126. Polyacrylonitrile:

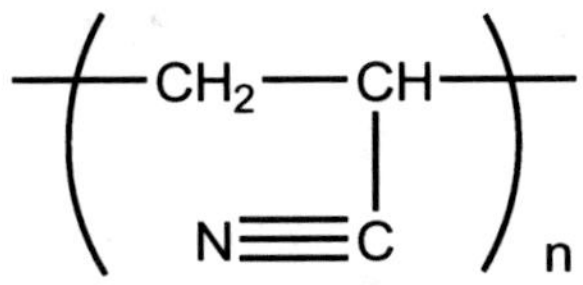

The CN triple bond is very strong and will not easily break in the combustion process. A likely combustion product is the toxic gas hydrogen cyanide, HCN(g).

127. a. The bond angles in the ring are about 60°. VSEPR predicts bond angles close to 109°. The bonding electrons are closer together than they prefer, resulting in strong electron-electron repulsions. Thus ethylene oxide is unstable (reactive).

b. The ring opens up during polymerization; the monomers link together through the formation of O–C bonds.

$$\left(O-CH_2CH_2-O-CH_2CH_2-O-CH_2CH_2 \right)_n$$

128.

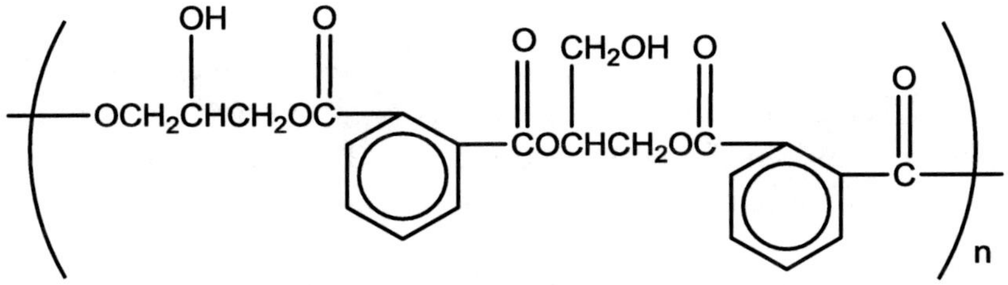

Two linkages are possible with glycerol. A possible repeating unit with both types of linkages is shown above. With either linkage, there are unreacted OH groups on the polymer chains. These unreacted OH groups on adjacent polymer chains can react with the acid groups of phthalic acid to form crosslinks (bonds) between various polymer chains.

129. Glutamic acid:

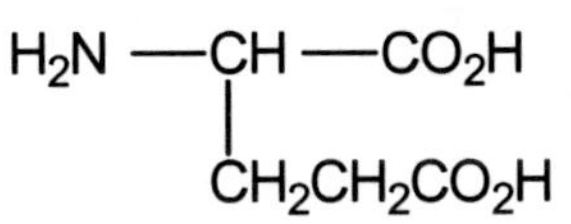

One of the two acidic protons in the carboxylic acid groups is lost to form MSG. Which proton is lost is impossible for you to predict.

Monosodium glutamate:

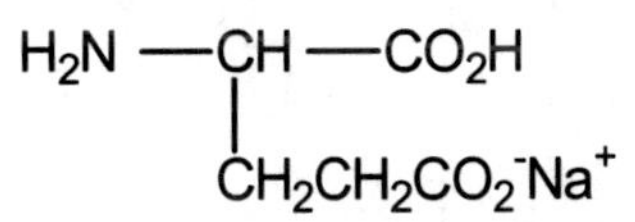

In MSG, the acidic proton from the carboxylic acid in the R group is lost, allowing formation of the ionic compound.

130. a.

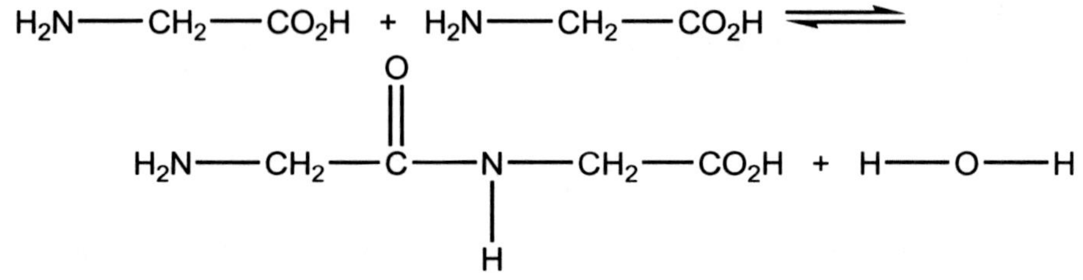

Bonds broken:

1 C–O (358 kJ/mol)

1 H–N (391 kJ/mol)

Bonds formed:

1 C–N (305 kJ/mol)

1 H–O (467 kJ/mol)

$\Delta H = 358 + 391 - (305 + 467) = -23$ kJ

b. ΔS for this process is negative (unfavorable) because the dipeptide has a more ordered structure (positional probability decreases).

c. ΔG = ΔH − TΔS; although ΔH is favorable (negative), the magnitude of ΔH is very small. In this process, the unfavorable entropy change will dominate, giving a positive ΔG value. The reaction is not spontaneous at normal temperatures.

131. ΔG = ΔH − TΔS; for the reaction, we break a P–O and O–H bond and form a P–O and O–H bond, so ΔH ≈ 0 based on bond dissociation energies. ΔS for this process is negative (unfavorable) because positional probability decreases. Thus, ΔG > 0 due to the unfavorable ΔS term, and the reaction is not expected to be spontaneous.

132. Both proteins and nucleic acids must form for life to exist. From the simple analysis, it looks as if life can't exist, an obviously incorrect assumption. A cell is not an isolated system. There is an external source of energy to drive the reactions. A photosynthetic plant uses sunlight, and animals use the carbohydrates produced by plants as sources of energy. When all processes are combined, ΔS_{univ} must be greater than zero, as is dictated by the second law of thermodynamics.

133. Alanine can be thought of as a diprotic acid. The first proton to leave comes from the carboxylic acid end with $K_a = 4.5 \times 10^{-3}$. The second proton to leave comes from the protonated amine end (K_a for $R{-}NH_3^+ = K_w/K_b = 1.0 \times 10^{-14}/7.4 \times 10^{-5} = 1.4 \times 10^{-10}$).

In 1.0 *M* H^+, both the carboxylic acid and the amine end will be protonated since H^+ is in excess. The protonated form of alanine is below. In 1.0 *M* OH^-, the dibasic form of alanine will be present because the excess OH^- will remove all acidic protons from alanine. The dibasic form of alanine follows.

1.0 *M* H^+ :

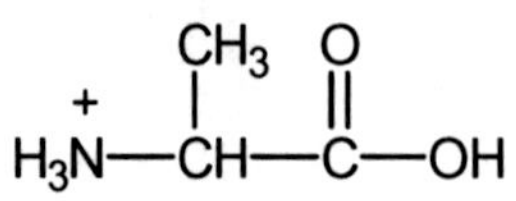

protonated form

1.0 *M* OH^- :

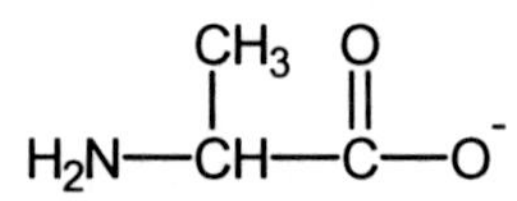

dibasic form

134. The number of approximate base pairs in a DNA molecule is:

$$\frac{4.5 \times 10^9 \text{ g/mol}}{600 \text{ g/mol}} = 8 \times 10^6 \text{ base pairs}$$

The approximate number of complete turns in a DNA molecule is:

$$8 \times 10^6 \text{ base pairs} \times \frac{0.34 \text{ nm}}{\text{base pair}} \times \frac{1 \text{ turn}}{3.4 \text{ nm}} = 8 \times 10^5 \text{ turns}$$

135. For denaturation, heat is added so it is an endothermic process. Because the highly ordered secondary structure is disrupted, positional probability increases, so entropy will increase. Thus ΔH and ΔS are both positive for protein denaturation.

136. a. $^+H_3NCH_2COO^- + H_2O \rightleftharpoons H_2NCH_2CO_2^- + H_3O^+$

$$K_{eq} = K_a(-NH_3^+) = \frac{K_w}{K_b(-NH_2)} = \frac{1.0 \times 10^{-14}}{6.0 \times 10^{-5}} = 1.7 \times 10^{-10}$$

b. $H_2NCH_2CO_2^- + H_2O \rightleftharpoons H_2NCH_2CO_2H + OH^-$

$$K_{eq} = K_b(-CO_2^-) = \frac{K_w}{K_a(-CO_2H)} = \frac{1.0 \times 10^{-14}}{4.3 \times 10^{-3}} = 2.3 \times 10^{-12}$$

c. $^+H_3NCH_2CO_2H \rightleftharpoons 2\ H^+ + H_2NCH_2CO_2^-$

$$K_{eq} = K_a(-CO_2H) \times K_a(-NH_3^+) = (4.3 \times 10^{-3})(1.7 \times 10^{-10}) = 7.3 \times 10^{-13}$$

ChemWork Problems

The answers to the problems 137-144 (or a variation to these problems) are found in OWL. These problems are also assignable in OWL.

Challenge Problems

145. For the reaction:

$$^+H_3NCH_2CO_2H \rightleftharpoons 2\ H^+ + H_2NCH_2CO_2^- \quad K_{eq} = 7.3 \times 10^{-13} = K_a(-CO_2H) \times K_a(-NH_3^+)$$

$$7.3 \times 10^{-13} = \frac{[H^+]^2[H_2NCH_2CO_2^-]}{[^+H_3NCH_2CO_2H]} = [H^+]^2,\ [H^+] = (7.3 \times 10^{-13})^{1/2}$$

$[H^+] = 8.5 \times 10^{-7}\ M$; $pH = -\log[H^+] = 6.07$ = isoelectric point

146. a. The new amino acid is most similar to methionine due to its $–CH_2CH_2SCH_3$ R group.

b. The new amino acid replaces methionine. The structure of the tetrapeptide is:

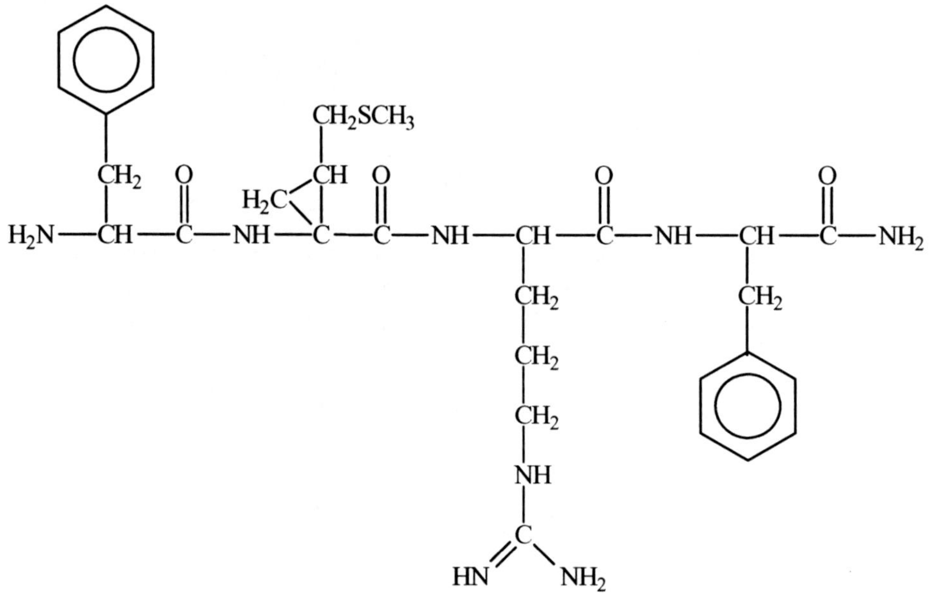

c. The chiral carbons are indicated with an asterisk.

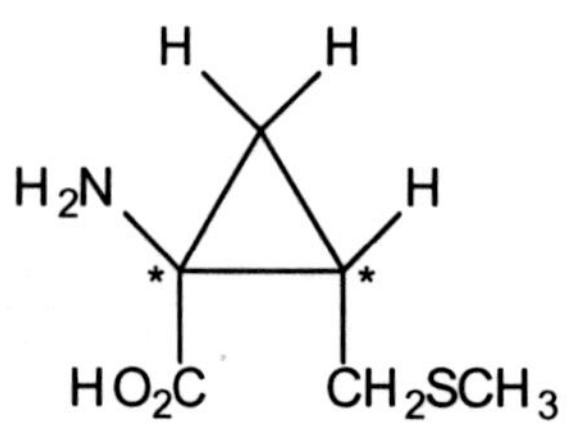

147. a. Even though this form of tartaric acid contains 2 chiral carbon atoms (see asterisks in the following structure), the mirror image of this form of tartaric acid is superimpossible. Therefore, it is not optically active. An easier way to identify optical activity in molecules with two or more chiral carbon atoms is to look for a plane of symmetry in the molecule. If a molecule has a plane of symmetry, then it is never optically active. A plane of symmetry is a plane that bisects the molecule where one side exactly reflects on the other side.

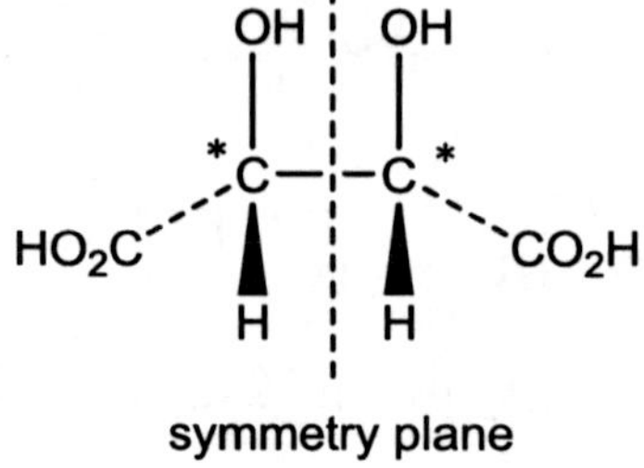

b. The optically active forms of tartaric acid have no plane of symmetry. The structures of the optically active forms of tartaric acid are:

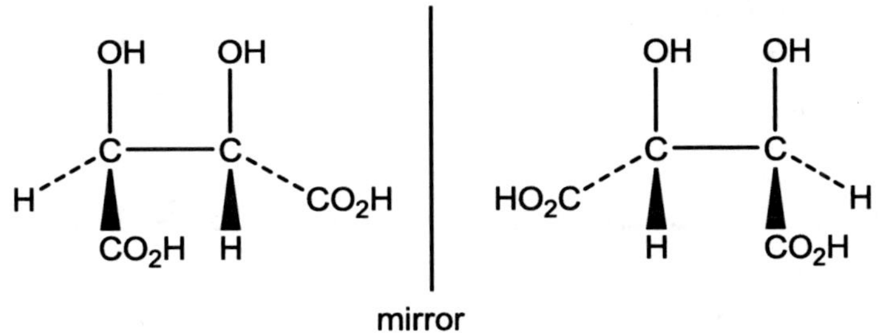

These two forms of tartaric acid are nonsuperimposable.

148. Assuming 1.000 L of the hydrocarbon (C_xH_y), then the volume of products will be 4.000 L, and the mass of products ($H_2O + CO_2$) will be:

$$1.391 \text{ g/L} \times 4.000 \text{ L} = 5.564 \text{ g products}$$

$$\text{Moles of } C_xH_y = n_{C_xH_y} = \frac{PV}{RT} = \frac{0.959 \text{ atm} \times 1.000 \text{ L}}{\frac{0.08206 \text{ L atm}}{\text{K mol}} \times 298 \text{ K}} = 0.0392 \text{ mol}$$

$$\text{Moles of products} = n_p = \frac{PV}{RT} = \frac{1.51 \text{ atm} \times 4.000 \text{ L}}{\frac{0.08206 \text{ L atm}}{\text{K mol}} \times 375 \text{ K}} = 0.196 \text{ mol}$$

C_xH_y + oxygen $\rightarrow x\ CO_2 + y/2\ H_2O$; setting up two equations:

$$0.0392x + 0.0392(y/2) = 0.196 \text{ (moles of products)}$$

$$0.0392x(44.01 \text{ g/mol}) + 0.0392(y/2)(18.02 \text{ g/mol}) = 5.564 \text{ g (mass of products)}$$

Solving: $x = 2$ and $y = 6$, so the formula of the hydrocarbon is C_2H_6, which is ethane.

149.

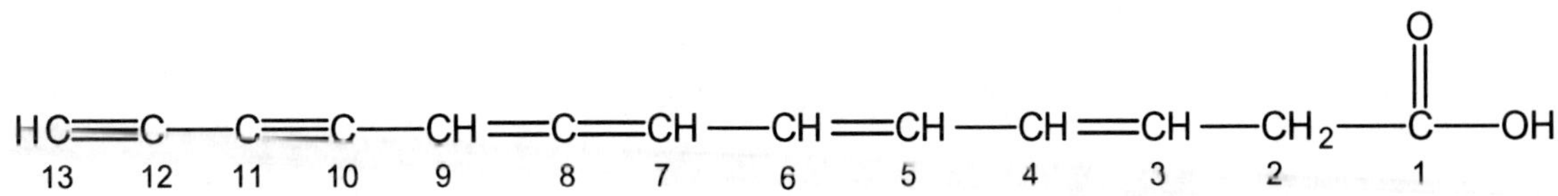

150.

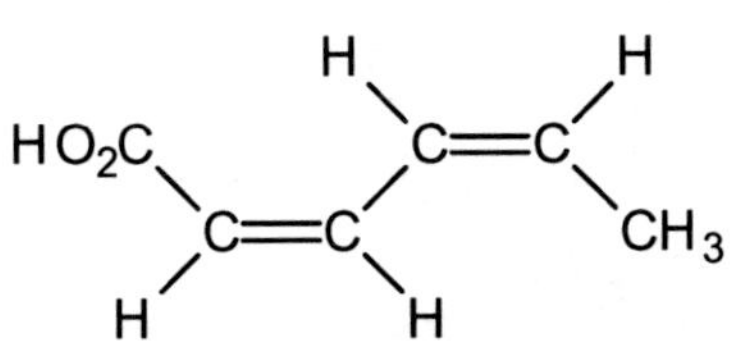

cis-2-cis-4-hexadienoic acid

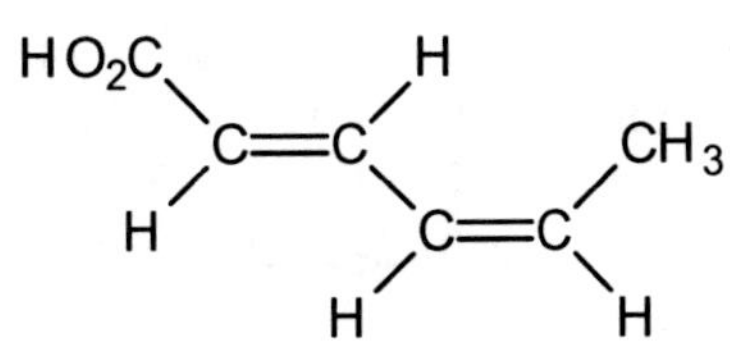

trans-2-cis-4-hexadienoic acid

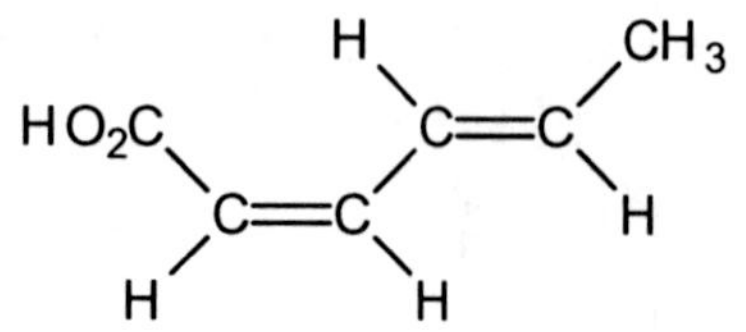

cis-2-trans-4-hexadienoic acid

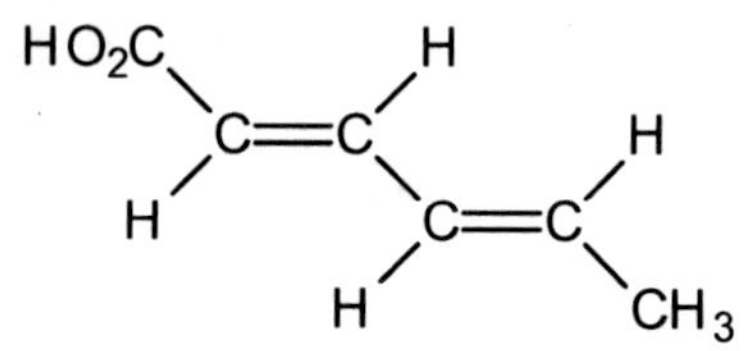

trans-2-trans-4-hexadienoic acid

151. a. The three structural isomers of C_5H_{12} are:

$CH_3CH_2CH_2CH_2CH_3$

n-pentane

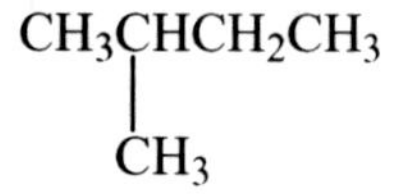

2-methylbutane

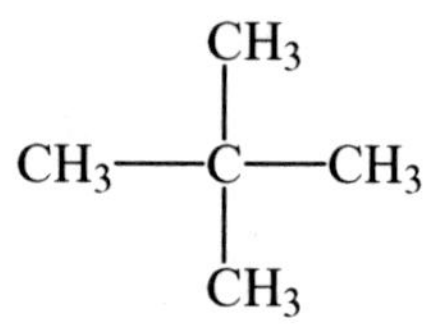

2,2-dimethylpropane

n-Pentane will form three different monochlorination products: 1-chloropentane, 2-chloropentane, and 3-chloropentane (the other possible monochlorination products differ by a simple rotation of the molecule; they are not different products from the ones listed). 2,2-Dimethylpropane will only form one monochlorination product: 1-chloro-2,2-dimethylpropane. 2-Methylbutane is the isomer of C_5H_{12} that forms four different monochlorination products: 1-chloro-2-methylbutane, 2-chloro-2-methyl-butane, 3-chloro-2-methylbutane (or we could name this compound 2-chloro-3-methylbutane), and 1-chloro-3-methylbutane.

b. The isomers of C_4H_8 are:

$CH_2{=}CHCH_2CH_3$

1-butene

$CH_3CH{=}CHCH_3$

2-butene

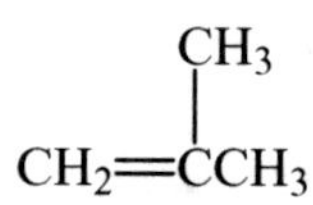

2-methyl-1-propene or 2-methylpropene

cyclobutane

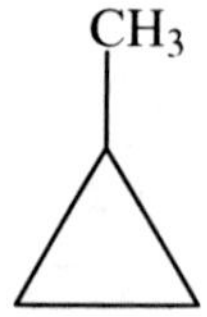

methylcyclopropane

The cyclic structures will not react with H_2O; only the alkenes will add H_2O to the double bond. From Exercise 62, the major product of the reaction of 1-butene and H_2O is 2-butanol (a 2° alcohol). 2-Butanol is also the major (and only) product when 2-butene and H_2O react. 2-Methylpropene forms 2-methyl-2-propanol as the major product when reacted with H_2O; this product is a tertiary alcohol. Therefore, the C_4H_8 isomer is 2-methylpropene.

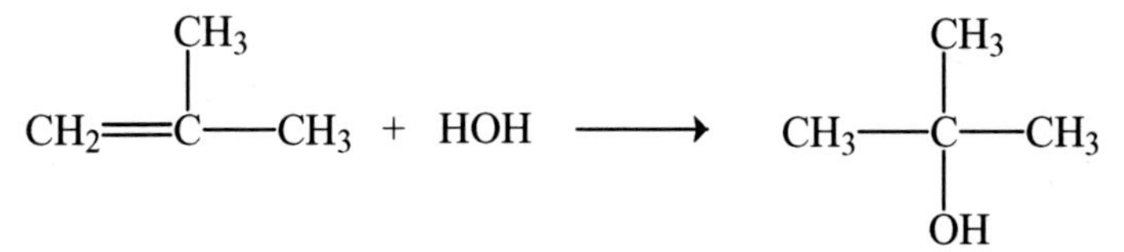

2-methyl-2-propanol (a 3° alcohol, 3 R groups)

c. The structure of 1-chloro-1-methylcyclohexane is:

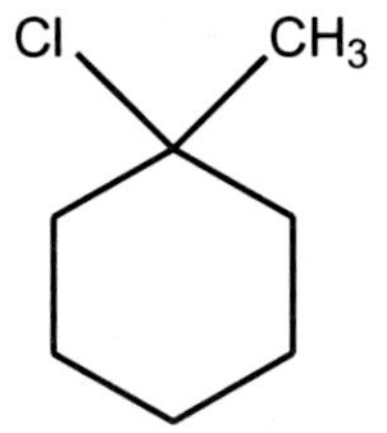

The addition reaction of HCl with an alkene is a likely choice for this reaction (see Exercise 62). The two isomers of C_7H_{12} that produce 1-chloro-1-methylcyclohexane as the major product are:

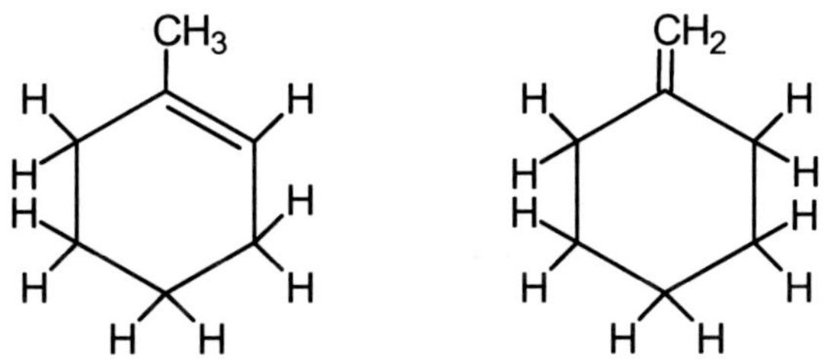

d. Working backwards, 2° alcohols produce ketones when they are oxidized (1° alcohols produce aldehydes, then carboxylic acids). The easiest way to produce the 2° alcohol from a hydrocarbon is to add H_2O to an alkene. The alkene reacted is 1-propene (or propene).

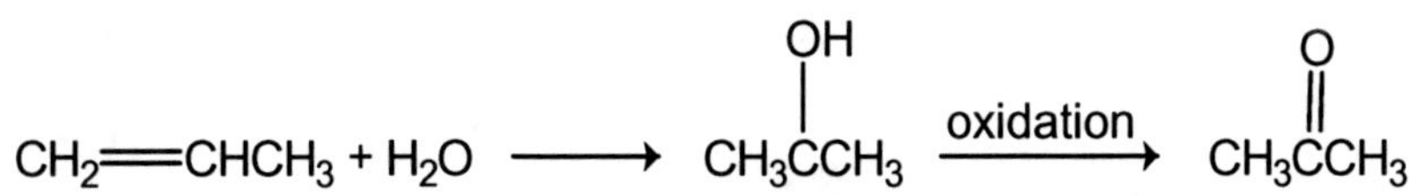

propene acetone

e. The $C_5H_{12}O$ formula has too many hydrogens to be anything other than an alcohol (or an unreactive ether). 1° Alcohols are first oxidized to aldehydes, then to carboxylic acids. Therefore, we want a 1° alcohol. The 1° alcohols with formula $C_5H_{12}O$ are:

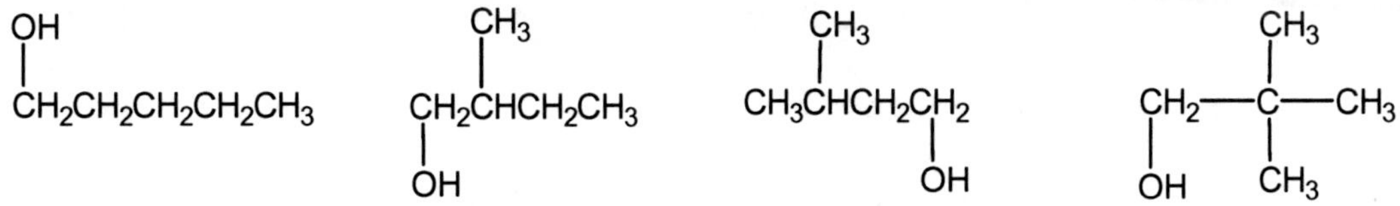

1-pentanol 2-methyl-1-butanol 3-methyl-1-butanol 2,2-dimethyl-1-propanol

There are other alcohols with formula $C_5H_{12}O$, but they are all 2° or 3° alcohols, which do not produce carboxylic acids when oxidized.

152. a.

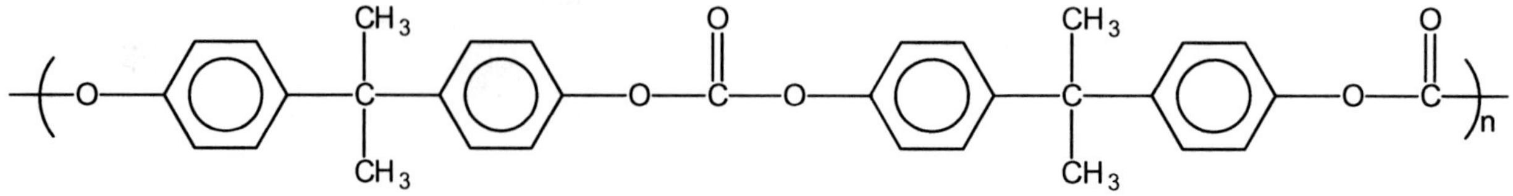

b. Condensation; HCl is eliminated when the polymer bonds form.

153.

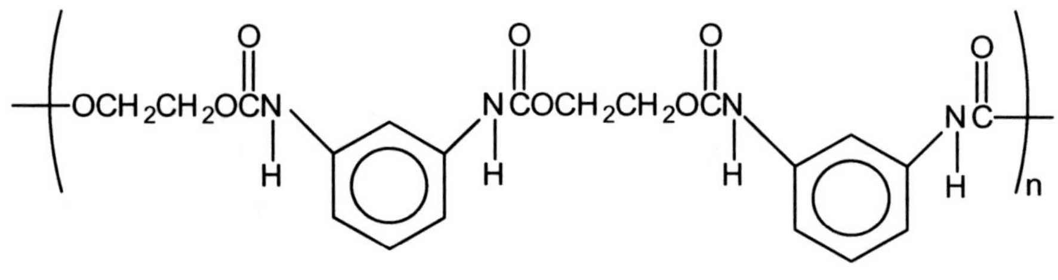

154. a.

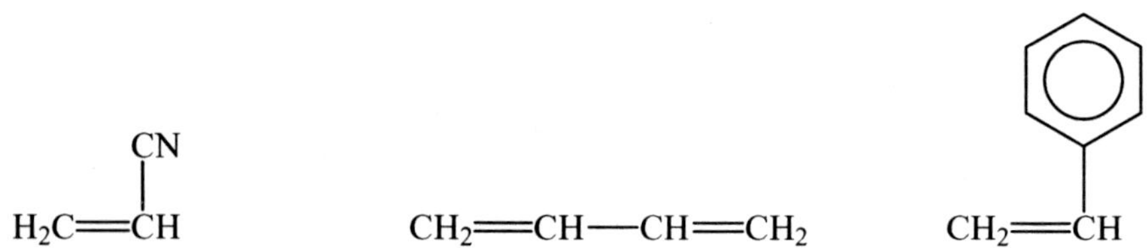

acrylonitrile butadiene styrene

The structure of ABS plastic assuming a 1 : 1 : 1 mole ratio is:

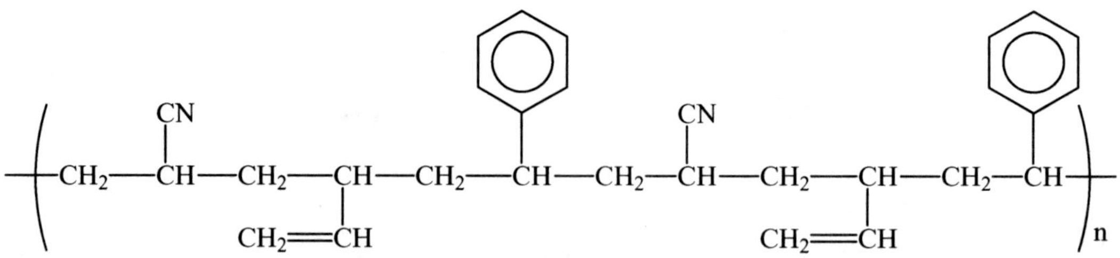

Note: Butadiene does not polymerize in a linear fashion in ABS plastic (unlike other butadiene polymers). There is no way for you to be able to predict this.

b. Only acrylonitrile contains nitrogen. If we have 100.00 g of polymer:

$$8.80\text{ g N} \times \frac{1\,\text{mol}\,C_3H_3N}{14.01\,\text{g N}} = \frac{53.06\,\text{g}\,C_3H_3N}{1\,\text{mol}\,C_3H_3\,N} = 33.3\text{ g }C_3H_3N$$

$$\text{Mass \% } C_3H_3N = \frac{33.3\text{ g } C_3H_3N}{100.00\text{ g polymer}} = 33.3\%\ C_3H_3N$$

Br_2 adds to double bonds of alkenes (benzene's delocalized π bonds in the styrene monomer will not react with Br_2 unless a special catalyst is present). Only butadiene in the polymer has a reactive double bond. From the polymer structure in part a, butadiene will react in a 1 : 1 mol ratio with Br_2.

$$0.605\text{ g } Br_2 \times \frac{1\text{ mol } Br_2}{159.8\text{ g } Br_2} \times \frac{1\text{ mol } C_4H_6}{\text{mol } Br_2} \times \frac{54.09\text{ g } C_4H_6}{\text{mol } C_4H_6} = 0.205\text{ g } C_4H_6$$

$$\text{Mass \% } C_4H_6 = \frac{0.205\text{ g}}{1.20\text{ g}} \times 100 = 17.1\%\ C_4H_6$$

Mass % styrene (C_8H_8) = 100.0 − 33.3 − 17.1 = 49.6% C_8H_8.

c. If we have 100.0 g of polymer:

$$33.3\text{ g } C_3H_3N \times \frac{1\text{ mol } C_3H_3N}{53.06\text{ g}} = 0.628\text{ mol } C_3H_3N$$

$$17.1\text{ g } C_4H_6 \times \frac{1\text{ mol } C_4H_6}{54.09\text{ g } C_4H_6} = 0.316\text{ mol } C_4H_6$$

$$49.6\text{ g } C_8H_8 \times \frac{1\text{ mol } C_8H_8}{104.14\text{ g } C_8H_8} = 0.476\text{ mol } C_8H_8$$

Dividing by 0.316: $\frac{0.628}{0.316} = 1.99$; $\frac{0.316}{0.316} = 1.00$; $\frac{0.476}{0.316} = 1.51$

This is close to a mole ratio of 4 : 2 : 3. Thus there are 4 acrylonitrile to 2 butadiene to 3 styrene molecules in this polymer sample, or $(A_4B_2S_3)_n$.

155. a. The temperature of the rubber band increases when it is stretched.

b. Exothermic because heat is released.

c. As the polymer chains that make up the rubber band are stretched, they line up more closely together, resulting in stronger London dispersion forces between the chains. Heat is released as the strength of the intermolecular forces increases.

d. Stretching is not spontaneous, so ΔG is positive. ΔG = ΔH − TΔS; since ΔH is negative, ΔS must be negative in order to give a positive ΔG.

e.

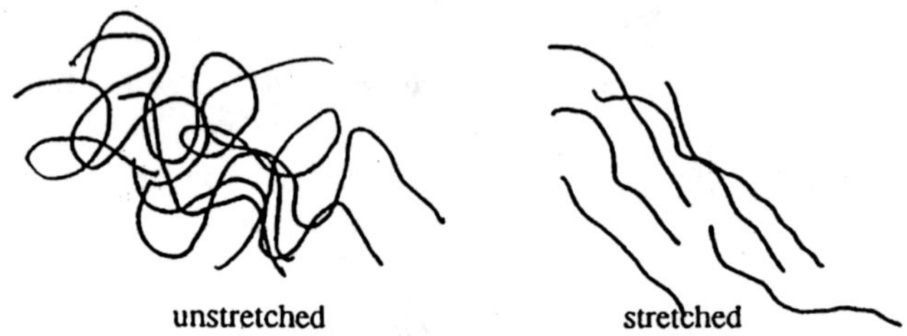

The structure of the stretched polymer chains is more ordered (has a smaller positional probability). Therefore, entropy decreases as the rubber band is stretched.

156. a.

Step 1: $\underset{\text{1-butanol}}{CH_2(OH)\text{—}CH(H)CH_2CH_3} \xrightarrow{H^+} \underset{\text{1-butene}}{CH_2{=}CHCH_2CH_3} + H_2O$

Step 2: $\underset{\text{1-butene}}{CH_2{=}CHCH_2CH_3} + H_2 \xrightarrow{Pt} \underset{\text{butane}}{CH_3CH_2CH_2CH_3}$

b.

Step 1: $\underset{\text{1-butanol}}{CH_2(OH)\text{—}CH(H)CH_2CH_3} \xrightarrow{H^+} \underset{\text{1-butene}}{CH_2{=}CHCH_2CH_3} + H_2O$

Step 2: $\underset{\text{1-butene}}{CH_2{=}CHCH_2CH_3} + H_2O \xrightarrow{H^+} \underset{\text{2-butanol (major product)}}{CH_2(H)\text{—}CH(OH)CH_2CH_3}$

Step 3: $\underset{\text{2-butanol}}{CH_3CH(OH)CH_2CH_3} \xrightarrow{\text{oxidation}} \underset{\text{2-butanone}}{CH_3C({=}O)CH_2CH_3}$

157. $4.2 \times 10^{-3}\ g\ K_2CrO_7 \times \frac{1\ mol\ K_2Cr_2O_7}{294.20\ g} \times \frac{1\ mol\ Cr_2O_7^{\ 2-}}{mol\ K_2Cr_2O_7} \times \frac{3\ mol\ C_2H_5OH}{2\ mol\ Cr_2O_7^{\ 2-}}$

$= 2.1 \times 10^{-5}$ mol C_2H_5OH

$$n_{breath} = \frac{PV}{RT} = \frac{\left(750.\ mm\ Hg \times \frac{1\ atm}{760\ mm\ Hg}\right) \times 0.500\ L}{\frac{0.08206\ L\ atm}{K\ mol} \times 303\ K} = 0.0198 \text{ mol breath}$$

$$\text{Mole \% } C_2H_5OH = \frac{2.1 \times 10^{-5}\text{ mol } C_2H_5OH}{0.0198\text{ mol total}} \times 100 = 0.11\%\text{ alcohol}$$

158. The five chiral carbons are marked with an asterisk.

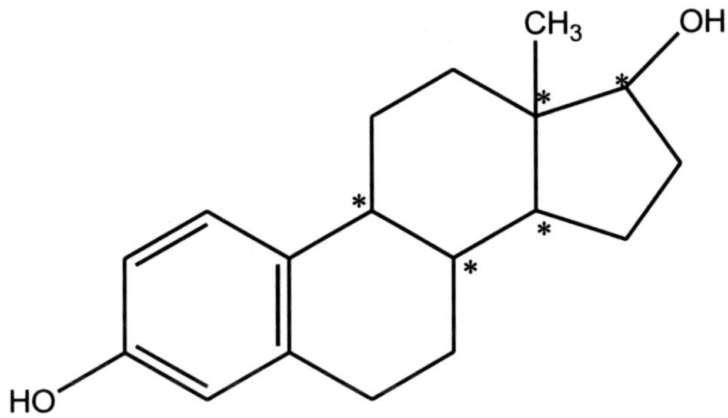

Each of these five carbons has four different groups bonded to it. The fourth bond that is not shown for any of the five chiral carbons is a C–H bond.

Integrative Problems

159. a. Zn^{2+} has the [Ar]$3d^{10}$ electron configuration, and zinc does form 2+ charged ions.

$$\text{Mass \% Zn} = \frac{\text{mass of 1 mol Zn}}{\text{mass of 1 mol } CH_3CH_2ZnBr} \times 100 = \frac{65.38\text{ g}}{174.34\text{ g}} \times 100 = 37.50\%\text{ Zn}$$

b. The reaction is:

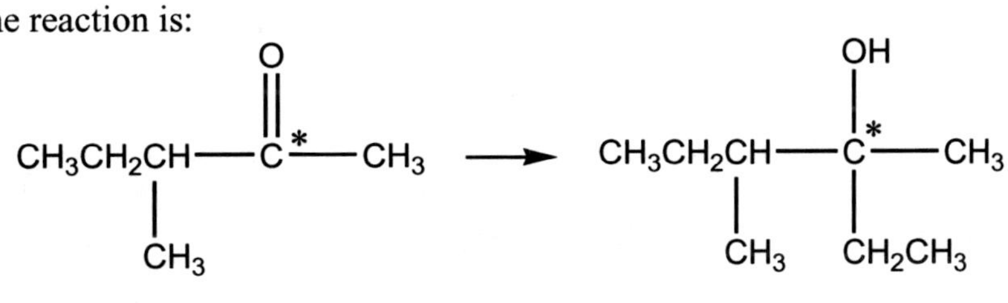

The hybridization changes from sp^2 to sp^3.

c. 3,4-dimethyl-3-hexanol

160. a. $0.5063\text{ g } CO_2 \times \frac{1\text{ mol } CO_2}{44.01\text{ g}} \times \frac{1\text{ mol C}}{\text{mol } CO_2} \times \frac{12.01\text{ g C}}{\text{mol C}} = 0.1382\text{ g C}$

$$\text{Mass \%C} = \frac{0.1382\text{ g C}}{0.1450\text{ g compound}} \times 100 = 95.31\%$$

Mass %H = 100.00 – 95.31 = 4.69%H

Assuming 100.00 g compound:

$$95.31 \text{ g C} \times \frac{1 \text{ mol C}}{12.01 \text{ g C}} = 7.936 \text{ mol C}/4.653 = 1.706 \text{ mol C}$$

$$4.69 \text{ g H} \times \frac{1 \text{ mol H}}{1.008 \text{ g H}} = 4.653 \text{ mol H}/4.653 = 1 \text{ mol H}$$

Multiplying by 10 gives the empirical formula $C_{17}H_{10}$.

b. Moles of helicene $= 0.0125 \text{ kg} \times \frac{0.0175 \text{ mol helicene}}{\text{kg solvent}} = 2.19 \times 10^{-4}$ mol helicene

$$\text{Molar mass} = \frac{0.0938 \text{ g}}{2.19 \times 10^{-4} \text{ mol}} = 428 \text{ g/mol}$$

Empirical formula mass $\approx 17(12) + 10(1) = 214$ g/mol

Because $\frac{428}{214} = 2.00$, the molecular formula is $(C_{17}H_{10}) \times 2 = C_{34}H_{20}$

c. $C_{34}H_{20}(s) + 39\ O_2(g) \rightarrow 34\ CO_2(g) + 10\ H_2O(l)$

Marathon Problems

161.
a. urea, ammonium cyanate b. saturated c. tetrahedral
d. straight-chain or normal e. bonds f. –ane
g. longest h. number i. combustion
j. substitution k. addition l. hydrogenation
m. aromatic n. functional o. primary
p. carbon monoxide q. fermentation r. carbonyl
s. oxidation t. carboxyl u. esters, alcohol

162.
a. statement (17) b. statement (13) c. statement (15)
d. statement (12) e. statement (8) f. statement (9)
g. statement (16) h. statement (2) i. statement (4)
j. statement (10) k. statement (11) l. statement (7)
m. statement (14) n. statement (3) o. statement (6)
p. statement (1) q. statement (5)

163.
a. deoxyribonucleic acid b. nucleotides c. ribose
d. ester e. complementary f. thymine, guanine
g. gene h. transfer, messenger i. DNA